MEDICAL
BIOCHEMISTRY

ELSEVIER *science & technology books*

ELSEVIER

Companion Web Site:

http://textbooks.elsevier.com/web/Manuals.aspx?isbn=9780128035504

Medical Biochemistry
Antonio Blanco and Gustavo Blanco, Authors

ELSEVIER

TOOLS FOR ALL YOUR TEACHING NEEDS
textbooks.elsevier.com

ACADEMIC PRESS

MEDICAL BIOCHEMISTRY

ANTONIO BLANCO
Emeritus Professor
National University of Cordoba
Cordoba, Argentina

GUSTAVO BLANCO
Professor and Kathleen M. Osborn Chair
Department of Molecular and Integrative Physiology
University of Kansas Medical Center
Kansas City, KS, USA

ACADEMIC PRESS

An imprint of Elsevier

Academic Press is an imprint of Elsevier
125 London Wall, London EC2Y 5AS, United Kingdom
525 B Street, Suite 1800, San Diego, CA 92101-4495, United States
50 Hampshire Street, 5th Floor, Cambridge, MA 02139, United States
The Boulevard, Langford Lane, Kidlington, Oxford OX5 1GB, United Kingdom

Notices
Knowledge and best practice in this field are constantly changing. As new research and experience broaden our understanding, changes in research methods, professional practices, or medical treatment may become necessary.

Practitioners and researchers must always rely on their own experience and knowledge in evaluating and using any information, methods, compounds, or experiments described herein. In using such information or methods they should be mindful of their own safety and the safety of others, including parties for whom they have a professional responsibility.

To the fullest extent of the law, neither the Publisher nor the authors, contributors, or editors, assume any liability for any injury and/or damage to persons or property as a matter of products liability, negligence or otherwise, or from any use or operation of any methods, products, instructions, or ideas contained in the material herein.

Library of Congress Cataloging-in-Publication Data
A catalog record for this book is available from the Library of Congress

British Library Cataloguing-in-Publication Data
A catalogue record for this book is available from the British Library

ISBN: 978-0-12-803550-4

For information on all Academic Press publications visit our website at
https://www.elsevier.com/books-and-journals

  Working together
to grow libraries in
developing countries

www.elsevier.com • www.bookaid.org

Publisher: Sara Tenney
Acquisition Editor: Linda Versteeg-Buschman
Editorial Project Manager: Fenton Coulthurst
Production Project Manager: Karen East and Kirsty Halterman
Designer: Greg Harris

Typeset by Thomson Digital

to Noemí

to Gladis

Contents

About the Authors   xiii
Preface   xv
Introduction   xvii

1. Chemical Composition of Living Beings

Biogenic Elements   1
Biological Compounds   2
Summary   3
Bibliography   3

2. Water

The Water Molecule is Polar   5
Hydrogen Bond   5
Water as Solvent   7
Water as an Electrolyte   8
Equilibrium Constant   9
Equilibrium of Water Ionization   10
Acids and Bases   11
Acids and Bases Strength   12
pH   12
Buffers   13
Titration Curve of Acids and Bases   15
Appendix   16
Summary   18
Bibliography   19

3. Proteins

Proteins are Macromolecules Formed
   by Amino Acids   21
Amino Acids   22
Peptides   30
Proteins   33
Collagen   47
Keratins   49

Hemoglobin   49
Heme   50
Globin   51
Hemoglobin Derivatives   56
Abnormal Hemoglobins   56
Blood Plasma Proteins   58
Main Proteins in Blood Plasma   60
Muscle Proteins   64
Muscle Contraction   67
Summary   69
Bibliography   71

4. Carbohydrates

Monosaccharides   73
Disaccharides   83
Polysaccharides   84
Summary   95
Bibliography   97

5. Lipids

Classification   99
Fatty Acids   99
Essential Fatty Acids   104
Simple Lipids   104
Complex Lipids   108
Substances Associated with Lipids   114
Summary   117
Bibliography   119

6. Nucleic Acids

Nucleotides   121
Nucleic Acids   123
Virus   136
Free Nucleotides   137
Summary   139
Bibliography   140

7. Elements of Thermodynamics and Biochemical Kinetics

Thermodynamics   141
Chemical Kinetics   148
Summary   151
Bibliography   152

8. Enzymes

Enzymes are Biological Catalysts   153
Nomenclature and Classification of
    Enzymes   153
Chemical Nature of Enzymes   156
Metalloenzymes   157
Enzymatic Catalysis   157
Active Site   158
Zymogens   160
Genetic Alterations that Affect Enzyme Function   160
Enzyme Distribution Within the Cell   160
Multienzyme Systems   161
Determination of Enzyme Activity   161
Factors that Modify Enzyme
    Activity   161
Enzyme Inhibitors   165
Regulation of Enzyme Activity   168
Constitutive and Inducible
    Enzymes   170
Enzymatic Processes in Cascade   171
Isozymes   172
Determination of Enzymes in the Clinical
    Laboratory   173
Summary   174
Bibliography   175

9. Biological Oxidations: Bioenergetics

The Energy-Rich Intermediate ATP   177
Oxidation–Reduction   178
Biological Oxidations   181
Respiratory Chain   184
Oxidative Phosphorylation   190
Phosphorylation at Substrate Level   201
Other Electron Transport Systems   202
Summary   202
Bibliography   203

10. Antioxidants

Reactive Species   205
Summary   213
Bibliography   213

11. Membranes

Structure   215
Transport Across Membranes   224
Passive Transport or Facilitated Diffusion   226
Active Transport   233
Summary   249
Bibliography   250

12. Digestion - Absorption

Saliva   251
Gastric Juice   254
Pancreatic Juice   258
Intestinal Mucosa   260
Bile   261
Summary of the Digestive Process   264
Absorption   267
Summary   272
Bibliography   273

13. Metabolism

Metabolic Pathways   275
Metabolic Studies   277
Regulation of Metabolism   279
Summary   281
Bibliography   281

14. Carbohydrate Metabolism

Glucose Uptake Into Cells   284
Glucose Phosphorylation   285
Metabolic Pathways for Glucose   287
Metabolism of Other Hexoses   311
Biosynthesis of the Oligosaccharides of
    Glycoproteins   316
Summary   320
Bibliography   323

15. Lipid Metabolism

Blood Lipids   326
Plasma Lipoproteins   326
Tissue Lipids   334
Fat Metabolism   334
Fatty Acid Catabolism   335
Fatty Acid Biosynthesis   343
Eicosanoid Biosynthesis   349
Triacylglycerol Biosynthesis   353
Biosynthesis of Phospholipids   354
Congenital Disorders of Complex
    Lipid Catabolism   356
Cholesterol Metabolism   358
Cholesterol Biosynthesis   358
Summary   363
Bibliography   365

16. Amino Acid Metabolism

Essential Amino Acids   370
Amino Acid Catabolism   373
Metabolic Pathways of Ammonia   376
Other General Mechanisms of Amino Acid
    Metabolism   384
Metabolic Pathways of Amino Acids   387
Summary   397
Bibliography   399

17. Heme Metabolism

Heme Biosynthesis   401
Catabolism of Heme   405
Summary   411
Bibliography   412

18. Purine and Pyrimidine Metabolism

Purine Biosynthesis   413
Purine Salvage Pathway   416
Purine Catabolism   416
Uric Acid   418
Gout   418
Pyrimidine Biosynthesis   419
Pyrimidine Catabolism   421
Di- and Triphosphate Nucleoside Biosynthesis   421

Deoxyribonucleotide Biosynthesis   422
Pharmacological Applications   422
Summary   423
Bibliography   423

19. Integration and Regulation of Metabolism

Metabolic Integration   425
Metabolic Regulation   427
Summary   444
Bibliography   445

20. Metabolism in Some Tissues

Liver Metabolism   447
Skeletal Muscle Metabolism   455
Heart Metabolism   459
Adipose Tissue Metabolism   460
Nervous Tissue Metabolism   461
Summary   462
Bibliography   463

21. The Genetic Information (I)

DNA Replication   465
Transcription   475
Methods Used in Molecular Biology   481
Summary   491
Bibliography   492

22. The Genetic Information (II)

Protein Biosynthesis   493
Mechanism of Protein Biosynthesis   502
Genetic Mutations   510
Oncogenes   518
Epigenetics   521
Summary   522
Bibliography   523

23. Regulation of Gene Expression

Gene Regulation in Eukaryotes   525
Structure of Gene Regulatory Proteins   527

Role of Noncoding RNAs   529
Small RNAs   529
RNA Interference   529
Long Noncoding RNA   531
Riboswitches   532
CRISPR   532
Summary   533
Bibliography   533

24. Posttranslational Protein Modifications

Protein Folding   535
Pathologies Caused by Misfolded Proteins   536
Posttranslational Protein Modifications   537
ADP-Ribosylation   538
Ubiquitination   541
Sumoylation   542
Nucleotidylation   542
Summary   545
Bibliography   545

25. Biochemical Basis of Endocrinology (I) Receptors and Signal Transduction

Receptors   548
Signal Transduction Systems   556
Neurotransmitter Receptors   568
Summary   571
Bibliography   572

26. Biochemical Bases of Endocrinology (II) Hormones and Other Chemical Intermediates

Hormones: Chemical Nature   573
Hypothalamic Hormones   576
Anterior Pituitary (Adenohypophysys)   578
Adrenocorticotropin Hormone   578
Thyroid-stimulating Hormone   580
Gonadotropins   580
Lactogenic Hormone or Prolactin   581
Growth Hormone or Somatotropin   582
Melanocyte Stimulatory Hormone   585
Placenta   585
Chorionic Gonadotropin Hormone   585
Placental Lactogen   585
Posterior Pituitary (Neurohypophysis)   586
Oxytocin   586

Arginine Vasopressin   587
Thyroid   588
Adrenal Gland   593
Adrenal Cortex Hormones   593
Adrenal Medulla Hormones   601
Pancreas   605
Insulin   605
Glucagon   611
Somatostatin   612
Glucose Homeostasis   613
Diabetes Mellitus   613
Testis Androgens   616
Ovary   620
Ovarian Hormones   620
Ovarian Peptide Hormones   624
Parathyroid Gland   624
Parathyroid Hormone   624
Calcitonin   627
Kidney   628
Renin   628
Erythropoietin   629
Bradykinin   629
Endothelin   630
Urodilatin   630
Calcitriol   630
Heart   630
Gastrointestinal Hormones   631
Gastrin   631
Secretin   631
Cholecystokinin   632
Vasoactive Intestinal Peptide   632
Enteroglucagon   632
Ghrelin   633
Amylin   633
Galanin   633
Pineal Gland   633
Eicosanoids   634
Growth Factors   636
Nervous System Chemical Intermediaries   637
Summary   640
Bibliography   644

27. Vitamins

Fat-soluble Vitamins   646
Soluble Vitamins   663
Summary   685
Bibliography   687

28. Water and Acid–Base Balance

Water Distribution in the Body   690
Water Balance   691
Ionic Composition of Body Fluids   691
Alterations of the Water Balance   697
H^+ Concentration in Body Fluids   699
Regulation of H^+ Concentration   700
Carbon Dioxide Transport   702
Regulatory Systems   704
Respiratory Regulation   704
Renal Regulation   705
Regulation of Intracellular pH   707
Acid–Base Balance Disorders   708
H^+ Excretion Regulation   710
Laboratory Studies in Acid–Base
   Disorders   710
Summary   711
Bibliography   713

29. Essential Minerals

Macrominerals   715
Trace Elements   735
Summary   741
Bibliography   743

30. Molecular Basis of Immunity

Immune System   745
Innate System   745
Adaptive System   746
Genetic Diversity of Immunoglobulins   753
Complement   760
Cellular Immunity   764
Cytokines   773
Summary   777
Bibliography   780

31. Hemostasis

Blood Coagulation   781
Summary   788
Bibliography   789

32. Apoptosis

Extrinsic Pathway   793
Intrinsic or Mitochondrial Pathway   793
Summary   795
Bibliography   796

Alphabetic Index   797

About the Authors

DR. ANTONIO BLANCO, MD, PhD

Dr. A. Blanco obtained his MD and PhD (in Medicine) degrees in the Faculty of Medical Sciences, National University of Córdoba, Córdoba, Argentina (UNC). During all his academic career, he worked in the Department of Biochemistry of that university, moving up in the ranks from Instructor, Assistant, Associate, and Full Professor to becoming Chairman of the department. He remained as Chairman at the UNC from 1965 to 1996, the year in which he retired. He was then named Emeritus Professor. Between 1961 and 1965, he obtained support from the National Institutes of Health to work as a Research Associate in the Johns Hopkins University Medical School in Baltimore, MD, USA. From 1971 to 2001, he was Researcher of the National Council of Scientific Investigations of Argentina in the highest category. He is currently a member of the National Academies of Sciences and of Natural and Physical Sciences, and of the Academy of Medicine of Córdoba, all from Argentina. He has published 5 textbooks for medical students and 126 original research articles in international journals. He has received numerous awards and distinctions for his research and teaching activities.

DR. GUSTAVO BLANCO, MD, PhD

Dr. G. Blanco obtained his MD and PhD (in Medicine) degrees in the Faculty of Medical Sciences, National University of Córdoba, Córdoba, Argentina (UNC). As a Fellow of the National Council of Scientific Investigations of Argentina, he completed his Doctoral Thesis dissertation, which received the Award of the UNC. He was Instructor in the Department of Biochemistry of the UNC from 1986 to 1990. Between 1990 and 1995, he performed his postdoctoral training, working as a Research Associate in Washington University School of Medicine in Saint Louis, MO, USA. In 2001, he was appointed Assistant Professor in the Department of Molecular and Integrative Physiology in University of Kansas Medical Center in Kansas City, KS, USA. He moved through the ranks as Associate Professor and Professor, and at present he serves as the Kathleen M. Osborn Chairman of the department. Dr. G. Blanco has published more than 60 original research articles in international journals. On account of his scientific investigations and teaching activities he has received numerous awards and distinctions.

Preface

This is the English version of the 10th Spanish edition of *Química Biológica*, published by El Ateneo, Buenos Aires, Argentina.

The book is primarily intended for premedical and medical students. It presents a comprehensive and updated view of the biochemical processes that occur in the human body in health and disease.

Preparing a Biochemistry textbook for students being interested in the medical field poses a great challenge. This requires the selection, from the enormous amount of information available, of the data that will be most relevant to the future physician; the presentation and discussion of the material in a didactic manner to facilitate the learning process; and an up-to-date revision of the topics that include the most important discoveries resulting from the rapid progress in the Biological Sciences and Biochemistry areas.

Following the experience that we have acquired by teaching biochemistry for medical students over the years, our approach to this book was to start describing elemental biochemical concepts, continue developing each topic incorporating more complex concepts and new discoveries with sound experimental basis, widely accepted by the scientific community, and emphasizing their relevance to different clinical scenarios.

Our purpose with this textbook has been to offer a useful tool that will help students learn the bases of Biochemistry and that will serve as a reference for more advanced biochemical concepts, which will cover the needs of the future professionals working in the health care area.

We hope that the textbook will also spark in the reader, their curiosity and interest for scientific research, which will help to continue expanding the frontiers of knowledge.

Introduction

Biological Chemistry, the science that seeks to understand life processes at the molecular level, comprises two broad areas: one studies the components of living organisms and is called *Static* or *Descriptive Biochemistry*; the other investigates the chemical transformations that occur in biological systems and is known as *Dynamic Biochemistry*. Both of these areas have witnessed an amazing development in the past years.

Descriptive Biochemistry. Due to the vast complexity of living organisms, studying the composition of living beings has been an extremely challenging undertaking that demands immense research efforts. Even the simplest unicellular organism contains a myriad of substances.

The study of the components of cells and tissues requires isolation, purification, structure determination, and characterization of their functional properties. Initially, biochemists directed their attention to simple compounds, which could be readily removed from animal or plant organisms, or compounds that could be easily obtained by degradation of more complex substances.

As in other disciplines, the advancement of Biological Chemistry has closely followed technological developments. The availability of new equipment and methods with increasing sensitivity and resolution has allowed researchers to study molecules of the highest complex organization. Isolation, purification, and analysis of macromolecules (proteins, nucleic acids, and polysaccharides), which were once unimaginable, have now become customary in biochemical laboratories.

The discovery of the structure of biological molecules has advanced the understanding of their functions and mechanisms of action, and has helped deciphering their roles.

Dynamic Biochemistry. Countless chemical reactions continuously take place in living organisms in what is known as *metabolism*. The study of these reactions is contemplated by dynamic biochemistry. The earliest goals in this area of biochemistry were directed to decipher the changes that occur to substances ingested with the diet and to identify the origin of waste products excreted by the body. Later, studies were focused to understand the biosynthesis of different endogenous components of the body.

The advances made in the area of metabolism have been, in part, a consequence of the progress of *enzymology*. The study of enzymes, catalysts of biochemical reactions, has been crucial in the interpretation of biological phenomena.

Most chemical conversions in living beings take place gradually, through a series of reactions or *metabolic pathways* which, through sequential steps, convert a compound into a final given product.

From the first observations in the second half of the 19th century, researchers have worked relentlessly to unravel the astounding diversity and complexity of chemical pathways. This resulted in the assembly of "metabolic maps," which illustrate the intricate pathways that compounds follow as they are chemically modified. Although these maps look like a chaotic network, each reaction proceeds in a highly ordered and coordinated manner. In a normal individual, metabolic reactions are exquisitely adapted to serve the needs of each cell and the organism as a whole. Studies on the regulation and cross talk between metabolic pathways have been particularly active in the last 60 years. These have uncovered numerous highly refined regulatory mechanisms which, through modulation of enzyme activity, precisely coordinate the ordered "traffic" of substances, to adjust the flow of compounds along specific metabolic pathways as required.

The nervous and endocrine systems play an essential role in coordinating and integrating the function of many pathways in multicellular organisms. These systems operate in response to a variety of stimuli by releasing intermediary substances or chemical messengers. These messengers include neurotransmitters, hormones, cytokines, and other factors that selectively target receptors to trigger signal transduction mechanisms, which will finally induce specific cell effects. Signal transduction mechanisms have been shown to exhibit a striking degree of complexity, displaying a series of mediator molecules that interact in different ways to activate or inhibit particular events in the cell.

The unity of the biological world. Despite their great diversity, living beings show a remarkable unity with respect to the structure and basic processes that govern their function. For example, although macromolecules, such as proteins and nucleic acids, differ from species to species, and even from individual to individual, they show an overall basic structure which follows the same general plan. All proteins are composed of long chains of the same 20 fundamental units (amino acids), and nucleic acids are constituted by long strands resulting from the assembly of 4 basic structures (nucleotides).

Likewise, the mechanisms underlying cell metabolism also show great similarity even across phylogenetically distant species. For example, the reactions by which human skeletal muscle fibers obtain energy for contraction mimic the transformations that yeast and other microorganisms perform during the process of fermentation. This unity in biology has allowed studying simpler organisms as an approach to understand the function of multicellular organisms of higher complexity.

The body as an energy converting machine. According to their energy requirements, living beings can be divided into *autotrophic* and *heterotrophic*. Plants are autotrophic, which means they are able to synthesize complex organic compounds (carbohydrates, fats, nucleic acids, and proteins) starting from simple inorganic substances (water, carbon dioxide, nitrogen, and phosphates). The energy for this synthesis comes from the sun (*photoautotrophs*) or from electron donors provided by inorganic chemical sources, such as ammonium, sulfur, and ferrous iron (*chemoautotrophs*).

In contrast, multicellular animals are heterotrophic, depending on the intake of compounds produced by other organisms. Normally, carbohydrates, lipids, proteins, vitamins, and minerals that animals obtain from the diet undergo chemical changes that serve two main purposes: (1) the transfer of the energy contained in these substances and (2) the supply of raw materials for synthesis of the organism's own molecules. This process of synthesis, like other cellular activities, requires energy.

The chemical energy obtained from components of the diet can be transformed to fuel the multiple activities occurring within cells. For this reason, living organisms can be considered true energy converting machines. Chemical reactions called *exergonic*, which are those accompanied by the release of energy, are frequently coupled with others that require energy (*endergonic*). Many of these reactions result in the production of new molecules that are able to trap and store the released energy until it is needed. Among the many molecules that play this role, *adenosine triphosphate (ATP)* is the most common carrier of readily usable chemical energy.

Reproduction capacity. A distinctive property of living beings is their ability to reproduce and create, over generations, new organisms that are similar to their predecessors in external features, internal structure, and physiological characteristics. This is possible through the transmission of heritable traits, from cell to cell and from parents to offspring, carried in the *genetic information* contained in deoxyribonucleic acid (DNA) found in chromosomes. This genetic information is ultimately expressed through the synthesis of proteins with unique characteristics for each species and individual. The "code" or "language" of the genetic information in DNA is universal and practically the same for all living beings. This is another example of the remarkable unity of the biological world, which points to the common origin of Earth's living matter.

Changes (*mutations*) that spontaneously appeared in the "genetic message" through millions years of evolution have created the diversity of the living world that we see today, and have shaped the particular characteristics of each species.

The progress made in recent decades on understanding the structure and function of DNA has been remarkable. This has allowed a better understanding of the mechanisms responsible for the transmission of genetic characteristics and the eventual modifications seen between an individual and its progeny. In addition, the understanding of DNA has opened unforeseen opportunities for applications in different fields. A relatively recent discipline, *Molecular Biology*, which emerged within Biochemistry, has made astonishing contributions to our knowledge of the complete sequence of the DNA in the genome of humans, plants, and animals. Also, this discipline has allowed the manipulation of genes; a technology that has direct application in biology and medicine, including the production of genetically modified organisms with specific desired properties and the development of genetic therapeutic approaches directed to treat a variety of diseases.

Along with the evident benefits brought about by the advancement of molecular biology, new ethical and social challenges arose, which require careful consideration by scientists and the general public.

Scope of Biological Chemistry. Advances in Biological Chemistry have opened new horizons and prompted the development of other disciplines, such as Cell Biology. The expansion of Biochemistry into areas, such as Molecular Biology and Cell Biology, has made the boundaries between these disciplines less defined. At present, any Biological Chemistry textbook needs to expand its scope "invading" into areas which were originally foreign to it.

Reductionism–complexity. Scientific research has provided rational explanations for the properties and functions of biomolecules. Today, biochemistry has a solid experimental basis, which resulted from a reductionist conception in the study and interpretation of phenomena. This strategy has proven highly successful in acquiring new knowledge. From studies carried out in isolated molecules to those performed in reconstituted in vitro systems, scientists have been able to develop models in an attempt to describe different in vivo processes.

As these models developed, it became apparent that biological systems are highly complex. Paradoxically, achievements of reductionism have shown its limitations. It is now clear that living organisms cannot just be defined as the sum of their components. The integrated operation of these components generates a functional intricacy that makes the analysis and understanding of living beings extremely difficult. The vast complexity of biological phenomena represents a major challenge for the human mind. This challenge is the stimulus that fuels scientific curiosity, drives the continuous search for understanding biological phenomena, and contributes to the constant expansion of the frontiers of knowledge.

Importance of Biological Chemistry. Without a doubt, the progress of Biochemistry has been one of the factors that have contributed the most to the development of the Biological Sciences. Medical disciplines have benefited tremendously from the advances of Biochemistry, and it is anticipated that this progress will continue at an even faster pace in the future.

Biologists and physicians need to acquire a solid background in basic sciences. In Biological Chemistry, not only will they find the grounds for a rational interpretation of many physiological and pathological phenomena, but also the stimulus for a permanent search of new knowledge.

1

Chemical Composition of Living Beings

BIOGENIC ELEMENTS

Life emerged on Earth many millions of years after the planet was first formed. Only a small number of elements within the inorganic matter of the Earth's crust and atmosphere were selected as the building blocks of all living organisms. These basic elements of life are called *biogenic elements*. Mammals, animals of great complexity, are composed of merely 20 elements, 4 of which (oxygen, carbon, hydrogen, and nitrogen) are the most abundant, comprising approximately 96% of the total body mass (Table 1.1).

All elements of the human body, with the exception of iodine (which has an atomic number of 53), are placed within the first 4 periods of the periodic table and possess atomic numbers lower than 34. Among the four most abundant ones, oxygen has the highest atomic number (8). While oxygen is relatively common on Earth, the other fundamental elements of living organisms are less abundant, suggesting that they have properties, which gave them a selective advantage in becoming the basic units of life. For example, carbon, and not silicon, has been the element around which life developed despite the fact that silicon is widespread and constitutes approximately 21% of the total Earth's weight.

Carbon belongs to the same group in the periodic table and shares many of the properties of silicon. However, carbon can form more stable chemical bonds, long branched chains, double and triple bonds, covalent bonds with different atoms, and adopts a variety of different spatial conformations. This gives carbon the unique potential to generate a variety of chemical combinations that are essential for the makeup of the molecules of living organisms.

The selection of the other elements that accompany carbon as components of the living matter depends on the size of these atoms and their ability to share electrons in covalent bonds. The smaller atomic size of these elements favors their capacity to establish more stable bonds and stronger molecular interactions.

Taking into consideration their relative amounts, biogenic elements can be classified into three main categories:

1. *Primary elements*. These elements include oxygen, carbon, hydrogen, nitrogen, and the less abundant, calcium and phosphorus. Together, these six elements account for more than 98% of the total body mass. Oxygen and hydrogen form water, the most abundant substance in the body. Carbon, oxygen, hydrogen, nitrogen, and phosphorus form part of a variety of essential organic molecules. Calcium, in combination with other substances, is mainly found in

TABLE 1.1   Elements of the Human Body and Their Relative Abundance

PRIMARY ELEMENTS

Oxygen	65.0	Nitrogen	3.0
Carbon	18.5	Calcium	1.5
Hydrogen	10.0	Phosphorus	1.0

SECONDARY ELEMENTS

Potassium	0.30	Chlorine	0.15
Sulfur	0.25	Magnesium	0.05
Sodium	0.20	Iron	0.005

OLIGOELEMENTS

Fluorine	0.001	Zinc	Traces
Cuprum	0.0002	Cobalt	Traces
Iodine	0.00004	Molybdenum	Traces
Manganese	0.00003	Selenium	Traces

Values are expressed as a percent of total body mass.

bone; and in its ionic state, is involved in numerous physiological processes.

2. *Secondary elements.* The elements in this group comprise potassium, sulfur, sodium, chlorine, magnesium, and iron. Found in much lower relative quantities than those of the previous group, these elements exist as salts, inorganic ions, and form part of organic molecules.

Na^+ and Cl^- are the main extracellular ions, while K^+ is the main intracellular ion. Mg^{2+} is indispensable for many reactions catalyzed by enzymes. Iron is an essential component of substances of high biological importance, including hemoglobin. Sulfur is present in almost all proteins and other molecules of biological interest.

3. *Trace elements.* These elements are also known as microconstituents or oligoelements because they are present in the body in very small quantities.

Iodine is a constituent of thyroid hormone. The other oligoelements (Cu, Mn, Co, Zn, Mo, and Se), while also scarce, are vital for

the normal function of the body and most are necessary factors for the proper activity of biological catalysts (enzymes).

BIOLOGICAL COMPOUNDS

The elements mentioned earlier are found as part of different inorganic or organic compounds. Among the *inorganic* compounds, water is of exceptional importance not only because it exists in large quantity (comprising 65% of the total body weight of an adult individual), but also due to its numerous biological roles. Also of great significance are the minerals deposited in hard tissues, such as calcium phosphate, which is the nonsoluble component of bones and teeth. The remaining inorganic components dissolved in body fluids and in the cytoplasm of cells play critical roles in many body functions as ions.

Within the *organic* compounds, carbon is a key component of most of the solid substances of the body, forming part of molecules of high biological relevance, such as proteins, carbohydrates, lipids, and nucleic acids. There are other compounds, such as vitamins, hormones, and pigments, that also have essential roles. Table 1.2 provides a general view of the relative chemical composition and approximate amounts of

TABLE 1.2   Chemical Composition of Human Tissues

	Muscle	Bone	Brain	Liver
Water	75.0	22.0	77.0	70.0
Carbohydrates	1.0	Scarce	0.1	5.0
Lipids	3.0	Scarce	12.0	9.0
Proteins	18.0	30.0	8.0	15.0
Other organic substances	1.0	Scarce	1.5	1.0
Other inorganic substances	1.0	45.0	1.0	Scarce

Values are expressed as a percent of total tissue mass.

inorganic and organic compounds in some human tissues.

The chapters that follow will describe in more detail the structure and properties of the main substances that compose the human body.

SUMMARY

Carbon has been the element around which life developed. This is due to the capacity of carbon to form stable chemical bonds, long branched chains, double and triple bonds, covalent bonds with different atoms, and a variety of different spatial conformations.

Biogenic elements include a variety of substances, which can be divided in the following groups:

1. *Primary elements* include oxygen, carbon, hydrogen, nitrogen, and the less abundant, calcium and phosphorus. They comprise ~98% of the total body mass and constitute all of the essential body molecules and water.

2. *Secondary elements* comprise potassium, sulfur, sodium, chlorine, magnesium, and iron. While in much lower amounts, they exist as salts and inorganic ions, and form part of some body molecules. Na^+ and Cl^- are the main extracellular ions, while K^+ is the main intracellular ion. Mg^{2+} is indispensable for many reactions catalyzed by enzymes. Iron is an essential component of substances, such as hemoglobin. Sulfur is present in almost all proteins.

3. *Trace elements*, microconstituents, or oligoelements, are present in the body in scarce quantities. Iodine is a constituent of thyroid hormone. Cu, Mn, Co, Zn, Mo, and Se are essential for the normal function of biological catalysts (enzymes).

Biological compounds include inorganic and organic substances. Among inorganic compounds is water, the solvent present in body fluids and tissues. It comprises 65% of the total body weight of an adult individual. Nonsoluble inorganic compounds have different roles, for example, calcium phosphate is an essential component of bone. Organic biological compounds include proteins, carbohydrates, lipids, and nucleic acids. Others, such as vitamins, hormones, and pigments have essential roles and all have carbon as a key component.

Bibliography

de Duve, C., 1995. The beginnings of life on Earth. Am. Sci. 83, 428–437.

Morowitz, H.J., 2002. The Emergence of Everything (How the World Became Complex). Oxford University Press, Oxford.

Water

Water is the most abundant component of the human body; it constitutes approximately 65% of the weight of an adult individual. All cells contain and are immersed in an aqueous medium. There is no biological process that occurs independent from the direct or indirect participation of water.

Water has exceptional properties. For example, its freezing (0°C or 32°F) and boiling temperatures (100°C or 212°F), and its heat of vaporization (40.71 kJ/mol) are significantly higher than those of other compounds of similar molecular mass. These, and other particular features of water, depend on the molecular structure of this compound.

In the water molecule (H_2O), oxygen (O) is bound by simple covalent bonds to two hydrogen (H) atoms. As O is more electronegative than H, the pair of electrons shared in each of the bonds is closer to the nucleus of the O atom. This creates a partial electronegative charge in the vicinity of the O nucleus and an electropositive charge around each H, which remains almost reduced to a "naked" proton. Individually, the O—H bonds are polar (covalent polar bond).

THE WATER MOLECULE IS POLAR

If the three atoms of a water molecule were distributed linearly (H—O—H), the resulting "center of gravity" of the positive charges would be located in the center of the molecule, coinciding with the position of the negative charge. In this case, the molecule would be nonpolar, as occurs with other compounds that have polar covalent bonds, such as carbon dioxide (CO_2) or carbon tetrachloride (CCl_4). However, in water the two O—H bonds are not in a straight line, but form an angle of 104.5° (Figs. 2.1 and 2.2). The negative charge is located around the vertex of the molecule, while the resultant of the positive charges is conceivably located at the midpoint of the line joining the two hydrogen nuclei. This creates a dipole, allowing the water molecule to behave as polar, although it is electrically neutral (Fig. 2.3).

The polarity of water molecules enables them to attract each other electrostatically. The partial positive charge of one H in a molecule is pulled toward the negatively charged O of another molecule, thereby establishing a hydrogen bond (Fig. 2.4).

HYDROGEN BOND

The hydrogen bond is not exclusive to water molecules; as it will be discussed in following chapters, it is also found in other compounds of great biological importance.

The hydrogen bond is easily established between an electronegative atom (typically O or N) and a H atom covalently bonded to another electronegative atom. The binding is most stable when the three components involved, the two electronegative atoms and the intermediate H, are on the same line (Fig. 2.4).

FIGURE 2.1   **Position of atoms in the water molecule.**

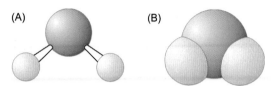

FIGURE 2.2   **Water molecule.** (A) Model of spheres and rods; and (B) a space filling spatial model, with oxygen in *red*.

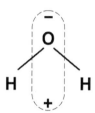

FIGURE 2.3   **Distribution of electric charges in the water molecule.**

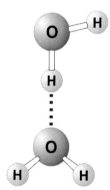

FIGURE 2.4   **Hydrogen bond between two molecules of water.**

The hydrogen bond allows the interaction among water molecules and explains the unique properties of this substance. In fact, the behavior of water does not correspond to that expected from a compound with the chemical formula H_2O, but to

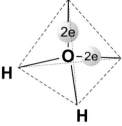

FIGURE 2.5   **Tetrahedral geometry of the water molecule.**

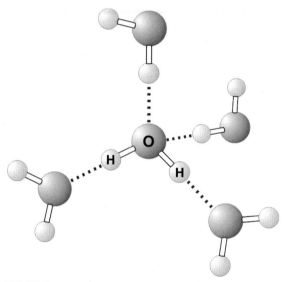

FIGURE 2.6   **Association of water molecules.** Hydrogen bonds *(red dotted line)* result from the tetrahedral arrangement of the water molecule.

a polymeric complex $(H_2O)_n$, which water adopts when it is in its solid and liquid states.

A water molecule can theoretically be conceived as a tetrahedron. In this configuration, the oxygen atom is in the center, the O—H bonds are directed toward two of the vertices of the tetrahedron, and the nonshared oxygen electrons are located in sp^3 hybrid orbitals, which are oriented toward the other two vertices of the tetrahedron (Fig. 2.5). This particular arrangement allows each water molecule to form hydrogen bonds with other four molecules of water (Fig. 2.6).

In a regular tetrahedron, the angle between shared pairs of electrons between O and H should be 109.5°. In the water molecule, the two pairs of electrons involved in covalent bonding with H are pushed away by the nonbonding electron pairs. These nonbonding electron pairs are closer to the oxygen atom and cause a slight distortion of the tetrahedron, resulting in an angle of 104.5° between the O—H bonds.

In its solid form (ice), water adopts the tetrahedral configuration, allowing it to produce a regular crystal lattice. The molecules remain at fixed distances from each other, which are determined by the length of the hydrogen bonds. The ice crystal lattice has relatively more void space than liquid water. In liquid water, the molecules not associated by hydrogen bonds have more freedom and can become closer to each other at a distance of 0.45 nm. This explains why ice is less dense than liquid water.

In liquid water, the crystal lattice order is lost. Although water molecules are associated with each other (on average, it is calculated that each water molecule is connected to other 3.4 water molecules), the hydrogen bonds that hold them together are very unstable, constantly forming and breaking apart. The molecular water clusters are therefore "fluctuating" or "oscillating." This dynamic interaction between water molecules accounts for the high heat of vaporization and high boiling temperature of water. These intermolecular attractions, which do not exist in other substances of similar molecular mass, are strong and a large amount of energy is required to break them. The hydrogen bonds among water molecules also explain the higher surface tension and viscosity of water compared to that of most other organic liquids. However, due to the continuous fluctuation of the hydrogen bonds, liquid water has more fluidity and lower viscosity than other polymeric molecules.

WATER AS SOLVENT

The polar property of water is responsible for the many interactions that it establishes with other substances. The type of interaction with other substances varies and depends on the nature of the other substance.

Ionic compounds are, in general, soluble in water. For example, NaCl crystals have electrostatic attractions between the Na^+ and Cl^- ions that maintain the highly ordered lattice of this salt. When NaCl crystals come in contact with water, the organization of the molecules of both compounds is altered. The attraction between the dipolar water molecules and the Na^+ and Cl^- ions has enough strength to dissociate the ions from the lattice they form, eventually separating and dispersing them into the solvent. The ions in aqueous solution become hydrated, or surrounded by water molecules, forming a layer or "halo" around them. Inorganic cations (Na^+, K^+, Ca^{2+}, and Mg^{2+}) and organic amine groups ($C—NH_3^+$) attract the negative charge of water molecules. Inorganic anions (Cl^-, HPO_4^{2-}, HCO_3^{3-}) and organic carboxylate ($—COO^-$) attract the positive charge of the water dipole (Fig. 2.7).

Nonionic polar compounds, such as alcohols, aldehydes, or ketones form, via their hydroxyl or carbonyl groups, hydrogen bonds with water (Fig. 2.8). This facilitates the solubility of nonionic polar compounds in water.

As ionic and nonionic polar compounds interact with water, they are hydrophilic and capable of forming stable water solutions.

Nonpolar compounds, such as hydrocarbons, are practically insoluble in water. This is due to the lack of attraction between these molecules and water. They are hydrophobic and, in general,

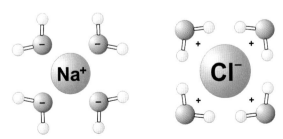

FIGURE 2.7 **Interaction of ions with water molecules.** Hydrated Na^+ is shown on the left and hydrated Cl^- on the right.

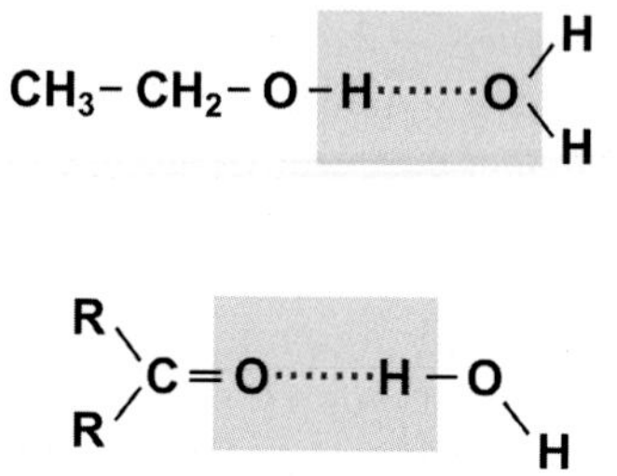

$$CH_3-CH_2-O-H\cdots\cdots O\begin{array}{c}H\\ \\H\end{array}$$

$$\begin{array}{c}R\\ \\R\end{array}C=O\cdots\cdots H-O\begin{array}{c}\\ \\H\end{array}$$

FIGURE 2.8 Hydrogen bonds (*red dotted line*) between water molecules and an alcohol (top), and a ketone (bottom).

they dissolve well in nonpolar or slightly polar organic solvents (benzene, carbon tetrachloride, and chloroform). Nonpolar molecules can establish mutual attractions through what is known as hydrophobic interactions.

Amphipathic compounds, such as phospholipids or monovalent metal salts of long chain fatty acids (sodium or potassium soaps), possess hydrophobic and hydrophilic groups in the same molecule, which makes them amphiphilic or amphipathic. Upon contact with water, they orient their hydrophilic moiety toward the water phase and their nonpolar portion away from the aqueous phase (Fig. 2.9B). When these molecules are in water, they can form spherical clusters called *micelles*. Nonpolar chains of the amphipathic compounds span toward the interior of the micelle, mutually attracted by hydrophobic interactions. In contrast, the hydrophilic ends of amphipathic molecules interact with the aqueous phase (Fig. 2.9C), preserving the stability of the micelles.

WATER AS AN ELECTROLYTE

Substances that dissociate into charged particles or ions in aqueous solution are called electrolytes, and the solutions they form allow the passage of an electric current through them.

Solutions containing equal concentrations of different electrolytes may have different capacity to conduct electrical current. Some are excellent, while others are poor conductors. According to this criterion, electrolytes can be divided into *strong* and *weak*.

For example, a 1 M solution of HCl is a much better conductor than a 1 M solution of acetic acid (CH_3COOH). The capacity to carry an electrical current depends on the availability of ions in the solution. As both the HCl and acetic acid

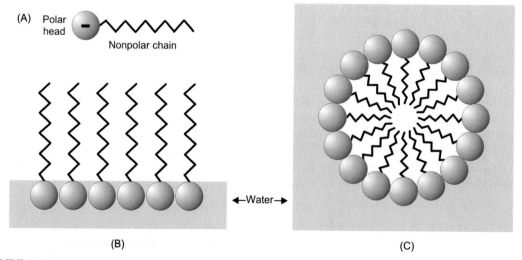

FIGURE 2.9 (A) Scheme of amphiphilic or amphipathic molecules. (B) Arrangement of amphiphilic molecules in the water–air interface. (C) Amphiphilic molecules forming a micelle in water.

1 M solutions have the same number of molecules per liter (1 mol of substance or $6.022 \cdot 10^{23}$ molecules), the higher conductivity of the HCl solution must be due to a greater number of ions in this solution than in the acetic acid solution. This shows that HCl ionizes more than acetic acid. HCl is a strong electrolyte, while acetic acid is a weak electrolyte, in which only a small part of the total molecules dissolved in water separate into ions.

EQUILIBRIUM CONSTANT

The ionization process for dissolved electrolytes can be compared to a chemical reaction. For strong electrolytes, it can be considered that almost all molecules dissociate into ions and the reaction proceeds in one direction:

$$HCl \rightarrow H^+ + Cl^-$$

For weak electrolytes, the reaction is reversible. If one considers the weak electrolyte molecule AB, this reactant will partially ionize into the products, A and B ions, as follows:

$$AB \rightleftharpoons A^+ + B^-$$

The dissociation of AB into ions continues until equilibrium is reached, in which whole molecules and ions coexist in the solution at relative constant amounts.

The concentration of ions and of entire molecules at equilibrium depends on the nature of the electrolyte, the initial concentration of the substance, and the temperature. For a system in chemical equilibrium at a given temperature, there is a constant numerical relationship between its components which is always satisfied regardless of the initial concentration of substance. This relationship is the *equilibrium constant* (K_{eq}) represented by the following equation:

$$K_{eq} = \frac{[A^+][B^-]}{[AB]}$$

$[A^+]$, $[B^-]$, and $[AB]$ represent concentrations of A^+, B^-, and AB, respectively.

The equilibrium of a weak electrolyte is dynamic, with molecules constantly ionizing and ions merging together to form the original molecules. In equilibrium, both processes occur at the same rate; therefore, the concentration of each component in the system remains unchanged.

In the ionization reaction of the weak electrolyte AB:

$$AB \underset{2}{\overset{1}{\rightleftharpoons}} A^+ + B^-$$

The velocity (V_1) of the ionization reaction (reaction 1) will be:

$$V_1 = k_1 [AB]$$

where k_1 is a constant value that is unique to each substance at a given temperature and $[AB]$ is the molal concentration of AB.

For the reverse reaction (2), the velocity V_2 at which A^+ and B^- associate to form AB is:

$$V_2 = k_2 [A^+][B^-]$$

In equilibrium, the rates of the forward and reverse reactions are equal ($V_1 = V_2$), therefore:

$$k_1[AB] = k_2[A^+][B^-]$$

then:

$$\frac{k_1}{k_2} = \frac{[A^+][B^-]}{[AB]}$$

k_1/k_2 is a ratio of two constants, which results in another constant, designated K or K_{eq}, therefore:

$$K_{eq} = \frac{[A^+][B^-]}{[AB]}$$

This constant gives an idea of the degree of ionization of an electrolyte. Weak electrolytes have a low K_{eq} value. A K_{eq} close to one indicates that the electrolyte is highly ionized. When K_{eq} is approximately 10, the electrolyte is practically completely ionized. In biological systems, an electrolyte can be considered strong when K_{eq} is greater than 0.0001 or 10^{-4}. A substance with a K_{eq} lower than 10^{-14} is not considered an electrolyte. Weak electrolytes have K_{eq} that vary between 10^{-4} and 10^{-14}.

EQUILIBRIUM OF WATER IONIZATION

Water ionizes very weakly generating hydrogen ions or protons and hydroxyl ions according to the reaction:

$$H_2O \rightleftharpoons H^+ + OH^-$$

The representation of H^+ as a free ion in water is not strictly correct, since protons rapidly react with nonionized water molecules. To depict a more accurate view of the state of H^+, many authors refer to it as hydronium ion (H_3O^+). Even more precise is the notation $[H(H_2O)_n]^+$. However, for practical purposes, we will continue using H^+ to represent the H ions, but recognizing that they do not exist independently.

The water ionization constant is expressed by the following equation:

$$K_{eq} = \frac{[H^+][OH^-]}{[H_2O]}$$

The K_{eq} value of pure water has been determined by measuring its electrical conductivity. By knowing the conductivity of the ions, it is possible to calculate the concentration of the free ions present in water. In pure water at 25°C, the $[H^+]$ is 0.0000001 or 10^{-7} M and the $[OH^-]$ is the same, since water ionization generates equal quantities of OH^- and H^+.

The K_{eq} value for pure water is very low and varies with temperature. For example, at 25°C, water K_{eq} is $1.8 \cdot 10^{-16}$ M, while at 37°C it is $4.3 \cdot 10^{-16}$ M. According to these figures, water should not be considered an electrolyte. Why so much attention is placed on the H^+ generated by

water ionization when its K_{eq} is so small? Changes in $[H^+]$ often have important biological effects. Hydrogen ions are very small (they consist of a proton) and have, in comparison to other ions, a large charge density, which creates a significant electrical gradient around them. This affects the hydrogen bonds that help maintain the structure and conformation of biologically important macromolecules. The ionization state of functional groups, critical for the function of certain compounds dissolved in water, is also affected.

Since the molecular mass of water is 18 Da (1 mol of water = 18 g), the molal concentration of water molecules in pure water is approximately 55.55 (in 1000 g of water there are $1000/18 = 55.55$ moles). Only 0.0000001 moles of the total 55.55 moles are ionized, which means that only 1 molecule every 555,500,000 is ionized. Then, the water ionization constant can be represented by the following equation:

$$K[H_2O] = [H^+][OH^-]$$

As the amount of ionized molecules is negligible compared to the total number of molecules, it can be considered, without making any appreciable error, that the ionization process does not modify the concentration of nonionized molecules $[H_2O]$. Therefore, the first term of the equation is the product of two constants (K and $[H_2O]$), which is a new constant, designated K_w:

$$K[H_2O] = K_w \quad \therefore \quad K_w = [H^+][OH^-]$$

The K_w constant is designated *ion product* for water; its value at 25°C is:

$$K_w = [H^+][OH^-] = 0.0000001 \times 0.0000001$$
$$= 10^{-7} \times 10^{-7} = 10^{-14}$$

In pure water, $[H^+] = [OH^-]$. So, $[OH^-]$ can be replaced by $[H^+]$ in the equation, or vice versa:

$$K_w = [H^+][H^+] \quad \therefore \quad [H^+] = \sqrt{K_w}$$
$$\text{or} \quad K_w = [OH^-][OH^-] \quad \therefore \quad [OH^-] = \sqrt{K_w}$$

K_w is a constant value in pure water and in any aqueous solution. Any increase in the concentration of one of the two ions of water will cause an immediate decrease in the concentration of the other ion, shifting the equilibrium toward the formation of whole water molecules; the ion product remains unchanged. If an electrolyte ionizes and provides hydrogen ions when dissolved in water, as seen with HCl (HCl $\rightarrow$ H$^+$ + Cl$^-$), the H$^+$ concentration will increase compared to that of pure water. The value of the ionic product will tend to be restored by increasing the rate of H$^+$ and OH$^-$ association to form whole water molecules. Consequently, the concentration of OH$^-$ ions decreases until equilibrium is reached when the value of K_w is 10^{-14} (at 25°C). Thus, if the HCl added increases the H$^+$ concentration to 10^{-1} M (1 million times higher than that of pure water at 25°C), the OH$^-$ must be reduced to a value of 10^{-13} (1 million times lower than that of pure water). Therefore, the ion product will remain constant:

$$K_w = 10^{-1} \cdot 10^{-13} = 10^{-14}$$

An analogous adjustment in the concentration of H$^+$ ions occurs if the electrolyte that is dissolved in water increases the concentration of OH$^-$ (NaOH $\rightarrow$ Na$^+$ + OH$^-$). The increase in [OH$^-$] will produce a decrease in the concentration of H$^+$ by forming whole water molecules (H$_2$O) to maintain the K_w value.

ACIDS AND BASES

A solution is neutral when the hydrogen ion concentration is equal to the hydroxide ion concentration. Pure water is neutral because [H$^+$] = [OH$^-$]. At 25°C, pure water has $1 \cdot 10^{-7}$ M [H$^+$] and $1 \cdot 10^{-7}$ M [OH$^-$], but this markedly changes with temperature. For example, at 0°C [H$^+$] or [OH$^-$] = $3.4 \cdot 10^{-8}$ and at 100°C, $8.8 \cdot 10^{-7}$.

When the concentration of hydrogen ions in a solution is greater than that of hydroxide ions, it is considered *acidic*. On the other hand, when the hydrogen ion concentration is lower than that of hydroxide ions it is deemed *basic* or *alkaline*.

These concepts lead to some practical definitions regarding acids and bases. Acids are substances that increase [H$^+$] when dissolved in water or aqueous solutions. Conversely, bases are substances which decrease [H$^+$] when dissolved in water or aqueous solutions.

According to the Brønsted and Lowry acid–base theory, acids are compounds or ions with the ability to release protons (H$^+$) into the medium, while bases are those which accept protons from the medium.

Here, the definition of acids is extended to ions, such as HSO$_4^-$ or NH$_4^+$, which can transfer one H$^+$ ion to the solution. Usually the hydroxides of alkaline metals are considered typical bases (i.e., NaOH), and strictly, according to the Brønsted–Lowry acid–base theory, the base is the OH$^-$ that those compounds release into the solution. When an OH$^-$ accepts H$^+$ from the solution, it will result in the formation of water molecules. All ions that can capture H$^+$, such as Cl$^-$ or CO$_3^{2-}$, are considered bases.

When a molecule or anion accepts H$^+$ (also known as a Brønsted–Lowry base), a "conjugated acid" of that molecule is formed.

Base		Proton		Conjugated acid
OH$^-$	+	H$^+$	$\rightarrow$	H$_2$O
NH$_3$	+	H$^+$	$\rightarrow$	NH$_4^+$
CO$_3^{2-}$	+	H$^+$	$\rightarrow$	HCO$_3^-$

When an acid loses H$^+$, its "conjugated base" is formed.

Acid		Proton		Conjugated acid
HCl	+	H$^+$	$\rightarrow$	Cl
H$_2$SO$_4$	+	H$^+$	$\rightarrow$	HSO$_4^-$
HNO$_3$	+	H$^+$	$\rightarrow$	NO$_3^-$

ACIDS AND BASES STRENGTH

The strength of an acid or a base is determined by its tendency to lose or gain protons. Just like electrolytes, acids can also be divided into strong (HCl, H_2SO_4, HNO_3) and weak ($H_2PO_4^-$, CH_3—COOH, H_2CO_3) acids. When dissolved in water, strong acids ionize almost completely, while weak acids only ionize in a small proportion. Consequently, the hydrogen concentration for the same acid concentration will be greater in the solution of a strong than a weak acid.

The ability of an acid to lose protons (acid strength) is expressed by its ionization constant (K_a). If we consider an acid HA:

$$K_a = \frac{[H^+][A^-]}{[HA]}$$

The larger the K_a, the easier the acid releases protons. In general, the K_a of strong acids reaches values that are very large (10^2 to 10^{10}) and for practical purposes it is not taken into account. Strong acids are considered to be completely ionized in diluted aqueous solutions. Weak acids are partially ionized and their K_a constants are very small.

Bases can also be divided into strong [NaOH, KOH, $Ca(OH)_2$, etc.] and weak (NH_3, trimethylamine, aniline, etc.). Strong bases ionize completely in a solution. Similar to weak acids, the ionization constants of weak bases (K_b) reflect their degree of ionization.

A useful generalization regarding the relative strength of the acid–base pair is: if an acid is strong, its conjugate base is weak and if a substance is a strong base, its conjugate acid is weak.

pH

As the ion product of water $[H^+] \cdot [OH^-]$ has a constant magnitude at a given temperature, it is sufficient to know the concentration of one of the ions to deduce the concentration of the other. Usually, the concentration of hydrogen ion is used. However, the small amount of free $[H^+]$ makes its use cumbersome; simpler ways to express $[H^+]$ have been proposed. In 1909 the Danish biochemist P.L. Sørensen proposed pH as the notation for the $[H^+]$ of a solution. The term pH received wide acceptance and is today the most commonly used expression to indicate the concentration of hydrogen ions.

To obtain the pH value of a solution, a double transformation is required. This includes taking the reciprocal of the $[H^+]$ and then, the logarithm of that value. In this manner, pH can be defined as the logarithm of the reciprocal of the hydrogen ion concentration or, in other words, the negative logarithm of the H^+ concentration.

$$pH = \log\frac{1}{[H^+]} = -\log[H^+]$$

In the case of pure water at 25°C $[H^+] = [OH^-] = 10^{-7}$ M, therefore:

$$pH = \log\frac{1}{[H^+]} = \log\frac{1}{10^{-7}} = \log 10^7 = 7$$

The same notation can be used for other parameters, including $[OH^-]$, K_w, and K_a, which will render pOH, pK_w, and pK_a, respectively.

$$pOH = \log\frac{1}{[OH^-]} \qquad pK_a = \log\frac{1}{K_a}$$

There are several points related to the pH notation to be taken into account:

1. No direct relationship exists between the magnitudes of $[H^+]$ and pH. When the value of $[H^+]$ increases, the pH decreases and vice versa. Moreover, since the pH scale is logarithmic (not arithmetic, such as $[H^+]$), any change in pH indicates a 10-fold change in $[H^+]$. For example, a change in 2 pH units indicates a 100-fold change in $[H^+]$, a pH change of 3 shows a 1000-fold change in $[H^+]$, and so on (Fig. 2.10).

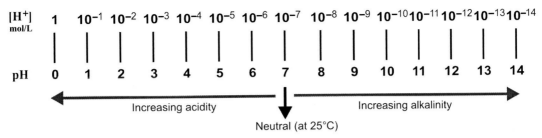

FIGURE 2.10 **Correlation between [H⁺] and pH values.**

2. A pH of 7 indicates neutrality only at 25°C. As the [H⁺] changes with temperature, so does the value of pH. Pure water (neutral) has a pH 7.5 at 0°C and 6.1 at 100°C. At 37°C, the human body temperature, neutral pH is 6.8.

3. Usually, the pH of a solution varies between 0 and 14. These limits of pH cover the range of [H⁺] from a 1 M strong acid solution (pH = 0) to a 1 M strong base solution (pH = 14). However, in theory, the pH scale can extend beyond these limits. The lowest value of [H⁺] that can be obtained in aqueous solutions is 10^{-15} M and the highest is 15 M, corresponding to pHs of 15 and -1.2, respectively.

BUFFERS

Buffers are systems that reduce the changes in hydrogen ion concentration in a solution when acid or alkaline electrolytes are added. In other words, buffers can minimize the pH deviations produced by acids or bases in a given medium.

In general, a buffer solution consists of a mixture of a weak electrolyte (acid or base) and the corresponding salt of that acid or base, which acts as a strong electrolyte. Examples of buffer systems include:

Carbonic acid–sodium bicarbonate ($H_2CO_3/NaHCO_3$)

Acetic acid–sodium acetate (CH_3COOH/CH_3COONa)

Monosodium phosphate–disodium phosphate (NaH_2PO_4/Na_2HPO_4)

Ammonia–ammonium chloride (NH_3/NH_4Cl)

Mechanism of buffers action. Carbonic acid in solution ionizes into bicarbonate (HCO_3^-) and hydrogen (H^+) ions according to the following reaction:

$$H_2CO_3 \rightleftharpoons HCO_3^- + H^-$$

The acid ionization constant (K_a) is given by the relationship:

$$K_a = \frac{[HCO_3^-][H^+]}{[H_2CO_3]}$$

Being a weak electrolyte, carbonic acid has a low capacity to ionize in water; it produces a very small number of ions compared to the total number of whole carbonic acid molecules. When a salt of the same acid is added to this solution, a buffer system is formed. The sodium salt is a strong electrolyte that completely ionizes into bicarbonate and sodium:

$$NaHCO_3 \rightarrow Na^+ + HCO_3^-$$

Both components of the system release bicarbonate ion when they ionize. As HCO_3^- is added to this system, the K_a value of the acid must be maintained. A shift in carbonic acid toward the formation of nonionized $NaHCO_3$ molecules takes place, decreasing the ionization of H_2CO_3 to the point that it can be considered practically in all the nonionized state. A new equilibrium is achieved that is characterized by a high concentration of bicarbonate ions and acid molecules, and by a low concentration of hydrogen ions. When a strong acid (HCl) is added to this solution, the increase in hydrogen ions shifts the equilibrium to the formation of whole carbonic acid molecules, and its concentration increases. Concomitantly, the amount of bicarbonate ions decreases.

$$HCl + HCO_3^- \rightarrow H_2CO_3 + Cl^-$$

In other words, a large portion of the hydrogen ions from the hydrochloric acid is taken up by the bicarbonate ion generating nonionized carbonic acid. Therefore, the increase in [H⁺] in the solution is very low and the pH remains almost unchanged.

If a base (e.g., NaOH) is added to a buffered system, the OH⁻ ions bind to the hydrogen ions in the solution to produce water. This causes a shift in the equation to the right, favoring carbonic acid ionization, generating new hydrogen ions that combine with OH⁻:

$$NaOH + H_2CO_3 \rightarrow HCO_3^- + Na^+ + H_2O$$

Consequently, the increase of OH⁻ concentration in the solution is small and the pH is altered only slightly.

pH of buffer solutions. In a system consisting of a weak acid and its salt, the pH can be calculated from the ionization constant of the acid and the initial concentration of the salt.

Let's consider a buffer consisting of a weak acid (HA) and its salt (NaA). The weak acid dissociates according to the equation:

$$HA \rightarrow A^- + H^+$$

The K_a of the system will be:

$$K_a = \frac{[A^-][H^+]}{[HA]}$$

The salt acts as a strong electrolyte and dissociates according to the equation:

$$NaA \rightarrow A^- + Na^+$$

The salt is completely dissociated in the solution and the concentration of anions (A⁻) and cations (Na⁺) is equal to the initial concentration of the salt. As the anion A⁻ is common to both acid and salt, its high concentration in the solution (from total ionization of the salt) causes a displacement of the acid ionization toward formation of nondissociated molecules. Virtually all of the acid in the solution is nonionized. Hence, the concentration of acid molecules can be considered equal to the initial concentration of the acid.

In the acid ionization equilibrium equation, [A⁻] can be replaced by the initial concentration of salt and the concentration of HA by the initial concentration of the acid:

$$K_a = \frac{[salt][H^+]}{[acid]}$$

Then, the [H⁺] is:

$$[H^+] = \frac{[acid]}{[salt]} \times K_a$$

The logarithm of the reciprocal of [H⁺] is:

$$\log \frac{1}{[H^+]} = \log \frac{[salt]}{[acid]} + \log \frac{1}{K_a}$$

As $\log 1/[H^+] = pH$ and $\log 1/K_a = pK_a$:

$$pH = pK_a + \log \frac{[salt]}{[acid]}$$

The pH of a buffer system is equal to the pK of the acid (pK_a), plus the logarithm of the ratio between the initial concentration of the salt and the initial concentration of acid. This relationship is known as the *Henderson–Hasselbalch equation.*

The buffering capacity of a buffer system varies depending on the relative concentrations of the acid and the salt within the system. It is maximal when the concentration of the acid is equal to that of the salt. In this case, the ratio [salt]/[acid] is equal to 1. Log1 is equal to zero, therefore, according to the Henderson–Hasselbalch equation, the pH equals the pK_a.

This can be conveyed as follows: the capacity of a buffer system to minimize the pH change produced in a medium by the addition of acids or bases is maximal when the pH of the buffer is equal to the pK_a of the acid.

TITRATION CURVE OF ACIDS AND BASES

Acid–base titration is a procedure used to determine the concentration or "titer" of an acid (or a base) solution, by adding a known and equivalent amount of a base (or acid) to the solution. During the titration process, a neutralization reaction is produced. When the same number of equivalents of acid and base combine, a point of equivalence is reached.

As the reaction proceeds, once the point of equivalence is surpassed, the medium shows progressive changes in pH. Any additional small amount of acid or base added to the solution causes a marked pH change. The pH values at each step of the titration are determined by potentiometric methods.

The titration process can be followed by using what is known as neutralization or titration curves. These represent the relationship between the changes in pH and the volume of the acid or base of known concentration added to the solution. The pH values are shown on the x axis and the volume of added acid or base on the y axis (Fig. 2.11). During the titration of a weak acid or base by a strong base or acid, a buffer system is formed, in which the concentration of its components changes as the titration proceeds. The pH changes that occur at different stages of neutralization provide data of interest related to the buffering capacity of the system.

As an example of the titration curve of a weak acid, we will analyze the neutralization of 10 mL of 0.1 N acetic acid with a 0.1 N NaOH solution.

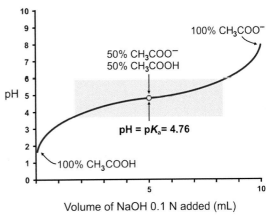

FIGURE 2.11 **Titration curve of acetic acid.** The *pink area* indicates the buffer zone.

Acetic acid is weakly ionized, giving H^+ and acetate ions (CH_3COO^-)

$$CH_3COOH \rightarrow H^+ + CH_3COO^-$$

The ionization constant of the acid can be represented by the relationship:

$$K_a = \frac{[H^+][CH_3-COO^-]}{[CH_3-COOH]}$$

The K_a of acetic acid at 25°C is $1.74 \cdot 10^{-5}$ M and the pK_a is 4.76.

Two equilibrium conditions, the ionization of acetic acid and the solvent (water, $K_w = [H^+][OH^-]$) are maintained in the system throughout the titration. For simplicity, only the acid ionization will be considered.

Before NaOH is added, the pH of the medium is equal to the pH of the 0.1 N acetic acid solution. While adding NaOH, the following reaction occurs:

$$CH_3COOH + NaOH \rightarrow CH_3COO^- + Na^+ + H_2O$$

The newly added OH^- combines with H^+ to form H_2O molecules. The decrease in $[H^+]$ is compensated by the ionization of the acid to form acetate (salt). The value of K_a is maintained constant. The Henderson–Hasselbalch equation can be used to calculate the pH from the buffer system formed by the nonneutralized weak acid and the acetate. When 50% neutralization of the starting acid is reached, equal amounts of acid and acetate (salt) are present in the medium. At this point, the pH value will be equal to the pK_a (4.76) (Fig. 2.11). If the titration is continued, the remaining acid is gradually converted to acetate. Eventually, the solution will become completely neutralized. This occurs when the number of base equivalents is equal to that of the acid. The curve will show a sharp upward inflection, showing that the equivalence point has been reached. The pH at this point does not correspond to neutrality, but is rather shifted to the alkaline side (8.72). This is due to hydrolysis of the acetate, which results in increased $[OH^-]$:

$$CH_3COO^- + H_2O \rightarrow CH_3COOH + OH^-$$

It is interesting to note that the curve flattens on either side of the titration midpoint (Fig. 2.11). Consecutive addition of base produces relatively less variation in pH in an area covering approximately one pH unit around the midpoint compared to other sections of the curve. This indicates that the buffering capacity of a buffer system is higher in the area near the midpoint and is dependent on the value of [salt]/[acid]. The [salt]/[acid] ratio where the buffer system is effective ranges from 1/10 to 10/1 and the buffer capacity is maximal when the ratio is equal to 1 (pH = pK_a). If one applies these values to the Henderson–Hasselbalch equation:

for [salt]/[acid] = 1/10:

$$pH = pK_a + \log\frac{1}{10} = pK_a + \log 10^{-1} = pK_a - 1$$

for [salt]/[acid] = 1:

$$pH = pK_a + \log 1 = pK_a + 0 = pK_a$$

for [salt]/[acid] = 10/1:

$$pH = pK_a + \log\frac{10}{1} = pK_a + \log 10^1 = pK_a + 1$$

This indicates that a buffer has greater buffering capacity in the pH range between ±1 + pK_a of the system.

APPENDIX

Expression of Concentrations

Concentration is the ratio between the amount of solute and solvent in a solution. Different types of expression are used to express the concentration of a solute.

Percent concentration. For the components of biological fluids it is common to indicate the amount of solute (weight in g, mg, µg, ng) dissolved in 100 parts of solvent (by volume, 100 mL or the equivalent, 10 dL, 1 dL = 100 mL).

Molar concentrations. A mol refers to the quantity of any element equal to the atomic mass of that element in grams (gram atom), which contains $6.022 \cdot 10^{23}$ atoms (Avogadro's number). For example, there is the same number of atoms in 1 g of ^1H than in 12 g of ^{12}C: $6.022 \cdot 10^{23}$. For any compound a mol is equal to its molecular weight in grams (gram molecule), which has $6.022 \cdot 10^{23}$ molecules. For example, 180 g of glucose ($C_6H_{12}O_6$), 60 g of urea $[CO(NH_2)_2]$, and 142 g of Na_2HPO_4 contain $6.022 \cdot 10^{23}$ molecules.

One *mol* can be defined as the quantity of matter containing one Avogadro's number of particles (electrons, ions, or molecules). Often a 1000-fold smaller unit, the millimole (mmol), is used. In certain cases it is necessary to use even smaller units, such as the micromole (µmol) equal to 10^{-6} mol, or the nanomole (nmol) equal to 10^{-9} mol of substance.

Molarity is the number of moles of solute existing in 1 L of solution and is indicated by the notation M (capital m). A 1 M solution contains 1 mol of solute dissolved in 1 L; a 0.5 M solution, 0.5 mol/L; and a 3 M, 3 mol/L.

Given a 0.2 M $CaCl_2$ solution, its concentration will be 0.2 mol of calcium chloride per liter of solution. Since this salt ionizes by the equation:

$$CaCl_2 \rightarrow Ca^{2+} + 2Cl^-$$

calcium ion concentration in the solution is 0.2 M because an equal number of calcium ions as that of molecules originally present in the solution (assuming full dissociation) are formed. Instead, each mol of $CaCl_2$ gives two chloride ions, so that the Cl^- concentration is 0.4 M.

Molarity corresponds to the ratio:

$$M = \frac{Moles}{Volume(L)}$$

The amount of moles in a given mass of substance is estimated by the relationship:

$$M = \frac{Mass(g)}{Molecular\ weight\,(g\,/\,mol)}$$

Given the low concentrations of certain ions or substances in biological fluids, it is sometimes convenient to express them in millimoles per liter (millimolar concentration or mM) or in micromoles per liter (micromolar concentration or μM).

To calculate the molarity of a solution when its concentration is known in grams per liter, the following formula is applied:

$$Molarity\,(mol/L) = \frac{Concentration\,(g/L)}{Molecular\ weight\,(g/mol)}$$

Generally the concentration of many substances in organic liquids is given in mg per 100 mL or dL. It is preferable to express the molarity in millimoles per liter (mM concentration); it is calculated from the concentration in mg per dL using the formula:

$$Molarity\,(mM\,/\,L) = \frac{Concentration\,(mg\,/\,dL) \times 10}{Molecular\ weight\,(g\,/\,mol)}$$

For example, the plasma calcium concentration is 10 mg/dL or 100 mL (Ca atomic mass 40). Its molarity in millimoles per liter (mM) will be:

$$Molarity\,(mM) = \frac{10 \times 10}{40} = 2.5\,mM$$

Molal concentrations. Molality refers to the moles of solute per 1000 g of solvent. This is indicated by the symbol m (lowercase m). In this notation, the concentration is not influenced by temperature and can be used to calculate boiling or freezing temperatures of solutions. In chemical practice; however, molar solutions are frequently used because it is common to use volume units. In this last case, it is important to mention the temperature at which the molar solution was prepared.

Equivalents and their use for concentrations. It is usual to express concentrations in terms of chemical equivalents (Eq.). One equivalent gram or equivalent weight is the mass in grams of an element, ion, or compound that can displace or combine with 1 g of hydrogen or 8 g of oxygen. In redox reactions, 1 Eq. is the amount of substance that gains or losses 1 mol of electrons. In acid–base reactions, 1 Eq. is the amount of acid or base that liberates or accepts 1 mol of protons.

The equivalent weight is actually a reactive unit. Expressing the concentration in Eq., it is possible to compare the number of chemical units that can combine. One Eq. of an oxidizing agent reacts exactly with 1 Eq. of a reducer. One Eq. of acid is precisely neutralized by 1 Eq. of base. One Eq. of Na exactly combines with 1 Eq. of Cl or bicarbonate.

The equivalent weight of an element is calculated by dividing its atomic weight by the number of electrons (*n*) that an atom gains or losses when the element reacts. For acids or bases, the mass of 1 mol is divided by the number of protons (*n*) that each molecule of acid or base yields or accepts. In the case of an ion, the mole weight is divided by the number of electric charges (*n*) it contains. Thus, one equivalent of the following substances will be: sodium, 23/1 = 23 g; chloride, 35.5/1 = 35.5 g; calcium, 40/2 = 20 g;

bicarbonate (HCO_3^-), $61/1 = 61$ g; and phosphate (PO_4^{3-}), $95/3 = 31.6$ g.

As the concentrations of electrolytes in body fluids are low, it is customary to express them in milliequivalents (mEq.) per liter. A mEq. is one thousandth of an equivalent.

Frequently, the concentration of a specific ion in biological fluids is given in mg per dL and this expression must be converted into milliequivalents per liter. The conversion formula used in this case is:

$$(mg / dL) \times 10 \times \frac{n}{Atomic\ mass} = mEq. / L$$

Examples: sodium concentration in plasma is 322 mg/dL. Expressed in mEq./L, the concentration will be:

$$322 \times 10 \times \frac{1}{23} = 140\,mEq. / L$$

The plasma calcium concentration is 10 mg/dL. Expressed in mEq./L as:

$$10 \times 10 \times \frac{2}{40} = 5\,mEq. / L$$

SUMMARY

Water is the most abundant component of the human body. Approximately 65% of the weight of an adult human is water. The function of all cells and tissues depends on it.

Water has unique properties. The particular molecular structure of water provides it with physical properties, such as melting and boiling temperatures, and heat of vaporization that are higher than those of other substances of similar weight.

Molecular structure of water. The elements in water (H—O—H) are arranged in a 104.5° angle, which gives the molecule its polar nature. The negative charge is displaced toward the O atom and the positive charges to the H atoms.

Hydrogen bond. The charge distribution within the water molecule allows the formation of hydrogen bonds between different water molecules (the positive charge of an H in a water molecule is attracted by the negative charge of another). In this manner, water can form polymeric complexes with the formula (H_2O)$_n$. These complexes are more common in solid (ice) or liquid water, than in water vapor. Each water molecule can form hydrogen bonds with another four molecules. In ice, a regular crystal lattice is formed, which has fixed distances between molecules. In liquid water, the H bonds easily form and break apart, giving the molecules the freedom to move closer; this explains why liquid water is denser than ice.

Polarity of water molecules influences its interaction with other substances. The electrostatic attractions of ionic and polar nonionic compounds favor their interactions with water. They are hydrophilic substances that can form stable water solutions. Nonpolar compounds are hydrophobic and do not dissolve in water. Amphipathic substances which exhibit hydrophilic and hydrophobic groups in the same molecule (phospholipids and soaps) can form micelles in water. These molecules have their polar groups oriented toward the polar aqueous medium and the hydrophobic groups toward the interior of the micelle, where they are mutually attracted by hydrophobic interactions.

Water is an electrolyte. Water weakly dissociates into hydrogen and hydroxide ions. In pure water at 25°C, the hydrogen ion concentration [H^+] and the hydroxyl ion concentration [OH^-] are the same, 0.0000001 M or 10^{-7} M.

Water ion product (K_w). The product [H^+] $\cdot$ [OH^-] is a constant. Its value at 25°C is: $10^{-7} \cdot 10^{-7} = 10^{-14}$. This remains constant in both pure water and aqueous solutions. For this reason, when a substance that increases [H^+] or [OH^-] is added to aqueous solutions, a concomitant decrease of [OH^-] or [H^+] immediately occurs, trying to maintain the value of the product [H^+] $\cdot$ [OH^-] constant at 10^{-14}.

Acids and bases. When [H^+] is equal to [OH^-], the solution is considered neutral. Any substance dissolved in water, which increases [H^+] is an acid; and any that decreases [H^+], is a base or alkali. According to Brønsted and Lowry, acids are compounds that yield H^+ ions or protons to the solution and bases are the compounds that can accept protons from the medium. According to their degree of ionization, acids and bases can be strong or weak.

pH. To simplify the expression of [H^+] in a solution, the pH notation is used. The pH corresponds to the logarithm of the reciprocal of the [H^+] or the negative logarithm of [H^+]. For pure water at 25°C, [H^+] = 10^{-7} M, the pH = $-\log 10^{-7} = 7$. At this temperature, acids have a pH below 7 and alkalis or bases, above 7.

Buffers. These are systems that minimize the changes in [H^+] produced by addition of acids or bases to a solution. A buffer solution is generally constituted by a mixture of a weak electrolyte (commonly a weak acid) and its salt, which functions as a strong electrolyte. The pH of the buffer solutions can be calculated by knowing the acid dissociation constant (K_a) and the initial concentrations of the acid and the salt. The relationship between these values is expressed in the Henderson–Hasselbalch equation:

$$pH = pK_a + \log\frac{[salt]}{[acid]}$$

Buffers display their highest buffering capacity at a pH range between $\pm 1 + pK_a$.

Titration curve of weak acids. This is obtained by plotting the changes in pH produced by neutralization of the acid with a strong base against the volume of the base used. The concentrations of the components of the buffer system formed during the course of the titration changes as the base is added. The buffering capacity of the system is maximal when the pH of the solution equals to the pK_a.

Bibliography

Chaplin, M., 2006. Do we underestimate the role of water in cell biology? Nat. Rev. Mol. Cell Biol. 7, 861–866.

Nicolls, P., 2000. Introduction to the biology of the water molecule. Cell. Mol. Life Sci. 57, 987–992.

Paladini, A.C., 1983. Agua. In: Torres, H.N., Carminatti, H., Cardini, C.E. (Eds.), Bioquímica General. El Ateneo, Buenos Aires, pp. 81–90.

Stewart, P.A., 1981. How to Understand Acid–Base: A Quantitative Acid–Base Primer for Biology and Medicine. Edward Arnold Ltd., London.

Proteins

Proteins are molecules of immense importance for living organisms. They are the most abundant organic compounds in vertebrates, accounting for approximately 50% of their tissue dry weight. Virtually all biological processes depend on the function of proteins. The following are only a few of the many examples of different protein types and the variety of roles that they perform: enzymes catalyze most chemical reactions in the body; hormones regulate many cellular activities; hemoglobin and transport proteins carry a series of substances in blood; antibodies defend the body against the attack of foreign agents; receptors trigger specific responses in cells, actin and myosin allow muscle contraction; and collagen forms the highly resistant fibers of connective tissue.

Major advances have been made in understanding the structure and function of proteins over the last 60 years. This has helped to explain the molecular basis underlying many biological processes. One of the most challenging problems that investigators were faced with in the area of protein research has been the isolation and purification of proteins from the extremely complex mixture of molecules that composes living matter. The development of methodologies and research tools for the separation of proteins in a pure form, and their crystallization, allowed understanding the structure and function of many proteins.

All proteins contain carbon, hydrogen, oxygen, and nitrogen and most also have sulfur. Although there are slight variations, the nitrogen content in proteins is approximately 16% of their total molecular mass. Thus, 6.25 g of protein contains 1 g of nitrogen. The value 6.25 is used as a factor to estimate the amount of protein in a sample with known N content.

PROTEINS ARE MACROMOLECULES FORMED BY AMINO ACIDS

Proteins are large size molecules (macromolecules), polymers of structural units called *amino acids.* A total of 20 different amino acids exist in proteins and hundreds to thousands of these amino acids are attached to each other in long chains to form a protein. Amino acids can be released from proteins by hydrolysis. (Hydrolysis is the cleavage of a covalent bond by addition of water in adequate conditions.)

Due to their large size, proteins obligatorily form colloids when they are dispersed in a suitable solvent. This property characteristically distinguishes proteins from solutions containing small size molecules.

Since amino acids are the "building blocks" for proteins, their structure and properties will be considered first.

AMINO ACIDS

Amino acids are compounds that have an acid or carboxyl (—COOH), and a basic or amine (—NH$_2$) groups, both of which are attached to a α carbon (αC) of an organic acid (the α carbon is that located next to the carboxyl group). This is why amino acids are called α-amino acids, and their general formula is

$$H_2N - \underset{\underset{COOH}{|}}{\overset{\overset{R}{|}}{C}} - H$$

R corresponds to the side chain, which varies for each of the 20 amino acids obtained by protein hydrolysis.

According to the general formula, the αC of all amino acids (except glycine, with H as side chain) is bound to four different functional groups. This allows for the existence of two optical isomers with different spatial arrangement for each amino acid.

Optical Isomerism

Isomers refer to compounds that have the same molecular formula but are structurally different. When isomers only differ in the spatial arrangement of their atoms, they constitute spatial isomers, or *stereoisomers*, a group that also includes the optical isomers. According to the tetrahedral configuration, the four bonds of the carbon atom are equivalent and oriented to the vertices of the tetrahedron. When each carbon valence is bound to different elements or atomic groups, the molecule becomes asymmetric and the carbon involved is known as *asymmetric carbon*.

For example, in the amino acid alanine:

$$H_2N - \underset{\underset{COOH}{|}}{\overset{\overset{CH_3}{|}}{C}} - H$$

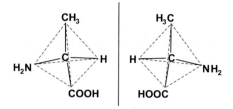

FIGURE 3.1   **Alanine enantiomers.**

The asymmetric carbon, in red, is bound to four different atomic groups (—H, —CH$_3$, —NH$_2$, and —COOH).

The groups attached to the central asymmetric carbon can be spatially arranged in two different ways. This results in two molecules, each of which is the mirror image of the other. As their relationship is analogous to that of the left and right hands, these compounds are called *chiral compounds* (from the Greek *chiros*: hand). They are also known as *optical isomers, enantiomorphs,* or *enantiomers*. Fig. 3.1 shows the two optical isomers of alanine.

Many of the chemical and physical properties of enantiomers are identical, except for the ability of these compounds to deflect polarized light.

Polarized light. A beam of ordinary light consists of waves vibrating in all planes intersecting the axis of beam propagation (Fig. 3.2). In contrast, polarized light consists of waves that vibrate in a single plane. Polarized light can be

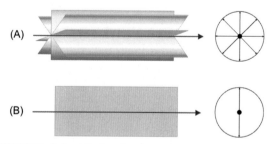

FIGURE 3.2   (A) Graphical representation of an ordinary light beam. The circle on the right represents the ideal cross section of the beam. Only some of the infinite planes of vibration that intersect at the axis of propagation of the beam are represented. (B) Polarized light; the wave vibrations occupy only one plane.

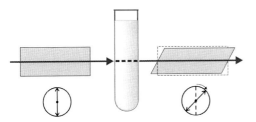

FIGURE 3.3 **Rotation of polarized light.** The plane of light vibration, after passing through a tube containing a solution of an optically active substance is deviated from its original position (in this case, a clockwise deviation is shown).

obtained by passing a beam of ordinary light through a Nicol prism (device constructed with calcite crystals), or through the synthetic material called polaroid.

Optical activity. If a beam of polarized light passes through a solution of a chiral compound, the plane of light vibration is rotated on its axis (Fig. 3.3). It is said that the substance is *optically active*. By convention, when the rotation is in the clockwise direction, it is considered to be right or positive (+); when the rotation takes place in the counter-clockwise direction, it is considered to be left or negative (−).

Compounds that deflect the plane of vibration of polarized light to the right are called *dextrorotatory* and those that rotate it to the left, are known as *levorotatory*.

The degree of rotation can be estimated by an instrument called polarimeter, which measures the angle of rotation of the polarized light plane when it passes through a solution of an optically active substance.

The angle of rotation of the polarized light plane depends on several conditions, including temperature, light wavelength, concentration, and thickness of the solution that is traversed by light. Moreover, if the conditions mentioned earlier are maintained constant, an optically active compound rotates the polarized light to a degree that is unique for that substance (*specific rotation*). Optical isomers of a same compound deviate the polarized light in the same manner;

however, while one turns polarized light to the right, the other turns it to the left.

Notation. Two optical isomers can be distinguished by the configuration of the four substituents around the chiral or asymmetric carbon, which can be determined by the use of X-ray diffraction. The compound that is taken as a reference is glyceraldehyde. Both of the optical isomers of this substance, one levorotatory and the other dextrorotatory, are designated L and D, respectively. Their chemical formulas are the following:

$$
\begin{array}{cc}
\begin{array}{c}
H \\
\diagup \\
C=O \\
| \\
HO-C-H \\
| \\
CH_2OH
\end{array}
&
\begin{array}{c}
H \\
\diagup \\
C=O \\
| \\
H-C-OH \\
| \\
CH_2OH
\end{array}
\\
\text{L-Glyceraldehyde} & \text{D-Glyceraldehyde}
\end{array}
$$

The L-isomer is represented by placing the hydroxyl group to the left of the asymmetric carbon and the D-isomer, with the carboxyl group to the right.

By convention, all compounds that have similar configuration to that of L-glyceraldehyde are called L, even if they are not levorotatory. Similarly, compounds that are analogous to the spatial arrangement of D-glyceraldehyde are called D, although they may not be dextrorotatory. Thus, the D and L denomination denote the configuration with respect to the αC and does not always correspond to the dextrorotatory and levorotatory activities of the compound. For this reason, the direction of polarized light rotation must be indicated with a (+) or (−) sign after the letter L or D. L-alanine, for example, has a specific rotation of +1.8° and its notation is therefore, L(+)-alanine.

$$
\begin{array}{cc}
\begin{array}{c}
CH_3 \\
| \\
H_2N-C-H \\
| \\
COOH
\end{array}
&
\begin{array}{c}
CH_3 \\
| \\
H-C-NH_2 \\
| \\
COOH
\end{array}
\\
\text{L-Alanine} & \text{D-Alanine}
\end{array}
$$

To avoid ambiguities with the D and L designation, another system called RS has been proposed. However, the notation DL continues being the one widely used. All amino acids, except glycine, offer D and L isomers. Isoleucine and threonine, amino acids that have a second asymmetric carbon in addition to the αC, give rise to four isomers. Only one of these isomers will participate in the formation of proteins.

Cells are able to distinguish stereoisomers with great efficiency. Only amino acids with the L-configuration are incorporated into the proteins of the human body. We will almost exclusively refer to them in the following pages. Thus, when no letter is included to the name, it is understood that we are referring to L-amino acids.

Classification of Amino Acids

Most of the 20 α-amino acids contain one acidic carboxyl group and one basic amine group, bound to the αC, which give them a neutral net charge. Two amino acids have an additional carboxyl group that gives the molecule acidic characteristics. Others, in contrast, have additional basic groups. Two amino acids contain sulfur. Finally, proline is the only amino acid in which the carbon adjacent to the carboxyl group forms part of a cyclic five-membered ring.

Amino acids will be presented dividing them into groups according to the characteristics of their side chains. Usually, each amino acid is indicated by abbreviated notations, one of which uses three letters and another uses only one letter. Both notations are shown below the name of the corresponding amino acid chemical formula. The side chains are shown in red.

Neutral Amino Acids With Nonpolar Aliphatic Chain

Glycine only has one hydrogen in its side chain and the polar groups, carboxyl and amine, have a predominant role in the molecule. Alanine, with a methyl group side chain, is more soluble

in water than those containing hydrophobic chains. *Valine, leucine,* and *isoleucine* have non-polar branched chains.

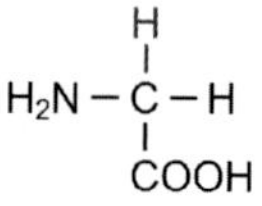

$$H_2N-\underset{\underset{COOH}{|}}{\overset{\overset{H}{|}}{C}}-H$$

Glycine
(Gly - G)

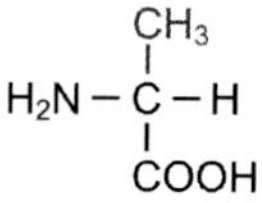

$$H_2N-\underset{\underset{COOH}{|}}{\overset{\overset{CH_3}{|}}{C}}-H$$

Alanine
(Ala - A)

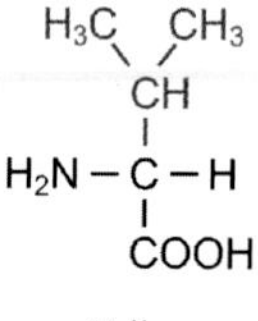

$$H_2N-\underset{\underset{COOH}{|}}{\overset{\overset{\overset{H_3C}{\diagdown}\overset{}{\underset{CH}{|}}\overset{CH_3}{\diagup}}{|}}{C}}-H$$

Valine
(Val - V)

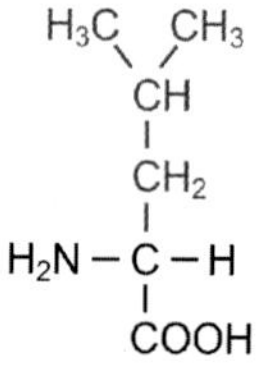

$$H_2N-\underset{\underset{COOH}{|}}{\overset{\overset{\overset{H_3C}{\diagdown}\overset{}{\underset{\underset{CH_2}{|}}{CH}}\overset{CH_3}{\diagup}}{|}}{C}}-H$$

Leucine
(Leu - L)

$$H_2N-\underset{\underset{COOH}{|}}{\overset{\overset{\overset{CH_3}{|}}{\underset{CH-CH_3}{|}}}{C}}-H$$

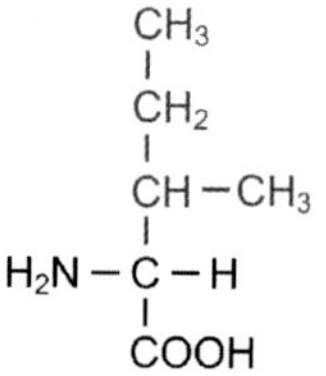

Isoleucine
(Ile - I)

Neutral Aliphatic Amino Acids With Nonionizable Polar Chain

Serine and *threonine* contain a hydroxyl functional group in their side chains, which gives these amino acids polar characteristics.

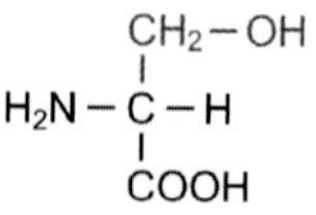

$$H_2N-\underset{\underset{COOH}{|}}{\overset{\overset{CH_2-OH}{|}}{C}}-H$$

Serine
(Ser - S)

$$H_2N-\underset{\underset{COOH}{|}}{\overset{\overset{\overset{CH_3}{|}}{\underset{CH-OH}{|}}}{C}}-H$$

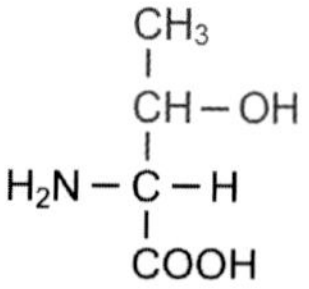

Threonine
(Thr - T)

Aromatic Neutral Amino Acids

Phenylalanine, which contains a benzene ring, and *tryptophan,* with an indole heterocyclic side chain, are both markedly nonpolar and hydrophobic. *Tyrosine* has a phenolic hydroxyl that adds to the polarity of this amino acid. At pH values above 10, tyrosine releases a proton and becomes negatively charged.

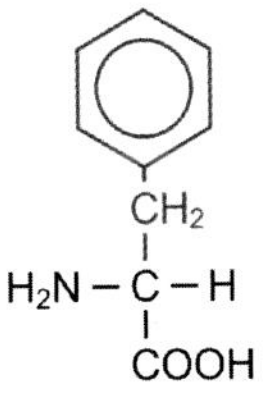

CH2
|
H2N – C – H
|
COOH

Phenylalanine
(Phen - F)

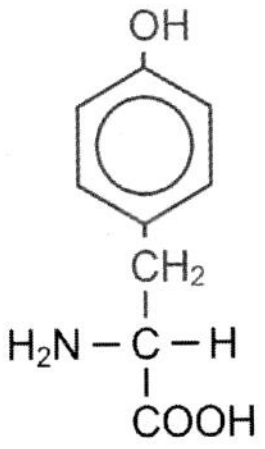

OH
|
CH2
|
H2N – C – H
|
COOH

Tyrosine
(Tyr - Y)

NH2
|
CH2 – C – COOH
|
H

N
|
H

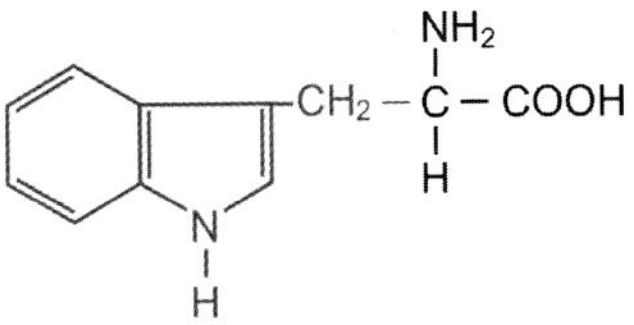

Tryptophane
(Trp - W)

These three amino acids with aromatic side chains strongly absorb light in the ultraviolet range of the light spectrum (280 nm). This property is used to detect the presence of proteins in a sample.

Amino Acids With Sulfur

Cysteine contains a sulfhydryl group (−SH) that is slightly polar. At a pH 9, it will release a proton. *Methionine* has a nonpolar side chain.

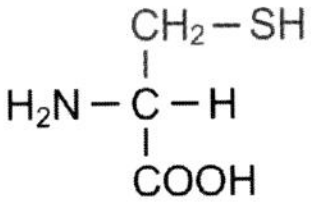

CH2 – SH
|
H2N – C – H
|
COOH

Cysteine
(Cys - C)

CH2 – S – CH3
|
CH2
|
H2N – C – H
|
COOH

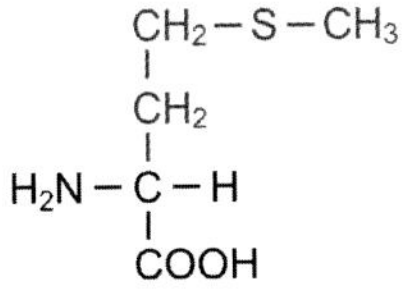

Methionine
(Met - M)

Acidic Amino Acids (Dicarboxylic)

Aspartic acid and *glutamic acid* are amino acids that have an additional carboxyl group that can release a proton and acquire a negative charge at the pH of body fluids. Often, these amino acids are designated with the name of their ionized form, aspartate and glutamate, respectively.

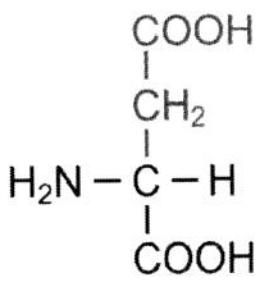

COOH
|
CH2
|
H2N – C – H
|
COOH

Aspartic acid
(Asp - D)

COOH
|
CH2
|
CH2
|
H2N – C – H
|
COOH

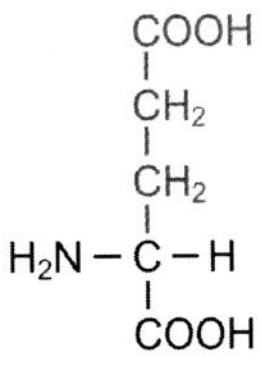

Glutamic acid
(Glu - E)

Asparagine and *glutamine* are derivatives of aspartic and glutamic acid, respectively; they possess an amide functional group in the carbon distal from the α carbon. Unlike their acidic analogs, the side chains of asparagine and glutamine have no electric charge; they *are polar.*

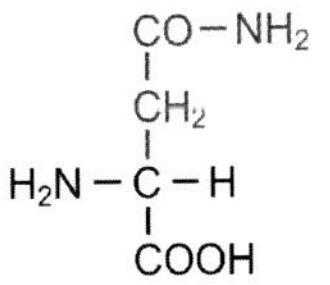

CO – NH2
|
CH2
|
H2N – C – H
|
COOH

Asparagine
(Asn - N)

CO – NH2
|
CH2
|
CH2
|
H2N – C – H
|
COOH

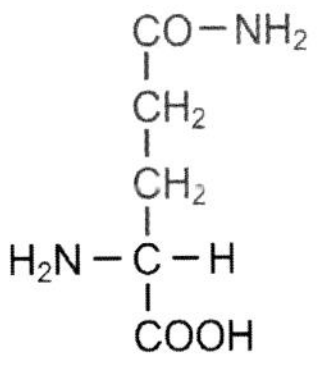

Glutamine
(Gln - Q)

Basic Amino Acids

Lysine has an additional amine functional group and *arginine* has a guanidine group, both of which can accept protons. At the pH of tissues the residues in these amino acids display a positive electrical charge.

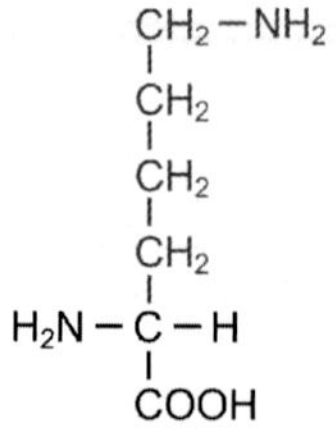

Lysine
(Lys - K)

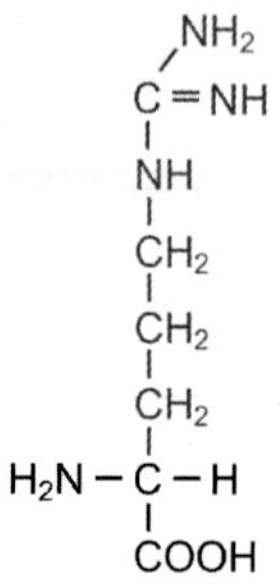

Arginine
(Arg - R)

Some authors consider that the nitrogen of proline forms an *imino* function (=NH) and, therefore, they call proline an imino acid rather than an amino acid. Having the α carbon and N in the ring gives the proline molecule greater stiffness than other amino acids. Some proteins contain a hydroxylated derivative of proline, the hydroxyproline.

The side chain of *histidine* is the heterocyclic imidazole side chain. One of the nitrogen atoms in imidazole can acquire a positive charge. The side chain of histidine has an ionization pK_a value around 6.0, so it can act as a base. The basic amino acids, lysine and histidine are highly polar.

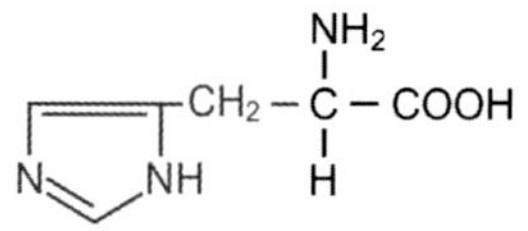

Other Amino Acids

Some of the amino acids presented earlier can be modified by covalent addition of different chemical groups. This includes, for example, phosphoserine, γ carboxyglutamic, and the already mentioned 4-hydroxyproline and 5-hydroxylysine.

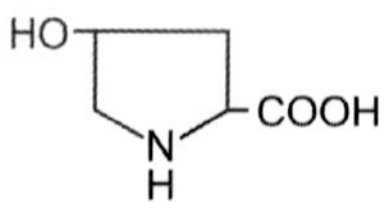

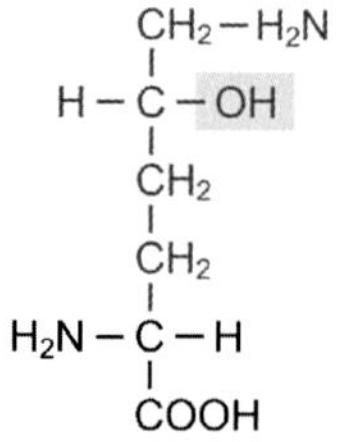

Histidine
(His - H)

The protein collagen contains hydroxylysine, which is a lysine derivative that has a hydroxyl group at carbon 5 (chain carbons are counted starting from the carboxyl group).

Proline

In the amino acid proline, the α carbon and the nitrogen bound to it are included in a pyrrolidine cycle. This chemical ring gives the amino acid an aliphatic character.

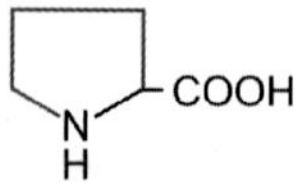

5-Hydroxylysine

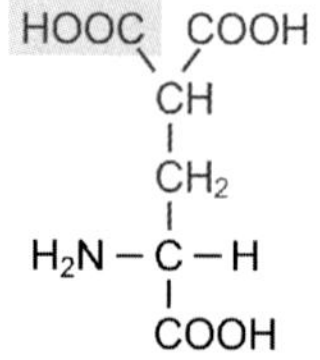

γ-Carboxyglutamic acid

Much less common than the amino acids mentioned earlier are selenomethionine and selenocysteine, in which the sulfur of cysteine and methionine is replaced by selenium. Approximately 15 proteins (selenoproteins) possessing these selenoamino acids have been described. Selenocysteine is preferably found in animal proteins, while selenomethionine is mainly found in plant proteins.

Other biologically important amino acids exist as free compounds or are part of nonprotein molecules. These include: β alanine, D-alanine, sarcosine, γ aminobutyric acid, D-glutamic acid, ornithine, homoserine, tyrosine, citrulline (p. 375) and homocysteine (p. 386).

Proline
(Pro - P)

4-Hydroxyproline

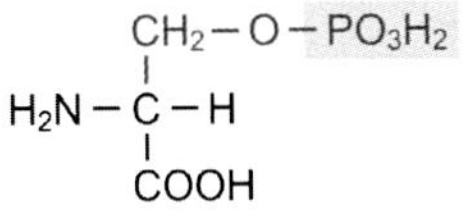

CH₂ — O — PO₃H₂
|
H₂N — C — H
|
COOH

O-phosphoserine

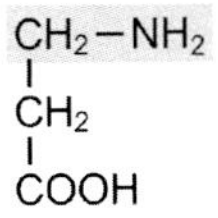

CH₃
|
NH
|
CH₂
|
COOH

Sarcosine

CH₂ — NH₂
|
CH₂
|
CH₂
|
H₂N — C — H
|
COOH

Ornithine

CH₂ — NH₂
|
CH₂
|
COOH

β-Alanine

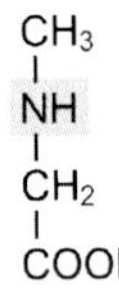

CH₂ — NH₂
|
CH₂
|
CH₂
|
COOH

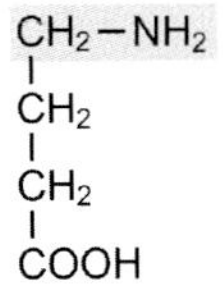

γ-Aminobutyric acid

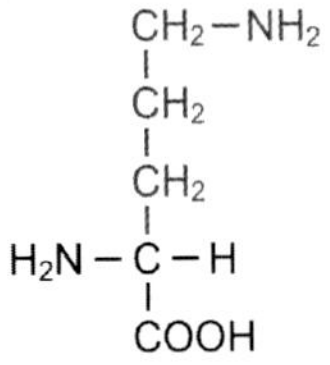

CH₂ — OH
|
CH₂
|
H₂N — C — H
|
COOH

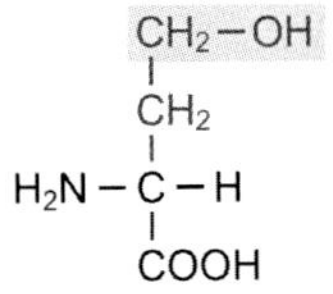

Homoserine

OH
|
[benzene ring]
|
O
|
[benzene ring]
|
CH₂
|
H₂N — C — H
|
COOH

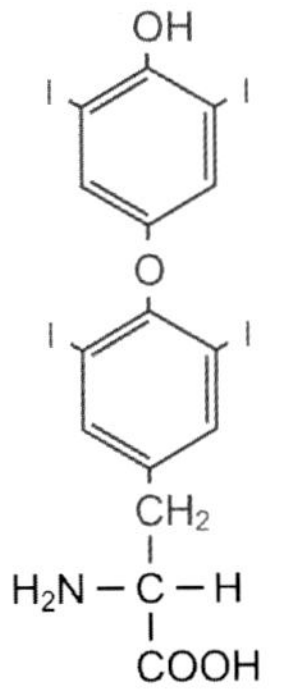

Tyroxine

Properties of Amino Acids

The properties of each amino acid chain predict their behavior. The sulfhydryl group of cysteine is highly reactive and easily combines with other sulfhydryl groups to form disulfide bonds (—S—S—). Two cysteines linked by this type of covalent bond form a compound known as *cystine*.

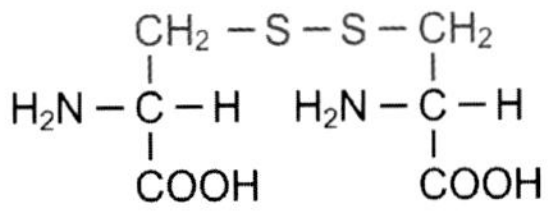

CH₂ — S — S — CH₂
| |
H₂N — C — H   H₂N — C — H
| |
COOH   COOH

Cystine

The additional carboxyl group of aspartic and glutamic acids not only give them acidic character, but also the ability to interact with basic substances to form salt-like bonds. The basic diaminated amino acids can also establish electrostatic linkages.

Amino acids can be grouped based on the polarity of their side chains:

Polar amino acids are glycine, serine, threonine, cysteine, tyrosine, aspartic acid, glutamic acid, asparagine, glutamine, lysine, histidine, and arginine.
Nonpolar amino acids are alanine, valine, leucine, isoleucine, methionine, phenylalanine, tryptophan, and proline.

Acid–Base Properties of Amino Acids

The existence of both an acidic and a basic group within the same molecule gives amino acids particular electrical properties. The carboxyl group behaves as an acid or proton donor:

$$—COOH \rightarrow —COO^- + H^+$$

The amine group accepts protons and acts as a base:

$$—NH_2 + H^+ \rightarrow —NH_3^+$$

In the formulas shown earlier, the amino acids are in a nonionized state, a situation that is not usual in biological media. In the crystalline state or in aqueous solutions, these compounds ionize, giving rise to both positive and negative charges on the same molecule. For this reason, amino acids are considered *dipolar ions*, or *ampholytes*. The German word *zwitterion* (hybrid ion) is also used to describe these types of polar molecules. Therefore, it is more correct to represent the α-amino acids as dipolar ions:

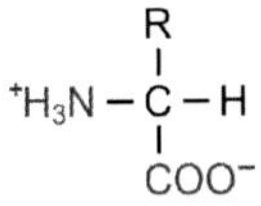

R
|
⁺H₃N — C — H
|
COO⁻

Ionized groups in red

The electrical charge of an amino acid depends on the pH of the medium in which it is dissolved. Increases in the hydrogen ion concentration in the medium diminish the dipolarity of the amino acid because the carboxylate ($-COO^-$) groups accept protons and act as bases. The amino acid then becomes a cation.

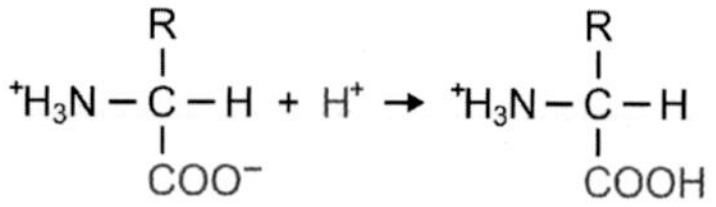

$$^+H_3N - \overset{\displaystyle R}{\underset{\displaystyle COO^-}{C}} - H + H^+ \rightarrow \ ^+H_3N - \overset{\displaystyle R}{\underset{\displaystyle COOH}{C}} - H$$

In contrast, when the amino acid is in an alkaline medium, the concentration of H^+ decreases and OH^- increases, the $-NH_3^+$ groups act as an acid, releasing H^+. The amino becomes negatively charged, it is an anion.

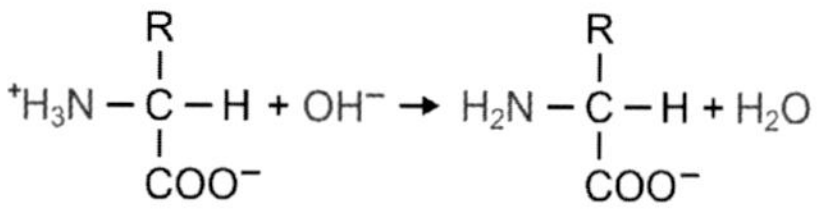

There is a pH value, characteristic for each amino acid, in which the ionization of positive and negative charges is equal and, therefore, the electric charge of the whole amino acid is zero. This pH value is called the *isoelectric point* (pHi or pI). At the pHi, an amino acid in solution subjected to an electrical field will not move toward any of the electrodes.

In dicarboxylic and diaminated amino acids, there is an additional ionizable group. For example, aspartic acid in a strong acid medium is fully protonated. In contrast, if the pH in the solution is increased by addition of a strong base (NaOH), aspartic acid will release protons to the medium. Aspartic acid forms successively different ionic species as the pH increases (Fig. 3.4).

For diaminated amino acids, such as lysine (Fig. 3.5):

$$^+H_3N - \overset{\displaystyle R}{\underset{\displaystyle COO^-}{C}} - H + OH^- \rightarrow H_2N - \overset{\displaystyle R}{\underset{\displaystyle COO^-}{C}} - H + H_2O$$

Amino Acids Titration Curve

The study of the titration curves of weak acids and bases can be applied to amino acids to

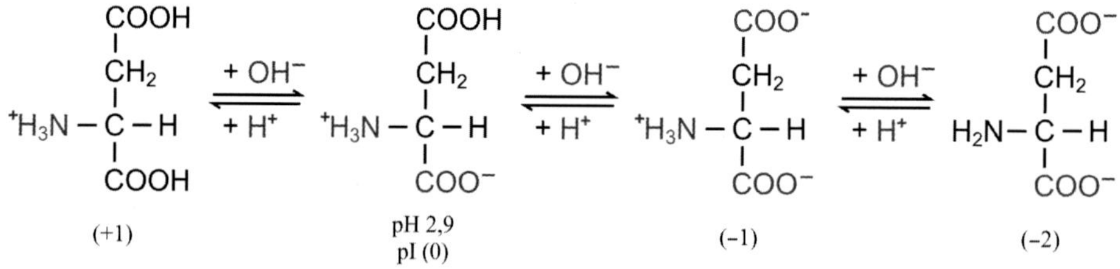

FIGURE 3.4 **Effect of pH on the electric charge of aspartic acid.** The value and sign of the net charge is shown in parenthesis.

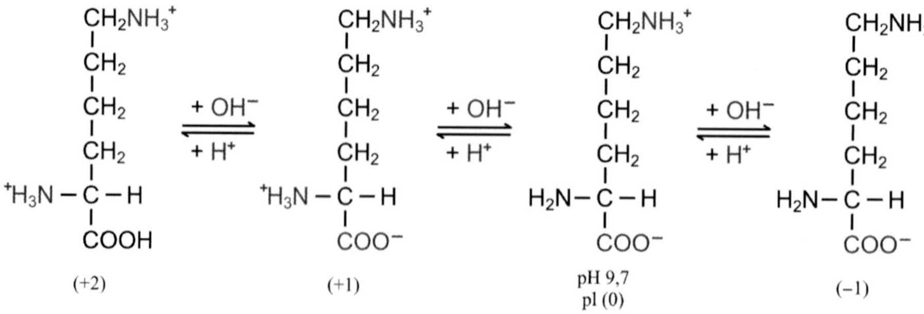

FIGURE 3.5 **Effect of pH on the electric charge of lysine.** The value and sign of the net charge is shown in parenthesis.

understand their acid–base behavior in solution. The titration curve corresponding to an amino acid with two ionizable groups, such as alanine ($^{+}H_3N$—CHR—COO^{-}, R stands for methyl in alanine), is analyzed in the following. The ionization equilibrium for each of the ionizable groups of alanine is, according to the following equations:

For the carboxyl:

$$^{+}H_3N\text{—CHR—COOH} \rightleftharpoons {}^{+}H_3N\text{—CHR—COO}^{-}+H^{+}$$

$$K_1 = \frac{\left[^{+}H_3N-CHR-COO^{-}\right]\left[H^{+}\right]}{\left[^{+}H_3N-CHR-COOH\right]}$$

For the amine:

$$^{+}H_3N\text{—CHR—COOH} \rightleftharpoons H_2N\text{—CHR—COO}^{-}+H^{+}$$

$$K_2 = \frac{\left[H_2N-CHR-COO^{-}\right]\left[H^{+}\right]}{\left[^{+}H_3N-CHR-COO^{-}\right]}$$

When alanine is dissolved in pure water, both its carboxylic and amine groups become ionized and the pH of the medium changes to the isoelectric point (pH = 6.02) of alanine. From this point, two separated titration phases can take place. One corresponds to that of the –COO^{-} group, which acts as a base accepting protons. The other depends on the —NH$_3^{+}$ group, which acts as a weak acid, releasing H^{+}. A solution of HCl can be used as a neutralization agent for the first titration and NaOH for the second. When plotting the pH changes produced during the titration against the volume of acid or base added, a biphasic curve is obtained (Fig. 3.6).

Before the addition of HCl, all of the amino acid molecules are in their ionic state (pH = pHi = 6.02). The addition of acid to the medium increases the concentration of H^{+} and these protons are accepted by the —COO^{-} groups. At a pH of 2.34, half of the —COOH groups are nonionized and there are equal concentrations of $^{+}H_3N$—CHR—COOH^{-} and $^{+}H_3N$—CHR—C

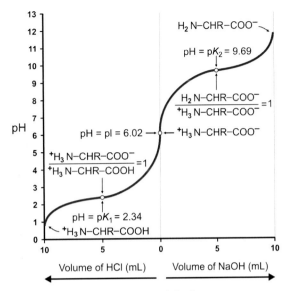

FIGURE 3.6 **Titration curve of alanine.**

OO^{-}. This pH value corresponds to the pK_a of the carboxyl group (pK_{a1}). If the addition of acid continues, a point is reached in which virtually all of the amino acid molecules will be protonated ($^{+}H_3N$—CHR—COOH), corresponding to the lower end of the graph.

The upper half of the titration curve results from the addition of NaOH. Starting from the pHi, the increase in the alkali augments the concentration of OH^{-} in the medium, which forms water with the H^{+} released by the NH$_3^{+}$ groups. The midpoint of this second phase of the curve is reached at pH 9.69. This corresponds to the pK_a of the —NH$_3^{+}$ of alanine (pK_{a2}), in which concentrations of ions $^{+}H_3N$—CHR—COO^{-} and H$_2$N—CHR—COO^{-} in the solution are the same. If addition of NaOH is continued until complete neutralization is reached, all NH$_3^{+}$ groups release their H^{+} and the alanine is completely deprotonated (H$_2$N—CHR—COO^{-}).

The biphasic titration curve shows the pK_a values of each ionizable group of alanine. The flattening of the curve that is seen around both pK_a values indicates the range of pH where the amino acid functions as a buffer.

All neutral amino acids, which have a non-ionizable side chain, such as alanine, produce similar titration curves. The pK_{a1} values of the carboxyl group next to the αC of different neutral amino acids are all close to 2. The pK_{a2} values of the amine group attached to the αC can vary from 9 to 10.

The amino acids with an additional ionizable group produce titration curves that have three components and three pK_a values. In addition to dicarboxylic and basic amino acids, this occurs with cysteine and tyrosine, which contain SH and phenol groups, respectively, that behave as weak acids. In general, the pK_a values of the ionizable groups are far from the normal pH in the body. An exception is histidine, which, at the normal pH of the body tissues, can accept a proton on one of the N of the imidazole ring. The pK_a of imidazole is 6.0, which allows histidine to be the only amino acid that functions as a buffer at physiological pH.

Chemical Properties of Amino Acids

Amino acids participate in a series of chemical reactions which may involve either the amine or carboxyl groups bound to the α carbon, or the side chains specific to each amino acid. The reactivity of these side chains is used to identify the presence of a particular amino acid in a sample.

A widely used reagent for the recognition of α-amino acids is ninhydrin. The α amine group reacts with this compound rendering a deep purple color. Proline, instead, gives a yellow product. Ninhydrin is used not only for the identification of amino acids, but also for the determination of their concentration. A colorimetric reaction, based on a ninhydrin reaction,

has been developed. This reaction is highly sensitive and can measure very small (nanomolar, 10^{-9} mol) amounts of amino acids. Even a higher sensitivity for amino acid detection can be achieved with the use of fluorescence techniques, which can identify amino acid concentrations lower than 10^{-15} molar. Amino acids also react with o-phthalaldehyde and can be identified by the indole derivative that this reaction produces.

A major problem in the study of proteins has been the isolation and identification of the amino acids from a complex mixture, such as that resulting from the total hydrolysis of a protein with a concentrated acid. The introduction of chromatographic methods, such as paper and thin-layer chromatography, and highly sensitive ion exchange columns have immensely advanced our capability of amino acid identification.

PEPTIDES

Peptide Bond

Amino acids can establish covalent bonds between the carboxyl group of one amino acid and the α amine group of another. This amide type link is called a *peptide bond* and it is accompanied by the loss of water (Fig. 3.7).

The product formed by linking two amino acids together is called a *dipeptide*. The subsequent binding of additional amino acid units to this dipeptide via peptide bonding generates tripeptides, tetrapeptides, pentapeptides, etc. The polymers formed by more than 10 amino acids linked by peptide bonds are designated *polypeptides*. A polypeptide chain is considered

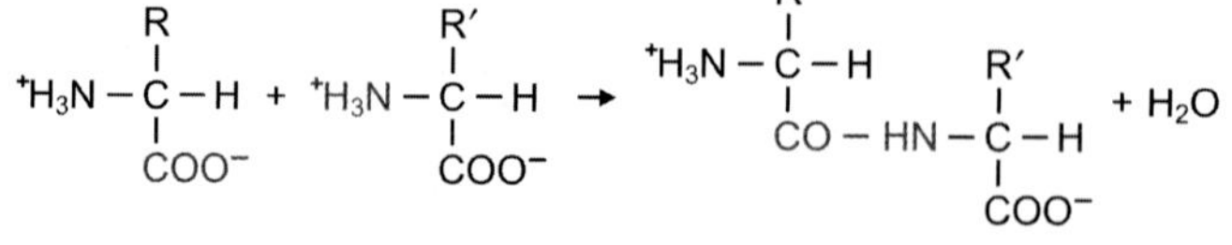

FIGURE 3.7 **Peptide bond.**

a *protein* when it has a molecular mass greater than 6000 Da (Daltons). Dalton (Da) is the unit of atomic mass; it is 1/12 the mass of one atom of ^{12}C. Frequently the expression kilodalton (kDa), 1000 daltons is used. Relative mass is the ratio between the molecular mass of a given substance and the mass of one atom of ^{12}C, which corresponds to the mass of a polymer of more than 50 amino acids. Below this mass, the compounds are designated *peptides*. There is no precise distinction between peptides and proteins; 6000 Da is arbitrary and was chosen because it is the approximate mass of insulin, a hormone produced in the pancreas, which was the first protein whose entire structure was deciphered.

At one end of every polypeptide chain there is an amino acid with a free α amine group. By convention, this end is considered the beginning of the chain and is called the amino-terminal or N-terminal portion of the polypeptide. The other extreme of the chain ends with a free carboxyl group and it is considered the C-terminal end of the polypeptide chain.

When integrated in the peptide or protein chain, amino acids lose the H of the amine group and the OH of the carboxyl group that are involved in the peptide binding. The amino acid units forming the polymer are referred to as *amino acid residues*.

Nomenclature

Peptides are named following the order of the amino acids that constitute them, beginning from the N-terminal of the chain. Amino acid residues are indicated by the root of its name followed by the suffix "yl." The last residue (with the free carboxyl group) is mentioned by its full name. For example, in the case of the hexapeptide composed of serine, aspartic acid, tyrosine, lysine, alanine, and cysteine (Fig. 3.8), the name is: seryl-tyrosyl-aspartyl-lysyl-alanyl-cysteine, which can be abbreviated: ser-asp-tyr-lys-ala-cys or SDYKAC.

Acid–Base Properties

Carboxyl and α amine groups involved in the peptide bonds lose OH and H, respectively, and cannot ionize. Therefore, the acid–base properties of peptides are determined by the terminal amine and carboxyl groups, as well as the ionizable groups in the side chains of the amino acid residues ($-COO^-$, $-NH_3^+$, $-SH$). The pH of the medium influences the magnitude and sign of the net charge of a peptide in a similar manner to that described for amino acids. Peptides also have an isoelectric point, the pH in which there is an equal number of positive and negative charges in the molecule.

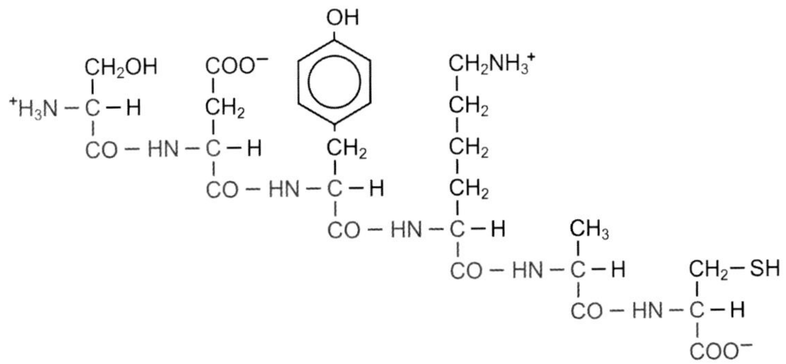

FIGURE 3.8   **Hexapeptide seryl-tyrosyl-aspartyl-lysyl-alanyl-cysteine.** The peptide bonds *(red)* form the backbone of the molecule.

Biologically Important Peptides

Peptides have important functions, both in plants and animals. Generally, peptides are chains of amino acids linked by peptide bonds; however, some of them have peculiar characteristics, such as atypical peptide bonds, unusual amino acids or amino acid derivatives, formation of cyclic structures, etc.

The number and type of amino acids, their sequence, spatial arrangement, and their biological activity are important determinants of the functional role of peptides. *Glutathione* is a peptide widely distributed in nature, found in bacteria, plants, and animals. This tripeptide is composed by glutamic acid, cysteine, and glycine. Glutamic acid is bound to the amine group of cysteine by an atypical peptide bond, which involves the distal or γ carboxyl group of glutamic acid.

When oxidized, glutathione forms a disulfide bridge (—S—S—) with another molecule of glutathione in a reaction that is reversible. Glutathione participates in enzymatic redox systems. It contributes to the prevention of oxidative damage in blood cells and tissues.

TABLE 3.1   Some Peptide Hormones

Name	Number of amino acids	Function
Angiotensin II	8	Hypertensor
Vasopressin	9	Water balance
Oxytocin	9	Uterine contraction
Bradikinin	9	Hypotensor
Kalllidin	10	Hypotensor
Gastrin I	17	Gastric secretion of HCl
Melanocyte stimulant (β MSH)	18	Increases melanin
Secretin	27	Pancreatic juice secretion
Glucagon	29	Hyperglycemic
Calcitonin	32	Reduces Ca level in blood
Cholecystokinin–pancreoenzymin	33	Contraction of gall bladder; secretion of pancreatic enzymes
Adrenocorticotrophin	39	Stimulate adrenal cortex

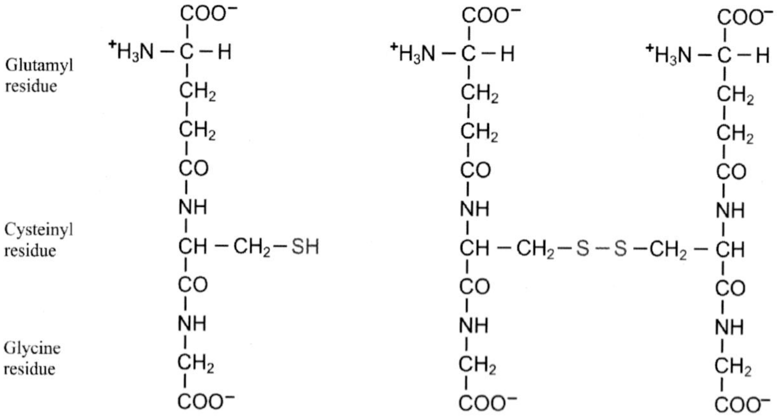

Reduced glutathione
γ-Glutamyl-cysteinyl-glycine

Oxydized glutathione

Many hormones are peptides. Table 3.1 presents a list of some of them. Another interesting group of peptides are the *enkephalins*, which are released by the central nervous system and produce analgesia by binding to specific brain receptors. Many antibiotics, substances synthesized by microorganisms that have toxic effects on other organisms, are peptides or contain a peptide as part of their molecule. Some of these antibiotics have a cyclic structure and often contain D-amino acids. Certain species of fungi produce peptides highly toxic to humans (α amanitin).

PROTEINS

Acid–Base Properties

Some of the properties of peptides also apply to proteins. In a polypeptide chain, the carboxyl and α-amino groups involved in a peptide bond are not ionizable and only the α-amine group at the N-terminus and the carboxyl group at the C-terminus of the protein are free to ionize. Those two terminal groups alone do not have major influence on the acid–base property of a macromolecule. Rather, the electric charge of a protein depends on the ionization of all ionizable groups of the amino acid side chains. The presence of many lysine, arginine, or histidine residues gives a protein molecule a basic character. Lysine and arginine have additional amine groups that can accept protons and become positively charged. If there is a predominance of aspartate and glutamate residues within a protein, it will have acidic properties, due to the additional carboxyl groups, which release hydrogen ions. The phenolic group of tyrosine and the sulfhydryl group of cysteine are weakly acidic. The presence of these two groups gives the protein buffering capacity, since they accept or release protons, according to the concentration of H^+ ions in the medium. However, at the pH range of cells (approximately between 6.0 and 8.0), only histidine residues significantly act as buffers, because the pK_a of the ionizable imidazole functional group ($pK_a = 6.0$) is close to the values of physiological pH.

In a strong acidic solution, most of the free amine groups accept hydrogen ions, while the majority of the acidic groups do not ionize. Under these conditions, the protein has a net positive charge. By contrast, in a strong alkaline solution, the free amine groups lose protons and are not ionized, while the carboxyl groups dissociate; the protein will exhibit an overall negative charge.

If alkali is added to an acid protein dispersion, the pH gradually rises and changes in the magnitude and sign of the net electric charge occur. The initial positive charge of the protein decreases because acidic groups (—COOH) release protons, causing the molecule to become electronegative. During the addition of alkali, a point is reached in which the number of positive and negative charges is the same; the total charge of the molecule is zero. The pH of the dispersion in which the net charge is zero is known as the *isoelectric point* and is indicated by the symbols pHi or pI. If the pH of the medium continues increasing, the weaker acidic groups (—SH, phenol, $—NH_3^+$) also release protons and the protein becomes progressively more electronegative.

Two proteins with different numbers of free acid and basic groups will have different net charges at a given pH. The magnitude of the electrical charge of a protein is proportional to the difference between the pH of the medium and the protein pHi. The protein will be more electropositive, as the acidity in the medium increases beyond the isoelectric point. On the other hand, the electronegative charge will be greater as the pH of the medium becomes more alkaline with respect to the protein pHi.

Electrophoresis

If a protein is subjected to an electric field in a medium with pH lower than that of its isoelectric point, the protein behaves as a cation (it has a positive charge) and moves toward the negative pole or cathode. When the pH is above the isoelectric point, the protein migrates toward the positive pole or anode (it will behave as an anion). At the isoelectric point, the protein has no electric charge and will not migrate in the dispersion. This migration by action of an electric field is known as *electrophoresis*.

In a mixture of two or more proteins of different pHi dissolved in a medium of a given pH, the differences between the pH and the pHi of each protein will be different and, therefore, the value of the net charge and the rate of migration in the electric field will be also different. This phenomenon is the basis of a widely used technique for protein separation, known as *electrophoretic fractionation*.

Electrofocusing is a variant of electrophoresis in which protein separation is further achieved by a gradual change in pH of the separation medium (there is a pH gradient). The migrating protein stops when it reaches the pH region corresponding to its pHi.

Molecular Mass

Proteins differ widely in shape, size, and molecular mass. Smaller proteins are approximately 6000 Da, while larger proteins can reach up to several million Daltons. The number of amino acids that constitutes a protein can be estimated by dividing the protein's molecular mass by 120, which is the average weight for an amino acid.

Determination of the molecular mass of proteins can be performed by various different methods. Ultracentrifugation requires equipment capable of developing centrifugal forces of at least 500,000 times the force of gravity through rotation of the sample. This procedure enables determination of the sedimentation velocity of a protein in dispersion. Under this strong centrifugal force, proteins tend to settle with a speed proportional to its mass. Other methods, such as chromatography, gel filtration, gel electrophoresis, density gradient centrifugation, are more accessible and all yield good information on the protein molecular mass.

Solubility

Many proteins disperse easily in water or aqueous media. Their stability in these dispersions is due to several factors. A very important feature is the ability of dispersed protein molecules to interact with water, or other polar solvents, by forming what is called a solvation layer (hydration layer when the solvent is water). Due to its high dielectric constant, water prevents protein aggregation and precipitation by separating oppositely charged groups in different proteins.

Ionized functional groups such as $—NH_3^+$, $—COO^-$, and other polar groups ($—OH$, $—SH$, $=NH$) in proteins attract water molecules. These water molecules orient around the protein, forming a hydration layer. The difference in solubility between different proteins depends on their degree of hydration, determined by the number of polar groups within the protein. The number and distribution of nonpolar hydrophobic groups (aromatic rings and aliphatic chains) affects the degree of solvation.

An important stabilizing factor for protein solubility is the net electrical charge of the molecule; if all the particles of a protein have the same sign, there is mutual rejection that prevents its clustering and precipitation. The net charge may be different for different proteins under similar environmental conditions, which influences their degree of solubility. Protein solubility varies with pH, temperature, and the presence of nonpolar solvents or inorganic salts in the medium. The behavior with respect to these factors differs in different proteins, a characteristic that is used to design protein fractionation methods. Thus, selective precipitation of proteins can be achieved by addition of salts or nonpolar solvents under various conditions of concentration, pH, and temperature.

Effect of pH. The pH of a medium is an important factor for protein solubility because the magnitude of the net electric charge of the molecule depends on it. The total electric charge of a protein molecule is zero at its isoelectric point and, therefore, the solubility will be minimal at that pH. Based on this property, proteins from a mixture can be separated by changing the pH to values in which one of the proteins tends to precipitate while the remaining, of different pHi, are not affected (isoelectric separation). At the pHi, the repulsive intermolecular forces disappear and precipitation can occur directly or after addition of agents that act on the solvation layer.

Effect of salts. At low concentrations, salts promote the solubility of many proteins since the inorganic ions interact with the ionized groups within the protein. The phenomenon is most notable in proteins with marked dipolar character, such as those in which the molecule electric charges of opposite sign are distributed asymmetrically. Addition of salts reduces the electrostatic attraction between the protein molecules and facilitates their dispersion. However, as the salt concentration increases, the salt ions attract water molecules and tend to remove them from the solvation layer, decreasing the solubility of the protein. When the salt concentration reaches a certain value, the effect on the solvation layer is strong enough to cause precipitation of the protein.

Ammonium, sodium, magnesium sulfates, and sodium thiosulfate are the most commonly used salts for selective protein precipitation. In fact, when one of these salts is added to protein mixtures, precipitation of different fractions can be obtained as different salt concentrations are attained in the medium; this separation method is called *salt fractionation*. These methods are convenient because the proteins are not altered and can be dispersed again without changes in their properties.

Effect of poorly polar solvents. The addition of poorly polar solvents (ethanol, acetone, etc.) decreases the solubility of proteins. Precipitation of the protein occurs when the concentration of the solvent reaches certain values that are variable for each protein. The poorly polar solvent affects protein structure (denaturation), unless one

operates at very low temperatures. If ethanol is added gradually to a complex protein mixture, different solubility conditions are created for different proteins and selective precipitation is obtained (alcoholic fractionation). Ethanol has a dielectric constant lower than that of water; therefore, it reduces the insulating power of the medium, allowing attraction of oppositely charged groups and promoting the production of molecular clusters, which tend to precipitate. To achieve fractional protein precipitation, careful adjustments of temperature and pH are required to avoid denaturation and to enhance neutralization of the protein charge.

Dialysis-Ultrafiltration

Most biological systems are complex mixtures containing low molecular mass solutes along with macromolecules. Through a process called *dialysis*, it is possible to separate small molecules. Porous membranes, such as cellophane and other synthetic materials, are used to allow the movement of water and low-weight molecules, while retaining macromolecules. These types of membranes are called *semipermeable* membranes.

If a cellophane bag containing blood serum is immersed in pure water, small mass solutes (inorganic ions, glucose, and urea) will pass through the membrane toward the water, while the larger proteins, which do not permeate the membrane, will remain within the dialysis bag. By repeated changes of the water surrounding the dialysis bag, it is possible to remove almost all of the diffusible solutes that were originally present in the serum sample.

A variant of this method is *ultrafiltration*, in which a filter system equipped with a porous plate acting as a semipermeable membrane is used. After adding a complex mixture to the filter, a pressure difference between both sides of the filter is applied, either by increasing the pressure on the mixture or performing a vacuum on the opposite side of the plate. Water and low mass solutes will pass through the filter. This method not only separates small molecules, but also concentrates macromolecules by removal of the solvent; this is useful to enrich proteins that are too diluted in the dispersion.

Molecular Shape

In its native condition, each protein has a characteristic molecular shape. Two general types of proteins can be distinguished: globular and fibrillary.

Globular proteins. These molecules fold and form a compact spheroid or ovoid with three axes of similar length. In general, they are functionally active proteins, such as enzymes, antibodies, hormones, and hemoglobin. They disperse well in aqueous media.

Fibrillary or fibrous proteins. These molecules are constituted by parallel polypeptide chains that form extended fibers or sheets, with the longitudinal axis markedly predominating over the transversal one. In general, these proteins do not disperse in water, and form part of structures that require great physical strength, such as the connective tissue fibers.

Molecular Structure

Protein structure is complex; proteins present different levels of organization:

Primary structure refers to the number and sequence of the amino acids in the polypeptide chain. The peptide bond can only form linear structures and proteins do not contain branching chains.

Secondary structure defines the regular spatial and repetitive arrangement of the polypeptide chain, generally held together by hydrogen bonds.

Tertiary structure describes the complete three-dimensional architecture of the protein.

Quaternary structure applies only to proteins formed by two or more polypeptide chains and refers to the spatial arrangement of the chains and the links established between them.

Primary Structure

When a protein is subjected to complete hydrolysis, the type and amount of amino acids that are released can be identified. For example, bovine insulin, a small protein with a mass close to 6000 Da, is constituted by the following amino acid residues: 3 alanines, 1 arginine, 3 asparagines, 6 cysteines, 3 phenylalanines, 4 glycines, 4 glutamic acids, 3 glutamines, 2 histidines, 1 isoleucine, 6 leucines, 1 lysine, 1 proline, 3 serines, 4 tyrosines, 1 threonine, and 5 valines. Certainly, the overall amino acid composition is of interest, but it provides no information on the arrangement or sequence of amino acids in the chain. Primary protein structure specifically refers to the protein amino acid sequence. Each protein is characterized by a defined amino acid composition and sequence. The amino acid assembly of proteins is carried out according to the information contained in the genome. The order of the units that make up the genetic material of a cell (DNA nucleotides) is a "coded message" that indicates the sequence in which amino acids must be inserted when the protein is synthesized (see p. 493). Twenty different amino acids linked by peptide bonds generate protein molecules of hundreds and even thousands of units; theoretically, the number of different possible associations is enormous. For example, the number of proteins that can be formed with a chain in which each amino acid only appears once is 2×10^{18}. This is for just an icosapeptide (20 amino acids), which is smaller than the average protein (over 300 amino acids). If the total number of amino acids is increased and repetition of amino acids occurs in different proportions, an almost infinite number of different polymers could be obtained. However, not all the possible combinations occur in nature because many of them are not functional. Nevertheless, the potential for protein molecular variation is enormous.

The determination of primary protein structure is a fundamental step in the study of a protein and great emphasis has been placed in identifying the amino acid sequence of proteins. The amino acid sequence of bovine insulin constituted a milestone of modern biochemistry. In the early 1950s, the English researcher Frederick Sanger developed a method for protein sequencing, based on the reaction with 1-fluoro-2,4-dinitrobenzene. In alkaline medium, this compound reacts with the terminal α-amino group of peptide chains. After acid hydrolysis of the peptide, the amino acid residue at the N-terminus is a dinitrophenyl derivative and can be identified by chromatography.

Other reagents used to identify N-terminal residues are dansyl chloride, dabsyl chloride, and phenylisothiocyanate. The latter was proposed by Pehr Edman. The technique, known as Edman degradation, has advantages over previously used methods. This is because the phenylisothiocyanate derivative of the amino acid at the N-terminus can be separated by mild acid hydrolysis, without affecting the peptide bonds within the rest of the chain. The method allows for identification, one by one, of the amino acid residues that make up a polypeptide. This technique is normally applied to protein fragments that have been obtained from degradation of the protein with chemical reagents and enzymes that produce selective hydrolysis of specific peptide bonds. Once the amino acid sequence of the protein segments is known, these segments can be arranged and the primary structure of the whole protein can be determined. There are sequencing machines that can automatically determine the primary structure of regular sized polypeptides. At present, indirect methods using genetic engineering techniques are used. For example, a gene can be isolated, its nucleotide sequence is determined, and the primary structure of the protein deduced. With the methods available today, it is easier to establish the nucleotide sequence of DNA than the amino acid sequence of a polypeptide. The amino acid sequence of millions of polypeptide molecules is known and the number continues increasing.

The synthesis of polypeptides can also be performed. To achieve this, the most effective method is solid phase synthesis. This technique consists of binding the C-terminal amino acid to a suitable support (synthetic resin) and then adding amino acids one by one via peptide bonds in the desired sequence. Robert B. Merrifield, author of this method, was able to synthesize a 124-residue protein, the enzyme ribonuclease.

Importance of the primary structure of proteins. The amino acid sequence of a protein is the most important determinant of its conformation, properties, and functional characteristics. The structural requirements for a protein to fulfill its physiological role are very stringent. Not only the number and type of constituent amino acids, but the exact position of each amino acid in the polypeptide chain must be maintained. Alterations in this order or substitutions can affect the functional capacity of the molecule and render it inactive.

Cells synthesize their proteins by assembling amino acids according to precise instructions. These instructions are contained in the DNA constituting the chromosomes. Modifications of this genetic material (mutations) can lead to "errors" in synthesis, producing abnormal proteins. Many inherited diseases have this origin.

In general, the primary structure of a protein that is responsible for a specific function is identical in all individuals of the same species and differs from the corresponding protein from other species. However, there are proteins of the same kind that differ even among individuals of the same species. These inter- and intraspecies differences are of medical interest. Humans and many animals are capable of detecting the presence of foreign proteins, which are different in primary structure to those of their own tissues, by initiating a response called the immune reaction. This reaction leads to the production of specific antibodies, capable of binding to the intruder protein to promote its destruction. This type of response is one of the natural defense mechanisms against foreign agents (see Chapter 30).

Proteins, such as hemoglobin, perform the same function in different animals and show important interspecies variability and similarities in its primary structure. For example, hemoglobin polypeptide chains have identical or nearly identical length; the same amino acid residues are present in many positions along the protein chain and its general shape is similar in all species. These similarities suggest that all hemoglobin molecules are derived from a common ancestral gene and that variation of the genetic material by successive mutations in the course of evolution has generated the existing protein variety.

The comparative analysis of the amino acid sequences of homologous proteins from organisms of different levels in the biological scale is of particular interest. For example, cytochrome c, a 104-amino acid protein, is widely distributed in nature. In many species, 27 of those amino acids are identical. These invariable sites are closely related to the function of the protein. The remaining residues exhibit differences which are more marked in species that are evolutionarily more distant in the phylogenetic tree. For example, horse and yeast cytochrome c differ by 48 amino acid residues, while chicken and duck cytochrome c homology is much higher, as they differ in only 2 amino acids. Knowing the sequence of amino acids of proteins from different species provides important data that allows one to build phylogenetic trees. This is important in evolution to estimate the time when a particular organism differentiated from a common ancestor.

Secondary Structure

The spatial arrangement that a polypeptide chain adopts depends on the orientation of the bonds between the backbone atoms —C—N—αC—. The peptide bond has intermediate

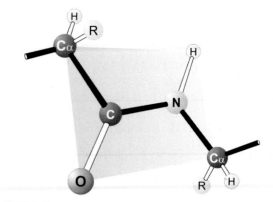

FIGURE 3.9 **Molecular model of a peptide bond.** The four atoms directly involved in the peptide bond (C, O, N, and H) are in the same plane, as the αC bonded to C and N. The αC rotates freely.

characteristics between a single and a double bond. This provides some rigidity to the C—N link, preventing free rotation of the involved atoms. This limitation of rotation has several consequences: (1) the four atoms linked in the peptide bond (C, O, N, and H) and the two carbons attached to the αC and N are in the same plane (Fig. 3.9). (2) The O of the carbonyl group (=CO), and the H bound to the nitrogen are in *trans* position. (3) The side chain (R) and the hydrogens linked to α carbons project out of the plane containing the other atoms.

The bonds of the α carbon (αC—C and N—αC) are simple (sigma) and allow free rotation (Fig. 3.10). Consequently, the orientation that such bonds adopt determines the spatial disposition of the chain.

A method that has immensely contributed to understanding protein structure is X-ray diffraction. Subjecting highly purified and crystallized proteins to X rays produces a diffraction pattern of rays that establishes the spatial arrangement of the atoms composing a molecule. Other techniques used in the structural analysis of proteins are optical rotatory dispersion, circular dichroism, nuclear magnetic resonance, and mass spectrometry. The description of these methods

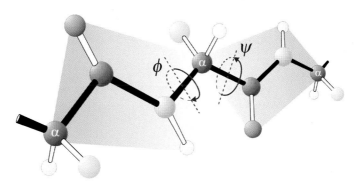

FIGURE 3.10 **Rotation of the peptide bonds.** Only N–αC and C–αC can rotate; rotation angles are designated φ (phi) and ψ (psi), respectively. The extended conformation, where φ and ψ have a value of +180° is shown. C: *Black*, O: *red*, N: *gray*, H: *white*, side chains: *pink*.

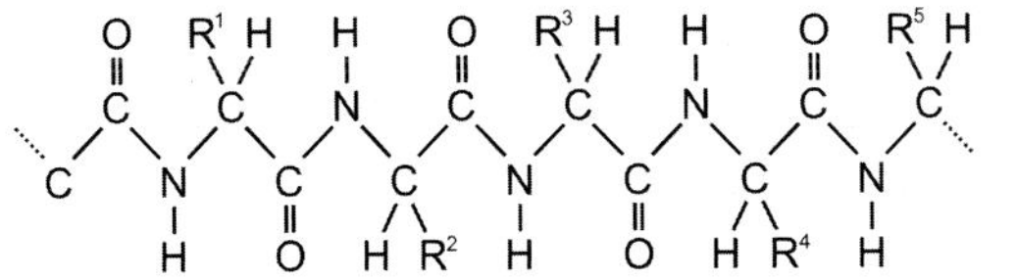

FIGURE 3.11 **Simplified diagram of a segment of a polypeptide. R^1, R^2, R^3, etc. indicate the side chains of amino acid residues.**

is beyond the scope of this textbook. A schematic representation of a polypeptide chain segment is shown in Figures 3.11 and 3.12.

In the late 1940s, Linus Pauling and Robert Corey determined the lengths of the bonds and the angles formed by these bonds. On the basis of X-ray diffraction studies, they confirmed their theoretical models and proposed two types of periodic structures for proteins: the *α helix* and the pleated β sheet (the α and β designations simply indicate the order in which both structures were proposed).

α helix. The polypeptide chain is arranged coiling on a central axis, as if the chain wrapped around a cylinder (Fig. 3.13). A complete turn of the helix covers a distance of 0.54 nm along the protein axis and encompasses 3.6 amino acid residues. This means that each residue spans a distance of 0.15 nm, covering an angle of 100° around the protein axis. The rotation of the chain is in the clockwise direction and it is said that the helix is in the right direction. This type of secondary structure is maintained by hydrogen bonds.

In the polypeptide chain, the hydrogen bound to the N of an amine moiety is attracted by the oxygen of a carbonyl group (Fig. 3.14).

In the helix, each =CO group forms a hydrogen bond with the =NH group of the amino acid that is situated four residues downstream in the chain; this causes the chain to come a full circle and to bring together both of those chemical groups (Figs. 3.13 and 3.15). The *trans* arrangement of these groups allows formation of an H bond. Although a hydrogen bond is weak, the existence of a large number of these bonds makes the helix very stable and compact.

FIGURE 3.12 **Schematic representation of a segment of an extended polypeptide chain, with the planes comprising all atoms involved in each peptide bond.** C: *Black*, O: *red*, N: *gray*, H: *white*, side chains: *pink*.

FIGURE 3.13 **Schematic representation of an α helix segment.** R indicates amino acid residue side chain. The *red dotted lines* represent the hydrogen bonds stabilizing the helix. Elements located behind the helix are not shown.

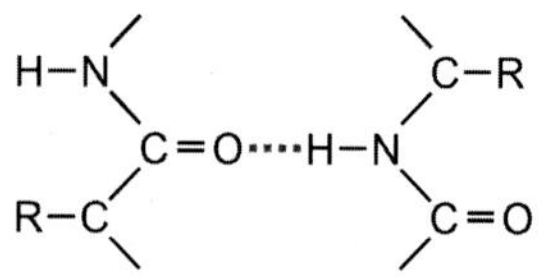

$$H-N \qquad\qquad \diagdown C-R$$
$$\diagdown C=O\cdots H-N \diagup$$
$$R-C \diagup \qquad\qquad \diagdown C=O$$

FIGURE 3.14 **Hydrogen bond** *(dotted red line)* **in a peptide chain.**

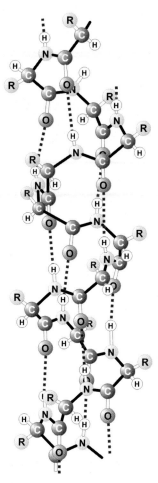

FIGURE 3.15 **Spheres and rods molecular model of a α helix.** R represents the side chains of amino acid residues. The *red dotted lines* indicate hydrogen bonds that stabilize the helix (Source: Imitated from Pauling and Corey).

For the α helix to be formed, the primary structure of the protein must meet certain conditions. For example, the presence of proline is incompatible with a α helix structure because it causes torsion in the chain which interrupts the regular orientation of bonds. The α carbon included in the pyrrolidine ring of proline cannot rotate, forcing a fixed position of the chain. Moreover, when bulky electrically charged side chains (such as those of arginine, lysine, and glutamic acid) are located close together, electrostatic forces that affect the spatial arrangement of the polypeptide prevent the formation of α helices.

β sheet. The polypeptide chains in this configuration are more extended than those in the α helix. Each amino acid residue covers a length of 0.35 nm instead of the 0.15 nm in the α helix. When two or more protein chains extended in this manner are paired, they can establish H bonds between =NH groups of one chain and =CO groups of the other, forming lamellar

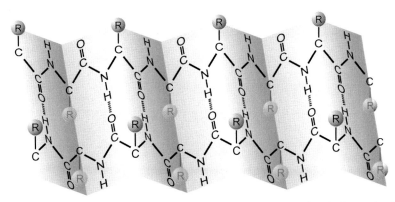

FIGURE 3.16 **Scheme of β sheet of two paired polypeptide chains.** Hydrogen bonds that maintain this structure are indicated in *dotted red lines*. Amino acid side chains alternate above and below the pleated sheet.

structures with a zigzag folding (pleated sheet structure) (Figs. 3.16 and 3.17).

If paired protein chains have the same direction (N-terminal→C-terminal) they are parallel, and if they progress in opposite directions they are antiparallel. Sometimes a chain can make a turn back on itself forming a hairpin loop in which the pairing will be antiparallel.

Other types of secondary structure are present in certain proteins. One of them is the collagen helix which will be considered later.

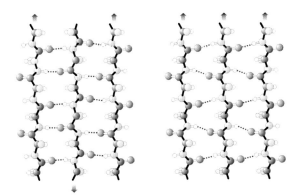

FIGURE 3.17 **Two types of β sheets.** Spheres and rods molecular model. Left, antiparallel strands; right, parallel. The *arrows* indicate the direction N-terminal → C-terminal. Carbon, *Black spheres*; side chains, *pink*; oxygen, *red*; nitrogen, *gray*; hydrogen, *white* (Source: Imitated from Pauling).

Random arrangement. Sometimes the polypeptide chain does not have a regular structure and exhibits a random coil disposition. However, this does not mean that the protein chain follows a completely random orientation; rather, it tends to adopt the thermodynamically most favorable spatial arrangement.

Usually different types of secondary structure are found in a molecule. A globular protein, for example, could not be formed in its entirety by a α helix because there would be a predominance of the longitudinal axis and the shape would correspond to a fibrous protein. α Helices are frequently found in various segments of a molecule connected by pieces of random coils. Globular proteins generally have α helical segments, β pleated sheets, and random coils (Fig. 3.20). Fibrous proteins have exclusively α helical or β sheet structures.

Tertiary Structure

The overall architecture of a protein determines its three-dimensional conformation. According to their final shape, they are classified in globular or fibrillary.

In the fibrous or fibrillary proteins, primary and secondary structures are sufficient to shape its conformation. The whole molecule can be arranged as α helix or β pleated sheet, allowing the prevalence of a longitudinal axis. Conversely, in globular proteins, even though there are portions of the chain that adopt α helical or β sheet structures, they contain segments that are randomly arranged, which provides the folds needed for the protein to achieve the overall spherical shape. The three-dimensional arrangement that the globular protein adopts is maintained through interactions between the side chains of the amino acids. In addition, thermodynamic factors should be considered; the most stable form of a protein is that of lowest free energy.

Forces responsible for maintaining the tertiary structure are of different types:

1. *Attraction/repulsion of electrostatic forces.* Electrically charged groups such as the —NH$_3^+$ of lysine and arginine form ionic, salt-type links when they come in contact with groups that have an opposite charge (such as the —COO$^-$ of aspartate and glutamate). This type of link has the ability to approximate and connect remote areas within the protein chain. If two groups having the same charge come into contact, the effect is instead a repulsion.

2. *Hydrogen bonds.* The carbonyl oxygen of a free carboxyl group in aspartate or glutamic acid residues and the imidazole nitrogen in histidine can attract the hydrogen in the OH group of serine, threonine, or tyrosine. This establishes H bonds, but it is important to note that these bridges are different from those that stabilize the α helix or β sheet secondary structures.

3. *Disulfide bridges.* When the sulfhydryl (—SH) groups of two cysteine residues come in close contact, they can establish a covalent —S—S— bond or disulfide bridge via oxidation. This type of bond is common and approximates two regions that are far apart in the protein sequence.

4. *Interactions of hydrophobic side chains.* Hydrophobic amino acid residues, such as leucine, isoleucine, valine, methionine, phenylalanine, and tryptophan are important determinants of the protein conformation in aqueous solutions. The nonpolar side chains of those amino acids tend to move away from the water environment and cause folds in the protein, clustering those residues in the interior of the molecule (hydrophobic interactions).

5. *Role of hydrophilic side chains.* Hydrophilic groups tend to lie toward the outside of the molecule, in contact with the polar solvent. When the protein is in an environment of nonpolar molecules, as when crossing lipid membranes, the arrangement is reversed, exhibiting the hydrophobic amino acid chains toward the medium.

6. *Van der Waals forces.* The electron cloud of every atom undergoes fluctuations that generate transient asymmetries in the distribution of electrical charges and dipoles. When two atoms not bonded to each other by covalent linkages approach to a sufficiently small distance (0.3–0.4 nm), one atom dipole induces a dipole on the other, determining their mutual attraction. These interactions are weak, but become significant when a large number of atoms are involved. Each atom has a characteristic Van der Waals force, depending on its radius, which determines the distance it will establish with another atom. At small distances, the repulsion between the negative charges of the electron orbitals predominates.

The diagram in Fig. 3.18 summarizes the forces contributing to maintain protein tertiary structure.

Quaternary Structure

So far, proteins constituted by a single polypeptide chain were presented. However, there are many proteins that consist of more than one polypeptide chain. These are called *oligomeric*

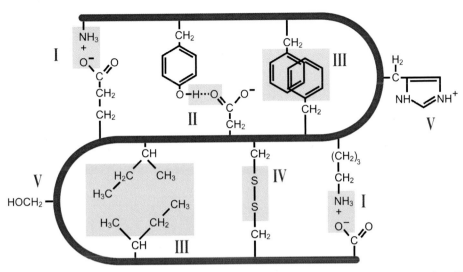

FIGURE 3.18 **Forces maintaining the tertiary structure of a polypeptide chain.** I. Electrostatic attractions. II. Hydrogen bond. III. Interactions of nonpolar groups. IV. Disulfide bond. V. Hydrophilic groups oriented outwards (Source: Modified from Anfinsen).

proteins, in which each chain constitutes a subunit of the protein complex. For example, insulin is composed of two subunits, while hemoglobin and lactate dehydrogenase are composed of four.

The quaternary structure refers to the spatial arrangement that the different polypeptide subunits adopt in the complex. Hydrogen bonds, electrostatic attractions, hydrophobic interactions, and disulfide bridges are the forces responsible for maintaining the position of the different subunits within the oligomeric protein. Obviously, only proteins that contain several polypeptide subunits can exhibit quaternary structure (Fig. 3.19).

Multimolecular Complexes

Another level of protein organization is represented by the interactions among proteins or between proteins and other type of compounds. These multimolecular complexes perform highly specialized functions. Some of these complexes represent "molecular machines," such as the F_1F_0 ATP synthase complex (p. 191), ribosomes (p. 135), and proteasomes (p. 369).

Considerations About Protein Structure

The primary structure of proteins plays a fundamental role in determining their conformation. The forces and interactions that maintain the secondary, tertiary, and quaternary structures result from the presence of certain amino acids at defined positions in the protein

FIGURE 3.19 **Quaternary structure.** A scheme of the arrangement of polypeptide subunits in an oligomeric protein is shown (tetrameric protein).

molecule. For this reason, small changes in sequence may cause large protein conformational changes.

When considering the primary structure of proteins, the possibility of obtaining an almost infinite number of different amino acid sequences through random assembly was discussed. However, not all theoretically possible sequences are feasible, since the formation of α helices or β sheets imposes limitations to the amino acid composition of the chain. If amino acids are arranged randomly not all possible polypeptides will be able to form secondary and tertiary structures compatible with a particular function.

Protein molecules are the result of the slow process of evolution, during which the most suitable sequences to fulfill a given physiological role were selected. In this manner, the collection of amino acid combinations, although vast, is less varied than that expected from their random combination.

Although there is great protein diversity, distant species express many proteins that are homologous. These proteins exhibit remarkable similarity in secondary and tertiary structures allowing them to have similar functions in each species. Clearly, during the course of evolution, certain amino acids in critical positions have been conserved. Other amino acids that did not modify physicochemical properties or the overall conformation of the protein molecule were changed.

Sequence identification of thousands of proteins has facilitated their comparative analysis and confirmed the existence of common structural elements. Particularly in globular proteins there are segments of protein molecules showing associations of α helices and β sheets (Figs. 3.20 and 3.21) in stable arrangements. These structural assemblies behave as a unit and are called protein *domains*. A domain is a section of the molecule with defined structure and folds, often related to specific functional tasks. Evolution has been conservative and has kept constant protein segments that have proven to be functionally efficient.

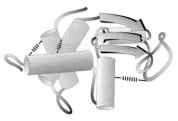

FIGURE 3.20 **Protein molecule formed by α helices and pleated β sheets joined by random coils.** The α helices are represented as *cylinders* and the β sheets as *flat arrows*. The diagram corresponds to a globular protein (lysozyme).

Protein families having common, but not identical, structural features and similar functions have been recognized within humans. Examples of these proteins are steroid receptors, immunoglobulins, and serine proteases. These all share a common genetic origin and, by successive duplications and subsequent independent differentiation steps of the ancestral gene, gave rise to the different proteins of each family.

Another structural characteristic of proteins is their flexibility and plasticity, which is essential for proper functional behavior.

As indicated, the primary structure is the main factor responsible for the final arrangement that

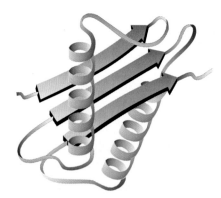

FIGURE 3.21 **Diagram of a protein domain formed by combination of α helices (*helical ribbons*) and β sheets (*flat arrows*).** The tips of the *arrows* indicate the direction of the polypeptide chain. These combinations are supersecondary structures.

a protein adopts. The secondary and tertiary structures are the result of multiple interactions between the functional groups of amino acid residues, located sometimes very far apart from each other in the sequence. Altogether, these interactions maintain the folds and characteristic three-dimensional arrangement of the "native" protein. The process by which a polypeptide chain reaches its correct conformation is carefully controlled and facilitated by auxiliary molecules known as *chaperones*. Defectively folded proteins can cause serious disorders and diseases (see p. 536). Fig. 3.22 shows two different graphical representations of a properly folded polypeptide chain.

Protein Denaturation

The correct functional performance of a protein requires proper conformation of its primary, secondary, tertiary, and quaternary structures. Protein conformation can be altered when proteins are subjected to physical agents, such as heat, radiation, repeated freezing, high pressure, or chemicals (including acids, alkalis, organic solvents, and concentrated urea solutions). Alterations of the structure of a protein leads to the loss of its properties and function. This process is called *protein denaturation* and involves the disruption of bonds and forces that hold the secondary, tertiary, and quaternary structures together. The result is unfolding of the polypeptide chain (Fig. 3.23).

In general, agents that cause denaturation do not directly affect the peptide bonds and, therefore, the primary structure of a denatured protein is maintained. Commonly, the denaturation process causes precipitation of the protein. A common example of protein denaturation is the distortion caused by heat on egg proteins. Denaturation is accompanied in this case by coagulation, an agglomeration of all the molecules into a solid mass. Often denaturation is an irreversible process, as is observed with the hardening of egg proteins, which cannot be returned to

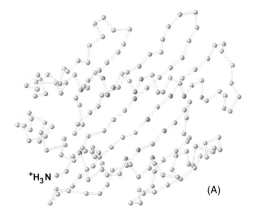

(A)

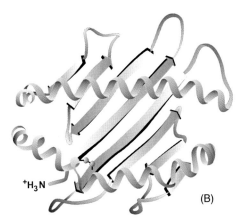

(B)

FIGURE 3.22 **Graphical representation of a polypeptide chain.** Both diagrams correspond to the same structure. (A) Shows the arrangement of the polypeptide chain with *small spheres* representing only the α carbons of the amino acids. (B) Shows the α helical segments as *helical ribbons* and β sheet pieces as *flat arrows*. These structures are linked by segments of random coils. Both figures represent the antigen-binding site of the major histocompatibility complex (MHC) class I. Both α helices delimit a grove in the protein formed by β sheets. The processed antigen is fixed within the protein grove. In (A), the *black spheres* correspond to the αC of amino acids that vary between different MHC proteins. For details about the function of this protein, see p. 766

their original state. In contrast, when the molecular disruption is not very intense, the process may be reversible, with the protein resuming its original conformation after removal of the denaturing agent.

FIGURE 3.23 **Schematic representation of a protein in native state** *(red)* **and the same protein denatured** *(black)*, **unfolded after the break of the forces that stabilize the tertiary and secondary structure of the protein.** The primary structure is not affected.

Protein Classification

Proteins can be classified into two groups: simple and conjugated.

Simple Proteins

This category includes proteins that, when broken down by total hydrolysis, give only amino acids. However, the availability of very sensitive methods has detected the presence of carbohydrates associated with most simple proteins. They are still classified in this group because of the highly predominant proportion of proteins on the total mass of these molecules.

Albumin. This protein easily disperses in water and can be separated from a dispersion by precipitation induced by addition of high concentrations of salts (e.g., saturated ammonium sulfate). The isoelectric point of albumin is approximately 4.7 and its molecular mass is between 60 and 70 kDa. Albumin is a globular protein found in animal and plant tissues. It has been given different names, depending on its origin: ovalbumin (from egg), lactalbumin (from milk), serum albumin (from plasma serum), and legumelin (from plants).

Globulins. These proteins are insoluble in pure water, but disperse in diluted salt solutions. They are less hydrophylic than albumin and precipitate in solution readily by the addition of salts, such as ammonium sulfate, at 50% saturation. Careful precipitation with ammonium sulfate allows proteins to return to their native state when the excess salt is removed by dialysis.

The molecular mass of different globulins is approximately 150 kDa. Usually the molecule comprises several polypeptide chains. Globulins are globular proteins of ovoid shape, which are found in blood plasma, egg (ovoglobulin), milk (lactoglobulin), and animal and plant tissues.

Histones are strongly basic proteins that contain a high proportion of lysine, arginine, and histidine. They make complexes with nucleic acids and are found in association with DNA in the cell nucleus.

Protamines are basic, relatively small proteins that associate with nucleic acids in sperm from different species.

Glutelin and *gliadin* are proteins found in wheat and cereals. Approximately half of the total protein in wheat flour is gliadin. Corn contains a protein of this type, called *zein*. In general, these proteins have a relatively poor nutritional value, because they lack essential amino acids (see further section, "Proteins in Nutrition"). There are pathological conditions that arise due to intolerance of these types of proteins (celiac disease).

Scleroproteins, or albuminoids, are insoluble fibrous proteins present only in animal tissues. They are important for forming part of the supporting tissues and structures of great physical strength. The most important members of this group are keratin, collagen, and elastin. *Keratin* is found in high amounts within the stratum corneum of epidermal tissue: hair, nails, wool, feathers, horns, and hooves. *Collagen* is a material of connective tissue fibers. *Elastins* form the elastic fibers of conjunctive tissue.

Conjugated Proteins

These include proteins that are associated with other compounds. The protein moiety is commonly called *apoprotein*, while the other component is known as *prosthetic group*. According to the prosthetic group bound, conjugated proteins are distinguished in different categories. These include the following:

Nucleoprotein. The protein portion is represented by a strongly basic protein, such as histones, linked to nucleic acids by salt-type links. Nucleic acids are large compounds with important biological functions. A whole chapter will be devoted to these molecules.

Chromoproteins consist of a protein associated with a colored prosthetic group. Many important proteins belong to this category, including hemoglobin, cytochromes, flavoproteins (involved in redox processes), and rhodopsin (present in the retina, plays an important role in the process of vision).

Glycoproteins. These are proteins linked to carbohydrates which may be mono-, oligo-, or polysaccharides. The hydrocarbons are attached to one or multiple sites in the polypeptide chain by *N*-glycosidic bonds (which involve the nitrogen of the amide group of asparagine residues) or *O*-glycosidic bonds (which involve the hydroxyl group of serine or threonine residues). Often the prosthetic group consists of heteropolysaccharides, which represent a significant portion of the molecule.

Phosphoproteins. These proteins act as reservoirs of phosphate in milk (casein) and egg yolk. Moreover, the reversible covalent bonding of phosphate to hydroxyl groups of serine, threonine, or tyrosine is a mechanism that regulates the function of many proteins.

Lipoproteins. In this type of proteins, the prosthetic group is represented by various kinds of lipids. Complexes between phospholipids and proteins are widely distributed in animal tissues. In blood plasma, lipoproteins carry insoluble lipids in the aqueous medium.

Metalloproteins. There is a large group of proteins that are conjugated to metals (Fe, Cu, Zn, Mg, and Mn) which are essential for the structure and function of those molecules.

Proteins in Nutrition

Proteins are very important components of the diet. Carbohydrates and lipids primarily function as energy suppliers, a role which is secondary for proteins. Proteins provide the building blocks (amino acids) required for the synthesis of endogenous proteins and the nitrogen needed for the formation of other compounds. The main function of proteins is structural, a role that cannot be replaced by any other components of the diet.

Within the existing amino acids, eight (phenylalanine, isoleucine, leucine, lysine, methionine, threonine, tryptophan, and valine) cannot be synthesized by humans. These amino acids are designated *essential* or *indispensable* and must be necessarily supplied by proteins present in foods. Two other amino acids, arginine and histidine, are synthesized in human tissues; however, they are produced at a rate that is insufficient to meet increased demands of the body, such as growth, pregnancy, and lactation. Under these circumstances, arginine and histidine also should be considered essential. From a nutritional point of view, the quality or "biological value" of dietary proteins depends on their content of essential amino acids. Animal proteins (from red and white meats, milk, and eggs) are generally of higher biological value than those from plants. Therefore, a correct diet shall include animal protein of high biological value.

Protein Structure and Function

The relationship between protein structure and function can be better understood by considering some specific proteins. These are described in the following sections.

COLLAGEN

Collagen is the most abundant protein in vertebrates (it makes up more than 25% of total proteins in the body). It is a structural component of connective tissue fibers, especially abundant in skin, bones, tendons, and cartilage. Collagen is water insoluble and cannot be digested by the enzymes of the gastrointestinal tract. When subjected to boiling water, it turns into gelatin, which makes this protein soluble and digestible.

The primary structure of collagen is unique. It has a high proportion of glycine and proline residues, with one out of three residues in its sequence being a glycine and one out of four residues being a proline. Many of the proline residues are hydroxylated (4- and 5-hydroxyproline). These two hydroxy derivatives, which are rare in other proteins, represent 25% of the total amino acid residues of collagen. Another distinctive feature of collagen is the presence of the sequence glycyl-prolyl-hydroxyprolyl, which is repeated many times along the chain. This sequence repetition is rare in other proteins. Tryptophan and cysteine are almost absent in the collagen molecule and essential amino acids are found in very small proportion. The polypeptide chains of collagen consist of approximately 1000 amino acids. Due to its primary structure rich in proline and hydroxyproline, collagen cannot form α helices because the pyrrolidine ring is incompatible with that conformation. The protein chain instead forms a more extended helix that rotates counter clockwise, comprising in each turn three amino acid residues. This type of helix is almost exclusive of this protein. Three of these collagen chains associate to form a coiled coil. All three helices are tightly wrapped around each other and, together, they coil around a central axis like strands of a rope (Fig. 3.24).

The three collagen chains are interconnected and held together by hydrogen bonds, which form transversal bridges between the protein chains. These bonds are different from the H bonds that maintain the α helix.

Since glycine is a small amino acid, without a side chain, it can easily accommodate within the collagen helix. This allows the collagen chains to become close to each other. In contrast, other bulkier amino acid residues project outward in the protein chain. No disulfide bridges are formed because there are virtually no cysteines in the chain.

Triple helices of collagen form structural units called *tropocollagen*, which are 300-nm long rods of 1.5 nm in diameter. Although all of the polypeptide chains of collagen have similar characteristics, there are several varieties of this protein that present slightly different primary structures. According to the class of chains involved, various types of collagen can be distinguished.

Tropocollagen units are arranged in rows and these in turn are packed into bundles which form fibers. All units in a fiber have the same orientation, with the N-terminal end of the polypeptide chains directed toward the same side. The ends of each unit are not in direct contact, leaving a space between them (Fig. 3.24). Tropocollagen fibers form the matrix where bone calcification takes place, and the spaces between the units in each row are the sites where the crystallization of the mineral component of bone (hydroxyapatite) is initiated.

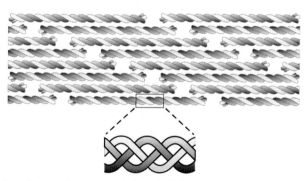

FIGURE 3.24 **Diagram of a collagen fiber segment.** The top panel shows the arrangement of tropocollagen units with the spaces between their ends in each row and the displacement of units in one row with respect to the adjacent one. All constituent molecules have the same orientation. The bottom panel shows and enlarged view of the triple helix of tropocollagen.

Each tropocollagen molecule in the row is shifted with respect to the one in the adjacent row by approximately one-quarter of its total length (Fig. 3.24). This displacement is maintained with great regularity in all rows and is responsible for the striations that collagen fibers show when viewed by electron microscope. Tropocollagen molecules of adjacent rows establish a special type of bond between them, which involves lysine or hydroxylysine side chains. This gives collagen a great resistance to traction. It is estimated that a fiber of 1-mm diameter of tropocollagen can withstand a weight of 10 kg. The number of cross-linkages between neighboring lysines of tropocollagen increases with age, and this causes the fibers to gradually become more rigid and fragile. This partly explains the increased susceptibility to bone fractures observed in the elderly.

There are at least 14 types of collagen that differ slightly from each other in primary structure. Those designated I, II, and III are the most abundant. Type I is found in skin, tendons, bone, and dentin; type II is found in cartilage and vitreous humor; type III is found in skin, muscle, and blood vessels; types IV and VII do not form fibers, but a two-dimensional lattice, which is the main constituent of the basal lamina; types IX and XII are found in cartilage, tendons, and ligaments.

Genetic defects can affect the structure of collagen. Different abnormalities of collagen are responsible for *Ehlers Danlos syndrome*. This syndrome appears in patients with defects in the gene that directs the synthesis of collagen type I, resulting in loose joints and a flaccid skin that can be easily stretched. Alterations in the gene for type III collagen cause defects in the arterial walls and vascular accidents. Most cases of *osteogenesis imperfecta*, or brittle bone disease, are due to mutations that affect the synthesis of collagen type I. Lastly, vitamin C plays an important role in the formation of collagen; deficiencies of this vitamin can lead to scurvy, a disease that is characterized by poor wound healing, subperiosteal and joint hemorrhage, anemia, and "corkscrew" hair.

KERATINS

Two main types of keratins can be identified, α and β. The α keratins are the major components of hair, nails, and skin. These filament-like proteins are arranged into α helices and consist of more than 300 amino acids in length, with cysteine being a predominant residue. Two keratin α helices usually wrap around each other forming a double helix. In hair, keratin commonly constitutes flexible fibers that have four polypeptide chains assembled as super helices. In other tissues, especially nails, the fibers have less flexibility due to the existence of a large number of disulfide bonds between the cysteines of different chains. The β keratins are characterized by presenting a β sheet structure; they are found in bird feathers and in the scales of reptiles.

HEMOGLOBIN

Hemoglobin is a conjugated protein whose prosthetic group, *heme*, gives it its typical intense red color. Among the heme proteins are the cytochromes, which are substances that act as electron carriers and enzymes, such as catalase and peroxidase; myoglobin, a molecule that carries and stores oxygen in muscle; and hemoglobin, which is responsible for the transport of oxygen in blood.

Both myoglobin and hemoglobin are constituted by the protein globin, which is rich in lysine, arginine, and histidine. Myoglobin, with a molecular mass of 16,700 Da, is composed of one globin chain bound to a heme group, while hemoglobin, with a mass of 64,500 Da, is a tetrameric molecule composed of four globin chains, each associated with a heme group. Myoglobin resembles one of the subunits of hemoglobin both in size and structure. Max Perutz and John

Kendrew, using X-ray crystallography, provided evidence on the structure and function of both of these molecules.

HEME

The prosthetic group of hemoglobin derives from the porphine molecule, a large ring or macrocycle of four pyrrole groups linked by methine bridges (=CH−) (Fig. 3.25). Pyrroles are numbered from I to IV and the methine bridges are designated by Greek letters, from α to δ. The numbers 1–8 indicate the positions of hydrogen atoms bonded to the carbons of the pyrrole groups.

All atom components of the porphine ring are located in the same plane. This structure possesses resonance; the double bonds in Fig. 3.25 are only intended to indicate that the carbon and nitrogen atoms of the macrocycle have all their valences saturated, but several resonance forms could be represented.

When the hydrogens in the positions 1–8 are replaced by carbon radicals, the porphine becomes porphyrin. Porphyrins have different names depending on the type of substituents. Protoporphyrin III (IX in Fisher's original notation) is the prosthetic group of hemoglobin, it

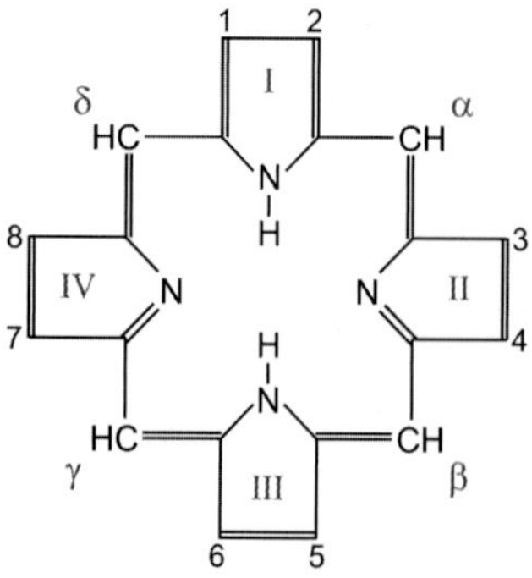

FIGURE 3.26 **Heme.**

FIGURE 3.25 **Porphine.** The pyrrole groups are indicated by Roman numerals; the methine bridges, by Greek letters. Numbers 1–8 indicate the positions that can be substituted by different carbon chains.

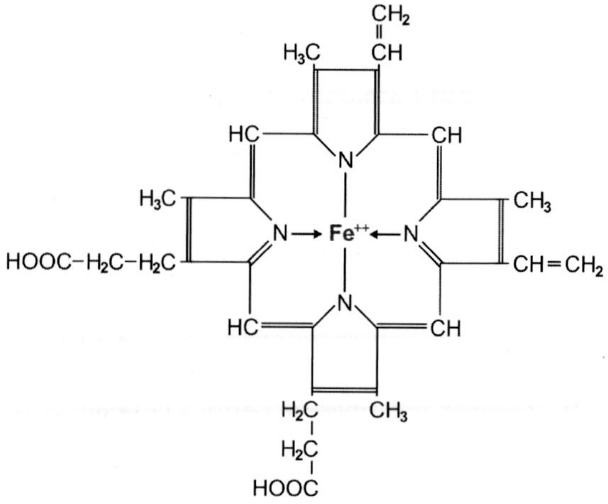

has methyl groups (—CH_3) at positions 1, 3, 5, and 8; vinyl (—CH=CH_2) at positions 2 and 4; and propionyl (—CH_2—CH_2—COOH) at positions 6 and 7.

One of the important properties of porphyrins is their ability to form complexes with metals. Heme consists of protoporphyrin III with an iron atom enclosed in the center of the tetrapyrrole ring (Fig. 3.26).

The heme iron of hemoglobin is bivalent or ferrous (Fe^{2+}). This Fe^{2+} forms six coordinate bonds, accepting unshared electron pairs from other ligands. (Ligand: atom or group of atoms that participate in a coordinated bond yielding two not-shared electrons.) Those electrons are located in six hybrid orbitals. When hemoglobin binds oxygen, these orbitals have an octahedral configuration. Four of these bonds coordinate the union of Fe^{2+} to the N atoms of the pyrroles. The fifth bond attaches heme to the N of imidazole in a histidine residue in globin. The sixth coordination position of Fe^{2+} can bind to an O_2 molecule (Fig. 3.27).

Hemoglobin can only fulfill its function of binding and transporting oxygen when the heme iron is in the ferrous state. If iron is oxidized to ferric (Fe^{3+}) heme is converted into hematin and

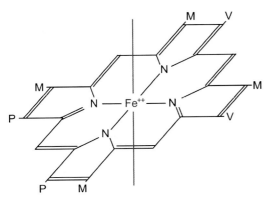

FIGURE 3.27 **Spatial arrangement of Fe²⁺ links in hemoglobin (red).** One is attached to a histidine of the globin chain, another to a molecule of oxygen when hemoglobin is converted into oxyhemoglobin.

hemoglobin becomes methemoglobin, which is unable to carry oxygen.

GLOBIN

Hemoglobin (Hb) is a tetrameric molecule. It is formed by the association of 2 polypeptide chains of 141 amino acids each (α or ζ) and 2 chains of 146 amino acids (β, δ, γ, or ε), which vary depending on the type of hemoglobin considered. Despite the differences in the sequences of these chains, they all have a similar shape.

Two different types of hemoglobin are normally found in a normal adult human: (1) Hemoglobin A_1 (HbA$_1$), which consist of two α and two β subunits (the notation for this molecule is $\alpha_2\beta_2$). This is the most abundant form, accounting for over 95% of the total hemoglobin in red cells. (2) Hemoglobin A_2 (HbA$_2$) consists of two α and two δ chains ($\alpha_2\delta_2$) and represents up to 3% of the total hemoglobin.

In contrast, the human newborn blood contains another type of hemoglobin, called fetal hemoglobin (HbF), which is composed of two α and two γ chains ($\alpha_2\gamma_2$). Hemoglobin F is predominant in the fetus during the last 6 months of intrauterine life. During this time, HbF is important for the growth and survival of the fetus due to its higher affinity for oxygen than HbA, which allows for the extraction of oxygen from the mother's blood as it passes through the placenta. At birth, approximately 80% of the hemoglobin in cord blood is HbF (the rest is HbA$_1$). Thereafter, a sustained decrease in HbF takes place, until it totally disappears at approximately 6 months of age. Then, the normal distribution of HbA$_1$ and HbA$_2$, typical of adulthood, is reached. During the earliest periods of intrauterine life (between the first and third months) other types of hemoglobin can be also identified. These are known as embryonic hemoglobins, Hb Gower 1, Gower 2, and Portland. The Hb Gower 1 consists of two zeta (ζ) and two epsilon (ε) subunits ($\zeta_2\varepsilon_2$). The ζ chains are later replaced by α chains and Gower 2 ($\alpha_2\varepsilon_2$) hemoglobin is formed. The Portland Hb is a $\zeta_2\gamma_2$ tetramer.

The various hemoglobin types can be distinguished by electrophoresis. Each hemoglobin chain differs in amino acid composition and total net charge at a particular pH; therefore, they have a particular mobility when subjected to an electrical field.

Studies of the structural and functional properties of hemoglobin have been primarily focused on HbA$_1$. These aspects are discussed in the following paragraphs.

Hemoglobin quaternary structure. The four chains of hemoglobin are assembled to produce a compact, almost spherical molecule of 5.5-nm diameter, with a cavity along its central axis (Fig. 3.28). Each of the chains forms a niche where the heme is housed. The four heme groups are apart from each other and oriented with the two propionyl chains outward. The links of the subunits are stronger between different chains (α–β) than between same class chains (α–α or β–β). The interactions in the α–β chains involve hydrophobic attractions and hydrogen bonds, whereas the contacts between subunits of the same class are scarce and limited to electrostatic attractions (salt bridges). This last type

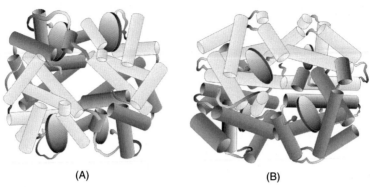

(A)   (B)

FIGURE 3.28 **Schematic representation of hemoglobin.** The α helices, constituting 80% of the molecule, are depicted as *cylinders*; α subunits are shown in *light gray* and β chains, in *dark gray*; hemes are represented by *red disks*. (A) Hb molecule upper view; note the empty space in the center. (B) Side view. The hemes occupy niches or pockets that open to the outside.

of link is flexible and can be modified, an event that is important for the function of hemoglobin.

Hemoglobin secondary and tertiary structures. Hemoglobin has features that are peculiar for a globular protein. Each subunit is formed by helix segments connected by random coils and almost 80% of the molecule has a helical structure (Fig. 3.29). The β chain has eight helices designated A–H, starting from the protein N-terminus. The α chain has seven helices, presenting the same regions of the β chain, with the exception of the D segment that is missing. Amino acids in hemoglobin are numbered according

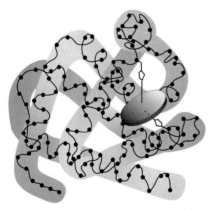

FIGURE 3.29 **Hemoglobin subunit β.** α helices and random coil segments are shown. Red disc represents the prosthetic heme group (Perutz).

to their position in the chain or to their order in the segment to which they belong to. For example, histidine in position 58 of the α subunit is the seventh residue of helix E; its notation is therefore His α58 or His αE7. In general, polar residues are placed toward the surface of the hemoglobin molecule, in contact with the aqueous medium, while the majority of the hydrophobic residues are located inside. These nonpolar residues help to stabilize the assembly of the four subunits and to form the niche where the heme is located. The pocket for the heme involves the space between helices E and F of each chain. The prosthetic group binds to globin by a coordinate bond between the Fe^{2+} and a nitrogen of the imidazole of histidine 87 (F8) in the α chains and histidine 92 (F8) in the β chains. The nonpolar environment around the heme is essential to maintain the Fe in ferrous state and thus, its ability to carry oxygen.

Functions. Hemoglobin reversibly binds oxygen to form oxyhemoglobin. Each Hb molecule reacts with four oxygen molecules (one per heme):

$$Hb + 4O_2 \rightleftarrows \underset{\text{Oxyhemoglobin}}{Hb(O_2)_4}$$

The course of the reaction is highly dependent on the partial pressure of oxygen (P_{O_2}) in the blood. An increase in O_2 partial pressure shifts the reaction toward the right (allowing formation of oxyhemoglobin) while a decrease of P_{O_2} shifts the reaction to the left. A partial pressure of 100 mmHg, or 100 Torr [Torr (from Torricelli), is the unit of pressure equivalent to 1 mm of Hg at 0°C. The International System (IS) unit is the Pascal (Pa). 1 kPa = 7.5 mm Hg; 1 mm Hg = 133.322 Pa], or 133,322 hectopascals in alveolar air almost completely saturates hemoglobin, with nearly 100% of Hb being converted into oxyhemoglobin. Under these conditions, each gram of Hb is able to fix 1.34 mL of oxygen. The blood of a normal adult contains about 15 g/dL of Hb; therefore, the total capacity for O_2 transport is approximately 20 mL/dL of blood (the solubility of O_2 in plasma is poor, the amount of gas dissolved in it is very small compared to that transported by Hb so, it is not taken into account in this calculation).

Anemia is the pathologic condition in which Hb content falls below normal values, reducing the ability to transport O_2 to tissues.

Oxygen dissociation curve of hemoglobin. The ratio between oxygen partial pressure and oxyhemoglobin formed is described by the oxygen dissociation curve of hemoglobin. To determine this, a suspension of hemoglobin is placed in contact with a gas mixture of known oxygen partial pressure until it reaches equilibrium. Then, the percent of hemoglobin transformed into oxyhemoglobin at different P_{O_2} levels is measured (Fig. 3.30). By plotting the P_{O_2} values on the X axis and the percentage of formed oxyhemoglobin on the y axis, a sigmoid curve is obtained. At low oxygen tension, there are only little ascent in the curve; however, at a P_{O_2} above 10 Torr (13.3 hPa), the curve rises sharply. Subsequent increases in P_{O_2} produce large increases in the percentage of oxyhemoglobin. When a P_{O_2} of 60 Torr (79.8 hPa) is attained, approximately 90% of oxyhemoglobin is saturated with O_2 and the curve reaches a plateau. At a P_{O_2} of 90 Torr (119.7 hPa), virtually all Hb is saturated with O_2. These changes in O_2 saturation at different P_{O_2} are important to understand the delivery of oxygen to tissues in the body. In the lungs, the P_{O_2} is high and allows the oxygen binding for transport to

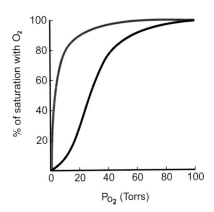

FIGURE 3.30 **Hemoglobin *(black)* and myoglobin *(red)* saturation curves.** The myoglobin curve is hyperbolic; it shows high affinity at low partial pressures. That of hemoglobin is sigmoid, it shows lower affinity at low partial pressures; this facilitates release of oxygen in tissues.

the rest of the body. Once the red blood cells reach the capillaries of an organ, the P_{O_2} drops dramatically as the P_{CO_2} increases and the oxygen is released in exchange for CO_2 to provide that organ with the oxygen it needs to function and survive.

If the same experiment is performed with myoglobin, the curve has a very different appearance (Fig. 3.30). Its shape is not sigmoid but instead hyperbolic, presenting a rapid raise at low P_{O_2} levels.

Comparison between the myoglobin and hemoglobin curves shows that myoglobin has a higher affinity for O_2 than hemoglobin. At a P_{O_2} of 10 Torr (13.3 hPa), more than 85% of myoglobin is oxygenated, whereas less than 5% of Hb is converted into oxyhemoglobin. This shows that myoglobin is not a useful carrier of oxygen in blood, because at the P_{O_2} existing in tissue capillaries (approximately 30 Torr, 39.9 hPa) it does not release the O_2.

The sigmoid shape of the Hb curve indicates that at very low O_2 pressures, this gas hardly binds to the heme, but when the pressure is increased, the formation of oxyhemoglobin is rapidly stimulated. Although this phenomenon was described in the early 20th century, it was understood many years later, when better knowledge

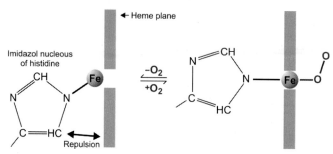

FIGURE 3.31 **Displacement of heme Fe^{2+} during oxygenation of Hb.** In deoxygenated Hb (left) the Fe atom is out of the heme plane because the imidazole ring of histidine to which Fe^{2+} binds is rejected by a nitrogen of heme. When Hb oxygenates (right) the Fe^{2+} is displaced toward the heme plane and drags the histidine modifying conformation of the Hb chain.

of the Hb structure was obtained. In its deoxygenated form, the four Hb chains remain closely associated.

Both α and both β chain pairs are joined by salt-type links giving the Hb a conformation called "tight" (T) because it prevents O_2 access to the heme. The Fe^{2+} is bound by its fifth coordination position to histidine F8 of the globin chains and is slightly displaced from the plane of the heme (Fig. 3.31). When an oxygen molecule accesses one of the pockets occupied by the heme and binds to the Fe^{2+} at its sixth coordination position; it displaces the Fe^{2+} toward the center of the ring (Fig. 3.31). This Fe^{2+} shift drags the histidine F8 residue, bringing the E and F helices together, which finally causes a conformational change in the whole polypeptide subunit. Due to the associations between the Hb chains, the change exerted on one of the chains is transmitted to the others. The relieving of the salt bridges between chains alters the arrangement of Hb subunits.

This conformational change of Hb facilitates the access of O_2 to the heme groups in the other subunits and Hb takes a "relaxed" conformation (R). The phenomenon by which O_2 binding to one Hb subunit modifies the shape of all Hb subunits to make them more receptive to O_2 is called heme–heme interaction or heme *cooperative effect*. The reverse heme–heme interaction also occurs; the exit of one of the oxygen molecules of oxyhemoglobin restores the formation of salt bridges, returning Hb to its T state,

to facilitate the release of O_2 from the remaining subunits. This type of cooperative effect has also been observed in other protein molecules, mainly enzymes, which are oligomeric molecules as the Hb. In contrast, monomeric molecules, such as myoglobin, do not show cooperative effect (Fig. 3.29). Factors as low pH, increase in partial CO_2 pressure or temperature, and presence of organic phosphorus containing compounds produce a rightward shift of the Hb dissociation curve (Fig. 3.32). The same factors have no effect on myoglobin. The action of pH and CO_2 on the

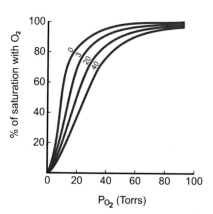

FIGURE 3.32 **Hemoglobin oxygen saturation curves.** Effect of CO_2 partial pressure. Numbers on each curve indicate CO_2 partial pressure (in Torr). P_{CO_2} increase decreases the affinity of Hb for O_2; the phenomenon (Bohr effect) is significant at low partial pressures of O_2; at pressures of 90 Torr or greater, the percent saturation is almost the same at any P_{CO_2}.

O_2 binding by Hb is called the *Bohr effect*. The increase in $[H^+]$ or CO_2 favors the return of the Hb molecule to the T state, decreasing its affinity for oxygen.

The Bohr effect has great physiological significance. Oxyhemoglobin releases its oxygen more readily at the tissue level, where the pressure of CO_2 is high and the pH is low. This represents a regulatory mechanism promoting faster release of oxygen where it is most needed.

2,3-bisphosphoglycerate. Present in substantial quantities in red blood cells, 2,3-bisphosphoglycerate (BPG) is linked to both β chains in the central cavity of Hb. BPG stabilizes Hb in the T form, hindering O_2 access to the heme and decreasing the affinity of Hb for oxygen at low P_{O_2} (Fig. 3.33). When Hb is oxygenated, salt bridges between the chains become weaker, the space in the central cavity narrows, and BPG is displaced causing Hb to return to the R form.

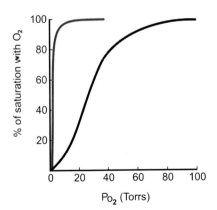

2,3-Bisphosphoglycerate

FIGURE 3.33 **Effect of 2.3-bisphosphoglycerate (BPG) on the saturation curve of hemoglobin.** In the presence of BPG *(red)* the affinity of Hb for O_2 at low partial pressures is significantly increased. In the absence of BPG *(black)* the affinity markedly decreases.

The BPG plays an important physiological role. The presence of BPG in red cells reduces the affinity of Hb for O_2 and promotes its release in tissues. Variations in the concentration of BPG have clinical interest:

1. Decreased affinity of Hb for O_2 is a mechanism of adaptation to hypoxic conditions (high altitude and severe cardiopulmonary conditions) favoring a more readily release of O_2 in the tissues. This adaptation is accomplished by increasing the BPG concentration in red cells.
2. Citrate–glucose anticoagulant solution is normally added to blood used in transfusions. This treatment causes the concentration of BPG to decrease with time. This can cause disorders in patients transfused with large volumes of blood that has been stored for long periods of time. Due to the low content of BPG, the Hb transfused does not release efficiently oxygen to the tissues and the recipient can suffer from hypoxia even when the blood volume has been restored to normal levels.
3. Fetal hemoglobin (HbF, $\alpha_2\beta_2$) has higher affinity for oxygen than HbA_1. This is an important physiological property, since it allows the transfer of O_2 in the placenta from maternal to fetal Hb. The difference in oxygen affinity between HbA_1 and HbF is due to the lower tendency of the latter to bind BPG.

The regulatory mechanisms described for Hb, in which the binding of a substance to an effector molecule cause conformational and functional changes, are called *allosteric* effects. Allosteric effects arise from cooperative actions among subunits of oligomeric proteins.

Transport of Carbon Dioxide

Hemoglobin is also involved in the transport of CO_2. Approximately 5% of the total CO_2 carried by blood and released in the lungs is transported as *carbamino* (bonded to amino-terminal groups of globin):

$$Hb—NH_2 + CO_2 \rightleftarrows Hb—NH—COOH$$

When the blood reaches the lungs, oxyhemoglobin is formed and this favors the release of CO_2 from the carbamino. This phenomenon (Haldane effect) is opposite to the Bohr effect.

HEMOGLOBIN DERIVATIVES

Carboxyhemoglobin

Carboxyhemoglobin is produced by the binding of carbon monoxide (CO) to hemoglobin. CO is generated during incomplete combustion of organic products and has toxic effect because it competes with oxygen for the same binding site in the Fe^{2+} of Hb. Hemoglobin bound to CO is unable to transport O_2. Hb has a 210-fold greater affinity for carbon monoxide than for oxygen, which explains the extraordinary toxicity of CO even when its concentration in the inspired air is relatively low. For example, breathing air containing 0.02% CO for 2 h causes symptoms, such as headache and nausea. A concentration of 0.1% produces loss of consciousness within 1 h and death by asphyxia in only 4 h. Car exhaust gases contain between 4% and 7% CO and toxic concentrations are readily achieved in poorly ventilated environments (garages). Carboxyhemoglobin has a cherry red color. For that reason, subjects intoxicated with CO show apparently "healthy" reddish lips and cheeks.

Methemoglobin

Methemoglobin is a type of Hb that has *hematin* or ferriheme as a prosthetic group instead of ferrous heme. The conversion of Fe^{2+} to Fe^{3+} prevents hemoglobin from transporting oxygen. The heme iron is maintained in the ferrous state by several factors. These include: (1) presence of nonpolar amino acid residues in the niches of the Hb molecule in which the heme is housed and (2) reductive systems in red blood cells that prevent the elevation of methemoglobin levels. Genetic defects can produce changes in the hydrophobic environment of the heme or affect the reducing system enzymes, generating methemoglobinemia. Moreover, agents, such as nitrophenols, aniline, or drugs can cause methemoglobinemia and variable degrees of tissue hypoxia. Methemoglobin has a dark brown color. Its increase in blood is clinically apparent by the presence of brownish cyanosis.

Hemoglobin A_{1c}

Besides hemoglobins A_1 and A_2, adults have a derivative of HbA_1, designated HbA_{1c}, which is produced by Hb glycosylation. HbA_{1c} can reach up to 3.5% of total hemoglobin in blood and it is slowly generated within the RBCs by a reaction between hemoglobin and glucose-6-phosphate, which produces a ketoamine (amino-l-desoxifructose) on the N-terminal end of the Hb subunits.

In poorly controlled diabetic patients, high amounts of HbA_{1c} are found. Levels of this glycosylated derivative can comprise up to 15% of the total Hb and are directly related to the concentration of blood glucose during the previous 2 or 3 months (the red blood cells remain in circulation no more than 120 days). The level of HbA_{1c} is used as a marker to determine if glycemia has been high during the period immediately prior to the determination.

ABNORMAL HEMOGLOBINS

The synthesis of the polypeptide chains (α, β, γ, δ, and ε) that form part of the different types of hemoglobin is controlled by different genes. (Diploid cells have two genes for each of the β and δ subunits, one inherited from each parent. α and γ have four genes each, two from each parent.) Alterations in the information contained in the genes (mutations) can generate defects or the inability to synthesize globin. Sometimes mechanisms regulating the production of chains are affected. These alterations cause a group of hereditary diseases, known as *hemoglobinopathies*. Currently, more than 650 genetic hemoglobin abnormalities have been described in humans. Initially, each variant found was designated with a capital letter or the initial letter of the most striking disease feature. Soon the number of variants exceeded the available letters; then the variants were named according to the place where they were first observed (Mexico, Zurich, and Bethesda), the name of the patient carrier of the defect (Sabine and Alexandra), or the ethnic group from which it was found (Babinga).

Several hemoglobinopathies have been described that include: (1) substitution of one or two amino acids (these are the most common type); (2) lack of one or more amino acids; (3) presence of "hybrid" chains, formed by pieces of two different polypeptide subunits (i.e., Hb Lepore or hybrid δβ); (4) presence of chains longer than normal; (5) decreased or lack of synthesis of a certain type of chain (thalassemia).

Many of the findings of hemoglobin variants have been casual because a change in Hb sequence does not always cause functional defects. However, sometimes just the replacement of a single amino acid for another can greatly alters Hb. Critical changes in Hb could lead to the following defects:

1. Instability of the hemoglobin with a tendency to precipitate in the erythrocytes. This causes premature destruction of red blood cells, with clinical symptoms of hemolytic anemia. This happens in *sickle cell disease* or sickle cell anemia, which contains *hemoglobin S* (HbS). HbS differs from HbA in only one of the amino acids within the β chains. The substitution in this disease includes the replacement of a glutamic acid in position 6 by valine and is designated by the notation $\alpha_2\beta_2^{6val}$. This amino acid exchange makes Hb highly unstable, especially when it is in its deoxygenated form. This causes Hb polymerization and precipitation, with deformities in red blood cell shape (sickle or crescent appearance). The defect occurs almost exclusively in individuals of African origin.

2. Modification of amino acids in the portion of Hb housing the heme, involving histidines linked to Fe^{2+}, can favor oxidation of Fe^{2+} to Fe^{3+}, generating methemoglobin that leads to cyanosis.

3. Amino acid changes in regions of contact between subunits can affect the cooperative or allosteric response of Hb to O_2. This results in Hb that exhibits greater affinity for O_2, or that does not respond to changes in pH or CO_2 concentration (lacks the Bohr effect). This abnormality affects the ability of Hb to release O_2 in tissues.

4. Amino acid replacements of residues involved in the binding of 2,3-bisphosphoglycerate can lead to the inability of Hb to bind this compound. Consequently Hb affinity for O_2 increases and the release of this gas in the tissues is impaired. In all cases the condition is most severe in individuals who have inherited defective genes from both parents (homozygous). In individuals heterozygous for the mutation, only one copy of the gene is altered producing a less severe disease.

An important group of hemoglobinopathies includes the *thalassemias*, genetic diseases in which the synthesis of a given subunit is reduced or completely missing. This defect is characterized by erythrocytes that are morphologically abnormal, presence of inclusion bodies (due to precipitation of unstable Hb), and premature destruction of red blood cells, leading to severe anemia.

In thalassemias the polypeptide chains of Hb have a primary structure identical to normal Hb. However, the abnormality consists in an alteration of the proportion of Hb types, or generation of Hb homotetramers. Thalassemias are more common in countries of the Mediterranean basin (Italy and Greece), Arabia, and other Asian countries. Along with HbS, they are the most common hemoglobinopathies.

When the disorder affects the synthesis of α chains, the defect is designated α thalassemia. In these cases hemoglobin lacks the α chains and is composed by a homotetramer of β chains, HbH (β_4), or more rarely δ chains, as in Bart Hb (γ_4). The β_4 and δ_4 homotetramers have higher affinity for O_2 than HbA, they do not present the Bohr effect, and they do not respond to BPG, rendering HbH and Bart Hb functionally incompetent. The total inability to produce α chains is a lethal condition.

The β thalassemias result from deficiencies in β chain production. These are accompanied by a compensatory increase in the synthesis of other subunits. Most frequent among these conditions is the increment of δ chains, which leads to a rise in HbA$_2$ ($\alpha_2\delta_2$). In other cases, the synthesis of γ chains may persist over time and thus, fetal Hb (HbF $\alpha_2\gamma_2$) is abnormally found in adults. Less common is the formation of α_4 homotetramers. Individuals that are homozygous for this defect suffer from a disease known as *major beta thalassemia*, which is characterized by severe anemia that appears shortly after birth. Heterozygotes carriers of this disease can synthesize some level of β chains, causing *minor beta thalassemia*, which usually does not require treatment.

In his description of HbS defect, Pauling proposed the name "molecular pathology" for this type of diseases. As for other genetic diseases, the development of molecular biology techniques and genetic engineering has immensely helped to understand the structure and control of Hb gene synthesis. This information raises hopes for finding a solution for these diseases.

BLOOD PLASMA PROTEINS

Blood plasma is the medium in which the blood cells (red cells, white cells, and platelets) are suspended. Proteins are the most abundant solid components of plasma, with total protein plasma concentration ranging from 6 to 8 g/dL (average 7.2 g/dL) in a normal adult human.

Early methods for protein fractionation, using salt precipitation, allowed the separation of two main plasma proteins: *albumin* and *globulins*, as well as a minor protein fraction of lower solubility, called *fibrinogen*, which participates in the process of blood coagulation.

To obtain full plasma, an anticoagulant agent must be added because if blood is allowed to clot, plasma fibrinogen is converted to fibrin, which forms the network that traps all formed elements of blood. After a clot is produced, it reorganizes releasing, a yellowish liquid, the serum, which consists of plasma devoid of fibrinogen.

Ammonium sulfate (at half saturation), or 27% sodium sulfate, selectively precipitates serum globulins, leaving only albumin in the dispersion. Determination of total protein in serum, before and after precipitation of globulins, allows for estimation of the concentration of albumin and globulins in plasma. In clinic, it is of interest to know the ratio between albumin and globulins (Alb/Glob). The normal average value for Alb/Glob ratio is 1.5. A lower ratio is a common finding in many disease states.

Edwin J. Cohn and coworkers developed a method for isolating plasma proteins based on their fractional precipitation with ethanol and pH changes in a medium with low salt concentration maintained at low temperature. This procedure yields six protein fractions, two of which correspond to albumin (fraction V) and γ globulin (fraction II) in a relatively pure state. The remaining fractions are complex protein mixtures. This method is used to process large amounts of plasma proteins that are used for therapeutic purposes.

Electrophoresis of serum proteins. At a given pH, the various plasma proteins exhibit distinct electric charges and they migrate with different velocity to one of the electrodes when subjected to an electric field. This is the basis of electrophoresis, a separation method used extensively in the study of proteins in different biological samples. Electrophoresis commonly employs filter paper, cellulose acetate, or semihydrolyzed starch, agar or polyacrylamide gels as support medium in which the migration of proteins takes place. Due to its simplicity the method of cellulose acetate has been the most commonly used in clinical laboratory practice.

Usually, electrophoresis on cellulose acetate allows the separation of five protein fractions from blood serum. The fractions can be identified as bands, after the support medium on which the proteins have migrated is stained with specific dyes. The intensity in color and the area of each band are proportional to the amount of protein contained in the sample (Fig. 3.34A). The use of densitometry measures the intensity of the stained bands. A scan of the stained supporting material provides a graph with a series of waves, which correspond to each of the protein bands in the sample (Fig. 3.34B). The area under each curve in the graph is proportional to the intensity and color of the original bands and provides a close estimate of the relative protein amount in each fraction. These types of graphs are usually known as *electrophoretic proteinograms.*

Usually protein separation is performed at a pH of 8.6. Under these conditions, all serum proteins have electronegative charge and, therefore, they all migrate toward the anode. The serum protein that shows the fastest electrophoretic migration is albumin. Following albumin are the globulins, which separate into four fractions, α_1, α_2, β, and γ, in order of decreasing motility. The normal serum protein values are presented in Table 3.2.

The only determinant of protein separation on cellulose acetate is the electrical charge. The five protein fractions obtained by using the electrophoretic techniques should not be interpreted as the existence of only five individual proteins in serum. Each protein fraction, especially the globulins, is composed of a mixture of different proteins that migrate together because they have the same net charge. The introduction of other electrophoretic methods, such as starch and polyacrylamide gels, which rely on electrical field separation as well as protein size and shape, significantly improved the method for protein discrimination. These techniques allow for the distinction of approximately 20 protein fractions.

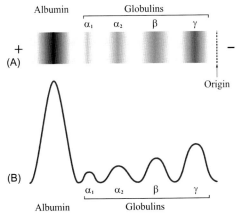

FIGURE 3.34 **Serum protein separation by electrophoresis on a cellulose acetate film.** (A) Substrate colored with a specific protein dye. (B) Densitometric graph of A.

TABLE 3.2 Serum Protein Fractions Separated by Electrophoresis in Cellulose Acetate

Fraction	Range (%)	Average (%)
Albumin	47–70	59.0
α_1 Globulin	3–7	4.5
α_2 Globulin	5–12	9.0
β Globulin	5–16	12.0
γ Globulin	10–20	15.5

Values indicate percentage of total proteins.

Immunoelectrophoresis. Another method of serum protein separation combines electrophoretic techniques with immunochemical methods. If an animal (rabbit or horse) is injected with human blood serum, it reacts against the foreign human macromolecules, producing a series of antibodies specific for each of the human serum proteins. These antibodies can be isolated from the animal's blood and can be used to selectively recognize human serum proteins in a serum sample. In immunoelectrophoresis, serum proteins are first separated in an agar gel. Once the migration of the proteins is complete, a groove parallel to the direction of sample migration is carved in the gel. This channel is then filled with the antihuman serum generated in animals. After incubation for a period of time, the antibodies contained in the grove slowly diffuse through the agar. When they encounter the separated plasma proteins, they specifically react with them and precipitate in the gel. This precipitate can be visualized as an arch, which becomes more apparent if the preparation is stained with a protein dye. A series of different arches are produced, which correspond to each of the immunologically distinct proteins that exist in human serum. This helps in the identification of numerous protein components in blood serum.

MAIN PROTEINS IN BLOOD PLASMA

Albumin is the most abundant plasma protein, present in concentrations of approximately 3.5–4.0 g/dL. Albumin is a globular protein with a mass of 69 kDa, composed of a polypeptide chain of 585 amino acids. Its isoelectric point of 4.7 explains its electronegative charge at a pH of 8.6. The presence of numerous reactive groups in albumin confers this molecule with the ability to bind a wide variety of substances, allowing it to function as a carrier for many compounds (fatty acids, bile pigments, steroids, hormones, and drugs). Due to its relatively low mass and high concentration, albumin contributes to approximately 80% of the total plasma colloid osmotic pressure (p. 695).

The concentration of albumin in plasma decreases in many conditions in which its synthesis is reduced (protein malnutrition or liver failure) or in which the protein is abnormally excreted (kidney disease).

Globulins. The plasma globulin fraction is a complex mixture of proteins. They are conjugated proteins, of two main types: the glycoproteins and the lipoproteins, which are linked to either carbohydrates or lipids, respectively. Only some of the main globulin components will be mentioned.

Glycoproteins. These types of globulins include the following:

α_1 *globulins.* They comprise:

α_1-*antiproteinase* or α_1-*antitrypsin* is a glycoprotein with a mass of 50 kDa and a total carbohydrate content of 12%. It is one of the main inhibitors of plasma serine proteases, protecting tissues (especially the lung) from the action of proteases released by polymorphonuclear granulocytes. Deficiency in this protein is responsible for some cases of pulmonary emphysema. Smoking promotes the oxidation of a methionine residue of α_1-antiproteinase which inactivates it. This is the reason why smokers who have low levels of this glycoprotein are predisposed to proteolytic destruction of lung tissue and the production of lung emphysema and pulmonary obstructive chronic disease (POCD).

Orosomucoid. Also called α_1 acid glycoprotein, has a high proportion of carbohydrates. The concentration of this protein increases during inflammatory processes, along with the C-reactive protein (so called because it reacts with the pneumococcal polysaccharide C).

Prothrombin. This is the precursor to thrombin, the enzyme that catalyzes the conversion of fibrinogen to fibrin, the final step of blood coagulation (Chapter 31).

Transcortin. This is the protein responsible for the transport of cortisol, a hormone produced in the adrenal gland cortex.

α_2 globulins. Include the following:

Ceruloplasmin. This is a protein with a mass of 151 kDa, which contains six copper atoms per molecule and has an intense blue color. It functions as a copper carrier and is an enzyme which exhibits ferroxidase activity. In Wilson's disease, an autosomal recessive genetic disorder, ceruloplasmin levels in plasma are decreased. The copper accumulates in the brain and liver causing neurological symptoms and liver disease.
Haptoglobin is a protein that binds to hemoglobin. Hb normally circulates in plasma in low amounts due to its release by eventual red blood cell hemolysis. This protein–protein association prevents free circulating Hb from being excreted in the urine (free hemoglobin in plasma can cross the kidney filter).
α_2 macroglobulin is a 720-kDa homotetramer that transports 10% of the total zinc present in plasma and functions as a protease inhibitor.
Erythropoietin is a hormone produced in the kidney that controls red blood cell production (p. 629).

β globulins. These include the following:

Transferrin. This protein binds two atoms of iron and serves as an iron carrier in plasma. In cases of iron deficiency or during normal pregnancy, transferrin levels increase significantly. It is decreased in pernicious anemia, chronic infections, and liver failure.
β_2 microglobulin. This is a small 11.7-kDa protein found in cell membranes. It is part of the major histocompatibility complex (MHC) class I (p. 766).

If electrophoresis is performed in blood plasma instead of serum, the presence of another protein can be observed that migrates between β and γ globulins. This is *fibrinogen*, a glycoprotein that participates in the final stage of coagulation. Fibrinogen represents 4%–6% of the total plasma protein; its concentration is 0.35 g/dL. It has a mass of 340 kDa and a shape with predominance of the longitudinal axis. Other structural data for this protein will be considered in Chapter 31. Fibrinogen is synthesized in the liver; therefore, in patients with severe hepatic failure levels of plasma fibrinogen are decreased. Hereditary diseases have been described in which fibrinogen is either diminished or absent, *fibrinogenopenia* and *afibrinogenemia*, respectively.

γ globulins or immunoglobulins. These glycoproteins are *antibodies*. Antibodies are generated in an individual when foreign substances or antigens enter the body. These could be pathogenic agents (such as bacteria, viruses, and fungi) or macromolecules different from those of the individual that receives them (heterologous proteins). These agents exogenous to the body trigger a specific response, called an immune reaction (see Chapter 30). During the immune reaction, an organism produces a series of antibodies capable of specifically binding to the invading antigen. This will cause the precipitation and destruction (lysis) of the foreign particle. After an infection with viral or bacterial pathogens, or after vaccination with a specific antigen, most individuals maintain the levels of antibodies against the foreign agents for variable periods of time. This, known as immunization, helps protect the individual from later reinfections. Antibodies belong to the γ globulin fraction of blood plasma proteins and are usually named *immunoglobulins* (Ig).

Studies using ultracentrifugation, electrophoresis, immunochemical, and other techniques, recognized the existence of various types of immunoglobulins, which were designated by the letters G, A, M, D, and E.

Immunoglobulin G (IgG) comprises approximately 75% of immunoglobulins circulating in blood plasma. Their molecular weight is 150 kDa and they have a sedimentation rate of 7S; for this reason they are called 7S γIgG (the S denomination refers to Svedberg units, which is related to the rate of sedimentation of a substance after ultracentrifugation). The carbohydrate content of immunoglobulin G represents 3% of the total weight of the molecule. There are four subclasses of this glycoprotein, designated IgG_1, IgG_2, IgG_3, and IgG_4.

Immunoglobulin A (IgA) represents 12%–15% of plasma immunoglobulins. They have a mass that ranges from 150 to 380 kDa, a sedimentation rate that ranges from 7S to 11S, and a carbohydrate content of approximately 12% of their total weight. Two subclasses, IgA_1 and IgA_2, have been identified.

Immunoglobulin M (IgM) is the largest molecule of this family of proteins, with a mass close to 1000 kDa and a sedimentation rate of 19S or more. They are also designated macroglobulins and constitute 8% of total γ globulins; about 12% of its weight is represented by carbohydrates.

Immunoglobulin D (IgD) represents approximately 1% of the total Ig content. Their mass is 180 kDa and 13% of their weight corresponds to carbohydrates.

Immunoglobulin E (IgE) has a mass of approximately 180 kDa and is the antibody found in lowest concentrations in plasma, accounting for less than 0.3% of the total immunoglobulins.

The structure, function, and genetic control of immunoglobulins will be considered in Chapter 30.

Circulating antibodies from an individual who has become immune to a particular infectious disease can be transferred to another individual by parenteral injection. This "passive" transfer of immunoglobulin confers the recipient temporary protection against the infection and is the basis for the preventive use of gamma globulin as an agent against the risk for infectious diseases.

Genetic defects have been identified in which there is a lack or reduced ability to synthesize immunoglobulins (agammaglobulinemia or hypogammaglobulinemia, respectively). These conditions have serious prognosis, since the patient's natural defenses against infection are reduced or absent.

In myeloma, a malignant tumoral process, there is a marked increase of immunoglobulins. All Ig secreted by a myeloma are of the same class and all will have identical antigen specificity.

Lipoproteins. These are globular proteins which are bound to lipids. As lipids are hydrophobic substances that cannot be freely dispersed in plasma, they need to bind to proteins to be transported in plasma. In lipoproteins, nonpolar lipids (triacylglycerols and cholesterol esters) are arranged inside of the particle, surrounded by phospholipids, free cholesterol, and proteins, with their hydrophobic portions oriented toward the nonpolar core and their hydrophilic regions facing the outside of the particle. This arrangement maintains the lipids stable in the aqueous medium. Several types of plasma lipoproteins have been identified and they all differ in their lipid and protein content, density, and electrophoretic mobility (p. 326).

Since lipids have a low density, the specific density of a lipoprotein is lower as the content of lipids increases. This property allows the separation of the different plasma lipoproteins by ultracentrifugation. When placed in a NaCl solution with a density of 1.063 subjected to centrifugation, lipoproteins tend to float or localize in the upper layers of the medium. The tendency to float in a medium of known density is expressed in Svedberg flotation units (Sf). This property can be used as a tool to estimate the content of lipids in the molecule. Thus, the larger the Sf value, the lower the density of a given lipoprotein (Table 3.3).

According to their density, four groups of plasma lipoproteins can be recognized:

1. *Chylomicrons*, with the lowest density: 0.96.
2. *Very low density lipoproteins* (VLDLs), which have a specific gravity between 0.96 and 1.006.

TABLE 3.3   Human Blood Plasma Lipoproteins

Fraction	Density	Sf	Mobility	% Proteins	% Total lipids	Composition % Total lipids Triacyl-glycerols	Phospho-lipids	Cholesterol
Chylomicrons	<0.96	>400	Origin	1	99	88	8	4
Very low density lipoproteins (VLDL)	0.96–1.006	20–400	Pre-β	7	93	56	20	23
Low density lipoproteins (LDL)	1.019–1.063	2–20	Beta	11–21	79–89	13–29	27	43–58
High density lipoproteins (HDL)	1.063–1.210	0–2	Alpha	33–57	43–67	13–16	45	35–41

3. *Low-density lipoprotein* (LDLs), exhibiting a density ranging from 1.019 to l. 063.

4. *High density lipoproteins* (HDLs), with a density from 1.063 to 1.210.

This classification, proposed by Fredrickson and coworkers, has been widely adopted and used in clinics.

Lipoproteins can also be separated by electrophoresis and there is close correlation between the fractions obtained by electrophoresis and those distinguished on the basis of their density. HDLs migrate as α globulins and are designated α lipoproteins. LDLs have the mobility of β globulins and are called β lipoproteins. The VLDLs move in front of β globulins and are named pre-β globulins. Chylomicrons do not migrate in the electric field.

The composition of different lipoproteins differs markedly. Those with higher specific weight have relatively higher protein content (Table 3.3). For example, VLDLs have a protein content of 7%, while HDLs contain between 33% and 57% of protein. Chylomicrons are particles of about 0.5 μm in diameter that mainly carry triacylglycerols from the intestine to the lymphatic system and contain very low protein

(nearly 1%). Proteins form a thin film on the surface of chylomicrons.

The relative lipid content of each lipoprotein type is shown in Table 3.3. The pre-β (VLDL) lipoproteins are particularly rich in triacylglycerols while β lipoproteins (LDL) are primarily responsible for the transport of cholesterol. Increased LDL is a common finding in cases of hypercholesterolemia.

The protein or apoprotein portion of lipoproteins is generally heterogeneous. There are several classes of apolipoproteins, designated by the letters A, B, C, D, and E, and all are commonly called apo A, apo B, etc. (p. 327). Several apolipoprotein subtypes have also been recognized. Thus, for apo A, three species are described (A-I, A-II, and A-IV). Apo AI is the major component protein of HDL and is found as a minor constituent in chylomicrons and VLDL. Apo AI is an activator of lecithin–cholesterol acyltransferase (LCAT) in plasma (see p. 330). The apo A-II is the second most abundant apoprotein and constitutes almost 20% of HDL. Apo A-IV is associated with chylomicrons and synthesized in intestinal mucosa.

Apo B is an apolipoprotein present in LDL, VLDL, and chylomicrons. It presents two subtypes, apo B-100 and apo B-48. Apo B-100 is the largest subtype and consists of a polypeptide chain of more than 4500 amino acids synthesized in liver. It is the most abundant protein of the LDL and VLDL fractions. Apo B-100 is recognized by the LDL receptors of various cell types throughout the body, which helps in the uptake of cholesterol by cells. LDL is commonly referred to as the "bad cholesterol". Apo B-48, which shares the same sequence as the N-terminal half of apo B-100, is found in chylomicrons and synthesized in the intestine.

Apo C is the apoprotein with the lowest mass and presents three subtypes, named C-I, C-II, and C-III. Apo C-II is an activator of lipoprotein lipase that hydrolyzes the triacylglycerols of chylomicrons and VLDL.

Apo D is found in HDL and is responsible for promoting the exchange of cholesterol esters and triacylglycerols between HDL and VLDL lipoproteins.

Apo E is a minor component of chylomicrons, VLDL, and HDL. It is a glycoprotein with four subtypes, named E-I–E-IV. They have an important role in the catabolism of cholesterol and participate in the receptor mediated recognition and uptake of lipoproteins by cells.

The β lipoproteins, with a molecular mass of about 1300 kDa, are among the largest plasma proteins. The α lipoproteins (HDL) are much smaller, with a mass of 200 kDa.

Synthesis of plasma proteins. The liver is the main organ involved in the production of albumin, fibrinogen, prothrombin, and most of the α and β globulins. The liver in the adult male synthesizes approximately 20 g of plasma proteins per day. The γ globulins are formed in plasma cells derived from B lymphocytes (Chapter 30).

Synthesis of plasma proteins is very active, since they are degraded and replaced at a fast rate. The average half-life of plasma proteins (time during which half the proteins are renewed) is approximately 10 days.

Due to the essential role of the liver in plasma protein synthesis, any important malfunction of the liver results in alterations of the protein plasma composition. Such alterations are known as dysproteinemias. Protein electrophoresis is an excellent method to reveal the changes in the relative content of the different plasma proteins. A reduction in the albumin fraction (hypoalbuminemia) is a typical finding in liver failure.

The synthesis of plasma proteins, like any other protein, requires the input of the necessary amino acids provided by a normal diet. There is a direct relationship between the quantity and quality of proteins ingested through the diet and the synthesis of plasma proteins. Protein deficiencies lead to a reduction of plasma proteins, specially albumin, and deficiency in the synthesis of antibodies. This explains the altered protection that malnourished individuals have against infections.

General functions of plasma proteins. A general important role of plasma proteins is maintenance of extracellular body fluid volume (see p. 696). In addition, plasma proteins exert a buffer action. At normal blood pH, the buffering capacity of proteins depends on the presence of histidine residues within the molecule. This depends on the capacity of the imidazole nitrogen in histidine to accept protons (pK_a 6.0).

MUSCLE PROTEINS

Muscle structure. Skeletal muscle comprises nearly 40% of the total body mass in a normal adult. This tissue is composed of multinucleated cells surrounded by the *sarcolemma*, an electrically excitable membrane surrounding the cells. Inside, the cell contains fibers formed by bundles of myofibrils bathed by the *sarcoplasm* (the cytoplasm of muscle cells). The muscle fibers are surrounded by a complex system of canaliculi, called the *sarcoplasmic reticulum*, which is equivalent to the endoplasmic reticulum of other cells. Transversal tubules emerge from the sarcolemma at regular intervals and make contact with the sarcoplasmic reticulum, forming what is known as the *T system*.

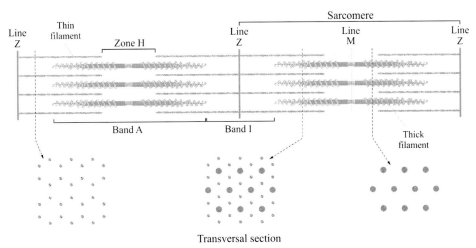

FIGURE 3.35 Schematic representation of a striated muscle fiber. Top: longitudinal section of relaxed muscle; below, transverse sections made at the levels shown by dashed lines. Note that I bands only contain thin filaments; the H area only has thick filaments and the denser portion of the A band comprises both types of filaments.

Under the light microscope, skeletal muscle fibers show alternating dark and light transversal bands, giving it a striated appearance. The dark regions are designated A bands (anisotropic, birefringent) and the light regions, I bands (isotropic). The central region of the A band is known as the H zone and is less dense than the rest of the band, except at its midpoint, where the H zone presents a line (M line). In the center of the I band, there is a very dense and fine segment, perpendicular to the longitudinal axis of the myofibril, designated the Z line (Fig. 3.35). The region of the myofibril between two Z lines corresponds to a functional unit or *sarcomere.*

Regular striations are produced by the arrangement adopted by two kinds of myofibril filaments: thick filaments and thin filaments. Thick filaments extend along the A band, are 1.6-μm long, and have a diameter of 16 nm. Thin filaments are 6 nm in diameter and extend from the Z line to the H zone. In cross sections, the I band only presents thin filaments, the H area only exhibits thick filaments, and the denser regions of the A band contain both thin and thick fibers. Within the A bands, each thick filament is surrounded by six thin filaments, forming a hexagonal arrangement, and each thin fiber is surrounded by three symmetrically distributed thick filaments (Fig. 3.35).

When a muscle contracts, there is a shortening of the sarcomere causing the I bands and H zones to decrease in length, reducing the distance between Z lines. The length of the A band, which corresponds to the length of the thick filaments, remains unchanged. Also, the distance between Z line and the H region remains unmodified, which indicates that the length of thin filaments does not vary. The shortening of the muscle has been explained by a sliding mechanism, in which the thin fibers penetrate into the spaces between the thick rods, shortening the sarcomere (Fig. 3.36). The maximum muscle contraction in vertebrates is able to shorten the length of the muscle fibers by 40%.

Muscle proteins. The major proteins contained in skeletal muscle myofibrils are myosin, actin, tropomyosin, and troponin.

Myosin represents 55% of the total muscle protein and forms the thick filaments. Its molecule is asymmetric, with a globular "head" and a fibrillary long "tail." Each myosin molecule is a hexamer of approximately 500 kDa. It has two

3. PROTEINS

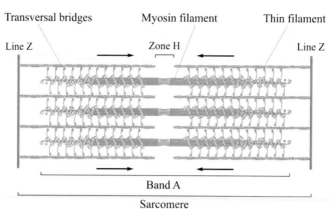

FIGURE 3.36 **Schematic representation of a cross section of a skeletal muscle myofibril during contraction.** Thin filaments penetrate between the thick filaments and occupy the space corresponding to the H zone in the relaxed muscle. This area and I band lengths decrease significantly.

large subunits of 200 kDa each, having a globular structure at one end. Globular domains of myosin have ATPase activity and catalyze the hydrolysis of ATP into ADP and P_i. The rest of the myosin chain is a long α helix that coils in a double helix with the homologous portion of the other large subunit (Fig. 3.37A).

Myosin fibrillary portions associate in bundles to form the thick filaments of the A band (Fig. 3.37B). All molecules in each half of these filaments are polarized and display the same orientation. Heads are directed toward the free end and tails point to the M line in the middle. Globular heads project out of the thick filaments that look like arms that can contact the thin filaments forming cross-bridges (Figs. 3.36 and 3.39).

The thin filaments are composed of three different proteins: actin, tropomyosin, and troponin.

Actin. Actin represents 25% of all muscle protein. Monomers of 45-kDa globular proteins, called G actin, polymerize to form F actin. This is a fibrous protein consisting of two strands of

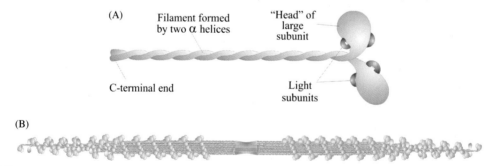

FIGURE 3.37 (A) Schematic representation of a myosin molecule. Each of the two major chains has a globular head and a long helical segment wrapped around the other strand to form a fibrillary tail. Two smaller subunits are attached to the head of the larger chain. (B) Outline of a thick filament. Fibrillary portions of myosin molecules are grouped into a bundle. The globular heads project out of the beam (arms or cross-bridges that can bind to the thin filaments). Note that the central portion of the filament is devoid of globular heads.

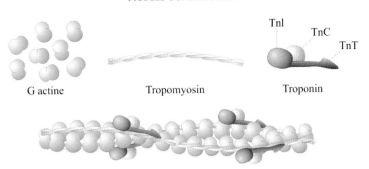

FIGURE 3.38 **Components of the thin filament proteins are shown in the top panel.** Below, a thin filament assembly is illustrated. G actin monomers are arranged into strands of F actin, which form a double helix. Tropomyosin molecules are situated in the grooves left between the actin helices. Troponin comprises three proteins (TnT, TnI, and TnC); it binds to the tropomyosin filament at regular intervals.

coiled G actin monomers (Fig. 3.38). F actin has no catalytic activity.

Tropomyosin. This is a fibrous protein of approximately 64 kDa that consists of two strands twisted around each other in a double helix. Tropomyosin molecules are arranged one after another, housed in the spaces between the actin filaments (Fig. 3.38).

Troponin. This is a complex molecule that contains three different proteins: troponin T, troponin C, and troponin I. These domains are noncovalently linked to each other. This protein complex binds to the thin filaments at regular intervals, with one troponin complex bound per tropomyosin molecule (Fig. 3.38). Troponin T facilitates binding to tropomyosin, whereas troponin I inhibits the interaction between actin and myosin. Troponin C is a Ca^{2+}-binding protein with a structure similar to that of calmodulin (p. 567).

A series of other proteins are present in muscle fibers. Only those proteins most directly involved in muscle contraction will be mentioned.

Titin. This is a large fibrillary protein (with a mass of 3000 kDa) that stretches like a spring from the M line to the Z disk. Titin holds myosin rods centered in the sarcomere and helps to maintain the tension of muscular fibers, allowing them to return to its rest position if muscle is stretched in excess.

Nebulin. This protein forms part of the thin fibers and is associated with actin. Nebulin appears to determine the length of actin filaments.

The sarcoplasm contains *myoglobin,* a hemoprotein that reversibly binds oxygen and functions as a molecule for storage and transport of O_2 within the muscle.

Phospholamban. This is a pentamer of identical subunits (20 kDa) that is especially important in cardiac muscle. Phospholamban is an inhibitor of the Ca^{2+}-ATPase of the sarcoplasmic reticulum. When phosphorylated by protein kinase A, cyclic AMP, or calmodulin-dependent kinases, phospholamban inhibits the Ca^{2+}-ATPase. Ca^{2+} from the cytosol will then return to the sarcoplasmic reticulum, promoting muscle relaxation.

MUSCLE CONTRACTION

Contraction of muscle myofibrils is the result of complex molecular interactions. The mechanism underlying sarcomere shortening involves the sliding of the thin fibrils between the thick filaments. The globular domains of myosin at the ends of the thick filaments protrude like arms; they have ATPase activity and a site that can bind to the actin of thin filaments. At rest, muscle has a high concentration of ATP and low levels of Ca^{2+} (approximately 10^{-7} M). This maintains myosin dissociated from actin. Each globular actin head binds one molecule of ATP and hydrolyzes it to ADP and P_i; these products remain bound to myosin. The energy released in the reaction is retained in the cross-arm, which adopts a "tense" conformation. To transform this energy into myofibril movement, myosin heads must contact the actin filaments, forming bridges between the thick and thin rods.

The arrival of a nerve stimulus to the motor endplate produces the depolarization of the sarcolemma. The stimulus travels through the T system and is rapidly transmitted into the fiber. Then, Ca^{2+} channels in the sarcoplasmic reticulum membrane open, releasing Ca^{2+} to the cytosol. Cytoplasmic Ca^{2+} concentration sharply increases from 10^{-7} to 10^{-5} M; the ion binds to troponin C and causes a conformational change in this protein as well as in troponin I, troponin T, and tropomyosin. These changes in protein shape cause a displacement of the tropomyosin strands in the helical groove between the actin filaments. In the resting position, tropomyosin blocks the myosin binding sites for actin and prevents the formation of cross-bridges. The increase in intracellular Ca^{2+} promotes the conformational change that exposes myosin binding sites. Some authors believe that this model of steric blocking and unblocking of actin by tropomyosin is somewhat simplistic and suggest the existence of more complex interactions between the acting proteins. In any case, the result is the binding of myosin heads to actin,

establishing a close association between thick and thin filaments.

Each globular domain of myosin is attached to the fibrillary portion of the molecule through a flexible arm, with at least two hinge sites that allow the angle of the cross-arm to change. The cocked myosin head tends to return to a position of lower energy. This force provides the "power stroke" that causes the sliding of the thin filament, displacing it toward the center of the sarcomere (Fig. 3.39). As the thick filament molecules are polarized, the direction of the forces developed on each half of the rod is opposite to that of the other half, causing an overall shortening of the sarcomere. The myosin head detaches from the actin and returns to its rest position, ready to start a new movement by binding to another site on the thin filament. This action is repeated while the concentration of Ca^{2+} is high.

The displacement of the transversal arms of myosin is the result of a cycle of reactions (Fig. 3.40).

1. The globular domain of myosin binds ATP and hydrolyzes it to ADP and P_i. One high energy myosin-ADP-P_i complex is formed.
2. Nerve stimulation that induces contraction produces a sudden elevation of $[Ca^{2+}]$ in the sarcoplasm and thereby a change in the troponin complex and tropomyosin which allows the binding of the myosin head to the actin filament. P_i is released and the power stroke occurs to achieve a position of lower energy. This movement drags the

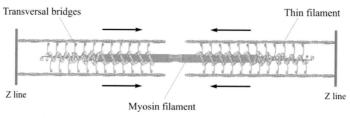

FIGURE 3.39 **Sliding mechanism of muscle contraction.** Cross-links of the myosin thick rods with actin thin filaments perform a power stroke that displaces actin toward the center of the sarcomere. As myosin molecules are polarized, cross-arms on each half of the rod will pull in opposite directions.

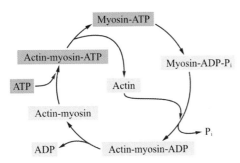

FIGURE 3.40   **Cycle of reactions in muscle contraction.**

thin filament and determines the myofibril shortening.

3. ADP is released and another molecule of ATP binds to myosin. This decreases its affinity for actin and causes the globular head to detach from F actin; myosin-ATP can start another cycle. When muscle is depleted of ATP, it remains contracted and the myosin–actin complex cannot be dissociated. This scenario explains the *rigor mortis* that occurs after death.

The cycles of thin fiber sliding between myosin continue while the calcium level is high in the sarcoplasm. When the stimulus that caused the contraction ceases, Ca^{2+} is sent back into the sarcoplasmic reticulum by the action of Ca^{2+}-ATPase or "calcium pump"; this pump transfers two Ca^{2+} ions per ATP hydrolyzed. When Ca^{2+} concentration in the cytoplasm is reduced, troponin C dissociates from Ca^{2+}, tropomyosin returns to its rest position, the transverse arms cannot keep myosin bound to actin and muscle relaxation occurs.

SUMMARY

Proteins are the most abundant organic compounds in animals, playing key roles in many biological processes. They are composed of carbon, hydrogen, oxygen, and nitrogen; and most of them also contain sulfur.

Amino acids (AA) are the building blocks of proteins. They have at least one functional group that is acid (carboxyl) and another that is basic (amine), both linked to the α carbon (α-amino acids). Twenty different AAs are obtained after complete hydrolysis of proteins, each differing in their carbon side chain. AA are classified according to the characteristics of their carbon chain:

1. aliphatic AA contain a neutral nonpolar chain (glycine, alanine, valine, leucine, and isoleucine),
2. neutral AA have a polar, aliphatic, nonionizable chain (serine and threonine),
3. neutral AA with aromatic chains (phenylalanine, tyrosine, and tryptophan),
4. sulfur containing AA (cysteine and methionine),
5. dicarboxylic AA (aspartic and glutamic acids),
6. amidic derivatives (asparagine and glutamine),
7. basic AA (lysine, arginine, and histidine), and
8. AA with an imino group (proline), in which the αC is included in a pyrrolidine ring.

AA exhibit acid–base properties and function as dipolar ions, or zwitterions, with acid and basic groups in the same molecule.

In neutral solutions, the acid (carboxyl) and basic (amine) functions of AA are ionized.

In strong acid solutions, the carboxyl group accepts a proton and the AA becomes positively charged.

In strong alkaline medium, the amine group releases a proton and the AA charge becomes negative.

When the dissociation of positive and negative groups in the molecule is the same, the overall net charge of the AA is zero. The pH at which this occurs is known as the *isoelectric point* of the AA (pI or pHi). At the pH of the body, only histidine functions as a buffer (the N in the imidazole ring can accept a proton because it has a pK_a 6.0).

Optical isomerism refers to the capacity of AA to rotate the polarized light that passed through them. This is a property of all AA in solution (except for glycine) that is due to the spatial configuration of the α carbon. This determines the existence of two optically active isomers, D or L. Animal proteins have AA that are all L isomers.

Peptidic bonds are formed between the carboxyl group of one AA and the amine group of the αC of another. These bonds allow AAs to form long chains. By convention, the AA that has its α amine group free is considered the first one in the chain; this is the N-*terminal* end of the protein. The AA with the free carboxyl group is considered the last one; this is the C-*terminal* portion of the protein.

Peptides are AA polymers with a total molecular mass of less than 6 kDa. The acid–base properties of AA are maintained in peptides, due to the ionizing capacity of the amine and carboxyl groups of the end and the side groups of the peptide.

Proteins are AA polymers of more than 6 kDa. They are charged molecules, whose electrical charge depends on the dissociation state of all ionizable functional groups in the side chains of the AAs. The pH of the medium determines the ionization state of the protein. In medium that is acid compared to the isoelectric point (pI) of the protein, the polypeptide has a positive charge, whereas in medium that is alkaline with respect to the protein pI, the molecule is negatively charged. The magnitude of the electric charge is proportional to difference between the pH of the medium and the pI of the protein. Due to this property, the stability of a protein in aqueous solution depends on the pH of the medium and on the presence of salts and nonpolar solvents in the medium.

Electrophoresis is a method that allows the separation of proteins based on their charge. It uses an electric current, which forces the migration of proteins through a proper medium, to either the anode or cathode, with a speed that is proportional to the protein's charge.

Dialysis is a method that allows the separation of proteins from other small molecules in a mixture, based on their differential movement across semipermeable membranes.

Protein Structure depends not only on the amount, but also the nature of its constitutive AAs. Protein *Primary structure* refers to the order or sequence of AA, which is determined by the genes. The C—N in the peptide bond is an intermediate between a single and a double bond, which does not allow free rotation. Consequently, these two atoms and the O and H bound to them remain in the same plane. The αC—N and N—αC bonds can rotate freely, so that the groups linked to the αC adopt different positions in space. These generate the *secondary structure* of the protein, which results in different spatial distributions. The most important are as follows:

1. α *helix*, in which the polypeptide chain coils around a central axis, with each turn of the helix comprising nearly four AA residues. This structure is maintained by H bonds between the N of an amine and the O of the carboxyl moiety involved in the peptide binding of AA residues located on adjacent turns of the helix.
2. β *sheet*, in which the chain is fully extended, forming pleated lamellar structures maintained by H bonds.
3. *Random coil*, the chain does not follow a repetitive pattern but adopts the most thermodynamically favorable configuration.

The three-dimensional shape of the protein gives it its *tertiary structure*. This depends on various forces that helps folding the polypeptide chain, including:

1. Electrostatic attraction and repulsion forces.
2. H bonds between functional groups of the AA side chains.
3. Interactions of hydrophobic or hydrophilic residues.
4. Disulfide bridges between cysteines.

As a result of these forces, certain protein segments present stable associations of α helices and/or β sheets, usually forming functional units within the molecule called protein *domains*. Proteins comprised by multiple subunits (oligomeric) have a *quaternary structure*, which refers to the spatial arrangement of the different polypeptide chains or subunits of the protein.

Based on their shape, proteins are classified as *globular* or *fibrous*. *Globular proteins* may present α helical portions, β sheets, or both, joined by segments of random coil, or simply be randomly arranged throughout.

Conjugated proteins are those which, besides AA (*apoprotein*), contain a nonprotein portion (*prosthetic group*).

Protein denaturation takes place when the forces that maintain the secondary, tertiary, and quaternary structures of proteins are disrupted by physical or chemical agents. Denaturation alters the native conformation and original properties of proteins.

Collagen is the protein that constitutes the fibers of the connective tissue. Collagen is rich in glycine, proline, and hydroxyl derivatives of these AA and its structure consists in helixes that extends for a longer distance than the α helix. Three of these chains associate in a coiled–coiled structure, forming *tropocollagen*. Tropocollagen units are regularly arranged in bundles to form collagen fibers of high mechanical strength.

Hemoglobin is a hemoprotein, with *heme* as prosthetic group. Its main function is the transport of O_2 from lungs to tissues. It is composed of four polypeptide chains (HbA$_1$, the most abundant in the adult, is a $\alpha_2\beta_2$ tetramer). Each of the hemoglobin subunits is associated with a tetrapyrrole with Fe^{2+} (*heme*). When O_2 binds to one of the deoxy-Hb hemes, Fe^{2+} produces a conformational change that is transmitted to the rest of the molecule and facilitates access of O_2 to the other subunits (cooperative effect). The increase in [H^+], P_{CO_2}, or the presence of phosphorous compounds, such as *2,3-bisphosphoglycerate*, decrease the affinity of Hb for O_2, favoring the release of O_2 from oxyHb in peripheral tissues. Cooperative actions are also observed in other oligomeric proteins (*allosteric proteins*). Certain genetic defects (mutations) result in changes in AA sequence in hemoglobin and the loss of its properties.

Plasma proteins circulate in the bloodstream at a concentration of 6–8 g/dL. Electrophoresis of blood serum on cellulose acetate allows the separation of several protein fractions. These are albumin and α_1, α_2, β, and γ globulins.

Albumin is the most abundant plasma protein fraction (3.5–4.0 g/dL). It is a globular protein of 69 kDa, composed of a 610 amino acids and a pHi of 4.7. Albumin carries in blood fatty acids, bile pigments, steroids, hormones, and drugs.

Glycoproteins include the following proteins:

α_1 *globulins* include α antiproteinase, orosomucoid, prothrombin, and transcortin.

α_2 *globulins* include ceruloplasmin, haptoglobin, α_2-macroglobulin, and erythropoietin.

β *globulins* include transferrin and β_2-microglobulin.

γ *globulins* are antibodies or *immunoglobulins* (Ig), of which there are five classes:

IgG has a molecular mass of 150 kDa and is the most abundant, constituting 75% of all Ig.

IgA, with a mass of 150–600 kDa accounts for 12%–15% of all Ig.

IgM is the largest with a mass of 1000 kDa and an abundance of 3%.

IgD has 180 kDa and makes up 1% of all Ig.

IgE has the lowest concentration and plays a role in allergies and parasitic infections.

Lipoproteins are carriers of lipids in plasma. They exist as four groups:

1. *Chylomicrons*, with a density <0.96, do not migrate when subjected to electrophoresis.
2. VLDLs have a density between 0.96 and 1.006 and migrate in front of β globulins (pre-β).
3. LDLs have a density between 1.019 and 1.063 and migrate as β globulins.
4. HDLs, with a density of 1.063–1.210, migrate as γ globulins.

The denser lipoproteins have proportionally more protein (HDL between 33% and 57%, VLDL 7%, and chylomicrons 1%). Chylomicrons and VLDL are rich in triacylglycerols; LDL has high proportion of cholesterol. The protein portion or apoprotein are designated apo A, apo B, apo C, apo D, and apo E; there are subtypes in each class. Apo A is the major protein component of HDL; apo B of LDL and VLDL.

Synthesis of plasma proteins occur in the liver for albumin and most globulins (approximately 20 g/day) and in plasma cells for immunoglobulins.

Functions of plasma proteins include the maintenance of blood volume; albumin is responsible for over 75% of the plasma oncotic pressure. Proteins also have a buffering function, which at the normal pH of tissues is exerted by histidine residues.

Muscle proteins include *myosin, actin, tropomyosin,* and *troponin*. Myosin thick filaments form the A band of the sarcomere. They have globular heads protruding from the filaments as side arms and possess ATPase activity. G actin polymerizes as F actin to form thin filaments. Tropomyosin and troponin (complex of T, C, and I troponin) associate to F actin.

Muscle contraction is the shortening of the muscle fibers produced by the sliding of fine filaments between the thick filaments. At rest, there is high ATP and low levels of Ca^{2+} in muscle, which maintain myosin and actin dissociated. The globular heads of myosin bind ATP and hydrolyze it to ADP and P_i. The myosin-ADP-P_i complex has high energy content and remains in a "tense" position. The arrival of a nerve stimulus determines a sharp increase in Ca^{2+} concentration in the sarcoplasm; Ca^{2+} binds to C troponin and produces a conformational change that is transmitted to tropomyosin. This allows the attachment of the transversal arms of myosin to actin filaments. As the myosin heads tend to return to a position of lesser energy, they drag the thin filament and this determines the shortening of the muscle fibers. This leads to the release of ADP and P_i. Another molecule of ATP binds to myosin and decreases the affinity for actin-myosin binding and these molecules are separated. Myosin-ATP repeats the cycle as long as the level of Ca^{2+} remains high.

Bibliography

Branden, C., Tooze, J., 1999. Introduction to Protein Structure, second ed. Garland Publishing Co., New York.

Creighton, T.E., 1993. Proteins, Structures and Molecular Properties, second ed. W.H. Reeman, New York.

Geeves, M.A., Holmes, K.C., 1999. Structrural mechanisms of muscle contraction. Ann. Rev. Biochem. 68, 687–728.

Hsia, C.C.W., 1998. Respiratory function of hemoglobin. New Engl. J. Med. 338, 239–247.

Molloy, J.E., Veigel, C., 2003. Myosin motors walk the walk. Science 300, 2045–2046.

Nigg, B.M., Herzog, W. (Eds.), 1998. Biomechanics of the Muscle-Skeletal System. second ed. John Wiley & Sons, Chichester.

Ponting, C.R., Russell, R.R., 2002. The natural history of protein domains. Ann. Rev. Biochem. 31, 45–71.

Prokod, D.J., Kivirikko, K.L., 1995. Collagens, molecular biology, diseases and potentials for therapy. Ann. Rev. Biochem. 64, 403–434.

4

Carbohydrates

Carbohydrates are another important component of living beings. They have a structural role, forming the fibrous components of plants and serve as nutrient reserve, stored in roots, seeds, and fruits. Carbohydrates are also widely distributed in animals, where they form molecules of diverse structural and functional relevance.

Plants synthesize carbohydrates from CO_2 and H_2O by capturing the energy from light in the process of *photosynthesis*. These carbohydrates are ingested by animals, and largely used as fuel. In humans, carbohydrates are the main source of energy. In a balanced diet, they provide 50%–60% of the total calories needed by an individual.

Carbohydrates are composed of carbon, hydrogen, and oxygen and are defined as polyhydroxy-aldehydes or polyhydroxy-ketones. They have an aldehyde or ketone and various alcoholic functions. Substances that render these polyhydroxy-aldehydes or polyhydroxy-ketones when subjected to hydrolysis are also considered carbohydrates.

Classification. Depending on their complexity, carbohydrates are classified into monosaccharides, oligosaccharides, or polysaccharides.

1. *Monosaccharides*, also known as simple sugars, they consist of only one polyhydroxy-aldehyde or polyhydroxy-ketone. They are obtained as water soluble white crystals and many of them are sweet. Glucose is the most important member of this group.

2. *Oligosaccharides* are polymers formed of 2–10 monosaccharides that can be separated by hydrolysis. According to the number of molecules that constitute them, they are designated disaccharides, trisaccharides, tetrasaccharides, etc. Representatives of greater interest within this group are the disaccharides. They are water soluble, can be obtained in crystalline state, and generally have a sweet taste.

3. *Polysaccharides* are large molecules, formed by the assembly of monosaccharides, arranged in linear or branched chains. In general, they are water insoluble, tasteless, and amorphous.

MONOSACCHARIDES

Simple sugars can be defined as polyhydroxy-aldehydes (polyols-aldehydes) or polyhydroxy-ketones (polyols-ketones). In general, carbohydrates are distinguished with the suffix "ose." When they have an aldehyde function, the monosaccharides are called *aldoses*; if they contain a ketone function, they are named *ketoses*. Usually, they are designated trioses, tetroses, pentoses, etc. depending on the number of carbons in the molecule. The monosaccharide can be described by indicating the number of carbons in the saccharide and its function. Thus, an aldohexose is a monosaccharide with aldehyde function and six carbons, while a ketopentose is a monosaccharide with ketone function and five carbons. The simplest monosaccharides are the trioses: the aldotriose glyceraldehyde and the ketotriose dihydroxyacetone.

Glyceraldehyde   Dihydroxiacetone
(aldotriose)   (ketotriose)

The carbohydrates with a higher number of carbons (tetroses, pentoses, hexoses, etc.) could be considered triose derivatives to which $=CH—OH$ groups have been added to the chain, between the aldehyde or ketone group and the adjacent alcohol function.

Monosaccharides are substances with reducing capacity, particularly in alkaline medium. This property depends on the aldehyde or ketone groups. Some reactions used to identify monosaccharides take advantage of this reducing property.

Isomerism

The second carbon of glyceraldehyde is asymmetric or chiral; all of its valences are saturated by different functional groups, which raises the possibility of two optical isomers. One of the isomers deviates polarized light clockwise, it is dextrorotatory and designated with the letter D before its name. The other is levorotatory (L). Both compounds are enantiomers, one being the mirror image of the other.

D(+)-Glyceraldehyde   L(−)-Glyceraldehyde
(dextrogyre)   (levogyre)

Glyceraldehyde with its asymmetric carbon shown in red.

By convention, D-glyceraldehyde is represented with the hydroxyl group in the asymmetric carbon placed to the right and the L-glyceraldehyde, with the hydroxyl group positioned to the left.

The aldotetroses may be considered derived from glyceraldehyde by addition of a $=CHOH$ group between the aldehyde and the next alcohol. This group originates a new chiral carbon, giving aldotetroses two asymmetric carbons. If another $=CH—OH$ group is added to an aldotetrose, an aldopentose is created, which has three chiral C. Addition of another secondary alcohol function to an aldopentose generates an aldohexose, which has four asymmetric carbons. The different isomers formed in each case are not mirror images of one another or *enantiomers*. These are called *diastereoisomers*.

The number of possible optical isomers is given by the formula 2^n, where n equals the number of asymmetric carbons. There are 4 aldotetroses (2^2) (2 diastereoisomers, each of which present 2 enantiomers), 8 aldopentoses (2^3), 16 aldohexoses (2^4). Optical isomers differ in their specific capacity to rotate polarized light.

As aldoses are considered derived from glyceraldehyde, there are two families of these monosaccharides: one related to D-glyceraldehyde and the other to L-glyceraldehyde. The configuration of the secondary alcohol that is farthest from the aldehyde function is, in all members of the D series, equal to that of D-glyceraldehyde. For each compound of the D series there is a corresponding enantiomer of the L series.

For ketoses there are also two series, D and L, depending on the configuration of the secondary carbon that is farthest from the ketone function.

The optical activity of a compound, which has several asymmetric carbon atoms, is the result of the effects of all of its asymmetric carbon atoms. For this reason, the D notation of a sugar with more than three carbons does not necessarily indicate that it is dextrorotatory. Therefore, the optical activity of the compound must be indicated with (+) or (−) following the D or L. Thus, D(+) aldohexose glucose indicates that it belongs to the D series and has dextrorotatory capacity. D-Fructose, a ketohexose, is strongly levorotatory and its notation is D(−) fructose.

Differentiation of carbohydrates in series or families has biological significance. Higher organisms metabolize and synthesize almost exclusively D carbohydrates and only very few L compounds are present in human tissues or body fluids.

Monosaccharides of Interest in Human Biochemistry

Only those monosaccharides important from the point of view of human biochemistry will be considered. These include: (1) the trioses glyceraldehyde and dihydroxyacetone, which are generated in the body by metabolic transformations of carbohydrates and other substances; (2) the aldopentose ribose; (3) the aldohexoses glucose, galactose, and mannose; (4) finally, fructose, the most important ketose.

Glucose

Glucose, also called dextrose because of its dextrorotatory properties, is the most abundant monosaccharide and is used by cells as fuel. It is present as free glucose in honey, ripe fruits, and in body fluids of vertebrates. Glucose also integrates disaccharides, including sucrose and lactose. Glucose polymerizes to form polysaccharide molecules, such as starch, cellulose, and glycogen.

Cyclic structure. Monosaccharides have been presented as aldehydes or ketones with a linear carbon chain. However, this structure does not explain some of the properties that these substances have. For example, most monosaccharides do not react immediately as aldehydes or ketones. In addition, some monosaccharides present two crystalline forms, which differ in

specific rotation. Glucose presents α and β forms, α-D-glucose rotates polarized light +112.2° and β-D-glucose, +18.7°. Both forms show the phenomenon of mutarotation, which consist of a spontaneous change in polarized light rotation. Thus, when an aqueous solution of α-D-glucose is prepared, it has an initial specific rotation of +112.2°. However, after some time, the specific rotation of the solution decreases until it stabilizes at +52.7°. On the other hand, a recently dissolved aqueous solution of β-D-glucose has a specific rotation of +18.7°, but in subsequent measurements, this value increases stabilizing at +52.7°.

The existence of α and β forms of a monosaccharide, and the abnormal reactivity of its aldehyde or ketone groups, is due to the formation of a cyclic structure. It depends on the orientation of the bonds between the glucose carbons, which allows the ends of the hexose chain to come close to each other. This brings together the aldehyde of the first carbon and the hydroxyl group of carbon 5, forming a hemiacetal or hemiketal type of bond (a hemiacetal bond results from the reaction between aldehyde and alcohol, whereas a hemiketal bond results from the reaction between a ketone and alcohol):

$$R-\overset{H}{\underset{}{C}}=O + HO-R' \rightarrow R-\overset{\overset{\textstyle H}{|}}{\underset{\underset{\textstyle OH}{|}}{C}}-O-R'$$

The interaction between the carbons in glucose generates a heterocyclic six-membered ring (Fig. 4.1). However, in certain cases, the hemiacetal bond of the monosaccharide takes place between carbons 1 and 4, giving rise to a five-membered ring composed of four carbons and oxygen. As these hexagonal and pentagonal rings are *pyran* and *furan* derivatives, monosaccharides that adopt these conformations are referred to as having *pyranose* or *furanose* forms. In solution, pyranose sugars are more stable, so

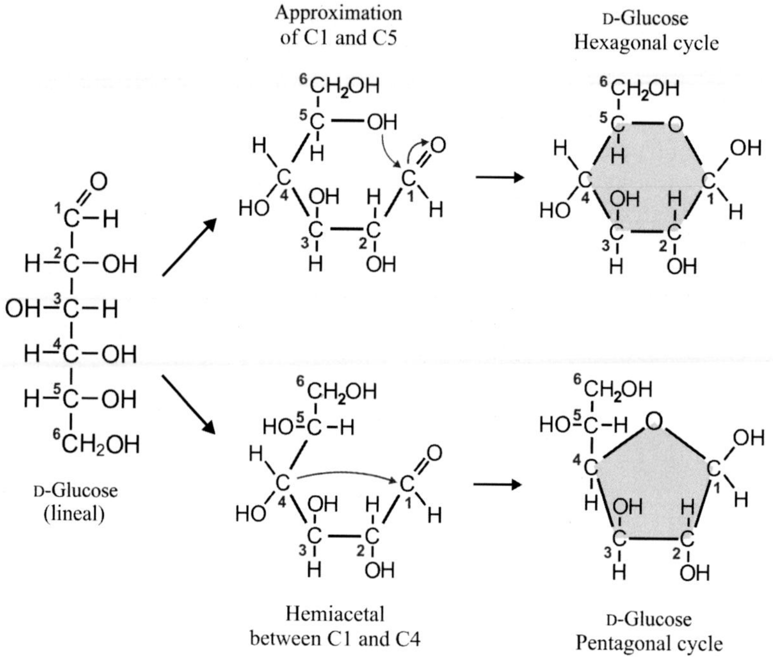

Approximation
of C1 and C5

D-Glucose
Hexagonal cycle

D-Glucose
(lineal)

Hemiacetal
between C1 and C4

D-Glucose
Pentagonal cycle

FIGURE 4.1   **Glucose.** Cycle formation.

they are more frequently found in nature than furanose sugars.

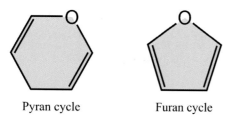

Pyran cycle   Furan cycle

When the monosaccharide adopts the cyclic structure, carbon 1 no longer displays the aldehyde function. However, this function can be exhibited when the ring structure is opened. That is the reason why the typical reactions of the aldehyde group occur more slowly in these sugars. It is said that the cyclic aldose has a "potential" aldehyde group, which is responsible for the reductive property of these compounds. The cyclic structure of monosaccharides also explains

the existence of α and β forms of sugars and the phenomenon of mutarotation. Carbon 1 in the cyclic form is asymmetric, allowing the possibility of two isomer configurations (Fig. 4.2). These types of isomers are known as *anomers* and the C in position 1 is the anomeric carbon.

Usually, the α form is represented with the OH linked to the anomeric C1 facing down, and the β form with the OH oriented upward.

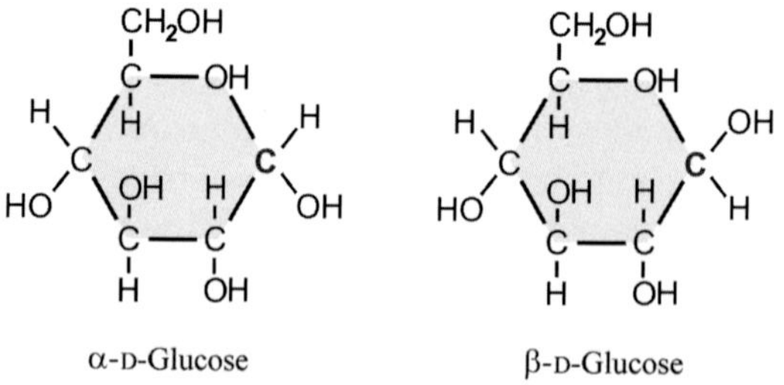

α-D-Glucose   β-D-Glucose

FIGURE 4.2   **Different forms of glucose.** The anomeric C1 is shown in *red*.

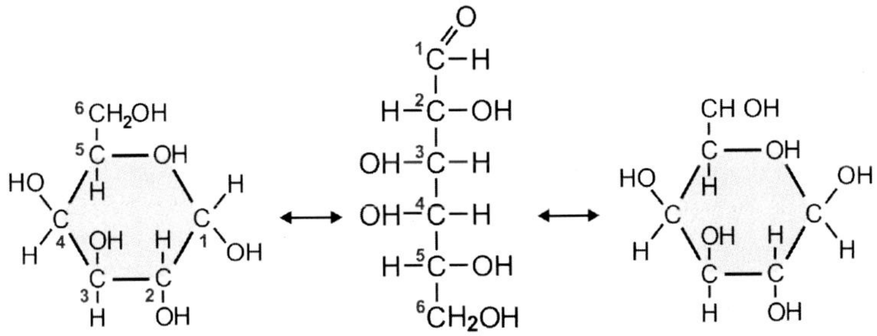

α-D-Galactose
α-D-Galactopyranose

D-Galactose
(lineal)

β-D-Galactose
β-D-Galactopyranose

FIGURE 4.3 **Different forms of galactose.**

When α glucose is dissolved in water, a portion of the molecules spontaneously convert into the β form and mutarotation takes place. Equilibrium is attained when two-third of the molecules in the solution are in the β form and one-third are in the α form. The mixture has a specific rotation of +52.7°. The same equilibrium is also reached if β glucose is originally dissolved.

Galactose

This sugar is commonly associated with other compounds to form complex molecules. Galactose is exceptionally found free in nature. With glucose, galactose forms the disaccharide *lactose*, which is present in milk. Galactose is less sweet than glucose. It is an epimer of glucose, differing in the configuration of C4.

Galactose is present as a cyclic pyranose form and, therefore, it has α and β anomers (Fig. 4.3).

Mannose

This aldohexose integrates oligosaccharides associated to glycoproteins in animals. It is also obtained by hydrolysis of plant polysaccharides known as mannans. Mannose is an epimer of glucose, differing in the configuration of C2 (Fig. 4.4).

Fructose

This sugar is a ketohexose, also called levulose because it is strongly levorotatory, with a specific polarized light rotation of −92.4°. It is

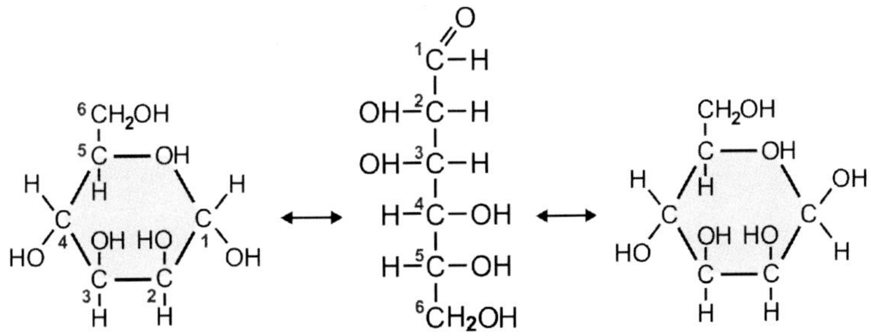

α-D-Mannose
α-D-Mannopyranose

D-Mannose
(lineal)

β-D-Mannose
β-D-Mannopyranose

FIGURE 4.4 **Different forms of mannose.**

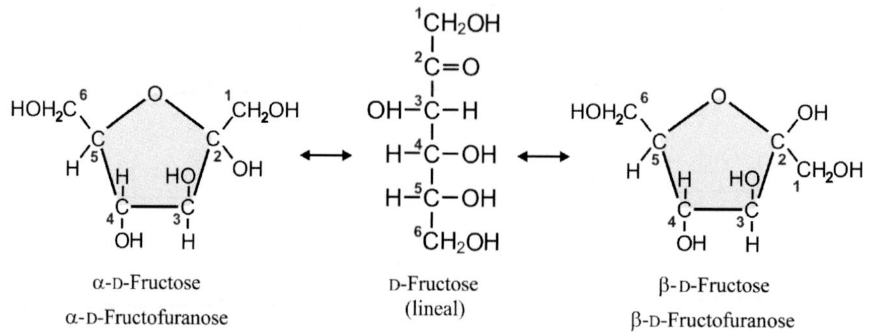

FIGURE 4.5   **Different forms of fructose.**

present as a free compound in ripe fruits, plant tissues, and in honey. When fructose is bound to glucose, it forms sucrose or sugar cane. Free fructose has higher sweetening power than sucrose and it is much sweeter than glucose. Thanks to this property, fructose is used in the manufacturing of soft drinks and candies. It is produced in large scale from maize starch after its hydrolysis to glucose and subsequent conversion into fructose by enzymatic isomerization.

In natural products containing fructose, this sugar adopts a cyclic configuration with a hemiketalic bond between C2′ and C5′, forming a five-membered ring similar to that of furan. Thus, fructose has a potential ketone group at C2′, which gives it delayed reducing properties. Fructose in its cyclic form has two possible configurations at carbon 2′: α and β (Fig. 4.5).

When forming complex molecules, fructose is mainly in its furanose form; however, when it is free in solution, it predominates in the pyranose form.

Pentoses

The most important pentose is aldopentose *D-ribose*, which is a component of ribonucleic acids (RNA) and other substances of great biological interest. It adopts the cyclic furanose form; therefore, it exhibits the α and β anomers (Fig. 4.6).

Haworth's Formulas

Haworth proposed the representation of the pyran and furan rings of monosaccharides in a

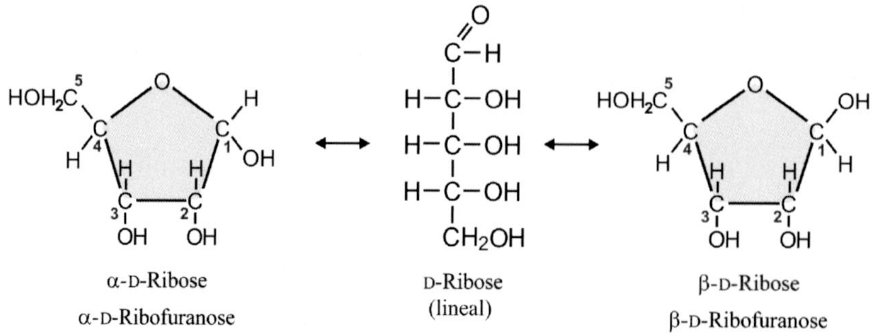

FIGURE 4.6   **Different forms of ribose.**

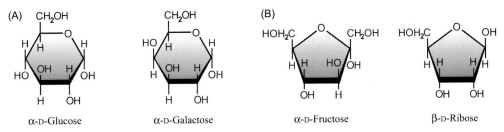

(A) CH$_2$OH   CH$_2$OH   (B)

α-D-Glucose   α-D-Galactose   α-D-Fructose   β-D-Ribose

FIGURE 4.7 **Haworth's formulas.** (A–B) Representation of the indicated monosaccharides according to Haworth.

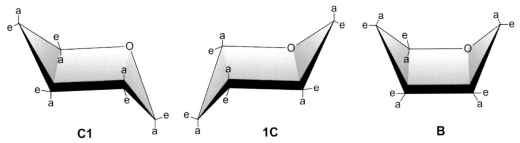

C1   1C   B

FIGURE 4.8 **Pyranose conformations.** C1 on the left, and 1C in the center correspond to the "chair" forms, while B on the right depicts the "boat" form. C1 is the most stable state. The direction of the bonds is shown, *a*: axial, *e*: equatorial.

same plane and situated the elements or functional groups bound to the carbon ring above or below that plane. In Haworth's formulas, the carbons of the ring are omitted and the bottom side of the hexagon or the pentagon is shown in bold, to emphasize that it is closer to the reader and give the molecule a three dimensional appearance (Fig. 4.7).

Haworth's representation is not entirely correct because the atoms included in the pyran ring are not located in the same plane. The molecule tends to adopt conformations of lower energy called the "chair conformation" and the "boat conformation" (Fig. 4.8).

The "chair" conformation (C1) is thermodynamically more stable. In both the chair and boat conformations, the bonds on the carbons forming the ring extend in two directions: one perpendicular to the plane of the cycle (axial) and the other in the same direction of the plane (equatorial). These are represented by the letters "a" and "e" in Fig. 4.8. The most common conformation for glucopyranose and other monosaccharides is the C1 form (Figs. 4.9 and 4.10).

The furanose form of monosaccharides is also not flat. One of the carbons in the cycle deviates from the plane where the other four carbons are located, giving what is called the "envelope configuration" due to its shape (Fig. 4.11). In ribose, which is a component of ribonucleic acid, C3' is outside of the plane where the rest of the molecule lays and it is oriented to the same side of the plane as C5'. Due to this, C3 is also called C3'-endo. In contrast, in the deoxyribose that forms part

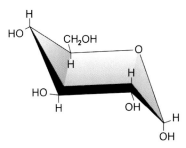

FIGURE 4.9 **α-D-glucopyranose ("chair" C1).**

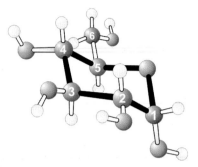

FIGURE 4.10 **α-D-glucopyranose ("chair" C1).** Spheres and rods model. C, *black*; O, *red*; H, *white*.

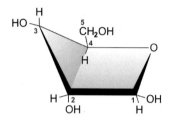

FIGURE 4.11 **Ribofuranose in its "envelope" form (C3 endo form).**

of deoxyribonucleic acid, C2 is the carbon that projects out of the plane (C2′-endo form).

Monosaccharide Derivatives

Glycosides

Carbon 1 or 2 of aldoses and ketoses, respectively, may react with another molecule to form a compound known as *glycoside*. For example, if methanol reacts with D-glucose in an acid medium, the carbon 1 of the sugar forms a bond with the alcohol and a molecule of water is released:

CH_2OH — O — H   H + HOCH_3 → CH_2OH — O — H   H + H_2O
HO   OH   H   OH   HO   OH   H   O-CH_3
     H   OH   H   OH

α-D-Glucose   Methanol   α-D-Methyl-glucoside

The carbon involved in glycosidic binding are the hemiacetalic carbon of aldoses or the hemiketalic carbon of ketoses. Depending on the configuration of the original monosaccharide, two types of glycosides are produced, α or β. The bond formed with the chemical group added is of the α or β glycosidic type, respectively. The α and β glycosides cannot spontaneously convert into each other; therefore, they do not exhibit the mutarotation phenomenon. Furthermore, the reactions typical of the aldehyde or ketone are no longer evident; glycosides lack the reducing capacity of sugars.

Glycosides have different names depending on the monosaccharides that form them. When the monosaccharide is glucose, these compounds are designated glucosides; if the sugar is galactose, they are called galactosides; if fructose is the monosaccharide present, they are named fructosides.

Monosaccharides can also establish glycosidic links between them, forming oligo- and polysaccharides. When the chemical group bound to the hemiacetal carbon of the monosaccharide is not a carbohydrate, it is called *aglycone*. The aglycone can be very simple, as in the case presented earlier (methyl), or more complex. Some glycosides are compounds of great medical interest and have been used to increase cardiac output in heart insufficiency due to their cardiotonic effects. Compounds in this group, including digitalis and ouabain, have a steroidal group forming the aglycone portion of the molecule.

Products Obtained From the Reduction of Hexoses

By reduction of the aldehyde or ketone group of sugars (under high hydrogen pressure and in the presence of a catalyst), the corresponding polyol is formed. Glucose generates a hexa alcohol called sorbitol and ribose produces ribitol, which makes up part of vitamin B_2 (riboflavin). Logically, these compounds cannot acquire cyclic form because they have lost the capacity of forming a hemiacetal bond.

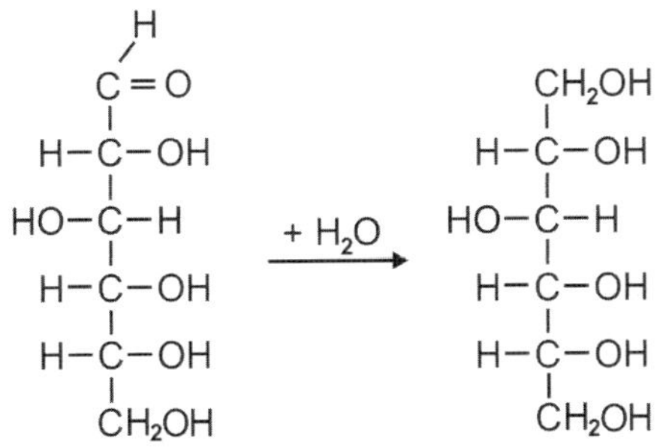

D-Glucose   Sorbitol

Deoxysugars

These compounds are monosaccharide derivatives, which are produced by loss of oxygen from one of the alcohol groups. The most abundant one in nature is 2-deoxyribose, which results from the removal of oxygen from carbon 2 of the aldopentose ribose. This compound is of great biological relevance, since it forms part of deoxyribonucleic acid (DNA).

Fucose is a deoxysugar and a L-galactose derivative, which lacks oxygen at carbon 6 (6-deoxy-L-galactose). Fucose forms part of complex molecules, such as glycoproteins of higher animals and bacterial cell walls.

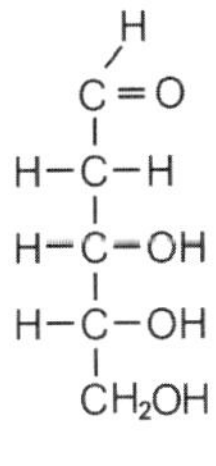

2-Deoxi-
D-ribose

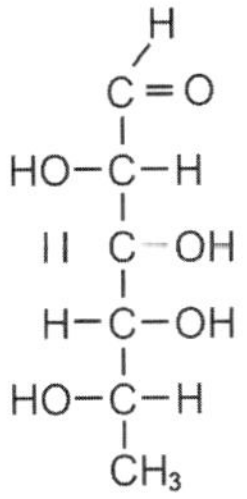

L-Fucose

Rhamnose is another deoxyhexose. It is an L-mannose derivative (6-deoxy-L-mannose), which is found in plant gums, mucilage, and animal glycoproteins.

Products of Oxidation of Aldoses

The aldehyde functional group of aldoses can be oxidized to carboxyl groups under the action of mild oxidants, forming *aldonic acids*. The aldonic acid formed by the oxidation of carbon 1 in glucose is designated *gluconic acid*. Stronger oxidation affects both aldose terminal carbons (C1 and C6), producing dicarboxylic acids. These diacids are called *aldaric acids*. The glucose derivative is *glucoaldaric acid*.

Under controlled conditions and protection of carbon 1, oxidation of only carbon 6 can be achieved. This originates *uronic acids*, which are part of complex polysaccharides. The uronic acid generated from glucose is *glucuronic acid*. Among all the derivatives produced by oxidation, obviously only the uronic acids can exist in cyclic form, since the reaction does not affect the hemiacetal bond.

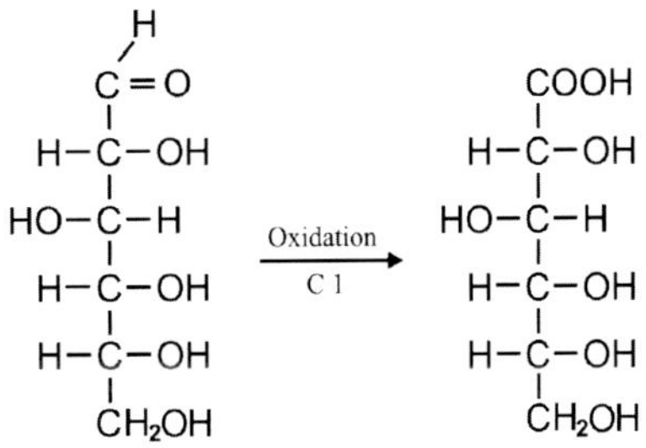

D-Glucose   Gluconic acid

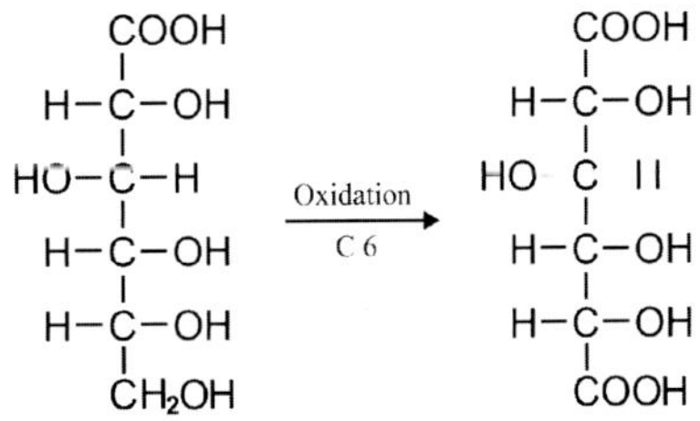

Gluconic acid   Glucaric acid

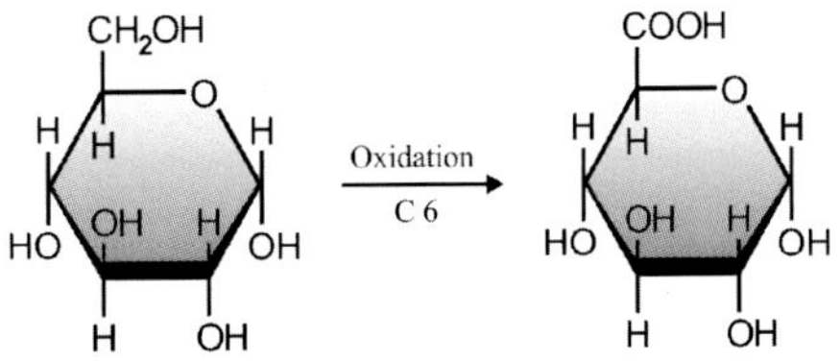

α-D-Glucose   α-D-Glucuronic acid

Phosphoric Esters

Monosaccharide esters with phosphoric acid (phosphorylation) are generated in many biological reactions. In general, this reaction is the first step in the metabolism of monosaccharides.

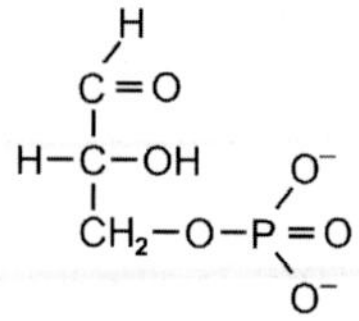

D-Glyceraldehyde-3-phosphate 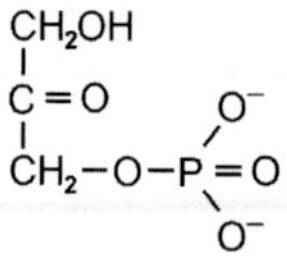   Dihydroxiacetone-phosphate

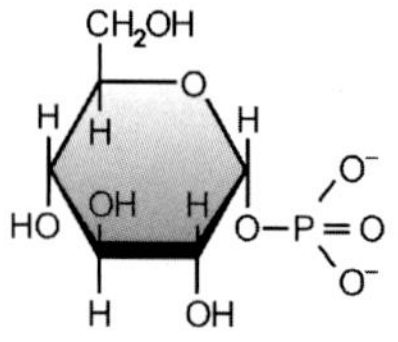

α-D-Glucose-1-phosphate 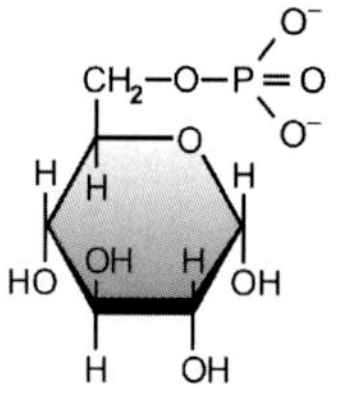   α-D-Glucose-6-phosphate

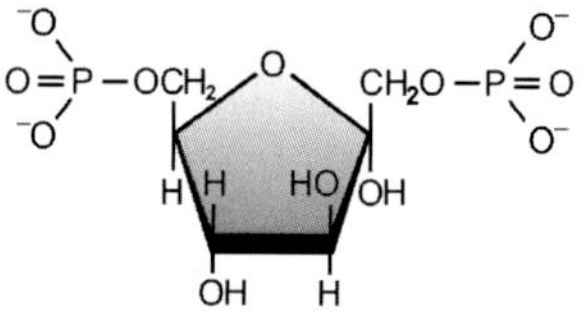

α-D-Fructose-1,6-bisphosphate 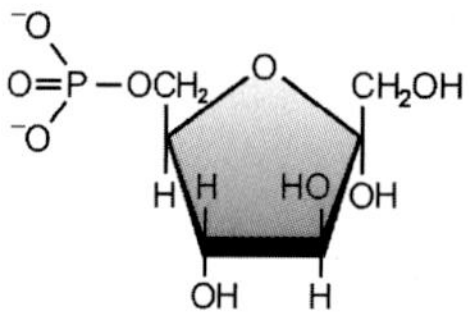   α-D-Fructose-6-phosphate

Amino Sugars

When one hydroxyl group of a monosaccharide is replaced by an amine group, an aminosugar is formed. Glucosamine and galactosamine, in which the amine group is attached to carbon 2, are the most common ones in nature. They constitute glycolipids and complex polysaccharides and are often acetylated on their amine group. An acetylated derivative of glucosamine is the basic component of chitin, a polysaccharide abundant in the exoskeleton of arthropods and insects, as well as in the cell wall of fungi.

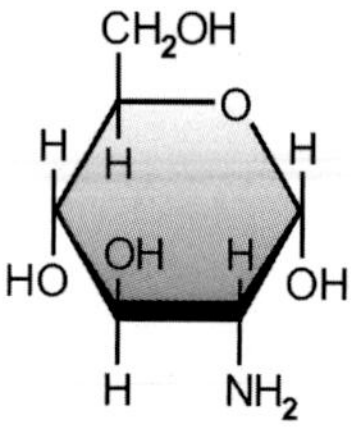

α-D-Glucosamine 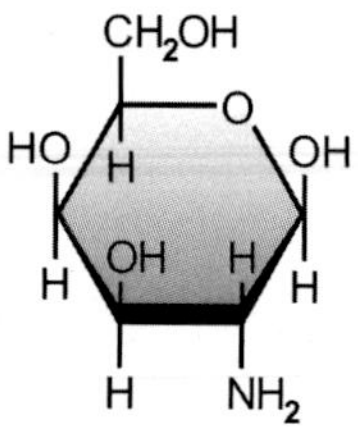   α-D-Galactosamine

Other nitrogen containing compounds related to hexoses are *neuraminic* and *muramic acids*. Neuraminic acid is an important component of polysaccharide chains in glycoproteins and glycolipids of cell membranes. This nine-carbon compound is formed by the amino-sugar mannosamine and pyruvic acid; generally the N is acylated, forming *sialic acids*. The most common sialic acid is N-acetyl-neuraminic, one of the strongest organic acids in living organisms ($pK_a = 2.6$).

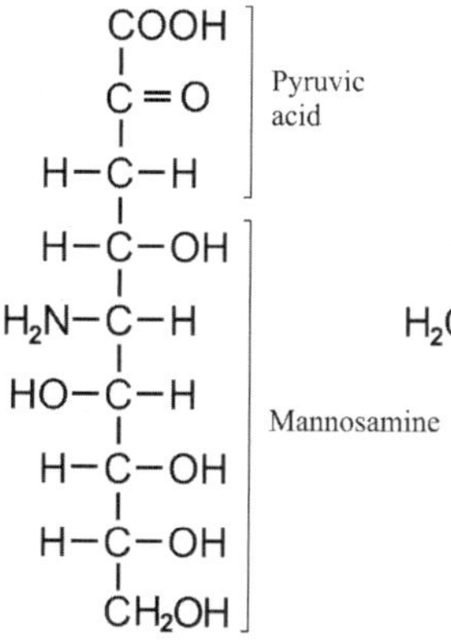

Neuraminic   N-acetyl neuraminic*
acid   (Sialic acid)

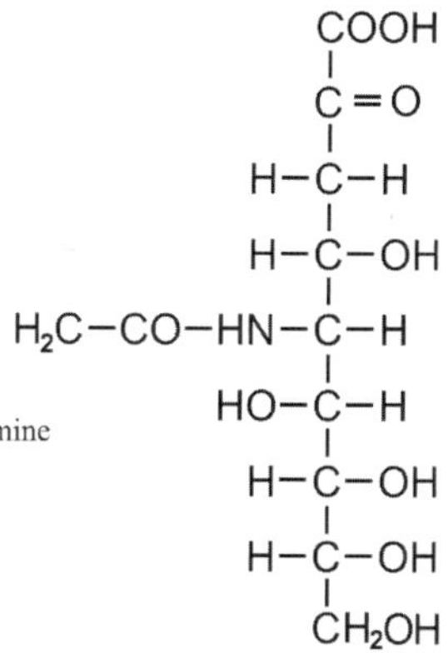

*N indicates that the acetyl moiety is attached to the nitrogen.

Muramic acid is formed by D-glucosamine with its C3 bonded to C2 of lactic acid (ether bond). An acetyl derivative of muramic acid, N-acetyl-muramic, is a component of the polysaccharide bacterial cell wall.

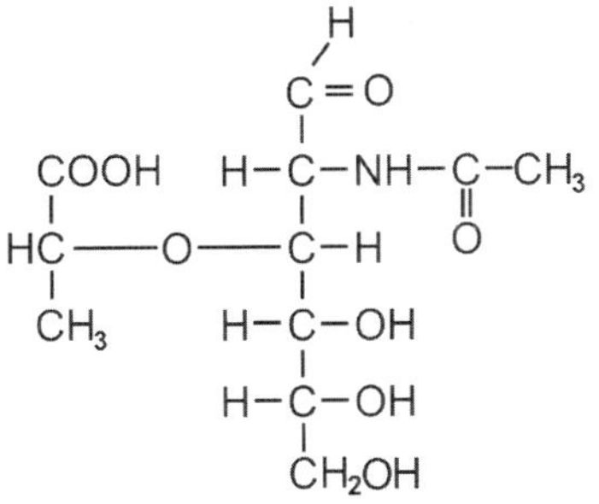

N-acetylemuramic acid*

*N indicates that the acetyl moiety is attached to the nitrogen.

DISACCHARIDES

Disaccharides are formed by the binding of two monosaccharides. This reaction produces a water molecule. Only the most important disaccharides in human biochemistry will be mentioned here.

Maltose

Malt sugar or maltose is a product of the hydrolysis of starch, catalyzed by the enzyme amylase. It is slightly sweet, very soluble in water, and results from the binding of carbon 1 of α-D-glucose (α-glycosidic bond) to carbon 4 of another D-glucose. Maltose is generated during brewing of beer and related beverages (malt beverages).

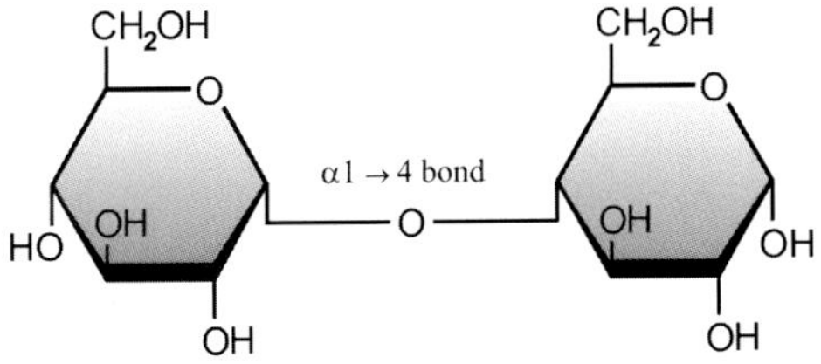

α1 → 4 bond

α-Maltose

O-α-D-Glucopyranosyl-(α1 → 4)-α-D-glucopyranose*

The aldehyde group of one of the glucoses remains free, giving the disaccharide its reducing properties and allowing it to have α and β forms.

Lactose

This disaccharide is found in milk. When hydrolyzed, galactose and glucose are released. Carbon 1 of β-D-galactose (β-glycosidic bond) is bound to carbon 4 of D-glucose. As carbon 1 of glucose remains free, lactose presents α and β forms and has reducing capacity.

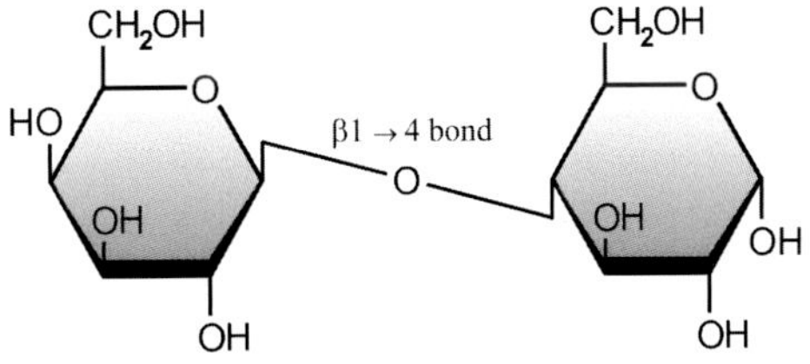

β1 → 4 bond

α-Lactose

O-β-D-Galactopyranosyl-(1 → 4)-α-D-glucopyranose*

*Name according to current nomenclature. O indicates glucose C1 oxygen bound to C4 of the other.

Saccharose

This sugar is commonly used as a sweetener in foods. It is obtained from sugar cane and beet. It consists of glucose and fructose, linked by a double glycosidic bond between carbon 1 of α glucose and carbon 2 of β-fructose. Both groups, aldehyde and ketone, are blocked and the disaccharide does not have reducing characteristics.

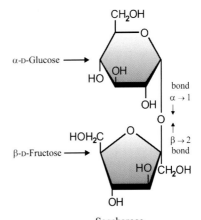

α-D-Glucose

bond α → 1

β → 2 bond

β-D-Fructose

Saccharose

β-D-Fructofuranosyl-(2 → 1)-α-D-glucopyranoside

Sucrose is dextrorotatory and subjected to hydrolysis produces an equimolar mixture of glucose and fructose, in which the levorotatory action of fructose predominates over the dextrorotatory activity of glucose. Due to this change in polarized light rotation, the mixture of glucose and fructose resulting from hydrolysis of sucrose is commonly known as "inverted sugar." Honey contains inverted sugar.

Cellobiose. This is a disaccharide that results from hydrolysis of cellulose. It is formed by two glucose units linked by a β-1→4 bond.

Trehalose. This is a nonreducing disaccharide composed of two α-D-glucose molecules linked by their anomeric hydroxyls (α-D-glucopyranosyl-(1→1)-α-D-glucopyranoside).

POLYSACCHARIDES

These compounds are more complex than the carbohydrates considered previously. They consist of many monosaccharide units, joined by glycosidic bonds. Some of these are polymers of one monosaccharide type and are called *homopolysaccharides*. Others have more than one class of monosaccharides and are called *heteropolysaccharides*. Generically they are all *glycans*. Most glycans are white, tasteless, amorphous compounds. They belong to the category of macromolecules, presenting a generally large molecular size. Some are insoluble in water, while others form colloidal dispersions in water.

Homopolysaccharides

These sugars are designated by adding the suffix *"an"* to the name of the monosaccharide

that constitutes them. For example, the homopolysaccharides composed by glucoses are called glucosans or glucans; those formed by mannose are mannans. The molecular mass of glycans varies within a wide range due to the constant addition or subtraction of monosaccharide residues to the polysaccharide chain. This synthesis or degradation of homopolysaccharides is regulated depending on the needs of the organism.

Starch

This sugar is a nutrient reserve in plants, where it is deposited in the cells as granules of size and shape that vary according to the plant of origin. Starch is the main carbohydrate in the human diet. It is found abundantly in cereals, potatoes, and some vegetables. Starch is composed of two different glucans, *amylose*, and *amylopectin*. Both are glucose polymers, but differ in structure and properties. Although the proportion varies in different plants, starch generally contains approximately 20% amylase and 80% amylopectin.

Amylose. This molecule is composed of 1000–5000 units of D-glucose and has a molecular mass between 180 and 900 kDa. The glucose molecules are associated by α-glycosidic bonds from the carbon 1 of one glucose to carbon 4 of the following (α-1→4 bond), forming long chains (Fig. 4.12).

This type of bond allows the chain to form a helical structure that coils around a central axis with six glucose molecules being contained within each turn of the helix (Fig. 4.13).The hydroxyl groups of the monosaccharide residues are arranged on the outer surface of the helix leaving the inside relatively hydrophobic.

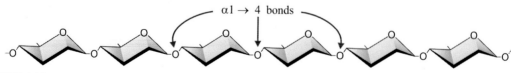

FIGURE 4.12　**Diagram of an amylose molecule segment.**

FIGURE 4.13 **Helical conformation of amylose.** Spheres and rods model. Only the elements of pyran rings are shown. C1 to C5, *black*; oxygen atoms of the glycosidic linkage α-1→4, *red*.

In aqueous suspensions amylase molecules tend to associate and precipitate. The reaction with iodine is used to recognize starch. Amylose gives a deep blue color when exposed to iodine. The internal diameter of the amylose helix is large enough to accommodate the iodine molecules. The blue color is the result of the amylose–iodine complex association.

Amylopectin. This molecule has higher molecular size than amylose and its mass can reach up to 100 million Da due to the polymerization of over 600,000 glucoses. The basic structure of amylopectin is similar to that of amylose, consisting of glucose molecules linked by α-glycosidic bonds (C1→C4). However, amylopectin differs from amylose because it presents branched chains. The ramifications are linear chains of approximately 24–26 glucose units, linked together by glycosidic bonds α-1→4 that are inserted onto the backbone chain by an α-glycosidic linkage from carbon 1 of the first glucose of the chain, to carbon 6 of a glucose molecule in the main chain (α-1→6). The branches are separated from each other by a distance of 10–15 glucose molecules in the main chain, where they are attached. There are also secondary and tertiary branches of 15–16 glucose units each. The scheme shown in Fig. 4.14 shows the structure of amylopectin.

When starch is heated in water, amylopectin forms highly viscous dispersions. The numerous hydroxyl groups on the surface of starch attract water molecules allowing it to make a stable gel (starch glue). The structural differences between amylose and amylopectin results in the formation of different complexes with iodine, which have a different color. An amylopectin–iodine complex produces a violet color, which is distinctly different from the blue color seen with amylose–iodine complexes. Starch has no reducing capacity because the glycosidic linkages in amylose and amylopectin block the potential aldehyde functions of glucose (except the one at the end of the main chain). Starch from food is hydrolyzed and degraded into free glucose molecules by digestive enzymes of animals. This allows glucose to be metabolized in tissues, since

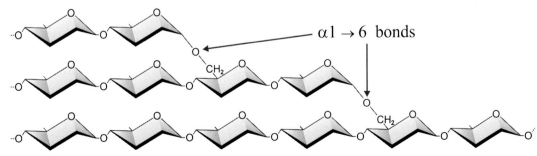

FIGURE 4.14 **Diagram of an amylopectin molecule segment.**

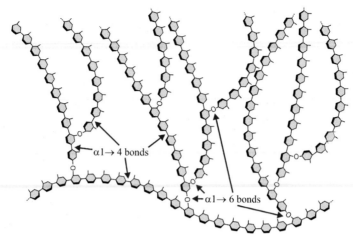

FIGURE 4.15   **Diagram of a glycogen molecule segment.**

only monosaccharides can be absorbed by the intestinal mucosa.

Glycogen

This is a polysaccharide present in the cytoplasm of most animal cells, mainly in liver and muscle. The size of glycogen molecules varies widely. They have an average diameter of approximately 25 nm, made of approximately 55,000 glucose units (10^7 Da). Twenty to forty glycogen molecules cluster to form rosettes that are visible under the microscope (β particles in muscle, and larger α particles in liver). Similar to amylopectin, glycogen is a polymer of α-D-glucose, presenting a branched structure with a main linear chain of glucose molecules linked by α-1→4 bonds and side linear α-1→4 chains inserted by α-1→6 links. Glycogen branches split from the main chain every 10 or less glucose molecules. The molecule has several layers of branches, while the internal branches have two sites of ramification, the external ones do not have ramifications. A diagram of a glycogen molecule segment is presented in Fig. 4.15. The structure of glycogen is very compact due to the proximity of the branches. This does not leave much space for water to accommodate

within the molecule and this is the reason why glycogen cannot form gels. This is in contrast to the branched structure of amylopectin, which is more open to accept water molecules and can easily form aqueous dispersions. However, glycogen can form aqueous dispersions that have opalescent appearance. Glycogen does not have reducing properties and gives a burgundy color when reacting with iodine. Although glycogen is a polymer of glucose, it contains traces of glucosamine and phosphate. Patients with a form of myoclonic epilepsy called *Lafora disease* suffer from an alteration in glycogen metabolism, which results in glycogen molecules excessively phosphorylated.

Dextrins

When starch is partially hydrolyzed by the action of acids or enzymes (amylases), it is degraded to maltose, maltotriose, and an oligosaccharide called *dextrin*. One type of dextrin, known as "limit dextrin" is one of the products after digestion with amylase. Since this enzyme catalyzes the hydrolysis of α-1→4 but does not affect links α-1→6, the digestive action of amylase stops at the starting points of the starch branches. These nonhydrolyzed sections of the starch molecule represent the limit of the action of amylase, giving them their name.

Dextrans

These polysaccharides are produced by certain microorganisms and are polymers of D-glucose. They have a branched structure with bonds that are different from those of amylopectin and glycogen. The main strands are glucose chains, linked by glycosidic α-1→6 bonds. The ramifications have α-1→2, α-1→3, or α-1→4 bonds, depending on the type of dextran. Dextrans of approximately 75 kDa produce highly viscous dispersions that are of clinical relevance. These are employed as emergency substitutes of plasma to restore blood volume in cases of acute loss of blood or plasma, until an appropriate transfusion therapy can be performed.

Inulin

Inulin is an energy reserve polysaccharide present in dahlia tubers and artichoke roots. It is a fructan formed by long chains of fructose molecules linked by β-2→1 glycosidic bonds; it is soluble in hot water. Inulin has been used in renal function tests, to estimate the kidney glomerular filtration rate (GFR).

Cellulose

This polysaccharide is the most abundant organic compound in nature. It is a glucan that

plays a structural role in plants and is one of the main components of plant cell walls. Cellulose is found in high proportion in bran, legumes, nuts, and cabbage. Wood pulp contains a high percentage of cellulose and cotton is almost pure cellulose. Industry processes more than 800 million tons of wood pulp per year for many different purposes.

Cellulose consists of over 10,000 glucose units linked by β-1→ 4 glycosidic bonds. It has a linear structure, with no branches. Although both amylose and cellulose are linear polymers of glucose, the difference in geometry of the α-1→ 4 and β-1→ 4 bonds gives these molecules a different conformation. In cellulose β-1→ 4 junctions, each glucose unit rotates 180° with respect to the preceding one (Fig. 4.16). This allows the formation of long straight chains, stabilized by hydrogen bonds. Instead, the α-1→ 4 bonds of amylose favor a helical conformation. Cellulose chains cluster in parallel strands that form strong microfibers. This structure is maintained by numerous hydrogen bonds, established between neighboring cellulose chains.

Human digestive secretions do not contain enzymes capable of catalyzing the hydrolysis of β-glycosidic linkages. Therefore, the cellulose ingested with plant foods cannot be modified during its transit through the gastrointestinal tract.

Chitin

Chitin is a polysaccharide abundant in nature, which constitutes the exoskeleton of arthropods,

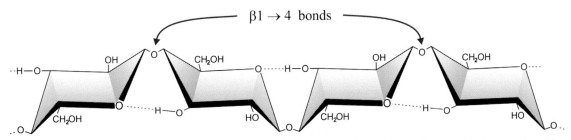

FIGURE 4.16   **Diagram of a cellulose molecule segment.** Each glucose unit turns 180° with respect to the preceding. Note the H bonds that stabilize the polymer strand.

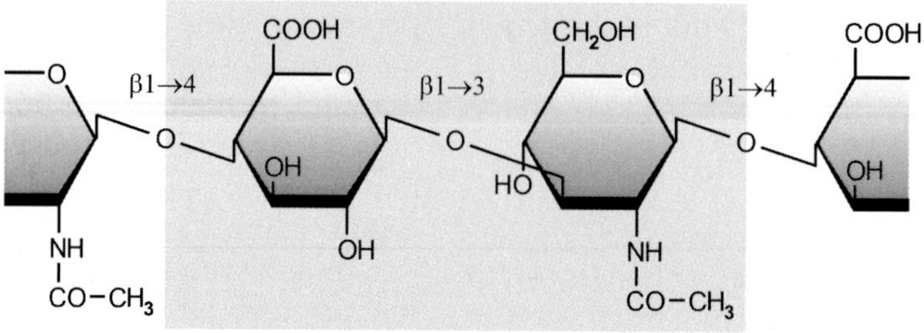

FIGURE 4.17   **Structural unit of hyaluronic acid (disaccharide on *pink box*).** D-Glucuronic acid-β-1→3-N-acetyl-D-glucosamine. At the body pH, the carboxyl function is ionized (COO⁻).

such as insects and crustaceans. It consists of units of N-acetyl-D-glucosamine, linked by β-1→4 glycosidic bonds.

Heteropolysaccharides

When hydrolyzed, these compounds give more than one type of monosaccharide or monosaccharide derivatives. Often they associate with proteins to form large molecular complexes.

Glycosaminoglycans

Formerly called mucopolysaccharides, these compounds are of great biological interest. They are linear polymers, formed by a succession of a disaccharide, generally composed of an uronic acid and a hexosamine. They usually contain sulfate groups. Glycosaminoglycans behave as polyanions thanks to the existence of many ionizable groups of uronic acid (—COO⁻) and sulfate (—SO₃⁻) in the molecule. Except for heparin, which is an intracellular compound, the other members of this group are found in the extracellular space, especially in the ground substance or extra fibrillar matrix of connective tissue. The structure of several types of glycosaminoglycans will be analyzed.

Hyaluronic acid. The structural unit of this compound is a disaccharide composed by D-glucuronic acid linked to N-acetyl-D-glucosamine by β-1→3 glycosidic bond. Each unit is attached to the next by a β-1→4 bond (Fig. 4.17).

Hyaluronic acid is the glycosaminoglycan of highest molecular weight (from 100,000 to

several million Da). It forms highly viscous solutions (gels) with lubricating properties. It is located in the intercellular substance of connective tissue, especially in skin, cartilage, the vitreous humor of the eye, Wharton's jelly in the umbilical cord, and synovial fluid.

Chondroitin sulfate. The disaccharide unit is similar to that of hyaluronic acid, but has N-acetyl-D-galactosamine instead of N-acetyl-D-glucosamine. The bonds are the same as those in hyaluronic acid. Chondroitin sulfate also differs from hyaluronic acid because it has a sulfate molecule (—SO₃⁻) forming an ester with C4 or C6 hydroxyl of galactosamine. According to the position of this group, two types of chondroitin are distinguished: chondroitin-4-sulfate or type B (Fig. 4.18) and chondroitin-6-sulfate or type C. The mass of these compounds ranges between 10 and 50 kDa. Both are important components of cartilage and bone.

Dermatan sulfate. This is a substance similar to chondroitin sulfate, except that it has L-iduronic

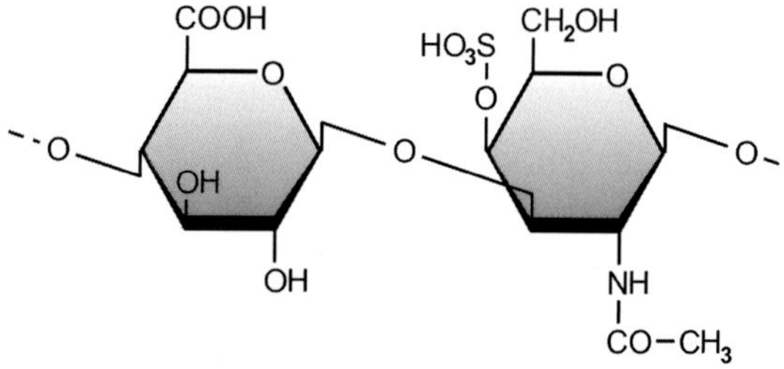

FIGURE 4.18   **A structural unit of B chondroitin sulfate (chondroitin-4-sulfate).** D-Glucuronic acid-β-1→3-N-acetyl-D-galactosamine-4-sulfate. At the pH of the body, the carboxyl and sulfide functions are ionized (-COO⁻, SO₃⁻).

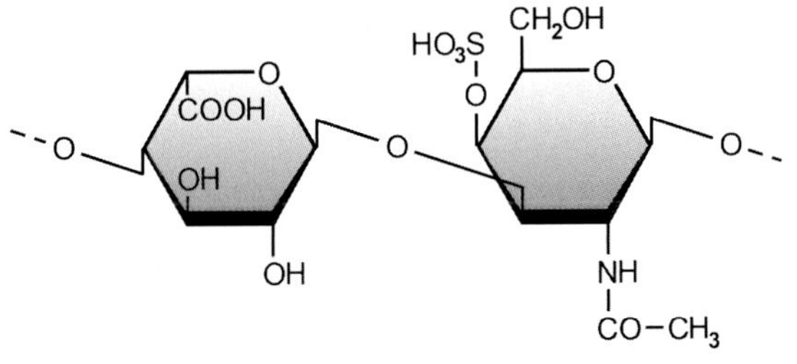

FIGURE 4.19 **Structural unit of dermatan sulfate.** L-Iduronic acid-β-1→3-acetyl-D-galactosamine-4-sulfate. At physiological pH the carboxyl and sulfate functions are ionized.

acid instead of glucuronic acid. This compound results from C5 epimerization of glucuronic acid (C6 carboxyl functional group is below the plane of the pyran ring). Sulfates are linked to C4 and/or C2 of galactosamine (Fig. 4.19). Originally, dermatan sulfate was called chondroitin sulfate B. It is found in skin and connective tissue of various organs.

Keratan sulfate. This glucosaminoglycan lacks uronic acid. The structural unit is composed of galactose and acetylated glucosamine, esterified by sulfate at C6. It is found in cornea and cartilage.

Heparin. The disaccharide unit of this compound is uronic acid and glucosamine linked by a β-bond. The uronic acids are glucuronic and iduronic acids, with the latter existing in greater proportion. Many glucosamine amine groups are sulfated, a few are acetylated. Sulfates are bound to C6 of the glucosamine and C2 of the uronic acid. The presence of so many sulfates gives this compound a strong acidic character. Heparin is the biomolecule with the highest density of negative charges. The glycosidic bonds

between disaccharides are α-1→4 (Fig. 4.20). Its molecular mass ranges between 8 and 20 kDa. It is found in granules contained in mast cells of connective tissue.

Due to its numerous negative charges, heparin has great tendency to interact with a variety of proteins, including enzymes, enzyme inhibitors, extracellular matrix proteins, cytokines, etc. Such interactions give heparin the ability to function as an anticoagulant, both in vitro and in vivo. This property has given heparin a frequent use in medicine. Another action of heparin is clearing plasma from fat after a meal high in lipids. Lipids are absorbed in the intestine and pass to the blood, forming particles called chylomicrons, which give plasma a milky aspect. Heparin accelerates the disappearance of chylomicrons circulating in blood.

Heparan sulfate. This is a compound similar to heparin, although with more sulfate groups and less iduronic acid groups. Both heparin and heparan sulfate show great variability in the sequence of monosaccharides that constitute them. Heparan sulfate is distributed on the surface of cells and in the extracellular matrix. The heparan sulfate–protein interaction is responsible for various physiological processes, including cell–cell adhesion, enzyme regulation, and cytokine action.

Dietary Fiber

Dietary fiber includes different plant components, which are not hydrolyzed by human gastrointestinal tract enzymes. Some are homopolysaccharides (cellulose), while others are heteropolysaccharides (hemicelluloses, pectins,

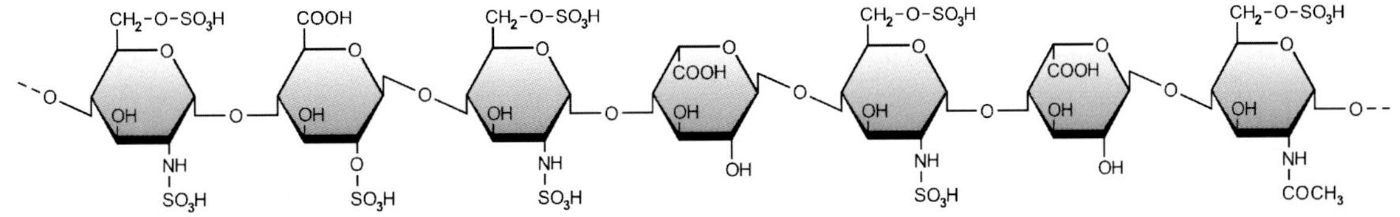

FIGURE 4.20 **Heparin molecule segment.** The units from left to right are: first, third, and fifth, α-D-glucosamine-2,6-sulfate; second, β-D-glucuronic acid-2-sulfate; fourth and sixth, α-L-iduronic acid; and seventh, N-acetyl-α-D-glucosamine-6-sulfate.

gums, and mucilages); finally, fibers such as lignin, are aromatic alcohols and not polysaccharides. Plant cell walls contain almost all of the dietary fiber, containing cellulose embedded in a matrix composed of heteropolysaccharides and fibrous proteins. The composition of this matrix varies in different vegetables and even in different parts of the same plant.

Hemicelluloses are heteropolysaccharides formed by a main chain of aldose molecules (glucose, galactose, xylose, or mannose) linked by β-1→4 bonds, with multiple branches or side chains of arabinose, galactose, and glucuronic acid. The main chain is stabilized by hydrogen bonds to the surface of the cellulose microfibers. The branches establish cross-links between the microfibers and other components of the cell matrix (pectins), forming a network responsible for the mechanical strength of plant cell walls. Hemicellulose represents 20%–30% of the plant components.

Pectins are a complex group of polysaccharides composed of galacturonic acid molecules joined into chains by α-1→4 links. The end of the chain is bound to short chains of monosaccharides (galactose, fucose, xylose, rhamnose, and arabinose). Pectins are found in fruits (apple and citrus), roots (beets and carrots), and plant stems. These molecules with numerous negative charges (carboxyl moieties of uronic acids) attract cations, particularly Ca^{2+} and water molecules. They are highly hydrated and tend to form gels, a property that is used in food (jellies and jams) and cosmetics industries.

Gums comprise a variety of monosaccharides, often galactose, and derivatives linked in a straight chain through β-1→3 and β-1→6 bonds, with branches of uronic acids, often galacturonic, arabinoses, and mannoses. Gums are secreted by plants (such as acacia and tragacanth) at sites of injury. They form highly viscous solutions and can also be used in the food industry.

Mucilages are substances with structure and properties similar to pectins. These are also able to form gels.

Lignin is not a polysaccharide; however, it is associated to polysaccharides of plant tissues, forming part of the dietary fiber. It is a highly branched and complex polymer of phenols, with strong intramolecular links. Lignin provides wood its typical resistance and density. It is also present in roots, wheat, and seeds.

Proteoglycans

Glycosaminoglycans are carbohydrates associated with proteins through glycosidic linkages between the polysaccharide chains and the hydroxyl of serine or threonine, or the nitrogen of asparagine residues in the protein. More than 100 glycosaminoglycan chains are linked to the protein. Several glycosaminoglycans attach to a central stem of hyaluronic acid, through their N-terminal ends. Association between this stem and the protein portion of proteoglycan takes place through another intermediary linking protein (Fig. 4.21). Up to 100 molecules of proteoglycan can bind to a hyaluronic acid backbone. The resulting large molecular aggregate, whose mass reaches tens of millions Daltons, are visible through electron microscopy.

Due to the polyanionic nature of glycosaminoglycans, these complexes interact with other macromolecules. In connective tissue, glycosaminoglycans bind to collagen by electrostatic forces. They also have great ability to attract water so, that much of the extracellular water in the body is fixed to the proteoglycans of connective tissue. The capacity of cartilage to serve as a cushion against compression depends on these highly hydrated polyanions. The cross-linked proteoglycans form a three-dimensional network that acts as a barrier to the extracellular transport of compounds. They contain chondroitin sulfate, dermatan, or keratan. The composition of the glycosaminoglycans of connective tissue from different organs is not exactly the same and also changes with age. For example, in newborns cartilage has a very low proportion of keratan sulfate. The length and relative amount of keratan sulfate increases with age, reaching, in the elderly, almost 50% of the total weight of all glycosaminoglycans.

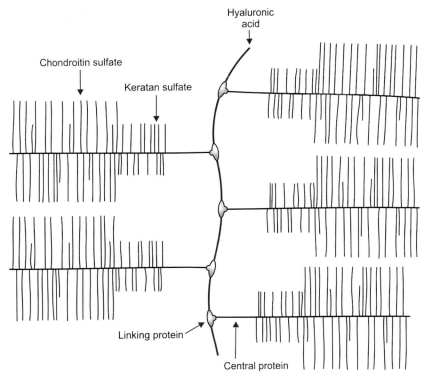

FIGURE 4.21   **Schematic representation of a proteoglycan segment.**

A cell membrane proteoglycan, *syndecan*, is composed by heparan and chondroitin sulfates; it binds to collagen and mediates cell adhesion to the extracellular connective tissue matrix. It also participates in signal transduction across the cell membrane to promote recruitment of ligands to the cell surface.

Peptidoglycans

Bacteria possess, outside of the cell membrane, a resilient wall which protects them from changes in the environment. This wall is formed by strands of a polysaccharide whose structural units are N-acetyl-D-glucosamine and N-acetylmuramic acid. The strands run parallel to each other and are interconnected by oligopeptides, which form a dense network that surrounds the entire bacteria. This structure is called *murein* and allows these organisms to take up and retain the Gram dye, rendering them Gram positive. One of the most useful antibiotics (penicillin) used to treat bacterial infections, inhibits the synthesis of murein. Gram-negative bacteria have yet another cover that is rich in lipids and hydrophobic proteins, which does not allow these organisms to retain the Gram stain as easily as the Gram-positive organisms do.

Glycoproteins

These compounds are conjugated proteins in which the prosthetic groups are carbohydrates; they also include proteoglycans. Glycoproteins differ from proteoglycans because their carbohydrate chains are shorter (oligosaccharides) and branched. Upon hydrolysis, they release more than two different types of monosaccharides.

In the oligosaccharide chains of glycoproteins there are D-galactose, D-mannose, L-fucose,

D-xylose, *N*-acetylglucosamine, *N*-acetylgalactosamine, glucuronic acid, iduronic, and sialic acids. The number of oligosaccharide chains in glycoproteins is highly variable; some have only one (ovalbumin) and others can have up to 800 chains (submandibular gland glycoprotein). The relative proportion of carbohydrates in glycoproteins ranges between 5% and 85%. Glycoproteins that are rich in sialic acids produce viscous solutions and act as lubricants.

Glycoproteins from part of: (1) almost all of the outer surface proteins of the plasma membrane of animal cells (the carbohydrate portion of these proteins form what is known as *glycocalyx*); (2) the majority of plasma proteins; (3) proteins excreted by mucous glands of the digestive, respiratory, and genital tracts; (4) some hormones;

and (5) many enzymes. In general, the export proteins synthesized by cells are glycoproteins.

Structural diversity of glycoprotein oligosaccharides. The oligosaccharides that constitute the side chains of glycoproteins have different structures; however, they also share certain common characteristics. For example, *N*-acetyl-glucosamine and galactose residues tend to be near the end attached to a protein, while sialic acids are at the opposite end. Sialic acid usually is located after a galactose.

The oligosaccharide binds through glycosidic bonds to hydroxyl groups of serine or threonine residues in the protein (O-glycosidic bond) (Fig. 4.22) or to the amide nitrogen of asparagines (N-glycosidic bond). In collagen, the oligosaccharides bind to the hydroxyl group of hydroxylysine, or hydroxyproline residues.

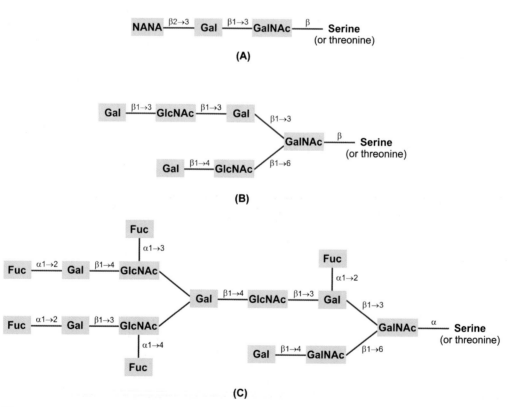

FIGURE 4.22 **Oligosaccharide structures of glycoproteins linked by O-glycosidic bond.** (A) Glycophorin (erythrocyte membrane glycoprotein) oligosaccharide. (B) Gastric mucin oligosaccharide. (C) Submandibular gland mucin oligosaccharide. *Fuc*, Fucose; *Gal*, galactose; *GalNAc*, N-acetyl-D-galactosamine; *GlcNAc*, N-acetyl-D-glucosamine; *NANA*, N-acetyl-neuraminic acid.

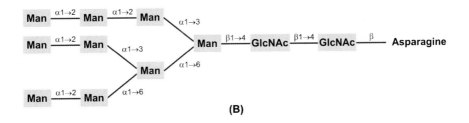

(A)

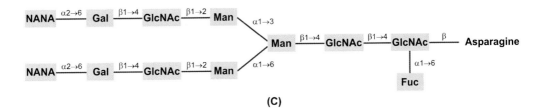

(B)

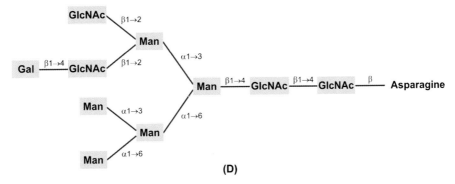

(C)

(D)

FIGURE 4.23 **Structure of oligosaccharides linked to protein by N-glycosidic bond.** (A) Common core for all oligosaccharides in N-glycosidic linkage. (B) High mannose type oligosaccharide. (C) Complex type. (D) Hybrid type. *Fuc,* Fucose; *Gal,* galactose; *GlcAAc,* N-acetyl-glucosamine; *Man,* mannose; *NANA,* N-acetyl-neuraminic acid.

The most frequent residue in O-glycosidic bonds is *N*-acetyl-D-galactosamine. In contrast, *N*-acetyl-D-glucosamine participates almost always in *N*-glycosidic bonds. Fig. 4.22 shows examples of oligosaccharide chains attached by O-glycosidic bonds to serine (or threonine) residues.

Oligosaccharides with N-glycosidic bonds contain a five-saccharide core, which is formed by two N-acetylglucosamines and three mannose residues (Fig. 4.23). Additional sugars are attached to this core structure, to originate a variety of carbohydrate patterns. The oligosaccharides linked to the amide N of asparagine residues are classified into three different types according to the amount of mannose units: (1) high mannose content; (2) complex oligosaccharides; and (3) hybrid oligosaccharides. All three types share the same basic structure, which consists of two N-acetyl-D-glucosamines and three mannose units (Fig. 4.23A). In the high mannose content, five to nine mannose molecules are added to the basic carbohydrate core (Fig. 4.23B). The complex types have varying amounts of other carbohydrates attached to the basic structure, which are different from mannose. These include N-acetyl-D-glucosamine, galactose, glucose, fucose, and sialic acid (Fig. 4.23C). In the hybrid or mixed type, on one of the distal mannoses of the common core only mannoses are added, while on the other, complex carbohydrate chains are added (Fig. 4.23D).

The multiplicity of possible chain combinations of oligosaccharides not only depends on the sequence of the component units, or "primary structure" as in polypeptides, but also on other peculiar features of this class of polymers: (1) The anomeric carbon (C1) can adopt two configurations (α or β), originating two different kinds of glycosidic linkages. (2) Glycosidic bonds involve any of the hydroxyl groups in each monosaccharide residue (C2, C3, C4, and C6). (3) Ramifications are frequent. All of these structural features offer a wide range of possibilities for building different oligosaccharides from a relatively small number of monosaccharides. For this reason, the amount of theoretically possible monosaccharide assemblies is larger than the amount of peptides that can be formed from the binding of the same number of amino acids.

Lectins. Proteins capable of recognizing and binding specific mono- or oligosaccharides with high affinity were discovered many years ago in different plants. These proteins, called *lectins*, are useful to study cell surface carbohydrates. Later, it was found that this type of protein is widely distributed in nature, being found not only in plants, but also in bacteria and animal tissues.

Blood groups. The surface of red blood cells and other cells contain glycoproteins and glycolipids that act as antigens. The antigenic determinant of these molecules lies in the carbohydrate portion of the glycoproteins and its structure is determined genetically (see Chapter 30). Depending on the oligosaccharide composition, it is possible to characterize different groups of individuals in a population. Several of these antigens are known in humans; the best studied is called AB0, which allows for the classification of individuals into four groups (A, B, AB, and 0).

Antigen 0 is an oligosaccharide composed by N-acetyl-glucosamine, galactose, and fucose and is bound to the galactose in a lactosyl residue attached to a ceramide of glycolipids or to a hydroxyl of an amino acid residue in glycoproteins. Antigens A and B have the same components than 0 antigen but they have in addition N-acetylgalactosamine (A group) or galactose (B group) linked to the terminal galactose (Fig. 4.24). Everyone can synthesize antigen 0, but to complete the A and B chains a particular enzyme is required to specifically catalyze the transfer of the additional monosaccharide in each group. People with group 0 lack both enzymes; group A individuals have inherited the gene that synthesizes the enzyme transferring N-acetylgalactosamine; in group B individuals, the enzyme that links galactose is present; people in group AB have both enzymes. Moreover, each individual produces antibodies against the antigen they lack. Thus, people in group 0 have antibodies against antigens A and B, while groups A and B produce antibodies against B and A, respectively; group AB does not produce antibodies against A or B antigens. In transfusions, the donor and recipient must be carefully recognized, since administration of the incompatible blood type causes severe reactions.

Glycoproteins as marker molecules. The complexity and diversity of carbohydrates in

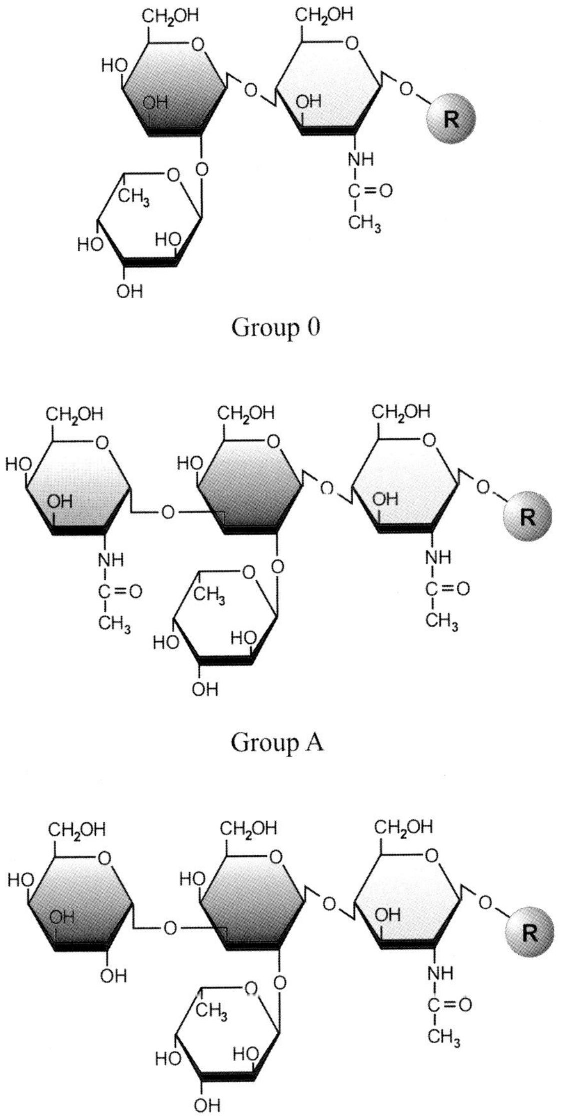

FIGURE 4.24 **Composition of oligosaccharides of AB0 blood group system.** Fucose, *White*; galactose, *red*; N-acetyl-galactosamine, *gray*; N-acetyl-glucosamine, *pink*.

glycoproteins makes them suitable molecules to contain information. The oligosaccharides on the cell surface represent "markers" or "signals" that serve for recognition. The following examples illustrate their role in cellular and molecular interactions.

A glycoprotein of the ovule, zona pellucida (ZP3) contains oligosaccharides recognized by receptors on the surface of sperm. This allows the interaction of female and male gametes prior to fertilization.

The adhesion of bacteria, viruses, and toxins to the cell surface requires the presence of specific carbohydrates in glycoproteins and glycolipids of cell membranes. For example, *Escherichia coli* and *Salmonella typhimurium* bind to mannose residues on the cell plasma membrane. An observation of interest related to this phenomenon is the loss of invasive ability of bacteria, viruses, or toxins when they are previously incubated with the carbohydrate that they selectively recognize. The added carbohydrate blocks the receptor site in the microorganism or toxin and renders it harmless.

Changes in surface oligosaccharides have been observed in malignant tumor cells. Some authors propose that such changes are the determinants of the anomalous behavior of neoplastic cells. The alteration of the signals that determine the "social relations" among cells contribute to the uncontrolled cell growth and multiplication.

These examples give an idea of the importance of these molecules. Advances in this field have opened new perspectives for the interpretation of many biological phenomena and provide opportunities for application in clinical practice.

SUMMARY

Carbohydrates are *polyhydroxyaldehydes, polyhydroxyketones,* and *polymers,* which, upon hydrolysis, can generate those compounds. They are classified into:

1. monosaccharides, or simple sugars;
2. oligosaccharides consisting of 2–10 monosaccharides;
3. polysaccharides, formed by a large number of monosaccharides.

Monosaccharides (MS) include *aldoses* or *ketoses*, which contain an aldehyde or ketone function, respectively. According to the number of carbon atoms in the molecule they can be trioses, tetroses, pentoses, etc. All have reducing capacity when placed in alkaline medium.

Optical isomerism is a property of MS; the simplest aldotriose, glyceraldehyde, has an asymmetric or chiral C, which makes it optically active. The dextrorotatory isomer is designated D and the levorotatory L. The capacity of a MS to produce isomers increases with the number of asymmetric C in the MS molecule. The number of possible optical isomers for any given MS is calculated by the formula 2^n, where *n* is the number of chiral C. All MS in which the configuration of the C containing the secondary alcohol (which is located farther away from the aldehyde or ketone function) is the same as that of the C2 of D-glyceraldehyde, belong to the D series, regardless of the sign of its optical activity. The human body synthesizes, almost exclusively, D carbohydrates.

Glucose is an aldohexose and the most important MS in humans. It is used as fuel by cells. Glucose adopts a cyclic structure (*pyran* derivative) by establishing a hemiacetal linkage between C1 and C5. Sometimes the hemiacetal bond is formed between C1 and C4 producing a *furan* cycle derivative. In the cyclic conformation, glucose has two isomers, α and β, that differ in specific optical rotation (α: +112.2°; β: +18.7°). In solution, both isomers convert into each other (*mutarotation*) until they are in equilibrium, which occurs when 2/3 of the molecules are in the β form and 1/3 are in the α form. This mixture has a specific rotation of +52.7°.

Galactose and *mannose* are aldohexoses that form part of complex molecules. They differ from glucose in the configuration of their C4 (galactose) or C2 (mannose). Both have reducing properties and present α and β forms.

Fructose is a ketohexose. D-Fructose is levorotatory (−92.4°). It is a reducing agent and has α and β forms.

Ribose is the most important aldopentose and a component of RNA.

Haworth's formulas represent pyran and furan rings as a plane. More realistic formulas are the "chair" and "boat" representations. The first is the most stable.

Glycosides are MS derivatives in which the hemiacetalic C of the MS binds to another compound. When this compound is not a carbohydrate, it is called *aglycone*. They are nonreducers and do not exhibit mutarotation. Glucose derivatives are called *glucosides*.

Polyalcohols are obtained by reduction of the aldehyde or ketone of MS; the glucose derivative is known as *sorbitol*.

Deoxysugars are produced by loss of oxygen from the alcohol group of a MS. The most abundant is *deoxyribose*, which is present in DNA. Fucose is another deoxysugar from animal and bacterial glycoproteins.

Oxidation products of MS include *aldonic acids*, result from mild oxidation of the C1 of aldoses to carboxyl. The glucose derivative of this oxidation is *gluconic acid*. Stronger oxidation of MS affects both C1 and C6, producing diacids called *saccharic* or *aldaric acids*. The glucose derivative of this type of oxidation is *glucaric acid*. Controlled oxidation, in which C1 is protected and only C6 is oxidized, produces *uronic acids*. The glucose derivative is called *glucuronic acid*.

Phosphate esters can be formed by phosphorylation of an MS and are commonly found as MS metabolic products.

Amino sugars generally have an amine group attached to C2. Examples of these compounds are *glucosamine* and *galactosamine*.

Neuraminic acid is a nine-carbon compound formed by mannosamine and pyruvic acid; it constitutes *sialic acid*, present in cell membranes.

Muramic acid is formed by *N*-acetyl-D-glucosamine, is a polysaccharide that forms part of bacterial walls.

Disaccharides include:

Maltose is a product of starch hydrolysis by amylase. Formed by two D-glucoses linked by glycosidic bond α-1→4; it is a reducer and has α and β forms.

Lactose is the main component of milk sugar, formed by D-galactose and D-glucose via β-1→4 glycosidic linkage. It is a reducer and has α and β forms.

Sucrose is used as a sweetener, formed by D-fructose and α-D-glucose linked through β-2→1. It does not reduce.

Polysaccharides or glycans are polymeric macromolecules, classified into:

Homopolysaccharides formed by glucose are glucans. *Starch* is the nutrient reserve of plants composed by *amylose* (± 20%) and *amylopectin* (± 80%). Amylose comprises between 1000 and 5000 D-glucose units joined linearly by α-1→4 glycosidic bonds. It forms a helix structure that when exposed to iodine produces a blue color. *Amylopectin* is a polymer of up to over 600,000 glucose units. It contains the basic structure of amylose plus branches formed by approximately 25 glucose residues inserted on the main chain by α-1→6 bonds. When exposed to iodine it gives off a violet color.

Glycogen is a polymer that serves as energy reserve polymer in animals. It is structurally similar to amylopectin, but with more branches. When exposed to iodine, it gives a burgundy color.

Dextrins are end products of partial hydrolysis of amylopectin by amylase.

Dextrans are the branched polymers of D-glucoses, with major chains formed by glucoses joined by α-1→6 bonds and ramifications that arise from α-1→2, α-1→3, or α-1→4 bonds.

Inulin is a polymer of fructose molecules bound via α-2→1.

Cellulose plays an important structural role in plants and is a linear polymer of glucose with β-1→4 bonds.

Chitin constitutes the exoskeleton of insects and crustaceans and it is a polymer of *N*-acetyl-D-glucosamine units, linked by β-1→4 bonds.

Heteropolysaccharides are constituted by more than one type of monosaccharide.

Glycosaminoglycans include:

Hyaluronic acid. Its structural unit is the disaccharide D-glucuronic acid-β-1→3-N-acetyl-D-glucosamine. Each of these units is attached to the next by β-1→4 bonds. Its molecular mass ranges from one hundred to thousands of kilodaltons.

Chondroitin sulfate is formed by units of the disaccharide D-glucuronic-β-1→3-N-acetyl-D-galactosamine, which is esterified with sulfate. It constitutes cartilage and bone tissues.

Dermatan sulfate is similar to condroitin sulfate, except that it has L-iduronic acid instead of glucuronic acid, linked by α-1→3-bonds to N-acetyl-D-galactosamine. It has sulfate on C4 of galactosamine and C2 of the iduronic acid. It is present in connective tissue and skin.

Keratan sulfate has no uronic acid and its structural unit consists of galactose and N-acetyl-D-glucosamine esterified with sulfate. It is abundant in cartilage.

Heparin is a repeat of the disaccharide D-glucosamine and uronic acids (iduronic and glucuronic). It is highly sulfated, which contributes to the acidic character of this compound. Heparin has a mass between 8 and 20 kDa. It is an anticoagulant and clears chylomicrons from plasma.

Heteropolysaccharides bound to other kind of molecules constitute proteoglycans, peptidoglycans, glycolipids (gangliosides), and glycoproteins. *Proteoglycans* result from the association of glycan chains (chondroitin sulfate, dermatan sulfate, and keratan) and proteins, bound via glycosidic bonds to the hydroxyl of serine or threonine residues (O-glycosidic bond), or to the N of asparagine residues (N-glycosidic bond). More than 100 glycosaminoglycan chains are attached to a polypeptide chain. This structure is in turn inserted through a binding protein, to a hyaluronic acid backbone chain. These large molecular complexes are arranged in three-dimensional networks in the extracellular space of connective tissue.

Peptidoglycans are the main component of bacterial cell walls. They consist of N-acetyl-D-glucosamine and N-acetyl-muramic acid connected by oligopeptide transversal bridges.

Glycoproteins are carbohydrates conjugated to proteins by O- or N-glycosidic linkages. The O-glycosidic link takes place at the hydroxyl group of serine or threonine residues on the protein. The N-glycosidic link occurs at the N of an asparagine residue. The chains in N-glycosidic link are: (1) high in mannose content; (2) complex, containing mannose, and other carbohydrates; and (3) hybrids, constituted by a mixture of mannose and complex carbohydrates.

Gangliosides and glycoproteins differ from proteoglycans because they have shorter carbohydrate chains (oligosaccharides) and produce more than two different MS when hydrolyzed. They play important roles, and the oligosaccharides that they contain function as markers for antigen/antibody recognition on the surface of cells.

Bibliography

Casu, B., Lindahl, U., 2001. Structure and biological interactions of heparin and heparin sulfate. Adv. Carbohydr. Chem. Biochem. 57, 159–206.

Elgavish, S., Shaanan, B., 1997. Lectin-carbohydrate interactions: Different folds, common recognition principles. Trends Biochem. Sci. 22, 462–467.

Freeze, H.H., Aebi, M., 2005. Altered glycan structures: the molecular basis of congenital disorders of glycosylation. Curr. Opin. Struct. Biol. 15, 490–498.

Greenberg, R.E., 1995. New dimensions in carbohydrates. Am. J. Clin. Nutr. 61 (Suppl.), 915S–1011S.

Lindhorst, T.K., 2003. Essentials of Carbohydrate Chemistry and Biochemistry, sixth ed. Prentice Hall, Upper Saddle River, NJ.

Sharon, N., Lis, H., 1995. Lectins–proteins with a sweet tooth: function in cell recognition. Essays Biochem. 30, 59–75.

Taylor, M.E., Drickamer, K., 2006. Introduction to Glycobiology, second ed. Oxford University Press, New York, NY.

Varki, A., Cummings, R., Esko, J., Freeze, H., Hart, G., Marth, J., 2002. Essentials of Glycobiology. Cold Spring Harbor Laboratory Press, New York, NY.

5

Lipids

Lipids, widely distributed in animals and plants, comprise a heterogeneous group of substances which share the common characteristic of having little or no solubility in water, but good solubility in nonpolar substances (a substance is soluble in solvents of similar nature; polar substances dissolve in polar solvents, nonpolar substances in nonpolar solvents). Different from polypeptides or polysaccharides, lipids do not form macromolecular polymers and do not have high molecular mass.

The study of lipids is important because (1) lipids are essential components of living beings, constituting a fundamental part of cell membranes; (2) they are the main energy reserve (neutral fats) in animals; (3) they are of essential nutritional value because lipids have a high-caloric content, they transport fat-soluble vitamins, and supply indispensable compounds [essential fatty acids (FAs)] that humans cannot synthesize; and (4) they comprise a number of substances of critical physiological activity, including hormones, certain vitamins, and bile acids.

CLASSIFICATION

According to the complexity of the molecule, lipids can be classified into two different types: *simple* and *complex*. In addition, there are other substances associated with lipids that share their solubility properties. Simple lipids include acylglycerols and waxes, while complex lipids include phospholipids, glycolipids, and lipoproteins. Among the compounds associated with lipids are sterols, terpenes, and vitamins that are soluble in organic solvents.

Lipid molecules contain monocarboxylic organic acids, normally referred to as FAs. Due to their biological relevance, they will be discussed first.

FATTY ACIDS

The FAs isolated from animal lipids are monocarboxylic and have a linear chain. Only very few of them are free; the vast majority form part of simple or complex lipids. FAs with a cyclic structure are only found in lipids from some microorganisms and seeds. FAs with branched chains are present in waxes. Animal lipids have, in general, an even number of carbon atoms (between 4 and 26 carbon atoms) and only very few FAs with an uneven number of carbon atoms have been isolated. The number of carbon atoms in a FA chain depends on the manner in which they are synthesized or degraded in animals, which is achieved through addition or subtraction of two carbon atoms at a time. FAs can be saturated, with the general chemical formula $CH_3—(CH_2)_n—COOH$, or unsaturated, with double bonds in the hydrocarbon chain. The unsaturated FAs can have one or multiple double bonds. When FAs have more than one double bond, generally they are not conjugated: —CH=CH—CH=CH—, but instead separated by a methylene group (—CH$_2$—)-: —CH=CH—CH$_2$—CH=CH—. In animals, the most common FAs have 16–18 carbon atoms. FAs with even carbon number linear chain are listed in Table 5.1.

TABLE 5.1   Fatty Acids (FAs) Common in Nature

Trivial name	No. of C	Systematic name	Fusion temp. (°C)	Formula
SATURATED FATS				
Butyric	4	Butanoic	−7.9	CH$_3$—(CH$_2$)$_2$—COOH
Caproic	6	Hexanoic	−3.4	CH$_3$—(CH$_2$)$_4$—COOH
Caprylic	8	Octanoic	16.3	CH$_3$—(CH$_2$)$_6$—COOH
Capric	10	Decanoic	31.2	CH$_3$—(CH$_2$)$_8$—COOH
Lauric	12	Dodecanoic	43.9	CH$_3$—(CH$_2$)$_{10}$—COOH
Myristic	14	Tetradecanoic	54.1	CH$_3$—(CH$_2$)$_{12}$—COOH
Palmitic	16	Hexadecanoic	62.7	CH$_3$—(CH$_2$)$_{14}$—COOH
Stearic	18	Octadecanoic	69.9	CH$_3$—(CH$_2$)$_{16}$—COOH
Arachidic	20	Eicosanoic	75.4	CH$_3$—(CH$_2$)$_{18}$—COOH
Behenic	22	Docosanoic	80.0	CH$_3$—(CH$_2$)$_{20}$—COOH
Lignoceric	24	Tetracosanoic	84.2	CH$_3$—(CH$_2$)$_{22}$—COOH
SATURATED HYDROXY ACID				
Cerebronic	24	2-Hydroxytetracosanoic	−100.0	CH$_3$—(CH$_2$)$_{21}$—CHOH—COOH
UNSATURATED FATS				
MONOETHYLENIC				
Palmitoleic	16	*cis* Δ9 Hexadecenoic ω7	0.5	CH$_3$(CH$_2$)$_5$CH=CH(CH$_2$)$_7$COOH
Oleic	18	*cis* Δ9 Octadecenoic ω9	13.4	CH$_3$(CH$_2$)$_7$CH=CH(CH$_2$)$_7$COOH
Erucic	22	*cis* Δ13 Docosenoic ω9		CH$_3$(CH$_2$)$_7$CH=CH(CH$_2$)$_{11}$COOH
Nervonic	24	*cis* Δ15 Tetracosenoic ω9		CH$_3$(CH$_2$)$_7$CH=CH(CH$_2$)$_{13}$COOH
DIETHYLENIC				
Linoleic	18	*cis* Δ9,12 Octadecadienoic ω6	−5.0	CH$_3$(CH$_2$)$_4$CH=CH—CH$_2$—CH =CH(CH$_2$)$_7$COOH
POLYETHYLENIC				
Linolenic	18	*cis* Δ9,12,15 Octadecatrienoic ω3	−10.0	
Arachidonic	20	*cis* Δ5,8,11,14 Eicosatetraenoic ω6	−49.5	
Timnodonic	20	*cis* Δ5,8,11,14,17 Eicosapentaenoic ω3		
Clupanodonic	22	*cis* Δ7,10,13,16,19 Docosapentaenoic ω3		
Cervonic	22	*cis* Δ4,7,10,13,16,19 Docosahexaenoic ω3		

The systematic name for the FAs results from adding the suffix "-oic" to the name of the hydrocarbon from which they derive, but commonly a trivial name is used (Table 5.1). FA carbons are numbered from the carbon that has the carboxyl group, which is considered C1. Greek letters are also used, α (or C2) is the carbon adjacent to the carboxyl, and carbons β, γ, etc. are the ones that follow. Carbon ω (omega) is always considered the last carbon, whatever be the number of carbons in the chain. To represent each FA, a simplified notation is used, which indicates the number of carbons and number of double bonds in the chain, separated by a colon. This notation allows for the differentiation between FAs that have the same number of carbon atoms, but different numbers of carbon double bonds. For example, the notation for stearic acid is 18:0, while the notation for linolenic acid is 18:3. For unsaturated FAs, the positions of the double bonds are indicated in parentheses, with the number of the carbon where the double bond begins. For example, oleic acid is 18:1(9) and arachidonic acid is 20:4(5,8,11,14), which indicates that the double bond is located between C9 and C10 in oleic acid and between C5–C6, C8–C9, C11–C12, and C14–C15 in arachidonic acid. The symbol Δ (delta) followed by the carbon number where the double bond begins is also used. For example, linolenic acid is 18:3Δ9,12,15.

Another notation is used which indicates the position of double bonds from the ωC. For example, oleic acid is represented as 18:1ω9 (or n9); linoleic, 18:2ω6 (or n6); linolenic, 18:3ω3 (or n3); arachidonic, 20:4ω6 (or n6). Since the double bonds are separated by methylene bridges ($-CH_2-$), knowing their number and position from the ωC, the location of the other double bonds can be deduced. The ω notation is useful when considering polyethylene acid biosynthesis.

Fatty Acids Properties

Physical Properties

Solubility. FAs possess a polar group (hydrophilic) represented by the carboxyl group and a nonpolar (hydrophobic) portion that includes the carbon chain. The water solubility of FAs decreases as the length of the carbon chain increases. FAs with more than six carbons are practically insoluble in water and soluble in organic solvents because the long hydrophobic chain prevails over the hydrophilic carboxyl group.

Melting and boiling temperatures. The melting temperature of FAs increases with the chain length (Table 5.1). Saturated FAs from two to eight carbons are in liquid state at 20°C, while those of longer chains are solid. The presence of double bonds decreases the melting temperature of FAs. Thus, stearic acid melts at 69.9°C and is solid at 20°C. If a double bond between C9 and C10 is inserted into stearic acid, it becomes oleic acid and the melting temperature is reduced to 13.4°C; unlike stearic acid, oleic acid is liquid at 20°C. Adding a second double bond to oleic acid, between C12 and C13 generates linoleic acid, which melts at 5°C. The boiling temperature of FAs increases with the chain length.

Geometrical isomerism. Saturated FAs adopt different spatial arrangements because the single bonds between the carbon atoms allow their free rotation. However, the extended conformation, forming a zigzag with angles of 109° between two successive links (Fig. 5.1), is more

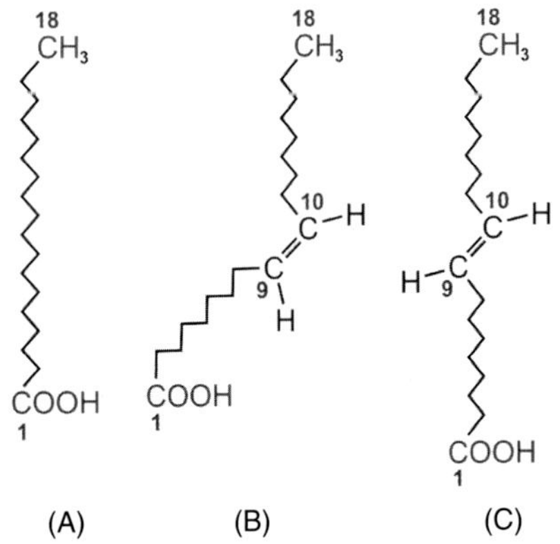

FIGURE 5.1 **Arrangement of fatty acid (FA) carbon chains.** (A) Saturated FA (stearic); (B) monoethylenic FA, *cis* configuration (oleic acid); and (C) monoethylenic FA, *trans* configuration (elaidic acid).

stable (with lower free energy) due to the hydrogen atoms linked to the carbon atoms of the chain.

The ethylenic or unsaturated FAs have a more rigid structure because the carbons joined by the double bond cannot rotate freely. The existence of a double bond creates the possibility of geometrical isomerism. According to the position of the substituents with respect to the plane of the double bond, *cis–trans* isomers can be formed:

$$CH_3-(CH_2)_7 \diagdown \diagup H$$
$$C$$
$$\|$$
$$C$$
$$HOOC-(CH_2)_7 \diagup \diagdown H$$

Oleic acid
(*cis*)

$$CH_3-(CH_2)_7 \diagdown \diagup H$$
$$C$$
$$\|$$
$$C$$
$$H \diagup \diagdown (CH_2)_7-COOH$$

Elaidic acid
(*trans*)

Almost all natural unsaturated FAs are *cis* isomers. The *cis* configuration produces a kink in the chain at each ethylenic bond, which causes the chain to adopt different layouts (Figs. 5.1 and 5.2), commonly exhibiting a U-shaped configuration (Fig. 5.3). In contrast, the *trans* isomers have an extended structure, similar to that of saturated chains (Fig. 5.1C). The *cis* form is less stable than the *trans* form and may be converted to the *trans* configuration by action of various agents, including heat. With an increase in temperature, oleic acid (18:1Δ9 *cis*) becomes elaidic acid (18:1Δ9 *trans*), which shows different properties.

Although natural products have great predominance of *cis* FAs, there are always a small proportion of *trans* isomers. For example, milk fat contains 4–8% of FAs in *trans* configuration. In contrast, fats subjected to hydrogenation, as

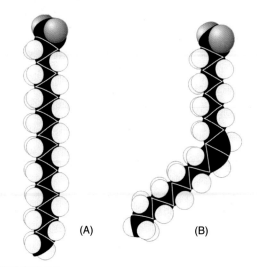

(A)   (B)

FIGURE 5.2   **Compact molecular models.** (A) Saturated FA and (B) monoethylenic FA (oleic). Carbon, *Black*; oxygen, *red*; hydrogen, *white*.

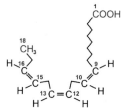

FIGURE 5.3   **Arrangement of the carbon chain of a *cis* triethylenic FA (linolenic).**

margarines, have significantly higher proportions of these isomers.

Chemical Properties

PROPERTIES THAT DEPEND ON THE CARBOXYL GROUP

Acidic character. The acidic property of a FA depends on the carboxyl group. In water soluble molecules like acetic acid, ionization occurs as follows:

$$CH_3COOH \longrightarrow CH_3COO^- + H^+$$

Increasing the number of carbons in the lipid chain reduces its water solubility and acidic character.

Formation of salts (soaps). When a FA reacts with a base, a salt is formed:

$$CH_3(CH_2)_{12}COOH + NaOH \longrightarrow$$

Myristic acid

$$\longrightarrow CH_3(CH_2)_{12}COONa + H_2O$$

Sodium myristate

Salts are designated by the name of the metal followed by the FA name with the suffix "-ate" (e.g., potassium stearate). These FA salts are called *soaps*. Soaps generated from alkaline metals (Na, K, etc.) are water soluble and act as emulsifiers or detergents. Salts formed with elements in Group II of the Periodic Table (Ca, Mg, Ba), or with any other heavy metal, are insoluble in water and organic solvents. Soaps containing calcium or magnesium precipitate and do not form foam.

Emulsifying action of soluble soaps. An *emulsion* is a heterogeneous mixture of two insoluble liquids causing one to disperse in small droplets within the other. A common example of emulsion is seen when mixing oil and water. Both liquids spontaneously separate, with the less dense oil forming a layer on top of the water. If the mixture is intensely shaken, the oil splits into small droplets, forming an emulsion. However, once the mixture is left unstirred, the oil droplets come together and reconstitute the initial oil top layer. If a soluble soap (sodium palmitate) is added before stirring the mixture, a stable emulsion is obtained and the oil will remain dispersed in fine droplets within the water. Soap molecules have a nonpolar carbon chain (CH_3—$(CH_2)_{14}$—) that is hydrophobic (soluble in oil and insoluble in water) and a —COONa group, which ionizes (—COO^- + Na^+) and is polar and hydrophilic (soluble in water and insoluble in oil). At the interface between oil droplets and water, soap ions are arranged with the alkyl group oriented toward the oil and the carboxylate ions positioned toward the water (Fig. 5.4). The surface of the droplets covered by the —COO^- groups is

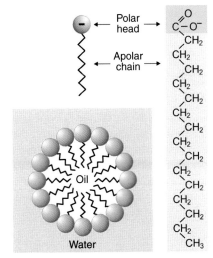

FIGURE 5.4 **Emulsifying action of soluble soaps.**

negatively charged, maintaining the droplets repelled from each other, which helps to stabilize the emulsion.

Ester formation. FAs form esters by reacting with alcohols:

$$CH_3-(CH_2)_{16}-COOH + CH_3-CH_2OH \xrightarrow{-H_2O}$$

Stearic acid   Ethanol

$$\longrightarrow CH_3-(CH_2)_{16}-CO-O-CH_2-CH_3$$

Ethyl stearate

PROPERTIES DEPENDENT ON THE CARBON CHAIN

Oxidation. Unsaturated FAs are easily oxidized. Atmospheric oxygen oxidizes oleic acid at the level of the double bond to form peroxides. This peroxide is susceptible to further oxidation, which produces the rupture of the double bond, breaking the carbon chain into different compounds (such as short-chain monocarboxylic and dicarboxylic acids and aldehydes), which are responsible for the odor and taste, typical of rancid fats.

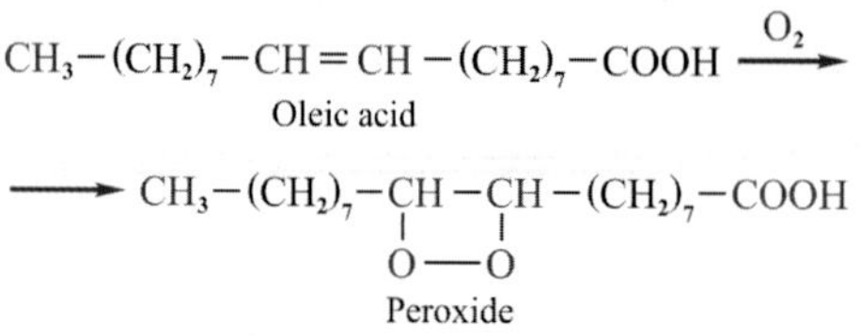

$$CH_3-(CH_2)_7-CH=CH-(CH_2)_7-COOH \xrightarrow{O_2}$$

Oleic acid

$$\longrightarrow CH_3-(CH_2)_7-\underset{O-O}{CH-CH}-(CH_2)_7-COOH$$

Peroxide

Hydrogenation. Unsaturated FAs are more abundant in nature than saturated FAs and their use in industry is important. Saturated FAs are obtained from unsaturated FAs by hydrogenation in the presence of a catalyst (Pt, Ni, or Pd). The hydrogen atoms bind to the carbon atoms that contain double bonds, transforming these into single bonds.

$$CH_3-(CH_2)_7-CH=CH-(CH_2)_7-COOH \xrightarrow[+H_2]{Ni}$$

Oleic acid (liquid at 20°C)

$$\longrightarrow CH_3-(CH_2)_{16}-COOH$$

Stearic acid (solid at 20°C)

Halogenation. The double bonds of FAs can easily incorporate halogens (F, Cl, Br, and I).

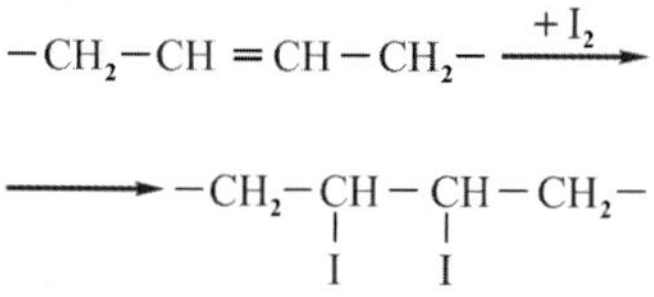

$$-CH_2-CH=CH-CH_2- \xrightarrow{+I_2}$$

$$\longrightarrow -CH_2-\underset{I}{CH}-\underset{I}{CH}-CH_2-$$

This property is used to determine the degree of unsaturation of FAs in a biological sample. Most commonly iodine is used for halogenation. Under controlled conditions, the quantity of iodine consumed by a certain amount of lipid is proportional to the number of double bonds within the FAs present in the sample. *Iodine number* is defined as the amount of iodine in grams required to halogenate 100 g of lipid material.

ESSENTIAL FATTY ACIDS

As it will be discussed in Chapter 15, animals produce FAs from the addition of two carbon atoms at a time. However, there are some FAs,

which cannot be synthesized by the body and must be supplied with the diet. These are called essential or indispensable FAs. Essential FAs include the polyethylenic or polyunsaturated *linoleic, linolenic,* and *arachidonic* FAs.

SIMPLE LIPIDS

Acylglycerols

Most of the FAs of the body form esters with different alcohols, preferably glycerol, generating *acylglycerols* or *acylglycerides*. Glycerol has three alcohol groups bound to each of its carbons. Glycerol carbons are designated by Arabic numerals or Greek letters. The primary carbons C1 and C3 are also named α and γ, while C2 is named β.

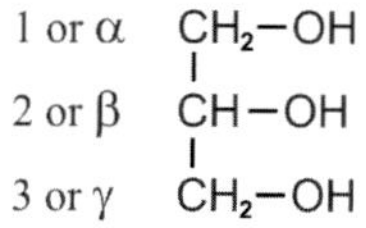

$$
\begin{aligned}
1 \text{ or } \alpha \quad & CH_2-OH \\
2 \text{ or } \beta \quad & CH-OH \\
3 \text{ or } \gamma \quad & CH_2-OH
\end{aligned}
$$

Depending on the number of alcohol groups esterified by FAs, acylglycerols are designated monoacylglycerols, diacylglycerols, or triacylglycerols. The names mono-, di-, and triglycerides, widely used, are not correct and should be abandoned. Triacylglycerols are commonly called *neutral fats*.

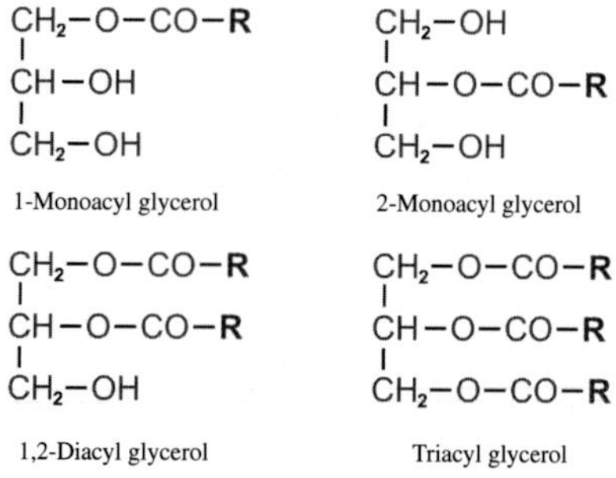

$$
\begin{array}{ll}
CH_2-O-CO-R & CH_2-OH \\
CH-OH & CH-O-CO-R \\
CH_2-OH & CH_2-OH \\
\text{1-Monoacyl glycerol} & \text{2-Monoacyl glycerol} \\
\\
CH_2-O-CO-R & CH_2-O-CO-R \\
CH-O-CO-R & CH-O-CO-R \\
CH_2-OH & CH_2-O-CO-R \\
\text{1,2-Diacyl glycerol} & \text{Triacyl glycerol}
\end{array}
$$

R: FA carbon chain

If the FAs are all the same, the di- and tria-cylglycerols are called homoacylglycerols; if they are different, the acylglycerols are named heteroacylglycerols. The more recommended designation for each acylglycerol includes the FA with the suffix "-oil" and a number that describes the position of the FA within the molecule. For example,

$$CH_2-O-CO-(CH_2)_{16}-CH_3$$
$$CH-O-CO-(CH_2)_{16}-CH_3$$
$$CH_2-O-CO-(CH_2)_{16}-CH_3$$

Tristearoylglycerol or tristearin
(homotriacylglycerol)

$$CH_2-O-CO-(CH_2)_{14}-CH_3$$
$$CH-O-CO-(CH_2)_7-CH=CH-(CH_2)_7-CH_3$$
$$CH_2-O-CO-(CH_2)_{14}-CH_3$$

1,3-Dipalmitoyl-2-oleyl-glycerol
(heterotriacylglycerol)

Trivial names for homotriacylglycerols, such as tripalmitin, tristearin, or triolein, are also commonly used.

In 1-monoacylglycerol, the C2 of glycerol is asymmetric or chiral, producing two stereoisomers: D and L. According to the convention adopted for glyceraldehyde (see p. 23), the D-form is represented with the hydroxyl in C2 to the right and the L form, with the hydroxyl to the left.

$$CH_2-O-CO-R \qquad CH_2-O-CO-R$$
$$H-C-OH \qquad HO-C-H$$
$$CH_2-OH \qquad CH_2-OH$$

D-Monoacylglycerol   L-Monoacylglycerol
(The asymmetric C is shown in *red*)

In 1,2-diacylglycerols and in di- and triacylglycerols, C2 is asymmetric when the C1 and C3 are esterified by different acyl residues. All natural compounds of this type belong to the L series.

To avoid confusions that may be created with the numbering of carbons (the primary carbons of glycerol are called either C1 or C3), the *stereospecific numbering* (*sn*) notation has been proposed. Glycerol is represented with the C2 hydroxyl to the left, the upper carbon is C1.

$$CH_2-O-CO-(CH_2)_{12}-CH_3$$
$$R-CO-O-C-H$$
$$CH_2-O-CO-(CH_2)_{16}-CH_3$$

R: $CH_3-(CH_2)_4-CH=CH-CH_2-CH=CH-(CH_2)_7-$

Systematic name:
1-Myristoyl-2-linoley-3-stearoyl-*sn*-glycerol
or 1-myristoyl-2-linoleyl-3-stearoyl-L-glycerol

Acylglycerols Properties

PHYSICAL PROPERTIES

Solubility. Acylglycerols are less dense than water and are water insoluble. In contrast, mono- and diacylglycerols are polar molecules due to their free hydroxyl groups; they have emulsifying power. Triacylglycerols are soluble in chloroform, ether, and hot alcohol. These are all solvents that are used to extract these triacylglycerols from tissues.

Melting temperature. The melting temperature of an acylglyceride depends on the FAs that compose them. Those containing FAs with long saturated chains melt at higher temperature, while those with FAs that are unsaturated or have short saturated chains melt at lower temperature. For example, tristearin melts at 71°C, while triolein at −17°C.

Heteroacylglycerols with unsaturated FAs are either liquid at room temperature or solids with a low melting temperature, depending on the amount of ethylenic FAs present in the molecule. Vegetable oils are rich in triacylglycerols containing long-chain unsaturated FAs.

Isomerism. Heteroacylglycerols exhibit structural and also optical isomerism.

CHEMICAL PROPERTIES

The chemical properties of acylglycerides depend mainly on their ester functions and the FA chains that constitute them.

Hydrolysis. Acylglycerols are hydrolyzed when heated in an aqueous acidic medium. This reaction releases FAs and glycerol.

$$CH_2-O-CO-(CH_2)_{14}-CH_3$$
$$CH-O-CO-(CH_2)_{14}-CH_3 \quad + 3\ H_2O \longrightarrow$$
$$CH_2-O-CO-(CH_2)_{14}-CH_3$$

Tripalmitin

$$CH_2-OH$$
$$\longrightarrow CH-OH + 3\ CH_3-(CH_2)_{14}-COOH$$
$$CH_2-OH$$

Glycerol   Palmitic acid

Acylglycerides are also readily cleaved when heated in the presence of a strong base (KOH or NaOH), leading to the release of glycerol and the corresponding FA salts (soaps). This process is called *saponification*.

$$CH_2-O-CO-(CH_2)_{16}-CH_3$$
$$CH-O-CO-(CH_2)_{16}-CH_3 \quad + 3\ KOH \longrightarrow$$
$$CH_2-O-CO-(CH_2)_{16}-CH_3$$

Tristearin

$$CH_2-OH$$
$$\longrightarrow CH-OH + 3\ CH_3-(CH_2)_{16}-COOK$$
$$CH_2-OH$$

Glycerol   Potassium stearate

Usually, in addition to triacylglycerols, non-ester substances exist in fats. These form the unsaponifiable fraction of fats and include hydrocarbons, free sterols, and pigments. After saponification, compounds with ester functions (acylglycerols) are converted into glycerol and soap, both of which are, unlike the original fat, soluble in water and insoluble in ether. The unsaponifiable fraction of fat remains soluble in ether and insoluble in water, a characteristic that allows for the separation of the two groups of substances.

Hydrogenation. Solid fats are obtained by hydrogenation of oils in the presence of nickel as catalyst. This process is used to produce margarine. The hydrogenation of unsaturated FA from the acylglycerols in oils is only partial, which gives these fats a consistency similar to that of butter. If the hydrogenation were complete, the fats obtained would be hard. The consistency of butter is due to short-chain FAs, such as butyric and caproic. Margarine has acylglycerols with partially hydrogenated, long-chain FAs and lacks the vitamins that are present in butter.

$$CH_2-O-CO-(CH_2)_7-CH=CH-(CH_2)_7-CH_3$$
$$CH-O-CO-(CH_2)_7-CH=CH-(CH_2)_7-CH_3 \ +$$
$$CH_2-O-CO-(CH_2)_7-CH=CH-(CH_2)_7-CH_3$$

Triolein (liquid at 20°C)

$$CH_2-O-CO-(CH_2)_{16}-CH_3$$
$$+ 3\ H_2 \longrightarrow CH-O-CO-(CH_2)_{16}-CH_3$$
$$CH_2-O-CO-(CH_2)_{16}-CH_3$$

Tristearin (solid at 20°C)

The hydrogenation process also produces isomerization of *cis* unsaturated chains, with part of the acylglycerides becoming *trans* isomers.

Oxidation. Acylglycerols can undergo oxidation on their ethylenic FAs, producing compounds that are responsible for the odor and flavor that stale (rancid) fats have.

Nutritional Importance of Fats

Lipids have a much higher caloric value than other components of a diet. One gram of fat provides 9.3 kcal (38.9 kJ), while the same amount of carbohydrates only offers 4.1 kcal (17.2 kJ). All animals store neutral fats as energy reserve. This reserve is more important than that of carbohydrates that are rapidly depleted during fasting. Triacylglycerols are efficient molecules for the storage of energy. Nearly all of the carbon atoms in triacylglycerols are relatively less oxidized than in carbohydrates; their complete oxidation to CO_2 and H_2O yields higher amounts of energy. Moreover, due to their hydrophobicity, fats practically do not retain water, unlike other reserve material, such as glycogen. This gives fat the possibility to store more energy relative to its weight.

The chemical composition of fat varies according to the site or organ where it is located within an animal. In general, fats that function as mechanical support predominantly have saturated long-chain FAs and are semisolid (perirenal fat). Fat reserves that are used as energy source are almost liquid at body temperature. The composition of reserve fat is influenced, in part, by the composition of dietary fat.

The predominant FA in animal fat reserve is oleic acid. Oleic acid is also the most abundant FA in plants; however, vegetable oils contain a high percentage of polyethylenic FAs. For example, corn oil is made up of 41.8% linoleic acid. Grape, sunflower, and peanut oils are also rich in polyunsaturated FAs. Olive oil is relatively low in these essential FAs. Fats from several species of fish contain polyunsaturated ω3 FAs of 20–22 carbon atoms in length.

It has been observed that the consumption of diets rich in polyethylenic *cis* FAs contributes to the reduction of cholesterol concentration in people with elevated blood cholesterol. This is a valuable preventive factor for atherosclerosis. In contrast, animal fats with a higher proportion of saturated FAs or foods with *trans* unsaturated FAs favor high levels of cholesterol.

Numerous studies have shown that linoleic acid (LA,18:2Δ9,12), ω3 (or n3), linolenic (LNA,18:3Δ9,12,15), eicosapentaenoic acid (EPA, 20:5Δ5,8,11,14,17), and docosahexaenoic acid (DHA, 22:6Δ4,7,10,13,16,19) exert a protective action against coronary heart disease. These acids, mainly LA, have a regulatory effect on the metabolism of plasma lipoproteins (see p. 348), diminishing the production of LDL and promoting its removal. Essential FAs tend to reduce hyperlipemic effects of other diet components, such as saturated FAs, *trans* unsaturated FAs, and cholesterol. EPA and DHA help to maintain normal vascular endothelium, blood pressure, levels of plasma triacylglycerides, and prevent platelet aggregation. An adequate diet should contain a FA ratio of ω6/ω3 (n6:n3) of 6/1. However, the most important factor is the total amount of essential FAs. It is recommended that essential FAs provide between 1% and 2% of the total calories needed by an individual.

Conjugated linoleic acids. These comprise dienoic FA isomers and stereoisomers of linoleic or octadecadienoic acid (18:2Δ9 *cis*, 12 *cis*). They are found in foods (meat, milk, and milk products) from bovine and ovine origin. The most abundant isomer in these foods is the 9 *cis*, 11 *trans* (also called rumenic acid), the 7 *trans*, 9 *cis*, and the 11 *cis*, 13 *trans* (Fig. 5.5).

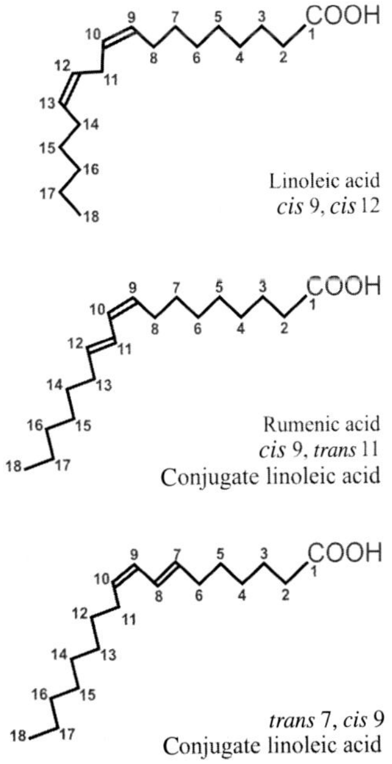

FIGURE 5.5 **trans 7, cis 9 conjugated linoleic acid.**

The inclusion of these acids in the diet of laboratory animals (rats and mice) showed beneficial effects, reducing body fat and weight and decreasing the incidence of malignant tumors, atherosclerosis, and diabetes. While some studies have reported similar effects in humans, there are others which present conflicting results. Until more convincing experimental evidence is obtained, ingestion of conjugated linoleic acids in amounts greater than 0.5% of the total dietary FA is not justified.

Waxes

These types of lipids are esters of long-chain monohydric alcohols and higher FAs. For example, one of the most important components in beeswax is the ester of a 30-carbon alcohol ($C_{30}H_{61}OH$) and palmitic acid. Waxes are solid at room temperature and insoluble in water. Usually they play lubricative and protective roles. For example, waxes help to lubricate the skin and provide waterproof protection to hair and feathers; bees use them to build the honeycombs. In vegetables, waxes form a protection coat over leaves and fruits. Plankton organisms are rich in waxes, and marine animals from cold regions consume them and accumulate waxes that they can store as energy reserve.

COMPLEX LIPIDS

Complex lipids have other components in addition to the alcohols and FAs present in simple lipids. Depending on these additional components, they are divided into *phospholipids* and *glycolipids*, which contain phosphoric acid and carbohydrates, respectively. Lipoproteins are also considered complex lipids.

Phospholipids

Phospholipids are complex lipids that have phosphoric acid linked by an ester bond. Some tissues are very rich in phospholipids. They represent up to 30% of brain dry weight, while they only comprise approximately 2% of muscle dry weight. Phospholipids are constituted by an alcohol, FAs, and phosphoric acid. They are subdivided into glycerophospholipids (when the alcohol is glycerol) and sphingophospholipids (when the alcohol is sphingosine).

Glycerophospholipids

Glycerophospholipids are the most abundant phospholipids. They are found in highest amounts in the membranes of all cells and are present in very small quantities in fat stores. In addition, glycerophospholipids are a source of physiologically active compounds. They commonly have arachidonate (at position C2), a FA that is released for the synthesis of eicosanoids (p. 349). Glycerophospholipids also participate in cell signaling systems and as an anchor for proteins in cell membranes.

Glycerophospholipids derive from *phosphatidic acids*, compounds formed by a molecule of glycerol with two of its hydroxyl groups esterified by FAs, and the third hydroxyl esterified by phosphoric acid. C2 of the glycerol moiety is asymmetric, producing stereoisomers. Natural glycerophospholipids have the L configuration. The numbering of the carbon atoms follows the rules of stereospecific numbering (*sn*). The carbon whose OH is esterified with phosphate is termed C3.

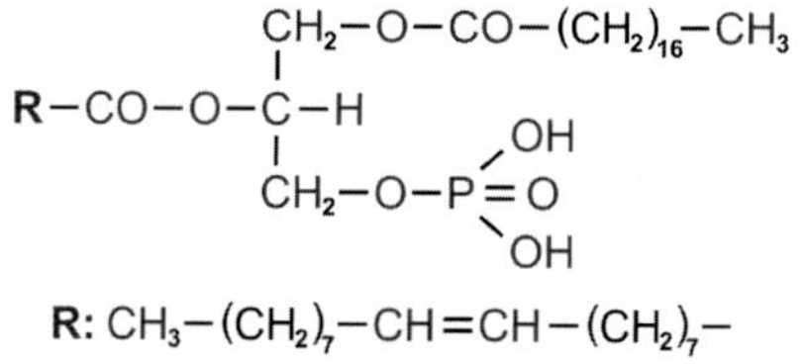

$$CH_2-O-CO-(CH_2)_{16}-CH_3$$
$$R-CO-O-C-H$$
$$CH_2-O-P=O$$

R: $CH_3-(CH_2)_7-CH=CH-(CH_2)_7-$

Phosphatidic acid
Systematic name: 1-stearoyl-2-oleyl-*sn*-glycerol-3-phosphate

Phosphatidic acids are produced in the body as intermediates in the synthesis of triacylglycerols and glycerophospholipids; however, because they are not stored in tissues, they exist in small amounts. Generally one of the free OH groups on the phosphate moiety is esterified with another component, generating different glycerophospholipids. When the added component is the amino alcohol *choline*, the phospholipid is *phosphatidylcholine*, also known as *lecithin* (Fig. 5.6). If the amino alcohol is ethanolamine, the resulting phospholipid is *phosphatidylethanolamine* or *cephalin*.

$$CH_2OH-CH_2-\overset{+}{N}\overset{CH_3}{\underset{CH_3}{-}}CH_3 \qquad CH_2OH-CH_2-\overset{+}{N}H_2$$

Choline   Ethanolamine

$$CH_2-O-CO-R_1$$
$$R_2-CO-O-C-H$$
$$CH_2-O-\overset{O^-}{\underset{}{P}}=O$$
$$O-CH_2-CH_2-\overset{+}{N}\overset{CH_3}{\underset{CH_3}{-}}CH_3$$

Phosphatidylcholine

Addition of the amino acid serine to phosphatidic acid produces *phosphatidylserine*, while addition of the cyclic polyol inositol gives *phosphatidylinositol*. Inositol is a cyclic hexalcohol; the most abundant isomer in nature is *meso*-inositol. When the six hydroxyls are esterified by phosphate, hexakisphosphate inositol or phytic acid is obtained.

$$CH_2-O-CO-R_1$$
$$R_2-CO-O-C-H$$
$$CH_2-O-\overset{O^-}{\underset{}{P}}=O$$
$$O-CH_2-CH_2-\overset{+}{N}H_3$$

Phosphatidylethanolamine

$$CH_2OH-CHNH_2-COOH$$

Serine

Inositol

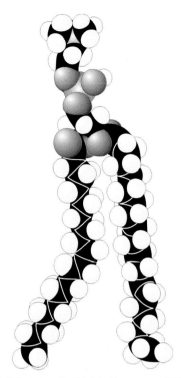

FIGURE 5.6 **Phosphatidylcholine, space filling model.** C, *Black*; H, *white*; O, *red*; P, *gray*; N, *pink*.

At physiological pH, phosphatidylcholine and phosphatidylethanolamine are neutral molecules that behave as dipolar ions or zwitterions. Phosphatidylserine and phosphatidylinositol have a net negative charge and behave as acids.

Phosphatidylinositol bisphosphate, unlike the phospholipids mentioned earlier, has three phosphate groups instead of one. Both of the additional phosphate groups are attached to the —OHs of the C4 and C5 of inositol. *Phosphatidylinositol bisphosphate* is found in cell membranes. In response to extracellular signals (hormones and chemical intermediates) *phosphatidylinositol bisphosphate* is hydrolyzed to diacylglycerol and inositol trisphosphate (1,4,5-trisphosphateinositol), both are products that function as "second messengers" in signal transduction systems (p. 560).

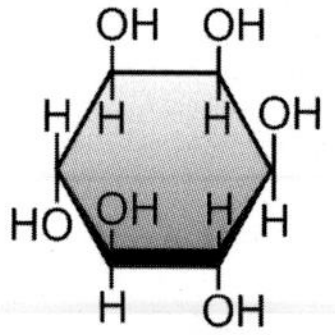

$$CH_2-O-CO-R_1$$
$$CH-O-CO-R_2$$
$$O-CH_2$$
$$CH_2-O-P=O$$
$$CH-OH \quad O^-$$
$$O^-$$
$$CH_2-O-P=O$$
$$O-CH_2$$
$$CH-O-CO-R_3$$
$$CH_2-O-CO-R_4$$

Cardiolipin

Meso-inositol
Most abundant isomer in nature

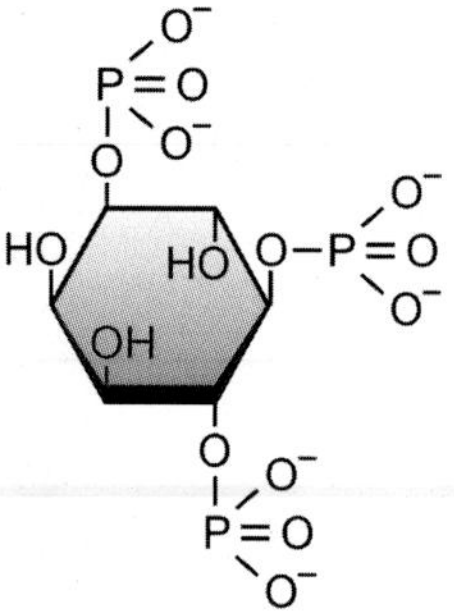

Inositol-1,4,5-trisphosphate

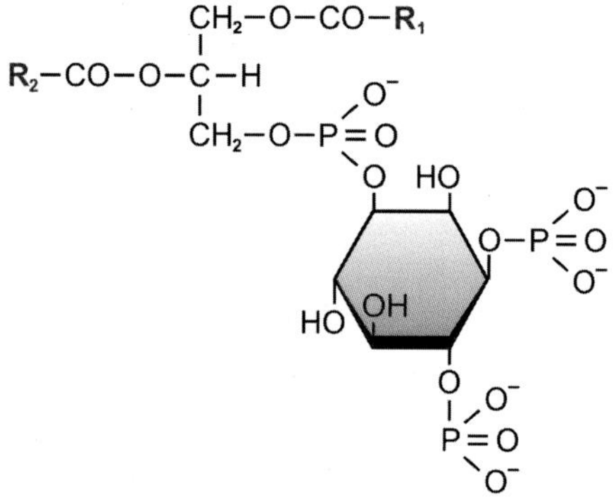

$$CH_2-O-CO-R_1$$
$$R_2-CO-O-C-H$$
$$CH_2-O-P=O$$

Phosphatidylinositol-4,5-bisphosphate

Plasmalogens. These compounds are glycerophospholipids that have glycerol, phosphoric acid, a nitrogenous base (choline or ethanolamine), and a FA. The difference with other phospholipids that these compounds have is, a long-chain *fatty aldehyde* linked by an ether bond to the C1 of glycerol. The fatty aldehyde easily acquires the enol form by transposition of a hydrogen and double bond change.

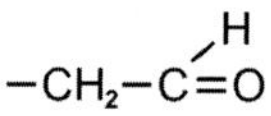

$$-CH_2-C=O \qquad -CH=C-OH$$

Aldehyde   Enol

Inositol derivatives with a high proportion of phosphate groups participate in regulation of various biological processes, such as gene expression, hormone signal transduction, and metabolism among others.

Phosphatidylglycerol, an acidic compound with a net negative electric charge, is usually found in the inner membrane of mitochondria and in the fluid covering the pulmonary alveolar epithelium. Another component of inner mitochondrial membranes and bacterial membranes is *cardiolipin,* formed by two phosphatidic acid molecules joined to one molecule of glycerol by phosphodiester bonds. Cardiolipin is acidic and has two negative electrical charges.

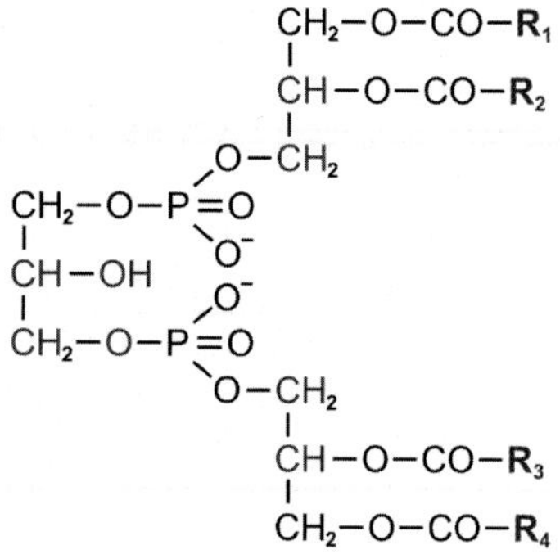

The enol form binds to the primary alcohol of glycerol with water loss. The fatty aldehyde can be palmital with 16 carbons, or any other aldehyde derived from FAs. Plasmalogens are found in cell membranes, especially in muscle and nerve cells.

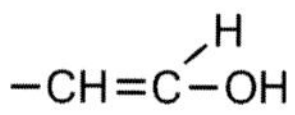

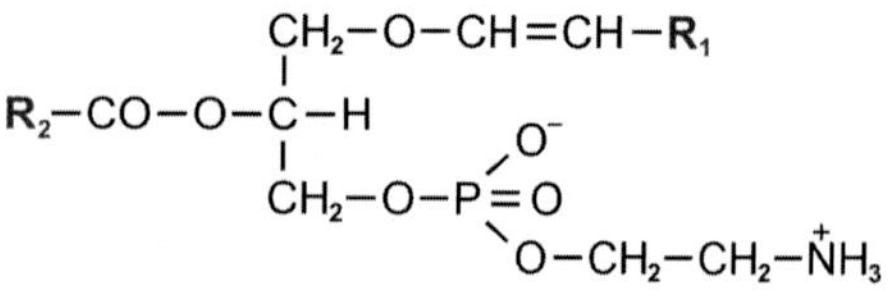

$$CH_2-O-CH=CH-R_1$$
$$R_2-CO-O-C-H$$
$$CH_2-O-P=O$$
$$O-CH_2-CH_2-\overset{+}{N}H_3$$

Plasmalogen

Properties of glycerophospholipids. Glycerophospholipids show great diversity in the composition of their FAs. Most glycerophospho-lipids have a saturated FA at *sn*–1 and another unsaturated FA at C2; however, others have two saturated or two unsaturated FAs. Phosphati-dylcholine often contains palmitoyl (16:0) or stearoyl (18:0) at *sn*–1 and unsaturated residues oleyl (18:1), linoleyl (18:2), or linolenoyl (18:3) at C2. Phosphatidylethanolamine has the same saturated FAs in *sn*–1, but frequently the un-saturated acid at *sn*–2 has a longer carbon chain (20–22C). Phosphatidylinositol presents almost exclusively stearic acid (18:0) at *sn*–1 and arachi-donic acid (20:4) at *sn*–2.

The glycerophospholipid molecule is am-phipathic or amphiphilic, with a polar head and a nonpolar tail. The polar head comprises the free —OH groups of the phosphoryl acid moi-ety and the basic nitrogen of the amino alcohol, while the nonpolar region comprises the tails or carbon chains of the FAs (Fig. 5.7).

The marked polarity of glycerophospholip-ids is an important property of these molecules and plays an essential role in the formation of cell membranes. These molecules arrange in a double layer, with their polar heads facing the aqueous medium (either the cytosol or the ex-ternal medium bathing the cell) and the nonpo-lar acyl chains oriented toward the membrane interior.

The venom of some snakes contains enzymes (phospholipases) that catalyze the hydrolysis of phosphoglycerides. One of these, phospholipase A_2, hydrolyzes the ester function of glycero-phospholipids located at position two, resulting in a structural change of the membrane and lysis

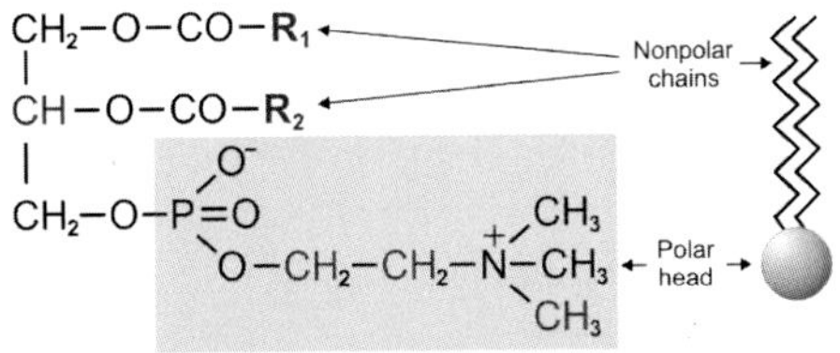

of the cell. The glycerophospholipids that result from the release of the FA at *sn*–2 by phospholi-pase A_2 are called lysoderivatives. The action of phospholipase A_2 on phosphatidylcholine pro-duces a free FA and *lysophosphatidylcholine*.

Glycerophospholipids are detergents and, as such, they reduce water surface tension and stabilize the dispersion of hydrophobic com-pounds (cholesterol and neutral fats) in aque-ous solutions. The capacity of phospholipids to function as detergents is important in bile, where they favor the solubility of cholesterol. In the lung, phospholipids prevent the collapse of the alveoli by decreasing surface tension. Type II pneumocytes of the lung secrete a substance called *surfactant* into the alveolar lumen. This is a lipoprotein complex, 80%–90% of which is composed of lipids and 10% is made by up of an 18–26 kDa proteins. Half of the surfactant lipids correspond to dipalmitoylphosphatidyl-choline, an unusual compound that contains the same saturated acid, palmitic acid (16:0), in the *sn*–1 and *sn*–2 positions. Surfactant also con-tains phosphatidylglycerol. The ability to syn-thesize and excrete surfactant develops during fetal life and reaches its normal level around the 8th month of gestation. Children that are born prematurely and are deficient in surfactant suf-fer severe respiratory disorders (respiratory dis-tress syndrome) due to reduced gas exchange in the alveoli. A method previously used to es-timate the degree of fetal lung maturity in po-tential premature infants is the determination of the amount of dipalmitoylphosphatidylcholine in amniotic fluid. The lethicin to sphingomyelin ratio is now more often used for this purpose. In the newborn, the determination of these lipids can be made in the gastric content because the fetus swallows amniotic fluid during its devel-opment in uterus.

Phosphatidylinositol and other phosphoglyc-erides, in addition to their role as structural com-ponents of cell membranes, function as a reservoir of arachidonic acid, whcih will be used for the future synthesis of prostaglandins, leukotrienes,

FIGURE 5.7 Schematic representation of a glycero-phospholipid.

and thromboxanes. Phosphatidylinositol also serves as an anchor for proteins, fixing them to the outer surface of the plasma membrane.

A special glycerophospholipid is the *platelet activating factor* (PAF), or 1-O-alkyl-2-acetyl-glyceryl-phosphoryl-choline. This lipid differs from phosphatidylcholine in a 16 carbon alkyl radical at *sn*–1 (linked via an ether instead of an ester bond) and an acetate moiety at *sn*–2. PAF exhibits several important actions, among which is its capacity to increase platelet aggregation, reduce blood pressure, and activate inflammatory processes.

Sphingophospholipids

The most abundant sphingolipid is *sphingomyelin*, which is composed of: (1) an alcohol called sphingosine, (2) a FA, (3) phosphoric acid, and (4) choline. Sphingosine has 18 carbon atoms, an alcohol functional group at C1, an amine at C2, a secondary alcohol at C3, and a double bond between C4 and C5. The rest of the molecule is a saturated hydrocarbon chain.

$$\overset{1}{CH_2OH}-\overset{2}{CHNH_2}-\overset{3}{CHOH}-\overset{4}{CH}=\overset{5}{CH}-(CH_2)_{12}-\overset{18}{CH_3}$$

Sphingosine

(Carbon numbering indicated in *red*)

FAs bind to sphingosine C2 amine via an amide function. This basic structure is called *ceramide* (Fig. 5.8).

$$
\begin{array}{c}
CH_2OH \\
| \\
R-CO-HN-CH \\
| \\
CH-OH \\
| \\
CH \\
\| \\
CH \\
| \\
(CH_2)_{12} \\
| \\
CH_3
\end{array}
$$

Ceramide

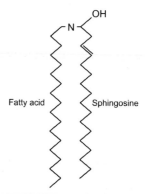

FIGURE 5.8   **Schematic representation of ceramide.**

The phosphoric acid esterifies the —OH at C1 of sphingosine and choline binds to the phosphate, just as it does in phosphatidylcholine.

$$
\begin{array}{c}
CH_2-O-P=O \\
| \qquad\quad O^- \\
R-CO-HN-CH \qquad O-CH_2-CH_2-\overset{+}{N}-CH_3 \\
| \qquad\qquad\qquad\qquad\qquad CH_3 \\
CH-OH \\
| \\
CH \\
\| \\
CH \\
| \\
(CH_2)_{12} \\
| \\
CH_3
\end{array}
$$

Sphingomyelin

Sphingomyelin is an important component of cell membranes; it forms the myelin sheaths of nervous tissue. Similar to other glycerophosphatides, sphingomyelin has a polar head (choline phosphate) and two nonpolar tails, represented by the hydrocarbon chains of sphingosine and the FA (Figs. 5.9 and 5.10).

Glycolipids

These lipids have carbohydrates instead of phosphates in their molecule. The most abundant in higher animals are glycosphingolipids, mainly *cerebrosides* and *gangliosides*. They are amphipathic compounds and are found in cell membranes.

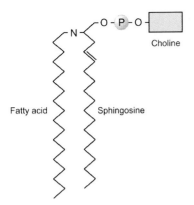

FIGURE 5.9 **Schematic representation of sphingomyelin.**

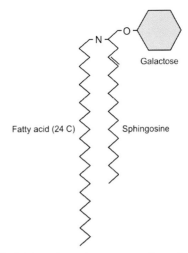

FIGURE 5.11 **Schematic representation of a cerebroside.**

FIGURE 5.10 **Space filling molecular model of sphingomyelin.** C, *Black*; H, *white*; O, *red*; P, *gray*; N, *pink*.

Cerebrosides

Cerebrosides are neutral compounds that consist of ceramide (sphingosine and FA) and a monosaccharide bound by a β-glycosidic bond to the C1 of esfingol. Often the carbohydrate is galactose (galactocerebroside) (Fig. 5.11). The most common FAs are lignoceric and hydroxylignoceric or cerebronic acid, both of which have 24 carbons. The cerebrosides containing lignoceric acid are called *kerasin*, while those having cerebronic acid are known as *phrenosin*.

Glucocerebrosides (glucose bound to ceramide) are found in very small proportions in the body, along with galactocerebrosides. Cerebrosides are abundant in brain white matter and nerve myelin sheaths and they are present in small quantity within the cell membranes of other tissues.

Brain white matter and, to a lesser extent, other tissues, also have lipids that contain sulfur. These compounds, formerly called sulfatides, are galactocerebrosides in which the monosaccharide is esterified with sulfate.

Glycosphingolipids with a more complex carbohydrate portion (di, tri, and tetrasaccharides instead of a monosaccharide) have been identified. Compounds of this type containing N-acetyl-galactosamine are called *globosides*.

Gangliosides

This is another important group of glycosphingolipids, whose basic structure is similar to that of cerebrosides, but the carbohydrate portion is of greater complexity. Linked to the ceramide, they contain an oligosaccharide composed of several hexoses and one to three acetylneuraminic acid (sialic acid) residues.

Many types of gangliosides have been recognized that differ in the number of hexoses and sialic acid residues and in the relative position of these residues. In virtually all gangliosides, the first hexose residue of the oligosaccharide attached to the ceramide is glucose, then galactose, N-acetyl-galactosamine, and another glucose or galactose, are subsequently attached by β-glycosidic bonds. Sialic acid is bound to one of the monosaccharides in the chain. According to the most commonly used notation, gangliosides are designated with the letter G followed by a subscript indicating the number of sialic acid residues existing in the molecule (G_M: mono-, G_D: di-, and G_T: trisialoganglioside). Another subscript indicates the order of migration of the compound in chromatography. The compound shown schematically in Fig. 5.12 corresponds to monosialoganglioside G_{M2}.

Gangliosides are not only a structural component of cell membranes. They also play a role as cell markers. For example, bacterial toxins, such as those of cholera, tetanus, botulism, and diphtheria selectively bind to specific cell surface gangliosides. If the toxin is first incubated with the specific ganglioside and then placed in contact with the cell, the binding site of the toxin is blocked and it cannot bind to the cell, becoming harmless. Surface gangliosides also serve as specific binding sites for other molecules, including interferons, which are potent antiviral agents.

Lipoproteins

Lipids are carried through the blood circulation in association with proteins, which allows them to be dispersed in the aqueous plasma medium. There are different types of plasma lipoproteins, which vary depending on the amount and composition of lipids that they have (pp. 62). In the lipoprotein complex, the hydrophobic lipids (triacylglycerol and cholesterol esters) are located in the interior of the molecule and the polar components (proteins, complex lipids, and free cholesterol) are arranged on the surface. Lipoproteins are also among the compounds that make up mitochondria, microsomes, and myelin membranes.

SUBSTANCES ASSOCIATED WITH LIPIDS

Terpenes

Polyisoprene or *terpenes* are hydrocarbon compounds derived from isoprene or 2-methyl-1,3-butadiene.

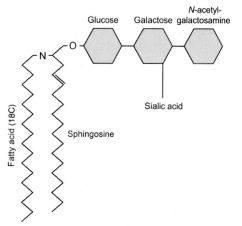

FIGURE 5.12 **Schematic representation of a ganglioside.**

$$\overset{1}{C}H_2=\overset{2}{C}-\overset{3}{C}H=\overset{4}{C}H_2$$
$$\underset{CH_3}{|}$$

Isoprene

Terpenes are formed by two or more isoprene units. The binding generally takes place between C4 of one isoprene and C1 of another.

Polyisoprenes can exhibit linear structure, such as *geraniol* (composed of two isoprene units), *farnesol* (containing three isoprene units), or *squalene* (with six isoprenes). They may also form cyclic structures in vitamin A, carotenes, lanosterol, and ubiquinone.

The polyprenols belong to the terpene group of compounds; among them are *dolichol*, which has a chain of 17–21 isoprene units (85–105 carbons). Some of the double bonds of dolichol are in the *trans* configuration; the initial isoprenyl moiety, with an alcohol functional group, is saturated. When esterified with phosphate (dolichol phosphate) dolichol participates in the biosynthesis of glycoproteins (p. 317).

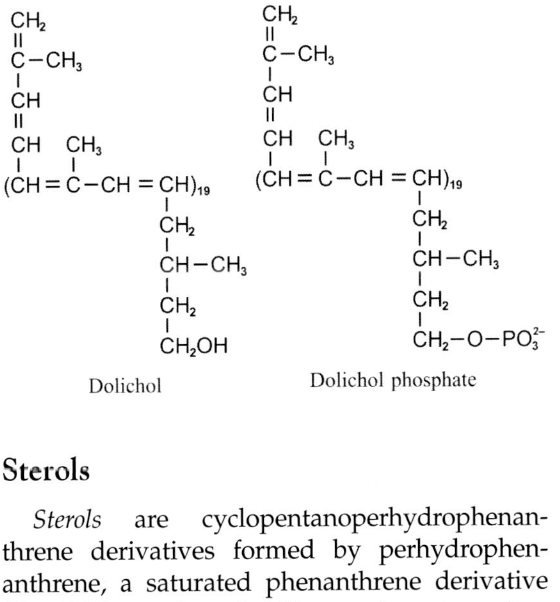

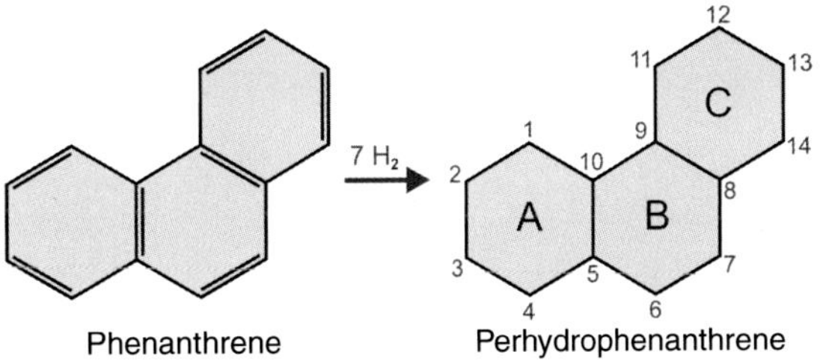

FIGURE 5.13　Carbon numbering in cyclopentanoperhydrophenanthrene.

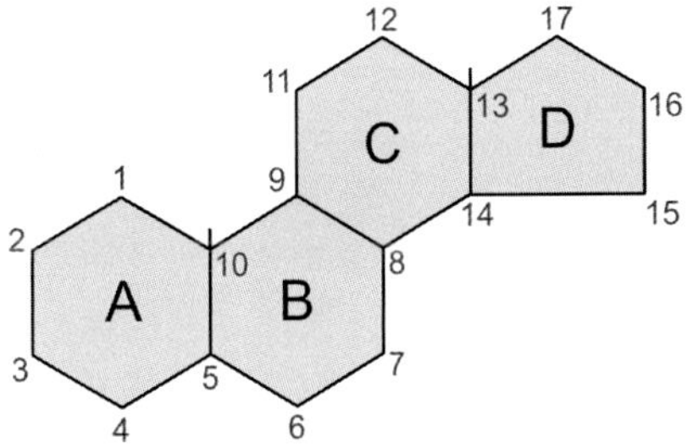

FIGURE 5.14　Cyclopentanoperhydrophenanthrene.

Dolichol

Dolichol phosphate

Sterols

Sterols are cyclopentanoperhydrophenanthrene derivatives formed by perhydrophenanthrene, a saturated phenanthrene derivative (Fig. 5.13), condensed with a cyclopentane ring. Cyclopentanoperhydrophenanthrene's rings are designated by letters and the carbons are numbered as shown in Fig. 5.13. All substances that have this chemical ring structure are called *steroids*.

Many biologically important compounds are derived from cyclopentanoperhydrophenanthrene, including sexual and adrenocortical hormones, bile acids, vitamin D, and sterols (Fig. 5.14).

All cyclopentanoperhydrophenanthrene carbons are located in a plane and the substituents attached to the carbons may be located on either side of the plane creating geometric (*cis–trans*) isomerism. There are six asymmetric centers in the cyclopentanoperhydrophenanthrene ring (carbons 5, 8, 9, 10, 13, and 14), which opens the possibility for the existence of many isomers of this compound. However, in nature only isomers in C5 occur, while substituents at the remaining asymmetric carbons have the same relative position in all compounds of biological interest.

In a flat cyclopentanoperhydrophenanthrene molecule, the hydrogen or side groups can be placed above or below the plane where the carbon atoms are located. Those placed above are called β and the bond to the carbon is represented by a solid line; those located below are designated α and a dashed line is drawn to represent the bond (Fig. 5.15).

In the majority of naturally occurring steroids, methyl groups are linked to C10 and C13 (the carbons of those methyl groups are numbered

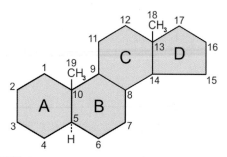

FIGURE 5.15 **Cyclopentanoperhydrophenanthrene with methyl molecules at C10 and C13.** Additional carbons numbered 18 and 19 are located above the molecular plane (indicated by a *solid line*). C5 hydrogen is below the molecular plane (shown by a *dashed line*).

19 and 18, respectively). The steroids of interest in human biochemistry have both C18 and C19 above the plane (β). Methyl C10 (C19) serves as a reference; the hydrogen of C5 or any other substituent is considered β (*cis*) when it is on the same side of the plane as C10, or α (*trans*) if it is on the opposite side.

In fact, the A, B, and C hexagonal rings of cyclopentanoperhydrophenanthrene do not lie in a plane; they adopt a more stable chair configuration. Fig. 5.16 shows the steroid conformation of the AB ring in the *trans* position.

If a branched hydrocarbon chain of eight carbons is inserted on C17 and a (—OH) hydroxyl group on C3 of cyclopentanoperhydrophenanthrene, the basic structure of a *sterol* is obtained. In these compounds, the addition of the hydroxyl group to C3 creates new *cis–trans*

isomers according to its position with respect to the methyl on C10. Sterols exist as free sterols or as esters of the C3 —OH with long-chain FAs. The most abundant sterol in animal tissues is cholesterol; it exists both free and esterified. The —OH on C3 of cholesterol is in the *cis* position (β) and there is a double bond between carbons five and six (Fig. 5.17). Cholesterol is insoluble in water, but easily soluble in organic solvents, such as chloroform and benzene. Increased cholesterol in blood plasma and deposition of this substance in vascular walls is commonly observed in atherosclerosis.

The spatial conformation adopted by cholesterol is shown in Fig. 5.18. The presence of the double bond between carbons five and six modifies the B ring arrangement. As there is no hydrogen at C5, there are no A/B (*cis–trans*) isomers.

Cholesterol is the raw material from which some tissues synthesize a series of compounds of high biological activity, including adrenocortical hormones, sex hormones, and bile acids. Cholesterol is present in animal fat. In plasma it is found free and esterified. It is particularly abundant in bile, where it may precipitate and form stones in the gallbladder or bile ducts (gallstones).

Another sterol present in animal organisms is 7-dehydrocholesterol (Fig. 5.19). The chemical formula of this compound is similar to that of cholesterol, except that it has an additional double bond between C7 and C8. The 7-dehydrocholesterol is a provitamin, which is converted into vitamin D_3 by ultraviolet light.

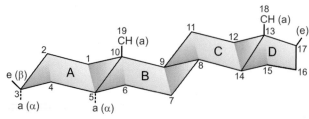

FIGURE 5.16 **Spatial conformation of cyclopentanoperhydrophenanthrene.** Hexagonal cycles adopt the chair configuration. This corresponds to the A/B *trans* isomer. The direction of the links is indicated by *a*, axial; *e*, equatorial.

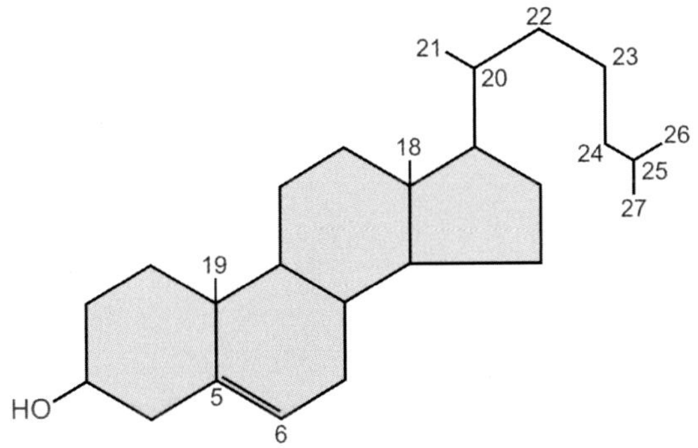

FIGURE 5.17   **Cholesterol.**

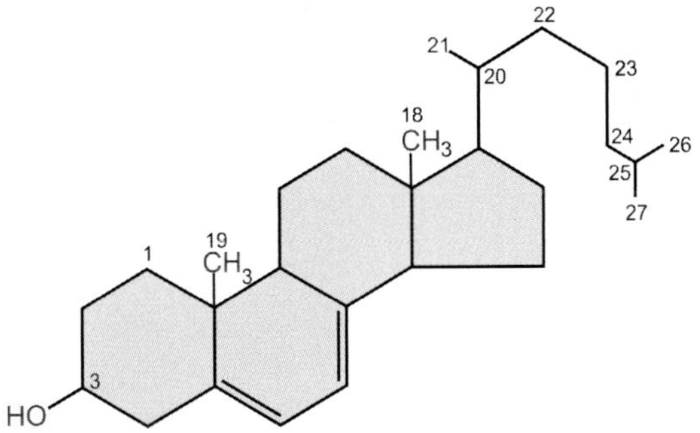

FIGURE 5.19   **7-Dehydrocholesterol.**

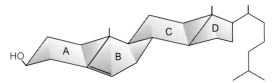

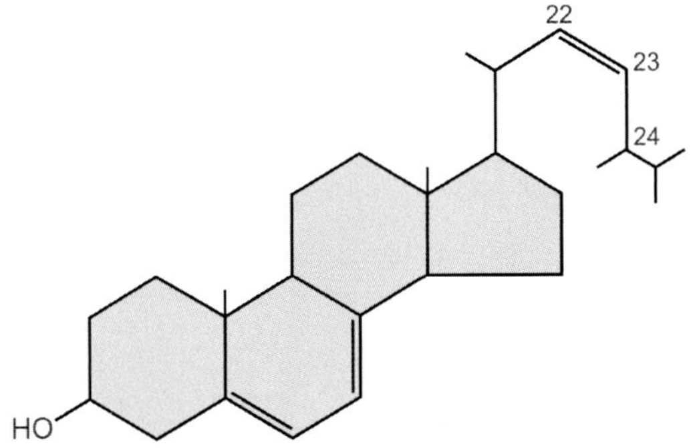

FIGURE 5.18   **Spatial conformation of cholesterol.** The double bond between C5 and C6 deforms the B ring.

Ergosterol is the most important plant sterol (Fig. 5.20). Its chemical structure is similar to that of 7-dehydrocholesterol, with an additional double bond between C22 and C23 and a methyl group at C24. This compound is also converted into vitamin D with exposure to sunlight.

FIGURE 5.20   **Ergosterol.**

SUMMARY

Lipids are a heterogeneous group of substances that share the property of being poorly soluble in water but soluble in organic solvents.

Fatty acids (FA) are monocarboxylic organic acids and essential components of lipids. Animal lipids contain FA composed of a linear carbon chain that has an even number of carbon atoms (4–26); the most common ones have 16–18C. FA can be *saturated* or *unsaturated* and mono- and polyethylenic.

The most common saturated FAs include: butyric (4C); caproic (6C); caprylic (8C); capric (10C); lauric (12C); myristic (14C); palmitic (16C); stearic (18C); arachidic (20C); and lignoceric (24C) acids. Unsaturated FA comprise: palmitoleic (16C) (double bond between C9 and C10, indicated by the notation 16:1Δ9); oleic (18C) (18:1Δ9),

linoleic (18C) (18:2Δ9,12),
linolenic (18:3Δ9,12,15), and
arachidonic (20:4Δ5,8,11,14) acids.

Another notation used for unsaturated FAs is based on the position of the double bond in relation to the carbon distal to the —COOH function (ωC). For example, oleic acid is 18:1ω9; linoleic, 18:2ω6; linolenic, 18:3ω3; arachidonic, 20:4ω6.

FA properties include the following:

1. Poor water solubility, which decreases as the carbon chain increases in length, due to the greater influence of the hydrophobic portion of the molecule over the polar —COOH group. FAs with more than six C are practically insoluble in water.

2. Melting and boiling temperatures are variable and increase with the FA carbon chain length. FAs with 1–8C are liquid at 20°C. Those with a higher number of C are solid. The presence of ethylenic bonds in FA decreases their melting temperature.

3. Geometrical isomerism, which is given by the rigidity of the ethylenic double bond of FAs. Almost all natural unsaturated FAs have a *cis* configuration. Double bonds produce angulation of the carbon chain.

4. Acidic character, which decreases as the number of C in the chain increases.
5. Capacity to form soaps when the —COOH group reacts with a base. Soaps derived from FA and alkaline metals (Na and K) are very soluble and act as emulsifiers or detergents.
6. FAs reactivity with alcohols allows them to form esters.
7. FAs can become oxidized and unsaturated ones are more readily oxidized, forming peroxides.
8. FA can be hydrogenated in the presence of a catalyst, with the ethylene FA becoming saturated.
9. Unsaturated FA easily incorporates halogens; the amount of halogen consumed by a certain amount of acylglycerol reflects the number of double bonds in their molecule (*iodine number*).

Lipids are classified as simple lipids, which include acylglycerols and waxes; and *complex lipids*, which comprise phospholipids, glycolipids, and lipoproteins.

Acylglycerols are esters of glycerol and FA. Depending on the number of esterified alcoholic functions, acylglycerols can be monoacylglycerols, diacylglycerols, or triacylglycerols. Triacylglycerols are also called *triacylglycerides* or neutral fats. If FAs in the triacylglycerides are all the same, they are called *homoacylglycerols*; if they are different, *heteroacylglycerols*. Many acylglycerols exhibit optical isomerism; the common ones found in nature belong to the L series. Neutral fats are the most abundant lipids in living beings; they represent energy reserve material. The animal adipose tissue also functions as mechanical protection and thermal insulation.

Properties of acylglycerols include a lack of solubility in water and a melting temperature that depends on the type of FA they contain. Those having long-chain saturated FAs show higher melting temperature. The presence of unsaturated or saturated short-chain FA allows natural fats to stay in liquid state at room temperature (oils).

Saponification is a process that results from heating a neutral fat in the presence of a strong base (KOH and NaOH). This generates free glycerol and soaps.

Hydrogenation of oils leads to the formation of a solid fat (margarine). Oxidation of unsaturated FAs results first in the formation of peroxides and second it causes a break in the FA chain, which generates compounds with the taste and odor of rancid fat.

Fats are important in nutrition. Their caloric value (38.9 kJ/g or 9.3 kcal/g) is much higher than that of other components of the diet. *Essential FAs* are linoleic, linolenic, and arachidonic acids (polyunsaturated); humans do not synthesize them, they must be provided with food.

Waxes are esters of long-chain monohydric alcohols and a FA.

Complex lipids include:

Phospholipids, which are composed by an alcohol, a FA, and phosphoric acid. Depending on the type of alcohol, they are divided into *glycerophospholipids* and *sphingophospholipids*.

Glycerophospholipids are main components of cell membranes; they are derivatives of *phosphatidic* acid, which is composed of glycerol esterified by FAs in C1 and C2 and phosphoric acid in C3. Generally, one of the —OH groups in the phosphate of phosphatidic acid is esterified by an amino alcohol. Depending on the amino alcohol, the glycerophospholipid could be *phosphatidylcholine*, *phosphatidylethanolamine*, *phosphatidylserine*, or *phosphatidylinositol*. Phosphatidylinositol-4,5-bisphosphate participates on cell signal transduction systems. Hydrolysis of the ester linking the FA to C2 produces a *lysoderivative*.

Plasmalogens are similar to glycerophospholipids, but they contain a fatty aldehyde instead of a FA.

Glycerophospholipids are amphipathic, with a polar head (which includes a —OH group of phosphate, the basic N of amino alcohols, and the —OH groups of serine and inositol) and a nonpolar chain (constituted by the FA carbon chains).

Sphingophospholipids. The most abundant is *sphingomyelin*, which is formed by:

1. A 18C alcohol (sphingosine).
2. FA (The FA does not form an ester, but an amide bond with the sphingosine —NH2 group of C2).
3. Phosphoric acid.
4. Choline.
5. FA and sphingosine form *ceramide*, which is the basic structure of various sphingolipids.

Glycolipids have no phosphate. They include cerebrosides, gangliosides, and sulfolipids.

Cerebrosides are formed by ceramide and a monosaccharide joined by β-glycosidic bond to the C1 of esfingosine. The monosaccharide of cerebrosides is commonly galactose and the FA usually is 24C long (lignoceric in *kerasin* and cerebronic in *phrenosin*). Cerebrosides are abundant in brain white matter and myelin sheaths.

Gangliosides are formed by ceramide, an oligosaccharide composed of several hexoses and 1–3 N-acetylneuraminic acid residues. Gangliosides function as marker molecules on the surface of membranes, which can be readily recognized by other molecules (toxins and interferon).

Sulfolipids or *sulfatides* are galactocerebrosides that contain sulfur.

Lipoproteins are complexes containing a hydrophobic lipid core (triacylglycerol and esterified cholesterol) and a polar surface, formed by proteins, complex lipids, and free cholesterol.

Substances associated to lipids refer to the following:

Terpenes, which are isoprene derivatives. Some have a linear molecule (geraniol, farnesol, squalene, and polyprenols), others present cyclic structures (vitamin A, carotenes, lanosterol, ubiquinone). A polyprenol of interest is *dolichol*, which has 17–21 isoprene units.

Sterols are derivatives of *cyclopentanoperhydrophenanthrene*. All carbons in this structure are in one plane, which allows the presence of geometric isomerism. When a chemical group on one of the C is on the same side of the plane as the methyl on C10, it is considered *cis* or β-isomer. In contrast, if the group is placed on the other side, the isomer is *trans* or α.

All substances in this group (sterols, sexual and adrenocortical hormones, bile acids, and vitamin D) are generically called *steroids*.

Cholesterol is a sterol with a branched chain of eight carbons bound to C17, a hydroxyl group on C3, and a double bond between C5 and C6. It is only found in animal tissues. Steroidal hormones and bile acids are synthesized from cholesterol. *7-Dehydrocholesterol* is a provitamin that can be activated by UV into vitamin D. In plants, the major sterol is *ergosterol*.

Bibliography

Angerer, P., von Schacky, C., 2000. Omega-3 polyunsaturated fatty acids and the cardiovascular system. Curr. Opin. Lipidol. 11, 57–63.

Belury, M.A., 2002. Dietary conjugated linoleic acid in health: physiological effects and mechanisms of action. Annu. Rev. Nutr. 22, 505–531.

Gurr, M.I., Harwood, J.L., Frayn, K.N., 2002. Lipid Biochemistry: An Introduction, fifth ed. Blackwell Science Ltd., Oxford.

Mozaffarian, D., Katan, M.B., Ascherio, P.H., Stampfer, M.J., Willet, W.C., 2006. *Trans* fatty acids and cardiovascular disease. N. Engl. J. Med. 354, 1601–1613.

Vance, D.E., Vance, J.E., 2002. Biochemistry of lipids, lipoproteins and membranes. New Comprehensive Biochemistry, vol. 36. Elsevier Science Publishing Co. Inc., New York, NY.

Wijendren, V., Hayes, K.C., 2004. Dietary *n*-6 and *n*-3 fatty acid balance in cardiovascular health. Annu. Rev. Nutr. 24, 597–615.

6

Nucleic Acids

Nucleic acids received their name because they were originally isolated from cell nuclei. They contain carbon, hydrogen, oxygen, nitrogen, and phosphorus; have acidic character; and are found in all living beings. They are linear macromolecules formed by the polymerization of units called *nucleotides*. Nucleic acids play important functions in the cell: (1) they are the repository of the genetic information responsible for the transmission of inherited characteristics from parents to children and from one cell to another; (2) they guide cell protein synthesis and are responsible for the correct assembly of amino acids in defined sequences. Ultimately, the morphological and functional uniqueness of each living being are determined by the information contained in its nucleic acids.

The structure of the nucleotides that constitute nucleic acids will be considered first.

NUCLEOTIDES

Nucleotides include: (1) a nitrogenous base, (2) a five-carbon monosaccharide (aldopentose), and (3) phosphoric acid.

Nitrogenous bases. Nucleotide hydrolysis produces two types of substances derived from the heterocyclic rings *purine* and *pyrimidine* known as the purine and pyrimidine bases. Fig. 6.1 shows their chemical structure and the numbering of the elements in the molecule. Purines

are derived from pyrimidines by addition of an imidazole group. Both purines and pyrimidines have all their atoms on the same plane.

Nucleic acids contain five different nucleotide bases. Three are pyrimidines and two purines. The pyrimidine bases are *thymine* (5-methyl-2,4-dioxipyrimidine), *cytosine* (2-oxo-4-aminopyrimidine), and *uracil* (2,4-dioxoypyrimidine) (Fig. 6.2).

Purine bases include *adenine* (6-aminopurine) and *guanine* (2-amino-6-oxypurine) (Fig. 6.3).

Pyrimidine and guanine bases in Figs. 6.2 and 6.3 correspond to the ketone or lactam forms of these nucleotides, which predominate in natural products. There are isomers (tautomers) that produce the enol or lactim form of these nucleotides, which exist in much lower proportion. These isomers are produced by displacement of the hydrogen atom bound to the neighboring nitrogen toward the oxygen. Eventually, nucleic acids may contain a small amount of other bases that derive from the main ones, such as 5-methyl-cytosine.

Due to their aromatic nature, purine and pyrimidine bases absorb radiation in the ultraviolet (UV) region of the spectrum, with a maximum at a wavelength of 260 nm. This property allows to identify nucleic acids in a sample and to estimate their concentration by spectrophotometry.

Aldopentoses. The monosaccharide that forms nucleic acids can be *D-ribose* or *D-2-deoxyribose*. According to the pentose present, two kinds

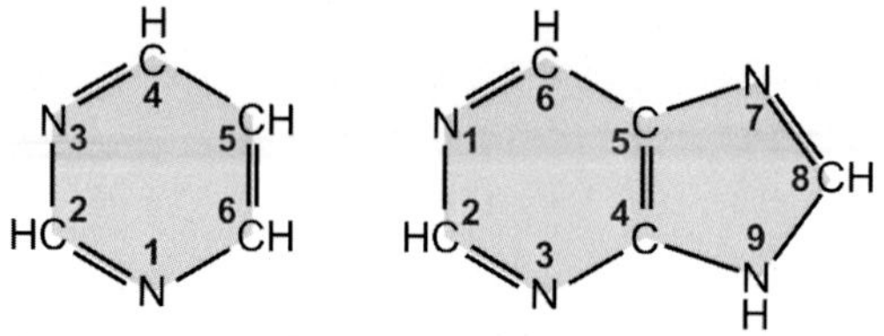

Pyrimidine

Purine

FIGURE 6.1 **Numbering of pyrimidine elements is different to that of purine.** Only carbons 2 and 5 have the same number in both cycles.

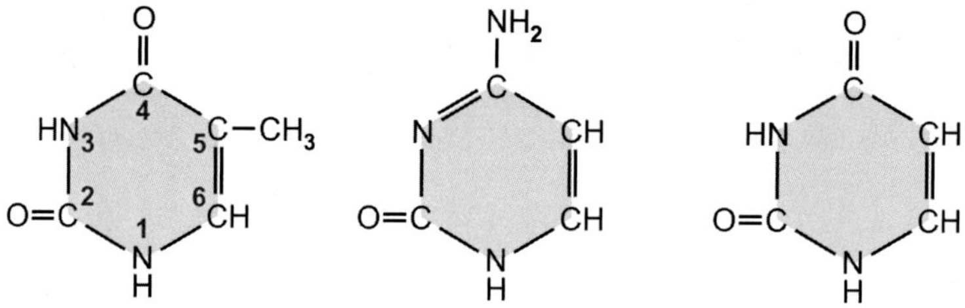

Thymine

Cytosine

Uracil

FIGURE 6.2 **Pyrimidine bases.**

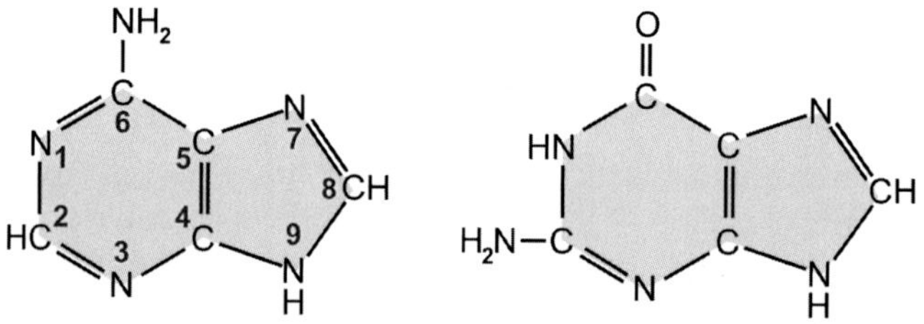

Adenine

Guanine

FIGURE 6.3 **Purine bases.**

of nucleic acids can be distinguished: ribonucleic acids (RNAs) and deoxyribonucleic acids (DNAs). The aldopentoses in nucleic acids adopt the furanose form (Fig. 6.4) (carbons of the pentose are distinguished from those of the base by adding a quotation mark, 1′, 2′, etc.).

Both ribose or deoxyribose, through their carbon 1′ are linked to nitrogen 9 of the

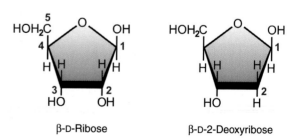

β-D-Ribose

β-D-2-Deoxyribose

FIGURE 6.4 **Aldoses present in nucleic acids.**

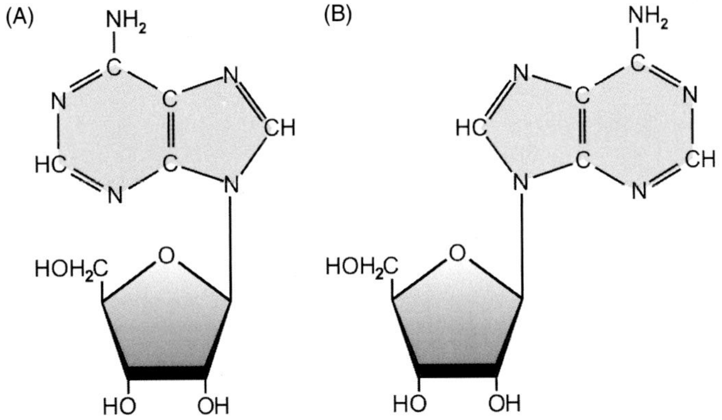

(A)

(B)

FIGURE 6.5 **Adenosine (nucleoside).** (A) *syn* form; (B) *anti* form.

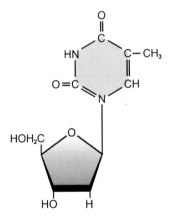

FIGURE 6.6 **Thymidine (nucleoside).**

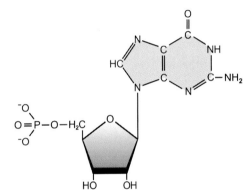

FIGURE 6.7 Guanylic acid or guanosine monophosphate (nucleotide, *anti* form).

purine or nitrogen 1 of the pyrimidine bases by a β-glycosidic bond, which allows their free rotation. The compound formed by a nitrogenous base, purine or pyrimidine and aldopentose is called *nucleoside*. The relative spatial arrangement of the nitrogenous base and the monosaccharide varies between the two main configurations shown in Fig. 6.5. These correspond to the *syn* and *anti* forms. The latter is thermodynamically more favorable (Fig. 6.6).

A nucleotide is formed by esterification with phosphoric acid of the hydroxyl group in carbon 5′ of the ribose or deoxyribose that forms part of the nucleoside (Fig. 6.7).

NUCLEIC ACIDS

To form nucleic acids, nucleotides link to each other by ester bonds, with the phosphate establishing a bridge between carbon 5′ of the pentose of one nucleotide and carbon 3′ of the pentose of another nucleotide. The type of link between nucleotides has been confirmed by studies that used agents (alkali and enzymes), which specifically cleave a particular type of bond.

The first nucleotide in the chain has its phosphate free, while the pentose of the last nucleotide has the hydroxyl group of C3′ free. These are considered the 5′ and 3′ ends of the chain (Fig. 6.8).

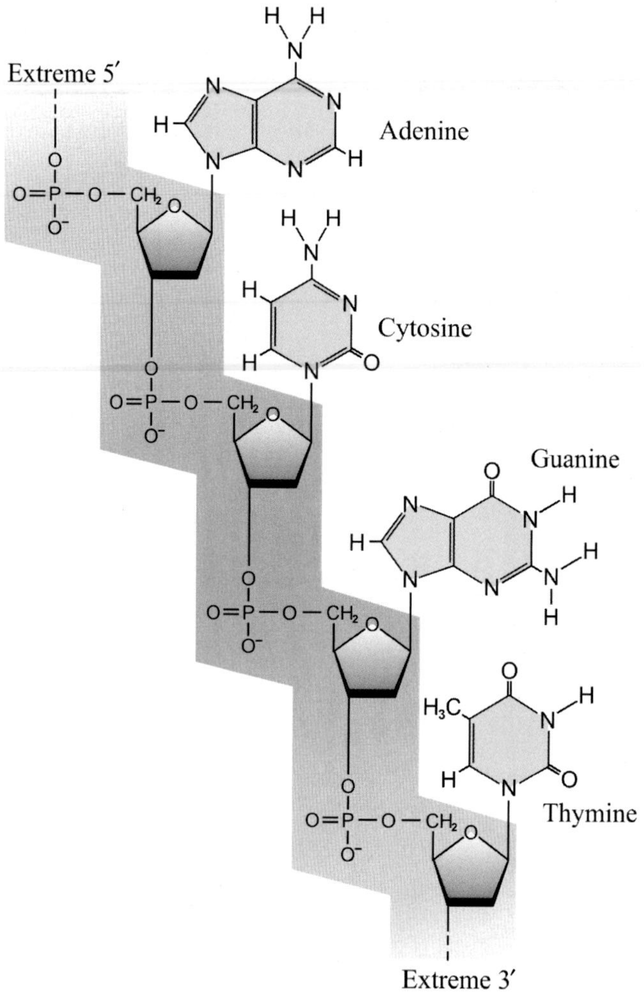

FIGURE 6.8 **Basic structure of a polynucleotide chain.** On *gray background*: pentoses and phosphates forming the backbone or continuous strand of the nucleic acid.

DNA and RNA are distinguished according to the pentose present. There are structural and functional differences between these two types of compounds and they will be described separately.

Deoxyribonucleic Acid

DNA is found almost entirely in the cell nucleus, more precisely, in the material forming the chromatin. In addition, a small amount of DNA is also found in mitochondria and chloroplasts.

All somatic cells of individuals of the same species have equal DNA content. In humans, the amount of DNA per cell is 6×10^{-6} µg (micrograms) or 6 pg (picograms). Male and female gametes have half of that amount. The DNA quantity remains constant with age and under different environmental or nutritional conditions.

DNA is a long linear molecule, which can sometimes reach several inches in length. In contrast, DNA cross section is only 2-nm thick, which makes the molecule susceptible to breaks during the procedures used for its separation and purification. DNA is densely packed in cells; the smallest human chromosome, which is 2 μm in length, contains a DNA molecule that is 30 million kDa and 1.4-cm long. If extended one after another, all DNA molecules of the 46 chromosomes of a human somatic cell reach a length of about 2 m.

Hydrolysis of DNA first releases the corresponding nucleotides, while further hydrolysis frees the purine and pyrimidine bases, deoxyribose, and phosphoric acid. DNA contains the purine bases *adenine* (A) and *guanine* (G) and the pyrimidines *thymine* (T) and *cytosine* (C). Commonly, the bases are named by the first letter of their name.

The proportion of bases in the DNA is characteristic for each species; however, the ratio between bases in all DNA samples is constant despite its origin. For example, the molar content of purine bases is always equal to that of pyrimidine bases or, in other words, the sum of adenine plus guanine molecules always equals that of cytosine plus thymine (A + G = T + C). In addition, the ratio of adenine/thymine and guanine/cytosine is always one, that is, the number of molecules is the same for adenine and thymine (A = T) and for guanine and cytosine (G = C).

These observations made by Erwin Chargaff, and X-ray diffraction analysis carried out by Maurice Wilkins and Rosalind Franklin, were the basis for James Watson and Francis Crick to propose, in 1953, the molecular model of DNA. The structure of DNA described by Watson and Crick was in perfect agreement with previous findings on the proportion of bases and X-ray images of DNA. Also, it provided a model that successfully explained the ability of DNA to duplicate during cell division and to serve as a repository of genetic information.

DNA Molecular Structure

The DNA molecule is a double helix consisting of two polynucleotide chains that coil around the same axis.

For each of the helices, a continuous strand is formed by a succession of deoxyriboses and phosphates that extend between C5′ of the pentose from one nucleotide to C3′ from another (the region of the molecule on gray in Fig. 6.8, which represents only one of the chains). In the double helix, phosphates and deoxypentoses, which are highly hydrophilic, are positioned on the outside of the molecule, in contact with the aqueous medium. Purine and pyrimidine bases, planar structures that are poorly polar, are oriented to the inside of the molecule and perpendicular to the central axis. Fig. 6.9 is a schematic representation of the double helix, based on the

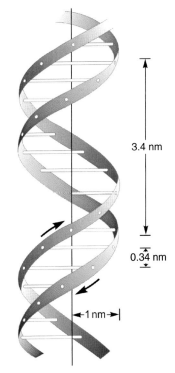

3.4 nm

0.34 nm

|←1 nm→|

FIGURE 6.9 Schematic representation of deoxyribonucleic acid (DNA) double helix (as viewed by Watson and Crick).

original representation from Watson and Crick. This shows two chains formed by the deoxyribose and phosphate ribbons, with the purine and pyrimidine bases perpendicular to the central axis and arranged like rungs of a spiral ladder.

The DNA helix has a clockwise rotation. Each turn of the helix is 3.4-nm long, and each base is at a distance of 0.34 nm from the following, which means that in a full turn, the helix contains 10 nitrogenous bases (Fig. 6.9). The cross section of the DNA molecule is 2-nm wide.

The DNA strands are *antiparallel*, that is, they run in opposite direction. While in one of the strands the phosphate bonds are established from carbon 5′ to C3′ of the pentoses involved, the other is oriented in the opposite direction (C3′ to C5′). At each end of the DNA molecule are the 5′ end of one chain and the 3′ terminal of the other (Fig. 6.10).

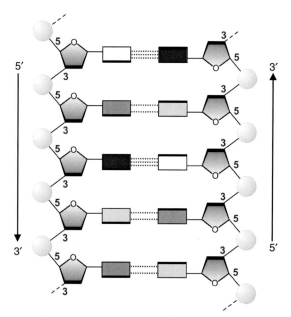

FIGURE 6.10 **Scheme of a DNA segment showing the chain pairing in the double helix.** Note how the orientation of the phosphates (represented by *pink circles*) and pentoses in one strand has an opposite direction than those of the other. *Red dotted lines* represent the hydrogen bonds holding paired bases (*rectangles* represent the bases: guanine, *pink*; cytosine, *red*; adenine, *white*; thymine, *gray*).

The primary structure of the DNA chains, corresponding to the order or sequence of nucleotides, is essential for the study of DNA. Given the enormous size of the DNA molecule, often comprising millions of units, the possibility of forming different arrangements with the four bases A, G, T, and C is virtually endless.

It is precisely the sequence of bases in DNA, which encodes the genetic information of each living being. As will be discussed later, the order of nucleotides in segments of DNA that correspond to some genes indicates the sequence in which the amino acids that constitute a given protein must be inserted.

The DNA sequence is usually indicated in an abbreviated form, with the initials of the corresponding nucleosides, for example, AGCTACT. Similar to proteins, in which the polypeptide chains are named from the N- to the C-terminus, the DNA sequences are named starting from the 5′ end.

Frederick Sanger, Allan Maxam, and Walter Gilbert developed efficient and practical methods for identifying the sequence of bases in nucleic acids. Later, other methods, faster and more economical, begun to be used.

The purine and pyrimidine bases in each DNA chain are arranged perpendicularly to the central axis of the molecule. The space between both DNA helices is not large enough to accommodate a pair of purine bases, and it is too wide to house two pyrimidine bases. The space between both DNA strands is adequate to exactly accommodate a purine–pyrimidine pair. The bases are maintained and held together by hydrogen bonds. Adenine and thymine establish two hydrogen bonds, while guanine and cytosine establish three hydrogen bonds between each other (Fig. 6.11). These are the only stable pairings possible between bases, with adenine and cytosine from one DNA strand always facing a thymine and a guanine from the other.

The double helix is very stable; thanks to the presence of the hydrogen bonds between bases. These forces, although individually weak, provide

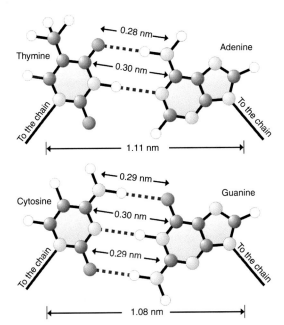

FIGURE 6.11 **Hydrogen bonds** *(dotted red lines)* **between DNA bases.** Adenine–thymine are always linked by two H bonds; guanine–cytosine, by three. C, *Black*; O, *red*; N, *pink*; H, *white* (Watson).

great strength to the molecule, when the large number of base pairs (bp) in the molecule are considered. There are other forces that contribute to maintain the structure of the DNA double helix. These are the van der Waals forces established among base pairs stacked inside the molecule.

Although the double helix forms a compact structure, it is flexible enough to bend and coil on itself or on other molecules. This physical property allows DNA to pack in very small volumes and to interact with other molecules.

Two aspects of the structure and function of DNA should be emphasized: (1) both DNA chains of a DNA molecule are not identical, but *complementary*. Base pairing requires that adenine and guanine from one chain is located opposed to thymine and cytosine in the other; (2) chains are antiparallel, while one runs in the 5′→3′ direction, the other one extends in the 3′→5′ direction. These characteristics are important for all processes in which DNA participates, including DNA replication, transcription, and translation, which will be described in Chapters 21 and 22.

The coiling of both DNA chains forms, on the surface of the molecule, two types of grooves parallel to the turns of the double helix. One, named the major groove, is wider than the other (minor groove) (Fig. 6.12). At the bottom of these grooves, atoms of the purine and pyrimidine bases are exposed and can establish interactions with proteins or other substances.

DNA conformations. The structure of DNA described by Watson and Crick and later confirmed by other authors, corresponds to the so-called B conformation, which is most commonly found in the cell (Fig. 6.12). There is another form, designated A, wider and shorter than helix B, in which each turn of the helix has 11 bp instead of 10 and covers a length of 2.8 nm instead of 3.4 nm. In addition, in the A conformation, the difference between the grooves disappears, with both grooves having almost equal depth. The A form is not found under physiological conditions, it occurs when DNA is poorly hydrated and is characteristic of duplexes formed by RNA or by RNA and DNA.

The advent of methods for the synthesis of chains of DNA allowed the study of double helices with defined sequences. This showed that chains formed by alternating guanine and cytosine (GCGC...) sequences adopt a conformation different from the B form, which is called Z. The coiling of the double helix changes its direction; it is counter clockwise instead of clockwise, the minor groove is deeper and the major groove practically disappears (Fig. 6.13). The molecule is elongated and thinner and has 12 bp/turn. The deoxyriboses and phosphates in the chain follow a zigzag line (hence the name Z). It is unknown whether the Z DNA has a functional role. However, these synthetic molecules show that certain nucleotide sequences change the conformation of the double helix and alter the access of the bases to the grooves, modifying their breadth and depth. This phenomenon may

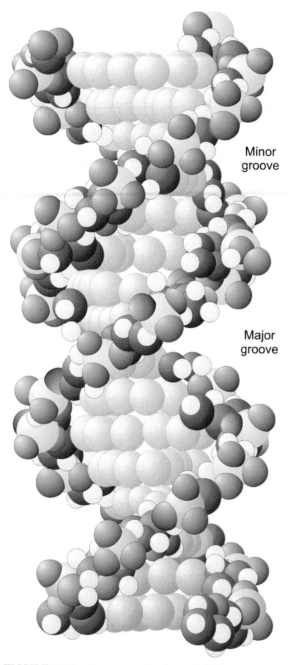

Minor
groove

Major
groove

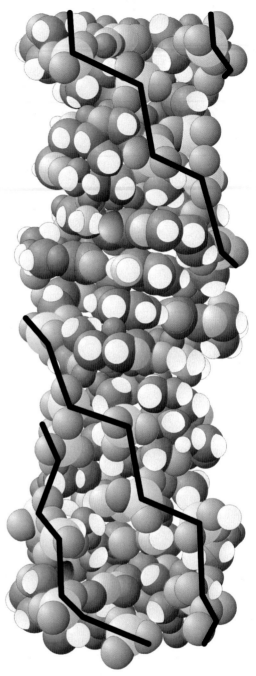

FIGURE 6.12 **Compact molecular model of DNA in its B configuration.** Note the major and minor grooves parallel to the turns of the double helix. Elements that form the continuous strand: C, *Black*; O, *red*; P, *pink*; H, *white*. Base pairs are shown in *gray*.

FIGURE 6.13 **Space filling model of DNA Z conformation.** The molecule is elongated and thinner than form B; the direction of rotation is to the left (counter clockwise) and the chain draws a broken, zigzag, line. C, *Black*; O, *red*; N, *gray*; P, *pink*.

be of biological relevance, since conformational changes facilitate or hinder DNA interaction with regulatory proteins that modulate DNA function.

Also relevant to DNA regulation is the presence of methylated bases in certain regions of the double helix. In this manner, within the DNA structure resides, not only the capacity to store genetic information, but also the signals to control its own functions.

DNA Denaturation

The compact structure of the double helix is maintained by the hydrogen bonds between base pairs and the van der Waals interactions between the stacked bases. Various agents (heat, strong alkalis, urea, and formamide) weaken such forces and promote the separation of the strands, in a process called *denaturation*. The resulting unwound polynucleotide strands adopt a random arrangement.

DNA denaturation can be followed spectrophotometrically by measuring the absorption of UV light at 260 nm. In its native state, DNA absorbs less UV light than the separate polynucleotide chains, a phenomenon that is called *hypochromicity*. If a DNA dispersion is slowly heated and its UV light absorption is followed, the temperature/absorbency relationship is an indicator of DNA denaturation (Fig. 6.14). A sigmoid curve is obtained and as the temperature augments, an increase of absorption (hyperchromic effect) occurs. When a given temperature is reached, the optical density does not further increase, which shows that DNA chains are completely separated and the molecules are maximally denatured.

The temperature at which half of the DNA is denatured (corresponding to the midpoint or inflection of the curve) is known as the DNA melting temperature (Mt). The Mt is characteristic for each DNA under defined conditions of pH and salt concentration, it ranges from 80 to 100°C for DNA isolated from different organisms. Determining the value of Mt is useful to estimate

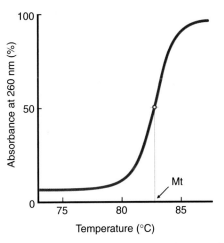

FIGURE 6.14 **DNA denaturation.** The change in absorbency at 260 nm with increasing temperature (hyperchromic effect) is shown. When both DNA helices are completely separated, absorbance does not further increase with temperature. *Mt*, Melting temperature corresponding to the temperature at which 50% of DNA is denatured.

the base composition of DNA. As the GC pair is maintained together by three hydrogen bonds and the AT pair by two, a higher GC content results in a higher resistance to denaturation and is reflected by a higher melting temperature.

Renaturation

The DNA denaturation process is reversible under controlled conditions of pH and ionic strength. If the temperature is slowly decreased in the solution where the DNA had been denatured, the DNA chains will spontaneously reanneal and the original double helix structure is restored. This process can be followed in a spectrophotometer at 260 nm and the temperature/absorbency relationship can be described by a curve that is the opposite of the denaturation curve shown in Fig. 6.14. The DNA renaturation resulting from slow cooling is called *reannealing*. When the complementary strands meet, they completely reconstitute the double helix.

The rate of renaturation depends on the structure of DNA. When a given DNA has segments with the same sequence (repetitive sequences), the annealing time is shorter because the chance that one chain meets a complementary one is greater. In contrast, DNA sections with unique sequences require a longer time to find its complementary strand to reform the double helix. In mammalian DNA preparations, fragments exist that are repeated many times (presenting hundreds of thousands to millions of copies), which are called *highly repetitive DNA* segments. Portions of highly repetitive DNA of more than 6 bp in length are also designated as *satellite DNAs*. These are located in the chromosomes at sites close to the centromere and in their ends or telomeres. Also, repeats of up to six nucleotides are frequently found and these are known as short tandem repeats or *microsatellites*. The repeated segments allow for a faster reassociation of DNA after denaturation. Another portion of DNA has *moderately repetitive* pieces, composed of hundreds to thousands of copies. These anneal at lower rate. Finally, a third fraction of the DNA that in mammals comprises about 60% of total DNA, corresponds to sequences only found in one to three copies (*single copy* DNA) that anneal very slowly. In bacteria, almost all the DNA exists as single copy DNA.

Hybridization. When mixed, in suitable conditions, denatured DNA from two different organisms, may have segments with complementary sequences that can pair forming hybrid double hybrid helices (with DNA chains from each organism). Such DNA hybridization is useful to estimate the evolutionary proximity between two different species. DNAs from phylogenetically close species are able to form hybrid molecules in higher frequency than that from species which are far apart. For example, human DNA can form hybrid duplexes in larger proportion with rhesus monkey DNA than with mouse DNA.

Chromatin

Nuclear DNA from eukaryotic cells, which have a nucleus surrounded by a membrane, is located in chromosomes, each containing a DNA molecule (two molecules immediately after replication, when sister chromatids are formed). These large molecules are densely packed in the nucleoprotein complexes that form chromatin. The organization of chromatin shows changes during the cell cycle. At the interphase, only an irregular reticle of extended chromatin is detected and no chromosomes are visible. At the beginning of mitosis, and more precisely at the end of prophase, the chromatin condenses into the chromosomes, which attain maximum density at metaphase.

Nucleoproteins in chromatin have DNA and a variety of nuclear proteins. Among the proteins, the most abundant ones are the basic histones. Over 20% of the amino acids that constitute histones are lysine and arginine. The free ionizable groups of these amino acids are positively charged, which provide electrostatic attractions with the negative phosphate groups of the DNA strands. These create interactions between polycationic and polyanionic molecules. Five different histones have been isolated in the cell nucleus: Hl, H2A, H2B, H3, and H4, with a molecular mass ranging between 11 and 21 kDa. They are only found in eukaryotic cells.

The other proteins associated to DNA are a heterogeneous group which includes polypeptides with structural functions, like the *high-mobility group* (HMG) that regulate gene activity (*Fos* and *Myc*) and enzymes necessary for the synthesis and processing of nucleic acids in the nucleus (DNA and RNA polymerases).

Studies using X-ray diffraction and electron microscopy have allowed to understand the arrangement of chromatin and to explain how the DNA molecule is packed into chromosomes. It is estimated that, on average, each human chromosome houses a DNA double helix of approximately 4-cm long).

At regular intervals, the DNA molecule makes two turns around a core composed by a histone octamer, containing two units of the histones H2A, H2B, H3, and H4. This type of

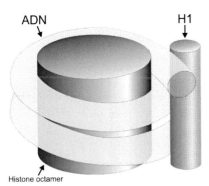

FIGURE 6.15 **Schematic representation of a nucleosome.** The DNA double helix coils two turns around a core consisting of a histone octamer *(central cylinder)*. The *thinner cylinder* at right, placed where the double helix enters and exits the nucleosome, is histone H1.

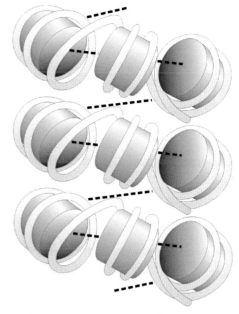

FIGURE 6.16 **DNA packing.** Nucleosomes are arranged in solenoid winding about a central axis; each turn comprises six nucleosomes. The *dashed line* indicates direction of winding. Only chromatosomes located forward are represented.

structure, in which a helix is wound on an axis, is called *superhelix*.

The DNA superhelix coiled around the spool formed by histones has a length of 146 bp. The core of histones and the superhelix form a particle called *nucleosome* (Fig. 6.15). A histone H1 unit is associated with the DNA double helix at the site of entry and exit of the nucleosome. The assembly of nucleosomes and H1 is called *chromatosome*. After making two turns around the histones, a piece of the DNA chain of approximately 50–60 bp (DNA spacer) extends free, until the DNA coils on to another histone octamer. The DNA segment between adjacent nucleosomes can be shorter, sometimes containing only 8 bp. The nucleosome assembly is visible by electron microscopy and it appears as beads of a rosary linked together by the DNA strand. This string of nucleosomes with a cross section of 10 nm appears extended during interphase. Upon initiation of mitosis, DNA condensation occurs. It has been proposed that the chromatosomes coil as in a solenoid, including 6 units/turn (Fig. 6.16); the fiber acquires now a cross section of 30 nm. Probably H1 histones play a role in this condensation; they are all located inside the solenoid. The long fiber with nucleosomes densely clustered in solenoids is folded into loops. For mitotic

chromosomes, this structure is maintained by cross-bindings in chromatin.

The common depiction of chromosomes is that of its most condensed form in the metaphase stage. Both of the sister chromatids resulting from cell duplication are joined at the centromere, on the place where the kinetochore is attached. The kinetochore is a protein complex to which the microtubules of the mitotic spindle assemble. The microtubules separate the chromatids to different sides of the cell during anaphase. The DNA of the centromere in yeast has an AT-rich sequence 88-bp long flanked by two short conserved regions. In mammalian cells, this sequence is longer and is flanked by a large segment containing DNA repeats, the *satellite* DNA. The telomeres, at the end of the chromosomes, contain the 5′ and 3′ terminals of the DNA molecule. These have hundreds of short repeated sequences (TTAGGG), which protect the ends of chromosomes from degradation.

During interphase, chromosomes adopt a fuzzy structure; however, part of the condensed chromatin remains visible under the microscope at the nucleus periphery. These portions of DNA are inactive and are known as *heterochromatin*. The sequence of heterochromatin usually contains satellite DNA repeated thousands of times, or DNA covalently modified by methylation of cytosine bases. Most heterochromatin is located near the centromeres and at the telomeres.

In mammalian female cells, one of the X chromosomes remains as heterochromatin (Barr corpuscle). The rest of the chromatin, not detectable as heterochromatin, is called *euchromatin*. It is generally accepted that the euchromatin comprises regions of active DNA, which is extended to allow transcription to take place (p. 477).

Circular DNA

Bacteria. The genetic information of prokaryotic organisms (without a nucleus) as bacteria is contained within a single chromosome comprising a DNA duplex that has no free ends, the molecule closes on itself to form a circular DNA. Apparently, bacterial DNA is not organized into nucleosomes, although it is common to observe compact clusters of genetic material. Initially it was thought that this DNA was not associated to proteins and was considered to be "naked" DNA. However, several small, histone-like proteins with the ability to form complexes with bacterial DNA have been isolated.

To give an idea of the size of bacterial DNA, we will describe the DNA of *Escherichia coli*, bacteria of the intestinal flora. The chromosome in *E. coli* has approximately 4 million bp or 4000 kilobases (the unit kilobase is commonly used for simplicity, which corresponds to 1000 bp). As the average mass of a base pair is 660 Da, the whole DNA molecule has a mass of $2.6 \cdot 10^9$ Da, its length is 1.36 mm, significantly smaller than that of higher animals.

If the circular DNA is extended on a plane (Fig. 6.17A), the DNA is considered to be in a

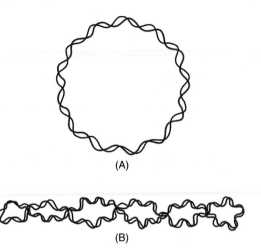

FIGURE 6.17 **Bacterial circular DNA.** (A) Relaxed; (B) coiled DNA.

"relaxed" state. Most frequently DNA is found forming supercoils (Fig. 6.17B). The existence of enzymes that specifically catalyze DNA supercoiling and the energy that is expended for the process, suggest that supercoiling has functional importance.

Plasmids. Besides the chromosome, where almost all of the genetic information is stored, many bacteria have other small size, circular DNA molecules (2–200 kb), which can duplicate independently and contain accessory genetic information. These small extrachromosomal DNA molecules are called *plasmids*. They contain the genes responsible for providing bacteria antibiotic resistance. They can be transferred from one bacterium to another through cell walls, which explains why the genes included in plasmids have spread so widely. Several techniques are available to isolate pure bacterial plasmids, free from chromosomal DNA.

Mitochondria. In mitochondria and chloroplasts, a DNA with characteristics similar to those of the bacterial chromosome has been isolated. These DNAs, of smaller size, are circular and contain approximately 15,000 bp (15 kb). The human mitochondrion has 16,569 bp and encodes information for the synthesis of RNA and some proteins of that organelle.

Mitochondria are not self-sufficient from a genetic viewpoint since the bulk of their proteins is synthesized from information contained in the cell nuclear DNA.

The existence of mitochondrial DNA, with characteristics similar to that of bacteria, supports the hypothesis that mitochondria evolved by an endosymbiotic process of bacteria with eukaryotic cells.

Genome

All individuals of a species have the same amount of DNA in every cell. In sexually reproducing diploid organisms, somatic cells contain twice the amount of DNA than the gametes (haploid cells). The total DNA in each cell is called the *genome*, which represents the "genetic reserve" of each individual.

The size of the genome varies with the complexity of the organism. For example, *E. coli* contain only slightly over 4 million $(4 \cdot 10^6)$ bp in their chromosome; the yeast *Saccharomyces cerevisiae* has $1.4 \cdot 10^7$ bp; the fruit fly *Drosophila melanogaster*, $1.7 \cdot 10^8$ bp; *Homo sapiens*, $6 \cdot 10^9$ bp. The amount of DNA in eukaryotic cells, compared to prokaryote cells is larger than that expected for the differences in complexity between these cells. Among eukaryotes, some striking facts are observed: there are fish and amphibian species whose genomes are much larger (up to 10^{11} bp) than those of mammals. This apparent DNA excess, based on the complexity difference between species, is due to the existence of functionally inactive large portions of the genome, carrying noncoding genetic information (previously considered "junk" DNA). Recent evidence suggests that many of these "inactive" regions play important functions.

The complete genome or DNA content of a human haploid cell (sperm or egg) is distributed in 23 chromosomes, 22 autosomal and 1 sexual (X or Y) and contains 3 billion bp. Diploid cells have 46 chromosomes; each of the 22 autosomal chromosomes has its homologous counterpart. The DNA molecules of homologous chromosomes are similar, but not necessarily identical. Each chromosome pair has one chromosome that comes from the father and the other from the mother. Each diploid cell also has two sex chromosomes, two Xs in women and one X and one Y in men.

Ribonucleic Acid

RNA is a polynucleotide whose main structural differences with DNA are: (1) the pentose is D-ribose in place of D-2-deoxyribose. (2) The pyrimidine base thymine is replaced by *uracil*. The remaining bases are the same as those of DNA (adenine, guanine, and cytosine). (3) RNA has a single polynucleotide chain and not two as in DNA. However, commonly RNA bends forming a hairpin and coils on itself in segments that mimic the double-stranded antiparallel strands of DNA. This obviously occurs at complementary sections of the molecule. RNA has greater conformational flexibility and ability to perform various functions.

The characteristic single chain of the RNA does not require maintaining the molar ratios of purines and pyrimidine bases that is observed in DNA. The amounts of guanine and adenine are not necessarily equal to those or cytosine and uracil, respectively.

In many cells there are several types of RNAs: messenger (mRNA), ribosomal (rRNA), transfer (tRNA), small nuclear RNA, microRNA, and others.

Messenger Ribonucleic Acid

Messenger ribonucleic acid (mRNA) represents about 5% of the total cell RNA. It is located in the nucleus and the cytoplasm. The molecular mass and base composition of mRNA is highly variable. The physiological role of mRNA is the transmission of the genetic information from nuclear DNA to the protein synthesis site in cytoplasm and to serve as guide for the assembly of amino acids in the correct sequence.

Nuclear RNA is heterogeneous; it attains large sizes of more than 10^6 Da. The RNA molecules synthesized in the nucleus are subjected to a process called *splicing*, in which the original RNA chains are cleaved in segments, some of which are rejoined and end up forming the final molecule strands that move to the cell cytoplasm. Approximately 20% of the original RNA is used to form mRNA, the rest is degraded. The large precursor molecules, the intermediate products of splicing, and the *mature mRNA*, which is much shorter than the precursor RNA are altogether called *heterogeneous RNA* (nhRNA). The nhRNA associated to proteins forms nuclear heterogeneous ribonucleoproteins (nhRNP).

Processed mRNA is modified at both ends; 7-methyl-guanosine triphosphate is attached at the 5′ terminus, which serves as marker for recognition of the mRNA by the protein synthesis system. In addition, this triphosphate nucleotide contributes to the stability of mRNA protecting the 5′ end against enzymatic hydrolysis. At the 3′ terminal end, a polymer of 100–250 units of adenilic acid or adenosine monophosphate (AMP), called *poly-A*, is added. This poly-A tail gives stability to the mRNA. mRNA is the most labile RNA; its half-life in mammalian cells is only approximately 6 h. The mRNA from bacteria has an even shorter half-life.

If denatured DNA and RNA are mixed in adequate conditions, associations in double helices can be produced when the chains possess complementary sequences; this DNA–RNA hybridization, involves the G–C, A–U, and T–A pairings.

Transfer Ribonucleic Acid

This class of RNA, also called soluble, has the smallest molecular size (25 kDa); its chain contains about 75 nucleotides. The transfer ribonucleic acid (tRNA) participates in the process of protein synthesis carrying amino acids

to the site of assembly. It acts as an adapter molecule, assuring the exact location of each amino acid. There are different kinds of tRNA in the cell that are specific for each of the different amino acids. The nucleotide sequence and primary structure of tRNA isolated from different organisms is known. In these tRNAs, nucleosides other than A, U, G, and C (such as dihydrouridine, pseudouridine, inosine, and methylated derivatives of A, U, G, and C) are frequently present.

Initial studies suggested that tRNA molecule has the general arrangement of a trilobed leaf (Fig. 6.18). The polynucleotide chain in tRNA possesses complementary segments that can pair with each other. Approximately half of the nucleotide residues of tRNA are paired in four double helix zones. There are also free segments of the chain that form the three lobes of the molecule. The central lobe contains a group

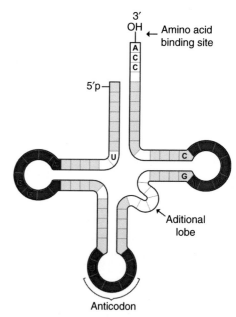

FIGURE 6.18 **Schematic representation of a tRNA molecule.** In *red*, the three lobes; *pink*, the double helix segments; 3′ terminal has the sequence CCA, where the carried amino acid is attached. *Letters* indicate the bases invariable for all tRNA.

of three bases, known as the *anticodon*, which is responsible for the amino acid specificity and the adapter function of tRNA. This "clover leaf" like structure has a stalk (one of the segments in double helix) where the molecule 5' and 3' ends are located. In the 5' terminal end there is a guanosine (G) or cytidine (C) residue, in the 3' end all tRNAs have the sequence CCA for the three last nucleotides. The amino acid carried by the tRNA is bound by an ester link between the carboxyl group of the amino acid and the C3' —OH of the last adenosine ribose. The stalk is called the *acceptor arm* because of its amino acid binding function. There is also an extra arm, the additional lobe of Fig. 6.18, which has variable length on the different tRNAs.

Studies by X-ray diffraction showed that the tRNA molecule has the shape of a letter L (Fig. 6.19); the acceptor site is on one of the ends of the L and on the other is the *anticodon* lobe. Although this is the real tRNA structure, the "clover leaf" image continues being a useful way to represent the molecule.

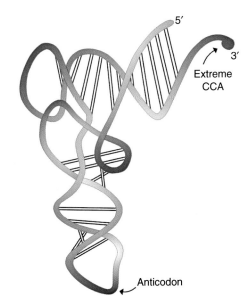

FIGURE 6.19 **More accurate scheme of tRNA.**

Ribosomal Ribonucleic Acid

Ribosomal ribonucleic acid (rRNA) is the most abundant, comprising approximately 80% of the total cell RNA. It constitutes the prosthetic group of nucleoproteins that forms the *ribosomes*, small granules located in the cytoplasm, either free or bound to the rough endoplasmic reticulum. More than 55% of a ribosome mass corresponds to RNA, the rest is constituted by proteins.

Eukaryotic ribosomes have a sedimentation coefficient of 80S (Svedberg units) and are composed of two main particles. The larger particle, of 60S and 2700 kDa, is integrated by 3 RNA molecules (28S, 5.8S, and 5S) and approximately 45 different proteins. The smaller particle, of 40S and 1300 kDa, has 1 RNA molecule (18S) and about 30 proteins. Bacteria possess 70S ribosomes with a major particle of 50S and a minor one of 30S (the values are not additive because S depends not only on the mass but also on the molecule shape). Although ribosomes contain more proteins than RNA, these last molecules are bigger and represent about two-third of the total ribosomal mass.

The rRNA molecules present many foldings and double helix segments. Each rRNA plays an important structural and functional role. If all the rRNA and the protein molecules of a particle are added separately to an adequate medium, they can assemble spontaneously. A tridimensional model of a ribosome shows that both of its particles have an irregular shape (Fig. 6.20).

Electron microscopy has allowed detecting groups of ribosomes linked together by a strand of mRNA as rosary beads, these are called *polysomes*.

Small Nuclear Nucleoproteins

There are other particles of RNA in the cells that integrate the *small nuclear nucleoproteins* (snRNP), also called *"snurps."* RNAs of snRNP have less than 300 nucleotides and are rich in uracil, which gives them their UsnRNP

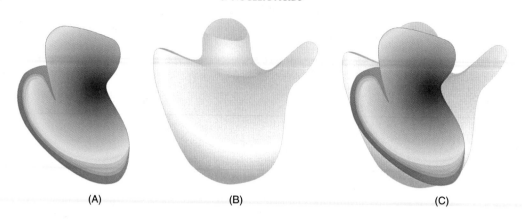

(A)   (B)   (C)

FIGURE 6.20   **Ribosome.** (A) Small particle, (B) large particle, and (C) assembly of the two particles into a whole ribosome.

designation. Several UsnRNP species have been isolated, which are designated by the letter U and a number (U1, U2, and U6). The snRNP participate in the processing of mRNA in the nucleus. There are also *small cytosolic ribonucleoproteins* (scRNP).

Other important RNAs in the cell are the *microRNA, small silencing RNA,* and *long noncoding RNA* which will be presented in Chapter 23.

VIRUS

Viruses are particles formed by nucleic acids surrounded by protein. They are agents that cause numerous diseases in animals, as well as plants. Some of them can induce tumors in the host they invade. Some viruses, called bacteriophages, are able to attack bacteria.

Basically, viruses consist of a *capsid* or protein cover that contains nucleic acids. The capsid is formed by association of polypeptide units or *capsomers* that are symmetrically arranged and frequently give the virus a geometric conformation. In most of the simplest viruses, the cover is formed by a single type of protein; in complex viruses, more than one class of protein is present and they can even contain lipids or carbohydrates.

The genetic material of the virus can be either DNA or RNA and it is all contained within the viral capsid. The complete viral structure, formed by DNA or RNA and its cover, is called *virion*. In contrast to prokaryote or eukaryote organisms that possess both types of nucleic acids, viruses only have one kind. Viruses that attack animals may possess DNA or RNA. These nucleic acids have, in some cases, one strand or form a double helix in others. Due to their relatively small size, it is easier to isolate intact DNA molecules from viruses than from more complex organisms, this has promoted their use for studies of DNA structure and function. One of the smallest viruses is bacteriophage ϕX174, with a single DNA strand that is 5386 nucleotides long. This was the first DNA molecule to be entirely sequenced.

Fig. 6.21 shows the structure of a plant virus, the mosaic of tobacco virus. It has a RNA helical chain covered by multiple polypeptide units; the whole structure is a cylinder. The RNA is 6400-nucleotides long and the capsid is composed by 2130 identical units disposed in a helix conformation.

Fig. 6.22 represents an *adenovirus*, with a cover formed by 252 capsomers disposed in icosahedral symmetry. Inside the viral capsid is a double helix DNA.

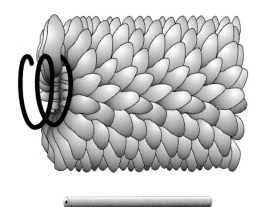

FIGURE 6.21 **Tobacco mosaic virus.** Polypeptide units form the cover, RNA helix inside *(black spiral)*. At the bottom, a complete virus *(cylindrical rod)* is shown.

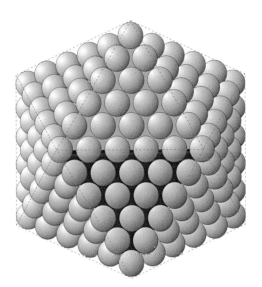

FIGURE 6.22 **Adenovirus (icosahedral symmetry).**

The T series bacteriophages are viruses with a more complex structure, the cover proteins form a polyhedral "head," able to contract, containing DNA with the genetic information. In addition, they have a hollow stalk made up of proteins and six filaments at its base that attach to the attacked bacteria (Fig. 6.23).

Nucleic acid within the capsid contains the information needed to synthesize the viral

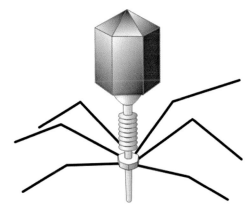

FIGURE 6.23 **T4 bacteriophage.**

proteins. Viruses multiply invading a cell, introducing their nucleic acid and forcing the cell to reproduce new viral particles. They behave as parasites that use the metabolic machinery and energy of the invaded cells for the synthesis of their own proteins and nucleic acids.

Viroids are infectious agents that cause diseases in plants. They lack a protein cover and are constituted by only one circular molecule of RNA of approximately 300 nucleotides.

FREE NUCLEOTIDES

There are nucleotide compounds that have important functions, not as components of nucleic acids, but free or integrating relatively small size molecules. These comprise di- and triphosphonucleosides derived from ribonucleotides (Table 6.1) by addition of one or two phosphoric acid residues. The most abundant in all living organism is *adenosine triphosphate* (ATP). This compound is the most important mediator in phosphoryl group and energy transfer in the cell. The hydrolysis of the bonds between the second (β) and third (γ) phosphate, and between the first (α) and second phosphate, has a marked negative ΔG, which makes ATP an "energy-rich" compound (see p. 147). After hydrolysis of the bond between the second and

TABLE 6.1   Most Common Nucleosides and Nucleotides

Base +	Aldopentose =	Nucleoside +	Phosphoric acid =	Nucleotide	Abbrev.
Adenine	Ribose	Adenosine	Phosphoric acid	Adenylic acid or adenosine monophosphate	AMP
Adenine	Deoxyribose	Deoxyadenosine or D-adenosine	Phosphoric acid	Deoxyadenylic or D-adenosine monophosphate	dAMP
Guanine	Ribose	Guanosine	Phosphoric acid	Guanylic acid or guanosine monophosphate	GMP
Guanine	Deoxyribose	Deoxyguanosine or D-guanosine	Phosphoric acid	Deoxyguanylic acid or D-guanosine monophosphate	dGMP
Thymine	Deoxyribose	Deoxythymidine or D-thymidine	Phosphoric acid	Deoxythymidylic acid or D-thymidine monophosphate	dTMP
Cytosine	Ribose	Cytidine	Phosphoric acid	Cytidylic acid or cytidine monophosphate	CMP
Cytosine	Deoxyribose	Deoxycytidine or D-cytidine	Phosphoric acid	Desoxycytidylic or D-cytidine monophosphate	dCMP
Uracil	Ribose	Uridine	Phosphoric acid	Uridylic acid or uridine monophosphate	UMP

third phosphate, ATP is converted into adenosine diphosphate (ADP); further release of another phosphate generates AMP (Fig. 6.24).

Guanosine triphosphate (GTP) is another physiologically important nucleotide. CTP and UTP also take part in phosphoryl transfer reactions. Ribonucleotide and deoxyribonucleotide triphosphates are basic compounds needed for the synthesis of RNA and DNA, respectively.

Nucleoside mono- and diphosphates are converted into triphosphates by the energy liberated during oxidation of compounds coming from the diet (oxidative phosphorylation) or by energy transfer from energy-rich metabolic intermediaries (substrate level phosphorylation) (see pp. 190 and 201).

Several nucleoside diphosphates participate in transfer of molecules for synthesis or

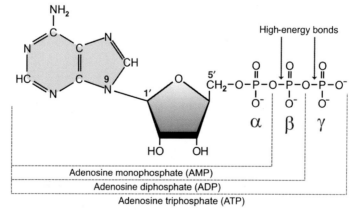

FIGURE 6.24   **Adenosine triphosphate (ATP).**

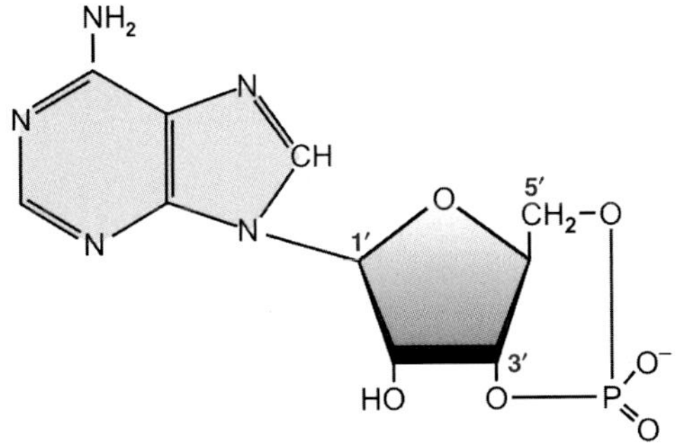

FIGURE 6.25 **3′,5′-Cyclic-adenosine monophosphate (cyclic AMP).**

conjugation processes; for example, uridine diphosphate glucose (UDPGlc), uridine diphosphate galactose (UDPGal), and uridine diphosphate glucuronic (UDPGlu) are important intermediaries in metabolism. Cytidine diphosphate choline (CDPcholine) transfers choline for phospholipid synthesis and phosphoadenosine phosphosulfate (PAPS) transfers sulfate groups.

Other nucleotides integrate factors essential for enzyme action. These factors, called coenzymes, frequently present nucleotide structure.

Some nucleotides take part in signal transduction systems (p. 556), for example, *cyclic-3′,5′-AMP* (Fig. 6.25). This compound is formed from ATP by *adenylate cyclase*, enzyme regulated by hormones. Cyclic AMP is a chemical messenger, intermediary of hormone actions.

The 3′,5′-cyclic-guanosine monophosphate is another nucleotide with important function as an intracellular messenger.

SUMMARY

Nucleic acids perform essential functions in the cell. They contain genetic information and play a key role in protein biosynthesis. They are macromolecules formed by *nucleotides*.

Nucleotides consist of a nitrogenous base, an aldopentose and phosphoric acid. The base can be of the *pyrimidine* and *purine* type. The first include *thymine* (T), *cytosine* (C), and *uracil* (U); the second, *adenine* (A) and *guanine* (G). The aldopentose is D-*ribose* in RNA or D-2-*deoxyribose* in

DNA. The pentose is linked by a β-glycosidic bond to the N1 of pyrimidine bases or N9 of purine bases.

Nucleic acids are polymers containing numerous nucleotides and bound through ester links established by phosphate between the —OH group in C3′ of the pentose of one nucleotide and the —OH of C3′ of the pentose from another nucleotide. One extreme of the chain has the 5′ of pentose free and is considered the *start* of the chain. The last nucleotide has the C3′ nonesterified and it is considered the 3′ *end* of the molecule.

DNA forms the chromatin of the cell nucleus, a small amount is in mitochondria. Individuals of the same species have equal amount of DNA in each cell (6 pg in humans), female and male gametes contain half of that amount. DNA molecules are long, with a length of several centimeters and a cross section of approximately 2 nm. After DNA hydrolysis, the purine bases adenine (A) and guanine (G) and the pyrimidine base thymine (T) and cytosine (C) are obtained. There is a defined ratio between purine and pyrimidine bases in DNA: A + G is always equals to T + C. A/T and G/C are always equal to 1.

DNA has a double helix structure with two polynucleotide strands wrapped around the same axis. Deoxyriboses and phosphates joining the C5′ and C3′ of pentoses form the continuous thread of the chain. Purine and pyrimidine bases are flat structures that project toward the interior of the molecule perpendicularly to the double helix axis. Each helix turn comprises 10 bases; it has a length of 3.4 nm, with a distance of 0.34 nm between bases. The DNA helix coils clockwise, both chains are antiparallel, one running 5′→3′ and the other 3′→5′. The double helix forms two grooves, one wider (major groove) than the other (minor groove).

The purine and pyrimidine bases are paired in a defined way: always an adenine from one chain is linked to a thymine from the other, the A–T pair is maintained by two H bonds. Guanine is paired with cytosine by three H bonds. The double helix is very stable not only because of the great number of H bonds, but also by the interactions between the bases. As a consequence of base pairing, the chains are not equal, they are *complementary*.

The nucleotide sequence of DNA chains is important since the order of bases encodes the genetic information.

DNA denaturation can be obtained by heat and some chemical reagents, which weaken the H bonds and produce the separation of the DNA chains. DNA absorbency at 260 nm increases as the chains separate (*hyperchromic effect*). The temperature at which half of the DNA molecules are denatured is the *melting temperature* or Mt. The determination of Mt is an estimate of the base composition of DNA; the greater the Mt, the greater the G–C content of DNA. By slowly cooling denatured DNA, reassociation of the chains can be obtained and the double helix can be formed again (*reannealing*). Reassociation of denatured DNA shows that in eukaryote cells there are highly repetitive, moderately repetitive, and single-copy DNAs. Denatured chains from different organisms can associate into hybrid molecules (*hybridization*) if they have adequate complementary zones.

Chromatin results in eukaryotes from the association of DNA to proteins. Proteins are basic, predominantly *histones*. In chromatin, the DNA molecule makes a double turn (DNA superhelix) at regular intervals on a core of eight histone units. Both DNA turns comprise 146 bp.

Nucleosomes are constituted by the histone core and the DNA superhelix. They are connected by segments of DNA of approximately 50 bp. Nucleosomes are extended during interphase; however, at the beginning of mitosis, they are packed in a solenoid-like structure of 6 nucleosomes per turn.

Heterochromatin is highly condensed chromatin. It is inactive.

Circular DNA is found in bacteria, plasmids, mitochondria, and chloroplasts.

RNA is another polynucleotide that differs from DNA in the following:

1. It contains ribose instead of deoxyribose.
2. It has uracil instead of thymine.
3. It is formed by a single chain instead of two.

Denatured DNA and RNA that have complementary segments can form hybrid double helices DNA/RNA.

Different types of RNA exist in cells; these include: mRNA, tRNA, rRNA, and other classes of RNA.

mRNA represents approximately 5% of the total RNA; it is found in the cell nucleus and cytoplasm. Its mass and composition are very variable. In the cell nucleus, it is processed to form mature mRNA that always has the 5′ terminal linked to 7-methyl-guanosine triphosphate and the 3′ end bound to a piece of 100–250 units of adenilic acid (*poly-A tail*). mRNA transmits genetic information from the DNA in the nucleus to the cytoplasm, where protein synthesis takes place. mRNA is the most labile of all RNA.

tRNA or soluble RNA are small molecules of ~75 nucleotides in length and a mass of ~25 kDa.

They carry amino acids to the site of protein synthesis. It frequently has unusual bases (dihydrouridine, pseudouridine, inosine, and methylated A, G, U, C. The tRNA molecule has an L shape, three lobes, and double helix segments. The terminal 5′ end contains G or C, the 3′ end always has the CCA sequence, which binds to the transported amino acid.

rRNA is the most abundant one, comprising ~80% of the total RNA of the cell. RNA is the prosthetic group of nucleoproteins of *ribosomes*, formed by a *larger particle* (60S), which contains 3 RNA molecules and approximately 45 proteins, and a *minor particle* (40S), with 1 RNA molecule and close to 30 proteins. Both portions present an irregular conformation that associate during the synthesis of polypeptides. At the electron microscopy level, groups of ribosomes are attached to an mRNA chain like beads in a rosary (*polysome*).

Other types of RNA include *small nuclear, small cytosolic RNA, microRNA, small silencing RNA*, and *long noncoding RNA*.

Virus are agents of disease in animals and plants. They reproduce using the metabolic machinery and energy of the invaded cell.

They are particles formed by nucleic acids surrounded by proteins. The protein cover is the viral *capsid*, composed of polypeptide units or *capsomers*. The capsid contains the viral genetic material, DNA or RNA, commonly single stranded.

Free nucleotides contain two or three phosphate residues. The most abundant one is ATP, the main energy carrier in living beings. GTP is another free nucleotide that plays important cellular functions. Nucleoside diphosphates (UDP, CDP, and PAPS) transfer molecules in synthesis. Other nucleotides form part of coenzymes. 3′,5′-Cyclic AMP and 3′,5′-cyclic GMP act as chemical messengers in signal transduction systems.

Bibliography

Fifty years of DNA, 2003. Papers on ADN. Nature 421, 395–453.

Krebs, J.E., Goldstein, E.S., Kilpatrick, S.T., 2013. Lewin's Genes XI. Jones & Bartlett Publishers Inc., Burlington, MA.

Turner, R., 2002. RNA. Nature 418, 213–258.

Van Holde, K.E., Zlatanonovai, J., 1995. Chromatin higher order structure: chasing mirage. J. Biol. Chem. 270, 8373–8376.

Watson, J.D., Baker, T.A., Bell, S.P., Gann, A., Levine, M., Losick, R., 2008. Molecular Biology of the Gene, sixth ed. Pearson/Benjamin Cummings, San Francisco, CA.

Elements of Thermodynamics and Biochemical Kinetics

The structure of the compounds present in living organisms is not static. They continuously change and are subjected to a diversity of chemical reactions, in a process known as *metabolism*. In normal individuals, these chemical reactions take place in an orderly and highly regulated manner, following pathways that developed through millions of years of evolution. Their study is the subject of an area of biochemistry known as dynamic biochemistry.

This chapter presents basic concepts regarding the general requirements and characteristics of chemical reactions, which will serve as an introduction to the Chapters 8 and 9 that follow.

THERMODYNAMICS

Besides the required reagents and the resulting products, another important factor for the development of chemical reactions is *energy*. Chemical transformations are accompanied by energy changes. Understanding these changes is important in biochemistry since it helps to recognize how metabolic processes proceed.

Thermodynamics is the branch of physics that deals with energy and its transformations. Its basic principles are applicable to biological processes. The fundamental principles of thermodynamics are expressed in the following two laws:

First law of thermodynamics: The total energy of the universe remains constant (all forms of energy are exchangeable; the energy is neither created nor destroyed).

Second law of thermodynamics: The entropy of the universe increases constantly (entropy is associated with disorder or randomness).

Energy

Commonly, energy is defined as the capacity to produce work. Energy exists in different modes: chemical, thermal, mechanical, electrical, and radiant, all of which can be converted into each other. Energy conversions occur frequently in biological processes. The development and growth of an organism and the continuous renewal of its structures involves a large number of chemical reactions, which are only possible if there is energy input. Similarly, maintenance of body temperature in warm-blooded animals; mechanical work in muscles, cilia, and flagella; generation of electrical impulses in the nervous system; and active transport of substances across membranes are all processes that demand energy.

The primary source of energy for all forms of life is solar radiation. This is captured and stored as chemical energy by photosynthetic organisms and transferred to other living beings through the feeding chain of biosphere. In aerobic organisms, energy is generated mainly by oxidation of substances incorporated with the food and transferred to compounds that retain it to be used when necessary.

Chemical energy plays an important role in biological processes. The chemical energy of a compound is represented by the movement and relative position of its atoms and particles and by bonds and attractions between its elements. When a chemical reaction occurs, frequently bonds are broken or formed and the energy content of the molecules changes. The course of a chemical reaction is ultimately determined by the energy content of the system under consideration and the energy exchange between it and the environment.

Energy Changes in Chemical Reactions

The term "system" is used to designate the portion of matter under study; other matter includes the medium surrounding the system.

The total energy content of the system before any reaction takes place is known as *initial state*, while the energy content after the chemical change occurs is the *final state*. During the transit from the initial to final states energy is released or gained. Thermodynamics concerns the study of the difference in energy that takes place between the initial and final states and not the mechanism by which the process proceeds. That difference is the same, independent of the steps involved in the reaction.

Measuring the energy content of a system can be difficult; it is easier to determine the energy change that takes place between the initial and final states (change is symbolized by the Greek capital letter delta, Δ).

The most common form of energy is heat. Almost all chemical processes are accompanied by heat consumption or production; the first are referred as *endothermic* and the second, as *exothermic* reactions.

According to the first law of thermodynamics, it is possible to determine the energy change in a reaction by measuring the gain or loss of heat in the system in conditions of constant temperature and pressure. This can be performed using a calorimetric pump, which determines the energy released by the reaction measuring the temperature of a known volume of water surrounding the calorimeter. Multiplying the temperature increase by the weight of water, the heat released can be estimated. This is usually measured in calories.

A *calorie* (cal) is the amount of heat required to raise the temperature of 1 g of water from 14.5 to 15.5°C. The kilocalorie (kcal or Cal), a 1000 times greater, is commonly used. The unit Cal has been widely used by biochemists. Currently the Joule or most often the kilojoule (kJ), units of the International System (IS), are preferred to express energy. (*Joule* = 10^7 ergs. *Erg*, unit of work, corresponds to the work done by 1 Dine along 1 cm. *Dine* is the force acting during 1 s on a 1 g mass, moving it with a speed of 1 cm/s.)

When a substance is oxidized, energy is produced and heat is released to the environment. The magnitude of this "heat of combustion" depends on the molecular structure of the substance and the remaining energy content in the products formed. If a mole (342 g) of common sugar (sucrose) is oxidized to CO_2 and H_2O, 1350 kcal (5648 kJ) are produced (the potential energy of sucrose is represented by the bonds and configuration of the molecule; the energy content of the products CO_2 and H_2O is much lower).

Heat from the combustion of a substance is the maximum energy that can be obtained from complete oxidation of the elements of that substance.

When the reaction takes place at a constant pressure, the change in heat produced is called *enthalpy* change and is symbolized by the notation ΔH. It has a negative sign when heat is released.

Enthalpy (H) is the caloric energy released or consumed in a system at constant temperature and pressure. When the volume is not changed, that is, when no work is done, the energy change (ΔE) equals the change in enthalpy (ΔH).

Heat is the simplest manner by which energy can be applied to produce work. Typical examples are represented by engines and machines. However, in biological systems which operate at constant temperature and pressure, heat change is not a source of energy.

Second Law of Thermodynamics

The first law of thermodynamics allows us to understand and determine the energy changes that occur during a chemical reaction, but cannot predict its direction. One might think that a reaction always proceeds in the direction in which ΔH is negative (exothermic) and does not occur when ΔH is positive (endothermic). But this is not the case because ΔH is not the only factor that determines the direction of a reaction. The second law of thermodynamics introduces another element that needs to be taken into account: *entropy.*

If a system with a given energy content is supplied with a certain amount of heat Q, the added heat, according to the principle of energy conservation, will appear as a change in energy content of the system (ΔE) and/or work (W) performed on the environment. This is represented by the equation $Q = \Delta E + W$, which can be transformed into $\Delta E = Q - W$, which is an expression of the first law of thermodynamics. The flow of heat between the system and its surroundings is accompanied by energy changes in the system and/or work performed; they exactly compensate each other. However, the conversion of heat to mechanical work is never complete.

For example, in a steam-powered machine, where the steam enters at a temperature T_1 and leaves after use at a lower temperature, T_2, the difference between T_1 and T_2 reveals the heat taken, Q. This heat is used and converted, as much as possible, in useful work. If W is called the maximum useful work output:

$$W = Q\frac{T_1 - T_2}{T_1} = Q - Q\frac{T_2}{T_1} \qquad (7.1)$$

where T_1 and T_2 are absolute temperatures.

Unless T_1 is infinite or T_2 equal to absolute zero ($-273°C$), conditions that are normally impossible to achieve, the useful work is always less than the total energy supplied by a factor equal to $Q(T_2/T_1) = (Q/T_1)T_2$.

The ratio Q/T_1 is known as *system entropy (S)*. The amount of energy unavailable for useful work $(Q/T_1)T_2$ or ST_2, is the energy lost in the transfer process.

Eq. (7.1) is similar to:

$$G = H - TS$$

where G (from Gibbs, mathematician and physicist who developed the application of thermodynamics to chemistry) is *free energy*, analogous to W (useful work); H (enthalpy) is analogous to Q (heat content of the system at constant pressure); and S (entropy) is analogous to Q/T_1, heat energy wasted.

Free Energy

Biochemical reactions in homeothermic organisms are carried out in a medium of constant temperature and pressure. Only a fraction of the energy released in the biochemical processes is available to do work of some kind; this is called *free energy.*

In a chemical reaction, if the reagents have a higher energy than that of the formed products, energy is released during the reaction. The energy of the reactants and products is difficult to measure and is of less interest than the change of energy that occurs during the reaction. The free energy change in the system is given by the equation:

$$\Delta G = \Delta H - T\Delta S$$

where ΔG is the free energy change, ΔH the enthalpy change, T the absolute temperature (°C + 273 = K = Kelvin degrees), and ΔS the entropy change.

Entropy is degraded energy, not usable for work. Only the change in free energy (ΔG) is available to perform work.

Direction of a Chemical Reaction

In chemical reactions in which there is reduction in free energy, it is possible to predict if they will occur spontaneously. When a system loses free energy during the reaction, the G of the system decreases in a defined amount and the ΔG is negative. If the free energy of the system increases, the change has a positive sign. For a reaction to occur spontaneously there must be reduction in free energy, which is represented as $-\Delta G$.

The tendency to increase entropy is the force that determines the direction of the process. Heat is released or absorbed by the system to allow the system and the medium to reach a state of higher entropy, following the second law of thermodynamics. Just as entropy tends to increase, free energy tends to decrease.

All chemical processes occur with a decline in free energy, until an equilibrium is reached in which free energy is minimal; simultaneously, there is an increase in entropy, which at equilibrium is maximal. It is possible to determine the change in free energy (ΔG) of a chemical reaction, and therefore to predict its direction.

Chemical Equilibrium

In the case of a reversible reaction in which two substances, A and B (reagents), react to form C and D (products) and vice versa, the direction of the reaction toward product formation is the direct reaction. This is named reaction 1 and indicated in the equation with an arrow going from left to right. C and D products can combine to reform A and B, this is the reverse reaction

(or reaction 2), indicated by the arrow going from right to left.

$$A + B \underset{2}{\overset{1}{\rightleftharpoons}} C + D$$

The rate at which the direct and reverse reactions occur follows the Guldberg and Waage's law (*Guldberg and Waage law*: The rate of a chemical reaction is proportional to the product of reagents masses):

$$\text{direct reaction}(1): v_1 = k_1[A][B]$$
$$\text{reverse reaction}(2): v_2 = k_2[C][D]$$

The brackets indicate reagent and product concentrations; v_1 and v_2 are the rates of reactions 1 and 2 respectively; k_1 and k_2 are the rate constants for each reaction at a given temperature.

At the beginning of the reaction, A and B combine to form C and D. As the reaction proceeds, the concentrations of A and B decrease as well as the rate of the reaction. Moreover, the C and D products react to reform A and B at a rate proportional to their concentrations. Eventually, a point is reached in which the rates of the forward and the reverse reactions are the same and no more changes in the concentrations of reactants and products occur; v_1 is equal to v_2. If $v_1 = v_2$, then:

$$k_1[A][B] = k_2[C][D] \tag{7.2}$$

The reaction has reached equilibrium. This is a dynamic equilibrium, since the A and B molecules continue reacting to form C and D and vice versa. The rates at which both forward and reverse reactions occur are equal, which explains why there are no changes in the concentration of the substances participating in the reaction. If terms are transposed in Eq. (7.2):

$$\frac{k_1}{k_2} = \frac{[C][D]}{[A][B]}$$

As the ratio k_1/k_2 is constant, the result is also a constant value, which is called the *equilibrium* constant (K_{eq}). The K_{eq} value is characteristic for each chemical reaction at a defined temperature.

$$K_{eq} = \frac{[C][D]}{[A][B]}$$

A similar reasoning was made in Chapter 2, when considering ionic dissociation (p. 9).

The conversion of reagents A and B into products C and D, and also during the reverse reaction, are accompanied by energy changes. The molecules involved in the reaction contain energy, trapped in the bonds and interactions of their atoms; when a chemical transformation occurs, the energy content of the reagents can be higher or lower than that of the products.

If K_{eq} is greater than 1, the reaction will take place predominantly to the right; if K_{eq} is less than 1, the reaction will preferably proceed from right to left. In other words, the equilibrium value is related to the tendency to reach the minimum of free energy and, therefore, it must be a function of the ΔG of the reaction. This relationship is expressed in the equation:

$$\Delta G° = -RT.\ln K_{eq} \qquad (7.3)$$

where $\Delta G°$ is *standard free energy change* [standard conditions correspond to 25°C (298°K) and 1 M concentration of reactants]; R, gas constant (8.314 J/mol degree or 1987 cal/mol degree); T, absolute temperature, and $\ln K_{eq}$, natural or neperian logarithm of the equilibrium constant; ln K_{eq} is converted into log K_{eq} (base 10) multiplying by the factor 2.303, thus the equation becomes:

$$\Delta G° = -RT(2.303)\log K_{eq}$$

Most biological reactions occur at pH close to 7, then, $\Delta G°$ is represented by $\Delta G°'$ (standard free energy change at pH 7).

The change in standard free energy corresponds to the maximum amount of work the reaction can accomplish under conditions in which the temperature is unchanged. This amount of work could be theoretically performed with a device that can prevent energy loss (useless frictions). Without such a device friction will produce heat, which will be released and not used to perform work.

The relationship between the value of the equilibrium constant and $\Delta G°$ indicates that when K_{eq} is high (greater than 1), $\Delta G°$ is negative, the reaction proceeds with a decrease in free energy. When K_{eq} is low (less than 1), $\Delta G°$ is positive and energy must be supplied for the reaction to occur (under standard conditions). When $\Delta G°$ is negative, the reaction will be spontaneous (however, as discussed later, this does not mean that the process will occur by itself or quickly. Combustion of sucrose is a spontaneous process because it has a negative $\Delta G°$, but sucrose can be years in contact with air at approximately 20°C without entering in combustion).

Reactions with a negative free energy change are *exergonic*. Reactions with positive $\Delta G°$ are not spontaneous and will not take place unless energy is supplied. Such processes are called *endergonic*.

The $\Delta G°$ value is determined by measuring the concentrations of reactants and products at equilibrium under standard conditions. The value of the standard free energy change is useful to predict whether or not a reaction will occur under standard conditions. If $\Delta G°$ is less than 0, the reaction is spontaneous; if $\Delta G°$ is greater than 0, the reaction will not proceed by itself.

In Eq. (7.3), $\Delta G°$ refers to a system in equilibrium, unable to produce useful work because there is no net change in the concentrations of reactants and products. Instead, a system that has not reached equilibrium produces work. The ΔG, useful for performing work at constant temperature and pressure, is related to $\Delta G°$ from the equation:

$$\Delta G = \Delta G° + RT\ln\frac{[products]}{[reactants]}$$

or, in decimal logarithms:

$$\Delta G = \Delta G° + 2.303\,RT\log\frac{[products]}{[reactants]} \qquad (7.4)$$

when the concentration of products is equal to that of reagents ([product]/[reactants] = 1), as log 1 = 0, then $\Delta G = \Delta G°$. Furthermore, at equilibrium, there is no change in free energy and $\Delta G = 0$.

The ΔG value determines if a reaction occurs under conditions that are different from the standard, such as those existing in the cell. Even if the reaction $\Delta G°$ is positive, the possibility of its spontaneous course is not excluded, since by changing the products [P] and/or reagents [R] concentrations, it is possible to obtain values of log [P]/[R] sufficiently negative to attain a negative ΔG value.

In the cell, the concentrations of reactants and/or products change and so does the value of ΔG [Eq. (7.4)], which determines whether the reaction will occur. For example, in the reaction dihydroxyacetone-P → glyceraldehyde-3-P (a step in glycolysis, the main route of glucose utilization), the $\Delta G°'$ equals +7.53 kJ/mol. The reaction will not occur spontaneously in the direction indicated by the arrow. However, in cells the reactant and product concentrations are far from 1 M; physiological values are $2 \cdot 10^{-4}$ M for dihydroxyacetone-P and $3 \cdot 10^{-6}$ M for glyceraldehyde-3-P. Substituting these values into Eq. (7.4) results in a $\Delta G = -2.93$ kJ/mol. The ΔG is negative and the reaction will proceed in the direction shown.

Depending on the concentrations of reactants and products, the ΔG of a reaction may be lower, higher, or equal to the $\Delta G°'$. The criterion to be considered for a reaction to proceed spontaneously is the value of ΔG, not of $\Delta G°'$. However, it is customary to give the value of $\Delta G°'$ when describing biological reactions. This only serves as an approximation to the possible behavior of the reaction.

It is certainly of interest to know the ΔG value for a system that is not in equilibrium, the most common situation in biological systems. This can be determined from $\Delta G°$ and the actual concentrations of products and reactants according to the Eq. (7.4). If ΔG is negative, in the conditions of the cell, the reaction will proceed to the formation of products. If ΔG is positive, the reaction will proceed in the opposite direction.

In biological systems, equilibrium state is the exception to such an extent, that life has been defined as the capacity to use an external source of energy to maintain the body's chemical reactions in a nonequilibrium state. Equilibrium for a cell is only achieved at death.

In the reaction A + B = C + D, equilibrium can be avoided by removing C and/or D as they are formed to produce other products, or by adding A and B as they are consumed. Chemical processes in cells show series of reactions in which a product is used as a reagent by a following reaction. This permanent removal of a product prevents the system reaching equilibrium and favors the reaction to proceed in a given direction.

Cell as an Open System

The principles of thermodynamics apply to closed systems that do not exchange matter with the environment and can reach a thermodynamic equilibrium. Living cells are open systems, which are in constant exchange of matter and energy with the environment. While the concentrations of many of the components of the cell may appear unchanged, they are in a dynamic steady state in which the rate of formation of a particular compound is balanced by its rate of removal. Biological processes appear to contradict the second law of thermodynamics, since they increase the free energy and order of the system. For example, the use of energy for the synthesis of highly ordered molecular structures means decreased entropy of the system. However, this is at the expense of increased disorder in the medium. Living organisms are open systems, they exist in a dynamic state; they increase the entropy of the universe and are therefore, irreversible.

High-Energy Compounds

From a thermodynamic viewpoint, endergonic processes would never occur in living beings without energy input. The approach used by cells to counteract the positive ΔG is to couple those processes with reactions in which "high-energy" compounds are produced. For example, glucose synthesis from CO_2 and H_2O is an endergonic process. In photosynthetic organisms, the energy for glucose synthesis is provided by solar light; a given number of photons are captured by pigments present in the cells. Analysis of this process, taking into consideration the reactants the photon-activated pigments, shows that the reactions proceed with a decrease in free energy.

In animals, synthetic processes undergo stages in which the reagents are "activated" by input of energy transferred from high-energy compounds. Thus, the biosynthetic reactions occur with release of free energy and, therefore, they are thermodynamically possible.

There are many energy-rich compounds involved in the operation of the metabolic machinery of the cell; they have bonds, which once ruptured, produce a significant decrease in free energy. When these compounds participate in reactions, the negative free energy change simultaneously enables the occurrence of endergonic reactions.

In biochemistry, a chemical bond is considered of high energy when the free energy change of the reaction in which that bond is involved is greater than 20 kJ/mol (ΔG from -20 to -60 kJ/mol). In chemical notation it is customary to indicate high energy bonds by the sign $\sim$. Table 7.1 presents examples of energy-rich compounds. Many contain phosphates (Ⓟ) and acetyl-coenzyme A is a thioester. Compounds having phosphate moieties with high energy of hydrolysis, some of which are listed in Table 7.1, participate in reactions of phosphoryl group transfer to other compounds of lower free energy. The ability of a substance to transfer a phosphate moiety to another is called *phosphoryl group transfer potential*. The value of this potential is expressed by the $\Delta G°$ with the sign changed. For phosphoenol-pyruvate, $\Delta G°$ is 14.8 kcal/mol, 61.9 (kJ/mol), for ATP, 7.3 kcal/mol (30.5 kJ/mol).

TABLE 7.1   Energy-Rich Compounds

Compound		$\Delta G°'$ kJ/mol	$\Delta G°'$ kcal/mol
Creatine-phosphate	HN H $\|\|$ $\|$ R — C — N $\sim$ Ⓟ	-43.9	-10.5
Phosphoenol-pyruvate	CH_2 $\|\|$ R' — C — O $\sim$ Ⓟ	-61.9	-14.8
Acetyl-phosphate	O $\|\|$ H_3C — C — O $\sim$ Ⓟ	-42.2	-10.1
Adenosine-diphosphate	O $\|\|$ R'' — O — P — O $\sim$ Ⓟ $\|$ OH	-30.5	-7.3
Acetyl-coenzyme A	H_3C — C $\sim$ S — CoA $\|\|$ O	-43.9	-10.5

The most important high energy compound is adenosine triphosphate (ATP). In ATP, the phosphoric anhydride bonds between the second (β) and the third (γ) and between the first (α) and second (β) phosphoryl moieties have high free energy of hydrolysis.

$$\text{ADENOSINE} - O - \underset{\underset{O^-}{|}}{\overset{\overset{O}{\|}}{P}} - O \sim \underset{\underset{O^-}{|}}{\overset{\overset{O}{\|}}{P}} - O \sim \underset{\underset{O^-}{|}}{\overset{\overset{O}{\|}}{P}} - O^-$$

The high-energy content of a chemical bond is not a property of a particular type of link, but rather depends on intramolecular tensions that disappear when the hydrolysis of the bond occurs.

At physiological pH, the phosphate residues of ATP are deprotonated; the ionic form is predominantly ATP^{4-}. The negative charges are close to each other and the electrostatic repulsion creates intramolecular tensions. The hydrolysis of ATP to ADP and phosphate, or to AMP and pyrophosphate produces an important reduction in free energy because the products have a lower tension level. Moreover, the phosphate is stabilized by formation of resonance hybrids, whereas in ATP this factor is not significant. In addition, ADP and phosphate dissolve in aqueous media better than ATP. The phenomenon of resonance, stabilization, and solvability of products also help to reduce the free energy and drive the reaction toward the hydrolysis of the compound.

The energy of phosphate bonds in ATP can be retained in the body until required, carried within the cell to the site of use, or transferred to form other high energy compounds, such as uridine triphosphate (UTP) or acyl-coenzyme A.

Energetically Coupled Reactions

A highly exergonic reaction can drive an endergonic one if both of them are coupled. Usually, coupled reactions share a common intermediate. For example, the exergonic reaction:

$$A + B \rightarrow C + D$$
$$\Delta G^{\circ\prime} = -8.0 \, kcal/mol \, (-33.5 \, kJ/mol) \quad (7.5)$$

and the endergonic reaction:

$$D + E \rightarrow F + G$$
$$\Delta G^{\circ\prime} = +3.0 \, kcal/mol \, (+12.5 \, kJ/mol) \quad (7.6)$$

Eqs. (7.5) and (7.6) share compound D, which is a product in the first and a reagent in the second. Both reactions can be coupled, in which case the overall reaction is:

$$A + B + E \rightarrow C + F + G \quad (7.7)$$

$\Delta G^{\circ\prime}$ of coupled reactions is equal to the sum of the $\Delta G^{\circ\prime}$ of the individual reactions. In the case of reaction (7.7), $\Delta G^{\circ\prime}$ will equal $-8 + 3 = -5 \, kcal/mol$ or $-21 \, kJ/mol$; that is, the total process becomes spontaneous.

ATP hydrolysis is an exergonic reaction, with $\Delta G^{\circ\prime}$ of $-7.3 \, kcal/mol \, (-30.5 \, kJ/mol)$ for each of the energy-rich bonds. This value corresponds to standard conditions (25°C, a concentration of reagents of 1 M and a pH of 7.0). In the cell conditions, ΔG is actually higher.

$$ATP + H_2O \rightarrow ADP + P_i + Energy$$
$$P_i \text{ indicates inorganic phosphate}$$

The formation of the ester glucose-6-phosphate is, in contrast, an endergonic reaction with $\Delta G^{\circ\prime}$ +3.3 kcal/mol (+13.8 kJ/mol).

$$Glucose + P_i + Energy \rightarrow Glucose\text{-}6\text{-}P + H_2O$$

This second reaction should not proceed spontaneously. However, it may occur if "coupled" to the first:

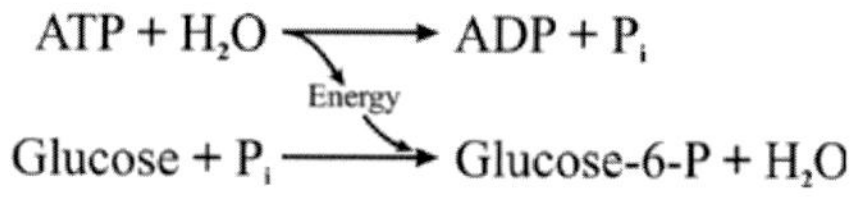

$$ATP + H_2O \searrow \atop Energy \quad ADP + P_i$$
$$Glucose + P_i \longrightarrow Glucose\text{-}6\text{-}P + H_2O$$

The resulting total reaction:

$$ATP + Glucose \rightarrow Glucose\text{-}6\text{-}P + ADP$$

has $\Delta G^{\circ\prime}$ of $(-30.5 + 13.8) = -16.7 \, kJ/mol$ $(-4.0 \, kcal/mol)$, which allows it to proceed in the indicated direction due to the coupling of an endergonic with an exergonic process. The final balance indicates a relatively small energy loss, released to the environment as heat.

In living beings, reactions that require energy are coupled directly or indirectly to ATP hydrolysis. Often the resulting "activated" compounds, as, for example, glucose-6-phosphate, are intermediate products in metabolic pathways of synthesis or degradation.

CHEMICAL KINETICS

Thermodynamics allows explaining the direction in which a given reaction will most probably occur. However, it does not provide information regarding how or through which steps the transformation of reactants into products occurs. *Chemical kinetics* studies the rate and mechanisms of chemical reactions.

In a chemical reaction, the bonds between atoms are broken or formed and interactions among elements in the participant molecules are modified. The reactive molecules must collide with enough energy and proper orientation; in other words, collisions must be effective. These effective collisions determine the rate at which a reaction takes place.

The rate of a reaction is expressed in terms of amount of reagent converted into product per unit time (mol/s).

As already mentioned, at chemical equilibrium, the rate of a reaction is given by the product of the reaction constant and the concentration of the reactants. In the reaction:

$$A + B \underset{2}{\overset{1}{\rightleftharpoons}} C + D$$

the rate of reaction 1 is:

$$v_1 = k_1[A] \cdot [B]$$

k_1 corresponds to the rate constant, which has a defined value for each system under defined conditions.

Reaction Order

The equation rate is more exactly expressed in the form:

$$v = k[A]^x \cdot [B]^y$$

The exponents x and y can be experimentally determined. These exponents commonly have values 1, 2, 0; however, they can also be fractional. Values of x and y indicate the order of reaction, namely, the relationship between the reaction rate and the reactant concentration. If $x = 1$, the reaction is first order with respect to A; if $y = 2$, the reaction is second order with respect to B. The overall order of the reaction is given by the sum of the exponents ($x + y$).

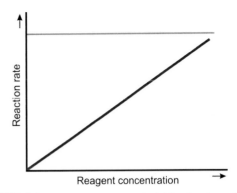

FIGURE 7.1 **Order zero chemical kinetics** *(gray line),* **first-order kinetics** *(red line).*

A reaction is zero order when the products are formed in constant amount, regardless of the concentration of reactants. The representation of the ratio between rate and reagent concentration, gives a horizontal line (Fig. 7.1).

Reactions involving more than one molecule may be of zero order with respect to one of the reactants, but not for all of them. This can be explained if one of the reactants is in limited amount and the other is in excess. It is possible that the reaction rate does not change beyond that determined by the limiting reagent. The rate, in this case, is independent of the nonlimiting reagents concentration; for these, reaction is of zero order.

A *first-order* reaction is that in which the rate of reaction is proportional to the concentration of the reactant. An increase of the reagent concentration proportionally increases the reaction rate (Fig. 7.1).

The concept of reaction order can be related to the number of molecules that must collide simultaneously. If in a first-order reaction, the process involves a single molecule, any molecule with enough free energy is spontaneously converted into product. A bimolecular reaction may also be of first order when one of the reactants is in excess with respect to the other. The reaction rate is proportional to the concentration of the reactant that is in lower amount.

A reaction is second order when the rate is related to the concentration of two reactants. In a second-order reaction, the rate depends on the concentration of two reactants or it may be second order with respect to one of them, in which case the rate is proportional to the square of its concentration. In the second-order reaction, both of the participating molecules need to have sufficient energy, but also must collide in the right direction to form one or several products.

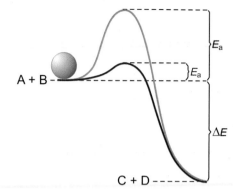

FIGURE 7.2 **Energy changes during the course of an uncatalyzed reaction** (*gray line*) **and a catalyzed one** (*red line*). *Ea*, Activation energy; *ΔE*, change in free energy of the reaction. The highest point reached by the curve corresponds to the energy of the activated state.

Activation Energy

If the thermodynamic data shows that a reaction can take place in a certain direction, that its ΔG is clearly negative and that it is spontaneous, why the reaction does not occur immediately and instead the reagents can remain long time together without reacting? (As occurs, e.g., with sucrose and oxygen.)

Regardless of ΔG, initiation of any reaction requires that energy is provided to the reactants. This can be kinetic energy of translation or rotation, which is needed for the reacting molecules to collide, electronic or vibrational energy, required for molecular reactions, in which a molecule eliminates or redistributes some of its atoms to form the product. In other words, molecules must reach a *transition* or *activation state* before the reaction can take place. The activated compound is an intermediate from which the reaction can spontaneously progress. In this intermediate state, links are rearranged to produce the new grouping of atoms and bonds that will lead to products.

The energy required to reach the activated state is called *activation energy* (Ea). This energy is something like an initial barrier that needs to be overcome before the reaction can proceed.

The rate at which this transition or activated intermediate is produced depends on several factors: (1) the energy difference between the initial state of the reactive molecules (basal or ground state) and that of the activated state;

this difference is the *activation energy*. A high activation energy generally corresponds to a low reaction rate; (2) number of effective collisions; when the frequency of collisions between reactant molecules is low, the rate is reduced, and (3) the need to properly orient the molecules involved in the formation of the activated state.

The course of a chemical reaction can be schematized using a mechanical analogy (Fig. 7.2). For example, a ball that must move from an A + B position in the side of a mountain to a lower level (C + D). A + B, located in a higher position, potentially have higher energy content than C + D and the ΔG from A + B to C + D has a negative sign. If one imagines that, to reach the bottom position (C + D), the ball must first overcome a terrain elevation (Fig. 7.2), the path cannot be satisfied if the ball is not supplied the necessary kinetic energy to overcome this initial barrier.

Energy changes in a chemical reaction can be represented by differences in the height of the terrain. The difference between A + B and C + D can be compared to the ΔG of the reaction; the difference between the level of A + B and the peak of the initial rise corresponds to the activation energy. The maximum height indicates the energy of the activated transition intermediary.

The magnitude of the activation energy varies for each chemical reaction. In the example given, it would be much greater for the reverse reaction (C + D → A + B).

A chemical reaction can be accelerated if energy is provided to the system, so a higher number of reactant molecules reach the activated state. This can be accomplished, for example, by supplying heat, which increases the internal energy of the reacting molecules and the possibility of effective collisions between them. This feature is commonly used in the laboratory, but it is not possible in biological media, which function under isothermal conditions.

Another manner to accelerate a reaction is to reduce the activation energy, so that a greater number of molecules in the system can reach the transition state. This is the effect produced by the catalysts, which increase the rate of the activated intermediate formation by reducing the activation energy. In the mechanical analogy, the action of a catalyst consists in decreasing the height of the initial rise in the ground. Then, a small impulse will be sufficient for the ball to start rolling. Fig. 7.2 shows curves for the energy changes in a reaction without and with the aid of a catalyst.

While a catalyst increases the reaction rate, it does not alter the net energy change; the value of ΔG and the equilibrium constant are unchanged whether the reaction is catalyzed or not.

SUMMARY

Thermodynamics is the branch of physics that deals with energy and its transformations.
Energy changes in biochemical reactions obey fundamental principles of thermodynamics, are expressed in two laws:
First law. The total energy of the universe remains constant.
Second law. The entropy of the universe increases.
Energy changes occur in chemical transformations.

Heat is one of the most common forms of energy and it is easy to measure.
At constant pressure and temperature, the change in heat is the change in *enthalpy* (ΔH).
Energy change (ΔE) equals ΔH when no work is produced in the system.
Free energy (G) is the fraction of released energy available to perform work.
Free energy change (ΔG) of a reaction is given by the equation: $\Delta G = \Delta H - T \cdot \Delta S$.
T is absolute temperature and ΔS, entropy change, corresponding to energy that is not available to perform work.
Open system. From a thermodynamic viewpoint, a living organism or cell is an open system, constantly exchanging matter and energy with its surroundings.
A reaction occurs spontaneously when its ΔG is negative.
The force that determines the direction of a reaction is the tendency to increase entropy, which is equated with increasing disorder.
All processes occur with free energy decrease until *equilibrium* is reached, in which G is minimal.
The ΔG for a reaction is related to the value of the equilibrium constant.
Under standard conditions (298°K, 1 M concentration of reactants), the free energy change ($\Delta G°$) is: $\Delta G° = -RT \cdot \ln K_{eq} = -RT (2.303) \log K_{eq}$.
When referring to the standard energy change at pH 7.0, the notation is $\Delta G°'$.
If $K_{eq} > 1$, $\Delta G°$ will be negative; if $K_{eq} < 1$, $\Delta G°$ will be positive. In the first case the reaction is spontaneous.
Reactions with negative ΔG are *exergonic*; those with ΔG positive are *endergonic*.
When the system is at equilibrium, $\Delta G = 0$.
Knowing the value of $\Delta G°$ of a reaction and the concentrations of reactants and products, it is possible to determine the direction in which the reaction will proceed, since: $\Delta G = \Delta G° + RT \cdot \ln[\text{Products}]/[\text{Reagents}] = \Delta G° + RT \cdot 2303 \cdot \log[\text{Products}]/[\text{Reagents}]$.
The reaction occurs spontaneously if ΔG is negative.
When $\Delta G°$ is positive, if the value of log [Products]/[Reagents] becomes sufficiently negative, it will have a negative ΔG.
When [Products] = [Reagents], $\Delta G = \Delta G°$.
Coupled reactions. Two successive reactions are coupled when the product of one of them is a reagent for the next. In this case, the overall process ΔG equals the sum of ΔG of individual reactions.
In biochemical transformations, endergonic reactions are possible by coupling with other sufficiently exergonic, so that the overall ΔG is negative.
ATP, an energy-rich compound, participates in coupled reactions; its $\Delta G°'$ of hydrolysis (-30.5 kJ/mol) makes ATP an excellent energy transfer mediator.

Chemical kinetics.

From a kinetic standpoint, reactions can be classified according to the relationship between the reaction rate and the concentration of reactants.

A reaction is *zero order* when the rate of reaction is constant regardless of the reagent concentration. It is *first order* when the rate is directly proportional to the reactant concentration.

Activation energy is the energy needed by the reactants to reach the *transition* or *activated state*, from which the reaction can proceed spontaneously.

Catalysts reduce the activation energy.

Bibliography

Alberty, R.A., 1994. Biochemical thermodynamics. Biochim. Biophys. Acta 1207, 1–11.

Hammes, G.G., 2000. Thermodynamics and Kinetics for the Biological Sciences. John Wiley and Sons, New York, NY.

Haynied, T., 2001. Biological Thermodiynamics. Cambridge University Press, Cambridge.

Newsholme, E.A., Leech, A.R., 1986. Bioquímica Médica. Interamericana, Mexico.

8

Enzymes

Countless chemical reactions take place at a given time in every living being. Many of them transform exogenous substances, which come with the diet, to obtain energy and the basic materials that will be used for the synthesis of endogenous molecules.

Biochemical transformations are performed at a remarkable fast rate and with great efficiency. To reproduce them in the laboratory, these reactions would need extreme changes in temperature, pH, or pressure to take place; these changes are not compatible with cell survival. Under normal physiological conditions (37°C for warm-blooded organisms, pH near neutrality, and constant pressure), most of the reactions would proceed very slowly or may not occur at all. It is the presence of catalysts that allow chemical reactions in living beings to occur with great speed and under the mild conditions that are compatible with life.

ENZYMES ARE BIOLOGICAL CATALYSTS

A catalyst is an agent capable of accelerating a chemical reaction without being part of the final products or being consumed in the process. In biological media, macromolecules called *enzymes* act as catalysts.

As any catalyst, enzymes work by lowering the reaction activation energy (A_e) (see p. 152).

Enzymes are more effective than most inorganic catalysts; moreover, enzymes show a greater specificity of effect. Usually inorganic catalysts function by accelerating a variety of chemical reactions, whereas enzymes catalyze only a specific chemical reaction. Some enzymes act on different substances, but generally, these are compounds with similar structural characteristics and the catalyzed reaction is always of the same type.

The substances that are modified by enzymes are called *substrates*.

The specificity of enzymes allows them to have high selectivity to distinguish among different substances and even between optical isomers of a compound. For example, glucokinase, an enzyme that catalyzes D-glucose phosphorylation, does not act on L-glucose.

NOMENCLATURE AND CLASSIFICATION OF ENZYMES

Enzymes are often named by adding the suffix *-ase* to the name of the substrate that they modify. For example, amylase, urease, and tyrosinase are enzymes that catalyze reactions involving starch, urea, and tyrosine, respectively. Enzymes are also designated by the type of reaction catalyzed, for example, dehydrogenases and decarboxylases catalyze hydrogen and carboxyl removal from different substrates, respectively.

Certain enzymes have arbitrary names; among them are saliva ptyalin, gastric pepsin, and pancreatic trypsin and chymotrypsin.

The confusion created by the use of names according to different criteria, led the International Union of Biochemistry and Molecular Biology (IUBMB) to propose a classification system, to assign each enzyme a descriptive name and a number that allows its unequivocal identification. In this classification, six main classes of enzymes are considered according to the type of reaction catalyzed. Each class is divided into subclasses and sub-subclasses. The numeric code used to identify the enzyme consists of four components: the first number corresponds to the main enzyme class, the second refers to the subclass, the third denotes the sub-subclass (these numbers are assigned taking into account the nature of the atom groups involved in the reaction), and the fourth represents the enzyme order number in its sub-subclass. Periodically the IUBMB publishes the nomenclature of enzymes giving their systematic name, type of substrates used, and reaction catalyzed; the common or recommended trivial name and the code number is also indicated. To accurately identify an enzyme, the code number must be mentioned. In this book, the trivial name recommended by the IUBMB will be used, except for cases, in which the use has imposed another denomination.

The six major groups of the international classification are:

1. *Oxidoreductases.* These enzymes catalyze redox reactions. They are associated with coenzymes (mentioned later) and include dehydrogenases, oxidases, peroxidases, and oxygenases.

    Dehydrogenases use hydrogen donors as the substrates and a coenzyme as the acceptor. They are designated with the name of the substrate they modify, followed by the word dehydrogenase. Less frequently, when the reverse reaction is considered, they are called *reductases*.

Oxidases are enzymes that catalyze reactions in which the hydrogen acceptor is molecular oxygen (O_2); oxygen atoms do not appear in the oxidized product. *Peroxidases* are enzymes utilizing H_2O_2 to oxidize a substrate, H_2O_2 is converted into H_2O.

Oxygenases incorporate oxygen to the substrate. They can be distinguished as *dioxygenases* and *monooxygenases*. Dioxygenases catalyze reactions in which both O_2 atoms are placed in the substrate. Monooxygenases only incorporate one atom from O_2 in the substrate, the other is reduced to form H_2O. They require a cosubstrate, which provides hydrogen atoms. Commonly a hydroxyl group is formed in the substrate, which gives them the alternative name of *hydroxylases*. Other enzymes in this group are the *mixed function oxidases*, which oxidize two substrates simultaneously.

An example of oxidoreductase is lactate dehydrogenase. Its systematic name is L-lactate:NAD oxidoreductase; its recommended trivial name is lactate dehydrogenase and the code number is 1.1.1.27. This enzyme catalyzes the oxidation of lactate to pyruvate, as well as the reverse reaction (reduction of pyruvate to lactate). The enzyme uses nicotinamide adenine dinucleotide (NAD) as coenzyme. The catalyzed reaction is:

$$
\begin{array}{ccc}
CH_3 & & CH_3 \\
| & & | \\
HOCH & + NAD^+ \rightleftharpoons & C=O \ + NADH + H^+ \\
| & & | \\
COO^- & & COO^- \\
\text{Lactate} & & \text{Pyruvate}
\end{array}
$$

2. *Transferases* are enzymes that catalyze the transfer of a group of atoms, such as amine, carboxyl, carbonyl, methyl, acyl, glycosyl, and phosphoryl from a donor substrate to an acceptor compound. An example includes

L-aspartate:2-oxoglutarate aminotransferase, or aspartate aminotransferase (code number, 2.6.1.1.). This is an aminotransferase or transaminase, which catalyzes the transfer of an amine group from one compound to another through the following reaction:

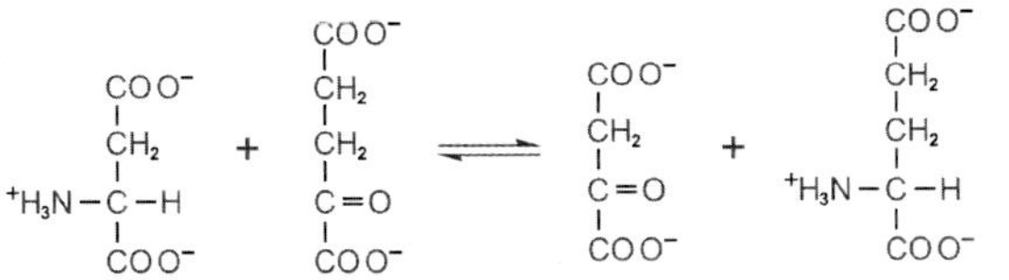

| L-Aspartate | 2-Oxoglutarate α-Ketoglutarate | Oxaloacetate | L-Glutamate |

3. *Hydrolases* are enzymes that catalyze the cleavage of C—O, C—N, C—S, and O—P bonds by addition of water. Their recommended designation results from the name of the substrate and the suffix "-ase." For example, acetylcholinesterase and ribonuclease belong to this group; they hydrolyze the ester bond between acetate and choline in acetylcholine and the links between nucleotides in ribonucleic acid (RNA), respectively. Another example is arginase or L-arginine amidino hydrolase (code number, 3.5.3.1), which catalyzes the hydrolysis of arginine to form urea:

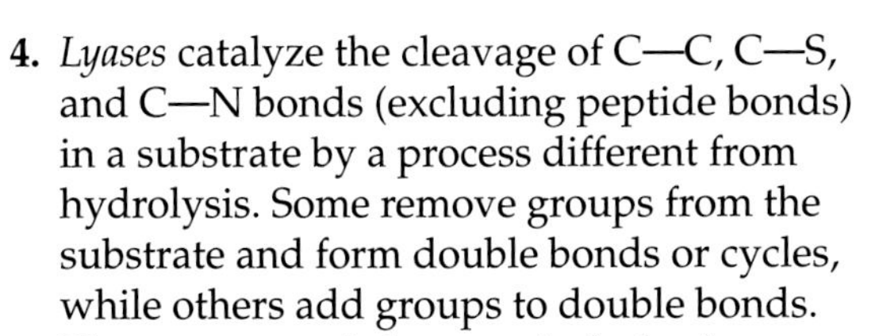

| L-Arginine | Urea | L-Ornithine |

4. *Lyases* catalyze the cleavage of C—C, C—S, and C—N bonds (excluding peptide bonds) in a substrate by a process different from hydrolysis. Some remove groups from the substrate and form double bonds or cycles, while others add groups to double bonds. The recommended names include, for example, *decarboxylases, dehydratases, aldolases*

when they remove CO_2, water, or aldehyde, respectively. When the reverse reaction (addition of a group) is performed, they are named *synthase* to stress the function of synthesis of these enzymes. Synthases do not require nucleoside triphosphate (ATP, GTP, etc.) as energy suppliers. An example of a lyase is aldolase, which cleaves fructose-1,6-bisphosphate into two trioses phosphate.

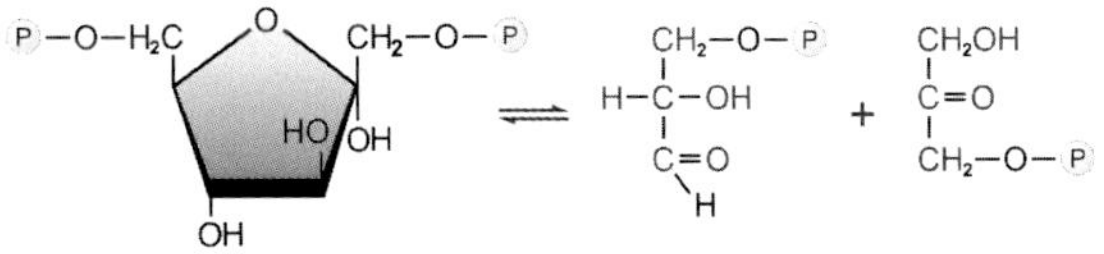

| Fructose-1,6-biphosphate | D-Glyceraldehyde-3-phosphate | Dihydroxyacetone phosphate |

The systematic name of aldolase is fructose-1,6-bisphosphate:D-glyceraldehyde-3-phosphatelyase; its recommended trivial name, fructose-bisphosphate aldolase; and its code number, 4.1.2.13.

5. *Isomerases* interconvert isomers of any type: optical, geometric, or positional. Some of them have trivial names, such as epimerase, racemase, *cis–trans* isomerases, cycloisomerase, and tautomerase. Within this group of enzymes is phosphoglucoisomerase, which catalyzes the interconversion of glucose-6-phosphate 1 to fructose-6-phosphate. Another isomerase is triosephosphate isomerase or D-glyceraldehyde-3-phosphate ketol isomerase (code number, 5.3.1.1), which catalyzes the following reaction:

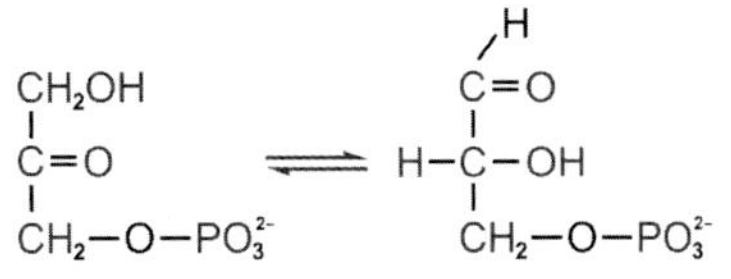

| Dihydroxyacetone phosphate | D-Glyceraldehyde-3-phosphate |

6. *Ligases* catalyze the binding of two molecules in a process that is coupled to the hydrolysis of a high-energy bond of a nucleoside

triphosphate. The resulting reactions form C—C, C—S, C—O, or C—N bonds. These enzymes are also usually designated synthetases; however, this name must be reserved for those enzymes that strictly meet the previous definition, and should not be confused with the synthases. To avoid errors, it is recommended that these enzymes be called ligases. An example is glutamine synthetase, which catalyzes the reaction that uses glutamic acid and ammonia to form glutamine. The energy required for the synthesis is provided by hydrolysis of ATP:

$$\text{COO}^- \quad \text{CH}_2 \quad \text{CH}_2 \quad {}^+\text{H}_3\text{N}-\text{C}-\text{H} \quad \text{COO}^- + NH_4^+ + ATP \rightarrow \text{CO}-\text{NH}_2 \quad \text{CH}_2 \quad \text{CH}_2 \quad {}^+\text{H}_3\text{N}-\text{C}-\text{H} \quad \text{COO}^- + ADP + P_i$$

L-Glutamate (Ammonium) → L-Glutamine

CHEMICAL NATURE OF ENZYMES

In 1926, James B. Sumner purified and crystallized an enzyme, *urease*, for the first time. Urease catalyzes the hydrolysis of urea to ammonia and carbon dioxide. Later, this enzyme was found to be a protein. Subsequently, many enzymes have been obtained in a pure state and their structures have been characterized. As a result of these initial studies, the concept that all enzymes are proteins developed. However, more recently, it was found that another type of molecule, RNA, also have catalytic activity. These molecules are called *ribozymes* (p. 501).

Current techniques have allowed us to understand the exact structure and build accurate models of many enzymes. This has shed light on the mechanism of enzyme action. Some enzymes are composed by only amino acids. Hydrolases, in general, are simple proteins. Other enzymes are formed by the association of various subunits or polypeptide chains, constituting oligomers. Often, the association between the subunits that form an enzyme is of functional significance.

Coenzyme. Many enzymes only perform their catalytic role when associated with another nonprotein molecule, of relatively small size, called a *coenzyme*. Coenzymes can be firmly attached to the enzyme by covalent or other strong bonds, forming complexes that are difficult to separate. Some authors prefer to call them *prosthetic groups* and reserve the name coenzyme for those chemical groups that are more loosely associated to the protein. In this textbook, the name coenzyme will be used indistinctly. Both the protein and nonprotein portions are essential for enzyme activity. The entire system is called *holoenzyme* and consists of the protein or *apoenzyme* (a macromolecule, thermolabile and nondialyzable) and the *coenzyme* (nonprotein molecule, thermostable and of relatively small size).

Holoenzyme = Apoenzyme + Coenzyme
Whole enzyme   Protein   Nonprotein
Thermolabile   Thermostable
Does not dialyze

Oxidoreductases, transferases, isomerases, and ligases require coenzymes. These are actively involved in the reaction, undergoing changes that compensate for the transformation undergone by the substrate. For example, the coenzymes in oxidoreductases accept or transfer the hydrogens or electrons subtracted from or donated to the substrate. Transferases have coenzymes, which accept or donate the group transferred in the reaction.

Many coenzymes present structures that resemble nucleotides. Moreover, coenzymes are related to *vitamins*, compounds that the body cannot synthesize and must be supplied with the diet. B complex vitamins function as coenzymes themselves or form part of the structure of coenzymes. This participation in enzymatic processes gives many vitamins their physiological importance. Table 8.1 lists different coenzymes and the vitamin they are related to. The functional role of vitamins will be described in Chapter 27.

Although they are always involved in a particular type of reaction, a coenzyme can associate

TABLE 8.1   Coenzymes

Name	Vitamin
Thiamine pyrophosphate	Thiamine
Pyridoxal phosphate	Pyridoxine
Biotin	Biotin
Flavin adenine mononucleotide (FMN)	Riboflavin
Flavin adenine dinucleotide (FAD)	Riboflavin
Nicotinamide adenine dinucleotide (NAD)	Niacin
Nicotinamide adenine dinucleotide phosphate (NADP)	Niacin
Coenzyme A	Pantothenic acid
Tetrahydrofolic acid	Folic acid
Coenzyme B_{12}	Vitamin B_{12}

to different apoenzymes and act on different substrates. For example, lactate dehydrogenase, malate dehydrogenase, glutamate dehydrogenase, and oxidoreductases use the same coenzyme, NAD. This coenzyme accepts hydrogens removed from the substrate. It is not the coenzyme but the apoenzyme that is responsible for the accurate recognition of the substrate, and which gives specificity to the holoenzyme.

METALLOENZYMES

Metal ions are essential for the catalytic action of some enzymes. Metal ions contribute to the catalytic process through their ability to attract or donate electrons. Some metals bind the substrate by coordination links. Others contribute to maintain the tertiary and quaternary structures of the enzyme molecule. In all metalloenzymes, removal of the metal component causes loss of activity. The following are examples of these types of enzymes:

Catalase, peroxidase, and cytochrome are hemoproteins in which iron is essential for enzymatic activity.

Tyrosinase, ascorbic acid oxidase, and cytochrome oxidase (which also has Fe) contain copper.

Carbonic anhydrase and alcohol dehydrogenase are enzymes that contain zinc as part of their molecule.

Xanthine oxidase (which also has Fe) and other oxidases and dehydrogenases contain molybdenum.

Magnesium is required by enzymes that use ATP as cofactor. The active form of ATP is an ATP–Mg^{2+} complex, where Mg^{2+} binds to the negatively charged phosphate group of ATP.

Manganese is essential for the action of acetyl-CoA carboxylase, DNAase, and other enzymes.

Glutathione peroxidase is covalently bound to selenium.

Many enzymes require Ca^{2+} or are activated by it.

The activity of some enzymes depends on the presence of certain ions in the medium, such as the cations Na^+ and K^+, and the anion Cl^-.

ENZYMATIC CATALYSIS

As mentioned in the previous chapter, enzymes increase the rate of a reaction by lowering the activation energy (A_e). Thus, more molecules reach the intermediate or transition state, and the chemical transformation is accelerated. Enzymes greatly increase the reaction rate and, like any catalyst, they do not modify the net energy change or the equilibrium constant of the reaction.

During the course of the reaction, an enzyme actually attaches to the substrates forming a transient complex. Eventual modifications of the molecule during such binding are ephemeral, meaning the enzyme appears unchanged at the end of catalysis.

If an enzyme E catalyzes the transformation of substrate S to product P, enzyme and substrate form a complex ES, which then dissociates into enzyme and product:

$$E + S \rightleftharpoons ES \longrightarrow E + P$$

Enzyme   Substrate   Enzyme–substrate   Enzyme   Product

Formation of the ES complex, initially proposed on theoretical grounds, has been demonstrated experimentally. In the course of the reaction, the enzyme effectively binds to the substrate. Finally, the enzyme is unchanged, and can again bind to another substrate molecule. This explains why very small amounts of enzyme greatly accelerate the reaction rate. The same molecule is reused many times.

ACTIVE SITE

To form the ES complex, the substrate is attached to a defined place on the enzyme. This region of the molecule has received the names of *active site, active center, catalytic site,* or *substrate site.* This is where the catalytic action of the enzyme is accomplished (Fig. 8.1).

The active site has binding and catalytic sites. The substrate is placed in the binding site in such a manner that the place that will be modified in the substrate is positioned exactly at the catalytic site. Both the binding and catalytic actions require a highly specific three-dimensional conformation at the level of the active site, where the side chains of amino acid residues

play an essential role. For example, the reactive side chains of cysteine, glutamate, aspartate, lysine, arginine, histidine, serine, threonine, and hydrophobic residues play an important role in substrate binding.

The active site is composed of a small number of amino acids, which are precisely distributed. This spatial distribution is maintained by the primary, secondary, tertiary, and quaternary structures of the enzyme. Amino acid residues that participate in the formation of the active site are sometimes located at distant positions in the polypeptide chain and converge to a restricted area in appropriate positions due to the folding and twisting of the polypeptide chain.

The importance of certain amino acid residues in the catalytic capacity of enzymes is highlighted by the observation that patients, who suffer from metabolic diseases, exhibit mutations that alter the primary structure of an enzyme (i.e., inborn errors of metabolism).

Binding of the substrate to the enzyme involves noncovalent bonds, such as hydrogen bonds, ionic attractions, hydrophobic bonds, and van der Waals interactions. The chemical groups at the active site are spatially arranged to face and interact with groups in the substrate, fixing them in the proper position. During the course of the reaction, transient covalent bonds between enzyme and substrate are also formed.

The substrate bound to the enzyme undergoes a conformational change, acquiring a tense

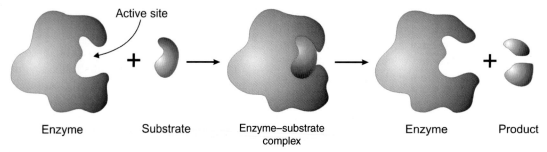

Enzyme   Substrate   Enzyme–substrate   Enzyme   Product
                                        complex

FIGURE 8.1   **Schematic representation of an enzymatic reaction.**

state, from which it easily passes to form the product(s). This state of tension, or intermediary transition, of the substrate is the mechanism by which the enzyme reduces the activation energy required to catalyze the reaction.

So far only reactions with one substrate have been mentioned. However, many chemical reactions catalyzed by enzymes involve two or more different substrates. In these cases, the active site provides a niche in which each substrate is located in the most favorable orientation to react, promoting formation of the transition state, reducing the activation energy, and increasing the reaction rate.

There is always a great difference in size between the enzyme and the substrate. Even if the substrate is a macromolecule, its binding site is small when compared to the enzyme as a whole. Many authors wonder why, throughout evolution, macromolecules (RNA and proteins), and not smaller molecules, have been selected to function as enzymes. The most satisfactory explanation to this question is that it would be very difficult for small molecules to obtain the three-dimensional configuration necessary to adapt a given substrate and create the precise niche in the enzyme for its modification. While the active site is a small portion of the molecule, the whole enzyme contributes to maintain a proper configuration at the catalytic site.

The coenzyme is also involved in ensuring the optimal conformation of the enzyme. It binds to the enzyme in specific places that are usually near the active site, or form part of the substrate site.

In the late 19th century, E. Fischer proposed a hypothesis to explain enzyme–substrate interaction by comparing it with the precise fit existing between a key and its lock. This requires a perfect structural complementarity for accurate assembly (Fig. 8.2).

The "lock and key" hypothesis is useful to explain the mechanism of action of enzymes that have strict specificity, but the model requires

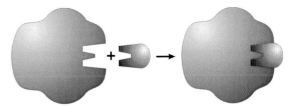

FIGURE 8.2 Schematic representation of the "lock and key" hypothesis for the formation of the enzyme–substrate complex.

a level of rigidity that is incompatible with the current knowledge of molecular structure and conformation of macromolecules. There are agents that cause conformational changes in the enzyme, increasing its catalytic activity. Currently, another hypothesis, proposed by Koshland, is more accepted. This "induced fit" hypothesis introduced the concept of *adaptive* or *induced fit*, which takes into consideration that the enzyme structure is not rigid, but plastic. This flexible model allows the enzyme to modify its conformation when in contact with the substrate, adapting to it and orienting essential residues to obtain the optimal conformation to form the ES complex. Only the appropriate substrate can induce the precise conformational change needed for catalysis. When the substrate binds to the enzyme, it produces conformational changes in the molecule; only then do the critical functional groups rearrange to ensure the most effective location (Fig. 8.3).

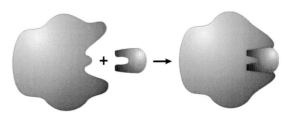

FIGURE 8.3 Schematic representation of the adaptation or "induced fit" hypothesis for the formation of the enzyme–substrate complex.

ZYMOGENS

Some enzymes are synthesized in the cell in an inactive state called *zymogens* or proenzymes. In most cases, these precursors are simple proteins that become catalytically active by hydrolysis. Specific agents, frequently hydrolytic enzymes, break the polypeptide chain of the zymogen, changing the conformation of the molecule, allowing it to acquire catalytic activity.

Some components of the digestive secretions are produced as zymogens and are activated when they reach the lumen of the gastrointestinal tract (p. 258). Early activation of these zymogens, prior to their arrival to the intestinal tract, may cause damage to the organs they are released from. Other zymogens are present in plasma and are precursors of proteolytic enzymes involved in the process of blood coagulation (Chapter 31). Any alterations in the activation of these proenzymes may lead to bleeding disorders or coagulopathies.

GENETIC ALTERATIONS THAT AFFECT ENZYME FUNCTION

Alterations of the amino acid sequence of an enzyme, either at the active site or at any critical position of the molecule, can alter its conformation and activity.

This is the cause of many genetic diseases, known as "inborn errors of metabolism." DNA mutations can lead to the synthesis of abnormal enzymes that are less efficient, completely inactive, or unable to respond to physiological demands. These abnormalities can interfere with a metabolic pathway in which the affected enzyme participates, usually producing serious disorders. Examples of such diseases will be mentioned in the chapters devoted to metabolism (Chapters 14–18).

ENZYME DISTRIBUTION WITHIN THE CELL

In general, enzymes are synthesized in the cell cytoplasm and are then exported to the place where they fulfill their mission. In addition, there are enzymes that are secreted and act outside the cell, such as those of the digestive system or those related to blood coagulation.

The vast majority of enzymes are intracellular. They are not distributed randomly, but localized to different intracellular compartments, where they efficiently exert their specific functions. The intracellular distribution of enzymes can be studied in cell fractions isolated by centrifugation. The difference in density between organelles produces their sedimentation at different centrifugal forces. This allows obtaining cell fractions comprised predominantly by nuclei, mitochondria, lysosomes, endoplasmic reticulum membranes, and components of the cytosol, which can then be used to study different enzymes. Histochemical methods are also used that apply specific stains to reveal the location of enzymes in tissue sections.

These techniques have helped understanding that: (1) many enzymes associated with the nucleus are involved in the maintenance and function of the genetic apparatus, (2) enzymes in mitochondria participate in oxidative reactions that supply energy, (3) hydrolases in lysosomes, which degrade molecules at the end of their useful life, work at a pH that is more acidic than that of other cell enzymes, (4) enzymes involved in protein synthesis are found in ribosomes, (5) enzymes in microsomal fractions, formed by fragments of the endoplasmic reticulum, are involved in the synthesis of complex lipids and in metabolism and inactivation of foreign substances, (6) enzymes in the Golgi complex, consisting of flattened sacs, catalyze the synthesis of oligosaccharides and the glycosylation of proteins and lipids, (7) enzymes in the cytosol are implicated in glycolysis, the main pathway of glucose utilization, and in the biosynthesis of fatty acids and other compounds; (8) many enzymes of the plasma membrane are involved in different mechanisms for the transport of solutes.

MULTIENZYME SYSTEMS

In some cases, organized complexes consisting of several different enzymes with complementary actions are formed. *Multienzyme systems* are arranged (usually within membranes) in such a way that the product of the reaction catalyzed by the first enzyme is received as a substrate by the second enzyme and from this to subsequent enzymes. The chemical transformations occur in the sequence determined by the spatial arrangement of the catalysts. An example of these systems is the electron transport or "respiratory chain" associated with the mitochondrial inner membrane (p. 185).

There are also *multifunctional enzymes*, which present several different catalytic sites in a single polypeptide chain. An example of this type of enzyme is fatty acid synthase.

DETERMINATION OF ENZYME ACTIVITY

The activity of an enzyme can be determined in a medium by measuring the amount of product formed or the amount of substrate consumed by the enzyme in a given time.

The activity of an enzyme directly depends on the total amount of enzyme present in the sample and is not significantly influenced by changes produced in the mixture during the reaction if activity is measured at *initial velocity*. Initial velocity refers to the time in the enzymatic reaction in which the amount of substrate used by the enzyme is still negligible compared to the total substrate existing in the mixture. It is accepted that enzyme activity is determined at initial velocity before the substrate consumption has reached 20% of the total originally present in the sample; however, it is preferable to set initial velocity at a lower limit (about 5%).

To define the activity of an enzyme in a preparation, various practical expressions are used. The amount of enzyme is usually specified in International Units (IU). The amount of enzyme present in a given volume of sample (the enzyme concentration), is commonly used and expressed in international units per milliliter of preparation. A unit of any enzyme is the amount of enzyme that catalyzes the conversion of one micromole (1 μmol = 10^{-6} mol) of substrate per minute under defined conditions of pH, temperature, and pressure.

Specific activity indicates the relative purity of the enzyme preparation. It correlates enzymatic activity, not to the volume, but to the total protein content present and it is indicated as units of enzyme per milligram of protein present in the sample. When the enzyme is in a pure state, its activity can be expressed per milligram of enzyme and, if its molecular weight is known, one can calculate the *molar activity*, *catalytic constant*, or *turnover number* of the enzyme. This corresponds to the number of substrate molecules converted into product by an enzyme molecule working under conditions of substrate saturation, in a unit time (second or minute). Examples of enzymes with high turnover number comprise catalase (which uses H_2O_2 as a substrate), with a molar activity of 40,000,000 s^{-1}, and carbonic anhydrase which catalyzes the reversible reaction between carbon dioxide and water to form carbonic acid, at a molar activity of 400,000 s^{-1}.

FACTORS THAT MODIFY ENZYME ACTIVITY

Enzyme concentration. When the initial velocity of an enzymatic reaction is determined at different concentrations of the enzyme, in the presence of saturating amounts of substrate and keeping constant all other conditions in the reaction medium, the relationship between enzyme amount and rate of reaction (equivalent to enzymatic activity) can be established. This

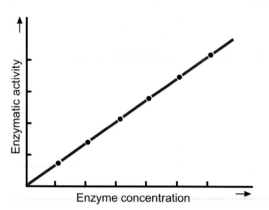

FIGURE 8.4 **Effect of enzyme concentration on the reaction rate or activity of an enzyme.**

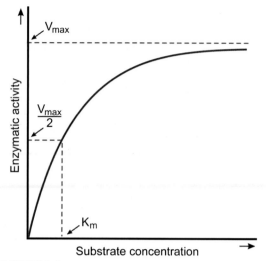

FIGURE 8.5 **Effect of the substrate concentration on enzymatic activity.**

relationship is maintained only if initial velocity is determined. The reaction should not be prolonged beyond the time required to consume more than 10% of the substrate originally present. If the results are plotted on a graph (enzyme concentration on the y-axis and enzyme activity on the x-axis), a curve similar to that shown in Fig. 8.4 is obtained. As indicated in the graph, enzyme activity is directly proportional to the enzyme concentration. Commonly used methods to determine the amount of enzyme present in a sample are based on this direct relationship.

Substrate concentration. If enzyme activity is determined at a single enzyme concentration, maintaining all other reaction conditions constant, but under varying substrate concentrations, a plot, such as the one shown in Fig. 8.5 is obtained. As shown, the relationship between substrate concentration and enzyme activity follows a hyperbola.

When the substrate concentration is low, enzyme activity increases linearly with the substrate concentration. This part of the curve reflects a first-order reaction and reflects a direct relationship between the enzyme reaction rate and the substrate concentration. As the substrate increases, the enzyme rate augments, reaching a point at which the activity does not further increase, even if the concentration of substrate continues to raise. The curve tends to become horizontal at a point that corresponds to

the maximum velocity (V_{max}) of the enzyme. The hyperbolic curve is characteristic of reactions in which a single substrate is involved in the reaction. When two substrates participate in the reaction, a hyperbolic curve can also be obtained for one of them if an excess concentration of the other substrate is used. The substrate–activity relationship is a well-known concept since the early 20th century, when Michaelis and Menten first described it. In the simplest case, the substrate rapidly binds to the enzyme in a reversible reaction. The complex formed dissociates more slowly than the first reaction and the enzyme releases the product:

$$E + S \rightleftharpoons ES \rightleftharpoons E + P$$

At very low substrate concentrations, most of the enzyme (E) molecules are free. As the substrate (S) increases, a greater number of enzyme molecules are involved in forming the enzyme–substrate complex (ES). If the substrate concentration continues increasing, a point is reached in which virtually all the enzyme molecules are occupied by the substrate (assuming the concentration of enzyme is constant). At this point, the enzyme becomes saturated with the substrate. If the concentration of substrate continues to raise and far exceeds the amount of enzyme, a *steady state* is achieved in which the reaction rate does not increase any further. Any additional increase in substrate can no longer produce increments in enzyme activity and the reaction behaves as *zero order*.

Theoretically, the speed of the reaction reaches a maximum only at an infinite concentration of substrate; the curve never completely achieves a horizontal line corresponding to V_{max}. Therefore, it is not possible to accurately predict the concentration of substrate at which V_{max} will be obtained. To establish a precise relationship between initial velocity and substrate concentration, Michaelis and Menten defined a constant called K_m (Michaelis constant). *K_m corresponds to the substrate concentration at which the reaction rate reaches a value equal to half the maximum.*

Under defined conditions of pH and temperature in the medium, K_m has a particular value for each enzyme, which allows to characterize it.

The hyperbolic enzyme–substrate saturation is described by the following equation derived by Michaelis and Menten:

$$v = \frac{V_{max}[S]}{K_m + [S]} \qquad (8.1)$$

where v corresponds to the initial rate with substrate concentration equal to [S], V_{max} represents the maximum rate, and K_m equals the Michaelis constant for the specific substrate.

From this equation, when [S] is below the K_m, the reaction rate depends on the concentration of substrate (initial portion of the curve where the reaction is *first order* with respect to [S]).

When [S] is much higher than the K_m value, the initial velocity is almost maximal (final portion of the curve, *zero-order* reaction).

When [S] is equal to K_m, the rate of reaction is equal to half the maximal velocity, by replacing, Eq. (8.1) becomes:

$$v = \frac{V_{max}[S]}{K_m + [S]} = \frac{V_{max}[S]}{[S] + [S]} = \frac{V_{max}[S]}{2[S]} = \frac{V_{max}}{2}$$

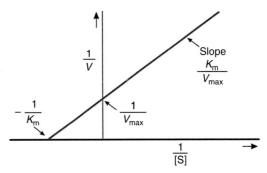

FIGURE 8.6 **Double reciprocal representation of enzyme activity (Lineweaver–Burk plot).** The experimental points are not shown.

The Michaelis–Menten equation can be transformed algebraically into equivalent equations useful for practical determination of the K_m value. Consider taking the inverse of Eq. (8.1):

$$\frac{1}{v} = \frac{K_m + [S]}{V_{max}[S]} = \frac{K_m}{V_{max}[S]} + \frac{[S]}{V_{max}[S]}$$

$$= \frac{K_m}{V_{max}} \cdot \frac{1}{[S]} + \frac{1}{V_{max}} \qquad (8.2)$$

This last equation is known as the Lineweaver–Burk equation and it corresponds to a linear plot.

Representation of the inverse of the initial velocity ($1/v$) versus inverse of substrate concentration ($1/[S]$) yields a straight line (Fig. 8.6). The slope of this line is equal to K_m/V_{max}, the intersection with the vertical axis corresponds to $1/V_{max}$ and intersection with the horizontal axis provides $-1/K_m$. Accordingly, the representation of double reciprocals ($1/v$ as a function of $1/[S]$) allows the determination of K_m and V_{max} values from measurements of enzyme activity at various substrate concentrations. There are other more powerful graphical methods to calculate V_{max} and K_m; however, the Lineweaver–Burk plot is more commonly used.

The K_m value is characteristic for each enzyme and each of its substrates when determined under the same conditions of temperature and pH. Values for different enzymes vary, for example, the K_m of catalase for hydrogen peroxide (H_2O_2)

is 25 mM, while the K_m of hexokinase (the enzyme that catalyzes the phosphorylation of glucose at carbon six) for D-glucose is 0.005 mM.

The K_m value of most enzymes is inversely related to the affinity of the enzyme for its substrate. In general, higher affinity to a substrate is indicated by a lower K_m value.

When an enzyme acts on several substrates, the K_m value is usually different for each of them. The substrate with the lower K_m value is usually considered the natural or "physiological" substrate for that enzyme.

Temperature. The rate of a chemical reaction increases, when temperature rises, as a result of the increase in kinetic energy in the system. Within certain limits, enzyme-catalyzed reactions follow this behavior and the speed of many biological reactions almost doubles for every 10°C increase in temperature. This activity–temperature increase in the reaction rate is called Q_{10} or *temperature coefficient*. If concentrations of enzyme, substrate, and other factors of the reaction medium are maintained constant and the reaction activity is determined at increasing temperatures, the curve shown in Fig. 8.7 is obtained.

Although enzyme activity increases with temperature, a maximum value is reached, which corresponds to the optimal temperature for the

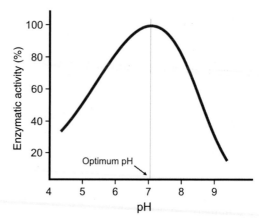

FIGURE 8.8 **Effect of pH on enzyme activity.** Enzyme activity is expressed as a percent of the maximum activity achieved. Experimental points are not shown.

catalytic activity of a given enzyme. Above this temperature, enzyme activity rapidly drops.

For the great majority of warm-blooded animal enzymes, the optimal temperature is approximately 37°C. Beyond that temperature, activity decreases, and approaching 60°C, most of the enzymes are completely inactivated. This inactivating effect of temperature, which occurs above 40°C, is explained by the denaturing action that heat has on the molecular structure of the enzyme.

pH. If enzyme activity is measured at different pH values, maintaining other conditions constant, the effect of the hydrogen ion concentration becomes apparent (Fig. 8.8).

For most enzymes, optimum activity is between pH 6 and 8. Below or above these values, the reaction rate drops more or less rapidly. However, exceptions include gastric pepsin which exhibits maximal activity at acidic pH (around 1.5). Acid phosphatase, abundant in the male prostate, exhibits its greatest activity at pH 5 and alkaline phosphatase from bone and other organs has optimal activity at a pH of 9.5.

The pH changes in the medium affect the state of ionization of functional groups on the enzyme molecule and the substrate. For the formation of the enzyme–substrate complex,

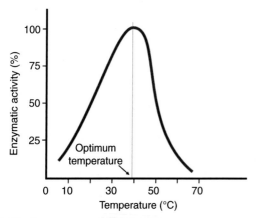

FIGURE 8.7 **Effect of temperature on enzyme activity.** Enzyme activity is expressed as percent of the maximal activity achieved. Experimental points are not shown.

proper distribution of electrical charges in both molecules is required. The optimal pH is that in which the state of dissociation of the essential groups is most appropriate for interaction of enzyme and substrate to form the ES complex. Extreme pH changes, above or below optimal, may cause denaturation of the enzyme and subsequent inactivation.

ENZYME INHIBITORS

There are chemical agents that inhibit the catalytic activity of enzymes. Some of them exert their action by binding to essential sites or functional groups in the enzyme. These substances are useful to study the mechanisms of action of enzymes, the functional groups involved in the catalytic process and the structural requirements for enzyme–substrate binding.

The reduction in enzyme activity caused by inhibitors can be reversible or irreversible.

Irreversible Inhibitors

These inhibitors produce permanent changes in the enzyme molecule causing definitive alteration of its catalytic capacity. Examples of these inhibitors are organophosphates, poisons used as insecticides. They produce irreversible inhibition of acetylcholinesterase, a very important enzyme in the nervous system. Patients intoxicated with organophosphates present with severe and excessive salivation, lacrimation, urination, diarrhea, and emesis.

Among the irreversible inhibitors are "suicide inhibitors," which, due to their structural similarity with the substrate, can occupy the active site on an enzyme and are transformed into products. The products establish covalent bonds that irreversibly block the active site of the enzyme, as if the enzyme had "committed suicide." Suicide inhibitors are very specific. An example is allopurinol, a xanthine oxidase inhibitor that is used in the treatment of gout (p. 419).

Reversible Inhibitors

There are three types of reversible inhibitors: competitive, noncompetitive, and uncompetitive.

Competitive inhibitors increase the value of the Michaelis constant (K_m), but do not modify the maximum velocity (V_{max}) of the enzyme. These effects are achieved by different mechanisms:

In some cases, the inhibitor has structural similarity with the substrate and competes for the active site in the enzyme. A well-known example is that of succinate dehydrogenase inhibition by malonate (ionized form of malonic acid). Succinate dehydrogenase catalyzes the oxidation of succinate, which loses two hydrogens to become fumarate.

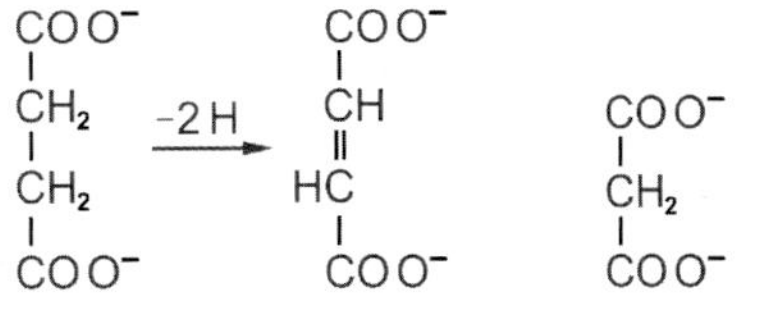

Succinate
Ionic form of
succinic acid

Fumarate
Ionic form of
fumaric acid

Malonate
Ionic form of
malonic acid

Malonic acid has structural similarity with succinic acid; it is a straight-chain dicarboxylic acid, but with one less carbon atom. In the active site of succinate dehydrogenase there are groups that bind the positively charged carboxylates of succinate. Other diacids with two separated carboxylic functions, such as in malonate and some other dicarboxylic acid, can bind to the enzyme's active site. Different from succinate, malonate cannot be dehydrogenated because its carbon chain is different from that of succinate and therefore the action of the enzyme is blocked.

Certain molecules can act as competitive inhibitors binding to the active site of an enzyme even if they do not possess structural similarity with the substrate. An example is salicylate, a competitive inhibitor of alcohol dehydrogenase and 3-phosphoglycerate kinase.

In other cases, inhibitors and substrates bind to different sites on the enzyme, but the binding

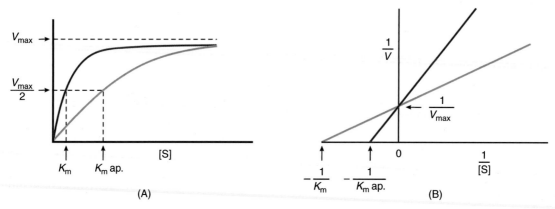

FIGURE 8.9 **Effect of a competitive inhibitor on enzyme's activity** (*red line*, **without inhibitor;** *gray line*, **with inhibitor).** (A) Curve of initial velocity versus substrate concentration and (B) double reciprocal representation.

of one of the molecules prevents the binding of the other, most likely by inducing protein conformational changes. Competitive inhibition can be reversed by increasing the substrate concentration. If the substrate predominates in the mixture, it will tend to displace the inhibitor bound to the enzyme.

An inhibitor's mechanism of action can be studied by determining enzyme activity at various concentrations of the substrate, in the absence and presence of a fixed concentration of inhibitor. The graphical representation of the results provides information on the type of inhibition. Fig. 8.9A shows the dependence of enzyme activity, in the presence of an inhibitor, with increasing concentrations of the substrate. The activity is reduced by the presence of the inhibitor (I) at low substrate concentrations. At very high substrate, the maximum velocity is the same than that reached by the noninhibited enzyme because the substrate displaces the inhibitor from the enzyme's active site. In double reciprocal representation (Fig. 8.9B), the line corresponding to values obtained in the presence of inhibitor intersects the vertical axis at the same point as the line of the noninhibited enzyme. The V_{max} is equal in both cases. In contrast, the intersections with the horizontal axis ($-1/K_m$ value)

are different, indicating that the inhibitor produced an increase of the enzyme's K_m.

As inhibitor and substrate can only bind to free enzyme, they exclude each other. The enzyme bound to the inhibitor is inactive. In the presence of a competitive inhibitor, higher concentrations of substrate are required to obtain the same rate as in its absence. There is an apparent reduction in affinity of the enzyme for the substrate (increased K_m). The reactions in the presence of a competitive inhibitor can be represented as follows:

$$E + S \rightleftharpoons ES \rightarrow E + P$$
$$+$$
$$I$$
$$\updownarrow$$
$$EI$$

Competitive inhibitors have pharmacological application. Some powerful antibacterial chemotherapeutic agents are competitive inhibitors. Many pathogenic microorganisms synthesize folic acid, an essential factor for their development, from *para*-aminobenzoic acid (PABA). Sulfanilamide, a structural analog of PABA, blocks the synthesis of folic acid by competitive inhibition of enzymes using PABA as a substrate. The resulting deficiency of folic acid is fatal for the bacteria.

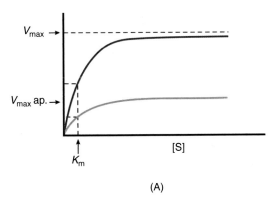
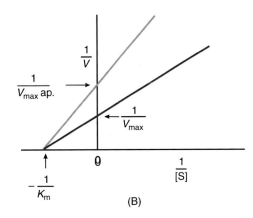

(A)

(B)

FIGURE 8.10 **Effect of a noncompetitive inhibitor on enzyme activity** (*red line*, no inhibitor; *gray line*, inhibitor). (A) Initial velocity curve against substrate concentration and (B) double reciprocal representation.

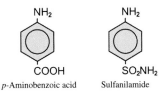

p-Aminobenzoic acid   Sulfanilamide

Folic acid is also a vitamin related to factors that participate in important metabolic processes (p. 675). In humans, folic acid is required by tissues containing cells that actively divide, such as hematopoietic cells. Malignant tumors, with great mitotic activity, also need folic acid. Administration of folic acid analogs, which act as competitive inhibitors of enzymes that utilize folic acid, affects tissues with high cell division rates more intensely. This is the rationale for the use of substances, such as aminopterin and amethopterin (antifolic agents) in the treatment of proliferative diseases. There are many other examples of these types of *metabolic antagonists* used for different therapeutic purposes.

Noncompetitive inhibitors bind to the enzyme at a site different from the active site and they decrease the maximum rate without modifying the Michaelis constant or K_m. This type of inhibition cannot be reversed by increasing the substrate concentration.

In the presence of a noncompetitive inhibitor, the initial velocity curve versus substrate concentration (Fig. 8.10A) shows lower activity at all substrate concentrations. In the double reciprocal representation (Fig. 8.10B), the intersection of the vertical axis with the line of values in the presence of inhibitor indicates decreased V_{max}. In contrast, the intersection of lines with the horizontal axis is the same in the absence and presence of the inhibitor, showing that inhibition does not change the K_m value.

The reactions can be represented by the following equations:

$$\begin{array}{ccccc}
E & + & S & \rightleftharpoons & ES & \rightarrow & E & + & P \\
+ & & & & + \\
I & & & & I \\
\updownarrow & & & & \updownarrow \\
EI & + & S & \rightleftharpoons & ESI
\end{array}$$

The binding of the substrate to the enzyme is not affected by the presence of the inhibitor; the inhibitor binds free enzymes as well as the ES complex. Once the enzyme–substrate–inhibitor (ESI) complex is formed, the enzyme is inactivated. The K_m value does not change, since the substrate binds to the enzyme as it would in the absence of the inhibitor. However, the amount of ES able to release product is reduced and the result is similar to that obtained if there was less enzyme present in the medium.

Reagents that reversibly bind to sulfhydryl groups (—SH) of cysteine residues also belong to this category. Metal ions, such as Cu^{2+}, Hg^{2+}, and Ag^+ inhibit enzymes combining with —SH groups. The binding of the metal ion causes conformational changes that inactivate the enzyme.

Other noncompetitive inhibitors bind to metal components of an enzyme to inhibit it. This is the mechanism of action of cyanide, a powerful compound that binds to iron molecules in cytochromes, catalases, and peroxidases, blocking their activity. Ethylenediamine tetraacetate (EDTA) is a "chelating" agent of divalent cations, which inhibits enzymes that require these ions for activity. Sometimes the mechanism of inhibition is more complex. This is the case of *mixed inhibition*, in which the inhibitor modifies both the K_m and V_{max}.

Anticompetitive or uncompetitive inhibitors represent another type of reversible inhibitors. The graphs of activity versus substrate concentration determined in the presence and absence of an inhibitor of this kind shows reduction in activity at all substrate concentrations (Fig. 8.11A). Double reciprocal representation gives a line parallel to that of the noninhibited enzyme (Fig. 8.11B). The intersection with the vertical and horizontal axes indicates that the inhibitor produces a decrease in both V_{max} and K_m.

The reactions can be represented as follows:

$$E + S \rightleftharpoons ES \rightarrow E + P$$
$$+$$
$$I$$
$$\updownarrow$$
$$ESI$$

The uncompetitive inhibitor binds to the ES complex and forms an inactive ESI complex. There are two reactions that consume ES, one leads to the formation of product and the other to ESI; reaction $E + S \rightleftharpoons ES$ is favored toward the right, appearing as if substrate affinity is increased (decrease in K_m). The activity is diminished because the ES complex bound to the inhibitor is ineffective.

This type of inhibition is observed in cases in which various substrates participate in the reaction. It cannot be reversed by increasing the substrate concentration.

REGULATION OF ENZYME ACTIVITY

The enzyme activity in cells changes constantly and it is adjusted to physiological requirements. Various mechanisms of enzyme regulation have been demonstrated. Enzyme

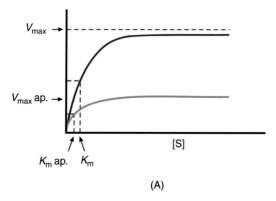

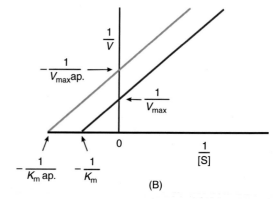

FIGURE 8.11 **Effect of an uncompetitive inhibitor on enzyme activity (*red line*, without inhibitor; *gray line*, with inhibitor).** (A) Plot of initial velocity versus substrate concentration and (B) double reciprocal representation.

activity is directly proportional to the level of substrate. Under physiological conditions, the substrate concentration is below or close to the enzyme's K_m. In this manner, the substrate level determines the degree of enzymatic activity. Increasing the substrate concentration in the cell accelerates its utilization and vice versa.

Transformation of a specific compound in the body is usually accomplished by a series of steps, each catalyzed by a different enzyme. At each step, a new product is formed, which is used as a substrate by the subsequent enzyme. In almost all of these reaction sequences (metabolic pathways) there are one or more enzymes that functions as regulators of substrate and product formation, adjusting the pathway to the needs of the cell. These regulatory enzymes not only fulfill their catalytic function but increase or decrease their activity in response to specific signals.

The enzyme that catalyzes the first step in a metabolic pathway is usually the regulator of the pathway. The following enzymes in the pathway adjust their activity to the availability of substrate regulated by the first reaction.

According to the type of signal to which they respond, regulatory enzymes can be distinguished as *allosteric* or regulated by *covalent modification*.

Allosteric enzymes. In some metabolic pathways, the enzyme that catalyzes the first step in a series of reactions is inhibited by the final product. When the amount of the final product exceeds the requirements, the operation of the pathway is slowed, reducing the activity of the regulatory enzyme. This is called *feedback inhibition*. For example, aspartate transcarbamylase, which catalyzes the first reaction in the biosynthesis of pyrimidine nucleotides, is inhibited by cytidine triphosphate (CTP), the end product of the metabolic pathway.

In other cases, the enzyme is stimulated by compounds that accumulate in the medium. The activator may be the enzyme's substrate. When there is an excess of substrate, the enzyme is activated.

Both inhibition and activation are reversible; when the concentration of the modifying substance diminishes, the enzyme activity is normalized. The modifying agent acts by binding to the enzyme at a site different to the catalytic center, hence the name of this type of regulation (Greek *allo*: other, *stereo*: site or place).

In addition to the catalytic site, allosteric enzymes have other sites where regulatory molecules that affect the catalytic activity are specifically bound. These agents are known as *modulators*, modifiers, or *allosteric effectors*. These effectors are said to be positive when they stimulate and negative when they depress enzyme activity. When the allosteric modulator is the substrate itself, it is known as *homotropic*. When it is different from the substrate, it is called *heterotropic*.

Allosteric enzymes catalyze unidirectional reactions, not reversible ones.

In some cases, several modulators act on the same enzyme, although with opposite effects. Each of the modulators has a binding site for the enzyme (allosteric site) with structural complementarity to ensure specificity (Fig. 8.12).

When the enzyme activity is plotted against increasing concentrations of the substrate, generally a hyperbolic curve (Fig. 8.5) is obtained according to the classical enzyme kinetics. In contrast, allosteric enzymes produce a *sigmoid curve* (Fig. 8.13), similar to that seen in the hemoglobin–oxygen saturation curve (p. 53). In hemoglobin, the bisphosphoglycerate behaves as a negative allosteric modulator. Oxygen is another example of an allosteric modulator. When binding to the first heme molecule within a red blood cell, oxygen causes a conformational change that facilitates the entry of O_2 to the remaining heme molecules. It produces a homotropic cooperative effect similar to that induced by the substrate in allosteric enzymes.

Allosteric enzymes, such as hemoglobin, consist of several polypeptide subunits, among which there is some communication. When a

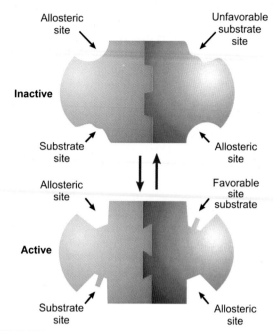

FIGURE 8.12 **Schematic representation of an allosteric enzyme.** It is a dimer that has an inactive and an active conformation, which are in equilibrium. In the active form, the catalytic site acquires a favorable configuration to receive the substrate.

modulator binds to one of the enzyme's subunit, a conformational change is transmitted to the other subunits, modifying the ability of the active site to bind substrate (Fig. 8.14).

Covalent modification. There are also enzymes regulated by covalent addition or subtraction of groups. An example of this is glycogen phosphorylase, the enzyme that initiates the glycogen degradation pathway. This enzyme is in a state of low activity, called phosphorylase *b*, which is converted to active phosphorylase *a* by addition of phosphate to the hydroxyl of serine residues in the enzyme. Phosphorylase *a* is deactivated by removal of phosphate and reverts to phosphorylase *b*.

Covalent regulation is performed in a number of enzymes by a bonding or removal process similar to that mentioned for phosphate. There are enzymes whose activity is modulated by the covalent insertion of other groups. The same enzyme can respond to more than one type of regulation. Phosphorylase, mentioned as an example of covalent regulation, is also an allosteric enzyme responsive to various modulators.

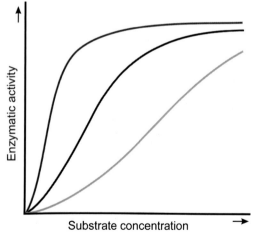

FIGURE 8.13 *Black*: Curve of an allosteric enzyme activity as a function of substrate concentration. *Red*: Effect of the presence of a positive modulator (activator). *Gray*: A negative (inhibitory) modulator.

CONSTITUTIVE AND INDUCIBLE ENZYMES

The absolute amount of a particular enzyme in a cell depends on the ratio between its synthesis and degradation. When both processes are maintained constant throughout the life of the cell, the amount of enzyme remains stable. This is the case of *constitutive* enzymes. For other enzymes, the synthesis is activated or depressed according to the requirements of the cell at a given time. For example, the synthesis of some enzymes used in glucose and amino acid metabolism is stimulated with the presence of their respective substrates. These enzymes are said to be *inducible*. Induction implies de novo synthesis of the enzyme; the synthesis can be activated by hormones (Chapter 26).

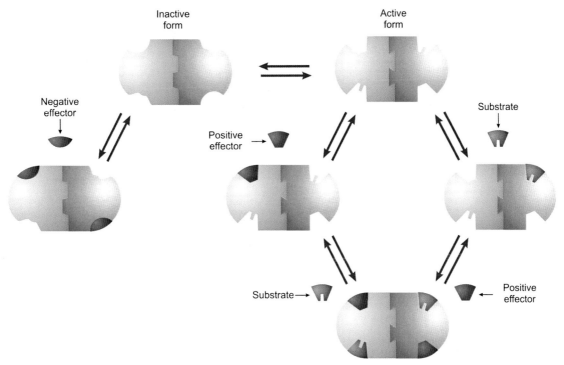

FIGURE 8.14 **Schematic representation of allosteric interactions in a dimeric enzyme.** Sites of the negative and positive effectors are arbitrarily drawn in the same area. On the left, binding of a negative modulator changes the enzyme to the unfavorable or inactive conformation. At the center, binding of a positive modulator promotes the favorable conformation to receive the substrate. On the right, the entry of the substrate into the active site of one of the monomers may also promote favorable conformation in the other catalytic sites of the enzyme (positive homotropic effect). The inactive and active forms are initially in equilibrium. If the positive effector or the substrate concentration increases in the medium, the displacement of the equilibrium toward the active conformation is favored. When the negative modulator increases in the medium, enzyme molecules are converted to the inactive conformation.

ENZYMATIC PROCESSES IN CASCADE

A number of zymogens or proenzymes participate in activation of one another in a chain reaction.

In a system consisting of the pro-A, pro-B, and pro-C zymogens, an initial stimulus triggers the activation of proenzyme A to give product A. Product A will then act as a substrate on pro-B to give B, which in turn catalyzes the conversion of pro-C to active C. In this example, if all three enzymes A, B, and C have a molar activity of 100, a single initial molecule can produce 100 B molecules in 1 min, which will activate 10,000 C molecules in the next minute. As this is a logarithmic progression, the end result is a vast increase in enzyme activity. Examples of these cascades are found in blood coagulation and signal transduction triggered by hormonal stimulation.

Regulatory mechanisms include allosteric effects, covalent modifications, and enzymatic cascades acting on preexisting molecules, all of which are very fast. In contrast, enzyme induction requires de novo protein synthesis, which provides a much slower response.

ISOZYMES

In an organism, and even in a cell, there may be different proteins with the same enzymatic activity. These different molecular forms of an enzyme are called *isoenzymes, isozymes,* or *isoforms*. A widely used method to demonstrate isozymes is gel electrophoresis, followed by specific staining after separation. Proteins with enzyme activity appear as colored bands on the surface of the gel.

A great amount of enzymes exist as multiple isoforms. Lactate dehydrogenase (LDH) is one of those enzymes. It has five isoenzymes in most animal tissues, numbered in accordance with their electrophoretic mobility. LDH 1 is the isoform that migrates faster toward the anode (Fig. 8.15). The relative distribution of enzyme activity between the five isozymes is characteristic for each tissue. Diaphragm muscle extracts show the five isoforms with approximately similar intensity. In contrast, heart predominantly exhibits isozymes LDH 1 and LDH

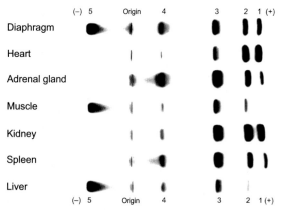

FIGURE 8.15 **Lactate dehydrogenase isozymes of adult human tissues aqueous extracts separated by starch gel electrophoresis and stained with a specific reagent.** The areas in which the dye is deposited correspond to the different molecular forms of the enzyme. Origin indicates where the sample was loaded. The numbers 1–5 indicate location of the corresponding isozyme. Note the predominance of isoforms 1 and 2 in heart and 5 in liver and muscle.

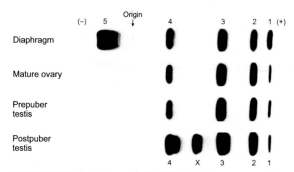

FIGURE 8.16 **Lactate dehydrogenase isozymes of human tissues, separated by starch gel electrophoresis and stained with a specific reagent.** Origin indicates where the sample was inserted. Numbers 1–5 indicate location of the corresponding isoforms. Note the presence of an additional band, between isoenzymes 3 and 4, in postpubertal testis. This fraction corresponds to the X or C4, sperm-specific isozyme.

2, whereas liver and voluntary muscle show LDH 5 as the most abundant form (Fig. 8.15).

All five LDH isoforms are tetramers formed by the associations of two different polypeptide chains, designated A and B or M and H. Isozyme LDH 1 is a homotetramer of B subunits (B_4) and isozyme LDH 5 is composed by four A chains (A_4). The remaining forms correspond to the following associations: LDH 2 = A_1B_3, LDH 3 = A_2B_2, and LDH 4 = A_3B_1.

In adult human testis, an additional isozyme whose electrophoretic mobility is intermediate between fractions LDH 3 and LDH 4 (indicated with an X in Fig. 8.16) is found. This additional form was originally designated as isozyme X, but currently, the name LDH C4 is preferred. LDH C4 is a homotetramer (C4) formed by C polypeptide units, which are different from subunits A and B. It appears in testes of many mammals and birds at the time of sexual maturation and is the main lactate dehydrogenase of sperm.

Although the six molecular forms are all lactate dehydrogenases, each isoenzyme has catalytic peculiarities that confer different functional capacity to different molecular forms. For this reason, the possibility of synthesizing various isozymes gives a tissue great physiological flexibility, as each organ produces the isoforms that are most suitable for its specific requirements.

Other enzymes, for example, aspartate aminotransferase and malate dehydrogenase, present two isoforms with different intracellular locations. One is free in the cytosol and the other is found within mitochondria.

Advances in the area of isozymes have provided valuable information not only for enzymologists, but also for general geneticists and biologists. The determination of molecular forms of enzymes has found application in the clinical laboratory.

DETERMINATION OF ENZYMES IN THE CLINICAL LABORATORY

The clinical biochemistry laboratory frequently uses the determination of enzymes in body fluids or tissue biopsies for diagnostic purposes. The origin of enzymes in plasma or serum will be briefly discussed since those are the fluids in which they are most commonly found.

Blood plasma enzymes. Enzymes circulating in the bloodstream may be commonly found in tissues or may be specific to plasma. Specific plasma enzymes include those which are normally within the bloodstream. Among them are thrombin and plasmin, both of which participate in blood coagulation and fibrinolysis. Ceruloplasmin, a ferroxidase, and cholinesterase are other examples. These enzymes are of clinical interest, especially when their activity is diminished.

Nonspecific plasma enzymes have no function in blood. Normally their concentration in circulation is negligible or zero. Within these enzymes, intracellular and extracellular enzymes can be distinguished.

Extracellular enzymes are normally produced by external secretion glands. Secreted enzymes, such as pancreatic amylase, lipase, and pepsinogen (zymogen of gastric pepsin), are found in plasma at very low concentrations. Their concentration in serum increases with alterations in the passage from the gland of origin to the interstitial space. For example, obstruction of the pancreatic duct or serious inflammatory processes within the duct can increase the levels of amylase and lipase in serum.

Intracellular enzymes are involved in metabolism and are distributed in different cell compartments. The normal plasma membrane does not allow the passage of enzymes, therefore, they remain within the cell and very small quantities are found in blood plasma and interstitial space. Disruption of the cell membrane causes the appearance of these enzymes in plasma. If the cell is altered by severe inflammatory processes or interruption of the oxygen and nutrient supply (lack of blood perfusion), the cells and their membranes deteriorate. More serious processes cause cellular destruction or necrosis. In these cases the cell's content, including enzymes, are released into the interstitial space and from there into the bloodstream. When the number of damaged cells is large, the level of intracellular enzymes in plasma increases markedly. For example, in myocardial infarction, a process in which blood flow to the heart muscle is blocked, tissue destruction and release of enzymes from the injured myocardial fibers to the circulation occurs. A sharp increase in enzymes, particularly aspartate aminotransferase (also called glutamic oxaloacetic transaminase), lactate dehydrogenase, and creatine kinase, is detected due to their abundance in heart muscle. The magnitude and persistence of the increase in these enzymes within the serum have diagnostic and prognostic value in patients with myocardial infarction. Another index of heart damage is the increase of troponin (not an enzyme) in plasma. In liver diseases accompanied by cellular damage, there is an increase in the levels of aminotransferase and lactate dehydrogenase in serum.

If an enzyme is found predominantly in one tissue or organ (such as alcohol and sorbitol dehydrogenases in liver and acid phosphatase in prostate), an increase in plasma level clearly indicates the organ of origin. In contrast, for other enzymes that are distributed across many different tissues, an increase in plasma is not an accurate measure to use in detecting the origin of tissue damage. In these cases, determination of the isoenzyme type helps identify the source of the circulating enzyme. For example, increased isozyme LDH 1 (B4) is characteristic of myocardial damage. On the other hand, liver alterations cause a rise in isoform LDH 5 (A4) in serum. The study of creatine kinase isoforms and isozymes of alkaline phosphatase is also useful in identifying damage to specific organs.

Determination of isoenzymes also helps to estimate the severity of cell damage in a particular disease process. The presence in plasma of isozymes that are normally localized in different subcellular compartments is an indicator of cell necrosis. Examples include malate dehydrogenase, aspartate, and other transaminases, all of which have isozymes which are found both in the cytoplasm and within mitochondria. These enzymes can be detected in plasma during episodes of myocardial infarction or serious hepatitis. In inflammatory processes that have not caused total cell destruction, cytosolic isoforms may be seen in plasma, but mitochondrial forms are rarely found. There are many diseases in which the increase in plasma enzymes is an index for diagnosis and prognosis. Their enumeration is beyond the scope of this book.

In some cases, enzymes are determined in other biological fluids, such as urine and cerebrospinal fluid. There is a group of diseases caused by the genetic inability to synthesize a particular enzyme. The study of enzyme activity in tissue biopsies or blood cells certifies diagnosis and helps to detect patients with these kinds of defects. Prenatal diagnosis can be ascertained by determinations in fetal cells obtained by amniocentesis.

The examples presented previously highlight the importance of the clinical determination of enzymes for diagnosis and prognosis of disease. For a more comprehensive consideration of this topic, the reader is referred to specific literature in this area.

SUMMARY

Enzymes are catalysts that, within the mild conditions of temperature, pH, and pressure of the cells, carry out chemical reactions at amazing high rate. They are characterized by a remarkable efficiency and specificity.

Substrates are the substances on which enzymes act.

Enzymes are named by adding the suffix *-ase* to the name of the substrate that they modify (i.e., urease and tyrosinase), or the type of reaction they catalyze (dehydrogenase, decarboxylase). Some have arbitrary names (pepsin and trypsin). The International Union of Biochemistry and Molecular Biology assigns each enzyme a name and a number to identify them.

Enzymes are classified into six categories according to the type of reaction catalyzed:

Oxidoreductases, transferases, hydrolases, lyases, ligases, and isomerases.

Structurally, the *vast majority of enzymes are proteins.* Also RNA molecules have catalytic activity (*ribozymes*).

Coenzymes are small nonprotein molecules that are associated to some enzymes. Many coenzymes are related to vitamins. Coenzymes and the protein portion with catalytic activity or *apoenzyme* form the *holoenzyme.* The apoenzyme is responsible for the enzyme's substrate specificity. Coenzymes undergo changes to compensate for the transformations occurring in the substrate.

Metalloenzymes are enzymes that contain metal ions.

The mechanism of action of enzymes depends on the ability of enzymes to accelerate the reaction rate by decreasing the *activation energy.* During the course of the reaction, the enzyme (E) binds to the substrate/s (S) and forms a transient *enzyme–substrate complex* (ES). At the end of the reaction, the product/s are formed, the enzyme remains unchanged, can bind another substrate and can be reused many times.

Active site or *catalytic site* is the specific place in the enzyme where the substrate binds. The structural complementarity between E and S allows an exact reciprocal fit. The enzyme adapts to the substrate via a conformational change known as *induced fit.* The presence in the active site of amino acids that bind functional groups in the substrate ensures adequate location of the substrate and formation of the *transition intermediary,* which will be subjected to catalysis.

Zymogens or *proenzymes* are inactive precursors of enzymes. They acquire activity after hydrolysis of a portion of their molecule.

Cellular location of enzymes varies, the majority being in different compartments of the cell, while others are extracellular.

Multienzyme systems are those composed of a series of enzymes or enzyme complexes. There are also *multifunctional enzymes* with several different catalytic sites in the same molecule.

Enzyme activity is determined by measuring the amount of product formed, or substrate consumed in a reaction in a given time. *Initial velocity* corresponds to the activity measured when the amount of consumed substrate is less than 20% of the total substrate originally present. One IU of enzyme catalyzes the conversion of 1 μmol of substrate per second under defined conditions of pH and temperature. *Specific activity* is the units of enzyme per milligram of protein present in the sample. Molar activity or turnover number are the substrate molecules converted into product per unit time per enzyme molecule, under conditions of substrate saturation.

The rate of the enzymatic reaction is directly proportional to the amount of enzyme rate present in the sample.

Also, at low [S] and under constant conditions of the medium, enzyme activity rapidly increases with the raise in [S]. At higher substrate levels, the activity increases slowly and tends to reach a maximum. *The effect follows a hyperbolic* function; at low [S] the reaction is first order; at high [S] the reaction is zero order with respect to the substrate.

K_m *or Michaelis constant* is the [S] at which the reaction rate reaches a value equal to half the maximum.

Under given conditions of pH and temperature, the K_m value is distinctive for each enzyme and is used to characterize it. For most enzymes, the K_m value is inversely related to the affinity of the enzyme for the substrate, the higher the affinity, the lower the K_m.

Temperature affects enzyme activity, increasing it to reach a peak, which corresponds to the optimal enzyme activity. Beyond this maximum, enzyme activity rapidly drops. The optimal temperature for most mammalian enzymes is around 37°C. The inactivating effect of temperatures above 40°C is due to protein denaturation.

pH affects enzyme activity, by influencing the state of dissociation of functional groups involved in the ES complex. Enzymes have an optimum pH and extreme values of pH cause enzyme denaturation.

Enzyme inhibitors can be classified as:

> *Irreversible*, which permanently inactivate the enzyme, and

> *Reversible*, which consist of the following inhibitors:
> *Competitive*: increase the K_m but not the V_{max}, its action is reversed by increasing [S]. Some have structural similarity to the substrate and compete with it for the active site.
> *Noncompetitive*: bind to the enzyme in a site different to the catalytic center. They decrease V_{max}, leave K_m unaffected, and are not influenced by [S].
> *Anticompetitive*: reduce K_m and V_{max}.

Enzymes are subjected to regulation, to adapt to the requirements of different cells. When the [S] in the cell is below the K_m, changes in [S] modify the activity. *Allosteric enzymes* are those modulated by agents that bind to them at a site different to the active center. The curve of initial velocity versus [S] for allosteric enzymes is not hyperbolic, but sigmoid. Enzyme activity is also changed by *covalent modification*, such as phosphorylation.

Constitutive enzymes are those whose levels remain constant throughout the life of the cell. *Inducible enzymes*, are those whose synthesis is activated as required.

Isozymes are different proteins that have the same enzyme activity.

Bibliography

Blanco, A., 1991. Functional significance of the testis and sperm-specific lactate dehydrogenase isozyme (LDH C$_4$). Miscelánea 84, 1–33.

Christensen, H.N., Palmer, G.A., 1980. Cinética Enzimática. Programmed Course for Medical Students of Medicine and Biological Sciences. Edit. Reverté, Barcelona.

Cornish-Bowden, A., 1995. Fundamentals of Enzyme Kinetics. Portland Press, London.

Frey, P.A., Hegeman, A.D., 2006. Enzymatic Reaction Mechanisms. Oxford University Press, New York, NY.

Lilley, D.M., 2003. The origins of RNA catalysis in ribozymes. Trends Biochem. Sci. 28, 495–501.

Biological Oxidations: Bioenergetics

The multiple activities of living beings can only take place with an adequate energy supply. Events, such as the synthesis of new cellular components, the active transport of solutes across membranes, muscle contraction, nerve transmission, and the movement of cilia and flagella would not be possible without the adequate energy supply.

Primarily, the energy for biological processes proceeds from the sun. Plants and some microorganisms capture the energy from light and transform it, mainly into chemical energy, through the process of photosynthesis. Chemical energy is then used for the synthesis of components of those organisms (carbohydrates, lipids, proteins, etc.), which are made from simple substances (CO_2, H_2O, N_2, NH_3, and NO_3^-), taken from the environment. These organisms are called *phototrophs*.

Other living beings obtain the chemical energy from molecules synthesized by photosynthetic organisms. They are called *chemotrophs*. For example, glucose is synthesized by phototrophic organisms from CO_2 and H_2O. The overall reaction of photosynthesis can be expressed by the following equation:

$$6 \text{ moles } CO_2 + 6 \text{ moles } H_2O + 686 \text{ kcal mol} \rightarrow 1 \text{ mol } C_6H_{12}O_6 + 6 \text{ moles } O_2$$

Carbon dioxide and water are converted into a compound of higher energy content, glucose, in a reaction that requires 686 kcal (2870 kJ)/mole, supplied by solar radiation. The 686 kcal incorporated in each mol of glucose can be released by organisms capable of degrading glucose, which will then use it for their different needs.

THE ENERGY-RICH INTERMEDIATE ATP

Both the conversion of light into chemical energy by phototrophs and the use of chemical energy by chemotrophs require special high-energy intermediate molecules. These compounds serve as reservoirs and carriers of energy that can later be used to perform work (chemical, osmotic, and mechanical) in the cell. The main energy-rich intermediate compound in living beings is adenosine triphosphate (ATP) (Fig. 9.1).

Aerobic organisms require oxygen to survive and exhibit a greater efficiency for the utilization of the chemical energy contained in molecules coming with food. The primary mechanism to release chemical energy from these molecules is oxidation. In the case of glucose, its complete oxidation in the body produces the same end products as its combustion in the laboratory:

$$C_6H_{12}O_6 + 6 O_2 \rightarrow 6 CO_2 + 6 H_2O$$

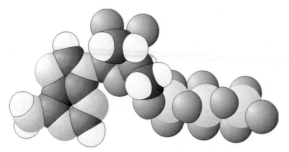

FIGURE 9.1 **Compact molecular model of adenosine triphosphate (ATP).** C, *Black*; N, *gray*; H, *white*; O, *red*; P, *pink*.

This reaction is strongly exergonic ($\Delta G° = -686$ kcal/mol or -2870 kJ/mol).

When combustion is performed in one step, chemical energy is abruptly released in the form of heat. The oxidation of glucose cannot be performed in the body in this manner. The resulting rise in temperature would be incompatible with cell survival and, even if cells could tolerate the temperature increase, the sudden release of free energy could not be used fast enough by cells. Instead, cells oxidize glucose and other substances through a number of chemical steps, which allows the gradual release of energy for its efficient utilization.

Due to the importance of oxidation as the main source of energy in aerobic beings, it is convenient to describe here some basic concepts of oxidation–reduction reactions.

OXIDATION–REDUCTION

Initially, *oxidation* was exclusively understood as the combination of an element or compound with oxygen. The inverse phenomenon, namely the loss of oxygen from a compound, was called *reduction*.

A piece of iron left outdoors slowly reacts with the atmospheric oxygen and forms ferric oxide. The iron has been oxidized and the reaction can be represented as follows:

$$4 \, Fe + 3 \, O_2 \rightarrow 2 \, Fe_2O_3 \qquad (9.1)$$

When carbon is burned in the presence of oxygen, carbon dioxide is produced:

$$C + O_2 \rightarrow CO_2 \qquad (9.2)$$

The carbon has been oxidized. This type of process, which proceeds rapidly, with large releases of energy (heat and light), is called combustion.

Cuprous oxide in the presence of oxygen becomes cupric oxide:

$$2 \, Cu_2O + O_2 \rightarrow 4 \, CuO \qquad (9.3)$$

The Cu/O ratio in the cuprous oxide is 2/1, while in the cupric oxide it is 1/1. The second compound has, relative to Cu, proportionally more oxygen than the first; that is, the copper has been oxidized.

Reduction, a process inverse to oxidation, was considered equivalent to the loss or decrease of the oxygen content of a compound. For example, iron oxide in the presence of hydrogen forms free iron and water:

$$Fe_2O_3 + 3 \, H_2 \rightarrow 2 \, Fe + 3 \, H_2O \qquad (9.4)$$

Iron has lost the oxygen, which was bound in ferric oxide; therefore, it has been reduced.

New experimental findings prompted scientists to widen the concept of oxidation. It was observed that an element or compound can reach the oxidized state not only by the addition of oxygen, but also by the subtraction of hydrogen:

$$3 \, H_2S + 2 \, HNO_3 \rightarrow 3 \, S + 2 \, NO + 4 \, H_2O \qquad (9.5)$$

In this reaction, sulfur in hydrogen sulfide becomes free sulfur; it has not incorporated oxygen, but has instead lost hydrogen. This reaction is equivalent to oxidation. Moreover, a compound can be reduced by the addition of hydrogen.

Oxidation Involves Electron Loss

The reactions described earlier can be analyzed from the perspective of electron exchanges between the participating elements. In this manner:

Reaction (9.1): Oxygen has six electrons in its highest energy level and to achieve a stable configuration it must gain two electrons. Iron can donate two or three electrons in its combinations. In ferric oxide, iron gives up three electrons that are captured by oxygen atoms that are more electronegative.

Reaction (9.2): Carbon is oxidized to carbon dioxide; polar covalent bonds are established. In these bonds, electrons are closer to the oxygen atom, which is more electronegative. There is a relative electron transfer from carbon to oxygen.

Reaction (9.3): Copper in cuprous oxide binds oxygen by a single polar covalent bond in which the pair of electrons is closer to the oxygen atom; copper has virtually donated an electron to oxygen. In cupric oxide, the copper atom gives two electrons to oxygen. While changing from cuprous to cupric, copper loses one more electron.

Reaction (9.4): The iron in ferric oxide donates three electrons to the oxygen molecule it is bound to and becomes reduced. To convert back into free iron, it must recover the three electrons.

Reaction (9.5): In hydrogen sulfide, sulfur is linked to hydrogen by polar covalent bonds, in which electrons are attracted by the sulfur that loses those electrons when it becomes free sulfur.

In all preceding reactions, the oxidized or reduced elements undergo electron changes. They can be generalized as follows: all the elements that lose electrons become oxidized and elements that gain electrons become reduced. From this arises a new definition of the oxidation and reduction processes:

Oxidation involves a total or partial loss of electrons, while reduction results from a gain of electrons.

Reaction of iron and chlorine produces ferric chloride:

$$2\ Fe + 3\ Cl_2 \rightarrow 2\ FeCl_3$$

In this case, a compound with ionic bonds is formed. Iron yields three electrons and becomes a ferric ion; the three electrons lost by the metal are captured by chlorine atoms, which are transformed into chloride ions. In this reaction, iron is oxidized and chlorine is reduced. This is an example of oxidation and reduction proceeding without participation of oxygen or hydrogen; instead, it involves only electron transfers from one element to another.

Oxidation and reduction are always coupled; in any reaction in which one element is oxidized, another element is simultaneously reduced. The electrons given by an element must be captured by another, as electrons cannot be free. Thus, these reactions are oxidation–reduction reactions, or *redox* reactions.

In a redox reaction, the oxidized element is the reducing agent and the reduced element is the oxidizing agent.

Reduction Potential

If a piece of metallic zinc is introduced into a cuvette containing a cupric sulfate solution (blue) and is gently heated, the solution loses its blue color. A reaction occurs in which zinc sulfate and a deposit of metallic copper are formed:

$$CuSO_4 + Zn \rightarrow ZnSO_4 + Cu$$

In cupric sulfate, the Cu is oxidized (as Cu^{2+}). When converted to its metallic form, copper gains two electrons (it has been reduced from its previous state). The metallic zinc is oxidized by displacing Cu^{2+} from sulfate and losing two electrons.

The reverse reaction will not occur; copper is able to donate electrons to zinc and to displace it from its combinations. Electrons can only flow from one element to another if the former has a greater tendency to lose electrons than the latter.

The ability of a substance or element to oxidize other elements depends on its avidity to accept electrons, which is expressed quantitatively by the *reduction potential* of the substance.

In the previous example, it is evident that copper has higher reduction potential than zinc, as electrons can be transferred from Zn to Cu^{2+}, but not in the reverse direction.

Reduction potential (E) is the tendency of an element, ion, or compound to gain electrons from another element, ion, or compound.

Half-reaction. A redox reaction may be considered as the sum of two half reactions or semireactions, one of which represents the oxidation step and the other, the reduction.

For example, in the reaction $2 Na + 2 Cl_2 \rightarrow 2$ NaCl, sodium is oxidized (loses an electron) while chlorine is reduced (gains an electron). The corresponding half-reactions are as follows:

Oxidation: $2 Na \rightarrow 2 Na^+ + 2 e^-$

Reduction: $Cl_2 + 2 e^- \rightarrow 2 Cl^-$

By convention, the redox reactions are written in the direction of reduction.

The oxidized and reduced forms (Na^+—Na, Cl—Cl^-) in each of these reactions constitute a redox couple or pair.

Reduction Potential Determination

The principle of electrochemical cells can be used to measure the reduction potential. This helps identifying which of two oxidants is stronger and in which direction a reaction will occur.

Electrochemical cell. A piece of zinc and a solution of cupric sulfate produce a redox reaction, simplified in the following equation:

$$Cu^{2+} + Zn \rightarrow Zn^{2+} + Cu$$

The same reaction occurs if the reagents are placed in an electrochemical cell system as shown in Fig. 9.2. The cuvette on the left, with a rod of pure zinc immersed in it, is filled with a zinc sulfate solution. The other cuvette, with

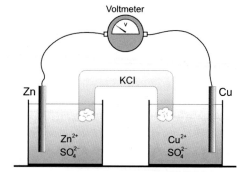

FIGURE 9.2   **Electrochemical cell.**

a copper bar in it is filled with a copper sulfate solution. Both cells are connected by a tube with a solution of KCl (salt bridge). If the two metal rods (electrodes) are connected by a wire to a voltmeter, a difference in potential between both electrodes is observed. Electrons flow from the zinc bar to the copper electrode. This system functions as a cell and each container with its electrode constitutes a hemicell. The flow of electrons through the conductor indicates that the Cu electrode attracts electrons. The zinc rod dissolves, releasing free Zn^{2+} ions into the solution. In the copper electrode, copper metal is deposited. The solution of zinc ions in the left container becomes more concentrated, whereas the solution of copper ions in the right container becomes more dilute.

With this system it is possible to determine, in electromotive force units, the difference in the tendency to gain electrons (reduction potential) of two redox couples. When the ion concentration in the cells is 1 mol/L (l molar solution) and the temperature is 25°C (standard conditions), the potential difference for the Zn–Cu cell system or battery is 1.10 V (volts).

Measurements of redox potential are performed by comparison with a reference system which is assigned a value of zero, in a device similar to the electrochemical cell. One hemicell is replaced by the normal hydrogen electrode, comprising a piece of platinum coated with hydrogen molecules after exposure to hydrogen gas at a pressure of 1 atmosphere. This electrode is then immersed in a 1 mol/L solution of H^+ ions (1 M, pH = 0). The redox pair in this case is $H^+/\frac{1}{2} H_2$, which has conventionally been assigned a potential of zero. The potential is measured by connecting the system under study with the hydrogen hemicell. If the determination is performed at an ion concentration of 1 M and at 25°C, the measured potential is the standard reduction potential, or E_0.

The Zn^{2+}/Zn hemicell connected to the standard hydrogen electrode shows a potential difference of 0.76 V, with the electrons flowing from the Zn to the H electrode. The voltage difference between the Cu^{2+}/Cu hemicell and the hydrogen electrode is 0.34 V, but in this case the electrons flow from the H to the Cu electrode.

A positive sign (+) is assigned to the redox couples with a reducing tendency that is higher than $H^+/\frac{1}{2}H_2$ and a negative sign (−) is assigned if their reducing tendency is lower.

Reduction potentials are determined against the same reference system. Knowing the potential of two redox couples, it is possible to predict, under standard conditions, that the system with higher potential will gain electrons and will become reduced, while the other will be oxidized. Electrons always flow from the element of lower reduction potential to that of higher reduction potential.

The Zn in the Zn^{2+}/Zn hemicell is capable of reducing H^+ ions, indicating that the reduction potential of $H^+/\frac{1}{2} H_2$ is greater than that of Zn^{2+}/Zn. Thus, the potential has negative sign.

$$Zn^{2+} + 2e^- \rightarrow Zn \qquad \Delta E_0 = -0.76\,V$$

The potential is higher for the Cu_2^{2+}/Cu hemicell, than for the $H^+/\frac{1}{2}H_2$ hemicell, therefore, it has a positive sign.

$$Cu^{2+} + 2e^- \rightarrow Cu \qquad \Delta E_0 = +0.34\,V$$

Differences in reduction potential between two redox pairs create differences of electromotive potential that determine the electron flow

and the release of usable energy. The ΔG will be more negative as the difference between the reduction potentials of the reacting substances increases. The change in standard free energy ($\Delta G°$) can be calculated, as E_0 and $\Delta G°$ are related by the following equation:

$$\Delta G° = -n \cdot F \cdot \Delta E_0$$

where n is the number of electrons transferred and F is a constant (Faraday caloric equivalent: 23,062 kcal or 96.49 kJ/V/mol). ΔE_0 is expressed in volts.

In the example of Zn^{2+}/Zn and Cu^{2+}/Cu redox pairs, ΔE_0 value is

$$+0.34 - (-0.76) = 1.10\,V$$

The $\Delta G°$ of reaction: $Cu^{2+} + Zn \rightarrow Zn^{2+} + Cu$ is the following:

$$\Delta G° = -2 \cdot 96.49 \cdot 1.10$$
$$= -50.736\,kcal/mol\,(-212.28\,kJ/mol)$$

Table 9.1 shows standard reduction potential values (at 25°C and 1 M concentration) for various redox pairs of biological interest. The values are adjusted to pH 7.0 (E_0'). At this pH, which is closer to physiological pH, the reduction potential for the $H^+/\frac{1}{2}H_2$ couple is not zero but $-0.42\,V$ (H^+ concentration is not 1 M but $10^{-7}\,M$).

The values in Table 9.1 correspond to systems in which the concentrations of reduced and oxidized forms are equal to each other (1 M). If the concentration of the reduced form is higher than the oxidized form, the reduction potential becomes more negative and vice versa.

BIOLOGICAL OXIDATIONS

Much of the substrates oxidized in the body undergo dehydrogenation. Hydrogens subtracted from the substrate finally bind molecular oxygen to form water:

$$\frac{1}{2}O_2 + 2H^+ + 2e^- \rightarrow H_2O$$

TABLE 9.1   Standard Reduction Potentials (E_0') of Some Biologically Important Redox Couples.

Semireaction	$E_0'(V)$
α-ketoglutarate + $2H^+$ + 2 e^- ⇌ Succinate + CO_2	−0.67
Acetate + $2H^+$ + 2 e^- ⇌ Acetaldehyde	−0.60
H^+ + e^- ⇌ ½ H_2	−0.42
NAD^+ + $2H^+$ + 2 e^- ⇌ NADH + H^+	−0.32
$NADP^+$ + $2H^+$ + 2 e^- ⇌ NADPH + H^+	−0.32
Pyruvate + $2H^+$ + 2 e^- ⇌ Lactate	−0.19
FAD + $2H^+$ + 2 e^- ⇌ $FADH_2$	−0.12
Fumarate + $2H^+$ + 2 e^- ⇌ Succinate	+0.03
Coenzyme Q + $2H^+$ + 2 e^- ⇌ Cytochrome b (Fe^{3+})	+0.04
Cytochrome b (Fe^{3+}) + e^- ⇌ Coenzyme QH_2	+0.07
Cytochrome c_1 (Fe^{3+}) + e^- ⇌ Cytochrome c_1 (Fe^{2+})	+0.22
Cytochrome c (Fe^{3+}) + e^- ⇌ Cytochrome c (Fe^{2+})	+0.25
Cytochrome a (Fe^{3+}) + e^- ⇌ Cytochrome a (Fe^{2+})	+0.29
Cytochrome a_3 (Fe^{3+}) + e^- ⇌ Cytochrome a_3 (Fe^{2+})	+0.55
½ O_2 + $2H^+$ + 2 e^- ⇌ H_2O	+0.82

The reduction potential (E_0') of this reaction is +0.82 V.

Dehydrogenation reactions are catalyzed by enzymes (dehydrogenases) specific for each substrate; hydrogens are accepted by the coenzyme, a nucleotide of nicotinamide (NAD or NADP) or a flavin (FAD).

For NAD, the semireaction can be represented as follows:

$$NAD^+ + 2\,H^+ + 2\,e^- ⇌ NADH + H^+$$

$$E_0' = -0.32\,V$$

The potential difference between ½O_2/H_2O and NAD^+/NADH+H^+ is +0.82 − (−0.32) = 1.14 V. Therefore, the $\Delta G^{o\prime}$ of the redox reaction between the two pairs will be

$$\Delta G^{o\prime} = -n \cdot F \cdot \Delta E_0 = -2 \cdot 96.49 \cdot 1.14$$
$$= -52.6\,kcal/mol(-220\,kJ/mol)$$

If a highly exergonic reaction occurs in a single stage, it will produce sudden release of energy (heat) that is not usable by the cell. Generally, in biological processes, hydrogen ions removed from the substrate are not directly oxidized by oxygen. Instead, they are transferred, in successive steps, to different acceptor substances of increasing reduction potential. In this manner, the energy is gradually released in smaller amounts that can be utilized by the cell. This process is illustrated in Fig. 9.3. A reduced substrate (substrate H_2) is oxidized by a compound A of higher reduction potential. The substrate donates hydrogens to A, which becomes reduced (AH_2). The release of energy in this reaction is proportional to the difference between the reduction potentials of the substrate and A. In a second step, AH_2 is oxidized by compound B, of greater potential, and again a moderate release of energy is produced. The situation is repeated

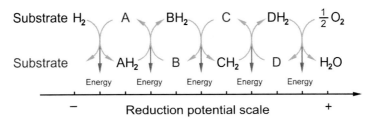

FIGURE 9.3 **Schematic representation of a stepwise oxidation process.** *Red,* Oxidized substrates; *black,* reduced substrates.

with BH_2 and the following compound (C) and then again, between CH_2 and compound D, whose reduction potential is closer to the terminal oxidant, molecular oxygen. The end result is the same as that of direct oxidation, oxygen captures two hydrogens to form water, but the release of energy has been divided into fractions that can be used to perform work.

Mitochondria

Mitochondria are organelles in which the orderly transfer of electrons and the capture of energy generated by the flow of electrons take place. The size, shape, and number of mitochondria vary from one tissue to another, but its basic structure is similar in all of them. Mitochondria possess an *outer membrane* in contact with the cell cytosol, which is permeable to ions and molecules that have a mass lower than 6 kDa. This permeability is due to the existence of pores, or channels, formed by transmembrane proteins called *porins*. Within the mitochondria, and under the outer membrane, there is a second membrane, the *inner membrane*, surrounding a central space, called the *mitochondrial matrix* (Fig. 9.4). The outer membrane and inner membrane are separated by an *intermembrane space*. The inner membrane has selective permeability; it can only be traversed by water, O_2, CO_2, NH_3, and some selected compounds via different transport mechanisms. It has invaginations called *crests*, most abundant in mitochondria of cells with intense respiratory activity. In the inner membrane, the side

facing the matrix has a large number of spheroidal formations (*submitochondrial particles*) attached by a short stalk.

Composition of the inner membrane has distinctive characteristics; it is extremely rich in proteins, which constitute nearly 80% of its overall components, and it contains *cardiolipin*, a lipid nonexistent in other membranes. This

(A)

(B)

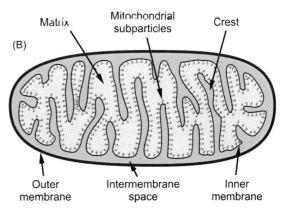

FIGURE 9.4 On top (A) Tridimensional diagram of a sectioned mitochondria. (B) Cross section.

phospholipid participates in various activities of the organelle and in the production of ATP.

The mitochondrial matrix contains many enzymes that participate in central pathways of oxidative metabolism (citric acid cycle and fatty acid oxidation pathway). The inner membrane contains transport systems, proteins of the respiratory chain, and the enzymes responsible for energy capture and synthesis of ATP.

In addition to its critical role in energy metabolism, mitochondria are involved in another important process: apoptosis or programmed cell death (see Chapter 32).

RESPIRATORY CHAIN

As indicated in biological oxidations, hydrogens removed from substrates are gradually transferred through acceptors that undergo reversible changes in their redox state. In the inner membrane of mitochondria, these acceptors are arranged according to a gradient of increasing reduction potential and are intimately associated with the enzymes that catalyze the transfer. The whole system is called the *respiratory chain* or *electron transport chain*.

The transfer of two hydrogens in a redox reaction encompasses two protons (H^+) and two electrons (e^-), they are often called *reducing equivalents*. The respiratory chain consists of a series of stages. The electrons flow through it naturally in the direction fixed by the reduction potential gradient of the acceptors. In a system, as outlined in Fig. 9.3, the electrons pass from A to B to C to D to oxygen sequentially, driven by the growing potential.

All proteins of the electron transport chain are located in the inner mitochondrial membrane, forming a highly ordered multienzyme system. Their inclusion in the membrane ensures proper spatial arrangement for optimum performance.

By analyzing the process of oxidation of a substrate and electron transfer, one might wonder why the substrate does not deliver its hydrogens directly to oxygen or other acceptors with higher reduction potential than NAD since these reactions are thermodynamically more favorable. This does not happen because chemical reactions, in cells, occur with the help of catalysts. The presence of specific enzymes ensures the organized movement of electrons through oxidation/reduction steps and the release of energy in a gradual and controlled way. Hydrogens do not pass directly to oxygen or any other acceptor if enzymes that catalyze the transfer are not present.

Transfer of Reduction Equivalents

Nicotinamide adenine dinucleotide. Numerous oxidizable substrates (pyruvate, α-ketoglutarate, malate, isocitrate, glutamate, and 3-OH-acyl-coenzyme A) are dehydrogenated within the mitochondrial matrix in reactions catalyzed by specific enzymes (dehydrogenases) that use a nucleotide of nicotinamide as coenzyme.

Nicotinamide, a derivative of pyridine, is related to niacin, a member of the vitamin B complex (p. 667). There are two nicotinamide coenzymes in the cells, nicotinamide adenine dinucleotide (NAD) and nicotinamide adenine dinucleotide phosphate (NADP) (Fig. 9.5).

In their reduced state these coenzymes are bound to a hydrogen ion and an electron. When a substrate transfers two hydrogens, one of the protons remains in the medium. The reactions are the following:

$$AH_2 + NAD^+ \rightarrow A + NADH + H^+$$
$$AH_2 + NADP^+ \rightarrow A + NADPH + H^+$$

Reduced   Oxidized
substrate   substrate

The nicotinamide portion of the molecule functions as the acceptor of hydrogen and electrons (Fig. 9.6). One hydrogen ion is bound to carbon 4 of the pyridine ring, while the electron of the other hydrogen is bound to the nitrogen in the cycle.

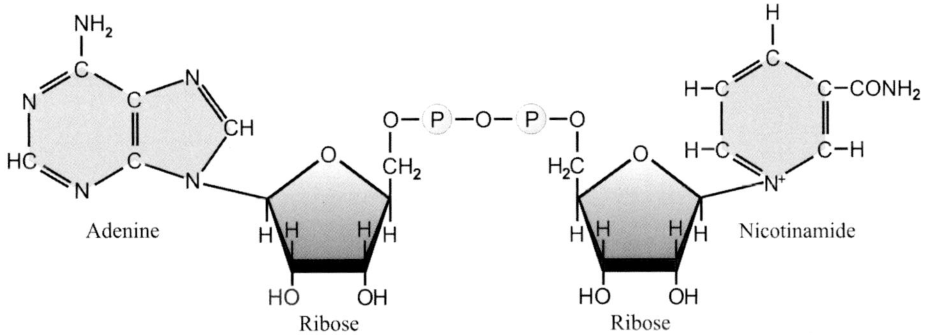

FIGURE 9.5 **Nicotinamide adenine dinucleotide (NAD).** In NADP, the −OH group (in *red*) is esterified with phosphate.

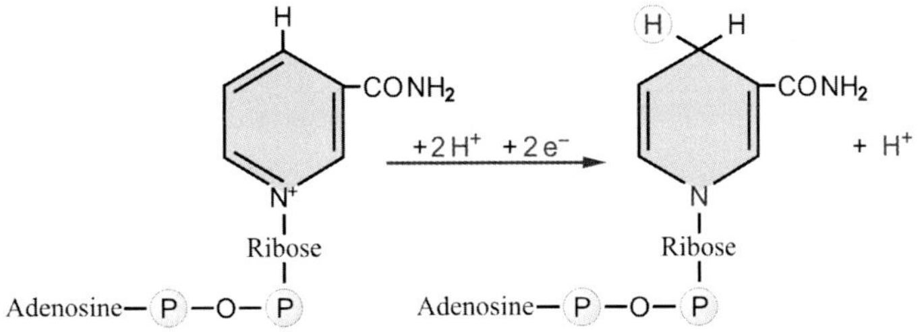

FIGURE 9.6 **The nicotinamide moiety of NAD is the H and electron acceptor.**

NAD or NADP bound dehydrogenases are not part of the respiratory chain; they are found within the mitochondrial matrix. The reduced coenzyme NAD donates reducing equivalents to the first acceptor of the electron transport chain and becomes oxidized.

There are also dehydrogenases bound to NAD and NADP in the cytosol, but due to the impermeability of the inner mitochondrial membrane to adenine nucleotides, NADH and NADPH formed in the cytosol cannot directly transfer their hydrogens to the chain. Oxidation is performed by "shuttle" systems, or switches, capable of transferring hydrogen from the cytosol to the respiratory chain within mitochondrial membranes. These systems use intermediary acceptors that can cross the inner membrane (see subsequent sections).

In general, the fate of reduction equivalents is dependent on whether they are captured by NAD or NADP. Reduced NAD in the mitochondrial matrix yields hydrogen to the respiratory chain to produce energy, while reduced NADP hydrogens are preferably used in the synthesis of various compounds. Eventually, the reduced NADP can transfer a hydrogen ion to NAD in a *transhydrogenase* catalyzed reaction:

$$NADPH + NAD^+ \rightarrow NADP^+ + NADH$$

Through this pathway, hydrogens originally accepted by NADP can be donated to the respiratory chain.

Respiratory Chain Components

The respiratory chain is a group of electron acceptors and reducing carriers imbedded in the inner membrane of mitochondria. Many of its components are clustered into multimolecular complexes that cross the entire thickness of the lipid bilayer. There are four complexes and two

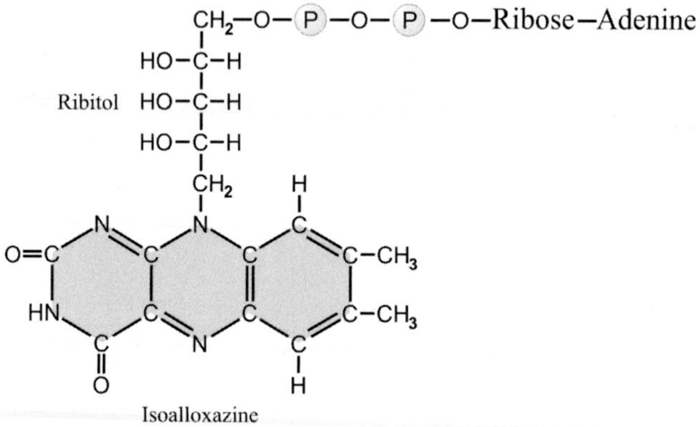

FIGURE 9.7   **Flavin adenine dinucleotide (FAD).**

other single components of the chain (coenzyme Q and cytochrome c) (Fig. 9.11).

Reduced NAD is oxidized by the first complex in the respiratory chain, called *NADH–ubiquinone reductase*. This is a large complex (>900 kDa) consisting of about 45 polypeptide chains, one prosthetic group called flavin mononucleotide (FMN), and 16–24 iron atoms in 7–9 iron–sulfur centers (see subsequent sections).

Electrons transferred from NADH to the NADH–ubiquinone reductase complex are initially accepted by FMN, which is reduced to $FMNH_2$. Then, electrons pass successively through different Fe atoms in the Fe–S centers, and finally are transferred to ubiquinone or coenzyme Q. The FMN and Fe atoms are reoxidized and coenzyme Q is reduced ($CoQH_2$).

Flavoproteins are firmly bound to the prosthetic group, flavin. NADH–ubiquinone reductase complex has FMN (formed by isoalloxazine), a penta alcohol derivative of ribose, ribitol, and orthophosphate.

Flavin adenine dinucleotide (FAD) is similar to FMN in that it is composed of isoalloxazine, ribitol, phosphate, plus another nucleotide, adenosine monophosphate (AMP). The AMP binds to the FMN phosphate through a pyrophosphate bond (Fig. 9.7).

Substrates in the mitochondrial matrix such as succinate, acyl coenzyme A, glycerol-3-phosphate, and others, are oxidized by flavin dehydrogenase using adenine dinucleotide (FAD) as a coenzyme. In the reaction, the FAD is reduced to $FADH_2$. The isoalloxazine core and ribitol, part of FMN and FAD molecules, are components of riboflavin, a complex B vitamin. The isoalloxazine in FMN and FAD is the portion of the molecule, which accepts both of the transferred hydrogens (Fig. 9.8).

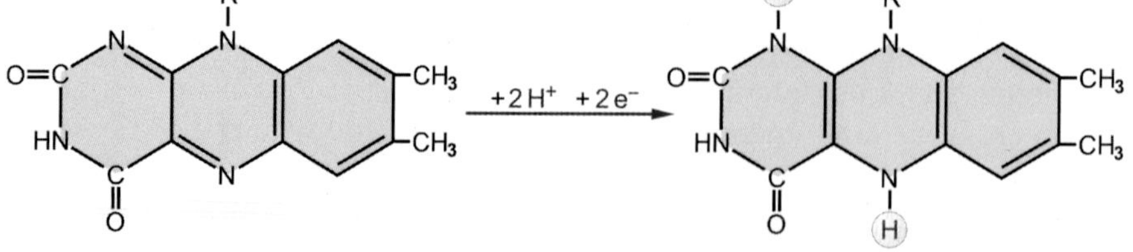

FIGURE 9.8   **Isoalloxazine of FMN and FAD as hydrogen acceptor.**

FeS Fe_2S_2 Fe_4S_4

FIGURE 9.9 **Iron–sulfur centers.** Fe, *Red*; inorganic S, *gray*; cysteine S, *black*; cysteine residues, *pink*.

A flavoprotein linked to FAD catalyzes the oxidation of succinate (succinate dehydrogenase). It integrates the complex called *succinate–ubiquinone reductase*, which comprises four polypeptide subunits, a prosthetic group constituted by FAD, and eight Fe atoms divided into three iron–sulfur centers. Electrons flow from succinate to FAD (which becomes $FADH_2$) and then flow through the Fe^{3+} atoms in the S–Fe centers, to be finally transferred to coenzyme Q. The $FADH_2$ is reoxidized to FAD. The free energy change that occurs during the passage of electrons through the succinate–ubiquinone reductase complex is lower than that of the NADH–ubiquinone reductase complex.

Fe–S centers have nonheme iron (iron not associated with heme) bound to sulfur. The simplest form has only one iron atom coordinated with four cysteine sulfhydryl residues in the protein. Some centers have two iron atoms attached to two inorganic sulfur atoms (Fe_2S_2), while others have four iron atoms bound to four inorganic sulfur atoms (Fc_4S_4). These sulfur atoms are easily separated by strong acid treatments, releasing hydrogen sulfide. In Fe_2S_2 and Fe_4S_4 centers,

the iron also binds to cysteine side chains in the protein (Fig. 9.9). Each iron atom captures an electron and passes reversibly from the ferric to the ferrous state ($Fe^{3+} \rightleftharpoons Fe^{2+}$). These Fe–S proteins are found in several respiratory chain complexes. The standard reduction potential of the Fe^{3+}/Fe^{2+} redox pair in different iron–sulfur proteins ranges between -0.24 and $+0.30$ V.

Coenzyme Q is also called ubiquinone because of its chemical nature and ubiquity in biological systems. It is a benzoquinone with a long chain of 10 isoprene units (a total of 50 carbons). The structures of the oxidized and reduced forms are shown in Fig. 9.10.

This is the only acceptor of the respiratory chain that is not bound to proteins. Its isoprenoid chain is hydrophobic and allows for its location to be in the nonpolar lipid bilayer area of the inner membrane; here it can move freely. It acts as a mobile carrier of electrons.

The reduction stages described earlier show the transfer of a pair of hydrogen atoms from the substrate to ubiquinone. For some substrates, the reducing equivalents are donated to NAD and FMN and then to the Fe–S centers of

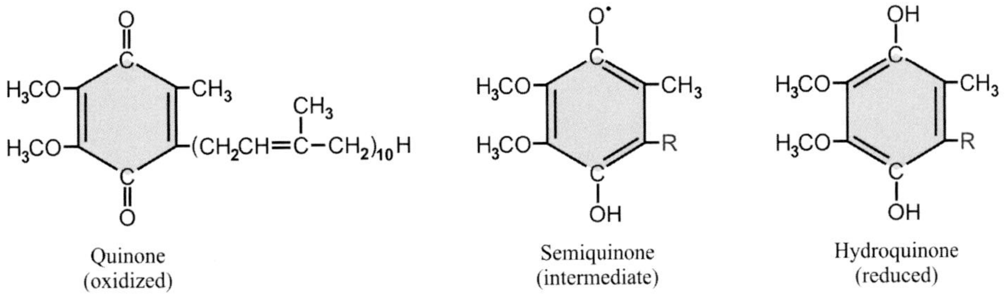

Quinone (oxidized)   Semiquinone (intermediate)   Hydroquinone (reduced)

FIGURE 9.10 **Coenzyme Q or ubiquinone.**

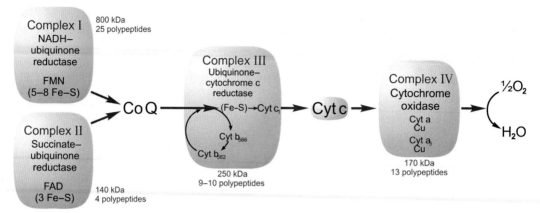

FIGURE 9.11   **Order of respiratory chain components.**

NADH–ubiquinone reductase complex. For other substrates, such as the succinate–ubiquinone reductase complex, the hydrogen is transferred to flavoprotein with FAD. In both cases they are accepted by ubiquinone. Thus, coenzyme Q receives hydrogen from different sources (Fig. 9.11). The decrease in free energy (ΔG) with the passage of reducing equivalents from the succinate–ubiquinone reductase complex is less than that seen with NADH–ubiquinone reductase.

In the next step, the reduced ubiquinone ($CoQH_2$) transfers two electrons to the following acceptors, the cytochromes.

Cytochromes are heme proteins that are able to accept electrons. The heme iron captures an electron, passing from the oxidized state (Fe^{3+}) to the reduced (Fe^{2+}). Several types of respiratory chain cytochromes have been identified: two types of cytochrome-a (a and a_3), two types of cytochrome-b (b_{566} and b_{562}), and two types of cytochrome-c (c and c_1). They differ in their reduction potential and light absorption spectrum. In the case of cytochromes-b, the subscript indicates the wavelength at which they have their maximum absorption. They can be ordered according to their increasing reduction potential: b_{566}-b_{562}-c_1-c-a-a_3.

Cytochromes-b and cytochromes-c possess a prosthetic group identical to that found in hemoglobin and myoglobin (the heme and iron are complexed with protoporphyrin IX). Cytochromes-a contain heme A, with two different side chains. Molecules of cytochrome-a and cytochrome-a_3 are equal, but have different reduction potentials because they are located in different environments within the complex. The cytochromes are associated with a copper ion, located near the heme A iron. Metals undergo redox reactions which aid in recycling between Fe^{3+}/Fe^{2+} and Cu^{2+}/Cu^+ states.

Cytochrome-b and cytochrome-c_1 form part of a complex called *ubiquinone–cytochrome c reductase*, which comprises 11 different polypeptide subunits, including the cytochromes. Also located in this complex is an iron–sulfur protein (Fe_2-S_2). From the ubiquinone–cytochrome c reductase complex, electrons pass to cytochrome-c.

Cytochrome-c is a heme protein made up of 104 amino acids (13 kDa) and a prosthetic group identical to the heme of hemoglobin and myoglobin. The amino acid sequence of cytochrome-c has been preserved with little change along evolution. It is a peripheral protein located on the outer side of the inner mitochondrial membrane. It is associated with the membrane through electrostatic attractions to the polar heads of phospholipids. Cytochrome-c can be easily separated from the membrane by treatment with saline solutions. Its molecule has various side chains of lysine residues surrounding the area in which the heme is housed. The positive charges on the lysine are important for recognition and binding to the complexes located immediately before and after in the respiratory chain of electron acceptors. Cytochrome-c is a mobile carrier which receives electrons from ubiquinone–cytochrome-c reductase and transfers them to the cytochrome oxidase complex, responsible for the last stage of the system.

Cytochrome oxidase is formed by 11–13 polypeptide subunits. This complex, just like the previous ones, spans the entire thickness of the membrane. It contains one cytochrome-a, one cytochrome-a_3, and two copper ions that form two heme–Cu centers. Cytochrome oxidase is the only component of the electron transport system with the capacity to directly react with oxygen. An oxygen molecule (O_2) captures four electrons to form two molecules of water. The overall reaction is

$$O_2 + 4\,e^- + 4\,H^+ \rightarrow 2\,H_2O$$

In this reaction, four electrons converge simultaneously. This poses a problem, since cytochrome-a_3, the last acceptor of the chain, carries only one electron at a time and the converging electrons cannot exist free in the medium. To explain this phenomenon, it has been proposed that complete reduction of oxygen is performed in a cycle of several stages, called the Q cycle, which will be considered later.

Respiratory Chain Complexes

Except for ubiquinone (free within the lipid bilayer) and cytochrome-c (attached to the outer face of the inner membrane), the remaining components of the respiratory chain are grouped into four complexes that span the membrane. These complexes are designated by Roman numerals.

Complex I is made up of the NADH ubiquinone reductase. Through NAD, it receives hydrogens from substrates oxidized by dehydrogenases associated with that coenzyme. It contains FMN and several Fe–S centers. It delivers hydrogens to ubiquinone.

Complex II corresponds to the succinate–ubiquinone reductase. It contains FAD as a prosthetic group and three Fe–S centers. It transfers reducing equivalents from succinate to coenzyme Q.

Complex III is the ubiquinone–cytochrome-c reductase. It contains cytochrome-b_{566}, cytochrome-b_{562}, cytochrome-c_1, and a Fe–S center. It transfers electrons from ubiquinone to cytochrome-c.

Complex IV is the cytochrome oxidase, that includes cytochrome-a and cytochrome-a_3 as well as two copper ions. It catalyzes the reduction of O_2 to H_2O.

Fig. 9.11 outlines the different complexes of the respiratory chain and their organization. This arrangement has been experimentally confirmed, based on the reduction potentials of the various redox pairs. E_0' values (Table 9.1) gradually increase in the direction of electron flow. It is important to note that the values of E_0' are determined in equilibrium, with the same concentrations (1 M) of the oxidized and reduced forms in each pair, and in conditions that do not reproduce normal physiological conditions (e.g., CoQ determinations must be performed in ethanol solutions given the insolubility of the compound in water). Therefore, there may be important differences between the E_0' values determined in vitro and those corresponding to the redox reaction in the cell. However, the standard values are used as an approximation to predict the most likely direction of electron movement. Distribution of components in the four complexes is perfectly compatible with the proposed system.

Electron Transport Inhibitors

A useful resource to ascertain the chain sequence has been the use of inhibitors acting at specific sites within the chain. The redox state of the various components can be determined directly through spectrophotometry of mitochondrial suspensions. When an inhibitor that blocks electron transfer at a specific point is added to the preparation, the components of the chain upstream to the blocked site appear reduced. Downstream to the blocked site, the chain components are oxidized because they do not receive electrons. Studies with different electron transport blockers support the proposed order of reactions in the respiratory chain. As an example, some frequently used inhibitors are mentioned in the subsequent sections.

Compounds that act at the level of complex I (NADH–ubiquinone reductase) include the following: the plant product rotenone, a fish poison, amytal, barbiturates, and anesthetics such as halothane. These substances prevent the delivery of hydrogens from substrates oxidized by NAD-bound dehydrogenases to CoQ. They do not affect the oxidation of succinate and other substrates that yield hydrogens to flavoprotein FAD, thus proving that they follow another path. The antibiotic antimycin inhibits electron transport at complex III. Cyanide (CN^-), carbon monoxide (CO), and azide (N^{3-}) specifically inhibit cytochrome oxidase (complex IV) and block the final step of oxygen activation (Fig. 9.12).

The use of inhibitors has not only helped to deduce the sequence of the respiratory chain, but has also allowed a better understanding of the mechanism of action of some drugs and poisons.

OXIDATIVE PHOSPHORYLATION

The steps of oxidation and flow of electrons within the respiratory chain is an exergonic process, which proceeds with an overall decrease in free energy ($-\Delta G$). This energy is used for the phosphoryl transfer in the synthesis of ATP from ADP and inorganic phosphate (P_i), to produce ATP and water:

$$ADP + P_i \rightarrow ATP + H_2O$$

This is an endergonic reaction that occurs when it is coupled to processes that provide the necessary energy. ATP production using the energy released during electron transport in the respiratory chain is called *oxidative phosphorylation*.

The $\Delta G^{o\prime}$ of the bond between the third (γ) and second (β) phosphate of ATP is -7.3 kcal/mol (-30.5 kJ/mol). Therefore, an equal or greater amount of energy is required to synthesize ATP from ADP and P_i.

When a substrate is oxidized in a reaction catalyzed by enzymes using NAD, it transfers the reducing agents to the first component of the respiratory chain; this is due to the action of NADH dehydrogenase. From there, they will move through all of the intermediate acceptors to finally produce a water molecule. According

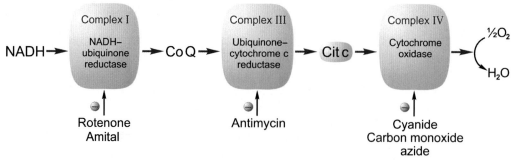

FIGURE 9.12 **Respiratory chain.** Sites of action of blocking inhibitors.

to the estimation already presented, this process proceeds with an overall decrease in free energy of -52.6 kcal or -220 kJ/mole, which is accomplished after a number of steps. Three of these steps release energy that can be coupled to the formation of a high energy phosphate bond: (1) transfer of hydrogen from NADH to CoQ, at the level of NADH–ubiquinone reductase or complex I, (2) transfer of electrons in the ubiquinone–cytochrome-c reductase or complex III, and (3) the reaction occurring at the level of cytochrome oxidase or complex IV.

The synthesis of ATP does not occur by direct coupling with these reactions. However, the understanding that the release of free energy takes place at these three sites has been useful for the study of the respiratory chain function.

P:O ratio. Experiments that measure oxygen and inorganic phosphate consumption by mitochondria in the presence of an oxidation substrate have allowed for the calculation of ATP molecules produced in the electron transport chain. Malate donates two hydrogens to NAD in the malate dehydrogenase catalyzed reaction. Through the use of malate, a ratio between phosphate and oxygen atoms (P:O) consumed can be determined; this ratio was found to be 3:1. This indicates that for each pair of electrons or hydrogens transferred along the respiratory chain, three phosphate molecules are bound to ADP. Therefore, the flow of an electron pair allows the synthesis of three ATP molecules. The same ratio has been calculated using other substrates capable of transferring hydrogen to NAD.

When using succinate, substrate oxidized by a succinate dehydrogenase catalyzed reaction, linked to FAD, the P:O ratio was 2. Succinate undergoes oxidation by succinate dehydrogenase within the respiratory chain. In this reaction, succinate donates two hydrogens to FAD and the P:O ratio for this substance has been shown to be 2:1, slightly lower than that of malate. Therefore, for every electron pair transferred, two molecules of ATP are produced. Other substrates that use FAD as a coenzyme also have been shown to

have a P:O of 2:1. These observations indicated that one of the sites of ATP production is associated with the NADH–ubiquinone reductase and when reducing equivalents enter through another enzyme (FAD $\rightarrow$ CoQ), the yield is lowered by one ATP.

Although these values have been accepted for a long time, new determinations of the amounts of ATP production in oxidative phosphorylation have shown that for each electron pair transferred from NADH to O_2, ~ 2.5 molecules of ATP are produced. When the electrons are derived from substrates which donate hydrogen ions to FAD (succinate, glycerol-3-phosphate), ~ 1.5 molecules of ATP are produced. The following section will return to this issue.

Mechanism of Oxidative Phosphorylation

The molecular mechanisms by which the energy produced through the electron transfer is used for ATP synthesis was a topic that intrigued researchers for many years. A breakthrough for this phenomenon was the discovery and isolation of molecules capable of converting ADP and P_i into ATP.

ATP synthase. It is estimated that an adult human consumes, under a normal day, approximately 40 kg of ATP. This enormous amount of ATP is continuously regenerated by the body. Most of the ATP is synthesized by the activity of an enzyme located in the inner mitochondrial membrane, *ATP synthase*, constituted by the association of two protein complexes called F_1 and F_0.

F_1 is part of the submitochondrial particles. F_1 is composed of at least nine subunits, including three α, three β, one γ, one δ, and one ε subunit, giving the enzyme a total mass of ~ 380 kDa. The α and β polypeptides form the "head" of the enzyme and are assembled into three dimers arranged like orange wedges (Fig. 9.13). Each β subunit contains a catalytic site. Polypeptide γ forms the stem of the particle, with its base implanted in the center of the F_0 complex and the

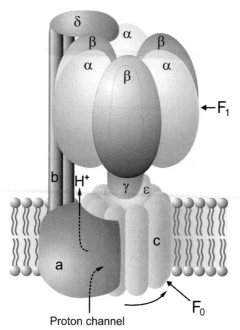

Proton channel

FIGURE 9.13 **Schematic representation of the F_1–F_0 complex (ATP synthase).** Conformation and arrangement of subunits is hypothetical. The F_0 portion is included into the mitochondrial inner membrane; F_1 protrudes to the matrix.

other end forming an axis that inserts into the $\alpha_3\beta_3$ hexamer. The ϵ subunit is firmly secured to the base of the γ subunit and the F_0 complex. The δ polypeptide adheres to the top of the particle (Fig. 9.13).

When separated from the mitochondrial membrane, the F_1 particles were found to only catalyze ATP hydrolysis, which initially lead scientists to call the complex adenosine triphosphatase (ATPase).

The F_0 complex, embedded in the membrane, has a mass of ~250 kDa and consists of one a, two b, and 10-14 c subunits. The c polypeptides have two transmembrane segments which are arranged to form a wheel, the γ and ϵ subunits of F_1 bind to its center. Adjacent to this wheel is the a subunit, which has eight transmembrane segments that form a channel (along with

the c subunits) for the passage of protons from one side of the mitochondrial membrane to the other. Both of the elongated b polypeptides form a stem which joins the a with δ subunit. In this manner, the a, b, and δ subunits and the $\alpha_3\beta_3$ hexamer are all held together (Fig. 9.13).

Hypothesis on the mechanism of oxidative phosphorylation. Some of the mechanisms that couple the electron transfer to the ATP synthesis, referred to as *energy transduction* by some authors, are still not completely understood. These include the transmission of energy produced during the electron transport to the site of phosphorylation and the use of that energy for ATP synthesis. To explain the phenomenon, several hypotheses have been postulated:

1. In 1953, Slater proposed the chemical hypothesis, which postulated the formation of an unstable high energy chemical intermediate between the respiratory chain and ATP synthesis. At present, the existence of such intermediary has not been demonstrated.

2. In 1964, Boyer presented a conformational hypothesis, which predicted that the energy transduction process from the redox systems to ATP synthesis occurs through proteins capable of undergoing conformational changes.

3. In 1961, Mitchell proposed the *chemioosmotic hypothesis*, which has gained more experimental support over years and is currently the most accepted hypothesis. This hypothesis proposes that proton transfer occurs from the matrix to the intermembrane space simultaneously with the process of electron transport in the respiratory chain. Thus, the electron transport coupled with the unidirectional pumping of protons produces a greater H^+ concentration in the outer side of the inner membrane, creating a transmembrane H^+ electrochemical gradient. The pH is lower in the intermembrane space compared to the mitochondrial matrix. There

is also a difference in electrical potential between both sides of the membrane, with the external side being relatively more positive than the inner side. The result is the creation of a proton-motive potential, which tends to drive the movement of protons back into the matrix. As the inner membrane is impermeable to H^+, this ion can only cross through the F_0 channel within the ATP synthase. The return of protons into the mitochondrial matrix is the driving force for ATP synthesis.

In conclusion, the respiratory chain is asymmetrically arranged in the inner mitochondrial membrane and it uses the energy released by the transfer of electrons to pump protons to the intermembrane space, creating a H^+ gradient. Protons return to the matrix through the only route available, offered by the F_0F_1 complex. This inward flow of H^+ provides the energy that generates ATP from ADP and Pi (Fig. 9.14).

Coupling of Electron Transport and Proton Translocation

Electron transport in the respiratory chain and its coupling with proton translocation will be examined according to the chemio-osmotic hypothesis.

The two electrons donated by NADH in the inner face of the NADH–ubiquinone reductase complex (complex I) bind to FMN, reducing it to $FMNH_2$. This transfers the electrons to Fe–S centers and finally to ubiquinone. At this stage, proton passage occurs from the matrix to the outside of the membrane through complex I. The number of translocated protons has been calculated to be between 2 and 4. The mechanism of electron transport coupling and the movement of protons in complex I is not well understood.

The unidirectional pumping of H^+ takes place through complexes I, III, and IV, which completely traverse the mitochondrial membrane. Moreover, the flow of electrons through each of

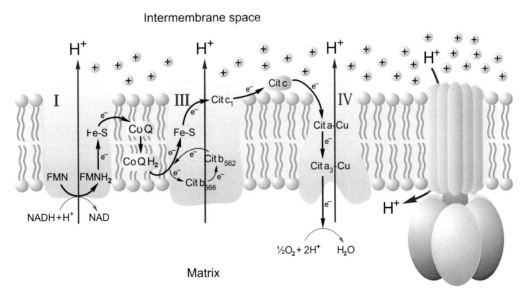

FIGURE 9.14 **Proton transfer according to the chemio-osmotic hypothesis.** Arrangement of the components of electron transport chain in the inner mitochondrial membrane. ATP synthase F_1F_0 complex is shown on the right.

these complexes is highly exergonic. The energy generated can be coupled to the transport of protons. The mechanism of proton translocation has been better explained for complex III, also called cycle Q.

Q Cycle. Coenzyme Q, a hydrogen (H^+ plus e^-) carrier, receives two electrons from NADH or $FADH_2$ through complexes I and II, respectively, and takes two protons from the matrix to become $CoQH_2$. This reduced ubiquinone migrates to complex III where it gives up an electron to the Fe–S center at a site next to the outer face of the inner membrane. This electron is then passed onto cytochrome c_1 and, finally, taken out of the complex to cytochrome c. The $CoQH_2$ that has transferred the electron releases two protons to the intermembrane space and is completely oxidized to CoQ. The remaining electron is donated to cytochromes b (successively b_{566} and b_{562}). CoQ becomes semiquinone (CoQ−) (Fig. 9.10). This completes the first phase of the cycle. In the second phase, another $CoQH_2$ molecule repeats the steps described earlier, transferring one electron to cytochrome-c, the other to cytochrome-b, and then pumps two protons to the intermembrane space. However, at this point, the electron received by cytochrome-b_{562} is donated to the semiquinone formed previously, which captures this second electron and two protons from the

matrix to form $CoQH_2$. The transport chain is now ready to start a new cycle (Fig. 9.15). Each cycle results in the oxidation of one molecule of $CoQH_2$, the pumping of four protons across the mitochondrial inner membrane, and the transfer of two electrons to two cytochrome-c molecules into the intermembrane space.

Final stage: reduction of O_2. The electrons are transported by cytochrome-c to the acceptor site of complex IV (cytochrome-c is a mobile carrier). The cytochrome oxidase complex moves electrons to the matrix and donates them to oxygen. Complex IV represents the third proton pumping site. It has been proposed that 2–4 protons are transported at this stage.

The overall process of oxygen reduction requires the overall transfer of four electrons. The transport by cytochromes involves the movement of one electron at a time. This poses the risk of releasing partial reduction products of O_2, which are toxic (see Chapter 10). To reduce this risk, the oxygen molecule is bound between the iron and copper atoms in the heme–Cu center of cytochrome-a_3 and remains there until it has received four electrons. The stages of this process are shown in Fig. 9.16.

1. Initially the Fe and Cu atoms of cytochrome-a_3 are oxidized (Fe^{3+} and Cu^{2+}).

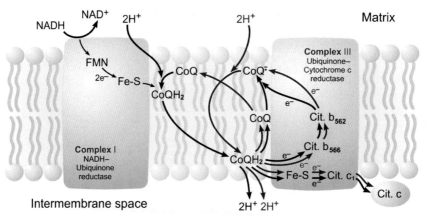

FIGURE 9.15   **Q cycle.** *Black arrows* indicate the electron and proton path during the first cycle phase. *Red arrows* indicate the electron and proton path taken the second stage.

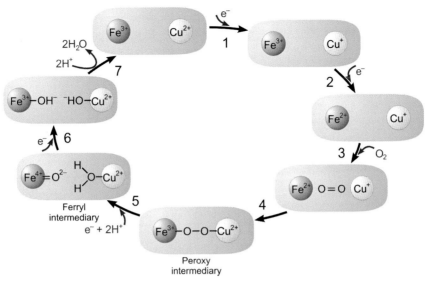

FIGURE 9.16   **O$_2$ reduction catalyzed by cytochrome oxidase.**

In the first stage, an electron enters and is captured by Cu^{2+} causing it to become Cu^+.

2. Another electron is taken by Fe^{3+}, which becomes Fe^{2+}.
3. The heme–Cu center binds an O_2 molecule.
4. Each Fe^{2+} and Cu^+ ions donates one electron to oxygen and a *peroxy intermediary* is formed.
5. This intermediary is cleaved and leaves H_2O bound to copper while the other oxygen atom binds iron, forming a *ferryl intermediary* (Fe^{4+}).
6. Next, the fourth electron and two protons enter the cycle.
7. Two water molecules are released along with oxidized Fe^{3+} and Cu^{2+}. The cycle is ready to be repeated.

The total number of protons translocated from the matrix to the intermembrane space for each pair of electrons transferred in the respiratory chain has been a matter of debate. The most accepted number is currently around 10 (2–4 in complex I, 4 in complex III, and 2–4 in complex IV).

As indicated, the pumping of protons across the membrane results in the creation of a H^+ gradient (pH gradient) and an electric potential difference. This proton-motive force drives the movement of protons back into the mitochondrial matrix. Due to the impermeability of the membrane to H^+ ions, the return flow can only occur through transporters that allow the passage of protons. These sites are found in the F_0 portion of ATP synthase. The proton channel is formed by the c and a subunits of the F_0 complex (Fig. 9.13). The entry of protons, propelled by the pH and membrane potential gradients, provides the energy necessary for ATP synthesis.

ATP Synthesis in the Complex F$_0$F$_1$

The most widely accepted model explaining the coupling mechanism of protons to the synthesis of ATP proposes that the F_0F_1 functions as a "rotary machine." The flow of protons that enters the F_0 complex from the intermembrane space provides the energy needed to generate a rotary motion of the c subunits. As the γ stem and polypeptide ε are fixed to the center of the ring, they rotate with it. Subunits α, β, δ, *a*, and *b* remain fixed.

When the stem in the center of the $\alpha_3\beta_3$ spheroid rotates, it allows for contacts that induce conformational changes and alter the affinity of the catalytic sites for its substrates and products. The active site in each β polypeptide passes successively through three states: "open" (O), "loose" (L), and "closed" (C). There is a cooperative effect among the subunits that constitute the hexamer, so that at any moment the three catalytic sites are in different conformations. While one of the β subunits is in a "loose" state, the following will be in the "closed" state, and the third in the "open" state. ADP and P_i enter into the "open" site and immediately the site acquires the "loose" conformation. A reaction occurs spontaneously without energy expenditure (ΔG equaling close to zero) and ATP is formed and firmly retained within the "closed" state. ATP is released into the medium when the complex returns to the "open" form (Fig. 9.17). Each full turn of γ produces three ATP molecules.

The rotation and conformational changes are driven by the energy generated by the H^+ flow; the cycle is repeated continuously, while protons enter through the F_0 channel.

ATP-ADP Transport

Most of the ATP generated in eukaryotic cells is synthesized within the mitochondria and consumed elsewhere. For this reason, most of the ATP formed must be moved outside of mitochondria, while the compounds needed to regenerate it (ADP and P_i) must be carried into the mitochondrial matrix.

All three adenylates have electric charge and do not diffuse through the membrane. A translocator or counter-transport system exchanges matrix ATP for external ADP (Fig. 9.18). ATP has four negative charges, while ADP has three. The exchange tends to neutralize the electrical gradient created through the pumping of protons. In

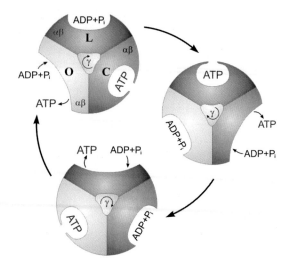

FIGURE 9.17　**Schematic representation (cross section) of ATP synthase rotary machine in the synthesis of ATP.** Rotation of the subunit assures that the F_1 complex catalytic site in each β unit successively passes through the three conformations: loose (L), closed (C), and open (O). In a complete turn, three molecules of ATP are released.

other words, the potential across the inner membrane favors the ATP^{4-}/ADP^{3-} exchange.

P_i transport takes place inwards through another inner membrane translocator, also driven by the proton gradient. In this case, P_i transport occurs in the same direction (toward the interior) of H^+, in exchange for OH^-. Usually the system is represented as the counter-transporter $H_2PO_4^-/OH^-$ (Fig. 9.18).

Approximately 25% of the energy generated by electron transfer in the respiratory chain is used to drive the ATP^{4-}/ADP^{3-} and P_i^-/OH^- exchangers.

ATP Yield of Oxidative Phosphorylation

As mentioned in a previous section, ATP production is considered to be coupled with highly exergonic reactions at three sites of the respiratory chain. These are approximately the same sites where protons are translocated. Synthesis of ATP can be related to the proton flow, and the

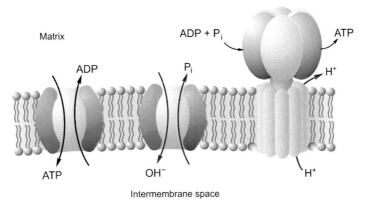

FIGURE 9.18 **The protons return through the F_0F_1 complex (right) and promotes synthesis of ATP.** On the left and center are the ADP/ATP and P_i/OH$^-$ exchangers, respectively.

coupling is performed in the F_0F_1 complex. A total of ten H$^+$ ions are pumped from the matrix to the intermembrane space per pair of electrons transported. The synthesis of a molecule of ATP requires the flow of three protons through the F_0F_1 complex. Taking into account the requirement of an additional proton for ATP translocation (exchange ATP^{4-}/ADP^{3-}) and another to exchange P_i/OH$^-$, each electron pair that passes through the entire chain from NADH to O_2 yields approximately 2.5 molecules of ATP (between 2.5 and 3). When electrons are donated by FADH$_2$, the yield is about 1.5 ATP (between 1.5 and 2). However, to simplify calculations, the estimated energy efficiency of metabolic reactions, which will be used in the following chapters, will consider three and two molecules of ATP per pair of electrons donated by NADH and FADH$_2$, respectively.

Inhibitors of Oxidative Phosphorylation

The electron transport inhibitors presented previously affect oxidative phosphorylation. Other agents do not block the flow of electrons, but dissociate the phosphorylation process. These agents are called mitochondrial *uncouplers*. One of these compounds is 2,4-dinitrophenol (DNP), which transfers hydrogen ions from the

outer side of the mitochondrion to the matrix and dissipates the proton gradient created by the respiratory chain. Compounds that carry ions across the membrane are called ionophores; DNP acts as a proton ionophore.

Another mitochondrial uncoupler is salicylate (aspirin is acetylsalicylic acid). When it traverses the outer mitochondrial membrane, salicylate encounters a low pH in the intermembrane space and acquires a proton; it becomes salicylic acid, uncharged, allowing it to easily cross the inner membrane. Once in the matrix, the pH is high and salicylic acid releases a proton. This tends to reduce the gradient across the inner membrane, decreasing proton transport and inhibiting ATP synthesis by the F_0F_1.

There are antibiotics of peptide nature, such as valinomycin and nigericin, which act as K$^+$ ionophores. K$^+$ transfer can eliminate or at least minimize the electrical potential gradient across the mitochondrial membrane, hindering ATP phosphorylation coupled to proton flow.

Another substance that interferes with oxidative phosphorylation is the antibiotic oligomycin, which specifically binds and inhibits one of the proteins of F_0 of the ATP synthesis complex.

Mitochondrial uncouplers do not interfere with electron transport. Normally, approximately

40% of the energy released by oxidations in the respiratory chain is used for the synthesis of ATP; the remaining 60% is released as heat to maintain body temperature. Uncoupling is manifested by hyperpyrexia, which is one of the symptoms of intoxication with high doses of salicylic acid or aspirin.

Reye's syndrome is a rare clinical condition in which there is a mitochondrial disturbance, with inhibition of oxidative phosphorylation and fatty acid oxidation. It may result, among various causes, from complication of common viral infections (influenza and varicella), or from intoxications (salicylates). It is characterized by fatty degeneration of the liver and encephalopathy with noninflammatory cerebral edema.

Production of ATP is not the only function of mitochondria. They also act in apoptosis (see Chapter 32) and cell Ca^{2+} homeostasis. Thereby, mitochondrial defects may also affect those processes.

Respiratory Control

Like other metabolic processes, oxidative phosphorylation requires the presence of adequate concentrations of the necessary substrates in the medium, including ADP, P_i, O_2, and a metabolite that by oxidation can donate electrons to NAD or FAD. When the availability of any of these four factors decreases, the synthesis of ATP is limited. The level of ADP in the mitochondrial matrix is very important as a regulator of oxidative phosphorylation. When there is no ADP, respiration stops. In the presence of oxygen, and an adequate supply of oxidation substrates, mitochondrial respiratory activity depends on the availability of ADP. This requires a close relationship between electron transport and phosphorylation and has been called *respiratory control*. It is clear that the synthesis of ATP depends on the continuous flow of electrons from a substrate of oxidation to O_2. Furthermore, in intact mitochondria, electron flow occurs only while ATP is synthesized. This is explained by the coupling between electron transport and proton pumping across the inner membrane. The proton-motive force is responsible for the synthesis of ATP. If external protons cannot return to the matrix through the F_0 channel piece, not only the production of ATP in the F_0F_1 complex will be blocked, but also the passage of electrons through the respiratory chain will stop. Proton accumulation in the outside of the membrane will increase the gradient, reducing, and eventually stopping, the transfer of any additional protons. This regulatory mechanism is intended to limit mitochondrial oxidation to adjust it to the physiological requirements of ATP of the cells, preventing wasteful consumption of substrates.

The respiratory chain is highly responsive and reacts very quickly to any demand. In general, the system maintains high levels of ATP in cells. Stimulation of ATP consuming activity, such as muscle contraction, immediately increases the production of ADP. Increasing levels of ADP in the cell activate substrate oxidation, electron transport, proton pumping, flow of H^+ via the F_0F_1 complex, and the production of ATP.

The ATP/ADP translocator is inhibited by the plant glucoside *atractyloside*. This compound blocks ADP entry, reducing the main stimulus of respiration, and stopping electron transport.

When mitochondrial phosphorylation is uncoupled by the action of agents like DNP, the respiratory control disappears. In these cases, because electron transport without ATP production continues, all the energy released is dissipated as heat.

Genetic Deficiencies of the Respiratory Chain

Most of the mitochondrial proteins are encoded by nuclear genes; only 3% of all proteins depend on genes found within the mitochondria. The father never transmits these organelles to his offspring, all the mitochondrial genome is inherited from the mother. Furthermore, not all mitochondria have identical DNA (homoplasmy), some may contain mutations (heteroplasmy). Thus, modes of respiratory chain proteins inheritance are varied.

Genetic defects of oxidative phosphorylation produce diverse clinical symptoms before the age of 2 years; renal compromise is frequent, characterized in some cases by proximal tubulopathies (de Toni-Debré-Fanconi Syndrome), in others, the glomeruli and distal tubules may be affected. Plasma lactate levels can be elevated.

Alterations of the respiratory chain complexes, specifically complex IV (cytochrome-c oxidase), may be manifested by the *Leigh syndrome*. This disease causes very severe neurological and motor disorders that lead to death within the first 2 years, mainly by respiratory failure. It is estimated that about 25% of Leigh syndrome cases are due to defects in mitochondrial proteins.

Brown Fat

In newborns and in many animals, especially those that experience periods of hibernation, there are deposits of tissue called *brown fat*. The fat is brown in color due to the large number of mitochondria within it. Interscapular deposits of brown fat have been found in human infants; these deposits are reduced or absent in adults. While the mitochondria in this tissue do not perform oxidative phosphorylation, they exhibit a particularly high electron flow. These mitochondria operate as if they were under the effect of uncouplers; this is due to the presence of *thermogenin* in the inner membrane. Thermogenin, or uncoupling protein (UCP), acts as an alternative pathway for proton transport. UCP maintains a flow of protons, not coupled to ATP synthesis, activated by free fatty acids. The energy produced by the electron transport in brown fat is released as heat to maintain body temperature.

Adipocyte groups that express UCP were found in the white adipose tissue of mice. They develop in response to various stimuli, and have been designated *beige* cells. These cells have not been demonstrated in humans.

The medical interest on brown and beige cells arises from the observations of mice capacity to counteract certain metabolic diseases, like obesity and type 2 diabetes.

Hydrogen Shuttle Systems

During operation of the respiratory chain, the hydrogens from the oxidized substrates, linked to NAD are transferred to the first complex of the electron transport system (NADH–ubiquinone reductase) and from there continues through a series of steps, to bind with oxygen and form water. The NAD involved in these reactions must be in the mitochondrial matrix to deliver the hydrogens to the respiratory chain, located in the inner membrane.

There are reactions catalyzed by NAD-dependent oxidoreductases which take place in the cytosol. The NADH formed in these reactions cannot directly transfer reducing equivalents to the respiratory chain, because the inner membrane of the mitochondrion is not permeable to nicotinamide nucleotides. The amount of NAD is limited in the cytosol, and if there were no mechanisms to reoxidize NADH resulting from the reactions in which the coenzyme is involved, the ability to oxidize some substrates would be greatly reduced. For example, during glycolysis, in the step of glyceraldehyde-3-phosphate oxidation by glyceraldehyde-3-phosphate dehydrogenase, NADH and H^+ are produced. In anaerobiosis, the NADH reoxidation reaction takes place by conversion of pyruvate to lactate by lactate dehydrogenase. Aerobically, however, pyruvate penetrates into the mitochondria to continue its oxidative path and NADH remains in the cytosol. Under these conditions, the degradation of glucose will be limited by the availability of oxidized NAD. If all existing cytosolic coenzyme is reduced, glycolysis cannot proceed.

The impermeability of the inner mitochondrial membrane for NADH requires the existence of indirect transfer systems, called shuttle systems or switches that can transport the reducing

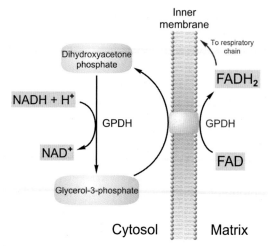

FIGURE 9.19 **Glycerophosphate hydrogen suttle system.** GPDH: glycerol-3-phosphate dehydrogenase. In *black*, cytosolic enzyme; in *red*, mitochondrial enzyme.

equivalents to the respiratory chain. The operation of shuttle systems requires: (1) presence of an NAD-dependent oxidoreductase in cytosol, which is able to reoxidize NADH, transferring hydrogen to an acceptor substrate, (2) the acceptor substrate must have a transport system that allows passage through the inner mitochondrial membrane, (3) the acceptor should yield reducing equivalents to the respiratory chain in the mitochondria, which requires the existence of an enzyme able to oxidize it.

Glycerophosphate shuttle. The first of these systems described in the literature was that of glycerophosphate. Present in skeletal muscle and brain, this shuttle system functions as shown in Fig. 9.19.

1. The hydrogen acceptor substrate is dihydroxyacetone. In the reaction catalyzed by glycerophosphate dehydrogenase (an NAD-linked cytosolic enzyme), NADH hydrogens are transferred to dihydroxyacetone to form glycerol-3-phosphate.
2. The outer side of the inner mitochondrial membrane has glycerophosphate dehydrogenase linked to FAD. Thanks to its

location, glycerol-3-phosphate formed in the cytosol does not need to penetrate into the matrix to be oxidized to dihydroxyacetone phosphate and transfer its hydrogens to FAD.
3. $FADH_2$ donates the reduction equivalents to ubiquinone.

In this system, the reduction equivalents enter the respiratory chain at the level of coenzyme Q, which yields two molecules of ATP per pair of electrons.

Aspartate–malate shuttle. Another hydrogen transfer system is the aspartate–malate shuttle (Fig. 9.20). This shuttle is very active in liver, kidney, and heart. Cytosolic and mitochondrial isozymes of aspartate aminotransferase and malate dehydrogenase make possible the following series of steps:

1. Transamination between L-aspartate and α-keto-glutarate catalyzed by cytosolic aspartate aminotransferase produces L-glutamate and oxaloacetate.

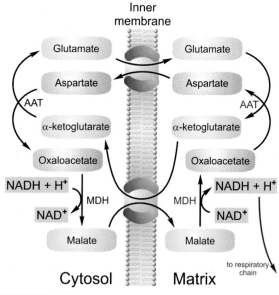

FIGURE 9.20 **Malate–aspartate shuttle system.** AAT, Aspartate aminotransferase; MDH, malate dehydrogenase. In *black*, cytosolic enzymes; in *red*, mitochondrial enzymes.

2. Oxaloacetate, in a reaction catalyzed by cytosolic malate dehydrogenase, accepts two hydrogens of cytoplasmic NADH. NAD$^+$ and malate are formed as products. Malate penetrates the matrix by crossing the inner membrane via a dicarboxylate transport system.

3. Within the mitochondria, malate is oxidized to oxaloacetate by mitochondrial malate dehydrogenase, yielding its hydrogens to NAD in the matrix. The NADH produced transfers the reducing equivalents to the electron transport system in the inner membrane.

4. Oxaloacetate and L-glutamate, in a reaction catalyzed by mitochondrial aspartate aminotransferase, form L-aspartate and α-ketoglutarate.

5. Aspartate and α-ketoglutarate have transport systems in the inner membrane that allow them to move out into the cytosol. In turn, the cytosolic L-glutamate can enter the matrix.

A cycle is in this manner closed, allowing for the transfer of hydrogens from NADH in the cytosol to the respiratory chain, with malate as an intermediary. In this case, the hydrogens are received by the mitochondrial matrix NAD; therefore, reduction equivalents will generate three molecules of ATP when transferred to the respiratory chain.

Besides the systems described, there are other shuttle systems, including the citrate–pyruvate shuttle. This system works in acetyl transfer from the mitochondrial matrix to the cytosol (p. 344). Shuttle systems exhibit some tissue specificity. Aspartate–malate is very active in liver, kidney, and heart, whereas glycerophosphate is very active in brain and skeletal muscle.

PHOSPHORYLATION AT SUBSTRATE LEVEL

The use of various substances in cells to obtain energy generally occurs through a series of chemical transformations. In the first of these reactions, an intermediate passes to the next stage to give an intermediate product, which then undergoes successive reactions until the final product is formed. This ordered sequence of chemical transformations is called a metabolic pathway. In some of these pathways, molecular changes in the original substrate lead to redistribution of the energy contained in it to create intermediates that have bonds with high hydrolytic energy. These compounds react directly or indirectly with ADP to form ATP. This type of energy transfer, without involvement of the respiratory chain, is called *phosphorylation at the substrate level*.

This type of phosphorylation produces direct transfer of a phosphoryl group to ADP. An example of this is seen at a stage of the glucose degradation in the metabolic pathway known as glycolysis. 2-Phosphoenolpyruvate, an intermediate of this pathway, is a compound high in energy. The standard free energy of the phosphate group hydrolysis in this compound is -14.8 kcal/mol (-61.9 kJ/mol). The hydrolysis is coupled to the phosphorylation of ADP to form ATP in an endergonic reaction with a $\Delta G^{\circ\prime}$ of -7.3 kcal/mol (-30.5 kJ/mol). The reaction is catalyzed by pyruvate kinase:

$$\begin{array}{ccc} CH_2 & & CH_3 \\ \| & & | \\ C{-}O{\sim}\textcircled{P} + ADP \longrightarrow & C{=}O + ATP \\ | & & | \\ COO^- & & COO^- \end{array}$$

Phosphoenol pyruvate   Pyruvate

$$\Delta G^{\circ\prime} = -7.5\,\text{kcal/mol}(-31.4\,\text{kJ/mol})$$

An indirect ADP energy transfer occurs in the metabolic cycle of citric acid, in which the products of carbohydrate, lipid, and amino acid degradation are converted into CO_2 and H_2O. In this cycle, one of the steps forms a high-energy intermediate, succinyl-coenzyme A. This metabolite reacts with guanosine diphosphate (GDP) and inorganic phosphate (P_i). The energy is transferred to form a high energy phosphate bond in guanosine triphosphate

(GTP) is synthesized. The reaction is catalyzed by succinate thiokinase.

$$\begin{array}{ccc}
\text{COO}^- & & \text{COO}^- \\
| & & | \\
\text{CH}_2 & & \text{CH}_2 \\
| & + \text{GDP} + \text{P}_i \rightleftharpoons \text{GTP} + & | \\
\text{CH}_2 & & \text{CH}_2 \\
| & & | \\
\text{CO}\sim\text{S}-\text{CoA} & & \text{COO}^- \\
\end{array} + \text{CoA}-\text{SH}$$

Succinyl-CoA   Succinate

GTP, in turn, can transfer phosphate to ADP to yield ATP:

$$\text{GTP} + \text{ADP} \rightleftharpoons \text{GDP} + \text{ATP}$$

Phosphorylation at a substrate level, in contrast to oxidative phosphorylation, does not require the presence of oxygen to form ATP.

OTHER ELECTRON TRANSPORT SYSTEMS

There are electron transport systems different from the respiratory chain, not involved in ATP synthesis, but in substrate hydroxylation or dehydrogenation reactions.

For hydroxylations, the systems require NADPH and O_2. One oxygen atom of the molecule is incorporated into the substrate and the other is reduced to H_2O. The process can be represented by the following equation (A = substrate):

$$\text{AH} + O_2 + \text{NADPH} + \text{H}^+$$
$$\rightarrow \text{AOH} + H_2O + \text{NADP}^+$$

The enzymes that catalyze these reactions are designated *monooxygenases* or *mixed function oxygenases*.

Introduction of a −OH group is a resource commonly used by liver cells as a mechanism of biotransformation of foreign substances. Hepatocytes have an electron transport system responsible for hydroxylations. It is attached to the endoplasmic reticulum and can be isolated with the microsomal fraction. The system contains a FAD flavoprotein called *NADPH–cytochrome P$_{450}$ reductase* and *cytochrome P$_{450}$*. Cytochrome P$_{450}$ is a heme protein of approximately 50 kDa, anchored to membranes, belongs to the cytochrome-b group, and has an absorption maximum at 450 nm when bound to CO. In humans there is a large family of genes encoding P$_{450}$ enzymes of different specificity.

NADPH donates reducing equivalents to NADPH–cytochrome P$_{450}$ reductase and this transfers electrons to cytochrome P$_{450}$. In the final step, cytochrome P$_{450}$ acts as a mixed function oxygenase catalyzing the addition of a hydroxyl group to the substrate. The system is inducible, meaning it increases markedly in hepatocytes when substrates are administered.

In adrenal cortex mitochondria, there is a hydroxylation system that receives NADPH hydrogens for steroid hormone synthesis from cholesterol. The electrons are transferred to a flavoprotein called *NADPH–adrenodoxin reductase* that transfers to a protein with no heme iron, designated *adrenodoxin*. This in turn reduces cytochrome P$_{450}$, which catalyzes the final activation of O_2 and hydroxylation of the substrate.

In the endoplasmic reticulum of hepatic cells there is another electron transport system responsible for fatty acid desaturation. It is composed of a flavoprotein with FAD, NADH–cytochrome b$_5$ reductase, cytochrome b$_5$, and desaturase. The reducing equivalents are provided by NADH and flow through the system to be finally transferred to O_2 along with the two hydrogen ions taken from the substrate. The result is the formation of two water molecules and the creation of a double bond in the carbon chain of the fatty acid.

SUMMARY

Biological oxidations do not take place by direct transfer of electrons (e$^-$) from substrate to oxygen. They are carried out in successive stages by different e$^-$ acceptors with increasing *reduction potential*. This allows for energy to be released stepwise, which can be captured and used by cells.

The respiratory chain or electron transport chain comprises hydrogen and/or electron acceptors, called *reduction equivalent* acceptors, which are arranged in an orderly fashion, from lower to higher reduction potential and associated with enzymes that catalyze the transfer of e$^-$. The respiratory chain is located in the inner membrane of mitochondria.

The components of the respiratory chain include NAD-linked dehydrogenases in the mitochondrial matrix that generate NADH + H$^+$. The hydrogens are transferred from the coenzyme to other components of the respiratory chain:

Complex I or *NADH–ubiquinone reductase* is composed of 45 polypeptides, FMN and seven to nine Fe–S centers. Coenzyme FMN of the complex captures reduction equivalents, and becomes FMNH$_2$. e$^-$ are successively transferred through Fe–S centers and are finally moved to coenzyme Q. As a result, FMNH$_2$ and Fe^{+2} are reoxidized and CoQ reduced to CoQH$_2$.

Complex II or *succinate–ubiquinone reductase*, composed of four polypeptides, FAD and three Fe–S centers, works in the movement of two H, which are taken from succinate and transferred to CoQ.

Ubiquinone or *coenzyme Q* (CoQH$_2$) is the only acceptor unbound to protein, located in the membrane lipid bilayer; it acts as an e$^-$ mobile carrier. It accepts hydrogens from complex I or II and donates the e$^-$ to complex III.

Complex III or *ubiquinone–cytochrome c reductase*, formed by cytochrome c$_1$, b$_{566}$, b$_{562}$ and Fe–S center. Cytochromes are hemoproteins, with Fe that reversibly captures an e$^-$ from ubiquinone–cytochrome c reductase, e$^-$ are transferred to *cytochrome c*.

Cytochrome c, located on the outer face of the membrane, donates e$^-$ to complex IV.

Complex IV or *cytochrome oxidase* includes cytochromes a and a$_3$, and two Cu atoms. It transfers 4 e$^-$ to O, which binds to 4 H$^+$ and forms H$_2$O.

Inhibitors can block e$^-$ transfer at different levels of the respiratory chain. Rotenone, amytal and other barbiturates act at the level of complex I; antimycin A, at complex III, and cyanide, carbon monoxide and azide, on complex IV.

Oxidative phosphorylation. The energy produced by the flow of electrons is coupled to phosphoryl transfer for the synthesis of ATP from ADP.

Each pair of e$^-$ from substrates dehydrogenated by enzymes bound to NAD generates three molecules of ATP. Two e$^-$ from substrates oxidized by FAD-dependent enzymes produce two molecules of ATP. In the first case, the P:O ratio is 3, in the second, 2.

The chemio-osmotic hypothesis explains the mechanism underlying oxidative phosphorylation. The energy generated by the flow of reducing equivalents is used to pump mitochondrial matrix protons outward of the inner membrane. Proton transport occurs at the complex I, III, and IV sites. This creates a proton gradient between the two faces of the mitochondrial inner membrane, with pH being lower and the electric potential being positive outside.

ATP synthesis is performed by the synthase complex in the submitochondrial particles. It is formed by the F$_1$ portion of the ATP synthase complex (containing nine polypeptide subunits), bound to the inner face of the membrane by a hollow stalk (segment F$_0$) that traverses the lipid bilayer. The [H$^+$] gradient and electrical potential created by the respiratory chain drives the movement of protons inwards to the mitochondrial intermembrane space. Due to the impermeability of the mitochondrial membrane, return of protons to the matrix can only occur through the F$_1$F$_0$ complex. This acts as ATP synthase and binds P$_i$ to ADP to form ATP, using the energy released by the backflow of H$^+$.

Inhibitors of oxidative phosphorylation include compounds such as *2,4-dinitrophenol* (DNP), which is a *proton ionophore*, and *nigericin* and *valinomycin*, which are K$^+$ ionophores. They suppress the mitochondrial electrical potential gradient, acting as uncoupling agents (e$^-$ flow continues to be active), but ATP is not synthesized and all energy is released as heat. Other inhibitors, such as *oligomycin* interfere with oxidative phosphorylation by binding to the F$_0$ segment of the F$_1$ F$_0$ ATPase.

Brown fat present in infants and hibernating animals has abundant mitochondria that function uncoupled due to the presence of thermogenin or UCP. They contribute to maintain body temperature.

Regulation of oxidative phosphorylation is mainly controlled by the level of ADP, absence of ADP stops oxidations.

Phosphorylation at the level of substrate is another pathway to generate ATP in reactions where the potential of phosphoryl transfer to ADP is provided directly by high energy metabolites.

Other electron transport systems do not form ATP, but they catalyze hydroxylations (*mixed function oxygenases* requiring NADPH and O$_2$). For example, a system constituted by NADPH–cytochrome P$_{450}$ reductase and cytochrome P$_{450}$ is present in the liver and the adrenal gland. Another hydroxylase system, composed of NADH–cytochrome b$_5$ and desaturase catalyzes fatty acid desaturation in hepatocytes.

Bibliography

Boyer, P.D., 1997. The ATP synthase. A splendid molecular machine. Ann. Rev. Biochem. 66, 717–749.

Cannon, B., Nedergaard, J., 2004. Brown adipose tissue: function and physiological significance. Physiol. Rev. 84, 277–359.

Capaldi, R.A., Aggeler, R., 2002. Mechanism of the F$_1$F$_0$-type synthase, a biological rotary motor. Trends Biochem. Sci. 27, 154–160.

Cechini, G., 2003. Function and structure of complex II of the respiratory Caín. Ann. Rev. Biochem. 72, 77–109.

Claypool, S.M., Koehler, C.M., 2012. The complexity of cardiolipin in health and disease. Trends Biochem. Sci. 37, 32–41.

Cross, R.L., 2004. Turning the ATP motor. Nature 427, 407–408.

Harms, M., Seale, P., 2013. Brown and beige fat: development, function and therapeutic potential. Nat. Med. 19, 1252–1263.

Hosler, J.P., Ferguson-Millers, S., Mills, S., Mills, D.S., 2006. Energy transduction: proton transfer through the respiratory complexes. Ann. Rev. Biochem. 75, 165–187.

Isaac, I.S., Dawson, J.H., 1999. Haem-iron containing peroxidases. Essays Biochem. 34, 51–69.

Ricquier, D., 1998. Neonatal brown adipose tissue, UCP1 and the novel uncoupling proteins. Biochem. Soc. Trans. 26, 120–123.

Roehm K.H., 2001. Electron Carriers: Proteins and Cofactors in Oxidative Phosphorylation. Encyclopedia of Life Sciences. John Wiley & Sons, Inc., Wiley InterScience, www.els.net.

Rötig, A., Munnich, A., 2003. Genetic features of mitochondrial respiratory chain disorders. J. Am. Soc. Nephrol. 14, 2995–3007.

Mitochondria dysfunction in human disease, 1999. Several articles. Biochim. Biophys. Acta 1410, 99–228.

Sligar, S.G., 1999. Nature's universal oxygenases: the cytochrome P450. Essays Biochem. 34, 71–83.

Antioxidants

Normal metabolism generates products with oxidizing activity which have toxic effects if they accumulate in cells. Oxidizing agents cause what has been called *oxidative stress*. To eliminate the deleterious action of oxidizing products, the organism has developed defense mechanisms that maintain a balance between prooxidant and antioxidant factors. This chapter gives an integrated view of the origin and harmful action of oxidizing agents and the protective mechanisms that are turned on in the body to remove or neutralize them.

REACTIVE SPECIES

Reactive oxygen species (ROS) and *reactive nitrogen species* (RNS) are compounds that can cause oxidative damage to cells. Both include free radicals and various other substances that are highly reactive and have oxidizing capacity.

A *free radical* is a group of atoms that has one or more unpaired electrons in its outer orbit. Free radicals are designated by a dot next to the symbol(s) of the element(s) that describes them ($\cdot$A or A$\cdot$). Some free radicals have a relatively long half-life because their reactivity is relatively low, while others are very unstable, exhibiting a lifespan of only microseconds.

Free radicals react to gain back the electron they lack. The molecules donating the electron turn into a new free radical, which can react with another molecule, propagating radical formation in a chain reaction.

For example, if a radical A$\cdot$ reacts with a molecule B:

$$A\cdot + B: \rightarrow A: + B\cdot$$

This B$\cdot$ radical then reacts with an electron donor and can generate another free radical as follows:

$$B\cdot + C: \rightarrow B: + C\cdot$$

C$\cdot$ can react with another molecule to continue propagating the generation of radicals.

The chain reaction stops when the newly formed free radical is poorly reactive or when two radicals react, which results in a nonreactive molecule:

$$A\cdot + B\cdot \rightarrow A:B$$

Reactive species are mainly generated in organelles, such as peroxisomes, mitochondria, and endoplasmic reticulum. In the respiratory chain, the reaction catalyzed by cytochrome oxidase, which incorporates electrons to oxygen, can produce partially reduced products and free radicals.

Another example are the ROS formed by invasion of bacteria, viruses, or foreign molecules into the body. This promotes a reaction that initiates the release of cytokines and factors capable of triggering an inflammatory response. The activated phagocytic cells (neutrophils, monocytes, and macrophages) attracted to the site of inflammation respond with a "respiratory burst," increasing oxygen consumption and forming ROS. These oxidizing radicals have the positive effect of destroying the invading agent, but can also damage the cells surrounding the inflammation process.

Reactive species can also be generated by exposure to external agents, including environmental pollutants (tobacco smoke, pesticides, combustion products, and industrial waste), drugs, ionizing radiations, ozone, and high oxygen concentrations.

Reactive Oxygen Species

ROS are normally formed in the body and they include free radicals, such as superoxide anion, peroxides, and hydroxyl radicals, as well as nonradical compounds, for example, singlet oxygen and hydrogen peroxide.

Singlet oxygen. The oxygen molecule (O_2) in its most common form or basal state, has two unpaired electrons with parallel spin, one in each of the two external orbitals. Basal oxygen is a biradical and is called triplet oxygen; it is indicated with the notation $^3O_2$. Unlike many free radicals, basal oxygen reacts in a very slow manner. To accept an electron pair from a substrate, one of the electrons must reverse its electron spin, which imposes a thermodynamic barrier. This explains why organic matter does not undergo spontaneous combustion on contact with oxygen. However, if enough energy is supplied to reverse the spin of one of the unpaired electrons, triplet oxygen converts to the singlet state, which is represented by the notation $^1O_2$. Singlet oxygen can be formed in some biological reactions, such as lipid peroxidation, respiratory burst in phagocytes, and also by absorption of energy. Although not a radical, $^1O_2$ is much more reactive than triplet oxygen and can cause damage. The excess energy of singlet oxygen is easily transferred to other molecules, which returns $^1O_2$ to the "relaxed" triplet or basal state.

Hydrogen peroxide (H_2O_2). Although not a radical, hydrogen peroxide is a reactive molecule that participates in reactions that result in additional reactive products. It readily crosses cell membranes and, therefore, it, can produce effects far from its site of origin. Hydrogen peroxide is commonly generated in reactions catalyzed by oxidases. Its production is particularly abundant in peroxisomes during degradation of long-chain fatty acids (greater than 20 carbons) and other compounds.

In phagocytic cells, myeloperoxidase activated during the respiratory or oxidative burst produces hydrogen peroxide. The respiratory burst refers to a process that involves the rapid release of reactive oxygen species from immune system or other cells, (which is often triggered by bacteria or fungi). The produced hydrogen peroxide can react with chloride ions to produce powerful oxidants and bactericides such as hypochlorous acid (hypochlorite, ClO^-) and hydroxyl radicals. This is the primary line of defense against foreign pathogens.

$$H_2O_2 + Cl^- \rightarrow HClO + {}^\cdot OH$$

Superoxide anion. This free radical is an anion that results from incomplete reduction of oxygen (the oxygen molecule acquires an electron) and is represented with the notation $^\cdot O_2^-$. This anion is not particularly harmful, but it is the precursor of other highly reactive oxidizing agents that are toxic. Due to its electronegative charge, $^\cdot O_2^-$ is not easily soluble in lipids and does not cross membranes. However, it can act far from the site of production due to the existence of anion transporters that allow its passage across cells.

The respiratory chain incidentally produces electrons that partially reduce oxygen. Complexes I and III are sites of superoxide generation. This can also occur in the final stage catalyzed by cytochrome oxidase, or by passage of an electron to oxygen from coenzyme Q of the semiquinone radical ($CoQH^\cdot$):

$$CoQH^\cdot + O_2 \rightarrow CoQ + H^+ + {}^\cdot O_2^-$$

Superoxide anion is generated in reactions between O_2 and different compounds, including catecholamines (dopamine and epinephrine) and tetrahydrofolate. Enzymes of the endoplasmic reticulum associated with cytochrome P_{450}, which catalyze the hydroxylation of various substrates (fatty acids, steroids, and xenobiotics), produce superoxide radicals. This free radical is formed abundantly in phagocytes during the respiratory burst, mainly by the action of NADPH oxidase:

$$NADPH + O_2 \rightarrow NADP^+ + H^+ + \dot{O}_2^-$$

The main toxic effect of superoxide anion and hydrogen peroxide is exerted by the hydroxyl radicals ($\dot{O}H$) formed in the Haber–Weiss reaction:

$$\dot{O}_2^- + H_2O_2 \rightarrow O_2 + OH^- + \dot{O}H$$

Superoxide anion also produces RNS (see subsequent sections).

Hydroxyl radical. This free radical can be generated in the Haber–Weiss reaction or in the Fenton reaction, when Fe^{2+} or other transition metals are present:

$$Fe^{2+} + H_2O_2 \rightarrow Fe^{3+} + OH^- + \dot{O}H$$

Hydroxyl radical is also formed in cells after γ-ray exposure. It can readily cross cell membranes but, in general, reacts near its place of origin, so that its site of action is often very limited. It is the most powerful and harmful of the reactive oxygen species, with great capacity to produce damage. Hydroxyl radicals react with various organic compounds, taking an electron; H_2O is formed and the substrate becomes a radical. This initiates a chain reaction with production of additional reactive products. Removal of these radicals is critical to prevent tissue damage.

Peroxide radicals. The previously mentioned radicals can react with various substances producing new radicals. For example, the hydroxyl radical is the main initiator of lipid peroxidation.

It reacts with polyunsaturated fatty acids to form lipid radicals (represented by the abbreviated notation $\dot{L}$) that causes further production of reactive oxidative species. By addition of O_2, lipoperoxyl radicals ($LOO\dot{}$) are formed and these generate new lipidic radicals:

1. $LH + \dot{O}H \rightarrow \dot{L} + H_2O$ (initiation)
2. $\dot{L} + O_2 \rightarrow LOO\dot{}$ (spread)
3. $LOO\dot{} + L'H \rightarrow LOOH + \dot{L'}$

The oxidative chain process stops when two radicals react with each other; however, this occurs much less frequently than the formation of new radicals. Due to the continuous formation of peroxide radicals, they need to be eliminated by regulatory systems of the body.

Reactive Nitrogen Species

RNS include radicals nitric oxide ($NO\dot{}$) , nitrogen dioxide (NO_2), and nonradicals nitrous acid (HNO_2), and peroxynitrite (ONO_2^-).

Nitric oxide. Nitric oxide is a free radical of high chemical reactivity, therefore, it has a brief half-life (less than 5 s). It diffuses easily through cell membranes. It is generated from arginine in a reaction catalyzed by nitric oxide synthase (NOS) with participation of NADPH, FMN, FAD coenzymes, and tetrahydrobiopterin. NOS contains a heme domain with Fe^{2+} and is activated by the Ca^{2+}–calmodulin complex. Activity of NOS produces as reaction products citrulline and nitric oxide (see p. 393).

Three isozymes of NOS have been recognized in mammals, each with different cellular distribution and regulatory properties. One is found in endothelial cells of blood vessels (eNOS or NOS I), another in macrophages (iNOS or NOS II), and the third in neurons (nNOS or NOS III). While there is a predominant expression of NOS in the indicated cells, these enzymes are also distributed in many others tissues. Isozymes I and III are constitutive, while the enzyme from macrophages (NOS II) is inducible.

Nitric oxide was found to be the *endothelium-derived vascular relaxing factor*, whose chemical structure had remained elusive for some time. Nitric oxide participates as a messenger in guanylate cyclase signal systems and produces vasodilation in different vascular territories. It can influence different molecules involved in signal transduction systems. There is also evidence of nitric oxide participation as a major neurotransmitter in the nervous system. In macrophages, and other phagocytic cells, NOS is induced by factors released in the inflammatory foci triggered by invasion of foreign agents. The sudden release of the free radical nitric oxide has toxic effects on bacteria, tumor cells, and other phagocytosed particles.

During the respiratory burst in macrophages, nitric oxide and superoxide radical can react and produce an active oxidizing molecule, peroxynitrous acid (peroxynitrite at neutral pH):

$$NO^{\cdot} + {}^{\cdot}O_2^- \rightarrow ONOO^-$$

Peroxynitrite reacts with cysteine and methionine residues in proteins. It also reacts with tyrosine residues, adding NO_2 to the benzene ring of this amino acid. Through protein modification, oxidative stress can affect signal transduction systems and regulate different cell processes.

Harmful Effects of Reactive Species

ROS and RNS can damage various components of the cell, including DNA, protein, lipid, and carbohydrate molecules.

In DNA, free radicals induce chemical modifications of purine and pyrimidine bases. For example, hydroxyl radicals generate 8-OH-guanine, thymine glycol, or 5-OH-uracil. The amount of these compounds that are excreted in urine can be taken as an indicator of the extent of oxidative damage. In addition, free radicals can react with the deoxyribose–phosphate links in DNA, rupturing DNA strands. Some of these modifications are mutagenic; accumulation of

mutations beyond the repair capacity of cells can lead to cancer.

In proteins, oxidation of sulfhydryl groups in cysteine residues results in formation of disulfide bridges or S-nitrosylated derivatives, which alter the structure of polypeptides. Other amino acid residues, including arginine, methionine, histidine, and proline, can also be modified by hydroxyl radicals. The alterations in protein oxidative damage range from breaks to conformational changes of the polypeptide chain. In the case of enzymes, these changes can reduce or suppress their activity.

In lipids, ROS particularly react with unsaturated lipids of the cell membrane. Also, blood plasma lipoproteins are sensitive to oxidation. Peroxides produce conformational changes and decrease the hydrophobicity of the fatty acid hydrocarbon chains in the cell membrane, which alters its structure. Erythrocytes, for example, easily undergo hemolysis. Peroxidation cleaves unsaturated fatty acid chains with three or more double bonds, producing compounds such as *malondialdehyde* (Fig. 10.1).

Determining the concentration of malondialdehyde in blood and urine provides an index of the damage caused by ROS.

On carbohydrates, hydroxyl radicals act by subtracting hydrogen ions and generating radicals at random. This results in the formation of peroxyl groups and breakdown of polysaccharides such as hyaluronic acid.

Pathologies associated with ROS effects. These include the following:

Cancer. The damage caused by free radicals in DNA bases and structure and alterations

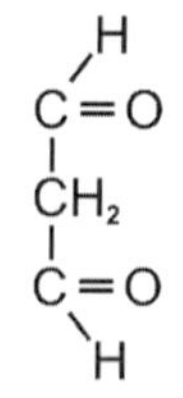

FIGURE 10.1 **Malondialdehyde.**

of proteins with genomic regulatory activity are important factors in the initiation and promotion of cell malignant transformation.

Atherosclerosis. Low density lipoproteins (LDL) penetrate the vascular intima through pores in the endothelium and bind to extracellular matrix proteoglycans. A direct correlation exists between the amounts of circulating LDL and its accumulation in the vessel wall. Lipoprotein deposits in the endothelium favors their oxidation by hydrogen peroxide, peroxide radicals, and hydroxyl radicals. This oxidation stimulates the expression of factors that attract leukocytes to the affected area and initiate an inflammatory reaction. Monocytes that arrive at the site where oxidized LDL is accumulated are converted into macrophages. These cells internalize and store LDL and the cytoplasm of macrophages becomes filled with oil droplets, giving them an aspect of foamy cells. Clusters of foamy cells form visible fatty streaks. The release of cytokines and growth factors at the site of inflammation promotes proliferation of smooth muscle cells and synthesis of collagen and elastin, which are responsible for vascular fibrosis and vessel stiffness. Foamy cells die and release cholesterol esters, which constitute the initiator of the atherosclerotic plaque. Therefore, the oxidation of LDL by reactive species is an important factor in the development of atherogenesis.

Reperfusion syndrome. In myocardial infarction, stroke, and other conditions that determine insufficient blood supply to tissues (ischemia), ROS, especially H_2O_2, are generated. This is due to: (1) in situ release of compounds that attract and activate neutrophils, with subsequent production of H_2O_2 and superoxide. (2) Disturbance in the respiratory chain, which allows electron release and production of superoxide. (3) Interruption of ATP production, which leaves an excess of ADP that is catabolized to hypoxanthine, xanthine, and uric acid. During these reactions H_2O_2 is produced, in a reaction catalyzed by xanthine oxidase.

In the treatment of ischemia, oxygen is administered and if the blood supply is restored, reperfusion could aggravate the damage, as xanthine oxidase continues contributing to the accumulation of hydrogen peroxide.

Rheumatoid arthritis. Generation of reactive species in the synovial fluid is a cause of damage in inflammatory processes of joints.

Cataracts. Lens proteins in the eye have numerous cysteine and methionine residues, both of which are very sensitive to oxidation. Light can contribute to oxidative stress of the lens and the resulting alterations of protein structure are responsible for the loss of lens transparency.

Lung and *brain* are particularly sensitive to oxidizing agents. Breathing high concentrations of oxygen for long periods of time favors the formation of ROS, which can cause severe alterations in the lung and nervous system.

Reactive species have also been implicated in the process of aging. Accumulation of effects of free radicals on cell membranes, DNA, and proteins contribute to acceleration of cellular aging.

Risk groups. Chances of oxidative damage are higher in certain individuals. For example, *preterm infants* have an immature, inadequate antioxidant system to face the challenge posed by high oxygen tensions in incubators or light therapy used for the treatment of hyperbilirubinemia.

Smokers are particularly exposed to free radicals and have increased risk of diseases induced by reactive species. Smokers suffer from an extent of damage in the lung's mitochondrial DNA that is sixfold greater than in nonsmokers. Active or passive exposure to tobacco smoke reduces the activity of the respiratory chain and energy generation in the alveolar cells. Furthermore, the carbon monoxide from tobacco inhibits cytochrome c oxidase.

Alcohol abusers develop an increased metabolic capacity to handle ethanol overloads and this causes increased generation of reactive species.

People who usually work in an environment containing volatile solvents or are exposed to exhaust fumes are also at risk for oxidative damage. Another condition that favors production of reactive species is increased exposure to ionizing radiations.

Oxidative damage has also been described in many diseases, including diabetes, emphysema, cataracts, acute renal failure, Down syndrome, and Parkinson's disease.

Cell Signaling Role of Reactive Species

While production of reactive species in relatively high amounts frequently result in tissue damage, at low concentrations, they have a beneficial effect. Controlled production of reactive species plays an important role in cell function. ROS are increasingly recognized as important second messengers, which participate in signal transduction systems (redox signaling). Redox signaling regulates the reversible and mild oxidation of transcription factors, protein kinases, phosphatases, receptors, and ion channels, to influence different events in the cell. For example, ROS and RNS generated by NADPH oxidase and NOS activate signaling pathways that control cellular differentiation and apoptosis. Mitochondria are the site where redox signaling mediated by lipid oxidation products is coordinated. Likewise, the production of superoxide and oxidized lipid derivatives is not always harmful, but responsible for posttranslational modifications. Some lipid derivatives or breakdown products from complex lipids, such as lysophosphatidylcholine, also act as signaling molecules.

Defense Mechanisms Against Reactive Oxygen Species

To counteract their toxicity, the body has efficient protective mechanisms against reactive species. Under normal conditions, at the O_2 tension of atmospheric air, the concentrations of these toxic agents remain at very low levels. This is due to the action of endogenous antioxidant systems and exogenous compounds provided by food.

Endogenous systems. These include:

Superoxide dismutase (SOD), an enzyme with extremely high activity and widely distributed in all tissues, which catalyzes the removal of superoxide anions from tissues. The reaction involves two superoxide anions and two protons:

$$\cdot O_2^- + \cdot O_2^- + 2H^+ \rightarrow H_2O_2 + O_2$$

The transfer of an electron from one superoxide anion to another corresponds to a dismutation reaction. Hydrogen peroxide is formed; it is also a toxic product and must be removed. Superoxide dismutase presents two isozymes, one that contains zinc and copper (located in the cytosol) and the other that has manganese (located in mitochondria). Zn, Cu, and Mn are essential components of SOD and indispensable for oxidative stress protection. *Catalase.* Decomposition of hydrogen peroxide is catalyzed by catalase, an active hemoprotein present in almost all cells, particularly liver, kidney, bone marrow, and blood cells. Phagocytes are rich in this enzyme. Catalase produces water and two oxygen atoms from two molecules of hydrogen peroxide:

$$2H_2O_2 \rightarrow 2H_2O + O_2$$

The enzyme is found predominantly in peroxisomes, small organelles containing oxidases that produce hydrogen peroxide. Although at very low levels, catalase can also be found in cytosol, mitochondria, and microsomes.

Little or no catalase activity has been described in patients with a genetic defect known as *acatalasemia*. Interestingly, this deficiency does not cause major clinical symptoms, likely due to compensation by other enzymes with the activity of catalase.
Glutathione peroxidase (GPx). This selenium-containing enzyme catalyzes the reduction of hydrogen peroxide and organic hydroperoxides (ROOH), using glutathione (GSH) as a hydrogen donor.

$$ROOH + 2GSH \rightarrow GSSG + ROH + H_2O$$

$$H_2O_2 + 2GSH \rightarrow GSSG + 2H_2O$$

In these reactions, each glutathione gives an electron from its —SH group, which becomes a thiyl radical (–S·). Two of these radicals form oxidized glutathione (GS—SG) with a disulfide bridge. Due to the role of selenium in the activity of glutathione peroxidase, this micronutrient mineral is considered an antioxidant. Glutathione peroxidase is localized in cytosol and mitochondria. It helps maintaining low concentrations of hydrogen peroxide in tissues, converts fatty acid peroxides to hydroxy derivatives, and reverses the oxidation of sulfhydryl groups in proteins and other compounds.
Oxidized glutathione is reduced by the action of *glutathione reductase*, an NADPH-dependent enzyme:

$$GS - SG + NADPH + H^+ \rightarrow 2GSH + NADP^+$$

Other peroxidases contribute to the removal of hydrogen peroxide. They are hemoproteins very common in plants and are also present in leukocytes and milk. They catalyze the reaction:

$$H_2O_2 + AH_2 \rightarrow 2H_2O + A$$

where AH_2 represents different hydrogen donor substrates used by peroxidases. The enzyme of neutrophil granulocytes uses NADPH as a hydrogen donor.

Other molecules synthesized in the body that possess thiol groups, and can act as electron donor/acceptor, are thioredoxin and lipoic acid.
Coenzyme Q. This compound, also designated Q_{10} or ubiquinol, functions as a reducer to eliminate free radicals, including hydroxyl radicals. The reduced form of CoQ ($CoQH_2$) is soluble in lipids and found in the inner mitochondrial membrane, where it forms part of the respiratory chain. It is also found in lipoproteins, protecting them against oxidation and interrupting the formation of peroxy radicals. In the process, $CoQH_2$ becomes oxidized to $CoQH_2$ semiquinone (CoQH).

Synthesis of these antioxidant agents is induced by the *erythroid nuclear factor* and related factors (Nfe, Nrf1, Nrf2, and Nrf3), which are transcription factors. When activated by oxidizing compounds, these cytosolic transcription factors move to the nucleus where they bind specific sites in DNA and promote production of components of the defense system.
Dihydrolipoic acid. This compound is the reduced form of lipoic acid, a powerful reducing agent that functions in aqueous, as well as in lipid media.
Metallothionein. This is a protein rich in cysteine and methionine residues that functions as a neutralizing agent for hydroxyl radicals.

Uric acid and plasma bilirubin also have antioxidant properties.

Nutrients with antioxidant properties. Vitamins E and C are the most important. The major difference between these vitamins is their solubility properties. Vitamin C, or ascorbic acid, is water soluble and is found in the intracellular and extracellular aqueous media. Vitamin E is highly lipophilic and localized in membranes and as part of lipoproteins.

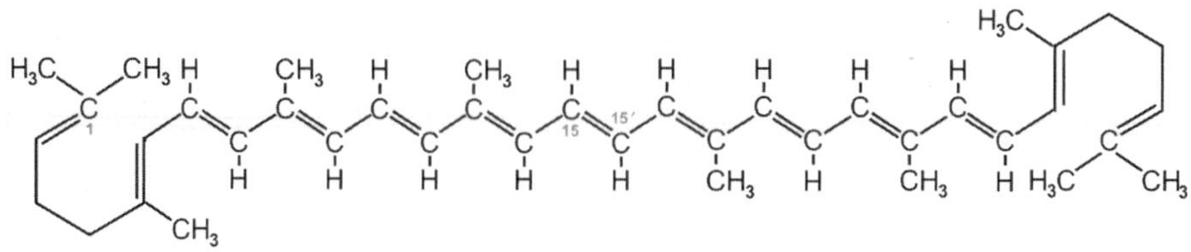

All-*trans*

FIGURE 10.2 **Lycopene.**

Carotenoids are another group of micronutrients that act as protectors against ROS.

Ascorbic acid (see p. 680). Vitamin C is involved in the defense against superoxide ions, hydrogen peroxide, hydroxyl radicals, and singlet oxygen. It also neutralizes RNS. Ascorbic acid ($AscH_2$) dissolved in blood and cytosol can act quickly before cell damage occurs. Ascorbate can reduce the hydroxyl radical producing water and ascorbyl radical ($AscH^.$), which is relatively stable and less active.

$$AscH_2 + {}^.OH \rightarrow H_2O + AscH^.$$

With superoxide anion ascorbic acid forms hydrogen peroxide. Initially ascorbate ($AscH_2$) becomes ascorbyl radical. Two ascorbyl groups react to form a molecule of ascorbic acid and another of dehydroascorbic.

Vitamin C also reacts with the hydrogen peroxide to produce water and dehydroascorbic acid (Asc):

$$AscH_2 + H_2O_2 \rightarrow 2H_2O + Asc$$

Vitamin E (see p. 656). This vitamin is usually located in cell membranes with the phytyl chain inserted in the double layer and the chromanol group facing the surface. It donates the hydrogen of the chroman ring —OH group, interrupting chain reactions of lipid and lipoperoxyls radicals. It is important to protect lipoproteins.

$$L^. + VitE.OH \rightarrow LH + VitE.O^.$$

$$LOO^. + VitE \, L^. + VitE.OH \rightarrow LH$$
$$+ VitE.O^..OH \rightarrow LOOH + VitE.O^.$$

Tocopheryl radical ($VitE.O^.$) is formed, which is a relatively low reactive semiquinone that is stabilized by resonance. Tocopheryl is not an antioxidant and should be recycled to its initial state to maintain its ability to neutralize membrane peroxyl groups. The most important of the eight vitamers of vitamin E is α tocopherol.

Carotenoids. These organic pigments are effective in the removal of singlet oxygen and peroxy radicals (see p. 648). For singlet oxygen, carotenoids in membranes and in solution are able to absorb the extra energy to form triplet or basal oxygen. The ability of carotenoids to neutralize singlet oxygen is related to the conjugated double bonds of its side chain. Carotenoids release energy to the environment as heat and do not need to be regenerated for further reaction. Lycopene (Fig. 10.2) is a carotenoid with antioxidant activity, abundant in tomatoes.

Flavonoids. These are polyphenols contained in various components of the diet. They have potent antioxidant action in vitro; however, it is not widely accepted that they have the same action in vivo in humans.

Regeneration of antioxidants. As a result of antioxidant reactions, radical and oxidized products are formed which, although being less active, still need to be permanently converted to their original state to regenerate the antioxidant defense system of the body. Ascorbic acid is recycled from ascorbyl to dehydroascorbic by hydrogen transfer from reduced glutathione (GSH) and NADPH. For vitamin E, regeneration must be very active and immediate to preserve its antioxidant capacity. It is estimated that there are only approximately 9 molecules of vitamin E per every 2000 molecules of unsaturated fatty acids present in cell membranes.

Ascorbate is one of the molecules that reconvert the tocopheryl radical (semiquinone) to its original form:

$$VitEO^{\cdot} + AscH_2 \rightarrow VitEOH + AscH^{\cdot}$$

Semiquinone is also reversed to vitamin E in reactions with dihydrolipoic acid or the reduced forms of thioredoxin, glutathione, and CoQ.

Supplementation of antioxidant micronutrients. ROS and RNS are constantly produced and are potentially harmful metabolic products. However, a healthy person, adequately nourished, has the appropriate mechanisms to repair or minimize the potential damage caused by reactive species. In cases in which the generation of free radicals increases either by infections, physical trauma, exposure to various oxidizing agents, or other causes beyond the body's ability to neutralize them, they can render irreversible changes and create conditions that favor the malignant transformation of cells (cancer) or lead to chronic processes (atherosclerosis and others).

In view of the properties shown by micronutrients with antioxidant capacity, it was thought that diet supplementation may have preventive effects on damage and diseases caused by reactive species. To test the effect of antioxidants, the incidence of diseases related to oxidative stress was studied on large number of people, who were administered antioxidant micronutrients, alone or combined, for long periods of time. At present, the results are inconclusive; many of the investigations reporting favorable effects have been the subject of reevaluation by other authors, who found deficiencies in experimental design. Other, more rigorously controlled studies, showed no positive action. Thus, the beneficial effects of antioxidant supplementation have not been sufficiently proven yet. On the other hand, it has been well established that individuals with an appropriate dietary intake of fresh vegetables and fruits have lower risk of chronic diseases than those who do not consume vegetables in sufficient amounts. It seems unquestionable that a diet rich in fresh fruits and vegetables is an essential part of a good nutrition, providing the antioxidant factors required to maintain health.

SUMMARY

Oxidant agents, including ROS and RNS are harmful molecules that can cause *oxidative stress* in tissues and organs. They are either endogenously produced during normal metabolism or they can come from the environment.

Reactive oxygen species (ROS) comprise *singlet oxygen, hydrogen peroxide,* and the *free radicals superoxide anion* and *hydroxyl.*

Nitrogen reactive species include the free radical *nitric oxide* and *peroxynitrite.*

Free radicals are groups of atoms that have one or more unpaired electron(s) in their external orbitals. They tend to react with other molecules to obtain their missing electron.

The deleterious actions of oxidants include damage to DNA, proteins, and lipids, promoting the development of malignant transformation of cells, atherosclerosis, rheumatoid arthritis, and cataracts. Oxidants are also responsible for the ischemia–reperfusion syndrome.

Defense mechanisms operate against oxidizing agents in the body, including endogenously produced molecules and factors that come with the diet.

Superoxide dismutase, catalase and *glutathione peroxidase, coenzyme Q, dihydrolipoic acid,* and *metallothionein* are endogenous factors that protect against oxidation.

Ascorbic acid, vitamin E, carotenoids, and *flavonoids* constitute exogenous antioxidants. A diet containing fresh vegetables and fruits are rich in antioxidants.

Bibliography

Bánhegyi, G., Benedetti, A., Csala, M., Mandl, J., 2007. Stress and redox. FEBS Lett. 581, 3634–3640.

Bonomini, F., Tengattini, S., Fabiano, A., Bianchi, R., Rezzani, R., 2008. Atherosclerosis and oxidative stress. Histol. Histopathol. 23, 381–390.

Buelna-Chontal, M., Zazueta, C., 2013. Redox activation of Nrf2 & NF-κB: a double end sword? Cell Signal. 25, 2546–2557.

Dröge, W., 2002. Free radicals in the physiological control of cell function. Physiol. Rev. 82, 47–95.

Galkin, A., Higgs, A., Moncada, S., 2007. Nitric oxide and hypoxia. Essays Biochem. 43, 29–42.

2005. The antioxidant nutrients, reactive species, and disease. In: Gropper, S.S., Smith, J.L., Grote, J.L. (Eds.), Advanced Nutrition and Human Metabolism. fourth ed. Thomson Wadsworth, Belmont, CA, pp. 368–377.

Gutiérrez, J., Ballinger, S.W., Darley-Usmar, V.M., Landar, A., 2006. Free radicals, mitochondria, and oxidized lipids. Circ. Res. 99, 924–932.

Halliwell, B., 2007. Biochemistry of oxidative stress. Biochem. Soc. Trans. 35, 1147–1150.

Navah, M., Ananthramaiah, G.M., Reddy, S.T., et al., 2004. The oxidation hypothesis of atherogenesis: the role of oxidized phospholipids and HDL. J. Lipid Res. 45, 993–1007.

Valko, M., Leibfritz, D., Moncol, J., Cronin, M.T., Mazur, M., Telser, J., 2007. Free radicals and antioxidants in normal physiological functions and human disease. Int. J. Biochem. Cell Biol. 39, 44–84.

Membranes

The surface of cells is covered by the *plasma membrane*, a very thin film (6–10 nm thick) consisting of lipids, proteins, and carbohydrates. Far from being just a shell, the plasma membrane is a structure with remarkable functional properties: (1) it is a selectively permeable barrier, which controls the passage of solutes and water and prevents the random mixture of components of the extracellular environment with those of the intracellular space or cytosol. Transport systems at the plasma membrane are largely responsible for maintaining the constancy of the cell cytosol. Also, the plasma membrane creates and maintains differences in concentration of many solutes and ions between the intracellular medium and the cell exterior. (2) It provides the proper environment for numerous enzymes that operate within it. (3) It has receptors that bind hormones, growth factors, neurotransmitters, and other chemical messengers, which can activate signaling systems and trigger specific responses in each cell. (4) It actively participates in the incorporation or secretion of particles and macromolecules (endocytosis and exocytosis). (5) It has molecules that function as recognition signals that are essential for the adhesion of cells between them, or to their substrate. (6) In addition, the plasma membrane contributes to maintain the cell shape.

Intracellular organelles are also surrounded by membranes whose basic structure is similar to that of the plasma membrane. Due to their membranes, subcellular organelles maintain their individual composition and function relatively independent from the rest of the cell.

STRUCTURE

All biological membranes are composed of lipids and proteins. The relative amount of these compounds varies considerably in different cell types and in organelles from the same cell. For example, the red blood cell membrane contains approximately 50% proteins and 50% lipids. Membranes of the neuronal myelin sheaths, which cover nerve fibers, have a lipid and protein amount of 80 and 20%, respectively. The inner mitochondrial membrane contains 80% protein. Lipids and proteins are often associated with carbohydrates, glycolipids, and glycoproteins.

Membrane Lipids

Lipids form the basic structure of all biological membranes. They include: (1) phospholipids (glycerophospholipids and sphingomyelin), (2) glycolipids (cerebrosides and gangliosides), and (3) cholesterol. The complex lipids, phospholipids and glycolipids, are amphipathic with a polar "head" and long nonpolar hydrocarbon chains or "tails." This dual polarity has great importance in the structure of membranes.

Lipid films. When amphipathic lipids are placed and extended over an orifice communicating two compartments of a container filled with aqueous solutions, a double lipid layer can be formed. In this layer, the lipids spontaneously orient their polar heads toward the aqueous solution of each compartment, while the nonpolar tails are directed toward the highly hydrophobic interior of the film (Fig. 11.1A).

If a small amount of amphipathic lipids are placed on the surface of an aqueous solution, a one molecule thick layer (monolayer) is formed in which the lipid polar heads are oriented toward the solution, attracted to water, and the hydrophobic chains are oriented away from the solution (Fig. 11.1A). If lysophospholipids or fatty acid salts (soaps) are mixed with an aqueous medium, they spontaneously form amphiphilic small spheres called *micelles* with the hydrophilic lipid heads in contact with the medium and the hydrophobic tails oriented toward the micelle's interior (Fig. 11.1B). The phospho- and glycolipids dispersed in aqueous solutions can also adopt other arrangements.

Liposomes. When higher amounts of amphipathic lipids are mixed in aqueous suspension, they normally organize into closed bilayers, generating vesicles. These vesicles are called liposomes and can be formed by a single bilayer film (monolamellar liposome) (Fig. 11.1B) or by several concentric bilayers arranged like layers in an onion (multilamellar vesicles).

Liposomes have been used in the study of the physicochemical properties of lipid bilayers, since they behave similar to biological membranes. In addition, proteins of the plasma membrane have been incorporated into liposomes of known lipid composition to study their function in a more controlled environment than that of the cell membrane. This has been important, for example, for understanding the function of ion transport proteins of the plasma membrane.

Liposomes have other interesting uses; the vesicles can trap small amounts of the solution in which they were prepared. These "loaded" liposomes can carry different substances (e.g., drugs or nucleic acids), bind and fuse with the plasma membrane of cells, and serve as a vehicle to deliver different compounds into cells.

Nanodiscs. Another membrane model system, which is used for the study of membrane-associated proteins are nanodiscs. These consist of a small discoidal portion of a lipid bilayer formed by amphipathic lipids, which are surrounded on their hydrophobic edge by a membrane scaffold protein (generally a lipoprotein, such as apolipoprotein A) that keeps the lipids in solution.

Aqueous solution

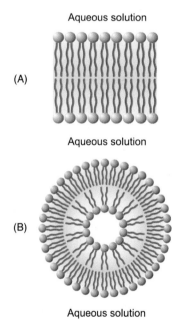

Aqueous solution

Aqueous solution

FIGURE 11.1 (A) Phospholipid bilayer. The lipid polar heads *(red spheres)* lie in contact with the aqueous solution. The hydrophobic chains, indicated by *wavy lines*, are oriented toward the interior of the film. (B) Cross section of a monolamellar liposome vesicle formed by a phospholipid bilayer. The polar heads are located on the outer and inner surfaces of the vesicle in contact with the aqueous solution.

The membrane protein of interest can be embedded in the nanodisc for the study of its structure and arrangement in the membrane. Nanodiscs are uniform in size and allow a native local lipidic environment that mimics cell membranes.

Cell membrane composition. Similar to liposomes, biological membranes are sealed structures surrounding the cell or subcellular organelles. The cell plasma membrane has a cytosolic side facing the cell interior and an extracellular side facing the medium. Membranes from organelles have the external side oriented toward the cell cytosol. Among the molecules that make up the membrane bilayer, the most abundant components are phospholipids, with nonpolar saturated or unsaturated hydrocarbon chains of 10–24 carbons in length (most commonly the chains contain 16–18 carbons).

The saturated carbon chains freely rotate around single bonds, oscillating between an expanded zigzag arrangement (Fig. 5.1), called *trans* or anti, and a contorted arrangement (called gauche). The *trans* configuration, perpendicular to the plane of the membrane, is more commonly found because it has lower free energy. The mutual interactions between these side chains (maintained by van der Waals forces) provide the structural stability needed to maintain the phospholipids grouped. Unsaturated fatty acids contain double bonds most commonly in *cis* configuration, which produces rigid angles that tend to separate the hydrocarbon tails (Fig. 5.1B); this gives them a more open arrangement.

Phosphatidylcholine is the predominant phospholipid in membranes; phosphatidylethanolamine, phosphatidylserine, sphingomyelin, and phosphatidylinositol are present in smaller amounts. Membranes contain glycerophospholipids. The hydrolysis of the ester linkage of the fatty acid at position two is catalyzed by phospholipase A; the reaction releases arachidonate, an eicosanoid precursor. Also, the plasma membrane contains phosphatidylinositol, phosphatidylinositol-4,5-bisphosphate, which is part of a signal transduction system activated by hormones; its hydrolysis leads to the formation of inositol-1,4,5-trisphosphate and diacylglycerol, both of which function as second messengers (p. 560).

Glycolipids, mainly cerebrosides and gangliosides form a small proportion of the membrane components. The carbohydrate portion of these molecules often functions as cell "recognition signals."

Cholesterol is quantitatively important, mainly in the plasma membrane. It has a hydroxyl group at C3; the rest of the molecule, with the steroid nucleus and the hydrocarbon chain on C17, is hydrophobic. Different from phospho- and glycolipids, cholesterol does not form bilayers. It is inserted in the membrane orienting its hydroxyl group toward the polar heads of amphipathic lipids, and the flat cyclic core and extended side chain toward the hydrophobic tails in the nonpolar region of the double layer.

The various lipid components described are kept in order within the bilayer due to interactions with the aqueous medium on one side and the hydrophobic chains of neighbor lipids on the other, but there are no covalent bonds between them.

The lipid bilayer is asymmetric. Each layer of the cell membrane has a different composition, which makes the membrane asymmetric. Generally, phosphatidylcholine and sphingomyelin predominate in the outer layer, while phosphatidylethanolamine and phosphatidylserine are more abundant in the inner layer (cytosolic for the plasmatic membrane).

The lipid membrane composition varies in cells from tissues of the same individual. For example, the molar ratio of phospholipid:cholesterol:glycolipids is 1:0.8:0.1 for erythrocyte membranes and 1:1.3:0.7 for myelin. Even greater differences can be found within the same cell, between the plasma membrane and membranes from organelles. For example, mitochondrial membranes have *cardiolipin*, a phospholipid that is absent from other membranes. Glycolipids are components almost exclusively present in the external face of the plasma membrane. Cholesterol is abundant in plasma membrane and very low in the inner mitochondrial membrane.

Membranes are fluid films. At physiological temperatures, the lipid bilayer behaves as a fluid structure. Fluidity is greater as the proportion of unsaturated fatty acids that constitute the membrane increases. Saturated fatty acid chains preferably adopt an extended conformation, forming compact groupings that confer stiffness to the membrane. As temperature increases, a larger number of C—C bonds in the hydrocarbon chains of phospholipids lose their extended state, adopting the gauche conformation. This alters the packing of the chains and increases the membrane's fluidity. Independent from temperature changes, increases in the amount of unsaturated fatty acids also disturbs the approximation of the chains and enhances the fluidity of the membrane. Cholesterol has different effects: at high temperature, it interferes with the movement of the hydrocarbon chains and tends to reduce the membrane's fluidity. In contrast, at low temperatures, cholesterol decreases the rigidity of the hydrocarbon chains.

The fluidity of the lipid bilayer allows its components to be displaced laterally with certain freedom and to rotate on their axis perpendicularly to the membrane. In addition to this ability to move within the plane of the plasma membrane, phospholipids can eventually jump from one membrane layer to the other in what is called the "flip-flop" movement. This shift of components between membrane's lipid leaflets is more restricted than the displacement within each layer.

Flip-flop movement of membrane lipids does not occur spontaneously; in the case of phospholipids, the polar portion of the molecule does not penetrate easily through the hydrophobic zone of the double layer. For this reason, the change from one leaflet to the other must be catalyzed by enzymes. *Flippases* are the enzymes that transport lipids from the external surface toward the cytosolic surface of the plasma membrane. They also move lipids from the internal side of organelles to the cytoplasmic side. This transport is carried out against the concentration gradient and requires energy, provided by hydrolysis of ATP. *Floppases* are lipid carriers that work in the opposite direction of flippases; that is, they carry lipids from the cytosolic layer to the other side of the membrane. These enzymes are also ATP-dependent. Other enzymes, called *scramblases*, do not require ATP and use the concentration gradient to facilitate flip-flop in both directions; they do not need energy supply.

Membrane Proteins

Proteins constitute an important component of biological membranes. They are associated to the components of the lipid bilayer by noncovalent interactions. Depending on the nature of these interactions, membrane proteins can be distinguished as *integral* or intrinsic, and *peripheral* or extrinsic.

Integral proteins are inserted or embedded into the lipid bilayer. Most of them completely traverse the membrane one or more times (Fig. 11.2), while others are partially inserted, reaching the middle hydrophobic region of the lipid bilayer. Often, oligosaccharides are covalently bound to protein domains that face the extracellular side of the plasma membrane.

Integral proteins show different characteristics in different areas of the membrane. Regions exposed to the aqueous medium are enriched in hydrophilic amino acid residues. Portions traversing the lipid hydrocarbon chains have greater proportion of nonpolar amino acids. Generally, the transmembrane domains adopt the α helical structure (Fig. 11.3) made of 20–25 nonpolar amino acid residues that expose a hydrophobic surface to the membrane lipids. Some proteins form cylindrical structures composed of the association of several α helices. Although most of the protein transmembrane domains have an α helix structure, others consist of β

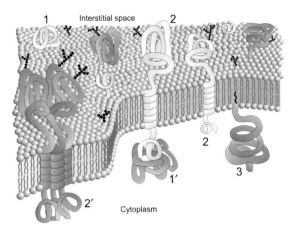

FIGURE 11.2 **Structure of a biological membrane (fluid mosaic model).** The basic lipid bilayer, associated proteins, and carbohydrates are shown. 1, Peripheral protein juxtaposed on the outer face; 1', peripheral protein linked to the cytoplasmic domain of an integral protein; 2, integral proteins embedded in the bilayer by a transmembrane α helix; 2', integral protein with multiple transmembrane segments; 3, peripheral membrane protein anchored by a lipid. Chains of *black circles* represent oligosaccharides linked to proteins or lipids on the outer surface of the membrane.

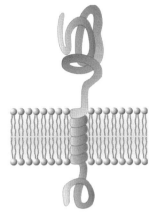

FIGURE 11.3 **Integral membrane protein with a transmembrane α helix and external and internal domains.**

sheets arranged in a barrel-like shape. Many of these structures, with multiple α helices or β sheets, form a tunnel spanning the membrane. These tunnels consist of an interior lined with polar groups, which creates pathways for the transport of different substances across the membrane.

For integral proteins that span the membrane more than once, the segments that connect the protein transmembrane domains form loops that emerge on both sides of the lipid bilayer (Figs. 11.2 and 11.3).

Extracting integral proteins from the membrane requires drastic procedures, such as the use of special detergents or solvents. Peripheral proteins do not reach the hydrophobic core of the lipid bilayer. They are simply juxtaposed on one side of the membrane, connected by noncovalent interactions between the polar side chains of amino acids in the peripheral proteins and the polar heads of lipids on the surface of the membrane (Fig. 11.2). These proteins can be easily removed from the membrane by mild treatment with saline solutions. Some peripheral proteins are attached to the membrane by anchoring structures. These may be a fatty acid, an isoprenoid chain, or complexes, such as the glycosyl-phosphatidyl-inositol (GPI) (Fig. 11.4).

The fatty acids used to anchor proteins to the membrane include myristic and palmitic acid, made of 14 and 16 carbon atoms, respectively. Myristic acid is bound by an amide link to the amine function of glycine in the N-terminal end of the polypeptide. Palmitic acid is generally bound to the sulfur of a cysteine residue near the C terminus of the polypeptide chain. The hydrocarbon chain of the fatty acids is inserted into the membrane where it interacts with the hydrophobic environment of the bilayer. Two types of isoprenoid chains serve to insert proteins into cell membranes: farnesyl, which has 15 carbon atoms, and geranylgeranyl, made of 20 carbon atoms. These chains are linked by a thioester bond to the cysteine near the polypeptide C-terminal end. Generally, proteins anchored by fatty acids or isoprenoid chains are in the inner side of the plasma membrane; some of them have important roles in cell signaling systems.

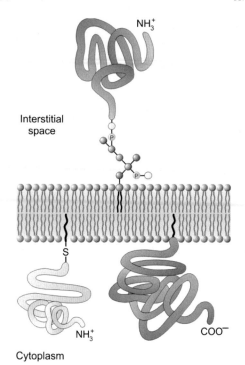

NH$_3^+$

Interstitial
space

S

NH$_3^+$ COO$^-$

Cytoplasm

FIGURE 11.4 **Peripheral proteins anchored in the membrane lipids.** A protein attached to the outer side by glycosyl-phosphatidyl-inositol (GPI) and two attached to the cytosolic side by lipid chains are shown.

Another protein anchor system is GPI. It is an oligomeric structure, whose basic skeleton consists of phosphatidylinositol, glucosamine, mannose, galactose, and phosphoethanolamine. The hydrophobic chains of the phosphatidyl portion of GPI are the segments in the molecule that serve as attachment sites. The phosphoethanolamine, at the other end of GPI, binds to the terminal carboxyl group of the polypeptide. GPI contributes to protein binding at the outer surface of the membrane. Alkaline phosphatase and acetylcholinesterase are examples of proteins that are fixed by GPI.

The proteins attached to membranes present greater asymmetry than lipids.

Proteins can displace laterally in the membrane or rotate on their axis perpendicular to membrane; they can be compared to icebergs floating in the fluid lipid bilayer. In 1972, the concept of the membrane as a *fluid mosaic* was proposed by Singer and Nicolson (Fig. 11.2). The ability to migrate within the membrane facilitates interactions between proteins and lipids, making it a dynamic and transient phenomenon. However, in most cases, protein–lipid associations are relatively stable. The lipids that surround a given protein (lipid annulus) generally maintain their contact with a specific protein. Proteins isolated from membranes commonly lose the properties they displayed when in the native environment of the membrane. These properties are sometimes recovered after addition of the corresponding lipids, indicating that the lipid annulus is important for proper protein conformation and function. These observations have led to a closer analysis of the lipids surrounding integral proteins. The membrane has "lipid domains" of different lipid composition, where particular proteins are embedded. These microenvironments in the membrane provide the appropriate medium for each protein to be fully active.

The membrane mobility of some proteins is restricted by their binding to the cytoskeleton, which localizes them to a certain position. For example, integral membrane proteins of erythrocytes, such as glycophorin and the anion transporter band 3, are immobilized in the membrane by association with the cytoskeleton (ankyrin, actin, spectrin). This mechanism forms a complex network that helps to maintain the red blood cell shape. In addition, association to peripheral proteins also maintains glycophorin and anion transporter band 3 in place.

Some cells (e.g., those of renal tubular epithelium and intestinal mucosa) are polarized; the plasma membrane shows two functionally distinct areas. The membrane domain located toward the tubular lumen corresponds to the *apical* membrane and the remaining membrane, in contact with neighboring cells and the interstitial space, corresponds to the *basolateral* membrane. Each of these membrane domains has a different protein composition; while proteins can freely diffuse within each membrane domain however, they cannot cross from the apical to the basolateral membrane due to the existence of *tight junctions*. Tight junctions are protein complexes that serve to maintain the adhesion between epithelial cells and also to create a barrier that limits the movement of lipids and proteins, confining them to the apical or the basolateral domains of the membrane.

Protein insertion in the membrane. Proteins that are synthesized in ribosomes bound to the rough endoplasmic reticulum (ER) are introduced into the lumen of this organelle or attached to the membrane by a channel or translocation system. The proteins to be inserted have sequences that serve as signals which indicate the cell protein synthesis machinery that they need to be attached to, or translocated across the membrane. This mechanism is outlined in Figs. 11.5 and 11.6.

In the N-terminal portion of the protein, there is a hydrophobic segment of 25–30 amino acids that serves as a signal which attaches the polypeptide chain to the channel or protein translocation complex. Transfer start segments promote the passage of the polypeptide through the channel as it is synthesized. If there are no additional signal sequences, the rest of the protein is transferred across the membrane and remains attached to it by the signal segment (Fig. 11.5). If the protein is destined to the ER lumen, the segment is then cleaved by action of a signal peptidase within the ER and the protein is released into the ER. Other proteins possess hydrophobic segments of about 25 amino acids in inner regions of the chain and bind to the channel to stop the passage of a polypeptide chain (stop transfer sequences). This allows for a portion of the polypeptide chain to be retained in the lipid bilayer. For some proteins, the N-terminal domain remains immersed in the cytosol and the C-terminus remains inside the ER; in other proteins, the opposite occurs. In these cases the resulting integral protein has a single transmembrane α helix segment (Fig. 11.3).

Frequently, the protein crosses the lipid bilayer several times, and has multiple transmembrane segments. This is determined by the presence of a series of transfer start and stop transfer signals alternating along the polypeptide chain. This establishes loops between transmembrane

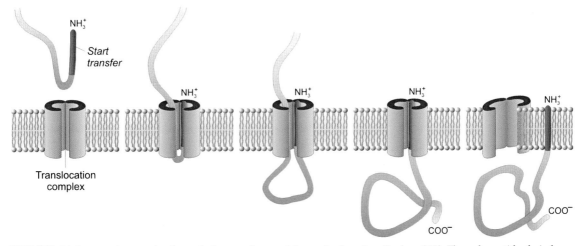

FIGURE 11.5   **Protein transfer through the membrane of the endoplasmic reticulum (ER).** The polypeptide chain has a start or start transfer *(black segment)* signal, which is fixed inside the translocation channel while the rest of the protein is transferred to the ER lumen. Once the protein passage is completed, the translocation channel opens and leaves the protein inserted in the membrane. Subsequently, the signal segment is cleaved by a peptidase and the protein is released into the lumen.

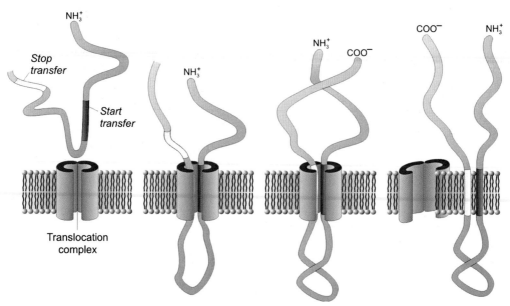

FIGURE 11.6 **Insertion of integral proteins in the membrane.** The polypeptide chains have internal start *(black)* and stop *(white)* signals. A loop with the start signal is initially inserted into the translocation channel, the chain is driven through the channel until the stop sign is reached, which is retained. The existence of multiple alternating start and stop signals along the protein chain allows it to repeatedly cross the double layer several times. Protein loops emerge on either side of the membrane. When there are multiple transmembrane segments, terminal ends may be on the same or on opposite sides of the membrane.

domains of the protein that emerge from both sides of the membrane (Figs. 11.2 and 11.10). The N- and C-terminal ends of the protein may be located in the same or opposite sides of the membrane. After reaching the correct position, the translocation complex channel opens laterally and the protein is inserted in the bilayer.

Many of the integral proteins inserted in the ER membrane are exported to other organelles or to the plasma membrane. To accomplish this, membrane portions containing the embedded proteins emerge from the ER as vesicles and are delivered to their final destination by the mechanisms that will be described in p. 246. The final arrangement of proteins in the membrane is in most cases asymmetric. Protein orientation is maintained at new sites of implantation (all portions of the protein facing the cytosol into the ER membrane will remain oriented to the cytosol in both the target organelle and plasma membrane).

Carbohydrates

The plasma membrane contains carbohydrates covalently linked to lipids (cerebrosides and gangliosides) or to proteins (glycoproteins). Carbohydrates in the outer face of the membrane form the cell *glycocalyx*.

Carbohydrates attached to lipids or proteins may be monosaccharides (glucose or galactose in cerebrosides) or oligosaccharides of different complexity; commonly, branched carbohydrates are found (gangliosides and glycoproteins).

Membrane carbohydrates are important for intercellular recognition (cell–cell interactions) or for binding of different ligands (messenger molecules, toxins, bacteria, viruses). An example of the role of carbohydrates in cellular interactions is the adhesion of leukocytes to the endothelium of blood vessels. This takes place prior to the migration of phagocytic cells from the blood to the sites of inflammation. A family of transmembrane proteins, known as *selectins*, is able to recognize and bind specific oligosaccharides on the surface of leukocytes.

Lectins are proteins found in plants, bacteria, and animals which recognize and selectively bind certain carbohydrates on the cell surface. They have been widely used for the identification and study of cell surface carbohydrate composition and function.

Subcellular Organelles

Internal membranes in cells maintain the identity of subcellular organelles. An exception is mature red blood cells, which do not have organelles and therefore lack internal membranes. While the basic structure of the lipid bilayer is the same for all organelles and the plasma membrane, the lipid and protein composition is significantly different. The type of molecules present in each membrane responds to the specific functions of the organelle.

The main intracellular compartments surrounded by membranes in animal cells are the following:

1. *Nucleus.* This structure is separated from the cytoplasm by a double membrane envelope with large pores that allow passage of macromolecules (RNA and proteins) between the nucleus and cytosol. The nucleus houses the chromosomes, which contain most of the genetic material of cells. It is the site of DNA and RNA synthesis. Messenger RNAs, ribosomal RNAs, transfer RNAs, and other types of RNAs pass to the cytosol through these membrane pores.

2. *Mitochondria.* Total oxidation of various metabolites and production of most of the chemical energy produced by the cells occur in these organelles. The mitochondrial external and inner membranes are lipid bilayers. The inner mitochondrial membrane contains the components of the respiratory chain (p. 189).

3. *ER.* This organelle is formed by a network of membranous ducts that occupy a significant portion of the cell's cytoplasm. Part of this system is associated with ribosomes, which gives it a rough appearance when viewed with the electron microscope, thus its designation as *rough* ER. Another section of the ER, the *smooth* ER, does not have ribosomes; it contains enzymes related to the synthesis of lipids, steroid hormones, and metabolism of foreign substances (drugs and poisons). Processing of membrane-associated proteins to be exported to different locations in the cell begins in the ER.

4. *Golgi complex.* This is a series of flattened membranous sacs that receives proteins synthesized in the ER. It is involved in completing the processing of proteins, which are then targeted to their final destination in the cell. In the Golgi, the oligosaccharide chains of glycoproteins are modified and elongated; also glycolipids are synthesized. Both the ER and Golgi are sealed membranes; the material exported to other destinations in the cell or out of the cell travel in vesicles formed by the membrane of these organelles.

5. *Lysosomes.* These are vesicles that contain hydrolytic enzymes capable of degrading different types of molecules that need to be removed from the cell. They have a pH of 5.0, which is more acid than that of the cytosol. The products of lysosomal digestion pass to the cytosol where they are metabolized or reused in the synthesis of new molecules.

6. *Peroxisomes.* These organelles metabolize compounds by oxidative reactions using oxygen and generating hydrogen peroxide, which is later removed by a catalase present in the same organelle.

TRANSPORT ACROSS MEMBRANES

The movement of ions and solutes between cells and the surrounding medium or between subcellular organelles and the cytoplasm, require their passage through membranes. The transmembrane exchange of substances is carried out by different membrane transport mechanisms, which will be discussed in the following sections.

Diffusion

When a solute is added to a suitable solvent, the solute tends to disperse in the whole volume of the solvent. The movement of these particles, called *diffusion*, takes place from the region with a higher solute concentration to that of a lower solute concentration in the solution. This occurs at a speed that is proportional to the difference in solute concentration or gradient. If the diffusing particle has electrical charge, the difference in electrical potential as well as the concentration gradient contribute to favor diffusion. In this case, the movement of the solute will depend on the *electrochemical potential gradient*.

All molecules dissolved in a liquid or gaseous solvent are constantly moving, following a random path (Brownian motion). When solute is added and its concentration in a region of the solution increases, the random motion of the solute continues. However, because the number of particles per volume unit is higher in one site of the solution than in another, more molecules move toward the site of lower concentration than in the reverse direction. The result is a net flux or diffusion of the solute toward the place where it is less concentrated. The direction and magnitude of diffusion are related to the gradient. In addition, the flow rate depends on the *diffusion coefficient* (*D*), the size of the molecules, and the temperature and viscosity of the solvent. Fick's first law defines the flow (*J*) through a given area (*A*) in aqueous solutions, using the following equation:

$$J = DA\left(-\frac{dC}{dx}\right)$$

where dC/dx represents the concentration difference over a distance x (solute concentration gradient).

After some time, a uniform distribution of solute in the solution is attained. An equilibrium is reached in which there is no net flow of substance, even though the molecules continue moving randomly.

The diffusion follows the solute's gradient in a passive or "downhill" process that proceeds spontaneously, with no need to provide energy to the system.

The displacement of molecules by passive diffusion can also take place between two aqueous compartments separated by a lipid bilayer. Of course, the presence of the membrane represents a barrier that limits free diffusion and the moving particles need to overcome the barrier represented by the bilayer. Lipid soluble substances diffuse more easily through the hydrophobic region of the membrane. There is a linear relationship between the rate of diffusion of a substance through biological membranes and lipid solubility (expressed by the *oil/water partition coefficient*) (*oil/water partition coefficient* of a substance is measured by mixing and stirring water and oil with the compound of interest. After a while, when the two phases separate, the concentration of the substance dissolved in each of the solvents is determined. The ratio of solute concentration in oil/solute concentration in water, gives the oil/water partition coefficient value).

A solute that enters or leaves cells and organelles by diffusion must pass through the lipid bilayer. The hydrophobic interior of the membrane has different solvent properties compared to those of water. The flux is dependent, among other factors, on the membrane permeability of the solute. Fick's law of simple diffusion can also be applied for the diffusion of a solute through membranes:

$$J = PA\ (C1 - C2)$$

where J is the flow; P, permeability coefficient; A, area considered; $C1$ and $C2$, solute concentrations on either side of the membrane.

Small nonpolar molecules, such as O_2, N_2, and CO_2 and larger liposoluble compounds, such as fatty acids and steroid hormones, freely diffuse through membranes. Water and urea, despite being polar, can still cross cell membranes because they are small and have no net charge. In contrast, the flow of larger polar molecules is limited. Hexoses, for example, diffuse with great difficulty. Although small, ions are unable to cross the lipid bilayer.

Simple diffusion takes place spontaneously at a rate directly proportional to the concentration difference (gradient) between one side of the membrane and the other. The flux rate of the solute depends on the permeability coefficient, which includes several factors: (1) the partition coefficient of the solute in lipids, (2) the mobility of the solute in the lipid bilayer (the inner side of the bilayer consists of the hydrocarbon chains and is a medium of higher viscosity than water), and (3) the thickness of the lipid bilayer, which is ~5 nm.

Molecules with low permeability coefficient due to their size or polar nature diffuse slowly through lipid bilayers. However, the flow of many of these molecules across biological membranes is greater than that expected when considering the Fick's equation. This is due to the existence of transport systems, which are integral proteins embeded in the membrane. Among these systems are carriers and channels, which are polypeptide chains with multiple transmembrane segments. These segments form a pore through which the polar solutes can pass, without coming in contact with the hydrophobic interior of the bilayer.

Different types of carriers can be distinguished: (1) *Uniporters*, which transfer one type of solute across the membrane (Fig. 11.7A). (2) *Symporters*, which simultaneously carry two different solutes across a membrane in the same direction. This movement is either from the

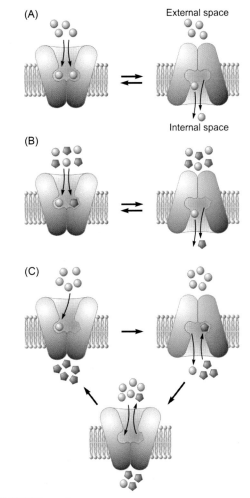

FIGURE 11.7 **Models of different transport system.** The carrier protein may exist in two conformational states. In one of them, the solute is bound to one side of the membrane and in the other the solute is exposed to the opposite side. (A) *Uniport*: the binding site specifically recognizes a particular solute. Upon solute binding, the carrier undergoes a conformational change, which opens to the other side of the membrane to release the solute. (B) *Cotransport*: the carrier has specific binding sites for two different solutes. On the left, each is bound to its respective site. Conformational change occurs and both solutes are released to the other side of the membrane. (C) *Countertransport*: the carrier has specific sites for two different solutes. In the initial position, one of the solutes is bound to its site, this triggers the conformational change of the transporter, moving it to the opposite side, from which the other solute is bound and transported in the reverse direction.

extracellular space into the cytoplasm or vice versa (Fig. 11.7B). (3) *Antiporters* or *exchangers*, which transport a solute in one direction while simultaneously transporting another solute in the opposite direction (countertransport) (Fig. 11.7C). Symports are similar to exchangers because the transfer of one solute is coupled to the other.

The transport through a number of channels and carriers is driven by the chemical or electrochemical gradient. This takes no energy expenditure and is called *passive transport* or *facilitated diffusion*. Other carriers are capable to transfer solutes through the membrane against their gradient. This process, called *active transport*, requires energy input, which in most cases comes from the hydrolysis of ATP.

PASSIVE TRANSPORT OR FACILITATED DIFFUSION

Unlike simple diffusion, facilitated diffusion exhibits specificity and saturability. The carriers form a carrier–solute complex that is similar to the enzyme–substrate complex. In facilitated transport (and also in active transport) the flow rate versus the solute concentration can be described by a hyperbolic curve (Fig. 11.8). This curve follows the one seen in Michaelis–Menten kinetics, which depicts enzyme activity as a function of substrate concentration (Fig. 8.5). The relationship between the flow rate and solute concentration indicates that the process is saturable. When all of the carriers in the membrane are filled with solute, maximum flow is attained and does not further increase, even when the solute concentration increases.

A constant, K_m or $K_{0.5}$, corresponds to the solute concentration at which half the maximum flow rate is reached. In almost all cases, the K_m value is inversely related to the affinity of the carrier for the solute. The lower the K_m value, the higher the affinity and vice versa.

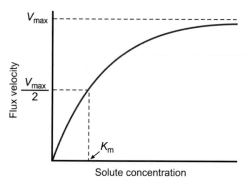

FIGURE 11.8 **Graphic representation of the flow rate depending on the concentration of solute in a facilitated diffusion process (same type of curve is obtained for active transport).** The curve is a hyperbola (Michaelis–Menten kinetics).

The flow rate of the solute in this type of system can be expressed by an equation similar to the Michaelis–Menten equation:

$$J = \frac{J_{max}[S]}{K_m + [S]}$$

where J, flow rate; J_{max}, maximum flow rate; [S], solute concentration; K_m, solute concentration at which the flow is equal to half the maximum.

The characteristic of saturability is the same for active and facilitated transport.

Similar to enzymes, substances with molecular structure analogous to that of the solute can bind to the carrier's specific site and produce competitive inhibition. Noncompetitive inhibition is also observed. The K_m value and type of inhibition can be determined experimentally (see pp. 163, 166).

The process of facilitated diffusion always follows the gradient. If the concentration or electrochemical gradient on either side of the membrane is reversed, the direction of flow is also reversed. The gradient is the driving force for facilitated diffusion, which requires no coupling to an energy source. From this point of view, facilitated diffusion is similar to simple diffusion, the difference lies in the carrier protein, which is not involved in simple diffusion.

Facilitated Diffusion Transporters

Numerous proteins that facilitate the diffusion of hydrophilic solutes in both directions through plasma and organelle membranes have been identified. Due to their similarity with enzymes, they are often called *permeases*. The following carriers are responsible for the transport of different organic and inorganic solutes.

Glucose transporters constitute a family of proteins widely distributed in the body, responsible for ensuring the supply of glucose and other hexoses to cells. They are very selective uniport systems, recognizing D, but not L isomers. Various classes of these transporters, which differ in kinetic properties, regulation, and tissue location have been described. They are designated with the acronym GLUT followed by a number (1–14). According to their primary structure, the GLUTs have been classified into three groups: class I includes GLUT 1-2-3-4 and 14, which were the first ones to be characterized. GLUT 1 is expressed in virtually all cells and carries basal glucose transport. GLUT 2 is responsible for glucose uptake in kidney and intestine and is the glucose sensor in pancreatic β cells. GLUT 3 is the neuronal form, vastly found throughout the nervous system. GLUT 4 is present in skeletal muscle and adipose tissue, its expression and function in the membrane are regulated by insulin. Class II is comprised of GLUT 5-7-9 and 11. GLUT 5 is a specific fructose carrier in the intestine. Finally, class III includes GLUT 6-8-10-12 and 13. This class of transporters is expressed in brain, muscle, liver, heart, and lung and they carry not only glucose, but also fructose and galactose.

While there are some differences in amino acid sequence among the different carriers, the overall structure is similar. GLUT 1 is a dimer. The others are either monomers or multimers of a five hundred amino acid subunit that has twelve transmembrane α helix segments, with amino and carboxyl terminals located in the cytoplasmic face, and a glycosylated outer loop connecting helices 1 and 2. The functional characteristics of these transporters will be considered in another chapter (p. 285).

Urea transporters. These transporters include proteins that facilitate the passive movement of urea through the plasma membrane. They are known as UT and include two types: the renal tubule transporters or UTA, with five different forms (UTA 1–5) and the erythrocyte transporter or UTB. UTA is involved in absorption and secretion of urea in the renal tubules as well as in the recycling of urea in the renal medulla. These transporters maintain the typical high urea levels of the renal interstitium, which is essential for the countercurrent mechanism and the control of urine concentration. UTB also facilitates the exchange of urea in the erythrocyte while traversing the renal medulla, which serves to maintain the osmotic balance of the cells in the hypertonic medium of the kidney.

UT is formed by two hydrophobic moieties, each possessing five transmembrane domains, connected by glycosylated extracellular segments. UTA 1 is the largest and consists of a duplication of that basic structure.

Cation and anion exchangers. These transporters mediate the uptake and release of organic substances with negative or positive charge (such as dopamine, adrenaline, serotonin, choline, histamine, prostaglandins) from cells. They also participate in the biotransformation of exogenous molecules of varied structure. Since these exchangers mobilize substances with electric charge, they are considered electrogenic. Cation exchangers include the organic cation transporters (OCT), while anion exchangers include the organic anion transporters (OAT). Like other transport systems, OCT and OAT have multiple molecular forms, which are primarily expressed in liver, kidney, and intestine.

Channels

Channels form hydrophilic pores or tunnels across the membrane. The movement of solutes through channels is always spontaneous, not requiring energy from ATP, but following a gradient.

Carriers carry a maximum of 10^5 particles per second, while the transport rate of channels is much higher (up to 10^8 per second). This enables channels to rapidly transmit signals in and between cells, including propagation of the nerve impulse, muscle contraction, and other biological responses. Channels differ from carriers in the way they recognize a solute. The carrier has a binding site, which can only be occupied by a specific substrate; channels discriminate solutes by their size and electric charge (Fig. 11.9).

Most of the known channels allow the passage of ions across the membrane. Many ion channels have been described in the plasma and organelle membranes of cells. They are highly selective for only one type of small ion (Na^+, K^+, Ca^{2+}, or Cl^-).

There are notable differences between concentrations of some solutes on either side of the plasma membrane due to the activity of active transport systems that use the energy of ATP hydrolysis to promote the passage of substances against a gradient. For example, K^+ is the predominant cation in the intracellular space and Na^+ is the most abundant one in the extracellular space. Anion distribution also shows differences on either side of the membrane (p. 692).

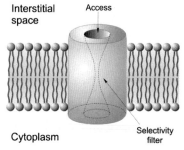

Interstitial space   Access

Cytoplasm   Selectivity filter

FIGURE 11.9 **Schematic representation of an ion channel.**

Selective ion transport and ion concentration gradients create a potential difference across the membrane of approximately −70 mV (millivolts), which depends on the cell type. The cell is negatively charged with respect to the outside medium. The membrane is therefore electrically polarized. The movement of charged particles generates an electric current; hence, ion channels play a fundamental role in the activity of excitable tissues, such as nerve and muscle. In those tissues, the entry of Na^+ or Ca^{2+} through the membrane reduces the negative intracellular charge and causes membrane depolarization. K^+ output or Cl^- input tends to reverse that effect and cause membrane repolarization.

Ion flow is driven by the electrochemical gradient resulting from the sum of the electric potential gradient and the ion concentration on either side of the membrane. In unstimulated cells, the electronegative interior favors the intake and hinders the escape of positive ions (the reverse is true for negative ions). In the case of K^+, for example, the electrical gradient opposes its release from the cell; however, the concentration gradient (with 140–150 mM inside and 5 mM in the extracellular space) promotes K^+ output. When these two forces are balanced, the electrochemical gradient is zero and there is no net ion flow. In most cells, the concentration of K^+ is close to equilibrium, with the membrane electric potential counteracting the concentration gradient. According to the Nernst equation, the equilibrium potential of an ion can be calculated by knowing its intra- and extracellular concentrations:

$$V = \frac{RT}{zF} \cdot \ln \frac{C_e}{C_i} = 2303 \frac{RT}{zF} \cdot \log \frac{C_e}{C_i}$$

where V, equilibrium potential in volts; R, gas constant [8.31 J (1987 calories)/mol · °K]; T, absolute temperature; F, Faraday constant (96 kJ/V); z, charge of ion; C_e and C_i, extra- and intracellular concentrations.

Knowledge of the structure and function of ion channels greatly advanced with the advent of molecular biology and genetic methods, as well as with the development of electrophysiological techniques, such as *patch clamp*. Patch clamp is used to isolate membrane portions of ~1 μm diameter inside a micropipette tip, which allows to analyze the function and properties of single channels.

In addition to being a pore, channels have other regions of great functional importance. A region within the channel forms the selectivity filter, which allows the movement of only one type of ion with the right size and charge. Channels also have a gate that controls the opening and closing of the pore. This gate is driven by specific stimuli. The opening of some channels occurs in response to a change in membrane potential; these are channels dependent on voltage (*voltage-gated channel*). In others, the binding of a signal substance or chemical intermediate opens the pore; these depend on ligands (*ligand-gated channel*). There are channels that attach ligands covalently to a specific site on the extracellular face of the channel, while others are activated by intracellular ligands. Some are sensitive to mechanical stimuli (i.e., those used for hearing in the inner ear).

Channels are constantly oscillating between the open and closed state. The study of ion currents through channels and the measure of time they remain open allows for the estimation of the *opening probability*, which defines the electrophysiological properties of the channel.

Channels with different specificity and properties have been described. Some of them are presented next.

Voltage-gated (K+, Na+, and Ca²+) channels. These channels are responsible for the electrical signals in excitable cells. They transport cations with high selectivity at a rate that is close to that of free diffusion; they are extraordinarily efficient. Regulated opening and closing of these channels determine the generation of action potentials in nerves and muscles. A short time (milliseconds) after opening, channels are inactivated and interrupt the solute flow. Thus, voltage-dependent cation channels may be in different states: (1) resting (closed), (2) activated (open), and (3) inactivated (closed). Voltage activation of Na^+ channels allows this cation to enter the cell and generate the fast initial phase of the action potential. The opening of voltage-dependent K^+ channels occurs during the depolarization phase; this facilitates the exit of K^+ from the cell and helps the membrane return to the resting membrane potential. Ion flux through Ca^{2+} channels induces entry into the cell and produces a sustained depolarization current that perpetuates action potentials in certain cells, such as myocardiocytes. In addition, the entry of Ca^{2+} through channels is the determinant of contraction in skeletal, cardiac, smooth muscle, and in many other processes in which this cation serves as an intracellular messenger.

Several voltage-gated K^+, Na^+, and Ca^{2+} channels have been identified. The structure is similar for all: they are polymers constituted by several polypeptide chains. The pore is formed by α subunits. Many have other subunits that are not directly involved in formation of the channel but contribute to their function.

In Na^+ and Ca^{2+} channels, there is one α subunit, a long polypeptide chain with intracellular N- and C-terminals, and four domains with six transmembrane segments each (Fig. 11.10). In contrast, K^+ channels possess four α subunits, each homologous to that of the domain of the Na^+ and Ca^{2+} channels (Fig. 11.11).

In all of these structures, the fourth transmembrane segment functions as a sensor of voltage changes. The loop connecting the fifth and sixth segments penetrates half way into the lipid bilayer and is responsible for ion discrimination. A portion of the protein complex functions as the pore cap responsible for opening and closing of the channel.

Nonvoltage-dependent K^+ channels. Another group of K^+ channels is known as the internal rectifier or K_{ir} channel. It is found predominantly

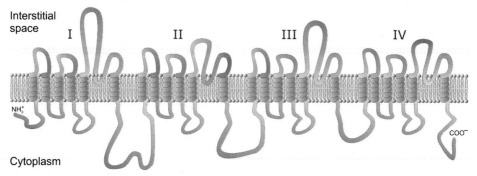

FIGURE 11.10 **Structure of voltage-dependent Na⁺ and Ca²⁺ channels.** One polypeptide chain with four repeated domains constitutes these channels. Each of these domains is similar to one of the four subunits of K⁺ channels.

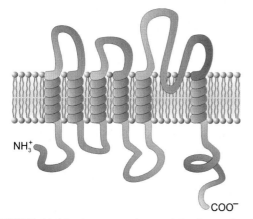

FIGURE 11.11 **Structure of one of the four subunits that form a voltage-dependent K⁺ channel.** Each subunit has six transmembrane α helices.

in skeletal and cardiac muscle cells, where it is responsible for maintaining K⁺ conductance at rest and the membrane potential equilibrium. This channel is a tetramer, consisting of units having two segments similar to transmembrane chains five and six of voltage-gated K⁺ channels. Both segments are bound on the extracellular side and form a conical structure with the vertex oriented toward the cytoplasm.

Extracellular ligand-dependent channels. This group of channels includes the acetylcholine nicotinic receptors and γ-aminobutyric acid, glycine, glutamate, serotonin type 3 (5-HT3) channels. These are all involved in synaptic

transmission. They are described in p. 241. Their overall structure is formed by five oligomeric complexes with four α-helix transmembrane segments each.

Cyclic nucleotide-gated channels (intracellular ligands). These transporters are sensitive to intracellular messengers. Among them are the cyclic GMP-activated channels present in heart, kidney, retina (rods and cones), and the olfactory organ. They have high permeability to Ca²⁺. Its structure is similar to that of voltage-dependent K⁺ channels, with a nucleotide-specific binding site at the C-terminus.

Amiloride sensitive Na⁺ channels. These channels are located in the apical membrane of epithelial cells and are responsible for NaCl absorption. The Na⁺ channels of the collecting tubules in the kidney are regulated by aldosterone. They are composed of three subunits (α, β, and γ), each with two transmembrane helices and a highly glycosylated external loop. Amiloride, a drug that binds to the channel and blocks it, prevents Na⁺ reabsorption, exerting a diuretic action.

Chloride channels. Besides Cl⁻, these channels allow the passage of other small anions. They are found in all cells and play an important role in epithelial Cl⁻ transport, cell volume regulation, acidification of organelles, and membrane potential stabilization. Ion channels are the target of a large number of pharmacological agents that have clinical application.

Several inherited disorders caused by mutations in genes encoding channel proteins have been identified. They have been grouped into a series of diseases called channelopathies.

Ionophores. There are compounds that can bind to and become incorporated into biological membranes, forming a pathway which increases the membrane permeability to certain ions. They are relatively small molecules with a hydrophobic surface, which allows them to enter the lipid bilayer. Two types of these molecules are known: the mobile carriers and the channel-forming ionophores.

Mobile carriers. These bind the ion on one side of the membrane, trap it inside the molecule, and move it across the other side of the membrane. The antibiotic *valinomycin*, a ring shaped peptide, is a mobile carrier that transfers K^+ from one side of the membrane to the other. Another ionophore of this class is A23187, which transfers Ca^{2+} and Mg^{2+} across the membrane. It is used experimentally to study the effect of rapid increase in intracellular Ca^{2+} concentration. *Channel-forming ionophores.* These molecules are inserted into the membrane generating a hydrophilic duct through which ions can pass. An example is *gramicidin A*, a peptide consisting of 15 amino acid residues. This antibiotic forms a transmembrane pore that is formed by the association of two molecules, which allow the movement of monovalent cations (H^+, K^+, Na^+).

Similar to the channels, ionophores allow ion flow only in the direction dictated by the electrochemical gradient.

Water Channels—Aquaporins

Water, a polar molecule, is practically insoluble in lipids. This property highly restricts water passage, which can only minimally flow across the cell membrane driven by the osmotic gradient. However, it was observed that some

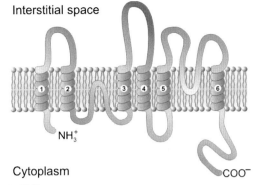

FIGURE 11.12 **Structure of aquaporin 1 (AQP1).**

cells had significantly higher water permeability than that expected from simple diffusion of water through the lipid bilayer. This observation led to propose the existence of pores or channels in the membrane for the passage of water. This hypothesis was confirmed by the isolation of water channel proteins. These channels constitute a family consisting of 13 isoforms, generically called *aquaporins*. They are designated with the acronym AQP and a number (0–12).

Hydrophobicity analysis of amino acids in the polypeptide chain of aquaporins indicates that they have six transmembrane α helices. The N- and C-terminals are intracellular (Fig. 11.12). The α helices 1, 2, and 6 on one side and α helices 3, 4, and 5 on the other are arranged symmetrically to form a pore so the water molecules can pass one at a time. The loops connecting α helices 2–3 and 5–6 are hydrophobic; they are inserted into the bilayer from the intracellular and extracellular side, respectively. This creates a structure that resembles that of an hourglass (Fig. 11.13). In the best known water channel (AQP1), four of these subunits are associated; only one is glycosylated. Each monomer functions as a separate channel. The channel is blocked by mercurial compounds that bind to a cysteine residue in the protein.

All aquaporins allow passage of water, which moves in the direction driven by the osmotic gradient. AQP0, 1, 2, 4, 5, and 6 carry only water,

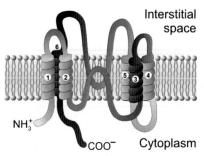

Interstitial space

NH_3^+

COO^- Cytoplasm

FIGURE 11.13 **Schematic representation of a unit of AQP1.** The channel is formed, in part, by the loop between helices 2 and 3 and the one that connects helices 5 and 6, both introduced into the membrane (structure in sandglass).

the rest are also permeable to other substances. For example, AQP3, 7, 9, and 10 allow the transmembrane transport of glycerol and are known as aquaglyceroporins. AQP7 and 8 are permeable to urea and AQP9 allows passage of solutes of greater mass, such as polyols and purines. The aquaporins are widely distributed in the body and in a specific location for each tissue:

Aquaporin 0 (AQP0). This AQP is located in the membrane of lens fibers. It has reduced water channel activity and it is thought that its main function is to maintain lens transparency. Mice with a mutation in the AQP0 gene suffer from opacity of the lens or cataracts.

Aquaporin 1 (AQP1). This was the first water channel identified and initially isolated from the red cell membrane. It is also present in nephron proximal tubes, the descending loop of Henle, the vasa recta, apical and basolateral cell membranes, capillary endothelial cells, choroid plexus microvilli, male reproductive tract, gallbladder epithelium, and spleen.

Aquaporin 2 (AQP2). This AQP is localized exclusively in the principal cells of renal collecting ducts. When these cells are not activated, most of the AQP2 is located in endosomal membrane vesicles, which results in low water permeability of the apical membrane. Antidiuretic hormone released from the neurohypophysis binds to specific receptors in principal renal cells, increases cyclic AMP, and promotes the delivery

and fusion of AQP2 containing endosomal vesicles to the apical membrane. This substantially increases water permeability in the cells. These channels are involved in the facultative, or regulated, reabsorption of up to 16 L of water per day. Hereditary *diabetes insipidus* can be caused by mutations in the AQP2 gene, which results in the inability to reabsorb water in the kidney, increasing urine excretion greatly (polyuria).

Aquaporin 3 (AQP3) is located in the basolateral membrane of renal collecting tubular cells. Together with AQP2, AQP3 is important in the process of urine concentration. AQP3 is also expressed in conjunctival epithelia, meninges, urinary, respiratory, and intestinal tract epithelium.

Aquaporin 4 (AQP4) is present in brain, retina, respiratory tract epithelium, and skeletal muscle.

Aquaporin 5 (AQP5) is located in salivary and lacrimal glands, corneal epithelium, and airway submucosal glands.

Aquaporin 6 (AQP6) is detected predominantly in intracellular compartments of intercalated cells in the renal collecting tubules.

Aquaporin 7 (AQP7) is abundant in the brush border of kidney proximal tubular cells and adipose tissue, where it mediates glycerol output.

Aquaporin 8 (AQP8) is expressed in liver, pancreas, small intestine, colon, placenta, and the testes.

Aquaporin 9 (AQP9) is mainly found in liver, lung, spleen, bone marrow, and leukocytes.

Aquaporin 10 (AQP10) is the intestinal form of aquaporin.

Aquaporin 11 and *12* (AQP11–12) are found in liver, brain, and ER of the pancreas.

It is clear that these channels represent an important homeostatic mechanism in the control of water balance at the cellular level; their alteration is associated with different clinical disorders.

Channels Between Cells

Channels exist between cells that are called *gap junctions* or *connexons*. These proteins form channels that connect the cytoplasms of adjacent cells. They allow the direct exchange of ions and small molecules, such as cyclic nucleotides, inositol trisphosphate, and glucose between neighboring cells without going through the interstitial space. They are found in nearly all cells, except for those that circulate freely, such as blood cells.

Gap junctions play an important role, helping to coordinate the response of cell groups to regulate tissue function in an integrated manner. Gap junctions are not very selective, being able to transport various ions, metabolites, and secondary messengers, such as Ca^{2+} and cyclic AMP. In excitable tissues, such as cardiac muscle cells, the direct passage of ions through these channels is synchronized to allow the coordinated contraction of muscle myofibrils.

Structurally, gap junctions are formed by two juxtaposed hemichannels, each of which belongs to one of the cells involved in the union. These hemichannels are formed by a hexamer of a protein called *connexin*. This protein with a mass of 30 kDa has four transmembrane segments with N- and C-terminals facing the cytoplasm. The six subunits form a cylinder with a pore, which allows for the passage of solutes (Fig. 11.14). The opening and closing of the channel is regulated by Ca^{2+} and is dependent on a conformational change of the subunits to produce its rotation in a plane perpendicular to the membrane.

Connexins encompass a family of homologous proteins. Twenty different forms have been identified, each of which are designated CX with a number indicating their molecular mass. Each connexon is made of the same type of connexins

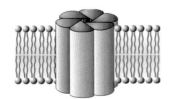

FIGURE 11.14 **Schematic representation of a connexon or nexus.**

or by several different types, but each class only assembles connexin channel subunits within the same tissue type. For example, an endothelial cell connexin cannot form a pore with connexins from muscle cells.

ACTIVE TRANSPORT

Simple and facilitated diffusion occurs following a gradient and has a negative ΔG. When the flow of a substance is in the direction opposite to a gradient, the ΔG is positive and the flow can only take place if energy is supplied. This constitutes *active transport*. Coupling of an exergonic reaction supplies the energy needed to drive the flow of solutes or ions in the thermodynamically unfavorable direction. In most cases, the transport is coupled to the hydrolysis of ATP. Approximately 50% of the ATP produced in cells is invested in active transport activities, which indicates the physiological relevance of these processes.

Active transport is mediated by carrier proteins that have the same characteristics of specificity and saturability described for facilitated diffusion. A main example of active transport is the maintenance of ion gradients across cell membranes, which depends on the function of specific carriers or "ion pumps." These ion transporters are grouped into three main classes, known as ATPases P, F, and V. The most important will be considered.

P-Type ATPases

The basic structure of all the ATP transporters is similar. They are composed of a subunit of approximately 100 kDa, which is responsible for the ATPase activity of the transporter. Specifically, the ATPase P contains a subunit with an ATP-binding site and becomes transiently phosphorylated during the transport process, giving these ATPases its P-type designation. Through cycles of phosphorylation/dephosphorylation, these transporters undergo conformational changes which allow the movement of the ions across the membrane. ATPases that drive the transport of more than one kind of ion have a second subunit called β, which is important during synthesis, assembly, and delivery of the transporter to the plasma membrane. Na⁺,K⁺-ATPase, Ca²⁺-ATPase, and H⁺,K⁺-ATPase are all examples of P-type ATPases.

Na⁺,K⁺-ATPase

One of the most widespread active transport systems is responsible for maintaining the difference in Na⁺ and K⁺ concentrations across a cell plasma membrane. It is called the Na,K-ATPase, or more commonly "sodium pump." It uses the free energy from the hydrolysis of ATP to move Na⁺ to the extracellular space in exchange for K⁺ which is transported into the cell cytoplasm (Fig. 11.15). Thus, it can be considered a countertransporter, antiport, or exchanger. The Na⁺ and K⁺ transmembrane gradients generated by the Na⁺,K⁺-ATPase contribute to many essential cell processes including: maintenance of cell osmolarity and volume, conservation of the cell resting membrane potential, and the transport of different solutes and water in and out of the cell.

The Na⁺,K⁺-ATPase is a complex formed by two major subunits, both of which are integral plasma membrane proteins. One of them, the α subunit, has a mass of approximately 100 kDa

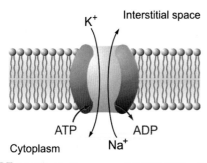

FIGURE 11.15   **Sodium pump (Na⁺,K⁺-ATPase).**

and traverses the membrane 10 times. The other subunit, called β, is a glycoprotein of approximately 45 kDa linked to oligosaccharide chains on its outer side; it traverses the membrane only once. The mass of the isolated complex indicates that the Na⁺,K⁺-ATPase is composed of at least two α and two β subunits. A third small hydrophobic subunit (approximately 8 kDa), called γ, has also been described as a component of the Na,K-ATPase. This polypeptide is expressed only in certain cells (i.e., kidney cells) and its function is that of a regulator of Na⁺,K⁺-ATPase activity. Families of hydrophobic polypeptides that share a particular amino acid motif (FXYD) with the γ subunit have been more recently identified. These FXYD proteins are differentially expressed in tissues and each have particular effects on the transport function of the Na,K-ATPase.

Obtaining Na⁺,K⁺-ATPase in crystal form has greatly contributed to understanding the structure of this transporter and that of other P-type ATPases. The α subunit is mostly placed on the cytoplasmic side of the membrane, where three domains can be distinguished: (1) the A or "acting" domain, including the N-terminal portion of the protein, (2) the P or "phosphorylation" domain, and (3) the N or "binding nucleotide" domain. The last two domains are located between transmembrane domains 4 and 5. The relative changes of these domains in a repeated and cyclic manner is induced by phosphorylation and dephosphorylation of the protein and allows for ion transport during the reaction catalyzed by the Na⁺,K⁺-ATPase.

Membrane lipids associated to the polypeptide chains are important in the operation of the sodium pump and their complete removal impairs the activity of the transporter.

The α subunit has binding sites with high affinity for Na⁺ on its internal (cytoplasmic) surface and high affinity for K⁺ on its outer face. Under normal conditions, the movement of Na⁺ out of the cell and the uptake of K⁺ into the cytoplasm are coupled; one cannot proceed without the other. The result of the pump operation is the exchange of intracellular Na$^+$ by extracellular K$^+$. The ion exchange is performed against the gradient for both Na$^+$ and K$^+$ and the necessary energy is obtained from the hydrolysis of ATP. Na$^+$,K$^+$-ATPase catalyzes the hydrolysis of ATP in a reaction that requires the presence of Na$^+$, K$^+$, and Mg$^+$. ATP binds to a specific site of the α subunit on the cytoplasmic face of the membrane and the hydrolysis is coupled to the transport of ions. Each mole of ATP hydrolyzed powers the export of 3 moles of Na$^+$ in exchange for the uptake of 2 moles of K$^+$. The result of the pump operation can be summarized by the following equation:

$$3Na_i^+ + 2K_e^+ + ATP \rightarrow 3Na_e^+ + 2K_i^+ + ADP + P_i$$

The subscripts i and e on the Na$^+$ and K$^+$ symbols indicate intracellular and extracellular location of the ions, respectively.

The direction of ion transport can be reversed by increasing the concentrations of Na$_e^+$, K$_i$, ADP, and P$_i$. In this case, the Na$^+$,K$^+$-ATPase functions as an ATP synthase.

Normally, the sodium pump operates according to the equation shown earlier, allowing the passage of three Na$^+$ out for every two K$^+$ that are brought into the cell, thus, creating a difference of potential across the membrane; therefore, the sodium pump is electrogenic.

Na$^+$,K$^+$-ATPase alternates between phosphorylated and dephosphorylated states of its α subunit, which determines a reversible conformational change and the translocation of ions. Experimental evidence has allowed to propose the following mechanism of action for Na,K-ATPase:

1. In a protein conformation known as E1, the α subunit binds ATP, Mg^{2+}, and three Na ions from the cytoplasmic side with high affinity. ATP hydrolysis occurs and the third phosphoryl moiety of ATP is transferred to the carboxyl of an aspartate residue at the α subunit.

2. Three Na ions are trapped by the phosphorylated enzyme, ADP is released, and a conformational change (to a conformation known as E2) is produced. This results in Na$^+$ translocation to the external side of the membrane and a decrease in the affinity of the Na$^+$-binding sites, which allows the release of three Na ions into the interstitial space.

3. Two K$^+$ ions bind to high affinity sites, accessible from the outer surface of the phosphorylated enzyme. K$^+$ binding promotes the release of phosphate linked to the enzyme.

4. Potassium ions are trapped by the dephosphorylated enzyme. ATP binding, now at low affinity sites, promotes a conformational change that returns the protein to the E1 conformation. The enzyme decreases its affinity for K$^+$, which is exposed to the interior of the cell, and releases the ions into the cytosol.

This mechanism is shown schematically in Fig. 11.16.

Both the α and β subunits of the Na$^+$,K$^+$-ATPase are present in different molecular forms, encoded by different genes. Four α polypeptides (α_1, α_2, α_3, and α_4) and three β subunits (β_1, β_2, and β_3) have been found in mammals, each of which has a cell-specific and developmentally regulated pattern of expression.

The Na$^+$,K$^+$-ATPase is the receptor for glycosides, such as ouabain and digoxin. These compounds, of steroid structure, are used clinically as cardiotonics. These substances bind to the outer surface of the α subunit, blocking Na$^+$ and K$^+$ transport. Inhibition of Na$^+$,K$^+$-ATPase by cardiac glycosides leads to increased intracellular Na$^+$ concentration, decreasing the gradient for this cation across the plasma membrane. Since the transmembrane Na$^+$ gradient drives secondary active transport via other transporters (such as the Na$^+$/Ca^{2+} countertransport), Na$^+$,K$^+$-ATPase inhibition by cardiac glycosides reduces calcium

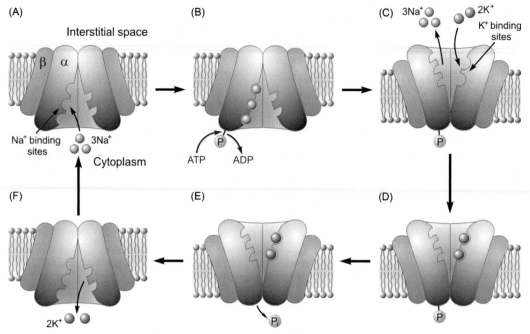

FIGURE 11.16 **Schematic representation of the reaction cycle of the Na⁺,K-ATPase.** (A) E1 conformation. Binding sites for Na⁺ are accessible from the intracellular side of the membrane, ATP–Mg²⁺ binds with high affinity. (B) Three Na⁺ are bound to their respective sites, ATP hydrolysis occurs, and α subunit is phosphorylated, which induces a conformational change. (C) Conformation E2. Three Na⁺ are released and K⁺ sites are exposed. (D) Two K⁺ ions bind to their sites. (E) Phosphoryl separates from the α subunit. (F) New conformational change that frees the two K⁺ ions into the cytoplasm and the ATPase returns to the initial conformation.

release from the myocardial cells. This increases the calcium reserves in the sarcoplasmic reticulum (SR), which upon release during systole, improves heart contractility. Glycosides were first found to be synthesized in plants; however, more recently, it was discovered that ouabain is a hormone endogenously synthesized in the adrenal gland of mammals and acts as a regulator of Na⁺,K⁺-ATPase function when released into the circulation.

Ca²⁺-ATPase

Calcium concentration in the cell cytoplasm is maintained at very low levels (approximately 10,000 times lower than those existing in the extracellular fluid) due to the existence of systems that expel the cytosolic Ca²⁺ out of the cell or deliver it to intracellular compartments. All cells possess Ca²⁺ pumps in the plasma membrane and membranes of the ER. The most studied and better understood calcium pump is the SR calcium pump of muscle cells. This ion transporter has a structure and function similar to that of Na⁺,K⁺-ATPase, except that it is a uniporter and not an antiporter. It has a single α subunit with 10 transmembrane segments, whose cytoplasmic face has specific high affinity binding sites for Ca²⁺ and ATP. Similar to Na⁺,K⁺-ATPase, the calcium pump requires Mg²⁺ and catalyzes the hydrolysis of ATP to ADP and P$_i$.

Plasma membrane Ca²⁺-ATPase (PMCA). This ATPase transfers one Ca²⁺ ion for each ATP molecule hydrolyzed and helps regulate cytosol Ca²⁺ levels. The increase in cytosolic concentration of free Ca²⁺ promotes binding to a protein called *calmodulin* (p. 567). The Ca–calmodulin complex interacts with a site in the Ca^{2+}-ATPase C-terminus and produces its allosteric activation. The calcium pump can also be activated by phosphatidylinositol bisphosphate (PIP_2) and covalent modification by protein kinase dependent on cyclic AMP.

There are four molecular forms of Ca^{2+}-ATPase (PMCA1–4). PMCA1 and 4 are expressed in most cells, whereas PMCA2 and 3 are found in brain, skeletal muscle, liver, and kidney.

Sarcoplasmic reticulum Ca^{2+}-ATPase (SERCA). This ATPase constitutes 80% of the total integral membrane proteins found in the SR and nearly cover one-third of its surface. It pumps Ca^{2+} from the cytosol into the SR cisternae, where it is stored, to be released during cell depolarization. The Ca^{2+} concentration in cytosol plays a critical role in the mechanisms of muscle contraction and relaxation (p. 67).

The SR Ca^{2+}-ATPase, unlike PMCA, carries two Ca^{2+} ions per molecule of hydrolyzed ATP. Its activity is regulated by *phospholamban* and *sarcolipin*, both of which are low molecular weight proteins found in the SR membrane.

As in the case of other carriers, there are three forms of SR Ca^{2+}-ATPase. SERCA1 is found in skeletal muscle while SERCA2 and 3 are expressed in heart.

H^+,K^+-ATPase

This enzyme has structural similarity to Na^+,K-ATPase, it has α and β subunits. It catalyzes the countertransport of H^+ and K^+ coupled to hydrolysis of ATP. Unlike the Na^+,K^+-ATPase, which is electrogenic, the H^+,K^+-ATPase is electroneutral, exchanging one H^+ out for each K^+ introduced into the cell. It is located in the parietal cells of gastric mucosa, where its action increases H^+ secretion into the stomach and increases K^+ concentration in the cell cytoplasm. Furthermore, H^+,K^+-ATPase is present in kidney collecting ducts, where it plays an important role in H^+ elimination from the body through the process of urine acidification.

F-Type ATPases

The most interesting F-type ATPase is present in the inner mitochondrial membrane (p. 191). The structure and mechanism of action of these ATPases are different to those of the P-type pumps. They are composed of a multi-subunit association (F_1–F_0) with a large protein extramembrane complex (F_1) which contains the catalytic site, and a transmembrane complex forming a proton channel (F_0). It functions as an ATP synthase, driven by the flow of protons. Proton translocation occurs in the inner membrane of mitochondria during electron transfer in the respiratory chain (p. 195). This generates an electrochemical potential used for the synthesis of ATP when the H^+ ions return to the mitochondrial matrix.

V-Type ATPase

Representatives of this group are found in lysosomal and endosomal membranes, they pump protons from the cytosol into these organelles, maintaining their pH at approximately 5.0. This means that the ATPase activity can maintain a concentration of H^+ within the organelle that is 100 times greater than that of the cytoplasm (where pH is closer to 7). This difference in pH is important to maintain the optimum medium for enzymes (hydrolases) involved in the degradation of proteins within lysosomes. Moreover, this ATPase creates the acidic environment in endosomes which promotes the dissociation of ligands from their receptors after their introduction into the cell.

The structure and mechanism of action of these ATPases is similar to those of type F ATPases. They consist of two parts, a cytosolic (V_1) which binds ATP and exerts the hydrolytic action. The other part is membrane bound (V_0) and forms the channel for the passage of protons. The V_1 sector is composed of eight subunits, designated by letters A–H. Three A subunits and three B subunits alternate to form a spherical particle, which is connected to V_0 by a stem consisting of subunits C, D, E, and F. The protein looks like a "mushroom" on the surface of the membrane. The V_0 segment has the shape of a ring, consisting of six proteins (a, d, e, c, c', and c").

The V-ATPase uses energy from ATP hydrolysis to produce a conformational change in the molecule that results in a rotational movement of the stem and V_0 portion as well as a translocation of protons from the cytoplasmic face to the endosomal lumen. Its function is regulated by the assembly and dissociation mechanisms of V_1 and V_0 subunits, which responds to cytosolic glucose levels.

Type V ATPases are also found in plasma membranes of kidney tubule cells and in osteoclasts. In the kidney they contribute to H^+ secretion in urine and in osteoclasts they aid in the bone resorption process. This proton pump expels protons and acidifies the medium, contributing to the dissolution of bone calcium phosphate crystals.

ABC Transporters

This group of active transporters constitutes a large family of proteins characterized by having an ATP-binding domain. They are named by the acronym for ATP-binding cassette (ABC transporter). The main physiological role of these transporters is to carry endogenous and exogenous substances of varied structures. Forty eight ABC transporters have been recognized in humans; almost half of them carry lipids or compounds related to lipids. They use the energy from ATP hydrolysis to remove lipids, carbohydrates, peptides, bile salts, and antibiotics from cells.

Examples of these transporters include: the *multidrug resistance protein (MRP)*, the *cystic fibrosis transmembrane conductance regulator (CFTR)*, and the *sulfonylurea receptor (SUR)*.

Multidrug resistance protein (MRP). In patients with some malignant tumors, it has been found that MRP transporters present resistance to the antineoplastic drugs used for treatment. The cause of this resistance was found to be an overexpression of a 170 kDa glycoprotein (P170 or P-glycoprotein) in cancer cells. This was the first transporter of this type to be discovered (also known as MRP1). It uses the energy of ATP hydrolysis to extrude hydrophobic compounds from the cell and also functions as a chloride channel. In cancer cells, particularly of hepatic origin (hepatoma), this protein is responsible for resistance to chemotherapy. It has an amino acid chain of 1280 amino acids in length with 2 identical domains that contain 6 transmembrane helices and an ATP-binding site at the terminal cytosolic loop.

Nine multiple drug resistance proteins have been discovered. MRP1 is virtually in all cells, while other forms (MRP2–MRP9) are expressed in liver, kidney, and intestine. At these locations, these types of MRPs promote excretion of toxic substances to the bile, urine, and feces. In addition to drugs export, the MRP1 is involved in cellular efflux of glutathione, cysteinyl leukotriene LTD_4, and other lipid-derived compounds.

Cystic fibrosis transmembrane conductance regulator (CFTR). This transporter has a structure similar to that of the P170 protein, with a total of 12 transmembrane segments. It differs from other ABC transporters in that only one of its ATP-binding sites has hydrolytic activity. The central cytoplasmic loop of CFTR exhibits phosphorylation sites, which are phosphorylated by the cyclic AMP–dependent protein kinase. CFTR is a chloride channel found in lung epithelial cells, sweat glands, pancreas, and other tissues apical membranes. It is activated by ATP and cyclic AMP.

Different mutations in the CFTR gene can alter the Cl^- transport function of this carrier and produce a lethal disease, *cystic fibrosis*. This disorder is one of the most common autosomal recessive genetic disorders in Caucasians. The disease is manifested by increased levels of Na^+ and Cl^- in sweat and thickening of respiratory, pancreatic, and intestinal tract secretions which produces obstructions and favors infections.

SUR (sulfonyl-urea regulated) transporter. Unlike other members of the ABC group, this transporter is not a carrier. Instead, it is a protein associated to K^+ channels to form a larger structure that functions as a K^+ channel regulated by ATP (K_{ATP}). It has the typical structure of ABC proteins with the addition of a region that spans the membrane 5 times at the N-terminus. It is found in pancreatic β cells, where it is important for insulin release. It regulates K^+ flow by responding to changes in ATP levels, which leads to changes in membrane potential and secretion of insulin. SUR is regulated by glucose and sulfonylurea, a drug commonly used to treat type 2 diabetes.

Secondary Active Transport

Transport systems, such as Na^+,K^+-ATPase, and Ca^{2+}-ATPase are considered primary active transport systems, since the transfer of ions is directly coupled to the hydrolysis of ATP.

Secondary active transport does not use the energy supplied by coupled exergonic reactions directly, instead it uses the electrochemical potential gradient created by the operation of a primary transport system. For example, the sodium pump generates a considerable difference in chemical (Na^+ gradient) and electric potential (the inside of the cell is negative with respect to outside) which favors the entrance of Na^+ ions from the extracellular space. This Na^+ entry has a negative ΔG and can drive the simultaneous flow of other ions or substances. The flow of these solutes, coupled to the inward movement of Na^+, can be performed in the same direction as that of the Na^+ ions (from the outside to the inside of the cell). This is completed by the action of cotransporters (symporters). The movement of a solute can be carried out in a direction opposite to that of Na^+ movement by countertransporters (antiporters) (Fig. 11.17).

The secondary transport systems are directly dependent on the activity of the primary system. For example, secondary transport associated

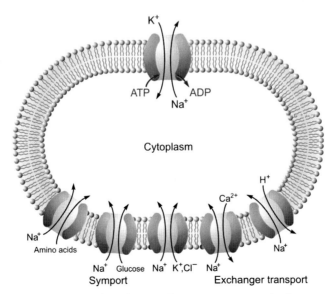

FIGURE 11.17 Schematic representation of a primary active transport system (sodium pump, top) and secondary active transport systems [secondary active cotransport (Na^+–amino acid, Na^+–glucose, and Na^+–K^+–Cl^-) and countertransport (Na^+–Ca^{2+} and Na^+–H^+), bottom].

with the Na^+ gradient halts if the Na^+,K^+-ATPase is inhibited, since the transmembrane Na^+ electrochemical gradient is dissipated.

Na^+-dependent transport systems include the following:

Na^+-dependent glucose transporter (SGLT). This cotransporter is responsible for mediating the entry of glucose into cells and can be found in the intestinal mucosa (SGLT1) and the renal tubules (SCLT2). Glucose is transported against its concentration gradient from the apical side of these cells. SGLT is a glycoprotein composed of a 664 amino acid chain, 14 transmembrane helices forming the channel, and specific sites to which Na^+ and glucose bind. Binding of the transported molecules produces a protein conformational change that allows Na^+ and glucose entry into the cell. The passage of Na^+ favored by the electrochemical gradient has a negative ΔG large enough to drive glucose movement against its chemical gradient. Two Na^+ ions are introduced for every glucose molecule transported. Once inside the cell, glucose is transported into the interstitial space and blood by facilitated transport (basolateral carrier GLUT2).

In renal tubular epithelial cells, there is also a Na^+-glucose cotransporter, called SGLT2. It is located in the initial portion of the nephron proximal tubule and transfers one Na^+ ion per molecule of glucose. In the final segment of proximal tubules, the SGLT1 carrier predominates, with a stoichiometry of two Na^+ per glucose cotransported.

Na^+-dependent amino acid transporters. Amino acids are absorbed in the intestine and renal tubules by Na^+-dependent transport mechanisms. There are also Na^+-independent amino acid transporters (see p. 373).

$Na^+/K^+/Cl^-$ cotransporter (NKCC). This system couples the uptake of Na^+ with that of K^+ and Cl^-. The transport has a stoichiometry of $1Na^+$:$1K^+$:$2Cl^-$ so, it is electroneutral. One form of the carrier (NKCC1) is expressed in epithelial cells across the body, while another form (NKCC2) is found in the loop of Henle of the kidney nephron. NKCC are proteins with 12 transmembrane domains and intracellular N- and C-terminals.

Another protein similar in structure to the NKCC cotransporter is the Na^+/Cl^- transporter, which is important for Na^+ and Cl^- reabsorption across the apical membrane of distal tubule cells of the kidney.

Na^+/Ca^{2+} exchanger. This antiport system introduces three Na into a cell for every one Ca^{2+} expelled into the extracellular space in an electrogenic process. This protein has nine transmembrane domains and is expressed in myocardial and arterial smooth muscle cells. It plays a fundamental role in the maintenance of intracellular Ca^{2+} levels and in regulation of muscle contractility.

Na^+/H^+ exchanger. This countertransporter couples the inward movement of Na^+ with the export of protons from a cell. It works with a stoichiometry of $1Na^+$:$1H^+$; it is found in almost all cells and in the apical membrane of epithelia, such as the renal proximal tubule. It has 9 different isoforms with a similar structure, with 12 transmembrane domains. The Na^+/H^+ exchanger is important in intracellular pH regulation.

Na^+-independent transporters. These transporters couple movement of a transported solute to ions other than Na^+. Among these systems are:

K^+/Cl^- cotransporter (KCC). This transport system utilizes the difference of chemical potential for K^+ generated by the sodium pump. The K^+ gradient favors its exit from the cell and is used to cotransport Cl^-. Both ions are removed from the cytosol in a ratio of one K^+ ion for every one Cl^- ion. It is found in almost all cells and has a structure similar to NKCC. It is involved in the maintenance of osmotic balance and cell volume.

$HCO_3^- - Cl^-$ *exchanger*. This countertransporter (antiport) exchanges chloride and bicarbonate ions with a $1HCO_3$:$1Cl^-$ stoichiometry. Since the transporter does not alter the charge balance on both sides of the membrane, the exchange is electroneutral. The anion gradient provides the direction of transport; in most cells, the extracellular Cl^- is exchanged with cytoplasmic HCO_3^-. Particularly in erythrocytes, ion flow is reversed depending on whether they are in the systemic or in pulmonary capillaries. Three molecular variants of this exchanger have been described and they are designated AE1–3 (anion exchanger). The AE1 is known as band 3 protein, due to its electrophoretic mobility; it is particularly abundant in red blood cell membranes and plays an important role in blood CO_2 transport. AE1 is also present in the basolateral membrane of renal intercalated tubule cells where it is involved in HCO_3^- reabsorption from the glomerular ultrafiltrate. AE2 is expressed in epithelial cells of gastric mucosa, colon, and kidney. AE3 predominates in excitable tissues, such as brain, heart, and smooth muscle. The AE directly contributes to mobilize HCO_3^- into the blood and, indirectly, to the secretion of H^+ and maintenance of intracellular pH. Cl^- transport through the AE is also involved in cell volume regulation.

AE is a polypeptide dimer with N- and C-terminals oriented toward the cytoplasm and 13 α-helix transmembrane domains. It is also glycosylated in extracellular domains 3 and 4.

H^+-dependent transporters. The electromotive potential created in the inner mitochondrial membrane due to proton pumping is also used in secondary active transport processes. Numerous ions and charged molecule carriers in the inner membrane are driven by the proton gradient. Examples include: (1) pyruvate, which enters the mitochondrial matrix countertransported with OH^- or cotransported with H^+, (2) the dicarboxylate carrier, which transports malate, succinate, and fumarate in exchange for P_i, (3) the tricarboxylate transporter, which exchanges citrate and H^+ for malate, and (4) the H^+/Ca^{2+} cotransport, which introduces calcium into the mitochondrial matrix. All of these transporters have similar structures, consisting of a polypeptide chain with three tandem repeats, each of which has two transmembrane helices.

Nerve Impulse

In nerve cells at rest, the intracellular space is electronegative with respect to the extracellular environment. This voltage difference, called the *resting membrane potential*, is due to the uneven distribution of ions across the cell membrane, mainly maintained by the concerted action of the Na^+,K^+-ATPase and K^+ channels. Normally, the resting potential has a value of -60 to -70 mV. When a stimulus reaches a neuron, it causes local membrane changes that shift the membrane potential to less negative values. If a threshold value of approximately -40 mV is reached, voltage-gated Na^+ channels (p. 229) in the vicinity of the stimulated area open. This causes an influx of Na^+ into the cell, favored by the intracellular negative electrochemical gradient. The entry of Na^+ moves membrane potential to positive values, a phenomenon called *depolarization*. Once the membrane potential reaches $+50$ mV, Na^+ entry stops because the balance for the ion is attained. Moreover, Na^+ channels are spontaneously closed after approximately 1 ms and remain closed for a few more milliseconds, during which they are insensitive to further stimuli. This time period is known as the *refractory period* of the cell. When the membrane potential returns to its resting value, the channels return to their closed but excitable condition.

At maximum depolarization ($+50$ mV), K^+ channels open, allowing the exit of this cation from the neuron, which lowers the membrane potential to reach the equilibrium value for K^+ (-75 mV). The action of the Na^+,K^+-ATPase restores the Na^+ and K^+ ion distribution and the potential returns to -60 mV.

The activation of Na^+ channels and the local plasma membrane depolarization spreads along the cell axon to transmit the *action potential*. During the period of spontaneous closure of Na^+ channels, the action potential is not able to propagate backward or permanently depolarize the membrane because the cells are in a refractory state.

The axon terminal contains multiple vesicles filled with different neurotransmitters depending on the neuronal type. The plasma membrane at the axon terminal has Ca^{2+} channels that are sensitive to voltage.

When the action potential reaches the terminal, the depolarization wave determines the opening of Ca^{2+} channels and the entry of this cation from the extracellular space driven by the inward electrochemical gradient. The abrupt increase in intracellular Ca^{2+} concentration induces the fusion of the vesicles with the plasma membrane. The neurotransmitter is released by exocytosis into the synaptic cleft, diffuses across the synaptic gap, and binds to specific receptors on the postsynaptic cell membrane. A neurotransmitter-receptor complex is formed, which generates a new stimulus in the postsynaptic neuron, target muscle, or gland cell. Often, the binding of a chemical intermediate to its receptor promotes opening of ion channels (ionotropic effect) and can cause: (1) income of Na^+ and membrane depolarization, whereby a new action potential is generated in the postsynaptic cell; this is the case of excitatory postsynaptic potentials. (2) Activation of anion channels (Cl^-) or K^+ channels; opening of Cl^- channels allows Cl^- entrance into the cell while opening of K^+ channels allows K^+ to escape from the intracellular space. In both cases, membrane potential becomes more negative (hyperpolarization) and prevents further development of the nerve impulse (inhibitory postsynaptic potential).

Formation of the transmitter-receptor complex is sometimes followed by chemical changes via activation of signaling systems; in this case the effect is metabotropic.

Endocytosis

Macromolecules and large particles penetrate into the cells by using mechanisms different to those mentioned in previous sections. The term *endocytosis* is used to designate processes in which the plasma membrane surrounds the material to be engulfed by a cell, forming a vesicle that is then introduced into the cell. Endocytosis is important for incorporation of nutrients into the cell, removal of undesired material from the interstitial space, stopping signaling systems activated on the cell surface by different ligands, and for the recycling of plasma membrane components. The greatest efficiency in the process of endocytosis is achieved when substances of the extracellular medium selectively bind to membrane receptors. This process, highly specific and regulated, begins with the formation of the ligand-receptor complex. This is followed by a series of steps aimed to introduce the ligand in the cytoplasm and finishes with the delivery of the endocytosed molecules to their final destination within the cell.

Two types of endocytosis are recognized: phagocytosis and pinocytosis.

Phagocytosis. This type of endocytosis occurs in some specialized cells, such as macrophages, neutrophils, leukocytes, and monocytes. Relatively large sized particles are attached to the cell surface of these cell types, which emit plasma membrane projections that surround a particle and finally engulf it in an intracellular vesicle. This process allows the destruction and elimination of pathogenic agents, such as bacteria, yeast, remains of dead cells, fatty deposits in arteries, and other foreign particles.

The process generally begins with the specific binding of the material to be removed to membrane receptors that activate signal transduction systems within the cell (see Chapter 25).

Initially, cell membrane extensions engulf the particle, forming projections that resemble a cup. The subsequent growth of these outgrowths requires assembly on the inner side of the membrane, of actin and myosin fibers, along with incorporation of new membrane, which is added at the base of the cup. The advancement of this process is driven by membrane receptors. The whole process requires a very organized series of steps regulated by numerous intracellular messengers. Activation of intracellular signals triggered by the formation of the ligand-receptor complex stimulates the inositolphosphate system and phosphatidyl kinases. Among effectors are GTPases that induce polymerization of actin, myosin contraction, and phagosome formation.

Pinocytosis. This process occurs in virtually all cells. Different types can be distinguished: macropinocytosis and endocytosis mediated by clathrin and caveolin, and endocytosis independent from clathrin and caveolin.

Macropinocytosis. Similar to phagocytosis, actin fibers inside the cell promote outward projections of the membrane, which then form large vesicles (from 0.2 to 10 μm). Different from phagocytosis, this process does not depend on the binding of a ligand to its receptor and the vesicles do not include a particular particle, they simply trap a portion of the liquid in contact with the outer surface of the cell. Molecules, ions, or small particles of any type dispersed in that fluid can be introduced into the cytoplasm. In some cells, macropinocytosis occurs spontaneously, in others it occurs after stimulation by growth factors.

Clathrin- or caveolin-mediated endocytosis. This type of endocytosis is highly specific. Membrane receptors selectively bind certain ligands in the extracellular medium. This ligand-receptor binding event has the characteristics of specificity and saturability following Michaelis–Menten kinetics. After ligand-receptor binding, a series of steps determine the invagination of the membrane in the area where the complex is located. Often, the complex is displaced in the plane of the membrane into an already formed depression of the membrane, which gathers various complexes. In all cases, the membrane invagination has a protein coat on its cytoplasmic face, which is why it is called a *coated pit*. The protein that lines the cytosolic face of the invagination can be *clathrin* or *caveolin*.

Clathrin-mediated endocytosis. This process is found in all mammalian cells and it is essential for the uptake of certain nutrients by the cell. For example, LDL particles, containing mostly cholesterol (p. 331), bind to specific receptors. Transferrin molecules, carrying iron (p. 732), bind to their own receptors. Clathrin-mediated endocytosis is also important in the regulation of the amount of receptors, ion channels, and transporters exposed in the plasma membrane, since their internalization in the cytoplasm can regulate their cell surface expression.

This type of endocytosis requires clathrin and adapter proteins. Clathrin consists of hexameric units (three heavy polypeptide chains linked to three light protein chains) called *triskelions*, which form a polygonal network on the inner invagination side of the membrane. Among the adapter proteins are heterotetrameric complexes formed by *adaptins*. The AP-2 (adapter protein 2) has been characterized in detail; it is essential for endosome formation. The adapter proteins direct the production of the desired curvature for the formation of the inward depression of the membrane coated by clathrin. This forms the coated pit which grows deeper and ends up forming a completely closed plasma membrane vesicle, covered by the clathrin mesh and adapter proteins (Figs. 11.18 and 11.19).

Another participant in the formation of vesicles is a small GTP-binding protein called *dynamin*. Dynamin forms a ring around the neck of the vesicle when it is deeply invaginated (Fig. 11.19). Hydrolysis of GTP to GDP and P_i determines the contraction of the ring, neck closure, and the release of the vesicle into the cytoplasm.

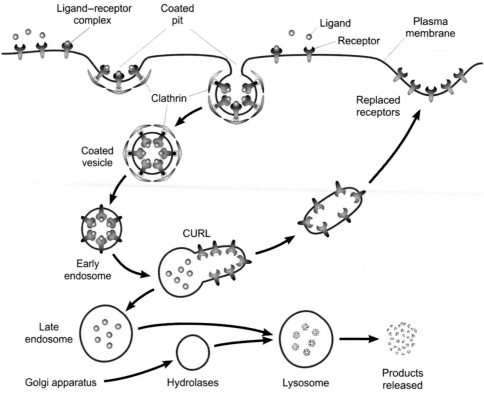

FIGURE 11.18   **Schematic representation of receptor mediated endocytosis.**

Once internalized in the cell, the vesicle with the ligand-receptor complexes loses the clathrin coating and is fused with other vesicles forming early endosomes. The interior of the endosome is progressively acidified, reaching pH values ~5.0. In many cases, this acidification promotes dissociation of the ligand-receptor complex.

Sometimes, the endosome undergoes changes. A portion of it adopts a tubular conformation in which the empty receptors, embedded in the membrane, are concentrated [*compartment of uncoupling of receptor and ligand* (CURL), Fig. 11.18]; the rest of the vesicle contains the free ligands. The vesicular and tubular portions of these compartments end up separated. The tubular segment, with the receptors, recycles back to the plasma membrane. In this manner, the same receptors can be reused repeatedly for new ligands. The vesicular portion, with the free ligands constitutes the late endosome, which fuses with another vesicle full of hydrolytic enzymes from the Golgi complex to form a *lysosome*. Hydrolases inside the fused vesicles cause degradation of the ligands.

Not all ligands are degraded. For example, transferrin is a plasma protein which transports iron; it binds with high affinity to its membrane receptor and is included in a coated vesicle. In the endosome containing the receptor-transferrin complexes, when the pH drops to 5.5, Fe^{3+} is liberated and transferred into the cytosol. Apotransferrin-receptor complexes that remain in the endosomal membrane are returned to the plasma membrane where apotransferrin is released into the extracellular space and returns to the circulation. In other cases (e.g., epidermal growth factor), the receptor does not return to the membrane and is instead degraded in the lysosome together with the ligand.

Caveolin-mediated endocytosis. Many cells, especially endothelial cells, contain invaginations of microdomains of the plasma membrane rich in cholesterol and sphingolipids. These are called *caveolae* and present a dimeric protein called *caveolin*, which maintains the organization and structure of these membrane invaginations.

Caveolin is inserted into the inner membrane hemilayer by binding to cholesterol and it associates to other molecules of the same class to form a layer of striated appearance that ends up completely coating the cytoplasmic face of the caveolae. The closure of the neck and the final release of the vesicle are promoted by dynamin.

This system generally depends on a phosphorylation cascade initiated by protein kinases, including the kinase *Src* which phosphorylates tyrosine residues of many proteins. *Src* kinase targets caveolin, which once phosphorylated dissociates from the membrane. *Src* also catalyzes the incorporation of phosphate into dynamin, activating it. Finally, *Src* phosphorylates actin and other cytoskeletal proteins. Altogether, these effects induce a fission process, which separates the endosome from the plasma membrane and releases the endocytic vesicles into the cell.

Caveolin-mediated endocytosis was originally described for the transport of serum proteins, particularly albumin, from the blood to the tissues through endothelial cells. A role of caveolin-mediated endocytosis in the regulation of signaling cascades has been described, and several molecular members of these systems are associated with caveolae. It is also believed that they participate in cholesterol traffic within cells and in the overall homeostasis of lipids.

Endocytosis is a complex, tightly regulated process in which many factors are involved. In addition to the mentioned proteins, there are additional accessory proteins involved, which are not mentioned to simplify this presentation.

In the endocytosis process, vesicle internalization represents a considerable removal of plasma membrane from the cell surface. There are mechanisms that return membrane to the cell surface continuously recycling them.

Exocytosis

Proteins destined for export are enclosed in membrane vesicles formed by the *trans* Golgi. They are directed to the plasma membrane, fuse with it, and release their contents into the extracellular space. This process is known as *exocytosis*. In epithelial cells, exocytosed vesicles can be selectively directed to the apical or basolateral membrane, which implies the existence of specific targeting mechanisms in the cell.

In many cells, exocytosis occurs continuously; it is a constitutive secretion pathway. However, in specialized secretory cells, exocytosis responds to different stimuli. Processed products (digestive enzymes, neurotransmitters, hormones) that remain in the cytoplasm are packaged in vesicles until the arrival of a signal (second messenger or membrane potential variation), which triggers adhesion and fusion of the vesicle to the plasma membrane with subsequent release of the vesicle's content. It is the regulated secretion pathway.

Exocytosis produces a continuous transfer of intracellular membrane from organelles to the plasma membrane. Membrane recycling via internalization occurs at a considerable magnitude to maintain the membrane balance. It is estimated that the plasma membrane of cells is completely replaced every 2 h.

Exosomes. A particular kind of vesicle, known as *exosome*, has been the focus of intense investigation during the last years. They are small vesicles (ranging from 30 to 100 nm) that are secreted to the extracellular medium by many cells, especially epithelial cells surrounding tubular structures of the body. Initially, exosomes are generated as intracellular vesicles, through

invagination of the plasma membrane, and included in late endosomes where multivesicular
bodies are formed. These bodies fuse with the
external membrane and are released as exosomes. They can be found in blood, urine, and
saliva. Exosomes contain lipids, proteins, and
RNAs. It is believed that they exert diverse actions, acting as messenger particles between
cells via the delivery of the molecules they carry. They play a role in intercellular communi-
cations, modulate immune responses, regulate
regenerative and degenerative processes, and
activate tumor proliferation. Due to their important functions, exosomes are promising can-
didates for clinical diagnosis, prognosis, and
even therapeutic uses.

Vesicular Transport

The processes of endocytosis and exocytosis
involve the transport of material enclosed or
inserted in membranes vesicles, formed either
by invagination of the plasma membrane (endocytosis) or evagination of organelle membranes
(exocytosis). Vesicular transport is a very important activity of molecular traffic between different compartments within the cell.

Vesicle formation. Organelle membrane evagination and plasma membrane invagination is
mediated by proteins that polymerize in a mesh
attached to the cytosolic side and direct the formation of the vesicle. Three types of coating proteins have been described: clathrin, COP I, and
COP II.

Clathrin covers plasma membrane vesicles in
endocytosis and the vesicles that move from the
Golgi apparatus to lysosomes. It is polymerized
in a network that resembles a poligonal mesh.

The COP proteins form complexes called
coatomers, which are not assembled in geometric arrangements like clathrin; rather, they are
arranged in a dense and diffuse layer. COP I-
coated vesicles emerge from the Golgi apparatus
and COP II-coated vesicles transport proteins
from the ER to the Golgi.

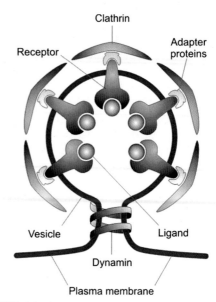

FIGURE 11.19 **Clathrin-coated vesicle.** Clathrin binding to the membrane is mediated by adapter proteins. Dynamin, bound to GTP, is arranged around the neck of the
vesicle. GTP hydrolysis provides the energy that allows
the neck closing by dynamin and the vesicle release.

Clathrin plays a purely structural role; it
helps to give the vesicle its shape. Between
clathrin and the membrane there are other accessory membrane proteins called *adapter* proteins (Fig. 11.19), which recognize the content of
the vesicle. There are other proteins involved in
providing specificity for the transport of vesicles
via different transport pathways. Dynamin participates in the release of vesicles.

Clustering and assembly of the coating molecules (clathrin and COP) are regulated by small
proteins with GTPase activity, designated by the
acronym *ARF* (ADP-ribosylation factor). *ARF*
binds to GTP (Fig. 11.20) and becomes hydrolyzed,
depolymerizing the protein coat on the membrane; the vesicle is now considered "naked."

Vesicle transport within the cell. Between the ER
and the *cis* Golgi, vesicles move by diffusion. In
other cases, they need to travel longer distances
within the cell (particularly notable is the axonal
transport from the body of the neurons to the

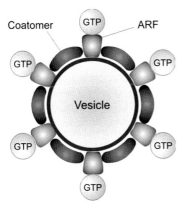

FIGURE 11.20 **COP protein (coatomer)–coated vesicle.** COP assembly requires protein *ARF* (ADP-ribosylation factor), which binds GTP.

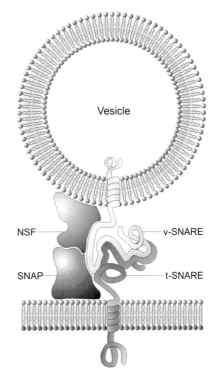

FIGURE 11.21 **Vesicle recognizes the target membrane by interactions between v-SNARE and t-SNARE.** The NSF–SNAP complex promotes SNARE dissociation.

synaptic terminals). This requires the presence of motor proteins, such as *kinesin* and *dynein*, which drive vesicles along cytoskeletal structures (microtubules).

Target site recognition and membrane fusion. A vesicle accurately recognizes its target site due to the interaction between its integral membrane proteins and its membrane receptor, designated SNARE (v-SNARE in the vesicle and t-SNARE in the membrane, the name SNARE comes from *SNAP receptor*). Vesicles and SNARE proteins have complementary structures and recognize each other with great specificity. SNARE proteins are expressed in all cells; there are about 36 different protein types. In neurons, the v-SNARE is *synaptobrevin* and the t-SNARE is *syntaxin*. The SNARE function is to juxtapose and closely approximate the vesicular and target membranes. Both juxtaposed membranes fuse in a process that requires Ca^{2+} and is promoted by other proteins incorporated into the docking site of the SNARE. *Rab* is among those proteins; it specifically requires GTP to bind to the SNARE complex (Fig. 11.21). There are different kinds of *Rab*, each associated with a specific type of vesicular fusion. Once the vesicle has been integrated into the host membrane, the SNARE complex is dissociated and it can be reused in subsequent fusions. Two other proteins are responsible for dissociation. One is *N*-methylmaleimide sensitive factor (NSF), a hexameric enzyme with ATPase activity, inhibited by *N*-methylmaleimide. The second is soluble NSF attachment protein (SNAP), which functions by binding to SNARE and favors NSF entry into the complex. ATP hydrolysis by NSF provides the energy needed for SNARE separation.

Cell Adhesion

Contacts between cells and with the extracellular matrix are largely dependent on the existence of integral plasma membrane proteins called *cell adhesion molecules* (CAM). Cell–cell and cell–extracellular matrix connections are essential in multicellular organisms. Some of these interactions are transient, as seen with immune cells and those directing leukocytes into sites of inflammation. Other more stable bonds are essential for the organization of tissues and organs.

Identical molecules from different cells can interact forming homophilic adhesion associations. The interaction of different molecules from two cells is a heterophilic adhesion.

Four major groups of CAM have been described: *selectins*, *integrins*, *cadherins*, and the *immunoglobulin super-family* (Ig). The first three groups are glycoproteins that require Ca^{2+} or Mg^{2+} to perform their function. The last group presents one or more domains with similar structure to that of the immunoglobulins (p. 747).

Selectins comprise a family of three types of Ca^{2+}-dependent lectins. All consist of a polypeptide chain having the following domains at the N-terminus: (1) a ligand-binding portion, similar to plant lectins, (2) a segment homologous to epidermal growth factor, (3) 2–9 tandem repeats similar to those in complement binding proteins (p. 760), (4) a transmembrane segment, and (5) an intracytoplasmic domain at the C-terminus.

The natural ligands of selectins are oligosaccharides with sialic acid or sulfate in mucin type glycoproteins. Selectins mediate transient adhesion and primary contacts of leukocytes with the vascular endothelium in the initial stage of deployment of these cells to inflammatory sites. These are transient adhesions.

Integrins are heterodimers resulting from the association of α and β integral proteins. Eighteen α and eight β forms have been identified constituting 24 different dimers. Both chains have a large extracellular domain, a transmembrane segment, and a small intracytoplasmic C-terminal piece.

Integrin ligands include cell surface polypeptides of the immunoglobulin superfamily or large molecules of the extracellular matrix, including *fibronectin* and *laminin*. The cytoplasmic face of integrins associates with another group of proteins, such as *vinculin*, which in turn binds

to actin and serves as a link between sites of cell adhesion and the cytoskeleton. This provides mechanical strength and stability to the cell contact points.

Integrins participate in cell–cell interactions, cell attachment to the extracellular matrix, and in a variety of adhesive functions. They can mediate both stable and transient cell–cell attachment.

Cadherins are Ca^{2+}-dependent adhesion molecules involved in stable cell junctions that are part of adherent junctions and desmosomes. They are glycoproteins that have a large extracellular domain with 5 tandem repeats of a 100 amino acid module in which the adhesive activity and binding sites for Ca^{2+} ions are located. They possess a transmembrane helix and a cytoplasmic domain generally connected to cytoskeletal proteins (actin and intermediate filaments) through binding molecules (α and β catenins). Different forms of cadherin are expressed in different cell types. E cadherins are found in epithelium and VE cadherins are found in endothelium, neurons, and mesenchymal cells. These molecules are essential for tissue formation and their decreased expression or loss in cancer cells correlates with the invasive characteristics of tumors.

Immunoglobulin superfamily. There are various classes of such molecules expressed in membranes after activation by cytokines (Chapter 30). They mostly act as integrin ligands. They are comprised of various proteins; including intracellular (ICAM), nerve (NCAM), vascular (VCAM), and mucosal cells (*addressins*) adhesion molecules.

The discovery of cell adhesion molecules has contributed in clarifying aspects of the immune response and to understand the mechanism of leukocyte migration across the capillary walls in inflammation. Moreover, evidence has indicated their involvement in cell malignant transformation and in the process of tumor metastasis. Knowledge in this field opens up new prospects for therapeutic applications.

SUMMARY

The double lipid layer is the basic structure of all biological membranes. All membranes are made of lipids and proteins, which often bind to carbohydrates.

Membrane lipids include phospholipids, which are the most abundant, glycolipids and cholesterol, which are present in smaller amounts. Lipids are asymmetrically distributed, with phosphatidylcholine and sphingomyelin predominating in the outer layer and phosphatidylethanolamine and phosphatidylserine facing the internal layer of the membrane. A property of cell membranes is to be fluid. This depends on the lipid content; fluidity is higher as the proportion of unsaturated fatty acids increases.

Membrane proteins differ in different cells and organelles. They can be *peripheral* sometimes loosely attached to the membrane, or *integral*, which cross the bilayer one or several times.

Transport across membranes is carried out via different mechanisms, including:

Simple diffusion is driven by concentration or electrical gradients; it does not require ATP. Nonpolar or slightly polar substances use this type of transport.

Passive transport or facilitated diffusion is performed following concentration or electrical gradients and it is mediated via *carriers* and *channels*. These are integral membrane proteins, with specificity and saturability for the transported solute.

Carriers can be divided into *uniporters*, which bind and transport only one type of solute; *cotransporters* or *symport*ers, which move two solutes in the same direction; *countertransporters*, *antiporters*, or *exchangers*, which transport solutes in opposite directions. Examples include the facilitated transport carriers of glucose (GLUT), urea (UT), organic cations, OCT and organic anions, OAT.

Channels are hydrophilic pores in the membrane. *Ion channels* allow the passage of ions (Na^+, K, Ca^{2+}, Cl^-) with high specificity. They contain a gate mechanism regulated by different stimuli. Depending on their regulation, they are distinguished in *voltage-gated* and *ligand-dependent* channels.

Ionophores are substances that when incorporated into the membrane form pores that increase the permeability to certain ions.

Aquaporins are water channels.

Active transporters carry out the movement of solutes against their electrochemical gradient; they require energy provided by the hydrolysis of ATP. Active transport is mediated by specific and saturable (Michaelis–Menten kinetics) carriers. They are grouped in several classes: P, V, F, and ABC.

Class P transporters include the Na^+,K^+-ATPase or sodium pump, Ca^{2+}-ATPases or calcium pumps, and H^+,K^+-ATPase.

Class V or proton pumps are found in lysosomes and the plasma membrane of some cells (osteoclasts).

Class F comprises the ATP synthase of the mitochondrial inner membrane. It synthesizes ATP from ADP and P_i using the energy provided by the flux of protons driven by the H^+ electrochemical gradient across the membrane to synthesized ATP.

Class ABC comprises a family of proteins with a protein domain that binds ATP. They move carbohydrates, lipids, peptides, biliary salts, and antibiotics out of the cell. Examples are the *multidrug resistance protein* (MRP), *cystic fibrosis transmembrane regulator* (CFTR), and the SUR.

Secondary active transporters use the electrochemical gradient created by primary active transport systems like the sodium pump. Examples include the Na^+-dependent cotransporter of glucose or amino acids in intestinal mucosa and renal tubules, the Na^+/Ca^{2+}, Na^+/H^+, and HCO_3^- / Cl^- exchangers, and the $Na^+/K^+/Cl^-$ cotransporter.

Nerve impulse. At rest, the interior of the neuron is electronegative with respect to the outside. A stimulus produces opening of Na^+ and K^+ channels. The membrane is depolarized and the change is propagated to the axon terminal as an action potential. In the terminal, the action potential produces Ca^{2+} channel opening and intracellular Ca^{2+} levels rise. Ca^{2+} stimulates fusion of synaptic vesicles with the plasma membrane and releases neurotransmitter into the synaptic space. The neurotransmitters bind to receptors on the postsynaptic neuron, target organ, or gland cell. The activity of the neurotransmitter is dependent on its conformation and the cell type receiving its stimulus.

Endocytosis (*phagocytosis* and *pinocytosis*) is the process by which substances are incorporated into the cell. This is mediated by vesicles which contain the material to be endocytosed surrounded by plasma membrane. Endocytosis can be *clathrin-* and *caveolin-*mediated. The substance to be internalized is first bound by a receptor in the outer surface of the plasma membrane. Invagination of the membrane forms a clathrin- or caveolin-coated vesicle that is internalized and emerges at the cytoplasmic side of the membrane. Fusion of several vesicles form *early endosomes*, whose interior is acidified. The ligand is separated from the receptor, which is recycled and returned to the plasma membrane. The remaining vesicle, with the free ligands, constitutes the *late endosome* which fuses with vesicles from the Golgi apparatus, loaded with hydrolytic enzymes, to form *lysosomes*. Finally, in these last organelles the endocytosed substance becomes degraded.

Exocytosis is the process by which cells transport substance to the medium. This requires that the molecules to be exported are enclosed in vesicles formed in the *trans* Golgi by membrane evagination. Vesicles then traffic to the plasma membrane, fuse with it, and release their content outside of the cell.

Bibliography

Beaugé, L., 2000. Distribución de material y mecanismos de transporte pasivo a través de membranas. In: Cingolani, H., Houssay, A.B. (Eds.), Fisiología Humana. seventh ed. El Ateneo, Buenos Aires.

Blanco, G., 2005. Na$^+$,K$^+$-ATPase subunit heterogeneity as a mechanism for tissue-specific ion regulation. Semin. Nephrol. 25, 292–303.

Borst, P., Elferink, R.O., 2002. Mammalian ABC transporters in health and disease. Annu. Rev. Biochem. 71, 537–592.

Cole, S.P.C., 2014. Multidrug resistance protein 1 (MRP1, ABCC1), a "multitasking" ATP-binding cassette (ABC) transporter. J. Biol. Chem. 289, 30880–30888.

Daleke, D.L., 2007. Phospholipid flippases. J. Biol. Chem. 282, 821–825.

Fujiyoshi, Y., Mitsuka, K., de Groot, B.L., Phillippsen, A., Grubmüller, H., Agre, P., Engel, A., 2002. Structural function of water channels. Curr. Opin. Struct. Biol. 12, 509–515.

Garrahan, P.J., 2000. Transporte activo. In: Cingolani, H., Houssay, A.B. (Eds.), Fisiología Humana. seventh ed. El Ateneo, Buenos Aires.

Gouaux, E., MacKinnon, R., 2005. Principles of selective ion transport in channels and pumps. Science 310, 1461–1465.

Haas, M., Forbush, III, B., 2000. The Na-K-Cl cotransporter of secretory epithelia. Annu. Rev. Physiol. 62, 515–534.

Jahn, R., Scheller, R.H., 2006. SNAREs—engines for membrane fusion. Nat. Rev. Mol. Cell Biol. 7, 631–643.

Janmey, P.A., Kunnunnen, P.K.J., 2006. Biophysical properties of lipids and dynamic. Trends Cell Biol. 16, 538–546.

Linton, K.J., 2007. Structure and function of ABC transporters. Physiology 22, 122–130.

Locher, K.P., 2009. Review. Structure and mechanism of ATP-binding cassette transporters. Philos. Trans. R. Soc. Lond. 364, 239–245.

Malsam, J., Kreye, S., Söllner, T.H., 2008. Membrane fusion: SNAREs and regulation. Cell. Mol. Life Sci. 65, 2814–2832.

Manolescu, A.R., Witkowska, K., Kinnaird, A., Cessford, T., Cheeseman, C., 2007. Facilitated hexose transporters: new perspectives on form and function. Physiology 22, 234–240.

Pauly, B.S., Drubin, D.G., 2007. Clathrin: an amazing multifunctional dreamcoat? Cell Host Microbe 15 (2), 288–290.

Pedersen, P.L., 2007. Transport ATPases into the year 2008. A brief overview related to types, structures, functions and roles in health and disease. J. Bioenerg. Biomembr. 39, 349–355.

Pokutta, S., Weis, W.I., 2007. Structure and mechanism of cadherins and catenins in cell–cell contacts. Annu. Rev. Cell. Dev. Biol. 23, 237–261.

Qin, J., Xu, Q., 2014. Function and application of exosomes. Acta Pol. Pharm. 71, 537–543.

Riordan, J.R., 2008. CFTR function and prospects for therapy. Annu. Rev. Biochem. 7, 701–726.

Takada, Y., Ye, X., Simon, S., 2007. The integrins. Genome Biol. 8, 215.

Vlassov, A.V., Magdaleno, S., Setterquisr, R., Conrad, R., 2012. Exosomes: current knowledge of their composition, biological functions, and diagnostic and therapeutic potentials. Biochim. Biophys. Acta 1820, 940–948.

Yeagley, P.L. (Ed.), 2004. The Structure of Biological Membranes. second ed. CRC Press, Inc., Boca Raton, FL.

Digestion - Absorption

Ingested water, inorganic salts, most vitamins, monosaccharides, and some lipids are absorbed unchanged in the intestinal mucosa. Other components of the diet must undergo degradation or digestion, during which complex molecules become broken down into smaller, simpler substances, which the body can then easily absorb and use. Digestion involves the hydrolysis of the dietary compounds, catalyzed by enzymes present in secretions produced at different levels along the digestive tract. Absorption of the digested substances is in some cases carried out by simple diffusion, but most frequently, requires selective transport systems that move them from the intestine lumen into the bloodstream. The composition and action of digestive secretions, as well as the absorption of substances in the intestine will be considered in the following sections.

SALIVA

The major salivary glands include the parotid, sublingual, and submaxillary glands. Other minor glands are also distributed in the mucosa of mouth and tongue (Ebner glands). In general, glands consist of acini formed by polygonal cells that secrete the initial saliva into the central cavity of the acini. From there, saliva moves along tubules lined by columnar epithelial cells, which secrete and absorb ions and other solutes,

modifying the final composition of saliva. The tubules merge in a duct that carries the final saliva to the oral cavity.

It is possible to obtain saliva from the parotid gland, or a mixture of submandibular–sublingual saliva, using special collection devices placed at the openings of the corresponding secretion ducts (partial saliva). The mixture of these fluids is known as mixed saliva. Whole saliva obtained directly from the mouth consists of a combination of all salivary gland secretions and other fluids, which come from small tubular glands of the oral mucosa (mucous) and serous secretions as gingival crevice fluid. Whole saliva is contaminated with particles from the oral environment (bacteria, desquamated epithelial cells, leukocytes, and food debris).

Saliva is a viscous, colorless liquid, with a pH of ~6.8 and a relative density ranging from 1.000 to 1.010. A normal adult produces approximately 1 L of saliva per day. Partial saliva is clear, while whole saliva is turbid due to particles within the suspension.

Composition

The chemical composition of partial saliva from the parotid and submandibular–sublingual glands shows that they are similar. Submandibular–sublingual saliva is more viscous than parotid gland saliva. This higher viscosity is due to the presence of mucus. Analysis of whole saliva samples does not provide an accurate view of its composition because although it can be easily collected, it contains several contaminants. While particles can be removed by filtration and centrifugation, these manipulations can modify the composition of saliva. In addition, many of the contaminants of saliva originate from the bacteria that are present in the mouth flora.

There are variations in the chemical composition of saliva from one individual to another, and even among samples collected from the same subject at different times or under different conditions (using or not stimulation, and depending on the type, intensity, and duration of the stimulus applied to collect saliva).

Human parotid saliva obtained without stimulation consists of 99.5% water, 0.24 g/dL of mineral components (mainly ions), and 0.26 g/dL of organic components. The composition of saliva is different from that of blood plasma because it is actively produced by the gland cells. Sodium and chloride ion concentrations are much lower than in plasma, while bicarbonate and potassium ions are several times higher. Saliva is the digestive juice richest in potassium. Almost all organic substances in saliva are proteins. These represent a complex mixture of different components. Electrophoretic studies show at least 21 different protein fractions. Salivary amylase, immunoglobulin A, and mucoproteins represent the major proteins in saliva.

Inorganic Components

The fluid originally secreted by the acinar cells has a ionic composition and molarity similar to that of blood plasma. The change in composition occurs during the passage through the ducts, where there is selective absorption and excretion of ions. The final saliva has lower Na^+ and Cl^- and higher K^+ and bicarbonate (HCO_3^-) concentrations than plasma. Different transport systems inserted in the apical (facing the glandular lumen) and basolateral (facing the interstitial

space) membranes of the acini and ducts are responsible for the final composition of saliva.

Acini. The operation of the sodium pump (Na^+, K^+-ATPase) in the basolateral membrane creates the electrochemical Na^+ gradient, which is used by the secondary Na^+/Cl^- cotransporter and the Na^+/H^+ exchanger to drive the influx of Na^+ and Cl^- and the secretion of H^+ across the basolateral membrane of the acinar cells. Increase in cell Cl^- concentration favors its exit to the lumen through apical membrane channels that also drive HCO_3^- release. Furthermore, Na^+ moves from the interstitial space into the lumen through the tight junctions that separate acinar cells (Fig. 12.1). This Na^+ is accompanied by water. K^+ passes to the interstitium and to the gland lumen through K^+ channels present in both membranes. As a result of these processes, the ion content and concentration of the acinar lumen is similar to that of plasma.

Ducts. The saliva produced in the acini circulates through the ducts before reaching the mouth. In the salivary ducts, Na^+ is reabsorbed and K^+ is secreted due to the ion gradients created by the basolateral membrane Na^+, K^+-ATPase

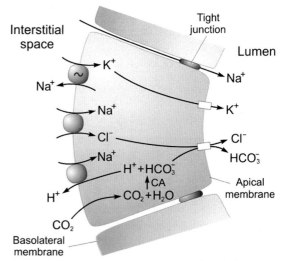

FIGURE 12.1 **Ion secretion in salivary acinar cells.** *CA,* Carbonic anhydrase; $\curvearrowright$, active transport; *white boxes,* ion channels.

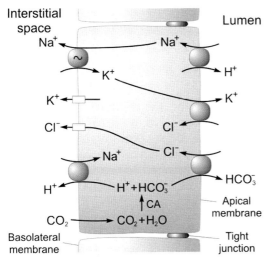

FIGURE 12.2 **Ion absorption and excretion in salivary duct cells.** *CA*, Carbonic anhydrase; $\sim$, active transport; *white boxes*, ion channels.

(Fig. 12.2). Na^+ is internalized at the apical membrane by the Na^+/H^+ exchanger and K^+ is secreted to the lumen via a K^+/Cl^- exchanger. Chloride ions in the lumen also return to plasma in exchange with HCO_3^-, via the Cl^-/HCO_3^- exchanger at the apical membrane of ductal cells and through Cl^- channels in the basolateral membrane.

In addition, the basolateral membrane of ductal cells contains the Na^+/H^+ exchanger which increases intracellular pH and activates the Cl^-/HCO_3^- exchanger in the apical membrane (Fig. 12.2). The tight junctions of ductal cells are less permeable than those in the acini, preventing the return of H_2O to plasma.

Digestive Action of Saliva

The acinar cells produce an enzyme called *ptyalin*, or salivary amylase, which is involved in the digestive process initiating the hydrolysis of starch present in food. The pH for optimal activity of ptyalin is ~ 7.0 and it requires the presence of Cl^-.

Salivary amylase belongs to the group of endoamylases, or α amylases, which catalyze the hydrolysis of internal α-1$\rightarrow$4 glycosidic bonds in starch. In contrast, plant amylases are β-amylases, or exoamylases, which catalyze starch hydrolysis from the chain ends. Starch consists of a linear component, amylase, and a branched component, amylopectin. Salivary amylase can completely degrade amylose. Its hydrolysis produces maltose and eventually maltotrioses (trisaccharides of glucose), when chains with an odd number of glucose molecules are digested. During digestion, amylopectin also produces *limit dextrins*, which are oligosaccharides of 5–10 residues containing α-1$\rightarrow$6 bonds, corresponding to the branching points of amylopectin. Amylase is unable to act on these molecules because it only catalyzes the hydrolysis of α-1$\rightarrow$4 bonds.

Under normal conditions, salivary amylase cannot produce complete degradation of starch molecules due to the rapid oral transit. Ptyalin continues functioning in the stomach, although briefly, since the gastric juice, which has a very low pH (about 1.5), completely inactivates it. This is why the role of ptyalin in starch digestion is limited.

Salivary ptyalin is similar to pancreatic amylase. Both are isozymes of amylase, controlled by different genes and presenting highly homologous sequences. Ptyalin absence does not lead to digestive alterations due to compensation by pancreatic amylase, which degrades the starch arriving at the second portion of duodenum.

Another digestive enzyme, salivary lipase, is secreted by the lingual Ebner glands. Lipase catalyzes the hydrolysis of ester bonds, at the *sn*–3 position of triacylglycerols, with short or medium sized fatty acids the hydrolysis produces 1,2-diacylglycerol and a free fatty acid;. After swallowing, it can exert some action in the stomach because it remains stable at low pH. While it plays an important digestive role in various species, its action in adult humans is not significant; it may have some relevance in infants.

Other Functions of Saliva

The presence of mucus gives saliva lubricating properties, which facilitate swallowing. Saliva also has protective action; it moderates the temperature changes brought by food and reduces the effect of acids, alkalis, and other substances through dilution. Another function of saliva is to provide defense against infections. It contains antibacterial agents, such as *lysozyme* and *lactoferrin*. Lysozyme exerts hydrolytic action on components of the bacterial cell wall. Lactoferrin is an iron chelator that inhibits the growth of microorganisms that depend on iron for development. Acinar cells produce the secretory component of immunoglobulin A (IgA) and excrete it into the glandular lumen. IgA of saliva contributes to the defense of the oral mucosa against viruses and bacteria (p. 752). Saliva also contains the antigens responsible for ABO blood groups.

GASTRIC JUICE

Gastric juice is secreted by glands of the stomach mucosa. The *principal* glands are the most important gastric glands. They have several types of cells with different functions: (1) *parietal* or *oxyntic* cells, (2) *principal* (zymogenic or peptic) cells, and (3) *mucous* cells. Approximately 80% of the principal glands are found in the proximal stomach mucosa (particularly in the fundus); the remaining 20% are located in the distal or pyloric antrum mucosa. Distally, gastrin secreting mucous and G glands can be found.

The human stomach secretes up to 2 L of gastric juice per day. The gastric secretion is a clear, pale, yellow liquid, which mainly consists of water (about 99% of total), hydrochloric acid, enzymes, and mucoproteins (mucin).

Hydrochloric Acid

The parietal cells of the principal stomach glands secrete hydrochloric acid into the gastric lumen allowing the stomach to reach a pH

of 0.87. Parietal cells are unique, since they have the capacity to make a strong inorganic acid from the components of blood plasma and interstitial fluid, which are neutral or slightly alkaline (pH 7.3–7.4). Both H^+ and Cl^- concentrations in gastric juice can reach values of up to 170 mM; in blood plasma, the H^+ and Cl^- concentrations are 0.00004 and 105 mM, respectively. In this manner, the H^+ concentration in the stomach is several million times higher than in plasma. The energy required to create this gradient is provided by the hydrolysis of ATP, generated in the numerous mitochondria that are present in parietal cells. The transport of protons or hydronium ions (H_3O^+) in parietal cells is performed by an active exchanger system, the P-type H^+,K^+-ATPase, or "proton pump" (p. 237). At rest, parietal cells contain numerous vesicles in their cytoplasm that contain H^+,K^+-ATPase. The cells have intracellular canaliculi, communicated with the gland lumen (Fig. 12.3). When

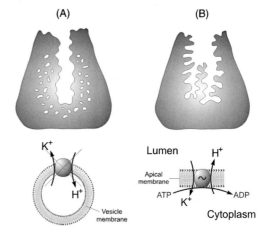

FIGURE 12.3   (A) Parietal cell of a principal gastric mucosa gland at rest (nonsecreting). Intracytoplasmic vesicles in the membrane have proton pumps (schematized at lower left). (B) Activated parietal cell. The vesicles have disappeared, because they were delivered and fused to the apical membrane. Also, intracellular canaliculi have expanded. The H^+,K^+-ATPase, now included in the apical membrane, transports H^+ into the lumen and introduces K^+ into the cell.

the cells are stimulated, they acquire full secretory activity, with a concomitant decrease in the number of vesicles, and expansion of the intracellular canaliculi (Fig. 12.3B). This is due to the delivery and fusion of the vesicles to the luminal plasma membrane of the cells.

The H^+,K^+-ATPase in intracytoplasmic vesicles is oriented with its H^+ ion releasing or "acid face" toward the interior of the vesicle. To operate, the proton pump requires K^+ ions at the luminal side. Due to the low K^+ content in parietal cell vesicles, H^+,K^+-ATPase remains inactive. After fusion and insertion in the plasma membrane, the acid face of H^+,K^+-ATPase is now oriented toward the gastric lumen. The K^+ ions present in the stomach, which come with the food, as well as from the cells of the gastric mucosa via K^+ channels, stimulate H^+,K^+-ATPase, initiating H_3O^+/K^+ exchange. At the same time, chloride moves into the stomach lumen through Cl^- channels in the apical side of the cells of the gastric mucosa. The net effect is HCl secretion and K^+ recycling back into the cell (Fig. 12.4).

The trafficking of vesicles containing H^+,K^+-ATPase is regulated by *histamine*. This potent activator of acid secretion is released by enterochromaffin cells in the fundus of the stomach, after stimulation by gastrin, acetylcholine, and ghrelin. Histamine binds to H_2 receptors in the basolateral membrane of parietal cells and initiates a series of intracellular signaling events that result in migration and fusion of the H^+,K^+-ATPase containing vesicles to the membrane of the intracellular canaliculi. Hydrochloric secretion is inhibited by somatostastin and interleukin-11. When the stomach content has reached a sufficiently low pH, the proton pumps at the plasma membrane are internalized by endocytosis and remain inactive until a new stimulus arrives to the cells to reinitiate their trafficking to the cell surface.

The origin of the protons secreted by parietal cells depends on the presence of *carbonic anhydrase* in the cells, which catalyzes the following reaction:

$$CO_2 + H_2O \rightleftharpoons H_2CO_3 \rightleftharpoons HCO_3^- + H^+$$

Some authors represent the reaction catalyzed by carbonic anhydrase as the combination of CO_2 with HO^- resulting from the dissociation of water, to form HCO_3^-. The end result is the same as that of the equation shown earlier. The CO_2 is generated in tissues as a result of metabolic processes and diffuses freely into cells. The carbonic acid formed dissociates into bicarbonate and hydrogen ions.

The H^+ are captured by the H^+,K^+-ATPase in the form of hydronium ions (H_3O^+) and transported to the gastric lumen in exchange for K^+. Bicarbonate passes into the blood through a basolateral membrane exchanger that simultaneously introduces Cl^- into the cell. For every proton released into the gastric lumen, one HCO_3^- appears in blood plasma. This produces the phenomenon called "alkaline tide," observed after meals. Various exchanger systems in the basolateral membrane of the parietal cell: HCO_3^-/Cl^-, Na^+/H^+, and the Na^+,K^+-ATPase (Fig. 12.4) help to maintain the appropriate intracellular ionic concentration.

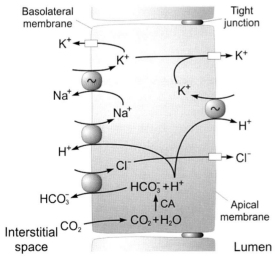

FIGURE 12.4 **Hydrochloric acid production in apical membranes of parietal cells of gastric principal glands.** *CA,* Carbonic anhydrase.

Hydrochloric acid in gastric juice ensures the proper pH for pepsin activity. It also has some direct action on food, making it more susceptible to digestion by hydrolytic enzymes. It also has an antiseptic effect, thus helping to prevent the development of bacteria in the stomach and avoiding fermentation of the gastric content. HCl also plays a role in iron absorption. This element is found in some foods as ferric hydroxide, clustered in colloidal micelles. Hydrochloric acid helps to solubilize ferric ions, which then can be easily reduced to ferrous ions. In its reduced state (Fe^{2+}), iron can be readily absorbed.

Digestive Action of Gastric Secretion

Pepsin. The main digestive action of gastric juice is exerted by *pepsin*, which catalyzes the partial hydrolysis of proteins. Pepsin is secreted in the state of pepsinogen by glands in the stomach's body and fundus.

Pepsinogen (42.5 kDa) is a proenzyme, or zymogen, activated by H^+ ions in gastric secretions. Its activity is further potentiated by its active form, pepsin. This mechanism, by which an enzyme activates its own zymogen, is called autocatalysis.

The activation of pepsin is accomplished by hydrolysis of the peptide bond between residues 42 and 43 of the zymogen, releasing a 42 amino acid segment from the N-terminus of the protein. Active pepsin has a mass of 35 kDa.

Pepsinogen secretion is stimulated by the same factors that activate HCl release: acetylcholine (vagal neurotransmitter), gastrin, and histamine. Approximately 99% of pepsinogen produced in the principal glands is secreted into the gastric lumen. The remaining 1% moves to the interstitial fluid and blood,

eventually reaching the kidney to be excreted into urine as uropepsin. Determination of uropepsin in urine serves as an index of stomach peptic activity.

Pepsin action. Pepsin acts on virtually all proteins except keratins, mucoproteins, and protamines. It catalyzes the hydrolysis of peptide bonds located in the interior of the protein chain. Due to this action, pepsin belongs to a family of enzymes known as *endopeptidases*. The product of protein hydrolysis catalyzed by pepsin are polypeptide fragments of high molecular weight, which were originally named proteoses and peptones. Although pepsin can hydrolyze virtually any peptide bond, it has certain preferences, selectively targeting bonds that contain the amine group of an aromatic amino acid (tryptophan, phenylalanine, and tyrosine). The optimum pH for pepsin activity of 1.0–2.0 is maintained in the stomach by HCl. When the pH of the medium increases to values greater than 3.0, pepsin is almost completely inactivated.

In young children, gastric acidity is usually higher than in normal adults. In the first few months of life the gastric pH is approximately 5.0. Proteolytic action does not depend on pepsin but on other proteinases, including *cathepsins*, which are present in the lysosomes of almost all cells and released from desquamated gastric mucosa cells. In addition, a proteinase capable to act at near neutral pH has also been described in gastric juice of children.

Lab ferment or rennin. This proteolytic enzyme, also known as chymosin, is secreted by the fourth stomach of ruminants at early stages of life. It produces milk coagulation, acting on the most abundant protein of milk, *casein*. Rennin transforms casein into paracasein in the presence of Ca^{2+} ions, forming a calcium paracaseinate precipitate. This change in casein conformation facilitates its digestion by other proteases.

In humans, pepsin catalyzes the same reaction at a pH of 4.0. Perhaps pepsin is responsible for milk coagulation in an infant's stomach.

$$Casein \xrightarrow{\text{Rennin}} Paracasein$$

$$Paracasein + Ca^{2+} \longrightarrow \underset{\text{Nonsoluble}}{Ca\ Paracaseinate}$$

Lipase. This enzyme is secreted by cells of the gastric fundus. Its optimum pH fluctuates between 3 and 6. It catalyzes the hydrolysis of ester bonds at positions 1 and 3 of triacylglycerides, especially those that have short or medium length chain fatty acids; their products are free fatty acids and diacylglycerols. Together with lingual lipase, which continues its action in the stomach, they contribute to the degradation of fats. These lipases are not essential and their absence does not produce clinical alterations because the pancreas secretes another lipase that is sufficient to meet the needs of digestion. Gastric and lingual lipases become important in neonates and young infants who have not yet fully developed their pancreatic function.

Mucus. This substance is secreted by cells of the principal gastric glands and mucosa. Mucus consists of glycoproteins; it cannot be digested by pepsin and helps protecting the gastric mucosa.

Parietal glands produce a glycoprotein with a molecular mass of 55 kDa, called *intrinsic factor*, which forms a complex with vitamin B_{12}. Formation of this complex is required for the absorption of vitamin B_{12} in the ileum. Vitamin B_{12} is an essential factor for normal erythropoiesis (p. 677). Lack of intrinsic factor abolishes vitamin B_{12} absorption and causes a type of anemia known as *pernicious anemia*.

Analysis of Gastric Secretion

The study of volume, acidity, enzyme content, and other gastric juice components is of clinical interest. This is achieved by gastric juice aspiration via nasogastric intubation, or through sample collection via endoscopic techniques. Acidity is determined by titration of the stomach fluid with sodium hydroxide. The secreted H^+ or *basal acid output* is expressed in mEq./h. While fasting, basal acid output is estimated in the aspirated liquid during four consecutive periods of 15 min each. The basal acid rate of a normal adult ranges from 0 to 11 mEq. of H^+ ions per hour. The maximum acid output is obtained by stimulating gastric secretion after injection of a secretagogue, usually pentagastrin. Histamine has also been employed. The maximal acid output in normal adults ranges from 10 to 63 mEq. of H^+ per hour. While determination of gastric pH is an adequate indicator of stomach acidity, it is not a measure of gastric secretion. These methods have the disadvantage of being invasive, uncomfortable, and time consuming.

Noninvasive methods have also been used to study gastric juice. Among them are urinary analysis, serum pepsinogen levels, alkaline tide determinations, scintillation techniques with ^{99}Tc, and impedance tomography. These methods are, at best, semiquantitative and their use is limited to detect significant decrease or absence of acid secretion.

Acid secretion alterations. Achlorhydria is a condition that consist in the absence of hydrochloric acid in the gastric juice. The reduction in HCl concentration is known as *hypochlorhydria*, while the increase in HCl concentration is called *hyperclorhydria*. Hyperchlorhydria occurs frequently in pathological processes, such as gastritis and peptic ulcer disease.

Knowledge of the mechanisms responsible for the secretion of HCl in the stomach has allowed for the development of effective drugs to control acid release. Initially, antagonists which block histamine H_2 receptors, such as *cimetidine*, were used. Later, specific H^+,K^+-ATPase inhibitors (*omeprazole*) were developed. Proton pump inhibitors are used in combination with antibiotics (usually clarithromycin and amoxicillin) to treat gastritis originated by the bacterium *Helicobacter pylori*. Protectors of the gastric mucosa, such as bismuth colloids, are also commonly included to this therapy regimen.

PANCREATIC JUICE

The pancreatic juice has some similarities with saliva. It contains small amounts of protein, hydrolytic enzymes, and inorganic components such as: Na^+, K^+, HCO_3^-, Ca^{2+}, and HPO_4^{2-}. It has a pH ranging from 7.5 to 8.0. A normal adult produces up to 1.5 L of pancreatic juice per day. Pancreas digestive secretions are produced by the exocrine pancreas, constituted by glandular acini, whose lumen continues with canals that converge into larger ducts and finish in a final conduit that drains into the duodenal portion of the intestine.

Inorganic Components

Pancreatic juice is isotonic with respect to the extracellular fluid. Most of the water and ions present in pancreatic secretions are secreted by cells lining the ducts next to the acini. A characteristic of these cells is their capacity to release high concentrations of bicarbonate (several times higher than that of plasma). This ion secretion is stimulated by secretin and potentiated by cholecystokinin and acetylcholine. The transport of HCO_3^- in the pancreatic duct cells, against its gradient, is a secondary active process. This is carried out by the apically located HCO_3^-/Cl^- exchanger, which depends on the operation of the basolateral Na^+,K^+-ATPase and the Na^+/H^+ exchanger. The diagram in Fig. 12.5 shows the ion exchange transporters involved in ion secretion of ductal cells.

Pancreatic ductal cells have carbonic anhydrase, which produces H_2CO_3 from CO_2 generated by body metabolism and water. Carbonic acid gives by its dissociation HCO_3^- and H^+. Protons are transported into the blood, where they are exchanged for Na^+ by the Na^+/H^+ antiport system. This exchanger is driven by the Na^+ gradient created by the Na^+,K^+-ATPase. In addition, the presence of a H^+-ATPase in the basolateral side of the ductal cells has been postulated. These transport mechanisms produce HCO_3^-

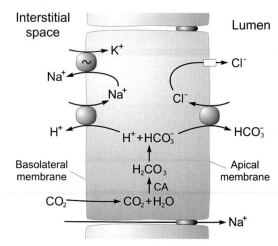

FIGURE 12.5   Ion exchanges in pancreatic ducts cells. *CA*, Carbonic anhydrase.

accumulation in the cytosol and its secondary excretion into the glandular lumen by the apical Cl^-/HCO_3^- exchanger. The rate of HCO_3^- excretion depends on availability of Cl^- in the lumen. Cl^- is recycled from the cell interior through a Cl^- channel. Na^+ is passively transported from the interstitium to the lumen through intercellular tight junctions (paracellular pathway) following the transcellular electrochemical gradient (Fig. 12.5).

Digestive Action of Pancreatic Juice

Pancreatic juice contains several powerful hydrolases synthesized by the acinar cells. Enzyme secretion is stimulated by endocrine factors (cholecystokinin, secretin, and, much more weakly, gastrin) and neuronal factors [acetylcholine and vasoactive intestinal polypeptide (VIP)]. The pancreas produces enzymes that act on starch, lipids, and proteins introduced with the diet. All pancreatic enzymes, except amylase and lipase, are secreted as zymogens and activated in the intestinal lumen. Proteolytic enzymes encompass three endopeptidases (trypsin, chymotrypsin, and elastase) and two exopeptidases (carboxypeptidases A and B) that separate terminal amino acids.

Endopeptidases are all neutral proteases, inactivated by acid pH. They share a similar structure and have an essential serine residue in the active site, which has given them their generic name of serine proteases. Carboxypeptidases A and B are also structurally related; both are metalloenzymes that contain Zn.

Trypsin is secreted by the pancreas as the zymogen trypsinogen, which is activated in the intestinal lumen by the enzyme *enterokinase* or *enteropeptidase* in the intestinal mucosa. Once activated, trypsin acts autocatalytically on trypsinogen.

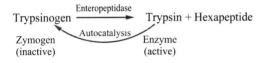

Trypsinogen has 229 amino acid residues, while trypsin contains 223 amino acid residues. Trypsinogen conversion to trypsin involves the separation of a hexapeptide from the zymogen's N-terminus. After removal of this segment, the catalytic site adopts the appropriate spatial arrangement for its operation. The enzyme has an optimal pH for activity between 8.0 and 8.5.

Trypsin is an endopeptidase, which catalyzes the hydrolysis of proteins on internal peptide bonds. It has selectivity for bonds containing the carboxyl group of diaminated amino acids (lysine and arginine). Their products are polypeptides with basic amino acids at the C-terminus. Trypsin activates all zymogens produced by the pancreas.

Chymotrypsin is secreted as chymotrypsinogen, a proenzyme that is activated by trypsin in intestine. Chymotrypsinogen is a polypeptide of 245 amino acid residues. Trypsin catalyzes the hydrolysis of the bond between amino acid residues 15 and 16, producing the active enzyme, β-chymotrypsin. β-Chymotrypsin has autocatalytic action; it converts other molecules of β-chymotrypsin into α-chymotrypsin by breaking a peptide bond distant from that hydrolyzed

by trypsin. The original chain, cleaved at two different sites, produces three segments that continue to be linked together by disulfide bridges. The change in the tridimensional structure of chymotrypsinogen produced by hydrolysis promotes activation of the enzyme. Chymotrypsin is also an endopeptidase. Although it targets several peptide bonds, it has preference for the carboxyl group containing aromatic amino acids (phenylalanine, tyrosine, and tryptophan).

Elastase is secreted as proelastase and activated by trypsin. It catalyzes primarily the hydrolysis of elastin in the elastic fibers of connective tissue and also acts on other proteins. Elastase promotes the breakdown of peptide bonds adjacent to aliphatic amino acids; the resulting products are peptides containing neutral amino acids at the C-terminus.

The hydrolytic action of the three endopeptidases on proteins and protein fragments produced by gastric digestion generates relatively small peptide segments.

Carboxypeptidases are secreted as procarboxypeptidases and are activated by trypsin in the intestinal lumen. They are exopeptidases, which catalyze the hydrolysis of peptide bonds adjacent to the C-terminal end of proteins, releasing the last amino acid residue. Carboxypeptidase A has preference for terminal peptide bonds with a neutral amino acid. Carboxypeptidase B acts on peptides with a basic C-terminal amino acid.

Ribonuclease and deoxyribonuclease catalyze the hydrolysis of nucleic acids, by hydrolysis of bonds between nucleotides.

Amylase has powerful hydrolytic action on starch. Its activity is identical to that described for salivary amylase. It requires Cl^- and cleaves only α-1→4 glycosidic bonds. The end products of starch digestion by pancreatic amylase are maltoses, maltotrioses, and limit dextrins.

Lipase catalyzes the hydrolysis of ester bonds in neutral fats. Its optimal pH is close to 8.0, but it remains active up to pH as low as 3.0. Below this pH, it is rapidly denatured. Pancreatic acini secrete, together with lipase, a polypeptide of

102–107 amino acid residues called procolipase. When this proenzyme arrives into the intestinal lumen, it is hydrolyzed by trypsin and converted into a polypeptide of 96 amino acid residues, colipase. Colipase then forms a complex with lipase and serves as an anchor for fixing lipase on micelles formed by bile acids and dietary lipids.

Lipase only affects esters in the primary carbon bonds of glycerol. The resulting products, in the first stage, are 2,3-diacylglycerol and a fatty acid. In a second stage, 2-monoacylglycerol and another fatty acid are formed. The action of an isomerase converts 2-monoacylglycerol into 1-monoacylglycerol (by transferring the fatty acid from the secondary alcohol to a primary carbon) so that the lipase can complete the digestion of triglycerides into glycerol and free fatty acids. Isomerase has modest activity, which explains the low rate and small proportion of dietary triacylglycerols being completely degraded by lipase. A large proportion of triacylglycerol molecules are degraded into 2-monoacylglycerols.

Cholesterolesterase catalyzes the hydrolysis of ester bonds of fatty acids, cholesterol, vitamins A, D, and E, and of acylglycerides. Cholesterolesterase is active against substrates incorporated into bile salt micelles. This enzyme is also present in the cells of the intestinal mucosa.

Phospholipase A_2 works on bonds formed between fatty acids and the glycerol hydroxyl carbon 2 of glycerophospholipids. This produces a free fatty acid and a lysophospholipid. Phospholipase A_2 is secreted as a proenzyme and activated by trypsin. The phospholipids in micelles are phospholipase A_2 substrates. The presence of bile acids is an absolute requirement for the activity of this enzyme.

INTESTINAL MUCOSA

The intestinal mucosa has numerous folds, villi, and crypts that greatly increase the absorptive surface area of the organ. Crypts and villi are lined by a columnar epithelium formed by enterocytes and mucous cells (goblet cells). Both these cell types originate from precursor cells located in the bottom of crypts that, upon differentiation, move from the lower portion of the crypt to the tip of the villi. When the mature cells reach the tip of the villus, they are shed into the intestinal lumen, where they are lysed. The lifetime of these cells is 3–6 days. The apical or luminal membranes of enterocytes present microvilli that greatly expand the area of the mucosa. The presence of microvilli gives a distinctive appearance to the apical face of the cells, which is known as the *brush border*. This brush border is composed of integral membrane proteins, hydrolytic enzymes, and transport systems.

Brush Border Enzymes

Endopeptidases. The first endopeptidase identified in intestinal mucosa was *enterokinase* or *enteropeptidase* (a name that is more appropriate). It catalyzes the hydrolysis of trypsinogen into trypsin, a reaction that initiates the activation of pancreatic zymogens in the intestinal lumen.

Exopeptidases. At least six exo-oligopeptidases have been described that have a wide range of substrate specificity. Three are aminopeptidases that catalyze the cleavage of the peptide bond adjacent to the oligopeptide N-terminus, releasing the first amino acid of the chain. Within this group of enzymes are also *dipeptidases*.

Disaccharidases. These enzymes are responsible for the final degradation of residues from starch digestion and of disaccharides present in the food. There are three of these enzymes, all of which have dual function and two different active sites within the same molecule. They include the following:

Sucrase–isomaltase is an integral glycoprotein with an N-terminal intracytoplasmic segment, a single transmembrane helix, and two catalytic sites projecting toward the intestinal lumen. The active site for isomaltase is close to the membrane, whereas the active site for sucrase is located distally. Isomaltase catalyzes the hydrolysis of the α-1→4 bonds of maltose and the α-1→6 bonds of limit dextrins and isomaltoses. Sucrase cleaves sucrose into glucose and fructose. This enzyme is also called invertase.

Lactase–phloridzin hydrolase is another dual enzyme, with its C-terminal end facing the cell interior. The portion located in the lumen has the lactase site which catalyzes the hydrolysis of lactose to glucose and galactose. The phloridzin hydrolase site produces rupture of β-glycosidic linkages (phloridzin is a compound having a bond of this type). It intervenes in digestion of glycolipids (e.g., glycosylated ceramides).

Maltase–glucoamylase acts on α-1→4 glycosidic bonds and, at a very small rate, on α-1→6 bonds. It accounts for approximately 20% of the total digested maltose; the remaining 80% is hydrolyzed by isomaltase.

Trehalase. Human intestinal mucosa contains small amounts of enzymes important in the digestion of food containing trehalose, such as yeast and fungi (trehalose is a nonreducing disaccharide formed by glucose in a α-1→1 double glycosidic bond).

Nucleases, phosphatases, and nucleosidases are enzymes that degrade nucleic acids. Nucleases degrade nucleic acids into nucleotides. Phosphatases also hydrolyze other phosphoric esters. Among the wide substrate spectrum of phosphatases is *intestinal alkaline phosphatase*, an enzyme with zinc that is anchored in the brush border membrane. Finally, nucleosidases complete the digestion of nucleosides resulting from the action of phosphatases; the hydrolysis produces pyrimidine or purine bases and pentose.

BILE

Bile is a fluid continuously produced by the liver and is stored in the gallbladder in periods between meals. The liver produces approximately 500–600 mL of bile per day. Bile is a liquid which has a golden yellow or slightly brownish color and viscous appearance with a bitter taste. The pH of bile ranges between 7.8 and 8.6 and its relative density is 1.010. Hepatic bile contains ~2.5–3.5% solid matter.

Composition

Bile is a complex fluid with high relative content of water, insoluble lipids (phosphatidylcholine and cholesterol) and compounds with detergent properties (bile acids). It also contains bile pigments, urea, varying amounts of proteins (mainly mucoproteins), and inorganic electrolytes in similar concentrations to those of plasma.

Bile acids are compounds structurally related to cyclopentanoperhydrophenanthrene. They constitute approximately 50% of the bile organic solutes. Originally, the liver synthesizes *primary bile acids* directly from cholesterol. This includes a series of stages initiated by cytochrome P_{450} monooxygenase. Then, α7-hydroxylase adds hydroxyl groups to cholesterol, shortens the side chain to 5 carbons, and oxidizes the C-terminal group to carboxyl. The most abundant of the primary bile acids is *cholic acid* (α3,α7,α12-trihydroxycholanic), shown in Fig. 12.6; less abundant is *chenodeoxycholic acid* (α3,α7-dihydroxycholanic), shown in Fig. 12.7.

Primary bile acids are converted into secondary bile acids in the intestine by bacteria of the enteric flora. The main ones are *deoxycholic*

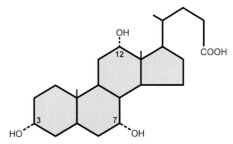

FIGURE 12.6   Cholic acid (3,7,12-trihydroxycholanic).

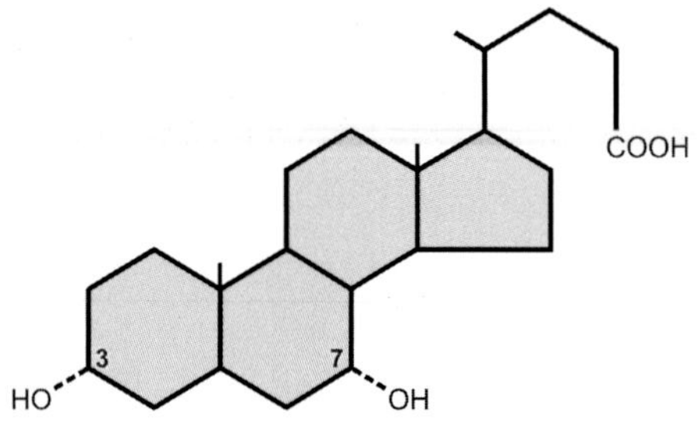

FIGURE 12.7 **Chenodeoxycholic acid (3,7-dihydroxy-cholanic).**

(α3,α12-dihydroxycholanic) and *lithocholic* (α3-monohydroxycholanic) acid.

The terminal bile acid carboxyl group is conjugated in the liver with glycine or taurine. One amide bond is formed and glycocholic and taurocholic acids are obtained (Fig. 12.8). These compounds are strongly hydrophilic and are more acidic (lower pK$_a$) than unconjugated bile acids. They are ionized over a wide range of pH and are neutralized mainly with Na$^+$ to form bile salts.

Bile salts are amphipathic or amphiphilic compounds. The hydroxyl groups are located on the same side of the molecule (all in the α position within common bile acids) and, along with the ionized groups (COO$^-$ and SO$_3^-$), interact with water. The other side of the steroid core is markedly hydrophobic. When a sufficient concentration of the bile salts (critical micelle concentration) is reached, bile salts tend to aggregate in micelles with the hydrophilic face oriented outward, in contact with the aqueous medium. These micelles contain other amphipathic molecules such as phospholipids and cholesterol; bile salts favor emulsification and stabilization of these substances in the bile.

During periods between meals, the bile secreted by hepatocytes into the canaliculi flows through the channels and is stored in the gallbladder. Here, there is intense absorption of inorganic ions, primarily Na$^+$, Cl$^-$, and HCO$_3^-$, as well as water absorption. This results in a slight acidification of the bile. The concentration of salt and bile pigment in the gallbladder's bile reach values 5–20 times greater than that of the

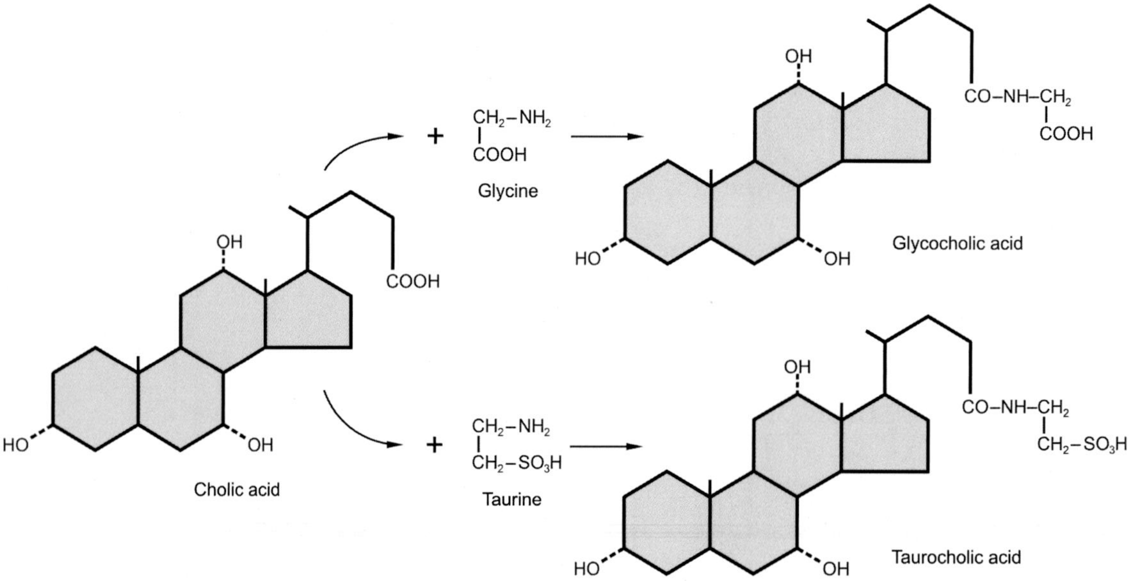

FIGURE 12.8 **Conjugation of cholic acid with glycine and taurine.**

original bile from hepatocytes. However, the bile in the gallbladder remains isotonic with respect to plasma because the osmotic activity of the micelles formed by bile acids, phospholipids, and cholesterol is minimal. The bile stored and concentrated in the gallbladder is greenish-yellow and sometimes can be a darker, olive green. Its pH ranges from 6.8 to 7.7 and its relative density reaches 1.040. It has high proportion of solids (up to 17%).

When the gastric content passes to the duodenum, components of the diet (mainly lipids) stimulate the release of cholecystokinin (CCK) by cells of the duodenal mucosa. The CCK transported in blood reaches the gallbladder and causes contraction of its muscle layer. CCK also relaxes the sphincter of Oddi, facilitating bile evacuation into the duodenum through the common bile duct.

In the intestine, bile salts participate in digestion and absorption of lipids and other related substances. Their action is due to their detergent properties, which facilitate the dispersion of lipids into fine droplets and increases the surface area of lipids exposed to hydrolytic enzymes (lipase, phospholipase, and cholesterolesterase); they also favor lipid absorption across the intestinal mucosa.

After exerting their function, bile salts continue along the small intestine, where they undergo the action of enteric bacteria, which generate secondary bile acids. Then, they are absorbed by the mucosa of the distal ileum and are sent back to the liver by the portal vein. These bile acids are taken up by hepatocytes and reexcreted, to return to the intestine. Normally, the bile contains primary and secondary bile acids (cholic, chenodeoxycholic, deoxycholic and, to a lesser extent, lithocholic) conjugated with glycine or taurine. This recycling of bile salts is called *enterohepatic circulation*. The total amount of these salts is 2–3 g/day. Twenty to thirty grams per day enter and leave the intestine, which indicates that each molecule is recycled approximately ten times. Normally, 0.5 g

of bile salts per day is eliminated from the body with feces. Synthesis in liver is regulated to replace this loss.

Bile salts have choleretic action, which refers to the stimulatory effect that salts reabsorbed via the portal vein have on the liver production of bile.

Phospholipids are the second most abundant organic compounds of bile, comprising 22% of the total solids. Phosphatidyl choline (lecithin) is the main phospholipid in bile. These are amphipathic molecules that associate with bile salts into micelles. Each mole of bile salts solubilizes approximately two moles of phospholipids. The association of bile salts with phospholipids has more capacity to emulsify other lipids, such as cholesterol, than bile salts alone.

Cholesterol, predominantly nonesterified, comprises ~4% of all solids in the bile. From this, only 10%–15% is esterified. Bile salts and phospholipids allow the incorporation of cholesterol esters into micelles and, depending on their relative proportion, they maintain cholesterol in suspension. From a standpoint of its metabolism, the bile is the major route for cholesterol excretion.

Bile pigments. These substances result from degradation of heme (p. 406). The most abundant is *bilirubin*, conjugated as soluble diglucuronide. The yellow color of the newly secreted bile is due to this pigment. Upon exposure to air, or even in the gallbladder, bilirubin is oxidized to *biliverdin*, which has a deep green color. Bile normally discharged into the intestine usually has a small proportion of biliverdin. Bile pigments are considered excretory products and have no known digestive function.

Gallstones. The stability of bile lipid components is ensured by an appropriate relationship among bile salts, phospholipids, and cholesterol concentrations. When there is a relative excess of cholesterol, it may precipitate. Precipitation of bile components produces solid masses of different sizes and shapes (gallstones), representing a common problem in clinics. Gallstones can be found in the gallbladder and throughout the biliary tract. Single gallstones are generally ovoid, but multiple ones present with facets, which are formed by the pressure or rubbing that they exert on each other. A cross section of gallstones shows a central core around which concentric layers of precipitated bile components are deposited. Gallstones are distinguished according to the prevailing substance that they contain. Most of them have 90%–98% of cholesterol, but they can also contain bile pigments, or calcium carbonate.

SUMMARY OF THE DIGESTIVE PROCESS

Carbohydrates

Approximately 50% of the energy provided by a normal diet corresponds to carbohydrates. These include polysaccharides, disaccharides (mainly sucrose and lactose), as well as the monosaccharides glucose, fructose, and galactose. Starch, composed by amylose and amylopectin, is the main carbohydrate in diets. It is present in many plant foods, such as grains, flour, fruits, tubers, and legumes.

Sucrose is typically the most abundant dietary disaccharide. It is present in fruits and plants, or in sweeteners, such as sugar, from cane and beet. The average daily intake of sucrose in an adult is 40 g. Lactose is another important disaccharide; its intake varies with age and dietary habits.

The most common monosaccharides in diets are glucose and fructose, which are present freely in fruits and honey. Fructose consumption has significantly increased in recent decades because it is the sweetest sugar and is widely used in the preparation of sweets and carbonated drinks.

Starch. Hydrolysis of polysaccharides, such as starch and glycogen, starts in the mouth and is catalyzed by salivary amylase (ptyalin). The role of this enzyme is only transient, since it is quickly inactivated as soon as it passes into the stomach due to the gastric acid content.

Starch digestion in the intestine is carried out mainly by the action of pancreatic amylase, which has catalytic properties identical to those of ptyalin. These enzymes are endoamylases, which catalyze hydrolysis of α-1→4 glycosidic bonds inside the molecules of amylase and amylopectin. They cannot cleave bonds at the end of the chains, amylopectin α-1→6 bonds, or the α-1→4 bond near the branched α-1→6 junctions. The final products of amylase action are maltoses, maltotrioses, and limit dextrins. These molecules are all hydrolyzed to free glucoses by the isomaltase activity of the sucrase–isomaltase enzyme complex and by both enzymes of maltase–glucoamylase complex.

Not all starch coming with food is digested completely. Part of it, called resistant starch, can escape hydrolysis. In some foods (raw vegetables and unripe bananas), starch grains usually adopt organized structures (pseudocrystalline) that protect them from amylase activity. In whole grains and in some vegetables, starch is physically protected by indigestible membranes that prevent the action of hydrolases. The amount of resistant starch coming into the body varies according to the type of food and the cooking process, which changes the starch status and makes it more accessible to enzymes. It is estimated that 2%–5% of the ingested starch escapes digestion.

Dietary fiber. Humans cannot digest cellulose, a polysaccharide abundant in plant foods, because they do not have enzymes with hydrolytic action on β-1→4 bonds between glucose molecules. In many animals, including ruminants, cellulose is degraded to absorbable products. This action is not due to the production of enzymes that can cleave cellulose in those animals, but rather to enzymes of the bacterial flora within their digestive tracts. This is an example of a symbiosis that allows the host organism to utilize cellulose from foods. While in humans there is also some bacterial fermentation of indigestible polysaccharides, this occurs at a much smaller degree.

Dietary fiber comprises cellulose, other polysaccharides in plant foods (hemicellulose, pectins, and gums), and lignin, which is not a polysaccharide. These compounds go through the small intestine without being modified, due to the lack of enzymes that can degrade them. Dietary soluble fiber includes pectins, gums, and mucilages; insoluble dietary fiber includes cellulose and hemicellulose (see p. 89). Resistant starch is also considered a type of dietary fiber. *Functional fiber* is a term used to denote those carbohydrates that cannot be digested, isolated from natural sources or synthesized. Their ingestion has favorable effects on the human digestive system.

Hemicellulose predominates in grains and cellulose is the main fiber found in legumes. Indigestible polymers increase the intestinal content volume and are an important factor for promoting peristaltic activity. Pectins, gums, and mucilage can be metabolized by bacteria in the colon and do not contribute to increase fecal volume.

In the large intestine, dietary fiber (especially pectin) is fermented by the normal bacterial flora, producing short-chain fatty acids (acetic, propionic, and butyric acid) and gases (methane and hydrogen). The fatty acids are used as an energy source by mucosal cells (colonocytes) and are also absorbed and sent to the liver by the portal vein. Cellulose, hemicellulose, and lignin are poorly fermented by the intestinal flora. The remaining fiber is eliminated unchanged in feces and helps to increase the mass of fecal matter.

Fiber binds lipids to reduce fat and cholesterol absorption. The acidic conditions produced by bacterial fermentation of fiber in the colon are considered favorable. For example, it has been proposed that they may play a role in the prevention of colon cancer. In conclusion, although fiber is not absorbed, it is a beneficial component of the diet and aids in stimulating peristalsis.

Disaccharides. All ingested sucrose is hydrolyzed into glucose and fructose by the enzyme sucrase present in the intestine brush border (part of the complex sucrase–isomaltase). Lactose is hydrolyzed into galactose and glucose by the action of lactase–phloridzin hydrolase within the mucosa.

Lactase activity is high in infants and declines during the course of life. In adults, especially in Asia and Africa, this enzyme commonly falls to very low levels, or even disappears. However, some people maintain intestinal lactase activity at relatively high levels throughout life. This is a genetic trait known as lactase persistence. The absence or deficiency of lactase from birth (lactose intolerance) is a rare inherited condition. Patients with this alteration do not tolerate milk containing lactose. The inability to hydrolyze lactose determines its accumulation in the intestine and leads to a series of severe gastrointestinal problems. Due to its osmotic effect, undigested lactose draws water into the lumen of the digestive tract causing diarrhea. In addition, bacterial flora can degrade lactose, generating products with irritating action on the intestinal mucosa, and resulting in enteritis, flatulence, and diarrhea. The treatment for this disorder consists in removing all foods containing lactose and supplying hydrolyzed milk, which contains free glucose and galactose that can be easily absorbed.

The end result of carbohydrate digestion is the production of monosaccharides, the only form of carbohydrates that can be absorbed and utilized by the body.

Lipids

Dietary lipids include triglycerides, phospholipids, cholesterol, cholesteryl esters, and fat-soluble vitamins (A, D, E, and K). The most abundant in common foods are triacylglycerols with long chain fatty acids (longer than 14 carbon atoms). Digestion of these nonwater soluble nutrients is favored by the presence of bile salts, which promote their incorporation into tiny micelles, keeping them suspended in aqueous environments and exposing them to the action of hydrolytic enzymes.

Triacylglycerols. In humans, salivary lipase secreted by lingual glands has negligible action on dietary lipids. The triacylglycerol degradation process starts in the stomach, catalyzed by gastric lipase. The enzyme is not affected by the proteolytic activity of pepsin, or by the gastric acidity. Lipase targets ester bonds of primary carbons in glycerol, producing free fatty acids, 1,2-diacylglycerol, and 2-monoacylglycerols. Its activity is responsible for the digestion of 10%–30% of the neutral fat contained in the diet. However, the presence of gastric lipase is not essential. Patients whose stomachs have been removed do not show deficiencies in fat digestion, due to the activity of pancreatic lipase in the small intestine. Gastric lipase is important in children during the first few months of life, when the levels of pancreatic lipase are low.

Human milk contains a lipase different from that found in the stomach or pancreas; it can hydrolyze any of the three ester bonds in triacylglycerol molecules. This lipase plays an important role in fat digestion in newborns.

When the stomach content passes into the duodenum, it is exposed to bile salts, which promote lipid emulsion. Bile salts form a layer around the hydrophobic molecules by their ability to reduce surface tension at the water–lipid interface; they allow for the dispersion of fat into fine particles and stabilize them in the aqueous medium. Presence of other amphipathic molecules, such as phospholipids, is instrumental in the formation of micelles containing nonpolar fat internally.

Although bile salts favor lipid emulsion, the layer formed on the surface of the micelles hinders the action of pancreatic lipase. This problem is solved by the presence of colipase, which binds to the micelle's surface, displaces bile salts, and allows attachment of pancreatic lipase for interaction with triacylglycerols.

The end products of fat digestion are free fatty acids, 2-monoacylglycerols, and small amounts of glycerol.

Phospholipids. The micelles formed by bile salts and phospholipids are good substrates for the action of phospholipase A_2, which catalyzes the hydrolysis of the ester function in position 2 to give free fatty acid and lysophospholipids. Phospholipase A_2 requires the presence of bile salts. Other esterases and phosphatases complete the degradation of lysophospholipids.

Cholesterol esters. These lipids are degraded into fatty acids and cholesterol by cholesterol esterase. This enzyme also hydrolyzes the three triacylglycerol ester functions and esters of vitamins A, D, and E. For this reason, cholesterol esterase is called nonspecific esterase. Its activity requires the presence of bile salts.

Proteins

Protein digestion depends on the type of protein and the manner in which it is prepared before being ingested. In general, plant proteins are more difficult to digest than animal proteins. Proline-rich polypeptides, such as gluten and casein, are relatively resistant to digestion. Cooking denatures proteins and makes them more accessible for hydrolases.

The recommended intake for proteins coming with the diet is approximately 70 g/day. There are also endogenous proteins that reach the intestinal lumen. Approximately 20–30 g of protein derives from digestive secretions and ~30 g of endogenous protein comes from desquamated mucosal enterocytes.

Saliva does not contain significant proteolytic enzymes. Protein hydrolysis begins in the stomach. The HCl in gastric juice helps to denature proteins and makes them more accessible to protease degradation. Pepsin, an endopeptidase, cleaves proteins into segments of high molecular weight. These polypeptides move on to the duodenum, where three powerful pancreatic endopeptidases (trypsin, chymotrypsin, and elastase) reduce them to small peptides. Up to this point, free amino acids have not been produced in substantial quantities; their release is catalyzed by exopeptidases, which act on the polypeptide extremes. Pancreatic carboxypeptidases separate amino acids from the C-terminal end of polypeptides, while intestinal aminopeptidases release them from the N-terminal side. Two brush border dipeptidases catalyze the hydrolysis of dipeptides. The end products of protein digestion are free amino acids, dipeptides, and tripeptides.

Nucleic Acids

Pancreatic and intestinal nucleases (ribo- and deoxyribonuclease) cleave nucleic acids in their corresponding nucleotides. These nucleotides finally undergo the action of intestinal phosphatases to produce phosphate and nucleosides, which can be easily absorbed or degraded by intestinal nucleosidases. Purines, pyrimidines, ribose, and deoxyribose are the final products.

ABSORPTION

Once the process of digestion is completed, nutrients are incorporated into the organism. Approximately 90% of nutrients are absorbed in the small intestine where they can follow two different routes: (1) move into blood and reach the liver through the portal vein system, or (2) pass into the lymphatic vessels of the intestine, draining to the thoracic duct to finally enter the general circulation.

Enterocytes play an important role in regulating the passage of substances from the intestinal lumen into the blood or lymph. The apical membrane of enterocytes is the main site for selection and control of the substances to be transported. However, before reaching the circulation, substances need to come across other barriers. These include: (1) the thin layer of liquid on the luminal surface of enterocytes, where solutes must

move into by diffusion; (2) the oligosaccharide glycocalyx or microvilli surface covering the enterocytes; (3) the apical plasma membrane, cytoplasm, and basolateral membrane of the enterocytes; (4) the interstitial space; (5) the basal lamina; and finally (6) the walls of the blood and lymph capillaries. The transport of nutrients through these structures includes passive diffusion, facilitated diffusion, and active transport processes.

Tight junctions between neighboring enterocytes generally represent an effective seal that prevents free exchange between the lumen and the interstitium. However, they occasionally exhibit some permeability and allow a limited flow of solutes and water (paracellular pathway).

Carbohydrates

The only carbohydrates which can be taken up by cells of the intestinal mucosa are monosaccharides. The most abundant simple sugars released by digestion in the intestine are glucose, fructose, and galactose. These compounds are highly hydrophilic, they cannot readily cross the membrane lipid bilayer, and their cell internalization by passive diffusion is negligible; it needs the help of membrane transport systems. Glucose and galactose share the same transport system in the brush border membrane. This consists of the SGLT1, a secondary active Na^+ dependent transporter (symporter), of 664 amino acid residues and 14–15 transmembrane helices (p. 240). SGLT1 function is driven by the Na^+ gradient created by the Na^+,K^+-ATPase located in the enterocyte basolateral membrane. SGTL1 transports glucose (or galactose) and Na^+ ions from the lumen into cells; glucose and galactose cannot bind the carrier if Na^+ has not previously bound to it. The stoichiometry of transport is two Na^+ ions for each monosaccharide that is internalized into the cells. This system allows for the uptake of glucose against its concentration gradient. Sodium is immediately pumped into the interstitial space by Na^+,K^+-ATPase

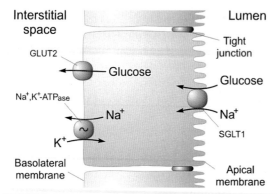

FIGURE 12.9 **Schematic representation of the glucose absorption process in the intestinal mucosa cells.**

(Fig. 12.9). The importance of SGTL1 is highlighted by cases of severe deficiency or absence, which is lethal.

After absorption at the apical side of the cells, the higher sugar concentrations in the enterocyte's cytosol drive its movement through the basolateral membrane via the GLUT2 facilitated transport system (pp. 227 and 284). A small amount of glucose is used by the intestinal cells to satisfy their needs, but most of it diffuses into the circulation, reaching the capillaries and the liver via the portal vein.

Fructose enters enterocytes through an apical membrane facilitated transport system. The protein responsible is a member of the monosaccharide transporter family called GLUT5, which is a specific fructose transporter. Fructose absorption increases in the presence of glucose. From the cell interior, fructose reaches the interstitial space and bloodstream through the basolateral membrane GLUT5 and GLUT2 transporters. GLUT2 is the same system used for the transport of glucose.

Lipids

Total hydrolysis is not an essential requirement for absorption of neutral fats. Most of the ingested fat is degraded to 2-monoacylglycerol, a compound that readily enters into enterocytes.

Bile salts have an important role in absorption of lipolysis products. When they reach a critical concentration, which is lower for conjugated (glyco- and taurocholate) than nonconjugated salts, they form molecular aggregates 3–10 nm in diameter. These micelles, with a hydrophilic surface and a hydrophobic interior, are stable in aqueous suspensions.

The final compounds produced by lipid digestion (monoacylglycerols, fatty acids, long-chain lysophospholipids, and cholesterol) as well as fat-soluble vitamins are contained in these micelles, which allows them to easily diffuse through the water layer covering the brush border. If they were free in the environment, these substances would find important limitations accessing the enterocyte apical membrane.

Due to the fact that lipid uptake is completed in the jejunum, while absorption of bile salt occurs in the distal ileum, it is believed that micelles are not incorporated as a whole. This is supported by the observation of different absorption rates for different products of lipolysis. While it is possible for these products to cross membranes by simple diffusion, transport systems in apical membranes of intestinal cells have been identified. These encompass proteins that bind and transfer long chain fatty acids to the enterocyte smooth endoplasmic reticulum, where they are processed. The protein FAT/CD36 has an important role in the uptake of fatty acids. Its synthesis is high in the intestine; its expression increases in the presence of dietary fat and in cases of obesity and diabetes mellitus.

A small amount of free glycerol and fatty acid chains of 10 or less carbons released in the intestinal lumen by the digestive action of enzymes is not incorporated into micelles. These compounds can passively diffuse through the membranes and capillaries of the portal system (Fig. 12.10).

Enterocytes are not only a pathway for the absorption of substances, but also a center of active and important metabolic activity. Some products absorbed in the intestinal mucosa are

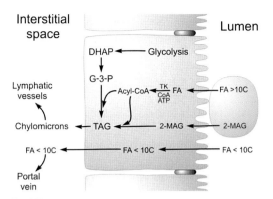

FIGURE 12.10 **Absorption of triacylglycerol degradation products in enterocytes.** *FA < 10C*, Fatty acids less than 10 carbons; *FA >10C*, fatty acids longer than 10 carbons; *G-3-P*, glycerol-3-phosphate; *2-MAG*, 2-monoacylglyceride; *TK*, thiokinase.

used for the synthesis of triacylglycerol. Triacylglycerol synthesis in intestinal cells starts with fatty acids and 2-monoacylglycerol. The fatty acids are linked to coenzyme A, to give the activated compound acyl coenzyme A. The reaction is catalyzed by an enzyme called *thiokinase* and the required energy is provided by hydrolysis of ATP to AMP and pyrophosphate:

$$\text{Fatty acid} + \text{CoA} \xrightarrow[\substack{\text{ATP} \qquad \text{AMP} + \text{PP}_i}]{\text{Thiokinase}} \text{Acyl CoA}$$

Acyl CoA transfers the fatty acid to form esters with the free hydroxyls of 2-monoacylglycerol. Another synthetic route of triacylglycerols and phospholipids is the phosphatidic acid pathway, which requires glycerol-3-phosphate. This is formed in the cell by glycerol phosphorylation catalyzed by *glycerokinase*, or by reduction of dihydroxyacetone catalyzed by *glycerophosphate dehydrogenase*. The dihydroxyacetone phosphate is derived from glycolysis (metabolic pathway of glucose degradation).

Newly formed triacylglycerols, together with small amounts of cholesterol, phospholipids, and proteins, form particles called *chylomicrons*, which pass to the lymphatic circulation. Proteins (apoproteins) and phospholipids (amphipathic compounds) form a hydrophilic surface film that allows the predominantly hydrophobic compounds in the interior of chylomicrons to be transported in the aqueous medium. Over 70% of the lipids incorporated by the intestinal mucosa follow the lymphatic way. Fig. 12.10 summarizes this process.

Cholesterol is absorbed in the intestine and is incorporated into chylomicrons. There is evidence suggesting the existence of cholesterol carriers, the principal ones are Nieman–Pick C1-like 1 (NPC1L1) and the ATP-binding cassettes ABCG5 and ABCG8. A great part of this cholesterol is esterified with fatty acids in mucosal cells and included in chylomicrons.

Large amounts of phospholipids are degraded and their products are incorporated into the cell.

When fats are not absorbed, they are eliminated with feces, leading to *steatorrhea*. Causes of steatorrhea include (1) lack of bile acids in the intestine, observed in cases of severe hepatic insufficiency or obstruction of the bile excretory duct, (2) absent or decreased pancreatic lipase due to severe pancreatic damage or obstruction of its excretory duct, (3) deficiencies in mucosal cells, with inability to synthesize triacylglycerols or apoproteins necessary for their transport.

Proteins

As a result of the action of gastric and pancreatic proteases, approximately 40% of ingested proteins are degraded into free amino acids; the remaining 60% are broken down into oligopeptides. These oligopeptides, in contact with the brush border peptidases, are hydrolyzed into free amino acids, dipeptides, and tripeptides (final products of digestion).

These compounds cross the apical membrane using different transport systems. Most of the free amino acids are cotransported with Na^+ in a system that is similar to that of glucose transport, dependent on the activity of Na^+,K^+-ATPase. A smaller proportion of amino acids enter the cell by facilitated diffusion.

Different transport systems with specificity for different groups of amino acids have been identified in the brush border:

1. *Na⁺ gradient dependent.* These include secondary active transporters. They are designated using capital letters: the B type transporters transport phenylalanine, tyrosine, tryptophan, isoleucine, leucine, valine, proline, and glycine; the Y_{AG} transporters carry basic amino acids; the $B^{o,+}$ transporters move cationic and neutral amino acids (system y^{+L}, ASC); the X_{AG} and X_c type are glutamate and aspartate transporters; the A type are glycine and methionine carriers, and the N type transport glutamine, asparagine, and histidine.
2. *Na⁺ gradient independent.* These are facilitated diffusion transporters. The different types are designated using lower case letters, except for types L and X_c. Type L systems, asc, $b^{o,+}$, and y^+ are neutral and cationic amino acids transporters; and type X_{AG} and X_c transporters move glutamate and cysteine through the cell membrane.

Dipeptides and tripeptides are carried by the enterocyte apical membrane transporter PEPT1, which acts as an electrogenic proton/peptide cotransport system (symport). Once in the enterocytes, di- and tripeptides are cleaved into amino acids by intracellular peptidases.

In patients with *Hartnup disease*, a hereditary condition in which the neutral amino acid transport system is lacking, dipeptides are absorbed normally, which demonstrates the existence of separate systems for the transport of free amino acids, dipeptides, and tripeptides.

From the cell interior, amino acids reach the basolateral membrane and pass to the portal vein capillary system by facilitated diffusion.

Under normal conditions, degradation of dietary protein is complete; that is, only free amino acids can be absorbed and reach the bloodstream. Therefore, there is no rationale for the administration of proteins by mouth for pharmacological purposes because they will not arrive intact to their final destination. For example, insulin, a hormone administered to diabetics, must be injected parenterally because when taken orally it rapidly degrades into its constituent amino acids by the digestive system.

Eventually, some small peptides can escape hydrolysis and reach the blood vessels. It is also possible, in certain physiological and pathological conditions, to absorb whole proteins or larger protein fragments. This is done by pinocytosis (endocytosis of small particles). During the first days of life, proteolytic activity is poor, which, coupled with an increased intestinal permeability, allows for the passage of polypeptides. This phenomenon is particularly noticeable in some ruminants and is a physiological mechanism that contributes to the transference of antibodies from mother to child. For example, in the newborn calf there are no immunoglobulins circulating in blood; however, immediately after they start ingesting milk, antibodies appear in the animal's plasma. *Colostrum*, a mammary gland fluid secreted by the mother during the first days after giving birth, is rich in immunoglobulins. These antibodies can move through the digestive tract of the newborn, providing protection against infections. In humans, the main transfer of immunoglobulins from mother to child occurs before birth through the placenta, but there is also antibody passage through the digestive tract. The breast-fed infants show greater defense against infections than those who are fed formula milk.

In *celiac disease*, there is a defect in the intestinal mucosa that enables absorption of polypeptides resulting from the partial digestion of wheat proteins (gluten). This leads to intolerance to a small protein of 88 amino acids called *gliadin*, which causes serious disorders. This is an example of pathological polypeptide absorption through intestine. Originally, the celiac disease designation was reserved for children, while for adults, the designation used was *nontropical sprue*.

Moreover, sensitivity expressed by allergic phenomena is relatively common for some proteins coming with food. Foreign polypeptides or proteins are antigenic. To induce immune response and production of antibodies, molecules must have a relatively large size. The abnormal absorption of poorly digested large polypeptides elicits protein susceptibility in some patients.

Water and Electrolytes

Approximately 8 L of water enter the digestive tract daily. Two liters come from food and beverages and the rest corresponds to secretions released in the gastrointestinal lumen. These include saliva (~1000 mL), gastric juice (~2000 mL), bile (~500 mL), pancreatic juice (~1500 mL), and intestinal secretions (~1000 mL). Most of this liquid is absorbed in the small intestine; less than 1 L reaches the colon. In the distal portions of the gut (colon and rectum), approximately 900 mL are absorbed and 100 mL are eliminated with feces. The movement of water from the lumen into the blood (and vice versa) follows osmotic and hydrostatic gradients (Starling forces) and is accompanied by the movement of solutes across membranes.

Concentrations of the most abundant electrolytes (Na^+, Cl^-, K^+, and HCO_3^-) are modified in the lumen as the intestinal content progresses along the digestive tract. Levels of Na^+ and Cl^+ decrease, while those of K^+ and HCO_3^- increase progressively from the duodenum to the colon.

There is passage of ions and water across the tight junctions between mucosal cells (paracellular way). The permeability of these junctions decreases from the proximal to the distal small intestine and is even lower in the colon. The majority of the water and electrolyte absorption is via the transcellular pathway through different transport systems in the cells' apical and basal membranes.

Sodium is absorbed across the apical membrane of enterocytes through four different mechanisms: (1) diffusion via Na^+ channels; (2) cotransport with organic solutes through transporters, including the Na^+/glucose and Na^+/amino acids transport systems; (3) cotransport with Cl^- ions; and (4) countertransport with H^+ ions. These last two mechanisms are electroneutral and all of the systems are driven by the Na^+ gradient generated by the basolateral membrane Na^+,K^+-ATPase. Activity of the Na^+ pump ensures a high intracellular concentration of K^+, creating a gradient that allows K^+ to exit by channels located in apical and basolateral membranes (Fig. 12.11).

The large intestine does not contain Na^+/glucose or Na^+/amino acid cotransport systems. Na^+ uptake at the apical membrane is accomplished by restricted diffusion through channels regulated by mineralocorticoid hormones and also by Na^+/H^+ and Na^+/Cl^- transporters, as seen in the small intestine. K^+ exits the cells through K^+ channels in the colon

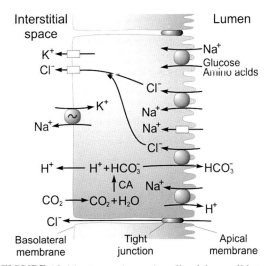

FIGURE 12.11   **Ion exchange in cells of the small intestine mucosa.** *CA,* Carbonic anhydrase.

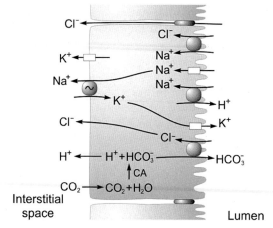

FIGURE 12.12 **Ion exchange in cells of the large intestine.** *CA,* Carbonic anhydrase.

apical mucosa (Fig. 12.12). Cl^- absorption in the small intestine is mainly coupled to Na^+ entry via the Na^+/Cl^- cotransporter. In the distal ileum, Cl^- transport is mediated through the HCO_3^-/Cl^- antiporter. The activity of these transporters increases in the colon and rectum, which explains the high HCO_3^- and pH of stool. HCO_3^- is produced from the CO_2 that results from cell metabolism by the action of carbonic anhydrase.

Cl^- entry into the cell against the electrochemical gradient is driven by energy derived from the action of the Na^+,K^+-ATPase and the output of HCO_3^- via the HCO_3^-/Cl^- transporter. The transepithelial potential difference, with the serosal side being positive with respect to the lumen, is higher across the colon than it is in the small intestine mucosa. This facilitates Cl^- transport in the later segments of the gut.

SUMMARY

Digestion is the process of degradation in the digestive tract of complex molecules that come with food. It is accomplished by hydrolytic reactions catalyzed by enzymes produced at different segments of the digestive system.

Saliva is a fluid with a pH of ~6.8, containing bicarbonate and digestive enzymes, including *ptyalin* or *salivary amylase*. This is an endoamylase that requires a pH of ~7.0 and Cl^- to function. It catalyzes the hydrolysis of α-1→4 glycosidic linkages in the interior portion of molecules. Amylase can completely degrade amylose producing maltose and maltotrioses, and it can partially digest amylopectin. However, it has no activity on α-1→6 bonds, leaving fragments that correspond to the branching points of amylopectin (*limit dextrins*). The action of salivary amylase is limited, as it is inhibited by the acid medium of the stomach. Saliva also has *lipase*, which catalyzes the hydrolysis of ester bonds in triacylglycerols.

Gastric juice is produced in glands of the stomach mucosa. The most important ones are the principal glands, where parietal cells excrete H^+ and Cl^-. Proton transport is performed in those cells by the H^+,K^+-ATPase. Protons are generated by *carbonic anhydrase*, which catalyzes the formation of carbonic acid from CO_2 and H_2O. The H_2CO_3 ionizes into bicarbonate and H^+. The HCO_3^- returns to plasma and produces the postprandial "alkaline tide." Cl^- moves into the stomach lumen from plasma through Cl channels. HCl concentration in gastric juice reaches concentrations of 0.17 M, acidifying the stomach lumen to a pH of 0.87. *Achlorhydria* refers to the lack of HCl; *hypo-* and *hyperclorhidria* are frequently observed clinical disorders.

The principal digestive action of gastric juice consists in the partial hydrolysis of proteins catalyzed by *pepsin*. This is an endopeptidase with an optimum pH of 1.0–2.0. It is secreted as pepsinogen by chief cells and activated by H^+ ions and by pepsin via autocatalysis. Stomach also possesses a *gastric lipase*, which catalyzes the hydrolysis of ester bonds of medium chain fatty acids in triacylglycerols. The parietal cells secrete *intrinsic factor*, a glycoprotein required for vitamin B_{12} absorption.

Pancreatic juice is produced by the exocrine pancreas; it is a fluid with an osmolarity similar to that of plasma and a pH between 7.5 and 8.0. It contains powerful hydrolases, including:

Trypsin is an endopeptidase that selectively hydrolyses carboxyl bonds of lysine and arginine. It is secreted as trypsinogen and becomes activated by enteropeptidase or via autocatalysis; activation is achieved by removing an N-terminal hexapeptide.

Chymotrypsin is also an endopeptidase, which acts on aromatic amino acid carboxyl bonds and is secreted as the zymogen chymotrypsinogen.

Carboxypeptidase is an exopeptidase that breaks the terminal peptide bond, releasing the C-terminal amino acid of proteins. It is secreted as procarboxypeptidase.

Elastase acts on elastin. All these proteases are activated by trypsin.

Amylase hydrolyzes starch by a mechanism identical to that of ptyalin.

Lipase hydrolyzes neutral fat ester bonds in primary alcohols of glycerol. Its action is facilitated by *colipase*.

Cholesterolesterase hydrolyzes cholesterol esters.

Phospholipase A$_2$ attacks bonds of the fatty acid chain attached to C2 in glycerophospholipids.

Ribo- and *deoxyribonuclease.*

Intestinal mucosa has a major role in absorption of nutrients; its surface is highly expanded by the presence of a brush border in the epithelium. It produces the following digestive enzymes:

Enterokinase, or *enteropeptidase*, which catalyzes the activation of trypsinogen.

Dipeptidases. Aminopeptidases that hydrolyze the peptide bond closest to the N-terminus of proteins.

Disaccharidases, which are responsible for the final degradation of starch residues and dissacharides. They include: *Sucrase–isomaltase, lactase–phloridzin hydrolase*, and *maltase–glucoamylase.*

Nucleotidases, phosphatases, and *nucleosidases* are enzymes that degrade nucleic acids into nucleotides.

Bile is produced in the liver and stored in the gallbladder. It is a complex solution with specific components including:

Bilirubin, the pigment resulting from heme degradation; it is excreted as glucuronide.

Cholic and *chenodeoxycholic* acids, which are synthesized in the liver from cholesterol. They are excreted conjugated with taurine or glycine (*glycocholic* and *taurocholic* acids). At physiological pH, they are ionized and neutralized by Na$^+$ to form *bile salts*. These have emulsifying action on fats and fat soluble vitamins favoring their absorption; they also stimulate liver production of bile.

Cholesterol is a normal component of bile and one of the main routes for its excretion. It can precipitate, and form gallstones.

Absorption of components of the diet after digestion follows different mechanisms as listed subsequently.

For carbohydrates, only monosacharides are absorbed by the intestine. D-glucose enters the enterocytes by a Na$^+$-glucose cotransporter (SGLT1) dependent on the sodium pump. Fructose is absorbed by facilitated transport (GLUT5). From the enterocyte to the interstitium, they pass using GLUT2 carriers.

For *lipids*, total hydrolysis of fats is not necessary for absorption. Products of partial lipid digestion are solubilized in micelles formed by biliary acids, phospholipids, and monoacylglycerides. Glycerol and fatty acids of less than 10 carbon atoms in length can directly move into the portal system. The cells of the intestinal mucosa cells resynthesize triacylglycerols that pass to the lymphatic vessels included in chylomicrons.

For *proteins*, only free amino acids, dipeptides, and tripeptides are absorbed. Amino acids use Na-amino acid transporters and other Na-independent transporters for absorption.

Bibliography

Bröer, S., 2008. Amino acid transport across mammalian intestinal and renal epithelia. Physiol. Rev. 88, 249–298.

Hannelore, D., 2004. Molecular and integrative physiology of intestinal peptide transport. Ann. Rev. Physiol. 66, 361–384.

Hersey, S.J., Sachs, G., 1995. Gastric acid secretion. Physiol. Rev. 75, 155–189.

Igbal, J., Hussein, M.M., 2009. Intestinal lipid absorption. Am. J. Physiol. Endocrinol., Metabol. 296, E1183–E1194.

Johnson, L.R., 1997. Gastrointestinal Physiol, fifth ed. Mosby, St. Louis.

Swallow, D.M., 2003. Genetics of lactase persistence and lactose intolerance. Ann. Rev. Genet. 37, 197–219.

Yamada, T., col., 2003. Textbook of Gastroenterology, fourth ed. Lippincott Co, Philadelphia.

Yao, X., 2003. Cell biology of acid secretion by the parietal cell. Ann. Rev. Physiol. 65, 103–131.

13

Metabolism

The chemical reactions that take place in living organisms encompass what is called *metabolism* or, more precisely, *intermediary metabolism*. The reactions that occur before substances are absorbed in the gastrointestinal tract are considered premetabolic stages.

Intermediary metabolism includes processes of varied nature, designed to fulfill the following purposes: (1) degradation of compounds absorbed in the intestine into simpler products that can be used as precursors for the synthesis of new endogenous molecules (proteins, nucleic acids, polysaccharides, and complex lipids), necessary for the structure and function of tissues; and (2) generation of the energy and reducing power needed for the synthesis of own compounds and different functions of the body.

Degradation processes correspond to what is known as *catabolism*, while biosynthesis events refer to *anabolism*.

METABOLIC PATHWAYS

Metabolic transformations, including both degradation and synthesis, generally occur through a series of enzyme catalyzed reactions arranged in a defined sequence. These series of reactions, known as *metabolic pathways*, convert a precursor or starting compound into a particular end product. Commonly, the initial substrate, under the action of a specific enzyme, is converted to a product that serves as a substrate for another enzyme in a following reaction; this continues until a final product is made in a linear sequence of reactions (Fig. 13.1).

Substance A (*initial substrate*) is transformed into product B by action of enzyme *a*, B is then converted into C by enzyme *b*, etc., until the *final product* E is obtained. This pathway involves four stages or reactions catalyzed by enzymes. The substances B, C, and D are considered intermediate products, or *metabolites*.

Some metabolic pathways are not just linear, but include branching points (Fig. 13.2).

Where the product (B) that results from the reaction that substrate (A) undergoes has two alternative routes. In the first, enzyme *b* catalyzes the conversion of B to C, and through a series of stages, the final product E is obtained. Alternatively, B can form P through the action of enzyme *b'* and the product R will be obtained after subsequent reactions.

If all reactions in a metabolic pathway are reversible, chemical transformations can proceed in both directions. For example, as shown in Fig. 13.3, it would be possible to obtain compound A as the final product when starting with E as an initial substrate, following the same reactions that lead from A to E, but in the reverse direction.

When one or more of the reactions are practically irreversible, the reverse reactions need to occur through a detour pathway (Fig. 13.4).

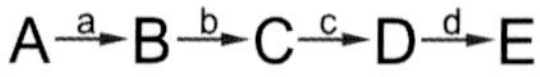

$$A \xrightarrow{a} B \xrightarrow{b} C \xrightarrow{c} D \xrightarrow{d} E$$

FIGURE 13.1   **Linear sequence of reactions.**

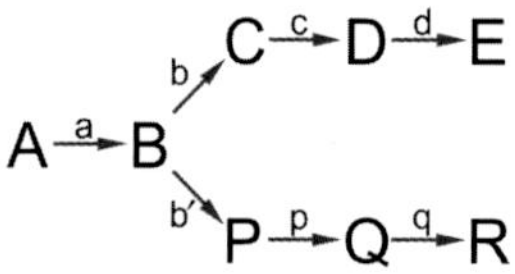

$$A \xrightarrow{a} B \begin{array}{c} \xrightarrow{b} C \xrightarrow{c} D \xrightarrow{d} E \\ \\ \xrightarrow{b'} P \xrightarrow{p} Q \xrightarrow{q} R \end{array}$$

FIGURE 13.2   **Branched reactions.**

When the reactions C→D and D→E are essentially irreversible, the reverse reactions between E and C require the formation of another intermediate, S.

In some cases, chemical transformations occur cyclically and are called *metabolic cycles*. These involve an ordered series of reactions, which results in the regeneration of the initial compound (Fig. 13.5).

In the example in Fig. 13.5, the cycle begins with the reaction between A and S to form substance B. The product B becomes C, which is eventually converted into D. Substance D then breaks down into two products, P and A. Substance P is released and compound A starts a new series of reactions by combining with molecule S. Substances contributed from the outside (S in this example) are called *feeders*; substances A, B, C, and D are *intermediate metabolites* generated during the cycle operation, and molecule P is the *released product*.

Metabolic cycles can be interconnected by one or more common intermediate metabolites (Fig. 13.6).

Cycles I and II share the intermediary C. Substances S and X are feeders and molecules

$$A \rightleftharpoons B \rightleftharpoons C \rightleftharpoons D \rightleftharpoons E$$

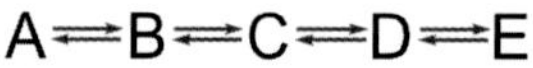

FIGURE 13.3   **Sequence of reversible reactions.**

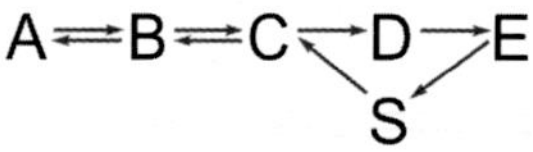

$$A \rightleftharpoons B \rightleftharpoons C \underset{S}{\overset{}{\rightleftharpoons}} D \rightharpoonup E$$

FIGURE 13.4   **Reversibility of a metabolic pathway through a detour route.**

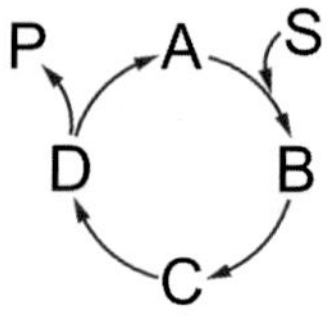

FIGURE 13.5   **Metabolic cycle.**

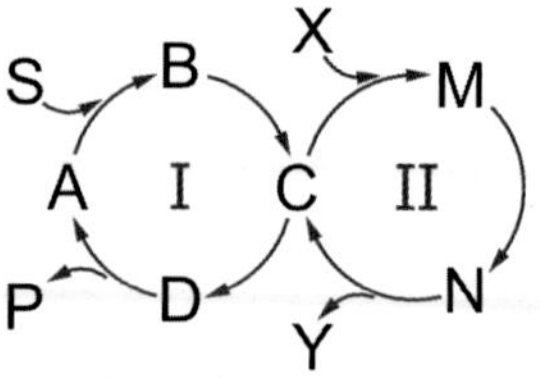

FIGURE 13.6   **Metabolic cycles interconnected by a common intermediate (C).**

P and Y are the released products. Compound C represents a common intermediary metabolite that determines the interdependent operation of both cycles.

In other metabolic processes, reactions follow a *"cascade"* or staggered arrangement of reactions (Fig. 13.7). Generally, these include reactions of enzyme activation. For example, a zymogen A is converted to the active enzyme B through catalysis by enzyme *a*. B catalyzes the conversion of inactive enzyme M into the active enzyme N, which, in turn, activates X to Y. This type of process allows a progressive amplification of the response. This is essential in cases in which rapid and abundant production of the final compound is required. An example of this type of reaction is the blood coagulation cascade (Chapter 31).

Metabolic pathways can be distinguished in three different groups: (1) catabolic, (2) anabolic, and (3) amphibolic reactions.

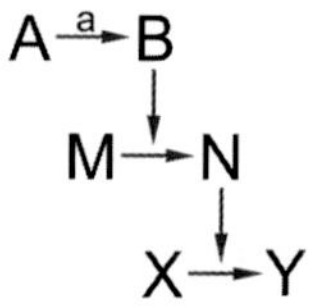

$$\begin{array}{c} A \xrightarrow{a} B \\ \downarrow \\ M \rightarrow N \\ \downarrow \\ X \rightarrow Y \end{array}$$

FIGURE 13.7   **Staggered or metabolic cascade sequence.**

Catabolic pathways. In these reactions, the original substrate is converted into simpler compounds. They include oxidative reactions that are involved in energy production; hence, they are exergonic $(-\Delta G)$. The chemical energy obtained from the substrates is trapped in the form of ATP and the reducing equivalents generated in the reaction are accepted by redox coenzymes, such as NAD^+. Examples of these type of reactions are the glycolytic and fatty acid oxidation pathways.

Anabolic pathways. These pathways form chemical products that are more complex than the initial substrates, following a series of endergonic reactions $(+\Delta G)$, which only proceed if they are coupled to exergonic reactions (commonly associated with ATP hydrolysis). They generally have reducing potential; the coenzyme NADPH is the main donor of hydrogen in anabolic processes. Examples include the fatty acid and cholesterol synthesis pathways.

Amphibolic pathways. These pathways can function as either anabolic or catabolic depending on the cell's needs. They degrade substrates by oxidation, but can also produce metabolites that are used for synthesis of new compounds. An example includes the tricarboxylic acid cycle, which oxidizes acetate to CO_2 and H_2O with energy production and generates intermediates that are substrates for the synthesis of various compounds.

The processes of biosynthesis and degradation operate continuously in living organisms. Chemical structures undergo permanent changes; anabolism and catabolism are in a *dynamic equilibrium.* As a result of this equilibrium, a balance between the organism and its environment is established. Food provides the energy and the elements necessary for the repair and maintenance of vital structures. In contrast, metabolic waste products and unused substances are excreted back to the medium. Maintenance of the balance between the substances that enter and are excreted from the body is a characteristic of a healthy adult and requires that appropriate nutrients are provided in adequate quantity and quality.

In contrast, in young growing organisms and in pregnant women there is predominance of anabolism, with net incorporation of matter. In older age, however, the balance shifts in favor of catabolism. In many pathological conditions, the balance between anabolism and catabolism can be altered in either direction.

METABOLIC STUDIES

The current state of knowledge of intermediary metabolism is the result of contributions from many researchers, who have solved, step by step, the intricate maze of biochemical transformations that take place in the body. This was not a simple task. The complete resolution of a metabolic pathway presents enormous challenges, mainly due to the extremely low concentration and short life of many of the intermediary metabolites. At present, it has been possible to precisely determine numerous metabolic pathways, including their interconnections, to establish and develop complex "metabolic maps," which provide an integrated view of metabolism.

How have scientists been able to decipher metabolism and its many pathways. A brief reference of the main tools and approaches used in this field follows.

Early studies related to chemical balance showed that certain precursors were converted by living organisms in a particular final product.

Observations on the fermentation process provided the basis of the material balance and stressed the importance of catalysis. For example, fermentation of glucose by yeast produces carbon dioxide and ethanol according to the following general equation:

$$C_6H_{12}O_6 \rightarrow 2\,CO_2 + 2\,C_2H_5OH$$

Carbon is not formed or lost in the process; the carbon atoms coming from glucose molecule are rearranged to form the final products.

Contributions by Claude Bernard are among the firsts in this field. It was known that glucose is secreted into the circulation by the liver. Bernard was able to demonstrate that a polymeric substance, glycogen, was the precursor of the glucose released to blood.

From this early evidence, the concepts of transformation of precursor molecules into products, chemical balance, and the fundamental role of catalysts emerged. This was followed by the characterization of the multiple reactions involved in different metabolic pathways.

Methods Used to Study Metabolism

The use of markers has been one of the most fruitful methodological resources used in metabolic studies. A precursor substance is labeled and its distribution and transformations are tracked in the body. One of the classic examples of this method is the experiment on body fatty acid oxidation by von Knoop in 1904. In this experiment, fatty acids were labeled by attaching to them a phenyl radical. When these acids were metabolized, their oxidized products remained linked to the phenyl radical, allowing for their identification.

The subsequent incorporation of isotopic elements as markers of a precursor substance, substantially improved metabolic studies. The body does not distinguish between artificial isotopes and natural elements; both undergo the same metabolic pathways and thus, radioisotopes allow following the distribution and fate of a substance in the body. The studies by Schoenheimer with deuterium and nitrogen isotopes (^2H and ^{15}N) are among the first that used this technique. He demonstrated that the molecules of living matter are in constant degradation and resynthesis. From those experiments arose the concept of body component "half-lives"; this is the time in which 50% of a given compound in a tissue is renewed.

The use of radioactive isotopes has allowed the dissection of many metabolic pathways and has become a valuable tool in biochemical research in general.

Another line of investigation is the direct study of the enzymes involved in metabolic pathways. The finding of specific inhibitors of those enzymes allowed understanding the properties of metabolic enzymes, the intermediate metabolites they modify or the compounds they produce, and the specific pathways where the enzymes exert their action. Thus, if a particular enzyme in a metabolic pathway is blocked with a specific inhibitor, its substrate, as well as other preceding metabolic intermediaries will accumulate, while the products of the reaction will decrease. In this case, measurement of the changes in substrates and products provides important information of that metabolic pathway.

Individuals who lack a particular enzyme can have alterations in specific metabolic pathways. These are commonly genetic defects and correspond to a class of diseases known as *inborn errors of metabolism*. The study of these disorders have unraveled important information on different metabolic pathways.

Techniques developed by molecular biology, which can remove or reduce the expression and activity of a particular enzyme in cells or even in a whole animal (e.g., knock-out mice, directed mutagenesis, RNA interference, see Chapters 22 and 23), have provided new essential tools in the area of metabolic research.

Model Systems Used in Metabolic Studies

The methods described previously have been applied to different biological systems. Each present different advantages and should be selected according to the specific objectives of the research and capability of the laboratory. These systems include:

1. *Intact animals* (in vivo). Experiments are performed in the whole organism.
2. *Intact organs* (in vivo or in vitro). An organ, either in situ or isolated, is used. The precursor substance is added by perfusion

through the arterial system and samples of blood exiting the venous system are obtained for the analysis of the products.

3. *Tissue slices* (in vitro). The experiment is carried out using thin sections of a tissue, in which the diffusion of substances between the sample and the medium is facilitated by the reduced thickness of the section.

4. *Whole cells.* The experiments are performed with a homogeneous cell population, either isolated from tissues (primary cell culture), or developed in culture as an immortal cell line.

5. *Homogenized tissues* (in vitro). This includes tissue preparations in which plasma membranes are disrupted by mechanical or physical means. If this is performed in isotonic solutions, with special precautions, intact subcellular organelles can be separated from the whole cell homogenate by fractional centrifugation (different particles have different densities). Homogeneous preparations of nuclei, mitochondria, lysosomes, and other organelles can be obtained in this manner.

6. *Tissue extracts.* Enzymes, factors, or metabolites can be isolated and purified from tissue extracts. Identification of these, even in very low concentrations, is possible today with the help of extremely sensitive methods.

Nuclear magnetic resonance spectroscopy (NMR). NMR has been used in recent years for the study of metabolism with great success. This technique allows studying chemical transformations, identifying metabolic intermediates, and determining the flow of metabolites through different cellular pathways in organs, and even the whole organism. It is a noninvasive technique that does not disturb normal physiology.

Metabolome

As a correlate of the genome and proteome studies (pp. 496 and 515), many laboratories began to characterize the metabolites resulting from chemical transformations. The array of metabolites represent what has been called the *metabolome*. It is estimated that the normal metabolome comprises at least 3,000 compounds; however, some authors claim that the number is much higher ($\sim$100,000). An accurate estimation of endogenous metabolites is difficult because there are multiple substances coming from the environment, as well as from the food ingested, which have the potential to generate new products derived from their own metabolism.

The metabolome has been considered by some authors as the best indicator of an individual's phenotype. A snapshot of the metabolic profile at a given time or under a particular condition provides important information regarding tissue and body function, being thus of great medical interest.

The advancement of metabolomics coincided with the development of highly resolute analytical techniques, especially capillary electrophoresis, gas chromatography, liquid chromatography (high-performance and ultra-high performance, HPLC and UPLC), mass spectrometry, and nuclear magnetic resonance (NMR).

REGULATION OF METABOLISM

The multiple metabolic pathways that exist in each cell and tissue need to be finely coordinated to ensure the proper operation of the organism as a whole. The supply and demand of a given compound varies at different times, which is why the flow of metabolites through catabolic or anabolic pathways must be constantly adjusted to the circumstances and needs that a cell or organism faces. Survival highly depends on the ability of an organism to regulate metabolic processes, to ensure the constancy of the intra- and extracellular media within narrow limits necessary to perform basic functions.

A number of metabolic regulatory processes have been studied at the molecular level especially in bacteria. This is due to the simplicity of microorganisms, compared to that of higher eukaryotes. Higher organisms need to integrate the metabolic function carried out in different organs to achieve a harmonious and integrated function of the whole body. This is accomplished mainly via the nervous and endocrine systems.

As all chemical reactions in a cell are catalyzed by enzymes, therefore, the control of their activity is essential to control the rate of specific reactions and ultimately of different metabolic pathways. The rate at which a particular reaction proceeds depends on both the absolute amount of enzyme present and its catalytic efficiency. The enzyme amount in a cell depends on the balance between its synthesis and its degradation. As will be seen in other chapters, protein biosynthesis is closely adjusted by induction and repression of its corresponding gene. The processes of specific protein degradation are also regulated.

Other mechanism to achieve metabolic regulation is the control of enzyme activity. This depends, in many cases, on the action of low molecular weight substances that function as *allosteric effectors*. Changes in the concentration of substrates, coenzymes, intermediates, or end products of metabolism may affect an enzyme's activity either by stimulating (positive effectors) or inhibiting (negative effectors) it.

A common type of regulation is the feedback mechanism.

In a metabolic sequence, such as the one shown in Fig. 13.8, the precursor A is converted in the final product D through the catalytic action of three enzymes (*a* to *c*). The final product (D) may act as an allosteric inhibitor of enzyme *a*. Thus, when a sufficient amount of D has accumulated, it will stop its own production by blocking

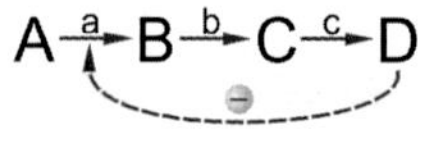

$$A \xrightarrow{a} B \xrightarrow{b} C \xrightarrow{c} D$$

FIGURE 13.8   **Feedback regulation in a linear metabolic pathway.**

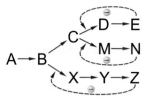

FIGURE 13.9   **Feedback regulation in a branched metabolic pathway.**

the conversion of new A into B molecules. This is an example of negative feedback, in which the end product of a metabolic pathway inhibits the enzyme responsible for the initial reaction.

In branched pathways, each of the end products can act as a negative allosteric effector of the enzyme at the branch point where the synthesis of that particular product begins. Fig. 13.9 shows a schematic representation of feedback inhibition sites within a branched metabolic pathway.

The covalent modification, by addition or removal of phosphate, adenylation, etc. is another mechanism used to modulate the activity of enzymes and regulate metabolic pathways. Specific examples of these types of control, integration mechanisms, and specific hormonal effects on metabolic pathways will be covered in more detail in other chapters.

Compartmentalization of Enzymes Involved in Metabolism

In eukaryotic cells, the enzymes acting in each metabolic pathway are usually confined to a defined cell compartment. This provides the appropriate spatial location and increases the efficiency of metabolic transformations. The compartmentalized distribution of enzymes allows them to function in sequential order, so the product of one reaction is released in the vicinity of the active site of the next enzyme in the metabolic pathway, facilitating its catalysis. The flux of metabolites in a given direction is called "channeling" or "tunneling." The mechanisms responsible for the proper spatial distribution of enzymes in a pathway are varied: (1) they can be ordered by insertion in a membrane (i.e., components of the electron transport chain in the inner membrane of mitochondria), (2) they can be organized into highly structured molecular complexes (i.e., the multienzyme pyruvate dehydrogenase complex in the mitochondrial matrix), and (3) they can interact even when they are in solution. The ordered distribution of enzymes in the cytosol limits the free diffusion of metabolites into the medium, facilitates enzyme–substrate encounters and increases the efficiency of reactions. An association of soluble enzymes with the structural elements of the cytoplasmic matrix has been proposed for enzymes of the glycolytic pathway.

Moreover, the existence of specific carriers that are specifically regulated in membranes at the cell surface and in organelles allows to control the traffic of substrates in different cell compartments and to regulate the function of metabolic pathways.

SUMMARY

Metabolism comprises all chemical reactions that occur within living organisms. The processes that involve degradation of compounds correspond to *catabolism* and those leading to synthesis of new molecules correspond to *anabolism*. In general, catabolism has an oxidative nature and uses coenzymes like NAD^+ as an electron acceptor. It is exergonic; the $-\Delta G$ is used to transfer phosphate to ADP for ATP synthesis. In contrast, anabolism involves an endergonic process that uses ATP as the main source of energy. In general, it involves reducing reactions and uses NADPH as the H^+ ion supplier. In the normal adult, both catabolism and anabolism are in equilibrium.

Metabolic pathways refer to the sequence of enzyme catalyzed reactions that lead to the conversion of a substance into a final product.

Metabolic cycles encompass a series of reactions in which the substrate is continuously reformed and the intermediate metabolites are continuously regenerated. Substances that enter the cycle are known as *feeders* of the cycle, while the molecules that leave the cycle are called *products*.

Cascade reactions are metabolic processes in which the product of one reaction activates the rate of a second reaction. This produces significant amplification of a response.

The study of metabolism has taken advantage of the use of biochemical markers to label a particular substance and follow its metabolic faith. *Isotopes* have been very useful markers in metabolic studies. *Nuclear magnetic resonance spectroscopy (NMR)* is another powerful resource to follow the chemical transformation of compounds. These type of techniques can be applied to intact animals or organs, cells, homogenates, and tissue extracts. In addition, studies of the function and properties of enzymes have greatly helped in understanding metabolism.

Metabolic regulation is essential to suit the needs of the organism. The mechanisms involved include modulation of enzyme activity (such as *allosteric effects* and *covalent modification*) and the up- or downregulation of the amount of enzyme molecules existing in cells, via *induction* or *repression* of their synthesis or degradation.

Compartmentalization in the cell allows the adequate spatial localization of the enzymes of metabolic pathways. The restriction in distribution and the sequential arrangement of enzymes in a cell compartment or organelle increases the efficiency at which chemical transformations can proceed.

Bibliography

Blow, N., 2008. Metabolomics: biochemistry's new look. Nature 455, 697–700.

Coffee, C.J., 1998. Metabolism. Fence Creek Publishing, Wisconsin.

Huang, X., Holden, H.M., Raushel, F.M., 2001. Channeling the substrates and intermediates in enzyme-catalyzed reactions. Annu. Rev. Biochem. 70, 149–188.

Shulman, R.G., Rothman, D.L., 2001. NMR of intermediary metabolism: implications for systemic physiology. Annu. Rev. Physiol. 63, 15–48.

14

Carbohydrate Metabolism

Carbohydrates, mainly coming as starch, represent a major component of the human diet. After ingested, they are hydrolyzed in the digestive tract and converted into monosaccharides. These simple molecules are the only carbohydrates that can be absorbed in the intestine and later metabolized in different cells.

Glucose is the most abundant monosaccharide in the body. Its main function is to serve as fuel, providing energy when oxidized. In addition, glucose is utilized as the raw material for the synthesis of several compounds. Fructose can also reach significant levels in the body when the diet is high in sucrose or free fructose. Galactose amounts become important when the main carbohydrate in the diet is lactose, as occurs in infants.

Monosaccharides absorbed in the intestine are transported via the portal vein to the liver. There, fructose and galactose are converted into metabolites identical to those derived from glucose. In this manner, the three monosaccharides end up converging in a common metabolic pathway.

The liver is the primary organ for glucose metabolism; it captures glucose from the circulation and incorporates it into a polymer (glycogen), to store it as energy reserve. Glycogen synthesis, also called *glycogenesis*, is an anabolic process that requires energy. All tissues receive a continuous supply of glucose. After a meal rich in carbohydrates, glucose levels increase; the liver cannot capture all of the coming glucose and the glucose

remaining in circulation is taken up by other tissues. These are also able to metabolize and store glucose as glycogen. The liver and muscle are the organs where glycogen synthesis is the greatest. Approximately one-third of the total body glycogen of an adult individual is found in liver and most of the rest is found in muscle.

Glycogen stored in the liver can be cleaved back into glucose and released to the general circulation. Degradation of glycogen to glucose, called *glycogenolysis*, functions according to the body's needs. Hepatic glycogenolysis is an important mechanism to maintain blood glucose levels constant during the intervals between meals. The maintenance of a constant glucose supply to tissues is vital, especially for the central nervous system, which depends almost exclusively on glucose as energy source. When blood glucose is determined in a normal individual under fasting conditions (over a period of 4 h after food intake), its level ranges from 70–110 mg/dL. After each meal, however, glucose blood levels transiently increase.

Muscle glycogen serves as an energy reserve for this tissue and it is intensely used when performing work. Unlike the liver, muscle does not release free glucose into the circulation. In muscle, glycogen breakdown produces pyruvate and lactate. As organic acids are generally ionized in the organism, the ionic form names (pyruvate for pyruvic acid and lactate for lactic acid) are commonly used.

The catabolism of glucose takes place through the following pathways:

1. *Glycolysis* or *Embden–Meyerhof pathway*, which produces pyruvate. Pyruvate is reduced to lactate if oxygen supply is insufficient. Lactate formation is particularly important in muscle, which can contract anaerobically using the ATP generated through glycolysis. Increased levels of lactate can be detected in blood and urine after intense exercise. This directly reflects the level of glycolytic activity of muscle. In red blood cells, glycolysis is the only pathway for energy production.

2. In the presence of oxygen, pyruvate generated during glycolysis is oxidized to CO_2 and H_2O. For this to occur, pyruvate is first subjected to decarboxylation, CO_2 is released, and acetate (a two carbon containing compound) remains as a product. Acetate moves into a metabolic cycle called the *citric acid cycle* [also known as the *tricarboxylic acid (TCA) cycle* or *Krebs* cycle], which is a very efficient energy producing system.

An anabolic pathway, called *gluconeogenesis*, allows the body to synthesize glucose from metabolites of noncarbohydrate origin. For example, lactate or pyruvate synthesized during the catabolism of certain amino acids or other substances can be converted to glucose by gluconeogenesis.

While noncarbohydrates can produce glucose or glycogen, glucose derivatives can be used to synthesize lipids or supply the carbon backbone for the synthesis of some amino acids. This use of common metabolites by different metabolic pathways highlights the elaborate integration of body metabolism. The separate study of the metabolic processes for carbohydrates, lipids, and amino acids is therefore only practical for a didactic purpose.

Fig. 14.1 shows an overall summary of glucose metabolism.

GLUCOSE UPTAKE INTO CELLS

As previously mentioned (p. 267), glucose uptake in the intestine takes place via the Na^+/glucose symporter (SGLT 1), placed at the apical membrane of the enterocytes. This secondary active transporter uses the Na^+ gradient created by the Na^+,K^+-ATPase, located in the basal side of the cells. SGLT1 allows glucose accumulation

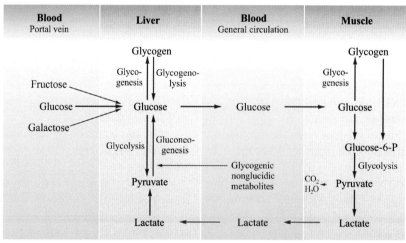

FIGURE 14.1   **Outline of carbohydrate metabolism.**

in the cell cytosol and its subsequent movement to the portal circulation by facilitated diffusion using the GLUT2 transporters located in the basolateral membrane. GLUT2 has low affinity but a large capacity for glucose transport, making it a very efficient mechanism for glucose absorption.

Once in blood, glucose enters cells via facilitated diffusion. Most cells (with the exception of intestinal mucosa and renal tubule cells) have a lower concentration of glucose than plasma and interstitial fluid. Facilitated uptake of glucose in cells is carried out by a family of integral membrane proteins, the glucose transporters or GLUTs. These proteins consist of a 500 amino acid polypeptide chain with 12 transmembrane segments that form the channel through which glucose enters the cell.

Glucose transport. There are two secondary active Na^+/glucose cotransport mechanisms. SGLT1 is found only in the apical membrane of small intestine epithelial cells and SGLT2 is present in renal tubules. SGLT1 transports glucose from the intestinal lumen against its concentration gradient. SGLT2 is involved in glucose reabsorption in the proximal tubules. Approximately 180 g of glucose are filtered in the glomeruli per day and all of it is efficiently reabsorbed.

The remaining glucose transport is performed via facilitated diffusion mediated by uniporters distributed in all cells. Fourteen members of this transporter family have been identified (GLUT1 to GLUT14) and they differ in tissue distribution, specificity, substrate affinity, and functional properties. GLUT1 is ubiquitously expressed in all fetal cells and it predominates in adult red blood cells, fibroblasts, blood vessel endothelial cells, and placental and brain barriers. It has high affinity for glucose (K_m, 1.4 mM), which allows it to work at a maximal rate, even when blood glucose levels are low. GLUT2 is expressed in the basolateral membrane of renal tubular cells, intestinal epithelium, hepatocytes, and Langerhans' islet β cells of the pancreas. GLUT2 has lower affinity ($K_m \sim 17$ mM) for glucose than GLUT1 and its transport rate is directly related to the glucose level. In the liver, where GLUT2 is abundant,

there is a net glucose flow into cells when the glucose level is high (after a meal). In contrast, if glucose level is low (during fasting periods) the glycogenolysis and gluconeogenesis processes are activated. This causes intracellular glucose to increase and this reverses glucose transport, which can now move to the interstitial space and the blood. GLUT2 performs glucose uptake in pancreatic β cells, depending on the glucose plasma concentrations. In these cells, GLUT2 function is important in stimulating insulin and glucagon secretion into the circulation. Galactose and fructose are also transported by GLUT2, which allows the liver uptake and metabolism of these hexoses. GLUT3 is the major glucose transporter in the brain and peripheral nerves. Its high affinity for glucose ensures a constant supply of glucose to the nervous tissue, protecting it from normal fluctuations in blood sugar. GLUT4 is expressed in adipose tissue, skeletal muscle, and cardiac muscle, where its activity is highly dependent on the presence of insulin. When blood glucose is low, most of GLUT4 transporters in adipose tissue and muscle remain in the cell cytoplasm (within vesicle membranes). Upon glucose increase and insulin secretion, GLUT4 moves from its intracellular store to the cell surface by transport and fusion of the vesicles to the plasma membrane. This rapidly activates glucose uptake into the cell. Different from other GLUTs, GLUT5 is a transporter specific for fructose present in the apical and basolateral membranes of enterocytes and in testis.

The remaining GLUTs are distributed in different tissues; some of their properties and functions are still unclear. For example, GLUT13 (also called HMIT) does not transport glucose, but is instead a H^+/myoinositol symporter.

GLUCOSE PHOSPHORYLATION

Phosphorylation is the initial step for the metabolic utilization of monosaccharides. For glucose, the first metabolic transformation is the esterification with phosphate to form glucose-6-phosphate (G-6-P). This reaction is catalyzed by *hexokinase*, an enzyme present in all cells.

D-Glucose   Glucose-6-phosphate

There are four types of hexokinase. Isoforms I, II, and III are found in varying proportions in many tissues and are rather nonspecific, catalyzing the incorporation of phosphate on carbon 6 of not only glucose but other hexoses as well. Their K_m for glucose ranges from 0.01 to 0.1 mM, which is well below the usual concentration of glucose in body fluids and cells (~5.0 mM). Consequently, at the physiological glucose level in tissues, these enzymes are working at their maximal rate and are not affected by changes in plasma glucose levels. Hexokinases I, II, and III are normally saturated and the reaction is zero order with respect to the substrate. They are allosterically inhibited by G-6-P, a product of the reaction. Isozyme IV, also known as *glucokinase*, is found exclusively in liver and pancreatic β cells. Glucokinase is highly specific, it utilizes only D-glucose as a substrate, but it has an affinity for glucose that is much lower than that of other hexokinases (K_m greater than 10 mM, which is above the glucose concentrations normally found in tissues). Glucokinase activity depends on the amount of glucose available. Since physiological glucose levels are within the range where activity of glucokinase is directly proportional to the substrate concentration (first-order reaction with respect to the substrate), its function is relatively low at plasma glucose fasting values. Unlike hexokinases, glucokinase is not inhibited by G-6-P. In pancreatic β cells, glucokinase acts as a glucose sensor, involved in regulating insulin secretion. Deficiencies in type IV hexokinase have been associated with diabetes mellitus.

The biological properties of each hexokinase are of great physiological relevance. The activity of isozymes I–III (with low hexose K_m) assures continued utilization of glucose and a permanent supply of energy to the cells, even if blood glucose levels vary. In contrast, glucokinase of hepatocytes and pancreatic β cells is only activated when blood levels are significantly increased, as is seen after a meal.

Hexokinases I–III are constitutively expressed in cells, while glucokinase synthesis is induced by insulin. All hexokinases have ATP-Mg as a substrate. ATP is used as the phosphate and energy donor and Mg^{2+} is the cofactor.

The reaction catalyzed by hexokinases comprises two coupled reactions, the endergonic synthesis of ester G-6-P and the exergonic hydrolysis of ATP. The algebraic sum of the $\Delta G°$'s of these reactions is −4.0 kcal/mol (−16.7 kJ/mol). Under physiological conditions, the reaction proceeds in the direction of glucose phosphorylation and is practically irreversible.

The formation of G-6-P is important for converting glucose into a more reactive compound, suitable for further transformations. In addition, because cell membranes are impermeable to G-6-P, glucose phosphorylation allows trapping of glucose into the cell, where it will follow different metabolic pathways. Moreover, rapid conversion of glucose to G-6-P maintains the intracellular glucose concentration low and facilitates the continued entry of glucose into the cell.

G-6-P is an important metabolite and represents a *metabolic crossroad* common to different pathways, including: glycogenesis, glycogenolysis, glycolysis, gluconeogenesis, and the pentose phosphate pathway.

METABOLIC PATHWAYS FOR GLUCOSE

The following are the main metabolic pathways that glucose can follow:

1. *Glycogenesis*: conversion of glucose into glycogen.
2. *Glycogenolysis*: formation of glucose from glycogen.
3. *Glycolysis* or *Embden–Meyerhof pathway*: degradation of glucose into pyruvate and lactate.
4. *Pyruvate oxidative decarboxylation*: conversion of pyruvate (formed in glycolysis) into acetate.
5. *Citric acid, tricarboxylic acid*, or *Krebs cycle*: oxidation of acetate to CO_2 and H_2O.
6. *Pentose phosphate or hexose monophosphate pathway*: alternative route for glucose oxidation.
7. *Gluconeogenesis*: glucose or glycogen production from noncarbohydrate sources, using glucogenic amino acids, lactate, and glycerol as substrates.

Glycogenesis

Glycogen synthesis from glucose takes place in many tissues, but it is particularly important in liver and muscle where its magnitude and functional relevance is more significant. In humans, approximately 8% of the liver's weight is glycogen, especially after a diet rich in carbohydrates. This amount is considerably reduced after prolonged fasting. In skeletal muscle, glycogen contains approximately 1% of its weight.

Glycogenesis is an anabolic process that requires energy. It consists of the following steps:

1. *Glucose phosphorylation*. The first step in glycogen synthesis is conversion of glucose to G-6-P. This reaction, catalyzed by hexokinases (*glucokinase* among them), was described in a previous section.
2. *Glucose-1-phosphate formation*. In the second stage, *phosphoglucomutase* catalyzes the transfer of the phosphate group from carbon 6 to carbon 1 on the glucose molecule. G-6-P is converted to glucose-1-phosphate. Phosphoglucomutase requires Mg^+ and glucose-1,6-bisphosphate as cofactors [the term *bisphosphate* indicates that there are two phosphates bound at different sites of the same molecule. When the phosphates are bound together by an anhydride bond, it is called diphosphate (i.e., ADP)]. The reaction is reversible.

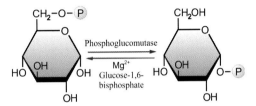

Glucose-6-phosphate   Glucose-1-phosphate

3. *Glucose activation*. Glucose-1-phosphate reacts with the high energy nucleotide uridine triphosphate (UTP) to give uridine diphosphate glucose (UDPG) and pyrophosphate (PP_i). The reaction is catalyzed by *uridine diphosphate-glucose pyrophosphorylase*, or *glucose-1-P uridyltransferase*.

$$\text{Glucose-1-P} + \text{UTP} \xrightarrow{\text{Glucose-1-P uridyltransferase}} \text{UDPG} + PP_i$$

Inorganic pyrophosphate is rapidly hydrolyzed by the action of pyrophosphatase. Immediate disappearance of pyrophosphate makes this reaction virtually irreversible.

Glucose inclusion in the nucleotide sugar (UDPG) gives it the necessary reactivity to participate in the synthesis of glycogen. Glucose is activated by binding to UDP. The metabolic role of nucleotide sugars, such as UDPG was discovered by Luis F. Leloir.

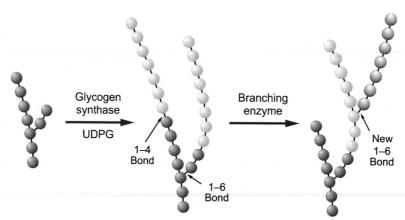

FIGURE 14.2 **Schematic representation of glycogen branching.** The *darker gray spheres* represent glucose molecules of the preexisting polymer. *Lighter gray spheres* indicate glucose residues added by glycogen synthase. *Red spheres* show the chain segment transferred by the branching enzyme.

4. *Glucose addition to the polymer backbone.* At this stage, the UDPG-activated glucose is transferred to preexisting glycogen. One glycosidic linkage to carbon 4 of a terminal glucose of a glycogen chain is formed. This reaction is catalyzed by *glycogen synthase*, a glucosyl transferase that requires the presence of the preexisting glycogen polymeric structure, which keeps adding glucose molecules via α1→4 bonds. The reaction is practically irreversible.

$$\text{UDPG} + (\text{Glucose})_n \xrightarrow{\text{Glycogen synthase}} (\text{Glucose})_{n+1} + \text{UDP}$$

Glycogen synthase can only form α1→4 bonds; therefore, its action can only achieve linear elongation of existing glycogen branches by successive addition of glucose molecules.

5. *Branch formation.* When glycogen synthase has built a glycogen chain of 10 or more glucose residues, another enzyme, *amylo-α(1,4)→α(1,6)-glucantransferase* (or *branching enzyme*), cuts a terminal segment of at least 6 glucose molecules and inserts it with an

α1→6 glycosidic bond on a neighboring chain (Fig. 14.2).

In this manner, glycogen is built by the joint action of glycogen synthase and the branching enzyme. Glycogen synthase is present in tissues in two interchangeable forms: *b* (inactive, phosphorylated) and *a* (active, dephosphorylated). Form *b* shows activity only in the presence of saturating concentrations of G-6-P, which acts as a positive allosteric effector. Form *a* is active, regardless of the existence of G-6-P. Synthase *b* is transformed into *a* by a phosphatase. The active *a* form is converted into *b* via phosphorylation. This shows the importance of phosphorylation as a mechanism to regulate glycogen synthase and the glycogen pathway.

Glycogenin. The activity of glycogen synthase requires preformed glycogen chains. Despite this, glycogen synthesis is possible in the total absence of glycogen. This depends on the function of an initiator protein called *glycogenin*, which acts as an acceptor of the first glucose molecule prior to the glycosidic linkage with a tyrosine protein. The process is autocatalytic, using UDPG as glucose donor. Glycogenin also catalyzes the successive addition of units to form a linear chain from six to seven glucose molecules bound by α1→4 bonds. Glycogen synthase and the branching enzyme continue acting on the chain anchored to glycogenin, giving the polymer its final characteristic structure. Each glycogen molecule is covalently linked to an initiator protein, which is why the number of particles of glycogen in the cell depends on the availability of glycogenin.

Energy Requirements for Glycogen Synthesis

Addition of glucose to glycogen is an endergonic process that requires energy. The first phosphorylation reaction (1), common to all pathways of glucose utilization, consumes one molecule of ATP. In the reaction of glucose activation (3), UTP (a compound with energy-rich bonds) is needed.

In the following reaction (4), UDP is released. UTP is regenerated from UDP in a reaction catalyzed by *nucleoside diphosphokinase*:

$$UDP + ATP \rightarrow UTP + ADP$$

Incorporation of one glucose molecule to glycogen consumes two ATPs. This energy expenditure to store glucose appears to be futile. It would be energetically more economical to accumulate G-6-P, since producing this metabolite requires less energy and it cannot leave the cell. However, using G-6-P as an energy store is not possible under physiological conditions. Osmotic pressure of a solution depends on the number of dispersed particles in a solution, not on their size. A glycogen molecule, consisting of thousands of glucose units, is equivalent (from an osmotic standpoint) to one molecule of G-6-P or any other solute. Therefore, accumulation of a number of G-6-P molecules would result in an increase in osmotic pressure, which could lead to cell swelling and lysis. Glycogen can be used to store a large amount of glucose molecules with little effects on cell osmolarity.

Glycogenolysis

Glycogenolysis is not simply the reverse process of glycogenesis. The existence of irreversible reactions in glycogenesis requires glycogen degradation to be performed using different enzymes that can catalyze the unidirectional steps of the pathway. The stages of glycogenolysis are:

1. *Glycogen phosphorolysis.* Glycogen degradation is initiated by the action of *phosphorylase*, a serine–threonine kinase which catalyzes the rupture of $\alpha1\rightarrow4$ glycosidic bonds by insertion of a phosphate at carbon 1. The phosphate employed in this reaction is obtained from the medium (P_i) and the hydrolysis of ATP is not necessary. Phosphorylase acts on the nonreducing ends of glycogen branches, releasing glucose-1-phosphate. Its action stops four glucose residues before an $\alpha1\rightarrow6$ junction (Fig. 14.3). At this point, another enzyme, *oligo-α(1,4)-α(1,4)-glucantransferase*, separates a trisaccharide from the terminal branch and transfers it to the end of a neighbor branch, where it is attached via a $\alpha1\rightarrow4$ bond. The branch is reduced to a single glucose with an $\alpha1\rightarrow6$ bond.

Phosphorylase contains pyridoxal phosphate covalently bound to it, which serves as an essential cofactor that promotes the hydrolysis of the glycosidic bond. It is an important regulatory enzyme of the pathway that responds to allosteric effectors and to phosphorylation–dephosphorylation modifications. Chapter 19 will expand on this topic.

2. *Hydrolysis of $\alpha1\rightarrow6$ glycosidic bonds.* This is catalyzed by *α1→6-glucosidase*, or *debranching enzyme*, which releases free glucose. The $\alpha1\rightarrow6$ glucosidase and oligo-α(1,4)→α(1,4)-glucantransferase activities depend on the action of a single protein with two active sites of different specificity. After the action of the debranching enzyme, phosphorylase continues releasing glucose-1-P until the enzyme reaches a distance that is four glucose molecules apart from the next $\alpha1\rightarrow6$ bond. At this point, the hydrolysis of $\alpha1\rightarrow6$ bonds is catalyzed by the enzymes

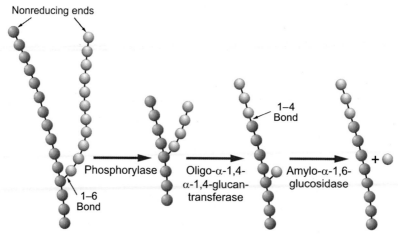

FIGURE 14.3 **Representation of the action of oligo-α(1,4)-α(1,4)-glucantransferase and α-(1,6)-glucosidase (debranching enzyme).** *Red spheres* represent degraded branch glucose residues.

mentioned previously. The concerted action of phosphorylase, oligo-α(1,4)→α(1,4)-glucantransferase, and α1→6-glucosidase releases glucose-1-P and some free glucose molecules (debranching enzyme causes hydrolysis and not phosphorolysis). Only the glucose in the branching position is released as free glucose. All other molecules are released as G-1-P. On average, a free glucose molecule is produced for every nine molecules of glucose-1-P, which indicates the degree of glycogen molecule branching.

3. *G-6-P formation.* Glucose-l-phosphate is converted into G-6-P by *phosphoglucomutase.* This is the reverse reaction that produces G-1-P in glycogenesis.

4. *Free glucose formation.* The last stage of glycogenolysis is the hydrolysis of G-6-P to glucose and inorganic phosphate, catalyzed by *G-6-P.*

$$\text{Glucose-6-P} \xrightarrow[\text{H}_2\text{O}]{\text{Glucose-6-phosphatase}} \text{Glucose} + \text{P}_i$$

The reaction from glucose to G-6-P, catalyzed by glucokinase in glycogenesis, is essentially irreversible. For this reason, the reverse process is accomplished by another enzyme, G-6-phosphatase. This enzyme is found in liver and kidney cell membranes and in intestinal cell endoplasmic reticulum, but not in muscle cells. This explains why the liver, kidney, and intestine can release glucose to the circulation and muscle cannot. G-6-phosphatase is a complex with five subunits; three of them carry G-1-P and G-6-P through the ER membrane. One of the subunits belongs to the GLUT transporter family (GLUT7) and transports glucose from the ER to the cytosol.

In muscle, glycogen degradation starts with stages similar to those described earlier. However, G-6-P cannot be hydrolyzed because there is no G-6-P in muscle; thus, G-6-P continues its catabolic pathway mainly through glycolysis.

Fig. 14.4 summarizes the pathways for glycogen synthesis and degradation in the liver.

Functional role of glycogen. In most tissues, glycogen is used as an energy reserve to obtain glucose during periods of fasting, hypoglycemia, or hypoxia. However, the role of glycogen is not the same in all organs. Particularly, glycogen function is significantly different in liver and muscle. The liver plays an important role in the control of glycemia, ensuring that all tissues receive a constant supply of glucose. Immediately

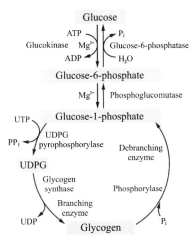

FIGURE 14.4 Glycogen synthesis and degradation pathways in the liver.

after a meal, blood glucose increases transiently. During this period, the liver clears glucose from the circulation and stores it as glycogen. Between meals, the liver breaks down glycogen and releases free glucose to the bloodstream. In contrast, glycogen serves as an energy reserve that can be rapidly mobilized to fuel contraction in muscle cells. Muscle does not release glucose to the circulation, its glycogen stores are exclusively used in that tissue.

Genetic Diseases Related to Glycogen Metabolism

Genetic alterations can lead to the decrease or absence of the enzymes involved in glycogen synthesis or degradation. Generally, these disorders are known as *glycogenosis*. They are characterized by the accumulation of abnormal levels of glycogen in tissues or the presence of abnormally structured glycogen molecules. At least 14 types of glycogenoses have been described, each due to deficiencies in different enzymes. The first one identified was *von Gierke's disease*, or glycogen storage disease type I, characterized by excessive glycogen deposition in the liver and kidney. In this disease, glycogen is normally synthesized, but there is a defect in

glycogenolysis that prevents the release of glucose from its stores. This is due to an absence or a decrease in G-6-phosphatase in liver and kidney cells.

In glycogen storage disease type II and types V–X, glycogenolysis is also the pathway affected and increased deposits of glycogen are found in different organs. McArdle's disease (glycogen storage disease type V) leads to increased glycogen in skeletal muscle, with a marked decrease in muscle working capacity and a lack of lactate increase in blood after exercise. This disease is caused by the deficiency or absence of phosphorylase activity in muscle. Liver phosphorylase is not altered, which shows the independent genetic control between the liver and muscle enzymes.

In all of the disorders mentioned previously, the structure of glycogen is normal. In other glycogenosis, it is abnormal. For example, glycogen storage disease type III is due to a deficiency of the debranching enzyme; the stored glycogen has very short branches. Type IV, or Andersen disease, is caused by the inability to synthesize the branching enzyme, which results in glycogen with excessively long and poorly branched chains.

In general, these pathological conditions are infrequent; however, it is important for general physicians to be aware of their existence so they can properly refer the patients to specialized centers. Some of the glycogen storage diseases are serious and have poor prognosis, leading to death at an early age. Table 14.1 presents a summary of the glycogenosis.

Glycolysis

The main pathway for glucose catabolism is *glycolysis*, or the *Embden–Meyerhof pathway*. In this metabolic pathway, a glucose molecule is cleaved into two molecules of pyruvate and energy is produced. The process can be accomplished in the absence of oxygen (under anaerobic conditions). Glycolysis is evolutionarily the oldest metabolic pathway, perhaps used by the first living

TABLE 14.1   Glycogen Storage Diseases

Type	Disease	Deficient enzyme	Symptoms
0		Glycogen synthase	Hypoglycemia, ketonemia, delayed growth, early death
Ia	Von Gierke	G-6-phosphatase	Glycogen accumulation in liver and kidney, hypoglycemia, ketonemia, lactic acidosis, hyperlipemia, hyperuricemia
Ib		Microsomal G-6-P translocase	Same as Type Ia with added susceptibility to infections
Ic		Microsomal P_i carrier	Same as Type Ia
II	Pompe	Lysosomal α-(1,4)-glucosidase	Glycogen accumulation in lysosomes, normal glycemia, mainly affected: skeletal and cardiac muscles
IIIa	Cori or Forbes	Debranching	Hepatomegaly, myopathy
IIIb		Debranching only in liver	Hepatomegaly
IV	Andersen or Amylopectinosis	Branching	Hepatosplenomegaly, myoglobinuria, early death
V	McArdle	Muscle phosphorylase	Glycogen accumulation in muscle, muscular weakness, cramps, diminished production of lactate during exercise
VI	Hers	Liver phosphorylase	Glycogen accumulation in liver, hepatomegaly, hypoglycemia, and light ketosis
VIb, VIII, IX		Phosphorylase kinase	Hepatomegaly
VII	Tarui	Muscle and erythrocytes phosphofructokinase	Same as Type V plus hemolysis
X		Liver protein kinase A	Hepatomegaly
XI	Fanconi–Bickel	Glucose carrier (GLUT2)	Hepatomegaly, delayed growth, rickets, kidney dysfunction
XII		Enolase	Myalgia, myoglobinuria, muscle weakness
XIII		Phosphoglucomutase	Muscle pain with effort, muscle necrosis, myoglobinuria

Note: All, except type 0, show glycogen accumulation. There are several subtypes of Hers disease (type VI) due to different mutations.

organisms that inhabited earth, when the atmosphere lacked oxygen. Glycolysis is also a remarkable example of the unity of the biological world, since this pathway exists in all living organisms. Many microorganisms metabolize glucose and other monosaccharides by this route, through a process called *fermentation*. However, the end products of fermentation vary in different organisms. Some produce lactate, while others generate ethanol or acetic acid and CO_2.

In aerobic organisms, glycolysis is the first part of glucose catabolism. The pyruvate produced continues oxidative degradation into CO_2 and H_2O. However, in aerobic organisms, when the oxygen supply in a tissue is insufficient (e.g., in skeletal muscle during intense exercise), pyruvate is converted to lactate, following the steps of lactic acid fermentation.

The chemical transformations during glycolysis include changes in the original substrate (glucose), with production of energy-rich metabolites for transfer of phosphoryl residues to ADP. This ability to generate ATP by substrate-level phosphorylation mechanisms, without participation of oxygen and the respiratory chain, makes glycolysis a process of high physiological relevance.

The series of reactions in the glycolytic pathway can be divided into two phases. In the first phase, glucose undergoes two phosphorylation steps and is divided into two triose-phosphate molecules. This is a preliminary phase during which energy is expended to form compounds that are unable to escape the cell; they are chemically more reactive than glucose and can more easily undergo subsequent transformations. The result of the first set of reactions is the breakdown of the initial six-carbon molecule into two molecules of three carbons each, glyceraldehyde-3-phosphate (G3P), and dihydroxyacetone phosphate (DHAP). DHAP is then converted to G3P; therefore, it can be considered that each glucose molecule entering glycolysis is broken down into two G3P molecules.

In the second part, glyceraldehyde-3-phosphate undergoes oxidation and redistribution of its atoms to form high energy intermediates, which are used in the synthesis of ATP through substrate-level phosphorylation. This is the energy producing stage of the pathway.

All the enzymes involved in glycolysis are found in the cytosol, so this metabolic pathway is fully completed in the cell cytoplasm.

First Phase of Glycolysis

1. *Formation of G-6-P*. A required initial stage for glucose utilization is its phosphorylation at carbon 6. The reactions necessary to obtain G-6-P are different depending on the starting material (glucose or glycogen). As previously indicated, glucose phosphorylation is catalyzed by hexokinase. Due to the isoforms with high affinity for the substrate (isoforms I–III), glucose utilization in tissues other than the liver is not affected by the usual fluctuations of blood glucose. In the liver, isozyme IV

(glucokinase) has low glucose affinity and works when glucose levels are high. The reaction of G-6-P formation is irreversible at the physiological conditions of the cell.

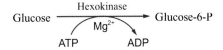

When glycolysis starts from glycogen, G-6-P formation is accomplished in two steps, catalyzed by phosphorylase and phosphoglucomutase.

$$\text{Glycogen} \xrightarrow{\text{Phosphorylase}} \text{Glucose-1-P}$$

$$\text{Glucose-1-P} \underset{\text{Phosphoglucomutase}}{\rightleftarrows} \text{Glucose-6-P}$$

The glycolytic pathway continues from G-6-P with the following reactions:

2. *Formation of fructose-6-phosphate (F-6-P)*. G-6-P is converted into F-6-P. The reaction is reversible and is catalyzed by *phosphoglucoisomerase* (PGI), which requires Mg^{2+} or Mn^{2+}.

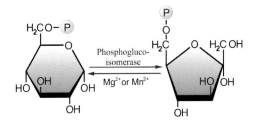

Glucose-6-phosphate   Fructose-6-phosphate

PGI is a multifunctional protein. It also acts as a cytokine and is a neurotrophic factor that promotes neurons differentiation.

Defects in the genes coding PGI are associated with hereditary spherocytosis, a type of hemolytic anemia.

3. *F-6-P phosphorylation*. F-6-P is phosphorylated at carbon 1 and converted into fructose-1,6-bisphosphate (F-1,6-bisP). The reaction requires the transfer of a phosphoryl group from ATP and is catalyzed by

phosphofructokinase (PFK) in the presence of Mg^{2+} ions.

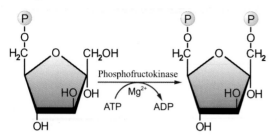

Fructose-6-phosphate   Fructose-1,6-bisphosphate

The coupling of this reaction to ATP hydrolysis enables the synthesis of the phosphoric ester at carbon 1. The reaction, essentially irreversible, is an important regulatory step of this pathway. Phosphofructokinase is an allosteric enzyme, whose activity is modulated by various effectors. It is activated by AMP, ADP, and fructose-2,6-bisphosphate. Conversely, it is inhibited by ATP and citrate.

Genetic defects of PFK results in Tarui disease (glycogen storage disease type VII).

4. *Triose-phosphates formation.* Fructose-1,6-bisphosphate is cleaved into two triose-phosphate molecules: glyceraldehyde-3-phosphate (G3P) and DHAP. The reaction is catalyzed in a reversible manner by *aldolase*, a lyase.

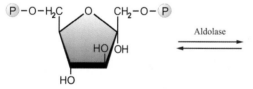

Fructose-1,6-bisphosphate

D-Glyceraldehyde-   Dihydroxyacetone-
3-phosphate   phosphate

Although cleavage of fructose-1,6-bisphosphate is endergonic, with a positive $\Delta G^{\circ\prime}$, it can easily proceed because the products (triose-phosphates) are rapidly eliminated by the following reactions.

There are three isozymes of aldolase. Isozyme A is expressed in muscle; isozyme B is found in the liver, stomach, intestine, and kidney; and isozyme C predominates in the heart, brain, and ovaries. Isozymes A and C are involved mainly in glycolysis, while isozyme B functions in fructose metabolism.

Mutation in the gene coding aldolase B is the cause of hereditary fructose intolerance.

5. *Triose-phosphate interconversion.* D-glyceraldehyde-3-phosphate is the only triose-phosphate produced in the previous reaction that continues through the metabolic pathway. While DHAP also follows the glycolytic pathway, it must be first transformed into G3P. This is possible due to the reversible conversion of DHAP into G3P, catalyzed by *triose-phosphate isomerase* (TPI).

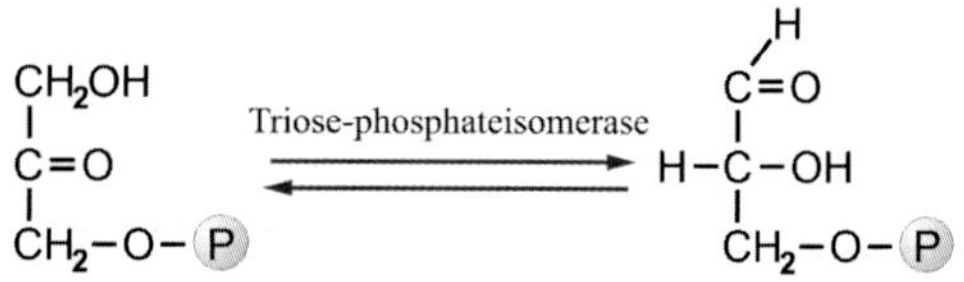

$$
\begin{array}{ccc}
CH_2OH & & H \\
| & & \diagup \\
C=O & \xrightarrow{\text{Triose-phosphate isomerase}} & C=O \\
| & & | \\
CH_2-O-\text{P} & & H-C-OH \\
& & | \\
& & CH_2-O-\text{P}
\end{array}
$$

Dihydroxyacetone phosphate   D-Glyceraldehyde 3-phosphate

While the reaction equilibrium is strongly shifted to the left, the reaction predominantly proceeds to the right in tissues due to the continuous removal of G3P by the next step of glycolysis.

Deficiencies of TPI cause hemolytic anemia and serious neurologic disorders.

Second Phase of Glycolysis

In the previous reaction, DHAP is converted into D-glyceraldehyde-3-phosphate. For this reason, it is considered that each glucose molecule generates two G3P molecules.

6. *Glyceraldehyde-3-phosphate oxidation and phosphorylation.* This is an important step of glycolysis, in which glyceraldehyde is dehydrogenated. The energy released is used to add phosphate (P_i) from the medium to form 1,3-bisphosphoglycerate. The reaction is catalyzed by *glyceraldehyde-3-phosphate dehydrogenase* (G3PDH), an oxidoreductase that uses NAD as coenzyme.

$$
\begin{array}{ccc}
\underset{\text{D-Glyceraldehyde-}\atop\text{3-phosphate}}{\begin{array}{c} \text{H} \\ \diagup \\ \text{C}=\text{O} \\ | \\ \text{H}-\text{C}-\text{OH} \\ | \\ \text{CH}_2-\text{O}-\text{\textcircled{P}} \end{array}} & \xrightarrow[\text{NAD}^+ \quad \text{NADH+H}^+]{\substack{\text{Glyceraldehyde-3P} \\ \text{dehydrogenase} \\ P_i}} & \underset{\text{1,3-Bisphosphoglycerate}}{\begin{array}{c} \text{CH}_2-\text{O}-\text{\textcircled{P}} \\ | \\ \text{CHOH} \\ | \\ \text{COO} \sim \text{\textcircled{P}} \end{array}}
\end{array}
$$

The mechanism of glyceraldehyde-3-phosphate dehydrogenase is well known. Glyceraldehyde-3-phosphate binds to an essential sulfhydryl group in the active site of the enzyme and is oxidized into 3-phospho-glyceric acid. A phosphate is then introduced and 1,3-bisphosphoglycerate, which has an acyl phosphate with high hydrolysis energy, is produced.

If G3PDH is translocated to the nucleus, it can interfere with the signal system of programmed cell death (apoptosis, see Chapter 32), particularly in neurons; thus, it may be involved in alterations leading to neurodegenerative diseases (such as Huntington's disease, Parkinson's disease, and Alzheimer's disease).

7. *Substrate-level phosphorylation.* High energy phosphate is transferred from 1,3-bisphosphoglycerate to ADP through a reaction catalyzed by *phosphoglycerate kinase*. The result is the production of 3-phosphoglycerate and ATP.

$$
\begin{array}{ccc}
\underset{\text{1,3-Bisphosphoglycerate}}{\begin{array}{c} \text{CH}_2-\text{O}-\text{\textcircled{P}} \\ | \\ \text{HC OH} \\ | \\ \text{COO} \sim \text{\textcircled{P}} \end{array}} & \xrightarrow[\underset{\text{ADP} \quad \text{ATP}}{\text{Mg}^{2+}}]{\substack{\text{Phosphoglycerate} \\ \text{kinase}}} & \underset{\text{3-Phosphoglycerate}}{\begin{array}{c} \text{CH}_2-\text{O}-\text{\textcircled{P}} \\ | \\ \text{HC OH} \\ | \\ \text{COO}^- \end{array}}
\end{array}
$$

Deficiency of phosphoglycerokinase is associated with hemolytic anemia, myopathy, central nervous system disorders, and growth retardation.

Together, the last two reactions (6 and 7) are reversible under normal cell conditions. The transfer of acyl phosphate phosphoryl allows the formation of ATP. This is a phosphorylation at the substrate-level and the first reaction within the glycolysis pathway where there is energy conservation.

8. *Phosphoglycerate formation.* 3-Phosphoglycerate is converted into 2-phosphoglycerate via an intramolecular phosphoryl transfer. This reaction is catalyzed, in both directions, by *phosphoglycerate mutase*, which requires Mg^{2+}.

$$
\begin{array}{ccc}
\underset{\text{3-Phosphoglycerate}}{\begin{array}{c} \text{CH}_2-\text{O}-\text{\textcircled{P}} \\ | \\ \text{HC OH} \\ | \\ \text{COO}^- \end{array}} & \underset{\text{Mg}^{2+}}{\overset{\text{Phosphoglyceromutase}}{\rightleftharpoons}} & \underset{\text{2-Phosphoglycerate}}{\begin{array}{c} \text{CH}_2\text{OH} \\ | \\ \text{HC O}-\text{\textcircled{P}} \\ | \\ \text{COO}^- \end{array}}
\end{array}
$$

Genetic defects of phosphoglycerate mutase produce muscle alterations with intolerance to exercise.

9. *Formation of phosphoenolpyruvate.* Dehydration and intramolecular redistribution in 2-phosphoglycerate generates an energy-rich compound, phosphoenolpyruvate (PEP). *Enolase*, which requires Mg^{2+} or Mn^{2+}, catalyzes the reversible reaction:

$$
\begin{array}{ccc}
\underset{\text{2-Phosphoglycerate}}{\begin{array}{c} \text{CH}_2\text{OH} \\ | \\ \text{CH}-\text{O} \sim \text{\textcircled{P}} \\ | \\ \text{COO}^- \end{array}} & \underset{\text{Mg}^{2+}}{\overset{\text{Enolase}}{\rightleftharpoons}} & \underset{\text{Phosphoenolpyruvate}}{\begin{array}{c} \text{CH}_2 \\ \| \\ \text{C}-\text{O} \sim \text{\textcircled{P}} \\ | \\ \text{COO}^- \end{array}} + \text{H}_2\text{O}
\end{array}
$$

Loss of a water molecule produces a redistribution of energy in 2-phosphoglycerate, generating phosphoenolpyruvate with a high energy phosphate bond.

Enolase presents three isoforms: α, β, and γ. Isoform α is found in almost all tissues, isoform β is found in muscle, and isoform γ is only found in nervous tissue. The γ isozyme can be released from the nervous system and appears in plasma in some neurologic diseases, tumors, strokes, and seizures.

Lack of enolase β is associated to glycogen storage disease type XII.

10. *Second substrate-level phosphorylation.* Phosphoenolpyruvate has enough potential to transfer a phosphate molecule to ADP, forming ATP. The reaction is catalyzed by *pyruvate kinase*, which requires Mg^{2+} or Mn^{2+} ions. Potassium ion is an activator of the enzyme. The product enolpyruvate is spontaneously transformed into pyruvate.

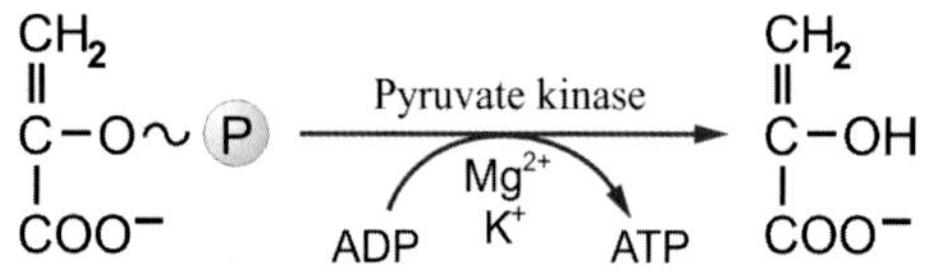

Phosphoenolpyruvate   Enolpyruvate

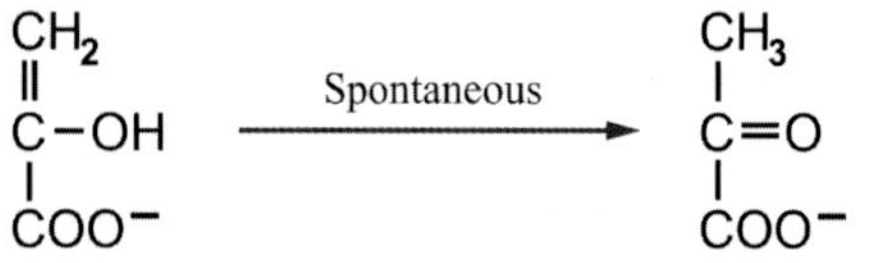

Enolpyruvate   Pyruvate

Another ATP molecule has been generated by substrate-level phosphorylation.

There are four isoenzymes of pyruvate kinase: isozyme L in liver; isozyme R in red blood cells; isozyme M1 in muscle, heart, and brain; and isozyme M2 in fetal tissues.

11. *Lactate formation.* The pyruvate resulting from glycolysis can follow different pathways. It can be degraded under aerobic conditions through a pathway that will be described in a later section. When oxygen is scarce or absent (anaerobic condition), pyruvate is reduced to lactate by *lactate dehydrogenase*, an enzyme that uses NAD as a coenzyme. The process is easily reversible.

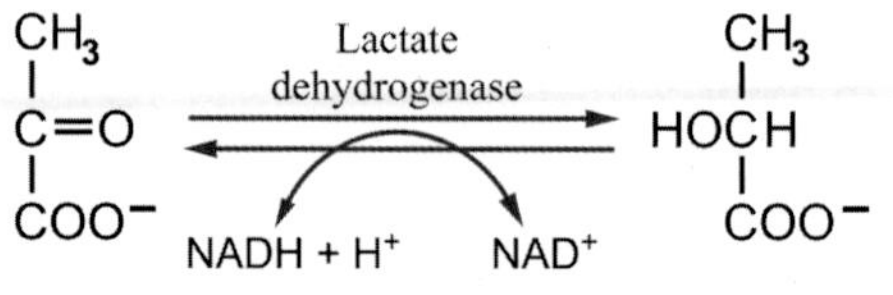

Pyruvate   Lactate

The reaction catalyzed by lactate dehydrogenase is of great functional relevance. In the absence or deficiency of oxygen, the NADH formed during oxidation of glyceraldehyde-3-phosphate (reaction 6) cannot be oxidized to NAD because the respiratory chain is not working (aerobically, the transfer of reducing equivalents is done indirectly, through hydrogen shuttles, p. 199). Glycolysis is limited by the availability of NAD in the cytoplasm; it stops when all NAD in the cytosol is reduced to NADH. Conversion of pyruvate to lactate is a mechanism by which reoxidation of NADH is ensured, which allows glycolysis to continue. This explains why lactate formation is the end product of glycolysis in tissues that work under relatively low oxygen conditions, such as skeletal muscle during intense exercise. Fig. 14.5 summarizes the reactions of the glycolytic pathway.

Energy Balance of Glycolysis

Each mole of glucose produces 2 moles of G3P and is finally converted into 2 moles of lactate. There are two steps in which ATP is consumed. One mole of glucose requires the energy

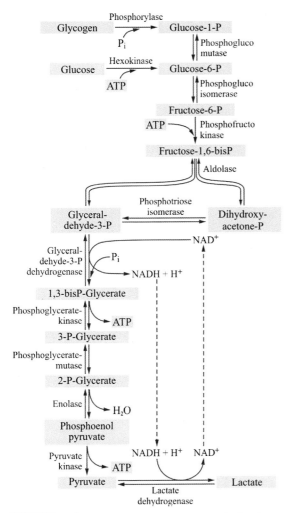

FIGURE 14.5 **Glycolysis or Embden–Meyerhof pathway.**

TABLE 14.2   Energy Yield of Glycolysis

ATP expenditure (per mole of glucose)	
Glucose → G-6-P	−1 mol ATP
F-6-P → fructose-1,6-bisP	−1 mol ATP
ATP yield (per mole of glucose)	
1,3-Bisphosphoglycerate → 3-bisphosphoglycerate	+2 mol ATP
Phosphoenolpyruvate → pyruvate	+2 mol ATP
Net balance	+2 mol ATP

mole of ATP from ADP. Another mole of ATP is produced from phosphoenolpyruvate. Therefore, since one molecule of glucose generates two triose-phosphate molecules, the ATP yield per mole of glucose is 4 moles of ATP. The final balance of glycolysis is a net gain of 2 moles of ATP per mole of glucose (Table 14.2).

Hydrolysis of 1 mole of ATP to ADP yields approximately 7.3 kcal (30.5 kJ) under standard conditions. The energy efficiency of glycolysis, per mole of glucose, is 14.6 kcal (61 kJ). Taking into consideration that the total energy content of glucose is 686 kcal/mol (2870 kJ/mol), the performance achieved by glycolysis is negligible. However, to measure the efficiency of this pathway, it is necessary to note that the two lactate molecules formed as the final product still contain more than 90% of the original energy of glucose, which can be extracted in part by full oxidation to CO_2 and H_2O. Glycolysis is a highly efficient mechanism to obtain energy independent from glucose oxidation. Due to its energy-producing capacity, muscle contraction and other processes can occur without significant consumption of oxygen.

Lactate generated in muscles during exercise moves to the blood and can be converted into glucose in the liver. It can also be completely oxidized in other tissues, producing the energy contained in the molecule, which is slightly lower than that in the original glucose molecule.

of 1 mole of ATP for the initial phosphorylation to form G-6-P, and another mole of ATP for the second phosphorylation step, which converts F-6-P into fructose-1,6-bisphosphate. When glycolysis starts from glycogen, only one ATP is consumed in the second phosphorylation.

In the second phase of glycolysis, two steps produce ATP by substrate-level phosphorylation. These are the reactions catalyzed by phosphoglycerate kinase and pyruvate kinase. Each mole of 1,3-bisphosphoglycerate generates 1

Functional Role of Glycolysis

While in most tissues glycolysis serves as the major initial pathway of glucose utilization, in others it has different specific roles.

Skeletal muscle uses glycolysis for generation of ATP during intense exercise. At rest or during light activity, muscle uses oxidative pathways, which yield higher ATP levels. During intense activity, glycolysis is the most important supplier of energy to the muscle.

Adipose tissue is specialized in storing triglycerides. A major function of glycolysis in adipocytes is to provide DHAP, a precursor used in triglyceride synthesis.

Red blood cells lack mitochondria and, therefore, cannot generate ATP by oxidative pathways. They depend entirely on glycolysis for ATP synthesis. 2,3-Bisphosphoglycerate, an important modulator of hemoglobin, is generated from 1,3-bisphosphoglycerate, a metabolite of glycolysis.

Exergonic and endergonic processes are coupled in glycolysis. The end result is conversion of one glucose molecule into two molecules of lactate:

$$\text{Glucose} \longrightarrow 2\ \text{Lactate} + 2\ H^+$$

The free energy change for this process is negative ($\Delta G^{\circ\prime} = -47.0$ kcal/mol or -196.6 kJ/mol). The net energy yield is formation of two high energy bonds, which can be expressed by the equation:

$$2\ ADP + 2\ P_i \longrightarrow 2\ ATP + 2\ H_2O$$

This is an endergonic process with $\Delta G^{\circ\prime}$ of +14.6 kcal/mol (+61 kJ/mol).

The sum of the previous reactions gives the total equation for glycolysis:

$$\text{Glucose} + 2\ ADP + 2\ P_i \longrightarrow$$
$$\longrightarrow 2\ \text{LACTATE} + 2\ H^+ + 2\ ATP + 2\ H_2O$$

The free energy change of this reaction is:

$$\Delta G^{\circ\prime} = -196.6\,\text{kJ/mol} + 61\,\text{kJ/mol}$$
$$= -135.6\,\text{kJ/mol}$$

The total process occurs with a marked decrease in free energy, which makes glycolysis an irreversible pathway under physiological conditions.

In the described sequence of reactions, three are markedly exergonic and determine the direction of reactions of the glycolytic pathway. They correspond to the reactions catalyzed by hexokinase, phosphofructokinase, and pyruvate kinase.

In tissues, such as the liver, which can transform lactate back into glucose, these irreversible steps are catalyzed in the opposite direction by enzymes different to those in glycolysis (this reverse process is called gluconeogenesis, p. 308).

Anabolic function of glycolysis. While glycolysis is essentially a catabolic pathway, it generates metabolites used for the biosynthesis of different molecules, including: (1) DHAP, from which glycerol-3-phosphate is formed in the synthesis of triacylglycerols and phospholipids, (2) 1,3-bisphosphoglycerate, a precursor of 2,3-bisphosphoglycerate (a hemoglobin modulator), and (3) pyruvate, which can be converted into the amino acid alanine by transamination.

Oxidative Decarboxylation of Pyruvate

When there is adequate oxygen supply, pyruvate produced in the glycolytic pathway is oxidized to carbon dioxide and water. Even the lactate formed in anaerobic conditions is converted into pyruvate by lactate dehydrogenase when oxygen is available. Thus, lactate resulting from intense muscular activity may be used as fuel.

Pyruvate formed in the cytosol as a result of glycolysis is degraded by oxidation within the mitochondria. This degradation requires pyruvate molecules to enter the mitochondrial matrix by traversing the inner mitochondrial membrane via carrier-mediated transport. The first step is oxidative decarboxylation, in which pyruvate loses the carboxyl group. CO_2 is released, and a two carbon compound (acetyl or acetate) is produced. The oxidative decarboxylation of pyruvate is catalyzed by a multienzyme system called *pyruvate dehydrogenase complex* (PDH). This molecular complex, of approximately 7000 kDa, is comprised of multiple copies of three enzymes: (1) *pyruvate decarboxylase* (E1), (2) *dihydrolipoyl transacetylase* (E2), and (3) *dihydrolipoyl dehydrogenase* (E3). The reaction intermediates remain bound to the enzyme complex, which is arranged in a manner which allows for the transfer of substrate from one enzyme to the next without release of partial products, making the whole system more efficient. This enzyme complex is an example of substrate channeling.

Five coenzymes are required for the activity of the PDH: (1) *thiamine pyrophosphate* (TPP), (2) *lipoic acid*, (3) *coenzyme A*, (4) *FAD*, and (5) *NAD*. Thiamine pyrophosphate (TPP) is derived from vitamin B_1 (thiamine) (p. 663). Lipoic acid (also called thioctic acid) is an eight carbon saturated fatty acid with sulfhydryl groups at carbons 6 and 8. Sulfhydryl groups are oxidized reversibly to form a disulfide (-S-S-) bridge. Presence of the sulfhydryl groups allows the lipoate to act as a hydrogen acceptor.

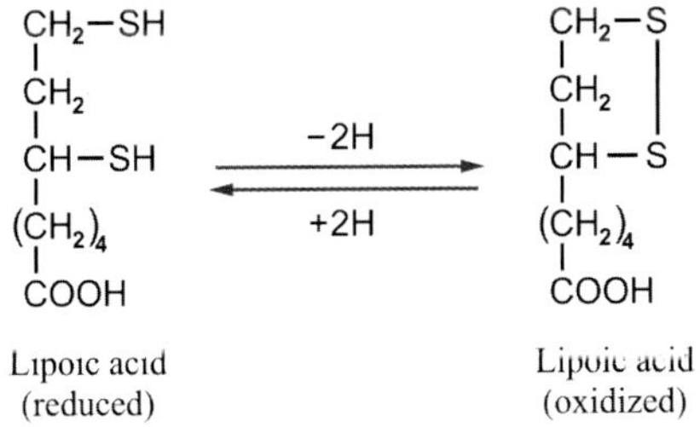

Lipoic acid (reduced)   Lipoic acid (oxidized)

Coenzyme A (CoA) has a nucleotide-based structure. It consists of adenine, ribose, two phosphate residues, pantothenic acid (vitamin B complex) (p. 671), and β-mercaptoethylamine. Coenzyme A functions as an acceptor and carrier of the acyl group, which is bound through a thioester bond with high free energy of hydrolysis.

Coenzymes NAD and FAD have been presented in Chapter 9.

Other components of the complex are protein kinase and a phosphoprotein phosphatase, which have regulatory roles.

The whole process can be summarized in the following equation:

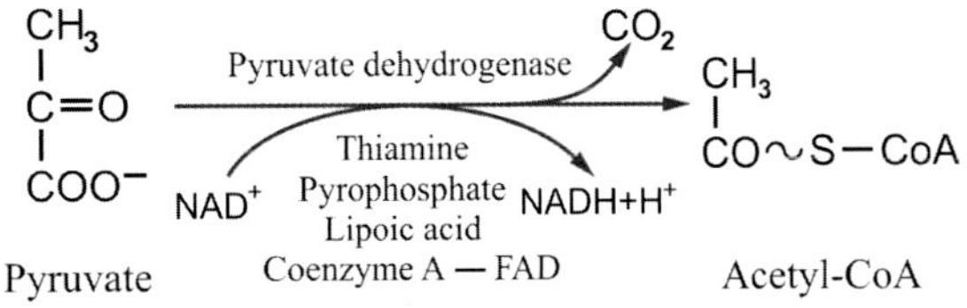

As shown, the reaction is complex, consisting of several steps (Fig. 14.6): (1) through the action of enzyme E1 (pyruvate decarboxylase), pyruvate loses its carboxyl group and CO_2 is released. Coenzyme TPP, firmly attached to E1, is the acceptor of the other two carbons. (2) The next couple of reactions are catalyzed by E2 (transacetylase dihydrolipoyl). Both carbon residues are oxidized to acetate by loss of two H^+ ions, which are accepted by lipoic acid. Then, acetate is transferred to CoA and acetyl-CoA is formed. (3) E3 (dihydrolipoyl dehydrogenase), an FAD-linked oxidoreductase, accepts dihydrolipoate hydrogens to regenerate lipoate. Finally, $FADH_2$ donates its reducing equivalents to NAD^+ resulting in the release of NADH + H^+ (Fig. 14.6). The process is irreversible under physiological conditions.

The pyruvate dehydrogenase complex is an important site of metabolic regulation. Its activity is modulated by different metabolites and by covalent modification (phosphorylation–dephosphorylation). Acetate originated from pyruvate decarboxylation is bound to coenzyme A by a high-energy thioester, to form acetyl-CoA (also called active acetate). The acetyl acquires high reactivity and can participate in further chemical transformations. The reduced NAD (NADH + H^+) generated in the reaction donates its hydrogen ions to the respiratory chain, and these eventually bind to oxygen to form water. Two hydrogen ions in the respiratory chain transferred from NAD generate 2.5–3 molecules of ATP from ADP (to simplify calculations, it can be considered that the oxidative decarboxylation of pyruvate produces 3 moles of ATP per mole of pyruvate).

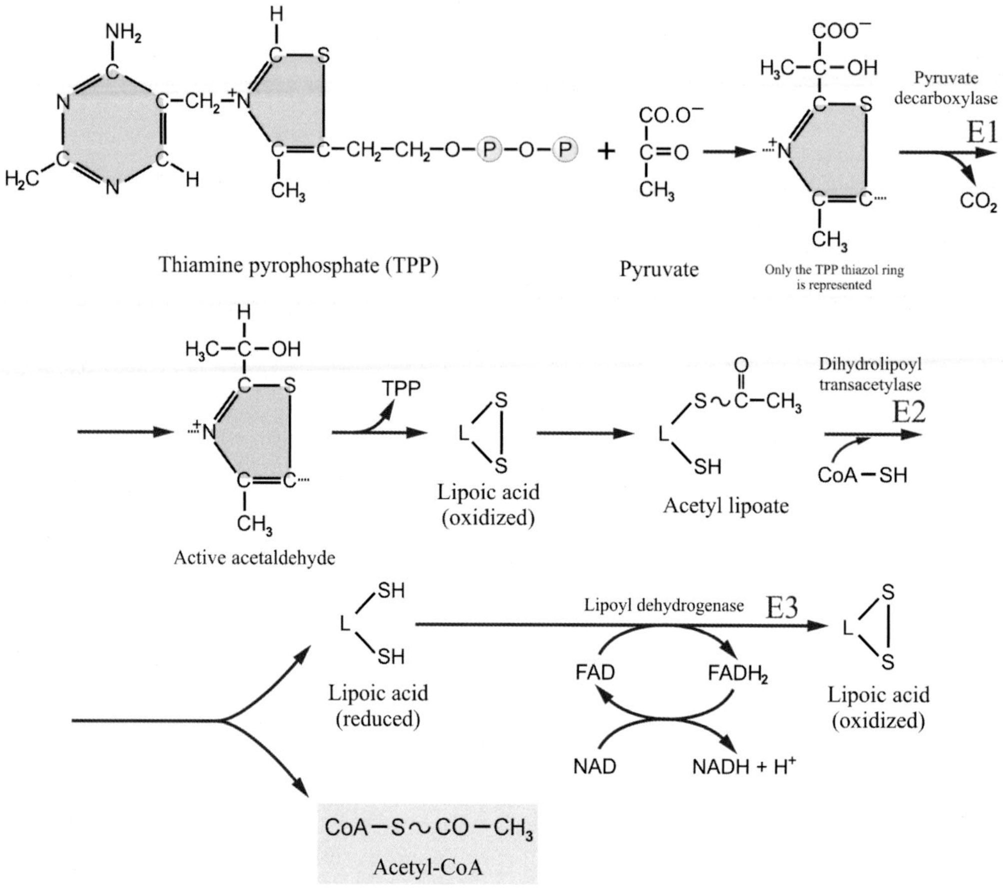

FIGURE 14.6   **Oxidative decarboxylation of pyruvate.**

Pyruvate is a molecule used in many different metabolic pathways. Its decarboxylation is the most important metabolic step under conditions of adequate oxygen supply. Activity of PDH is reduced or absent in certain conditions. Genetic deficiencies affect the synthesis of the system components. PDH-acquired deficiencies are observed in chronic alcoholism, vitamin B_1 deficiency (thiamine deficiency), and arsenite or mercury poisoning, which produce PDH inhibition. In all cases, the result is the overaccumulation of pyruvate, which is diverted to other pathways and mainly reduced to lactate or converted to alanine via transamination. Increased lactate production produces *lactic acidosis*, a type of metabolic acidosis characterized by low body pH, nausea, vomiting, hyperventilation, abdominal pain, and lethargy. Lactic acidosis can also be observed when oxygen supply is inadequate (hypoxia).

There is an E1 deficiency due to a chromosome X mutation. When there is a total absence of the enzyme, acetyl-CoA is not formed and the patients do not survive past the neonatal period. If the enzymatic activity is 25% or higher, the condition is less severe and results in reduction of ATP synthesis and lactic acid accumulation; the most conspicuous symptoms are psychomotor disturbances. In this case, a palliative treatment is a ketogenic diet, high in fats and proteins. The acetyl-CoA formed from these substrates can be oxidized in the citric acid cycle to produce ATP.

A genetic defect in the E3 component of the complex affects not only pyruvate dehydrogenase, but also other α-keto acid dehydrogenases (α-keto glutarate and α-keto acids derived from branched chain amino acids).

Citric Acid, Tricarboxylic Acid, or Krebs Cycle

Acetyl-coenzyme A is essential for oxidative metabolism and synthesis of many cell constituents. This intermediate is formed not only by decarboxylation of pyruvate, but also by oxidation of fatty acids and amino acid carbon chains. Furthermore, the remaining two carbons are used for synthesis of cholesterol, fatty acids, and other compounds. Acetyl-CoA is another important metabolic crossroad where many pathways converge.

The acetyl group is oxidized into CO_2 and H_2O via a metabolic cycle proposed in the 1930s by Hans Krebs, one of the most prominent biochemists of the 20th century. This is why the citric acid cycle, or TCA cycle, is also known as the Krebs cycle. This metabolic pathway is fully accomplished within the mitochondria and encompasses a series of reactions that lead to the total oxidation of acetate coming from different origins (carbohydrates, lipids, amino acids). Acetyl-CoA acts as a feeder compound of the cycle and initiates the reactions by combining with oxaloacetate. Oxaloacetate functions catalytically by oxidating the acetyl radical into two CO_2 molecules (products of the cycle). Oxaloacetate is regenerated at the end of the cycle, allowing the cycle to continue.

Citric Acid Cycle Reactions

1. *Citric acid formation.* Condensation of the acetyl-coenzyme A with the four carbon dicarboxylic acid intermediary oxaloacetate produces citrate (ionic form of citric acid), which is composed of six carbons and three carboxyl groups. This reaction is catalyzed by the condensing enzyme, *citrate synthase.* The activated state of the acetyl allows for the formation of a carbon—carbon bond between the acetate methyl and oxaloacetate carbonyl.

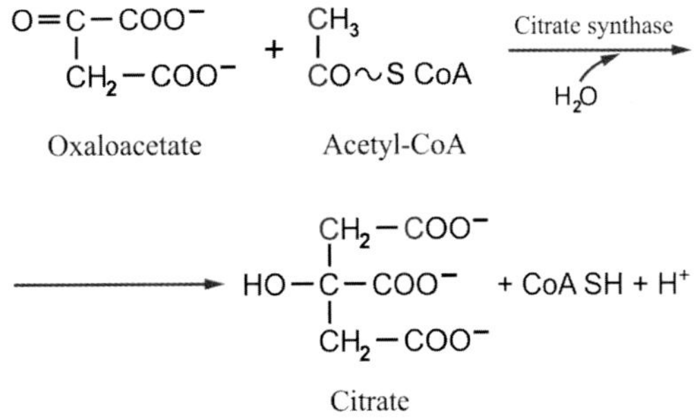

Oxaloacetate   Acetyl-CoA

Citrate

Carbon atoms of the acetyl group in red.

Due to the exergonic hydrolysis of the acetyl-CoA thioester bond, the reaction is strongly shifted toward citrate formation and is essentially irreversible. Citrate synthase is a regulatory step of the cycle, being inhibited by ATP.

2. *Isocitrate formation.* Citrate is converted into isocitrate by a two-step isomerization process. First it is dehydrated to *cis*-aconitate and then water is recovered to form isocitrate. Both steps are catalyzed by the same enzyme, *aconitase.*

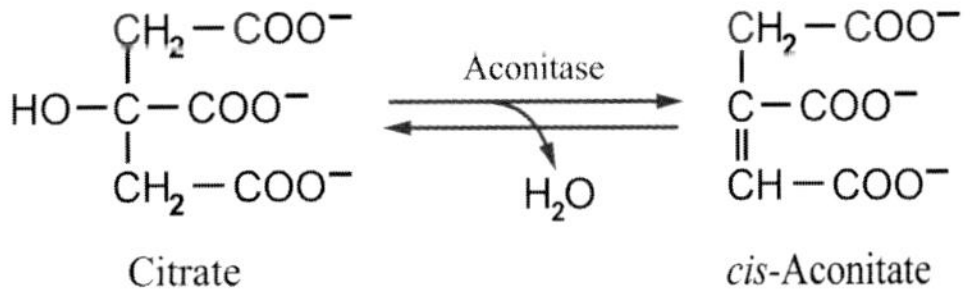

Citrate   cis-Aconitate

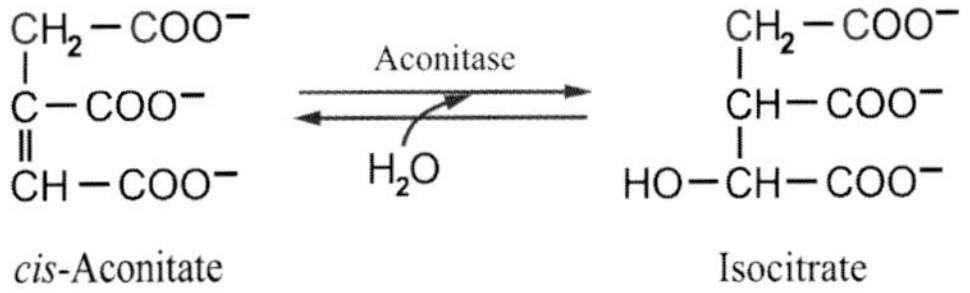

cis-Aconitate   Isocitrate

Aconitase has a dual role, functioning as an enzyme and as a modulator of gene activity. It has an iron–sulfur core, which upon release of the Fe molecule, allows it to function as a regulatory protein (IRP1). Once activated, aconitase upregulates the synthesis of proteins involved in iron homeostasis.

3. *Isocitrate oxidation.* Isocitrate undergoes dehydrogenation to become oxalosuccinate. *Isocitrate dehydrogenase* catalyzes this reaction. This enzyme is an oxidoreductase that uses NAD as a coenzyme and also requires Mg^{2+} or Mn^{2+}. Isocitrate dehydrogenase is an allosteric enzyme; its affinity for the substrate is increased by ADP and inhibited by ATP and NADH. This stage is considered the primary site for regulation of the cycle.

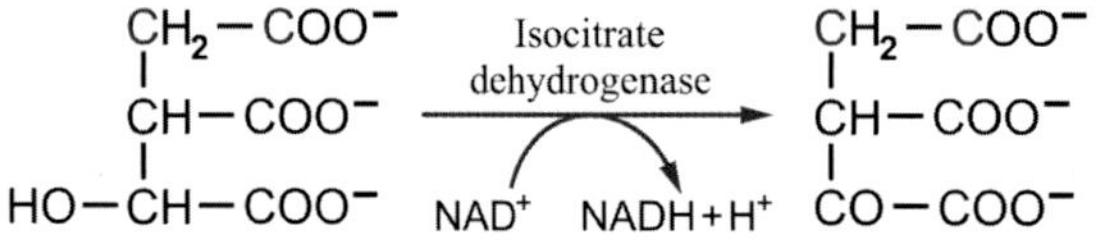

Isocitrate   Oxalosuccinate

In the cytosol, and in the mitochondrial matrix, there is another isocitrate dehydrogenase which is NAD independent, but instead regulated by NADP. This enzyme does not depend on NAD and is regulated differently.

Up to this point of the cycle, the intermediary metabolites have three carboxyl molecules, this gives this pathway its alternative name of TCA cycle.

4. *Oxalosuccinate decarboxylation.* The same enzyme responsible for the previous reaction, *isocitrate dehydrogenase*, catalyzes the decarboxylation of oxalosuccinate to give α-ketoglutarate.

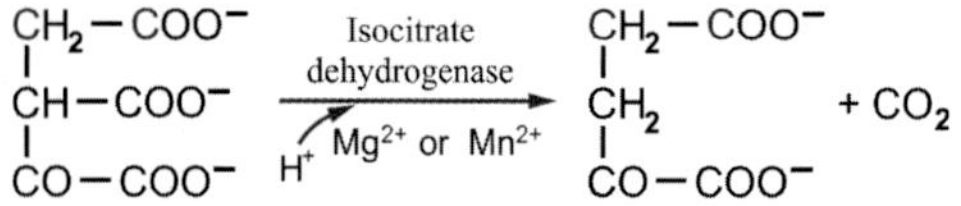

Oxalosuccinate   α-Ketoglutarate

The carbon molecule in the CO_2 released by the reaction does not correspond to any of the acetyl admitted in the first reaction

(shown in red). At this stage, the first carbon dioxide molecule is released and a five carbon dicarboxylic acid intermediary is generated.

5. *Oxidative decarboxylation of α-ketoglutarate.* This reaction is similar to the one described for pyruvate. The process is catalyzed by a multienzyme system called *α-ketoglutarate dehydrogenase complex*, which is composed of three enzymes that require thiamine pyrophosphate, lipoic acid, coenzyme A, FAD, and NAD as coenzymes. The mechanism of the reaction is similar to that of pyruvate decarboxylation. The reaction products are CO_2, NADH, H^+, and succinyl-CoA. This last compound is a four-carbon dicarboxylic residue bound to CoA by a high-energy thioester link. Briefly, the reaction can be represented as follows:

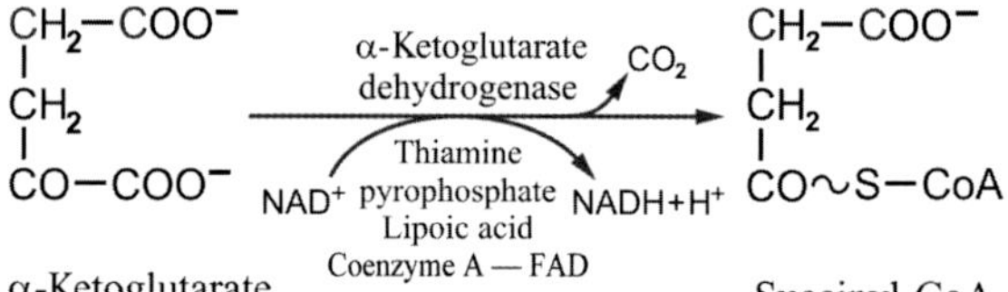

α-Ketoglutarate   Succinyl-CoA

The carbon in the CO_2 released in the reaction does not correspond to any of the acetyl group entered in the first reaction of the cycle (shown in red).

The reaction is strongly exergonic and practically irreversible.

6. *Succinate formation.* Succinyl-coenzyme A is converted to succinate and free CoA by action of *succinate thioquinase.* This reaction requires guanosine diphosphate (GDP) and inorganic phosphate (P_i). The energy in the thioester bond is used to transfer phosphate to GDP forming GTP.

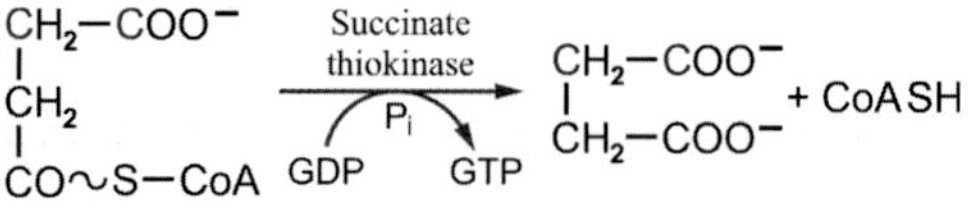

Succinyl-CoA   Succinate

This is the only stage of the cycle in which a high energy phosphate bond is generated at the substrate level. From GTP, ATP can be formed in a reaction catalyzed by *nucleoside diphosphate kinase*:

$$GTP + ADP \rightarrow GDP + ATP$$

7. *Succinate dehydrogenation.* Succinate is oxidized to fumarate by *succinate dehydrogenase*, a flavoprotein which uses FAD as a hydrogen acceptor. The enzyme has high specificity and produces only the *trans* isomer (fumarate). No maleate (*cis* configuration) is formed.

Succinate   Fumarate

Unlike other enzymes involved in the cycle, all of which are located in the mitochondrial matrix, succinate dehydrogenase is firmly attached to the inner membrane of the organelle. This enzyme is competitively inhibited by oxaloacetate.

8. *Fumarate hydration.* By addition of water, fumarate is converted to malate. The reaction is catalyzed by lyase *fumarate hydratase*, also called *fumarase*.

Fumarate   L-Malate

9. *Malate oxidation.* Malate loses two hydrogens and is transformed into oxaloacetate by the enzyme *malate dehydrogenase*, which is NAD-dependent.

L-Malate   Oxaloacetate

Carbons belonging to acetyl entered in the first reaction of the cycle (in red) are conserved in oxaloacetate.

This is an endergonic reaction with a positive $\Delta G^{\circ\prime}$. However, under physiological conditions, the continuous use of oxaloacetate drives the reaction to the right.

The cycle ends with the regeneration of oxaloacetate, the final and initial intermediary. During one complete cycle, two CO_2 molecules and eight hydrogen atoms are released. Three pairs of these hydrogens are transferred to NAD, and the remaining pair is transferred to FAD. In the respiratory chain, the four hydrogen pairs form four molecules of water by binding to oxygen.

Radioisotope studies have shown that the two carbons released as CO_2 do not belong to the acetate introduced into the cycle, but to the oxaloacetate. In subsequent cycles, the carbon atoms entered in the first cycle with the acetate will be released. Despite their origin, the two carbons released maintain the balance of the cycle and, for all practical purposes, they are considered the product of acetate oxidation.

Fig. 14.7 shows the stages of the cycle. The reactions are summarized in Table 14.3.

The resulting global equation is:

$$Acetyl\text{-}SCoA + 3NAD^+ + FAD + 2H_2O$$
$$+ GDP + P_i \rightarrow 2CO_2 + 3NADH + 3H^+ + FADH_2$$
$$+ CoASH + GTP$$

Genetic defects of citric acid cycle enzymes are rare. Mutations in any of the genes coding for aconitase, α-ketoglutarate dehydrogenase (α-KGDH), succinate dehydrogenase (SDH), or fumarase produce mainly neurological symptoms which may or may not be accompanied by muscular alterations (encephalopathy and Leigh syndrome, see p. 199) and lactic acidosis. SDH deficiency is known to cause cardiomyopathy. Patients with fumarase disturbances present with development of leiomyomas, tumors of the uterine smooth muscle. Symptoms of fumarase and α-KGDH defects appear before the first year of life. In contrast, SDH deficiency symptoms are observed after several years or in adulthood.

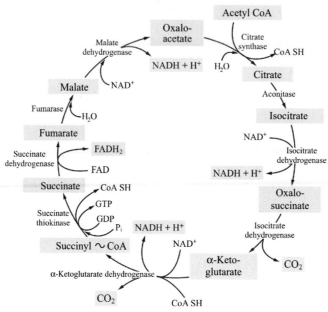

FIGURE 14.7   **Citric acid cycle.**

Alteration of the succinate dehydrogenase gene leads to adrenal tumors (pheochromocytomas).

General Considerations About the Citric Acid Cycle

Malate oxidation reaction (reaction 9) is endergonic; however, the rapid consumption of oxaloacetate in the first reaction of the cycle (1), shifts the reaction equilibrium to the right, determining the direction of metabolic flow.

The algebraic sum of all the citric acid cycle steps yields a $\Delta G^{\circ\prime}$ that is negative. The cycle's thermodynamic design favors its unidirectional operation. Reactions 1 (formation of citrate) and 5 (oxidative decarboxylation of α-ketoglutarate) are both strongly exergonic and are primarily responsible for the final result.

The intermediate metabolites of the cycle are continuously regenerated. The pathway is therefore autocatalytic, with the ability to generate its own substrates. The overall series of reactions shows the participation of a number of intermediate compounds, such as: acetyl-SCoA, NAD, FAD, GDP, P_i, and H_2O. The supply of these metabolites is critical for the proper function of the cycle.

Under normal conditions, there are no limitations in H_2O, P_i, and acetyl-SCoA supply in the cell. Acetyl-SCoA is not only formed from pyruvate but also from oxidative degradation of fatty acids, some amino acids, and other substances.

TABLE 14.3   Reactions of Citric Acid Cycle

1. Acetyl-SCoA + oxaloacetate + H_2O →
   citrate + CoASH + H^+

2. Citrate → *cis*-aconitate + H_2O → isocitrate

3. Isocitrate + NAD^+ → oxalosuccinate + NADH + H^+

4. Oxalosuccinate + H^+ → α-ketoglutarate + CO_2

5. α-Ketoglutarate + NAD^+ + CoASH →
   succinyl-SCoA + CO_2 + NAD + H^+

6. Succinyl-SCoA + GDP + P_i →
   succinate + GTP + CoASH

7. Succinate + FAD → fumarate + $FADH_2$

8. Fumarate + H_2O → L-malate

9. L-Malate + NAD^+ → oxaloacetate + NADH + H^+

The coenzymes NAD$^+$ and FAD are reduced in reactions 3, 5, 7, and 9; they must be reoxidized to maintain the cycle operational. The transfer of reducing equivalents from NADH and FADH$_2$ to the respiratory chain is also essential for normal cycle operation. In conclusion, this is a purely aerobic metabolic pathway that does not work in anaerobiosis. The enzymes used in the cycle are arranged conveniently in the mitochondria; this allows the coenzymes to easily deliver their hydrogen atoms to the electron transport chain included in the inner mitochondrial membrane.

The other compound needed by the cycle is GDP. This diphosphate nucleoside is generated within mitochondria in a reaction catalyzed by *nucleoside diphosphate kinase*:

$$GTP + ADP \rightarrow GDP + ATP$$

The enzymes of the citric acid cycle are found in the mitochondrial matrix (with the exception of succinate dehydrogenase, which is inserted in the inner mitochondrial membrane). However, they are not dispersed in free solution; in fact, evidence suggests that they form a multienzyme complex that ensures passage of an enzyme product from one enzyme to the next allowing for a more efficient operation of the system. This is another example of substrate channeling.

Citric Acid Cycle Functional Role

The description of the TCA cycle in this chapter does not mean it is an exclusive metabolic pathway for the final oxidation of carbohydrates. In fact, all the two carbon acetyl-CoAs that result either from carbohydrates, fatty acids, amino acids, or other substances are degraded into CO$_2$ and H$_2$O in this cycle. Therefore, the citric acid cycle is a major catabolic pathway and the *final common route* for oxidation of activated acetates, regardless of their origin. The citric acid cycle can also function as an anabolic pathway, with some of its intermediate metabolites participating in the syntheses of various products. Due to the importance of the intermediates of the citric acid cycle for anabolic purposes, the reactions responsible for replenishing those intermediates are essential. An example includes the conversion of pyruvate into oxaloacetate, which is catalyzed by *pyruvate carboxylase*. The reaction requires energy from the hydrolysis of ATP and the coenzyme *biotin*, a member of the vitamin B complex.

Pyruvate   Oxaloacetate

This is an important reaction feeding the citric acid cycle. Pyruvate carboxylase is activated allosterically by acetyl-CoA. The accumulation of active acetate promotes formation of oxaloacetate and stimulates the operation of the cycle. All the pathways that function as feeders of the citric acid cycle are called *anaplerotic*. Furthermore, there are *cataplerotic* reactions, such as that catalyzed by *phosphoenolpyruvate carboxykinase* (p. 309) which consumes oxaloacetate produced in the cycle.

Citric Acid Cycle Energy Balance

Oxidation of acetate in the citric acid cycle has high phosphoryl transfer potential and yields significant chemical energy. The oxidative reactions of the cycle donate the reduced hydrogen atoms from coenzymes to the respiratory chain. There, the flow of electrons is coupled to the pumping of protons from the mitochondrial matrix to the intermembrane space creating an electrochemical H$^+$ gradient. Return of H$^+$ through the ATP synthase channel provides the energy for ATP synthesis from ADP and P$_i$. Each pair of hydrogen ions transferred from NAD generates 2.5–3 molecules of ATP (in order to simplify calculations, factor 3 is used). The hydrogen atoms transferred from flavoproteins (FAD) produce 1.5–2 ATP (to simplify calculations, factor 2 is used). The total energy balance per mole of metabolized acetate of the citric acid cycle is summarized in Table 14.4.

TABLE 14.4   Energy Yield of Acetate Oxidation in the Citric Acid Cycle

Isocitrate → oxalosuccinate	(NADH)	3 mol ATP
α-Ketoglutarate → succinyl-CoA	(NADH)	3 mol ATP
Succinyl-CoA → succinate		1 mol ATP
Succinate → fumarate	(FADH₂)	2 mol ATP
Malate → oxaloacetate	(NADH)	3 mol ATP
Total per mole of acetate		*12 mol ATP*

TABLE 14.5   Energy Yield by Oxidation of 1 Mole of Glucose

Glycolysis	6–8 mol ATP[a]
Pyruvate oxidative decarboxylation	6 mol ATP
Citric acid cycle	24 mol ATP
Total yield	*36–38 mol ATP*

[a]According to the shuttle used to transfer H from cytosol to mitochondria.

Glucose Oxidation Energy Balance

Glycolysis. Anaerobically, each mole of glucose produces 2 moles of ATP. When there is adequate supply of oxygen, NAD reduced during oxidation of glyceraldehyde-3-phosphate transfers reducing equivalents from the cytosol to the respiratory chain by one of the shuttle systems (p. 199). Through this mechanism, the energy yield is either two (glycerophosphate shuttle) or three ATP (malate–aspartate shuttle). Two molecules of triose-phosphate produced per molecule of glucose yields 4–6 ATP. These, in addition to the 2 ATP made from glycolysis, gives a total of 6–8 molecules of ATP per glucose molecule.

Decarboxylation of pyruvate. Three ATPs are generated in the respiratory chain by transfer of reducing equivalents from reduced NAD. Each glucose molecule generates two molecules of pyruvate; thus ATP gain is 6 moles per mole of glucose.

Citric acid cycle. Oxidation of acetate yields a total of 12 ATP. One mole of glucose results in 2 moles of acetate, yielding a total of 24 moles of ATP.

Table 14.5 summarizes the total ATP production of 1 mole of glucose in oxidative catabolism.

Free energy of ATP hydrolysis is 7.3 kcal/mol (30.5 kJ/mol); therefore, the total energy captured in the form of ATP per mole of glucose is around 277 kcal (1159 kJ).

Combustion of 1 mole of glucose releases 686 kcal (2870 kJ). The efficiency of the oxidative pathway (percentage of the energy contained in the fuel utilized for work) in terms of energy obtained from glucose is approximately 40%. This is a remarkable difference when compared to a common combustion engine. Most engines barely reach 30% efficiency and lose energy as heat or useless frictions. This highlights how efficient living organisms are in utilizing fuel (Fig. 14.8).

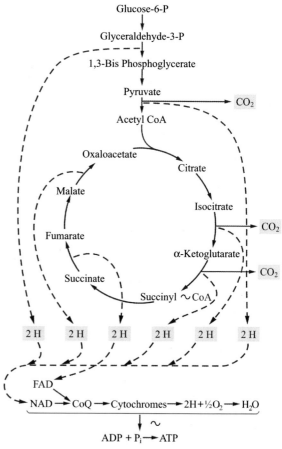

FIGURE 14.8   **Glucose oxidative catabolism.**

Hexose Monophosphate or Pentose Phosphate Pathway

In most tissues, 80% or more of glucose catabolism initially follows glycolysis. The rest enters an alternative pathway called the hexose monophosphate or pentose phosphate pathway. This pathway has two main functions: (1) to generate reduced nicotinamide adenine dinucleotide phosphate (NADPH), for future hydrogen ion donations in different syntheses, and (2) to produce pentose phosphate for the synthesis of nucleotides and nucleic acids.

The pentose phosphate pathway involves a series of reactions closely connected with glycolysis, as both processes have common intermediates (Fig. 14.9). It can be divided into two phases. In the first, G-6-P undergoes two oxidations, decarboxylation, and it is transformed into a pentose phosphate, ribulose-5-phosphate (CO_2 is released). These three reactions constitute the oxidative phase that is irreversible, producing all the NADPH generated by the pathway.

The second phase is nonoxidative, comprising a series of reversible reactions, in which aldoses and ketoses of 3–7 carbons are formed. Ribulose-5-phosphate gives two isomers: ribose-5-phosphate and xylulose-5-phosphate. These two combine and produce a triose-phosphate and a heptose-phosphate, which, in turn, generate hexosephosphate and tetrosephosphate. A new redistribution forms glyceraldehyde-3-phosphate and F-6-P, both of which are intermediates of glycolysis. All enzymes of this pathway are found in the cytosol.

Only the reactions of the first phase will be described.

1. *G-6-P oxidation.* G-6-P dehydrogenation is catalyzed by *G-6-P dehydrogenase*, which depends on NADP as hydrogen acceptor; it produces 6-phosphogluconolactone.

The enzyme is inhibited by NADPH, the regulatory mechanism adapting reduced coenzyme production to the demand.

2. *6-Phosphogluconate formation.* 6-Phosphogluconolactone is converted into 6-phosphogluconate by *lactone hydrolase* or *lactonase.*

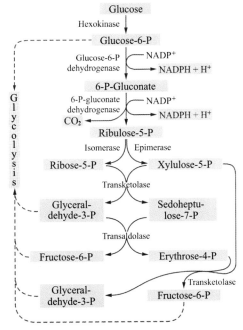

FIGURE 14.9 **Pentose phosphate pathway.**

3. *Phosphogluconate oxidation.* The second oxidative step is catalyzed by *6-phosphogluconate dehydrogenase* and is NADP-dependent.

The reaction products are ribulose-5-phosphate and carbon dioxide. The process takes two steps, with the formation of 3-keto-6-phosphogluconate as an intermediary.

$$
\begin{array}{ccc}
\text{COO}^- & & \text{COO}^- \\
| & & | \\
\text{H}-\text{C}-\text{OH} & \xrightarrow[\text{dehydrogenase}]{\text{6-Phosphogluconate}} & \text{H}-\text{C}-\text{OH} \\
| & & | \\
\text{HO}-\text{C}-\text{H} & & \text{C}=\text{O} \\
| & \text{NADP} \quad \text{NADPH}+\text{H}^+ & | \\
\text{H}-\text{C}-\text{OH} & & \text{H}-\text{C}-\text{OH} \\
| & & | \\
\text{H}-\text{C}-\text{OH} & & \text{H}-\text{C}-\text{OH} \\
| & & | \\
\text{CH}_2-\text{O}-\text{P} & & \text{CH}_2-\text{O}-\text{P}
\end{array}
$$

6-Phosphogluconate   3-Keto-6-phosphogluconate

$$
\begin{array}{ccc}
\text{COO}^- & & \text{CH}_2\text{OH} \\
| & & | \\
\text{H}-\text{C}-\text{OH} & \xrightarrow{\text{CO}_2} & \text{C}=\text{O} \\
| & & | \\
\text{C}=\text{O} & & \text{H}-\text{C}-\text{OH} \\
| & & | \\
\text{H}-\text{C}-\text{OH} & & \text{H}-\text{C}-\text{OH} \\
| & & | \\
\text{H}-\text{C}-\text{OH} & & \text{CH}_2-\text{O}-\text{P} \\
| & & \\
\text{CH}_2-\text{O}-\text{P} & &
\end{array}
$$

3-Keto-6-phosphogluconate   Ribulose-5-phosphate

Functional Significance of the Pentose Pathway

Although the amount of glucose metabolized by the pentose phosphate pathway is relatively small compared with that entering glycolysis, the operation of this alternative route is very important.

NADP hydrogens captured by the first phase steps are used in various processes: (1) fatty acid synthesis in liver, adipose tissue, and lactating mammary gland, (2) cholesterol and bile acid synthesis in liver, (3) steroid hormone synthesis in adrenal cortex, ovaries, and testes, and (4) cytochrome P_{450}-dependent biotransformation processes in the liver. The pentose phosphate pathway is highly active in all of the tissues mentioned earlier.

Theoretically, the NADPH can transfer reducing equivalents to NAD (reaction catalyzed by a transhydrogenase), which in turn are transferred to the respiratory chain to generate energy (under normal conditions), but preferably hydrogens from NADPH are derived to biosynthetic pathways.

The erythrocyte pentose phosphate pathway plays an important role. NADPH has a protective action against oxidizing agents. It helps to maintain the concentration of reduced glutathione near normal values (5 mM) and lower levels of methemoglobin. In neutrophils, and other phagocytes, reactive oxygen species are used to kill bacteria. One of these species, superoxide, is formed by reduction of oxygen catalyzed by *NADPH oxidase*, which requires NADPH generated in the pentose phosphate pathway.

Another function of the hexose monophosphate pathway is the production of ribose-5-phosphate, which is used for the synthesis of nucleotides and nucleic acids.

G-6-P dehydrogenase deficiency is the most common genetic enzymopathy in humans. It is frequent in African American individuals and citizens of Mediterranean basin countries. It is an inherited X-linked disorder characterized by red blood cell fragility and episodes of hemolysis when patients take certain drugs (primaquine derivatives, sulfonamides, nitrofurans, aspirin, and others) or foods (fava beans). Lack of NADPH significantly decreases the stability of erythrocytes and causes hemolytic anemia.

Other diseases produced by genetic defects in enzymes of the pentose phosphate pathway are very rare. The deficiency of transaldolase produces liver failure, the deficiency of ribose-5-P isomerase causes leukoencephalopathy and development delay, and the deficiency of xylitol dehydrogenase is considered benign; it is related to essential pentosuria.

Gluconeogenesis

Gluconeogenesis is the process of glucose and glycogen biosynthesis from noncarbohydrate sources. It is important to produce glucose when dietary carbohydrates are insufficient. Some tissues obtain energy from either carbohydrates or lipids, but they still require a basal supply of glucose. In contrast, nervous system tissue and red blood cells only use glucose. Under anaerobic conditions, glucose is the only energy source for skeletal muscle. Therefore, glucose production is essential for the body. In humans, when the external supply of glucose is insufficient, gluconeogenesis takes place mainly in the liver and kidney.

Gluconeogenesis is not simply the reverse path to glycolysis. In the pathway, there are irreversible reactions that do not allow using the same metabolic route backward. In gluconeogenesis, these reactions are carried out by different metabolic detour routes. These include:

1. *Pyruvate to phosphoenolpyruvate.* This conversion is performed by a detour route in which the following reactions are involved:

   a. Pyruvate is converted to oxaloacetate by *pyruvate carboxylase*, which requires biotin (a member of the vitamin B complex) and ATP as cofactors (p. 673). CO_2 is introduced to form a carboxyl. Oxaloacetate is an intermediate in the citric acid cycle, a reason why this is an important anaplerotic reaction that feeds that cycle. Pyruvate carboxylase is an allosteric enzyme, activated by acetyl-CoA.

   b. Oxaloacetate is converted to phosphoenolpyruvate by *phosphoenolpyruvate carboxykinase* with the release of CO_2; GTP is the phosphate and energy donor.

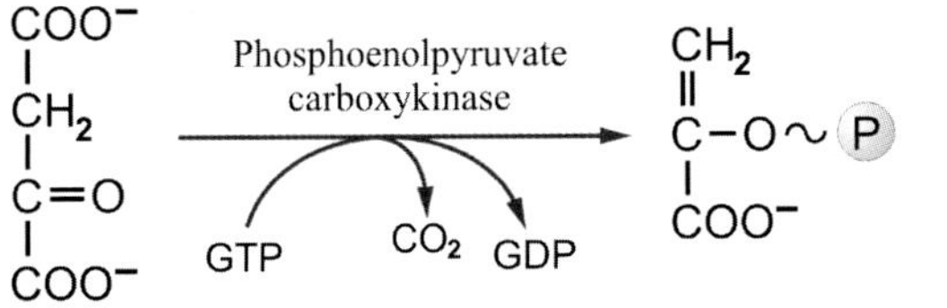

There are two isoforms of phosphoenolpyruvate carboxykinase. While many species have one of these forms localized in mitochondria, all cells have a cytosolic isozyme. Oxaloacetate is not able to cross the inner mitochondrial membrane; however, malate can permeate it, which influences the manner in which gluconeogenesis takes place. The reaction steps are the following: (1) pyruvate is converted into oxaloacetate (via pyruvate carboxylase). (2) Oxaloacetate is reduced to malate (by mitochondrial malate dehydrogenase). (3) Malate passes to the cytoplasm and is oxidized to oxaloacetate by the cytosolic isozyme of malate dehydrogenase. (4) Oxaloacetate is converted to phosphoenolpyruvate by phosphoenolpyruvate carboxykinase (Fig. 14.10). The first two reactions occur in the mitochondrial matrix, whereas the following reactions take place in the cytosol.

Oxaloacetate is a metabolic intermediate common to gluconeogenesis and the citric acid cycle. For this reason, any metabolite that acts as a feeder of the cycle can contribute to oxaloacetate formation and to the gluconeogenic pathway.

Lactate is a major gluconeogenic metabolite produced in anaerobic conditions; it is converted to pyruvate by lactate dehydrogenase.

Phosphoenolpyruvate carboxykinase participates in gluconeogenesis and also feeds into other pathways to produce glycerol from oxaloacetate (glyceroneogenesis), which is used in the synthesis of triacylglycerols and other molecules (e.g., serine). These pathways that utilize intermediates of the citric acid cycle are *cataplerotic*.

2. *Fructose-1,6-bisphosphate to F-6-P.* Fructose-1,6-bisphosphate is hydrolyzed, releasing the phosphate at carbon 1. The reaction is catalyzed by *bisphosphofructose phosphatase* and produces inorganic phosphate (P_i).

Oxaloacetate → Phosphoenolpyruvate
(Phosphoenolpyruvate carboxykinase; GTP, CO_2, GDP)

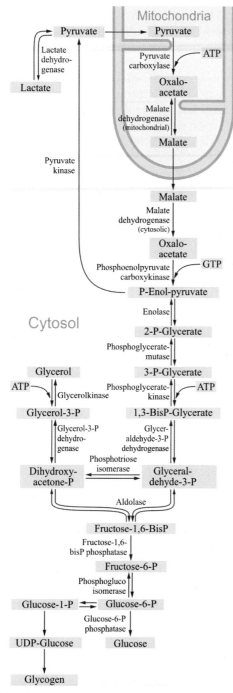

FIGURE 14.10 **Stages of gluconeogenesis.**

$$\text{Fructose-1,6-bis P} + H_2O \xrightarrow{\substack{\text{Bis-P-fructose} \\ \text{phosphatase}}}$$

$$\xrightarrow{\hspace{2cm}} \text{Fructose-6-P} + P_i$$

Bisphosphofructose phosphatase is an important regulatory enzyme.

3. *G-6-P to glucose.* This reaction is catalyzed by G-6-phosphatase, an enzyme found in the endoplasmic reticulum of liver, kidney, and intestinal cells. In these organs, G-6-P (formed in the cytosol) enters the endoplasmic reticulum through a specific carrier (T1) and is hydrolyzed by *G-6-*phosphatase. The products, glucose and P_i, return to the cytosol through two different carriers (T2 and T3, respectively). Glucose moves into the circulation by GLUT2 transporters. The specific location of G-6-P in the endoplasmic reticulum spatially isolates G6P from the glycolysis and pentose pathways. If the enzyme were in the cytosol, all the G-6-P would be hydrolyzed and these would interfere with the glycolytic and pentose pathways. Muscle and adipose tissue do not have G-6-P; this prevents these tissues from releasing glucose into the circulation.

The reactions of gluconeogenesis are presented in Fig. 14.10. In the conditions existing in the cell, the process is irreversible.

Deficiencies of genetic origin for the four enzymes of the gluconeogenic pathway (pyruvate carboxylase, phosphoenolpyruvate carboxykinase, fructose-1,6-bisPase, and G-6-P) are clinically characterized by hypoglycemia and lactic acidosis, which can be very severe and seriously affect the nervous system.

General Considerations About Gluconeogenesis

Lactate formed during anaerobic glycolysis enters the gluconeogenic pathway after oxidation to pyruvate by lactate dehydrogenase. After intense exercise, the lactate produced diffuses from the muscle into the blood and is taken up by the liver to be converted into glucose and glycogen.

Oxaloacetate is a common intermediary in the first reactions of gluconeogenesis and the citric acid cycle. All cycle intermediates and any compound producing it may become a glucose precursor. The carbon chains of some amino acids originate α-ketoglutarate, others produce succinate, fumarate, oxaloacetate, or pyruvate (p. 383) and can contribute to glucose formation.

Acetyl-CoA is not glucogenic. Practically, each acetate moiety entering the citric acid cycle is completely oxidized. Therefore, fatty acids degraded to acetyl-CoA in the organism are nonglucogenic. However, glycerol, another lipid component, is glucogenic. In liver tissue, for example, glycerol can be phosphorylated to glycerol-3-phosphate, which is subsequently oxidized to DHAP, and then oxidized. The triose-phosphate has two metabolic choices: (1) to follow the gluconeogenesis pathway by binding to glyceraldehyde-3-phosphate to yield fructose-1,6-bisphosphate or (2) to enter glycolysis to become glyceraldehyde-3-phosphate and 1,3-bisphosphoglycerate. The final destination is determined by the cell needs.

Gluconeogenesis Energy Cost

The formation of one glucose molecule from two of pyruvate or lactate is an endergonic process that requires energy. One ATP molecule is consumed for each molecule of pyruvate in the first reaction of gluconeogenesis, catalyzed by pyruvate carboxylase, one GTP is used in the phosphoenolpyruvate carboxykinase stage, and another ATP for the reverse reaction catalyzed by 3-phosphoglycerate kinase. In sum, synthesis of one molecule of glucose from two of pyruvate is coupled with the conversion of six molecules of ATP to ADP.

In the liver, lactate is converted to glucose, the energy is provided by oxidation of lactate to pyruvate. This reaction, catalyzed by lactate dehydrogenase, generates $NADH + H^+$. Hydrogen

transfer to the respiratory chain produces three ATP. Complete oxidation of 1 mole of lactate produces 18 moles of ATP (3 in the oxidation reaction by lactate dehydrogenase, 3 in the oxidative decarboxylation of pyruvate, and 12 in the oxidation of acetyl-SCoA in the citric acid cycle); this is sufficient energy for the synthesis of 3 moles of glucose.

Much of the oxygen consumed during the period following brief and intense activity (e.g., by an athlete in a 100 or 200 m race), is used in the synthesis of glycogen from glucose and lactate. The "oxygen debt" is paid after anaerobic exercise is performed. The liver then delivers glucose to the circulation. In muscle, this glucose reconstitutes the glycogen stores.

Cori's Cycle

Pyruvate formed in muscle by degradation of glycogen or glucose is oxidized into CO_2 and H_2O in the muscle itself when oxygen supply is sufficient. However, under conditions of intense contractile activity, oxygen supply falls short to support the oxidation needs of the tissue. Much of the pyruvate is reduced to lactate, which enters the blood and is taken up by the liver, where it is converted into glucose and glycogen. When blood sugar drops, liver breaks glycogen down and the produced glucose is released into the circulation, from where it is taken up by muscle to meet its needs and restore its glycogen stores. This closed metabolic cycle, called Cori's cycle, is summarized in Fig. 14.11.

METABOLISM OF OTHER HEXOSES

Fructose and galactose are components of commonly consumed foods. Normally, almost all the fructose and galactose that arrive via the portal vein from the intestine are taken up by the liver. After a moderate intake, very little of these monosaccharides is in the general circulation. Both undergo transformations in liver, generating metabolites equal to those produced from

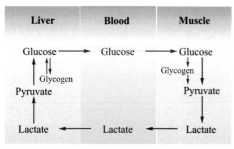

FIGURE 14.11   **Cori's cycle.**

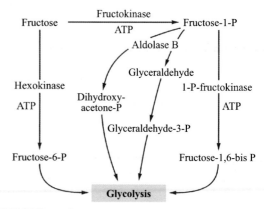

FIGURE 14.12   **Fructose metabolism.**

glucose; therefore the final destination of the three hexoses is the same.

Fructose Metabolism

For all monosaccharides, phosphorylation is the reaction previous to any further processing.

The principal utilization pathway begins with fructose phosphorylation at carbon 1 catalyzed by *fructokinase*, a very specific liver enzyme that transfers phosphoryl from ATP.

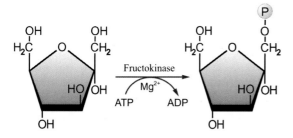

Fructose   Fructose-1-phosphate

Fructose-1-phosphate is cleaved between carbons 3 and 4 to give D-glyceraldehyde and DHAP. The reaction is catalyzed by *aldolase B* or *fructose-1-P aldolase*. This enzyme is different to aldolase A, located in muscle, which acts on fructose-1,6-bisphosphate.

Glyceraldehyde is phosphorylated to glyceraldehyde-3-phosphate (G3P) by an ATP-dependent *triosekinase.*

The three stages mentioned are catalyzed successively by *fructokinase, aldolase B,* and

triosekinase; they convert fructose into the same triose-phosphates formed in glycolysis. They can follow this pathway and finally be oxidized to CO_2 and H_2O to provide energy, or can be derived to gluconeogenesis and form glucose or glycogen (Fig. 14.12).

An alternative minor route for fructose is its phosphorylation at carbon 6, catalyzed by *hexokinase*. This enzyme has low affinity for fructose. F-6-P can then be incorporated into the glycolytic pathway.

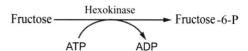

Fructose has synergistic effect with glucose metabolism. It increases affinity of glucokinase for glucose and stimulates its utilization in liver.

Fructose entry into the glycolytic pathway bypasses the limiting reactions of glycolysis (formation of G-6-P and F-1,6-bisP). Besides, F-1-P is an allosteric pyruvate kinase activator. This explains why lactate formation is faster from fructose than from glucose. An excessive intake of fructose may cause lactic acidosis. On the other hand, an overload of dietary fructose saturates the glycolytic pathway, which derives glyceraldehyde toward triacylglycerol formation and can consume ATP in hepatocytes, reducing other biosynthetic processes. Also, high fructose intake causes increase of uric acid in blood and urine.

There is fructose in human semen, which is used as an energy source for spermatozoa. Fructose is produced from glucose in the seminal vesicles. The process is accomplished in two steps: first, catalyzed by *aldol reductase* (NADPH dependent), which reduces carbon 1 of glucose and forms sorbitol. In the second, NAD-linked *sorbitol dehydrogenase* oxidizes the hydroxyl group on sorbitol carbon 2 and produces fructose. Both enzymes of this pathway are also found in the liver.

Several genetic diseases related to fructose metabolism have been described. One of them, called *fructose intolerance*, is due to deficiency of aldolase B. This disorder, which is inherited in an autosomal recessive manner, causes accumulation of F-1-P and fall of intracellular ATP and P_i concentrations. Oxidative phosphorylation and all metabolic processes that require energy are depressed. In addition, F-1-P inhibits phosphorylase and phosphoglucomutase, thus preventing glycogenolysis. Patients present with vomiting, failure to thrive, liver dysfunction, and hypoglycemia, even when there are glycogen stores in the liver. Symptoms occur when the infant stops breastfeeding and begins eating foods containing fructose or sucrose.

Deficiency in *aldolase A* has also been observed. This causes a delay in physical and mental development.

Other genetic defects that alter synthesis of fructose-1,6-bisphosphatase cause hypoglycemia, acidosis, and ketonuria. *Essential fructosuria* is an inherited disease due to lack of fructokinase. Patients eliminate fructose in urine; it generally has a benign course. Also, fructose malabsorption of genetic origin has been reported.

The treatment of these disorders consists in the elimination of fructose and sucrose from the diet.

Galactose Metabolism

Galactose is one of the products of lactose (milk sugar) hydrolysis in the intestine. The following reactions, that take place in the liver, transform galactose into metabolites identical to those of glucose:

1. *Galactose-l-phosphate formation.* The initial phosphorylation of galactose is catalyzed by *galactokinase*, an enzyme that uses ATP as a phosphate donor and energy source for the reaction. Phosphorylation takes place by phosphate esterification of galactose carbon 1.

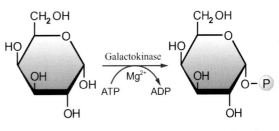

Galactose   Galactose-1-phosphate

2. *UDP-galactose formation.* Galactose-1-phosphate reacts with uridine diphosphate glucose (UDPGlc) to form UDP-galactose (UDP-Gal) and glucose-l-phosphate. In this reaction, catalyzed by *galactose-1-phosphate uridyltransferase*, galactose replaces glucose binding to UDP.

$$\text{Galactose-1-P} + \text{UDPGlc} \xrightarrow{\text{Gal-1-P-uridyl transferase}} \text{UDP-Galactose} + \text{G-1-P}$$

3. *UDP-glucose formation.* UDP-galactose is converted into UDP-glucose by an epimerase (*UDP-galactose 4-epimerase*), NAD-linked. The reaction comprises oxidation of the ketone at carbon 4, and then reduction to reform the hydroxyl with an inverted configuration.

$$\text{UDP-Galactose} \xrightarrow{\text{Epimerase}} \text{UDP-Glucose}$$

UDP-glucose is the intermediary for glycogen synthesis, sugar conversion, and glycosylation reactions. The mammary gland during lactation synthesizes galactose from UDP glucose, following reactions that are reversed to that of the pathway described.

There is an autosomally recessive inherited disease called *galactosemia*, characterized by inability to metabolize galactose. The most common cause of galactosemia is lack of galactose-l-phosphate uridyltransferase activity. Accumulation of galactose-1-P in the cells and increased galactose in blood and tissues causes serious damage in the nervous system, liver, and other organs. Excess galactose also produces lens opacification in eyes (cataracts). Galactosemia can also be caused by lack of UDP-galactose epimerase.

Another genetic disease related to galactose metabolism is produced by galactokinase deficiency.

To prevent the serious disorders caused by these diseases, milk and dairy products need to be suppressed from the diet. Early diagnosis, before the damage is irreversible, is therefore essential.

Blood Glucose

Glucose levels in venous blood of normal individuals during fasting periods are between 70 and 110 mg/dL or 100 mL. Values are practically the same when measured in plasma, serum, or whole blood. Blood glucose (glycemia) transiently increases in the postprandial period to maximum levels half an hour after food ingestion. Two or three hours later, glucose levels return to the fasting baseline. To detect possible disturbances of hexose metabolism, the glucose tolerance test is performed. For this, the patient is given a glucose load (50 g) after fasting for 8 h, the evolution of glycemia is determined during the following 3–4 h. In normal individuals, the increase in blood glucose concentration reaches a maximum (no more than 170 mg/dL) between 30 and 60 min. After 2 h, it goes down to its initial values.

Glucose is filtered in the renal glomeruli and is fully reabsorbed in the tubules by secondary active transport (via SGLT-2); the tubules reabsorb up to 350 mg of glucose per minute (glucose T_m). Normal urine does not contain detectable amounts of glucose; all filtered glucose is reabsorbed back to the blood. When blood glucose levels exceed 160–170 mg/dL, renal reabsorption capacity is exceeded and glucose appears in urine (*glucosuria*). A glycemia of 160–170 mg/dL corresponds to the renal threshold for glucose.

Blood glucose homeostasis. The constancy of glucose plasma levels reveals the existence of regulatory mechanisms that ensure its maintenance within narrow limits. Glucose homeostasis depends on the balance between the processes that release glucose into blood and those removing it from blood.

Among the metabolic processes that maintain blood glucose constancy are the following:

1. *Glycogenesis and glycogenolysis regulation in the liver.* When there is an excess of glucose in blood, glycogenesis is stimulated, glycogen storage is favored, and glucose is removed from circulation. When blood sugars drop below normal levels, hepatic glycogenolysis is activated and glucose is released to the general circulation.
2. *Glycogenesis and glucose utilization in muscle and other tissues.* Glycogen formation in muscle tissue and glycolysis are processes that tend to lower blood glucose.
3. *Conversion of glucose into other substances.* Glucose transformation, mainly in fat, reduces blood glucose levels.
4. *Gluconeogenesis.* Glucose formation from amino acids and other compounds provide glucose to the circulation.

All of these processes require the action of a number of hormones. Their mode of action will be discussed in Chapter 26. Insulin secreted by the pancreas exerts mechanisms that decrease blood glucose levels. Other hormones produced in the anterior pituitary, adrenal cortex and medulla, thyroid, and glucagon from the pancreas exert actions that increase blood glucose. Blood glucose concentration above the normal values (greater than 120 mg/dL) constitutes a symptom called *hyperglycemia*. Values less than 70 mg/dL correspond to *hypoglycemia*.

Glycemia disturbances. When blood glucose drops below normal levels, glucose receptors in the hypothalamus trigger signals that stimulate secretion of hormones that act to restore glucose levels. Conversely, hyperglycemia leads to increased insulin release.

Hypoglycemia occurs in prolonged fasting, or excessive administration of insulin or drugs, such as sulfonylureas, which promote insulin secretion. Hypoglycemia is also caused by excessive alcohol ingestion and by certain diseases, such as tumors of the Langerhans' islet β cells of the pancreas (insulinomas).

Hypoglycemia produces a number of symptoms. When it is not very pronounced, it is manifested by weakness, sweating, palpitations, tachycardia, and tremor. Most severe conditions, with blood glucose levels below 50 mg/dL, are accompanied by neuroglycemic symptoms, including: headache, hypothermia, visual disturbances, mental depression, loss of consciousness, and seizures.

Hyperglycemia is a symptom that characterizes diabetes mellitus.

Glycemic index. The relationship between increase in blood glucose produced by ingestion of a certain amount of carbohydrates and that caused by oral administration of an equivalent amount of glucose, gives a value called glycemic index (GI). This index can range from 0 to 100, indicating the increase in blood sugar levels in comparison to the amount of carbohydrate ingested. Values near zero indicate no glucose absorption from the carbohydrate ingested with the diet. Values closer to 100 indicate full absorption. In medical practice, glycemic index is measured by determining the area above the baseline of the curve obtained by plotting blood glucose levels measured every 30 min during 3 h after ingestion of a meal containing 50 g of carbohydrate in the Y axis, versus time in the X axis. This area, considered the test meal, is compared with that obtained after ingestion of 50 g glucose (reference meal). The surface area corresponding to the test meal is divided by that of glucose reference and multiplied by 100. Galactose/maltose and lactose/trehalose disaccharides have a GI of 100. Fructose and sucrose give values below 100. For starch, the value varies according to its state in the incoming food, which influences its capacity of being digested. If most of the starch is digested, the GI approaches 100; however, if it is in a digestion-resistant state (p. 264), GI is lower than 100. For polysaccharides that cannot be digested, the GI value is zero.

More recently, another parameter to measure the effect of a food on blood glucose levels is used. This is called *glycemic impact*. Glycemic impact is the amount of glucose (in grams) that produces a response equal to that induced by a given quantity of food.

The glycemic index reflects the effect of carbohydrates contained in food on blood glucose levels compared to that produced by an equal quantity of glucose, it is a relative value. Glycemic impact is expressed as the grams of glucose that cause a glycemic increase equivalent to a given ingested amount of food and not to their carbohydrates.

Nucleotide Sugars

In glycogenesis and galactose metabolism, nucleoside diphosphate sugar (UDP-glucose and UDP-galactose) formation was mentioned. Leloir and coworkers were the first to recognize the functional relevance of these types of compounds. Since their discovery in 1949, abundant evidence demonstrating the role of nucleotide sugars in metabolic processes has accumulated. Many compounds of this class have been isolated. Carbon 1 of the sugar is esterified with the nucleotide terminal phosphate (Fig. 14.13). Except for cytidine-monophosphate-acetylneuraminic (CMP-NANA), all other known nucleotide sugars are nucleoside diphosphates.

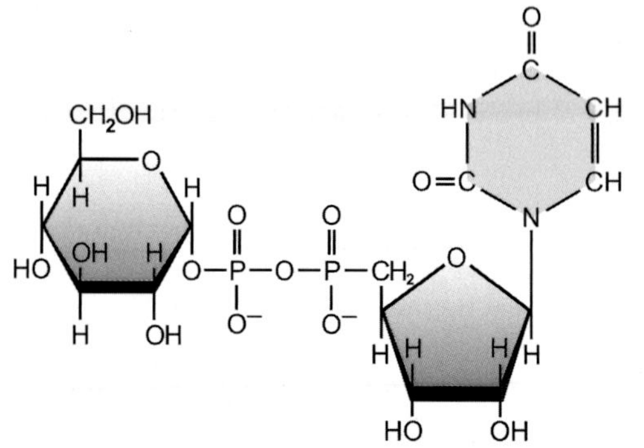

FIGURE 14.13 **Uridinediphosphate glucose (sugar nucleotide).**

Nucleotide sugars are mainly involved in the transfer of monosaccharides and glycosidic moieties. These substances are activated forms of monosaccharides; the carbohydrate moiety acquires high chemical reactivity, which enables it to intervene in various reactions. Examples of these reactions are: (1) epimerization of the galactosyl residue of UDP-galactose to give UDP-glucose, (2) oxidation of carbon 6 of the glycosyl residue of UDP-glucose to form UDP-glucuronate, and (3) conversion of GDP-mannose into GDP-fucose.

The role of nucleotide sugars in monosaccharide transfer is important for the polymerization of carbohydrates, such as poly- and oligosaccharides.

Active glucuronate, as UDP-glucuronate, is used to conjugate steroid hormones, bilirubin, many drugs, and foreign substances. These reactions are catalyzed by *glucuronyl transferase.*

The mammary gland synthesizes lactose by binding glucose to the galactosyl moiety of UDP-galactose.

BIOSYNTHESIS OF THE OLIGOSACCHARIDES OF GLYCOPROTEINS

The oligosaccharide portion of glycoproteins are bound to the apoprotein by two types of links: (1) *O-glycosidic bond*, in which carbon 1 of N-acetylated galactosamine binds with the O of the hydroxyl function of serine (Ser) or threonine (Thr) residues in the polypeptide chain; and (2) *N-glycosidic bond*, where C1 of N-acetylated glucosamine is attached to the *N*-amide of an asparagine residue (Asn) in the protein. Biosynthesis of each type of oligosaccharide follows different pathways. Both synthetic routes for glycoproteins need glycosyl transferase enzymes to catalyze the transfer of the sugar moieties from a donor to an acceptor molecule. The donor is usually a nucleoside diphosphate sugar. For example, UDP is used to carry glucose (Glc), galactose (Gal), N-acetylated glucosamine (GlcNAc), and N-acetylated galactosamine (GalNAc); GDP is a mannose (Man) and fucose (Fuc) carrier; and CMP transfers *N*-acetyl neuraminate (NAN). During synthesis of *N*-oligosaccharides, phosphorylated polyprenol, dolichol phosphate (Dol-P) (p. 115), also participates as a monosaccharide carrier. Glycosyl transferases have remarkable specificity for the sugar, the substrate on which the carbohydrate will be inserted, and the type of bond to be formed. Each of the glycosidic linkages in oligosaccharides requires a different enzyme. The N-linked oligosaccharide chains undergo processing involving the removal of sugar moieties catalyzed by glycosidases. Most glycosidases and glycosyl transferases are bound to membranes of the endoplasmic reticulum and Golgi apparatus.

Biosynthesis of Oligosaccharides Linked to Serine or Threonine (O)

Examples of the structure of O-linked oligosaccharides are shown in Fig. 4.22. Their synthesis is achieved by the orderly addition of monosaccharides, which is determined by the specificity of the glycosyl transferases involved. The first residue, generally N-acetylated galactosamine, is attached to the hydroxyl of serine or threonine. Then, the other carbohydrate molecules, such as galactose and fucose, which are donated by UDP-Gal and GDP-Fuc, respectively, are successively added. Bonds are α or β glycosidic and join C1 of a monosaccharide to C2, C3, C4, or C6 of the previous unit. When N-acetyl neuraminic acid (NANA) is inserted, monosaccharide addition ceases. NANA appears to be the signal to end glycosylation. In most cells, the O-linked glycosylation is carried out in the Golgi complex.

Biosynthesis of Asparagine-Linked Oligosaccharides (N-Oligosaccharides)

The process of N-glycosylation is more complex than that of O-glycosylation. A first series of reactions are required in which a precursor chain is assembled on a lipid support, *dolichol phosphate*, which is anchored to the endoplasmic reticulum membranes. Once the precursor is formed, the sugar is transferred from dolichol-P to an asparagine residue of the protein. From this step, the precursor undergoes a series of changes that comprise the removal of certain groups and the addition of others. Finally, the chain acquires the distinctive characteristics of the different types of N-oligosaccharides (rich in mannose, complex, and hybrid carbohydrates).

1. *Precursor synthesis.* The carrier molecule is dolichol phosphate. Synthesis is initiated with the addition of GlcNAc-P to dolichol phosphate. A pyrophosphate (GlcNAc-P-P-Dol) is formed. Another GlcNAc and five mannoses, transferred from UDP-GlcNAc and GDPMan, respectively, are then added. In successive stages, four mannose and three glucose residues, donated by dolichol-P-Man and dolichol-P-Glc are added. A branched oligosaccharide, composed of $(Gluc)_3$-$(Man)_9$-$(GlcNAc)_2$ is assembled and remains attached to dolichol-pyrophosphate (Fig. 14.13, see types of bonds in Fig. 4.21). This chain is then transferred from dolichol-

P-P to a protein asparagine residue. The reaction is catalyzed by *oligosaccharidyl transferase*. To function as an oligosaccharide receptor, the asparagine residue in the protein must be part of a sequence Asn-X-Ser/Thr, where X can be any amino acid.

Dolichol phosphate is released and can be reused (Dol-P cycle, Fig. 14.14).

2. *Precursor processing.* The $(Glc)_3$-$(Man)_9$-$(GlcNAc)_2$- chain attached to the N of an asparagine amide residue undergoes various modifications. The three glucose residues are removed by specific glucosidases, and an α-mannosidase removes a mannose, leaving the $(Man)_8$-$(GlcNAc)_2$-oligosaccharide. Up to this point all steps occur in the endoplasmic reticulum. From here the precursor is transferred in vesicles to the Golgi complex, where the final changes are carried out. After loss of the three glucose residues from the precursor, high mannose oligosaccharides undergo the removal of a variable number of mannose residues (up to four); the end product can have five to nine mannose residues (Fig. 14.14). Hydrolysis of mannoses is catalyzed by *mannosidase*.

Complex and hybrid oligosaccharides result from different processing of the intermediate precursor. For complex chains, mannosidase removes mannose residues and leaves the basic structure $(Man)_3$-$(GlcNAc)_2$, common to all types of N-linked oligosaccharides (Fig. 4.21). Other sugars are added to it, branches are elongated and sometimes new branches are created. Insertion of each unit is catalyzed by a specific glycosyl transferase and monosaccharide donors are the respective nucleotide sugars. This mechanism produces a wide variety of oligosaccharides. The synthesized carbohydrate strand is characteristic for each glycoprotein. Fig. 14.15 shows the different stages of the glycosylation process.

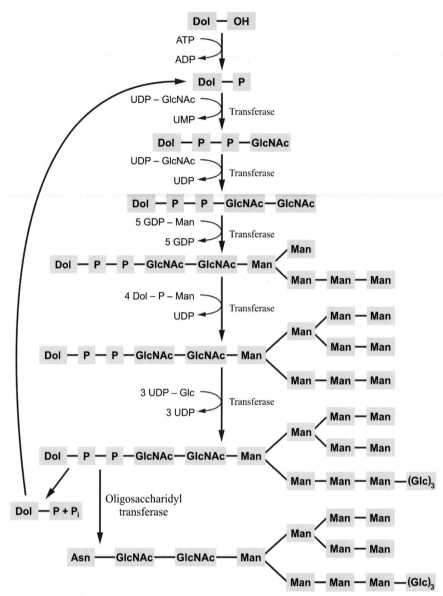

FIGURE 14.14 **Steps of oligosaccharide precursors linked to *N*-amide of asparagine in glycoproteins biosynthesis.** *Dol-OH*, Dolichol; *GlcNAc*, N-acetyl glucosamine; *Man*, mannose; *Glc*, glucose; *Asn*, asparagine; *UDP*, uridine diphosphate; *UMP*, uridine monophosphate; *GDP*, guanosine diphosphate.

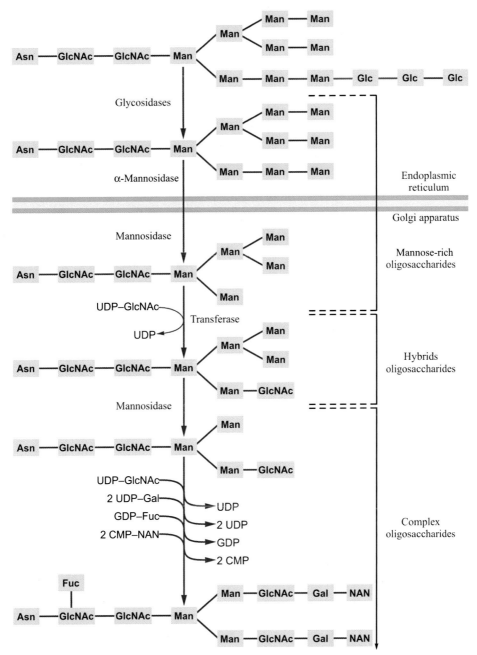

FIGURE 14.15   **Simplified representation of the asparagine-linked polysaccharides precursor processing.** See abbreviations in Fig. 14.14 legend. *Gal*, galactose; *CMP*, cytidine monophosphate; *NAN*, N-acetyl-neuraminate.

TABLE 14.6   Mucopolysaccharidoses

Type	Disease	Deficient enzyme	Symptoms
IH	Hurler	α-ʟ-Iduronidase	Severe mental deficiency, skeletal malformations, severe somatic changes, corneal opacity, dermatan sulfate, and heparan sulfate in tissue and urine.
IS	Scheie	α-ʟ-Iduronidase	Mild skeletal alterations, normal mental status, corneal opacity, and dermatan sulfate in urine.
II	Hunter (two types)	Iduronate sulfatase	Moderate mental deficiency, severe skeletal malformations, deafness, somatic changes, dermatan sulfate, and heparan sulfate in tissue and urine.
III	Sanfilippo (four types)	1. Heparan *N*-sulfatase 2. *N*-acetylglucose aminidase 3. *N*-acetyltransferase 4. *N*-acetylglucose amine-6-sulfatase	Abnormal glycogen (long chains with very little branches), early death due to cardiac or liver insufficiency.
IV	Morquio (two types)	1. Galactose-6-sulfatase 2. β-Galactosidase	Severe skeletal malformations, normal mental status, keratan sulfate in urine.
VI	Maroteaux–Lamy	*N*-acetylgalactose-amine-4-sulfatase (arylsulfatase B)	Severe skeletal malformations, corneal opacity, normal mental status, dermatan sulfate in urine.
VII	Sly	β-Glucuronidase	Mental deficiency, dermatan sulfate, and heparan sulfate in urine.

Degradation of Proteoglycans and Glycoproteins

Having reached the end of their useful life, proteoglycans and glycoproteins are incorporated into the cells by endocytosis. Endocytic vesicles fuse with lysosomes that contain various hydrolases, which degrade the internalized material. The protein is reduced to amino acids by *cathepsin* and the carbohydrate moiety is subjected to the action of *endo- and exoglycosidases*. The endoglycosidase (e.g., *hyaluronidase*) cleaves inside of oligosaccharide chains and subsequently exoglycosidases, specific for each type of bond, remove terminal residues from the nonreducing end, one at a time.

There is a group of diseases, known as *mucopolysaccharidoses*, which are due to genetic defects in glycosidase synthesis. The inability to degrade heteropolysaccharides in these conditions determines their accumulation in tissues. These alterations are included within the lysosomal storage disorders. Some of them are presented in Table 14.6.

SUMMARY

Glucose absorption in the intestine is mediated via the Na⁺-glucose cotransporter (SGLT-1), which is dependent on the sodium gradient generated by Na⁺,K⁺-ATPase. Renal tubules also absorb glucose from the filtrate by a secondary active transport system (SGLT-2). All other glucose transport in cells takes place by facilitated diffusion mediated by GLUT family proteins, with each member having selective tissue distribution and functional properties.

Glucose phosphorylation is the initial step in all hexose utilization pathways. The phosphate is esterified at C6 producing G-6-P in a reaction catalyzed by *hexokinases*. Hexokinases I–III have a K_m for glucose between 0.01 and 0.1 mM, which is below the usual levels of glucose in blood and tissues (~5 mM). This ensures continued activity of the enzyme and utilization of glucose by cells. Hexokinases I–III are inhibited by their product (G-6-P). Hexokinase IV or *glucokinase* is in liver and pancreatic β cells. Its K_m for glucose is 10 mM, its activity is adjusted to the amount of glucose available and phosphorylates glucose only when blood levels increase, such as after a meal. It is not inhibited by G-6-P. Hexokinases require ATP and Mg^{2+}. Under physiological conditions, the reaction catalyzed by hexokinases is irreversible.

Glycogenesis is the process by which glycogen is formed. This occurs in many tissues, but mainly in liver and muscle. It is accomplished through several stages:

1. *Glucose phosphorylation.* Hexokinase catalyzes conversion of glucose in G-6-P.
2. *Glucose-1-phosphate formation: phosphoglucomutase* converts G-6-P into G-1-P. It requires Mg^{2+} and cofactor G-1,6-bisP. This is a reversible reaction.
3. *Uridine diphosphate glucose formation: UDP-glucose pyrophosphorylase* catalyzes the formation of UDP-glucose from UTP and G-1-P. This irreversible reaction allows glucose to be activated.
4. *Glucose addition to the polymer: Glycogen synthase* or *glucosyl transferase* and preexisting glycogen are required. Glucoses are bound by α1→4 glycosidic linkages forming linear chains in an irreversible reaction.
5. *Branch formation: amylo-α(1,4)→α(1,6)-glucantransferase* or *branching enzyme.* It transfers straight chain segments of ~6 glucose molecules long and inserts them into a neighbor chain via α1→6 binding.

When there are no previous glycogen stores, *glycogenin* (a synthesis initiator protein) acts autocatalytically to form a linear chain of 6–7 glucose molecules that are bound by α1→4 bonds, anchored to a protein tyrosine residue. This initial chain is the substrate for the action of glycogen synthase and the branching enzyme.

Glycogenolysis, the process of glycogen breakdown, is not simply performed by reversal of the glycogenesis reactions. It is accomplished through the following stages:

1. *Glycogen phosphorolysis. Phosphorylase* catalyzes the cleavage of α1→4 glycosidic linkages introducing phosphate at C1 of glucose residues. It releases G-1-P from the chain up to approximately 4 units before an α1→6 bond.
2. *Oligo-α(1,4)→α(1,4)-glucantransferase* releases the terminal trisaccharide of the four glucoses branch and transfers it to the end of a neighbor branch by a α1→4 bond.
3. *Hydrolysis of α1→6 bonds.* Catalyzed by *oligo-α1→6-glucosidase* or *debranching enzyme,* which releases free glucose.
4. *G-6-P formation. Phosphoglucomutase* converts G-1-P in G-6-P.

5. *Glucose formation. G-6-P,* in an irreversible reaction, catalyzes the hydrolysis of G-6-P into glucose and phosphate. This enzyme is found in the endoplasmic reticulum of liver, kidney, and intestinal tissues; it is not expressed in muscle.

Glycolysis is the primary pathway for glucose catabolism, also called Embden–Meyerhof pathway. All enzymes of this pathway are found in the cytosol.

Its reactions include the following:

1. *G-6-P formation.* Catalyzed through an irreversible reaction by *hexokinase.* When glycolysis starts from glycogen, phosphorylase and phosphoglucomutase are required to obtain G-6-P.
2. *F-6-P formation. Phosphoglucoisomerase* converts G-6-P to F-6-P, requires Mg^{2+} or Mn^{2+} in a reversible reaction.
3. *F-6-P phosphorylation. Phosphofructokinase* catalyzes phosphate addition to F-6-P to give F-1,6-bisP, requires ATP and Mg^{2+}. This irreversible reaction constitutes an important regulatory step.
4. *Triose-phosphate formation. Aldolase,* in a reversible reaction, catalyzes cleavage of F-1,6-bisP into glyceraldehyde-3-P (G3P) and DHAP.
5. *Triose-phosphates interconversion. Triose-phosphate isomerase* converts DHAP to G3P in a reversible reaction.
6. *Oxidation and phosphorylation of glyceraldehyde-3-P. Glyceraldehyde-3-P dehydrogenase* uses NAD and requires P_i. It forms 1,3-bisP-glycerate, an energy-rich compound.
7. *3-Phosphoglycerate formation, phosphoglycerate kinase.* Phosphoryl is transferred to ADP to form ATP. It is a substrate-level phosphorylation. Together, reactions 6 and 7 are reversible.
8. *2-Phosphoglycerate formation. Phosphoglycerate mutase,* 3-phosphoglycerate is converted into 2-phosphoglycerate. It requires Mg^{2+}. This step is also reversible.
9. *Phosphoenolpyruvate formation. Enolase* catalyzes dehydration and conversion of 2-phosphoglycerate into phosphoenolpyruvate, an energy rich compound. The enzyme requires Mg^{2+} or Mn^{2+} and the reaction that it catalyzes is reversible.
10. *Pyruvate formation. Pyruvate kinase* catalyzes phosphate transfer from phosphoenolpyruvate to ADP. ATP is formed. It requires Mg^{2+} or Mn^{2+} and K^+. This irreversible reaction is the second substrate-level phosphorylation in the pathway. Enolpyruvate spontaneously transforms into pyruvate.
11. *Lactate formation.* Under anaerobic conditions, pyruvate is reduced to lactate in a reaction catalyzed by *lactate dehydrogenase.* NADH reoxidation allows maintenance of glycolysis.

The overall result of glycolysis is the conversion of one glucose molecule into two lactate plus 2 H^+. The final net energy balance is 2 moles of ATP per mole of glucose utilized.

Oxidative decarboxylation of pyruvate is a process that occurs in mitochondria. It is catalyzed by *pyruvate dehydrogenase*, the multienzyme system constituted by three enzymes: *pyruvate decarboxylase, dihydrolipoyl transacetylase,* and *dihydrolipoyl dehydrogenase.* It requires five coenzymes: thiamine pyrophosphate, lipoic acid, coenzyme A, FAD, and NAD. Reaction products are CO_2, acetyl-CoA, or active acetate and NADH + H^+.

Citric acid cycle also called tricarboxylic or Krebs cycle is the oxidation pathway of the acetyl groups produced by degradation of various compounds, including glucose, fatty acids, amino acid carbon chains, or others.

The steps of the cycle are the following:

1. *Citric acid formation. Citrate synthase* catalyzes, in an irreversible reaction, the condensation of the acetyl-CoA two carbon residue with oxaloacetate to form citrate. This is a regulatory step of the cycle.

2. *Isocitrate formation. Aconitase* catalyzes first conversion of citrate to *cis*-aconitate and then to isocitrate.

3 and 4. *Isocitrate oxidation and decarboxylation.* Both catalyzed by *isocitrate dehydrogenase.* Isocitrate is oxidized to oxalosuccinate and this is decarboxylated to α-ketoglutarate. NAD is the coenzyme, CO_2 is released. This is a regulatory step.

5. *α-Ketoglutarate decarboxylation.* Catalyzed by a multienzyme system, *α-ketoglutarate dehydrogenase,* which is similar to the pyruvate dehydrogenase complex, formed by three enzymes, and the same five coenzymes. This irreversible reaction produces CO_2.

6. *Succinate formation.* Succinyl-CoA formed in the preceding step is converted to succinate and free CoA by *succinate thiokinase.* It requires GDP and P_i. GTP donates a phosphoryl to ADP to give ATP. This is the only stage of the cycle in which there is substrate-level phosphorylation.

7. *Succinate oxidation. Succinate dehydrogenase* converts succinate to fumarate. It uses FAD, it is firmly attached to the inner mitochondrial membrane. This is another regulatory step of the cycle.

8. *Fumarate hydration. Fumarase* or *fumarate hydratase* catalyzes the addition of water to form malate.

9. *Malate oxidation. Malate dehydrogenase* uses NAD, it oxidizes malate to oxaloacetate; the cycle is then complete.

During one complete cycle turn, two CO_2 molecules are released and four H pairs are transferred, three to NAD and one to FAD.

The citric acid cycle is the oxidation pathway of acetates from any source. It also plays an anabolic role, providing intermediates for syntheses of various compounds.

The pathways that feed the citric acid cycle are called *anaplerotic* pathways.

Operation of the cycle produces 12 moles of ATP per mole of oxidized acetate.

The net energy yield from total oxidation of glucose, considering glycolysis, pyruvate decarboxylation, and the citric acid cycle, is 36–38 moles of ATP per mole of glucose. This means a total of 277 kcal/mol (1159 kJ/mol). This corresponds to 40% energy contained in 1 mole of glucose (686 kcal or 2870 kJ).

The pentose phosphate pathway is an alternative route for the metabolism of glucose. It includes the following reactions:

1. *G-6-P oxidation: G-6-P dehydrogenase,* NADP-linked, catalyzes the formation of 6-phosphogluconolactone.

2. *6-Phosphogluconate formation:* 6-P-gluconolactone is converted to 6-phosphogluconate by *6-P-gluconolactone hydrolase.*

3. *6-Phosphogluconate oxidation:* 6-phosphogluconate *dehydrogenase,* which uses NADP, produces ribulose-5-P and CO_2..

In subsequent stages, ribose-5-P and other metabolites, including glycolysis intermediates are formed.

This pathway produces NADPH and H, which are used for different synthesis processes, and ribose-5-P, which is a precursor of nucleotide and nucleic acid synthesis.

Gluconeogenesis involves the formation of glucose and glycogen from noncarbohydrate compounds. The reactions in this pathway are not just a reversal of the steps of glycolysis, requiring alternative pathways for those reactions that are irreversible. The reactions are:

1. *Pyruvate to phosphoenolpyruvate,* which includes two stages:
   a. Pyruvate is converted to oxaloacetate by *pyruvate carboxylase,* an allosteric enzyme. It requires biotin and ATP.
   b. Oxaloacetate is converted to phosphoenolpyruvate by *phosphoenolpyruvate carboxykinase.* It requires GTP.

2. *Fructose-1,6-bisphosphate to F-6-P: 1,6-bisphosphofructose phosphatase* catalyzes the hydrolysis. This is a regulatory enzyme in the pathway.

3. *G-6-P to glucose: G-6-P* releases free glucose. It is found in liver, kidney, and intestinal tissue; it is not present in muscle.

4. *Glucose-1-phosphate to glycogen:* requires *UDP glucose pyrophosphorylase, glycogen synthase,* and *branching enzyme.*

Lactate enters gluconeogenesis with previous oxidation to pyruvate. Any substance capable of becoming one of the citric acid cycle intermediates is potentially glucogenic. Acetyl-CoA is not glucogenic.

Synthesis of 1 mole of glucose from two of pyruvate requires 6 moles of ATP.

Fructose metabolism follows a main pathway that begins with fructose phosphorylation at C1. This is catalyzed by *fructokinase* and requires ATP. F-1-P is cleaved into glyceraldehyde and DHAP by *aldolase B.* Glyceraldehyde is phosphorylated to glyceraldehyde-3-P, which continues the glycolytic pathway. F-1-P can also be phosphorylated at C6 to give F-1,6-bisP, another intermediate of glycolysis.

Another pathway for fructose is its phosphorylation in C6 to give F-6-P, this is a minor pathway catalyzed by *hexokinase.*

Galactose metabolism follows the following stages:

1. *Galactose-1-phosphate formation: galactokinase* requires ATP.
2. *UDP-galactose formation: galactose-1-phosphate uridyl transferase.*
3. *UDP-glucose formation: epimerase* converts UDP-galactose to UDP-glucose.

Glycemia or glucose levels in blood are maintained during fasting periods between 70 and 110 mg/dL. After meals, a transient increase occurs. Maintaining these values requires complex regulation of the various pathways for glucose in different tissues.

Nucleotide sugars, such as UDP-glucose and UDP-galactose are active forms of monosaccharides. These are involved in interconversion of monosaccharides and in the transfer of glycosidic residues.

Glycoprotein synthesis involves the formation of O-glycosidic linkages by sequential addition of monosaccharides directly to OH⁻ groups of serine or threonine residues in the protein. The first sugar is commonly GalNAc. Specific glycosyl transferases are involved in each addition. Donor monosaccharides are nucleotide sugars.

Carbohydrates, attached to proteins by *N*-glycosidic bonds are synthesized by a more complex process. This is achieved by assembling a precursor attached to a lipid carrier, *dolichol phosphate.* The first carbohydrate is generally GlcNAc-P, others are added to form a branched structure [(Glc)$_3$-(Man)$_9$-(GlcNAc)$_2$-]. Donors of the first seven carbohydrate molecules are nucleotide sugars and the last seven, dolichol-P-sugars. The chain precursor is transferred to the protein and attached to an *N*-amide group in asparagine. Removal of the three glucose residues and some mannose generates oligosaccharides rich in mannose. Synthesis of complex and hybrid oligosaccharides requires removal of more mannoses and subsequent addition of different sugars. This process is accomplished in the endoplasmic reticulum and Golgi complex.

Bibliography

Alonso, M.D., Lomako, J., Lomako, W.M., Whelan, W.J., 1995. A new look at the biogenesis of glycogen. FASEB J. 9, 1126–1137.

Blanco, A., Kremer, R., Taleisnik, S., 2000. Inherited Metabolic Defects. Academia Nacional de Ciencias, Córdoba.

Bollen, M., Keppens, S., Stalmans, W., 1998. Specific features of glycogen metabolism in the liver. Biochem. J. 336, 19–31.

Mann, G.E., Yadilevich, D.L., Sobrevia, L., 2001. Regulation of amino acid and glucose transporters in endothelial and smooth muscle cells. Physiol. Rev. 83, 183–252.

Munro, J.A., Shaw, M., 2008. Glycemic impact, glycemic glucose equivalents, glycemic index, and glycemic load: definitions, distinctions, and implications. Am. J. Clin. Nutr. 87 (Suppl.), 237S–243S.

Rustin, P., Bourgeron, T., Parfait, B., Chretien, D., Munnich, A., Rötig, A., 1997. Inborn errors of the Krebs cycle: a group of unusual mitochondrial diseases in human. Biochim. Biophys. Acta 1361, 185–197.

Scriver, C.R., Beaudet, A.L., Sly, W.S., Velle, D. (Eds.), 2001. Metabolic and Molecular Bases of Inherited Disease. eighth ed. McGraw Hill, New York, NY.

Van den Berghe, G., 1994. Inborn errors of fructose metabolism. Annu. Rev. Nutr. 14, 41–58.

Van den Berghe, G., 1996. Disorders of gluconeogenesis. J. Inherit. Metab. Dis. 19, 470–477.

Wamelink, M.M.C., Struys, E.A., Jakobs, C., 2008. The biochemistry, metabolism and inherited defects of the pentose phosphate pathway: a review. J. Inherit. Metab. Dis. 31, 703–717.

15

Lipid Metabolism

Triacylglycerols (TAGs). Neutral fats or TAGs are the predominant lipids in the human diet. Their catabolism in the body yields an important amount of energy.

TAGs are synthesized in enterocytes from fatty acids (FAs) and monoacylglycerols, which are absorbed in the intestine and are the result of the degradation of fats in the digestive tract. TAGs are packed together with small amounts of cholesterol and proteins in particles known as chylomicrons. These particles transport all of the exogenous lipids that come with the diet. The liver is the other organ with active TAG synthesis. It incorporates TAGs to lipoprotein particles different from those of the intestine, these are the very low density lipoproteins (VLDLs), which are in charge of transporting endogenous lipids in blood.

Fat from chylomicrons and VLDL undergo almost complete hydrolysis in capillaries, releasing glycerol and FAs to the cells. Glycerol is taken up and metabolized by tissues capable of phosphorylating it. On the other hand, FAs are oxidized in most tissues by a process that generates two carbon residues bound to coenzyme A (acetyl-CoA).

Acetyl-CoA or active acetate is at a major metabolic crossroad, where various pathways converge. Acetate can follow several metabolic alternatives, including the citric acid cycle, FA production, and cholesterol synthesis. The production of FAs from two carbon segments is

particularly active in liver, adipose tissue, mammary glands, and brain tissue; these tissues also have the metabolic machinery capable of synthesizing TAGs.

TAGs constitute the majority of lipids in fat deposits and represent the main energy reserve of the body. The caloric value of fats is more than double (9 kcal or 37.6 kJ/g for fat) of that of carbohydrates or proteins (4 kcal or 16.7 kJ/g). Moreover, water content of fatty deposits is significantly less than that of glycogen. Therefore, fats constitute the most concentrated form of chemical energy stored in the body.

For a long time, lipid deposits were considered rather static, and only used when food intake did not cover the daily caloric needs. However, it is now clear that this is not the case. Body fat stores are very dynamic and constantly subjected to degradation and resynthesis. Fig. 15.1 presents a summary of triacylglycerol metabolism.

From an energetic standpoint, it is theoretically possible to replace dietary fats for carbohydrates; however, the presence of lipid in the diet is necessary to provide essential polyunsaturated FAs and liposoluble vitamins, which the body cannot synthesize. In addition, polyunsaturated FAs of 20 carbons in length are used for the synthesis of eicosanoids, which are compounds involved in the regulation of different cellular processes.

Glycerophospholipids and sphingolipids. The major role of these amphipathic substances is

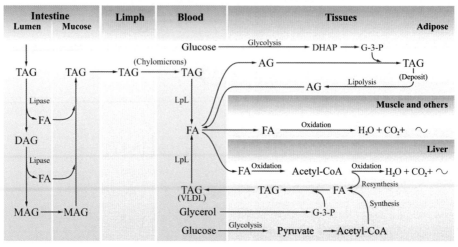

FIGURE 15.1 **Triacylglycerols (TAG) metabolism.** DAG, diacylglycerol; MAG, monoacylglycerol; FA, fatty acid; DHAP, dihydroxyacetonephosphate; LpL, lipoprotein lipase; VLDL, very low density lipoprotein.

structural; they participate in the formation of cell membranes. They also help to stabilize and maintain hydrophobic lipids in suspension in aqueous media, allowing lipid transport in blood or facilitating lipid digestion and absorption in the intestine.

Cholesterol. This lipid is an essential component of cell membranes and serves as the precursor of steroidal hormones and bile acids.

Table 15.1 lists the normal values for lipids circulating in human plasma after 12 h of fasting.

TABLE 15.1 Normal Serum Lipids Concentration (in mg/dL)

	Average	Range
Total lipids	540	315–670
Triacylglyicerols	140	80–170
Phospholipids	220	125–290
Cholesterol[a]	180	110–220

[a]*70% of cholesterol is sterified with fatty acids (FAs) and 30% is free.*

BLOOD LIPIDS

In plasma there are also free (nonesterified) FAs, which circulate at an average concentration of 20 mg/dL. They are transported primarily bound to albumin, in a ratio of 8–9 FA molecules per protein molecule. The half-life of FAs in plasma is only 2–3 min, indicating that their turnover is very active. Most of these FAs are generated by hydrolysis of fat deposits mobilized for use in tissues.

PLASMA LIPOPROTEINS

All of the lipids in plasma are associated with proteins. These lipoproteins consist of an external hydrophilic region and an internal hydrophobic core. The peripheral layer is constituted by amphipathic molecules (proteins, phospholipids, and nonsterified cholesterol), oriented with their hydrophilic groups toward the medium. This allows lipids to stay in suspension and to interact with enzymes and receptors on the surface of cells. The lipoprotein complex interior is hydrophobic, containing TAGs, and cholesterol esters.

Plasma carries several types of lipoproteins, which vary in size, density, and composition (p. 62). The external cover of all lipoproteins is approximately 2 nm wide, while the hydrophobic core, which accounts for most of the mass of these particles, is variable in size. As the neutral lipid content of lipoproteins increases, their diameter raises and their density decreases.

Based on their density, five major lipoprotein groups can be distinguished. These include (from low to high density): (1) chylomicrons, (2) VLDLs, (3) IDLs, (4) LDLs, and (5) HDLs.

In general, chylomicrons carry exogenous lipids from the intestine to the rest of the body. The remaining lipoproteins are involved in transporting endogenous lipids, synthesized in different organs. TAGs predominate in chylomicrons and VLDL, while cholesterol is present mainly in LDL and HDL.

Apolipoproteins. Ten different protein components of lipoproteins have been described. They are named with the prefix "apo," followed by a letter and a number, which identifies different members within the same apolipoprotein group (Table 15.2). The most important

apolipoproteins from a clinical perspective are apo A-I, B-48, B-100, C-II, E, and apo (a). Apolipoprotein A constitutes the main protein in HDL, which also contain apos C and E. Apo B-48 is found in chylomicrons and apo B-100 in VLDL, IDL, and LDL. Apo B-48 and B-100 are encoded by the same gene (p. 499). Apo B-48 is identical to 48% of the N-terminal portion of Apo B-100. In plasma, there is transfer of Apo C and E from HDL to chylomicrons and VLDL. In contrast, apo A, B-48, and B-100 are not exchangeable and are confined to their original particles. Apoproteins play a structural role in the biosynthesis and remodeling of lipoproteins. Apo A-I is essential for the synthesis and secretion of HDL. All apolipoproteins, except B-48, are produced in the liver. B-48 and a small proportion of apo A are made in the intestine. Apo B-48 and B-100 are necessary for the secretion of TAG rich lipoproteins (chylomicrons and VLDL, respectively).

Some apolipoproteins function as cofactors or activators of enzymes involved in lipoprotein metabolism. Particularly important is the role of apo C-II, which activates *lipoprotein lipase* (LpL). This enzyme is responsible for the intravascular hydrolysis of TAG contained in chylomicrons and VLDL. Apo A-I stimulates the activity of

TABLE 15.2   Blood Plasma Lipoproteins

Class	Main lipids	Apolipoproteins	Density	Diameter (nm)
Chylomicrons	Triacylglycerides Cholesterol esters (from the diet)	AI, AII, AIV, B-48, CI, CII, CIII, E	<0.95	100–500
Chylomicrons remnants	Cholesterol esters (from the diet)	B-48, E	<1.006	>30
VLDL	Triacylglycerides (endogenous)	B-100, CI, CII, CIII, E	<1.006	30–80
IDL	Triacylglycerides Cholesterol esters (endogenous)	B-100, E	1.006–1.019	25–35
LDL	Cholesterol esters (endogenous)	B-100	1.019–1.063	18–28
HDL$_2$	Cholesterol esters Phospholipids (endogenous)	AI, AII, CI, CII, CIII, E	1.063–1.125	9–12
HDL$_3$	Cholesterol esters Phospholipids (endogenous)	AI, AII, CI, CII, CIII, E	1.125–1.210	5–9

lecithin-cholesterol acyltransferase (LCAT), which catalyzes cholesterol esterification in HDL.

Apolipoproteins also are ligands capable of binding to cellular receptors. Apo B-100 is involved in binding LDL to its receptor in all cells and apo E binds to specific receptors in hepatic cells.

Lipoprotein Metabolism

Chylomicrons

Cells of the intestinal mucosa pack together triacylglycerides and a small amount of cholesterol into chylomicrons in an organized arrangement that includes a layer of phospholipids, free cholesterol, and apo B-48. The process requires a microsomal triacylglyceride transfer protein (MTTP), which is involved in the formation and secretion of chylomicrons. Lipids comprise 98%–99% of the total weight of chylomicrons, with proteins comprising the remaining 1%–2%. Within the lipids, TAGs represent 88%, phospholipids 8%, cholesteryl esters 3%, and free cholesterol 1%. Liposoluble vitamins absorbed in the intestinal lumen are also incorporated in the chylomicron core. Nascent chylomicrons are carried in vesicles from the cell Golgi complex to the basolateral membrane of enterocytes, where they are secreted by exocytosis into the interstitial space. Lymphatic capillaries carry them to the thoracic duct and the subclavian vein to reach the general circulation. Chylomicrons, with a diameter of 100–500 nm, appear in blood 1 h after fat ingestion and persist in circulation for more than 8 h after fasting. The presence of chylomicrons in blood upon fat absorption after a meal (absorptive lipemia) gives the plasma a cloudy or milky appearance. Depending on the lipid intake, chylomicrons carry approximately 100 g of triacylglycerol and 0.5–1 g of cholesterol per day.

Once they enter the circulation, chylomicrons receive apoproteins C and E transferred from HDL. Chylomicrons are processed in the blood vessel endothelium, where they interact with LpL, which is the enzyme that is activated by apo C-II and catalyzes the hydrolysis of triacylglycerol contained in the interior of the particle. The released FAs are rapidly taken up by the underlying cells. Main sites of LpL activity are the capillaries of adipose tissue, myocardium, skeletal muscle, and lactating mammary glands. Fatty acids are used in adipose tissue to synthesize and store TAG. In skeletal and cardiac muscles, FAs are oxidized for energy production. Lastly, in mammary glands they are used to synthesize and secrete TAGs. Glycerol, the other product of hydrolysis, is released into plasma and is primarily taken up by hepatocytes, where it is metabolized.

LpL is bound to the walls of capillaries through glycosaminoglycans of the heparan sulfate type. Administration of heparin releases LpL from its heparin-sulfate anchoring, which rapidly clears the milky plasma produced after lipid absorption. Lipolysis significantly reduces the size of chylomicrons and increases the relative proportion of cholesterol. The excess surface phospholipids are transferred to HDL and this receives apo C-II, Apo CI, and LpL activating proteins, all of which prevent premature interaction of chylomicrons with hepatic receptors. As a result of these changes, the particles become *chylomicron remnants*.

After lipolysis, chylomicron remnants dissociate from the capillary endothelium and recirculate back to the liver. Particles relatively enriched in cholesterol esters have only apo B-48 and apo E proteins. Hepatocytes possess apo E receptors [LDL receptor-related protein (LRP)] which bind the remnants and internalize them by endocytosis. Apo B-48, necessary for the secretion of chylomicrons in the intestine, is not involved in their removal. Although B-48 and B-100 are products of the same gene, B-48 lacks the B-100 C-terminal half, where the LDL receptor binding domain is located. The half-life of chylomicrons in circulation is less than an hour. Fig. 15.2 summarizes the described steps.

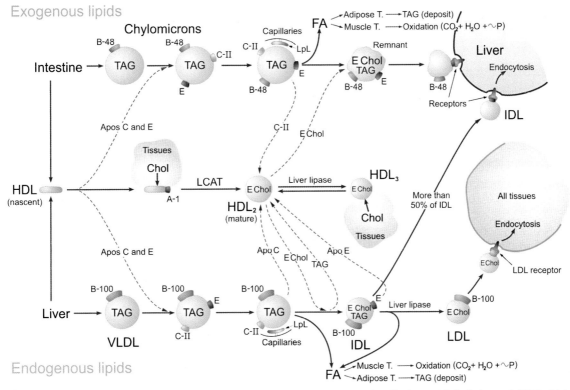

FIGURE 15.2 **Metabolism of plasma lipoproteins.** Chol, Free cholesterol; EChol, esterified cholesterol; HDL, high density lipoproteins; IDL, intermediate density lipoproteins; LDL, low density lipoproteins; LCAT, lecithin-cholesterol acyltransferase; LpL, lipoprotein lipase; TAG, triacylglycerol; VLDL, very low density.

Within liver cells, the endocytic vesicles bind to lysosomes where the chylomicron remnants are degraded. The products formed (cholesterol, FAs, and amino acids) are released into the cytosol. Cholesterol is used for bile acid synthesis or excreted as such in the bile, and can also be reexported to blood in VLDL particles. Fatty acids are oxidized to obtain energy, or used in the synthesis of TAG.

Very Low Density Lipoproteins (VLDLs)

TAGs synthesized in hepatocytes are incorporated into VLDL particles, along with cholesterol esters, which represent approximately 10% of the hydrophobic core mass of the nascent VLDL. The main protein in the amphipathic cover of VLDL is apo B-100. The complete particle is secreted by exocytosis into the hepatic perisinusoidal spaces and reaches the sinusoidal capillaries to enter the bloodstream. There, VLDL molecules undergo changes similar to those described for the chylomicrons. First, VLDLs receive apolipoproteins C and E from HDL, then, they are subjected to the action of LpL (activated by apo C-II), in extrahepatic tissues. Besides hydrolysis of TAG by LpL, the particles exchange TAG for cholesterol esters from HDLs. VLDL particles lose much of their TAG and are enriched in cholesterol; their size decreases and surface phospholipids are transferred to HDL. After these changes, apo C-II returns to HDL, and the action of LpL is interrupted. The modifications of VLDL particles transform them into IDLs. The half-life of VLDL is of approximately 4 h.

IDLs are particles rich in cholesterol, mostly esterified, and with small amounts of TAG. Their apoproteins are B-100 and E. Receptors in hepatocytes (LRP) that bind apo E capture more than half of the IDL particles from the circulation and internalize them by endocytosis. As IDLs lack apo C-II, LpL cannot hydrolyze the TAG present in IDL. Instead, TAG from IDL is still hydrolyzed by a liver lipase, which is located in the wall of sinusoidal liver capillaries and is not sensitive to apo C-II regulation; apo E is returned to HDL. All these changes transform IDL into LDL. IDL only remain in the bloodstream for a period of approximately 2–5 h. As the amount of IDL captured by the liver increases, the amount of LDL remaining declines.

Low Density Lipoproteins

LDLs contain almost only esterified cholesterol within the central core and only apo B-100 on their surface. They represent the final product of VLDL. The half-life of LDL is approximately 2.5 days. The average normal concentration in plasma is 100 mg/dL (2.6 mmol/L).

LDLs bind to receptors that are specific for apo B-100 (LDL receptors), which are expressed on the surface of all cells. In cells, LDLs become internalized by endocytosis, and are hydrolyzed by lysosomal enzymes (Fig. 15.2). The resulting products (amino acids, cholesterol, and FAs) move to the cytosol. Cholesterol is added to the cell membranes and in some cells (adrenal cortex, ovarian, and testicular), it is used in the synthesis of steroidal hormones. The excess is esterified again in a reaction catalyzed by *acyl-CoA cholesterol acyltransferase* (ACAT) and stored in the cell. The LDL receptor is recycled to the plasma membrane.

High Density Lipoproteins (HDLs)

HDLs are synthesized in the liver and, to a lesser extent, in the intestine. The nascent particles, of discoid shape, are complexes of apolipoproteins (A, C, and E) and phospholipids, predominantly phosphatidylcholine. HDLs perform several roles:

Apolipoprotein transfer. HDL contains apos A, C, and E. Apo A is the main HDL protein and always remains attached to the particles. Apos C and E are transferred to chylomicrons and VLDL. Apo C-II is an activator of LpL, which after fulfilling its mission, returns to HDL. Apo E plays an important role; it regulates cholesterol, triacylglyceride, and phospholipid metabolism in the blood and brain. Apo E transferred to VLDL returns to HDL from IDL, when these have lost almost all their TAGs.

Reverse cholesterol transport. HDL interacts with the plasma membrane of extrahepatic cells through a process in which apo A-I, and probably other types of apo A, are involved. The existence of specific HDL receptors has been proposed. Once in the cell cytoplasm, cholesterol is mobilized to the cell surface and transferred to the HDL particle. This free cholesterol is quickly esterified by LCAT, which is activated by apo A-I. The FA in position 2 of phosphatidylcholine (lecithin) is transferred to the C3 hydroxyl of cholesterol. A cholesterol ester and lysophospholipid are formed.

$$\text{Cholesterol} + \underset{\text{(Lecithin)}}{\text{Phosphatidylcholine}} \xrightarrow{\text{LCAT}}$$

$$\xrightarrow{\quad} \text{Cholesterol ester} + \underset{\text{(Lysolecithin)}}{\text{Lysophosphatidylcholine}}$$

Cholesterol esters increase the particle size of HDL and change their shape from oblong to spherical. Cholesterol esters incorporated by HDL can be transferred to triacylglycerol-rich lipoproteins; the VLDL and chylomicron remnants are then taken up by hepatic receptors and retrieved from circulation. The whole process is called *reverse cholesterol transport.*

The transfer of cholesterol esters from HDL to VLDL and chylomicrons is mediated by the cholesterol ester transport protein (CETP). In exchange for the esterified cholesterol transferred to lipoproteins rich in TAG, VLDLs transfer TAGs to HDL. Cholesterol esters are finally taken up by the liver, along with chylomicron remnants and IDL, to complete the reverse transport of cholesterol process. This pathway results in the net transfer of free cholesterol from peripheral tissues to the liver, where cholesterol can be processed for excretion. Three proteins are involved in this process: LCAT, apo A-I (activates LCAT), and CETP.

On the other hand, HDL provides cholesterol to steroidogenic tissues (e.g., adrenal cortex and gonads). To do this, HDL binds to special receptors and transfers cholesterol to the cells. The process does not involve endocytosis, only cholesterol transfer.

HDL₂ and HDL₃. Two main types of HDL can be distinguished, HDL_2 and HDL_3; they are different in size and cholesterol content. HDL_2 particles are larger and richer in cholesterol than HDL_3. Due to the transfer of cholesteryl esters to VLDL and chylomicrons, and subsequent hydrolysis of TAG by the action of hepatic lipase, HDL_2 becomes HDL_3. HDL_3 are smaller particles with capacity to acquire more cholesterol in peripheral cells, they are also able to regenerate HDL_2.

Desirable plasma concentrations of HDL are 60 mg/dL (1.6 mmol/L) or above, low HDL levels are an index of high risk of cardiovascular accidents. Low levels of HDL cholesterol have been observed in clinical cases of hypertriacylglyceridemia with high levels of VLDL. This may depend on an increased exchange of cholesterol esters and TAG between HDL and TAG-rich lipoproteins.

HDL removal. A small proportion of HDL is captured and removed from circulation by LRP (LDL receptor related protein) and by LDL receptors on hepatocytes, which also recognize the apo E. Cholesterol from HDL enters the liver and is finally excreted as such or forming part of bile acids.

Lipoprotein (a)

Besides those already described, another lipoprotein fraction exists in human plasma that has a different protein composition than LDL. This is lipoprotein (a) [lp(a)], composed of the apo B-100 present in LDL and apo (a), a glycoprotein with a molecular mass between 300 and 800 kDa. The glycoprotein binds to apo B-100 by a disulfide bridge. Lipoprotein (a) contains higher amounts of protein than LDL, which gives them greater density. The primary structure of apo (a) is homologous to plasminogen, a member of the serine proteases family (see blood clotting, Chapter 31). The concentration of lipoprotein (a) in plasma varies depending on the individual. A close relationship between high levels of lipoprotein (a) and incidence of atherosclerosis and cardiovascular accidents has been observed. This prompted the use of lipoprotein (a) levels as a clinically relevant index; values greater than 40 mg/dL are associated with coronary heart disease risk, especially if associated with increased LDL and total cholesterol.

Receptors

LDL receptor. This cell surface protein recognizes apolipoprotein B-100, present in VLDL, IDL, LDL, and probably apo-E. It is found in almost all cells. The LDL receptor is an integral membrane protein of 115 kDa, with five functionally distinct segments. Domain 1 comprises the N-terminal (280 amino acids) and is projected outside the cell; it is the binding site for apo B-100. This segment is rich in cysteines that form disulfide bonds, which gives the molecule stability, and has several glutamate and aspartate residues that allow its interaction with a positively charged site on the C-terminal portion of B-100. Domain 2 is homologous to the epidermal growth factor precursor, having two N-oligosaccharide chains. Domain 3 is rich in glycosylated serine and threonine. Domain 4 is a 22 amino acid α helix, predominantly hydrophobic, that traverses the plasma membrane lipid bilayer. Domain 5, which includes the C-terminal (50 amino acids), faces the cytosol. This segment gives the receptor its capacity to move laterally in the membrane and to cluster into the coated invaginations to be endocytosed. The synthesis of LDL receptor is highly regulated; when the intracellular cholesterol concentration increases, the LDL receptor production is inhibited (down regulation).

Receptor of remnants. This is also designated as lipoprotein receptor-related protein (LRP). Its main ligand is apo E. Apo CI inhibits binding and premature uptake of chylomicrons and VLDL. It is abundant in liver, brain, and placental tissues. Its synthesis is not affected by intracellular cholesterol levels.

Scavenger receptor. This receptor is less specific than the preceding ones. It is different from the LDL receptors, it binds to chemically modified LDL, and it plays an important role in the uptake of altered lipoproteins by macrophages. Its synthesis is not regulated like that of the LDL receptor.

HDL receptor. This protein is present in adipose cells, vascular endothelium, fibroblasts, and steroidogenic tissues. It binds apoproteins A-I, A-II, and A-IV. HDL interaction with the receptor initiates a signal transduction cascade which promotes the transfer of intracellular cholesterol to the plasma membrane and to HDL (in steroidogenic tissues, cholesterol transfer is reversed). HDL receptor function is critical in the reverse cholesterol transport, which removes excess cholesterol from extrahepatic cells.

Lipoproteins and Atherosclerosis

The study of lipoprotein metabolism is of primary medical interest, as it correlates with the development of atherosclerosis. This pathological condition is currently the leading cause of death in humans. Atherosclerosis is characterized by formation of plaques (atheromas) in the arterial walls, resulting from the accumulation of cholesterol, cellular debris, smooth muscle cells, and connective tissue fibers, which decrease the elasticity of the arterial wall and reduce the arterial lumen. Progression of the lesion favors the

production of intravascular clots that eventually block blood flow in the artery, leading to severe consequences, such as ischemia. This causes marked negative effects in the myocardium and brain, where it leads to heart infarcts, stroke, and brain damage.

The primary cause of atherosclerosis is not fully understood; however, risk factors have been identified that strongly contribute to the genesis and progression of the disease. Hypercholesterolemia, hypertension, smoking, and diabetes, are considered main predisposing causes of the disease development. Secondary factors include stress, a diet rich in animal fats (high in saturated FAs and cholesterol), obesity, and lack of exercise.

Development of atherosclerosis is common in individuals with plasma lipid alterations (dyslipidemia). Elevated levels of LDL and cholesterol, and an increase in triacylglyceride-rich lipoproteins enhance the risk for the development of atheromas. Arterial plaque formation is also favored by mechanical or toxic factors, such as high blood pressure and the exposure to tobacco derivatives. Genetic factors also play a major role.

The development of the atherosclerotic plaque is a complex process that can be summarized as follows: the disease can be triggered by an alteration of the vascular endothelium caused by mechanical stress (e.g., hypertension) or chemicals that stimulate expression of certain proteins. Among these proteins is vascular cell adhesion molecule-1 (VCAM-1), which promotes the recruitment and adhesion of monocytes, T lymphocytes, and platelets to the affected area. LDL enters the arterial intima layer, passing through pores in the vascular endothelium, and binds to extracellular matrix proteoglycans. The greater the amount of circulating LDL, the greater it becomes accumulated within the artery. This infiltration of lipoproteins alters the endothelium, causing the release of reactive oxygen species (superoxide ion, hydrogen peroxide, hydroxyl radicals, see p. 205) and cell oxidation. Oxidizing agents can activate monocytes and lymphocytes to release cytokines that initiate an inflammatory response. This inflammatory process is believed to be a pivotal factor in the pathogenesis of atherosclerosis.

Monocytes are attracted to the site where the LDL are accumulated and altered by oxidation; there, monocytes are converted into macrophages. These cells bind lipoproteins via scavenger receptors, LDL is internalized and degraded; the released cholesterol is esterified by ACAT and stored in cells. Scavenger receptors are not regulated by intracellular cholesterol levels, thus, LDL uptake continues as long as it is present in the medium. The macrophage cytoplasm, full of cholesterol ester droplets, acquires a foamy appearance. These cells group in clusters, which are visible as fatty streaks. At the same time, platelets release growth factors that stimulate the proliferation of smooth muscle cells and the deposition of collagen and elastin in the medial arterial layer. This last event is responsible for the fibrosis and stiffness of the vessel. Eventually, the cells that have accumulated lipids die and release their cholesterol esters content to form an oily mass. Platelet aggregation favors blood clotting at the site of injury, causing blockage of blood flow.

Numerous studies have been conducted to understand the relationship between lipoproteins and atherosclerosis. The current evidence indicates that increased plasma levels of LDL, total cholesterol, triacylglycerides, lipoprotein (a), and diminished HDL cholesterol are important risk factors. Cholesterol contained in HDL is commonly called "good" cholesterol, as opposed to "bad" LDL cholesterol.

LDL concentration in blood is related to the number of apo B-100 (LDL receptors) in cells. Plasma LDL increases when there is primary or secondary reduction of specific receptors. Primary receptor reduction due to genetic factors causes familial hypercholesterolemia.

Familial hypercholesterolemia. Numerous mutations have been identified in the gene controlling the synthesis of LDL receptors. These mutations may determine deficiencies in removal of LDL from the circulation by (1) absence or reduced production of receptors. (2) Failure in the mechanisms involved in intracellular vesicular transport and insertion of receptors in the plasma membrane. (3) Altered LDL receptor function, with decreased capacity to bind LDL or to become endocytosed.

In homozygous forms of the disease, with total absence of LDL receptors, LDL (and cholesterol) concentration in blood is greatly increased (up to six times the normal levels). In heterozygous forms, the amount of LDL receptors decreases and levels of plasma LDL and cholesterol are two fold higher than normal. These patients suffer from atherosclerosis at an early age and have a high incidence of vascular accidents. Studies in those patients were instrumental in proving the importance of the receptor for the removal of LDL from the circulation and the maintenance of normal LDL concentration.

Even in people without a genetic defect, the level of LDL (and cholesterol) in blood increases with age. This reflects an increase in the production of LDL and/or reduction of LDL receptors, which reduces the capacity to remove circulating LDL.

A direct relationship between lipoprotein (a) level and incidence of atherosclerosis has been proven. A high lipoprotein (a) concentration has been shown to be predictive of coronary disease. Instead, there is an inverse relationship between HDL cholesterol concentration and atherosclerosis. When the level of HDL is high (particularly HDL_2), the risk of vascular accidents decreases. This may be due to the fact that HDL plays a role in removing cholesterol from extrahepatic cells (inverted transport).

Hypertriacylglyceridemia. This is observed in people with the inability to hydrolyze TAG contained in chylomicrons and VLDL due to genetic defects that affect the synthesis of LpL or apo C-II, a factor required for LpL activity.

Hypercholesterolemia and hypertriacylglyceridemia secondary to failure of capturing chylomicron remnants and IDL have also been observed. The most common cause of this type of hyperlipidemia is usually apo E deficiency and the accumulation of VLDL.

Less frequent are primary hypolipemias, including abetalipoproteinemia due to failures in the synthesis of B-100 and B-48. In these cases, plasma lipid levels are very low and chylomicrons and VLDL cannot be formed. IDL and LDL concentrations are also reduced.

There is another inherited condition, called Tangier disease, in which there is a marked decrease in HDL with cholesterol accumulation in the cells, increased TAG in plasma, low levels of apos A-I and E, and reduced capture of chylomicron remnants and IDL.

TISSUE LIPIDS

Body fats can be distinguished according to their tissue distribution and functions in (1) stored lipids and (2) lipids contained in organs and tissues.

1. *Lipid storage.* These lipids are found mainly in subcutaneous adipose tissue and surrounding certain organs. They contain approximately 90% of neutral fats and very small amounts of cholesterol and complex lipids. The most abundant FAs in the TAGs of adipose tissue are oleic, palmitic, linoleic, stearic, and myristic acids.

The main function of stored lipids is to serve as an energy reserve. When the food intake exceeds caloric needs, the excess is deposited as fat. Only TAGs can be stored in large quantities. Carbohydrates are also deposited in the form of glycogen, but the storage capacity is very low comparatively.

Stored fat is deposited and degraded depending on the energy needs of the body. A hormonally regulated intracellular lipase in adipose tissue catalyzes the hydrolysis of TAG into glycerol and FAs. Fatty acids released into the circulation

reach the tissues to be used. Synthesis and degradation of stored fat are dynamic processes. It is estimated that the total reserve of TAG is renewed every 2–3 weeks.

Fat deposits also function as thermal insulators and provide a protective cover to some organs, such as the kidneys.

2. *Constituent lipids.* They are represented almost entirely by complex lipids and cholesterol. They do not include TAGs and are part of cell membranes and other cell structures. Under normal conditions, phospholipids, glycolipids, and cholesterol remain at relatively constant levels in the body and are not accumulated.

FAT METABOLISM

TAGs must be totally hydrolyzed before they can be used by tissues. Much of this hydrolysis affects the fat stored in adipose tissue. Thus, there is a permanent degradation, or lipolysis, of the stored TAGs that is catalyzed by intracellular lipases, whose activity is regulated to suit the body needs. The products formed (FAs and glycerol) are released into the plasma, where FAs bind to albumin.

Exogenous TAGs transported by chylomicrons and the endogenous ones transported by VLDLs are hydrolyzed by LpL in capillaries and the released FAs enter the cells where they are used. Hydrolyzed fats, either from stored TAG or transported by lipoproteins, also release free glycerol, which is taken up by cells that can metabolize it.

Glycerol Metabolism

Before it can be used by cells, glycerol requires its previous activation by phosphorylation, which is why only tissues possessing *glycerokinase* are able to metabolize free glycerol. Glycerokinase is found in liver, kidney, intestine, and lactating mammary gland tissues. It catalyzes the conversion of glycerol into L-glycerol-3-phosphate, by transferring a phosphate group from ATP. The reaction is practically irreversible.

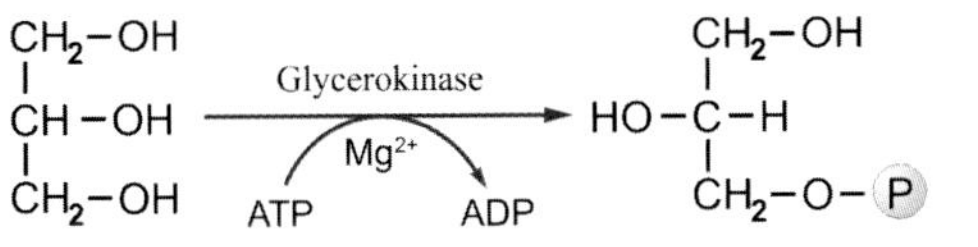

Glycerol   L-glycerol-3-phosphate

Glycerol-3-phosphate is converted into dihydroxyacetone, a triose phosphate of the glycolysis pathway, by the action of the NAD-linked *glycerophosphate dehydrogenase*. As noted when considering the stages of glycolysis, dihydroxyacetone is subsequently converted into glyceraldehyde-3-phosphate in the reaction catalyzed by phosphotriose isomerase.

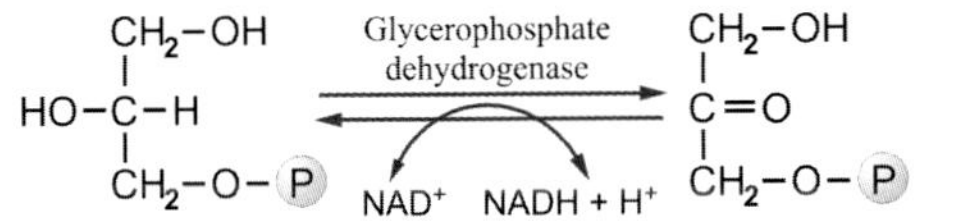

L-glycerol-3-phosphate   Dihydroxyacetone phosphate

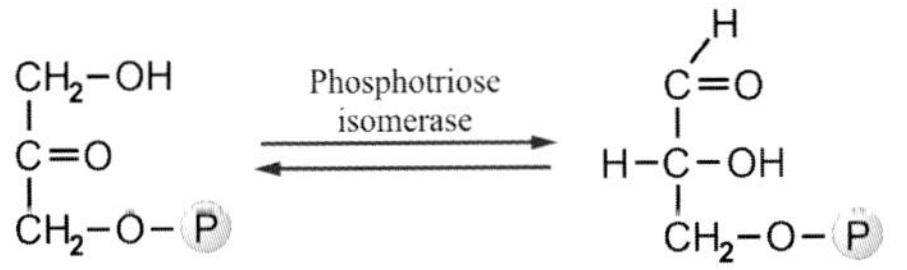

Dihydroxyacetone phosphate   D-glyceraldehyde-3-phosphate

Both reactions are reversible and through the same steps, glycerol-3-phosphate can be obtained from triose phosphate. The formation of triose phosphate provides a pathway that leads to the complete degradation of glycerol in the glycolysis and the citric acid cycles. Moreover, it can also follow the gluconeogenic pathway to form glucose or glycogen. Glycerol-3-phosphate is an important metabolite in TAGs and glycerophospholipids synthesis.

FATTY ACID CATABOLISM

Many organs, especially liver, muscle, myocardium, kidney, and adipose tissue, can oxidize long-chain FAs. The observation that most natural FAs have an even number of carbons led to the assumption that FAs must be synthesized or degraded in the body by addition or subtraction of two carbon residues at a time. Later studies fully confirmed this notion. The main degradation process comprises the oxidation of the FA β carbon (hence the name of β-oxidation to this process). Enzymes localized in the mitochondrial matrix are involved in β-oxidation. The proximity of the respiratory chain facilitates the transfer of reducing equivalents and ATP production by oxidative phosphorylation.

Before starting the oxidation process, two preparatory steps are necessary: (1) activation of the FA and (2) transport into mitochondria.

Fatty acid activation. The initial step consists in the formation of a highly reactive compound, able to participate in subsequent transformations. The reaction is catalyzed by *thiokinase*, or *acyl-CoA synthetase*, in the presence of coenzyme A, ATP, and Mg^{2+}. ATP is hydrolyzed at its second phosphate bond producing AMP and inorganic pyrophosphate (PP_i); the chemical energy of both high-energy bonds of ATP are consumed. The FA is attached to coenzyme A by an energy-rich thioester bond. Acyl-CoA or active FA is formed.

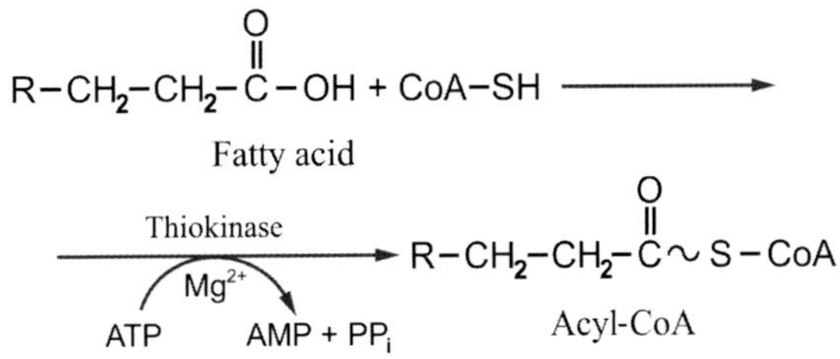

Pyrophosphate is rapidly hydrolyzed by *pyrophosphatase*, the FA activation is an irreversible process. In all reactions producing inorganic

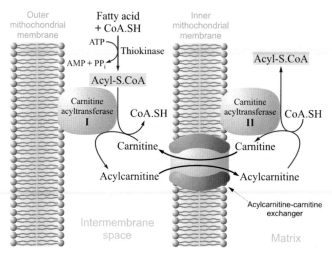

FIGURE 15.3 **Transfer system of acyl-CoA from the cytosol to the mitochondrial matrix (role of carnitine).**

pyrophosphate, its rapid removal by hydrolysis determines the irreversibility of the main reaction.

The activation of FAs takes place in the cytosol, while the oxidation occurs within the mitochondria. As the inner mitochondrial membrane is impermeable to acyl-CoA, a transfer mechanism is required.

Transfer of acyl-CoA from the cytosol to the mitochondrial matrix. Acyl-CoA is transferred to carnitine or β-hydroxy-γ-trimethylammonium butyrate (Fig. 15.3), a compound synthesized in the liver and kidneys from lysine. Acyl-carnitine is formed, which is transported to the mitochondrial matrix.

The system comprises two enzymes, *carnitine acyltransferase I*, located in the outer membrane of mitochondria, and *carnitine acyltransferase II*, located in the inner membrane, as well as an acylcarnitine/carnitine exchanger (Fig. 15.3). Acyl-CoA and carnitine enter the intermembrane space through pores in the mitochondrial outer membrane.

Carnitine acyltransferase I (CAT I) catalyzes the reaction:

$$\text{Acyl-SCoA} + \text{Carnitine} \underset{}{\overset{\text{CAT I}}{\rightleftharpoons}} \text{Acyl-Carnitine} + \text{CoA-SH}$$

The acyl moiety is linked to the carnitine β carbon hydroxyl by an ester bond. This is a high energy bond, such as the thioester bond of acyl—CoA. The acylcarnitine passes to the matrix, where it transfers the acyl to CoA-SH to regenerate acyl-CoA. The reaction is catalyzed by carnitine acyltransferase II (CAT II):

$$\text{Acyl-Carnitine} + \text{CoA-SH} \underset{}{\overset{\text{CAT II}}{\rightleftharpoons}} \text{Acyl-SCoA} + \text{Carnitine}$$

In the mitochondrial inner membrane, an exchanger transports acylcarnitine to the mitochondrial matrix and moves carnitine to the cytosol. Fatty acids of less than 12C enter the mitochondrial matrix without being transferred to carnitine.

Carnitine deficiency. This alteration is observed in preterm infants who have not developed the capacity to synthesize carnitine and in patients with organic aciduria (organic acids in urine) or renal disease that leads to loss of carnitine in urine. In these cases, FA oxidation, ketogenesis, and gluconeogenesis are altered and the plasma levels of FAs are elevated, causing hypoglycemia.

β-Oxidation

Acyl-CoA in the mitochondrial matrix initiates the oxidation process. This comprises four reactions which release acetyl-SCoA and an acyl chain that is two carbon residues shorter. Degradation cycles are repeated until the entire chain is reduced to compounds containing two carbon residues. The reactions are the following:

1. *First oxidation.* In the initial step, acyl coenzyme A loses two hydrogen atoms from its α and β carbon atoms (C2 and C3). This dehydrogenation is catalyzed by *acyl-CoA dehydrogenase* that uses FAD as a hydrogen acceptor. The product is an unsaturated α-β (or 2–3) acyl-CoA derivative in *trans* configuration.

$$R-CH_2-CH_2-\overset{\overset{O}{\|}}{C}\sim S-CoA \xrightarrow[\text{FAD} \quad \text{FADH}_2]{\text{Acyl-CoA dehydrogenase}}$$

Acyl-CoA

$$\longrightarrow R-\overset{\overset{H}{|}}{C}=\overset{\underset{H}{|}}{C}-\overset{\overset{O}{\|}}{C}\sim S-CoA$$

Acyl-CoA 2,3-unsaturated (*trans*)
trans-Δ2-enoyl-CoA

There are three isozymes of acyl-CoA dehydrogenase. One uses long chain FAs (of more than 12C) as substrate, the other acts on FAs of 6–12C in length, a third one catalyze the dehydrogenation of FAs of 4–6C in length. Complete oxidation of a long FA chain requires the action of all three isoforms.

2. *Hydration.* Water is added to saturate the double bond and to form β-hydroxyacyl-CoA (or 3-hydroxyacyl-CoA). The reaction is catalyzed by *enoyl hydratase* also called *crotonase*.

$$R-\overset{\overset{H}{|}}{C}=\overset{\underset{H}{|}}{C}-\overset{\overset{O}{\|}}{C}\sim S-CoA + H_2O \xrightleftharpoons{\text{Enoyl hydratase}}$$

trans-Δ2-enoyl-CoA

$$\xrightleftharpoons{} R-\overset{\overset{HO}{|}}{\underset{\underset{H}{|}}{C}}-\overset{\overset{H}{|}}{\underset{\underset{H}{|}}{C}}-\overset{\overset{O}{\|}}{C}\sim S-CoA$$

L-β-hydroxyacyl-CoA
L-3-hydroxyacyl-CoA

3. *Second oxidation.* The β-hydroxy derivative undergoes a new β carbon dehydrogenation to form the corresponding β-keto-acyl-CoA. The *β-hydroxyacyl-CoA dehydrogenase* is responsible for this reaction, with NAD acting as a hydrogen acceptor.

$$R-\overset{\overset{HO}{|}}{\underset{\underset{H}{|}}{C}}-\overset{\overset{H}{|}}{\underset{\underset{H}{|}}{C}}-\overset{\overset{O}{\|}}{C}\sim S-CoA \xrightleftharpoons[\text{NAD}^+ \quad \text{NADH} + \text{H}^+]{\substack{\beta\text{-hydroxyacyl-CoA} \\ \text{dehydrogenase}}}$$

L-β-hydroxyacyl-CoA

$$\xrightleftharpoons{} R-\overset{\overset{O}{\|}}{C}-CH_2-\overset{\overset{O}{\|}}{C}\sim S-CoA$$

β-ketoacyl-CoA

4. *Chain rupture and acetyl-CoA release.* Finally, the β-ketoacyl-CoA is cleaved at the junction between α and β carbons by the action of *thiolase (ketothiolase)*. This thiolytic reaction requires the presence of another molecule of coenzyme A. The products formed are acetyl-CoA and acyl-CoA, which are two carbons smaller than the original compound.

$$R-\overset{\overset{O}{\|}}{C}-CH_2-\overset{\overset{O}{\|}}{C}\sim S-CoA \xrightleftharpoons[\text{CoA}-\text{SH}]{\text{Thiolase}}$$

β-ketoacyl-CoA

$$\xrightleftharpoons{} R-\overset{\overset{O}{\|}}{C}\sim S-CoA + CH_3-\overset{\overset{O}{\|}}{C}\sim S-CoA$$

Acyl-CoA \qquad\qquad Acetyl-CoA

Fatty acids of 12 carbon residues or more are subjected to the enzymes that catalyze the last three stages (reactions 2, 3, and 4), which are located in the same trifunctional protein.

The series of reactions included in β-oxidation are summarized in Fig. 15.4.

The oxidation is repeated to completely degrade FAs into active acetates. At every step, the starting substrate is two carbons shorter than the substrate from the preceding round of reactions. Finally, the last step is initiated by a 4-carbon acyl-CoA.

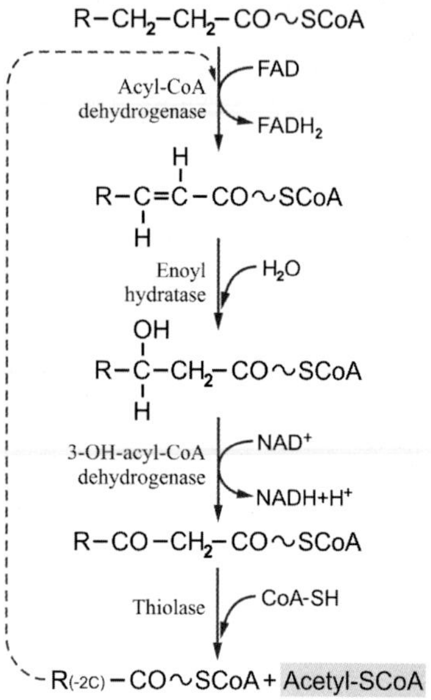

$$R-CH_2-CH_2-CO\sim SCoA$$

Acyl-CoA dehydrogenase — FAD → FADH$_2$

$$\underset{\underset{H}{|}}{\overset{\overset{H}{|}}{R-C=C}}-CO\sim SCoA$$

Enoyl hydratase — H$_2$O

$$R-\underset{\underset{H}{|}}{\overset{\overset{OH}{|}}{C}}-CH_2-CO\sim SCoA$$

3-OH-acyl-CoA dehydrogenase — NAD$^+$ → NADH+H$^+$

$$R-CO-CH_2-CO\sim SCoA$$

Thiolase — CoA-SH

$$\sim R_{(-2C)}-CO\sim SCoA + \boxed{Acetyl\text{-}SCoA}$$

FIGURE 15.4 **Fatty acid β-oxidation stages (R indicates the alkyl group, the FA has been activated by binding CoA-SH).** The four stages render acetyl-CoA and acyl-CoA that are 2C residues shorter than the starting compound. The process is repeated as many times as required, until the acyl chain is degraded in 2C segments. On the last step, a 4-carbon acyl substrate undergoes the stages of β-oxidation, producing 2 acetyl-CoA.

bound to FAD, with formation of a *trans* configuration unsaturated product. This compound is then hydrated and again oxidized to a ketoacyl, this time using NAD as a coenzyme (Fig. 15.5). The fourth stage of β-oxidation is not maintained in the Krebs cycle.

Fatty acid chains with an odd number of carbons are also subjected to β-oxidation. In the last cycle, the 5-carbon acyl-CoA produced is metabolized into acetyl-CoA and propionyl-CoA. Propionyl-CoA is the only product of FA catabolism that can enter gluconeogenesis.

β-oxidation is inhibited in Reyes' syndrome (see p. 198).

Genetic defects. A mutation in the gene encoding medium chain acyl-CoA dehydrogenase results in a failure in FA catabolism. Homozygous individuals are unable to oxidize FAs of 6–12 carbons in length. The disease is more prevalent in northern European countries. It is characterized by a pathological condition that presents with accumulation of fat in liver, high levels of octanoic acid in blood, high urinary excretion of dicarboxylic acids of 6–10 carbons in length, and reduced levels of ketones. The main symptoms include drowsiness, vomiting, hypoglycemia,

For example, β-oxidation of an eight-carbon fatty acid (caprylic acid) produces four acetyl CoA and requires three β-oxidation steps:

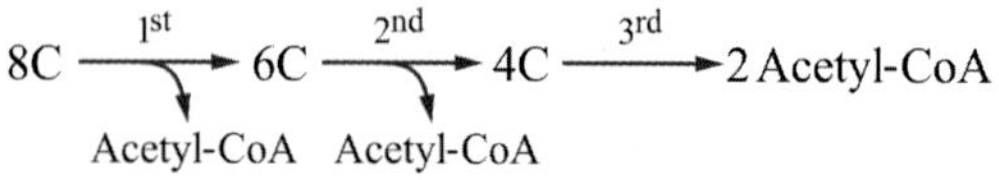

$$8C \xrightarrow[\text{Acetyl-CoA}]{1^{st}} 6C \xrightarrow[\text{Acetyl-CoA}]{2^{nd}} 4C \xrightarrow{3^{rd}} 2\,Acetyl\text{-}CoA$$

A 16-carbon FA (palmitic acid) is degraded into eight active acetates after seven β-oxidation steps.

The acetyl-CoA produced in the oxidative degradation of FAs enters the citric acid cycle for oxidation to CO$_2$ and H$_2$O.

The first three stages of FA β oxidation are similar to the last three stages of the citric acid cycle (from succinate to oxaloacetate). Both consist in the oxidation catalyzed by an enzyme

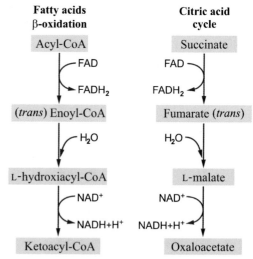

FIGURE 15.5 **Similarity between the metabolic reactions of FA β-oxidation and the citric acid cycle.**

hyperammonemia, and eventually coma. Without treatment, 30%–60% of these patients die at an early age (2–3 years). Prognosis is greatly improved if an early diagnosis is made and if a diet low in fat and high in carbohydrate content is followed, along with maintaining short intervals between meals to prevent the use of fat reserves as an energy source.

Much less common are other defects, such as loss of the activity of β-hydroxyacyl-CoA dehydrogenase.

Other Oxidation Pathways

Fatty acid oxidation in peroxisomes. Long chain FAs do not initiate their oxidation in mitochondria but in organelles called *peroxisomes*. In these organelles, a β-oxidation process different from that described before takes place. The enzymes involved are different from those of mitochondria and are encoded by other genes. The important differences are as follows: (1) the acyl-CoA that enters peroxisomes do not need the carnitine shuttle. (2) Fatty acids are subjected to oxidation and shortened to 8–10C in length. The remaining acyls undergo complete degradation in mitochondria. (3) FADH$_2$ is formed and hydrogens are not derived to the respiratory chain, but directly to molecular oxygen to form H$_2$O$_2$. This is converted to water and O$_2$ by peroxisome catalase. (4) There is no formation of high-energy bonds.

Peroxisomes are the only site where oxidation of long chain and branched chain (e.g., phytanic acid) FAs is carried out. They also contain enzymes for plasmalogens and the synthesis of bile acids.

Several alterations caused by genetic defects of peroxisomes have been described. Some affect the biogenesis of these organelles, due to failure of protein import mechanisms. This group of diseases includes various hereditary conditions, from the severe *Zellweger syndrome* to the milder infantile *Refsum disease*. In the first, patients suffer serious neurological damage that often determines death shortly after birth. Peroxisomal enzyme systems are absent, which causes an increase in very long chain FAs, phytanic acid in plasma, and deficiencies in plasmalogen synthesis. Other peroxisomal enzymes, among them catalase, are found free in the cytosol.

There are other genetic defects, such as the X-linked *adrenoleukodystrophy*, which lack the ABC FAs transporter in lysosomes.

α- and ω-oxidation. These are minor pathways in FA catabolism that catalyze the oxidation and removal of carbons from the ends of the FA chains. While these are not major pathways, genetic failures that lead to absence of α-oxidation enzymes determine neurological disorders (Refsum disease).

Unsaturated Fatty Acids Oxidation

The described β-oxidation process applies to saturated FAs. The unsaturated FAs undergo the same steps, with successive cycles of β-oxidation to release acetyl-CoA. However, as the natural unsaturated FAs have a *cis* configuration in their double bonds and the β-oxidation intermediate (enoyl-CoA) is *trans*, when the process reaches carbons joined by the double bond, additional enzymes are required to modify the steric arrangement. Monoethylenic FAs need the intervention of an *isomerase* (Δ^3,Δ^2-3-enoyl-CoA isomerase) to change the double bond position and configuration. Polyunsaturated FAs, in addition to the enoyl-CoA isomerase, require a *2,4-dienoyl-CoA reductase*.

Fatty Acid Oxidation: Energy Balance

Each β oxidation step has two stages (see reactions 1 and 3). The first consists of a hydrogen transfer; the hydrogen acceptor is FAD. In the other, the hydrogen acceptor is NAD. The transport of an electron pair from the first reaction produces, by oxidative phosphorylation, 1.5–2 high energy bonds and from the third reaction, 2.5–3 (to simplify, factors 2 and 3, respectively, will be used).

TABLE 15.3   Energy Balance of Fatty Acid Total Oxidation

Caprilic acid (8 carbons)	P high energy bonds
Initial activation	−2
Yield of β oxidation Three turns (5 × 3)	+15
Yield of oxidation in the citric acid cycle (4 acetyl-CoA) (12 × 4)	+48
Total	61

One mol of caprilic acid yields 61 moles of ATP.

Palmitic acid (16 carbons)	
Initial activation	−2
Yield of β oxidation Seven turns (5 × 7)	+35
Yield of oxidation in the citric acid cycle (8 acetyl-CoA) (12 × 8)	+96
Total	129

One mol of palmitic acid yields 129 moles of ATP.

For example, caprylic acid requires three rounds of β oxidation to achieve its complete degradation to acetyl-CoA. This produces 15 moles of ATP per mole of acid (5 × 3). Two high-energy bonds are used in the initial activation, therefore, the energetic balance is 15 − 2 = 13 moles of ATP.

Each of the four moles of the acetyl-CoA formed produces 12 ATP in the citric acid cycle (p. 301).

Table 15.3 shows the balance for the total oxidation of caprylic acid (8 carbons) and palmitic acid (16 carbons). One mole of caprylic acid generates 61 moles of ATP and one mole of palmitic acid generates 129 moles of ATP.

Considering that the hydrolysis of the ATP $\sim$P bond releases −7.3 kcal/mol (−30.5 kJ/mol), the total energy produced by the oxidation of one mole of caprylic acid is 445.3 kcal (1863 kJ) while that from palmitic acid is 941.7 kcal (3940 kJ).

The energy released during the complete combustion of palmitic acid in a calorimeter

reaches 2340 kcal/mol (9790 kJ/mol). The efficiency of oxidative degradation in the body is approximately 40%, similar to that of glucose.

The aforementioned figures indicate the importance of FAs as a biological fuel. In liver, heart, and skeletal muscle at rest, the total FA oxidation provides more than 50% of the body energy requirements.

Ketogenesis

Ketogenesis, or formation of ketone bodies, is an alternative catabolic pathway for active acetates. The amount of ketone bodies is small in normal individuals, but their levels become important in certain metabolic conditions. Acetoacetate, 3-hydroxybutyrate, and acetone are all ketone bodies.

Acetoacetate   D-3-hydroxybutyrate   Acetone

The synthesis of these compounds is performed in liver mitochondria from acetyl-CoA. The process comprises several stages:

1. *Acetoacetyl-CoA formation.* In a reaction catalyzed by thiolase, two acetyl-CoA molecules are bound to form acetoacetyl-CoA:

Acetoacetyl-CoA is also the resulting product, with acetyl-CoA, of the second to last FA β-oxidation round.

2. *3-OH-3-methylglutaryl-CoA formation.* Acetoacetyl-CoA reacts with acetyl-CoA to give 3-hydroxy-3-methylglutaryl-CoA (HMGCoA), also an intermediate

in cholesterol biosynthesis. This step is catalyzed by 3-OH-3-methylglutaryl-CoA synthase.

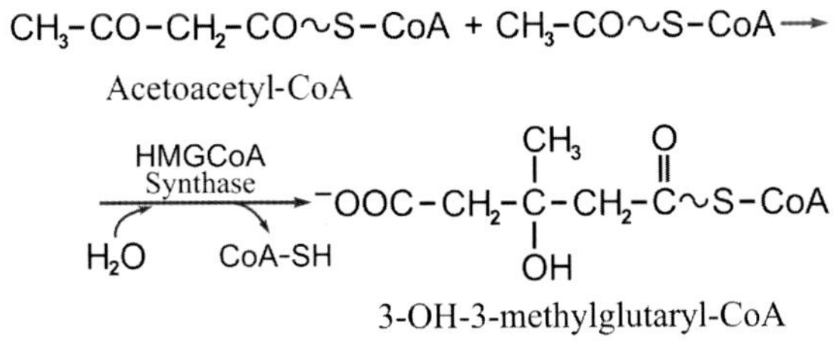

$$CH_3-CO-CH_2-CO\sim S-CoA + CH_3-CO\sim S-CoA \longrightarrow$$

Acetoacetyl-CoA

3-OH-3-methylglutaryl-CoA

3. *Acetoacetate formation.* 3-OH-3-methylglutaryl-CoA is cleaved to acetoacetate and acetyl-CoA in a reaction catalyzed by 3-OH-3-methylglutaryl-CoA lyase. This is the major pathway for the generation of acetoacetate in the liver.

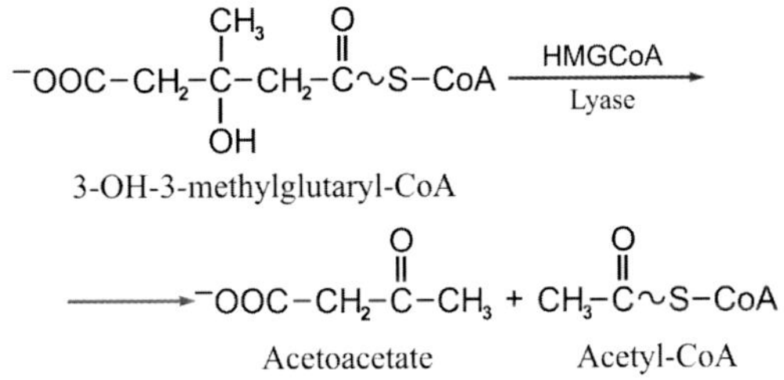

3-OH-3-methylglutaryl-CoA

Acetoacetate   Acetyl-CoA

Another minor pathway is the hydrolysis of acetoacetyl-CoA. The thioester (reaction 1) can be hydrolyzed by acetoacetyl-deacylase.

$$CH_3-CO-CH_2-CO\sim S-CoA + H_2O \xrightarrow{\text{Deacylase}}$$

Acetoacetyl-CoA

$$\longrightarrow CH_3-CO-CH_2-COO^- + CoA-SH + H^+$$

Acetoacetate

Acetoacetate is reduced by 3-hydroxy-butyrate dehydrogenase, an NAD-linked enzyme, to form D-3-hydroxybutyrate, another important ketone body. By decarboxylation, the acetoacetate produces acetone. The production of this compound is small compared to that of 3-OH-butyrate.

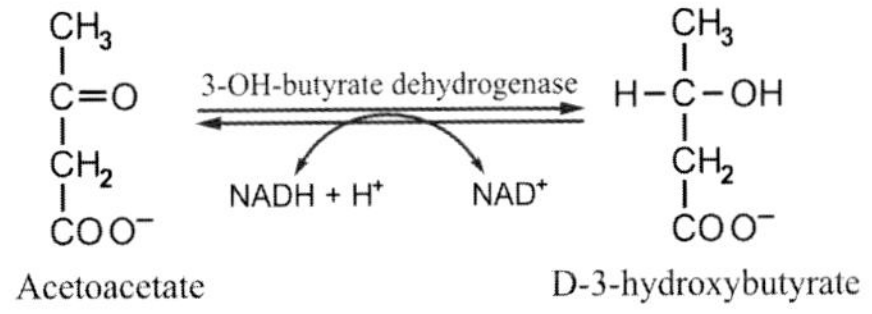

Acetoacetate   D-3-hydroxybutyrate

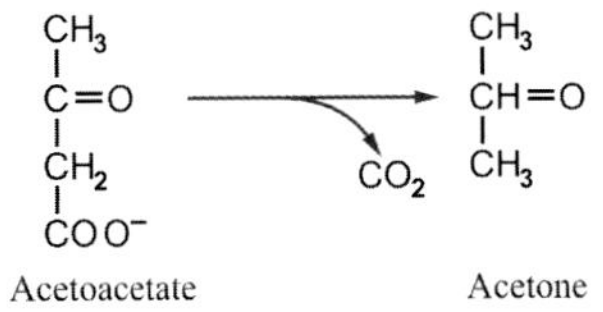

Acetoacetate   Acetone

The liver has all the enzymes necessary for the synthesis of ketones and is the main organ for the production of ketone bodies. However, it is unable to use them for energetic purposes. The ketone bodies pass from its place of origin in the hepatocyte mitochondria to the general circulation, where they are taken up by peripheral tissues.

Under normal conditions, the brain does not use ketone bodies. It is only after prolonged periods of fasting that the nervous system is able to oxidize ketone bodies as an alternative energy source. Skeletal muscle, heart muscle, and other tissues can readily metabolize ketone bodies. For this, 3-OH-butyrate is oxidized to acetoacetate in a reaction catalyzed by *3-OH-butyrate dehydrogenase*.

The tissues containing enzymes capable of activating acetoacetate can use this metabolite by two mechanisms:

1. Transfer of CoA from succinyl-CoA, a reaction catalyzed by *succinyl-CoA-3-ketoacid-CoA transferase*, also called *thiophorase*. This reaction is important in muscle.

$$\begin{array}{c}\text{Succinyl-}\\\text{CoA}\end{array} + \begin{array}{c}\text{Aceto-}\\\text{acetate}\end{array} \underset{}{\overset{\text{Thiophorase}}{\rightleftharpoons}} \text{Succinate} + \begin{array}{c}\text{Aceto-}\\\text{acetyl-CoA}\end{array}$$

2. Formation of acetoacetyl-CoA, catalyzed by acetoacetyl thiokinase, which requires ATP.

$$CH_3-CO-CH_2-COO^- + CoA-SH + ATP \xrightarrow{\text{Thiokinase}}$$

Acetoacetate

$$\longrightarrow CH_3-CO-CH_2-CO\sim S-CoA + AMP + PP_i$$

Acetoacetyl-CoA

For its final oxidation, acetoacetyl-CoA is separated into two acetyl-CoAs by action of *thiolase* (catalyzes the reaction in both directions). In individuals with an adequate diet, the brain does not possess thiophorase. However, when ketones in plasma reach high levels (2–3 mM), the central nervous system upregulates the synthesis of thiophorase.

Ketosis

In the normal adult, there is a balance between production and consumption of ketone bodies and their concentration in blood is low (1 mg/dL, or less than 0.2 mM). Occasionally, if a slight excess occurs, it is rapidly eliminated from the body by the kidneys. In normal individuals, the urinary excretion of ketone bodies is always less than 100 mg/day. However, in some abnormal conditions ketogenesis is abnormally increased, leading to ketosis.

When the absorption capacity of the renal tubules is exceeded, acetoacetate and 3-OH-butyrate are excreted in the urine (ketonuria). Acetone is removed from the body by the lungs and the exhaled air acquires a characteristic odor, easily detectable in patients with ketosis.

To understand the origin of this alteration, some metabolic characteristics of FAs and carbohydrates need to be considered.

The main catabolic pathway for active acetates, produced by FA β-oxidation, is the citric acid cycle. This depends on the availability of sufficient oxaloacetate to form citrate in the first stage. Oxaloacetate, regenerated in each turn of the cycle, is mainly provided via the anaplerotic reaction catalyzed by pyruvate carboxylase (p. 305). This takes place when there is abundant supply of pyruvate. The main supplier of pyruvate is glucose after its transformation in glycolysis. For this reason, the oxidization of acetyl-CoA, derived from FA oxidation in the tricarboxylic acid cycle, requires a balance between acetyl-CoA and glucose catabolism. There are situations in which the balance is disturbed, as is seen in diabetes mellitus. In this disease, glucose cannot enter the cells to be metabolized and energy is mainly obtained from the oxidation of FAs, which becomes more active than in normal individuals. The increased FA catabolism results in enhanced acetyl-CoA production, while the pyruvate supply is reduced, due to lack of glucose utilization. Furthermore, in diabetes mellitus, increased gluconeogenesis consumes oxaloacetate from the citric acid cycle. This decreases the levels of oxaloacetate and reduces the flow of metabolites through the citric acid cycle. The inability to oxidize acetate causes its accumulation and diversion to the formation of ketone bodies. The increase in acetoacetate and 3-OH-butyrate causes metabolic acidosis. This anion excess is eliminated by urine neutralized by Na^+ and K^+, which considerably increases the loss of cations. Ketosis, common in uncontrolled diabetics, produces serious imbalances in these patients.

Prolonged fasting mimics what happens during diabetes mellitus. Fasting requires the body to draw from its reserves to meet the tissues energy requirements. Glycogen is rapidly consumed and the fat deposits start being used, increasing FA oxidation. The decrease in carbohydrates as a source of energy upregulates gluconeogenic activity and oxaloacetate is consumed to synthesize glucose, reducing the activity of the tricarboxylic acid cycle. This, in addition to the increase in the supply of acetyl-CoA, exacerbates the production of ketone bodies.

Glucose Formation From Fats

Glycerol is the only component of triacylglycerides that is potentially glucogenic. It is phosphorylated into glycerol-3-phosphate which subsequently becomes converted by oxidation to dihydroxyacetone. This last compound is one of the triose phosphates of glycolysis and as such it has the possibility of following gluconeogenesis to synthesize glucose and glycogen.

Fatty acids, the other component of triacylglycerides, do not provide intermediates that can be derived to glucose formation. In animals, acetyl-CoA produced in β-oxidation does not form oxaloacetate and pyruvate, or any other glucogenic compound. Only FAs with an odd number of carbons generate a glucogenic product, propionyl-CoA.

FATTY ACID BIOSYNTHESIS

Fatty acids are synthesized from acetate residues (acetyl-CoA) through successive addition of two carbon fragments to the carboxyl at the end of the growing acyl chain. A system located in the cytosol is responsible for the synthesis of FAs up to 16 carbon residues in length (palmitate). In smooth endoplasmic reticulum membranes, there is another enzymatic system capable of elongating already formed FAs.

When the diet exceeds the caloric needs of the body, the excess acetyl-CoA is derived to the synthesis of acyls and these are incorporated into TAGs, contributing to an increase in body fat deposits.

De Novo Cytoplasmic Synthesis

The complete synthesis of saturated FAs from active acetate is catalyzed by multicatalytic proteins located in liver, kidney, brain, adipose tissue, lung, and mammary glands. The main product formed is free palmitate, the ionic form of palmitic acid. The main FA synthesized in mammary glands is decanoic, or capric acid. Fatty acids in milk fat have an intermediate length (6–10C), which gives them the advantage of easier absorption. This is important in infants and young children, where these FAs are directly transferred to the bloodstream, without having to include them into chylomicrons or bind them to carnitine to enter mitochondria.

Acetyl-CoA used for synthesis of FAs derives predominantly from the oxidative decarboxylation of pyruvate. Regulatory mechanisms (see p. 438) limit the availability of acetates from acyl β-oxidation.

Fatty acids are synthesized in the cytosol from acetyl-CoA that is mainly generated in the mitochondrial matrix. This requires acetyl-CoA to be transferred to the cytoplasm. The inner mitochondrial membrane is not permeable to acetyl-CoA and the carnitine transport system preferably operates with long chain acyls. Consequently, another transfer mechanism is required.

Citrate shuttle. The main system used in mammals involves citrate exit from mitochondria. Citrate is formed in the first step of the tricarboxylic acid cycle through the condensation of acetyl CoA and oxaloacetate, catalyzed by citrate synthase. When the ATP level in mitochondria is high citrate accumulates because ATP inhibits isocitrate dehydrogenase and reduces operation of the tricarboxylic acid cycle. Citrate passes through the inner membrane by a tricarboxylate or citrate/malate exchanger, which transports citrate out to the cytoplasm and moves malate into the mitochondrial matrix. Once in the cytosol, the citrate is cleaved by citrate lyase with participation of coenzyme A and ATP:

$$\text{Citrate} + \text{CoA-SH} \xrightarrow[\text{ATP} \quad \text{ADP} + P_i]{\text{Citrate lyase}}$$
$$\longrightarrow \text{Acetyl} \sim S - \text{CoA} + \text{Oxaloacetate}$$

Acetyl-CoA and oxaloacetate are regenerated. The two carbon residues produced are used for FA synthesis. Oxaloacetate cannot return to the mitochondrial matrix because the inner mitochondrial membrane is impermeable to it. In contrast, pyruvate and malate are translocated across the mitochondrial membrane by specific transporters. Oxaloacetate is reduced to malate by cytosolic malate dehydrogenase:

$$\text{Oxaloacetate} \underset{\text{NADH} + H^+ \quad \text{NAD}^+}{\overset{\text{Malate dehydrogenase}}{\rightleftharpoons}} \text{L-malate}$$

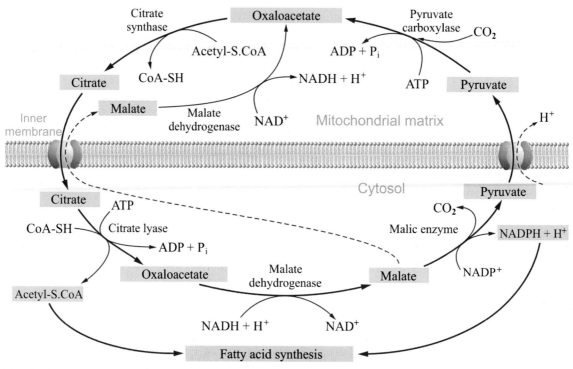

FIGURE 15.6   **Acetyl transfer from mitochondria to cytosol (role of citrate).**

In a second step, malate is decarboxylated to pyruvate. The reaction is catalyzed by malic enzyme, which is bound to NADP:

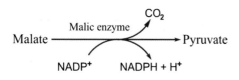

Once in the mitochondrial matrix, pyruvate has several metabolic alternatives. One is carboxylation to form oxaloacetate. The reaction is catalyzed by pyruvate carboxylase, which requires biotin and ATP.

$$\text{Pyruvate} + CO_2 + H_2O \xrightarrow[\substack{\text{Biotin} \\ ATP \qquad ADP + P_i}]{\text{Pyruvate carboxylase}} \text{Oxalo-acetate}$$

The acetyl transfer cycle from the mitochondrial matrix to the cytosol is completed with this

step, and NADPH is generated in the reaction catalyzed by malic enzyme. Together, with the NADPH produced in the pentose phosphate pathway, it provides H^+ ions for FA synthesis. Fig. 15.6 summarizes this process.

Steps of Fatty Acid Synthesis

Malonyl-CoA Formation

The first step in FA synthesis involves a process of carboxylation, which uses bicarbonate (HCO_3^-) as a CO_2 source. The acetyl-CoA is converted to malonyl-CoA in a reaction catalyzed by *acetyl-CoA carboxylase*, along with the coenzyme biotin. Biotin is a member of the vitamin B complex that functions as a HCO_3^- carrier. The reaction is coupled to the hydrolysis of ATP. This irreversible stage is the primary site of FA biosynthesis regulation. CO_2 is attached to the end of the acetate methyl to form a free carboxyl.

$$CH_3-\overset{\overset{\displaystyle O}{\|}}{C}\sim S-CoA+CO_2 \xrightarrow[\text{ATP} \quad \text{ADP}+P_i]{\overset{\text{Acetyl-CoA}}{\text{carboxylase}}} ^-OOC-\overset{\overset{\displaystyle O}{\|}}{CH_2}-\overset{\overset{\displaystyle O}{\|}}{C}\sim S-CoA$$

Acetyl-CoA   ATP   ADP+P_i   Malonyl-CoA

The C added from CO_2 is shown in red

Part of the energy released by the hydrolysis of ATP is stored in the C—C bond and serves to promote the following elongation reaction.

Fatty Acid Synthase

From malonyl-CoA formation, the synthesis of FAs up to 16C (palmitate) in length is catalyzed by a multienzyme FA synthase system. In animals, the system comprises two identical subunits which operate in close association. Each of the subunits is a multifunctional protein that clusters, in the same polypeptide chain, all the enzymes necessary for FA synthesis and also an acyl carrier protein (ACP). This system is active only when the complex is assembled into a homodimer. Each subunit has a mass of approximately 270 kDa, consisting of three globular domains linked by random chain segments. The first domain from the N-terminal side is the site of entry and condensation of the substrates. It contains three enzymes: *acetyltransferase*, *malonyltransferase*, and the condensing enzyme *ketoacyl synthase*. The latter enzyme has, at the active site, a cysteine residue which is essential for its function. The second domain is the reduction unit. It has three catalytic sites: *3-ketoacyl reductase*, *3-hydroxyacyl dehydratase*, and *enoyl reductase*. The ACP is located in this domain. ACP has 4'-phosphopantetheine as a prosthetic group, consisting of mercaptoethanolamine (with its —SH free), pantothenic acid (a component of the vitamin B complex), and a phosphate moiety through which it binds to a protein serine residue. This group is also found in coenzyme A (see p. 671). The ACP phosphopantethein binds acyls by a thioester bond to the SH group. Acyls remain anchored in the phosphopantetheine and ACP acts as a swinging arm

that supports the acyl and travels from the active site of an enzyme to the next, according to the reaction sequence. The third domain is the site at which the synthesized FA is liberated. It contains a *thiolesterase*, or *deacylase*. The segment located between domains I and II does not contain catalytic sites, but plays an important role in formation of the dimer. The subunits connect at their middle region, forming an X. Initially, it was proposed that the monomers were oriented in an antiparallel manner, with the head of one subunit in front of the tail of the other, but new evidence indicates a head to head orientation.

The complex catalyzes the sequential addition of 2C to the carboxyl end of the growing acyl. The addition of each carbon atom requires malonyl-CoA and releases carbon dioxide. This decarboxylation provides energy to form the new C—C bond.

The sequence of reactions is the following:

1. *Acetate transfer.* A molecule of acetyl-CoA reaches the income site in domain 1 and the acetyl transferase binds the acetyl residue by a thioester bond to the —SH group at the active site of the condensing enzyme. CoA-SH is released.
2. *Malonyl transfer.* Malonyl-CoA formed in the reaction catalyzed by the acetyl CoA carboxylase enters to the same domain 1 that received the acetyl radical. By action of malonyl transferase or malonyl-CoA-CTP transacylase, the malonyl residue is transferred to the —SH of the ACP phosphopantethein in domain 2 of the neighboring subunit.

$$^-OOC-CH_2-\overset{\overset{\displaystyle O}{\|}}{C}\sim S-CoA + HS-ACP \xrightarrow{\overset{\text{Malonyl}}{\text{transferase}}}$$

$$\xrightarrow{} {}^-OOC-CH_2-\overset{\overset{\displaystyle O}{\|}}{C}\sim S-ACP + CoA-SH$$

The malonyl group, linked by a thioester bond to ACP, comes very close to the acetyl bound to the other unit condensing enzyme.

After these preparatory steps, four successive reactions, condensation, reduction, dehydration, and further reduction, follow. In both reductions, NADPH is the hydrogen donor.

3. *Acetyl with malonyl condensation.* The free carboxyl group of malonyl is separated as CO_2. Immediately, acetyl is bound to the site previously occupied by the lost carboxyl. Acetyl is released from the active site of the condensing enzyme, leaving its —SH free. Both of the carbons from acetate will become the last two carbon residues of the FA. The methyl group of acetyl-CH_3 will constitute the FA terminal carbon (ω C).

The CO_2 released by malonyl is the same entered in the initial carboxylation reaction. The result after condensation can be viewed as the addition of two acetates to the molecule. The carboxylation of acetyl-CoA at the beginning of the process, and the subsequent removal of the added CO_2, appears to be useless. However, this reaction is justified because the malonyl decarboxylation reaction produces a marked decrease of free energy, which allows for the condensation of the acyl moieties.

Acetyl bonding with decarboxylated malonyl to form acetoacetyl-ACP is catalyzed by the *ketoacyl synthase*, also known as the condensing enzyme.

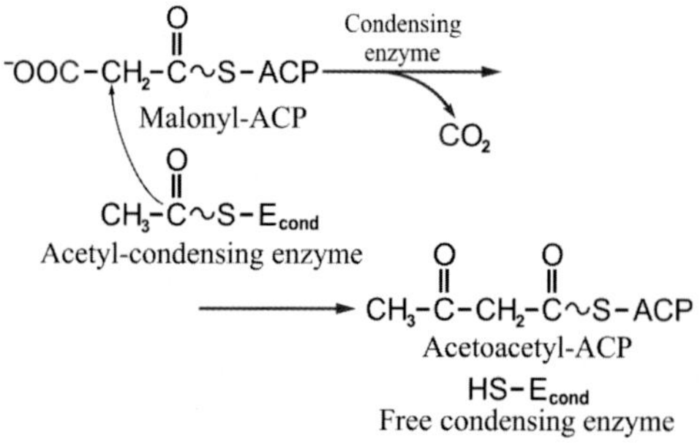

The acetoacetyl remains bound to the —SH group of phosphopantetheine, which allows its translocation to the next enzyme.

4. *First reduction reaction.* Acetoacetyl-ACP receives two hydrogens in a reaction catalyzed by *3-ketoacyl reductase,* or *3-ketoacyl-ACP reductase,* which transfers reduced NADP hydrogens to form 3-hydroxybutyryl-ACP.

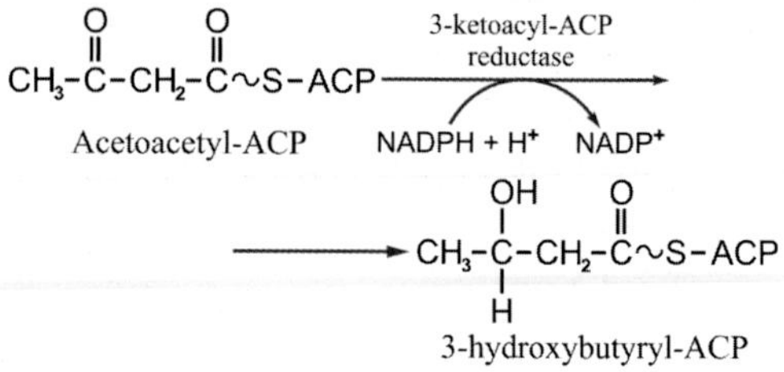

5. *Dehydration.* 3-hydroxyacyl-ACP loses a water molecule in a reaction catalyzed by *3-hydroxyacyl dehydratase.* An acyl, unsaturated between carbons 2 and 3 (Δ-2-enoyl-PTA), is formed.

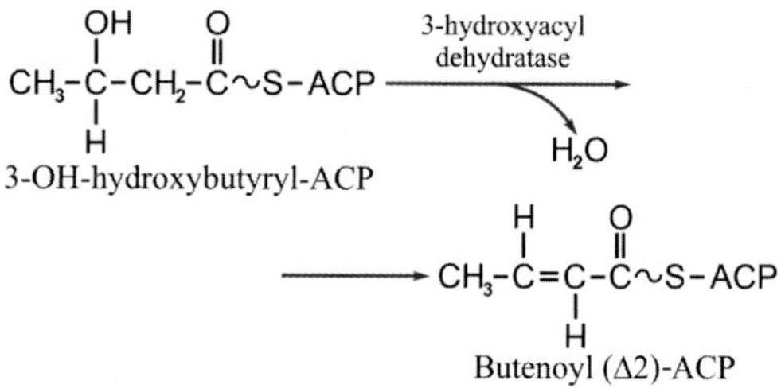

6. *Second reduction reaction.* The unsaturated compound is hydrogenated by the action of *enoyl reductase,* the sixth enzyme of the complex; reduced NADP functions as the hydrogen donor. ACP-butyryl is formed.

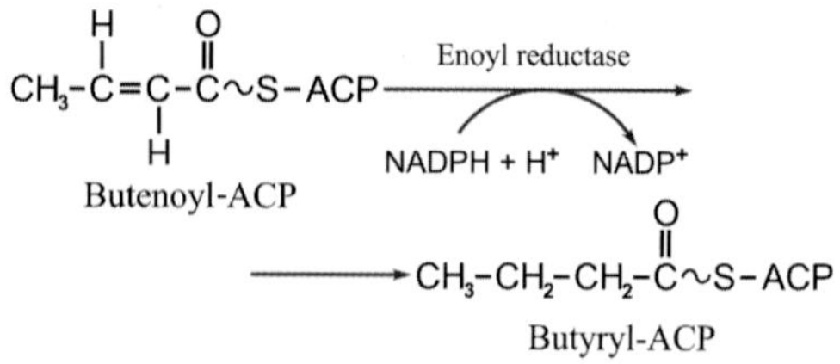

At this stage, the process has produced a saturated 4-carbon acyl. The system does not stop at short chain acids, but instead continues adding two carbon segments at a time to produce 16C (palmitate) acyls. Addition of carbons to the acyl moiety is initiated by the transfer of the 4-carbon acyl to the —SH group of a cysteine residue in the condensing enzyme. The ACP phosphopantetheine —SH group is now freed and binds another malonyl. The malonyl group enters the admission site of domain 1 facing the receiving ACP. *Malonyltransferase* transfers it to the phosphopantetheine —SH group. A new cycle starts to incorporate another two C to the FA chain. Malonyl loses its free carboxyl group, which is released as CO_2. Both C residues bind to the 4-carbon acyl molecules from the previous cycle and 3-ketocaproyl-ACP is formed. The cysteine —SH group in the condensing enzyme is free once again. The series of reduction–dehydration–reduction reactions is repeated to produce a 6-carbon saturated acyl caproyl-ACP, which is transferred to the condensing enzyme —SH. This translocation process of the newly formed chain, from one binding site to another, is similar to the synthesis of polypeptide chains in ribosomes (p. 502). Another cycle with a new malonyl adds two additional carbon residues to the chain. This process continues until a 16-carbon acyl is obtained.

Hydrolysis catalyzed by *thiolesterase*, or deacylase, releases palmitate that needs to be activated to acyl-CoA before it can follow other metabolic pathways.

The mammary gland is an exception because it contains a *decanoyl deacylase* that liberates the acyl chain from ACP when it reaches 10 carbons in length.

The complete synthesis of palmitic acid requires the cycle to be repeated seven times. The overall equation for this synthesis is

$$Acetyl\text{-}CoA + 7\,malonyl\text{-}CoA + 14\,NADPH + 14\,H^+ \longrightarrow$$

$$\rightarrow CH_3\text{-}(CH_2)_{14}\text{-}COO^- + 7\,CO_2 + 8\,CoA\text{-}SH + 14\,NADP^+ + 6\,H_2O$$
Palmitate

The 14 hydrogen pairs required for the reduction steps are donated by NADPH; abundant supply of the reduced coenzyme is needed. The main metabolic pathway producing reduced NADP is the pentose phosphate, which explains the high activity of this pathway in tissues with active FA synthesis, such as the liver, lactating mammary glands, and adipose tissue. The exchange of NADP-NADPH between the FA and pentose pathways is facilitated by the location of both systems in the cytosol. Another reaction that supplies NADPH is that catalyzed by malic enzyme, involved in the transfer of acetyl to cytosol.

Normally, free FAs do not accumulate in the cell because their production rate is regulated according to the requirement of acylglycerol and phospholipid synthesis. Acetyl-CoA carboxylase, which catalyzes the first reaction, is the major regulatory enzyme of the pathway. Long chain FAs have inhibitory action on this enzyme (negative feedback). Moreover, citrate allosterically activates the acetyl-CoA carboxylase.

Fatty Acid Elongation

The FA synthase system mainly produces palmitate. Longer chain FAs (18 or more carbons in length) are synthesized from palmitic acid by successive addition of two carbons at a time. The first step in this elongation process is the acyl activation by acyl thiokinase to form palmitoyl-CoA. This is performed by two systems; the most important (microsomal system) is located on the cytosolic face of the endoplasmic reticulum membrane. Malonyl-CoA residues provide the carbons, whereas NADPH provides the hydrogen ions required for the synthesis.

A second, less important elongation system operates within mitochondria. It uses acetyl-CoA and NADH or NADPH as a donor of reduction equivalents.

The stages of elongation are similar to those described for the synthesis, but the chain is not attached to ACP and the enzymes are controlled by different genes.

Unsaturated Fatty Acids Biosynthesis

Monoethylenic

Monounsaturated oleic and palmitoleic acids are synthesized in the smooth endoplasmic reticulum from stearic and palmitic acid, respectively. The initial step consists in the addition of a double bond to the activated acyl between carbons 9 and 10. The reaction is catalyzed by an electron transport system consisting of the flavoprotein *NADH-cytochrome b$_5$ reductase*, cytochrome b$_5$, and *9-desaturase*. The electrons are successively transferred from NADH to the cytochrome b$_5$ reductase FAD, then to the nonheme Fe (FeS center) of desaturase. Finally, desaturase interacts with O$_2$ and saturated acyl-CoA. A double bond between C 9 and 10 is formed, and two water molecules are released. The overall reaction can be represented as follows:

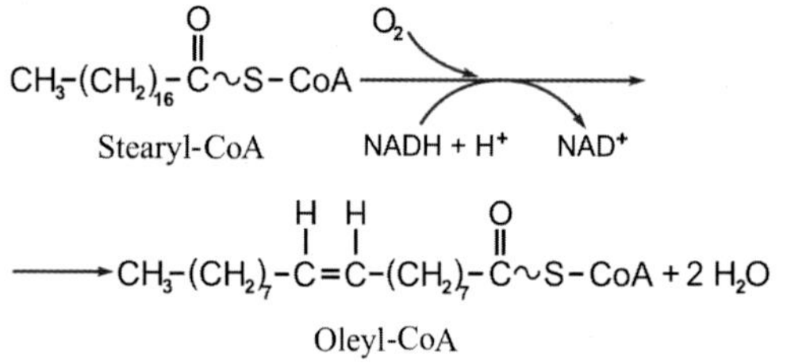

Oleic acid (18:1Δ9) is the main product of Δ9 desaturase and the most abundant FA in the TAGs of mammalian adipose tissue.

Polyethylenic

Polyunsaturated acids are formed from a monounsaturated acid synthesized as described earlier. When the first double bond between carbons 9 and 10 has been introduced, the subsequent double bonds are separated by a methylene group. There is substantial difference in the synthesis of polyethylenic FAs between higher plants and animals. Plants can introduce additional double bonds which are located between carbons 9 and 10 and the terminal segment of the chain, toward the ω methyl carbon. In contrast, mammals introduce the additional double bond in the proximal portion of the chain, toward the carboxyl group. Besides Δ9, humans have Δ6 and Δ5 desaturases. Position ω9 is the closest site to the ω end in which a double bond can be inserted. For example, octadecadienoic acid (18:2Δ6,9) can be produced from oleic acid (18:1Δ9). Linoleic (18:2Δ9,12) and linolenic (18:3Δ9,12,15), ω6, and ω3 FAs, respectively, cannot be synthesized by mammals. This inability of humans to introduce double bonds in ω3 or ω6 positions makes these FAs essential, or indispensable; they need to be provided with the diet to avoid deficiencies.

Arachidonic acid (20:4Δ5,8,11,14) is a partially essential FA, because animals can synthesize it from linoleic acid by additional elongation and desaturation.

Physiological role of polyunsaturated FAs. Essential FAs have different functions: (1) they form part of structural lipids in the cells, mainly in membranes, generally found in the sn2 position of glycerophospholipids. (2) They are the precursors of prostaglandin, thromboxane, and leukotriene synthesis. (3) They participate in the formation of cholesterol esters in blood plasma, affecting plasma cholesterol transport and level.

Polyunsaturated FAs and atherogenesis. Numerous studies have shown that continued consumption of an appropriate proportion of polyunsaturated FAs reduces elevated levels of plasma cholesterol and LDL, which are recognized risk factors for atherosclerosis. On the other hand, excessive intake of these FAs, which are susceptible to peroxidation, may predispose to lipid changes in LDL. LDL oxidation has been identified as an atherogenic factor.

Some observations have shown that ω−3 (*n*−3) polyunsaturated FAs, especially those contained in marine fish, have greater effect than ω−6 (*n*−6) in preventing atherogenesis. Linolenic acid, eicosapentaenoic (EPA, 20:5 *n*−3), and docosahexaenoic acid (DHA, 20:6 *n*−3) (18:3 *n*−3), interfere with platelet aggregation and the release of proinflammatory cytokines involved in atherosclerotic plaque formation. However, other evidence indicates that both $n-3$ and $n-6$ FAs have similar antiatherogenic properties.

More recently, it has been shown that a diet rich in monounsaturated FAs (i.e., oleic acid, 18:1 $n-9$) reduces LDL cholesterol and TAGs and increases HDL in a manner similar to that of diets containing polyunsaturated FAs.

Besides its role as an antiatherogenic agent, $\omega-3$ FAs (especially docosahexaenoic) are important components of brain membrane phospholipids and are essential for neuronal function. Deficiencies of these acids in the diet are a cause of disorders in neural system development and alterations in cognitive capacity and memory.

The *trans* isomers of polyunsaturated FAs present in small proportion in natural fats significantly increase during hydrogenation of vegetable oils, a process used for the production of margarines. Diets containing large amounts of these *trans* FAs, as well as those rich in saturated FAs and cholesterol (animal fats), favor the increase in LDL and cholesterol.

The long-term consumption of saturated FAs, particularly palmitic acids and shorter FA chains, is frequently accompanied by elevated LDL. The mechanism by which saturated FAs exert their hypercholesterolemic effect has not been clearly defined. It has been proposed that they have an inhibitory effect on the secretion of bile acids, increase cholesterol synthesis by interfering with the regulatory enzyme hydroxymethyl-glutaryl CoA reductase, depress the activity of LCAT (see p. 328), and act on the expression of factors involved in cholesterol biosynthesis.

Interestingly, stearic acid (a 18C saturated FA) does not increase cholesterol levels.

It is worthwhile noticing that although it is a risk factor that contributes to aggravating hypercholesterolemia, dietary cholesterol only has a minor effect on plasma LDL levels, which is otherwise dependent on cholesterol biosynthesis and metabolism. Moreover, beyond the proven effect of dietary lipids, there is great individual variability in the response to high intakes of saturated or *trans* polyunsaturated FAs. It is clear that genetic factors have a decisive influence in determining predisposition to cardiovascular disease and to establish the severity of risk factors.

EICOSANOID BIOSYNTHESIS

Polyunsaturated FAs of 20 carbon residues in length (eicosanoic) are precursors of a family of substances called the *eicosanoids*. They comprise compounds of great functional interest, including *prostaglandins, thromboxanes,* and *leukotrienes.*

Prostaglandins are unsaturated hydroxyacids with a cyclopentane ring. They are designated with the letters PG followed by a third letter (A–I). This third letter corresponds to different types of compounds that differ structurally in the position of hydroxyl and ketone functions. The most important are the PGE and PGF series; PGE has a ketone function at C9 whereas PGF has a hydroxyl group in that position. Both series have hydroxyls at carbons 11 and 15. A number included as a subscript is used to indicate the number of double bonds (PGE_2 and PGF_2 are the most common). Finally, another subscript (a Greek letter) is added to define the spatial configuration of the carbon chain linked to C8 in the cyclopentane. For example, $PGF_{2\alpha}$ is a prostaglandin with: (1) a hydroxyl group at C9, C11, and C15, (2), two double bonds between C5-6 and C13-14, and (3), the chains at C8 and C12 located in a different plane (α configuration).

Thromboxanes have a similar structure, but are characterized by a hexagonal instead of pentagonal ring. The most important is TXA_2. Prostaglandin PGI_2, also called prostacyclin, is a compound with two pentagonal cycles (Fig. 15.11). $PGF_{2\alpha}$, PGI_2, and TXA_2 are unstable under physiological conditions; they become inactivated quickly.

Leukotrienes are FAs with four double bonds ($\Delta7,9,11,14$).

Prostaglandins are ubiquitous substances. They are found in all cells, except erythrocytes. Thromboxanes are synthesized within platelets, prostacyclin is formed in blood vessel endothelium, and leukotrienes are made in leukocytes. In general, they act near the site where they are synthesized (paracrine action); they are not transported by the blood to act at distant places of their origin.

Series 2 prostaglandins, thromboxanes, and leukotrienes are synthesized from arachidonate. This FA is not commonly free in cells, but associated with membrane phospholipids. For this reason, the initial step in the synthesis of arachidonate derived compounds is the hydrolysis of a glycerophospholipid, such as phosphatidylcholine by phospholipase A_2 action. Lysophosphatidylcholine is formed and the FA at position 2, often arachidonate, is released. Arachidonate can follow two pathways: prostaglandin and thromboxane synthesis, or leukotriene formation.

Prostaglandins and thromboxane synthesis. This process begins with the conversion of arachidonic acid into cyclic endoperoxide (PGG_2) by action of *prostaglandin endoperoxide synthase* in the endoplasmic reticulum. PGG_2 is then transformed into PGH_2 by a peroxidase. The initial two steps (cyclization and peroxidation) are catalyzed by a bifunctional enzyme called *cyclooxygenase* (COX), which is expressed as two isozymes designated (COX-1 and COX-2). COX-1 is a constitutive enzyme, expressed in most cells. In contrast, COX-2 is inducible by cytokines and growth factors, and has a more limited cellular distribution. COX activity markedly increases in tissues affected by inflammatory processes and the increase is due to induction of COX-2 synthesis.

Starting from PGH_2, various chemical modifications lead to the formation of PGE_2, $PGF_{2\alpha}$, or thromboxane TXA_2 and prostacyclin (PGI_2) (Fig. 15.7).

Leukotriene synthesis. A specific route for synthesis of leukotrienes is initiated by the action of cytosolic enzymes, *lipo-oxygenases*, which catalyze the conversion of arachidonate into hydroperoxy-eicosatetraenoic acid (HPETE). Different lipo-oxygenases are expressed that are cell specific and can be differentiated by the site of the carbon chain where the hydroperoxy group is inserted. For example, leukocytes contain 5-lipoxygenase and 12-lipoxygenase, to produce 5-HPETE and 12-HPETE, respectively, while platelets have only 12-lipoxygenase. 5-lipoxygenase is the most important and is involved in the synthesis of leukotrienes.

5-HPETE is converted to an unstable epoxide leukotriene A_4 (LTA_4), which is converted into leukotriene B_4 (LTB_4) by the action of a hydrolase. Alternatively, carbon 6 of A_4 can bind glutathione (tripeptide γ glutamyl-cysteinyl-glycine) to form leukotriene C_4 (LTC_4).

LTC_4 glutamate is lost by hydrolysis, producing LTD_4, which is converted to LTE_4 by removing the glycine residue (Fig. 15.8). The mixture of LTC_4 and LTD_4 is called *slow reacting substance of anaphylaxis* (SRS-A), which participates in immunological reactions known long before the discovery of leukotrienes.

Functional role of eicosanoids. Prostaglandins have great biological activity. Concentrations around 1 ng/mL produce intense smooth muscle contraction. However, the action of prostaglandins can vary depending on each type, sometimes having antagonistic effects. For example, while $PGF_{2\alpha}$ has potent bronchoconstrictor and vasocontrictor action, PGE_2 causes bronchial, as well as heart, kidney, skeletal muscle, and mesentery vessels dilatation. In general, prostaglandins cause contraction of smooth muscle in the uterus and the gastrointestinal tract, inhibition of lipolysis in adipose tissue, and inhibition of the HCl secretion in stomach.

Thromboxanes have platelet aggregating properties and have vasoconstrictor effects. Prostacyclin functions as a platelet antiaggregation factor and has vasodilating effects. PGE_2

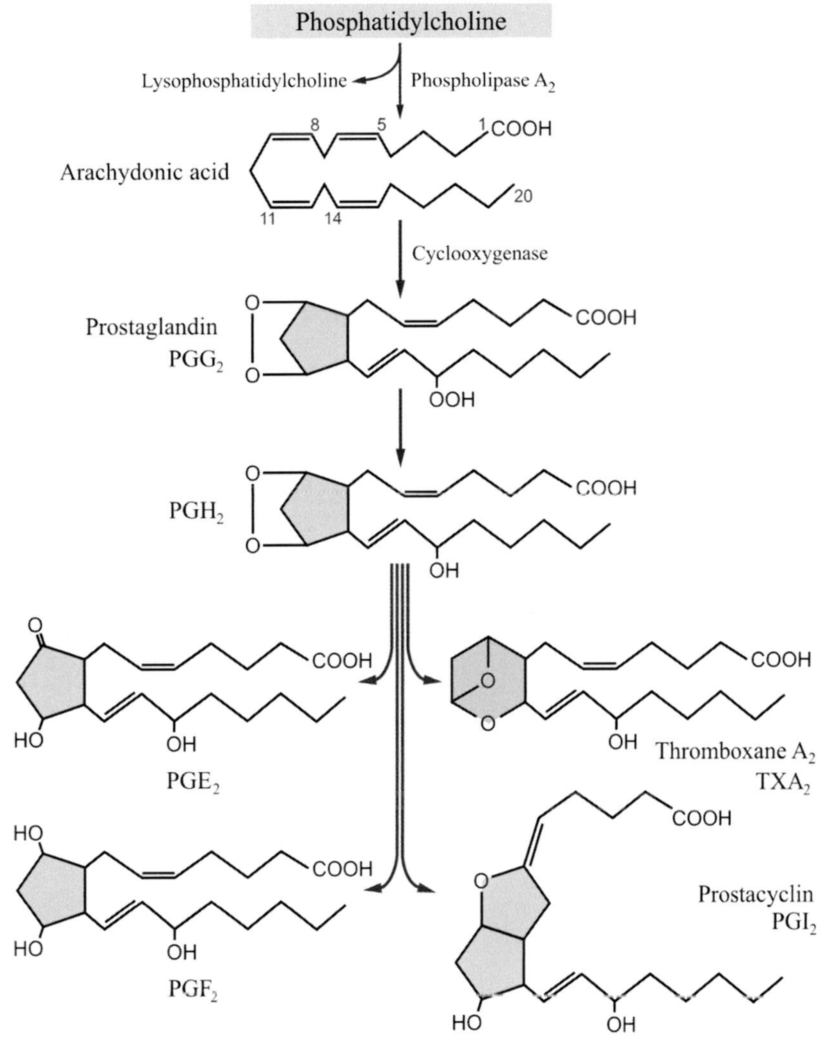

FIGURE 15.7 **Pathway for the synthesis of prostaglandin, thromboxane, and prostacyclin.**

and PGI$_2$ (prostacyclin) increase capillary permeability and contribute to the vascular inflammation phase, with production of edema. Leukotrienes are potent bronchial smooth muscle constrictors, eliciting an effect that is 1000 times more potent than histamine. They are small arteriole vasoconstrictors and increase capillary permeability more than histamine. Through this action, leukotrienes contribute to the edema that accompanies inflammation. Leukotrienes,

like prostaglandins, are inflammatory mediators and are associated with abnormal immune responses, such as allergies. Discovery of these compounds has generated great interest.

The ability to pharmacologically interfere with the production of eicosanoids is of great medical interest. Aspirin, indomethacin, and other nonsteroidal antiinflammatory drugs (NSAIDs) block prostaglandin synthesis by their inhibitory action on both cyclo-oxygenases.

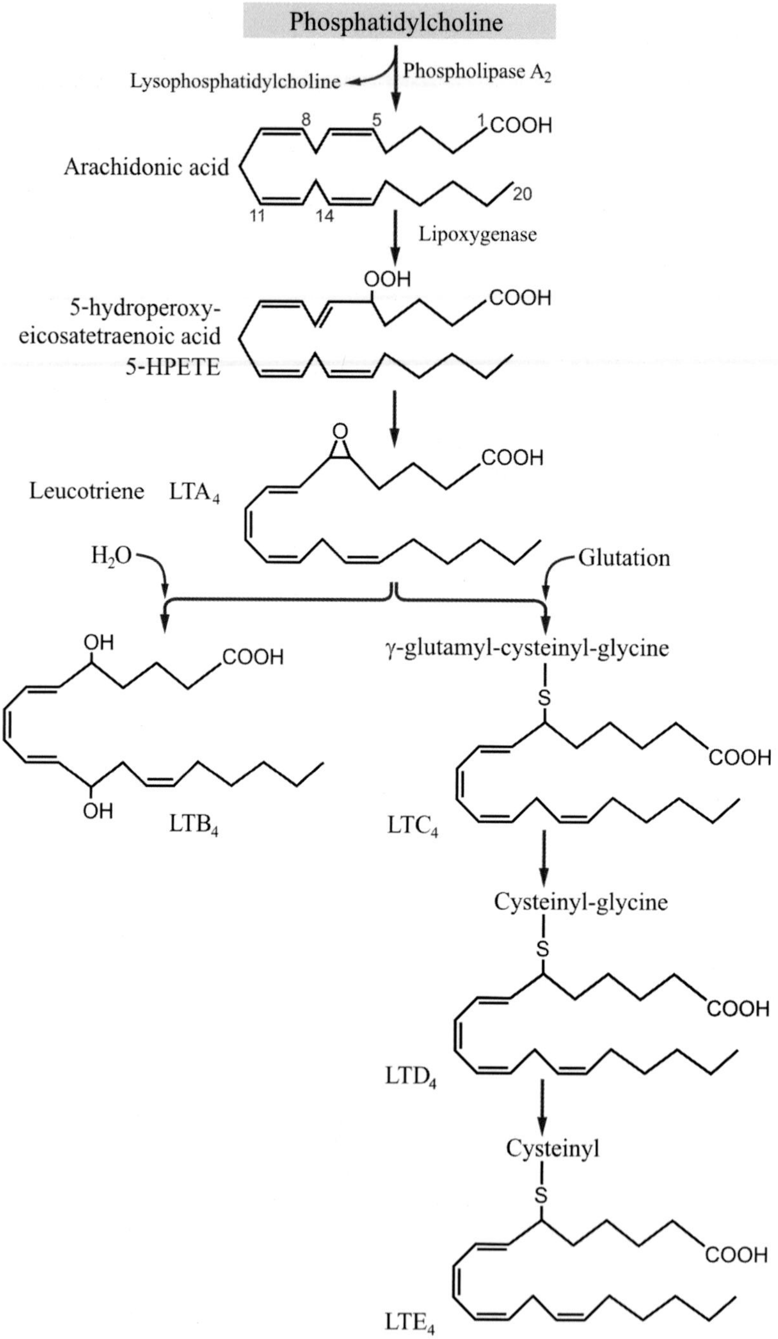

FIGURE 15.8   **Pathway for the synthesis of leukotrienes.**

These drugs, in addition to exerting antiinflammatory action, have undesirable side effects, such as gastroduodenal ulceration and bleeding.

Drugs that are designed to selectively inhibit COX-2, which is related more to prostaglandin production at sites of inflammation, have antiinflammatory activity without the side effects of COX-1 inhibitors.

Other drugs inhibit phospholipase A_2, decreasing the release of arachidonate and reducing the activity of eicosanoid synthesis. There are also compounds that inhibit some steps in the synthesis of leukotrienes.

TRIACYLGLYCEROL BIOSYNTHESIS

The first steps of triacylglycerol synthesis are the activation of glycerol to glycerol-3-phosphate and activation of FAs to acyl-CoA. In both cases, ATP and the corresponding kinase are required.

Glycerokinase catalyzes glycerol activation in liver, intestine, mammary glands, and kidneys. The enzyme is absent in muscle and adipose tissue and, in well nourished people, glycerol-3-phosphate is derived from dihydroxyacetonephosphate, a metabolic intermediary of glycolisis.

Glycerophosphate may also be generated in liver and adipose tissue, especially in fasting conditions, by *glyceroneogenesis*. This starts with the production of phosphoenolpyruvate, from oxaloacetate in a reaction catalyzed by *phosphoenolpyruvate carboxykinase*, and continues to form triose phosphate through the stages of gluconeogenesis (see Fig. 14.11). Dihydroxyacetone is converted to glycerol-3-phosphate by *glycerophosphate dehydrogenase*.

Fatty acids are activated to acyl-CoA by *thiokinase* utilizing ATP and CoA.

Glycerol-3-phosphate is esterified at carbons 1 and 2 by two acyls transferred from acyl-CoA to form 1,2-diacylglycerolphosphate, also called phosphatidic acid. The reaction is catalyzed by *glycerophosphate acyltransferase*.

$$\text{Glycerol-3-phosphate} + 2\,\text{Acyl-CoA} \xrightarrow{\text{Acyl transferase}} \text{Phosphatidic acid} + 2\,\text{CoA-SH}$$

Phosphatidic acid is hydrolyzed to 1,2-diacylglycerol by *phosphatidic acid phosphohydrolase*.

$$\text{Phosphatidic acid} \xrightarrow{\text{Phosphatase}} \text{1,2 diacylglycerol} + P_i$$

A new acyl-CoA molecule transfers another acyl to 1,2-diacylglycerol and triacylglycerol is produced. The enzyme involved is *diacylglycerol acyltransferase*. All the enzymes that participate in triacylglycerol synthesis are in the cellular microsomal fraction (endoplasmic reticulum). In the intestinal mucosa, the absorbed monoacylglycerols contribute, along with acyls from acyl-CoA, to form TAGs without previous activation to phosphatidic acid.

Obesity

Adipose tissue is the major site for body fat storage. The size of this deposit depends on the total number of existing adipocytes and the amount of TAGs per cell. The increase in the fat deposit is a characteristic of obesity, a frequent condition that is associated with different health problems in humans. It is estimated that there are over one billion people suffering from obesity of varying degrees of severity in the world.

Obesity is characterized by an energetic imbalance, with a higher intake than consumption of energy. The excess fat saturates the store capacity of adipose tissue and fat deposits develop in other organs, such as the liver, muscle, and heart, where they cause functional alterations.

Fat overload influences the levels of lipoproteins. For example, the increased flow of FAs from adipose tissue to the liver stimulates production of VLDL, diminishes lipolysis of triacylglyceride-rich lipoproteins by LpL, and increases HDL catabolism. Frequently, obesity is accompanied by insensitivity to insulin, which may result in hyperglycemia.

Multiple factors (genetic, neurological, psychological, hormonal, and environmental) contribute to this pathology. In many cases, the alteration primarily lies in the mechanisms that regulate appetite. The sensation of hunger is controlled by a complex system of hormones and peptides secreted by the hypothalamus, which influence the behavior of food intake. The most important of these factors are *α-melanocyte stimulating hormone* (αMSH), an appetite depressant with anorexic action, as well as *neuropeptide Y* and the *agouti-related protein*, which are orexigenic or appetite stimulators. Release of these factors is modulated by agents produced in the digestive tract and in adipose tissue. The stomach secretes *ghrelin*, an activator of the neurons that release neuropeptide Y. Its secretion increases during periods between meals, when the gastric lumen is empty. If food enters the stomach, ghrelin production drops rapidly. *Cholecystokinin* from the intestine functions as an antagonist of ghrelin. Its secretion increases immediately after a meal. Another appetite suppressant is the intestinal hormone PYY, which has a prolonged effect, lasting up to 12 h after food intake.

Adipose tissue is not only a reservoir of fat, but an active endocrine organ. It produces protein hormones, such as *leptin, adiponectin, resistin,* and proinflammatory cytokines.

Leptin is a 16 kDa protein secreted by adipocytes, its plasma levels in humans increase after about 3–4 days of excessive food intake and decreases after several hours of fasting. Leptin released into circulation traverses the blood brain barrier, binds to specific receptors on neurons in the arcuate nucleus of the hypothalamus, and triggers a series of actions. These include: inhibition of orexigenic, activation of anorexigenic neuropeptides, and stimulation of general catabolism. It has important effects on energy balance. As a result of these actions, food intake

decreases, body weight is reduced, and fat oxidation and energy expenditure increases. The absence of leptin, its inability to reach the brain, or its inability to interact with its receptors, lead to obesity. Lack of leptin due to genetic defects has been demonstrated in a strain of obese mice. In obese humans, leptin deficiency has not been detected, but its levels in plasma are found to be increased. This suggests that there is a lack of response to leptin dependent on an alteration of its receptor. Clinical tests show that leptin administration has no effect on body weight of obese individuals. However, leptin and other hormones have important actions on adipose tissue, such as regulating energy metabolism and improving tissue insulin sensitivity (see p. 615).

Insulin is a pancreatic hormone that also helps maintaining the sensation of satiety. It reduces the tendency to eat foods, either directly, by suppressing the release of neuropeptide Y in the hypothalamus, or indirectly, by stimulating adipogenesis from glucose. Different from other factors, leptin and insulin exert their effects over long periods of time.

BIOSYNTHESIS OF PHOSPHOLIPIDS

The most abundant phospholipids in higher animals are synthesized from phosphatidic acid, the synthesis of which has been considered in the previous section. Phosphatidic acid is an important metabolic intermediate that gives origin to glycerophospholipids and TAGs. The enzymes involved in phospholipid synthesis are embedded in the membranes of the endoplasmic reticulum.

Phosphatidic acid reacts with cytidine nucleoside triphosphate (CTP) to form cytidine diphosphate diacylglycerol (CDP-diacylglycerol), the activated form of phosphatidic acid. The reaction is catalyzed by *CDP-phosphatidic acid-cytidyltransferase.*

$$\text{Phosphatidic acid} + \text{CTP} \xrightarrow{\text{Transferase}} \text{CDP-diacylglycerol} + \text{PP}_i$$

Phosphatidylinositol and phosphatidylserine biosynthesis. Another specific transferase (*CDP-diacylglycerol-inositoltransferase*) catalyzes phosphatidylinositol formation. This compound is phosphorylated by action of kinases (phosphoinositide 3 kinases) that transfer phosphate from ATP, to render phosphatidylinositol-monophosphate, phosphatidylinositol-bisphosphate, and phosphatidylinositol triphosphate. Phosphatidylinositols are compounds of biological interest (see p. 560). Upon stimulation by hormones and other agonists, phosphatidylinositol-bisphosphate is hydrolyzed, releasing inositoltrisphosphate, a compound that acts as a "second messenger" in cells.

CDP-
diacylglycerol + Inositol $\xrightarrow{\text{Transferase}}$ Phospafatidyl-
inositol + CMP

Phosphatidylinositol $\xrightarrow[\substack{\diagup \quad \diagdown \\ \text{ATP} \quad \text{ADP}}]{\text{Kinase}}$ Phosphatidylinositol-monophosphate

Phosphatidylinositol-
monophosphate $\xrightarrow[\substack{\diagup \quad \diagdown \\ \text{ATP} \quad \text{ADP}}]{\text{Kinase}}$ Phosphatidylinositol-bisphosphate

Phosphatidylserine is a minor component of cell membranes in higher animals. It is also formed by transfer of serine to the activated intermediate CDP-diacylglycerol.

Phosphatidylcholine and phosphatidylethanolamine biosynthesis. Phosphatidylcholine and phosphatidylethanolamine are the most abundant phospholipids of cell membranes in higher animals. Their synthesis requires phosphatidic acid, which is hydrolyzed to 1,2-diacylglycerol and phosphate. Choline and ethanolamine react with ATP to form phosphorylcholine and phosphorylethanolamine. Subsequently, these compounds are combined with CTP to give cytidine diphosphate choline (CDP-choline) and citidinediphosphate ethanolamine (CDP-ethanolamine). The first reaction is catalyzed by kinases and the second by specific transferases. Cytidine nucleotides are high energy intermediates involved in the synthesis of these lipids.

Finally, CDP-choline or CDP-ethanolamine react with 1,2-diacylglycerol to form phosphatidylcholine or phosphatidylethanolamine, respectively.

The series of reactions is the following:

1. Phosphatidic acid $\xrightarrow{\text{Phosphatase}}$ 1,2-diacylglycerol + P$_i$

2. Choline + ATP $\xrightarrow{\text{Kinase}}$ Phosphorylcholine + ADP

3. Phosphorylcholine + CTP $\xrightarrow{\text{Transferase}}$ CDP-choline + PP$_i$

4. $\substack{\text{CDP-} \\ \text{choline}}$ + $\substack{\text{1,2-diacyl-} \\ \text{glycerol}}$ $\xrightarrow{\text{Transferase}}$ $\substack{\text{Phosphatidyl-} \\ \text{choline}}$ + CMP

Sphingomyelin biosynthesis. Sphingomyelin consists of ceramide linked to phosphorylcholine. Ceramide results from the binding of the amino alcohol sphingosine to a long chain FA, through an amide bond. Sphingosine is biosynthesized from serine and palmitoyl-CoA. Then, an acyl from acyl-CoA is transferred to N-sphingosine to form ceramide. Finally, another transferase inserts a phosphorylcholine moiety from CDP-choline to the hydroxyl group of the primary carbon of ceramide via an ester bond.

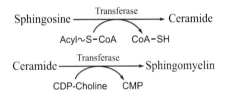

The synthesis of these compounds occurs in the smooth endoplasmic reticulum.

Cerebrosides and gangliosides biosynthesis. Cerebrosides are glycolipids composed of ceramide (sphingosine + FA) linked to a glycoside residue (galactose or glucose). For the synthesis of cerebrosides (the most abundant are galactosylcerebrosides), the UDP-galactose galactosyl residue is transferred to the primary hydroxyl of ceramide by *galactosyltransferase.*

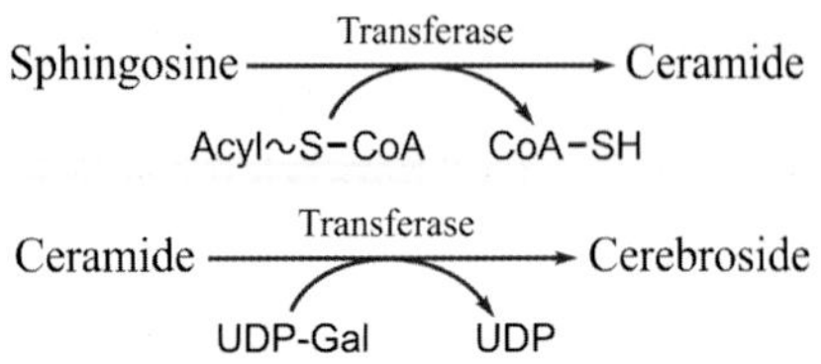

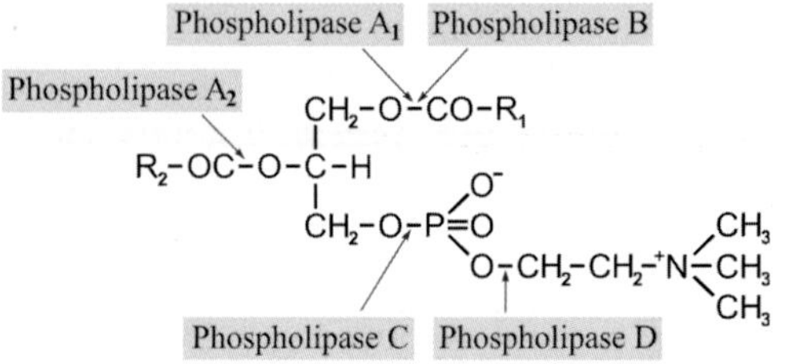

FIGURE 15.9　**Sites of action of the phospholipases on a glycerophospholipid.**

Gangliosides have an oligosaccharide chain attached to carbon 1 of sphingosine in the ceramide. The monosaccharides are transferred from the activated form of nucleotide-sugars (UDP-Gal, UDP-Glc, UDP-GalNAc, UDP-GlcNAc, GDP-fucose). The active sialic acid is CMP-NeuAc (cytidine monophosphate-N-acetyl neuraminic acid).

For the synthesis of sulfolipids (sulfatides), sulfate residues are transferred from the active form, phosphoadenosine phosphosulfate (PAPS).

Assembly of the oligosaccharide chains takes place mainly in the Golgi complex.

Disturbances of FA, phospholipid, and sphingolipid biosynthesis. Fourteen clinical conditions due to defects in the long chain FA, phospholipid, and sphingolipid synthesis have been described. The predominant symptomatology of these disorders can be grouped into three categories: (1) central nervous system alterations, (2) peripheral neuropathy, and (3) effects of heart and muscle, and eventually myoglobinuria.

Complex Lipid Degradation

Complex lipids in cell membranes undergo permanent turnover. The degradation is catalyzed by specific enzymes. Glycerophospholipids, for example, require several enzymes for their total degradation. *Phospholipase A$_2$* catalyzes hydrolysis of the ester bond of FAs at position 2, and forms lysophospholipid and a free FA (generally unsaturated). The FA at position 1 of the lysophospholipid is released by the hydrolysis catalyzed by phospholipase B (*lysophospholipase*). Subsequently, a hydrolase separates choline or ethanol amine and leaves glycerol-3-phosphate (Fig. 15.9).

Other enzymes act on the whole phospholipid: *phospholipase A$_1$* catalyzes the hydrolysis of the ester bond of FAs at position 1, *phospholipase C* acts on the phosphate ester bond in position 3, and *phospholipase D* acts on the binding of the phosphate.

In digestive juices, there are phospholipases involved in the degradation of phospholipids. Other intracellular phospholipases participate in the synthesis of substances with high physiological activity, some of which are intracellular messengers in cell signaling systems. For example, phospholipase A$_2$ releases arachidonate, a precursor of prostaglandins and leukotrienes, and phospholipase C, which is involved in the phosphoinositide cascade (see p. 561).

Other complex phospholipids are degraded by lysosomal enzymes that can specifically cleave particular chemical links. These consist of sphingomyelinase, α-fucosidase, hexosaminidase, galactocerebrosidase, sulfatidase, and ceramidase.

CONGENITAL DISORDERS OF COMPLEX LIPID CATABOLISM

There is a group of hereditary metabolic diseases that result in the accumulation of complex lipids in various tissues. In general, they are evident during childhood and seriously compromise the patient's development and survival. Among these genetic inborn errors of metabolism, a variety of diseases labeled generically as *sphingolipidoses* have been described. They are all characterized by abnormalities in the synthesis of enzymes involved in the degradation of sphingolipids (sphingomyelin, cerebrosides, gangliosides, and sulfatides). A few of these alterations are presented in the subsequent sections.

Niemann–Pick disease. These patients have significant enlargement of the liver and spleen (hepatosplenomegaly). Pathologically, the liver and spleen cells have a frothy aspect (foam cells). Large amounts of sphingomyelin accumulate in tissues. The primary defect is lack of *sphingomyelinase*, the enzyme which catalyzes the hydrolysis that separates ceramide and phosphorylcholine and constitutes the first step in the catabolism of sphingomyelin.

Gaucher disease. Individuals affected with this disease have hepatosplenomegaly and characteristic lipid-filled cells in bone marrow and other tissues. In the most common type of this disease, the nervous system is not compromised. The alteration consists in the lack of *glucocerebrosidase*, which leads to accumulation of glucocerebroside in cells.

Tay–Sachs disease. In this alteration, neuronal cells (particularly in the cerebral cortex) are full of gangliosides, mostly G_{M2}, which are very scarce in normal individuals. The biochemical defect is an absence of the *hexosaminidase*, which catalyzes the hydrolysis of the acetylgalactosamine in G_{M2} ganglioside [Glc-Gal-Ceramide (NeuAc)-GalNAc]. Accumulation of G_{M2} causes mental retardation, blindness, and muscle weakness.

Sandhoff disease. The genetic defect results in the absence of *hexosaminidases A* and *B*. G_{M2} ganglioside and globosides accumulate, causing serious neurological abnormalities.

The diseases mentioned, along with others consisting of defective proteoglycan degradation, have also been designated lysosomal diseases due to the lysosomal localization of the altered hydrolases. Table 15.4 lists some of these alterations.

TABLE 15.4   Sphingolipidoses

Disease	Deficient enzyme	Accumulated substance	Symptoms
Niemann–Pick	Sphingomyelinase	Sphingomyelin	Hepatosplenomegaly, mental deficiency, foam cells in bone marrow.
Gaucher	β-glucocerebrosidase	Glucocerebroside	Hepatosplenomegaly, long bone and pelvic bone erosions.
Tay–Sachs	Hexosaminidase A	Ganglioside G_{M_2}	Mental deficiency, blindness, early death.
Sandhoff	Hexosaminidase A–B	Ganglioside G_{M_2}	Mental deficiency, blindness, early death.
Fabry	α-galactosidase A	Ceramide trihexoside	X-linked. Pain in legs, skin lessions, and adrenal alterations.
Krabbe	Galactocerebrosidase	Galactocerebroside	Mental deficiency, demyelinization, and early death.
Farber	Ceramidase	Ceramide	Mental deficiency, skeletal alterations, dermatitis, and early death.
Leucodistrophia metachromatic	Arylsulfatase	Sulfatide	Mental deficiency, demyelinization, and progressive paralysis.
Generalized Gangliosidosis	β-galactosidase	Ganglioside G_{M_1}	Mental deficiency, hepatomegaly, and skeletal alterations.
Fucosidosis	α-fucosidase	Cer-Glc-Gal-GalNAc-Gal-Fuc	Degenerative lesions in brain, muscular spasticity, and skin thickening.

These diseases are rare, but the general practitioner must know their existence to diagnose, or at least be able to suspect them. This will allow the correct referral to a specialist.

CHOLESTEROL METABOLISM

The usual diet provides approximately 300 mg of cholesterol per day. This substance is absorbed free in the intestine and is esterified within intestinal cells, mainly with oleic acid, in a reaction catalyzed by ACAT. Cholesterol esters are incorporated, together with TAGs, into chylomicrons sent to the bloodstream, where they are subjected to the action of LpL. The remnant particles, with esterified cholesterol, are captured by the liver for final degradation.

An organism is not dependent on exogenous cholesterol for its needs, as almost all tissues are able to synthesize it. A normal adult produces approximately 1 g of cholesterol/day, with almost half of it being produced in the liver and the remainder mostly in the intestines, gonads, adrenal glands, skin, muscle, and adipose tissue.

CHOLESTEROL BIOSYNTHESIS

Biosynthesis of cholesterol is carried out through a series of stages, which were identified by Bloch and his coworkers. Early findings indicated that all carbon residues in cholesterol derive from acetate residues. When laboratory animals were administered ^{14}C labeled acetate, the radioisotope appeared incorporated into the cholesterol molecule.

Later, mevalonic acid (mevalonate) was identified as an important intermediate in the pathway of cholesterol synthesis. Mevalonic acid loses its carboxyl group and creates five carbon isoprenoid molecules, which, by successive condensations, forms squalene, a terpene hydrocarbon that contains 30 carbon atoms. By cyclization and other changes, this compound

will finally result in cholesterol. Three phases will be considered in the biosynthetic process: (1) acetate to mevalonate, (2) mevalonate to squalene, and (3) squalene to cholesterol.

1. *Conversion of acetate to mevalonate.* This stage is carried out in three reactions. In the first, catalyzed by *thiolase*, two acetyl-CoA molecules form acetoacetyl-CoA and one CoA molecule is released:

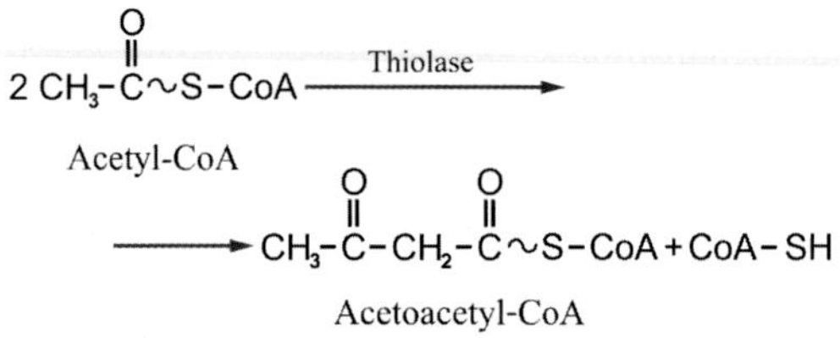

Then, the acetoacetyl-CoA reacts with another molecule of acetyl-CoA and generates 3-hydroxy-3-methylglutaryl-CoA (HMGCoA). The enzyme responsible for this reaction is *3-hydroxy-3-methylglutaryl-CoA synthase* (HMG-CoA synthase):

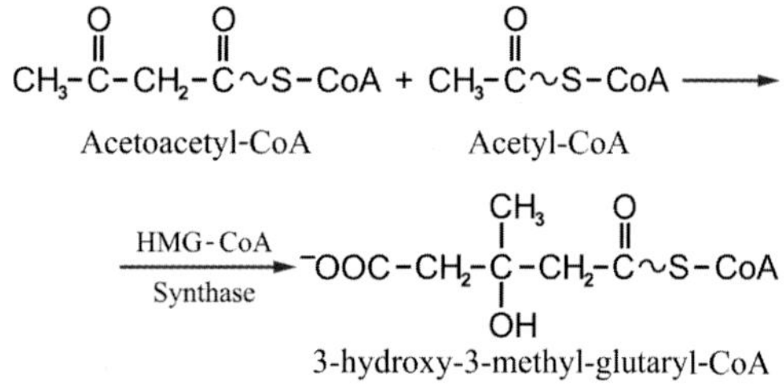

The HMGCoA is at a metabolic crossroad. It is an intermediate in the biosynthesis of cholesterol and ketones. In both pathways, HMG-CoA synthesis is performed through the same reaction, but the enzymes involved are not the same. The isozyme participating in the synthesis of cholesterol is localized in the cell cytoplasm while the enzyme forming ketone bodies is located in mitochondria.

In the pathway to synthesize cholesterol, one of the HMG-CoA carboxyl undergoes reduction to alcohol, releases CoA, and forms mevalonate, a six carbon compound:

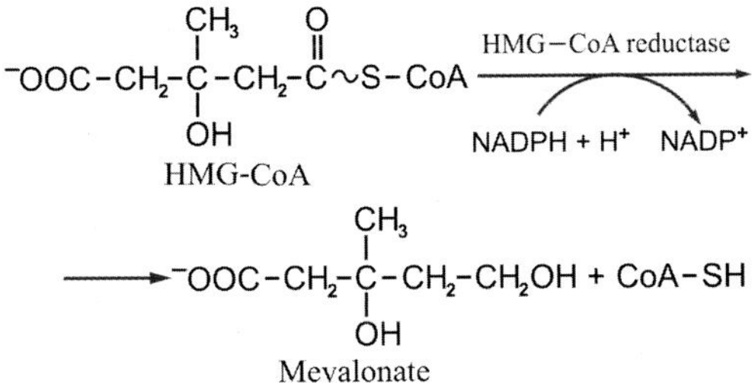

$$\text{HMG-CoA} \xrightarrow[\text{NADPH + H}^+ \quad \text{NADP}^+]{\text{HMG-CoA reductase}}$$

Mevalonate

This reaction is catalyzed by *hydroxy-methylglutaryl-CoA reductase,* enzyme which uses reduced NADP as a hydrogen donor. The enzyme is subjected to regulation; its activity is inhibited in the presence of exogenous cholesterol and bile acids.

2. *Conversion of mevalonate to squalene (Fig. 15.10).* The mevalonate receives a phosphoryl group from ATP to form

5-phosphomevalonate. This compound accepts another phosphate to render mevalonate-5-pyrophosphate. After a third phosphorylation, an unstable compound (not shown in Fig. 15.10) is formed, which is decarboxylated, loses water, and originates isopentenyl pyrophosphate.

The three ATP molecules necessary for the synthesis of cholesterol are used in the stages from mevalonate to isopentenyl pyrophosphate. This compound is transformed by an *isomerase* into *dimethylallyl pyrophosphate* with the double bond located in a different position.

Isopentenyl pyrophosphate and dimethylallyl pyrophosphate are condensed to form *geranyl pyrophosphate,* which contains 10 carbons. Geranyl pyrophosphate reacts with another molecule of isopentenyl pyrophosphate and produces *farnesyl pyrophosphate,* a compound with 15 carbons. Each of the isoprenoid condensation reactions

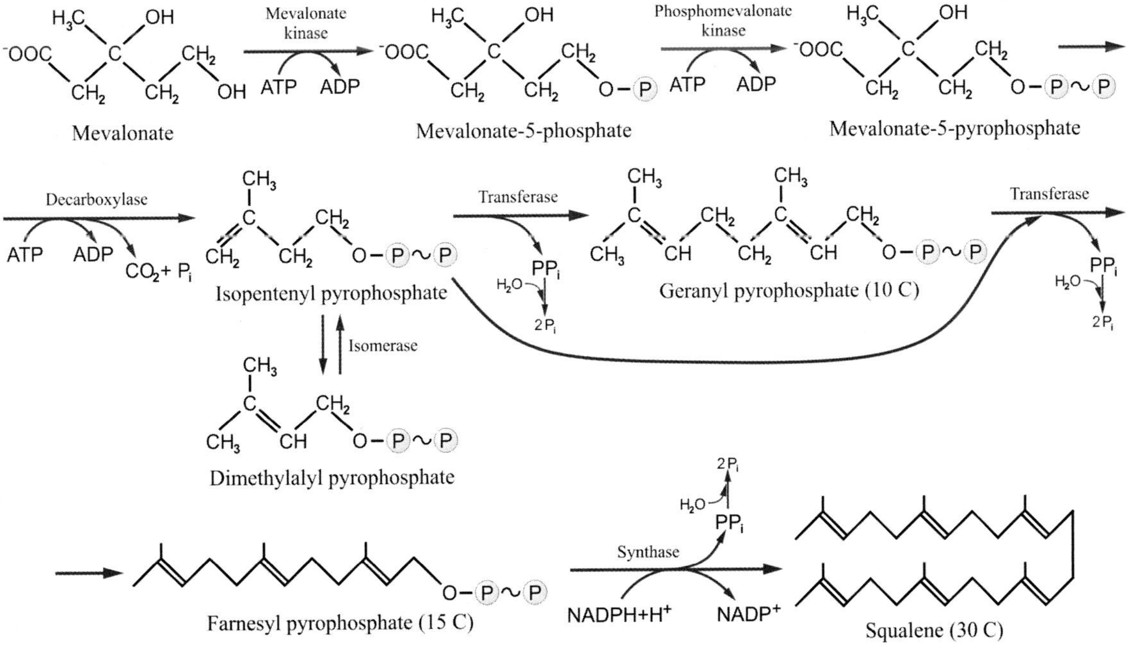

FIGURE 15.10 Conversion of mevalonic acid into squalene.

liberates pyrophosphate, which is immediately hydrolyzed. The energy generated is used to form the C—C links. Binding of two farnesyl pyrophosphate molecules forms squalene. In this reaction, the NADPH coenzyme acts as a hydrogen donor. Again, the hydrolysis of pyrophosphate liberated in the condensation reactions generates energy that allows the formation of the C—C bonds. The stages of this second phase in the cholesterol synthesis pathway are shown in Fig. 15.11.

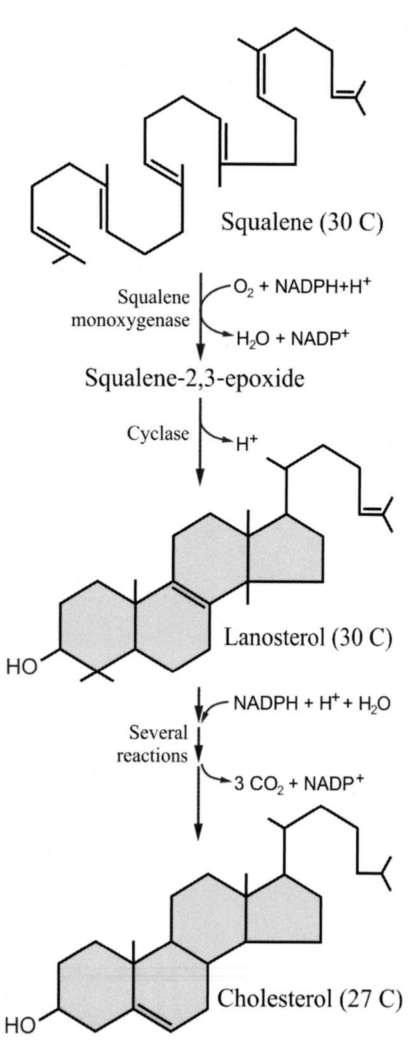

FIGURE 15.11   **Conversion of squalene into cholesterol.**

Activated isoprenoids are used not only for the biosynthesis of cholesterol, but also for other important compounds, including coenzyme Q, carotenoids, dolichol, and isoprenyl groups that are important in anchoring proteins to membranes.

3. *Conversion of squalene into cholesterol.* The structure of squalene is similar to that of steroids (Fig. 15.11). A process that comprises the closure of four rings, displacement of methyl groups, and saturation of double bonds, forms *lanosterol* (a molecule with 30 carbon atoms and two double bonds).

   After the loss of three methyl groups, released as CO_2, displacement of the side chain double bond to another position (C5–6) converts lanosterol into *cholesterol*.

   All enzymes catalyzing the synthesis of cholesterol from squalene are inserted in the smooth endoplasmic reticulum membrane and intermediate products remain bound to a sterol carrier protein. Fig. 15.11 shows the structures for squalene, lanosterol, and cholesterol. Other intermediary metabolites of this final stage in the biosynthesis are omitted.

Plasma cholesterol. The cholesterol that reaches the circulation is bound to lipoproteins, but most are associated with LDL. Two-thirds of the total circulating cholesterol are esterified. The esterification is catalyzed by LCAT which transfers an unsaturated acyl residue (commonly linoleate) from position 2 of phosphatidylcholine (lecithin) to hydroxyl 3 of cholesterol. This reaction is accomplished primarily in HDLs that receive free cholesterol from extrahepatic tissues. Once esterified, cholesterol is transferred, in exchange with TAGs, mainly to VLDL and chylomicrons. Esterified cholesterol transport protein (ECTP) is involved in the exchange (reverse cholesterol transport).

LDL is quantitatively the major carrier of cholesterol. Thus, there is a direct correlation between the levels of total blood cholesterol and LDL. Serum cholesterol is maintained within normal limits by its constant removal by all tissues. For this, cells have specific receptors (LDL receptors) that capture and internalize circulating LDL by endocytosis. Cholesterol esters are degraded in lysosomes and the products are released into the cytosol. The level of intracellular cholesterol exerts regulatory action: (1) increased cell cholesterol inhibits HMG-CoA reductase, depressing cholesterol synthesis. With inhibition of the pathway generating endogenous cholesterol, the cell only has the cholesterol incorporated by its LDL receptors. (2) The excess cholesterol, not incorporated to membranes or used in synthesis, is stored in the cell as cholesterol esters [the esterification is catalyzed by *ACAT* (ACAT), activated by high levels of free cholesterol]. (3) Cholesterol accumulation inhibits the LDL receptor synthesis. Thus, cells adjust the number of LDL receptors according to physiological needs.

Role of the liver in the metabolism of cholesterol. The cholesterol content in liver depends on: (1) dietary cholesterol incorporated by intestinal cells into chylomicrons and taken up with remnant particles by liver cells. (2) The transfer of cholesterol to HDL in extrahepatic tissues, where it is esterified by LCAT and then transferred to VLDL and chylomicrons. The remnants of these particles are finally captured by hepatocytes. (3) Synthesis in the liver, from acetyl-CoA, derived mainly from dietary carbohydrates.

Cholesterol catabolism and excretion. The human body does not have enzymes to degrade the cyclopentanoperhydrophenanthrene, so, it is excreted intact. The liver is the organ of cholesterol elimination. Most of it is converted into bile acids, of which 300–500 mg are produced per day. The synthesis of bile acids requires the participation of 17 enzymes. The synthetic routes include: (1) 7α hydroxylation of steroid precursors, (2) modification in the structure of the ring, (3) shortening and oxidation of the side chain, and (4) bile acid conjugation with amino acids (see p. 262).

Both bile acids and cholesterol are excreted into the intestine through the bile. In the intestine, these compounds are partly reabsorbed and returned to the liver to complete the enterohepatic circulation. Nonabsorbed cholesterol and bile acids undergo the action of the normal intestinal bacteria flora. Reduction of the cholesterol double bond produces coprostanol and cholestanol, isomers that are the principal sterols in feces. Bile acids appear in feces at levels of approximately 500 mg/day. Their synthesis and excretion are balanced, which indicates that they are tightly regulated. The excretion of steroidal hormone products in the urine represents only a small proportion of the total cholesterol excreted.

Regulation. The hepatic synthesis of cholesterol and bile acids is modulated by the level of free cholesterol in cells. Free cholesterol inhibits *3-hydroxy-3-methylglutaryl CoA (HMG-CoA) reductase*, the main site of regulation. In addition, it also activates *acyl-CoA-cholesterol acyltransferase* (Fig. 15.12) and decreases the synthesis of LDL receptor by reducing its mRNA levels.

Bile acids produced in hepatocytes, and those recycled by the enterohepatic circulation, depress the activity of HMG-CoA reductase. In addition, another enzyme involved in the biosynthesis of bile acids (7α hydroxylase) is also modulated by bile acid levels.

The close relationship between cholesterol and atherosclerosis has fueled the interest to study how cholesterol is metabolized. There is a direct correlation between levels of total plasma cholesterol and incidence of coronary accidents (heart attacks). Cholesterol values above 240 mg/dL (6.2 mmol/L), along with other risk factors, considerably increase the risk of such accidents. HDL levels of less than 35 mg/dL (0.9 mmol/L) are another significant risk factor for coronary heart disease.

Knowledge of cholesterol metabolism has allowed the design of pharmacological compounds to reduce its levels in blood. There are drugs which, by acting on HMG-CoA reductase,

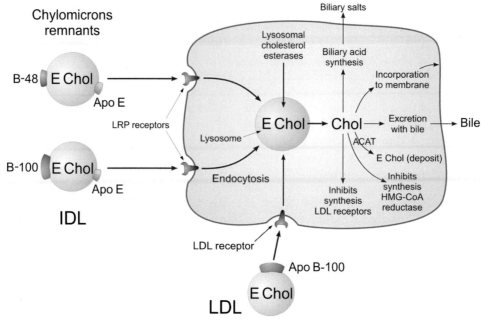

FIGURE 15.12   **Fate of cholesterol in hepatocytes.** The hepatocyte is the only cell with both LRP and LDL receptors (most cells have only LDL receptors). Chylomicron remnants and IDL are captured by LRP receptors. LDL receptors bind LDL. All particles enter the cell by endocytosis and are degraded in lysosomes. Cholesterol esters (CholE) are hydrolyzed and provide free cholesterol (Chol), which pass into the cytosol. Cholesterol is used for the synthesis of bile acids, incorporated into membranes, or excreted unchanged with bile. It is esterified by acyl-CoA acyltransferase (ACAT) and stored. Cholesterol levels in the cell have regulatory effects: increased cholesterol depresses its synthesis by inhibition of HMG-CoA reductase and inhibits the synthesis of LDL receptors.

inhibit cholesterol synthesis, decrease the intracellular cholesterol concentration, and stimulate synthesis of LDL receptors. This promotes removal of LDL from the circulation. Furthermore, the use of substances that bind bile acids in the intestine and form nonabsorbable complexes, block the return of bile acids to the liver and increase their elimination via the feces. Thus, resins which bind bile acids and prevent their absorption are also used to control cholesterol levels. Reduction of bile acids returning to the liver stimulates their synthesis, thereby decreasing the intracellular concentration of cholesterol and increasing the production of LDL receptors that subtract cholesterol from the circulation.

The combination of decreased synthesis and increased excretion of bile acids can reduce cholesterol plasma levels in patients with abnormally high values of the steroid.

For years, research has been intense in this field, hoping to reduce the burden that atherosclerosis causes to human life.

Recently a number of genetic disorders of the biosynthesis of cholesterol have been described. These include the relatively common (1 in 50,000 births), Smith–Lemli–Opitz syndrome produced by lack of the enzyme that catalyzes the last reaction of the synthetic route, the *7-dehydrocholesterol reductase*. Cholesterol plasma levels in these patients are extremely low and they suffer multiple malformations and neurological disorders. Another very rare disorder is the deficiency of *mevalonate kinase*, which causes *mevalonic aciduria*.

SUMMARY

TAGs are the predominant lipids in the human diet and an important energy source in the body.

Glycerophospholipids and sphingolipids mainly have a structural role in the formation of cell membranes and they help on the transport of lipids in blood.

Cholesterol is an essential component of cell membranes and is the precursor of steroidal hormones and bile acids.

Lipoproteins carry lipids in plasma. They consist of an external hydrophilic region (containing proteins, phospholipids, and nonsterified cholesterol), and an internal hydrophobic core, containing TAGs and cholesterol esters. Based on their density, five major lipoprotein groups can be distinguished in blood. These include the following:

Chylomicrons are constituted by TAGs with a small amount of cholesterol. TAGs have been absorbed and resynthesized in the intestinal mucosa, and move into the bloodstream from the lymph. Their protein portion contains predominantly apo B-48. Chylomicrons receive apo C, apo E, and esterified cholesterol (EChol) from HDL. In capillary endothelial cells, a *LpL* activated by apo C-II hydrolyzes TAGs. The free FAs enter the cells, where they are oxidized (muscle) or stored (adipose tissue). After almost all TAGs have been hydrolyzed, apo C is returned to HDL. The chylomicron is reduced to a particle formed by residual B-48, apo E, EChol, phospholipids, and little TAG (chylomicron remnants). Chylomicrons remain in circulation for less than an hour. Hepatocytes have receptors binding these particles (LRP), which aid in their clearance from circulation.

VLDLs are synthesized in hepatocytes. The protein portion is represented by apo B-100; they are rich in TAG. Unlike lipids in chylomicrons, which are exogenous, the VLDLs are endogenous. In circulation, VLDL receives apos C and E from HDL. In the capillaries, TAGs are hydrolyzed by LpL, activated by apo C-II. Apo C-II is returned to HDL, and this exchanges EChol for TAG with VLDL. The half-life of VLDL is 4 h. VLDLs are converted to IDL, composed of apos B-100 and E, Chol, mostly Ecol, and low proportion of TAG.

Hepatocytes have receptors for IDL that remove more than 50% of the circulating particles. The rest undergoes further hydrolysis of TAG by hepatic lipase, returns apo E to HDL, and becomes LDL. The half-life of IDL is 2–5 h. LDL, rich in EChol, contains only apo B-100. Their half-life is 2.5 days. In all cells, there are specific receptors to apo B-100 (LDL receptor), which capture LDL and internalize them into the cells by endocytosis.

HDLs are synthesized in the liver and intestine. When recently formed, they are oblong in shape and contain: apos A, C, and E, lecithin-cholesterol acyltransferase (LCAT, activated by apoA-I), and phospholipids. In blood, HDL donates apos C and E to VLDL and chylomicrons. In extrahepatic tissues, HDL binds to the capillary endothelium and receives Chol from the interior of the cells. The particle enlarges and becomes spherical. The Chol received is esterified by LCAT and transferred to VLDL and chylomicrons by the EChol transport protein. Remnants of these particles are taken up by the liver. The transport of cholesterol from peripheral tissues to liver mediated by HDL is called *reverse cholesterol transport*.

Lipoproteins (a) are similar to LDL but also contain apo (a). Its rise in plasma is a risk factor for atherosclerosis.

Receptors for LDL are in the plasma membrane of all cells. Their ligand is apo B-100. LRP is mainly found in liver and it recognizes and binds particles with apo E. *Scavenger receptors* are predominantly in macrophages, they are less specific. They bind modified LDL (oxidized).

Fat metabolism starts with the hydrolysis of TAG into glycerol (Glyc) and FA, by the action of intracellular lipase and LpL.

Glycerol metabolism starts with the incorporation of phosphate from ATP to glycerol (Glyc), to form Glyc-3-P (*glycerokinase*). Glyc-3-P is then converted to dihydroxyacetone-P (*glycerophosphate dehydrogenase* and NAD). Lastly, dihydroxyacetone-P is converted to glyceraldehyde-3-P (*phosphotriose isomerase*). The ultimate fate of these compounds is glycolysis or gluconeogenesis.

FA catabolism occurs in many tissues and they are used as energy source. FA degradation is accomplished in mitochondria by a process of *β-oxidation* after two preparatory steps:

1. *Activation* of acyl-CoA by *thiokinase* or *acyl-CoA synthetase*, which requires CoA, ATP, and Mg^{2+}). ATP is hydrolyzed to AMP and PP_i.

2. *Transfer* of acyl-CoA from cytosol to mitochondria: in the cytosol, acyl-CoA transfers the acyl to carnitine (*carnitine-acyltransferase I*). Acylcarnitine traverses the inner membrane by the acylcarnitine/carnitine exchanger. In the matrix, acyl is transferred back to CoA (*carnitine-acyltransferase II*). Carnitine returns to the cytosol. The acyl-CoA initiates the process of *β-oxidation*.

β-oxidation follows a series of steps:

1. *Oxidation.* The acyl-CoA loses 2H from C2 and 3 (α and β) (by *acyl-CoA dehydrogenase*, FAD). The acyl-CoA is unsaturated or *trans*.

2. *Hydratation.* Trans-Δ2-enoyl-CoA is converted into 3-hydroxyacyl-CoA (by *enoyl hydratase*).

3. *Second oxidation.* 3-hydroxyacyl-CoA is converted into 3-ketoacyl-CoA (by *3-hydroxyacyl-CoA dehydrogenase*, NAD).

4. *Release of acetyl-CoA*: 3-ketoacyl-CoA is cleaved into acetyl-CoA and acyl-CoA, 2C shorter than the original

(*thiolase, CoA*). The reaction sequence is repeated with the acyl-CoA formed.

A FA of 16C in length (palmitate) requires seven rounds of β-oxidation to be degraded and produces eight active acetates. Acetyl-CoA can enter the citric acid cycle for complete oxidation to CO_2 and H_2O. Energy balance is as follows: The initial activation step consumes 2 ~P. Each β-oxidation round produces five ATP, two in the dehydrogenation step with FAD and three in the step with NAD. Each acetyl CoA formed produces 12 ATP in the citric acid cycle.

Ketogenesis refers to the formation of ketone bodies from acetoacetate, 3-OH butyrate, and acetone. The synthesis stages are as follows:

1. Formation of acetoacetyl-CoA from two acetyl-CoAs (by thiolase).
2. Formation of 3-OH-3-methylglutaryl-CoA (by HMG-CoA). Acetoacetyl-CoA plus acetyl-CoA (by HMG-CoA synthase).
3. Formation of acetoacetate: HMG-CoA is cleaved to acetoacetate and acetyl-CoA (by HMG-CoA lyase).
4. Formation of 3-OH-butyrate: acetoacetate is reduced (by 3-OH-butyrate dehydrogenase, NAD).
5. Acetone: acetoacetate is decarboxylated.

Muscle, heart, and other tissues are capable of activating acetoacetate and can use ketone bodies as fuel, the liver cannot.

These involve two mechanisms:

1. Transfer of CoA from succinyl-CoA (by succinyl-CoA-3-ketoacid-CoA transferase or thiophorase).
2. Acetoacetate binds CoA (by thiokinase, ATP). For final oxidation, acetoacetyl-CoA is hydrolyzed into two acetyl-CoA (by thiolase).

FA biosynthesis takes place from acetyl-CoA, generated mainly by degradation of glucose and amino acid carbon chains. This is catalyzed by the *FA synthase multienzyme system*, a multifunctional protein. First, the acetyl-CoA is transferred from mitochondria to cytosol. In the mitochondrial matrix, acetyl-CoA reacts with oxaloacetate to form citrate (by *citrate synthase*). Citrate is transported into the cytosol, where it is cleaved into acetyl-CoA and oxaloacetate (by *citrate lyase* ATP). Acetyl-CoA is used in the synthesis of FA, and oxaloacetate is reduced to malate (by *malate dehydrogenase*, NAD). Malate is decarboxylated oxidatively to pyruvate (by *malic enzyme*, NADP). Pyruvate enters the matrix, where it is transformed into oxaloacetate (by *pyruvate carboxylase*). Oxaloacetate binds to acetyl-CoA and forms citrate, closing the cycle.

The stages of FA synthesis include the following:

Formation of malonyl-CoA: acetyl-CoA is carboxylated (by *acetyl CoA carboxylase*, biotin, ATP).

1. Acetate transfer. An acetyl residue is transferred from acetyl-CoA to the —SH of the active site of the condensing enzyme (by *acetyl transferase*).
2. Malonyl transfer. Malonyl is transferred to the —SH phosphopantetheine of ACP.
3. Condensation: malonyl is decarboxylated and acetyl remains bound to ACP (by *condensing enzyme* or *3-acetoacyl synthetase* ACP). CO_2 is released, the —SH group of the condensing enzyme is unoccupied and the acetoacetil-SH is attached to the ACP phosphopantetheine.
4. First reduction: acetoacetyl-ACP accepts 2H ions from NADPH to produce 3-OH-butyryl-ACP (*3-ketoacyl-reductase* ACP).
5. Dehydration produces 2-enoyl-ACP (by *3-hydroxyacyl dehydratase*).
6. Second reduction: 2H are transferred from NADPH and butyryl-ACP is formed (by *enoyl reductase*).

2C residues are added to this 4C acyl in successive cycles similar to that described, resulting in the formation of palmitate (16C). The addition of acyl starts with the 4C acyl transfer to the —SH group of the condensing enzyme.

The phosphopantetheine —SH group is free and joins a malonyl moiety. This is decarboxylated, and receives the acyl from the condensing enzyme. After seven rounds, a 16C acyl is formed, which is released by the action of thiolesterase. Longer FAs are formed by elongation into two systems, mitochondrial and microsomal.

Synthesis of unsaturated FA is performed by introduction of a double bond between C9 and 10. Desaturation system consists of the flavoprotein *NADH-cytochrome b$_5$ reductase*, cytochrome b$_5$, and *Δ9-desaturase*. Animals cannot introduce double bonds beyond C9. This is why linoleic and linolenic acids must be obtained from the diet (they are *essential FAs*). Arachidonic acid can be synthesized from linoleic acid.

Eicosanoid biosynthesis uses as precursors 20C FA (eicosanoic). Prostaglandins and thromboxanes are synthesized from arachidonate; the initial reaction is catalyzed by *cyclooxygenase* (COX1 and COX2 isozymes). Leukotrienes are also formed from arachidonate (lipoxygenase).

Biosynthesis of TAG.

Glycerol is activated to glycerol-3-P, and the acyl to acyl-CoA. In both cases, ATP is required. Liver, intestine, kidney, and mammary gland tissues have *glycerokinase*, muscle and adipose tissues do not. These tissues use glycerol-3-P formed from dihydroxyacetone-P. Glycerol-3-P is esterified at C1 and 2 to form 1,2-diacylglycerolphosphate, or phosphatidic acid (*glycero-P acyltransferase*). *Phosphatidic acid* is hydrolyzed (*phosphatase*) and transformed into 1,2-diacylglycerol, to which a third acyl gets transferred (*diacylglycerol acyltransferase*).

Bibliography

Barter, P.J., Rye, K.A., 1996. Molecular mechanisms of reverse cholesterol transport. Curr. Opin. Lipidol. 4, 171–176.

Blanco, A., Kremer, R.D., Taleisnik, S. (Eds.), 2000. Inherited Metabolic Defects. Academia Nacional de Ciencias, Córdoba.

Dessi, M., Noce, A., Bertucci, P., Manca de Villahermosa, S., Zenobi, R., Castagnola, V., Adessi, E., Di Daniele, N., 2013. Atherosclerosis, dyslipidemia, and inflammation: the significant role of polyunsaturated fatty acids. ISRN Inflamm. 2013, 1–13.

Genest, J., Marci, P.M., Denis, M., Yu, L., 1999. High density lipoproteins in health and disease. J. Invest. Med. 47, 31–42.

Hegele, R.A., 2009. Plasma lipoproteins: genetic influences and clinical implications. Nat. Rev. Genet. 10, 109–121.

Jensen-Urstad, A.P.L., Semenkovich, C.F., 2012. Fatty acid synthase and liver triglyceride metabolism: housekeeper or messenger? Biochim. Biophys. Acta 1821, 747–753.

Lamari, F., Sedel, M.F., Saudubray, J.M., 2013. Disorders of phospholipids, sphingolipids and fatty acids biosynthesis: toward a new category of inherited metabolic diseases. J. Inherit. Metab. Dis. 36, 411–425.

Maccioni, H.J.F., Daniotti, J.L., Martina, J.A., 1999. Organization of ganglioside synthesis in the Golgi apparatus. Biochim. Biophys. Acta 1437, 101–118.

Rinaldi, P., Matern, D., Bennett, M.J., 2002. Fatty acid oxidation disorders. Ann. Rev. Physiol. 64, 477–502.

Rosen, E.D., Spiegelman, B.M., 2014. What we talk about when we talk about fat. Cell 156, 20–44.

Rötig, A., Munnich, A., 2003. Genetic features of mitochondrial respiratory chain disorders. J. Am. Soc. Nephrol. 14, 2995–3007, Special focus on atherosclerosis. Article Series, 2002. Nat. Med. 8, 1209–1262.

Amino Acid Metabolism

Amino acids are commonly used in the body as the building blocks for the synthesis of proteins and a variety of physiologically active nitrogenous compounds (hormones, enzymes, and other functionally important substances). This structural function of amino acids is unique and irreplaceable. They are the main source of nitrogen in the body. While amino acids can be also utilized as fuel, this role is secondary and replaceable by carbohydrates and fats.

Unlike carbohydrates and fats, amino acids are not stored in the body. Their amounts depend on amino acid anabolism and catabolism in the body, or nitrogen balance, which is directly related to the rate of synthesis and degradation of proteins.

In normal adults, there is a balance between nitrogen intake from the diet and nitrogen elimination through urine and feces. During growth and pregnancy, nitrogen consumption exceeds its excretion. It is said that the nitrogen balance is positive; the excess of nitrogen is used in the synthesis of new tissue structures. In contrast, in cases of protein malnutrition, severe febrile conditions, cancer, or uncontrolled diabetes, nitrogen excretion exceeds its intake and the nitrogen balance is negative.

Proteins coming with foods are digested and hydrolyzed into their constituent amino acids, which are easily absorbed in the intestine and, via the circulation, are distributed to all tissues. Amino acids are taken into the cells by carrier systems (pp. 372). Once in the cytoplasm, they can follow different metabolic alternatives: (1) remain unmodified and be used for synthesis of specific proteins, (2) undergo transformation into physiologically important nonprotein compounds, or (3) become degraded for energetic purposes.

Different from glycogen, in which the molecule is subjected to permanent elongation and shortening, proteins undergo continuous renewal, being degraded and resynthesized at different rates, depending on their particular half-lives.

Amino acids released by degradation of endogenous proteins, along with those coming from proteins in the diet, form a common amino acid pool in the body from which new protein and other nitrogen containing compounds are synthesized.

A normal adult degrades approximately 400 g of protein per day. From the amino acids released, approximately 75% are reused in the synthesis of new proteins. The rest enter different metabolic pathways, including gluconeogenesis, ketogenesis, and synthesis of a variety of other nonprotein compounds.

Protein synthesis. Protein levels in the cell are controlled by regulating their synthesis and degradation. New proteins are built by attaching amino acids, one after the other, via peptide bonds, following a sequence dictated by the cell genome. Protein biosynthesis is a complex process that will be described in Chapter 22.

Half-life of proteins. After a period of time, cell proteins become degraded, with a turnover time that is variable depending on the protein considered. Proteins produced for cell export, such as digestive enzymes, hormones, and antibodies have a half-life of a few hours or days, after which they are rapidly degraded. Even shorter half-lives have been reported for proteins that have regulatory functions, such as enzymes, which catalyze limiting step reactions in metabolic pathways, cell cycle proteins, and transcription factors, which control cell division. In contrast, proteins with a structural role, such as collagen, are more stable and have an average half-life of several months.

The half-life of a protein can be influenced by different agents. In particular, reactive oxygen species can affect the structure, spatial conformation, and biological activity of proteins. Altered proteins are rapidly degraded in the cell.

Protein degradation. The cell has several systems involved in the elimination of proteins that have reached the end of their life. These include the following:

Lysosomal degradation. Lysosomes contain hydrolases, which degrade proteins, nucleic acids, lipids, and carbohydrates. Lysosomal proteases include a family of enzymes known as the *cathepsins*, which specifically function in acidic medium. Lysosomes catalyze the hydrolysis of extracellular proteins, that the cell captures via endocytosis, or intracellular soluble and organelle associated proteins, which have long half-lives, via a process called *autophagy*. Endocytosis has already been described (p. 242).

Besides cathepsins, other enzymes are involved in protein degradation. These include *calpains* (cytosolic cysteine proteases activated by Ca^{2+}) and *caspases* (proteases involved in the process of *programmed cell death* or *apoptosis*) (Chapter 32).

Autophagy is a process for the degradation of long-living proteins, carbohydrates, lipids, and organelles (mitochondria, peroxisomes, ribosomes, endoplasmic reticulum and nucleus), particularly in conditions of cell starvation and stress. Three different types of autophagy have

been described: (1) macroautophagy, (2) microautophagy, and (3) autophagy mediated by chaperones (see p. 535).

1. *Macroautophagy.* The material to be degraded is engulfed in vesicles derived from the endoplasmic reticulum membrane. These vesicles are directed and fuse to a lysosome, forming a new vesicle called *autophagosome*. The hydrolases contained within the lysosome degrade the imported material.
2. *Microautophagy.* The material to be degraded (mainly cytoplasmic proteins) is directly internalized and digested by lysosomes.
3. *Autophagy mediated by chaperones.* The polypeptide to be degraded is bound by a chaperone protein, which delivers it to the lysosome, where it will be hydrolyzed.

During periods of cell starvation, the cells hydrolyze cytosolic proteins that carry the signal sequence (Lys-Phe-Glu-Arg-Gln). These are, in general, proteins with a long half-life, functionally nonessential for the cell, but, such as the heat shock protein of 70 kD important as a source of energy and amino acids. These proteins enter the lysosomal compartment and are unfolded by association with chaperon proteins (Hsp70) (p. 535).

Ubiquitin–proteasome degradation. This is the main pathway for selective hydrolysis of, in general, short half-life proteins. Molecules degraded in this pathway are first labeled by insertion of a 76 amino acid polypeptide called *ubiquitin* (Ub), which is widely distributed in all cells. Ubiquitin has a mass of 8.5 kDa, seven lysine residues and a glycine residue at the C terminal end.

Ub binding to proteins or ubiquitination is preceded by a series of stages. These include:

1. *Ub activation.* A thioester is formed between the carboxyl of the terminal glycine residue of Ub and a cysteine residue of the *activating enzyme*, or E1. The energy that fuels this reaction is provided by hydrolysis of ATP.
2. *Conjugation to E2.* Activated Ub is transferred to a cysteine residue of a *conjugating enzyme*, or E2.

3. *Binding of Ub to E3.* A new transesterification attaches Ub to the enzyme *ubiquitin ligase,* or E3. Most cells have one type of E1, but different forms of E2 and E3. Different kinds of E2 and E3 recognize different proteins as substrate.

Insertion of Ub in the protein. Ub is bound by its terminal carboxyl to the ε amine group of a lysine residue in the protein targeted for degradation. The reaction is catalyzed by ubiquitin ligase. This step is repeated several times, until several Ub molecules are bound to form a poly-Ub chain (Fig. 16.1). A lysine residue (K48) is the main amino acid involved in linking the Ub molecules. Polymerization of Ub enables the protein to enter the *proteasome.*

Proteasomes are structures of 2000 kDa and 26S (S = standard sedimentation coefficient, a measure of the sedimentation rate of a particle subjected to ultracentrifugation) each formed by a variety of different proteins. They have a core that consists of a 20 S hollow cylinder formed by the association of four rings of seven polypeptide subunits each. The inner surface of this tubular structure is covered by numerous proteases. Both ends of the cylinder function as gates or caps and contain proteins with ATPase ubiquitin hydrolase activity. The cap receives the ubiquitinated polypeptide, separates Ub from the protein, and returns Ub to the cytoplasm, where it will be reused for the labeling of other proteins targeted for degradation. Once in the proteasome, the protein, now devoid of Ub, enters the cylinder cavity, is subjected to the action of the proteases of the inner wall of the proteasome, and is hydrolyzed to smaller peptides. These peptides are freed to the cytosol where they are further degraded by peptidases (Fig. 16.1).

The ubiquitin–proteasome pathway is responsible for the elimination of numerous regulatory proteins, such as cell cycle modulators, including cyclins, p27, kinases, the p53 tumor suppressor protein, transcription factors, proteins of the DNA repair system the regulatory subunit of cAMP-dependent protein kinase

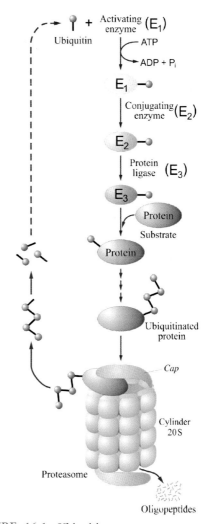

FIGURE 16.1 Ubiquitin–proteasome system. In the first step, one molecule of ubiquitin (Ub) associates with the activating enzyme (E1). Then, Ub is transferred to the conjugating enzyme (E2). Ligase (E3) catalyzes the binding of Ub to the substrate protein. This step may be accomplished by direct transfer of Ub from E2 to the substrate or by forming an E3–Ub intermediate as shown. These steps are repeated to form a polyUb chain. The substrate is captured by the proteasome, the polyUb chain is hydrolyzed, and its subunits are recycled for future use. The protein substrate is degraded to oligopeptides in the cavity of the proteasome.

(p. 560), and antigens (p. 749). Ubiquitin-mediated protein degradation plays an important role in cell growth, DNA repair, and immune response. Its dysfunction contributes to the development of cancer, neurodegenerative diseases, and immune system disorders.

Several factors influence protein half-life. These include, for example, protein structure. When the N-terminal amino acid is serine, the protein lasts more than 24 h; however, if aspartate is the N-terminal amino acid, the protein has a half-life of a few minutes. Also, those proteins that show the sequence proline-glutamate-serine-threonine have a short life. In addition, defects in protein structure or ubiquitination make them prone to rapid degradation.

The ubiquitin–proteasome system, as well as autophagy, contributes to maintain protein homeostasis and quality. In addition, autophagy degrades other nutrients, such as carbohydrates, lipids, and protein carrying minerals (iron).

Ubiquitination promotes protein degradation, but it can also exert other actions on the substrate protein (see Chapter 24).

ESSENTIAL AMINO ACIDS

Nutritional experiments led to the finding that there are two types of amino acids, essential and non-essential. Essential or indispensable amino acids are those that must be supplied in adequate quantity in the diet, to maintain normal growth in children and young individuals and proper nitrogen balance in adults. Their exogenous supply is required because humans do not have metabolic pathways to synthesize them.

In contrast, nonessential or dispensable amino acids are synthesized in the body, their administration in the diet is not required. However, the presence of nonessential amino acids in the diet decreases the need for essential amino acids. If the intake of nonessential amino acids are reduced, the body will synthesize them from essential amino acids.

TABLE 16.1 Minimal Daily Requirement of Essential Amino Acids (Normal Adult Human)

Phenylalanine	1.1 g	Methionine	1.1 g
Isoleucine	0.7 g	Threonine	0.5 g
Leucine	1.1 g	Tryptophan	0.25 g
Lysine	0.8 g	Valine	0.8 g

Table 16.1 indicates the minimum daily requirements of essential amino acids in an adult when the additional input of nitrogen and carbon for synthesis of nonessential amino acids is covered.

Arginine and *histidine* should be added to the eight essential amino acids, because while they are synthesized in the body, they are produced in insufficient quantities to meet the demands during growth, pregnancy, and lactation.

In certain situations, a nonessential amino acid may become essential. For example, in patients with phenylketonuria, a genetic defect in which there is an inability to convert phenylalanine to tyrosine, the amino acid tyrosine results essential. Inborn errors in the synthesis of cysteine from methionine require that cysteine be provided as an essential amino acid.

In premature infants, their functional immaturity precludes the synthesis of some nonessential amino acids, such as cysteine or proline. In patients with severe hepatic insufficiency, the ability to produce tyrosine or cysteine is affected.

When α-keto acids with the same carbon chains than leucine, isoleucine, valine, tryptophan, methionine, or phenylalanine are provided, these amino acids can be synthesized by the addition of an amine group catalyzed by aminotransferases. Lysine, threonine, and histidine do not participate in transamination reactions (see subsequent sections) and are not replaceable by their keto acids.

Protein Requirements

Proteins are an essential component of any normal diet, providing the amino acids necessary to synthesize body proteins, other nitrogenous compounds, and to maintain the proper nitrogen balance of the body. The simultaneous presence of adequate amounts of all amino acids in the diet is necessary. Protein synthesis decreases when a single amino acid is deficient, even if all the others are available. The recommended daily intake of proteins in adults is 0.8 g/kg of body weight. In women, 30 g/day should be added during pregnancy time, and 20 g/day should be provided during lactation to meet the needs of milk protein synthesis. The requirement for infants up to 1 year of age is 2 g/kg/day, for children 1–10 years old, 1.2 g/kg/day, and for adolescents, 1 g/kg/day. In all age groups, the need increases during processes that enhance protein catabolism (sepsis, trauma, surgery).

An adequate diet must contain not only the right quantity of protein, but also must have the appropriate protein quality.

Biological value. The biological value refers to an estimation of the proportion of absorbed protein present in a particular food, which is incorporated into the endogenous proteins of the body. The biological value of a protein is directly related to its digestibility, or amount of amino acids in the ingested protein that are absorbed, and essential amino acid content. It can be quantified by comparing the amount of nitrogen retained in the body after ingestion of a certain quantity of a protein in relation to the same amount of a reference protein (normally egg protein is used). A biological value of 100 corresponds to the egg protein and indicates that 100% of the absorbed nitrogen is incorporated.

Animal proteins are 90%–99% digestible, while proteins from vegetables are 70%–80% digestible. The high-quality proteins, also called complete proteins, contain all the essential amino acids in the amounts required by humans. Proteins from animal origin (milk, white, and red meats, eggs) are examples of complete proteins. An exception is gelatin, obtained by boiling collagen, which does not contain tryptophan. The majority of plant (legumes, cereals, vegetables) proteins

lack or have very low amounts of one or more essential amino acids. An exception is soy, which contains all amino acids. In cereals, the amount of lysine, threonine, or tryptophan is low, while legumes are very deficient in methionine.

To ensure an adequate supply of all essential amino acids, a diet should include animal proteins. In the case of a vegetarian diet, it is important to include beans and grains, which complement each other with respect to their amino acid content.

The quality of a protein is estimated by taking the ratio between its content in essential amino acids and the essential amino acids present in egg protein, multiplied by 100.

The *limiting amino acid* is the essential amino acid present in the lowest amount in a protein when compared to the content of that same amino acid in a reference protein, usually egg white. For example, if tryptophan is the limiting amino acid being present in a given protein at levels that are 50% of those in egg white, the protein has a score of 50 with respect to that amino acid. The optimum value is 100. The term *limiting amino acid* is also applied to those essential amino acids in a particular food, which falls short of meeting the amount required by humans.

The energy needs of the body are primarily met with carbohydrates and fats. If carbohydrate and fat in the diet are low, amino acids will be used as an energy source, which increases the amount of protein required in the diet.

A diet low in protein is the most common cause of malnutrition and one of the primary human health and social problems. Malnutrition is most common in developing countries, where it is associated to poverty and has the highest incidence among children. The most serious conditions of protein malnutrition are kwashiorkor and marasmus. *Kwashiorkor*, a word of African origin, is observed in infants with diets poor in proteins and rich in carbohydrates. It is characterized by marked growth retardation, a swollen abdomen, edema, decreased plasma albumin, anemia, and hepatomegaly. *Marasmus* is a clinical condition of severe wasting, caused by a chronic deficiency in the diet of both proteins and calories. It causes loss of total body fat and muscle mass.

Fate of Amino Acids in the Body

Amino acids can follow different destinations in the body.

1. Most amino acids are used, unmodified, for the synthesis of new protein. Protein synthesis will be discussed in Chapter 22.
2. Certain amino acids enter specific metabolic pathways that convert them in nonprotein nitrogenous compounds of important physiological relevance.
3. Amino acids can also be degraded and finally oxidized for energy production in a process that involves the separation and removal of the amine group.

Fig. 16.2 shows a summary of the metabolic pathways that amino acids are subjected to.

Amino Acid Transport

Amino acids cross cell membranes carried by specific transport systems which recognize only the L-isomers. Several transporters have been characterized; they are grouped into two categories: (1) *secondary active transporters*, which use the electrochemical Na^+ gradient created by the Na,K$^+$-ATPase to cotransport amino acids into the cell. (2) *Facilitated diffusion*, which uses uniport systems that are Na^+ independent and allow amino acid transport following their concentration gradient.

Among the Na^+-dependent carriers are the following:

1. *Neutral amino acid transporters*. These exhibit broad amino acid specificity. Some have a preference for small aliphatic amino acids (alanine, serine, or threonine). Others recognize hydrophobic, aromatic, and aliphatic amino acids (phenylalanine, tyrosine, methionine, valine, leucine, and isoleucine). These include the y^{+L}, ASC, and B$^{o,+}$ systems.

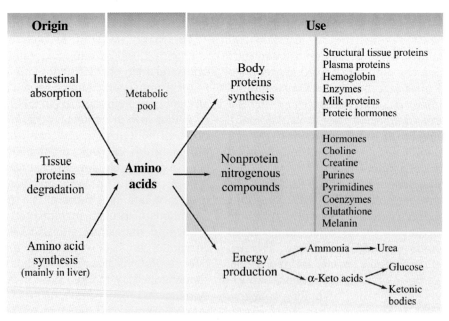

FIGURE 16.2 **Summary of amino acid metabolism.**

2. *Basic amino acid transporters.* These are carriers for lysine, arginine, and ornithine (i.e., y^{+L}, $B^{o,+}$).

3. *Acidic amino acid transporters.* These are selective carriers for aspartate and glutamate (i.e., X_{AG}).

4. *Iminoacid and glycine transporters.* These carriers transport proline, hydroxyproline, and glycine (Pro and Gly carriers).

5. *Glutamine, asparagine,* and *histidine transporters.* This subgroup of carriers includes the N transporters.

These amino acid transporters are found in many tissues, especially in the brush border of the intestinal mucosa and the renal tubular epithelium.

Among the Na^+ independent facilitated diffusion carriers, there are different types according to their specificity for the transported substrate:

1. *Neutral amino acids* (L, asc, $b^{o,+}$ systems).
2. *Cationic amino acids* (y^+, $b^{o,+}$ carriers).
3. *Glutamate* (X_c, x_{AG} transporters).

γ-Glutamyl cycle. This cycle transports amino acids in various organs, especially liver, kidneys, and intestines (p. 396).

Genetic defects in amino acid transport. Diseases have been described in which amino acid transport systems are altered. These include the following:

Hartnup disease. In this disease, the defect is in the hydrophobic aromatic and aliphatic neutral amino acid transporters. It is transmitted with an autosomal recessive trait. Patients with this alteration are unable to absorb neutral amino acids in the intestine and renal tubules, which leads to low plasma levels of these amino acids in plasma and their increased excretion in the urine (aminoaciduria). The disease can cause vitamin deficiency (pellagra) due to lack of tryptophan, which is the precursor of nicotinic acid, a B complex vitamin. Despite these alterations, patients do not suffer severe essential amino acid deficiency because di- and tripeptide transporters in the intestine can compensate for the

uptake of amino acids for which individual's intake is blocked.

Cystinuria. This condition results from mutations in the basic amino acid and cystine transporters. It is inherited in an autosomal recessive manner and is the most common genetic defect related to amino acid absorption (1 in 7000 births). The levels of cystine, lysine, arginine, and ornithine in urine are 20–30 times above normal. The main problem in these patients is the formation of kidney stones containing cystine due to the low solubility of this amino acid in acidic urine.

AMINO ACID CATABOLISM

Amino acids initiate their degradation process by the separation of their α-amino group. Specific metabolic pathways deal with the nitrogen group, which include transfer (transamination) and separation (deamination) reactions.

Transamination

This process refers to the transfer of the amino acid α-amine group to a α-keto acid. The amino acid becomes a keto acid, and the α-keto acid acceptor of the amine group is converted into the corresponding amino acid.

The general equation is:

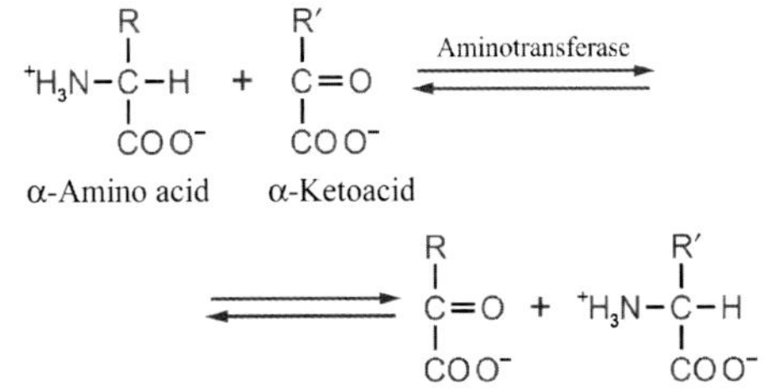

$$
\begin{array}{ccc}
\overset{\displaystyle R}{\underset{\displaystyle COO^-}{^+H_3N-\overset{|}{\underset{|}{C}}-H}} & + & \overset{\displaystyle R'}{\underset{\displaystyle COO^-}{\overset{|}{\underset{|}{C}}=O}} \quad \xrightarrow{\text{Aminotransferase}} \\
\text{α-Amino acid} & & \text{α-Ketoacid}
\end{array}
$$

$$
\xrightarrow{} \overset{\displaystyle R}{\underset{\displaystyle COO^-}{\overset{|}{\underset{|}{C}}=O}} \; + \; \overset{\displaystyle R'}{\underset{\displaystyle COO^-}{^+H_3N-\overset{|}{\underset{|}{C}}-H}}
$$

This reaction is readily reversible; it is catalyzed by *transaminase* or *aminotransferase*, an enzyme that uses pyridoxal phosphate as a coenzyme, which is tightly bound to the enzyme.

Pyridoxal phosphate is derived from pyridoxine, a B complex vitamin. It participates in

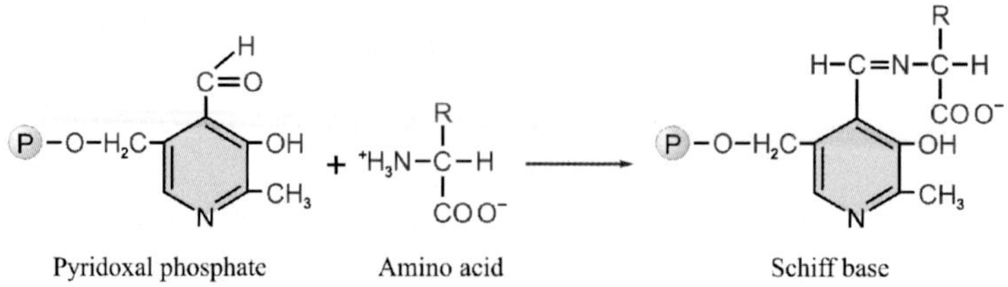

FIGURE 16.3   **Schiff base formation (amino acid-pyridoxal phosphate).**

numerous reactions forming with the amino acid a Schiff base intermediate compound (Fig. 16.3). This reaction allows that all α carbon bonds in amino acids become more labile, facilitating subsequent reactions (transfer of the amine group, decarboxylation, and others). The enzyme is responsible for guiding the direction of the reaction and ensuring the nature of the change. Aminotransferases catalyze the separation and transfer of the amine group attached to the α-carbon. Pyridoxal phosphate serves as an acceptor and transporter of the amine group.

Transamination is a bimolecular reaction and its mechanism is well known. First, the amino acid is bound to the active site to form a Schiff base with pyridoxal phosphate. Then, the α-amine group is separated by hydrolysis and an α-keto acid, derived from the original amino acid, is formed and released. The prosthetic group of the enzyme is converted into pyridoxamine phosphate. Subsequently, another α-keto acid enters the catalytic site as a second substrate, forming a Schiff base with pyridoxamine phosphate. The amine group is transferred to the keto acid, pyridoxal phosphate is regenerated, and the newly formed amino acid is released. Both substrates bind successively and independently to the enzyme and the first product is removed before the second substrate is bound. Pyridoxal phosphate acts as a transient acceptor of the amine group. Commonly, the α-keto acid is α-ketoglutarate; the enzyme receives its name from the amino acid donor of the amine group. For example, aspartate aminotransferase catalyzes the following reversible reaction:

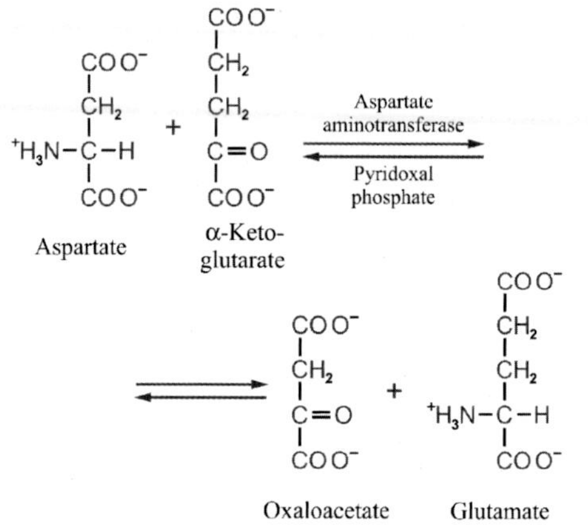

This reaction is particularly important in the liver. In the reverse reaction, oxaloacetate acts as an acceptor of the amine group donated by glutamate.

Alanine aminotransferase is responsible for the reaction:

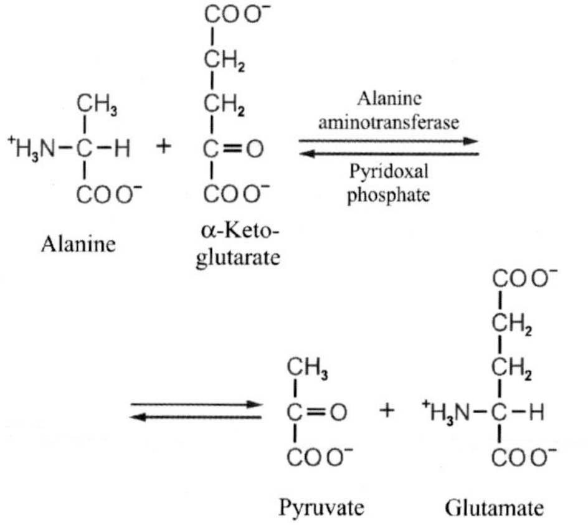

One of the substrates/products of this reaction, alanine, is an important amine carrier. In muscle, the amine groups are transferred from amino acids other than α-ketoglutarate to produce glutamate and eventually pyruvate. Alanine, which enters the circulation, is taken up by tissues, mainly the liver, where it transaminates again to regenerate glutamate and pyruvate.

Aspartate aminotransferase and alanine aminotransferase are the names recommended for these enzymes by the IUBMB; however, the initials GOT and GPT (glutamic-oxaloacetic and glutamic-pyruvic acid transaminases) are widely used in clinics. Both of these aminotransferases are particularly abundant in the liver and heart, which explains the increase of these enzymes in plasma when there are pathological processes of these organs (i.e., hepatitis and myocardial infarction). Hence, determination of these enzymes in plasma is often used for diagnostic and prognostic purposes.

Other examples of transamination reactions include the following:

$$\text{L-Phenylalanine} + \alpha\text{-Ketoglutarate} \xrightleftharpoons[\text{Aminotransferase}]{\text{Phenylalanine}}$$

$$\xrightleftharpoons{} \text{Phenylpyruvate} + \text{L-Glutamate}$$

$$\text{L-Leucine} + \alpha\text{-Ketoglutarate} \xrightleftharpoons[\text{Aminotransferase}]{\text{Leucine}}$$

$$\xrightleftharpoons{} \alpha\text{-Ketoisocaproate} + \text{L-Glutamate}$$

These reactions are all reversible. Certain aminotransferases are expressed as two isozymes with different intracellular localization, the cytosol and the mitochondrial matrix.

With the exception of lysine and threonine, all amino acids are involved in transamination reactions with the α-keto acids pyruvate, oxaloacetate, or α-ketoglutarate, which are converted to alanine, aspartate, or glutamate, respectively. The original amino acids form the corresponding α-keto acids. In turn, alanine and aspartate, produced by transamination from pyruvate, and oxaloacetate react with α-ketoglutarate. The amine groups are used for the formation of glutamate (Fig. 16.4).

In transaminations, the amine group of the amino acid is not eliminated, but transferred to a keto acid to form another amino acid. For this reason, the reaction is not only the first step in the degradation of the amino acids carbon chain, but also the last step in the synthesis of amino acids. Through transaminations, a given amino acid can be generated if the corresponding α-keto acid is available.

Glutamate Deamination

The substrate most frequently involved in transaminations is α-ketoglutarate. Amine groups from almost all amino acids converge to form glutamate. The nitrogen group of

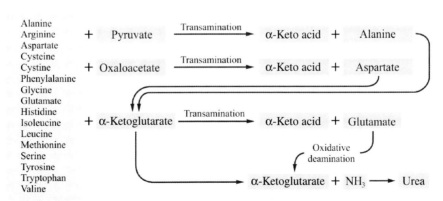

FIGURE 16.4 **Fate of the amine group of amino acids.**

glutamate can be removed by oxidative deamination catalyzed by *glutamate dehydrogenase*. This enzyme, active in most mammalian tissues, uses NAD and NADP as coenzymes. In the forward reaction, NAD^+ usually participates and α-ketoglutarate and ammonia are formed.

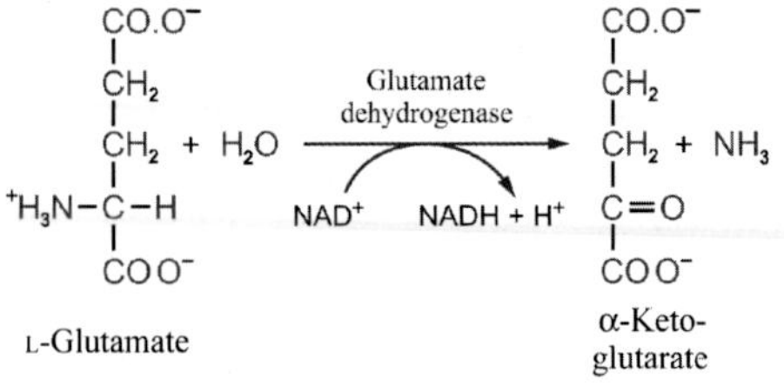

Most of the ammonia produced in tissues is generated by this reaction. At physiological pH, ammonia (NH_3) captures a proton and becomes ammonium ion (NH_4^+).

Glutamate dehydrogenase is found in the mitochondrial matrix. It is an allosteric enzyme, activated by ADP and GDP and inhibited by ATP and GTP. When the ADP level in the cell is high, the enzyme is activated. Increased production of α-ketoglutarate, a feeder of the Krebs cycle, enhances the operation of this pathway, generating ATP. When the cell has abundant ATP and GTP (the latter produced in the reaction catalyzed by succinate thiokinase), glutamate dehydrogenase is inhibited, the supply of α-ketoglutarate is reduced, and the cycle activity is depressed.

The reaction is reversible; ammonia can bind to α-ketoglutarate to form glutamate. Apparently, while the direct reaction preferably uses coenzyme NAD, the reverse reaction involves the reduction of NADP.

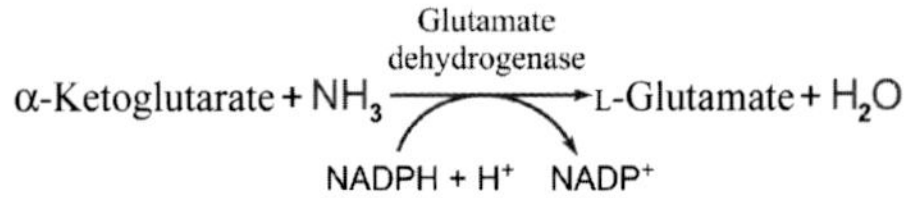

Utilization of different coenzymes depending on the direction of the reaction allows the

independent regulation of the deamination and amination events. Due to the reversibility of the reaction, glutamate dehydrogenase acts in catabolism, as well as in the synthesis of glutamate.

There are other enzymes that catalyze the oxidative deamination of amino acids, these are flavoproteins called *amino oxidases*. Their role in human tissues is not important.

Deamidation. The amide groups of asparagine and glutamine are released as ammonia by hydrolysis, catalyzed by *asparaginase* and *glutaminase* respectively, which produce aspartate and glutamate; the ammonia is protonated to give NH_4^+.

METABOLIC PATHWAYS OF AMMONIA

The main source of ammonia in the body is the oxidative deamination of glutamate in various tissues. Also, significant amounts of ammonia are produced from nitrogenous food by bacteria in the intestinal flora. This ammonia is absorbed via the portal circulation. In normal blood, ammonia levels remain low (10–50 μg/dL or 5–30 μM), indicating the high efficiency of the mechanisms responsible for its removal. Maintaining ammonia low is important because of the toxicity of this compound, particularly at the level of the central nervous system. The liver is the main organ involved in ammonia removal; in cases of severe liver failure, blood ammonia rises and this leads to encephalopathy, coma, and even death. The major route of ammonia elimination in humans is through the synthesis of urea. Another mechanism is the formation of glutamine.

Glutamine Formation

Glutamine is produced by *glutamine synthetase*, a mitochondrial enzyme that catalyzes the formation of the amide bond between ammonia and glutamate. This reaction requires energy released by hydrolysis of ATP into ADP and P_i. The reaction is practically irreversible.

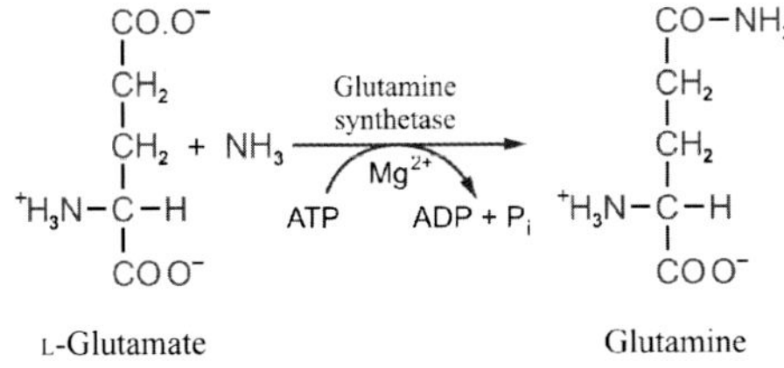

L-Glutamate   Glutamine

Glutamine synthesis is an important mechanism for elimination of ammonia in various tissues. In the liver, it primarily takes place in hepatocytes surrounding the central vein of the liver lobes. Glutamine synthetase activity is also important in muscle, kidney, and brain tissues. The brain is particularly sensitive to the presence of ammonia and, thus actively produces glutamine from ammonia.

Glutamine is hydrolyzed to glutamic acid and ammonia by action of the *glutaminase*. Since the glutamine synthesis reaction is irreversible, its hydrolysis is not performed by reversing the same process, but by a different mechanism. Glutaminase is expressed in periportal hepatocytes and renal tubules, where the production of ammonia and its excretion in urine is one of the mechanisms regulating acid–base balance and cation conservation (p. 707).

A similar reaction is catalyzed by asparaginase that hydrolyzes asparagine to aspartate and ammonia. Some tumors require high amounts of glutamine and asparagine for their development. For this reason, glutaminase and asparaginase have been tried as antitumor agents.

Urea Formation

Almost all ammonia produced by deamination is converted into urea in the liver, the only organ that possesses all the enzymes necessary for this conversion.

The synthesis of urea takes place mainly in hepatocytes surrounding portal vessels by a cyclic mechanism, originally described by Krebs and Henseleit in 1932. The cycle involves five enzymes; ammonia, carbon dioxide, and aspartate (amine donor) are the feeders of the cycle. The process consumes four high-energy phosphate bonds per molecule of urea. It includes the following reactions:

1. *Synthesis of carbamoyl phosphate.* This includes the condensation of ammonia, carbon dioxide, and phosphate (from ATP) to form carbamoyl phosphate. The reaction is catalyzed by *carbamoyl phosphate synthetase 1* (CPS-1), present in liver mitochondria.

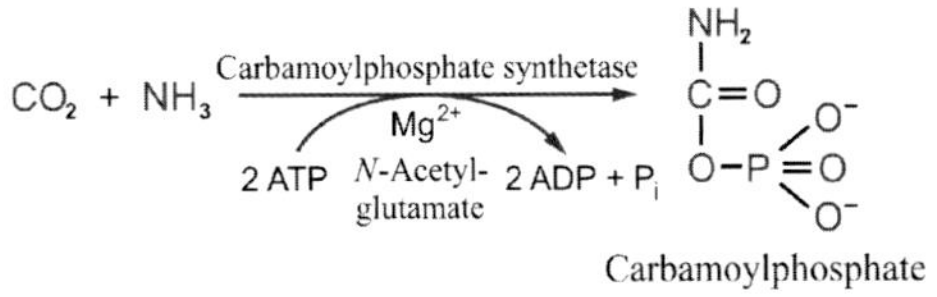

Carbamoylphosphate

Two molecules of ATP are hydrolyzed. The enzyme requires Mg^{2+} and *N*-acetyl glutamate, which acts as an allosteric activator.

There are two isoforms of carbamoyl phosphate synthetase; CPS-2, located in the cytosol of most cells, is involved in the synthesis of pyrimidine nucleotides.

2. *Synthesis of citrulline.* The carbamoyl portion is transferred from carbamoyl phosphate to ornithine (the first cycle intermediate). Citrulline is formed and P_i is released. The reaction is catalyzed by *ornithine transcarbamoylase*, an enzyme of the mitochondrial matrix.

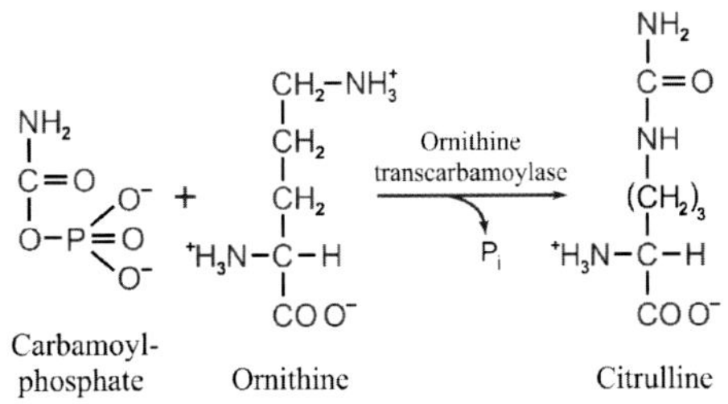

Carbamoyl-phosphate   Ornithine   Citrulline

The reaction equilibrium is strongly shifted toward the right. The following steps occur in the cytosol and citrulline must leave mitochondria via an exchange system.

3. *Argininosuccinate synthesis.* In this reaction, aspartate enters the cycle, which binds citrulline to form argininosuccinate. The responsible enzyme is *argininosuccinate synthetase.* ATP is required, hydrolyzed to AMP and inorganic pyrophosphate (PP$_i$). The process is practically irreversible due to the rapid hydrolysis of pyrophosphate.

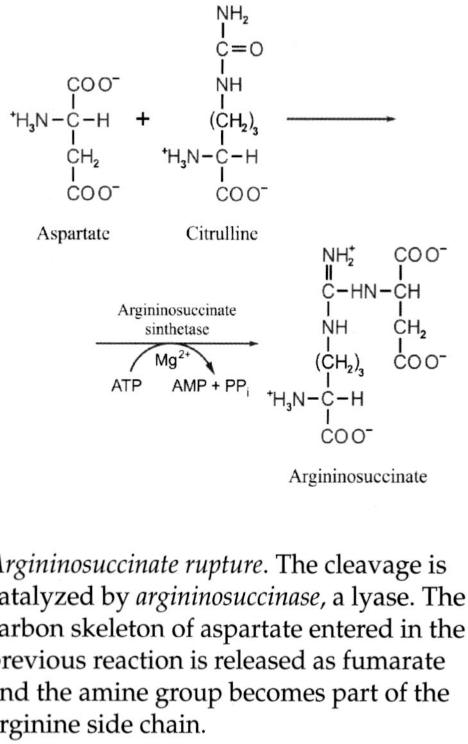

Aspartate   Citrulline

Argininosuccinate
sinthetase
→
Mg^{2+}
ATP   AMP + PP$_i$

Argininosuccinate

4. *Argininosuccinate rupture.* The cleavage is catalyzed by *argininosuccinase*, a lyase. The carbon skeleton of aspartate entered in the previous reaction is released as fumarate and the amine group becomes part of the arginine side chain.

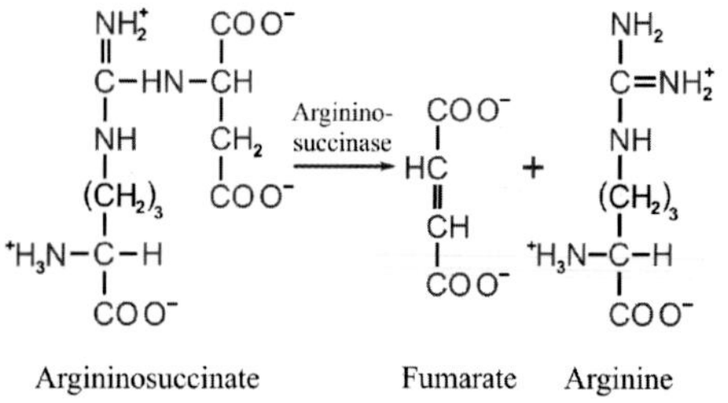

Argininosuccinate   Fumarate   Arginine

5. *Arginine hydrolysis.* This is the last stage of the cycle, where the guanidine group of arginine is hydrolyzed and forms urea and ornithine. Ornithine, the first cycle intermediate, is regenerated. *Arginase* is the enzyme responsible for the reaction.

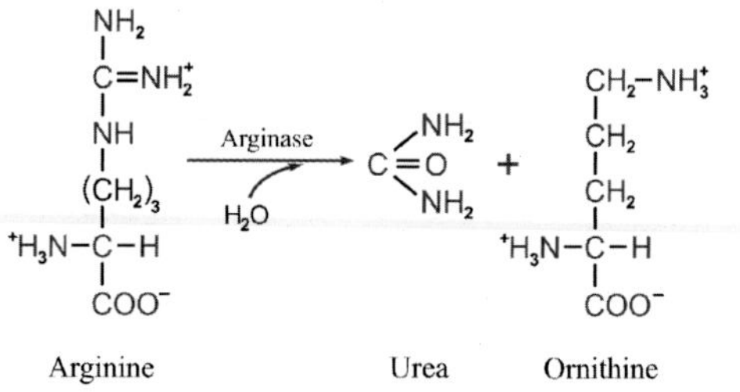

Arginine   Urea   Ornithine

Urea, the end product released after each cycle of reactions, diffuses from the liver to the systemic circulation. The kidney is the main organ for its excretion; approximately 75% of the formed urea is eliminated by the urine. The remaining urea passes to the colon, where it is hydrolyzed by urease of the normal bacterial flora and the ammonia produced returns to the liver via the portal vein.

Ornithine initiates another series of reactions joining a carbamoyl moiety. For this, it needs to enter the mitochondria using the citrulline/ornithine exchange system (antiporter) in the inner mitochondrial membrane.

Fig. 16.5 shows the stages of the urea cycle and the cellular location of reactions.

Considerations About the Urea Cycle

The total equation of the urea cycle is:

$$CO_2 + NH_3 + 3\,ATP + Aspartate + H_2O \rightarrow$$

$$\rightarrow Urea + 2\,ADP + 2\,Pi + AMP + PP_i + Fumarate$$

The first two steps of this pathway occur within the mitochondria. The carbamoyl that participates in the formation of urea is synthesized in the mitochondrial matrix, where carbamoyl phosphate synthetase isozyme CPS-1 is located.

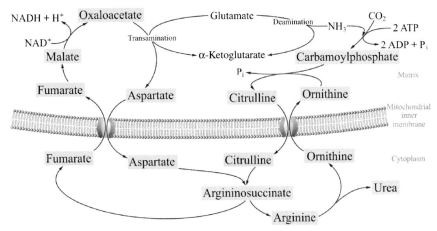

FIGURE 16.5   **The Urea Krebs–Henseleit cycle.** Cycle stages and their cell compartmentalization.

Synthesis of pyrimidine bases also requires carbamoyl; however, it is produced in the cytosol in a reaction catalyzed by the cytosolic isozyme carbamoyl phosphate synthetase CPS-2, which is different from CPS-1. This isozyme compartmentation is important to maintain the functional independence of metabolic pathways, avoiding interference in the use of substrates and products and allowing their separate regulation.

Both nitrogens in urea are derived from amino acids involved in transaminations. Ammonia entered in the first reaction comes primarily from the oxidative deamination of glutamate formed by amine transfer from another amino acid to α-ketoglutarate. The second nitrogen is donated by aspartate and derives from transaminations with oxaloacetate. Thus, nitrogenous residues from almost all catabolized amino acids converge into the cycle. The final product, urea, is a nontoxic compound that can be easily excreted.

Fumarate released in reaction 4 is a citric acid cycle intermediate. In this cycle, it is hydrated to malate and then subsequently oxidized to oxaloacetate. Oxaloacetate has the following metabolic alternatives: (1) condensation with acetyl-CoA to form citrate (first step of the citric acid cycle), (2) conversion to phosphoenolpyruvate in the reaction catalyzed by phosphoenolpyruvate carboxykinase (gluconeogenesis), (3) formation of aspartate by transamination, fueling the urea cycle. Oxaloacetate and fumarate both connect the urea and citric acid cycles in such a manner, that operation of the one cycle drives the function of the other (Fig. 16.6). Urea synthesis requires four P high energy bonds; the oxidation of fumarate to aspartate (oxaloacetate → malate) generates three ATP which help to sustain the overall process.

The efficient functioning of the urea cycle requires additional enzymes than those involved in the cycle. These include hepatic *glutaminase* and *N-acetyl glutamate synthetase*, as well as the ornithine/citrulline and aspartate/fumarate exchangers embedded in the inner mitochondrial membrane.

A normal adult, with a balanced diet, eliminates 25–30 g of urea in the urine every day, which corresponds to 90% of the total nitrogen excreted by this route. The amount of urea eliminated is related to protein intake. The levels of other nitrogenous substances, mainly uric acid, creatinine, ammonia, and amino acids, remain more or less constant and relatively independent of the amount of nitrogenous food ingested.

Urea is soluble, easily diffusible through cell membranes and nontoxic. It is found in circulating blood at concentrations of 20–30 mg/dL

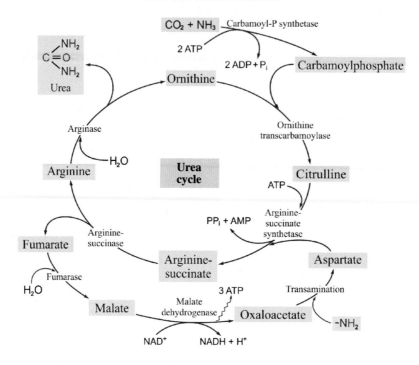

FIGURE 16.6 **Crosstalk of the urea and the citric acid cycles.** The nitrogen molecules entering the cycle to form part of the urea molecule are shown in *red*.

(average 0.4 mM). Its level increases in cases of renal insufficiency in a condition known as uremia. The toxicity associated with uremia is not directly related to increased urea in blood and tissues, but to the increase of other harmful catabolites that the kidney is not able to eliminate. Besides producing urea, the cycle serves as a pathway for arginine synthesis. In this manner, arginine is not essential for nitrogen balance in adults and should only be supplemented in the

diet during situations of increased requirement (growth, pregnancy, or lactation).

Inborn Errors of Urea Cycle

Diseases due to genetic alterations that affect the synthesis of enzymes of the urea cycle have been reported (Table 16.2). Severe deficiencies of carbamoyl phosphate synthetase, ornithine transcarbamoylase, argininosuccinate

TABLE 16.2 Hereditary Diseases Related to the Urea Cycle

Disease	Deficient enzyme	Accumulated substance
Type-I hyper-ammmoniemia	Carbamoylphosphate synthetase 1	Ammonia, glutamine, and alanine
Type II hyper-ammoniemia (linked to X)	Ornithine transcarbamoylase	Ammonia, glutamine, and orotic acid
Citrullinemia	Arginine succinate synthetase	Citrulline
Arginine succinic aciduria	Arginine succinate liase	Arginine and succinate
Argininemia	Arginase	Arginine

synthetase, or argininosuccinase block the synthesis of urea and produce marked increases in the concentration of ammonia in blood and tissues. The absence of ornithine transcarbamoylase, an X-linked condition, is relatively frequent. The other alterations follow an autosomal recessive pattern of inheritance. Since most of the reactions of the cycle are irreversible, determination of urea metabolic intermediate levels in blood and/or urine allows for identification of enzyme deficiency.

Complete lack of any of the enzymes of the cycle is incompatible with life. Affected children appear normal at birth, but 24–48 h later, they present with hypothermia, lethargy, apnea, and a clinical picture of severe encephalopathy, which leads to death.

Partial deficiencies of these enzymes may not be fatal, but can cause mental retardation and other disturbances. In the deficiency of arginase, hyperammonemia is not as severe. There is progressive mental retardation. In milder forms of these genetic diseases, a low protein diet, tending to reduce blood ammonia levels, helps to control the disease.

Ammonia Toxicity

The encephalopathy associated with severe defects of the urea cycle is due to the increase of ammonia in blood and tissues. In patients with severe hepatic impairment, hyperammonemia is primarily responsible for encephalopathy and coma.

At physiological pH, almost all of the ammonia (NH_3) is converted to ammonium ion (NH_4^+), the NH_4^+/NH_3 ratio is 100/1. Ammonia, an electrically neutral molecule, freely crosses cell membranes, while the ammonium ion does not. In the brain, normal levels of ammonia are approximately 0.18 mM, values reaching 0.5 mM are pathological and levels, as high as 1.0 mM can be associated with seizures and coma.

The mechanisms determining ammonia toxicity are not precisely known; however, some possible causes include the following:

1. *Accumulation of glutamine.* Glutamine is a major product of ammonia metabolism. The levels of this substance in blood, tissue, and cerebrospinal fluid, are markedly increased in cases of high ammonia. Glutamine accumulation in brain, especially in astrocytes, produces swelling, increased intracranial pressure, and cerebral hypoxia through its osmotic effects.
2. *Malate-aspartate shuttle inhibition.* The synthesis of glutamine reduces glutamate levels and inhibits the operation of the malate–aspartate shuttle. This leads to increased lactate and pH reduction in the brain.
3. *Glycolysis activation.* Ammonia stimulates phosphofructokinase and glycolytic activity. It increases the lactate/pyruvate and $NADH/NAD^+$ ratios.
4. *Citric acid cycle inhibition.* Ammonia increase diverts the reaction catalyzed by glutamate dehydrogenase toward the amination of α-ketoglutarate and production of glutamate. This drains a cycle intermediate and depresses the activity of this final oxidation pathway.

All the factors mentioned previously affect brain energy metabolism. In addition, they have been shown to alter the function of neurotransmitters and their receptors and, therefore the electrophysiological properties of neurons.

Role of Various Organs in Amino Acid Metabolism

The small intestines, liver, kidneys, and muscle are organs that play an essential role in amino acid metabolism. The main role of each is given as follows.

Intestine. Amino acids from protein digestion are absorbed in the small intestine. Intestine preferably uses glutamine and asparagine as energy suppliers. The products formed, together with the remaining amino acids in the diet, are sent to the liver via the portal vein. During fasting periods, the intestine oxidizes glutamine that is released into the circulation by muscle.

Liver. The catabolism of amino acids, except those with branched chains, starts in the liver. The amine group is separated and incorporated into urea. The carbon skeletons can be oxidized to CO_2 and H_2O or used for gluconeogenesis and ketogenesis.

The liver is very efficient in the removal of ammonia. However, not all hepatocytes are equally involved in this function. Liver cells located around the portal system vessels (portal hepatocytes) receive blood directly from the intestine and are rich in glutaminase, glutamate dehydrogenase, and all the enzymes of the urea cycle. NH_3 produced in reactions catalyzed by glutaminase and glutamate dehydrogenase is used for the synthesis of carbamoyl phosphate in the first step of the cycle. Hepatocytes located next to the cava vein system (venous hepatocytes) are rich in glutamine synthetase. Here, the NH_3 is mainly transferred to glutamate to form glutamine.

Muscle. The degradation of branched chain amino acids mainly starts in skeletal muscle. The amine groups are transferred to pyruvate to form alanine. More than half of the muscle amino acids released to the circulation are alanine and glutamine. Both act as carriers of amines from other tissues.

Kidney. This organ captures glutamine released from muscles. Reactions catalyzed by glutaminase and glutamate dehydrogenase produce ammonia, which is converted to ammonium ion and excreted in urine, neutralizing anions. The ammoniagenesis is one of the mechanisms used by the kidneys to maintain the body's acid–base balance (p. 707).

Fate of the Carbon Skeleton of Amino Acids

Studies using radioactive isotopes confirmed the concept of the dynamic exchange of carbon atoms between carbohydrates, fats, and proteins. In diabetic animals with negative nitrogen balance, administration of certain amino acids increases the excretion of glucose in urine, while others cause an increase in ketone bodies. This observation prompted the classification of amino acids into *glucogenic* and *ketogenic*. Almost all nonessential amino acids are glucogenic, indicating that the conversion of these amino acids to glucose is a reversible process; their carbon skeletons can be synthetized from intermediates of glucose metabolism.

Moreover, almost all the ketogenic amino acids are essential. They can be converted into ketone bodies, but do not serve as precursors of amino acids.

Glucogenic amino acids. Alanine, arginine, aspartate, cysteine, glycine, glutamate, histidine, hydroxyproline, proline, methionine, serine, threonine, and valine are glucogenic amino acids. The catabolism of these amino acids generates one of the following intermediates: pyruvate, oxaloacetate, fumarate, succinyl-CoA, or α-ketoglutarate.

Ketogenic amino acids. These include leucine and lysine.

Glucogenic and ketogenic amino acids. Some amino acids, such as phenylalanine, isoleucine, tyrosine, and tryptophan are both glucogenic and ketogenic.

Fig. 16.7 summarizes the fate of amino acids and their relation to the citric acid cycle.

Intermediate Metabolites Formed by the Catabolism of Amino Acids

Degradation pathways of deaminated amino acid carbon chains vary according to the nature of those chains. Intermediary products produced are related to metabolic pathways of carbohydrates or fats. The intermediate metabolites and amino acids from which they can be generated are indicated as follows.

Oxaloacetate is formed from aspartate and its amide, asparagine. Asparagine is hydrolyzed by asparaginase into aspartate and ammonia.

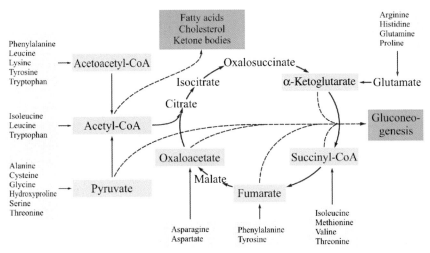

FIGURE 16.7 **Fate of amino acid carbon chains.**

Aspartate, by transamination, is converted to oxaloacetate, an intermediate in the citric acid cycle.

α-Ketoglutarate. Through different pathways, arginine, histidine, proline, and hydroxyproline are converted into glutamate. Glutamate, by transamination or oxidative deamination, produces α-ketoglutarate, an intermediate of the citric acid cycle.

Succinyl-CoA can be originated from methionine, isoleucine, and valine.

Pyruvate can be derived from alanine, serine, cysteine, and cystine. Glycine also forms pyruvate, but must be first converted into serine. Threonine loses the β and γ carbons to give acetyl-CoA and glycine, which contributes to form pyruvate, prior conversion into serine.

Acetyl-CoA. Under an adequate diet, all amino acids that convert into pyruvate render acetyl-CoA, which is oxidized to CO_2 and H_2O when the citric acid cycle operates normally (pyruvate, besides forming acetyl-CoA feeds the cycle by the anaplerotic reaction from pyruvate to oxaloacetate). During prolonged fasting, the pyruvate derived from amino acids, is used in gluconeogenesis instead of forming acetyl-CoA. Phenylalanine, tyrosine, tryptophan, leucine, and lysine do not form pyruvate first, but yield acetyl-CoA, which under fasting conditions or poor metabolism of carbohydrates produces acetoacetate, one of the ketone bodies.

Fumarate. Besides acetoacetate, phenylalanine and tyrosine generate fumarate, an intermediate of the citric acid cycle. Therefore, these amino acids are both gluco- and ketogenic.

Final catabolism. The metabolites mentioned previously continue their total degradation to CO_2 and H_2O in the citric acid cycle. The cycle intermediates and pyruvate can enter the gluconeogenesis to form glucose or glycogen. The acetyl-CoA has numerous metabolic possibilities, including the synthesis of fatty acids.

Metabolic Fate of Branched Chain Amino Acids

The catabolism of valine, leucine, and isoleucine takes place in skeletal muscle, where transamination of these amino acids is very active. The amine groups are primarily transferred to pyruvate with alanine production. The α-keto acids generated are used as fuel in muscle and liver tissue; they are decarboxylated by the branched-chain keto acid α-dehydrogenase, a multienzyme complex with similar structure, mechanism, and action, to pyruvate and α-ketoglutarate dehydrogenases. This enzyme complex is located in mitochondria, requires five coenzymes (thiamine pyrophosphate, lipoic acid, coenzyme A, NAD, and FAD), and is allosterically inhibited by the end products NADH and CoA. Branched-chain keto acid α-dehydrogenase is also regulated by phosphorylation–dephosphorylation; covalent binding of phosphate to the complex inactivates it.

Each acyl-CoA resulting from the reaction is independently degraded by oxidation. The acyl-CoA derived from valine produces propionyl CoA, that coming from leucine generates acetoacetyl-CoA and acyl-CoA, and the one from isoleucine, produces propionyl-CoA and acetyl-CoA. Propionyl-CoA is methylated to methylmalonyl-CoA, a reaction catalyzed by *methylmalonyl-CoA mutase*, which is vitamin B_{12} dependent. A redistribution of atoms in the molecule takes place to form succinyl-CoA, an intermediate of the citric acid cycle.

Genetic defects affecting the synthesis of enzymes in the branched-chain keto acid α-dehydrogenase complex are the cause of *maple syrup disease*. The urine of these patients smells like burnt sugar (maple syrup). The patients suffer brain damage and, in severe cases, they die during the first year of life. The disease is inherited as an autosomal recessive trait, with a frequency of 1:185,000 births.

Genetic alterations of propionyl CoA carboxylase and methylmalonyl CoA mutase synthesis produce metabolic acidosis, with an increase in the excretion of organic acids in urine (organic aciduria).

Amino Acid Biosynthesis

Human beings cannot synthesize essential amino acids; all others can be synthesized. In general, if the corresponding α-keto acid can be synthesized, production of the amino acid by transamination is possible.

OTHER GENERAL MECHANISMS OF AMINO ACID METABOLISM

Decarboxylation

Biosynthesis of Biological Amines

Some bacteria possess the capacity to decarboxylate amino acids. The reaction is one of the stages of protein putrefaction by bacteria; as a result, biogenic amines (physiologically active substances) are produced. Lysine and ornithine generate *cadaverine* and *putrescine*, respectively, through decarboxylation.

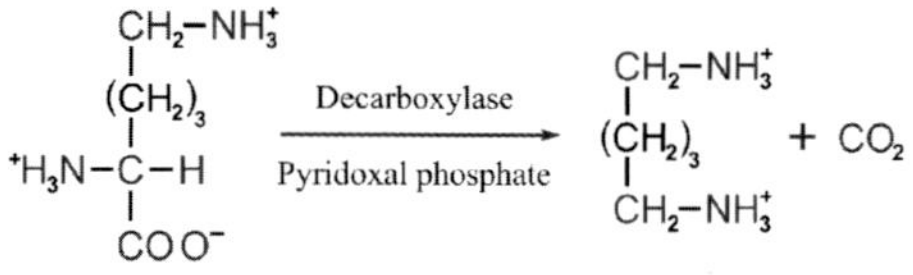

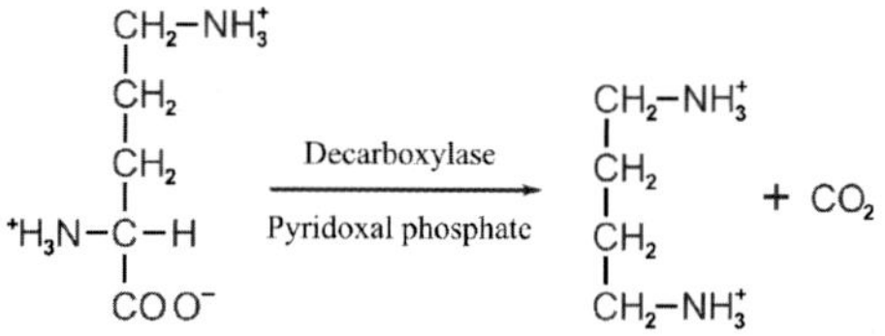

Animal tissues also have enzymes that catalyze decarboxylation of amino acids. The coenzyme pyridoxal phosphate is used by these decarboxylases. Many of the biological amines formed by decarboxylation of amino acids are substances of great functional importance.

Histamine. This biogenic amine is produced by decarboxylation of histidine (Fig. 16.8). The reaction is catalyzed by *histidine decarboxylase* and by an *aromatic amino acid decarboxylase*, which also uses phenylalanine, tryptophan, and tyrosine as substrates. Both enzymes require the coenzyme pyridoxal phosphate. Histamine has important physiological activity. It is a chemical messenger in many cellular responses, a vasodilator that contributes to lower blood pressure and, in large doses, can produce vascular

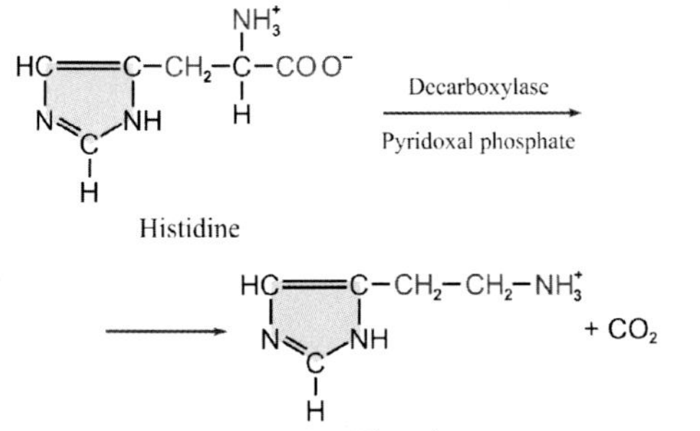

FIGURE 16.8 **Decarboxylation of histidine.**

collapse. It causes bronchiolar constriction in the lung and stimulates the secretion of hydrochloric acid and pepsin in the stomach.

Histamine is stored in mast cells, from where it is released abruptly in response to exogenous sensitizing agents that enter the body and cause allergic or inflammatory reactions of various kinds. Antihistamine compounds are used to treat allergic reactions. They are substances with a chemical structure similar to histamine and behave as competitive antagonists.

Given the intense activity of histamine, the body must degrade it quickly after it is produced. This process is performed by *histaminase,* an enzyme which catalyzes the oxidation of histamine and converts it into an inactive product.

Tyramine and tryptamine. The decarboxylation of tyrosine produces tyramine, while the decarboxylation of tryptophan generates tryptamine. Both these biogenic amines have vasoconstrictor action.

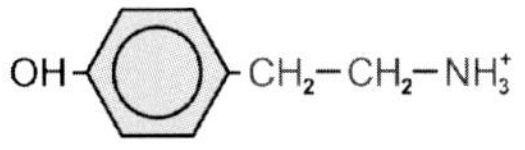

Tyramine

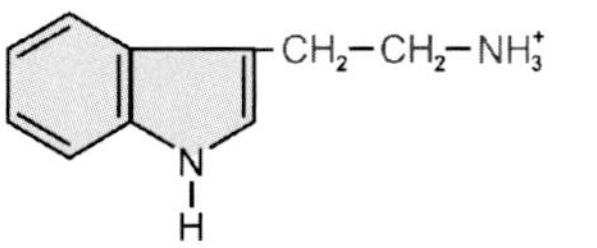

Tryptamine

γ-Aminobutyric acid (GABA). This compound is formed by decarboxylation of glutamic acid. The enzyme that catalyzes this reaction predominates

in the gray matter of the central nervous system and requires pyridoxal phosphate.

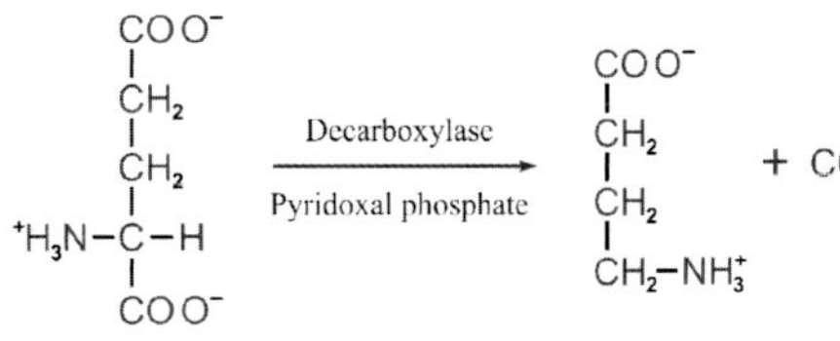

GABA plays an important functional role as a chemical regulator of neuronal activity. It depresses nerve impulse transmission.

Huntington's disease is an inherited, progressive, and fatal condition characterized by abrupt, involuntary movements (known as chorea). Patients suffer degeneration of GABAergic neurons, which leads to decreased levels of GABA.

Polyamines. Putrescine is generated from the decarboxylation of ornithine. It reacts with methionine decarboxylase to form the polyamines spermidine and spermine.

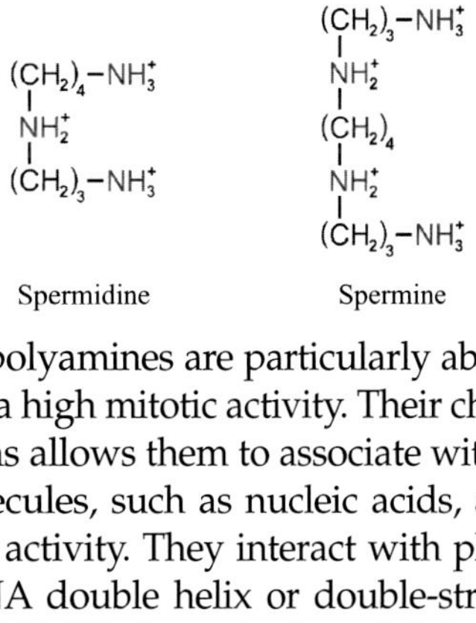

Spermidine · · · · · · · · · · · · Spermine

These polyamines are particularly abundant in cells with a high mitotic activity. Their character of polycations allows them to associate with polyanionic molecules, such as nucleic acids, and influence their activity. They interact with phosphates in the DNA double helix or double-stranded regions of RNA. The bacteriophage T4, for example, has approximately 40% of its negative DNA charges neutralized by polyamines. In some species, transfer RNA is bound to spermidine and spermine. There are proteins that have polyamines covalently linked to the carboxyl of glutamate.

Polyamines may play a role in cell cycle regulation. Difluoromethylornithine, an ornithine decarboxylase inhibitor, stops the progression of the cell cycle.

Transfer of Monocarbon Groups

In transamination reactions, amino acids can donate their amine functional group to an acceptor molecule. Some amino acids also participate in reactions in which other groups, used in the synthesis of functional important substances, are transferred. For example, the amidine group of arginine, or the amide group of glutamine and asparagine, are employed in the synthesis of various compounds.

Monocarbon groups are important molecular fragments frequently transferred in synthesis processes of more or less complex chemical structures. They may exhibit different degrees of oxidation, such as methyl ($—CH_3$), hydroxymethyl ($—CH_2OH$), formyl ($—COH$), and carbon dioxide (CO_2), which is also considered a monocarbon residue.

The main methyl donor is methionine; its methyl group is used in the synthesis of numerous substances such as choline, creatine, adrenaline, carnitine, and methylated RNA. This is possible only if the methionine is activated, for which it must react with ATP to form S-adenosyl methionine (SAM). The sulfur atom of methionine is attached to C5′ of adenosine. Two of the ATP phosphate groups are released as pyrophosphate and the third as an inorganic phosphate.

SAM is the active form of methionine. The bond between methyl and sulfur is of high energy, which explains the ability of the methyl group to participate in transfer reactions. The reactions in which SAM transfers methyl groups to form different compounds are catalyzed by methyltransferases that are specific for each compound.

Homocysteine. S-adenosyl-homocysteine, a product of transmethylations, is hydrolyzed to adenosine and homocysteine. Cysteine or methionine is synthesized from this compound. There are genetic defects that alter the metabolism of homocysteine (homocysteinuria and cisteinuria).

Homocysteine   Homocystine

Increased plasma homocysteine is observed in deficiencies of vitamin B_{12} and folic acid. Hyperhomocysteinemia is a risk factor for atherosclerosis of coronary, cerebral, and peripheral vessels.

Another transport agent of monocarbons [methyl, methylene ($—CH_2—$), methenyl ($=CH—$), and formyl] is tetrahydrofolic acid (THF), a derivative of folic acid, which is a member of the vitamin B complex. Methylcobalamin, related to vitamin B_{12}, is another essential factor in one-carbon transfer reactions. Most of these groups are donated to THF by metabolites produced by the degradation of the amino acids serine, glycine, histidine, or tryptophan and transferred for the synthesis of purine, thymine, and methionine.

Carbon dioxide (CO_2) produced in the citric acid cycle or amino acid decarboxylation reaction is transferred in carboxylase catalyzed reactions. The coenzyme is the vitamin B biotin. Carboxylations require ATP and form important metabolic intermediates. The principal three carboxylases using biotin are: (1) pyruvate carboxylase, which catalyzes the first step of gluconeogenesis (also anaplerotic reaction of the citric

S-Adenosylmethionine (SAM)
(Active methionine)

acid cycle), (2) acetyl CoA carboxylase, involved in the first step of fatty acid synthesis, and (3) propionyl-CoA carboxylase, a key enzyme in the catabolism of valine and isoleucine.

Folic acid, vitamin B_{12}, and biotin are further considered in Chapter 27.

METABOLIC PATHWAYS OF AMINO ACIDS

In addition to the general metabolic processes mentioned, each amino acid has specific pathways that lead to different products. Amino acid metabolism is one of the most complex in the study of biochemical transformations. It is not the purpose of this text to cover the comprehensive analysis of this subject. Only some examples will be presented, giving an idea of the multiplicity of possibilities of amino acid metabolism.

Metabolism of Phenylalanine and Tyrosine

Due to their structural relationship, these amino acids are considered together.

Humans cannot synthesize the benzene ring and this group needs to be exogenously provided from phenylalanine and tyrosine residues introduced with the diet.

Phenylalanine is converted to tyrosine mainly in the liver. There is no enzyme to catalyze the reverse reaction (formation of phenylalanine from tyrosine). For this reason, phenylalanine is an essential amino acid but tyrosine is not if there is an adequate amount of phenylalanine present.

The catabolic pathway of these amino acids leads to formation of fumarate and acetoacetate, metabolic intermediates related to carbohydrate and fatty acid metabolism, respectively. For this reason, phenylalanine and tyrosine are both glucogenic and ketogenic.

The catabolic reaction sequence of phenylalanine and tyrosine is shown in Fig. 16.9. The first step is the conversion of phenylalanine to tyrosine, a reaction that is catalyzed by a system consisting of two enzymes: *phenylalanine hydroxylase* and *dihydrobiopterin reductase*. These enzymes introduce a hydroxyl group in the *para* position on the benzene ring. Molecular oxygen and the hydrogen donor coenzyme tetrahydrobiopterin

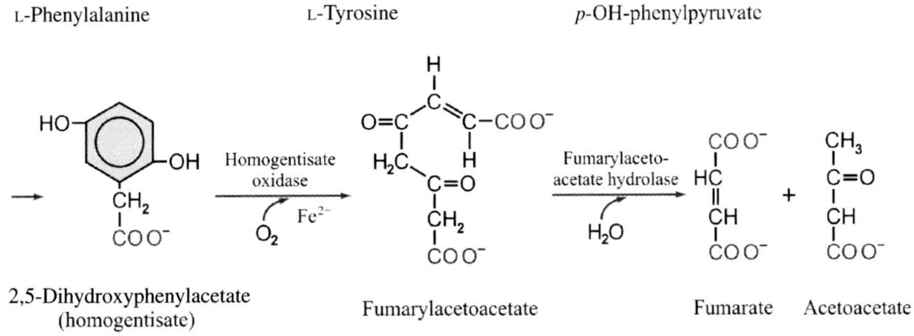

FIGURE 16.9 **Phenylalanine and tyrosine catabolism.**

(THB) are required. The reaction is irreversible. Dihydrobiopterin reductase regenerates THB from dihydrobiopterin and NADPH.

In a second stage, a transamination reaction generates the corresponding α-keto acid, p-hydroxyphenylpyruvic acid. The third step is a complex reaction that produces oxidation, displacement of the side chain to another position on the benzene ring, and decarboxylation to form homogentisic acid. This compound is oxidized to maleyl-acetoacetate (not shown in the sequence of Fig. 16.9), which is isomerized to fumaryl-acetoacetate. Finally, fumaryl-acetoacetate is hydrolyzed to fumarate and acetoacetate. Fumarate is an intermediate in the citric acid cycle, while acetoacetate is a ketone body. Both can be oxidized to carbon dioxide and water.

All the enzymes involved in this pathway can be found in the liver.

Most of phenylalanine catabolism follows the described path. However, there is a minor amount that follows an alternative route, which starts with a transamination reaction to give phenylpyruvic acid, which is then decarboxylated to phenylacetic acid or reduced to phenyl-lactic acid (Fig. 16.10). Normally, phenylpyruvic and phenyl-lactic acid are converted back to phenylalanine and metabolized by the pathway leading to tyrosine and homogentisic acid. Phenylacetate is excreted in urine.

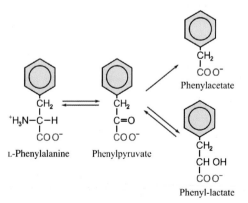

FIGURE 16.10 **Alternative metabolic pathway of phenylalanine.**

Catecholamine Biosynthesis

Another metabolic pathway of phenylalanine and tyrosine leads to the production of compounds with high physiological activity, such as the catecholamines *dopamine, norepinephrine,* and *epinephrine.* These compounds are synthesized in chromaffin cells from the nervous system and the adrenal medulla. The designation of chromaffin derives from the brownish granules that these cells acquire when labeled with potassium dichromate.

The reaction sequence for catecholamine synthesis from tyrosine is the following:

1. *DOPA formation. Tyrosine hydroxylase* introduces a second hydroxyl group on the benzene ring of tyrosine to form 3,4-dihydroxyphenylalanine, also known by its acronym DOPA. The reaction requires oxygen and THB. To maintain the THB level, the presence of dihydrobiopterin THB reductase and NADPH is required.

   Tyrosine hydroxylase is a regulatory enzyme. It is inhibited by dopamine and norepinephrine and activated by cAMP-dependent phosphorylation.

2. *Dopamine formation. DOPA decarboxylase* uses pyridoxal phosphate as coenzyme to catalyze the formation of this biogenic amine. The enzyme has broad substrate specificity, acting on aromatic amino acids and derivatives. Dopamine belongs to the catecholamine family and is a neurotransmitter (chemical intermediate) in the nervous system. In some neurons of the central nervous system, the metabolic pathway ends in dopamine production; in other neurons, it continues further to produce other catecholamines.

   Dopamine content in neuronal centers, such as the caudate nucleus, putamen, and pallidum, is very low in patients with Parkinson's disease. This finding is of great clinical and pharmacological interest, and has immediate clinical application.

Administration of DOPA in high doses is used in the treatment of this disease. The compound moves from the blood into the nervous center, where it is converted into dopamine. Dopamine cannot be used in the treatment of Parkinson's disease because the catecholamine cannot cross the "brain barrier"; therefore, the diffusible precursor DOPA is used.

3. *Norepinephrine formation.* Dopamine can be hydroxylated by *dopamine-β-hydroxylase* to produce norepinephrine (also called noradrenaline). Molecular oxygen, ascorbic acid (vitamin C), and copper are required for the reaction. Most of the norepinephrine synthesized in the body is produced in the sympathetic nervous system.

4. *Epinephrine or adrenaline.* Norepinephrine can be converted into epinephrine (also called adrenaline) by transmethylation. The reaction is catalyzed by *phenylethanolamine-N-methyltransferase* (PNMT) and requires SAM as a methyl donor. The enzyme is found in the adrenal medulla. Adrenaline is the main catecholamine synthesized in this gland.

PNMT synthesis is induced by hormone cortisol from the adrenal cortex.

Fig. 16.11 summarizes the pathway.

Catecholamines are chemical transmitters of the adrenergic system. Their action is varied; they have vasoconstrictive actions in some vascular regions and vasodilator actions in others. They increase heart rate and cardiac output, have relaxing effects on bronchial muscles, and stimulate muscle glycogenolysis and lipolysis in adipose tissue. The type of response depends on the type of adrenergic receptors expressed in the target organs. For example, α-receptors are involved in peripheral vasoconstriction, while the β-receptors are associated with heart stimulation and vasodilation in some areas. Norepinephrine acts mainly on α-receptors, so the vasopressor action predominates. Adrenaline has effect on both types of receptors.

Chromaffin tissue tumors from the adrenal medulla or extra adrenal tissue, called pheochromocytomas, secrete large amounts of catecholamines, mainly norepinephrine. Patients with pheocromocytomas suffer from severe hypertension, elevated heart rate and palpitations,

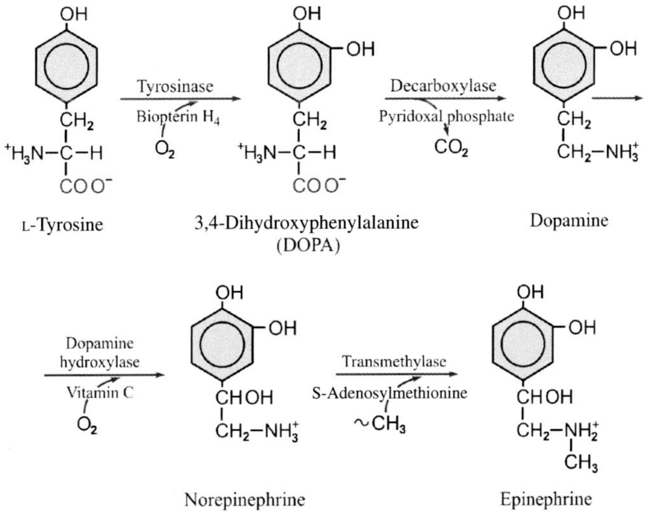

FIGURE 16.11 **Catecholamine biosynthesis.**

weight loss, high glucose levels, headaches, increased perspiration, pallor, and anxiety.

As a substance of great biological activity, catecholamines are rapidly degraded and eliminated from the body; their half-life is 15–30 s. The inactivation depends on the action of *monoamine oxidase* (MAO) and *catechol-O-methyl transferase* (COMT). MAO is responsible for the oxidative deamination of catecholamines and other biological amines, such as tyramine, tryptamine, and serotonin. COMT catalyzes the methylation of one of the hydroxyl groups of catecholamines, generally, the OH bound to the benzene ring in the *meta* position. Further details on the ways of catecholamine inactivation are given in the section on adrenal medullary hormones (p. 602).

Several MAO inhibitors are known. Its administration causes increase of catecholamine levels in the nervous system, causing stimulatory effects. Some of these drugs are used to treat nervous depressions. Reserpine is a drug that has effects opposite to MAO inhibitors. They determine reduction of adrenaline deposits accumulated in nerve endings.

Melanin Synthesis

Melanin, the pigment that gives color to the skin and hair, is derived from tyrosine, which, after conversion to DOPA, undergoes a series of transformations that result in the formation of melanin. Hydroxylation of *tyrosine* to DOPA in melanocytes is catalyzed by *tyrosinase*, a copper-containing enzyme with a function of tyrosine hydroxylase different to that involved in the pathway of catecholamine synthesis in nervous system and adrenal medulla.

Activity of tyrosinase also promotes the oxidation of DOPA to dopaquinone, which after a series of nonenzymatic oxidations is polymerized into brown and black granules that serve as pigments.

Thyroid Hormone Synthesis

Thyroxine and triiodothyronine, which are hormones of the thyroid gland, are synthesized from tyrosine. The corresponding pathway will be described in p. 589.

Inborn Errors of Tyrosine and Phenylalanine Metabolism

Several genetic diseases related to the metabolism of tyrosine and phenylalanine have been described. Only the most common ones will be mentioned here.

Phenylketonuria or phenylpyruvic oligophrenia. This hereditary disease is caused by the inability of cells to convert phenylalanine to tyrosine, which can be caused by lack of phenylalanine hydroxylase, biopterin reductase, or dihydro- and tetrahydro-biopterin. In these patients, tyrosine becomes an essential amino acid and its metabolism is not affected. Phenylalanine, which cannot follow the main pathway of degradation, is shuttled into alternative metabolic routes to form phenylpyruvate, phenylacetate, and phenyl-lactate (Figs. 16.10 and 16.12). Phenylalanine, along with these intermediates, accumulates in tissues and blood and is excreted in urine.

This defect occurs with a frequency of one case per 10,000 births. In children suffering from this disease, the central nervous system is affected and they present with severe mental retardation. Administration of a diet low in phenylalanine early in life prevents the mental defects. Hence, early diagnosis is critical. The presence of phenylpyruvic acid in urine of newborns should be systematically investigated. A safer index for a rapid diagnosis is the level of phenylalanine in plasma.

Alcaptonuria. In this disease, there is an inability to catabolize phenylalanine and tyrosine. The metabolic defect is due to lack of homogentisate oxidase. Homogentisic acid increases, which is excreted in urine. By oxidation, this substance becomes a brown blackish pigment called alcaptone. The darkening of urine exposed to air is precisely the most striking symptom of the disease. Generalized connective tissue pigmentation and arthritis can occur in adult patients.

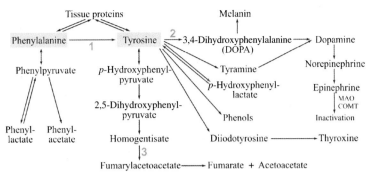

FIGURE 16.12 **Summary of phenylalanine and tyrosine metabolic pathways.** *1* indicates the step blocked in phenylketonuria, *2* in albinism, and *3* in alkaptonuria.

Albinism. This genetic disorder is caused by various inherited defects that compromise the production of melanin. Some forms of albinism are caused by lack of tyrosinase. These patients cannot synthesize melanin. There is lack of pigmentation in skin and hair, high sensitivity to solar radiation, and tendency to develop skin carcinomas.

Tryptophan Metabolism

Some pathways of tryptophan metabolism include the following:

Serotonin Biosynthesis

One of the pathways for serotonin synthesis comprises the hydroxylation of tryptophan at carbon 5 to form 5-hydroxy-tryptophan. This reaction is catalyzed by *tryptophan hydroxylase*, which requires oxygen and THB. In a second step, the compound is decarboxylated to 5-hydroxytryptamine (5HT), also known as *serotonin*, trombocitine, or trombotonine (Fig. 16.13). This compound is synthesized and stored in brain and intestinal cells. Platelets also contain serotonin that they receive from plasma.

Serotonin functions as a neurotransmitter and exerts multiple regulatory functions in the nervous system. It is involved in mechanisms such as sleep, appetite, thermoregulation, pain perception, and the anterior pituitary secretion control. 5HT is produced within the brain; it does not cross the blood–brain barrier. Decreased levels of serotonin in the central nervous system have a depressant effect on brain activity. 5HT stimulates contraction of smooth muscle, it is a powerful vasoconstrictor.

There is a type of malignant neoplasia that generates by transformation of enterochromaffin cells that synthesize serotonin (argentaffinoma). Among other symptoms, this pathology causes the carcinoid syndrome, which is characterized

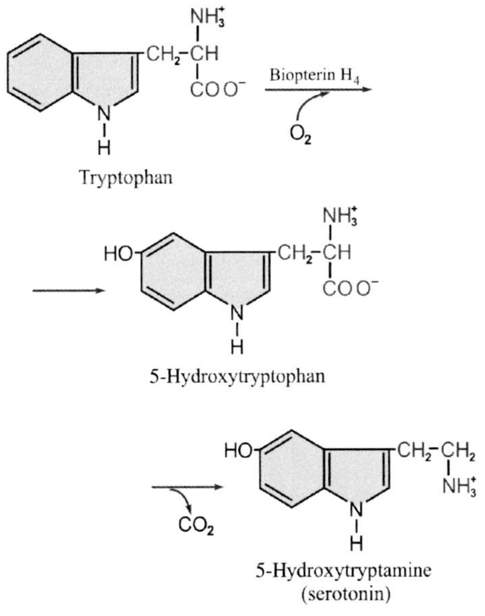

FIGURE 16.13 **Serotonin biosynthesis.**

by flushing, diarrhea, muscle cramps, and less frequently, bronchial spasm and restrictive cardiopathy.

Most of the released serotonin is subjected to oxidative deamination (MAO) to give 5-OH-indoleacetic, an inactive compound. MAO inhibitors produce psychic stimulation because they prolong serotonin action.

Lysergic acid diethylamide (LSD) competes with serotonin. LSD poisoning can be treated by administration of serotonin.

Melatonin Synthesis

Melatonin is a hormone released by the pineal gland and it is present in peripheral nerves in humans. It blocks the action of melanocyte stimulating and adrenocorticotrophic hormones. It is synthesized from tryptophan, which must first be converted into serotonin. This is then acetylated and methylated (Fig. 16.14).

Melatonin synthesis is regulated by light–dark photoperiods. This hormone is closely related to the production of circadian rhythm and reproductive function. Melatonin has been used as a sleep aid in sleep disorders.

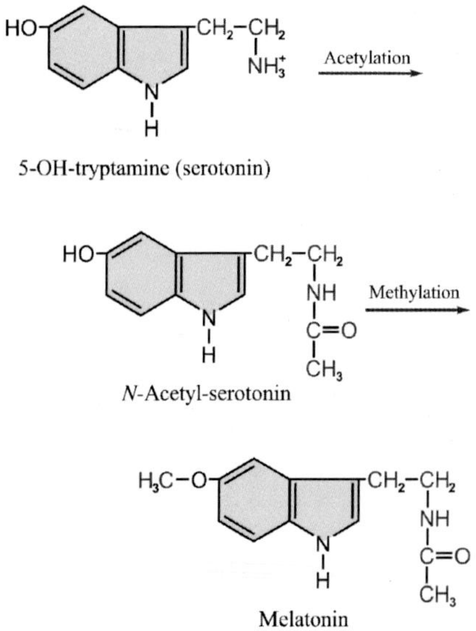

5-OH-tryptamine (serotonin)

N-Acetyl-serotonin

Melatonin

FIGURE 16.14 **Melanotonin biosynthesis.**

Nicotinic Acid Synthesis

Nicotinic acid is a vitamin of the B complex. One of its derivatives, nicotinamide, forms part of coenzymes NAD and NADP. Lack of this vitamin in the diet produces a very severe disease known as *pellagra* (p. 669). Consuming foods rich in tryptophan improves the symptoms caused by pellagra, even when no nicotinic acid is supplied. This is because nicotinic acid is produced in one of the metabolic pathways of tryptophan.

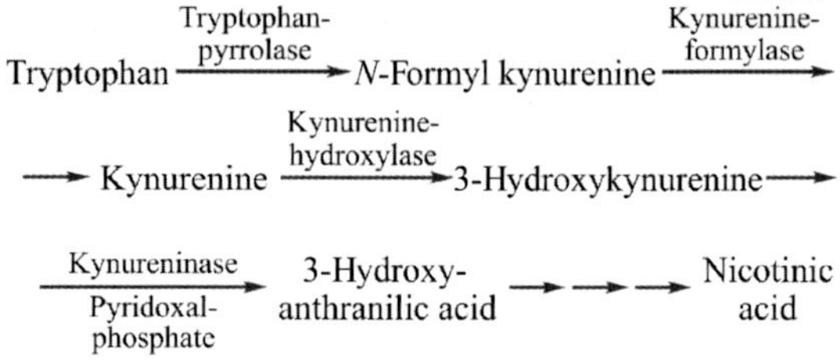

Bacterial Putrefaction

Bacteria from the intestinal flora act on tryptophan residues present in the digestive tract content and produce a series of substances. These include indole, indolacetate, skatole, and indoxyl. All these products can be excreted with the feces or absorbed and eliminated in urine. Indoxyl conjugated with sulfate is found in urine of people with high-protein diets. The potassium salt of indoxyl sulfate is called urinary *indican*.

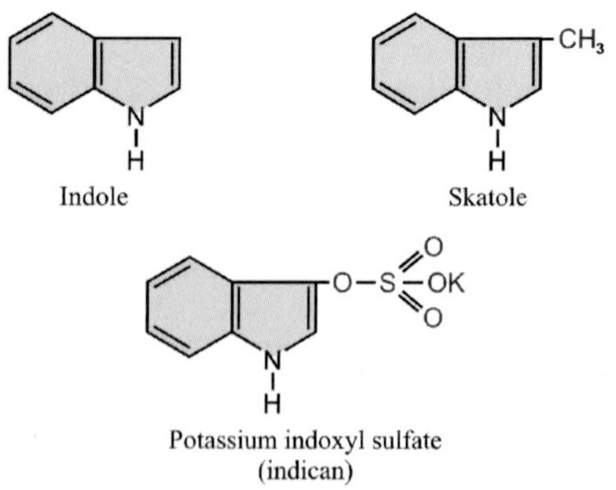

Indole

Skatole

Potassium indoxyl sulfate (indican)

Nitric Oxide Synthesis

Nitric oxide ($NO^•$) is a gas generated from arginine in a reaction catalyzed by *nitric oxide synthase* (NOS). Coenzymes NADPH, FMN, FAD, and THB participate in the reactions that generate NO. NOS contains heme Fe^{2+} and is activated by the Ca^{2+}–calmodulin complex. Citrulline and nitric oxide are produced.

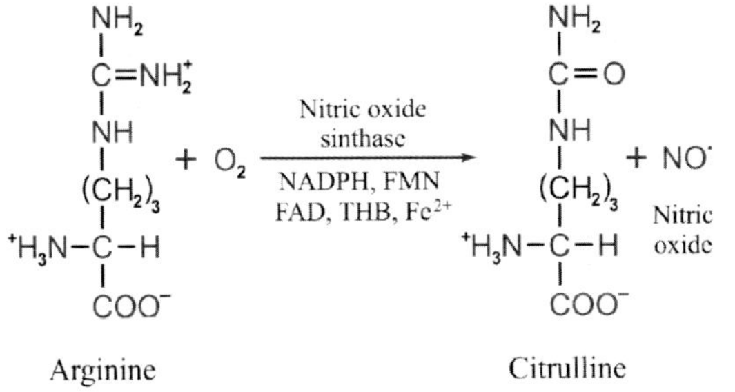

$$
\begin{array}{c}
NH_2 \\
| \\
C=NH_2^+ \\
| \\
NH \\
| \quad\quad + O_2 \xrightarrow[\substack{NADPH,\ FMN \\ FAD,\ THB,\ Fe^{2+}}]{\text{Nitric oxide sinthase}} \\
(CH_2)_3 \\
| \\
{}^+H_3N-C-H \\
| \\
COO^-
\end{array}
\qquad
\begin{array}{c}
NH_2 \\
| \\
C=O \\
| \\
NH \\
| \quad\quad + NO^• \\
(CH_2)_3 \quad \text{Nitric} \\
| \quad\quad \text{oxide} \\
{}^+H_3N-C-H \\
| \\
COO^-
\end{array}
$$

Arginine   Citrulline

NADPH and dihydrobiopterin are required to regenerate THB. Nitric oxide is a free radical of high chemical reactivity. For this reason, it has a very short half-life of less than 5 s. Thanks to its small size, $NO^•$ rapidly diffuses across membranes. $NO^•$ is an important regulator and multifunctional messenger.

Mammals express three isozymes of NOS, with different cellular distribution and regulatory properties. One is found in the endothelial cells of blood vessels (eNOS or I), another is found in macrophages (iNOS or II), and a third isozyme is present in neuronal cells (nNOS or III). While the mentioned tissues are the main location of NOS, other tissues also have NOS, albeit at lower levels. Isozymes I and III are constitutive. In contrast, the macrophage isoform (iNOS or II) is inducible.

$NO^•$ is the *vascular relaxing factor* derived from vascular endothelium. It was described before knowing its chemical nature. It participates as a messenger of a cell signal system that includes guanylate cyclase (p. 562). The effect of $NO^•$ is vasodilation in different vascular areas, including the corpora cavernosa of the penis, where it contributes to penis erection.

During many years, nitroglycerin and organic nitrates were used as vasodilators in the treatment of coronary disease. Today, it is known that these compounds act by releasing $NO^•$.

Nitric oxide is produced in different areas of the brain. Although it is a very atypical neurotransmitter, there is evidence of its involvement in important functions of the nervous system.

In macrophages, NOS is induced by stimuli and factors released in inflammatory foci or after the entry of foreign agents. The sudden release of $NO^•$ generates free radicals with toxic action on bacteria, tumor cells, and other phagocytosed materials.

Use of Amino Acids in Detoxification or Biotransformation Reactions

Some amino acids, or substances derived from them, are involved in reactions that facilitate the removal of toxic substances in the body. For example, glycine forms nontoxic complexes with various aromatic substances. When combined with benzoic acid, it forms *hippuric acid* (hippurate), which is excreted in the urine (Fig. 16.15). Hippuric acid synthesis takes place mainly in the liver, thus, its determination in urine after benzoate administration can be used as a liver function test.

Cysteine participates in biotransformation reactions of organic compounds. The amino acid

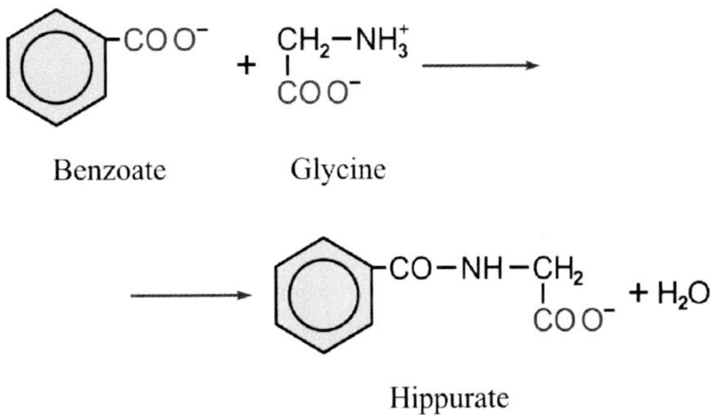

Benzoate   Glycine

Hippurate

FIGURE 16.15 **Hippuric acid biosynthesis.**

is converted to taurine after several metabolic steps. Taurine is involved in conjugations (e.g., see formation of taurocholate, p. 262). The cysteine residue of glutathione is used in the formation of *mercapturic acids*, soluble compounds that are easily eliminated in the urine, resulting from the detoxification of halogenated compounds, organophosphates, and others.

Sulfoconjugation is a biotransformation process in which sulfate participates after activation by reaction with ATP. Adenosine-3′-phosphate-5′-phosphosulfate and phosphoadenosine-adenosine-phosphosulfate (PAPS), or active sulfates, are able to transfer sulfate to an acceptor substance.

Much of the sulfate from the sulfur oxidation of cysteine and methionine is eliminated by urine as inorganic sulfate. A small part combines with organic compounds derived from phenol, indole, or skatole. Indican is the potassium salt of indoxylsulfate. The amount of this compound in urine reflects the magnitude of bacterial putrefaction in the intestine.

Creatine Synthesis

Creatine, free or bound to phosphate (phosphocreatine or creatinephosphate), is present in skeletal muscle, myocardium, and brain tissues. Three amino acids, arginine, glycine, and methionine are involved in the biosynthesis of creatine. The first reaction takes place in the kidneys and consists in the binding of the amidine moiety of arginine to glycine, to form guanidoacetic acid. Creatine synthesis is completed in the liver by methylation of guanidoacetic acid (Fig. 16.16). The methyl donor is S-adenosyl methionine.

From the liver, creatine enters the circulation and is taken up by skeletal muscle, myocardium, and brain tissues where it reacts with ATP to form creatinephosphate. The phosphorylation is catalyzed by *creatinephosphate kinase*, which is located in the intermembrane space of mitochondria.

Creatinephosphate kinase isozymes. The enzyme is a dimer formed by the possible associations

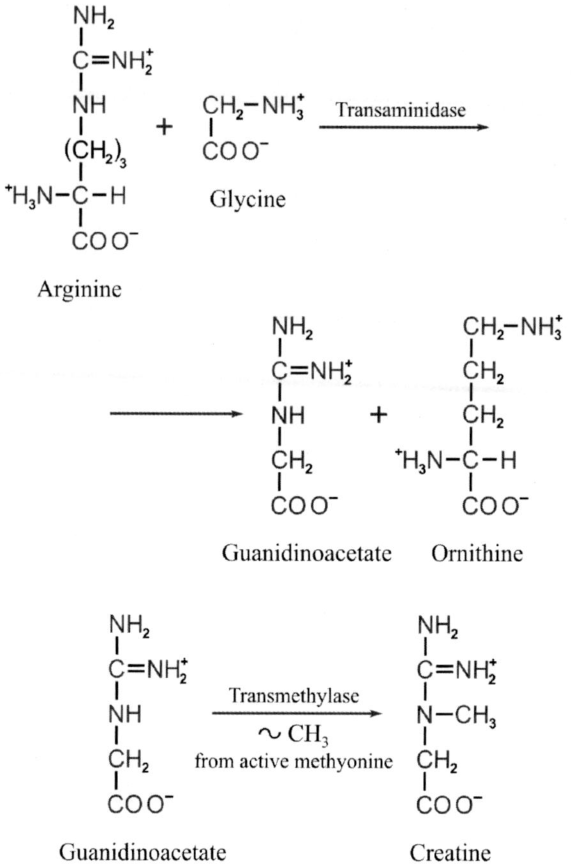

FIGURE 16.16 **Creatine biosynthesis.**

between subunits M (muscle) and B (brain), which results in the existence of three isoforms: MM, MB, and BB. The brain isozyme is BB, skeletal muscle is almost entirely MM, and the heart contains about 15% MB and the rest is MM. Determination of creatinephosphate kinase activity in blood plasma is useful in the diagnosis of myocardial infarction. The enzyme level increases in plasma 6–8 h after a myocardial infarct is produced; it reaches maximum levels at 24–48 h postinfarct. Detection of plasma isozyme MB allows for identification of heart damage.

The creatinephosphate has a high-energy phosphate bond. It is an energy reserve compound used to maintain the intracellular level of ATP in muscle during brief periods of intense contractile activity. The reaction is reversible; when ATP is required, it can be generated from creatinephosphate and ADP.

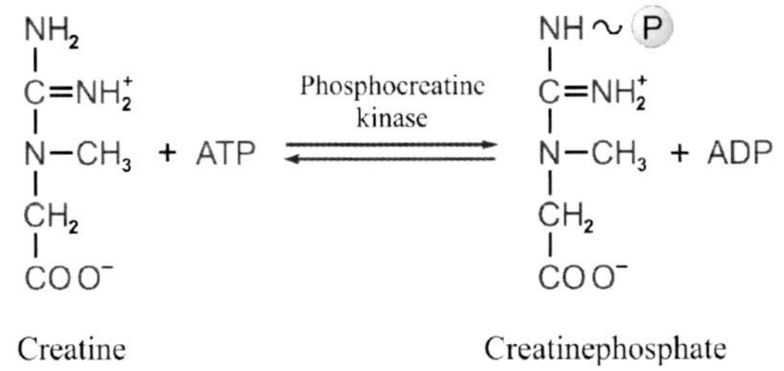

Creatine   Creatinephosphate

Creatinephosphate is an unstable compound, it cyclizes spontaneously and irreversibly to form *creatinine* and free phosphate.

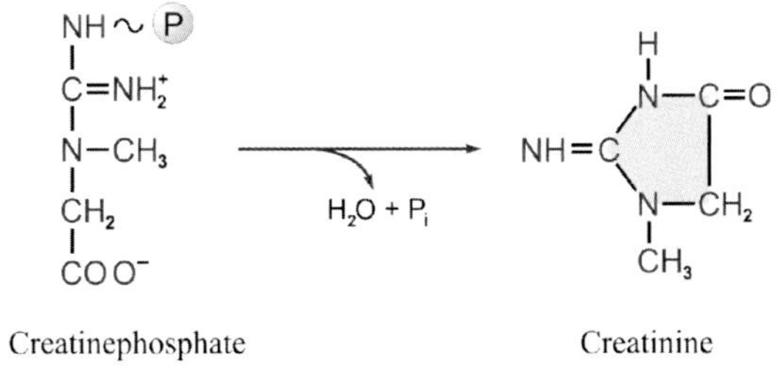

Creatinephosphate   Creatinine

Creatinine is excreted in urine; the amount excreted in 24 h is relatively constant for each individual and is dependent on the individual's muscle mass. The plasma concentration of creatinine in normal adults is 0.9–1.4 mg/dL in men and 0.8–1.2 mg/dL in women. Its levels are commonly used as an indicator of renal function.

Glutathione

All cells contain γ-glutamyl-cysteinyl-glycine (p. 32) at a concentration of ~5 mM. The main function of this compound depends on its reducing power. Glutathione helps to maintain the reduced state of protein sulfhydryl groups and is a mechanism of defense against the action of reactive oxygen species (p. 207). In red blood cells, glutathione decreases the formation of methemoglobin and prevents oxidative damage of the cell membrane. It is also required for the synthesis of eicosanoids and the biotransformation of substances that are foreign to the body.

Glutathione peroxidase, an enzyme containing selenium, is localized in the cytosol and mitochondria; it catalyzes the reduction of organic hydroperoxides and hydrogen peroxide. In this reaction, glutathione is oxidized (GSSG) and must be reduced to maintain the GSH/GSSG ratio within the normal value, which is close to 100. Reduction of GSSG is catalyzed by the enzyme *glutathione reductase* (NADPH-dependent), responsible for GSH regeneration (p. 211).

Glutathione synthesis. This metabolic pathway comprises two stages. In the first, an amide bond between the γ-carboxyl of glutamate and α-amine group of cysteine is formed. The reaction is catalyzed by *γ-glutamyl-cysteine synthetase*, which uses energy provided by the hydrolysis of ATP. The next step is catalyzed by *glutathione synthetase*, which forms a peptide bond between the carboxyl group of the dipeptide cysteine residue generated in the first reaction and the amine group of glycine. The reaction is coupled to the breakdown of ATP to ADP and P_i.

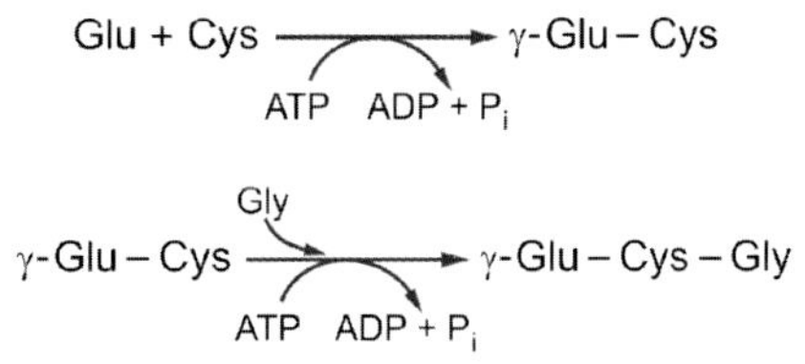

There are genetic defects that reduce the ability to synthesize glutathione (deficiency of γ-glutamyl cysteine and glutathione synthetase). Clinical conditions associated with these inherited defects present different symptoms, including hemolytic anemia and neurological and mental abnormalities.

γ-Glutamyl cycle. Another function of glutathione is to act as an intermediary in a membrane-bound amino acid transport system. It is a cyclic

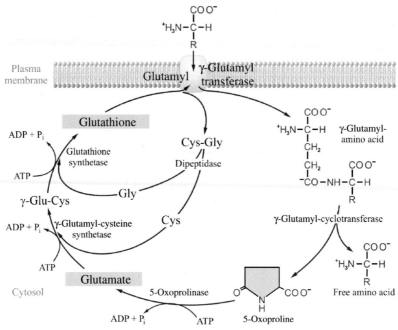

FIGURE 16.17 **γ-Glutamyl cycle.**

mechanism present in various organs (liver, kidney, and intestine). The process is summarized in Fig. 16.17. All amino acids, except proline, can use this mechanism to enter cells.

The steps of the cycle are the following: first, the amino acid to be transported is bound to a specific site on the outer face of the plasma membrane, where the *γ-glutamyl transferase* or transpeptidase is inserted. This enzyme catalyzes the separation of the γ-glutamyl moiety of cytosolic glutathione. The γ-glutamyl binds to the amino acid that is attached to the membrane. The products of the reaction are the dipeptides, γ-glutamyl-amino acid and cysteinyl-glycine. The latter is hydrolyzed to free cysteine and glycine by a dipeptidase. The dipeptide γ-glutamyl-amino acid is transferred into the cell and cleaved by *γ-glutamyl cyclotransferase*. The net result of this reaction is the incorporation of an amino acid to the cytoplasm and the release of glutamate.

The same γ-glutamyl cyclotransferase converts glutamate in a stable cyclic compound,

5-oxoproline, also called pyroglutamic acid. 5-Oxoprolinase promotes the conversion of 5-oxoproline in glutamate, in a reaction that requires the hydrolysis of ATP. In the remaining two phases, catalyzed by γ-glutamyl-cysteine synthetase and glutathione synthetase, glutathione is regenerated.

With the exception of the enzyme that catalyzes the first step of the cycle, which is an integral membrane protein, the other enzymes are all located in the cell cytosol. This transport system is energetically costly; each amino acid intake demands three high-energy bonds of ATP.

The levels of γ-glutamyl transferase are determined in clinical laboratories for diagnosis. Its activity in plasma increases in patients with biliary obstruction, liver damage from alcohol consumption, disseminated neoplasia, and prostate carcinoma.

Inborn Errors of Amino Acid Metabolism

These conditions usually result from mutations affecting enzyme synthesis. Sometimes the defect involves important sites or critical amino acid residues in the molecule, leading to alteration of its catalytic efficiency; others completely block synthesis. Any defect that causes loss of an enzyme's activity in a metabolic pathway causes the disruption of the flow of substrates and the accumulation of intermediary metabolites of stages that are before the affected site. In general, increased concentrations of metabolites have deleterious effects, especially on the central nervous system of infants, leading to frequent mental retardation and neurological disorders.

In some cases, it is possible to reduce the risk of severe injury by diets that eliminate or reduce the amino acid whose metabolism is affected. Therefore, it is imperative to diagnose these deficiencies before irreversible damage has occurred.

In most cases, human genetic diseases related to amino acid metabolism involve enzymes of catabolic pathways or synthesis of derivatives.

In well-nourished children, deficiencies of nonessential amino acids caused by alterations in their synthesis have not been observed.

In the course of this chapter, only some of the inherited diseases of amino acid metabolism have been mentioned. Nowadays, a great number of them are known. It is not within the scope of this book to make a comprehensive presentation of inborn errors, but to simply emphasize the existence of these disorders, which need to be identified, diagnosed early, and treated.

SUMMARY

Amino acids (AA) coming with the diet mix with those generated by the degradation of endogenous protein to form a *common metabolic pool* in the body.

Nitrogen balance is the equilibrium that exists in normal well-fed adults, between the intake of nitrogen (represented mainly by the protein content of the diet) and the nitrogen excreted via the urine and feces. The protein level in a cell is the result of the balance between protein synthesis and degradation.

Protein half-life is an indicator of the turnover time of a protein; it normally varies between hours and several months. Once the life of a protein has reached its end, it is hydrolyzed into its constituent amino acids. The main systems responsible for this function are:

1. *Lysosomes*, which contain proteases called *cathepsins* and hydrolases of lipids and carbohydrates. These hydrolyze extracellular proteins ingressed by endocytosis or cytosolic molecules and organelles by a process named *autophagy*.
2. *Ubiquitin–proteasome*, which degrades proteins marked by tandem insertion of multiple units of *ubiquitin*. The process is catalyzed by activating (E1), conjugating (E2), and ligase (E3) enzymes. Once ubiquitinated, proteins are fed to the proteasome, which hydrolyzes it to oligopeptides. The proteasome is a hollow cylinder formed by multiple subunits; on its internal surface it has proteolytic sites.

Essential amino acids are those which cannot be synthesized by human cells and need to be supplied with the diet. They include: phenylanine, isoleucine, leucine, lysine, methionine, threonine, tryptophan, and valine. Arginine and histidine are relatively essential.

Amino acid catabolism occurs when AA are not used in the synthesis of proteins or other nitrogenous compounds, but instead undergo degradation for energy purposes. The amine group is separated and follows metabolic pathways independent of those of the carbon chain. The processes related to the fate of the amine group include *transamination* and *deamination*.

Transamination is the reaction by which the α-amino group of AA is transferred to an α-keto acid. The reaction is catalyzed by *transaminases* or *aminotransferases*, using *pyridoxal phosphate* as a coenzyme. Pyridoxal-P serves as acceptor and transporter of the amine group, becoming reversibly pyridoxamine-P. Except for lysine and threonine, all AA participate in transamination reactions with pyruvate, oxaloacetate, or α-ketoglutarate to form alanine, aspartate, or glutamate, respectively, and the α-keto acids corresponding to the original AA. In turn, alanine and aspartate react with α ketoglutarate. Consequently, the amine groups of all AA converge to form glutamate.

Deamination of glutamate is catalyzed by *glutamate dehydrogenase*, which uses NAD or NADP. The products of the reaction are α-ketoglutarate and ammonia, which at the pH of cells, 99% of the ammonia is converted to ammonium ion (NH_4^+). NH_3 produced by deaminations in tissues, and also generated by the intestinal bacteria flora, reaches the blood and must be removed because it is toxic. One mechanism of removal is the formation of glutamine.

Glutamine is formed from NH_3 and glutamate in a reaction catalyzed by *glutamine synthetase*; it requires ATP. Glutamine is hydrolyzed to glutamate and NH_3 by *glutaminase*. In the kidneys, ammonia production is important as a mechanism to maintain body acid–base balance. In mammals, the main process for the removal of ammonia is the *urea cycle*.

Urea formation occurs in the liver by the following reactions:

1. *Carbamoyl synthesis*, by condensation of NH_3, CO_2, and phosphate (by *carbamoyl phosphate synthetase*). The reaction occurs in mitochondria, two ATPs are hydrolyzed and *N*-acetyl glutamate is required.
2. *Synthesis of citrulline*. The carbamoyl is transferred to ornithine (*ornithine transcarbamoylase*). Citrulline leaves the mitochondria and continues the following steps in the cell cytosol.
3. *Synthesis of argininosuccinate*. Citrulline is joined to aspartate (*argininosuccinate synthetase*). ATP is hydrolyzed to AMP and PP_i.
4. *Cleavage of argininosuccinate*. Argininosuccinate is separated into fumarate and arginine (by *arginino succinase*, a lyase).
5. *Hydrolysis of arginine*: Urea and ornithine are formed (by *arginase*).

To start another cycle, ornithine needs to move into the mitochondria.

One of the urea nitrogens comes from NH_3 produced by oxidative deamination of glutamate and the other from aspartate, which acquires its amine group by transamination with oxaloacetate.

Fumarate released in reaction 4 is an intermediate of the citric acid cycle. In this cycle, fumarate is converted successively to malate and oxaloacetate, which produces aspartate by transamination. This close association in the operation of the urea and the citric acid cycles contributes to the formation of urea (in the oxidation of malate, three ATPs are generated. Four ATPs are necessary for the production of one molecule of urea). Urea is excreted in urine.

Fate of the carbon skeleton of amino acids depends on the AA being *glucogenic* or *ketogenic*. Glucogenic AA are metabolized to pyruvate or intermediates of the citric acid cycle. Ketogenic AA generate acetyl-CoA or acetoacetate.

General metabolic mechanisms for AA degradation include the following:

Decarboxylation leads to the loss of the AA carboxyl group, which results in formation of biological, or *biogenic amines*, of intense physiological action. The reaction is catalyzed by *decarboxylase* (coenzyme pyridoxal phosphate). By decarboxylation, lysine forms: cadaverine; ornithine, putrescine; histidine, histamine; tyrosine, tyramine; tryptophan, tryptamine; and glutamic acid, GABA. *Polyamines*, such as spermidine and spermine are formed from putrescine.

Transfer of monocarbon groups (*methyl, OH-methyl, formyl*, and CO_2) are used in various processes of syntheses, catalyzed by specific *methyltransferases*. The main methyl donor is active methionine (*S-adenosyl-methionine*). Other transport agents of one C groups are *THF* and *methylcobalamin*, both derived from complex B vitamins. CO_2 is

transferred in *carboxylase* catalyzed reactions with coenzyme biotin (a B complex vitamin).

Phenylalanine and tyrosine follow the following metabolic pathways:

1. Degradation:
   a. Phenylalanine is converted to tyrosine (by *phenylalanine hydroxylase* and biopterin reductase, using O_2, THB, and NADPH).
   b. Transamination (by *tyrosine aminotransferase*). Forms *p*-OH-phenylpyruvic.
   c. Oxidation and decarboxylation (by *p-OH-phenylpyruvate hydroxylase*). Forms homogentisic acid.
   d. *Homogentisate oxidase* catalyzes the formation of maleylacetoacetate, which is isomerized to fumarylacetoacetate.
   e. Formation of fumarate and acetoacetate (by *fumarylacetoacetate hydrolase*).
2. Synthesis of catecholamines.
   a. Formation of *DOPA*: tyrosine is converted to 3,4-dihydroxyphenylalanine (*tyrosine hydroxylase*, O_2, *THB*).
   b. Formation of *dopamine* (by *DOPA decarboxylase*).
   c. Formation of *norepinephrine* (noradrenalin): dopamine is hydroxylated (by *dopamine-hydroxylase*, requiring O_2 and ascorbic acid).
   d. Formation of *epinephrine* (adrenalin): transmethylation from 5-adenosyl methionine (by *methyltransferase*).

Catecholamines act as chemical transmitters of the adrenergic system. Inactivation is performed by *MAO* and *COMT*.

Melanin synthesis is obtained by first formation of DOPA, then several reaction steps produce *melanin*.

Inborn errors of AA metabolism due to lack of *phenylalanine hydroxylase* or *dihydrobiopterin reductase* lead to *phenylketonuria*, or fenylpyruvate oligophrenia. Lack of homogentisic oxidase causes *alcaptonuria* and absence of *tyrosinase* in melanocytes causes *albinism*.

Serotonin is synthesized by hydroxylation of tryotophan in C5. The formed 5-OH-tryptophan is converted to 5-OH-tryptamine, or *serotonin* when decarboxylated. *MAO* inactivates serotonin.

Melatonin synthesis involves the acetylatilation and methylation of serotonin. *Melatonin* is a hormone secreted by the pineal gland.

Nicotinic acid (vitamin B complex) is synthesized from tryptophan.

Tryptophan forms indole, indoleacetate, skatole, indoxyl, and potassium salt of indoxylsulphate (*indican*) in the intestine as a result of bacterial putrefaction.

Nitric oxide synthesis.

NO• gas is a free radical that acts as an important chemical messenger. It is formed from arginine by the action of *NOS*.

Biotransformation reactions in which AA are involved serve as a mechanism of detoxification in the body. Some examples are the following:

Glycine reacts with benzoic acid to form *hippuric acid*.

Cysteine is converted to taurine which binds to bile acids (taurocholate).

Oxidation of sulfur from methionine and cysteine releases sulfate.

In sulfoconjugations, active sulfate (phosphoadenosine–phosphosulfate, PAPS) is used.

Creatine synthesis requires arginine, glycine, and methionine. The amidine group of arginine is transferred to glycine to form guanidoacetic acid. This is methylated and gives *creatine*, the methyl donor is SAM. Creatine forms creatinephosphate (*phosphocreatine kinase*, ATP). The creatinephosphate is a muscle energy reserve. Via dehydration, creatine becomes *creatinine*, which is excreted in urine.

Glutathione helps to maintain the reduced state of protein sulfhydryl groups and is a mechanism of defense against the action of reactive oxygen species. In addition, glutathione functions as an intermediary in a membrane-bound amino acid transport system, which brings AA into the cell (*γ-glutamyl cycle*). Glutathione synthesis requires γ-glutamyl-cysteine synthetase and glutathione synthetase.

Bibliography

Glickman, M.H., Ciechanover, A., 2002. The ubiquitin-proteasome proteolytic pathway. Destruction for the sake of construction. Physiol. Rev. 82, 373–428.

Hershko, A., Ciechanover, A., 1998. The ubiquitin system. Annu. Rev. Biochem. 67, 425–479.

Mann, G.E., Yudilevich, D.L., Sobrevia, L., 2003. Regulation of amino acid and glucose transporters in endothelial and smooth muscle cells. Physiol. Rev. 83, 183–252.

Morris, S.M., 2002. Regulation of enzymes of the urea cycle and arginine metabolism. Annu. Rev. Nutr. 22, 97–105.

Nath, D., Shadan, S., 2009. The ubiquitin system. Nature 458, 421.

Scriver, C.R., Beaudet, A.l., Sly, W.S., Valle, D. (Eds.), 2001. Metabolic and Molecular Bases of Inherited Disease. eighth ed. McGraw Hill, New York, NY.

Watford, M., 1991. The urea cycle: a two-compartment system. Essays Biochem. 26, 49–58.

17

Heme Metabolism

Heme is a ferroporphyrin and the prosthetic group of molecules of great functional importance called hemoproteins. *Hemoglobin*, responsible for O_2 transport in blood; *myoglobin*, which carries and stores O_2 in muscle; cytochromes, which are involved in electron transfer; and various enzymes (catalase, peroxidase, tryptophan pyrrolase) belong to this group of compounds.

Heme is constituted by a tetrapyrrole ring derived from protoporphyrin III (or IX in the original notation of Fischer). It contains methyl groups in positions 1, 3, 5, and 8; vinyl groups in positions 2 and 4; and propionate residues in positions 6 and 7. A ferrous atom binds to the nitrogen residues of the pyrrole (Fig. 17.1).

The average adult produces approximately 300 mg of heme per day to meet the requirements for renewal of hemoproteins in the body. Eighty-five percent or more of the synthesis is performed in erythrocyte precursor cells in the bone marrow and the rest in other tissues, primarily hepatic.

HEME BIOSYNTHESIS

The precursor compound, protoporphyrin III is synthesized from glycine and succinyl-CoA in three steps: (1) synthesis of δ-aminolevulinic acid (ALA), (2) formation of porphobilinogen, and (3) synthesis of protoporphyrin. Heme is obtained by adding an atom of ferrous iron to protoporphyrin.

1. *Synthesis of δ-aminolevulinic acid*. Succinyl-CoA and glycine form δ-aminolevulinic acid or 5-aminolevulinic acid (ALA). Succinyl-CoA is an intermediary of the citric acid cycle, generated by decarboxylation of α-ketoglutarate. Glycine comes from the pool of free amino acids. The reaction is catalyzed by *δ-aminolevulinate synthase*, a mitochondrial enzyme dependent on pyridoxal phosphate and Mg^{2+}. Initially, α-amino-β-ketoadipate is formed, which is rapidly decarboxylated to δ-aminolevulinate (Fig. 17.2). The CO_2 released corresponds to the carboxyl of glycine.

   The enzyme δ-aminolevulinate synthase is the major site of control of porphyrin synthesis. The final product, heme, exerts allosteric inhibitory action on this enzyme. Moreover, heme, hemoglobin, and other heme proteins act as repressors of δ-aminolevulinate synthase synthesis, which, due to the short half-life of this enzyme (~1 h in the liver), rapidly reduces its levels and action. In contrast, hypoxia, erythropoietin, certain steroid hormones, drugs (barbiturate), and alcohol induce the synthesis of δ-aminolevulinate synthase (the mechanisms of induction and repression

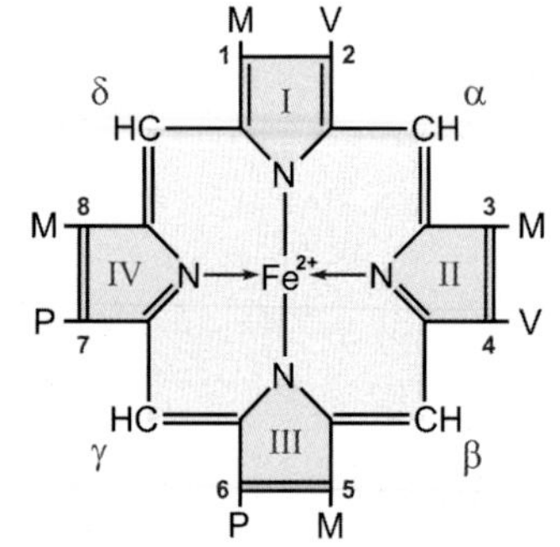

FIGURE 17.1 **Heme.** *M*, Methyl; *P*, propionate; *V*, vinyl.

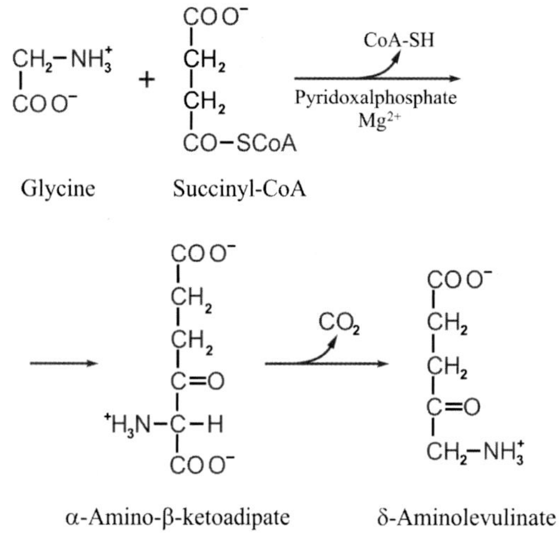

Glycine   Succinyl-CoA

α-Amino-β-ketoadipate   δ-Aminolevulinate

FIGURE 17.2 **δ-Aminolevulinate biosynthesis.**

2 Molecules of δ-aminolevulinate

Porphobilinogen

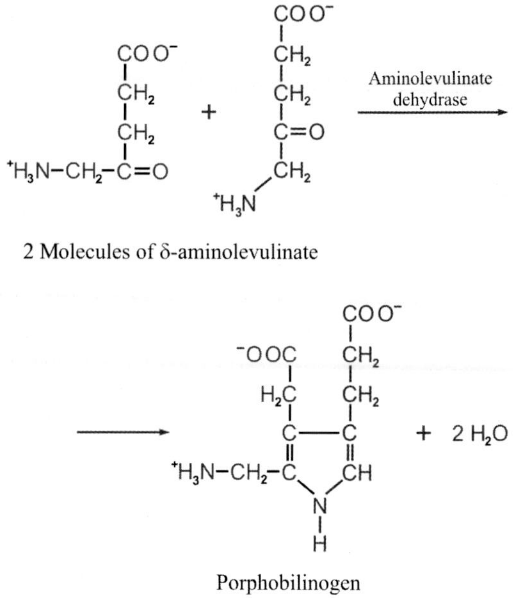

FIGURE 17.3 **Biosynthesis of porphobilinogen.** One of the δ-aminolevulinates is in *black* and the other in *red* to show the contribution of each molecule to the formation of the pyrrole group, the acetate, and the propionate side chains.

of protein synthesis will be considered in Chapter 23).

2. *Porphobilinogen synthesis.* The δ-aminolevulinate, formed in the mitochondria, needs to be transferred to the cytosol for the second step, which is catalyzed by *δ-aminolevulinate dehydrase*, also called *porphobilinogen synthase*. This cytosolic enzyme, which contains Zn^{2+} and sulfhydryl groups that are essential for its activity, is sensitive to reagents that react with SH groups, such as heavy metals (lead). Heme acts as an inhibitor of porphobilinogen synthase and also represses the synthesis of this enzyme.

Two molecules of δ-aminolevulinic acid react, water is released and porphobilinogen (PBG), which contains the pyrrole ring, is produced (Fig. 17.3).

Inhibition of PBG synthase results in increase in δ-aminolevulinic acid (ALA); this is secondary to the failure of heme synthesis, which upregulates the transcription of the ALA synthase gene. Elevation of ALA levels is believed to cause some of the neurologic effects of Pb poisoning.

3. *Porphyrin formation.* At this stage, four molecules of porphobilinogen form the tetrapyrrole ring of uroporphyrinogen. Fig. 17.4 shows the origin of the chemical groups in porphyrin.

The enzyme porphobilinogen deaminase removes the terminal amine group from the methylamine side chain of porphobilinogen and methylene bridges (—CH_2—) are formed between the pyrrol

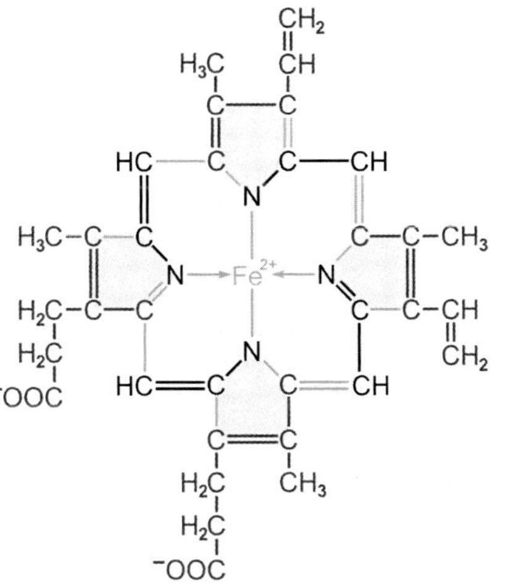

FIGURE 17.4 **Origin of the chemical groups of the porphyrin ring.** Elements in *red* are from the succinyl-CoA molecule, while those in *black* are from glycine.

groups. Uroporphyrinogens I and III are obtained (Fig. 17.5); in isomer I, the side chains acetate (A) and propionate (P) are arranged symmetrically, while isomer III is asymmetric (with A and P switching positions in pyrrole IV). The most abundant isomer formed, uroporphyrinogen III, is the precursor of heme. The uroporphyrinogen I and its derivatives, uroporphyrin I and coproporphyrinogen I, have no known function; they are eliminated in urine and feces, respectively. The uroporphyrinogen III is oxidized in small proportion to uroporphyrin III, which is also excreted.

Formation of tetrapyrrole results from the interaction of two enzymes: *uroporphyrinogen I synthase* and *uroporphyrinogen III cosynthase*. The reactions that follow modify the side chains and oxidate the molecule, decreasing the degree of saturation of the ring. The uroporphyrinogen III forms coproporphyrinogen III by decarboxylation of acetates in positions 1, 3, 5, and 8, which are converted to methyl residues. The reaction is catalyzed by *uroporphyrinogen decarboxylase*.

The remaining transformations, until heme is generated, occur within the mitochondria; therefore, coproporphyrinogen III must be transported into the matrix. Decarboxylation and oxidation convert propionate residues 2 and 4 into vinyl groups ($-CH=CH_2$) and form protoporphyrin III (Fig. 17.5).

4. *Fe addition.* The final stage of heme synthesis is the incorporation of ferrous iron (Fe^+) to protoporphyrin. The reaction is catalyzed by *ferrochelatase*, also called *heme synthase*, located in mitochondria.

Eighty-five percent of the heme synthesized is destined to hemoglobin formation, 10% to myoglobin, and the rest to other hemoproteins (i.e., cytochromes, catalase).

Porphyrias

There is a group of diseases characterized by increased excretion of porphyrins or their precursors. These diseases, called porphyrias, are caused by defects in the heme synthetic pathway. These disorders can be inherited or acquired. Porphyrias are characterized by accumulation of heme precursors from reactions in the synthetic pathway that are upstream from that catalyzed by the affected enzyme. Concurrently, there is a reduction of metabolites generated in downstream reactions. Heme levels are depressed, which stimulate the synthesis of δ-aminolevulinate synthase and exacerbates the accumulation of heme intermediates.

Hereditary porphyrias are distinguished as erythropoietic or hepatic, depending on the organ in which the defect is predominantly expressed. A description of some of these diseases follows.

Erythropoietic porphyria. This disorder is due to a deficiency of uroporphyrinogen III cosynthase, which results in an imbalance in the production of uroporphyrinogen isomers, causing an excess of isomer I over isomer III. Uroporphyrinogen I do not participate in the synthesis of heme and has no physiological role. Its levels in blood and

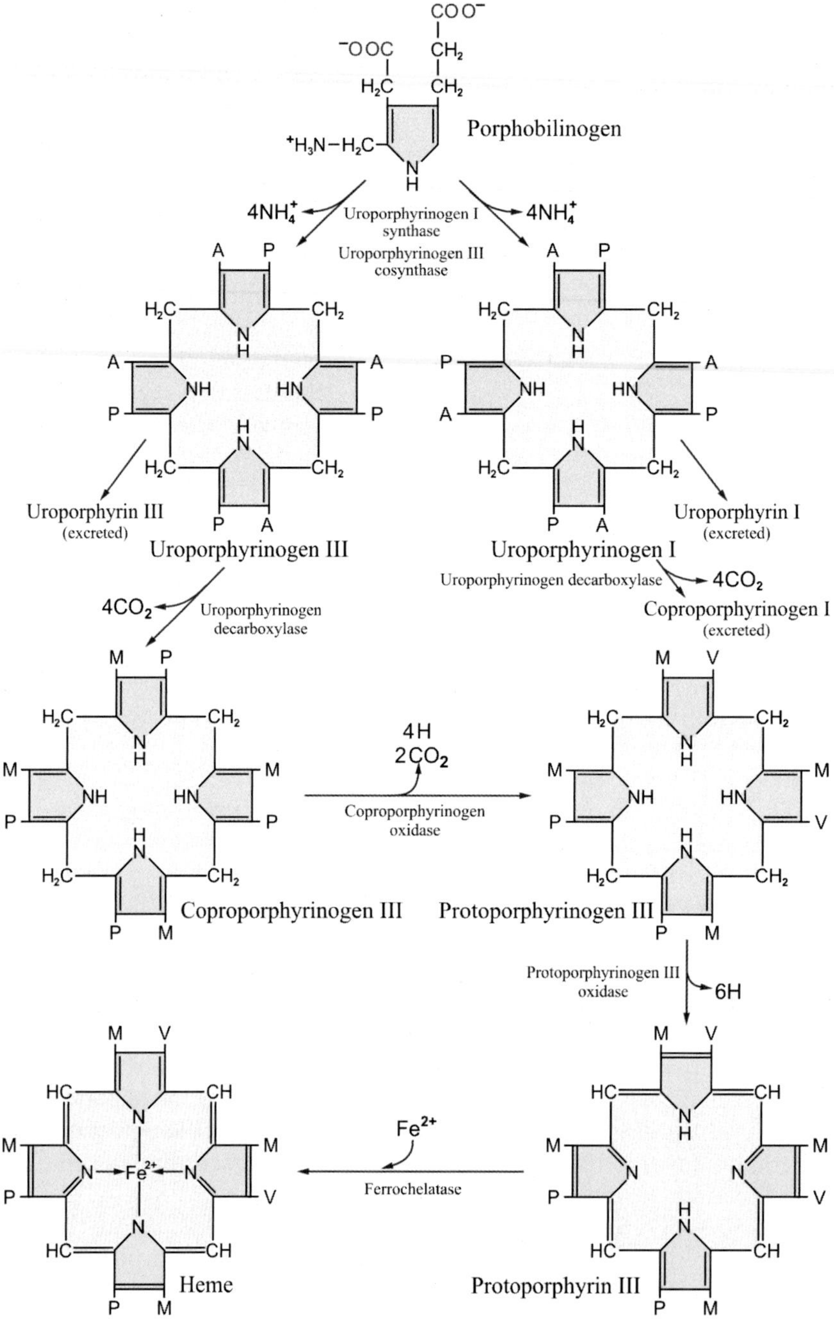

FIGURE 17.5 **Stages of heme synthesis from porphobilinogen.** *A*, Acetate; *M*, methyl; *P*, propionate; *V*, vinyl.

bone marrow rise sharply, as well as the levels of its derivatives, uroporphyrin I and coproporphyrinogen I, which are excreted, in large amounts, in urine and feces. These substances are colored, causing the patient's urine to turn red. The most important symptom of the disease is cutaneous photosensitivity due to porphyrin's capacity to absorb radiation at wavelengths of 400 nm. This causes porphyrin to acquire a reactive (excited) state that transmits energy to molecular oxygen, producing free radicals. These free radicals, acting on capillaries in the dermis, are responsible for the production of erythema, edema, blisters, and ulcerations on the skin, leaving scars and severe deformations. The disease is inherited in an autosomal recessive manner.

Acute intermittent porphyria (hepatic). This disease is due to defects in the synthesis of uroporphyrinogen I synthase. It is a codominant character. Heterozygous individuals, with defects in only one of the alleles in the enzyme's gene, produce half of the uroporphyrinogen I synthase levels of normal individuals. These patients excrete large amount of porphobilinogen and δ-aminolevulinate in urine. These compounds are colorless and do not cause photosensitivity. The disease is characterized by abdominal cramps, vomiting, and neuropsychiatric symptoms. On exposure to air and light, porphobilinogen forms colored compounds. The urine from these patients, originally of normal color, darkens after exposure to air and light. A consistent finding is the increased activity of δ-aminolevulinate synthase due to the reduction of heme production. The normal negative feedback mechanism for enzyme synthesis (caused by the end product) does not operate. Administration of hematin (ferriheme) intravenously to these patients has been proposed to attenuate the "derepression" of aminolevulinate synthase and to decrease the production of aminolevulinate and porphobilinogen.

In some cases, the symptoms are triggered by ethanol consumption or by drugs (barbiturates) that induce the expression of the cytochrome P_{450} oxidative system. Synthesis of cytochrome P_{450} consumes heme, which stimulates the production of δ-amino synthase and the accumulation of porphobilinogen.

Porphyria cutanea tarda (hepatic and erythropoietic). This is an autosomal recessive disorder and the most common among this group of diseases. It is due to a partial deficiency of uroporphyrinogen decarboxylase. The liver of patients affected by this disease contains large amounts of porphyrins. The main symptom is skin photosensitivity. Just a slight exposure to light triggers the formation of blisters and other skin lesions. The urine color varies from red to brown when viewed under natural light, and pink to red when viewed under fluorescent light. Sometimes, the symptoms appear after exposure to sunlight, viral infections (HIV, hepatitis B or C), or iron overload.

Acquired porphyrias. These are the result of agents, such as lead and other heavy metals and compounds, such as hexachlorobenzene and the antibiotic griseofulvin. These inhibit enzymes of the heme synthesis pathway, especially aminolevulinate dehydrase, uroporphyrinogen synthase, and ferrochelatase. Increased excretion of δ-aminolevulinate occurs. The determination of this compound in urine and the activity of δ-aminolevulinate dehydrase in blood can be used as indicators of the magnitude of intoxication.

CATABOLISM OF HEME

The average life span of red blood cells is approximately 120 days, after which they are destroyed by the *mononuclear phagocyte system (MPS)*, previously known as the reticuloendothelial system.

A normal man of 70 kg weight has ~750 g of hemoglobin circulating in blood, of which 6.25 g are degraded every day. The globin portion and the heme prosthetic group follow separate metabolic pathways.

Globin, like all proteins that have finished their cycle in the body, is hydrolyzed to its amino acids, which are reused for the syntheses of new proteins. Heme undergoes specific transformations in a multistage process that takes place in the MPS, in the liver, and intestine.

Stages in MPS. The initial phase is carried out in cells of the MPS, mainly in the liver, spleen, and bone marrow. A heme oxygenase multienzyme system of smooth endoplasmic reticulum catalyzes oxidations that convert ferrous to ferric iron, and the carbon of the α-methine bridge (between pyrroles I and II) into carbon monoxide (CO), causing the opening of the tetrapyrrole. Molecular oxygen and NADPH participate in these reactions (Fig. 17.6). The iron is separated immediately. Each atom is preserved and reused in the synthesis of new molecules. The product of these transformations is a green pigment called *biliverdin*, which consists of a chain of four pyrrole groups without iron.

Biliverdin is reduced in the methine bridge between pyrroles III and IV by *biliverdin reductase* (NADPH-dependent). The product formed with a methylene bridge (—CH₂—) between pyrroles III and IV, is the yellow-orange pigment *bilirubin*. This compound is insoluble in water and soluble in lipids, a property that allows it to readily diffuse through cell membranes. Bilirubin is toxic; it interferes with metabolic functions of the cells through mechanisms that are unknown.

When degraded, each gram of hemoglobin produces 35 mg of bilirubin. The daily production of this compound is ~300 mg; 80% of this amount comes from hemoglobin and the rest from other hemoproteins. Once produced, bilirubin moves from the tissues to the circulation.

Bilirubin transport. Since bilirubin is insoluble in aqueous media, it is transported by plasma proteins (mainly albumin), forming a macromolecular complex that cannot cross cell membranes or the barrier of the renal glomerulus. Thus, it is not excreted in urine.

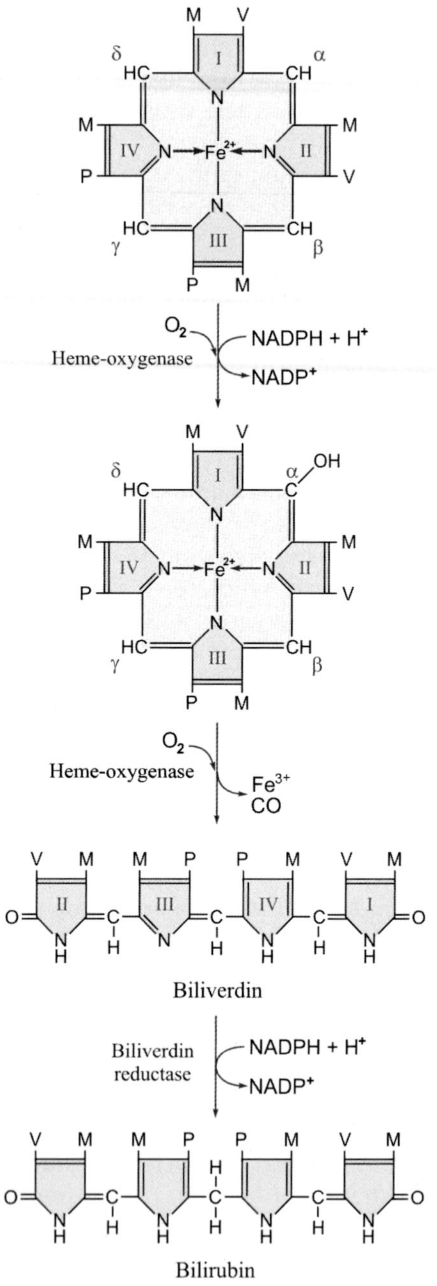

FIGURE 17.6 **Catabolism of heme in mononuclear phagocyte system (MPS).**

Albumin has a high and a low affinity site for bilirubin binding. Approximately 25 mg of bilirubin are bound to the high affinity sites and carried by 4 g of albumin, amount that is present in 1 dL of plasma. When bilirubin level surpasses the capacity of the high affinity sites, the excess is loosely bound to albumin's lower affinity sites. If the concentration of bilirubin rises even more, the binding sites in albumin become saturated and the extra bilirubin is carried weakly bound to other plasma proteins, which can easily release it to the cells.

The liver of the newborn, still immature, is unable to metabolize bilirubin efficiently. The concentration of bilirubin in blood increases during the first 5 days of life in about half of all neonates. When hyperbilirubinemia reaches levels above the capacity of albumin to fix all the pigment, it leaks into the cells and can cause toxicity. The problem is exacerbated when intense hemolysis occurs, which further increases heme degradation and bilirubin production, as is the case in erythroblastosis. In cases of very high hyperbilirubinemia (20 mg/dL or more), there is a great risk of a serious clinical condition called *kernicterus*, in which the central nervous system is affected.

Hepatic stage. Once it reaches the liver, bilirubin is separated from albumin and enters cells by facilitated diffusion mediated by a plasma membrane transporter that some authors call *bilitranslocase*. Inside the hepatocyte, pigment acceptor proteins bind and trap bilirubin in the cytosol. Several bilirubin-binding proteins have been isolated in liver cells, the best characterized is called *ligandin (or Y)*; it is currently identified as *glutathione-S-transferase* (GST). This protein has enzymatic activity and is also able to bind ligands that are not its own substrates.

In the cell, bilirubin is conjugated with polar molecules and converted into a water-soluble product that can be excreted and transported in the bile. The conjugation process is conducted in the smooth endoplasmic reticulum; it consists of the addition of glucuronic acid residues to

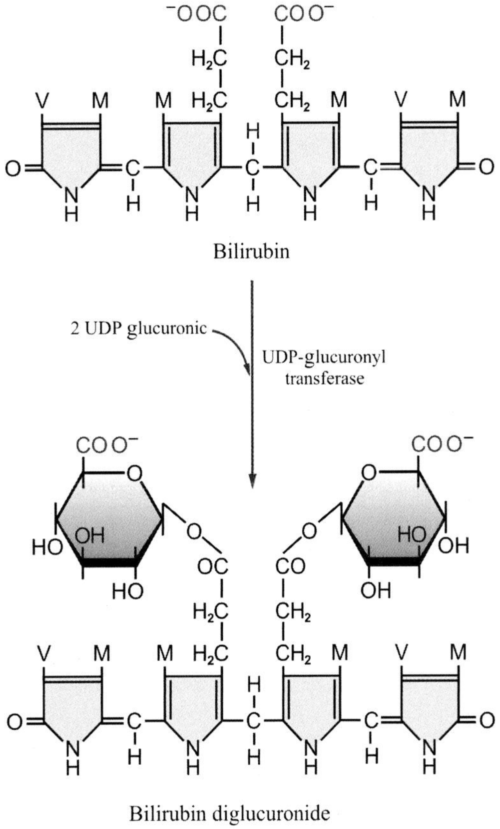

FIGURE 17.7 **Conjugation of bilirubin and glucuronic acid.**

propionic acid chains in bilirubin. The reaction is catalyzed by *bilirubin-glucuronyl transferase* (BGT); the donor of glucuronic acid is uridine-diphosphate-glucuronate (UDP-glucuronic). Mono- and diglucuronide bilirubin (Fig. 17.7) are formed, which are secreted into the bile ducts by an active transport mechanism. This carrier is a multidrug resistance protein 2 (MRP-2), also called multispecific organic anion transporter (MOAT), an ABC type transporter (p. 238).

The activity of bilirubin glucuronyl transferase and the transport system is increased by drugs, such as barbiturates.

Under physiological conditions all bilirubin secreted in bile is conjugated. In the normal adult 70%–90% corresponds to bilirubin diglucuronide and 7%–27% corresponds to bilirubin monoglucuronide. In the newborn, BGT activity is reduced and bile contains a lower proportion of diglucuronide.

Direct and indirect bilirubins. In 1913, van den Bergh distinguished two forms of bilirubin by a colorimetric reaction with diazotized sulfanilic acid (diazo reagent). One of those forms is called *direct bilirubin* because it immediately reacts with the diazo reagent. The other, reacts only after adding alcohol to the mixture and is designated *indirect bilirubin*. Direct bilirubin is bilirubin diglucuronide, namely, the water-soluble product formed in the liver. Indirect bilirubin corresponds to the pigment formed in the MPS, not yet conjugated with glucuronate.

In normal blood plasma there is a small amount of bilirubin, almost entirely of the indirect type, insoluble in aqueous medium and transported bound to albumin. The normal concentration is less than 1.0 mg/dL. Values greater than 1.5 mg/dL are considered abnormal (hyperbilirubinemia).

Intestinal stage. Once in the intestinal tract, bilirubin glucuronide is hydrolyzed and subjected to the reducing action of enzymatic systems of the anaerobic bacterial flora. The first compound formed is mesobilirubinogen and, after a series of chemical transformations, it is converted to stercobilinogen, the end product. This is a colorless compound, which is partially eliminated with feces.

In contact with air, stercobilinogen oxidizes and becomes stercobilin, or urobilin, a brownish pigment that contributes to the normal color of feces. Urobilin and stercobilin are similar compounds, but not necessarily identical. Due to the existence of several double bonds, these molecules generate, by oxidoreduction, a family of products called urobilinoids.

Enterohepatic cycle. Part of the reduction products of bilirubin in the intestine are reabsorbed and returned to the liver via the portal vein. The liver oxidizes them and regenerates bilirubin

glucuronides, which are excreted again with the bile into the intestine. This represents bilirubin's enterohepatic cycle.

Urinary pigments. Not all pigments are reabsorbed, processed, and reexcreted by the liver into the intestine; some pass to the general circulation and are eliminated by the kidney (1–4 mg/day). In urine, the compound is called urobilinogen because it produces the urobilin pigments (urobilinoids) when oxidized. Stercobilinoids and urobilinoids belong to the same family of compounds; their different name only indicates the way by which they are excreted. Fig. 17.8 shows the stages of heme catabolism.

Jaundice. Various pathological conditions can alter the catabolism of heme. A phenomenon of common clinical observation is the increase in the concentration of bilirubin in blood. When

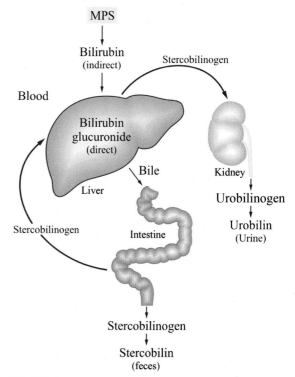

FIGURE 17.8 **Schematic representation of the changes in pigments derived from heme.**

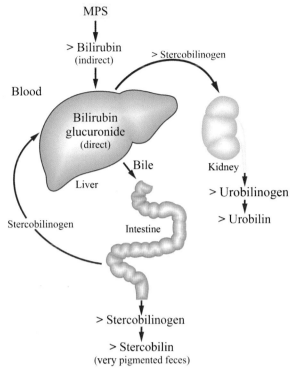

MPS

> Bilirubin
(indirect)

> Stercobilinogen

Blood

Bilirubin
glucuronide
(direct)

Bile

Kidney

Liver

> Urobilinogen

> Urobilin

Stercobilinogen

Intestine

> Stercobilinogen

> Stercobilin
(very pigmented feces)

FIGURE 17.9 **Alterations in hemolytic jaundice (scheme).** The excessive destruction of red blood cells leads to a greater amount of bilirubin; indirect bilirubin passes to the blood (hyperbilirubinemia), where it is transported bound to albumin, forming a macromolecular complex that is not filtered in the kidneys. There is no bilirubin in urine despite its increase in circulating blood. There is excess supply of bilirubin to the liver, an organ in which glucuronidation occurs, to produce direct bilirubin, which will be excreted in the bile in higher than normal amounts. Increased bilirubin glucuronide in the intestines determines greater production of stercobilinogen and increased pigments that will give the feces an intense color. Also, the reabsorption of stercobilinogen through the portal vein system increases. Stercobilinogen moves into the general circulation and is excreted in urine. Renal elimination of urobilinoids is increased.

bilirubinemia exceeds 2–2.5 mg/dL, the pigment moves into tissues. This is especially visible in the skin and the sclera, which turn a yellow hue; this sign is called *jaundice*.

According to the mechanism of production, jaundice can result from: (1) excessive hemolysis, (2) bile duct obstruction, and (3) secondary to functional impairment of the liver.

Determining the concentration of direct and total bilirubin in blood and the amount of urobilinoids excreted in 24 h in urine provides useful information for establishing the diagnosis of the type and cause of jaundice.

Figs. 17.9–17.11 present the most important aspects of the various types of jaundice. It should be noted, however, that in clinics the situation is not always as clear; often, mixed states displaying signs from more than one type of jaundice may appear.

As mentioned earlier, in the newborn there is potential for the appearance of hyperbilirubinemia, mainly due to the liver's inability to capture and conjugate bilirubin. In this case, indirect, or unconjugated bilirubin increases, which can have toxic effects, particularly for the brain (kernicterus). Unconjugated bilirubin affects

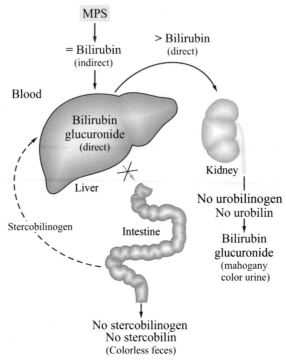

FIGURE 17.10 **Alterations in jaundice caused by bile duct obstruction.** Bilirubin production in MPS and the level of indirect bilirubin in blood plasma are normal. The liver forms bilirubin glucuronides but biliary obstruction does not allow its excretion into the intestine. Accumulation in bile ducts causes changes in the structure of liver lobules that let bile move to the capillaries. The level of direct bilirubin in blood increases (and can filter through the renal glomeruli). The excretion of bilirubin glucuronides in the urine gives it a mahogany color. As bile does not reach the intestine, stercobilinogen is not formed. The feces does not have a normal color, but a light gray that looks like clay. There is no reabsorption through the enterohepatic cycle, urobilinoids are absent in urine.

astrocytes and neurons, damaging mitochondria (which leads to energy metabolism disorders and apoptosis) and plasma membranes (which causes neurotransmitter transport disruptions).

When indirect bilirubin concentration in plasma exceeds dangerous limits (20 mg/dL), therapeutic procedures need to be applied. Phototherapy is a widely used method of treatment, based on the photosensitivity of bilirubin. When bilirubin circulating through the skin capillaries is subjected to UV light, in the presence of oxygen, it decomposes into water-soluble products that can be readily removed from the body. The child is exposed to intense UV light during a variable amount of time, necessary to reduce bilirubin levels.

There are genetic alterations that result from defects in the gene encoding bilirubin-uridylglucuronyl transferase, which cause a nonhemolytic increase of nonconjugated bilirubinemia. These diseases are inherited as an autosomal recessive trait. One of them is the *Crigler–Najjar syndrome type I*, with total absence of the conjugating enzyme in the liver. The values of plasma unconjugated bilirubin rise between 20 and 25 mg/dL. Without treatment, most of the patients die of kernicterus within the first 18 months of life. If they survive until puberty, they die from encephalopathy. The only possible cure for those cases is liver transplant. The *Crigler–Najjar syndrome type II* is a less severe variant; the deficiency of

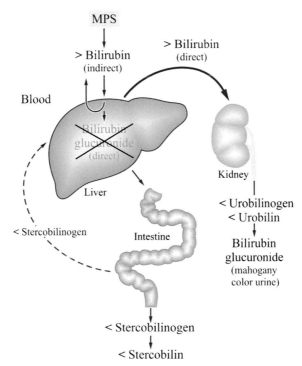

FIGURE 17.11 Alterations in jaundice caused by impairment of hepatic function. Production of indirect bilirubin in MPS is normal. The liver, with decreased functional capacity, cannot process all the bilirubin that it receives and only part of it is glucuronidated. The remaining unmodified bilirubin returns to the blood, explaining the increase of indirect bilirubin in plasma. Moreover, under conditions which produce this type of jaundice, the existence of alterations in hepatic parenchyma is common, with reflux of bile into blood capillaries. This explains the possible increase of direct bilirubin (glucuronide) in plasma of these patients. Direct bilirubin is eliminated by the kidneys and urine has a dark color. There is decreased production and excretion of bile into the intestine, which determines reduction of urobilinoids in feces and urine.

the enzyme is not total. Plasma unconjugated bilirubin varies between 7 and 20 mg/mL.

Another defect is *Gilbert's syndrome*, which is characterized by a decrease of bilirubin-uridylglucuronyl transferase to less than 30% of its normal level. Plasma unconjugated bilirubin values are not higher than 3 mg/dL; besides jaundice, these patients may manifest with fatigue, abdominal discomfort, and anxiety. This condition is relatively benign and normally causes little disturbance to everyday life.

Alterations in bile excretion can cause an increase of conjugated bilirubin in plasma. The *Dubin–Johnson syndrome* is an inherited disease produced by intrahepatic cholestasis due to

disturbance in the canalicular excretion of bile. Values of plasma unconjugated bilirubin are between 2 and 5 mg/dL. This rare disease is more frequent in Middle Oriental Jews. *Rotor syndrome* also shows an increase of conjugated bilirubin in plasma; it is an autosomal recessive disease. *Alagille syndrome*, is an autosomal dominant disease that affects the development of the biliary ducts.

SUMMARY

Heme biosynthesis starts with the formation of protoporphyrin III or IX, through the following steps:

    1. *Synthesis of δ-aminolevulinic acid* by *δ-aminolevulinate synthase*, which catalyzes the binding of succinyl-CoA

and glycine, and releasing CO_2. This is a regulatory step in the pathway, inhibited by heme.

2. *Porphobilinogen synthesis* by *δ-aminolevulinate dehydrase* or *porphobilinogen synthase*, which catalyzes the binding of two δ-aminolevulinate molecules, releasing two molecules of water. This step is inhibited by heavy metals (lead).

3. *Formation of porphyrin* by *uroporphyrinogen I synthase* and *uroporphyrinogen III cosynthase*, which join four molecules of porphobilinogen to form uroporphyrinogen, a tetrapyrrole ring. Isomers I and III are produced, with the latter one being formed in greater amounts.

   The uroporphyrinogen III is converted to coproporphyrinogen III by decarboxylation of the acetate chains in positions 1, 3, 5, and 8 (catalyzed by *uroporphyrinogen decarboxylase*).

   After oxidations and decarboxylations, protoporphyrin III is formed.

4. *Formation of heme* is achieved after incorporation of Fe^{2+} into the molecule by action of *heme synthase* or *ferrochelatase*.

   Eighty-five percent of the total heme synthesized is destined to Hb formation, 10% to myoglobin, and the rest to other hemoproteins (cytochromes, catalase, etc.).

Porphyrias are diseases produced by defects in the heme synthesis pathway.

Heme catabolism takes place when red blood cells are destroyed and degraded, after reaching their life span of 120 days. *Globin* is hydrolyzed to amino acids and heme undergoes the following transformations:

In the monocyte phagocitic system there is a *heme oxygenase multienzyme* that uses O_2 and NADPH.

Fe^{2+} is oxidized to Fe^{3+} and the carbon of the methine bridge is converted to CO.

The Fe^{3+} is immediately separated; *biliverdin* is formed and reduced to *bilirubin* by *biliverdin reductase*, which uses NADPH.

Bilirubin is transported in plasma bound to albumin in a macromolecular complex that does not filter in the renal glomeruli. In the liver, hepatocytes capture bilirubin from the blood via a plasma membrane carrier and bind it to proteins in the cytoplasm (*protein Y* or *ligandin* and *glutathione-S-transferase*).

In the hepatocytes, bilirubin is conjugated with glucuronic acid (transferred from UDP-glucuronic) by *BGT*, to form mono- and diglucuronides bilirubin, soluble compounds that pass into bile by active transport. Conjugated bilirubin is called *direct*; while unconjugated is known as *indirect*. Normally, there is only indirect bilirubin in blood (at levels less than 1 mg/dL); it is not excreted in urine. Under pathological situations, conjugated or direct bilirubin may appear in plasma and urine.

Bilirubin glucuronides are sent with bile to the intestines, where they are hydrolyzed and bilirubin is submitted to reductions. *Mesobilinogen* and *stercobilinogen* are formed, both of which are eliminated with the feces. In contact with air, *stercobilinogen* is oxidized to *stercobilin*, a pigment that gives stool its color. Part of the reduction products of bilirubin in the intestine are reabsorbed and returned to the liver, where bilirubin glucuronide is regenerated and reexcreted in the bile (*enterohepatic cycle*). Pigments reabsorbed in the intestines pass to the general circulation and are eliminated in part by the kidneys as *urobilinogen*, which is oxidized to *urobilin* or other *urobilinoids* (in amounts of 1–4 mg/day).

Jaundice refers to the increase of bilirubin in the blood and it is characterized by the yellowish tint of the teguments.

Bibliography

Erlinger, S., Arias, I.M., Dhumeneaux, D., 2014. New insights into molecular mechanisms and consequences. Gastroenterology 146, 1625–1638.

Fevery, J., 2008. Bilirubin in clinical practice: a review. Liver Int. 28, 592–605.

Gourlay, G.R., 1997. Bilirubin metabolism and kernicterus. Adv. Pediatr. 44, 173–229.

Scriver, C.R., Beaudet, A.L., Sly, W.S., Valle, D., 2001. Metabolic and Molecular Bases of Inherited Disease, eighth ed. McGraw Hill, New York, NY.

Yamada, T., et al., 2009. Textbook of Gastroenterology, fifth ed. Lippincott Co., Philadelphia, PA.

Zakim, D., Boyer, T.D., 2012. Hepatology, sixth ed. W.S. Saunders Co., Philadelphia, PA.

Purine and Pyrimidine Metabolism

Nucleic acids ingested with foods are degraded in the intestine; first they are broken down into free nucleotides and then into nucleosides and phosphate. The nucleosides are absorbed in the intestine as such or hydrolyzed by nucleosidases that separate the nitrogenous base from the pentose. Some of the bases released into the intestinal lumen are degraded by the bacteria of the intestinal flora. The remaining bases are absorbed and move into the portal circulation.

Purines and pyrimidines are not dietary requirements; they can be synthesized with high efficiency in the body. Most of the nitrogenous bases that come with the diet are degraded and the final products excreted; only a small fraction of the exogenous adenine is used in the production of new molecules. The synthesis of new nucleotides and polynucleotide molecules uses purines and pyrimidines that are endogenously generated in the body. The different route for the metabolism of exogenous and endogenous purine and pyrimidine establishes two metabolically independent pools for them.

Nucleic acids in the body are in permanent turnover. Some of the bases released during degradation processes are reused in the synthesis of nucleotides and nucleic acids through recovery or salvage pathways. The rest is catabolized and the end products are excreted (Fig. 18.1).

PURINE BIOSYNTHESIS

The purine ring is synthesized in cells from small molecules or molecular moieties coming from other compounds. The main initial component is the amino acid glycine, the remaining are residues with a single carbon or the —NH$_2$ group transferred from different sources. Studies with isotopic elements have established the origin of each of the atoms that form the purine ring (Fig. 18.2). Carbons 4 and 5 and nitrogen 7 of the purine rings are derived from glycine, nitrogen 3 and 9 are derived from the amide group of glutamine, nitrogen 1 is derived from aspartate, and carbons 2 and 8 are derived from formyl residues transported by tetrahydrofolic acid. The carbon atom at position 6 comes from CO_2, which is transferred in a reaction in which biotin is used as a coenzyme.

The molecular assembly of diverse segments of the purine ring is performed in a series of reactions in which ribose-5-phosphate plays an essential role. Ribose-5-phosphate is present throughout the synthesis process and remains attached to purine. In this manner, a nucleotide is produced.

The ribose-5-phosphate portion of the nucleotide is generated from glucose in the pentose phosphate pathway (p. 307). To be used in the

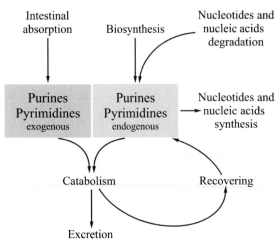

FIGURE 18.1 **Summary of purine and pyrimidine metabolism.**

FIGURE 18.2 **Biosynthesis of the purine nucleus.** Origin of the carbon and nitrogen residues.

biosynthesis of nucleotides, ribose-5-phosphate needs to be activated by transfer of pyrophosphate from ATP to carbon 1 of the pentose. The reaction is catalyzed by *phosphoribosylpyrophosphate synthetase* (or *ATP-ribose-5-phosphatepyrophosphate ligase*), a peculiar kinase that transfers pyrophosphate instead of phosphate. The compound 5-phosphoribosyl-1-pyrophosphate (PRPP), with α configuration of carbon 1 is formed.

Phosphoribosylpyrophosphate is an important metabolite involved in purine and pyrimidine synthesis and in the purine salvage pathway. PRPP synthetase is subjected to feedback control; it is activated by phosphate and inhibited by purine and pyrimidine nucleotides.

Assembly of the molecular fragments into the purine ring begins with the transfer of the amide group of glutamine to PRPP, catalyzed by *glutamine:PRPP amidotransferase*. The amide group displaces pyrophosphate from carbon 1 of ribose and forms 5-phosphoribosylamine. The nitrogen finally occupies position 9 in the purine ring. The spatial arrangement of carbon 1 follows the β configuration, which is the common configuration seen in natural nucleotides.

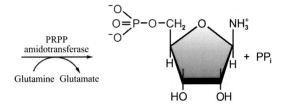

Immediate hydrolysis of the released pyro-phosphate makes this reaction virtually irre-versible. The most important regulatory factor in this metabolic step is PRPP. The concentra-tion of this compound in the cell is 10–100 times lower than the K_m of the enzyme. For this reason, changes in PRPP levels can rapidly affect the ac-tivity of this transferase.

5-Phosphoribosylamine reacts with glycine and ATP to form phosphoribosylglycinamide. This transfer is catalyzed by *phosphoribosylglyc-inamide synthetase.*

$$\text{5-phosphoribosylamine} + \text{Glycine} + \text{ATP} \rightarrow$$
$$\rightarrow \text{5-phosphoribosylglycinamide} + \text{ADP} + \text{P}_i$$

When glycine is incorporated in this reaction, it provides carbons 4, 5, and nitrogen 7 of the pu-rine ring. The remaining atoms are added in suc-cessive stages to form a ribonucleotide; ribose-5-phosphate remains attached to the N9 during the whole process. The resulting compound, ino-sinic acid or inosine monophosphate (IMP), con-tains the nitrogenous base hypoxanthine. Carbon 6 of hypoxanthine in IMP is aminated by transfer of the α-amino group of aspartate to form AMP; the energy that drives this reaction comes from the hydrolysis of a GTP phosphate bond.

An alternative route involves the oxidation of hypoxanthine to xanthine and then amination at C2 through the transfer of a glutamine amide group; the product is guanylic acid or guano-sine monophosphate (GMP). The hydrolysis of ATP to AMP and PP$_i$ supplies the energy for the transfer of the —NH$_2$ group.

Regulation. The synthesis of purine nucleo-tides is regulated by a feedback mechanism at several levels: (1) Formation of PRPP. Phos-phoribosylpyrophosphate synthetase is inhib-ited by the end products of the pathway (IMP, AMP, and GMP). Thus, when the level of these nucleotides increases, the production of PRPP is depressed. (2) The reaction from PRPP to phosphoribosylamine is the main control site of purine nucleotide synthesis. AMP, GMP,

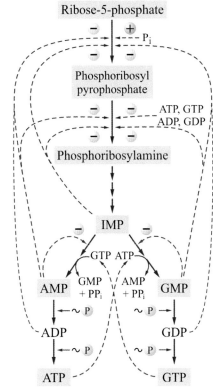

FIGURE 18.3 **Regulation of the pathway for biosynthe-sis of purines.**

and IMP act as negative effectors of glutamine PRPP amidotransferase. ATP, ADP, GTP, GDP, ITP, and IDP are also inhibitors. Each group of nucleotides occupy different allosteric sites on phosphoribosylpyrophosphate synthetase, making their effects additive. (3) From IMP, the reactions follow two different pathways. One leads to AMP and the other to GMP. The enzymes involved in the oxidation and ami-nation of IMP to generate GMP are inhibited by GMP, while the amination of IMP to form AMP is depressed by AMP (Fig. 18.3). (4) The synthesis of AMP from IMP uses GTP as en-ergy source, while the synthesis of GMP uses ATP. The levels of GTP and ATP regulate the two end branches of the pathway. An excess of GTP favors adenine nucleotide production, whereas a high level of ATP promotes GMP formation.

PURINE SALVAGE PATHWAY

The energy cost of purine synthesis is very high (six high-energy phosphate bonds are required per molecule of IMP). However, there is a more economical pathway, called the salvage pathway, which produces purine nucleotides from the degradation of preexistent nucleic acids in tissues or from those coming with the diet and absorbed in the intestine. This shows that both endogenous and exogenous bases can be used in a common pathway and that there is no absolute metabolic separation between both sources of bases.

Adenine is converted to adenylic acid or adenosine monophosphate (AMP) in a reaction that involves PRPP and the enzyme *adenine phosphoribosyl transferase* (APRT). The ribose-5-phosphate moiety is attached by a N-glycosidic bond to N9 of adenine to form AMP.

$$\text{Adenine} + \text{PRPP} \xrightarrow[\text{transferase}]{\text{Adenine phosphoribosyl-}} \text{Adenilic acid (AMP)} + \text{PP}_i$$

In a similar reaction, hypoxanthine and guanine are transformed into their corresponding nucleotides, inosinic acid or IMP and guanylic acid or GMP. This occurs by action of *hypoxanthine guanine phosphoribosyl transferase* (HGPRT).

$$\genfrac{}{}{0pt}{}{\text{Hypoxanthine}}{\text{Guanine}} + \text{PRPP} \xrightarrow[\text{phosphoribosyl transferase}]{\text{Hypoxanthine-guanine}}$$
$$\xrightarrow{} \genfrac{}{}{0pt}{}{\text{Inosinic acid (IMP)}}{\text{Guanilic acid (GMP)}} + \text{PP}_i$$

In most cells, HGPRT activity is significantly higher than that of APRT.

The energy cost for the recovery of purine molecules is 1 mole of ATP per mole of

nucleotide. ATP is consumed in the generation of PRPP. Compared with the energy cost of the de novo synthesis pathway, the salvage pathway "saves" 5 moles of ATP.

A hereditary X-linked disease has been described in which HGPRT is missing. This alteration, known as the *Lesch–Nyhan syndrome*, is characterized by the overproduction of uric acid, hyperuricemia, hyperuricosuria, associated with gout and kidney alterations. This disorder is accompanied by mental retardation, muscle spasticity, and self-mutilating behavior, characterized by lip and finger biting. Usually, symptoms appear in the first year of life.

PURINE CATABOLISM

Degradation of DNA and RNA in the cells releases the nucleotides of deoxyadenosine, deoxyguanosine, adenosine, and guanosine. These nucleotides are further hydrolyzed by nucleotidases or phosphatases, which free the nucleosides adenosine and guanosine; guanosine is then degraded into guanine and pentose.

Both adenine and guanine initially undergo hydrolytic deamination; however, adenine is deaminated to adenosine while still attached to the pentose, whereas guanine is deaminated once it is free.

Adenosine is converted to inosine hypoxanthine nucleoside by adenosine deaminase (Fig. 18.4). Subsequently, a phosphorylation reaction catalyzed by nucleoside phosphorylase separates inosine into hypoxanthine and pentose phosphate. In a following step, hypoxanthine is converted to xanthine by oxidation at carbon 2 in a reaction catalyzed by *xanthine oxidase*. This enzyme contains molybdenum, utilizes molecular oxygen, and releases hydrogen peroxide.

Guanine begins its catabolism with hydrolytic deamination catalyzed by *guanase*, an enzyme that produces xanthine (Fig. 18.4). Thus,

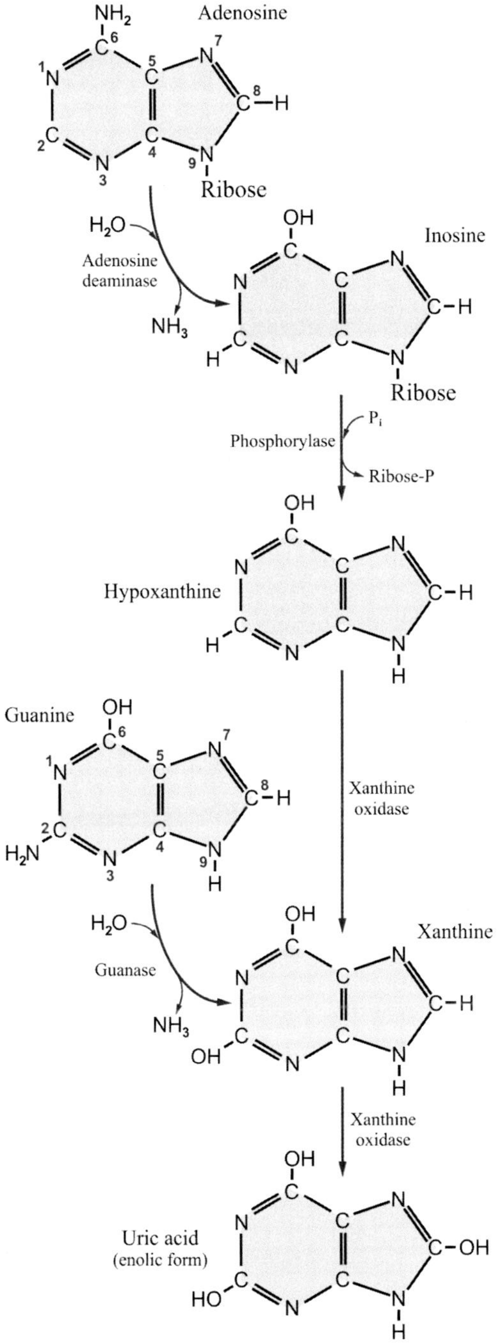

FIGURE 18.4 **Catabolism of purine bases.**

the catabolic pathways of adenine and guanine converge in the formation of a common intermediate, xanthine.

Finally, xanthine is oxidized into uric acid, an eight carbon compound, by action of *xanthine oxidase*, the same enzyme that catalyzes the oxidation of hypoxanthine. Uric acid, the end product of purine catabolism in humans, is poorly soluble in water; it is mainly excreted in the urine. In other species, there is an enzyme, uricase, which converts uric acid into a more soluble product.

A genetic disease that affects the synthesis of adenosine deaminase is the cause of approximately 15% of *severe combined immunodeficiency* (SCID) or alymphocytosis cases. The disorder is inherited in an autosomal recessive manner in 55% of cases. Normally, adenosine deaminase is found in all tissues, but it is particularly abundant in T and B lymphocytes. Therefore, these cells are affected when there are alterations in adenosine deaminase. Forty five percent of SCID cases are due to a defect linked to chromosome X, resulting in loss of T-cell receptor γ chains (see Chapter 30).

When adenosine deaminase is missing, there is accumulation of adenosine, which may eventually become ATP and dATP by the action of extracellular kinases. The increased level of dATP inhibits ribonucleotide reductase, which reduces the production of other deoxyribonucleotides and DNA. SCID patients have serious defects in their immune system and are prone to infectious diseases early in life. The need to maintain these children in a sterile environment has also given this disease the name of "bubble boy disease." Usually, these children do not survive beyond the second year of age. This condition was the first in which gene therapy was performed. Some of the treated patients have shown good results.

Another hereditary disease associated with the purine catabolic pathway is the absence of xanthine oxidase. The alteration, called xanthinuria, is characterized by high levels of xanthine in blood and in urine. This last results in the formation of urinary tract stones.

URIC ACID

In humans, the end product of purine catabolism is uric acid. A normal adult produces approximately 500 mg of uric acid per day. Approximately 80% of this amount is excreted in the urine; the rest is degraded and is eliminated as CO_2 and NH_3, or urea.

In normal plasma, uric acid concentration reaches 4–6 mg/dL. Males have blood levels that are 1 mg/dL higher than women; this difference disappears after 45–50 years of age.

The pK of the —OH group attached to carbon 8 of uric acid is 5.75. At the normal blood pH, most uric acid releases a proton and ionizes to urate. At pH 5.75, there are equal amounts of the protonated and deprotonated forms (uric acid and urate, respectively), while at lower pH, uric acid (not ionized) predominates. When urine is acidified, the proportion of uric acid increases and tends to precipitate; if urine is more alkaline, urate, which is 17 times more soluble in water predominates.

The filtered urate in renal glomeruli is partially reabsorbed in the proximal and distal tubules. It is also secreted in the proximal tubule and the loop of Henle. Usually, an average of 400 mg of uric acid is excreted every 24 h.

Several medications can block renal tubular urate reabsorption and, therefore, they increase urate excretion in the urine. Salicylate, cinchophen, and probenecid are uricosuric drugs. Urinary excretion of urate is also increased by adrenocortical hormones.

When purine ingestion increases, the production of uric acid and excretion of urate in urine increases. Meat, glandular tissues, meat extract, legumes, mushrooms, and spinach are rich in nucleoproteins. Coffee, cocoa, tea, and cola carbonated drinks contain caffeine, a methylated purine.

GOUT

The maximal amount of urate that can be dissolved in plasma at 37°C is 7 mg/dL; when this level is exceeded, urate tends to precipitate.

Gout is a disease characterized by elevated uric acid levels in blood (hyperuricemia) and urine (uricosuria). Urate crystals precipitate in joints synovial fluid. Urate deposits attract neutrophils to the affected area and initiate an inflammatory process, which results in release of cytokines and reactive oxygen species. The inflammation also increases the local production of lactate that reduces the pH in the tissue and promotes additional deposition of uric acid. The nodules formed at the area of inflammation surrounding the uric acid crystals are called tophi.

Interphalangeal and metatarsus joints are preferentially affected, causing painful arthritis. Typical location of the tophi is the big toe. Urate precipitates also occur in cartilages (pinna).

Primary gout is a metabolic disorder of genetic origin. There is excessive production or deficient urinary excretion of uric acid. Several mutations in the X-linked gene PRPP synthase synthesis have been identified; these mutations produce an enzyme with higher activity. Some cases present high V_{max} or low K_m for ribose-5-phosphate, or insensitivity of PRPP synthase to feedback inhibition. Any of these defects leads to an excess of purine production and an increase in the levels of blood uric acid. Hormonal factors also appear to influence the disease, since the frequency of primary gout is lower in females than in males, occurs in women after menopause, and is very rare in children and adolescents.

In other cases, the cause of gout is not metabolic, but is instead a defect in the transport system that secretes urate in the renal tubules. This is called *renal gout.*

Secondary gout refers to the hyperuricemia secondary to another underlying cause. Normally, these causes include chronic renal insufficiency, leukemias, and other cell proliferative processes treated with chemotherapy. It is also often produced in cases of excessive alcohol intake. Hyperuricemia is also observed when there is excessive production of lactate and ketone bodies (acetoacetate and 3-hydroxybutyrate), which compete with urate for renal excretion.

One of the most effective drugs used for reducing uric acid levels is *allopurinol*. In the body, this compound is converted into oxypurine, which is structurally similar to hypoxanthine and can inhibit xanthine oxidase. Oxypurine binds to the active site of xanthine oxidase and is oxidized to oxopurinol or alloxantin. This compound remains bound to the enzyme and blocks it. In this way, oxopurinol is considered a "suicide inhibitor" which helps decrease the production of xanthine and uric acid. Hypoxanthine accumulates, but it is more soluble and, therefore, less likely to precipitate and initiate inflammation. Allopurinol also reduces purine synthesis. Another compound inhibitor of xanthine oxidase is febuxostat.

OH

Allopurinol

Probenecid and sulfampyrazone are compounds that increase the excretion of urate and help reduce its levels in blood.

While hyperuricemia in gout is not due to an excess of purine ingested with diet, it is advisable in hyperuricemic patients to reduce the consumption of foods high in purines to avoid additional exogenous purine to contribute to the generation of uric acid.

Marked hyperuricemia in Lesch–Nyhan syndrome has already been mentioned. In these cases, the lack of hypoxanthine guanine phosphoribosyl PRPP transferase determines the increase

in uric acid levels within cells. As PRPP cannot be used in the salvage pathway, it is diverted to the de novo synthesis of bases. Increased production of purines increases the formation of uric acid.

PYRIMIDINE BIOSYNTHESIS

Similar to purine synthesis, pyrimidine bases are formed from relatively simple precursors (aspartate and carbamoyl). Fig. 18.5 shows the contribution of both of these compounds to the structure of the pyrimidine ring.

The first step is the formation of carbamoyl phosphate catalyzed by *carbamoyl phosphate synthetase 2* (CPS 2). A —NH_2 group derived from the deamination of glutamine is incorporated.

$$CO_2 + \text{Glutamine} + 2\,ATP \longrightarrow$$

$$\longrightarrow \underset{\text{Carbamoylphosphate}}{\overset{NH_2}{\underset{O-P=O}{\overset{|}{\underset{|}{C=O}}}}O^- \;+\; \text{Glutamate} + 2\,ADP + P_i}$$

The reaction is similar to the first step of the urea cycle. However, there are important differences between them. In cells, particularly hepatocytes, there are two separate pools of carbamoyl phosphate, one is in mitochondria and the other in the cytosol. The first participates in the

FIGURE 18.5 **Formation of the pyrimidine ring.** *Black,* elements are derived from carbamoyl; *red,* elements are derived from aspartic acid.

synthesis of urea, and the second in the synthesis of pyrimidine. These metabolic processes require the action of carbamoyl phosphate synthetase (CPS) isozymes with different subcellular localization. The mitochondrial isoform (isozyme 1), mainly expressed in hepatocytes, participates in the synthesis of urea. It requires *N*-acetyl glutamate as an allosteric activator and uses NH_3 as a substrate (most commonly derived from the oxidative deamination of glutamate). CPS 2, the cytosolic isoform involved in pyrimidine synthesis, is present in all cells. This enzyme is allosterically activated by ATP and PRPP, and inhibited by UTP. It uses only the amide group of glutamine molecules as a nitrogen donor. The existence of different enzymes for the synthesis of an intermediate metabolite common to two different pathways and their compartmentalization allows for independent regulation.

Carbamoyl combines with aspartate to form carbamoylaspartate. The reaction is catalyzed by *aspartate transcarbamoylase*, an allosteric enzyme and main regulatory site of this pathway.

Carbamoylaspartate is cyclized by the action of *dihydro-orotase* and is subsequently converted into orotic acid. The first three enzymes of this pathway (CPS 2, aspartate transcarbamoylase, and dihydro-orotase) are in different catalytic domains of a single multifunctional protein. Orotic acid reacts with PRPP to form the nucleotide structure orotate monophosphate (OMP). Pyrophosphate is removed, leaving ribose-5-phosphate bound by a N-glycosidic bond to N1 of the pyrimidine ring. OMP is then transformed by decarboxylation into uridylic acid or uridine monophosphate (UMP). Both of the enzymatic activities that convert orotate to UMP (orotate phosphoribosyl transferase and OMP decarboxylase) are located in the same bifunctional protein.

Transfer of two high-energy phosphate molecules allows forming UTP from UMP. Carbon 4 of uracil is aminated by the transfer of a glutamine amide group to form cytidine triphosphate (CTP).

Another pathway leads to the reduction of the 2′ carbon of UDP ribose, generating

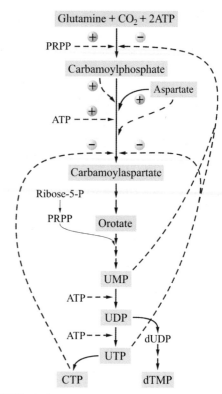

FIGURE 18.6   **Regulation of the pathway for biosynthesis of pyrimidines.**

deoxyuridine diphosphate (dUDP). This molecule is dephosphorylated to dUMP and methylated in carbon 5 of the pyrimidine ring to give deoxithymidylic acid or deoxythymidine monophosphate (dTMP). Tetrahydrofolate participates in the methyl transfer. This is the only reaction in the synthesis of pyrimidine that is linked to tetrahydrofolate. Therefore, dTMP formation is sensitive to folic acid antagonists.

Aspartate transcarbamoylase is one of the first enzymes in which allosteric regulation was described. Substrates and purine nucleotides act as positive modulators, while pyrimidine nucleotides behave as negative effectors (Fig. 18.6). The action of purine nucleotides on pyrimidine formation is important because it establishes a balance between the production of both types of nucleotides, especially for DNA synthesis.

There are similarities and differences between the processes of synthesis of purines and pyrimidines. The similarities include the following: (1) both bases require glutamine amide for their synthesis; (2) an amino acid is incorporated as the "core" of the purine and pyrimidine base to be synthesized. In the formation of the purine ring, glycine provides two carbon atoms and one nitrogen atom. In pyrimidine formation, aspartate provides three carbon atoms and one nitrogen atom; (3) as occurs with purines, there are salvage pathways to recycle or recover pyrimidines from nucleic acid degradation; finally, (4) the synthesis of both purines and pyrimidines is very costly in terms of high-energy bonds. Each molecule of UMP requires a total of five ATP molecules. Regarding the differences in metabolic pathways, for purine synthesis, the assembly of fragments is done from the beginning with bound ribosyl phosphate. In pyrimidines, ribosyl phosphate is incorporated after the heterocyclic ring has been formed.

Orotic aciduria. This is a genetic disease affecting enzymes of pyrimidine synthesis or of the urea cycle. In some cases, the deficiency is in the bifunctional protein responsible for orotate conversion in UMP. Patients excrete orotic acid in urine, causing megaloblastic anemia and growth retardation. Orotic aciduria also occurs when there is a deficiency of ornithine transcarbamoylase. In addition, this alteration also presents hyperammoniemia.

PYRIMIDINE CATABOLISM

Pyrimidine bases are degraded to soluble products that can be easily used or eliminated by cells.

Cytosine is converted, by deamination, to uracil and receives two hydrogen atoms, donated by NADPH, to form dihydrouracil. Subsequently, hydrolysis and rupture of the pirimidic ring, release β-alanine, CO_2, and NH_3 as end products.

Thymine is hydrogenated to dihydrothymine in a reaction linked to NADPH. Then, the cycle opens and β-aminoisobutyrate, CO_2, and NH_3 are produced. Eventually, the β-aminoisobutyrate becomes succinyl-CoA, an intermediary of the citric acid cycle.

In leukemic patients and those undergoing treatment with X-rays, DNA degradation causes an exaggerated increase in β-aminoisobutyrate elimination.

In individuals of Chinese or Japanese origin, an increased excretion of β-amino-isobutyrate has been observed. This is a genetically transmitted character, which does not produce clinical disorders.

DI- AND TRIPHOSPHATE NUCLEOSIDE BIOSYNTHESIS

Once a nucleoside monophosphate has been synthesized, phosphoryl groups are added by transfer from a nucleoside triphosphate. These reactions are reversible and catalyzed by nucleoside monophosphate and nucleoside diphosphate kinases.

$$GMP + ATP \underset{\substack{\text{Nucleoside monophosphate} \\ \text{kinase, } Mg^{2+}}}{\rightleftharpoons} GDP + ADP$$

Nucleoside diphosphate reacts with another ATP molecule to form nucleoside triphosphate.

$$GDP + ATP \underset{\substack{\text{Nucleoside diphosphate} \\ \text{kinase, } Mg^{2+}}}{\rightleftharpoons} GTP + ADP$$

The nucleoside triphosphates participate in the synthesis of nucleic acids (Fig. 18.7). While the reverse reactions can generate ATP from ADP, oxidative phosphorylation (p. 190) is the most important pathway for ATP production.

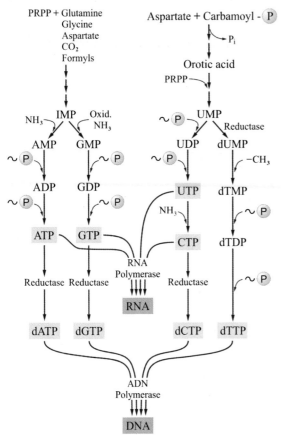

FIGURE 18.7 **Synthesis of nucleic acids.**

DEOXYRIBONUCLEOTIDE BIOSYNTHESIS

Deoxyribonucleotides are obtained by reduction of ribose already incorporated into nucleotides. Nucleoside diphosphate molecules (ADP, GDP) are used as substrates. The reaction is catalyzed by *ribonucleoside diphosphate reductase*, a multienzyme complex that is present in all cells and is active only when DNA is synthesized. The reducing agent for the reaction is a low molecular weight protein, *thioredoxin*, which has two adjacent cysteine residues, whose sulfhydryl groups yield a hydrogen atom. Another enzyme complex, thioredoxin reductase (NADPH dependent), regenerates reduced thioredoxin.

The synthesis of deoxyribonucleotide is regulated by ATP, which functions as an allosteric activator; dATP, dGTP, dTTP, and dCTP depress their own synthesis via negative feedback regulation.

PHARMACOLOGICAL APPLICATIONS

The better understanding of the metabolic pathways of purine and pyrimidine has allowed to develop compounds to treat hyperuricemia. The use of allopurinol, a xanthine oxidase inhibitor, has been mentioned earlier. Other compounds are used in the treatment of malignant tumors. One of the characteristics of neoplastic tissues is their uncontrolled cell proliferation, which requires increased purine and pyrimidine synthesis to produce nucleic acids. Compounds structurally analogous to intermediate metabolites of purines and pyrimidine pathways function as competitive inhibitors, suppressing cell mitotic activity. Among these drugs are glutamine analogs, such as asazerine and 6-diazo-5-oxo-L-norleucine. Glutamine is an important nitrogen donor in the synthesis of purine and pyrimidine bases, and in the formation of CTP from UTP and GMP from XMP. Administration of glutamine antagonists block the metabolic steps in which this amino acid participates.

The antifolic agents, such as methotrexate and aminopterin, are competitive inhibitors of dihydrofolate reductase, which catalyzes the formation of tetrahydrofolate. This agent has a very important role in the synthesis of purines, transferring single carbon groups; it is also needed to convert dUMP to dTMP.

6-Mercaptopurine, 8-azaguanidine, 5-fluorouracil, and other analogs of purine and pyrimidine bases are metabolic antagonists which inhibit cell replication. These pharmacological agents are used in the treatment of certain types of malignant tumors.

Azidothymidine (AZT) has structural similarity with thymidine nucleosides; it blocks the synthesis of DNA by reverse transcriptase. It is used in the treatment of acquired immunodeficiency syndrome (AIDS) syndrome.

Tissues with high mitotic activity, particularly the bone marrow, are often affected by treatment with these drugs. For this reason, their use should be carefully monitored.

SUMMARY

Humans produce nitrogenous bases endogenously and are not dependent on dietary intake of purines and pyrimidines.

Purine biosynthesis involves the formation of the purine ring from residues of different origins. C4, C5, and N7 are derived from glycine; N3 and N9 are derived from the amide group of glutamine; N1 is derived from aspartate; C2 and C8 come from formyl residues donated by formyl tetrahydrofolate; C6 is derived from CO_2. The molecular assembly is performed with ribose-5-P bound to it. First, PRPP is formed through a reaction catalyzed by *phosphoribosylpyrophosphate synthetase*, an enzyme inhibited by the end products, AMP, GMP, IMP. Finally a nucleotide is obtained.

Salvage pathway for purine synthesis requires the activity of *APRT* and *hypoxanthine-guanine phosphoribosyl transferase*.

Purine catabolism starts with the degradation of nucleic acids into nucleosides and nucleotides. Adenosine is deaminated (catalyzed by *adenosine deaminase*). Inosine formed is cleaved by phosphorylation (catalyzed by *nucleoside phosphorylase*) to produce hypoxanthine and ribose-P.

Then, hypoxanthine is oxidized to xanthine (catalyzed by *xanthine oxidase*). Guanosine is hydrolyzed to guanine and ribose. Guanine is deaminated to xanthine (catalyzed by *guanase*). Xanthine, formed from both adenine and guanine, is oxidized into *uric acid* (catalyzed by *xanthine oxidase*).

Uric acid is the end product of purine catabolism in humans. It is poorly soluble and is mainly excreted in the urine. The concentration of uric acid in normal plasma is 4–6 mg/dL. In some pathological conditions this value increases.

Gout is a disease characterized by elevated levels of urate in the blood and urine. Urate precipitates causing arthritis and kidney stones.

Pyrimidine biosynthesis requires the binding of aspartate and carbamoyl phosphate. Carbamoyl phosphate is synthesized from the amide group of glutamine and CO_2 (catalyzed by *CPS 2*). The reaction of carbamoyl-phosphate and aspartate forms *carbamoylaspartate* (catalyzed by aspartate transcarbamoylase), which is cyclized forming *orotic acid*. *Aspartate transcarbamoylase* is the main regulatory site of the pathway, it is inhibited by the end products (UTP, CTP).

Pyrimidine catabolism renders soluble compounds, which can be easily removed or used.

Degradation of cytosine produces β-alanine, CO_2, and NH_3. Thymine produces β-aminoisobutyrate, CO_2, and NH_3. β-Aminoisobutyrate is converted into succinyl-CoA.

Biosynthesis of nucleoside di-and triphosphate are obtained from nucleoside monophosphate by phosphoryl transfer from other nucleoside triphosphates (catalyzed by *nucleoside kinase*).

Deoxyribonucleotide biosynthesis is obtained by reduction of ribose already bound to the nucleotide by *ribonucleotide reductase*. NADPH and thioredoxin are required.

Bibliography

Blanco, A., Kremer, R.D., Taleisnik, S. (Eds.), 2000. Inherited Metabolic Defects. Academia Nacional de Ciencias, Córdoba.

Pang, B., et al., 2012. Defects of purine nucleotide metabolism lead to substantial incorporation of xanthine and hypoxanthine into DNA and RNA. Proc. Natl. Acad. Sci. USA 109, 2319–2324.

Scriver, C.R., Beaudet, A.L., Sly, W.S., Valle, D., 2001. Metabolic and Molecular Bases of Inherited Disease, eighth ed. McGraw Hill, New York, NY.

Integration and Regulation of Metabolism

METABOLIC INTEGRATION

The study of the metabolism of carbohydrates, lipids, amino acids, heme, purines, and pyrimidines in separate chapters, as was presented, could lead to the assumption that the chemical reactions that these compounds undergo in the body is independent from each other. However, this is not the case; despite their different structure, all of those compounds undergo chemical transformations that are closely related to each other and are interconnected. Substrates of varied origin and nature converge on a common metabolic pathway or generate end products that are identical, and conversely, products produced from metabolism of carbohydrates, lipids, or amino acids can be used as the raw material for the synthesis of many different compounds.

The catabolic pathways for carbohydrates, fats, and proteins will be described as a general example of metabolic convergence (Fig. 19.1). Digestive hydrolysis of the food ingested, generates simpler substances, which are absorbed in the intestine and enter the circulation. Once they are taken up by the cells, these substances are degraded to obtain from them energy; this process may result in the formation of common intermediates.

Monosaccharides and glycerol, through glycolysis, produce pyruvate, which is transformed into acetyl coenzyme A (acetyl-CoA) or active acetate. Fatty acids undergoing β-oxidation and the carbon skeletons of some amino acids also generate acetyl-CoA. This metabolite, shared by many pathways, is eventually oxidized in the citric acid cycle. Carbon residues of amino acids that end up in acetyl-CoA, as well as those converted into citric acid cycle intermediates, are oxidized to CO_2. Acetyl-CoA can also be used for the synthesis of fatty acids, cholesterol, and other substances.

The reducing equivalents (protons and electrons) removed from different substrates during their oxidation enter the respiratory chain, the final common pathway where hydrogen ions of diverse origin end up. H_2O is formed and much of the energy released is trapped in the γ and β phosphate bonds of ATP.

Interconversion of Carbohydrates, Lipids, and Amino Acids

Different compounds are interconverted in the body. Among the many examples are the following: carbohydrates can be converted into triacylglycerols (TAG) because, during their metabolism, they generate products that are used

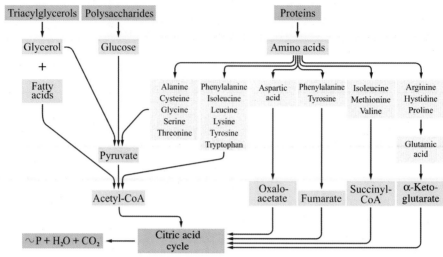

FIGURE 19.1   **Common metabolic fate of different compounds.**

in the synthesis of fats. Glucose produces acetyl-CoA, the main precursor in the synthesis of fatty acids. Glycerol is formed from an intermediate of glycolysis, dihydroxyacetone. Glucose metabolism produces α-ketoacids, easily convertible into amino acids by transamination. For example, pyruvate is converted to alanine, oxaloacetate produces aspartate, and α-ketoglutarate generates glutamate.

Following deamination, the carbon chains of amino acids can be converted into glucose or ketones; due to these possibilities, amino acids are classified as glucogenic or ketogenic.

Glycerol is the only lipid that is potentially glucogenic. Fatty acids are oxidized to acetyl-CoA, which is nonglucogenic.

Fig. 19.2 is a simplified scheme of the cross talk between metabolic pathways. As shown, some metabolites, such as glucose-6-phosphate (G-6-P), pyruvate, and acetyl-CoA, are at metabolic crossroads, where different metabolic pathways converge or start.

Metabolic alternatives. Examples of the various metabolic pathways that some substrates can follow include:

G-6-P can follow the following pathways: (1) glycolysis, (2) hydrolysis to become free glucose, (3) incorporation into glycogen, and (4) to enter the pentose phosphate pathway. Some stages of these pathways provide additional alternatives of chemical transformation.

Pyruvate can undergo: (1) decarboxylation to acetyl-CoA and all the metabolic possibilities for this compound, (2) glucose or glycogen formation in the gluconeogenic pathway, (3) transamination to give alanine, and (4) reduction to become lactate.

Acetyl-CoA follows: (1) oxidation in the citric acid cycle to produce energy, (2) fatty acid synthesis, (3) cholesterol synthesis, (4) formation of ketone bodies, and (5) incorporation into complex molecules (acetylation).

Tyrosine is used for the synthesis of: (1) proteins, (2) thyroid hormone, (3) catecholamines (dopamine, norepinephrine, epinephrine), (4) melanin, (5) phenols, or it can be (6) oxidized to CO_2 and H_2O, with energy production, (7) used for glucose

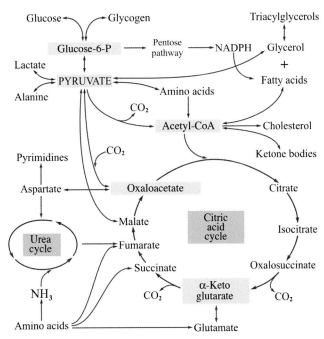

FIGURE 19.2 **Examples of interactions among metabolic pathways.**

formation (gluconeogenesis), and (8) transformed to ketone bodies (ketogenesis). *Glycine* can be used as substrate for the synthesis of: (1) proteins, (2) creatine, (3) glutathione, (4) heme, (5) purines; or can undergo (6) conjugation with bile acids (glycocholic acid), (7) biotransformation reactions, such as the formation of hippuric acid from benzoic acid, (8) production of one-carbon moieties (formyl) involved in numerous processes of synthesis, (9) conversion into serine, (10) formation of pyruvate, followed by all the pathways related to this compound, and (11) transamination.

Given the complexity of the metabolic networks and the numerous possibilities of transformation for certain substrates, it is really surprising that, under normal conditions, the system operates smoothly, producing the corresponding final products in amounts that are specifically tailored to the needs of the organism. It is evident that remarkably efficient regulatory mechanisms operate to control the flow, direction, and function of each metabolic pathway.

METABOLIC REGULATION

Operation of a metabolic pathway requires that cells contain all of the enzymes catalyzing the respective reactions and the needed substrates. The activity of enzymes must be regulated to maintain the levels of intermediate metabolites (usually at scales ranging from micromolar to millimolar) relatively constant. Moreover, the metabolic flow in each pathway is controlled to adapt them to fluctuations in the substrate supply or in the product demand.

The activity of enzymes can be regulated by two mechanisms: (1) modulating the catalytic activity of the preexisting enzyme, and (2) modulating the number of enzyme molecules.

1. *Modification of the activity of preexisting enzymes*
   a. One of the simplest mechanisms to modify an enzyme's activity is the change in the amount of substrate in the medium. When the substrate level is near or below the K_m value of the enzyme, the reaction rate is directly proportional to the substrate concentration.
   b. Another mechanism consists of the regulation by metabolites, allosteric modifiers, modulators, or effectors (p. 168). These molecules bind to the enzyme and produce changes in substrate affinity. Positive regulators are those that increase the enzyme's affinity for its substrate and activity, while negative effectors reduce enzyme's catalytic function.
   c. The activity of some enzymes can be covalently modified by the action of other enzymes, for example, the addition of a phosphate group by kinases.

   All these regulatory actions induce a rapid response. While the mechanisms listed under (a) and (b) occur almost immediately (no more than 2–3 s), covalent modification (c) occurs in a few minutes.

2. *Increase or decrease in the number of enzyme molecules.* The activity of an enzyme is directly proportional to the amount of enzymes present. The number of enzyme molecules in the cell at a given time depends on the balance between its synthesis and degradation.
   a. *Synthesis of protein.* Regulation of enzyme synthesis is exerted mainly at the transcriptional level, by regulating the production of messenger RNA coding for that enzyme from DNA (pp. 475 and 526). Another regulatory mechanism takes place at the translational level, to control the conversion of the message contained in mRNA into protein (p. 502). Regulation of protein synthesis is slower than the mechanism that directly modifies enzyme's activity.
   b. *Protein degradation.* The concentration of a given protein in the cell depends not only on the magnitude of its synthesis, but also on its degradation (p. 368). Enzymes are subjected to continual renewal, according to their half-lives (time by which 50% of the enzyme is degraded), which ranges from minutes to days, and the specific needs of the cell.

Regulatory mechanisms. The information regarding the mechanisms of metabolic regulation was originally based on studies performed on isolated enzymes or in preparations in vitro, which do not exactly resemble the complexity that metabolic systems present under in vivo conditions. These studies generated the basic knowledge that dominated this area of biochemistry for a long time. The main findings can be summarized as follows:

a. There are particular reactions in metabolic pathways which are considered limiting, since they regulate the rate at which metabolites flow along the entire pathway.
b. It is common that the first reaction and those at branching points in a metabolic pathway, as well as the reactions that are markedly shifted from equilibrium, play a decisive regulatory role in the pathway.
c. The end product often exerts feedback actions on one or more "key" enzyme of the pathway to control the magnitude of its own production.
d. Isozymes with different catalytic properties, or subcellular localization, catalyze the same reaction in different metabolic pathways, allowing independent regulation.

While some of these conclusions have been confirmed, others have being questioned and revised. For example, evidence obtained by quantification of metabolite flow in complex systems indicate that, although some enzymes have greater influence than others in the overall regulation of a pathway, there are several enzymes that together contribute to control a pathway. In general, these enzymes catalyze reactions that are far from equilibrium.

The mechanisms of regulation of some metabolic pathways are described in the following sections.

Examples of Metabolic Regulation

For glycogen, glucose, and fatty acids, synthesis and degradation follow different pathways. Some of the reactions are practically irreversible and require that different enzymes catalyze the forward or reverse reactions.

The differential control of anabolic and catabolic pathways is essential to control the direction of metabolite flow. Usually, factors that stimulate an anabolic pathway simultaneously depress the corresponding catabolic pathway or vice versa. This not only allows for regulation of metabolism according to the needs of the body, but ensures the most efficient use of energy resources. In the absence of these controls, *futile metabolic cycles* with energy wastage would occur. There are, however, exceptions to this; the operation of futile cycles sometimes plays an important functional role, as will be discussed in subsequent sections.

The following are examples that illustrate the operation of regulatory systems.

Regulation of Glycogen Synthesis and Degradation

With glucose-1-phosphate as the initial substrate, glycogen synthesis consumes 1 mole of ATP per mole of glucose incorporated into glycogen. ATP is needed to regenerate UTP that is produced from UDP during glycogen synthesis. Glycogenolysis produces glucose-1-phosphate and no energy. If there is no control of the enzymes needed for synthesis and degradation of glycerol, a futile cycle will be established, in which the glucose-1-phosphate to glycogen conversion in one direction will be continuously offset by the reaction that brings glycogen back to glucose-1-phosphate. This will result in a nonproductive expenditure of energy, as each turn would require hydrolysis of UTP to UDP and P_i.

Under normal conditions the operation of these reactions is tightly regulated. Glycogen synthesis is activated when there is abundant supply of glucose and sufficient energy in the form of ATP. Degradation is stimulated in the liver by decreases in blood glucose. Generally, conditions which promote glycogen synthesis inhibit its breakdown and vice versa.

Glycogenolysis

The most important regulatory enzyme in the glycogen degradation pathway is *glycogen phosphorylase* (or just *phosphorylase*).

In liver and muscle, which contain the largest reserves of glycogen in the body, the enzyme is found in two forms, inactive (*b*) and active (*a*).

Muscle glycogen phosphorylase is different than that found in the liver; they are distinct isozymes synthesized under the control of different genes. Their properties are not exactly the same, but the regulatory mechanisms are similar; both are susceptible to covalent modification and allosteric effectors.

Phosphorylase *b* in both muscle and liver is stimulated by addition of a phosphate ester bond to the hydroxyl group of a serine residue on each of its two subunits. This phosphorylation is catalyzed by *phosphorylase kinase*, which transfers phosphate from ATP and forms glycogen phosphorylase *a*.

$$\text{Phosphorylase } b + 2\text{ATP} \xrightarrow{\text{Phosphorylase kinase}} \text{Phosphorylase } a + 2\text{ADP}$$
(Inactive)   (Active)

Phosphorylase inactivation occurs by hydrolysis of the phosphate ester bond between serine and the phosphoryl group, which is catalyzed by *protein phosphatase 1* (PP1). Glycogen phosphorylase *a* is converted into *b*.

$$\text{Phosphorylase } a \xrightarrow{\text{Phosphatase}} \text{Phosphorylase } b + 2\text{P}_i$$

Activation of glycogen phosphorylase is an example of covalent modification and it is the result of a cascade of reactions in which the product of each reaction acts as the activator for the next. Fig. 19.3 shows the series of reactions involved in the activation of glycogen phosphorylase. The process is initiated by hormones, epinephrine in muscle and glucagon in liver. The hormone binds to specific cell membrane receptors and activates an enzyme called *adenylate cyclase* (p. 551) (Chapter 25 will revisit this issue and the mechanism of action of hormones).

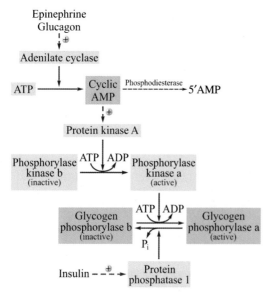

FIGURE 19.3   **Cascade mechanism for the activation of glycogen phosphorylase.**

Adenylate cyclase catalyzes the conversion of intracellular ATP into 3′,5′-cyclic adenosine monophosphate (cyclic AMP, p. 139). This cyclic-AMP functions as a regulator of many hormones; it is the "second messenger" in the system of chemical signals initiated by the arrival of the hormone, considered the "first messenger."

The increased level of 3′,5′-cyclic-AMP in the cytosol activates protein kinase A (for details see p. 559). Protein kinase A phosphorylates the —OH of a serine residue on *phosphorylase kinase*. This enzyme is also found in two forms, *a* and *b*, in the dephosphorylated state (*b*) it is inactive. Activated phosphorylase kinase (*a*), in turn, catalyzes the transfer of phosphate from ATP to glycogen phosphorylase *b* converting it into *a*. This initiates the degradation of glycogen to glucose-*l*-phosphate (Fig. 19.3).

Cascades of reactions of this type produce large amplifications of a response. A few molecules of hormone promote the production of significant amounts of intermediary molecules, such as cyclic-AMP and the desired downstream effects in the cells.

Due to its physiological activity, cyclic AMP must be rapidly degraded. A *phosphodiesterase* in the cytosol catalyzes the hydrolysis of the ester bond between the phosphate and hydroxyl group of ribose C3′, converting 3′,5′-cyclic AMP to 5′-AMP, with no activity.

Phosphorylase *a* is inactivated by a *phosphorylase phosphatase*, also called *phosphoprotein phosphatase 1* (PP1). This enzyme separates the phosphoryl group to convert phosphorylase *a* into its inactive form, phosphorylase *b*.

In addition to covalent modification, glycogen phosphorylase is subjected to allosteric modulation.

During a sudden and intense exercise, the intracellular concentration of ATP decreases rapidly and ADP and AMP increase. AMP acts as an allosteric effector of glycogen phosphorylase *b*, changing its conformation and activating it. ATP and G-6-P are negative effectors of the enzyme. In resting muscle ATP concentration is high;

most of the glycogen phosphorylase is inactive. Phosphorylase *a* (phosphorylated) is active irrespective of AMP, ATP, or G-6-P levels.

Hepatic glycogen phosphorylase behaves differently from that of muscle and it is not sensitive to variations in the concentration of AMP. This isoform is inhibited by high levels of glucose. Glucose in hepatocytes binds to an inhibitory allosteric site in phosphorylase *a*. A conformational change in the enzyme exposes bound phosphoryl serine residues to the action of the PP1. Thus, it acts as a sensor of glucose levels.

The difference between muscle and liver is related to the role of glycogenolysis in each of these tissues. In muscle, glycogen breakdown is activated to meet the energy needs of muscle. In the liver, glucose is not the main source of energy of the tissue. Hepatic glycogenolysis releases glucose to the bloodstream when blood glucose is low, which then is carried to all body tissues.

Phosphorylase kinase *b* is also allosterically activated by high levels of Ca^{2+}. One of the subunits of the enzyme is *calmodulin*, a protein that binds calcium and promotes activation. This mechanism of regulation is physiologically relevant in muscle, in which contraction is initiated by an abrupt rise in cytosol Ca^{2+} concentration. Thus, the increase of Ca^{2+}, which is the signal that triggers the contraction, favors the degradation of glycogen to provide the necessary energy to support muscle work.

When the muscle is at rest, another enzyme, PP1, cleaves the phosphoryl groups and converts phosphorylase *a* into its inactive form, phosphorylase *b*.

Glycogenesis

Glycogen synthesis is primarily regulated by modulating the activity of *glycogen synthase*. This enzyme exists in two forms, dephosphorylated (active or *a*) and phosphorylated (inactive or *b*). It is regulated by covalent modification, in an inverse direction to that of glycogen phosphorylase.

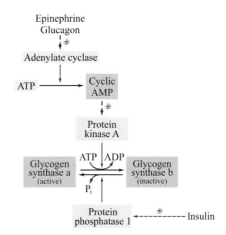

FIGURE 19.4 **Activation–inactivation mechanism of glycogen synthase.**

PP1 catalyzes the conversion of glycogen synthase *b* into *a*, and the inactivation of glycogen phosphorylase and phosphorylase kinase. In turn, PP1 is active when it is dephosphorylated.

Inactivation of glycogen synthase is produced by a cascade of reactions initiated by hormones, similar to the activation of glycogen phosphorylase. Adrenaline and glucagon binding to specific receptors on the membrane of muscle or liver cells, respectively, causes activation of adenylate cyclase, which catalyzes formation of 3′,5′-cyclic-AMP from ATP. Cyclic AMP activates protein kinase A, which phosphorylates and inactivates glycogen synthase and PP1 (Fig. 19.4). There is also an indirect regulatory mechanism involving inhibition of the PP1, catalyzed by protein kinase A; this involves the activating phosphorylation of an endogenous inhibitor (I-1) of PP1.

Another enzyme called *glycogen synthase kinase 3* (GSK-3), as well as kinases activated by Ca^{2+} also phosphorylate glycogen synthase. An insulin stimulated kinase, PKB or AKT, inhibits GSK-3 and helps maintain glycogen synthase active. G-6-P acts as a positive allosteric effector of the normally inactive phosphorylated form of the enzyme. Moreover, glycogen, the final product of the pathway, exerts direct inhibitory action on synthase *a* (Fig. 19.5).

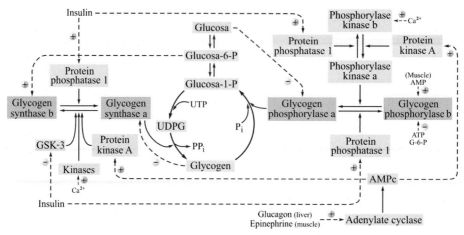

FIGURE 19.5 **Regulatory actions on glycogen synthesis and degradation pathways.** + and − indicate activating and inhibitory actions.

The role of glucokinase in the regulation of liver glycogenesis must be mentioned. This enzyme catalyzes the first step of glucose phosphorylation to form G-6-P; it has a high K_m for glucose. Therefore, its activity reaches physiologically significant values only when the supply of glucose is abundant enough, for example, after a meal rich in carbohydrates. G-6-P allosterically activates the phosphorylated form of glycogen synthase, which is normally inactive.

During fasting, most glucokinase remains "sequestered" in the nucleus of hepatocytes, linked to a regulatory protein that maintains it in the inactive state. After meals, arrival of glucose increases and glucokinase is separated from the regulatory protein so it can be transferred to the cytoplasm. Glucokinase activation increases the concentration of G-6-P. This metabolite, acting synergistically with glucose, promotes dephosphorylation (inactivation) of glycogen phosphorylase and activation (also by dephosphorylation) of glycogen synthase.

Regulation of Glycolysis and Gluconeogenesis

There are three strongly exergonic reaction steps in glycolysis and they are physiologically irreversible. In the opposite direction (gluconeogenesis), these steps are catalyzed by different enzymes from those used in glycolysis. Regulatory actions of both pathways are exercised on those three steps.

Furthermore, the intensity of glycolysis is dependent on the availability of substrate and the redox state of the cell. Glucose, ADP, P_i, and NAD^+ are required. The general process is controlled in part by the $NADH/NAD^+$ and the lactate/pyruvate ratios which, in turn, depend on the oxygen supply and function of the respiratory chain.

Stage glucose-G-6-P. If not controlled, the phosphorylation reaction of glucose and the hydrolysis of G-6-P establishes the following substrate cycle:

$$Glucose + ATP \xrightarrow{\text{Hexokinase}} Glucose\text{-}6\text{-}P + ADP$$

$$Glucose\text{-}6\text{-}P + H_2O \xrightarrow{\text{Glucose-6-P phosphatase}} Glucose + P_i$$

The net result of this substrate cycle is:

$$ATP + H_2O \longrightarrow ADP + P_i$$

There is a loss of one high-energy phosphate bond.

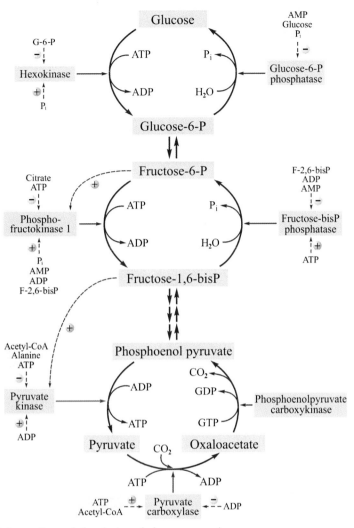

FIGURE 19.6 **Regulatory actions of glycolysis and gluconeogenesis.**

For this reason, these cycles are called *futile*. However, some of these cycles fulfill a functional role. Small changes in activity in one or both enzymes produce marked effect on the flux of metabolites in either direction.

Hexokinases, responsible for the formation of G-6-P in the vast majority of tissues, have high affinity for glucose. At normal glucose concentrations, these enzymes work practically at their V_{max}. Hexokinases are regulated by product, with G-6-P inhibiting their activity. This inhibitory action of G-6-P is counteracted by ATP (Fig. 19.6).

Glucokinase, or hexokinase IV, catalyzes the reaction in liver and pancreatic β cells. It has a high K_m for glucose, so concentrations of glucose, such as those found in the cell are not enough to saturate it. Therefore, the glucokinase activity is regulated by blood glucose levels. Glucokinase is not down-regulated by its product and may continue functioning under concentrations of G-6-P that completely inhibit the activity of the other hexokinases. Glucokinase is inhibited by the binding of a specific regulatory protein (glucokinase regulatory protein, GKRP) that "sequesters" it into the nucleus. The binding of this protein to the glucokinase is enhanced by the presence of fructose-6-P.

When blood glucose increases, the function of the membrane transporter GLUT2 quickly uptakes glucose from the blood into the cytosol. Glucose easily passes through the nuclear membrane and, in the nucleus, it competes with fructose-6-P; this action decreases the binding of the inhibitory protein to glucokinase, which is released and returns, active, to the cytosol.

Another regulatory mechanism for of glucokinase involves factors that promote the use of glucose, such as insulin and a decreased ATP/AMP ratio.

G-6-P phosphatase, present in gluconeogenic tissues, including liver, kidney, and intestine, is regulated by its products; glucose and P_i function as inhibitors (Fig. 19.6). Its synthesis is stimulated by conditions that require an increase in glucose supply, for example, hypoglycemia.

The hormonal control of the production of these enzymes will be discussed in Chapter 26.

Stage fructose-6-phosphate-fructose-1,6-bisphosphate. G-6-P has several metabolic possibilities, including entering glycolysis, the pentose phosphate pathway, and glycogenesis. The irreversible reaction catalyzed by phosphofructokinase 1 is the step that ultimately decides the entry of G-6-P into glycolysis.

If the reactions catalyzed by phosphofructokinase 1 and fructose-1,6-bisphosphate phosphatase were not controlled, a futile cycle would occur, with loss of energy.

$$\text{Fructose-6-P} + \text{ATP} \xrightarrow{\substack{\text{Phosphofructo-}\\\text{kinase}}} \text{Fructose-1,6-bisP} + \text{ADP}$$

$$\text{Fructose-1,6-bisP} + H_2O \xrightarrow{\substack{\text{Fructose-1,6-bisP}\\\text{phosphatase}}} \text{Fructose-6-P} + P_i$$

The sum of these reactions gives:

$$\text{ATP} + H_2O \longrightarrow \text{ADP} + P_i$$

Enzymes included in this step play a very important role in the regulation of glycolysis and gluconeogenesis.

Muscle and brain glycolysis is regulated according to the energy demand of cells, primarily through allosteric effectors of *phosphofructokinase 1*, which modify the value of its K_m for fructose-6-phosphate. This enzyme is the major site of glycolysis modulation. It is activated by AMP and P_i, while ATP and citrate act as negative effectors. Inhibition of the enzyme produces accumulation of metabolites produced in the previous steps in the pathway, including G-6-P, which depresses the activity of hexokinase and reduces the use of glucose via different pathways.

Gluconeogenesis practically does not work in muscle and brain, whereas it is active in liver and kidney. Therefore, the modulation of this step simultaneously in both directions is almost exclusively exerted in these tissues.

Fructose-1,6-bisphosphate phosphatase is allosterically inhibited by AMP and ADP.

In liver and other tissues, a compound with potent action on glycolysis and gluconeogenesis was isolated. This molecule is fructose-2,6-bisphosphate, formed from fructose 6-phosphate by the action of phosphofructokinase 2 (PFK2) and hydrolyzed back to F-6-P by fructose-2,6-bisphosphate phosphatase 2 (FBPase2). PFK2 and FBPase2 are different from PFK1 and FBPase1; they are bifunctional proteins with two active sites regulated by phosphorylation/dephosphorylation. Dephosphorylation activates PFK2 and inhibits FBPase2.

Fructose-2,6-bisphosphate is a powerful activator of phosphofructokinase 1 and inhibitor of fructose-1,6-bisP phosphatase 1 (Fig. 19.6). It modulates the activity of phosphofructokinase 1 by various actions that alter the kinetic properties of the enzyme: (1) it decreases the K_m for fructose-6-phosphate, (2) increases the association constant (K_a) of AMP, so it becomes a more efficient activator, and (3) increases the inhibition constant (K_i) for ATP (i.e., decreases the ability of this effector as enzyme depressor).

Glucagon, a hormone secreted by the pancreas when blood glucose decreases, induces in the liver (by a 3′,5′-cyclic AMP–dependent mechanism) a reduction of fructose-2,6-bisphosphate, which depresses glycolysis and stimulates gluconeogenesis. Insulin has the opposite effect.

Increased fructose-2,6-bisphosphate is an intracellular signal indicator of high blood glucose levels. The appropriate response to this situation is to increase glucose utilization (glycolysis) and to stop production (gluconeogenesis).

Activation of phosphofructokinase 1 and inhibition of fructose-1,6-bisP phosphatase by AMP and fructose-2,6-bisP establish directional control of the pathway simultaneously, stimulating glycolysis and inhibiting gluconeogenesis.

Another mechanism that controls the level of F-2,6-bisP depends on xylulose-5-P, an intermediate of the pentose phosphate pathway. Xylulose-5-P, whose concentration increases as more G-6-P enters glycolysis and the pentose phosphate pathways, activates phosphoprotein phosphatase 2 that dephosphorylates the bifunctional enzyme PFK2/FBPase2.

An apparently futile cycle has been demonstrated in some insects for the muscles involved in flight. The hydrolysis of ATP generates heat and ensures the proper temperature for contractile activity.

Stage phosphoenolpyruvate-pyruvate. Without regulation, the reactions in this stage would determine a substrate cycle:

$$\text{Phosphoenol pyruvate} + \text{ADP} \xrightarrow{\text{Pyruvate-kinase}} \text{Pyruvate} + \text{ATP}$$

$$\text{Pyruvate} + CO_2 + \text{ATP} \xrightarrow{\text{Pyruvate-carboxylase}} \text{Oxaloacetate} + \text{ADP} + P_i$$

$$\text{Oxaloacetate} + \text{GTP} \xrightarrow{\text{P-enolpyruvate-carboxykinase}}$$

$$\xrightarrow{} \text{Phosphoenol pyruvate} + CO_2 + \text{GDP}$$

The total balance is:

$$\text{GTP} + H_2O \xrightarrow{} \text{GDP} + P_i$$

While *pyruvate kinase* is a glycolytic enzyme, *pyruvate carboxylase* and *phosphoenolpyruvate carboxykinase* are involved in gluconeogenesis.

Pyruvate carboxylase is subjected to control by metabolites. In particular, the requirement of this enzyme for acetyl-CoA functions as a positive allosteric effector. The active acetate, product of fatty acid β-oxidation, promotes gluconeogenesis.

Oxaloacetate, formed in the reaction catalyzed by pyruvate carboxylase, is a metabolite intermediate of the citric acid cycle. Its condensation with acetyl-CoA starts the sequence of reactions. When the concentration of oxaloacetate decreases, acetyl-CoA tends to accumulate. This stimulates synthesis of oxaloacetate from pyruvate and CO_2, allowing the citric acid cycle to resume its rate of operation. In this respect, the action of acetyl-CoA carboxylase on pyruvate not only promotes gluconeogenesis, but also controls the function of the citric acid cycle.

Pyruvate carboxylase is inhibited by ADP. The enzyme is probably regulated in vivo by the relative concentrations of ATP and ADP. The increase in the ATP/ADP ratio stimulates the enzyme, while the decrease inhibits the enzyme (Fig. 19.6).

Pyruvate kinase is depressed by ATP, acetyl-CoA, and alanine, the effect of ATP is counteracted by ADP. The enzyme is sensitive to the ATP/ADP ratio in a way that is the opposite to that seen with pyruvate carboxylase. The fructose-1,6-bisphosphate activates pyruvate kinase. In vitro, the enzyme is inhibited by NADH and free fatty acids.

The liver isozyme of pyruvate kinase is also regulated by covalent modification (phosphorylation–dephosphorylation) dependent on cyclic AMP. The increase in 3′,5′-cyclic-AMP concentration activates protein kinase A, which catalyzes phosphorylation of pyruvate kinase and inactivates it.

Oxaloacetate formed by pyruvate carboxylase is converted to phosphoenolpyruvate by phosphoenolpyruvate carboxykinase. This enzyme is regulated primarily at the level of its synthesis and degradation. Glucagon increases the production of PEP carboxykinase and insulin depresses it.

Fig. 19.6 shows the steps of regulatory factors in the pathways of glycolysis and gluconeogenesis.

According to the role of the glycolytic pathway in each tissue, there are differences in regulatory factors. In muscle, AMP and ATP are the most important allosteric effectors, because the main role of glycolysis is to provide energy for contraction. However, in the liver, control mechanisms tend to maintain a constant supply of glucose to extrahepatic tissues. When blood glucose is elevated, insulin is secreted, which activates the glycolysis pathway. Under these conditions, metabolites, such as acetyl-CoA and phosphodihydroxyacetone are preferably diverted to the synthesis of fatty acids and TAG. If the level of blood glucose is low, glucagon secretion is stimulated and gluconeogenesis is activated. A very important factor coordinating degradation or production of glucose is fructose-2,6-bisphosphate, its presence accelerates glycolysis and depresses gluconeogenesis.

Fatty acids have stimulatory actions on gluconeogenesis. In general, the free fatty acids inhibit the kinases which catalyze irreversible stages of glycolysis (hexokinase, phosphofructokinase 1, and pyruvate kinase). They also inhibit the oxidative pentose phosphate cycle reactions, that is, those catalyzed by G-6-P and 6-phosphogluconate dehydrogenase.

Besides allosteric effects, actions at the level of enzyme synthesis are also exerted (see p. 525). For example, there is regulation of the synthesis of enzymes that catalyze irreversible reactions of glycolysis and of the enzymes responsible for the corresponding steps of gluconeogenesis. Factors that induce or repress their synthesis act simultaneously and in the same direction on all enzymes in each of these pathways.

Pasteur Effect

During his studies on fermentation, Pasteur observed a marked decrease in glucose consumption when bacteria were moved from an anaerobic medium to another where the supply of oxygen was abundant. He concluded that the presence of oxygen inhibited glucose utilization. The phenomenon, called the *Pasteur effect*, remained unexplained for a long time. Later, the better understanding of enzyme allosteric regulation explained the phenomenon and shed light on the mechanisms that adjust glycolysis to cell energy requirements.

The production of ATP from glucose is 19 times greater when oxygen is available than when it is absent. In anaerobiosis, more glucose needs to be consumed to obtain the same amount of ATP that can be obtained in the presence of oxygen. The Pasteur effect is explained as follows: (1) an increase in aerobic ATP and citrate production inhibits phosphofructokinase 1; (2) decreased activity of phosphofructokinase 1 causes accumulation of metabolites formed by previous stages, including G-6-P. This last substance inhibits the activity of hexokinase. These feedback effects contribute to decrease glycolytic activity in the presence of oxygen. Fructose-2,6-bisphosphate is not involved in the production of the Pasteur effect.

Warburg effect. Cancer cells have a remarkable difference in glucose utilization compared to normal cells. They consume less oxygen and more glucose than normal. Much of the glucose in cancer cells is converted to lactate, even when oxygen is fully available. This indicates a failure in the metabolic regulatory mechanisms; there is no Pasteur effect.

In aerobic organisms, cells normally catabolize glucose until it is finally oxidized to CO_2 and H_2O, which renders important level of ATP by oxidative phosphorylation. In neoplastic cells, the production of ATP from glucose is limited to that coming from glycolysis. This is not due to the lack of factors necessary for the complete

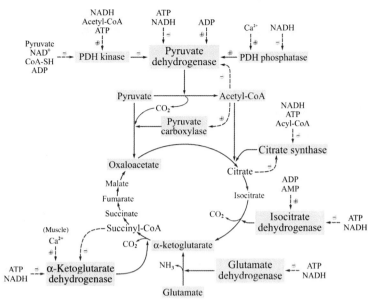

FIGURE 19.7 **Regulation of pyruvate dehydrogenase *(PDH)* and the citric acid cycle.**

metabolism of glucose, since the enzymes of glycolysis, pyruvate dehydrogenase (PDH), and citric cycle enzymes are present. This phenomenon, called aerobic glycolysis, or Warburg effect (named after the German biochemist who first described the effect), has not been clearly explained yet. However, it has been proposed that in tissues with high mitotic activity, instead of being directed to ATP production, metabolism is adapted to promote the uptake of nutrients, necessary to synthesize the components (nucleotides, amino acids, lipids) of the proliferating cells.

Oxidative Decarboxylation of Pyruvate

Oxidative decarboxylation of pyruvate is an irreversible reaction. Once pyruvate is converted into acetyl-CoA, it is not possible to use it to form glucose. The PDH multienzyme complex is regulated allosterically by covalent modification. It is directly inhibited by its products, NADH and acetyl-CoA, and by ATP. Also, it is

inactivated by phosphorylation catalyzed by PDH kinase. In contrast, the complex is stimulated by the action of PDH phosphatase. Both kinase and phosphatase are part of the PDH complex.

PDH kinase, in turn, is activated by acetyl-CoA, ATP, and NADH; it is inhibited by pyruvate, CoA-SH, ADP, and NAD$^+$. Thus, the increase in acetyl-CoA and NADH, depresses the activity of the PDH complex, while abundant supply of substrates stimulates it (Fig. 19.7).

The mechanism of PDH phosphorylation–dephosphorylation is not dependent on the levels of cyclic AMP. Insulin promotes the activation of PDH phosphatase. The enzyme is stimulated by increased levels of intramitochondrial Ca^{2+} and inhibited by NADH.

PDH activity in muscle tissue increases during intense exercise. This effect is mainly caused by the elevated levels of ADP and pyruvate-inhibiting PDH kinase, and by the increase of Ca^{2+}, which stimulates the PDH phosphatase. These factors contribute to maintain the PDH complex in the dephosphorylated (active) state.

PDH responds to the energy state of the cell. When the ATP level is high, glycolysis decreases and PDH activity is inhibited. It is also sensitive to the redox state, indicated by the value of the NADH/NAD$^+$ ratio. The reaction, catalyzed by PDH, controls the rate at which pyruvate originated from carbohydrates and from carbon skeletons of some amino acids is converted to acetyl-CoA. Products of fatty acid oxidation (acetyl-CoA, NADH, and ATP) inhibit PDH and save glucose and amino acids. In tissues that utilize both glucose and fatty acids as energy sources, this effect is important, since it allows for the saving of glucose and reserves it for tissues (such as nervous tissue) which rely almost exclusively on glucose as fuel. Amino acids, which are not stored as energy reserve, have a limited catabolism when other oxidizable substrates are available.

Regulation of the Citric Acid Cycle

Oxidation of acetyl-CoA and the simultaneous reduction of NAD$^+$ and FAD in the citric acid cycle provide energy for ATP synthesis in mitochondria. The intensity of ATP use is the main regulator of this cycle. The energy state of the cell is reflected by the adenylate ratio and the redox state is indicated by the NADH/NAD$^+$ ratio. These relationships have a key role in controlling the cycle activity.

There are three stages in which the most significant regulatory actions are exerted. The most important one is that catalyzed by isocitrate dehydrogenase, an allosteric enzyme. Two other control sites include the reactions catalyzed by α-ketoglutarate dehydrogenase and citrate synthase, which are not allosteric.

Isocitrate dehydrogenase is dependent on NAD, it is activated by ADP and inhibited by NADH. Binding of ADP to the enzyme causes a conformational change that reduces its K_m for the substrate. ATP is a negative effector. Increased values of the NADH/NAD$^+$ ratio decrease the activity of the enzyme. The NADP-linked isocitrate dehydrogenase does not have the regulatory properties of the NAD-dependent enzyme; it is neither stimulated by ADP nor inhibited by ATP and NADH.

The α-ketoglutarate dehydrogenase complex, while not allosteric, also responds to ADP, ATP, and NADH changes. It is inhibited by the product succinyl-CoA. The α-ketoglutarate dehydrogenase complex in myocardium and skeletal muscle is activated by increased Ca^{2+} levels.

Citrate synthase is not an allosteric enzyme. Its activity is modulated by the concentrations of substrate and product; citrate acts as an inhibitor. When isocitrate dehydrogenase is activated, citrate is consumed and the inhibition by product diminishes.

The substrate oxaloacetate is produced in the last reaction (malate to oxaloacetate) of the cycle. The equilibrium is largely in favor of malate, so that the concentration of oxaloacetate in mitochondria is very low. When the NADH/NAD$^+$ ratio decreases, the malate/oxaloacetate ratio is also reduced and the concentration of oxaloacetate and citrate synthase activity increases. The NADH/NAD$^+$ ratio indicates the path followed by the acetyl-CoA, especially in liver tissue. If the value of the ratio increases, citrate synthase is depressed. Under these conditions, acetyl-CoA preferably follows the pathway of ketone body synthesis.

It is also important to mention the modulatory actions on the reactions that are not part of the cycle, but that are involved in feeding it with metabolic intermediaries. Glutamate dehydrogenase, the enzyme that catalyzes oxidative deamination of glutamic acid, connects the amino acid metabolism with the citric acid cycle. This enzyme used as coenzyme, either NAD or NADP. The reaction with NAD provides electrons to the respiratory chain, while transference of hydrogens from NADP is preferably diverted to synthesis processes (fatty acids, steroids). ATP acts as a negative effector, inhibiting the oxidation reaction when NAD participates; yet, it lacks regulatory action on the reaction with

NADP. The effect is similar to that described for isocitrate dehydrogenase. Fig. 19.7 summarizes the actions described.

Regulation of Fatty Acid Metabolism

Lipolysis. Adipose tissue is the largest reserve of fat in the body. This reserve is mobilized to supply fuel to tissues, such as the liver, muscle, heart, kidneys, and others able to oxidize fatty acids.

The fatty acids from stored TAG are released by action of the adipose tissue-specific lipase, an enzyme sensitive to hormones. It is found in adipocytes in two forms, inactive (dephosphorylated) and active (phosphorylated). Its activation is achieved by a cascade process, similar to that described for glycogen phosphorylase.

Due to the intense activity of this enzyme, fats from adipose tissue are in permanent removal. The TAGs are completely renewed every 2 or 3 weeks.

If there is no demand from the tissues using fatty acids, lipase and *perilipins* (proteins surrounding fat droplets in each adipocyte) are dephosphorylated. In this state, perilipins do not enable lipase to contact TAG. Activation of protein kinase A stimulates phosphate addition to lipase and perilipins and triggers the lipolysis.

Catecholamines (epinephrine and norepinephrine), glucagon, and adrenocorticotropin (ACTH) bind to receptors in the plasma membrane of adipocytes and activate adenylate cyclase. Cyclic AMP levels in the cytosol increases, and protein kinase A (the enzyme that catalyzes phosphorylation of the lipase and perilipins) is stimulated.

Triiodothyronine (T_3), and prolactin hormones also increase lipolytic activity in adipocytes. Growth hormone and glucocorticoids stimulate lipase synthesis.

Insulin, some prostaglandins, and nicotinic acid (a B complex vitamin) act as antilipolytic factors because they decrease the intracellular concentration of cyclic AMP. Insulin also stimulates

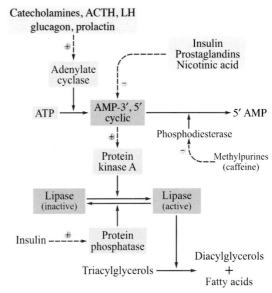

FIGURE 19.8   **Regulation of lipolysis.**

protein phosphatase, which dephosphorylates lipase and perilipins to inactivate them.

While they are not physiological mechanisms of regulation, the effect of caffeine and theophylline (methylated purines) inhibit the phosphodiesterase enzyme that catalyzes hydrolysis of 3′,5′-cyclic-AMP to 5′-AMP. These substances maintain the intracellular cAMP level elevated and, thus, the lipolytic activity in adipose tissue. Fig. 19.8 summarizes the actions described.

Along with the inhibition of lipolysis, insulin promotes fat deposition through multiple actions that determine increased supply of substrates for the synthesis of TAG. It stimulates the synthesis and secretion of lipoprotein lipase (LpL) in capillary endothelium. TAG hydrolysis of chylomicrons and VLDL provides fatty acids to the adipocytes (apo C-II, a LpL activator, transferred to chylomicrons and VLDL by HDL, contributes to this action).

Furthermore, insulin increases the number of GLUT4 transporters in the plasma membrane of adipocytes and thus increases glucose uptake. Glycolysis is activated and one of its intermediaries, dihydroxyacetone phosphate, is converted to glycerophosphate, a precursor in the synthesis of TAG.

The glucagon/insulin ratio in blood, which decreases when the glucose is high (and vice versa), is an important factor regulating lipolysis.

Under conditions of prolonged fasting, glucagon secretion is stimulated and that of insulin is depressed. In the liver, fatty acids are oxidized to acetyl-CoA, but this does not enter the citric acid cycle under these conditions, it is preferably converted into ketone bodies, offered to the tissues as an energy source. ATP generated in the liver by the oxidation of fatty acids is used to activate gluconeogenesis.

Blood levels of epinephrine, norepinephrine, ACTH, and hormones that activate lipolysis increase during intense exercise and stress. Rapid release of fatty acids ensures fuel supply to muscle, heart, liver, and kidneys.

Under conditions of appropriate food intake, tissues receive fatty acids released from the TAG of chylomicrons and VLDL by lipoprotein lipase of capillary endothelium (LPL). The muscle enzyme, particularly heart LPL, has a high affinity (low K_m) for both lipoproteins and functions even at low concentrations of chylomicrons and VLDL. LPL from the capillaries of adipose tissue has a higher K_m and is more active when these lipoprotein levels are high in blood.

β-Oxidation and fatty acid synthesis. Factors that promote fatty acid synthesis, inhibit oxidation (and vice versa). These exist to prevent the operation of futile cycles. The main regulatory enzyme in the pathway of fatty acid oxidation is carnitine acyltransferase I, which is an intermediate of fatty acid synthesis that is allosterically inhibited by malonyl-CoA. Thus, an increase of malonyl-CoA blocks the entry of fatty acids into mitochondria, where β-oxidation takes place.

After a meal rich in carbohydrates, liver fatty acid synthesis is activated and oxidation is depressed.

Acetyl-CoA carboxylase, which catalyzes the conversion of acetyl-CoA to malonyl-CoA, is the most important regulatory site of the synthetic pathway. Its activity is modulated by phosphorylation–dephosphorylation reactions. Phosphate addition, catalyzed by the protein kinase dependent on cyclic AMP or AMP (AMPK), reduces its activity. The activation is catalyzed by protein phosphatase, stimulated by insulin that, on the other hand, induces synthesis of acetyl-CoA carboxylase. This enzyme is also activated allosterically by citrate, an effect that is reversed by the presence of long-chain acyl-CoA (palmitoyl-CoA).

Resting muscles use fatty acids as fuel. When the substrate supply is plentiful, glucose oxidation is inhibited. β-Oxidation produces acetyl-CoA and NADH, both of which are PDH inhibitors. When the production of ATP by oxidation of fatty acids is sufficient to meet the needs of the muscle, the glycolytic pathway is inhibited.

When muscle performs intense work, there is a decrease of ATP and an increase of AMP. This inhibits the acetyl-CoA carboxylase, phosphorylated by AMPK. Subsequent reduction of malonyl-CoA tends to stimulate acylcarnitine transport into mitochondria and promotes β-oxidation, with consequent ATP production.

In well-fed individuals (low glucagon/insulin ratio), acetyl-CoA carboxylase in liver is active and malonyl-CoA levels are high. Under these conditions, the synthesized fatty acids are preferably incorporated into TAG instead of being oxidized in mitochondria, or converted into ketone bodies.

Other insulin actions include the induction of fatty acid synthase synthesis and synthesis of NADPH generating enzymes: malic enzyme, G-6-P, and 6-phosphogluconate dehydrogenases.

Regulation of Cholesterol Biosynthesis

Some regulatory steps in cholesterol synthesis have been mentioned (p. 361). The level of cholesterol in cells and several hormones, including insulin, glucagon, triiodothyronine (T_3), and cortisol, are factors that regulate cholesterol synthesis.

The primary mechanisms involved are covalent modification and control of gene transcription of different enzymes involved in cholesterol metabolic pathways. The most important step in the regulation of cholesterol synthesis is the conversion of 3-hydroxy-3-methylglutaryl-CoA to mevalonate, catalyzed by *3-hydroxy-3-methylglutaryl-CoA reductase* (HMG-CoA reductase). The activity of this enzyme is modulated by covalent modification (phosphorylation–dephosphorylation), with phosphorylation depressing the enzyme. Glucagon promotes inactivation of the HMG-CoA reductase through kinase-dependent phosphorylation. Instead, insulin, via dephosphorylation catalyzed by a phosphatase, enhances the activity of this reductase. As this hormone stimulates glycolysis, PDH, and the pentose phosphate pathway, it increases acetyl-CoA and NADPH, augmenting cholesterol synthesis.

Furthermore, cholesterol regulates HMG-CoA reductase activity in the cell. Increases in cholesterol depress the activity of the enzyme and its transcription. The gene corresponding to HMG-CoA reductase is controlled by a family of proteins called *separator of regulatory element-binding proteins* (SREBP). When intracellular cholesterol levels are high, SREBP are inactive; if cholesterol decreases, the N-terminal domain of SREBP is cleaved by hydrolysis and this induces the production of HMG-CoA reductase. In addition, thyroid hormone induces and cortisol represses the synthesis of this enzyme. HMG-CoA reductase has a short half-life (2 h); therefore, the actions of the enzyme are exerted for a brief period of time. Cholesterol levels in the cell also control the synthesis of LDL receptors. When the intracellular cholesterol rises, the presence of receptors in the membrane and rate of LDL uptake diminishes.

Other regulatory actions on cholesterol metabolism are exerted posttranslationally (see Chapter 24), particularly via ubiquitination, which marks proteins for its degradation. Key enzymes in cholesterol metabolism, such as [3-hydroxy-3-methylglutaryl-CoA reductase (HMGCR) and squalene monooxygenase (SM)], proteins involved in LDL (LDL receptors) and cholesterol uptake (ABC transporters) into cells, are regulated by ubiquitination and proteosomal hydrolysis (p. 368).

When the cholesterol level is low, HMGCR, SM, and LDLR are deubiquitinated to increase the synthesis and uptake of cholesterol; ABCA 1 and ABCG 1 are degraded when cholesterol levels are low. Instead, when cholesterol is high, ABCA 1 and ABCG 1 are stabilized to activate cellular efflux of cholesterol, while HMGCR, SM, and LDLR are ubiquitinated and hydrolyzed.

Regulation of the Metabolism of Nitrogenous Compounds

Activity of the urea cycle is closely related to diet. With a diet rich in protein, the carbon skeletons of amino acids are a major source of energy. Similarly, in states of starvation, there is degradation of endogenous proteins, especially those constituting muscle. In both cases, there is stimulation of synthesis of all the enzymes involved in the formation of urea, which is generated in large quantities. The activity of these enzymes is higher in people with starvation or hyperproteic diets than in individuals with a balanced diet in which carbohydrates and fats are the main energy suppliers. Rapid adjustment of the urea cycle operation is accomplished by regulation of a key enzyme, carbamoyl phosphate synthetase 1, which is allosterically activated by *N*-acetylglutamate, synthesized by *N*-acetylglutamate synthase.

Regarding amino acid metabolism, there are usually frequent examples of enzyme inhibition by the end product in the synthetic pathways. This has primarily been studied in bacterial systems, and the same control was later shown in mammalian tissues.

As presented in the respective chapters, heme, purine, and pyrimidine biosynthesis present examples of feedback regulation. In vitro studies indicate that the final product inhibits the enzyme that catalyzes the first step of the pathway. In the body, interactions are much more complex and there are probably other control mechanisms which are yet unknown.

Metabolic Interactions

The "cycles" of glucose–fatty acid and glucose–alanine will be described as examples of the cross talk that exist between metabolic pathways.

Glucose–fatty acid cycle. This, also called *Randle cycle*, refers to the significant reduction in the uptake and utilization of glucose that occurs in muscle when fatty acid oxidation is intense. This phenomenon is explained as follows: the oxidation of fatty acids yields acetyl-CoA from which citrate is generated by action of citrate synthase. High values of acetyl-CoA/CoA and NADH/NAD$^+$ ratios stimulate PDH kinase, which phosphorylates and inactivates PDH. ATP and citrate inhibit phosphofructokinase, with accumulation of substrates from previous stages, including G-6-P, which has an inhibitory effect on hexokinase. This depresses glycolysis.

This glucose–fatty acid cross talk is also observed in adipose tissue. When blood glucose is high, the pancreas secretes insulin. This hormone depresses lipolysis and stimulates lipogenesis, reducing the concentration of free fatty acids in plasma. After a meal, when glucose and insulin levels are elevated, the muscle tends to predominantly use glucose instead of fatty acids. In contrast, during periods between meals, blood glucose is present at baseline levels, insulin secretion decreases, and the concentration of free fatty acids in plasma increases. This mechanism allows preserving glucose for tissues that cannot use fatty acids and depend on glucose for energy.

Glucose–alanine cycle. This cycle, which operates between muscle and liver, is important to maintain blood glucose levels during the periods between meals. In muscle, glucose yields pyruvate through the glycolytic pathway. This compound may receive an amine group by transamination with glutamate to form alanine,

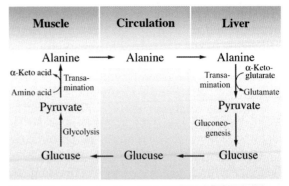

FIGURE 19.9   **Glucose–alanine cycle.**

which enters the bloodstream and is taken up by the liver. In this organ, alanine is converted to pyruvate by transamination with α-ketoglutarate, from which glucose can be generated through gluconeogenesis. Glucose formed in the liver returns to the circulation, where it is taken up by the tissues, including muscle, to close the cycle (Fig. 19.9).

Regulation of Cellular Oxidations

The electrons donated by the oxidized substrates are channeled to the respiratory chain, comprised by a series of acceptors in the inner mitochondrial membrane (p. 184). The transfer of electrons in this chain is accompanied by the pumping of protons out of the mitochondrial matrix. The return of the protons to the mitochondrial matrix through the F_0F_1 complex generates the energy needed to produce ATP from ADP.

In mitochondria, oxidation and phosphorylation are coupled in such a way that the transport of electrons cannot occur without concomitant phosphorylation of ADP. The presence of ADP is a prerequisite for cellular oxidations. Their activity is highly dependent on the relative concentration of adenylates. Within each cell, the sum of adenylates (AMP + ADP + ATP) is almost constant. Endergonic processes are generally performed at the expense of ATP hydrolysis and generate ADP, AMP, and P$_i$. ADP and AMP are then rephosphorylated to regenerate ATP. In some tissues, there is an adenylate kinase that catalyzes the following reaction in both directions:

$$\text{AMP} + \text{ATP} \xrightleftharpoons[]{\text{Adenylate kinase}} 2\,\text{ADP}$$

This reaction can transfer a high-energy phosphate bond from ADP to AMP or obtain one molecule of ATP from two molecules of ADP.

During phosphorylation coupled to oxidation in the respiratory chain, ADP is converted to ATP. If all of the adenylate molecules in the cell were as ATP, the operation of the electron transport system would eventually stop. The constant hydrolysis of ATP produces ADP. ADP stimulates oxidation, which in turn contributes to the regeneration of ATP.

Regulatory Role of Adenylates

The amount of energy stored in the cell is directly proportional to the number of high-energy phosphate bonds in adenylate molecules.

If the cell was compared to an electrical cell, the total number of high-energy phosphate bonds corresponds to the charge of the battery in a given moment; it is at a maximum when all the adenylate molecules exist as ATP, and at a minimum if they all become AMP.

The energy charge is expressed by the value of the following ratio between the level of adenylate molecules:

$$\text{Energetic charge} = \frac{[\text{ATP}] + 1/2[\text{ADP}]}{[\text{ATP}] + [\text{ADP}] + [\text{AMP}]}$$

Under normal conditions, the energy charge value fluctuates around 0.9, indicating that a great majority of adenine nucleotides are at the state of ATP.

In addition to influencing cell respiration, adenine nucleotides function as modulators of important regulatory enzymes in various metabolic pathways, as already mentioned. This adjusts

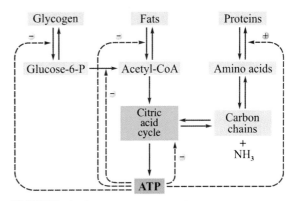

FIGURE 19.10 **Regulatory role of ATP.**

metabolism to the energy level of the cell. Metabolic systems that generate or consume ATP are sensitive to the cell energy charge (the total balance of adenylates). For an energy producer system, the activity increases as the charge decreases and becomes depressed when the charge is high. Metabolic pathways that use energy are activated when the charge is high and inhibited when it is low.

The relative increase in ATP has a negative effect on energy generating processes (glycolysis, citric acid cycle) and modulates positively energy-consuming pathways (amino acid synthesis, gluconeogenesis). Fig. 19.10 summarizes these actions.

An important effector of energy balance is an enzyme called AMP kinase (AMPK). It is activated by the reduction of cell ATP concentration, more precisely, by the increase in the ratio [AMP]/[ATP]. When activated, AMPK starts a cascade of phosphorylating events that trigger the expression of a variety of genes. As a result, metabolic pathways that consume ATP are inhibited, and those that generate ATP are activated.

In Chapter 26, when considering mechanisms of hormone action, other examples of regulatory actions on metabolism will be discussed. It is possible to infer how the concerted action of multiple factors ensures the operation of the "metabolic machinery" in harmony with the requirements of the body.

SUMMARY

Metabolic integration refers to the cross talk that exists between different metabolic pathways, which ensures the integrated function of metabolism in the body. Compounds of very diverse origin and nature may form common metabolites and products. Hexoses, glycerol, fatty acids, and some amino acids generate acetyl-CoA. Its final fate could be oxidation in the citric acid cycle to CO_2 and H_2O. Amino acids can also produce intermediates of this cycle. The respiratory chain is a common final destination of the electrons that come from many different substrates. Carbohydrates generate TAG through: (1) glucose production of acetyl-CoA, from which fatty acids are generated, and (2) the conversion of dihydroxyacetone phosphate, a glycolysis intermediate, into glycerol. Glucose also produces α-ketoacids that generate amino acids by transamination. After deamination, amino acids can form carbohydrates or ketone bodies, making them *glucogenic* or *ketogenic*. Glycerol is the only component of lipids potentially glucogenic, fatty acids are not. Some metabolites are true *"crossroads"*; different pathways converge into them and depart from them (e.g., G-6-P, pyruvate, acetyl-CoA).

Metabolic regulation is achieved via the following mechanisms:

1. *Modification of the activity of preexisting enzymes*, which is accomplished by:
   a. Changes in substrate levels [S]; at concentrations lower than the enzyme K_m, activity is directly proportional to [S].
   b. Regulatory metabolites, allosteric effectors.
   c. Covalent modification.
2. *Regulation of the number of enzyme molecules.*
   a. Control of enzyme synthesis mainly at the level of transcription.
   b. Control of enzyme rate of degradation.

Glycogenolysis is regulated through the control of *glycogen phosphorylase*. The inactive form (*b*) is converted to active (*a*) by phosphorylation (*phosphorylase kinase*). The reverse reaction or inactivation is catalyzed by *protein phosphatase*. Activation is accomplished by a reaction cascade initiated, in liver and muscle, by adrenaline and glucagon. These hormones activate *adenylate cyclase*, which produces 3′,5′-cyclic-AMP (cAMP), activates *protein kinase A*, and phosphorylates glycogen phosphorylase. The activation is abrogated by the hydrolysis of cyclic AMP to AMP by *phosphodiesterase*. In muscle, phosphorylase can be activated through another mechanism, that is, Ca^{2+}-dependent.

Another point of regulation is at the level of *phosphorylase b*, which is allosterically activated by AMP. ATP and G-6-P are negative modifiers. Phosphorylase *a* is active independently of the concentrations of ATP, AMP, or G-6-P.

Glycogenesis is mainly regulated modulating the activity of *glycogen synthase*. The inactive form of this enzyme (*b*) is phosphorylated, while the active form (*a*) is dephosphorylated. Stimulation of glycogen synthase is catalyzed by *protein phosphatase 1*; the same enzyme inactivates phosphorylase and phosphorylase kinase. Inactivation of glycogen synthase is produced by a reaction cascade, such as that activating phosphorylase. In the liver, stimulation of glycogen synthase is promoted by high levels of glucose and insulin. G-6-P and Ca^{2+} act as positive and negative effectors of GS respectively.

Glycolysis regulation is obtained by targeting *hexokinase*, which is inhibited by G-6-P, *and phosphofructokinase 1*, which is inhibited by ATP and citrate and stimulated by AMP, P_i, and F-2,6-bisP. Another enzyme susceptible of regulation is *pyruvate kinase*, which is inactivated by cAMP-dependent phosphorylation. It is inhibited by ATP and stimulated by F-1,6-bisP.

Gluconeogenesis is modulated at the level of: *G-6-P phosphatase*, which is inhibited by glucose and P_i, of *Fructose-1,6-bisP phosphatase*, which is inhibited by AMP, ADP, and F-2,6-bisP, and *pyruvate carboxylase*, activated by acetyl-CoA. In addition, expression of these enzymes is controlled by factors that, simultaneously induce glycolysis and repress gluconeogenic enzymes, or vice versa.

The Pasteur effect describes a phenomenon consisting in the decrease in glucose consumption in the presence of oxygen. Allosteric regulation of phosphofructokinase is responsible for this effect.

Oxidative decarboxylation of pyruvate is modulated via the PDH multienzyme complex. This is inactivated by phosphorylation not dependent on cAMP, and it is stimulated by a *phosphatase* activated by Ca^{2+}. Also, activation of *PDH kinase* by acetyl-CoA, ATP, and NADH regulates oxidative decarboxylation of pyruvate.

Citric acid cycle is regulated at various levels and through the following enzymes: *citrate synthase*, which is inhibited by ATP; *isocitrate dehydrogenase* (NAD dependent), which is stimulated by ADP and inhibited by ATP and NADH; *α-ketoglutarate dehydrogenase*, which is inhibited by succinyl-CoA, NADH, and ATP; *glutamate dehydrogenase*, that is, negatively regulated by ATP; and the $NADH/NAD^+$ ratio in the mitochondrial matrix, which when increased depresses the activity of NAD-dependent dehydrogenases.

Fatty acid metabolism is controlled via fatty acid metabolism, catalyzed by *lipase*. This step is activated by a cascade of events initiated by hormones (catecholamines, ACTH, glucagon) that activate *adenylate cyclase*, increases cAMP, and stimulates *protein kinase A*. Regulation of *β-oxidation* is accomplished via malonyl-CoA, which inhibits acyl transport to mitochondria acting on *acyl carnitine acyltransferase 1*. Also, via inhibition of *3-OH-acyl-CoA dehydrogenase* by NADH, and inhibition of *thiolase* by acetyl-CoA.

Fatty acid biosynthesis is mainly regulated at the level of *acetyl-CoA carboxylase*. This enzyme is inactivated by cAMP-dependent phosphorylation and stimulated by a *phosphatase*. Citrate is a positive allosteric effector and long-chain acyl-CoA act as inhibitors.

Cholesterol biosynthesis is regulated by controlling *3-OH-3-methylglutaryl-CoA reductase*. This enzyme is inactivated by phosphorylation and activated by a dephosphorylation. Increased cholesterol acts as an inhibitor.

Metabolism of nitrogenous compounds, such as the synthesis of amino acids, purines, and pyrimidines are regulated by the final product.

Cellular oxidations are adjusted by the content of nucleotides. The presence of ADP is an indispensable requirement for oxidations.

The energy charge of cells is expressed by the ratio:

$$\frac{[ATP]+1/2[ADP]}{[ATP]+[ADP]+[AMP]}$$

When this ratio value is high, energy-consuming metabolic pathways are stimulated, whereas those producing ATP are inhibited. AMP kinase is an important effector in this regulation. It is activated when the value of the ratio of adenylates decreases.

Bibliography

Agius, L., 2008. Glucokinase and molecular aspects of liver glycogen metabolism. Biochem. J. 414, 1–18.

Fell, D., 1997. Understanding the Control of Metabolism. Portland Press, London.

Jensen, T.E., Richter, E.A., 2012. Regulation of glucose and glycogen metabolism pathways during and after exercise. J. Physiol. 590, 1069–1076.

Morris, S.M., 2002. Regulation of enzymes of the urea cycle and arginine metabolism. Annu. Rev. Nutr. 22, 97–105.

Pilkis, S.J., Granner, D., 1992. Molecular physiology of the regulation of hepatic gluconeogenesis and glycolysis. Annu. Rev. Physiol. 54, 885–909.

Salter, M., Knowles, R.G., Pogson, C.I., 1994. Metabolic control. Essays Biochem. 28, 1–12.

Semenkovich, C.F., 1997. Regulation of fatty acid synthase (FAS). Prog. Lipid Res. 36, 43–53.

Sharpe, L.J., Cook, E.C.L., Zelcer, N., Brown, A.J., 2014. The UPS and downs of cholesterol homeostasis. Trends Biochem. Sci. 39, 527–536.

Van der Heiden, M.G., Cantley, L.C., Thompson, C.B., 2009. Understanding the Warburg effect: the metabolic requirements of cell proliferation. Science 324, 895–899.

Metabolism in Some Tissues

The metabolic routes described in previous chapters take place in almost all cells of the body; however, due to their functional specialization, each tissue exhibits differences in nutrient utilization and metabolism. This chapter discusses the specific metabolic pathways or "metabolic profile" of various tissues and organs.

LIVER METABOLISM

Most of the digestion products absorbed in the intestines move to the portal vein and are first received by the liver, which is responsible for the initial metabolism of nutrients. To handle the discontinuous supply of nutrients, which also vary in quantity and quality, the liver is endowed with high metabolic plasticity. This is necessary to maintain a variety of metabolites at constant levels in blood, and their steady supply to all tissues.

To cope with changes in its glycogen and protein content, the activity of specific enzymes in the liver is upregulated to adapt different metabolic pathways to the availability of substrates. The following sections summarize the major metabolic transformations that nutrients and other substances undergo in the liver.

Carbohydrates

Approximately two-thirds of all monosaccharides absorbed in the intestine are captured by the liver, the remaining one-third continues in the general circulation. The glucose arriving via the portal vein enters the hepatocytes by a facilitated diffusion process. Most of the glucose carriers in liver are of the GLUT2 type, characterized by low affinity for glucose and not influenced by insulin, which allows the steady uptake of this sugar following the gradient across the hepatocyte membrane. Once in the cell, glucose is converted to glucose-6-phosphate by glucokinase. This enzyme specifically uses glucose as a substrate, for which it has a high K_m. So its activity is significantly upregulated when glucose levels rise above baseline immediately after a meal rich in carbohydrates. Conversion into glucose-6-P "traps" glucose in the cell because the membrane is impermeable to this phosphorylated metabolite. Since the reaction consumes free glucose, its gradient is maintained and allows for the continuous entry of glucose into the cell.

Fructose and galactose are absorbed in the intestine, metabolized in the liver, and converted into the same intermediate products as those formed in the glucose utilization pathways (i.e., fructose is phosphorylated into fructose-6-P, which is then converted into glucose-6-P by phosphoglucomutase or into fructose-l-P, which can then be split into glyceraldehyde and dihydroxyacetone phosphate. These metabolites enter the glycolytic pathway or are used for synthesis of glucose through gluconeogenesis).

Galactose is converted to UDP-glucose and is used in the synthesis of glycogen.

Glucose-6-P is an intermediate compound at a metabolic crossroad and can follow several different pathways. These include:

Glycogenesis. Glucose-6-P is stored as glycogen in the liver after conversion into glucose-1-P and UDP-glucose. Glycogen reserves in well-nourished individuals are 5%–10% (75–150 g) of the total organ weight.

Degradation. Glucose-6-P is converted to pyruvate by entering the glycolytic pathway. Pyruvate is decarboxylated to acetyl-CoA, which is oxidized to CO_2 and H_2O in the citric acid cycle. This oxidative pathway generates ATP. However, glucose is not the most important source of energy in the liver; fatty acids are the main fuel.

Pentose phosphate pathway. This pathway is active in the liver. It generates NADPH necessary for the synthesis of fatty acids, cholesterol, and ribose-5-P, used in the synthesis of nucleotides.

Glucose release. The glycogen stored in the liver is degraded to glucose-6-P. Glycogenolysis and gluconeogenesis are both processes by which the liver generates glucose-6-P. These pathways are under tight regulation and are adjusted to the body's needs (pp. 289 and 308).

The existence of glucose-6-phosphatase in liver allows for the hydrolysis of glucose-6-P and production of free glucose, which can then enter the circulation. This makes the liver the main supplier of glucose to the body, which is essential during fasting periods between meals.

Gluconeogenesis. Lactate is produced mainly by degradation of glucose in muscle during intense exercise (anaerobic conditions). It is released to the bloodstream and taken up by the liver, where it is oxidized to pyruvate and may be converted to glucose through gluconeogenesis. Conversion of lactate to glucose requires 6 moles of ATP per mole of glucose. The glucose formed is released into the bloodstream and eventually returns to the muscle, completing the Cori's cycle (p. 311).

Lipogenesis. The glucose surplus not stored as glycogen and not diverted into other pathways is used in the synthesis of fatty acids, triacylglycerols, complex lipids, and cholesterol after conversion to acetyl-CoA.

The glycerol-3-phosphate required for the synthesis of phosphatidic acid, an intermediate in the process of triacylglycerol and glycerophospholipid formation, is generated by hydrogenation of dihydroxyacetone phosphate, a metabolite of the glycolytic pathway. The reaction is catalyzed by glycerol phosphate dehydrogenase. Fig. 20.1 summarizes the described processes.

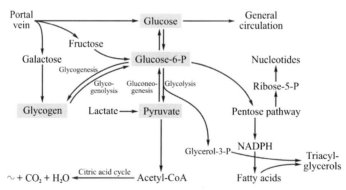

FIGURE 20.1 **Metabolic pathways of carbohydrates in the liver.**

Lipids

Different from glycerol and fatty acids of less than 10 carbon atoms in length, which arrive to the liver by the portal venous system, simple and complex lipids reach the liver via the systemic circulation.

Triacylglycerol (synthesized in cells of the intestinal mucosa from fatty acids) and mono- and diacylglycerols (resulting from digestion of neutral fats), are incorporated into chylomicrons and released first to the lymphatic system and then to general circulation. Once in the blood, the lipid particles lose most of their triacylglycerol content in peripheral vessels by the action of lipoprotein lipase. Chylomicron remnants are taken up and metabolized by the liver.

Intermediate density lipoproteins (IDL) and low density lipoproteins (LDL), rich in cholesterol, bind to specific receptors in the plasma membrane of hepatocytes and are internalized by endocytosis. In the cells, the cholesterol released is used for the synthesis of bile acids or is incorporated into membranes. The excess cholesterol is esterified and stored in hepatocytes.

The main metabolic pathways for lipids in the liver are as follows:

Fatty acid oxidation. Fatty acids are the main source of energy in the liver. Degradation of fatty acids into acetyl-CoA through β-oxidation and final conversion into CO_2 and H_2O in the citric acid cycle yields a substantial amount of ATP by oxidative phosphorylation.

Biosynthesis of fatty acids. The fatty acids present in liver originate from: (1) hydrolysis of triacylglycerols that come from chylomicron remnants taken up by the hepatocytes, (2) free fatty acids that arrive via blood bound to albumin, and (3) synthesis in the liver from acetyl-CoA, supplied by the degradation of fatty acids, glucose, and amino acid carbon chains.

Fatty acids, activated as acyl-CoA, are transferred to positions l and 2 of glycerol-3-P to form phosphatidic acid. This compound is a precursor of triacylglycerol and glycerophospholipids.

Glycerol-3-phosphate is formed in the liver by reduction of dihydroxyacetone phosphate (a glycolytic intermediate) and by phosphorylation of the glycerol generated by hydrolysis of glycerolipids. It is also provided by the circulation via absorption in the intestine or by the action of lipoprotein lipase.

Triacylglycerols, together with apoprotein B100 synthesized in liver, are incorporated in very low density lipoproteins (VLDL) and sent to the general circulation.

Acetyl-CoA generated by degradation of fatty acids and glucose is used in the synthesis of cholesterol. Cholesterol is the precursor of bile acids, which play an important role in digestion and absorption of lipids in the intestine.

The liver also synthesizes phospholipids, which are used for the membrane of hepatocytes and for export, along with free cholesterol, and the proteins apo A, and apo C forming the discoid clusters of nascent high density lipoproteins (HDL).

Ketone body formation. Acetyl-CoA not oxidized in the citric acid cycle, or not used in the synthesis of other compounds, forms ketone bodies, mainly acetoacetate and 3-hydroxybutyrate. These enter the bloodstream and are used by extrahepatic tissues as fuel. While the liver produces ketone bodies, it cannot use them as an energy source because it lacks the enzyme needed to activate acetoacetate, the succinyl-CoA-ketoacid-CoA transferase (thiophorase). Fig. 20.2 summarizes the mentioned processes.

Amino Acids

Amino acids absorbed in the intestine are taken up by the liver from the portal vein. Unlike glucose and triacylglycerols, amino acids are not stored in the liver.

Protein biosynthesis. Free amino acids are used for protein synthesis in the liver. The process is very active not only because of the rapid

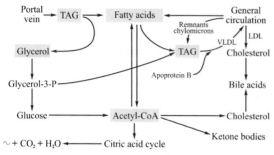

FIGURE 20.2 **Metabolic pathways of lipids in the liver.** *TAG,* Triacylglycerols; *LDL,* low density lipoprotein; *VLDL,* very low density lipoproteins.

turnover of the proteins in this organ, but also because the liver synthesizes most of the plasma proteins.

Amino acid degradation. Amino acids not used for the synthesis of proteins and other nitrogenous products (purines, pyrimidines) are deaminated. Enzymes involved in the separation of the amine group, such as transaminases and glutamate dehydrogenase, exhibit high activity in the liver. They are induced by the presence of substrate or hormones, such as glucocorticoids.

The carbon chains of amino acids, mainly of those unbranched, are used by the liver as an energy source. These chains are converted to acetyl-CoA or intermediates of the citric acid cycle.

The liver has great gluconeogenic activity. Glucose generated from amino acids in the liver is released into the circulation and is an important factor in the maintenance of blood glucose when the carbohydrate intake is low.

Glucose-alanine cycle. This metabolic cycle is established between muscle and liver and helps to maintain blood glucose in the periods between meals (p. 442).

Urea formation. The amine groups transferred in transamination reactions and released as ammonia (NH_3) in deaminations, mainly from glutamate, enter the urea cycle. The liver is the only organ that has all the necessary enzymes to convert the toxic product NH_3 into urea, a nontoxic, highly water-soluble compound that can be excreted easily by the kidneys. A normal adult produces 20–30 g of urea per day, most of which is excreted by urine. Fig. 20.3 summarizes the pathways involved in amino acid metabolism in liver.

Biotransformation

An organism is exposed to a number of compounds with structures that are different to that of usual nutrients. These include drugs administered for therapeutical purposes or environmental and food contaminants, such as waste chemicals produced by industries (pesticides, herbicides, food preservatives, and coloring compounds). Some of these compounds possess carcinogenic potential. Combustion products of tobacco, for example, are harmful and should be included in this group of exogenous substances.

The compounds that the body cannot use as energy source metabolites or for the synthesis of own components, are called *xenobiotics.* Generally, the body modifies and eliminates substances

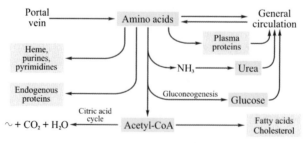

FIGURE 20.3 **Metabolic pathways of amino acids in the liver.**

that have toxic effects. These detoxifying mechanisms produce, in some cases, products that are equally, or more, toxic than the original compound. Therefore, it is better to refer to these processes as *biotransformation*.

The liver is the main site where biotransformation occurs. The reactions involved in the metabolism of foreign substances are catalyzed by enzyme systems associated with the endoplasmic reticulum of hepatocytes. In general, biotransformation of drugs and other foreign compounds tends to inactivate them, rendering products with little or no biological activity, which do not harm the cells. Moreover, biotransformation generates products that are more polar, hardly enter cells, and are more efficiently excreted by the kidneys or intestine than the original compound.

In spite of their wide structural variety, xenobiotics are subjected to the same general reactions. These include oxidation, conjugation, reduction, and hydrolysis. A substance may undergo one or more of these changes.

Oxidation. This is the most common and usually the first reaction that foreign substances undergo in the body. Many aliphatic compounds are completely oxidized and their carbon atoms are eliminated as CO_2. Aromatic substances and steroids are initially hydroxylated. These reactions constitute what is known as phase I of xenobiotic metabolism.

Oxidations are catalyzed by nonspecific oxidases that are members of an enzyme system associated to the endoplasmic reticulum known as the *microsomal mixed function oxidative system*, or *oxidizing system*. This system requires NADPH and molecular oxygen. The diatomic oxygen molecule is divided by the introduction of two electrons from NADPH; one oxygen atom is incorporated into the substrate and the other forms water. The primary source of electrons is NADPH via *NADPH-cytochrome P_{450} reductase* (a flavoprotein with FAD and FMN), which drives the movement of electrons, one at a time, to cytochrome P_{450}. Cytochrome P_{450} is a heme protein of 45 kDa (P stands for pigment and 450 refers

to the wavelength in nanometers at which the absorbency of the compound is the greatest). In humans, numerous genes encoding cytochrome P_{450} have been identified.

The mixed function oxidizing system is inducible. Its activity markedly increases in the presence of substrates. The upregulation of activity is due to stimulation of the synthesis of cytochrome P_{450} and other members of the system. Barbiturates are an example of drugs that induce synthesis of enzymes of the oxidizing system. In patients chronically treated with barbiturates, the dose must be increased as the treatment continues, to get the same effect. The patients also acquire greater ability to inactivate other drugs.

Alcohol is another foreign compound partially metabolized by the microsomal system. Chronic alcoholics have a greater microsomal oxidative system activity and metabolize certain drugs more rapidly than individuals that do not consume alcohol. Higher doses of sedatives or anesthetics are required to obtain the same effect in alcoholics than in normal individuals.

Moreover, different drugs that act as substrates of the cytochrome P_{450} oxidating system show competitive interactions. This phenomenon is important and clinically relevant. When several compounds are administered simultaneously, one of them can compete for the oxidation system with another. Thus, the metabolism of each individual compound is decreased, delaying the overall detoxification process. This event explains why the risk of toxicity is greatly increased when an individual who has been taking alcohol ingests drugs. Many fatal accidents occur by the consumption of sedatives or other drugs after heavy alcohol ingestion.

Conjugation. A xenobiotic which already has been modified by oxidation or hydrolysis is usually combined with a natural compound. The compounds most frequently used as conjugation agents are glucuronic acid, glycine, cysteine, ornithine, glutamate, acetate, sulfate, and methyl. This conjugation step is known as phase II of xenobiotic metabolism.

Glucuronic acid participates in conjugation reactions in its active form, UDP-glucuronic acid. Binding to other substances is catalyzed by *UDP-glucuronyl transferase*, an inducible enzyme.

Conjugation with glucuronic acid is not only a reaction for the transformation of foreign substances, but also a frequent reaction in the metabolism of endogenous compounds, such as bilirubin, steroid hormones, and others.

Conjugation with sulfate is made by transfer from phosphoadenosine phosphosulfate (PAPS), catalyzed by sulfokinase. Acetylation, another biotransformation process, uses active acetate (acetyl-CoA).

Conjugation with glycine has been mentioned in previous chapters (pp. 262 and 393).

Reduction and hydrolysis. These are two additional reactions performed by the liver. However, they are less common than oxidation and conjugation.

Excretion. The final stage by which the cells eliminate the products of the metabolized xenobiotics is known as phase III of xenobiotic metabolism. In general, the hydrophilicity of the biotransformed metabolites is higher than that of the original compounds; because of this, they require carriers to cross membranes. One of these carriers is the *multiple drug transporter protein*, an ABC membrane transporter (p. 238).

Individual variability. There are genetic variants of enzymes, transporters, and receptors involved in biotransformation. For this reason, not all individuals respond to treatment with a particular drug in the same manner, and can even present with adverse effects to the drug. *Pharmacogenetics* is the area that studies this metabolic variability among individuals.

Ethanol Metabolism

Alcohol consumption is a common practice in humans. Normal individuals have the ability to metabolize up to 100 mg of ethanol per kilogram of body weight and convert it into harmless products. For this reason, an amount of ≤25 g/day of alcohol (equivalent to two small glasses of wine) can normally be properly handled by the body. In contrast, acute or chronic consumption of higher quantities of ethanol can cause severe alterations. Understanding the metabolism of ethanol helps interpreting the harmful effects of alcoholism.

Complete oxidation of ethanol to CO_2 and H_2O produces 7.1 kcal (29.7 kJ) of energy per gram, a much higher energy efficiency than carbohydrates and proteins (about 4 kcal or 17 kJ/g). However, this does not make alcohol an adequate natural nutrient.

Absorption. Most of the ingested ethanol is absorbed in the duodenum and jejunum; a lesser quantity is also absorbed in the gastric mucosa, and other portions of the intestine. This takes place by passive diffusion. Alcohol is a small molecule, soluble in both water and lipids, which allows it to readily cross membranes and to distribute in all body compartments. Absorption is complete within 2 h after ingestion; the presence of food in the stomach delays gastric emptying and slows ethanol absorption.

Excretion. Only 2%–10% of the absorbed ethanol is eliminated unchanged in urine, sweat, and expired air, the rest is metabolized.

Metabolism. The liver is the main organ responsible for ethanol metabolism. Alcohol is metabolized by several steps of oxidation. In the first two stages, acetaldehyde and acetate are produced.

First stage. Three metabolic alternatives are possible, which involve the activity of: (1) alcohol dehydrogenase, (2) the oxidative microsomal system, and (3) catalase. All these systems give acetaldehyde as a product.

1. *Alcohol dehydrogenase*, a NAD^+-linked oxidoreductase, catalyzes the reaction:

$$CH_3-CH_2OH + NAD^+ \rightleftharpoons CH_3-C{\overset{O}{\underset{H}{}}} + NADH + H^+$$

Ethanol   Acetaldehyde

Alcohol dehydrogenase (ADH), found in the hepatocyte cytosol, is primarily responsible for the initial oxidation of ethanol in humans. In normal individuals, over 90% of the absorbed ethanol follows this pathway. ADH is a zinc metalloenzyme, which is composed of two polypeptide chains of 40 kDa each. The active enzyme exists as multiple isoforms that result from a homo- or heterodimeric association of seven different subunits, which are controlled by different genes and have different kinetic properties. Genetic variants of some of the subunits are found with different frequencies in different racial groups. ADH has broad substrate specificity; it oxidizes a variety of alcohol molecules, including aliphatic primary, secondary, tertiary, and some cyclic compounds.

There is also ADH activity in the stomach. Oxidation of ethanol in gastric mucosa is considered the "first step"; when small amounts of alcohol are ingested, this provides a protective barrier that reduces the amount of alcohol that arrives the liver.

2. *Ethanol microsomal oxidizing system* (EMOS). This is a monooxygenase or mixed function oxidase. It uses NADPH, cytochrome P_{450}, and O_2. The EMOS cytochrome P_{450} is called CYP2E1 or, more simply, 2E1. This enzyme catalyzes not only the conversion of ethanol to acetaldehyde, but also uses many other hepatotoxic agents as substrates. CYP2E1 is an inducible system located in the endoplasmic reticulum membrane of hepatocytes. In individuals who consume ethanol chronically, its activity is upregulated, being 4–10 times higher than in nonalcohol consuming individuals. In normal individuals, the EMOS contribution to alcohol oxidation is low (less than 10%). In contrast, it may

account for ~30% or more of the total ethanol metabolized in chronic alcoholics. Induction of EMOS explains the increased ability of alcoholics to oxidize various drugs.

3. *Catalase.* This enzyme, located in peroxisomes, is capable of oxidizing ethanol if a system that generates hydrogen peroxide is available. The peroxidase reaction is represented by the following equation:

$$CH_3-CH_2OH + H_2O_2 \rightleftharpoons CH_3-C\!\!\begin{array}{c}O\\H\end{array} + 2\,H_2O$$

Ethanol   Acetaldehyde

Under physiological conditions, this reaction does not play a significant role in ethanol metabolism.

Second stage. Acetaldehyde, a product of ethanol oxidation, is highly toxic. Normally the liver converts it rapidly to acetate by the action of *acetaldehyde dehydrogenase* (ALDH), which is NAD^+ dependent:

$$CH_3-C\!\!\begin{array}{c}O\\H\end{array} + NAD^+ + H_2O \longrightarrow$$

$$\longrightarrow CH_3-C\!\!\begin{array}{c}O\\O^-\end{array} + NADH + H^+$$

There are two isozymes of acetaldehyde dehydrogenase (ALDH), one localized in mitochondria and the other in the cytosol. The first, with higher affinity for acetaldehyde, is the most active.

A genetic defect responsible for the lack of activity of mitochondrial ALDH has been described. It is relatively common in individuals of Mongolian race. Patients are intolerant to ethanol and ingestion of even little amounts of alcohol triggers symptoms of acetaldehyde accumulation.

Final acetate oxidation. The acetate produced in the reaction catalyzed by ALDH enters the blood to be taken up by different tissues, mainly muscle, and is activated by binding to coenzyme A. This product can then follow different pathways, including oxidation to CO_2 and H_2O in the citric acid cycle.

Alternative metabolism. Another pathway for the metabolism of ethanol, which does not involve oxidation, has been described. This involves the esterification of the alcohol with carboxylic acids to form ethyl-fatty acid esters, a reaction catalyzed by *ethyl-fatty acid ester synthetase*, which is particularly active in liver and pancreas.

Metabolic consequences of high ethanol consumption. Excessive consumption of ethanol produces an increase in reducing equivalents in hepatocytes, primarily NADH. This causes a shift in redox equilibrium, with an increase in the $NADH/NAD^+$ and lactate/pyruvate ratios. The major metabolic effects of this change include the following: (1) *increase of lactic acid* and *ketone bodies* (specially β-hydroxybutyrate), which leads to acidosis. (2) *Hyperuricemia*, due to the competition of lactic acid with uric acid for its excretion in the kidney. (3) *Reduction of the citric acid cycle activity*, secondary to the lower availability of NAD^+. (4) *Accumulation of fatty acids*, due to inhibition of fatty acid oxidation in mitochondria, secondary to the overload of ethanol-derived reducing equivalents. On the other hand, the reoxidation of the excess NADH in mitochondria determines the generation of reactive oxygen species and oxidative stress. (5) *Increased hepatic lipogenesis*, which is secondary to the increase in the production of glycerol-3-phosphate from dihydroxyacetone phosphate, induced by the high NADH levels. The increased supply of glycerol-3-phosphate and fatty acids promotes the synthesis of triacylglycerols. (6) *Overload of shuttle systems.* The excess of cytosolic NADH from ADH and ALDH reactions saturates the mitochondrial hydrogen transfer systems. Among other effects, glycolysis is depressed. (7) *Production of fatty liver.* In early stages, while

the liver is not affected, the synthesis of VLDL is increased and triacylglycerides are exported. When liver function is altered, fats begin to accumulate. (8) *Effects on blood sugar.* In chronic alcoholics with a normal diet, hyperglycemia is usually observed. This is mainly due to reduced glucose utilization and stimulated glycogenolysis, caused by increased catecholamine release. In contrast, severe hypoglycemia sometimes occurs in acute alcohol poisoning, particularly in malnourished people without glycogen stores. The inhibition of gluconeogenesis also contributes to hypoglycemia. (9) *Effects on the nervous system.* Chronic consumption of ethanol perturbs some neurochemical functions and may produce structural damage. Main effects are exerted on white matter, producing brain atrophy, especially of the frontal lobes, which explains the alterations of cognitive functions that patients suffer.

Acetaldehyde is the main metabolite that interferes with neurotransmitters like glutamate, γ-amino butyrate, and others. The behavioral effects are in part due to disturbance of signal transduction pathways and ion channels. Acute alcohol intoxication may produce cerebral edema and vascular congestion, and, occasionally, focal subarachnoid hemorrhage.

Acetaldehyde toxicity. The deleterious action of ethanol is responsible, in part, to the product formed in the first oxidation stage, which complexes with proteins. Acetaldehyde inactivates enzymes, decreases DNA repair, and promotes formation of antibodies against the complexed proteins. Also, it favors reduction of glutathione and facilitates the action of free radicals and lipid peroxidation.

Competitive inhibition of acetaldehyde dehydrogenase by disulfiram (*antabuse*) has been used in the treatment of alcoholics. The administration of disulfiram produces increased levels of acetaldehyde after ethanol intake and thus elicits the unpleasant symptoms induced by this metabolite, including: peripheral vasodilation, skin flushing, nausea, headache, vomiting. The purpose of this treatment is to induce alcohol aversion in the subject.

Ethanol tolerance. In chronic alcoholics, ethanol tolerance has been attributed to adaptations of the central nervous system (CNS). In addition, the induction of EMOS activity is important and explains the development of increased tolerance to many drugs. Since ethanol and some drugs compete for the same system, the main effect of an acute ethanol load is the inhibition of drug metabolism, decreasing the drug disposal rate, prolonging its permanence in the body, and enhancing its action.

Cytochrome P_{450} 2E1 converts xenobiotics into toxic metabolites. Industrial solvents (bromobenzene, vinylidene chloride), anesthetics (enflurane), drugs (isoniazid, phenylbutazone, cocaine), and over-the-counter analgesics (acetaminophen) are substrates and inducers of 2E1, which metabolizes them into toxic products. Due to increased activity of EMOS, chronic alcoholics are at increased risk of liver damage due to accumulation of these xenobiotics. The simultaneous administration of these products and ethanol, in addition to nutrient deficiencies common in alcoholics, potentiate the effects. Decreased levels of reduced glutathione (GSH) and increased production of free radicals are common. Chronic ethanol consumption increases oxidative stress and impairs liver mitochondria. The reactive oxygen species formed promote enzyme inactivation and lipid peroxidation, which are factors that contribute to fibrosis and development of liver cirrhosis. Alcohol abuse has also been associated with increased incidence of upper respiratory and digestive tract cancer. The increased activity of cytochrome P_{450} 2E1 is responsible for the production of carcinogens. Another contributing factor is the lack of vitamin A.

The nonoxidative pathway produces ethyl fatty acid esters (EFA) that, in chronic alcoholics, accumulate in membranes (plasma and organelles) and destabilize them.

The EFA and toxic metabolites generated by the activity of CYP2E1 may activate some *MAP* kinase systems, such as *ERK* (extracellularly regulated kinase), *JNK* (N-terminal c-Jun kinase), and the *κB* nuclear factor (*NFκB*). They may also involve the peroxisome proliferator–activated receptor (PPAR) (Chapter 25). These actions affect transcription factors and eventually cause disruption of gene activity that some authors have related to the production of liver inflammation and steatosis.

Methanol poisoning. Methanol is a compound that causes very severe poisoning, particularly by ingestion of drinks adulterated with this substance. Methanol is oxidized by the liver alcohol dehydrogenase, which catalyzes the first step of ethanol metabolism. The product formed is formaldehyde, which is much more toxic than acetaldehyde. It causes serious damage of the nervous system and leads to changes in the optic nerve resulting in blindness. One of the therapeutic approaches for methanol intoxication at its early stages is ethanol administration. Ethanol competes with methanol for ADH, reducing the amount of formaldehyde formed and replacing it with acetaldehyde (less toxic).

SKELETAL MUSCLE METABOLISM

Muscle is a tissue with the capacity to convert chemical energy into mechanical energy. Its metabolism is predominantly geared to generate ATP, needed to produce work. ATP is not only needed for muscle contraction, but also for muscle relaxation. From the total ATP consumed in the body, approximately 30% is used by skeletal muscle at rest. During intense exercise, the total ATP consumed can raise up to ~90%. Unlike other tissues, skeletal muscle is characterized by its discontinuous activity; it can rapidly switch from rest to full contraction during exercise. To cope with this discontinuous action, muscle metabolism is adapted to provide the amounts of ATP required by the tissue.

Exercise not only involves the increase in activity in muscle, but also in other organs. For example, the heart and respiratory system increase their function to sustain the blood perfusion and oxygen demands of muscle; moreover, the liver and adipose tissue provide the substrates to be used for energy production.

The resting muscle receives fatty acids bound to serum albumin, glucose, and ketone bodies from the circulation. These substrates are oxidized to meet basal energy requirements (resting) and to maintain reserves of ATP and creatine phosphate (p. 395). Glucose is stored as glycogen, which comprises up to 1% of the total weight of the muscle. Also, a moderate reserve of triacylglycerols is deposited in muscle.

During work, muscle metabolism varies depending on the type of exercise, which can be: (1) brief and very intense exercise (maximum work cannot be sustained for more than 3 min), or (2) submaximal, which can be maintained for long periods (hours). In the first case, the O_2 supply and oxidative phosphorylation are not sufficient to sustain the high consumption of ATP caused by muscle contraction. Consequently, intense exercise requires anaerobic ATP production and the use of the muscle's own glycogen reserves. In contrast, during submaximal exercise,

contraction is maintained for a longer time because ATP is generated aerobically, which is more efficient to sustain muscle contraction. Thus, athletes competing in 100–200 m races, which require maximum effort during brief periods of time, almost exclusively generate ATP anaerobically. In contrast, during long races, as seen in marathons (42.2 km), aerobic production of ATP is favored.

Although ATP is predominantly generated either anaerobically or aerobically depending on the type of work performed, the operation of each pathway is never exclusive; both mechanisms function simultaneously to provide the energy needed for muscle contraction. The metabolic changes that occur during maximal (anaerobic) and submaximal (aerobic) muscular work are indicated in Fig. 20.4.

Anaerobic work. Upon intense exercise, the ATP demand can increase 100–1000 times from basal levels. This energy requirement cannot be provided by oxidative phosphorylation, since O_2 and fuel supply do not increase proportionally to satisfy the muscle needs.

In the initial stages of muscle work, ATP is generated from the reserves of creatine phosphate, a

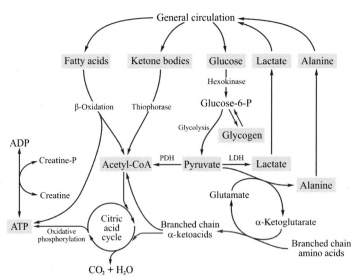

FIGURE 20.4 **Metabolic pathways in skeletal muscle.**

high energy compound whose hydrolysis has a ΔG of -10.3 kcal/mol (-43.1 kJ/mol) and can transfer phosphate to ADP to form ATP in a reaction catalyzed by *creatine phosphate kinase*:

$$\text{Creatine-P} + \text{ADP} \rightleftharpoons \text{Creatine} + \text{ATP}$$

Another reaction that contributes to the regeneration of ATP in muscle is catalyzed by *adenylate kinase*:

$$2\,\text{ADP} \rightleftharpoons \text{ATP} + \text{AMP}$$

The creatine phosphate and ATP reserves in the muscle are limited and can only provide energy for a very short time. Muscle glycogen degradation is the main, easily accessible source of substrate under anaerobic conditions. Ca^{2+} plays a main role in the coordination between contraction and glycogenolysis. Thus, the increase in intracellular Ca^{2+} levels that triggers muscle contraction by binding to troponin C, also binds to calmodulin, which stimulates phosphorylase activity and initiates glycogenolysis. In this manner, Ca^{2+} elicits muscle contraction and provides the substrate needed to support that process.

Secretion of catecholamines (adrenaline and noradrenaline) accompanying the start of a physical effort, also participates in the activation of glycogenolysis. The action of catecholamines in muscle cells is mediated by an increase in cyclic AMP levels.

Muscle contraction causes rapid decrease in ATP and increase in ADP and AMP concentrations. These changes can allosterically activate phosphorylase *b* and phosphofructokinase, which results in enhanced glycolysis.

Since during maximum effort the ATP requirements in muscle greatly exceed the rate of ATP regeneration via oxidative phosphorylation, it is necessary to utilize glucose from the muscle's own reserves. Through glycolysis, this produces two ATP and two lactate molecules per glucose molecule degraded.

Glycolysis could reach levels high enough to consume all the muscle glycogen stores. This explains why intense exercise cannot continue for more than a few minutes. Yet, the main limiting factor for contraction is the accumulation of lactate in muscle. Although blood flow increases during exercise, this is not enough to remove all the lactate formed. Local pH in the muscle drops to values close to 6.6. At this pH, phosphofructokinase is more sensitive to inhibition by allosteric effectors and glycolytic activity is reduced. In addition, the formed H^+ competes with Ca^{2+} for its binding to troponin C, displacing Ca^{2+} and decreasing contraction capacity.

Oxygen debt. During intense exercise, heart rate, muscle blood flow, pulmonary ventilation, and oxygen consumption are increased. The increase in oxygen uptake by muscle is, however, insufficient to fully maintain oxidative metabolism and satisfy the demands under strenuous exercise. To complete the oxidation of metabolites produced, an increase in oxygen consumption needs to continue after work. This sustained postexercise oxygen consumption is needed to "pay" what is known as the "oxygen debt" that the body has contracted. Most of the oxygen is used to oxidize the produced lactate. This metabolite diffuses from muscle into the bloodstream, is taken up by the liver to be oxidized to pyruvate, and finally converted into glucose (gluconeogenesis). Glucose is released by the liver into the circulation to return to the muscle, where it is used to regenerate glycogen (Cori cycle).

During exercise, there is an increase in catecholamines that promotes glycogenolysis, not only in muscle but also in liver. The amounts of these hormones return to baseline levels approximately 5–6 min after completion of work; during that time, the release of glucose from the liver continues.

Once at rest, oxidation of fatty acids, glucose, and ketone bodies (predominantly fatty acids) generates the ATP necessary to restore creatine phosphate and glycogen muscle deposits. Liver glycogen storage is replenished in part by using the lactate that arrives from muscle.

Aerobic work. When the intensity of exercise is low, the O_2 supply and oxidative phosphorylation generates sufficient ATP for muscle work. The maximum work intensity that can be maintained by the aerobic production of ATP depends on the tissue oxidation capacity. In addition, the aerobic work capacity is limited by the maximum amount of oxygen that blood can provide to the tissue and the amount that muscle can use per unit time. The maximum rate of oxygen uptake ($V_{O_2\,max}$) in young adults is normally between 40 and 50 mL of O_2/min/kg of body weight. Any exercise requiring an amount of oxygen greater than the $V_{O_2\,max}$ induces, at least partially, the generation of ATP by the anaerobic pathway, which involves consumption of muscle glycogen.

The fuel stored in muscle is limited and insufficient to sustain a prolonged work even under the highly efficient aerobic conditions. During sustained aerobic exercise, the muscle uses substrates coming from the bloodstream, such as glucose, ketone bodies (produced in the liver), or fatty acids (mobilized from adipose tissue). However, muscle also uses its own reserves of glycogen and triacylglycerols. Glucose enters the muscle mainly through GLUT4 transporters, which are increased in the plasma membrane by insulin.

From the oxygen consumption standpoint, the most efficient fuel is the one that generates the largest amount of ATP per oxygen molecule consumed. The comparison between glucose and fatty acid oxidation is shown here.

$$C_6H_{12}O_6 + 6O_2 \longrightarrow 6CO_2 + 6H_2O + 38\,ATP$$
Glucose

$$C_{18}H_{36}O_2 + 26O_2 \longrightarrow 18\,CO_2 + 18\,H_2O + 146\,ATP$$
Stearic acid

Energy generation per oxygen molecule is approximately 11% greater for glucose than for fatty acid utilization. For this reason, when aerobic exercise reaches levels close to the maximum rate of oxygen uptake (>90% $V_{O_2\,max}$), glucose is preferably used and muscle glycogen is consumed. On the other hand, if the ATP yield per fuel weight unit is considered, the fatty acids are 30%–50% more efficient than glucose.

During exercise in which O_2 consumption is 60% or less of the $V_{O_2\,max}$ value, fatty acids are preferentially used. They provide greater fuel efficiency, longer sustained work, and their reserves are significantly more abundant than those of glycogen. As oxygen consumption approaches the $V_{O_2\,max}$ the proportion of ATP generated from glucose oxidation progressively increases to enhance oxygen consumption efficiency.

The fuel supply to the muscle during exercise is regulated by the action of several hormones. When muscle effort reaches certain intensity, catecholamines are released and their increase is proportional to the intensity and duration of exercise. Catecholamines activate muscle glycogenolysis and lipolysis in adipose tissue, and glycogenolysis in the liver. Also, insulin decreases in plasma, which reduces the synthesis of glycogen, fatty acids, and triacylglycerols. During prolonged exercise, or upon short but very intense work, glucagon, cortisol, and somatotropin are released. These hormones stimulate glycogenolysis and gluconeogenesis in liver as well as lipolysis in adipose tissue.

Hepatic gluconeogenesis uses mainly lactate and alanine as starting substrates, both derived from muscle pyruvate. Pyruvate oxidation in muscle is depressed when fatty acids and ketones are used because the acetyl-CoA formed during degradation of these substrates inhibits pyruvate dehydrogenase.

The α-keto acids derived from branched chain amino acids are preferentially degraded and used as fuel in muscle.

Alanine is produced in muscle by transamination between pyruvate and glutamate. Once released in circulation, alanine is taken up by the liver, where the glucose-alanine cycle (p. 442) is completed. The other product of this reaction, α-ketoglutarate, undergoes transamination with amino acids.

During prolonged activity (submaximal O_2 consumption), as occurs in marathon runners, muscle use their own glycogen and triacylglycerol stores. In addition, free fatty acids (produced mainly by hydrolysis of triacylglycerols in adipose tissue), glucose, and ketone bodies released into the blood by the liver are utilized.

In the early phases of work, glucose oxidation in muscle and glycogen breakdown in liver predominate. The early stimulation of lipolysis leads to increased intake and consumption of fatty acids, which allows for a decrease in glucose oxidation. When exercise continues for more than 2 h, fatty acid and ketone body oxidation markedly predominates and glucose utilization is reduced. After submaximal work prolonged for more than 6 h, it was found that there is still a significant proportion of glycogen in muscle, which highlights the protecting effect that fatty acid oxidation has on glucose consumption. The availability of glycogen in muscle is one of the factors that determines the duration of muscle activity. When glycogen stores are depleted, exercise intensity is reduced or stopped.

Effect of training. In a physically trained subject, protein synthesis and vascularization in skeletal and cardiac muscles is increased. Myocardial hypertrophy improves blood supply to peripheral tissues. The number of mitochondria in skeletal muscle increases 2–3 times, and the ability to oxidize substrates is significantly incremented. Myoglobin content is higher and hence the capacity to store and transport O_2 in muscle is enhanced. Due to these changes, the values of the trained muscle $V_{O_2 max}$ are over twofold higher than that of nontrained individuals.

Trained muscle is able to more efficiently utilize fatty acids and ketone bodies, which reduces consumption of muscle glycogen and lactate production, delaying the onset of fatigue. Improved circulation not only increases O_2 delivery and fuel to muscle, but also allows a more efficient removal of lactate and its transport to the liver for conversion into glucose. All these result in increased strength and faster recovery.

The adaptation of skeletal muscle to exercise involves molecular and metabolic changes, which are secondary to transcriptional and translational regulation of protein synthesis.

According to their physiological characteristics (speed of contraction) and metabolic mode (oxidative capacity), muscle fibers can be distinguished in three different types: I, IIA, and IIB. Type I fibers are slow contracting and their metabolism is predominantly aerobic. The IIB fibers contract fast and have high glycolytic capacity. Type IIA fibers show properties that are intermediate between those of type I and IIB.

Some animals have muscles composed almost exclusively of a single type of fiber. Muscles that contain mostly type IIB fibers are called white muscles and include, for example, the pectoral muscle of chickens and abdominal muscles of fish. These muscles contract strongly, but for only short periods, presenting rapid fatigue. They have very few mitochondria and obtain energy from glycolysis and degradation of its own glycogen. In contrast, red muscles are predominantly composed of type I fibers. Their color depends on the high content of mitochondria, myoglobin, and cytochromes. Energy is supplied mainly by the oxidation of fatty acids and they can sustain long lasting activity. In humans, muscles contain a mixture of all three types of fibers, although the relative amount of each varies depending on the individual and the specific muscle considered.

HEART METABOLISM

The structure and contraction mechanism of the cardiac muscle are similar to those of skeletal muscle. However, the heart is characterized by its continuous activity, which is fueled by energy generated aerobically. The large number of mitochondria in cardiac muscle shows the ability of this tissue to oxidize substrates and to produce ATP through oxidative phosphorylation. The ATP content in heart is relatively low and, even at rest, its hydrolysis is high. Therefore, ATP regeneration in heart must be very active.

Only in situations of very intense physical activity, the heart relies on anaerobic glycolysis as a source of additional ATP. The cardiac tissue has a small reserve of glycogen that, in those circumstances, is degraded to produce lactate. At rest or during moderate exercise, the myocardium uses fatty acids as the main fuel. In addition, it oxidizes glucose, lactate, and ketone bodies that come from the bloodstream. These substrates are degraded into CO_2 and H_2O in the citric acid cycle.

Different disease conditions lead to metabolic disturbances in the heart. For example, excess of fatty acids and intense β-oxidation occurs in obesity and diabetes. Poor oxidation is found in cardiac ischemia.

Glucose enters myocardial cells mainly through GLUT4 transporters, whose insertion in the plasma membrane is stimulated by insulin. When exercise is increased, cardiac activity and glucose consumption is also increased. As previously mentioned, this allows for greater oxygen energy efficiency. Aerobic glucose consumption in the heart requires the operation of shuttle systems to transfer reducing equivalents from the sarcoplasm to the mitochondria. In the heart, the aspartate–malate shuttle (p. 200) is the most active of those systems. Glutamate, necessary for its operation, is formed by transamination with α-ketoglutarate, an intermediate of the citric acid cycle.

Amino acid metabolism is more intense in heart than in skeletal muscle. It also has a higher activity of protein synthesis. Methodic and continuous muscular work and training produces increased protein synthesis and muscle hypertrophy. New protein synthesis of muscle tissue has also been shown to occur during regeneration of the heart after an infarct, a type of myocardial injury due

to a lack of blood flow in an area of the heart. The complete blockage of vessels of the coronary system quickly leads to irreversible damage and death (necrosis) of the cardiac tissue perfused by those arteries. This is due to the high dependence of the heart on an aerobic metabolism, which does not tolerate the absence of oxygen.

ADIPOSE TISSUE METABOLISM

The adipose tissue is widely distributed, being particularly abundant in the subcutaneous tissue and surrounding organs (perivisceral fat); it represents approximately 25% of the body weight of a normal adult, with the percent amount being higher in women than in men.

Adipocytes contain a mixture of different triacylglycerols that occupy almost all of the cytoplasm of those cells. The triacylglycerols stored in adipose tissue account for ~15 kg of an adult individual. This is the largest energy reserve in the body and is equivalent to 140,000 kcal or 585,000 kJ (9.3 kcal or 39 kJ/g of fat). The importance of the fat deposit is better appreciated when it is compared to glycogen, which comprises ~225 g, divided mainly between liver and muscle, and has an energy equivalent of 900 kcal or 3760 kJ (4 kcal or 16.7 kJ/g).

Fats present several advantages over carbohydrates as energy reserve material. They have more than twice the caloric value per unit weight. Triacylglycerols, due to their low polarity, can be stored in cells unaccompanied by water, while glycogen, being hydrophilic, attracts an amount of water that duplicates its weight. If the energy content of 15 kg of fat were deposited as glycogen, the body weight would increase by approximately 60 kg.

Adipose tissue is a dynamic reservoir; it has an intense metabolic activity. The permanent replacement of accumulated fat contributes to supply fuel to the remaining tissues (excluding the CNS and red blood cells, which do not use fatty acids).

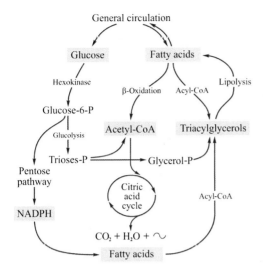

FIGURE 20.5 **Metabolic pathways in adipose tissue.**

Here, we will only discuss the major metabolic pathways functioning in adipocytes (Fig. 20.5). Glucose enters the fat cell from the blood by facilitated diffusion stimulated by insulin (GLUT4 transporters). Within the cell, hexokinase catalyzes the formation of glucose-6-P. The two main alternatives of this intermediary molecule include entering glycolysis or the pentose phosphate pathway. In the glycolytic pathway, glucose-6-P is degraded to pyruvate, decarboxylated to acetyl-CoA, and oxidized to CO_2 and H_2O in the citric acid cycle to produce energy. When glucose supply is abundant, it is also used to synthesize fatty acids. The pentose phosphate pathway generates the NADPH necessary for the synthesis of fatty acids. Fatty acids from chylomicrons, VLDL, and triacylglycerols hydrolyzed by lipoprotein lipase in capillaries are activated in adipocytes to form acyl-CoA, which is used for the synthesis of triacylglycerols or are degraded by β-oxidation and the citric acid cycle to produce ATP. Fatty acids, provided by the circulation and synthesized in the cell from glucose, are used for the synthesis of triacylglycerols, which are stored. For this, glycerol-3-phosphate plus acyl-CoA

are required. The adipocyte has no capacity to phosphorylate glycerol, and cannot use the glycerol released by the hydrolysis of fats from chylomicrons, VLDL, and adipose tissue itself. Glycerol-3-phosphate is generated by reduction of dihydroxyacetone phosphate, catalyzed by glycerolphosphate dehydrogenase. Thus, glucose contributes to lipogenesis not only providing acetyl-CoA for the synthesis of fatty acids, but also with glycerol-3-P.

Triacylglycerol stores are dynamic and rapidly renewed, having a half-life of approximately 5 days. When tissues require fuel, mobilization of the fat reserves takes place. A specific hormone-regulated lipase in adipose tissue catalyzes the hydrolysis of triacylglycerols. The released fatty acids enter the blood, and are bound and transported by albumin. The lipase from adipocytes is activated by phosphorylation catalyzed by protein kinase A (PKA), which is dependent on cyclic AMP. PKA also phosphorylates the perilipins, located on the surface of each adipocyte lipid droplet. When these proteins are not phosphorylated, they form a barrier that protects the triacylglycerols from the action of the lipase. Kinase action on lipase and perilipins is required to start lipolysis.

Norepinephrine, epinephrine, glucagon, glucocorticoids, somatropin, and adrenocorticotrophin bind specific receptors associated to G proteins in the fat cell membrane, increasing cyclic AMP levels and activating the lipase. Insulin and prostaglandin E1 depress it. Fig. 20.5 summarizes the concepts presented.

NERVOUS TISSUE METABOLISM

The central nervous system (CNS) has remarkable metabolic characteristics. Despite comprising less than 2.5% of the body weight of a normal adult, the CNS consumes a fifth (20%) of the total oxygen used by the body at rest. Moreover, in normal, well-nourished individuals, the CNS uses only glucose as fuel, as indicated by the value of 1 of the respiratory exchange ratio.

The *respiratory exchange ratio*, also referred as *respiratory quotient*, is the ratio between the moles of CO_2 produced and the O_2 moles consumed (moles CO_2/moles O_2). The value of this coefficient depends on the type of nutrient metabolized; for glucose, it is 1, for fat it is 0.7, for protein the average is 0.8. The combustion of food provided by a mixed diet with carbohydrates, fats, and proteins, in suitable proportions, gives an average respiratory quotient of 0.82.

Glucose enters nerve cells predominantly through GLUT3 transporters that have high glucose affinity to ensure provision of substrate even at low blood glucose levels. GLUT3 is not regulated by insulin. The action of hexokinase forms glucose-6-P, which is degraded to pyruvate in the glycolytic pathway. Pyruvate is decarboxylated to acetyl-CoA and is oxidized to CO_2 and H_2O in the tricarboxylic acid cycle, with the consequent production of ATP through oxidative phosphorylation.

While glucose is virtually the sole source of energy in the nervous system, under conditions of starvation or prolonged fasting, the brain develops the ability to use also ketone bodies (α-hydroxybutyrate).

Glucose consumption by the CNS does not show significant changes during the day, remaining at relatively constant levels during sleep, alertness, or intellectual activity.

Since the brain has no glycogen stores, it depends at all times on the supply of glucose from the circulation. With blood glucose levels of 80 mg/dL or more, the CNS has an adequate supply of fuel. The liver, through the processes of glycogenolysis and gluconeogenesis, is the main organ responsible for maintaining glucose homeostasis. Several hormones are involved in the regulation of constant blood glucose levels (p. 612).

Pronounced decreases in blood glucose affect the nervous system. A blood glucose of 40 mg/dL causes neurological symptoms (dizziness, fainting), while lower values can lead to coma and irreversible nervous damage.

In addition to glucose, the blood provides the oxygen levels that are necessary for the catabolism of glucose to CO_2 and H_2O. Given the great dependence of the CNS on blood supply, a vascular blockage of even a few minutes can result in permanent brain alterations.

Approximately two-thirds of the total ATP generated by the nervous system in its oxidative metabolism is employed in maintaining the membrane potential of neurons, the rest is used in the synthesis of neurotransmitters and proteins.

SUMMARY

Liver metabolism

Carbohydrate metabolism: glucose is converted to G-6-P by glucokinase. This reaction is significantly activated when glucose levels are elevated above baseline. Glycogenesis, glycogenolysis, glycolysis, the pentose phosphate pathway, and gluconeogenesis are very active in liver.

Lipid metabolism: β-oxidation of fatty acids is a main pathway for energy production by the liver. The liver synthesizes triacylglycerols, phospholipids, and cholesterol. It also forms ketone bodies but cannot use them because the liver lacks thiophorase.

Amino acid metabolism: the liver is active in protein synthesis. Liver is rich in aminotransferases and glutamate dehydrogenase. The liver is the organ of urea formation. Also, it participates in the *glucose-alanine cycle*, along with the muscle. In addition, the liver has intense gluconeogenic activity.

Biotransformation refers to the detoxification of foreign or *xenobiotic* substances by the liver.

The main reactions to detoxify foreign substance involve enzyme systems associated with the endoplasmic reticulum (microsomal fraction). They comprise:

1. *Oxidation* or *phase I* is an oxidizing system or *microsomal mixed function oxidative system*. It requires NADPH and O_2 along with FAD and cytochrome P_{450}.
2. *Conjugation* or *phase II* is performed by UDP-glucuronyl transferase, sulfokinase (PAPS), transacetylases (acetyl-CoA), or enzymes that catalyze conjugation with glycine.
3. *Reduction*.
4. *Hydrolysis*.
5. *Excretion* or *phase III*.

Ethanol metabolism takes place in the liver. Most of the ingested ethanol is absorbed in the duodenum and jejunum. Less than 10% is eliminated unchanged in urine, breath, and sweat. In a *first step*, ethanol is oxidized to acetaldehyde. The reaction is catalyzed by: *alcohol dehydrogenase, oxidant microsomal system*, and *catalase*. In normal individuals, *alcohol dehydrogenase*, which depends on NAD, is responsible for the oxidation of more than 90% of the absorbed ethanol. The *oxidizing microsomal system* uses NADPH and cytochrome P_{450} (or CYP2E1). This is an inducible, nonspecific system, which is able to oxidize other compounds besides ethanol. In chronic alcoholic, this system oxidizes over 30% of the total ethanol metabolized. In a *second step*, acetaldehyde, a toxic product, is oxidized to acetate by *acetaldehyde dehydrogenase*, which is NAD dependent. The acetate is joined to CoA and forms *active acetate*, which can follow several pathways, including final oxidation to CO_2 and H_2O in the citric acid cycle. Heavy consumption of ethanol produces an excess of reducing equivalents. There is hyperlactacidemia, reduced activity of the citric acid cycle, fatty acid accumulation, increased lipogenesis, shuttle system overload, and fatty liver disease.

Skeletal muscle varies its metabolism dependent on the type of work performed. Anaerobic metabolism occurs during intense muscle work; the amount of ATP needed by the muscle is much higher than that which can be provided by oxidative phosphorylation. Intense exercise can be maintained for a very short time by utilization of the muscle's own ATP and creatine phosphate reserves. It may be prolonged to not more than a few minutes generating ATP by anaerobic degradation of its own glycogen.

Muscle engages in aerobic metabolism when exercise intensity is low; supply of O_2 may be sufficient to generate the needed ATP by oxidative phosphorylation. The maximum rate of oxygen uptake ($V_{O_2 max}$) is the maximum amount of O_2 released from blood that muscles can use per minute. Energy generation per mole of O_2 is 11% greater when carbohydrates are metabolized than when fatty acids are oxidized. For this reason, if the intensity of aerobic exercise is close to $V_{O_2 max}$, glucose is preferably used and muscle glycogen is consumed. If O_2 consumption is 60% or less of the $V_{O_2 max}$, fatty acids are preferably used.

The muscle uses α-keto acids derived from branched chain amino acids as energy source.

Heart has continuous activity, which is accomplished with energy generated aerobically. At rest or moderate exercise, fatty acids are the main fuel. Glucose, ketone bodies, and lactate are also oxidized. Aerobic consumption of glucose requires H shuttles. The most active shuttle in myocardium is the malate–aspartate shuttle.

Adipose tissue represents the major energy reserve of the body, which mainly depends on the storage of triacylglycerols (TAG). Glycolysis and pentose phosphate are the main pathways for G-6-P in the adipocyte. Fatty acids provided by the blood and synthesized in adipocytes are used for the synthesis of TAG. Adipose cells do not phosphorylate glycerol, they use glycerol-3-P derived from dihydroxyacetone phosphate from the glycolytic pathway. The half-life of stored TAG is approximately 5 days. Lipolysis is catalyzed by cAMP-activated *lipase*.

Nervous tissue consumes 20% of the O_2 used by the whole body at rest.

The CNS uses only glucose to obtain energy. It depends on the supply of O_2 and glucose through blood. Under conditions of prolonged fasting, the brain can also oxidize ketone bodies. Of the total ATP generated by the CNS, approximately 2/3 is used to maintain the membrane potential. The remaining is used in synthesis, primarily of neurotransmitters.

Bibliography

Bollen, M., Keppens, S., Stalmans, W., 1998. Specific features of glycogen metabolism in the liver. Biochem. J. 336, 19–31.

Depré, C., Rider, M.H., Hue, L., 1998. Mechanism of control of heart glycolysis. Eur. J. Biochem. 258, 277–290.

Egan, B., Zierath, J.R., 2013. Exercise metabolism and the molecular regulation of skeletal muscle adaptation. Cell Metab. 17, 162–184.

Horowitz, J.F., 2002. Fatty acid mobilization from adipose tissue during exercise. Trends Endocrinol. Metab. 14, 386–392.

Lieber, C.S., 1997. Ethanol metabolism, cirrhosis and alcoholism. Clin. Chim. Acta 257, 59–84.

Lieber, C.S., 2004. CYP2E1: from ASH to NASH. Hepatol. Res. 28, 1–11.

Lopaschuk, G.D., Ussher, J.R., Folmes, C.D.L., Jaswal, J.S., Stanley, N.C., 2010. Myocardial fatty acid metabolism in health and disease. Physiol. Rev. 90, 207–258.

Nagy, L.E., 2004. Molecular aspects of alcohol metabolism. Annu. Rev. Nutr. 24, 55–78.

Zakim, D., Boyer, T.D., 2012. Hepatology, sixth ed. W.S. Saunders Co., Philadelphia, PA.

21

The Genetic Information (I)

The genetic content of the cell, or *genome*, is contained in the DNA, located in chromosomes and mitochondria of animals, or in chromosomes and chloroplasts of plants. Specific regions in the DNA direct the synthesis of the different classes of RNA, including ribosomal (rRNA), transfer RNA (tRNA), messenger RNA (mRNA), and other types of RNA, such as small and long noncoding RNAs.

Normally, during cell division, daughter cells receive the genetic material from the progenitor cell. This requires duplication, or *replication*, of the parental DNA.

The genetic information is encoded in the nucleotide sequence of DNA and is transferred to RNA molecules by a process known as *transcription*.

The sequence of mRNA carries the information that dictates the order in which amino acids will be assembled to produce proteins. The synthesis of polypeptide chains requires the *translation* of the message contained in the mRNA.

All the processes mentioned, especially in eukaryotes, involve multiple steps and a variety of regulatory factors, which working together via different mechanisms, tightly control gene activity. The extraordinary level of complexity of these processes has only been appreciated recently by the progress made in this area of research in the last several years. This chapter and the following provide a simplified overview of the mechanisms of gene replication, transcription, and translation.

DNA REPLICATION

All somatic cells of an organism, generated by successive divisions of a single primordial cell (egg), contain DNA that is identical in quantity and primary structure (nucleotide sequence). This reveals a remarkable property of living beings: the ability to transmit the information contained in the nuclear DNA without modification, from one cell generation to the next.

DNA Replication is Semiconservative

Each DNA strand of a progenitor cell serves as a template for the synthesis of a new complementary polynucleotide chain that is identical to that of the original cell. This process is known as *DNA replication*. The DNA received by each daughter cell contains one DNA strand that is newly synthesized at replication, and another strand that is directly received from the parental DNA. For this reason, the replication process is referred to as semiconservative (Fig. 21.1). DNA replication takes place before mitosis, during a limited period of the cell cycle, called S phase.

Cell Cycle

Eukaryotic cells undergo a cell cycle that consists of four steps. The first three (G_1, S, and G_2) are included in the interphase period. Cells spend most of their time in interphase, during

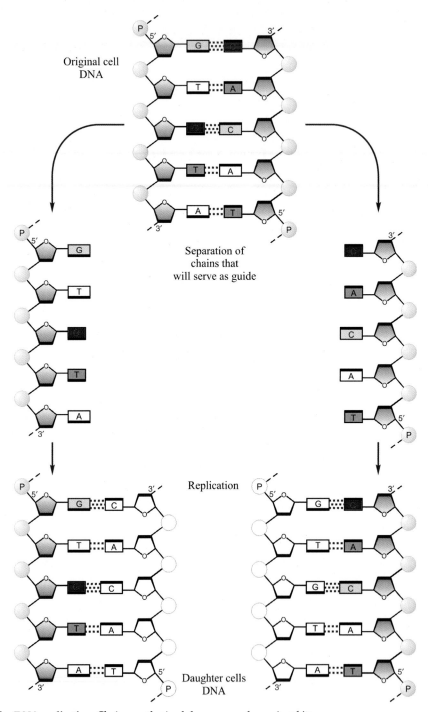

FIGURE 21.1   **DNA replication.** Chains synthesized de novo are shown in *white*.

which they perform their normal metabolic functions. The fourth phase, called M (mitosis), is shorter. In this phase, cell division and chromatid separation takes place to form two new cells.

The first phase (G_1) is the preparation period, previous to DNA replication. During this phase, the precursor nucleotides and all of the required enzymes and factors needed for replication are synthesized. In the second phase (S phase), DNA replication takes place. Nucleosomes disassociate as synthesis proceeds and production of histones and other proteins, closely associated with DNA, increases significantly. The amount of DNA and histones is duplicated in this phase. The third phase (G_2) is a step in which the cell prepares for mitosis. Tubulin, a protein that forms the microfibrils of the mitotic spindle, is synthesized. Finally, cell division occurs in the M phase.

The cell cycle of human cells in culture is of approximately 24 h; which include ~9 h for S phase, ~7 h for phase G_1, ~6 h for phase G_2, and 1–2 h for phase M. In vivo, the duration of the cycle varies greatly from one cell type to another. The G_1 and G_2 phases are very short in cells with high mitotic activity. In contrast, there are cells that do not divide and stay quiescent for long periods of time, such as liver cells, which divide once every 1 or 2 years.

The cell cycle is regulated by a set of proteins known as the cyclins and cyclin-dependent kinases. These proteins are activated by external agents (growth factors), which by binding to specific receptors in the plasma membrane, activate intracellular signaling cascades in the cell that target the various cyclins.

After mitosis, cells enter the G_1 phase to start a new cycle. Without stimulation, cells remain in a nonproliferative state, called G_0, in which all of the metabolic functions are performed but the cell remains undivided. Alterations in cell cycle regulation can increase cell mitotic activity and lead to different proliferative diseases.

Multiple Enzymes and Factors Are Involved in the DNA Replication Process

The replication process is exquisitely regulated to ensure that the DNA in each cell is duplicated only once per cell cycle, during the S phase. Cyclins and cyclin-dependent kinases are involved in this regulation. Although the basic mechanism is similar, replication is more complex in eukaryotes than in prokaryotes.

The initial step of replication is the separation of both DNA strands. This is necessary to copy the parental DNA, which functions as a guide for the assembly of the new complementary strand.

Replication begins at a defined point of origin, or initiation site, where specific base sequences serve as recognition signals for the enzymes and factors that initiate replication. Many initiation sites contain sequences rich in A-T pairs, which are easier to separate than G-C pairs. The first step is the formation of the origin recognition complex (ORC), composed of six subunits. Then, the mini-chromosome maintenance complex (MCM), also formed by six subunits, is added.

Prokaryotic chromosomes, mitochondrial DNA, and circular viral DNA have a single site of origin. In contrast, linear eukaryotic DNA molecules in chromosomes begin to replicate at multiple sites. In humans there are between 30,000–50,000 initiation sites. Separation of the DNA strands is initiated simultaneously in all chromosomes and at many different points along the molecule. As a result, the double helices form "bubbles" at the separation zones. These bubbles or replication units are called *replicons*. In the nucleus of every human cell about 50,000 "replication bubbles" can be formed. This simultaneous unwinding of DNA in many different sites is completed faster than if performed progressively from one end to another of the very long double helix of each chromosome.

The site where both strands are separated is called the *replication fork*. In each separation area,

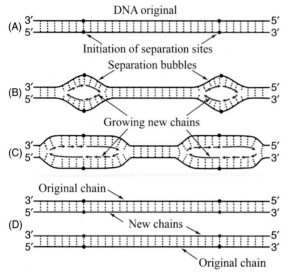

FIGURE 21.2 **Scheme of DNA replication from two initiation sites.** (A) Original molecule (double helix). (B) Separation and simultaneous replication of both DNA helix starts. (C) The same process at a more advanced stage. (D) Replication is complete and two double helices, identical to the original DNA, are formed.

two forks are formed, which progress in opposite directions, away from the point of origin (Fig. 21.2).

Helicase. Also called the *unwinding enzyme*, helicase catalyzes the separation of the DNA strands. The energy required for the process is provided by the hydrolysis of ATP.

Topoisomerases. In bacterial circular DNA, the separation of the DNA strands at a single site produces supercoiling in other regions of the double helix. In short linear DNA molecules, the separation does not create tensions; these are easily relieved by rotation of the free ends. In contrast, due to the enormous length and multiple interactions in the nucleosomes, the linear DNA molecules of eukaryotes do not rotate freely, and torsions frequently occur downstream of the replication fork. In both bacteria and eukaryotes, supercoiling is resolved by periodical cuts in the chain, in areas where torsions occur. These cuts are catalyzed by *topoisomerases*, of

which two types (I and II) are known. Isomerase type I cuts one strand of the double helix and relieves tension within the molecule by rotation of the noncut end on the other strand. No energy is required as the passage of the supercoiled to the relaxed state of DNA has a negative ΔG. The type II enzyme, also called *gyrase*, cuts both DNA strands. It is dependent on the hydrolysis of ATP. Both enzymes also have the capacity to bind back the cut ends of the chains and to reestablish the double helix once the relaxed state is attained.

Bacterial gyrase is inhibited by antibiotics, such as nalidixic acid and other substances. Eukaryotic topoisomerase II is unaffected by nalidixic acid, so, this compound is useful for the treatment of human infections caused by bacteria that are susceptible to this type of antibiotic.

Specific proteins maintain the DNA strands separated. As the double helix strands are separated, they bind to *single-strand DNA binding proteins*. These proteins in bacteria are designated with the acronym SSB. They stabilize and prevent chain reannealing. The binding of the SSB does not disturb the "copying" process of a complementary strand synthesis. In eukaryotes, this function is performed by *replication protein A* (RPA).

Helicase and binding proteins move along the double helix leaving behind two separate chains, ready to serve as a template for the synthesis of new complementary strands. Synthesis is initiated simultaneously at all sites of unwinding, before the original double helix is fully separated (Fig. 21.2).

DNA polymerases. Formation of the new chain is performed by assembly of deoxyribonucleotides catalyzed by *DNA polymerases*. In bacteria, three of these enzymes have been isolated, each designated with Roman numerals (I, II, III). In eukaryotes, a higher number of polymerases have been identified. They are named by Greek letters; the most important ones include the α, β, γ, δ, and ε polymerases. All DNA polymerases are molecular complexes formed by association of different subunits. They differ in their properties and functions; however, they all share the following characteristics:

1. Their substrates are the four deoxyribonucleoside triphosphates dATP, dGTP, dCTP, and dTTP. These molecules not only provide the "raw material" for the synthesis, but also the energy required for their assembly.
2. They need a free strand of DNA that serves as a guide or template. They only work on a single DNA strand to insert, one by one, nucleotides that are complementary to those in the template strand.
3. They are unable to bind free nucleotides and start a new strand. They can only extend a preexisting strand (the initiator or primer strand) correctly paired by hydrogen bonds with the template DNA bases.
4. All DNA polymerases catalyze the binding of deoxyribonucleotides to form chains in the 5′ to 3′ direction. This involves establishing diester connections from the 3′ carbon hydroxyl in the terminal nucleotide deoxyribose of the initiator-chain to the 5′ carbon phosphate of the entering nucleotide. However, the helices are antiparallel, so, there are polymerases that act on the chain that serves as a template in the 3′→5′ direction, which is opposite to the synthesis of the new strand.

Replisome. All proteins involved in DNA replication form a multienzyme complex that functions as a "replication machine"; this is called the *replisome*.

Mechanism of DNA Replication

DNA polymerases require a preexisting chain. Therefore, the process begins with the formation of a piece of RNA of approximately 10 nucleotides long, which is called the initiator RNA or *primer*. Its synthesis is catalyzed by an RNA polymerase called *primase*, firmly associated

with αDNA polymerase in eukaryotes. It adds ribonucleotides in the 5′→3′ direction, forming a segment of about 10 nucleotides complementary to the template DNA. The complementary nucleotides are as follows: guanine is always bound to cytosine and vice versa, thymine is bound to adenine, and uracil is bound to adenine (Fig. 21.3).

Then, the action of DNA α polymerase adds approximately 20 deoxyribonucleotides to the initiator RNA. When a fragment of 30 nucleotides (including the 10 of the RNA primer) is formed, the complex α polymerase–primase is released and the chain elongation is continued by δ polymerase, which successively adds the deoxynucleotides necessary to form a strand complementary to the original DNA template.

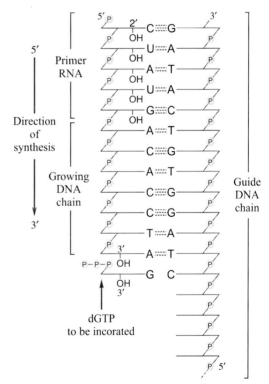

FIGURE 21.3 **Schematic representation of the DNA replication process.** After the initiator RNA, deoxyribonucleotides are inserted to form the new DNA strand complementary to the guide strand.

In the reaction that catalyzes the incorporation of each deoxynucleotide, pyrophosphate (PP_i) is released. The total synthesis process can be expressed by the equation:

$$n\ dATP + n\ dGTP + n\ dTTP + n\ dCTP \longrightarrow$$

$$\xrightarrow[\text{Primer RNA, guide chain}]{\text{DNA polymerase, Mg}^{2+}}$$

$$\longrightarrow \text{Polymer 4n dRibonucleotides} + 4n\ PP_i$$

The process is irreversible; pyrophosphate is hydrolyzed immediately by pyrophosphatase in a highly exergonic reaction:

$$PP_i + H_2O \xrightarrow{\text{Pyrophosphatase}} 2\ P_i$$

Both strands of the original DNA are replicated simultaneously but in opposite direction. They are antiparallel, presenting opposite directions. The synthesis advances from 5′ to 3′ in the newly formed chain, which is complementary and antiparallel to the template strand. This implies that DNA polymerase moves on the template chain in the 3′→5′ direction (Figs. 21.3 and 21.4). The most efficient polymerases can add up to 1000 nucleotides per second.

From the initial site of DNA strand separation, one strand will be in the 3′→5′ sense, ready for direct "reading" by the polymerase. On this chain, synthesis of the new complementary strand continues without interruption from the formation of the RNA–DNA primer (Figs. 21.2 and 21.3); this is the *leading* or *advanced strand*. The other strand cannot be assembled into a continuous chain, as the direction of the template strand is opposite to the progression of the DNA polymerase. For this reason, the synthesis is carried out in segments, after a sufficiently large segment of the template strand has been separated (Fig. 21.4). Similar to the leading strand, the synthesis of these pieces requires previous formation of a RNA primer of about 10 nucleotide units, catalyzed by the primase. Deoxyribonucleotides are bound to the 3′ end of this RNA according to complementarity with the template strand. The 20 first deoxynucleotides are inserted by α polymerase and then δ polymerase continues adding bases until it synthesizes a segment of about 200 nucleotides long (1000 bases or more for prokaryotes). These new segments of DNA are not yet connected to one another and are designated *Okazaki fragments*.

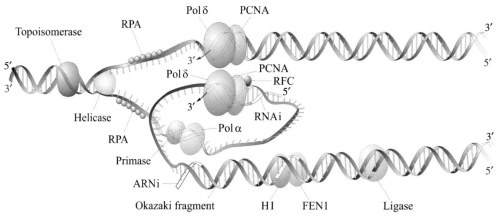

FIGURE 21.4   **Schematic representation of the DNA replication process at the replication fork.** RPA, replication protein; Polα, α DNA polymerase; RFC, replication factor C; PCNA, proliferating cell nuclear antigen; Polδ, δ DNA polymerase; RNAp, primer RNA; HI, ribonuclease; FEN1, nuclease.

These fragments have a length equivalent to the DNA comprised within a nucleosome, which suggests that DNA is released for replication from one nucleosome at a time. The chain synthesized in short segments is called the *delayed* or *lagged strand*. It has been proposed that the lagged strand forms a "loop" to face the replisome in the right direction (Fig. 21.4). The pieces of primer RNA of each Okazaki fragment are quickly eliminated, DNA is synthesized to cover the gap, and the free 3' and 5' ends are joined together to provide a continuous chain.

In their advance, two replication forks initiated in different parts of the same molecule approach each other and end up merging the "replication bubbles" (Fig. 21.2). Finally, there are two complete double helices, identical to the original one.

The process takes place through the following stages:

1. During the G_1 phase of the cell cycle, the initiation sequence in the template strand is recognized and a multiprotein complex forms at each of the sites where the synthesis will begin. The assembly of the complex is performed in an orderly manner. First, the ORC is bound to the double-strand and then, the MCM fixes the helicase that will unwind the DNA strands. The helicase is inactive during the G_1 phase and is stimulated to enter the S phase of the cycle by action of two protein kinases, one of which is a cyclin kinase. This step is regulated to ensure that replication will occur only once in each cycle.

   When the DNA strands are separated, SSB proteins, called RPA in eukaryotes, bind to each of the chains and prevent the reassociation of the strands into the double helix.

2. Activation of the complex promotes binding of other proteins, including primase and α, β, and ϵ polymerases. These proteins form the replication machinery, or *replisome*.

   Primase catalyzes the synthesis of the ~10 ribonucleotide RNA primer, complementary

to the parent strand. Then, the α polymerase, associated with primase, adds ~20 deoxyribonucleotides. Immediately, a protein called replication factor C (RFC) binds to the 3' end of the RNA–DNA primer and displaces the primase and α polymerase. RFC functions as a DNA-dependent ATPase. Its main role is to load the proliferating cell nuclear antigen (PCNA) onto the DNA being synthesized.

3. PCNA is a complex ring surrounding the DNA; it places δ polymerase on the template strand and ensures its continued action. The complex δ polymerase–PCNA moves on the template strand in the direction of synthesis. The ability of the polymerase to function continuously along a large number of nucleotides is called *processivity*. The δ polymerase is highly "processive," while the α polymerase is not.

   The described mechanism is identical for the leading and the lagging strands. On the leading strand, δ polymerase completes the synthesis from a single RNA primer to the end of the template strand. On the lagging strand, the δ polymerase–PCNA complex must be released after each Okazaki fragment is completed, and is reinserted again to elongate the next segment after the primase–α polymerase complex has assembled the RNA–DNA primer.

4. While synthesizing the DNA strand, δ polymerase detects eventual errors in base pairing. It also has *exonuclease* 3'→5' activity, which allows correcting the mistakes (see next section).

5. Immediately, the RNA primer is degraded. This function is performed by ribonucleases that catalyze the hydrolysis of 3'–5' phosphodiester bonds. The gap produced by the removal of RNA is covered with DNA synthesized by DNA δ or ϵ polymerases.

6. Finally, phosphodiester bonds are formed between the 3' and 5' of the synthesized neighbor segments. This action is carried out

by the DNA ligase that requires energy and acts only if pairing between the bases of the newly formed chain and the template strand is correct.

In conclusion, all of the DNA molecules in the progenitor cell are replicated. In each molecule, one of the strands has been synthesized de novo; the other strand is from the original cell and served as a template.

Functions of the other DNA polymerases. γ DNA polymerase is involved in the replication of circular mitochondrial DNA. The β DNA polymerase participates in DNA repair processes. The role of ε polymerase is similar to that of δ; it is also involved in DNA repair.

Table 21.1 presents some of the enzymes and factors involved in the replication process in eukaryotic cells.

DNA Repair

Under normal conditions, the genetic information stored in DNA passes unchanged from one cell division to the next with remarkable accuracy. A single base inserted incorrectly could create alterations to the cell, or may even be lethal. Therefore, in addition to its high efficiency, replication requires extreme precision. This is achieved through the existence of correction systems that repair the errors that could arise during synthesis.

The first quality control mechanism is DNA δ polymerase. This enzyme, while inserting nucleotides complementary to the template strand, detects eventual errors (in a proof-reading like manner). Polymerases δ, γ, and ε have $3' \rightarrow 5'$ exonuclease activity, removing the last base incorporated if it is not properly paired.

TABLE 21.1   Enzymes and Factors Involved in Eukaryotic DNA Replication

Protein	Function
Helicase (unwinding protein)	Unwinds the two helices by rupturing the hydrogen bonds that maintain the base pairs.
Topoisomerases I and II	Cut the double helix chains ahead of the replication fork to relieve the unwinding torsional stress.
Replication protein A (RPA)	Binds to the unwound DNA chains avoiding reassociation.
Primase (associated with polymerase α)	Catalyzes synthesis of primer RNA of about 10 nucleotides in length.
DNA polymerase α	Catalyzes synthesis of small segments of DNA of about 20 nucleotides following the primer RNA.
Replication factor C (RFC)	Displaces the polymerase (-primase complex. Binds nuclear antigen.
Proliferant cells (PCNA) nuclear antigen	Ring structure that fixes polymerase δ and allows for continuous synthesis.
DNA polymerase δ	Catalyzes DNA synthesis following the primer produced by polymerase (-primase. Acts as exonuclease in the $3' \rightarrow 5'$ direction.
RNAase HI Nuclease FEN1	Removes the pieces of primer RNA.
DNA ligase	Forms the phosphodiester $3' \rightarrow 5'$ bridges binding the Okasaki segments (newly formed DNA).
DNA polymerase ε	DNA synthesis and repair. Acts as an exonuclease in the $3' \rightarrow 5'$ direction.
DNA polymerase γ	Synthesizes circular mitochondrial DNA. Acts as exonuclease in the $3' \rightarrow 5'$ direction.
DNA polymerase β	Participates in DNA repair systems.

Despite this self-correcting ability of the polymerase, the number of bases to be inserted is so large, that some errors are inevitable. In eukaryotes, errors can occur as often as once every 109 bases or more. In addition, DNA is exposed to chemical and physical agents that can alter its structure. For example, ultraviolet light determines formation of covalent bonds between adjacent thymine residues, which causes the secondary structure of DNA to be disturbed, blocking replication. Repairing DNA errors or damage comprises different mechanisms: (1) mismatch repair, (2) base excision repair, (3) nucleotide excision repair, and (4) repair of DNA ruptures.

Each of these processes requires the presence of multiple proteins assembled in different complexes that perform the following repair actions: (1) separation by endo- and exonucleases of a base or chain segment that is wrongly paired or defective, (2) incorporation of the correct base, catalyzed by polymerases, and (3) binding, catalyzed by ligase, of the repaired DNA piece with the free ends of the main DNA strand.

Breakdown of DNA can be caused by various agents, particularly reactive oxygen species and ionizing radiations. There are several mechanisms for repairing these breaks in the DNA. One of them includes the exchange of equivalent regions of the double helix of homologous chromosomes. This process, called *recombination*, also occurs in germ cells during meiosis. Existing information in the homologous unmodified chromosome is used to replenish the damaged site in the other double helix. Participation of a multiprotein complex is also required.

Fanconi anemia, ataxia telangiectasia (a motor degenerative condition due to inability to repair oxidative damages in cerebellum), and *xeroderma pigmentosum* (characterized by hypersensitivity to UV light and appearance of pigmented areas) are diseases characterized by abnormal DNA repair. Patients have a high incidence of neoplasias and premature aging. A common type of hereditary colorectal cancer is due to genetic defects in one of the DNA repair systems.

Redistribution of genes in gametes. In ovules and spermatozoa, DNA *recombination* of homologous chromosomes occurs during meiosis with exchange of segments between two double helices. This results in a redistribution of the genes contained in the exchanged pieces. The mechanism of recombination will not be described here, it will only be pointed out that the process, similar to the repair mechanisms, requires energy, provided by the hydrolysis of ATP, and a group of proteins, including Rad51. Recombination contributes to increase the genetic variation in organisms with sexual reproduction.

Telomerase

Linear DNA replication in eukaryotic chromosomes poses a problem. Upon completion of the duplication process, a piece of RNA primer remains in the 3′ end of the template strand. These segments are removed by RNase, but cannot be replaced by DNA as DNA polymerases cannot direct DNA synthesis without a primer. Therefore, the 3′ end of each template chain cannot be duplicated. In theory, this would lead to progressively shorter chromosomes, with irreparable loss of genetic information, after successive cell divisions. This problem is delayed by the existence of DNA segments that do not contain information at the ends of chromosomes (telomeres); however, these segments get gradually shorter with successive divisions.

In humans, the disposable telomeric DNA is represented by hundreds of repetitions of a short sequence of bases (TTAGGG). In cells with high mitotic activity, additional segments are inserted at the 3′ end of preexisting telomeric DNA, catalyzed by a ribonucleoprotein enzyme called *telomerase*. This enzyme contains a piece of RNA used as a template to assemble the portions of the DNA chain to be elongated (the synthesis of a DNA strand on a template of RNA is similar to that performed by reverse transcriptase, see

p. 480). On the newly formed chain, a DNA polymerase synthesizes a complementary strand to form the double stranded telomeric DNA.

Telomerases compensate for the shortening produced by consecutive divisions. In somatic cells that divide infrequently, no telomerase activity is detected, or it is poor. It is assumed that the life of these cells is limited by the length of telomeres of the original cells, which become shorter with each replication up until the telomeric DNA is exhausted (cellular senescence). When this happens, the cell can no longer continue reproducing and they die. Cancer cells with high telomerase activity are immortal; they continue their reproduction indefinitely, because the DNA ends are replaced in each division.

Restriction Endonucleases

Restriction endonucleases are also called restriction enzymes or *restrictases*, they were first discovered in different species of bacteria. They catalyze the hydrolysis of the phosphodiester bonds between deoxyribonucleotides at specific sites of the double helix. Their function in the bacterial cell is to protect them from invasion of foreign DNA. The bacterial DNA has methylated bases in the endonuclease cleavage sites, hiding possible points of attack and preventing self-destruction of the genetic material.

Thousands of these enzymes have been isolated and purified. In general, they are named with the first letters of organism's name in which they are found. For example, Eco designates the endonucleases from *Escherichia coli*, Hin from *Haemophilus influenza*, Hae from *Haemophilus aegyptius*, Alu from *Arthrobacter luteus*, etc. Sometimes a fourth letter is added, indicating the strain, and a Roman numeral is added to differentiate each type.

Restriction endonucleases specifically recognize defined sequences. Many of the recognition sites are the same for different enzymes, which is why the number of those sites is lower (several hundreds) than the number of restriction enzymes. The recognition site is composed of four nucleotide base pairs for some enzymes and up to seven for others; frequently, the segment is self-complementary if rotated 180 degree. In other words, the segments in each strand have the same sequence in both the 3′→5′ and 5′→3′ directions. For example:

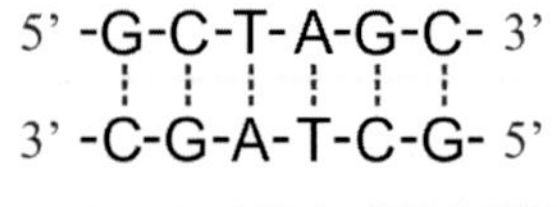

$$5' \text{ -G-C-T-A-G-C- } 3'$$
$$3' \text{ -C-G-A-T-C-G- } 5'$$

Fragments of this type are called *palindromic*.

The following diagram shows the specific sites of action (arrows) for the restriction enzyme EcoRI, isolated from *E. coli*.

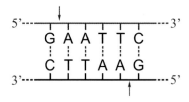

Frequently, the hydrolyzed diester linkages do not occur in the same position in both DNA strands, leaving overhanging segments in each cut single strand:

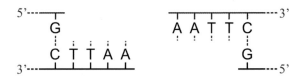

The 5′ end of one of the sectioned strands has a phosphate residue, and the 3′ of the other end has a hydroxyl group. These ends are known as "adhesive" because they tend to pair with a complementary end originated by the same enzyme in other DNA molecule. Fig. 21.5 shows examples of sequences recognized by various endonucleases. Some of these enzymes cut both

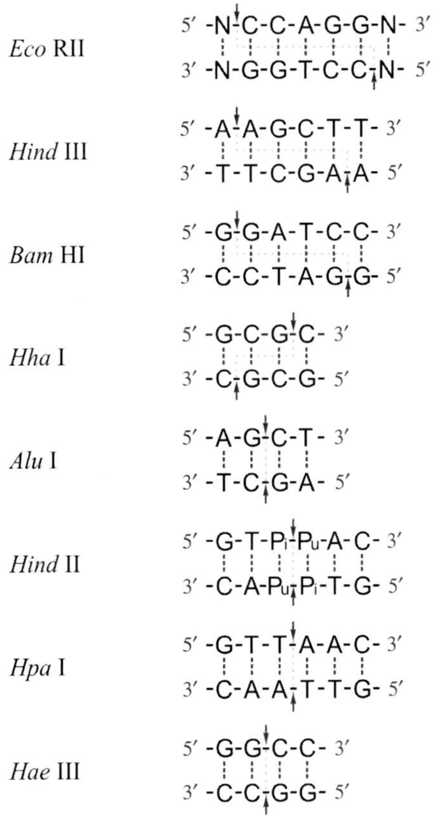

Eco RII

5′ -N$\overset{\downarrow}{-}$C-C-A-G-G-N- 3′
3′ -N-G-G-T-C-C$\underset{\uparrow}{-}$N- 5′

Hind III

5′ -A$\overset{\downarrow}{-}$A-G-C-T-T- 3′
3′ -T-T-C-G-A$\underset{\uparrow}{-}$A- 5′

Bam HI

5′ -G$\overset{\downarrow}{-}$G-A-T-C-C- 3′
3′ -C-C-T-A-G$\underset{\uparrow}{-}$G- 5′

Hha I

5′ -G-C-G$\overset{\downarrow}{-}$C- 3′
3′ -C$\underset{\uparrow}{-}$G-C-G- 5′

Alu I

5′ -A-G$\overset{\downarrow}{-}$C-T- 3′
3′ -T-C$\underset{\uparrow}{-}$G-A- 5′

Hind II

5′ -G-T-P$_\text{i}$$\overset{\downarrow}{-}P_\text{u}$-A-C- 3′
3′ -C-A-P$_\text{u}$$\underset{\uparrow}{-}P_\text{i}$-T-G- 5′

Hpa I

5′ -G-T-T$\overset{\downarrow}{-}$A-A-C- 3′
3′ -C-A-A$\underset{\uparrow}{-}$T-T-G- 5′

Hae III

5′ -G-G$\overset{\downarrow}{-}$C-C- 3′
3′ -C-C$\underset{\uparrow}{-}$G-G- 5′

FIGURE 21.5 **Sequences recognized by restriction enzymes.** The first four produce a staggered cut that generates adhesive ends. The last four cut the double helix transversely and the terminals are not single-stranded segments.

strands at exactly opposite sites and generate blunt ends, with no overhangs.

Restriction enzymes are used to cut DNA molecules at defined points. The endonuclease sections the DNA molecule at all sites where there is a sequence that serves a specific substrate, generating multiple segments. If the recognition site has a length of 6 bp, it is possible that this sequence will be repeated every 4096 bp; if the restriction site is 4 bp long, it can be estimated that the enzyme will cut the DNA double strand in 256 bp pieces. These fragments, treated with another restriction enzyme, are cleaved into

smaller pieces if they have the corresponding sequence in some segment of the double helix. By using restrictases it is possible to perform a controlled dissection of DNA. These enzymes have many valuable applications in molecular biology.

TRANSCRIPTION

RNA synthesis is similar to DNA synthesis. In both cases a polymerase catalyzes the assembly of nucleotides into a complementary strand of the template DNA. Both processes use nucleoside triphosphate with the logical differences required by the different nucleotide composition of RNA. The chain formed replicates the information contained in the piece of DNA used as a template.

As RNA is synthesized on a DNA chain, the transcription is said to be asymmetric. The template DNA strand on which the complementary RNA is assembled is called "antisense." The other DNA strand, not transcribed, has the same sequence (with T instead of U) as the synthesized RNA, this is the "sense" or "coding" strand. Nucleotide binding is catalyzed by DNA-dependent RNA polymerases, which require the presence of a DNA template to synthesize RNA.

All RNA molecules are generated in the nucleus; after synthesized, they are directed to the cell cytosol where they will perform their function.

Transcription in Prokaryotes

In bacteria, RNA polymerase is composed of an oligomeric complex of different subunits. One of these subunits contains the catalytic site responsible for the formation of phosphodiester bonds in the 5′→3′ direction, while another is responsible for attaching the holoenzyme to the template strand. A subunit called sigma (σ) recognizes the place where transcription will be started. All different monomers must be present for the holoenzyme to display synthesizing activity.

The DNA sites recognized by the σ subunit are segments with a defined sequence, called *promoters*. They are located at some distance from the place at which transcription begins. The location of a specific site on a given DNA strand is defined with respect to a reference point. Thus, the site is referred to as "upstream" if located toward the 5′ end of the reference site, or "downstream" if placed toward the 3′ end with respect to a particular position in the strand. The exact location is further specified by the number of nucleotides existing between the site and the reference point, preceded by a − or + sign, corresponding to an "upstream" or "downstream" site, respectively.

In bacteria, the promoter has two modules located at 10 and 35 bases upstream from the segment to be transcribed. These sites are specific sequences called boxes. The module in position −10 has a sequence rich in adenine and thymine (TATAAT) and is called the *Pribnow box*. The other, at position −35, has the sequence TTGACA. There are minor differences among the sequence of these boxes when different species are studied. Segments like these are *consensus sequences* and correspond to sequences in which each position is indicated by the nucleotide residue most frequently present when many sequences with the same function in different cells and organisms are compared.

RNA polymerase binds to the DNA and unwinds the DNA; the enzyme uses only one of the DNA strands as a template. The separation of both strands covers a stretch of 17 base pairs, just over one and a half turn of the double helix. The same RNA polymerase acts by separating the DNA strands and also restoring the double helix after transcription.

In the presence of all ribonucleoside triphosphate molecules (ATP, GTP, CTP, and UTP), the enzyme begins polymerization inserting a purine base nucleotide (ATP or GTP) as the first unit in the chain. This nucleotide retains its three phosphates, which serves as a signal of the initiation site. Unlike DNA polymerase, RNA polymerase does not require a primer segment. Nucleotide binding sequence is dictated by the DNA template, to form stable pairings. Thus, adenine is paired with uracil, guanine with cytosine, thymine with adenine, and cytosine with guanine. The pyrophosphate released is cleaved by pyrophosphatase when each nucleotide is bound.

$$n\,\text{ATP} + n\,\text{CTP} + n\,\text{GTP} + n\,\text{UTP} \xrightarrow[\text{Mg}^{2+}\ \text{DNA guide}]{\text{RNA polymerase}} \text{ARN} + 4n\,\text{PP}_i$$

Polymerization proceeds in the 5′→3′ direction of the newly formed strand; it is antiparallel to the template chain. When approximately 10 nucleotides are bound, the polymerase σ subunit is released. Once the polymerase has advanced enough to free the initial attachment place, another holoenzyme can bind to it. For this reason, on a template strand, it is possible to have simultaneous assembly of multiple RNA molecules using the same information. Each of them has, at any given moment, different lengths according to the distance traveled by the RNA polymerase along the DNA.

The end of a synthesized chain is indicated by a specific DNA sequence, which acts as a termination signal. Shortly before the end of the chain transcription there are GC repeats. The newly synthesized chain has a self-complementary piece and can turn around on itself to form a hairpin near the completion site. This slows or stops the progression of RNA polymerase. Moreover, proteins have been isolated which promote the release of the newly synthesized RNA chain, one of these proteins is called rho (ρ). The existence of initiation and termination signals at the beginning and end of a gene or set of genes is important because they ensure complete copying of the information contained in DNA. Otherwise, the transcript could start in the middle of a gene and end at any place.

Transcription in Eukaryotes

In eukaryotes, the transcription process is similar to that described for prokaryotes, although somewhat more complex. The cells of higher organisms have at least three polymerases (I, II, and III) in their nucleus, each responsible for the synthesis of different classes of RNA. RNA polymerase I (pol I), localized in the nucleolus, catalyzes the synthesis of the three major types of ribosomal RNA (28S, 18S, and 5.8S). RNA polymerase II (pol II) is in the nucleoplasm and it is responsible for the synthesis of the precursor mRNA and other RNAs, including small or short RNAs and long noncoding RNAs. RNA polymerase III (pol III), also in the nucleus, is responsible for the synthesis of tRNAs, ribosomal 5S RNA, and small RNA molecules (snRNA and scRNA). A fourth RNA polymerase in mitochondria is similar to the one in bacteria and transcribes the DNA of these organelles. All these enzymes are composed of 12 or more subunits each. Similar to DNA polymerases, RNA polymerases synthesize the polynucleotide strands in the 5′→3′ direction.

RNA polymerases, unlike those catalyzing DNA replication, cannot correct errors in base pairing. Hence, although transcription has a high degree of accuracy, it does not have the great reliability that the DNA replication process has.

The three polymerases can be distinguished by their sensitivity to α amanitin, a toxin from the fungus *Amanita phalloides*. RNA polymerase I is insensitive to the poison, pol II is completely inhibited by α amanitin (even at very low concentrations), and pol III is only affected by high concentrations of the toxin.

Transcription factors. The bacterial RNA polymerase initiates transcription without additional proteins; in contrast, eukaryotic cells require the assistance of proteins to initiate the process. These proteins are called *basal* or *general transcription factors* and they bind to DNA and place the polymerase in the correct position on the promoter to assist the separation of the double helix two strands and promote transcription initiation.

Transcription factors play a similar role to the prokaryotic σ subunit, but they do not form part of the RNA polymerase. They are denoted by the initials TF followed by Roman numerals I, II, or III according to the polymerase they act with.

Each polymerase interacts with different proteins that collaborate in recognizing promoters, but there is a common factor present in the initiation complex of the three polymerases, it is the protein binding to the TATA box (TBP).

DNA transcription and packing. This is another aspect of transcription that is unique to eukaryotes. While the DNA being transcribed is extended (euchromatin), the inactive (heterochromatin) is supercoiled in nucleosomes, which may hide the transcription initiation site if the promoter is covered.

In this case, histones should be displaced from their position to free promoter areas, allowing access of the transcription complex. Once the synthesis of the RNA chain is started, it continues smoothly. The presence of nucleosomes does not block the transcription.

DNA methylation and histone acetylation significantly influence chromatin structure and the activity of chromosomes. Highly methylated DNA is found in greater proportion in inactive genes. In contrast, when acetylated, histone chromatin "relaxes" and becomes looser, promoting transcription.

Synthesis of Precursor Messenger RNA

RNA polymerase II catalyzes the synthesis of precursor mRNA. In eukaryotes, this RNA is generally longer than the final or "mature" mRNA, whose molecule is used as a template for protein synthesis.

Promoter. The promoter comprises three sites. One of them is at position −25 relative to the initiation site and it is known as the *TATA box* or *Goldberg–Hogness* box. This region is a seven nucleotide consensus sequence formed by thymine and adenine residues (equivalent to bacterial Pribnow box). Most of the TATA boxes are flanked by GC-rich sequences. Other components of the promoter modules are found approximately at −40 and −110 bp, they are known as the *CAAT* and *GC boxes*. Such promoter sequences, called *cis-acting genetic elements*, are on the same DNA strand that is transcribed. The binding of transcription factors to these sites is essential to start the synthesis at the right place; this explains their location at a fixed distance from the initiation site. Single base changes in the promoter can significantly affect the activity of synthesis.

Activators and repressors. Transcription in eukaryotes is also influenced by regulatory proteins (activators and repressors) that bind to specific DNA sequences located sometimes thousands of pairs away from the promoter. The regions in the DNA that function as activators are called *enhancers*. These regulatory sequences can be located upstream, downstream, or even within the piece of the transcribed DNA. Several regulatory sequences may exist for a given gene. It is intriguing that transcriptional activators and repressors act by binding to DNA at distant sites from the promoter. The folding of DNA allows for the approximation of remote areas of the double helix to help place the activator protein, attached to the enhancer, close to the initiation complex and the promoter.

Other transcription factors. In addition to the general transcription factors mentioned, numerous proteins that bind with high affinity to specific sequences in promoters have been identified. These proteins are called upstream regulatory elements (URE) and enhancers. The control of the transcriptional machinery of the RNA polymerase II is mediated by a network of transcription factors, many of which are activated by hormones. In general, the sectors to which transcription factors bind are designated *response elements*. Chapters 23 and 24 will further discuss this topic.

Complex formation. The transcription factor IID (TFIID) initially binds to the TATA box. TFIID comprises multiple subunits, one of which is the TATA box binding protein, TBP. The remaining subunits are polypeptides called *TBP associated factors*, identified with the acronym TAF (Fig. 21.6).

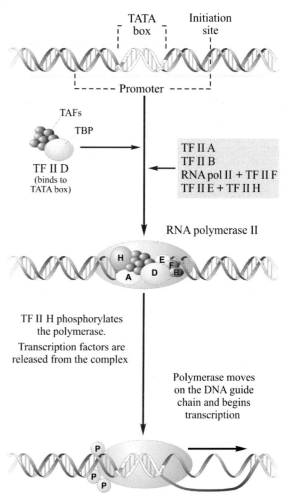

FIGURE 21.6 **Assembly of the transcription initiation complex (schematic).** First, TFIID binds to the DNA double helix in the TATA box of the promoter site, then the other factors attach approximately in the indicated order. The polymerase complex locates the polymerase in the correct position. As the enzyme moves, separation of the DNA helix occurs in a region involving approximately one and a half turns of the DNA.

Other factors are added (TFIIA, TFIIB, TFIIF, TFIIE, TFIIH) and pol II, together with the mediator (see subsequent sections), form the *transcription preinitiation complex*. TFIIH has helicase and protein kinase activity; the protein kinase catalyzes polymerase II phosphorylation at multiple sites.

The phosphorylated pol II detaches from the complex and starts transcription, moving along the DNA template strand. Transcription factors dissociate, separate from the promoter, and are available to start a new round with another pol II.

Cap formation. In eukaryotes, the 5′ end of the RNA chain is rapidly modified by attachment of a molecule of GTP through a 5′–5′ link (note that it is not the 3′–5′ bond commonly seen in polynucleotide chains). Immediately, the nitrogen 7 on guanine is methylated. Thus, the 5′ terminus of the chain is 7-methyl guanosine triphosphate, which functions as a "cap." This distinctive cap is used to identify the mRNA 5′ end by the complex that initiates translation. Also, the cap gives stability to the mRNA, protecting it from the action of phosphatases and exonucleases.

Insertion of the poly A tail. Another transformation of RNA after transcription is the addition of a segment of 100–200 adenine nucleotides (poly A tail) to the 3′ end of the chain. This addition occurs in most eukaryotic mRNAs; only a few, including histone mRNAs, are not polyadenylated.

The poly A tail insertion signal is a sequence of six bases (AAUAAA), located 11–30 bases before the final end. Some RNA chains have more than one insertion signal and more than one site where the poly A chain can be inserted. A specific endonuclease cuts the RNA strand at this site and then a poly A polymerase synthesizes an additional segment of 100–200 adenine nucleotides without a template, using ATP as a donor of adenosine units. The terminal 3′ segment or poly A tail contributes to the efficiency of translation and gives stability to the RNA.

The RNAs synthesized by pol II are released into the nucleoplasm where they form part of the heterogeneous nuclear RNA (hnRNA), which also includes "mature" mRNA. This is formed by processing of the original RNA after transcription. The process, called *splicing*, encompasses the removal of internal pieces of the RNA molecule and splicing of the cut ends. More details on this process will be given in the next chapter.

Mediators are multisubunit complexes which regulate transcription initiation and elongation, expression of RNA polymerase II transcripts (including those from genes coding proteins and noncoding ARNs), and also influences mARN processing.

Synthesis of Transfer RNA

The RNA polymerase III is responsible for the transcription of genes coding for tRNA.

Promoter. The tRNA genes have two promoters (A and B boxes) located "downstream" of the initiation site of the genes coding for the corresponding tRNA.

Complex formation. One of the transcription factors (TFIIIC) binds to the A and B promoter sequences. It acts as an assembly factor that allows for the correct association of another factor (TFIIIB) 50 bp upstream of box A. TFIIIB contains the TBP subunit; it favors the binding and positioning of the pol III to initiate transcription.

Precursor tRNA processing. Initially, a precursor RNA is synthesized; after being processed in the nucleus, it moves to the cytoplasm. The modifications include posttranscriptional methylation, removal of chain segments, joining of the fragments to form the final molecule, and addition of the CCA sequence characteristic to all tRNAs 3′-terminal ends. The addition of CCA is carried out one nucleotide at a time, catalyzed by nucleotidyl transferase. The tRNA has no 7-methylguanosine triphosphate cap or poly A tail. The 5′ end of the tRNA is generated by the action of ribonuclease P, a ribozyme.

Synthesis of Ribosomal RNA

Ribosomes of eukaryotic organisms are composed of a large portion (60S), consisting of three RNA molecules (5S, 5.8S, and 28S) and ~45 different proteins, and a minor portion (40E) comprising a RNA molecule (18S) and ~30 proteins (p. 136). There are thousands of copies of the genes encoding the three rRNA pertaining to the larger portion of the ribosome grouped in the nucleolus. The genes corresponding to the 5S fraction are located in another area of nuclear DNA.

Promoters. The promoters of the rRNA genes in the nucleolus extend ~50 base pairs, starting at position −100. This site is called the *upstream control element* (UCE).

Complex formation. Two factors bind to the promoter sequences, the upstream binding factor (UBF) and the selectivity factor 1 (SL1), which contains the TBP subunit. Eventually, pol I and other initiation protein complex bind to those two factors.

Processing of precursor rRNA. RNA polymerase I synthesizes a long chain of 45S that comprises the three major RNA (28S, 18S, and 5.8S) molecules. This precursor RNA undergoes cuts to generate the "mature" rRNAs. This processing includes methylation of ribose molecules at C2'. Small nuclear ribonucleoprotein particles (snRNA) are involved in these methylations.

Transcription of 5S rRNA. The 5S rRNA genes are transcribed by pol III. In humans there are approximately 2000 copies of the gene arranged in tandem. The promoters have a control region (C box) located downstream from the transcription start site (from position +81 to +99) and the A box (from position +50 to +65). After factor TFIIIA joins the C box, a complex is formed with TFIIIC and TFIIIB. The latter contains a TBP subunit. The pol III binds to the complex and initiates transcription.

Once released, the rRNA molecules are associated with the corresponding proteins. The 18S portion, bound to proteins, forms the ribosomal small particle. 5S RNA, transcribed by RNA polymerase III of extra-nucleolar genes, joins the 28S and 5.8S RNA (synthesized by pol I in the nucleolus) and a group of proteins to form the larger particle. Strict coordination of synthesis of the various components of ribosomes ensures production of equimolar amounts of each.

Reverse Transcriptase

As described in previous sections, the DNA serves as a template for replication by DNA polymerase and for the RNA synthesis during transcription catalyzed by DNA-dependent RNA polymerases. This led to the proposal that the transfer of information takes place only from DNA to RNA. However, subsequent findings demonstrated the existence of a reverse transcription process where DNA is synthesized on an RNA template.

This is catalyzed by a RNA-dependent DNA polymerase, designated *reverse transcriptase.* This enzyme is found in retroviruses causing tumors and other diseases (e.g., Rous sarcoma and acquired immunodeficiency syndrome, or AIDS).

When a retrovirus infects a cell, it forces the cell to synthesize DNA that reproduces the genetic information of the viral RNA. The virus has a reverse transcriptase which catalyzes the incorporation of deoxynucleotides in the 5'→3' direction to assemble a DNA strand complementary to the RNA strand of the intruder. The ribonuclease activity of the same enzyme promotes the hydrolysis of the original RNA template strand and releases the newly formed DNA strand. This serves as a template to build another chain of complementary DNA and to obtain a double helix with the genetic information of the viral RNA (Fig. 21.7). This double helix, called proviral DNA, has sequences at both ends called long terminal repeats containing signals for integration into the host cell DNA, and subsequent transcription of RNA identical to that of the virus. Azidothymidine (AZT) is an antiviral compound used in the treatment of AIDS. Its mechanism of action involves the inhibition of the reverse transcriptase.

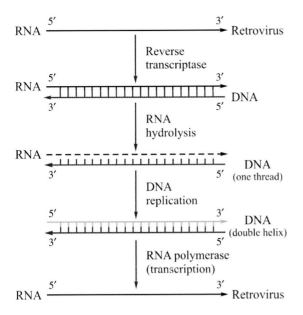

FIGURE 21.7 **Simplified scheme of reverse transcriptase action.** An RNA strand is the "template" for the synthesis of a complementary DNA strand. Subsequently, RNA is hydrolyzed and the DNA serves as a guide to assemble a complementary chain (replication). The double helix is incorporated into the genome of the host cell. This DNA is transcribed to reproduce the viral RNA.

Initially, it was thought that reverse transcription was exclusive to the retrovirus. However, the process also occurs in bacteria and eukaryotic cells. An example is telomerase, which has reverse transcriptase activity.

METHODS USED IN MOLECULAR BIOLOGY

A detailed description of the methodologies used in molecular biology is not the purpose of this textbook. However, a brief overview of the main procedures that are widely used will be presented, since they have allowed for the manipulation of genetic material and the advancement of molecular biology, cellular biology, biochemistry, and medicine.

Base sequence determination. Currently, it is possible to accurately determine the base sequence in a DNA sample. The first method was devised by Sanger. Another technique was proposed by Maxam and Gilbert. Now, these methods have been superceded by the so-called *next generation* technology. Development of automatic equipment and methodological refinements have greatly reduced the time and cost required for DNA sequencing. This has enabled large scale sequencing of wide portions of the genome, which has significantly advanced the field of genomics. The complete DNA sequence of the genome of an organism can be determined at a single time in what has been called whole genome sequencing. Also, the sequence of only the regions of DNA that will be expressed in a cell (exons) can be determined. This is known as whole exome sequencing.

Polynucleotide synthesis. Chemical methods and programmable instruments have allowed the synthesis of oligonucleotides of a defined sequence and relatively short length. As will be discussed later, these DNA fragments can be used for different purposes in molecular biology. Among their uses synthetic oligonucleotides can, for example, serve as primers for DNA sequencing and amplification, as probes (after labeling) for the detection of DNA and RNA, or for the introduction of mutations in a desired gene.

DNA probes. A probe is a segment of single stranded DNA whose sequence is complementary and can specifically pair (hybridize) with regions of single stranded DNA or RNA. They serve to identify, isolate, or amplify a specific region of DNA. Probes can be synthesized using molecular biology techniques, such as reverse transcriptase to generate a segment of RNA, or restriction enzyme digestion to separate a fragment of genomic DNA. Probes are labeled by incorporation of a radioactive isotope element (^{32}P) or an identifiable fluorescent chemical group. This allows for tracing of the probe to detect hybridization sites.

DNA Electrophoresis

When subjected to the action of a restriction endonuclease, a DNA sample is sliced at specific sites that have the sequence recognized by the enzyme. The presence of several sites for a particular restriction enzyme distributed along the DNA results in the release of different fragments that vary in number and length according to the endonuclease involved and the substrate DNA.

The pieces obtained after digestion can be separated by electrophoresis in a support medium of polyacrylamide and agarose (seaweed polysaccharide). If an electric field is established across the gel, the DNA will move toward the anode because it has negatively charged phosphate groups. The polymers forming the gels have a porous matrix, which acts as a sieve, allowing shorter DNA segments to move more rapidly than larger ones. The migration velocity of the pieces of DNA in the gel is inversely proportional to the logarithm of the number of bases contained in each fragment. Polyacrylamide gels can separate DNA segments that differ in only few base pairs and, thus, the method has high resolution.

To visualize the different fragments after migration, any of the following method can be used: (1) Staining with *ethidium bromide,* a fluorescent compound that binds to double stranded DNA. The stained DNA fragments appear as red–orange fluorescent bands when the gel is illuminated with ultraviolet light. Ethidium bromide can be carcinogenic, a reason why the use of another fluorescent stain, *SYBER green,* is preferred. (2) *Autoradiography.* A radioisotope (^{32}P) is incorporated into DNA segments before electrophoresis. Once the migration is complete, the gel is placed on a radiographic film, which is exposed to the radiation of the isotope. This autoradiography of the gel shows bands corresponding to the different DNA segments.

Comparisons of the band digestion pattern in the gel between different DNA samples treated with the same restriction enzymes help identify differences in the structure of DNA due to mutations or genetic polymorphisms. This type of analysis, called *restriction fragment length polymorphism* (RFLP), can be applied to the diagnosis of genetic diseases.

Recombinant DNA

The artificial recombination of DNA from different organisms has been one of the most remarkable advances in molecular biology. Examples of DNA recombination achievements include: (1) chromosome "mapping" to determine the exact location of genes, (2) gene cloning to isolate and control replication of a particular gene, (3) transmission of genetic information from one species to another, and (4) amplification of genome segments. The discovery of restriction endonucleases and mechanism of action and replication of plasmids and viruses has been instrumental in the development of these experimental techniques. These methods have wide application, because the DNA of all living beings has the same general pattern of organization.

Plasmids. Bacteria (prokaryotic) have a chromosome consisting of a circular DNA molecule containing almost all the genetic information necessary for their normal functions. In addition, many bacteria have other circular DNA molecules called *plasmids* (p. 133). Generally, compared to chromosomes, plasmids contain less genetic information presenting only a few genes instead of several thousand. Plasmids can replicate independently from chromosomal DNA and the information they contain can be transmitted across the outer membrane to the following bacterial generation and also to other strains of bacteria. There are usually several plasmids in bacteria and some contain genes that confer antibiotic resistance. Bacteria containing a plasmid conferring resistance to a given antibiotic can be isolated from a sample by growing them in a medium that contains that antibiotic. Only the resistant bacteria develop, those without the plasmid cannot survive. *E. coli,* human intestinal saprophytic bacteria, have been one of the most

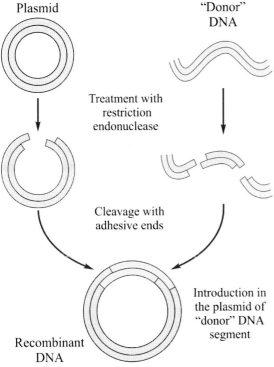

Plasmid

"Donor" DNA

Treatment with restriction endonuclease

Cleavage with adhesive ends

Introduction in the plasmid of "donor" DNA segment

Recombinant DNA

FIGURE 21.8 **Schematic diagram of recombinant DNA preparation.** A plasmid is used as a receptor of the donor DNA. Both are cut by the same restriction endonuclease to produce compatible adhesive ends.

extensively used organisms in studies of recombinant plasmids and DNA.

Formation of recombinant DNA. Recombinant DNA can be produced between DNA from a plasmid and that from any other organism. The plasmid or vector is treated with a restriction endonuclease that only sections the plasmid at a single site, this linearizes the DNA and creates adhesive ends with a particular sequence corresponding to the restriction site used (Fig. 21.8). The DNA from a donor organism is submitted to the same restriction enzyme, which cuts the DNA at all sites where there is the specific base sequence recognized by the endonuclease. Multiple fragments with adhesive ends are generated. One of the methods for DNA recombination uses all the donor DNA fragments formed. In other cases, a particular

piece of DNA obtained by "cloning" (see subsequent sections) is used. Both the digested donor and open plasmid DNA are mixed and DNA ligases are added to attach the DNA strands that have complementary "sticky" ends (Fig. 21.8). A recombinant DNA is produced with donor DNA sequences contained in the plasmid DNA.

The new recombinant DNA can be added to a culture of *E. coli* in the presence of $CaCl_2$ which makes the bacterial wall more permeable. The bacteria capture the new genetic information, replicate it, and synthesize proteins encoded in the piece of DNA introduced. At this stage, the bacteria are transformed. DNA segments of up to 3000 base pairs (3 kb) can be easily inserted in the plasmids. Viruses can also be utilized as vectors of genetic information from one organism to another.

If bacteriophage DNA (p. 137) is used as the vector, the recombinant DNA can be packed into the phage where it is mixed with the viral proteins. These assemble spontaneously in vitro and generate complete phage particles, which are able to infect and introduce the recombinant DNA into *E. coli*. Pieces of 23 kb of foreign DNA can be inserted into the phage. *Cosmids* are hybrids between plasmids and phage lambda (λ) and they contain only a small portion of the phage DNA. These particles are used when segments are larger; it is possible to introduce fragments up to 40 kb into the bacterium. Larger genomic portions (100–1000 kb) have been incorporated in *Saccharomyces cerevisiae* by assembling artificial yeast chromosomes that replicate in each division cycle.

In addition to restriction endonucleases, another widely used enzyme to produce recombinant DNA is *terminal transferase*. This enzyme catalyzes the sequential addition of nucleotides by phosphodiester link to the 3' end of the DNA chains. It indistinctly uses dATP, dGTP, dTTP, or dCTP, and does not require a template DNA for nucleotide assembly. Terminal transferase synthesizes the "sticky" ends of the donor DNA with the ends of the host DNA strands cut at the same level (blunt ends). When a single type of nucleotide is added to the reaction, the 3' ends of

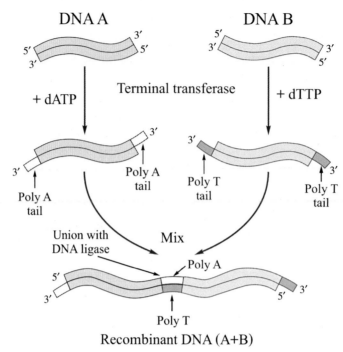

FIGURE 21.9 **Schematic representation of recombinant DNA formation with DNA segments from different sources (A and B) in which both strands of the double helix has been cut at the same level and then ligated by terminal transferase.**

the chains will solely consist of that nucleotide. Thus, if only adenine nucleotides are introduced, a poly A tail will be formed. If the terminal transferase is provided with dTTP as a precursor, it will form a poly T tail on the 3′ end (Fig. 21.9). These reactions, performed on donor and vector DNA, generate sticky poly A and T ends that pair when mixed. Finally, the binding between both DNAs is achieved by using DNA ligase. Deoxyribonucleic acids from many sources can be ligated when applying this technique; for example, mammalian DNA can be combined with DNA from viruses, bacteria, or any other organism. The introduction of foreign genetic material in eukaryotic cells is called *transfection*.

DNA recombination had a great impact on the development of the discipline known as genetic engineering. This has been one of the areas with most active progress in molecular biology. Recombinant DNA methods have numerous applications. Genetic information can be introduced by DNA transfer into a cell, giving it new properties. For example, using DNA transfer, it is possible to obtain bacteria that synthesize therapeutically useful proteins or to enhance the capacity of animals or plants to produce food.

Southern Blotting

Even when derived from different cells or organisms, single DNA strands with complementary sequence spontaneously associate or hybridize to form a double helix when they come in contact in a medium with the appropriate composition and the right temperature. The same occurs between DNA and RNA chains when they have the sufficient base pair complementarity.

This phenomenon of hybridization is used in several techniques to recognize specific sequences in a nucleic acid sample. A piece of DNA of known sequence (probe), complementary to that being investigated, is needed.

Probes have different applications, such as Southern blotting (named after its inventor, Edwin Southern). In this technique, a DNA sample is cut with a restriction endonuclease and the fragments are separated by agarose gel electrophoresis, which is then immersed in an alkaline solution (NaOH) to denature the DNA. Then, DNA is transferred from the gel to another support, such as a sheet of nitrocellulose, which allows for easier manipulation. This is achieved by placing several layers of absorbent paper soaked in a concentrated salt solution inside a cuvette. The gel block containing the separated DNA is placed over the paper block and a nitrocellulose sheet is arranged at the top. Finally, several layers of dry absorbent paper are added (Fig. 21.10). The solution from the lower stack of absorbent paper rises by capillary action to the gel and moves through the nitrocellulose to the top absorbent paper. The fluid moving up carries the DNA in the gel to the nitrocellulose, which will become a replica of the gel. The nitrocellulose sheet is removed and heated in an oven to fix and immobilize the DNA onto this substrate.

The presence of specific DNA sequences in the filter can be detected using a complementary ^{32}P labeled single-stranded segment that serves as probe. This probe will hybridize with its complementary DNA on the nitrocellulose. The unhybridized DNA is removed by washing the sheet. Finally, placing a photographic film on the nitrocellulose will produce an autoradiography showing bands of the DNA in which hybridization with the radioactive probe has occurred (Fig. 21.10).

Methods have also been developed for the transfer from the gel to a nitrocellulose sheet of RNA (Northern blot) and of proteins (Western blot).

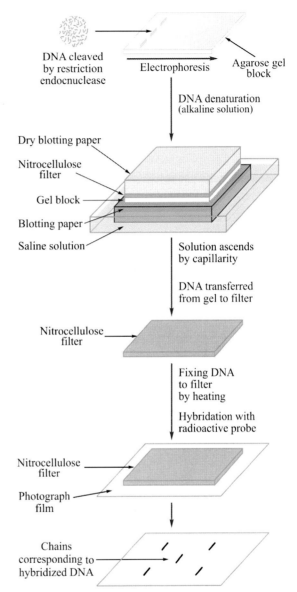

FIGURE 21.10 **Schematic representation of the Southern blot method.**

Amplification of DNA Sequences, Polymerase Chain Reaction (PCR)

Frequently, when the DNA present in a sample is scarce, it can be amplified. One of the methods uses cloning (see subsequent sections). Another widely used method is designated *polymerase chain reaction* (PCR), which can amplify any given DNA segment in a sample. To apply this method, it is necessary to know the sequences flanking the segment of DNA of interest, and to synthesize oligonucleotides complementary to each of the flanking regions. These oligonucleotides serve as primers (Fig. 21.11). The DNA sample is mixed with an excess of primers and deoxyribonucleoside triphosphate molecules (A,T,G,C), plus DNA polymerase. The sample is heated until complete denaturation of the DNA occurs. Upon cooling, the primers anneal with the complementary segments of each of the strands, forming the start site of replication. DNA polymerase begins the assembly of nucleotides from the 3′ end of the primers. Replication of the desired DNA fragment is produced, as this is the only segment capable of binding the specific primer. At the end of this stage, the DNA segment has been duplicated.

A second cycle is then initiated by heating the mixture and the DNA strands will separate again. Primer annealing to the complementary sites and polymerase elongation occurs to complete the chain. At this point, the initial amount of DNA of the selected pieces has quadrupled (Fig. 21.11).

The need to denature the DNA in each step by increasing temperature inactivates most DNA polymerases, which would require providing new enzyme after each stage. To avoid this, DNA polymerase isolated from the thermophilic bacterium *Thermus aquaticus*, which is resistant to high temperatures, is used.

Through successive cycles of PCR, the amount of DNA, specifically comprised within the sites of primer annealing in the mixture, is amplified exponentially. The process can be repeated as many times as necessary. After approximately 25 cycles, almost all the DNA in the preparation corresponds to the segment of interest, providing adequate amounts for subsequent studies. Currently there is equipment (thermal cyclers) that performs the process automatically.

To amplify mRNA, a method known as *reverse transcription* followed by *polymerase chain reaction* (RT-PCR) is used. This technique consists of two stages. In the first, cellular mRNA transcription

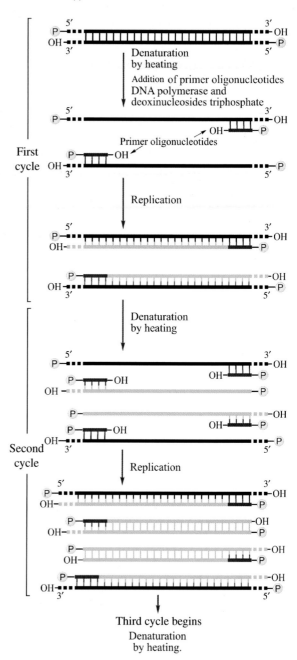

FIGURE 21.11 Schematic representation of the polymerase chain reaction.

from its corresponding DNA is performed in vitro. This is done using a primer with oligonucleotides complementary to sequences flanking the area of interest in the RNA and a reverse transcriptase which acts as a RNA-dependent DNA polymerase. In a second stage, the transcribed DNA, or complementary DNA (cDNA), is amplified by conventional PCR. The level of cDNA synthesized is proportional to the initial amount of RNA, so that the method is semiquantitative. Detection of the DNA synthesized is performed by electrophoresis in agarose gels. Although PCR enables the study of DNA, the reverse transcription PCR detects and estimates the amount and quality of specific mRNA expressed in different cells. This is of great value for the identification of active genes in a given tissue or cell type and the study of gene expression changes that take place under different physiological and pathological conditions. It is also used to study the genome of RNA viruses.

Real-time PCR. Another type of PCR has been developed not only to amplify, but also to determine the relative amount of a specific DNA that has been copied. This is known as *quantitative real time polymerase chain reaction*, or simply *real time PCR* (Q-PCR). The technique is based on the general principle of PCR, but in this case, the method determines simultaneously in each cycle, or "real time," the amount of DNA formed. Unlike traditional PCR, DNA estimation is performed in the initial stages of reaction during the exponential phase of synthesis and not at the end of it (toward the end the reaction is limited by consumption of the components in the assay medium). This method has high sensitivity and allows for relatively accurate determination of the DNA formed. Quantification of DNA is based on labeling the newly synthesized chains with fluorophores. Two different fluorescent molecules are used. In one case, the fluorophore intercalates with the double-stranded DNA formed. In the other, fluorescent oligonucleotides or probes are utilized, which hybridize with defined sequences of the synthesized

DNA. While in the former case any type of DNA is recognized, in the latter, more specific, only determined sequences of DNA can be identified. In both, the fluorescence levels depend on the amount of DNA produced, which in turn is proportional to the DNA molecules initially present in the sample under study. Often, the amount of DNA is expressed in relative terms, comparing it with PCR performed in parallel of known DNA dilutions. Real-time PCR can be performed to determine not only the DNA and products of PCR, but also those of RT-PCR. The method is known as RT-PCR. Applications of Q-PCR are numerous and include the study of expression levels of a given gene in different cell types at different stages of development and cell differentiation, or to determine the activator or repressor effect of agents that affect gene transcription. In addition, it is important in the detection of infectious disease levels and cancer marker genes.

DNA Fingerprinting

Variable number tandem repeats (VNTR), also called *mini-satellites*, are among the families of repetitive DNA dispersed in the genome. Each repeating unit comprises a sequence of 16–64 base pairs. By using restriction enzymes specific for sites flanking the VNTR, fragments of variable lengths in different individuals can be obtained because each subject has a different number of repeats. Fragments can also be obtained and multiplied by PCR using primers complementary to the VNTR flanking sequences.

The repeating units in each group have conserved sequences and probes can be prepared to recognize them. Mini satellites from an individual, analyzed by Southern blotting, give a characteristic banding pattern. Except for identical twins, it is highly unlikely that two individuals with the same pattern will be found because it is an individual characteristic, much like fingerprints. The method, called *DNA fingerprinting*, is used to identify a particular person in forensic cases, or to establish parenthood. Mini satellites have Mendelian inheritance, so that some of the bands detected in an individual will be common with those of the mother and others with those of the father. With PCR, minute amounts of DNA in a sample of blood, semen, or a hair are enough to establish DNA fingerprinting.

Gene Cloning

Cloning refers to the procedures used to obtain genetically identical cells or individuals. A cell population derived from a single progenitor cell is a clone. Applied to genes, the term has somewhat a different meaning. The terms "cloning genes" or "molecular cloning" are used to denote the methods used to produce multiple copies of a defined segment of the genome of an organism and their replication into the DNA of a host cell.

Isolation of a gene in a mammalian cell and its selective multiplication outside its native environment was unthinkable some years ago. Today, it is possible due to the development of genetic engineering.

Isolation of the gene. This represents the first step of cloning.

1. *Preparation of recombinant DNA.* One of the techniques used to isolate a gene is known as "shotgun." The DNA isolated from the organism to which the gene of interest belongs is sectioned into segments appropriate for recombination (between 2,000 and 20,000 base pairs in length) using a restriction endonuclease. The human genome, for example, can be divided into hundreds of thousands of pieces. Subsequently, these fragments are inserted into appropriate vectors (phages) using recombinant DNA methods. In this manner, all the genetic information of the donor cell is divided into the phages, each of which will receive a genome piece; some of these fragments should contain the sought gene.

The phages with recombinant DNA are introduced into bacteria (*E. coli*), where it is replicated. Taking some measures, it can be ensured that only one phage penetrates each bacterium. The bacteria are seeded in culture medium that is sufficiently diluted to allow them to multiply and form a separate colony. Distributed on the plates, many colonies can be easily recognized. Each is a clonal population, derived from a single stem cell; therefore, all bacteria of a colony possess and replicate the same piece of foreign DNA (Fig. 21.12).

The set of clones which are distributed among all the fragments of donor DNA constitutes a *genomic library*.

The next step is to identify the clonal population containing the gene of interest. This operation is simplified by having a piece of identical DNA to the one in the gene. DNA segments with those characteristics, called copy DNA can be used as a probe to find which colony population contains the desired gene.

2. *Preparation of the cDNA probe.* Complementary DNA (cDNA) probes can be synthesized from cells rich in the protein coded by the desired gene (i.e., for the globin gene, precursor cells of erythrocytes; for the proinsulin gene, β cells of pancreas Langerhans' islets). In these cells, there is a predominance of the mRNA that directs the assembly of the protein, making it easier to isolate it in pure form. Once the mRNA for the desired gene has been obtained, it is used as a template to synthesize a cDNA strand by reverse transcription.

cDNA libraries. Genomic libraries, especially those from eukaryotes with very large genomes, are inefficient for finding a particular gene, because a large part of the total DNA is noncoding DNA. DNA libraries prepared from mature total RNA from cells or tissues are more effective. This mRNA is used as a template for synthesis of copy DNA with reverse transcriptase. The obtained cDNA, complementary to

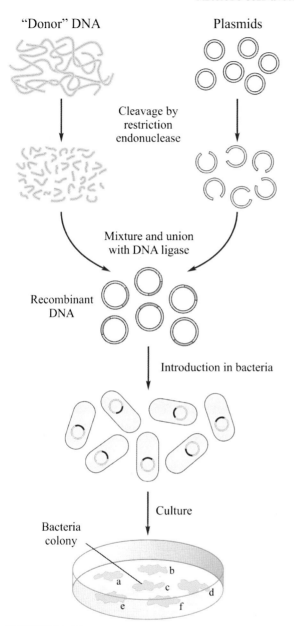

"Donor" DNA   Plasmids

Cleavage by
restriction
endonuclease

Mixture and union
with DNA ligase

Recombinant
DNA

Introduction in bacteria

Culture

Bacteria
colony

FIGURE 21.12   **Gene cloning.** Each bacterial colony constitutes a clone with a distinct piece of the donor genome. Together, they represent a "genomic library."

(see next Chapter) or other noncoding portions of the genome.

To obtain a radioactive cDNA probe, the deoxyribonucleosides triphosphorate dATP, dTTP, dGTP, and dCTP, labeled with ^{32}P in the first phosphate, are added. By action of reverse transcriptase, these nucleotides are assembled on the guide mRNA, forming a complementary DNA strand from which double stranded DNA can be prepared, using DNA polymerase (Fig. 21.13).

There are various methods, other than that mentioned earlier that are used to obtain cDNA probes. Selecting the most suitable procedure aids in the preparation of cDNA from any gene, even if it is in very low concentrations in cells.

3. *Identification of the clone containing the gene.* To achieve this, a disc of cellulose nitrate is applied on the culture, so that it contacts with the colonies developed on the plate. Part of each of the colonies is transferred to the disk, leaving a print of the culture. The bacteria attached to cellulose nitrate (Fig. 21.14) are lysed and their DNA is fixed to the paper on the site of the original colony. Then, DNA is denatured by heating. A cDNA solution (also denatured) is applied over the

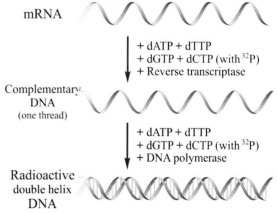

mRNA

+ dATP + dTTP
+ dGTP + dCTP (with ^{32}P)
+ Reverse transcriptase

Complementary
DNA
(one thread)

+ dATP + dTTP
+ dGTP + dCTP (with ^{32}P)
+ DNA polymerase

Radioactive
double helix
DNA

FIGURE 21.13   **Preparation of a copy DNA probe from mRNA of the gene of interest.**

the mature mRNA, possesses only the information for synthesis of proteins that the cell produces and contains no introns disc. After slow cooling, the annealing of complementary strands occurs. Radioactive cDNA hybridizes with the DNA on the paper. Excess cDNA is removed by washing the cellulose nitrate disc and it is placed on a radiographic film. In sites where there is hybrid DNA, the film is impressed (autoradiography), identifying the colony containing the gene of interest.

Once identified, the bacterial colony is seeded in a culture medium in which it will multiply separated from the rest of colonies. The gene has been cloned, now there will be millions of cells that continue replicating it, isolated from the rest of the genome.

4. *Expression of cloned genes.* A foreign gene introduced into the genome of a host organism cannot always be expressed. Often it continues replicating, but is not transcribed into mRNA and, therefore, the encoded protein is not synthesized. In this case, the gene is recovered from the cell in which it was cloned and inserted into another, more efficient vector. If insertion is done in a host genome at a site close to a promoter region, the gene may become expressed and have even greater activity than that in its cell of origin.

Recombinant DNA Techniques Applications

Production of therapeutic proteins. Introduction of human genetic information into bacteria to produce human proteins constitutes an important development. This has allowed for the synthesis of proteins in quantities that are adequate for therapeutic use, including insulin, growth hormone, interferon, erythropoietin, factor VIII, interleukins, and growth stimulating factors.

Production of DNA vaccines. The immune system responds to foreign proteins (antigens) with antibodies or cells that will interfere with the exogenous agent (Chapter 30). Traditional vaccination consists of introducing dead or attenuated infectious agents into the organism to

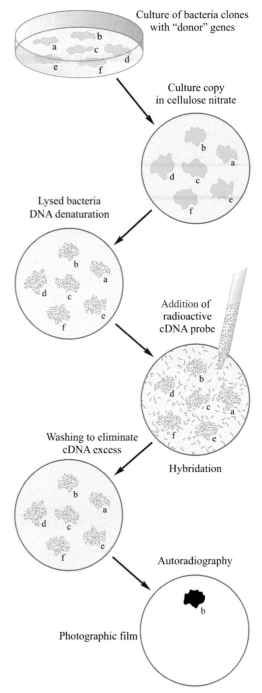

FIGURE 21.14 **Identification of the colony containing a particular gene by radioactive cDNA.** As shown, colony b has the gene of interest.

be immunized, which triggers the synthesis of specific antibodies by acting as antigens. Investigation of DNA vaccines processed through genetic engineering began approximately 25 years ago. One of the techniques used was dermal, subcutaneous, or intramuscular administration of plasmids grafted with the DNA sequence encoding an antigenic bacterial or viral protein. Plasmids penetrate the nuclei of cells (keratinocytes, myocytes, or antigen presenting cells) of the injected tissue and, after incorporation of this genetic information, the corresponding protein is expressed. The body recognizes the protein as foreign and triggers a humoral and cellular immune response with antibody production and T cell activation (Chapter 30).

DNA vaccination has less risk than traditional vaccines, mainly because no germs (either attenuated or dead) capable of causing undesirable reactions are introduced.

DNA vaccines are currently being used in clinical trials in humans with different pathological processes (AIDS, cancer). The results are encouraging, especially with regard to the safety of the procedure. In animals, they have shown to be effective in various diseases and some have been already approved for veterinary use.

Gene therapy. Introducing normal genes in cells with defective genes is now possible by recombinant DNA techniques. For example, it has been possible to incorporate, in vitro, the gene encoding hypoxanthine guanine phosphoribosyl transferase in cells of the connective tissue of patients with Lesch–Nyhan syndrome, which lack this enzyme, restoring their function. This opens new perspectives for genetic alterations that so far have no solution. Gene therapy has been performed, with good results, on patients with severe combined immunodeficiency, in which the normal adenosine deaminase gene was introduced to bone marrow cells.

Gene transfer into somatic cells only corrects the defect in the treated tissue, germ cells are not modified. Although clinical cure could eventually be achieved, the individuals continue to transmit the alteration to their offspring. Genetic manipulation with human germ cells presents challenging and ethical problems; currently, it is not allowed in many countries by government agencies supervising scientific research.

SUMMARY

DNA biosynthesis occurs when a cell divides and it involves the separation of the DNA double helix, with subsequent synthesis of complementary strands. The process is called *duplication* or *replication* and is said to be semiconservative. In *eukaryotic cells*, the separation takes place in the S phase of the cell cycle and occurs simultaneously in all chromosomes, at several sites. When both strands start separating, two "replication forks" move in opposite directions, forming a "bubble." The unwinding enzyme *helicase* separates the DNA strands. Single stranded DNA binding proteins (SSB or RPA) bind to the strands and prevent reassociation. An enzyme known as *topoisomerase* relieves tensions originated by the unwinding. The new strand of DNA is assembled by addition of complementary nucleotides on each chain of the original DNA, which serves as a template, by forming phosphodiester bonds in the $5' \rightarrow 3'$ direction (*DNA polymerases*). In bacteria, three DNA polymerases (I, II, and III) have been demonstrated and in eukaryotes five have been described (α, β, γ, δ, and ε). The chain grows in a $5' \rightarrow 3'$ direction. Structural units used for synthesis enter as deoxyribonucleoside triphosphates (dATP, dGTP, dTTP, and dCTP).

A primer or RNA fragment (10 bases long) is synthesized prior to assembly of the DNA chain by α *primase*, which is firmly attached to α polymerase. DNA α polymerase binds 20 deoxyribonucleotide molecules at the 3' end of the RNA primer, after which the α *primase–polymerase* complex is displaced by the RFC, which also binds PCNA.

Elongation of DNA is accomplished by δ *polymerase*, which moves along the template strand and is held together by *PCNA*. The strand is read in the $3' \rightarrow 5'$ direction (antiparallel to the synthesis of the new strand), thus, one of the strands can be assembled continuously (*leading strand*).

Okasaki fragments are the pieces of DNA being synthesized along the other DNA template strand; each piece is preceded by a primer RNA. The chain synthesized in segments is called the *delayed* or *lagging strand*. RNA primer pieces are removed by *RNase H1* and *FEN1*, and the vacant spaces are covered by DNA segments synthesized by δ *polymerase*. Binding of the adjacent ends of the resulting DNA fragments is catalyzed by *ligase*.

Telomerase is the enzyme that adds repeated sequences to the ends of chromosomes (telomeres) to replenish the loss that occurs in every replication due to the elimination of the initiator RNA in the 5′ ends. Cells that divide infrequently have poor or no telomerase activity; their life is limited by the length of telomeres. Cancer cells with high telomerase activity are immortal.

DNA repair is a mechanism able to correct errors during the process of DNA synthesis. *DNA δ, β, and ε polymerases* act in the repair system. The mechanisms for DNA restoration can repair the following errors in DNA:

(1) mismatches, (2) base excisions, (3) nucleotide excisions, (4) double stranded DNA breaks. This is catalyzed by multiprotein complexes.

DNA recombination is an event that takes place in the gametes during meiosis. DNA recombines when homologous chromosomes interchange segments between the two double helices.

Restriction endonucleases are a series of enzymes, which digest phosphodiester bonds at specific sites in the DNA double helix. They recognize sites formed by segments of 4–7 base pairs long, which are self-complementary when rotated 180 degrees (*palindromic fragments*). The digestion is frequently performed at different levels of each chain. The resulting ends have single stranded adhesives ends.

RNA biosynthesis is called *transcription* and involves the assembly of a complementary RNA using DNA as a template. It is catalyzed by *RNA polymerases*. In *bacteria*, RNA polymerase is an oligomeric complex. The *σ subunit* recognizes the *promoter* site. The promoter has two "boxes" located at positions −10 and −35 from the promoter. When the RNA polymerase binds to the double helix, it is unwound. Only one strand of DNA serves as a template. Ribonucleoside triphosphate molecules (ATP, GTP, CTP, and UTP) are used. The polymerization proceeds in the 5′→3′ direction.

In *eukaryotes* three RNA polymerases are present:

RNA polymerase I: localized in the nucleolus and catalyzes 5.8S, 18S, and 28S rRNA synthesis.

RNA polymerase II: located in the nucleoplasm and is responsible for the synthesis of precursor mRNA.

RNA polymerase III: located in the nucleoplasm and synthesizes tRNA and other small RNA molecules.

Transcription factors are responsible for promoter-enzyme interaction. The general or basal transcription factors are designated with the initials TF followed by the Roman numeral corresponding to the polymerase on which they act and a letter (A–H).

mRNA synthesis starts binding to a *promoter*, typically comprising three sites: the TATA box (in position −25), the CAAT box (in position −40), and GC (in position −110). In addition to promoters, *enhancer* sequences are necessary. mRNA transcription is initiated by the binding of TFIID to the TATA box. Then, it forms a complex with other factors that fix the polymerase II in the correct position. The enzyme detaches from the complex and initiates transcription. The 5′ end of the newly synthesized RNA strand gets a "cap" of triphosphate 7-methylguanosine, which provides stability. 100–200 adenine nucleotides are added to the 3′ end (*poly A tail*). Specific sequences located shortly before the end portion serve as polyadenylation signals. Subsequently, RNA precursors are processed in the nucleus to reach their "maturity."

Reverse transcriptase catalyzes the process that allows synthesis of DNA using RNA as a template. It is catalyzed by the *reverse transcriptase*, an enzyme first isolated from retroviruses.

Bibliography

Alberts B., Jonhson A., Lewis J., Raff M., Roberts K., Walter P., 2008. Molecular Biology of the Cell, fifth ed Garland Science, Taylor & Francis Group, New York.

Baker, T.A., Bell, S.P., 1998. Polymerases and the replisome: machines within machines. Cell 92, 295–306.

Collins, F.S., Ginsburg, D., Waga, S., Stilman, B., 1998. The DNA replication fork in eukaryotic cells. Ann. Rev. Biochem. 67, 721–751.

Cooper, G.M., Hausman, R.E., 2007. The Cell. A Molecular Approach, fourth ed Sinauer Associates, Inc., Sunderland.

Ishmael, F.T., Stellato, C., 2008. Principles and applications of polymerase chain reaction: basic science for the practicing physician. Ann. Allergy Asthma Immunol. 101, 437–443.

Kornberg, R.D., 2005. Mediator and the mechanism of transcriptional activation. Trends Biochem. Sci. 30, 235–239.

Krebs, J.E., Goldstein, J.E., Kilpatrick, S.T., 2013. Lewin's Genes XI. Jones & Bartlett Publishers Inc, Burlington.

Lodish, H., Berk, A., Kaiser, C.A., Kriger, M., 2007. Molecular Cell Biology, sixth ed W.E. Freeman, New York.

Morita, R., et al., 2010. Molecular mechanisms of the whole repair system: a comparison of bacterial and eukaryotic systems. J. Nucl. Acids 2010, 1–32.

Murphy, J., Bustin, S.A., 2009. Reliability of real-time reverse-transcription PCR in clinical diagnostics: gold standard or substandard? Expert Rev. Mol. Diagn. 9, 187–197.

Poss, Z.C., Ebmeier, C.C., Taatjes, D.J., 2013. The mediator complex and transcription regulation. Crit. Rev. Biochem. Mol. Biol. 48, 575–608.

Turner, R. (Ed.) 2002. RNA. Nature 418, 213–258.

Watson, J.D., Baker, T.A., Bell, S.P., Gann, A., Levine, M., Losick, R., 2008. Molecular Biology of the Gen, sixth ed Pearson Benjamin Cummings, Cold Spring Harbor Laboratory Press, San Francisco.

Weaber, R.F., 2008. Molecular Biology, fourth ed McGraw Hill, New York.

22

The Genetic Information (II)

PROTEIN BIOSYNTHESIS

The genetic information transmitted from parents to progeny and from cell to cell is determined by the primary structure of DNA. During cell division, DNA has the capacity to self-replicate and to generate exact copies of the original molecule, ensuring that all cells of an individual, derived from successive divisions of the egg, possess the same DNA.

The nucleotide sequence of specific regions of DNA, corresponding to genes, contains "encoded" information to guide the synthesis of RNA and proteins. Messenger RNA (mRNA), ribosomal RNA (rRNA), transfer RNA (tRNA), small nuclear RNA, small and long noncoding RNAs are all produced from the message contained in DNA. The mRNA is the only RNA that carries the code from DNA in the nucleus to the cytoplasm, where it will be used to build proteins. The other RNA types play additional functions that are also essential for the protein synthesis process.

The genetic message encoded in the DNA base sequence (A, G, C, and T) is first transcribed into another "language," represented by a sequence of RNA bases (A, G, C, and U). This RNA sequence is used as guide or "template" from which amino acids will be assembled into proteins. While genes provide information for the synthesis of all RNA types, this chapter will particularly focus on the production of mRNA, which is used as a "template" for the assembly of polypeptide chains.

The process of protein synthesis can be viewed as the translation between 2 "languages," one represented by the particular arrangement of 4 nucleotides in DNA and RNA and the other corresponding to the specific sequence of the 20 amino acids.

This process can be represented as follows:

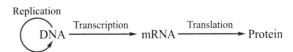

Aside from referring to the transfer of genetic message from RNA to proteins, the term *translation* in medicine has been recently used with another meaning.

Translational medicine denotes the application in clinical practice of new advances of basic research from different fields, such as genomics, general and molecular biology, biochemistry, and biophysics. This includes in vitro studies (using cell or tissue cultures) or in vivo studies (using animal models), which can improve the diagnosis or treatment of different clinical conditions.

Genetic Code

The sequence of bases in mRNA provides the template that dictates the order in which amino acids need to be placed to synthesize the proper polypeptide chain.

How is it possible that only 4 bases can indicate the sequence in which 20 amino acids need to be placed in a protein? If each base corresponded to a particular amino acid, the four bases would only contain information to accommodate four amino acids. If instead, 2 bases defined an amino acid, the various arrangements of the 4 bases would allow to encode information for 16 (4^2) different amino acids. This number is still insufficient to encode for the whole 20 amino acids. This puzzling problem was resolved by the demonstration that an amino acid is specified by a set of three consecutive nucleotide bases, also known as a triplet. The combination of 4 bases into triplets allows the generation of a total of 64 different triplets (4^3), which is more than sufficient to direct information for the existing number of amino acids.

TABLE 22.1   Genetic Code

Aspartic acid	GAU–GAC
Glutamic acid	GAA–GAG
Alanine	GCU–GCC–GCA–GCG
Arginine	CGU–CGC–CGA–CGG–AGA–AGG
Asparagine	AAU–AAC
Cysteine	UGU–UGC
Phenylalanine	UUU–UUC
Glycine	GGU–GGC–GGA–GGG
Glutamine	CAA–CAG
Hystidine	CAU–CAC
Isoleucine	AUU–AUC–AUA
Leucine	UUA–UUG–CUU–CUC–CUA–CUG
Lysine	AAA–AAG
Methionine	AUG
Proline	CCU–CCC–CCA–CCG
Serine	UCU–UCC–UCA–UCG–AGU–AGC
Tyrosine	UAU–UAC
Threonine	ACU–ACC–ACA–ACG
Tryptophan	UGG
Valine	GUU–GUC–GUA–GUG
Termination codons	UAA–UAG–UGA

A, Adenine; C, cytosine; G, guanine; U, uracil.

Ingenious experiments were conducted to decipher the "meaning" or translation of each triplet into an amino acid. The decoded "key" in the RNA message is known as the *genetic code*. Each base triplet is known as a *codon*. Table 22.1 shows the amino acids encoded by the 64 different codons.

Some amino acids are encoded by more than one base triplet. For example, arginine, leucine, and serine are represented by six different triplets each; five amino acids are represented by four; and nine are represented by two codons. Only methionine and tryptophan are specified by a single codon (AUG and UGG, respectively). The various triplets corresponding to the same amino acid are called synonyms. The existence of several triplets to indicate the same amino acid is known as degeneracy or redundancy of the genetic code. It is important to note that there are no ambiguities; each codon codes for only a specific amino acid.

In general, the third base in a codon is the less critical one; it can be changed without affecting the amino acid coded (Table 22.1).

The UAA, UAG, and UGA triplets do not correspond to amino acids (originally they were called nonsense triplets); they indicate the termination of the polypeptide chain. The following analogy can be made: while the triplets that encode for each amino acid represent the "words" of the genetic language, the termination codons symbolize the punctuation marks.

The genetic code is universal; codons have the same meaning in all living beings. However, there are some exceptions; the mitochondrial genome uses a slightly different nuclear code. Variations have also been observed among organisms. For example, the AGA codon, which codes for arginine in the standard code, indicates termination in mitochondria of some animals and serine in others. The triplet UGA, which indicates termination in the common code, corresponds to tryptophan in mammalian mitochondria and yeast. The triplet CUA indicates leucine in the general code, but codes for threonine in yeast mitochondria. Lastly, the triplets UGA and UAG that generally correspond to termination, code for glutamine in protozoa.

Nuclear DNA

Nuclear DNA forms the genes that are transcribed into mRNA and directs the order of amino acids in the synthesized polypeptides. Sets of 3 consecutive bases in the mRNA form *codons*; 61 of the 64 possible codons specify amino acids, the other 3 are termination codons, which indicate the end of the polypeptide chain. The codon sequence corresponds to the amino acid sequence in the corresponding protein. The set of codons with complete information for the synthesis of a polypeptide chain is called *cistron*. The gene coding for a protein, that is, 300 amino acids long (average size), corresponds to a segment of DNA of 900 nucleotides, which is transcribed into an mRNA of equal extension.

In bacteria, in general, each cistron is a continuous piece of DNA, whose base sequence corresponds, in the vast majority of genes, to the codons in the mRNA transcript and with the amino acid sequence of the resulting synthesized protein.

The bacterial genome contains, on average, more than 4 million base pairs. This number is expected due to the amount of DNA encoding the proteins that the bacteria produce. By contrast, in higher animals, there is a relative large excess of DNA in relation to the proteins to be synthesized; it is estimated that less than 2% of the total DNA is protein-coding DNA. In humans, the haploid genome comprises approximately 3 billion of base pairs, far more than those necessary for the polypeptide chains that need to be synthesized. In fact, only approximately 21,000 genes direct the production of mRNA in humans.

Unlike bacterial genes, eukaryotic genes are not organized into continuous pieces of DNA; instead, they are divided into segments called *exons*. Between exons are portions of DNA, which do not contain genetic information, these

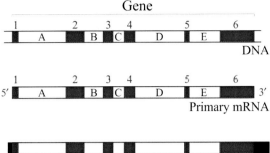

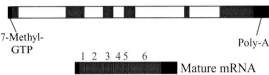

FIGURE 22.1 **Schematic representation of an eukaryotic gene, comprising six exons and five introns.** The precursor RNA, or the primary transcript, is processed to form "mature" RNA. The *numbers* indicate exons *(red)* and the *letters* introns *(white)*.

segments are called *introns*. In general, the length of introns far exceeds that of the exons (Fig. 22.1).

For example, the chicken ovalbumin gene is a DNA fragment that is 7700 bases long. Of these, only 1158 bases encode for protein segments and they are divided into 8 exons. The length of the intercalated sequences, or introns, is almost 6 times larger than that corresponding to the actual genetic information. In humans, the globin gene contains two interspersed sequences. The albumin gene has14 exons and the collagen gene has more than 40 exons. An extreme case is the gene coding for factor VIII, a protein involved in the process of blood coagulation (Chapter 31). It has a length of 186 kb (kilobases) and 26 exons, separated by 25 introns. The 25 introns correspond to over 175 kb, with the coding portion being only about 5% of the total gene. There are few mammalian genes that do not contain introns, these include those coding for proteins, such as histones and interferons.

Repeated genes. In addition to introns and other noncoding sequences, the size of the eukaryotic genome is also increased by the existence of repeated genes. In some cases in which large amounts of the gene product, such as rRNA and histones are required, multiple copies of the corresponding coding portions of DNA are present in the genome.

The multiple copies of a particular gene constitute a gene family. Sometimes the various members in a gene family are not exactly the same and are differently transcribed in diverse tissues, or during development. This is the case for subunits of globin, which are encoded by two gene families (α and β); ζ, α$_1$, and α$_2$ belong to the α family, while ε, γ^G, γ^A, δ, and β belong to the β family. Some are expressed in the embryo, others in the fetus or after birth (p. 51). The α family of globin genes are located on chromosome 16 and the β family genes are on chromosome 11. In other cases, members of the same family are scattered on different chromosomes.

According to the most accepted hypothesis, repeated genes have arisen in the course of evolution by successive duplications of an ancestral gene. Duplicate genes can mutate (see subsequent section) and differentiate themselves independently. As a result, after millions of years, a group of genes encoding homologous proteins with different functional properties were produced. For example, the γ subunits of globin form fetal hemoglobin with higher affinity to oxygen (p. 55). The different gene products resulting from variability in structure and function of a same protein are known as isoforms.

Pseudogenes. Not all mutations result in production of functional genes. In some gene copies, the change may suppress the ability of the gene to encode a useful molecule. For example, within each of the α and β families of globin there are two genes, which present mutations that make them nonfunctional. These inactive copies of genes are called *pseudogenes.* They can be considered evolutionary relics, or genetic "junk" that increases the size of the genome without providing any apparent function.

Human Genome Project

In 1990, an ambitious project started that was aimed to determine the full sequence of the ~3 billion base pairs of the human haploid genome. This represented the largest undertaking in the field of life sciences, involving an International Consortium composed of laboratories from the United States, Britain, France, Germany, Japan, and China. It was estimated that the project would be accomplished in 15 years. In 1998, Celera Genomics, a private company, started sequencing the human genome using different methodologies.

The improvement and automation of techniques accelerated the pace at which DNA was sequenced. A "draft," which described more than 90% of the human genome, was first reported in June 2000. The data were published in February 2001; in April 2003, the complete sequence of the human genome was announced and made available to the public. This impressive achievement had a huge impact in the life sciences in general, and medicine in particular. The results provided surprising information: (1) there are fewer genes than previously supposed. It is currently estimated that there are a total of approximately 21,000 genes. (2) Only less than 2% of the total number of bases in DNA are exons, 24% are introns and other sequences that do not encode for proteins. The majority of these DNA sequences serve as a guide for the synthesis of noncoding RNAs and small or long RNAs that have regulatory functions.

The results also posed new challenges for researchers, who needed to identify all functioning genes and the mechanisms that regulate their activity within the entire genome. Knowledge of the human genome is essential in medicine for the identification of genes associated with pathological conditions, the early detection of genetic predisposition to disease, and for possible gene therapy.

Along with these advances, ethical, and legal issues arose, which bring along risk of discriminating individuals based on the finding of altered genes.

Mitochondrial DNA

Mitochondria contain their own genome. In humans, mitochondria are composed of a total of 16,569 nucleotide base pairs. Sequence analysis has shown that mitochondrial DNA includes 37 genes, which code for 2 rRNAs, 22 tRNAs, and 13 components of polypeptide subunits in the respiratory chain. This genome is not self-sufficient; most mitochondrial proteins are encoded by nuclear genes, synthesized in the cytoplasm, and transferred to the organelle by a specific import system.

Mitochondrial DNA has similar characteristics to those of prokaryotic DNA and differs from nuclear DNA in several aspects: (1) mitochondrial DNA virtually does not possess introns and intergenic sequences; (2) mitochondrial DNA from an individual is only inherited from the mother, the mitochondria of fertilized eggs come from the ovule and those from the sperm are eliminated; (3) the mitochondrial genome divides independently of the cell cycle and it is not replicated synchronously with cell division; (4) four codons in the mitochondrial genetic code differ from the "universal" code. In the standard code, UGA is a termination codon, which in mitochondria encodes for tryptophan. AGA and AGG, which normally encode for arginine, are termination codons in mitochondria. Lastly, AUA, which encodes for isoleucine, represents methionine in mitochondria; (5) the mitochondrial genome has a higher mutation rate (10–100 times greater) than nuclear DNA. This is due to the lack of a mitochondrial DNA repair system.

On the other hand, the processes of replication, transcription, and translation in mitochondria are similar to those of prokaryotes.

Various pathological conditions caused by alterations of mitochondrial DNA have been described. These include: mitochondrial encephalopathy, ocular myopathy, Leber hereditary optic neuropathy, maternal retinitis pigmentosa, and myoclonic epilepsy.

Messenger RNA

During transcription, mRNA is synthesized containing a nucleotide sequence complementary to that of the DNA guide strand. In eukaryotes, DNA-dependent RNA polymerase II catalyzes the synthesis of mRNA.

Genes copied by RNA polymerase can be found in any of the two strands of DNA; with the antisense strand being the only one transcribed. The other DNA strand, called the coding or sense strand, has a sequence equal to that of the synthesized RNA (only differing by having thymine residues instead of uracil residues).

Each gene is transcribed in its entirety, including not only the coding portions (exons) but also the noncoding segments (introns). This primary precursor mRNA, forms part of the heterogeneous nuclear RNA (hnRNA) and, after being synthesized, it undergoes major changes in the nucleus.

In eukaryotes, the existence of the nuclear membrane separates the DNA transcription process that occurs in nucleus, from the RNA translation events that takes place in the cytoplasm. The mRNA produced in the nucleus must pass through nuclear membrane pores, where it becomes exposed to degradation catalyzed by nucleases. To prevent hydrolysis, the mRNA binds to proteins that protect it. The transport from nucleus to cytoplasm requires that the mRNA from eukaryotes have a greater stability than that of prokaryotes (half-life of hours or days, compared to minutes). The half-life of a specific mRNA depends on its poly-A tail, which also influences the regulation of mRNA expression.

Precursor mRNA processing. The posttranscriptional processing of mRNA does not end with the addition of the 7-metilguanosine triphosphate "cap" at the 5' end and the poly-A tail at the 3' end. Most of mRNA primary chains contain introns which must be eliminated. These noncoding segments are sectioned and exons are spliced in the correct order to form the "mature" mRNA strand, which will serve as the template for protein synthesis. This process is called *splicing*.

Intron excision and orderly joining of exons is normally performed with great precision. Defective splicing leads to significant protein synthesis alterations.

snRNP. In the nucleus, there is a group of particles called *small nuclear ribonucleoproteins* (snRNP), which play an important role in splicing. These are commonly known as "snurps." The snRNP proteins, designated with the letter U plus a number (U1, U2, U3, etc.), are attached to ribonucleic acid chains that are 100–300 nucleotides long and are usually rich in uracil.

Splicing. Introns vary in length, from 60 to 10,000 nucleotide bases. All introns start with GU (5' end) and finish with AG (3' end). These nucleotides are part of consensus sequences that span a relatively short stretch at each intron's extreme. Besides these, there is another consensus sequence located between positions −20 and −50, relative to the intron's 3' end, which is called the *branching site* (in yeast, this sequence is UACUAAC). With the exception of these sequences, the remaining regions show no homology among different introns.

Splicing starts with the formation of a complex, consisting of several snRNP and a precursor RNA bound to a protein. This assembly is designated *spliceosome*. A snRNP, called U1, has an RNA stretch with a sequence that is complementary to that found in the vicinity of the 5' end of introns. This allows the pairing of U1 to the cleavage site at the 5' end of the intron in the precursor RNA. Through a similar mechanism, a U2 snRNP particle joins the branching sequence. Then, U4 and U6, forming a single block, and U5 attach to the complex. The precursor RNA is sectioned at the 5' end of the intron, not by hydrolysis but by a transesterification reaction; a new phosphodiester bond is established between the 5' end of the intron and the OH on C2 of an adenine nucleotide at the branch site. As a result, a loop or lariat is formed. Then, the intron is released by excision at the 3' end. A new diester transesterification forms a 3'–5' bridge between the 3' end of one exon (exon 1, situated on

the left in Fig. 22.2), and the 5' end of the second exon (exon 2, located on the right in Fig. 22.2). The lariat intron is released and degraded by hydrolysis in the nucleus.

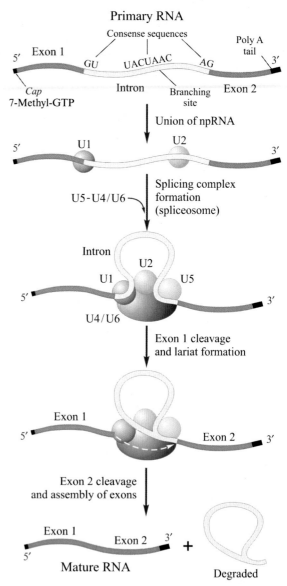

FIGURE 22.2 **Schematic representation of the mRNA splicing process.**

In general, exons encode for "functional domains" of proteins. For example, immunoglobulin heavy chains have three domains in the constant portion (CH1, CH2, and CH3), each of which is encoded by a separate exon (see Chapter 30).

Differential processing of mRNA. A precursor mRNA may generate different mature mRNAs. The splicing of mRNA can sometimes generate different proteins from the same gene. In the primary transcript, there is often more than one site where the poly-A tail can be inserted; depending on its location, the poly-A tail can give rise to different mature mRNAs, which will encode different proteins. An example is the mRNA that directs the synthesis of the thyroid hormone, calcitonin. In the pituitary gland, the same primary transcript is processed into mRNA that synthesizes a neuropeptide. This depends on the differential use of two different polyadenylation sites in the transcript.

It is also possible that splicing will either include all, or only some of the exons present in the primary transcript (alternative splicing). In this manner, the same precursor RNA can generate different proteins. Alternative splicing is an important process that modifies the genetic information from transcription to translation.

Another mechanism for RNA modification consists in editing the primary mRNA transcript. This is performed by insertion or deletion of bases, which creates a splicing site, or alters the structure of the RNA. A remarkable example of "differential RNA editing" is that of apolipoprotein B100. In the liver, mRNA transcribes, without modification, all of the exons encoding a polypeptide chain of 512 kDa (apolipoprotein B100). In intestinal tissue, the mRNA is edited by changing a cytosine residue to a uracil residue, which creates a stop codon. The resulting protein, apo B48, is shorter than apo B100 and has a primary structure equal to the initial 48% of apo B100. Genes that can originate more than one mature mRNA are called *complex transcription units*.

Diseases caused by splicing alterations. DNA mutations that affect the splicing process can lead to changes in the precursor mRNA, the spliceosome complex that recognizes the sites of exon cleavage, the splicing itself, and the binding of regulator factors. All these defects can produce different clinical conditions. Some are characterized by muscle and nervous alterations, including myotonic distrophia, spinal muscle atrophy, and ataxia telangiectasia, all of which were already mentioned when considering disturbances in DNA repair. Other cases present predisposition to malignant transformation, autoimmune diseases, or premature aging.

Thalassemias are anemias derived from different genetic failures that affect globin synthesis; some of them are due to splicing defects. In patients with the autoimmune disease lupus erythematosus, specific antibodies against various ribonucleoproteins, including U1 and snRNP are produced; these antibodies inhibit the splicing process.

Mature mRNA. Once the posttranscriptional modifications are completed, the terminal mature mRNA is exported to the cytoplasm through pores in the nuclear membrane. The mature mRNA contains a 7-methyl-GTP cap at the 5' end, followed by a noncoding sequence of approximately 100 bases in length. This is followed by the AUG codon, which encodes for the initiating methionine. Next, there is a coding portion indicating the amino acid sequence of the protein to be synthesized. This is followed by the termination codon (UAA, UAG, or UGA) and a segment of variable length (trailer sequence). Finally, most mRNAs contain the poly-A tail, which is approximately 200 adenine nucleotide residues in length. These last two segments correspond to nontranslated regions. The RNA segment that contains the nucleotide triplets encoding for a whole polypeptide, from the initiation to end, is called the *open reading frame*.

In bacteria, genes rarely have introns. Furthermore, prokaryotes commonly have polycistronic mRNA, which carry information to synthesize a series of polypeptide chains; several genes are transcribed one after another in the same mRNA molecule. In eukaryotes, mRNA is always monocistronic, encoding just a single polypeptide chain.

Transcriptome

Transcription is a key process for gene expression, requiring the synthesis of various types of RNA. At any given time, only a small portion of the genes encoding proteins are active. The analysis of mRNA provides a snapshot of the cell's transcriptional activity; this is functional data that the study of the genome per se cannot provide. The whole content of mRNA in a cell, tissue, or organism has been called *transcriptome* and the area of molecular biology that studies it is known as *transcriptomics*.

There is a marked difference in genome utilization between species. For example, almost all of the yeast genome encodes for proteins, whereas in mammals, only 2% or less corresponds to protein coding sequences. This suggests that the larger nonprotein coding transcriptome is related to the greater complexity of higher eukaryotes.

Although all cells of an organism have the same DNA, the mRNA repertoire or transcriptome of an individual varies from one cell type to another, and within the same cell, depending on the time at which it is determined. This is due to the influence that factors, such as hormones and other endogenous or exogenous agents, have on the transcriptome. From a medical point of view, studying the transcriptome is of great value to understand the changes that occur during disease. One of the most efficient methods for the characterization of mRNA in a tissue sample is the microarray assay (see subsequent sections).

Ribosomal RNA

The ribonucleic acid associated to proteins forms ribosomes, the cell particles involved in protein synthesis. Bacteria such as *Escherichia coli* contain ~20,000 ribosomes; in contrast, mammalian cells have approximately 10 million ribosomal particles. Due to this difference, mammalian cells have multiple repeats of the genes coding for rRNA. The synthesis of the major rRNAs (28S, 18S, and 5.8S) is catalyzed by RNA polymerase I in the nucleolus, where a precursor of 45S containing all three rRNAs is formed. From this rRNA, the 28S, 18S, and 5.8S rRNA species are released by cleavage of the primary rRNA. Sometimes, intercalated segments or introns are present and this requires processing that is similar to that described for mRNA. The 5S rRNA is assembled outside the nucleolus, by the action of RNA polymerase III. The "mature" rRNA binds to proteins to form the subunits of ribosomes (Fig. 6.21). When both ribosomal portions are joined, a cleft is left between them, through which the mRNA slides during protein synthesis. Two adjacent sites in the ribosome, named A (aminoacyl) and P (peptidyl), are involved in binding the tRNA loaded with an amino acid. Bacteria also have a third ribosomal position, called the E (exit) site. Ribosomes constitute the structure that provides support for the various components involved in protein synthesis.

Transfer RNA

The synthesis of tRNA is catalyzed by RNA polymerase III in the nucleus. Once produced, the tRNA precursor must be processed to remove an intron and to modify bases in its structure. This type of RNA, also called soluble, is responsible for binding free amino acids in the cytosol and transporting them to the site of assembly into new protein chains.

There are various tRNAs, which are specific for each amino acid. The 3' terminal sequence, CCA, is identical for all tRNAs and it binds to the amino acid. The amino acid specificity of tRNA lies in the central loop of the molecule (Figs. 6.19 and 6.20), where there is an anticodon triplet complementary to the codon corresponding to the amino acid carried by the tRNA. For example, the tRNA for tyrosine possesses the triplet GUA in a segment of the central loop, which is the anticodon of the tyrosine codon (UAC). For alanine, tRNA has the anticodon GGC (corresponding to the alanine codon, GCC). This codon–anticodon pairing is antiparallel and the sequence is always read in a 5′→3′ direction.

tRNA functions as an intermediary or adapter molecule, which recognizes a specific nucleotide sequence in the mRNA and locates the amino acid in the corresponding position.

Due to the redundancy of the genetic code (most amino acids are specified by more than one codon), there is more than one tRNA for some amino acids. Among the 64 possible codons, 61 encode amino acids (3 are termination signals). Similarly, in the bacterium *E. coli*, there are approximately 40 different tRNAs, indicating that some tRNAs recognize more than one codon corresponding to the same amino acid. The pairing of the third base of the codon with the first in the anticodon is usually less strict than that for the other two bases, allowing the pairing of a tRNA with more than one triplet.

Ribonucleic Enzymes or Ribozymes

Sumner discovered that urease, the enzyme that cleaves urea, is a protein. This discovery led to the dogma of biochemistry that states "all enzymes are proteins." The finding that some RNAs also contain catalytic ability was quite surprising. Ribonucleic acids behave as enzymes in many organisms; they have all the properties of biological catalysts and are called *ribozymes.*

According to their function, most ribozymes are hydrolytic or participate in the process of RNA splicing. They are composed of 40–200 nucleotides in length and present different secondary structures, which give them particular three-dimensional conformations. Hydrolytic ribozymes cleave the phosphodiester bonds that are placed between nucleotides in the RNA chains. Some have autocatalytic function, causing their own cleavage; others, such as RNase P, function as endonucleases in the processing of tRNA precursors. Ribozymes are involved in splicing, removing introns, and attaching adjacent exons to make mature RNA. In this process, introns themselves assist the enzymatic action.

There are other functions in which ribozymes are also involved. Formation of peptide bonds in the protein chain during synthesis is catalyzed by *peptidyl transferase*, a ribozyme in the larger particle of the ribosome. Some observations indicate that RNA molecules are involved in the reactions catalyzed by aminoacyl-tRNA synthetase (see next section).

From an evolutionary point of view, the discovery of catalytic RNAs led to the hypothesis that, during evolution, RNA appeared before DNA or proteins; they were needed to perform functions that later were assumed by these molecules. RNA contains information encoded in its base sequence and it duplicates by mechanisms that are similar to those of DNA. Furthermore, RNA forms secondary structures and can adopt various conformations that resemble those of proteins. According to some scholars, sometime during the development of life on Earth (about 4 billion years ago), RNA might have been the fundamental molecule, in what they called the "RNA world." The later appearance of proteins and DNA offered new molecules that were more suitable for a variety of cell functions. Proteins have the advantage of exhibiting greater chemical and conformational diversity, and are more flexible and efficient as catalysts than RNA. Similarly, DNA has advantages over RNA due to its higher chemical stability. Therefore, evolution and the forces of natural selection might have favored the replacement of RNA for DNA and proteins for activities in which the latter molecules are more advantageous. On the other hand, RNA was conserved for its highly efficient role in translation of the genetic information.

MECHANISM OF PROTEIN BIOSYNTHESIS

Protein biosynthesis comprises the translation of the message contained in mRNA and the assembly of amino acids in the order dictated by mRNA. Three types of RNA (mRNA, ribosomal, and tRNA), as well as a series of proteins participate in this process.

Ribosomal particles slide over the mRNA, reading from one codon to the next, and inserting tRNAs one at a time on their respective complementary codon. The amino acids carried by each tRNA are linked by peptidic bonds following the sequence indicated by the mRNA, to form the polypeptide chain. The incorporation of amino acids in the exact sequence requires the correct binding of specific tRNAs to the mRNA guide by codon–anticodon pairing. The mRNA is translated in the 5′→3′ direction and the polypeptide chain is assembled from the N- to the C-terminus.

When protein synthesis starts, both the major and minor subunits of the ribosome associate with each other, and with the mRNA; the complex disassembles soon after the polypeptide chain synthesis is finished. This process of association and dissociation of the ribosome is called the *ribosomal cycle*. Each ribosomal particle can simultaneously bind two tRNAs. One is the tRNA loaded with an amino acid (aminoacyl-tRNA), which binds to ribosomal site A; the other, which carries the nascent polypeptide chain in formation (peptidyl-tRNA), binds to ribosomal site P. Finally, the tRNA that has released the finished polypeptide is freed. In prokaryotes this occurs from a third binding site in the ribosome, called the exit site (E).

Protein biosynthesis involves four main steps (1) amino acid activation, (2) initiation, (3) elongation, and (4) termination of the polypeptide chain.

1. *Activation of the amino acids.* This stage requires free amino acids, aminoacyl-tRNA

synthetases (activating enzymes), tRNA, ATP, and Mg^{2+}. In a first step, the amino acid reacts with ATP to form an aminoacyl-AMP-enzyme complex. PP_i is released and hydrolyzed by pyrophosphatase.

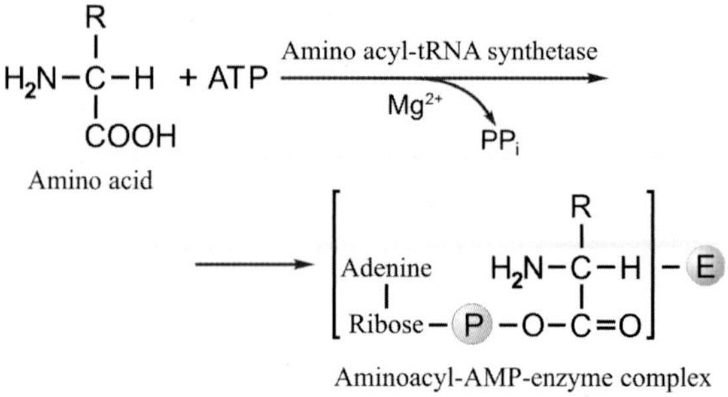

Aminoacyl-AMP-enzyme complex

The second step comprises the transfer of the activated amino acid to the 3′ end of tRNA. The specificity of the aminoacyl-tRNA synthetase allows it to recognize not only the amino acid but also the tRNA corresponding to that amino acid. There are enzymes specific for each amino acid.

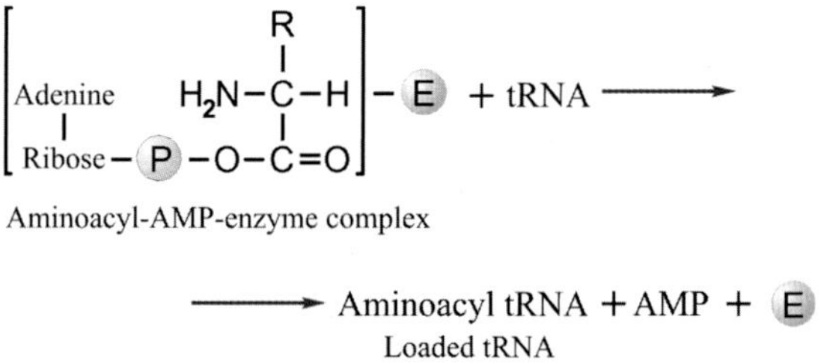

Aminoacyl-AMP-enzyme complex

⟶ Aminoacyl tRNA + AMP + Ⓔ

Loaded tRNA

The aminoacyl–tRNA complex is directed toward the site of synthesis, where the loaded tRNA, the mRNA, the ribosomes, and the factors that initiate the assembly of the chain gather together.

2. *Initiation of the polypeptide chain.* This step shows some differences between prokaryotes and eukaryotes. In prokaryotes, the chain begins with the insertion of formylmethionine, while in eukaryotes

the first amino acid is methionine, both indicated by the same AUG codon. In bacteria, a specific sequence in the mRNA, preceding the start codon, aligns the mRNA on the ribosome (it pairs with a complementary segment on the rRNA). This mRNA sequence is known as the Shine–Dalgarno sequence. As a result, the bacterial ribosomes can initiate translation not only at the 5′ end of the mRNA, but also at internal initiation sites across polycistronic mRNA. In eukaryotes, initiation of the protein chain starts by recognition of the 5′ end of the mRNA by the 7-methyl-guanosine triphosphate cap; the start codon, in most cases is the first AUG.

Initiation factors. Besides ribosomes, a special tRNA called initiator tRNA (itRNA) and proteins called initiation factors are required for the initiation of protein synthesis. In prokaryotes, three factors are necessary, which are designated by the initials IF and a number; in eukaryotes, at least nine factors, indicated with the letters eIF are present. The itRNA binds methionine (met-itRNA). This itRNA is different from the methionine tRNA that inserts methionine in internal AUG codons of the mRNA and it is only used to start protein synthesis. In bacteria, the initiator methionine is formylated (*N*-formyl-methionine).

In eukaryotic cells, the process can be summarized as follows.

Initially, the ribosomal subunits are separated by the dissociating factors eIF-3 and eIF-6, which bind to the ribosomal 40S and 60S subunits, respectively. Then, the following steps take place:

a. A complex called IF4F, which includes proteins eIF4E, eIF4G, and IF4A, binds to the 5′ cap at the end of the mRNA. eIF4E recognizes the cap and attaches to it; eIF4G binds to the poly-A tail at the 3′ end and maintains the 5′ and 3′ extremes close together. This arrangement facilitates regulation of translation. eIF4A, associated to eIFeF3, has helicase activity and unwinds the mRNA chain in complementary sectors that form double helical hairpins. All these actions require energy, which is provided by the hydrolysis of ATP (Fig. 22.3).

b. The factor eIF-2, bound to GTP, attaches to the initiator tRNA loaded with methionine (met-itRNA). This ternary complex is directed to the 40S subunit of a ribosome (bound to eIF-3). The aggregate of the 40S subunit, eIF-3, eIF-4C, and eIF-meth-itRNA-2 meets at the 5′ end of the mRNA with its accompanying factors. The small ribosomal subunit and the meth-itRNA are situated on the mRNA strand. This constitutes the *preinitiation complex*. The eIF-1 factor integrates this complex and assists in identifying the start codon.

c. The preinitiation complex begins to slide over the mRNA strand in the 5′→3′ direction to find the first AUG codon, flanked by a purine at position −3 and a guanosine at position +1; itRNA pairs to this codon. The energy to move the complex is provided by the hydrolysis of ATP.

d. The eIF-5, a GTPase, hydrolyzes the GTP bound to eIF-2 and all associated factors are released, including eIF-2-GDP.

e. A 60S ribosomal subunit detaches from eIF6 and associates to the 40S subunit, now free of factors, to form the complete 80S particle that is threaded on the mRNA at the cleft between the two subunits. The meth-itRNA adheres to the P site of the ribosome. The *initiation complex* has been formed (Fig. 22.3). The eIF-2 is an important site of regulation, as one of its three subunits is covalently phosphorylated by a protein kinase.

3. *Elongation.* In the initiation complex, the met-itRNA is placed at the ribosomal P site with the anticodon paired at the start codon

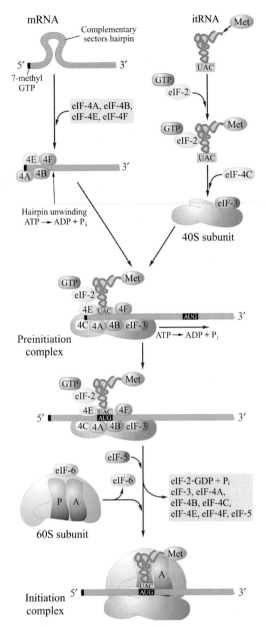

FIGURE 22.3 **Schematic representation of the initiation step in the biosynthesis of a polypeptide chain in eukaryotes.**

on the mRNA strand. Site A, located at the next codon, is empty.

The addition of amino acids requires participation of three elongation factors. These are eEF-α, eEF-βγ, and eEF-2 in eukaryotes and EF-Tu, EF-Ts, and EF-G in prokaryotes.

Elongation is performed through a cycle, which is repeated with each amino acid that is added to the newly synthesized polypeptide chain (Fig. 22.4).

a. *First stage of elongation.* An aminoacyl-tRNA with anticodon sequence complementary to the codon adjacent to the one occupied by meth-itRNA is driven to the A site of the ribosome by elongation factor eEF-1α, associated with GTP. The aminoacyl-tRNA binds to site A only if the P site is occupied; it pairs immediately with its anticodon in the second codon of the mRNA chain. The energy required is provided by the hydrolysis of GTP occurring in the complex. The eEF-1α, now bound to GDP, is released.

As indicated previously, there are several codons that code for the same amino acid (synonymous codons). The number of different tRNAs in cells is less than the number of codons, indicating that a tRNA is able to bind to more than one of the codons encoding the same amino acid. The third base of the triplet is the least specific one, and in many cases can be changed without altering the meaning of the codon. This is possible because, when the tRNA is paired to its anticodon on the mRNA, it has a rocking or wobbling movement, which makes the binding on the third base looser. This gives lower stringency to the complementarity of the base in the third position. Apparently, this mechanism ensures rapid dissociation of tRNA and mRNA, and increases the speed of the protein synthesis process. Another factor contributing to the pairing of an anticodon with different codons is the conversion, by deamination, of adenosine to inosine. Inosine can interact with more bases than adenosine.

The released eEF-1α-GDP is inactive, unable to bind aminoacyl-tRNA. Regeneration of active eEF-1α-GTP takes place in the cytosol by means of eEF-1βγ. This factor stimulates the exchange of GDP bound to eEF-1α by GTP.

$$eEF\text{-}1\alpha\text{-}GDP + eEF\text{-}1\beta\gamma \xrightarrow[\text{GDP}]{\text{GTP}}$$

Inactive

$$\longrightarrow eEF\text{-}1\alpha\text{-}GTP + eEF\text{-}1\beta\gamma$$

Active

b. *Formation of the peptide bond.* The carboxyl group of methionine attached to the itRNA in the P site forms a peptide bond with the amine α-amino acid attached to tRNA at site A. The reaction is catalyzed by *peptidyl transferase*, a ribozyme located in the ribosome's large subunit. A dipeptidyl is formed, which is bound to the tRNA that entered after the itRNA, and is located in site A. The itRNA on the P site, free from its methionine, is released. In prokaryotes, itRNA moves to the ribosomal E site before being released.

c. *Translocation.* This comprises the movement of peptidyl-tRNA from the A to the P site of the ribosome. It requires GTP and the translocation elongation factor eEF-2 (in prokaryotes, EF-G) (Fig. 22.4). Due to the displacement of peptidyl-tRNA, the A site becomes free. The energy for translocation is provided by hydrolysis of GTP to GDP and P_i. At the same time, the ribosome advances one codon on the mRNA, the new codon of the mRNA is now next to the vacant A site.

d. *New elongation cycles.* The steps described for the first elongation cycle are repeated for each amino acid added to the emerging polypeptide. When several cycles have elapsed, the tRNA in the P site has a peptide chain attached to its 3' end that begins with methionine and has a number of additional amino acids equal to the number of elongation steps accomplished. A new aminoacyl-tRNA-eEF-lα-GTP complex enters the A site, provided the tRNA anticodon is complementary to the next codon on the mRNA. GTP is hydrolyzed and eEF-lα-GDP is released. Peptidyltransferase catalyzes the formation of another peptide bond between the carboxyl terminus of the peptide bound to the tRNA in the P site and the amino group of the aminoacyl tRNA with the new aminoacyl-tRNA at the A site; the peptide chain is transferred to it. The P site tRNA, already freed of the peptide, is released. Then, eEF-2-GTP enters and promotes translocation of peptidyl-tRNA from the A site to the P site. Once the eEF-2 and GDP are released, a new cycle begins (Fig. 22.4).

The eEF-2 is the site of action of the toxin produced by *Corynebacterium diphtheriae*, an agent of diphtheria causing a disease virtually eradicated by vaccination. The lethal action of the diphtheria toxin is due to its capacity to block protein synthesis. It inserts adenosine-diphosphateribose to eEF-2 and inactivates it. In this reaction, called ADP-ribosylation, NAD is used as cofactor, it loses its nicotinamide portion, and the remaining adenosine-diphosphateribose is bound to eEF-2 (see p. 538).

In the elongation process, the ribosome moves on the mRNA in the 5'→3' direction, advancing one codon per cycle; the peptide chain grows one amino acid at a time, from the amino terminus to the carboxyl terminus.

The energy cost of a peptide bond formation is four high-energy phosphate bonds. Aminoacyl-tRNA production requires hydrolysis of ATP to AMP, the initial binding of aminoacyl-tRNA to the ribosome

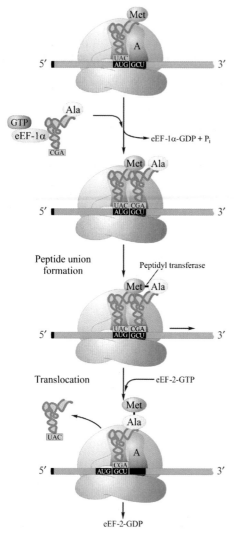

FIGURE 22.4 **Scheme of the first elongation cycle in the synthesis of a polypeptide chain.**

consumes one GTP, and translocation of peptidyl-tRNA from the A site to the P site requires another GTP→GDP + P$_i$ hydrolysis. The GTP and ATP molecules consumed in the formation of the initiation complex are not taken into account, since this is negligible compared to the total energy expenditure needed to assemble a protein.

4. *Termination of the polypeptide chain.* When the addition of all amino acids is finished, the tRNA bound to the ribosomal P site has the complete polypeptide chain attached to its 3′ end. The open reading frame of the mRNA used for protein synthesis spans from the initial AUG codon to the completion signal (a UAA, UAG, or UGA termination codon). This codon is not recognized by the tRNA, but instead by proteins called *releasing factors* that catalyze the hydrolysis of the bond between the polypeptide chain and tRNA. In prokaryotes, there are three releasing factors (RF-1, RF-2, and RF-3), in eukaryotes there is only one (eRF) and it is associated with GTP.

Releasing factors act on the ribosomal A site and require the P site containing the peptidyl-tRNA. The completed polypeptide chain is separated from the tRNA and liberated from the ribosome, which in turn releases the mRNA. If eIF-3 and eIF-6 are present, the ribosomal 40S and 60S subunits dissociate, and remain ready for the synthesis of another protein (Fig. 22.5).

Table 22.2 presents the factors involved in the different stages of protein synthesis.

Mutations in nuclear or mitochondrial genes, which encode the multiple factors involved in the process of protein synthesis, cause alterations in diverse organs. Many syndromes have been described, and the number continues to grow. It is impossible to describe them all here. It is sufficient to mention that they include deficiencies in ribosomes, tRNAs, and enzymes related to the activation of amino acids; in factors involved in the initiation, elongation, and termination steps of protein synthesis; and in their regulation. These mutations determine a variety of disorders, frequently of the central nervous system and muscles.

Polysomes. One mRNA strand can simultaneously direct the synthesis of several molecules of the same protein. The initiation complex covers

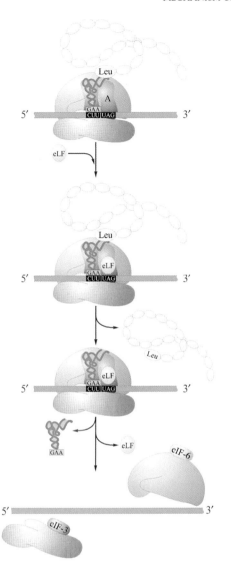

FIGURE 22.5 **Scheme of the steps for termination of the polypeptide chain.**

TABLE 22.2 Eukaryote Factors Involved in Protein Synthesis

Factor	Function
Initiation	
FIe 1	Participates in the search of initiator codon
FIe 2	Binds Met-itRNA to the ribosome
FIe 2A	Binds Met-itRNA to the ribosome
FIe 2B	Activates IF2 to replace GDP/GTP
FIe 2C	Stabilizes the complex
FIe 3	Binds to subunit 40S and inhibits its reassociation with the 60S
FIe 4A	Unwinds the secondary structure of mRNA (helicase activity)
FIe 4B	Assists FI4
FIe 4E	Recognizes the cap of mRNA and allows binding of subunit 40S to the 5′ end of mRNA
FIe 5	Promotes hydrolysis of GTP and liberation of initiation factors
FIe 6	Binds to the subunit 60S and prevents its reassociation with the 40S
Elongation	
FEe 1α	Binds amino acyl-tRNA and GTP
FEe 1βγ	Assists in the exchange of GTP/GDP in FEe 1α
FEe 2	Translocation of ribosome along the mRNA; hydrolyzes GTP Inhibited by ADP-ribosylation. Catalyzed by diphteric toxin
Termination	
FLe	Promotes hydrolysis of polypeptidyl-tRNA to free the polypeptide chain and tRNA Binds and hydrolyzes GTP

approximately the first 30 codons of the mRNA template. After moving through ~80 nucleotides, another ribosome can be inserted to initiate the synthesis of a new polypeptide chain. A single open reading frame of relatively small length (i.e., 1000 bases) can bind more than 10 ribosomes spaced along the mRNA strand. The whole structure, called a *polysome*, has the appearance of beads on a rosary (Fig. 22.6). Each of the ribosomes in a polysome is at a different place in its "journey" along the mRNA molecule. The closer the ribosome is to the termination point, the longer the polypeptide chain attached to it.

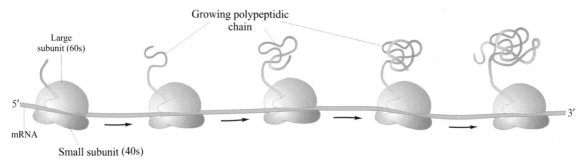

FIGURE 22.6 Schematic representation of a polysome. Five ribosomal particles move from the 5′ to 3′ ends of the mRNA strand. Each is at a different level of the polypeptide chain synthesis. As the ribosome progresses, the polypeptide chain length increases.

The synthesis of polypeptide chains is accomplished faster in prokaryotic than eukaryotic cells, with up to 20 amino acids per second being assembled in prokaryotes and 2–4 per second in eukaryotes. This difference is explained, in part, because the transcription and translation in bacteria are carried out in the same cellular compartment, while in eukaryotes transcription occurs in the nucleus and translation occurs in the cytoplasm.

Eukaryotic cells have the ability to degrade incomplete mRNAs or mRNAs with incorrectly located termination codons.

Antibiotic Action on DNA and Protein Synthesis

Antibiotics are substances produced by microorganisms, which have the property of inhibiting the growth of other microorganisms. Many of them exert their action by interfering with a particular stage in the synthesis of nucleic acids and proteins. They are widely used in medicine and are also valuable tools for the study of protein synthesis.

Antibiotics That Block Replication

Novobiocin is produced by *Streptomyces*. It inhibits binding of ATP to topoisomerases. It is used in the treatment of human infections.

Ciprofloxacin is a fluoroquinolone that blocks the final stage of the topoisomerase reaction and the ligation of both ends of DNA strands. It is a broad spectrum antibiotic that acts on all bacterial topoisomerases. Only when used at concentrations that are several orders of magnitude above its therapeutic level, ciprofloxacin acts on eukaryotic topoisomerases.

Antibiotics That Block Transcription

Actinomycin D is a molecule that consists of a fenoxazone ring attached to two identical cyclic pentapeptides. It blocks RNA transcription by firmly attaching to DNA, preferably at sites with guanine residues. This prevents DNA from acting as a template for RNA synthesis. Actinomycin D does not bind to single stranded DNA or RNA and, at low concentrations, it does not affect DNA replication. As this antibiotic does not directly affect the translation process, protein synthesis can continue from the preexisting mRNA. Actinomycin D works both in prokaryotes and eukaryotes. It has been administered to control the growth of malignant tumors, but its clinical use is limited due to its toxic effects. Thus, it is used for experimental purposes.

Rifamycin, rifampicin. Both of these antibiotics exert a similar action; however, while rifamycin is a natural product, produced by *Streptomyces*, rifampicin is a semisynthetic derivative. Both inhibit transcription, blocking the initiation of

RNA synthesis. They act on RNA polymerase and prevent the formation of the first phosphodiester 5′→3′ bond of RNA. They do not interfere with the synthesis of already initiated RNA chains. Only prokaryotic RNA polymerases are sensitive to these antibiotics. Rifamycin can inhibit RNA synthesis in human mitochondria, but at much higher doses than those used for bacterial infections.

Antibiotics That Block Translation

Puromycin acts on both prokaryotic and eukaryotic cells. It has a structure similar to that of the 3′ end of the aminoacyl-tRNA carrier of tyrosine or phenylalanine. For this reason, it occupies ribosomal site A and binds to the polypeptide chain synthesized by peptidyl transferase, blocking the entrance of the next aminoacyl-tRNA. Due to the weak binding of puromycin to ribosomal site A, the formed polypeptide chain is released and protein synthesis is prematurely interrupted. It was based on the effects of puromycin that the existence of the A and P sites in ribosomes was identified.

Paromycin binds at ribosomal site A, where the codon and aminoacyl-tRNA anticodon interact; preventing their adequate pairing and causing translation errors.

Hygromycin B is effective in both prokaryotes and eukaryotes. It binds to the A site of the ribosomal 30S subunit and prevents translocation of peptidyl-tRNA.

Streptomycin is an aminoglycoside (trisaccharide) that interferes with translation initiation in prokaryotes. It disrupts interactions of tRNA with the ribosome and mRNA. Streptomycin binds to initiating factors and to the 16S rRNA of the bacterial 30S ribosomal subunit. It acts on the ribosomal small subunit, preventing the binding of N-formylmethionyl-tRNA and promoting the accumulation of abnormal initiation complexes. When the synthesis of the polypeptide chain has been started, streptomycin causes errors in translation. At relatively high concentrations, it inhibits the initiation step.

Neomycin and *gentamicin* are aminoglycosides that interact with the ribosomal small subunit at sites different from those of streptomycin.

Tetracyclines inhibit protein synthesis by binding to the ribosomal small subunit (30S) of prokaryotes and blocking the aminoacyl-tRNA binding to site A. This effect is reversible; when administration of the antibiotic is discontinued, the bacteria resume their development. Tetracyclines have been used as a food additive to prevent infections in animals whose meat is consumed by humans. This increased exposure to these antibiotics unfortunately led to the development of bacterial strains with resistance to these antibiotics.

Chloramphenicol binds to the ribosomal large subunit (50S) and prevents the process of protein elongation beyond the first peptide bond, blocking the peptidyl chain transfer. It is a peptidyl transferase inhibitor. It affects prokaryotes and may also block the synthesis of proteins in eukaryotic mitochondria.

Erythromycin binds to the 50S ribosomal subunit of bacteria, near the site where chloramphenicol binds. It prevents the translocation of peptidyl-tRNA in the ribosome from site A to P.

Cycloheximide blocks protein synthesis in eukaryotic, but not prokaryotic cells. Its mechanism of action is similar to that of chloramphenicol, inhibiting peptidyl transferase.

Kirromycin blocks the function of elongation factor EF-Tu, avoiding conformational changes associated to the GTP hydrolysis. When the antibiotic binds to EF-Tu, it locks the aminoacyl-tRNA to the ribosomal A site; the complex EF-Tu-GDP is not released from the ribosome and the peptide bond is not formed. As a result, the ribosome is stopped along the mRNA and protein synthesis ceases.

Fusidic acid is a steroidal antibiotic that slows the ribosome at the posttranslocation step. It fixes to the EF-G and inhibits the conformational change that occurs upon hydrolysis of GTP. It stabilizes the ribosome-EF-G-GDP complex, so it is not released. The ribosome cannot bind a new aminoacyl-tRNA and protein synthesis is stopped. Fusidic acid acts on both bacteria and eukaryotes.

GENETIC MUTATIONS

Mutations, in a broad sense, are changes in the DNA sequence. According to the extent of the change, they can be: (1) point mutations, affecting only one base pair, or (2) structural changes of chromosomes that include larger sections of the DNA molecule.

1. *Point mutations* consist in the replacement of one base for another. The substitution of a base is called *transition* when a purine or a pyrimidine base is changed to another base of the same type. In contrast, *transversion* is the replacement of a purine base for a pyrimidine base or vice versa. The replacement of one base for another can produce a change in the meaning of the codon affected. For example, if adenine is replaced by guanine in the aspartic acid codon (GAU), the codon will now encode for glycine (GGU).

    Due to the redundancy of the code, mutations do not always result in alterations of the message. Frequently, the change of the third base of a codon does not modify its meaning; in this case the mutation is called "silent." Moreover, even when a change in the codon determines the substitution of an amino acid, this does not always causes functional disorders in the synthesized protein. For the mutation to affect the protein, the base replacement must produce some conformational alterations, loss of critical functional groups, or modify "key" sites in the molecule (such as active site of enzymes). Mutations which do not cause functional changes in the encoded protein are called *neutral* mutations.

Hemoglobin S (p. 57) is an abnormal protein resulting from a point mutation (transversion) in the gene encoding for the globin β chain. An adenine is substituted by thymine, resulting in a change in the glutamic acid (GAA) codon, for valine (GUA). In this case, a single amino acid change in the protein causes a functional alteration in the protein, which is lethal to individual homozygous for the abnormal gene. This type of mutation is known as a *missense* mutation.

Point mutations may also encompass the addition or loss of one base pair in the DNA (insertion or deletion). This type of mutation usually produces serious disturbances, because it completely changes the genetic message. The genetic information depends on the sequence of each codon and is orderly translated in a 5'→3' direction. Therefore, if a base is deleted or added, a shift in the sequence occurs, altering all subsequent codons and causing a total disruption of the open reading frame of the gene. Insertions and deletions create a new open reading frame, changing the amino acid sequence to render a protein that will be functionally useless. These are considered *frameshift* mutations.

If the change, addition, or subtraction of a base produces a termination codon, protein synthesis stops and an incomplete or truncated polypeptide chain will be formed. This is called a *nonsense* mutation.

Two examples follow which describe the effects of a base insertion or deletion on protein synthesis. Given the following piece of mRNA constituted by nine codons:

...-GUC-ACG-CUA-UGG-AUG-CAC-GCU-UAC-CGU-...
- Val - Thr - Leu - Trp - Met - His - Ala - Tyr - Arg -

If adenine is deleted in the second codon, the resulting shift will completely alter the message and a very different amino acid sequence will be produced:

-GUC- A CGC-UAU-GGA-UGC-ACG-CUU-ACC-GU..-
- Val -   Arg - Ile - Gly - Cys - Thr - Leu - Thr - Val -

If adenine is added between the first and second codons, the modification of the message will also result in a totally new polypeptide:

-GUC- AAC-GCU-AUG-GAU-GCA-CGC-UUA-CCG-U..-
- Val - Asp - Ala - Met - Asp - Ala - Arg - Leu - Prol -

2. *Changes in chromosome structure* affect a fairly large portion of DNA. Among such mutations are: (1) *Deletion*, when A DNA segment is lost. (2) *Duplication*, when a piece of DNA is repeated. (3) *Inversion*, when a region within the same DNA molecule is reversed. (4) *Translocation*, when the relative position of a fragment of DNA changes in the same chromosome or is incorporated into another. Some of these alterations may result in the lack, reduction, or an increase in the protein(s) encoded by the affected gene(s).
   Somatic and germline mutations. Mutations occur randomly anywhere in the genome and in any cell of an organism. The mutations affecting somatic cells are transmitted through successive divisions of the mutated cell. If the mutation leads to alterations, these will take place only in the tissue in which these cells are located. In organisms that reproduce by sexual cross, the somatic mutation disappears with the death of the organism. In contrast, mutations in germ cells are transmitted to the offspring, and they are perpetuated. These mutations originate genetic variability in populations. Alternative forms of the same gene are called *alleles*. In diploid organisms, the descendants receive a maternal and a paternal allele for genes that occupy the same locus on a pair

of homologous chromosomes. If both are different, the new individual is heterozygous for that character, and if the same allele is received from both parents, the individual is homozygous for the gene.
Frequency of mutations. In any organism, mutations occur either by errors in the synthesis of nucleic acids and proteins, or by environmental factors. Due to mechanisms operating in normal individuals that identify and correct for mutations, their incidence is low. However, there is a basal level of spontaneous mutations. In bacteria, the frequency of mutations that can produce gene alterations is around 10^{-6} per locus per generation. There are no precise estimates of the mutation rate in eukaryotes, but it is apparently less than in bacteria.

The basal mutation rate can be increased by factors that directly modify the DNA. These are induced mutations, and the factors involved in their production are known as *mutagens*. The effectiveness of a mutagenic agent is measured by its capacity to increase the basal level of mutations.
Mutagenic agents. Numerous physical and chemical agents act as mutagens. Radiations of higher energy, such as *ionizing* radiations; X-rays; α, β, and γ radiations emitted by radioactive isotopes; cosmic rays; and *nonionizing* radiations (ultraviolet light) have marked mutagenic effects.

Ionizing radiations have great tissue penetrating power. When they collide with the atoms in tissues, they release electrons and produce free radicals or ions. In contrast, ultraviolet light (UV light) has little penetrating power, mainly affecting surface tissues (skin). In general, the energy of UV light is not intense enough to cause ionization, but can promote the displacement of electrons to higher energy levels and render atoms in an "excited" state. Molecules with ionized groups have a higher chemical reactivity. The mutagenic effects of radiations are due to the higher reactivity that DNA acquires when subjected to these physical agents.

Chemical mutagens include a variety of compounds that act by different molecular mechanisms. Some directly modify DNA, either by introducing alkyl groups (e.g., mustard gas used during World War I) or by deamination (nitrous acid). Others, affect DNA by replacing a nitrogenous base (base analogs, such as 5-bromouracyl and 2-aminopurine) or by inserting themselves between the base pairs in the double helix (acridines). Combustion products of tobacco and many other compounds, especially of aromatic character, are also mutagenic.

There is a direct correlation between the mutagenicity of an agent and its efficacy to induce malignant transformation of cells.

Experimentally, it is possible to make changes in a gene sequence in a specifically selected site to study the effects of the modification. This is known as *directed mutagenesis*.

Technological development has improved the standards of human living; however, it has also contributed to create severe changes in the environment, which have potential carcinogenic effects. In an attempt to avoid pollution of our habitat, different countries have developed strict policies to control the release of mutagens to the environment. Still, despite the regulations, the growing number of mutagenic agents represents a constant threat to life in Earth.

Transposons

DNA sequence can also be altered by insertion of DNA segments that have the ability to "jump" or move from one place of the genome to another, either within the same chromosome, between different chromosomes, or into extra-chromosomal DNA (i.e., plasmids in bacteria). These genetic elements are known as *transposons*.

The existence of mobile genes was suggested many years ago, but only more recently have they been identified through studies in bacteria. The simplest, known as *insertion sequences* (IS) are portions of DNA that carry the information needed to synthesize the enzymes that catalyze DNA transposition. These DNA units are flanked at both extremes by short regions with sequences that have the same bases, although reversed in order (inverted repeats). In some cases, when the DNA region between two IS sequences is separated, they carry genes that confer resistance to antibiotics and mercurial antiseptics. This plasmid can be transferred from one plasmid to another and exchanged between bacteria, even from different species.

Transposons have been detected not only in bacteria but also in the genome of different plant and animal species, including humans.

The general mechanism for DNA transposition consists in first cutting both DNA strands at the specific site and creating a longer end or overhang, similar to that formed by some restriction endonucleases. The transposon is then inserted into the excised site, establishing a phosphodiester $5' \rightarrow 3'$ bond with the terminal single strand pieces of the DNA. On each side, the portion corresponding to the DNA, left as a single strand after the cut, needs to be filled with the missing complementary segment, to restore the double helix at both sides of the transposon. In this manner, two pieces of identical direct repeated sequences are originated. These pieces are always found at the sites where the insertion has occurred.

Several types of transposition can be distinguished. When the transposed element leaves its place and is housed in a new receptor site, the process is called *conservative transposition*. In other cases, such as those described in bacteria, the transposon previously undergoes replication, one copy remains at the original site and the other is inserted into the receptor site. This is called *replicator transposition*. A third type of DNA transposition consists in introduction to the double helix of DNA sequences produced by reverse transcription from RNA. These elements are called *retroposons*. The best one known are those that introduce genetic information of retroviruses into the genomes of bacteria and eukaryotic cells.

The mammalian genome contains a large number of repeated sequences and a significant number of them are retroposons. Two classes of retroposons can be distinguished: LINE (for long interspersed nucleotide element) and SINE (for short interspersed nucleotide element). LINE corresponds to long sequences (~6.5 kb), of which there are up to 50,000 copies. On the other hand, SINE includes short sequences. The most abundant in humans belong to the *Alu* family, which constitutes much of the moderate repetitive DNA. They are segments of ~300 bp with more than 300,000 copies in the haploid genome, intercalated with nonrepetitive DNA. They are called Alu because they are sectioned by the restriction enzyme Alu I. The genes coding for some small nuclear RNAs possibly originated as retroposons.

The possibility of DNA transposition is not unlimited. Different molecular mechanisms restrict the total number of DNA copies that can be inserted into the receiver.

In prokaryotes, when a mobile gene is inserted between a promoter and the structural gene, or within a polycistronic operon, it can inhibit transcription of the message "downstream" of the transposon. It acts as a regulator for the expression of other genes. In eukaryotes, the insertion of a transposon can activate or repress neighboring genes, or can favor the production of structural alterations. Thus, transposition becomes a phenomenon capable of producing genome modifications, which is an important determinant of evolutionary changes.

Variome

The complete genome of an increasing number of animals and plants is known. This knowledge has had an enormous impact in different areas of biology. The completion of the Human Genome Project has provided a more functional view of the genetic material, which has an obvious translation into clinics.

The genome project was followed by other ventures, such as the transcriptome, the proteome, and the metabolome, which are considered elsewhere in this text. A few years ago, an additional project was started. This is called the *human variome*, which refers to the variability of the genome of *Homo sapiens*. This study focuses on DNA mutations, directing its attention to those changes that have clinical implications and affect genes or noncoding regions of DNA with regulatory functions.

The Project Human Variome proposes to record, worldwide, all the mutations responsible for diseases and to create a large database with this information for free access. This will facilitate the detection of different anomalies.

Transgenesis

Transgenesis refers to the process of introducing an exogenous or modified gene (transgene) into a recipient organism of the same or different species from which the gene is derived. The transgene becomes incorporated into the genome of the host organism and can be transmitted to the offspring. Using this approach, genetic engineering has made it possible to produce genetically modified plants or animals, which are known as transgenic organisms. Due to its rapid reproduction, ease of maintenance, and simplicity of its genome, the fruit fly *Drosophila melanogaster* has been a species used with great success for transgenesis. The method has also been extended to vertebrates, with the mouse being one of the most common species used. Recent advances in genome sequencing have contributed to the expansion of transgenic animal production.

Animal gene manipulation involves inserting a previously existing or a new gene. In the first case, the total expression of the preexisting gene will be increased in the acceptor organism. The expression of a gene can also be suppressed or its structure can be modified to inactivate it. This results in what is known as *knockout* animals.

The suppression of a given gene can also be obtained by replacing it with another, to produce what is called *knock-in* animals. The phenotype resulting from the loss or gain of function of a gene in an organism helps to reveal the physiological role of that gene.

Sometimes, alterations of a gene can be compensated by upregulation of another functionally related gene, which suggests that there is redundancy for that particular gene. In other cases, due to the essential role of a gene, its alteration is lethal to the embryo or affects unspecifically the whole organism; in this case, the functional characterization of the gene is not possible using global knockouts. To avoid this problem, modification of the gene can be limited to a particular tissue, or its expression can be postponed to later stages of development. This is accomplished by introducing the gene of interest at specific DNA sites, controlled by defined promoters and transcription factors. Thus, the gene changes are induced only in certain cell types and during certain stages of tissue maturation. The temporal–spatial regulation of the incorporated gene results in *inducible* or *conditional transgenic* animals.

In addition to helping to understand gene function, transgenic animals are useful to determine the activity of promoter regions in DNA. Known genes whose expression can be easily detected by their intrinsic fluorescence or by its biochemical properties (reporter genes) are introduced in sites controlled by the promoter under study. This allows for the study of gene regulation in different tissues or during the development of the organism.

The methodology for generating transgenic animals includes different procedures. DNA recombination can be performed randomly, or in a region of homology between the introduced DNA and the acceptor genome. The exogenous DNA is introduced into the pronucleus of fertilized eggs, by direct microinjection, or by using vector viruses. Then, the eggs are implanted in the uterus of receptive animals. Thus, the DNA incorporated into the genome propagates through successive cell divisions, and is finally expressed in the developed animal. Another method is to use embryonic cells from blastocysts (stem cells). The desired DNA is first incorporated into these pluripotent cells by transfection in culture. Then, the cells are injected into blastocysts, where they divide and differentiate, forming part of the future embryo. The animal generated is a *chimera*, which expresses the particular gene and, if the new gene is incorporated into the germ cells, it will be transmitted to the offspring, generating a transgenic line. Normally, a number of crosses are required to obtain animals homozygous for the desired transgene. The aforementioned methods are widely used in the production of transgenic mice. Other procedures have also been used, such as gene transfer mediated by sperm. In this case, particular genes are first introduced in male gametes and these are used for insemination of eggs, which will produce embryos containing the new DNA. Nuclear transfer involves the replacement of the nucleus of an ovule by that of a somatic cell, which had previously been engineered genetically. This generates transgenic animals that express the modified genome. More recently, clustered regularly interspaced short palindromic repeats (CRISPR), along with the CRISPR-associated system (Cas) endonuclease has became a simpler and more accurate method to alter specific genes (p. 532).

The use of transgenic animals has been one of the greatest achievements of molecular biology. It allows (1) to study the function of a particular gene and its regulation in the whole animal, (2) to reproduce genetic diseases of humans in animals, creating experimental models for the study of these conditions, (3) obtaining the synthesis of human proteins in animals (such as hormones) to be used for therapeutic purposes, and (4) to make changes that will improve the quality and quantity of animal products for human consumption.

DNA Microarrays

DNA microarray is a technique used to detect genes that are expressed in a cell or tissue at a given moment. The method comprises the following steps:

1. Single-stranded DNA from different genes, or gene segments, are placed in an ordered manner on a grid or solid surface (plate or chip) of glass or other suitable material. Thousands of copies of different DNAs can be accommodated on a plate for testing.

2. The sample to be tested consists of mRNA extracted from the cells or tissue under study. This mRNA represents the transcriptome or transcribed genes of cells or tissues at the time of sampling. DNA complementary to the mRNA strand (cDNA) is synthesized by reverse transcription. This piece of DNA obtained are copies of all the genetic information that is being expressed in the sample of study. cDNA chains are labeled with a marker, commonly a fluorescent compound.

3. A suspension of labeled cDNA is added onto the microarray plate to cover it. The cDNA in the sample will specifically bind to the sites on the grid where the complementary DNA strands are located.

4. After washing away unspecifically bound DNA, the plate is taken to a detection system which can sense the fluorescently labeled DNA duplexes formed. A computer program produces a pattern of colored dots corresponding to different genes being expressed in the sample. Even a quantitative estimate of genes can be obtained, depending on the intensity of fluorescence. This provides information of the expression of many genes on a single plate.

The method has many applications and provides valuable information. For example, DNA microarrays are important for the determination of gene activity in different cells or tissues under healthy and diseased conditions, the analysis of the nature and evolution of neoplastic cells, the effects of drugs and their development, and the clinical diagnosis and prognosis of different diseases.

Proteome

According to the data obtained from the human genome project, there are approximately 21,000 genes encoding for proteins. However, the actual amount of proteins synthesized is much higher. This is because information in a gene can be used to produce more than one polypeptide, through mechanisms that include alternative splicing, mRNA editing, and selection of different polyadenylation sites within mRNA molecules. The discrepancy between existing genes and proteins synthesized fueled the interest to identify and characterize all existing proteins in an individual. The term *proteome* was coined to describe the complete set of proteins produced by a cell, tissue, or organism.

While the genome is virtually the same in all cells of an individual, the proteins synthesized differ significantly in different cell types. Even in the same cell, only a portion of the genome is active at any given time, which provides the cell with a characteristic temporally dependent protein repertoire. Thus, protein expression can vary at different stages of development, under different physiological or pathological conditions, or in response to environmental changes. The characterization of all proteins in a cell, tissue, or organism is a new research field called *proteomics*.

The detection of open reading frame sequences of DNA allows for the prediction of proteins that are synthesized by a cell, but does not directly indicate the regions of genome being transcribed at a given time. The investigation of the mRNA or *transcriptome* does not always exactly correspond with the proteins being synthesized, due to differential mRNA processing, or posttranslational modifications. The proteome, therefore, provides data that cannot be obtained by genomic or transcriptomic studies.

Proteomics includes not only the identification and determination of the proteins synthesized in cells, tissues, and organisms, but also the study of their biological activity, subcellular localization, interactions, and formation of molecular complexes.

The basic methodology, originally used in proteomics, was two-dimensional polyacrylamide gel electrophoresis. This method allows the proteins of a sample to be separated first according to their electric charge. In a second step, the sample is subjected to an additional current that is passed in a direction perpendicular to that of the previous step, in a medium which separates the proteins contained in each fraction based on their molecular size. The sensitivity of this technique was greatly enhanced with the application of mass spectrometry to analyze each of the fractions obtained. This approach can resolve thousands of proteins and peptides in biological samples.

The further use of fluorescent markers (i.e., green fluorescent protein, GFP) and successive technical refinements provides opportunities for increased resolution. Currently, the best equipped laboratories can simultaneously identify more than 5000 different proteins in a tissue sample.

Microarrays have also been added to the collection of approaches used in proteomics; they are useful for the identification of proteins. In microarrays, a large number of specific protein ligands [such as antibodies or synthetic RNA *aptamers* (*aptamers* are short chain RNA molecules and peptides that have a secondary and tertiary structure, which allows their mutual adaptation to other molecules in a selective manner)] are adsorbed on a grid in a plate or chip (see previous section). After incubating the sample with these chips, it is possible to follow the binding of proteins to the ligands on the chip and identify them. Even posttranslational modification of proteins can be identified using specific antibodies. For example, the set of phosphorylated proteins can be determined to decipher what is known as the phosphoproteome.

Research in the field of proteomics is currently very intense. Many diseases cause changes in the pattern of proteins expressed in the affected tissues. The proteome can unveil abnormal protein profiles, contributing to a better understanding of the pathogenic mechanisms involved in a particular disease. Moreover, given that most of the drug targets are proteins, knowledge of disease related proteins can serve for the development of more selective and effective compounds to treat pathologies. In this manner, proteomics has a direct impact on medicine, improving the chances for diagnosis and treatment of pathological conditions.

Personalized Medicine

Studies of the genome, transcriptome, proteome and metabolome have revealed the great variability that exists among individuals in the human population, both under physiological and pathological conditions. Among other parameters, this individual variability is expressed, for example, in the dissimilar response that patients have to treatment with certain drugs. Some patients respond very well to treatment with a compound, others do it to a lesser extent or do not respond at all, or suffer a variety of adverse effects.

Pharmacogenomics is an area that studies this phenomenon and has demonstrated a remarkable polymorphism of enzymes and transporters involved in the metabolism of drugs. For example, the cytochrome P_{450} enzymes that catalyze oxidation of xenobiotics constitute a family of isoenzymes and allelic variants, with different functional capabilities. Depending on the efficiency of cytochrome P_{450}, some individuals can metabolize compounds better than others.

The improvement of genomics and proteomics, and the reduction in costs, are making these techniques more accessible and widespread. Theoretically, the information obtained would allow tailoring the treatment schemes to the particular characteristics of each patient. This personalized medicine, will help provide the appropriate drugs, at the optimal dose, and at the right time, to obtain the most beneficial effects in each individual. An extension of personalized medicine is *precision medicine*, which has emerged as a broader term, referring to the overall medical strategies to be applied to a patient, according to the particular characteristics of each patient.

Mechanism of Virus Action

Viruses are pathogenic agents composed of one or more nucleic acid strands (DNA or RNA) enclosed in a protein cover called the *capsid* (p. 137). The nucleic acid contains the genetic information necessary to synthesize all the molecules that form the viral particle. The simplest viruses possess 3 genes; the most complex contain about 250 genes.

Viral particles cannot multiply by themselves because they lack the machinery and energy producing mechanisms necessary for the synthesis of their own structures. To reproduce, they invade a cell and use the biosynthetic apparatus of that cell.

In general, viruses enter cells by mechanisms similar to those used by other particles and macromolecules. The viral capsid proteins, in many cases glycoproteins, bind to other proteins which function as receptors on the plasma membrane. An invagination of the membrane internalizes the viral particle into the cell by endocytosis (p. 242). Once inside the cell, the viral particle loses its protein coat and its DNA or RNA is released into the cytoplasm. Viruses have factors that slow the synthesis of nucleic acids and proteins of the invaded cell and activate the production of the viral molecules. In other words, the virus forces the metabolic machinery of the cell to work for its own multiplication. This is an extreme case of parasitism.

In DNA containing viruses, replication and transcription of the viral DNA is initiated using polymerases and other enzymes of the infected cell, which also provides the primary material (nucleosides triphosphate and amino acids) and the energy required for the synthesis of new viral particles.

Bacteriophages, or simply phages, are viruses that infect bacteria. They do not enter cells by endocytosis. Instead, they attach to the bacterial wall by their base or foot and perforate it (p. 138).

Immediately, the viral cover contracts and injects the DNA contained in the phage head into the body of the invaded bacterium. Only the viral DNA penetrates the host cell, with the protein coat remaining out of the cell, functioning as a disposable syringe (Fig. 22.7). Few minutes after invasion, synthesis of bacterial DNA, RNA, and proteins cease and virus-specific molecules are produced; they assemble to form complete viral particles. Twenty minutes after infection, there is a large number of new phages within the bacterium. Finally, the bacterium is lysed and many bacteriophages are released, each of which is able to attack another bacterium.

Not all viruses lead to the destruction of the parasitized cell (lytic pathway). In some cases, the viral DNA is integrated into the genome of the host and continues replicating without production of viral particles (lysogenic pathway). These are called attenuated viruses; which only begin to multiply within the cell and eventually lyse it when activating factors are present. Among attenuated viruses, one of the most studied is the lambda phage (λ).

The cells infected by RNA viruses do not have enzymes capable of replicating RNA. The virus provides not only the genetic information, but also the required polymerase.

The mechanism of synthesis varies depending on the relationship between the viral RNA and the mRNA required for the synthesis of viral proteins. Arbitrarily, a virus is designated (+)RNA when their RNA is identical to the mRNA and (−)RNA when its sequence is complementary. According to the characteristics of their nucleic acid, viruses can be divided into four types. *Class 1.* (+)RNA: after infection, a (−)RNA strand complementary is synthesized, which serves as a template to assemble the mRNA (e.g., poliomielytis virus). *Class 2.* (−)RNA: serves directly as the guide for the synthesis of mRNA (e.g., rabies virus). *Class 3.* Double-stranded (±)RNA: the mRNA is assembled on the (−) strand of the duplex (i.e., reovirus). The aforementioned viral types have

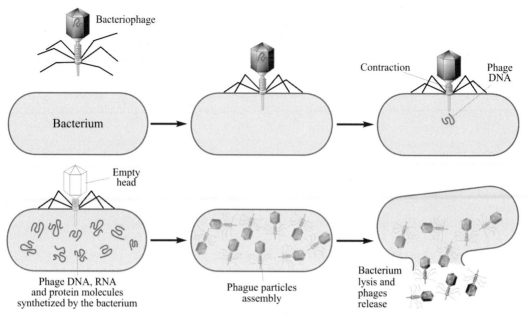

FIGURE 22.7 **Schematic representation of bacteriophage multiplication in bacteria and the production of the phage's nucleic acids and proteins (lytic pathway).**

an RNA-dependent RNA polymerase, called *RNA replicase*, which catalyzes the synthesis of new RNA. *Class 4*. Retrovirus: possess (+)RNA that serves as a template for DNA synthesis. The flow of information DNA→RNA→protein proposed by the "genetic dogma" is altered here. In this case, the order is reversed, and transcription goes from RNA to DNA, and then, it follows the usual direction. DNA synthesis is catalyzed by an RNA-dependent DNA polymerase, or *reverse transcriptase*, provided by the virus itself. The enzyme catalyzes the assembly of a DNA strand complementary to the viral RNA. Then, a complementary DNA chain is synthesized to form the double helix. The resulting DNA contains all of the virus's genetic information and is incorporated into the genome of the host cell, where it will be transcribed in the future.

Due to the simplicity of its genome, viruses have been widely used to study the function of nucleic acids as repositories and transmitters of information.

ONCOGENES

Cancer or malignant neoplastic processes are characterized by a series of transformations, the most conspicuous one being uncontrolled and invasive cell proliferation. In addition to their abnormal mitotic activity, cancer cells also present significant morphological and metabolic alterations.

Different possible causes for the malignant degeneration of cells have been described. Viruses, chemicals, and radiation are among the possible causes that trigger tumor formation. However, in most cancers, the causal agent cannot be detected.

The multifaceted pathophysiology of cancer has precluded finding a common mechanism for the induction of malignancies. However, studies of the molecular bases of cancer have provided evidence that neoplasias obey to genetic alterations, regardless of their apparent cause or characteristics.

A group of genes called *oncogenes* are responsible for carcinogenic transformation.

Initial observations came from the field of virology, which uncovered the existence of oncogenic viruses. Generally, these viruses belong to the retrovirus family and contain RNA as their genetic material. The cell infected with these viruses copies the information of the viral RNA into double-stranded DNA, due to the reverse transcriptase contained in the virus. The DNA is incorporated into the cell genome and is transcribed to synthesize viral RNA and protein.

A retrovirus widely used in these studies is the Rous sarcoma virus, a malignant tumor producing sarcoma in chickens. Genetic engineering techniques demonstrated that only a portion of the viral RNA is responsible for its tumorigenic capacity. The DNA segment, transcribed from the viral RNA, encodes for the synthesis of a protein capable of converting the affected cell into a cancerous cell. This segment of DNA is an oncogene, designated with the acronym *src* (for sarcoma).

Shortly after isolation of the *src* oncogene, a surprising finding came about. It was discovered that the *src* gene is not of viral origin, but it derives from the genome of the host animal. Thus, cells in normal uninfected chickens have genes almost identical to those of the *src* gene. How is it that the RNA virus possesses a piece of information equal to that in the host animal? Once they infect cells, many retroviruses are able to incorporate pieces of the host genetic information into their genome, such as the *src* gene. In subsequent infections, this segment is inserted in a normal cell with the rest of the viral RNA, now converted into an oncogene. The piece of normal DNA precursor of the oncogene is termed *proto-oncogene*. At some point in evolution, the proto-oncogene incorporated into the viral genome becomes modified after subsequent viral generations and acquires its oncogenic potential.

Neoplasias of viral origin, especially in humans, comprise only a very small fraction of the known tumors. New research has identified oncogenes causing cancer in nonviral tumors.

Transfer of DNA from human malignant tumors to normal cells, or to cells transformed by carcinogenic agents, can induce malignant transformation. Transmission of malignancy via DNA from one cell to another, involving only a DNA segment corresponding to the oncogene, has been demonstrated. These are called *cellular oncogenes* and are designated by adding the letter *c* before the three-letter acronym for the oncogene (i.e., *c-ras*), to distinguish them from viral (*v-src*).

Cellular and viral oncogenes are copies of proto-oncogenes present in normal cells. Proto-oncogenes can acquire new properties that lead to cancerous degeneration in the transformed cells, and also when incorporated into the genome of a retrovirus; in these cases, the proto-oncogene has been "activated."

The existence of proto-oncogenes raised the question of their role in normal cells. It is striking to find them, with almost identical structure, in many species, even in yeast. Cells have maintained these genes almost unchanged throughout evolution; in this respect, the conservation of proto-oncogenes appears as a biologic absurd, since they represent the seeds for the cell's own destruction. However, the high conservation of a gene across species suggests that it plays an important role. Proto-oncogenes encode proteins involved in the regulation of cell development. Proto-oncogenes control normal cell growth by encoding growth factors, growth factor receptors, transcription factors, and other growth-related cellular proteins. The *ras* proto-oncogenes encode proteins that bind GTP. Chapter 24 will describe the role of these proteins as mediators in signaling systems. Another proto-oncogene, named *myc*, directs the synthesis of a transcription factor normally stimulated by growth factors.

In summary, proto-oncogenes direct the synthesis of key proteins that control cell growth and development and, therefore, their role in cell development is of great importance. However, they can also confer malignancy to the cell by varying mechanisms. These include:

1. *Mutation.* Base changes in the DNA produced by different causes, either chemicals, radiations, or others, may determine the conversion of a proto-oncogene into an oncogene. Generally, the addition of several mutations is required for a normal cell to become malignant.
2. *Gene amplification.* A proto-oncogene can undergo several replications and produce multiple copies of itself. For example, existence of thirty copies of the *ras* oncogene has been found in human tumor cells.
3. *Translocation or transposition.* A piece of chromosome in which a proto-oncogene resides, is separated and placed on another chromosome. For example, Burkitt's lymphoma is a malignant tumor of lymphoid tissue in humans which is produced by this mechanism. In this case, the transformed B lymphocytes undergo a chromosomal rearrangement, which involves the translocation of a small end segment of a chromosome containing the *myc* oncogene to another chromosome, next to the gene encoding an immunoglobulin. Due to its proximity, the proto-oncogene is under the control of the same regulators controlling the immunoglobulin gene and is therefore actively expressed.
4. *Activation by retrovirus.* The information carried by a proto-oncogene is included in the genetic material of a retrovirus. When it infects a cell, the proto-oncogene introduced acquires oncogenic capacity because it is expressed under the control of the viral promoter, which has very high activity.

All mechanisms mentioned previously cause different levels of alteration in the synthesis of proteins normally encoded by the proto-oncogene. Mutations and the increase in the activity of the proto-oncogenes stimulate the synthesis of the encoded protein, leading to malignant transformation of the cell. Some authors have proposed that the proteins encoded by proto-oncogenes are functionally necessary at physiological concentrations, but promote malignant transformation when they are overexpressed in the cell.

The finding of oncogene's mechanisms of action has immensely contributed to elucidate, at the molecular level, the mechanism of cancerous degeneration.

Tumor Suppressor Genes

Besides the role of oncogenes in malignant degeneration, other genetic alterations are also responsible for tumorigenesis.

RBl. Studies in patients with retinoblastoma, a retinal tumor, indicated that tumor development is related to the loss of a gene. The predisposition to retinoblastoma is hereditary. There is also a sporadic, noninherited form of the disease. In the first case, patients have inherited a mutation due to a deletion on chromosome 13, where the RBl gene (retinoblastoma) is located. If a new mutation occurs in the retinal cells and inactivates the other RBl allele, both copies of the gene are lost, and a tumor develops. In sporadic forms of retinoblastoma, the chromosomes inherited from both parents are normal, but somatic mutations determine the inactivity of both the RBl alleles. Only the loss of both RBl alleles produces a complete loss of gene function and tumor development. This is in contrast to oncogenes, in which one abnormal allele is sufficient to cause tumor development. The RBl gene encodes a nuclear phosphoprotein of 110 kDa which binds to DNA and inhibits transcription of oncogenes, such as *myc* and *fos*.

Loss of the RBl gene allows for the expression of cancer-related genes. Its absence is also the cause of lung, breast, and colon cancers.

p53. Another tumor suppressor gene located on the short arm of chromosome 17 encodes for the p53 protein. The p53 mRNA is approximately

2.5-kb long and contains information for the synthesis of a nuclear protein of about 53 kDa that acts as a transcription factor. Normally, p53 is found in very low concentrations. When DNA damage occurs, the amount of p53 increases, and a signaling pathway is triggered that activate protein kinases. These kinases act on p53. Phosphorylated p53 functions as a regulator of gene activity. By binding to specific nucleotides in the grooves of DNA, p53 protects against cell damage and carcinogenesis.

Exposure to hypoxia, oxidizing agents, toxic chemicals, ultraviolet or gamma radiation, high activity of oncogenes, such as *Ras* and *Myc*, and excessive cell proliferation are all conditions that increase p53 levels. This has multiple actions: (1) blocks the cell cycle in the G1 phase (p. 465), (2) inhibits cyclin-dependent kinases, and (3) stops the cell cycle, allowing time for the DNA repair systems to correct the damage. If DNA cannot be repaired, it will induce cell death (apoptosis, Chapter 32). p53 also functions to inhibit angiogenesis, which is essential for the growth of tumors, and protects the cell's genetic stability. For example, when telomeres are very short, p53 inhibits cell division by a mechanism called *cellular replicative senescence*.

The gene encoding p53 is mutated in almost 50% of all human cancers. When p53 is lacking, neoplastic cells continue proliferating, even when the DNA is damaged. This allows for the accumulation of mutations, which increases tumor malignancy and depresses apoptosis, which prevents destruction of the abnormal cells. Currently, p53 is considered one of the most important tumor suppressor genes.

EPIGENETICS

Heritable changes in gene expression that occur without a change in the sequence or primary structure of DNA are called *epigenetic* changes. These occur mainly in germ cells during early stages of embryogenesis and persist through successive cell divisions or even across several generations.

The main epigenetic changes consist in (1) DNA methylation, (2) covalent histone modifications, and (3) actions of noncodifying RNAs.

1. *DNA methylation* includes insertion of methyl groups at position 5' of cytosines; the reaction is catalyzed by *methyl transferases*. Methylated cytosines are those adjacent to guanines in areas rich in the dinucleotide CG, called "CpG islands," usually located close to gene promoters (sites of DNA transcription start and regulation). Methylation tends to silence the expression of genes. The reverse process, demethylation, catalyzed by *DNA demethylases*, increases the expression of genes and is particularly critical in gametes and during the early embryo development.

2. *Histone modifications* include acetylation, methylation, phosphorylation, ubiquitination (p. 368), sumoylation (p. 542), ADP ribosylation (p. 638), and glycosylation. These histone changes modify the conformation and accessibility of chromatin, changing its degree of compactness, which influences DNA transcription.

   Some regions of the chromatin or entire chromosomes are very compact (heterochromatin) and the genes comprised in them cannot be expressed. In other portions or chromosomes the chromatin is relaxed (euchromatin) and permit free access to transcription factors, enzymes (polymerases), and the substrates needed for RNAs synthesis.

   The most common modifications are acetylation and methylation of lysine residues at the N-terminal of histone 3 (H3) and histone 4 (H4); the reactions are catalyzed by *histone acetyl transferases*. Increased acetylation reduces the compactness of chromatin and induces activation of transcription. Instead, deacetylation, catalyzed by *histone deacetylases*, represses it. Methylation is associated with activation or repression, according to the location of the methylated lysine residues.

3. *Actions of noncoding RNAs* (microRNAs and long noncoding RNAs) will be described in the next chapter.

All cells of an individual have the same genome, but epigenetic modifications may cause variability by changing the expression of genes even in cells of the same tissue.

Epigenetic changes related to hormones, malignancies, environmental factors, diet, and others have been demonstrated. For example, offsprings of mothers fed diets poor in proteins during pregnancy or lactancy have reduced expression of a nuclear transcription factor and a tendency to suffer obesity, hypertension, type 2 diabetes, and cardiovascular diseases during adulthood.

This phenomenon has been observed in bees. The queen bee, fed with royal jelly, lays eggs and lives years; instead, the worker bee is sterile and last only weeks. It has been demonstrated that the royal jelly promotes methylation of many genes.

Alcoholism, smoking, and environment contaminants may alter epigenetic gene expression.

The relationship between epigenetics and cancer has been intensely studied. In general, the patterns of DNA acetylation and methylation have been found to be altered in malignant tumors. The repression of genes coding for tumor suppressor proteins has also been demonstrated.

Imprinting. Genomic imprinting is considered an epigenetic event that determines the expression of only one of the parental alleles in the descendants. It generally occurs in the gametes. One locus or a cluster of genes is silenced, commonly by methylation, in the mother or the father gamete; then, only the father or the mother allele is expressed in the offspring. Imprinted

genes regulate fetal growth and influence newborn development, feeding, maternal care, and metabolism regulation.

Chromosomal defects may contribute to the production of diseases. For example, some rare cases of a deletion in chromosome 15 of spermatozoa will result in the embryo receiving only the imprinting pattern for this chromosome segment from the mother; the offspring will present with Prader–Willis syndrome, which is characterized by growth and mental retardation, obesity, hypogonadism, and muscle weakness. The inverse defect (the fetus receives the imprinting from the father) produces Angelman syndrome, which is characterized by mental deficit, epilepsy, and tremors.

SUMMARY

The genetic information is contained in DNA.

DNA replication ensures that all cells of an individual, by successive divisions of the egg cell, receive DNA of equal structure and quantity.

The sequence of nucleotides in DNA, transcribed to mRNA, is a "coded message" (genetic code) that indicates the amino acid sequence of the protein to be synthesized.

mRNA serves as a template upon which the assembly of amino acids that make up a protein is made. The message of mRNA, expressed with the bases A, G, C, and U, is translated into amino acid sequence. The unit of information in mRNA is a triple base unit or triplet called a *codon*. With the existing 4 bases, it is possible to produce 64 different codons. Each amino acid is represented by one or more triplets (synonymous codons). UAA, UAG, and UGA indicate *termination* of the polypeptide chain. The genetic code has universal validity, because the meaning of codons is the same in all living beings (a few exceptions can be found in mitochondrial DNA and in some unicellular organisms).

Cistron is the sequence of codons containing the complete information for the synthesis of a protein. In bacteria, each cistron is a continuous piece of DNA. In eukaryotes, the coding regions are separated by intercalated noncoding regions.

Exons and introns in eukaryotes refer to the coding and noncoding regions of DNA, respectively. Introns are almost always larger than exons.

During transcription, RNA is synthesized from a DNA template, called the *antisense* or *noncoding strand*. The other

DNA strand, not transcribed, is the *sense* or *coding strand* and has the same sequence as the synthesized RNA. In eukaryotes, introns and exons are transcribed to form the *precursor mRNA*, which is part of the heterogeneous nuclear RNA.

Splicing is the process by which the precursor mRNA undergoes major changes in the cell nucleus. A 7-methyl-GTP cap at the 5′ end and a poly-A tail at the 3′ are added, introns are removed. Each intron has a portion where *small nuclear RNAs* (snRNA) bind at its end and at its center. A spliceosome complex is formed, which separates the precursor RNA strand precisely at a site between the intron and the flanking exons. Then, exons are joined end to end in the correct order. The end product is *mature mRNA*, which is exported to the cytoplasm. In eukaryotes, the mRNA molecule contains information for the synthesis of one polypeptide chain, it is *monocistronic*. In bacteria, the mRNA can be *polycistronic*.

Ribozymes are RNA molecules that behave like enzymes. In many cases, RNA catalyzes the cleavage of introns. RNAase P is a ribozyme involved in tRNA processing.

Protein biosynthesis follows these stages:

1. *Activation of amino acids*. This requires free amino acids, *aminoacyl-tRNA synthetases*, tRNA, ATP, and Mg^{2+}. The amino acid reacts with ATP and forms an aminoacyl–AMP–enzyme complex. The activated amino acid is transferred to the specific tRNA.

2. *Initiation*. Various *initiation factors* (eIF) interact with the cap at the 5′ end of the mRNA. These bind to the smaller ribosomal (40S) subunit and methionyl itARN with eIF-2 bound to GTP to form the *preinitiation complex*. This complex moves on the mRNA, in a 5′→3′ direction, to locate the initiation codon (AUG), to which the itARN anticodon pairs. The larger ribosomal subunit is added and methionyl itARN adheres to the P site. An 80S *initiation complex* is formed.

3. *Elongation* takes place in a cycle that is repeated with each added amino acid. The aminoacyl-tRNA anticodon matching the corresponding codon in the mRNA binds to the A site.
   Peptide bond formation and translocation. The carboxyl group of the itRNA methionine in the P site binds to the amine function of the tRNA amino acid in site A; the reaction is catalyzed by peptidyl transferase. A dipeptidyl is now bound to the tRNA located at site A. The itARN, at the P site, unloads their methionine and is released. The dipeptidyl-tRNA is translocated from site A to site P. The ribosome advances on the mRNA to the next codon in the 5′→3′ direction. After several cycles, the peptide chain attached to tRNA in the P site has grown. The protein is assembled from the N-terminus to the C-terminus. The chain is transferred to the tRNA in site A. The energy expenditure to form a peptide bond is four high-energy phosphate bonds.

4. *Termination*. Once the chain is completed, there is a stop codon (UAA, UAG, or UGA) recognized by the *liberation factor* (eLF) associated with GTP. The polypetidic chain is separated from the tRNA by hydrolysis. Unloaded tRNA and ribosomal subunits are released. One mRNA chain directs simultaneously the synthesis of multiple protein molecules. A group of eight or more ribosomes on an mRNA is called a *polysome*.

Some antibiotics block the transcription process (actinomycin D, rifamycin, rifampin), others affect translation (puromycin, streptomycin, neomycin, gentamycin, tetracycline, chloramphenicol, erythromycin, cyclohexymide kirromycin, fusidic acid).

Mutations are changes in the DNA sequence. They may affect only a base pair or comprise more extensive structural changes in the chromosomes. Many diseases are due to mutations in the genetic material that result in the production of abnormal proteins or prevent their synthesis. Mutations in germ cells are transmitted to the offspring. Normally, the incidence of "spontaneous" mutations is very low. There are *mutagenic factors* (ionizing radiations, ultraviolet rays, various chemical compounds) that increase the frequency of mutations.

Transposons are DNA sequences capable of being inserted in different locations of the genome. *Retroposons* are introduced into the genome by reverse transcription from retroviruses.

The proteome refers to the total number of proteins synthesized by an organism.

Oncogenes are genes responsible for carcinogenic transformation of cells, inducing malignant proliferation. *Proto-oncogenes* that code for proteins related to cell development and multiplication normally exist in cells. They can be converted to oncogenes by exaggerated increases in their activity produced by mutations, amplifications, transpositions, or activation by retroviruses. There are also genes that function as *tumor suppressor genes* (RBl, p53).

Epigenetics includes the modification of gene expression without change in the DNA sequence.

Bibliography

Afshari, S.A., 2002. Microarray technology, seeing more than spots. Endocrinology 143, 1983–1989.

Alberts, B., Jonson, A., Lewis, J., Raff, M., Roberts, K., Walter, P., 2008. Molecular Biology of the Cell, Garland Science, fifth ed. Taylor & Francis Group, New York, NY.

Cooper, G.M., Hausman, R.E., 2007. The Cell. A Molecular Approach, fourth ed. Sinauer Associates, Inc., Sunderland.

Eichler, E.C., Clark, R.A., She, X., 2004. An assessment of the sequence gaps: unfinished business in a finished human genome. Nat. Rev. Genet. 5, 345–354.

Kornblihtt, A.R., Schor, I.E., Alló, M., Dujardin, G., Petrillo, E., Muñoz, M.J., 2013. Alternative splicing: a pivotal step between eukaryotic transcription and translation. Nat. Rev. Mol. Cell Biol. 14, 153–165.

Krebs, J.E., Goldstein, J.E., Kilpatrick, S.T., 2013. Lewin's Genes XI. Jones & Bartlett Publishers Inc., Burlington.

Larsson, N.G., Clayton, D.A., 1995. Molecular aspects of human mitochondrial disorders. Annu. Rev. Genet. 29, 151–178.

Lodish, H., Berk, A., Kaiser, C.A., Kriger, M., 2007. Molecular Cell Biology, sixth ed. W.E. Freeman, New York, NY.

Marte, B. (Ed.) 2003. Proteomics. Nature 422, 191–237.

Melo, E.O., Canavessi, A.M., Franco, M.M., Rumpf, R., 2007. Animal transgenesis: state of the art and applications. J. Appl. Genet. 48, 47–61.

Ohtsuka, M., Kimura, M., Tanaka, M., Inoko, H., 2009. Recombinant DNA technologies for construction of precisely designed transgene constructs. Curr. Pharmacol. Biotechnol. 10, 244–251.

Peters, J., 2014. The role of genomic imprinting in biology and disease: an expanding view. Nat. Rev. Genet. 15, 517–530.

Scheper, G.C., van der Knaap, M.S., Proud, C.G., 2007. Translation matters: protein synthesis defects in inherited disease. Nat. Rev. Genet. 8, 711–723.

Sherrat, E.J., Thomas, A.W., Akolado, J.C., 1997. Mitochondrial DNA defects. A widening clinical spectrum of disorders. Clin. Sci. 92, 225–235.

Shore, D., 1997. Telomere length regulation. Getting the measure of chromosome ends. J. Biol. Chem. 378, 591–598.

Solari, A.J., 2004. Genética Humana, third ed. Editorial Médica Panamericana, Buenos Aires.

Watson, J.D., Baker, T.A., Bell, S.P., Gann, A., Levine, M., Losick, R., 2008. Molecular Biology of the Gen, sixth ed. Pearson Benjamin Cummings, Cold Spring Harbor Laboratory Press, San Francisco.

Weaber, R.F., 2008. Molecular Biology, fourth ed. McGraw Hill, New York, NY.

23

Regulation of Gene Expression

The activity of the genome is a finely regulated process, which is under the control of numerous factors. Both in prokaryote and eukaryotes, gene expression is modulated, primarily, at the transcriptional level; however, the mechanisms operating are different. Bacteria regulate gene expression by controlling genetic units called *operons*.

An operon consists of a set of genes that encode functionally related proteins, all of which are under the control of a common promoter. Different regulatory factors bind to a gene promoter to either stimulate or inhibit the binding of RNA polymerase to DNA, which affects the transcriptional activity of the whole operon, located downstream in the same DNA molecule.

Eukaryotes have gene regulatory mechanisms that are more complex than those of prokaryotes. They need to adjust gene expression depending on cell type, developmental stage, and the particular physiological status of each cell type. In general, only a small portion of the eukaryotic genome is active; most of it is silenced. Eukaryotes do not have genes assembled into operons; however, multiple factors can regulate the genome not only at the level of an individual gene, but also by modulating groups of genes.

Each tissue in a eukaryotic organism has cells that markedly differ in morphology and function. This cell specialization is achieved during development, through the process of differentiation, which depends on the selective regulation of gene activity in the different cell types. For example, all human cells contain the genes encoding for globin; however, only the precursor of erythrocyte produces this protein. Neurons, fibroblasts, adipocytes, despite having the genetic information, do not synthesize globin. Even in erythroblasts, the genes for the different types of globin chains are activated at defined times during development; the γ units are only synthesized during fetal life and are silenced after birth and during the rest of the individual's life. This shows the high specificity of the systems that control gene activity.

GENE REGULATION IN EUKARYOTES

While gene expression is mainly exerted at the transcriptional level, it can also take place at other stages along the process of protein synthesis, through a variety of mechanisms. These include the following:

1. Changes in gene number and structure
   Loss of genes. Total or partial gene deletion prevents RNA production and the synthesis of the proteins encoded by those genes. An example of complete gene suppression is found in erythrocytes, where the whole genome of the cells is lost during cell differentiation.

Gene amplification. The generation of multiple copies of a gene is a mechanism that increases the production of the proteins, which are required in large quantities by the cell. This type of gene regulation is found for proteins, such as histones and ribosomal RNA.

Gene rearrangement. Genes can be transferred from one site of the genome to another. The genes encoding antibodies in B lymphocytes are regulated in this manner (Chapter 30).

Chemical modification of genes. Methylation of cytosine residues at promoter sites of the DNA can markedly reduce gene transcription or completely repress it. For example, globin genes are extensively methylated in cells that do not synthesize hemoglobin. Changes in methylation are also involved in the control of malignant cell transformation. Genes, such as those of the growth hormone and some proto-oncogenes, are less methylated in lung and colon tumor cells than in normal tissues.

2. Regulation of transcription

In general, this is the most frequent mechanism used to modulate protein synthesis. Transcription can be controlled via the following events:

Chromatin condensation. Transcription requires a close interaction of DNA with a set of proteins and other molecules. Therefore, the accessibility of DNA to these interacting effectors is essential for gene expression. The degree of "packing" of the nuclear chromatin is a key factor in this respect. DNA included in highly condensed accumulations, characteristic of heterochromatin, cannot be expressed. It is believed that only the portions of the DNA molecule comprised in the euchromatin can be transcribed. In other words, the tertiary structure of DNA (supercoiled state) greatly influences transcription. In eukaryotes, DNA is associated with histones to form nucleosomes

(p. 131). Only the DNA segments extended from one nucleosome to another ("naked" DNA) are accessible to regulatory proteins that can act as silencing or activating genes. It has been suggested that the displacement of nucleosomes exposes segments corresponding to promoter regions of the genome involved in transcription regulation. Methylation and acetylation of lysine residues and other covalent modifications (phosphorylation, ubiquitination, and acetylation) of histones, have an important role in the regulation of nucleosomes.

3. Modification of gene activity

Specific sequences in the genome are recognized by regulatory proteins, which can induce or repress transcription. These regions are known as *response elements*. Among a series of regulators, many hormones act as modulating response elements in DNA. In the case of repressor proteins, binding to DNA is usually sufficient to silence a gene. For activator proteins, in addition to binding to DNA, the regulator must interact with RNA polymerase and other transcription factors.

The targeted sequences in DNA, called *cis-acting* sequences, include promoters, such as TATA boxes, CAT, GC, and others, generally located upstream the start site of the coding portion of the gene. These sequences are targeted, among others, by the TATA box binding protein and a series of transcription factors.

Gene expression can be activated or suppressed by other *cis*-acting sequences, like the *enhancers*, located in any area of the DNA strand, sometimes far from the transcription initiation site. Some groups of genes share regulatory sequences and can be modulated simultaneously by the same factors.

4. Post-transcriptional regulation

This event takes place during the processing of the primary transcript (precursor RNA) and during transport of mRNA from the nucleus to the cytoplasm. Each of these steps are highly regulated.

5. Translational regulation

The synthesis of proteins is controlled by RNA binding proteins at the initiation and elongation stages of polypeptide synthesis. Most of the control mechanisms affect protein synthesis initiation by targeting, for example, the initiation factor eIF-2, which is inactivated by phosphorylation.

The amount of synthesized protein can also be controlled by varying the duration or "half-life" of mRNA. The presence of multiple copies of the AUUUA sequence in the 3' terminal noncoding region labels the mRNA for rapid degradation, reducing the time during which translation can operate. Another type of control consists of the direct binding to RNA of polypeptides that block the activity of factors involved in protein translation.

6. Posttranslational regulation

Proteins present variable half-lives, from hours to days for some and months or years for others. At the end of its life, the protein is degraded by lysosomal or cytosolic proteases (see p. 368). The activity of this proteolytic process also helps regulating the amount of a specific protein in the cell.

STRUCTURE OF GENE REGULATORY PROTEINS

The proteins that regulate gene activity in eukaryotes are characterized by a specific structure closely related to its function. The most important include the following:

1. *Helix-turn-helix*. This protein domain is present in both prokaryotic and eukaryotic regulatory proteins. It consists of two α helices joined by a short random coiled

segment (Figs. 23.1 and 23.2A). The axis of each helix in the helix-turn-helix is roughly perpendicular to the other. One of the helices accommodates in the major groove of DNA and binds to specific base sequences. The other helix lies at right angles with respect to the direction of the groove (Fig. 23.1). These proteins, commonly dimers, possess two "helix-turn-helix" domains that bind on the same side of the DNA molecule. The domains are separated by a distance equivalent to one or two turns of the double helix (Fig. 23.1).

The interaction of a regulatory protein and DNA most probably depends on the establishment of hydrogen bonds and van der Waals forces between side chains of the protein amino acid residues and the chemical groups in the base pairs accessible from the major groove space. Upon binding to DNA, repressors interfere, and inducers favor the activity of the DNA-dependent RNA polymerase. The mutual adaptation of regulatory proteins and DNA is critical for this action, which highlights the importance of conformational changes that can take place in the DNA molecule.

2. *Zinc fingers*. This is a ~30 amino acid motif in regulatory proteins in which a zinc atom is bound to two cysteine and two histidine residues, to give the polypeptide segment between those amino acid residues the shape of a finger (Fig. 23.2B). Usually from 2 to 13 "fingers" are found in a gene regulatory protein. Through zinc finger domains, the protein binds to the grooves of DNA. The steroid receptors belong to this class of transcription regulatory proteins.

3. *Leucine zipper*. This is a motif in some dimeric proteins which involves two helical structures. The helix in one of the dimers has a segment with seven leucine residues, oriented toward the same side of the helix. These leucine residues interdigitate with

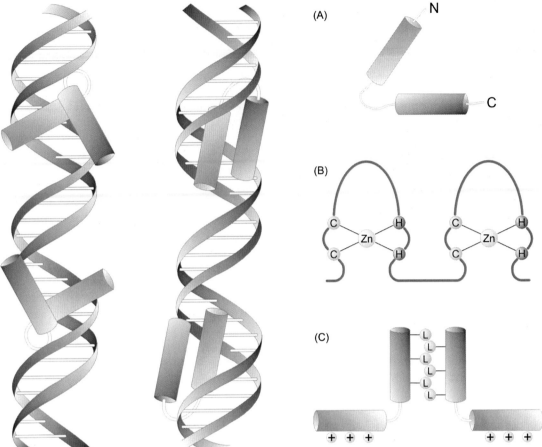

FIGURE 23.1 **Schematic representation of the "helix-turn-helix" binding domains.** *Gray cylinders* are the α helices, which interact with the major groove of DNA. Only the "helix-turn-helix" domains are shown, the remaining portions of the regulatory protein have been omitted. The left side of the figure presents a front view, while the right one shows a side view of the protein helix-turn-helix DNA domains.

FIGURE 23.2 **Common structural motifs present in transcriptional regulatory proteins.** (A) Helix-turn-helix. (B) Zinc fingers (Zn, zinc; C, cysteine; H, histidine). (C) Leucine zipper (L. leucine side chain; only three of the seven leucine residues in each helix are represented). The *gray cylinders* in A and C correspond to α helices that bind to DNA.

leucine residues present in the homologous helix of the other dimer of the protein.

The second helix is rich in positively charged residues and interacts with DNA. Each subunit can readily dimerize because one leucine-rich helix associates with the other by hydrophobic interactions. The name of this protein domain derives from the similarity of the structure to a zipper (Fig. 23.2C).

Mediator

This is a large multiprotein complex (consisting of 30 subunits in humans), which plays an important role in the regulation of transcription in eukaryotes. The mediator complex functions as an adaptor that connects the various components of the transcriptional machinery (enhancers or repressors-RNA polymerase

II-transcription factors-promoters) and ensures proper communication among them.

ROLE OF NONCODING RNAs

Besides mRNA, rRNA, and tRNA, the transcriptome includes other RNA molecules. These can be divided into two classes according to their size: small RNAs that consist in molecules of less than 200 nucleotides, and *long noncoding RNA* (lncARNs), which are of higher size. Small RNAs include the following: *micro RNA* (miRNA), *small interfering RNA* (siRNA), *small nuclear RNAs* (snRNA), PIWI interacting RNA (piRNA), *small RNA associated to promoters, RNA initiators of transcription*, and *small circular RNAs* which interact with miRNA.

Long noncoding RNAs include those synthesized by RNA polymerase II. It is estimated that more than 9,000 small RNAs, and between 10,000 and 30,000 long noncodifying RNAs exist in a cell.

Many RNA transcripts compete for binding to miRNA. They are known as *competing endogenous RNAs* (ceARN). Knowledge of these interactions can lead to a better understanding of the regulatory networks that are involved in the control of gene expression in normal and pathological states.

SMALL RNAs

The miRNAs are small noncoding RNA molecules of approximately 22 nucleotides in length. More than 2000 miRNAs have been identified and this number is likely to continue increasing. They function on posttranscriptional and translational regulation of many genes. It is estimated that 60% of protein coding genes are modulated by miRNAs. So, alterations in miRNAs can cause multiple disturbances.

The siRNA functions as a silencer of foreign genes or transposons. The *piARNs* are single-stranded RNA, 23–31 nucleotides long, which bind to Argonaut proteins called PIWI. One of their main actions is to silence transposons in germinal cells. The snRNA are involved in the process of RNA splicing.

RNA Interference

In plants and animals a mechanism was described, which regulates gene activity by selectively silencing gene expression, either by degrading mRNA, or inhibiting DNA translation. This process, called RNA interference (RNAi), was initially thought to represent a cell defense mechanism against the introduction of anomalous DNA in the genome, which may occur by viral infection or transposon insertion. Later, it was found that RNAi is a mechanism involved in the regulation of many genes.

RNAi is highly conserved across diverse organisms (protozoa, fungus, plants, and animals, including humans), which suggests that its appearance occurred early in the course of evolution.

Cells generate RNAi by producing small double-stranded RNA that has similar sequence (except that T is replaced by U) than the segment of the gene to be silenced. The process is different from the general antiviral reaction triggered by double-strand RNA that promotes release of interferon. RNAi production is a physiological process initiated by endogenous double-stranded RNA, synthesized in the same cell.

Two types of RNA, siRNA and miRNA are essential for RNAi.

Biogenesis of siRNA

siRNAs are generated from lineal double-chain RNA produced in the cell by mobile genetic elements (transposons) or by endogenous transcripts that can pair to other double-chain RNAs artificially introduced in the cell. This RNA is processed in the cytoplasm by a ribonuclease III enzyme called *Dicer* that cleaves it into segments of 21–25 nucleotide in length, leaving the 3′ end of one strand two bases longer than

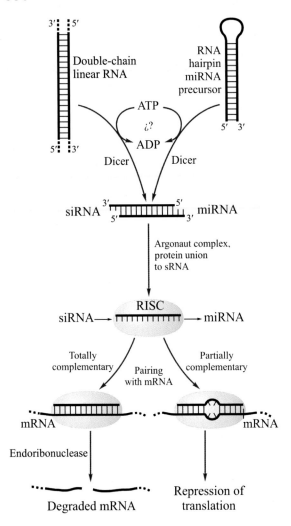

FIGURE 23.3   **RNA interference.**

there are genes responsible for the synthesis of 100–200 nucleotide chains with autocomplementary segments that allow them to form "hairpin" structures. These are the primary miARNs (pri-miRNA) (Fig. 23.3).

The processing of miRNA comprises a first stage in the nucleus with a protein complex containing an endoribonuclease, called *Drosha*, which cleaves the pri-miRNA (this step is not represented in Fig. 23.3). This precursor miRNA (pre-miRNA) moves to the cytoplasm, where, similar to siRNA, it is cleaved by the *dicer* RNA-ase in linear fragments 21–30 bp long with two free bases at the 3′ end. The sequence of this RNA may not be entirely identical to the segment of the target gene, but it has high homology with it.

Although the final function of siRNAs and miRNAs is to inhibit gene expression, there are differences in the origin and functions of these molecules. The siRNAs silence foreign genes; while miRNAs function by regulating genes involved in cell proliferation, hematopoiesis, and apoptosis (Chapter 32); in general, miRNAs are associated with the regulation of growth and development.

Formation of RISC

Both, siRNA and miRNA, form a complex with proteins, which is called RNA-induced silencing complex (RISC). Incorporation to the protein complex requires the presence on the small RNA of a phosphate group and a two nucleotide overhang at the 5′ and 3′ ends. This small RNA is cleaved by Dicer, rendering a double-stranded RNA of 22 base pairs that subsequently associates to Argonaut proteins; the miRNA can associate with proteins different from Argonaut. One strand of the RNA, totally or partially complementary to a segment of the target messenger RNA, is separated by a helicase; the other strand is eliminated. This produces the mature miRNA or siRNA. The specificity of the siRNA depends on its

the 5′ end of the other strand. Thus, the 3′ end has two unpaired bases (Fig. 23.3). The 5′ end is esterified by a phosphate. One of these segments has identical sequence to that of the target gene.

Biogenesis of miRNA

miRNAs are encoded in the genome and are produced by transcription, catalyzed by RNA polymerase II or III, from noncoding DNA regions localized in intron sequences. In humans, complementarity with the mRNA transcribed from the target gene.

RISC accommodates the miRNA or siRNA on the segment of the mRNA that is partially or totally complementary with it. If the sequences are exactly complementary, as in the case of siRNA, the pairing with mRNA is complete and the endoribonuclease function of the Argonaut protein is activated. This cleaves mRNA, preventing its translation. When the miRNA sequence is partially complementary to the target mRNA, the pairing is not complete. Under these conditions, hydrolysis of mRNA does not occur; instead, changes in the mRNA are produced including, for example, removal of the polyA tail, thus preventing ribosomal translation.

Although in most cases miRNAs inhibit gene expression, in some cases they can activate it.

The two-step hydrolysis of pri-miRNA by Drosha and pre-miRNA by Dicer, major sites of miRNA biogenesis, are closely regulated by different factors. Alteration of these regulatory processes may cause changes in the miRNA, affecting its activity.

It has been shown that miRNAs, through their action on gene expression, participate in the modulation of a broad spectrum of physiological processes, including cell proliferation, migration, differentiation, apoptosis, senescence, immune response, metabolism, and gametogenesis. They also contribute to maintain genome stability and the response of cells to stress. Therefore, dysregulation of miRNA biogenesis has been recognized as a cause of pathological conditions that can affect many organs and systems. Numerous observations show the existence of a correlation between miRNAs production and cancer, metastasis, or angiogenesis.

Techniques using RNAi have great impact on basic research. Thus, RNAi can be artificially generated by introducing synthetic siRNA into a cell. In theory, using siRNA with a defined sequence, it is possible to silence any gene of interest. This has made RNAi an attractive approach to block the expression of a gene to understand its function. To achieve this, a siRNA can be artificially introduced in cells and target tissues following various strategies: (1) Chemically synthesized small RNA can be associated to polycations, cholesterol, substrate surface membrane receptors, or included in nanoparticles, which facilitate their penetration into the cell. (2) Viral vectors can be used to carry hairpin RNA that will generate, by transcription, miRNA in the target cell.

Also, RNAi offers numerous possible uses in medicine. Current techniques can detect the presence of abnormal miRNA, and this is valuable for clinical diagnosis and prognosis. RNAi is being tested as a therapeutic approach in many pathological conditions, including Parkinson's disease, macular degeneration related to age, respiratory diseases, obesity, rheumatoid arthritis, hypercholesterolemia, type 2 diabetes, HIV infection, and some types of malignancies. The results in laboratory animals are very promising and in some cases miRNA have been approved for human clinical trials.

LONG NONCODING RNA

Less than 2% of the human genome corresponds to genes encoding proteins, part of the remaining DNA is actively transcribed into noncoding RNAs, including small and long noncoding RNAs (lncARN).

Although its mechanism of action is not exactly known yet, there is evidence that lncARNs are important regulators of gene activity. They influence histone modification, which results in chromatin condensation and inhibition of gene expression. A lncARN has been identified that participates in the inactivation of one X chromosome in female mammalian cells. Others are involved in the cell cycle, apoptosis, immune response regulation, and in various pathological processes, including oncogenesis.

Competing Endogenous RNA

Recently ceRNA has been discovered; it competes with miRNA for binding to response elements, reducing the inhibitory effect of miRNA on translation. Thus, ceRNA plays an important role in gene regulation, modifying the actions of miRNA.

RIBOSWITCHES

Riboswitches are segments of noncoding mRNA, ranging from 35 to 200 nucleotides, which change their conformation upon binding to metabolites, coenzymes, small molecules, or ions, to regulate gene expression. Riboswitches function as *aptamers* that serve as specific receptors for a given ligand, binding it when its concentration in the medium reaches a certain level. Often, another domain near the aptamer, called *expression platform*, is able to undergo conformational changes that enable the riboswitch to interact with DNA elements that control genes involved in the production of the ligand that bound the riboswitch. The regulatory effect can be exerted on transcription, splicing, mRNA stability, and translation. Most of them have repressive action (negative feedback effect), but some can stimulate gene expression.

Initially, riboswitches were demonstrated in bacteria and were thought to be unique to prokaryotes, for example, in *Escherichia coli* they regulate the production of enzymes involved in thiamine (vitamin B_1) biosynthesis. More recently, riboswitches were also found in eukaryotes.

CRISPR

CRISPR (Clustered Regularly Interspaced Short Palindromic Repeats) was first discovered in bacteria and consists of a repetition of short segments of DNA separated by spacer sequences. These sequences, along with an endonuclease CRISPR-associated system (Cas), correspond to an adaptive immune system that protects bacteria against the invasion of virus or plasmids. Currently, it has been widely used as a molecular tool for the targeted editing and modification of the genome of different cell types, including human cells.

CRISPR-Cas9 has opened a new era in genetic engineering. It can be used for the mutation, addition, or disruption of DNA sequences to alter specific genes with greater accuracy, simplicity, and lower cost than that provided by other methods.

The system is comprised by a guide RNA that has sequence complementary to regions of the DNA to be modified. Once introduced into a cell, this RNA binds to the target gene in the precise site, according to its sequence, and guides Cas9, which cleaves the DNA at one or both chains. This, similar to the RNA interference system of eukaryotes, silences the gene of interest. Alternatively, the gene can be just edited by only replacing one or more base pairs or longer DNA segments.

The system, has already allowed introducing gene modifications in mice, to produce animal models that resemble human genetic diseases, or for gene therapy to correct genetic abnormalities. For example, this technique has been shown to be successful for the treatment of tyrosinemia and hereditary cataracts in mice.

The CRISPR-Cas system is also being used in plants for the development of new varieties that are advantageous with respect to their growth and food production.

In humans, the experiments are at present limited to cultured cells. Theoretically, it will be possible to "cure" cells affected by mutations of only one pair of bases, like Hb S anemia or cystic fibrosis, or to replace longer regions of DNA, for the treatment of thalassemia and Huntington's disease. It is used in the treatment of viral infections and of other conditions (cancer, diabetes, Down syndrome, and others) is under study.

In conclusion, CRISPR-Cas technology is a powerful genetic tool that will have important medical applications in the near future.

SUMMARY

Regulation of gene expression takes place most commonly at the transcriptional level. An important factor for gene expression is the accessibility of DNA, which needs to interact with specific regulatory proteins. These recognize particular sequences in promoter and enhancer sites of DNA. Important structural motifs that allow gene regulatory proteins to interact with the DNA are *helix-turn-helix*, *zinc fingers*, and *leucine zipper domains*. Gene activity can also be regulated at other levels than transcription. Molecules, such as *small* and *long RNAs* are also involved in gene regulation.

RNA interference is a process that selectively silences gene expression. It is activated by the presence of double-stranded linear RNA with sequence similar to that of the gene to be silenced. An enzyme called *dicer* sections RNA into pieces of approximately 22 nucleotides, the miRNA and siRNA bind to the RISC and to the complementary mRNA to degrade it or to prevent its translation. Other regulators of gene expression are *long noncoding RNAs* and *riboswitches*, which are segments of noncoding RNA that upon binding to a small molecule can regulate gene expression.

CRISPR-Cas is a system that allows silencing, editing, or modifying genes with great accuracy. It is based on the delivery of a specific RNA sequence into cells, which binds to the targeted DNA in the genome and modifies it after specific cleavage by the endonuclease Cas and replacement of the DNA with the new desired sequence.

Bibliography

Charpentier, E., 2015. CRISPR-Cas 9 system: how research on a bacterial RNA-guided mechanism opened new perspective in biotechnology and biomedicine. EMBO Mol. Med. 7, 353–365.

Conaway, R.C., Sato, S., Chieri, T.S., Tingtin, Y., Conaway, J.W., 2005. The mammalian mediator complex and its role in transcriptional regulation. Trends Biochem. Sci. 30, 250–255.

Eggleston, A., Eggleston, A.K., 2004. RNA interference. Nature 431, 337–378.

Eggleston, A.K., 2009. RNA silencing. Nature 457, 395–434.

Hsu, P.D., Lander, E.S., Zhang, F., 2014. Development and application of CRISPR-Cas 9 for genome engineering. Cell 157, 1262–1278.

Kong, J., Lasko, P., 2012. Translational control in cellular and developmental processes. Nat. Rev. Genet. 13, 383–394.

Kung, J.T., Colognori, D., Lee, J.T., 2013. Long noncoding RNAs: past, present and future. Genetics 193, 651–669.

Mandal, M., Breaker, R.R., 2004. Gene regulation by riboswitches. Nat. Rev. Mol. Cell Biol. 5, 451–463.

Sander, J.D., Joung, J.K., 2014. CRISPR-Cas system for editing, regulating and targeting genome. Nat. Biotechnol. 32, 347–355.

Serganov, A., Patel, D.J., 2007. Ribozymes, riboswitches and beyond: regulation of gene expression without proteins. Nat. Rev. Genet. 8, 776–790.

Sunk, K., Lai, E.C., 2013. Adult specific functions of animal micro-RNAs. Nat. Rev. Genet. 14, 535–548.

Tay, Y., Rinn, J., Pandolfi, P.P., 2014. The multilayered complexity of ceRNA crosstalk and competition. Nature 505, 344–352.

Posttranslational Protein Modifications

PROTEIN FOLDING

The relationship between protein structural conformation and function was discussed in Chapter 3. Newly synthesized polypeptides need to fold and acquire the proper conformation to display their normal function. The manner in which proteins fold is predetermined in the primary protein structure. In other words, the genetic information, transcribed from the mRNA, dictates the three-dimensional arrangement that a protein will finally acquire. This arrangement, or "native" state, almost always corresponds to the protein structure that has the lowest energy and, therefore, is stable from a thermodynamic viewpoint. In some cases, proteins reach their final conformation only after they undergo posttranslational modifications.

In general, protein folding is initiated by interactions between hydrophobic and polar groups, which create a condensation core. It is over this core that the rest of the protein will be arranged until its final shape is acquired.

Attaining the correct three-dimensional structure is critical for proteins. Generally, defects in their folding determine functional disorders (see subsequent sections). For this reason, there are quality control systems, which detect defective proteins and rapidly label them for degradation in the proteasome (p. 368).

Chaperone proteins. Protein folding can be a spontaneous process, achieved by the sequential interaction between the side groups of the amino acid that integrate the protein. However, numerous steps are required until the proper spatial arrangement of the protein is accomplished, making the process slow and inefficient. In cells, protein folding is accelerated by the action of *chaperone proteins.* These proteins are key in facilitating protein folding.

Chaperone proteins can bind to the nascent polypeptide chain while it is still being synthesized in ribosomes. They stabilize intermediate conformations of the polypeptide before it reaches its final state and prevent unproductive interactions, formation of abnormal protein structures, and insoluble protein aggregates. Chaperone proteins also stabilize unfolded chains during transport from the cytosol to its final destination and are involved in the assembly of subunits of oligomeric proteins.

Originally, many chaperone proteins were identified as *heat shock proteins* (HSP). These proteins are expressed both in prokaryotes and eukaryotes in larger amounts when cells are subjected to higher than normal temperatures or other environmental stressors. Several families of HSP (Hsp 10, Hsp 60, Hsp 70, and Hsp 90) have been described. They bind to unfolded polypeptide chains during synthesis and stabilize them. When protein synthesis is completed, they bind to short segments of the polypeptide and keep it organized while transported within the cell.

Hsp 70 is a dimer, widely distributed in all tissues. It binds to its substrates via hydrophobic interactions in an ATP-regulated manner. When bound to ATP, the affinity of Hsp 70 for binding proteins is low; ATP hydrolysis increases its affinity. The chaperone-protein association of Hsp 70 is also regulated by proteins that stimulate ATP hydrolysis or increase ADP levels.

Hsp 60 forms multimeric complexes associated to Hsp 10. These complexes are known as *chaperonins*, of which two types has been described. GroEL is a chaperone of bacteria that is associated with GroES (Hsp 10) into a chaperonin complex. It is also found in eukaryote cells (as the TCP1 chaperonin) in chloroplasts, mitochondria, and in the cytosol. The GroEL (Hsp 60) complex is composed of two rings of seven subunits each. Both rings are attached to form a tube; GroES (Hsp 10) is another polymer of seven subunits. Chaperonins present in eukaryotic cytoplasm (TCP1) are formed by two rings of eight subunits each. The association of GroEL and GroES depends on ATP hydrolysis.

Chaperonins belong to the same class of protein macrocomplexes that proteasomes belong to (p. 368). They form hollow structures within which the substrate polypeptide chains remain sequestered while undergoing folding (in the case of chaperonins) or degradation into oligopeptides (in the case of proteasomes).

PATHOLOGIES CAUSED BY MISFOLDED PROTEINS

Misfolded proteins that escape quality control, or are beyond the capacity of proteasome degradation, may form intracellular or extracellular aggregates that have harmful effects. In general, these aggregates have fibrillary structure, with configurations predominantly made of β sheets.

Extracellular protein deposits are called *amyloid*; the pathological condition caused by amyloid accumulation is called *amyloidosis*. There are systemic types of amyloidosis which simultaneously affect various tissues, and local forms, which are limited to a single organ. Many examples of both systemic and local amyloidosis have been described which affect the central nervous system and result in neurodegenerative diseases.

In *Alzheimer's disease*, there is extracellular accumulation of β amyloid, a 4.3 kDa peptide produced by hydrolysis of the *amyloid precursor protein* in brain. Plaques and fibrillary bundles that compromise the function and viability of neurons are formed. Intracellular aggregates of a protein called *tau* are also observed.

In *Parkinson's disease* there are deposits of α synuclein, a filamentous protein (seen as Lewy bodies on pathological reports), in dopaminergic and noradrenergic neurons.

Another group of neurodegenerative diseases caused by protein misfolding is due to mutations that determine tandem repeats of glutamines, which leads to the synthesis of elongated proteins and their aggregation in the nucleus and cytoplasm of neuronal cells. This results in diseases, such as *Huntington's disease* and several *familiar spinocerebellar ataxias*.

Other interesting examples of neurodegenerative diseases are caused by *prions*, which are misfolded proteins. Among these alterations are *Creutzfeldt–Jakob disease* in humans, *scrapie* in sheep and goats, and *bovine spongiform encephalopathy* or "mad cow" disease.

Prion. The *prion* protein (PrP^c) is a sialoglycoprotein, whose structure is predominantly composed of α helices. PrP^c is expressed in the external surface of cell membranes, anchored by glycosylphosphatidylinositol. It is present in many cell types, but is most abundant in pre- and postsynaptic neuronal cell membranes. It participates in neural transmission and signal transduction in the nervous system. It also plays a role in redox status and calcium ion flux in other cells.

PrP^c can undergo a change in structure, becoming the altered PrP^{sc} protein. This abnormal form of the protein is predominantly made of β sheets and is resistant to the action of proteases. PrP^{sc} tends to form aggregates in the brain and determines a progressive, fatal degeneration. The conversion of PrP^c into PrP^{sc} rarely occurs spontaneously; however, it is favored by the existence of certain types of mutations. For this reason, some of the cases related to PrP^{sc} diseases are likely to be hereditary.

The most striking fact regarding prion proteins is that PrP^{sc} functions as an infectious agent. It is transmitted in epidemic form in cattle through ingestion of food contaminated with the abnormal protein. Theoretically, humans can also acquire the disease in this way. Prions are the unique pathogens known without nucleic acid, and the first known mammalian protein to behave as infectious agents.

It has been proposed that introduction of exogenous PrP^{sc} to a tissue may induce normal PrP^c molecules to adopt abnormal conformation. The process spreads, leads to the formation of intracellular deposits responsible for neuronal degeneration.

POSTTRANSLATIONAL PROTEIN MODIFICATIONS

Besides chaperone proteins, other proteins help to determine the final structural conformation of a polypeptide. Proteins can undergo a variety of modifications after being synthesized. Most of this modifications, especially those related to histones, can be included as part of cell epigenetic changes.

Disulfide bridge formation. The formation of disulfide bridges between cysteines, catalyzed by *disulfide isomerase*, is highly relevant for the final structure of a protein. When a polypeptide chain folds, distant cysteine residues may become close together. Oxidation by disulfide isomerase establishes covalent —S—S— bonds between cysteines of the same chain (intrachain). This largely contributes to maintain the tertiary structure of the protein. Disulfide bonds are also formed between cysteines from different polypeptide chains (interchain) in oligomeric proteins.

Isomerization of peptide bonds. Another important factor that determines protein folding and the secondary structure of a protein is the spatial configuration of the peptide bonds. In almost all proteins, the *trans* isomer is the most favorable form, except for those bonds containing the amino acid proline. *Peptidyl-prolyl isomerase* catalyzes the *cis–trans* isomerization of peptide bonds containing proline.

Hydrolysis of the polypeptide chain. In eukaryotes, all newly synthesized polypeptides have methionine as the first amino acid; in prokaryotes, the first amino acid is *N*-formyl methionine. This residue, and sometimes the next two or three amino acids, are removed by hydrolysis catalyzed by peptidases. In some cases, amino acid residues at the C-terminal end are also cleaved.

In molecules transported through membranes, the N-terminal segments that serve as membrane translocation signals are often removed by hydrolysis. This cleavage is also produced in precursor (inactive) proteins to obtain the final (active) product. For example, two segments from inactive pre-proinsulin are removed to produce insulin with hormonal activity (p. 606). Many zymogens, like trypsinogen and chymotrypsinogen, are converted to an active enzyme after hydrolysis.

Some mRNAs encode a polypeptide chain that is cleaved by specific proteases after synthesis, generating several different proteins. The parent polypeptide is designated *polyprotein* (e.g., *proopiomelanocortin*, p. 579).

Covalent modification. Following translation, proteins are often subjected to addition or removal of functional groups (hydroxyl, carboxyl, acetyl, methyl, amide, phosphate, ADP-ribose) on specific amino acid side chains. These changes are important for the function of the modified protein.

Hydroxylation. In collagen, for example, numerous proline and lysine residues on the original polypeptide are subjected to hydroxylation. Hydroxyprolines and hydroxylysines are very important in determining the shape and properties of a protein.

Carboxylation. A carboxyl can be added to glutamyl residues (γ-carboxylation) of a protein, such as the proteins involved in the process of blood coagulation (Chapter 31). An example of this protein modification is the γ-glutamyl carboxylation of *gla* proteins, Ca^{2+} chelators, which are essential for blood coagulation. This modification is catalyzed by an enzyme which requires vitamin K (p. 660).

Acetylation. Acetylation reactions, catalyzed by acetylases, have a great influence on the function of many proteins. It is estimated that 50% of eukaryotic proteins are acetylated posttranslationally at the N-terminus.

Methylation. The addition of methyl groups is another posttranslational change that proteins can undergo. In some muscle proteins and cytochromes there are monomethyl- and dimethyl-lysine residues. Calmodulin (p. 567) contains a trimethyl-lysine residue which is critical for its function. In other proteins, the glutamate carboxyl groups are methylated, which makes them lose their negative charge.

Phosphorylation. Phosphate addition to the hydroxyl group of serine, threonine, or tyrosine is catalyzed by protein kinases that transfer phosphate molecules from ATP. This modification adds a negative charge to the polypeptide; its functional significance varies from one protein to another. Phosphorylation plays an essential role in regulating the activity of numerous enzymes. The process is reversible; there are phosphatases that remove the phosphoryl groups. Casein in milk has many serine phosphorylated groups that bind Ca^{2+}. The protein is therefore an excellent source of not only amino acids, but also calcium and phosphorus.

Addition of carbohydrates. Glycoproteins are formed by addition of carbohydrates (in general, oligosaccharides) to the side chains of asparagine, serine, or threonine in proteins (p. 316).

Addition of lipids. In many cases, insertion of lipids to a polypeptide chain serves to anchor the protein to the plasma membrane. Myristoylation, palmitoylation, and isoprenylation are common in proteins associated with the cytosolic surface of the membrane. The addition of glycolipids (glycosylphosphatidylinositol) is important in proteins associated with the outer side of the plasma membrane. Interestingly, it has been observed that the isoprenylation of the ras oncogene abolishes its carcinogenic potential.

Addition of prosthetic groups. Prosthetic groups can be added after a protein is synthesized. For example, the heme group in hemoproteins and the biotin in carboxylase are added after synthesis and total release of the main protein from the ribosome.

ADP-RIBOSYLATION

For a long time, the function of the coenzyme NAD was believed to be related to its role in redox reactions. However, in the last 20 years, experimental evidence suggested that NAD participates in ADP-ribosylation events.

In NAD, the cleavage of the glycosidic bond between the C1′ of ribose and N11 of nicotinamide releases nicotinamide and ADP-ribosyl (Fig. 24.1). This can be attached to a variety of acceptor molecules. Numerous NAD-dependent reactions for the transfer of ADP-ribose (ADP-ribosylation) are known; all are of great functional importance.

There are different variants of ADP-ribosylation: (1) mono-ADP-ribosylation, (2) poly-ADP-ribosylation, (3) NAD-dependent deacetylations, (4) formation of cyclic ADP-ribose, and (5) NAADP formation.

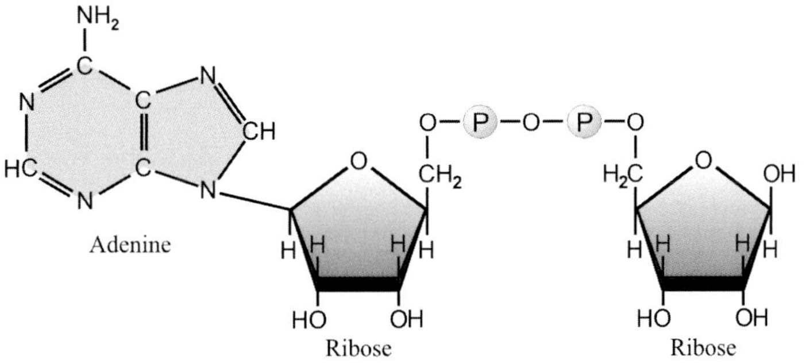

FIGURE 24.1 **ADP-ribose.**

Mono-ADP-ribosylation. In this posttranslational modification, the ADP ribosyl from NAD is transferred to an aminoacyl residue (arginine, cysteine, asparagine, or histidine) of an acceptor protein. It should be noted that binding of the ADP-ribosyl nicotinamide in NAD is a high energy bond; its rupture provides the energy that makes the reaction possible. *Mono-adenosinediphosphate-ribosyl transferase* (ART), initially described in bacteria toxins and later in eukaryotic cells, catalyzes the reaction.

Cholera toxin promotes the transfer of mono-ADP-ribosyl to the α subunit of G_s protein and activates it. This leads to stimulation of adenylate cyclase, increase in cyclic AMP levels, and higher function of ion transport channels in the luminal membrane of enterocytes. This causes severe diarrhea, a characteristic symptom of cholera toxin infection. The pertussis toxin (produced by the bacteria that cause whooping cough) determines the ADP-ribosylation of cysteinyl residues and uncouples G protein from its receptor. Diphtheria toxin and *Pseudomonas* exotoxin stop protein synthesis by ADP-ribosylation of elongation factor 2 (EF2). *Clostridium* toxin ADP-ribosylates actin molecules and prevents its polymerization. These actions show that the mono-ADP-ribosylation markedly influences the function of the modified protein.

In humans, several ADP-ribosylation enzymes have been recognized. Some are anchored by glycosyl-phosphatidylinositol to the external surface of the plasma membrane (ectoenzymes) and others are within the cell (endoenzymes).

The finding of ectoenzymes that act on NAD located within the cells was striking. It is believed that these enzymes use the NAD released into the interstitial space by lysed cells; alternatively, the existence of channels that allow the exit of NAD through the plasma membrane has been proposed. The ectoenzymes are functionally associated with the modulation of myocyte differentiation and other processes associated with immune and inflammatory responses, such as chemotaxis, recruitment of neutrophils, inhibition of T-cell cytotoxicity, and cell adhesion.

Intracellular ARTs are involved in the regulation of signal transduction systems in which G proteins are involved and serve as ART substrates. The mono-ADP-ribosylation can affect signaling and promote various cellular effects. Inhibition of protein translation, regulation of the Golgi apparatus, and cytoskeleton function are a result of these posttranslational modifications.

Poly-ADP-ribosylation. This is another type of posttranslational modification catalyzed by *poly-adenosinediphosphate-ribosyl polymerases* (PARP). Eighteen PARP genes have been identified, but not all the enzymes encoded by these genes have been characterized. PARP initially binds ADP-ribosyl to glutamyl or aspartyl residues in the acceptor protein. Then, it continues inserting ADP-ribosyl molecules, attaching them linearly by glycosidic 1'—2' bonds. At every 40–50 units, branch points are created in the main chain, inserting 1'—3' bonds. PARP can also undergo auto-poly-ADP-ribosylation; one of the major substrates of PARP is PARP itself.

ADP-ribose polymers are highly electronegative and affect the properties of the modified protein. The increase in the protein negative charge augments the repulsion of the ADP-ribosylated protein with other polyanions, such as DNA; or attracts positively charged molecules, such as histones.

Among the known ADP-ribosyl polymerases, some are located in the nucleus. PARP-1 and PARP-2 are activated by the presence of cleaved sites in the DNA strands, to which PARP binds. These cleaved sites usually occur during DNA replication and repair, or can be caused by external agents. PARP-3, is often associated with the centrosome and PARP-4 is associated with ribonucleoprotein particles. PARP-7 and PARP-10 are involved in histone ribosylation. TNKS and TNKS-2 are also poly-ADP-ribosyl polymerases and they are associated to telomeres.

The modification by poly-ADP-ribosylation of basic proteins, such as histones, alters DNA-histone interactions as well as intra- and inter-nucleosome attractions, promoting a looser chromatin structure. This facilitates access to DNA of enzymes involved in the processes of replication and repair, including helicase, topoisomerase, polymerase, and ligase. The PARP-ADP-auto-polyribosylate tends to repel nearby DNA strands as its electronegative charge increases, and finally separates.

PARP associated to telomeres promotes telomerase activity in chromosome elongation. Furthermore, their presence serves to repel other DNA strands and prevent abnormalities, such as translocations and end-to-end or deleterious recombination fusions.

Poly-ADP-ribosylation of some enzymes can modify their activity; for example, it can stimulate DNA ligase and inhibit endonuclease, preventing DNA degradation.

PARP is involved in the regulation of chromatin structure, transcription, replication, repair, maintenance of DNA integrity, and stimulation of DNA ligase. The absence or decrease in PARP activity leads to genome instability.

PARP associated to centrosomes contributes to an orderly separation of chromosomes during mitosis.

PARP is also involved in cell differentiation and protein degradation during programmed cell death (apoptosis, Chapter 32). The mechanisms of action are not yet clearly understood. PARP mediates apoptosis through proinflammatory signals. It controls the release of the apoptosis inducing factor (AIF) from mitochondria. Recent studies have also shown a relationship between poly-ADP-ribosylation and polyubiquitination in labeling proteins for degradation.

In cases of cell stress, overactivation of PARP can lead to depletion of NAD and ATP, with devastating consequences for the cell that end in cell necrosis.

Studies in laboratory animals have shown that ischemic conditions, in brain and heart, under septic shock, or severe inflammatory processes improve when PARP is inhibited. It is likely that these observations will have clinical application; however, the challenge of reduced ADP-ribose polymerase activity leading to genome destabilization, accumulation of mutations, and eventually to malignant transformation (carcinogenesis), must be first solved.

The ADP-ribose polymers are degraded by the poly-ADP-ribose glycohydrolase, which releases free ADP-ribose. An ADP-ribose lyase releases the first unit attached to the protein. Pyrophosphatase separates AMP and ribose phosphate.

NAD-dependent deacetylation. There is a protein family of *NAD-dependent deacetylases* called *sirtuins*, which release the nicotinamide group from NAD and utilize acetate separated from proteins as an ADP-ribosyl acceptor to form 2'-O-acetyl-ADP-ribose. This activity was first observed in yeast and was designated with the acronym SIR (*silent information regulator*); later it was also demonstrated in the nematode *Caenorhabditis elegans* and in *Drosophila melanogaster*. Their action increases the lifespan of these organisms, particularly in conditions where nutrients in the medium are restricted.

Sirtuins (SIRT) comprise a family of seven protein members (SIRT-1 to SIRT-7), which can be considered a variant of ADP-ribosylases. They use diverse substrates, such as histones, p53 protein, transcription factors, nuclear factor κB, and others. For example, histone deacetylation induces a more compact structure of chromatin, which promotes gene silencing and protects critical areas of chromosomes, such as telomeres and centrosomes. Deacetylation of protein p53 is important for genomic stability; it controls the cell cycle, DNA repair, and apoptosis. Deacetylation of p53 apparently increases its stability by blocking ubiquitination.

SIRT 1 is involved in energetic metabolism, oxidative stress response, cell senescence, and protection of vascular endothelia and nerves in different pathologic conditions.

Nicotinamide released from NAD-dependent deacetylations acts as a potent inhibitor of the activity of sirtuin, contributing to its regulation.

Since the action of sirtuins contributes to the prolongation of life in some organisms, it was thought to be a general prolongevity factor. However, there is not enough evidence yet to extrapolate these results to higher animals.

It is believed that the maintenance of normal levels of poly-PARP polymerases and sirtuins in the cells may prevent or delay carcinogenesis and aging-related disorders.

The product 2′-O-acetyl-ADP-ribose, resulting from sirtuin activity, functions as a second messenger.

Reactions that are unrelated to protein posttranslational modification, but can be considered variants of ADP-ribosylation, generate compounds with important physiological action: cyclic ADP-ribose and nicotinic acid-ADP-ribose-phosphate (NAADP) (see p. 669).

UBIQUITINATION

An important protein modification is protein ubiquitination. Ubiquitin (Ub) has lysine (K) residues at positions 6, 11, 27, 29, 33, 48 y 63, which participate in the association of Ub molecules. As described in p. 368, Ub linked together via their lysine 48 residues, form a polyubiquitin chain that binds to proteins and directs them for degradation in the proteasome.

In addition, Ub K48 chain does not only label and target proteins for hydrolysis. It has been shown that binding of Ub K48 chain to a transcription factor from yeast (M-4) does not promote its hydrolysis, but instead, inhibits its function. Therefore, K48 Ub has a role in gene regulation.

Polyubiquitin chains bound at lysine 63 (K63) modify the activity of a variety of proteins; for example, those involved in DNA repair and translation, mRNA splicing, and the transcription factor NFκB. Also, monoubiquitination may occur at selected lysine residues of a protein, (multi-monoubiquitination). This can modify the activity and localization of the protein, or affect protein–protein interactions of the targeted polypeptide.

Although less frequent, the binding between Ub units occurs not only at K48 or K63 but at any of the other five K residues that Ub has; even branched chains can be produced, or more than one chain can be attached to a protein.

Ubiquitination affects a variety of cellular processes, including DNA transcription and repair, cellular division, differentiation and development, neuronal morphogenesis, immune response, viral infection, and apoptosis.

Alteration of genes encoding Ub or related proteins, particularly enzymes E3, are associated to different pathologies, including neurodegenerative diseases (Parkinson and Alzheimer diseases), cardiovascular diseases, Fanconi anemia, and 3-M syndrome, characterized by development retardation.

SUMOYLATION

Sumoylation is a posttranslational modification, which consists in the addition of a small ubiquitin-like protein, designated SUMO (small Ub-like modifier), into lysine residues of numerous proteins. This results in significant changes of their properties.

SUMO has a molecular mass of about 10 kDa and a tertiary structure similar to that of Ub, but it contains less than 20% homology with Ub. In humans, four genes encoding SUMO proteins have been identified, all of which are expressed across many tissues. They are synthesized as an immature product, whose C-terminus needs to be cleaved to acquire the capacity to bind to target proteins.

Sumoylation involves the formation of a peptidic bond between the C-terminal glycine of the modified protein and the ε-amine group of a lysine residue in the acceptor protein. Like ubiquitination (p. 368), the process involves an enzymatic cascade involving three types of catalysts: activator (E1), conjugating (E2), and ligase (E3). In most cases, addition of only one SUMO molecule occurs, but formation of poly-SUMO chains have also been observed. The reaction is reversible; SUMO can be removed by a hydrolase.

Although there are similarities between the ubiquitination and sumoylation processes, the consequences of both protein modifications are different. Ub primarily labels substrates for degradation, while sumoylation results in different functional changes in proteins.

The proteins undergoing SUMO addition locate at different sites of the cell, including the nucleus, cytoplasm, endoplasmic reticulum (ER), and plasma membranes. The basic effect of sumoylation involves changes in the surface of the protein, or alters the conformation of the protein, affecting its interaction with other macromolecules. It can also modify the subcellular localization of the protein, suppress or facilitate its degradation, inhibit or activate enzymes, affect genomic or chromosomal stability, alter chromatin organization, modify DNA repair, and influence signal transduction systems. The result of these actions depends on the specific modified protein. Many transcription factors are common SUMO acceptors, some are activated and others inhibited. At times, the effect is influenced by other processes, such as phosphorylation, acetylation, methylation, etc.

Sumoylation may play a role in the regulation of basic metabolic functions; it is possibly involved in the adaptation of cells to stressful environments, such as hypoxia. The likely relationship to developmental defects and various diseases, such as cancer, neurodegenerative disorders, and others is the subject of ongoing studies.

NUCLEOTIDYLATION

This is another type of posttranslational modification, catalyzed by enzymes designated with the acronim *Fic* (filamentation induced by cAMP); these proteins has been conserved along evolution and can be found in bacteria and also in humans. Fic adds AMP, UMP, and phosphocholine to hydroxyl groups of amino acid lateral chains in many proteins, modulating their function. Among the Fic substrates are GTPases of the Rho family, which play an important role in signal transduction (see Chapter 25). Pathogenic bacteria employ this enzyme to disrupt host signaling pathways during infection.

Protein Traffic in the Cell

Protein synthesis takes place in the cytoplasm (except for some mitochondrial proteins). From the site of synthesis, each protein must be directed to their final destination, where they will perform their function. Some proteins remain in the cytosol, but most traffic to different organelles, membranes, or are secreted into the extracellular space. The protein transit from the place of origin to their final destination requires a signaling system capable of ensuring that each molecule finds its correct delivery pathway.

Often the signals are part of the structure of the molecules in transit. A segment of a polypeptide chain of variable length (10–60 amino acids) contains a special sequence indicating the target site, it is designated the *signal peptide*. This sequence is normally located at the N-terminus, at internal positions or, more rarely, at the C-terminal end.

The synthesis always starts on free ribosomes in the cytoplasm and can follow two main pathways: (1) the ribosomes continue in the cytoplasm and once the polypeptide chain is finished it is released into the cytosol; (2) the ribosomes move to the membrane of the ER and the polypeptide chain is transferred to the ER cisternae while they are synthesized.

1. The first path includes cytosolic proteins and proteins destined to the nucleus, mitochondria, and peroxisomes. Proteins without specific signals remain in the cytosol; those designed to specific organelles possess a signal peptide. For the proteins exported to the nucleus, the recognition signal is a short piece of 4–8 residues, rich in positively charged amino acids (lysine and arginine), located anywhere in the chain. This portion enables the protein to enter the pores of the nuclear membrane. The polypeptides transferred to mitochondria have, at the N-terminus, a signal peptide of 20–60 amino acids rich in positively charged residues. In each of the two mitochondrial membranes (outer and inner), protein complexes are responsible for the translocation of polypeptides. The passage through the channel formed by these complexes requires energy (ATP) and an electrical gradient. The chain must remain unfolded. Once inside the mitochondria, the signal sequence is cleaved by a peptidase. The protein can remain in the matrix, be inserted in the inner membrane, or return to the intermembrane space. In both of the latter cases, an internal hydrophobic segment directs the reintegration of the chain into the inner membrane. The protein can be attached to it as an integral protein or pass to the intermembrane space. When the polypeptide chain has reached its final destination, it undergoes the folds that will give it its normal conformation. This process involves chaperones of the Hsp70 family.

2. The second alternative includes the proteins initially introduced in the ER, which may remain in the lumen of the ER cisternae or move to the Golgi apparatus and then to lysosomes, plasma membrane, or outside of the cell (secretion). All have a signal peptide rich in hydrophobic amino acids at the N-terminus, also called the *leader peptide*. Shortly after initiation of synthesis, when this sequence emerges from the ribosome, it is recognized by cytoplasmic particles of ribonucleoprotein nature, designated *recognition signal particles* (RSP). The RSP lead the complex to the ER membrane, where there are specific receptors for RSP. The ribosome binds to the ER membrane on transmembrane protein complexes, forming the translocation system. Binding of a large number of ribosomes gives the membrane its characteristic *rough* appearance, as opposed to the smooth ER, which has no ribosomes attached.

While the RSP is attached to the leader peptide, the elongation of the peptide chain is slowed or stopped. When the ribosome binds to the ER membrane, RSP separates and the synthesis resumes its initial activity. The leader peptide then enters into the translocation system channel and attaches to it, the rest of the growing chain is driven into the ER lumen (Figs. 11.5 and 24.2), and a signal peptidase cleaves the leader sequence, even before completion of the polypeptide chain. After this initial processing, the protein is released into the ER lumen and may stay there or continue its way to the Golgi apparatus. The pathway followed depends on existing marker segments in different sectors of the molecule. Many proteins initiate the

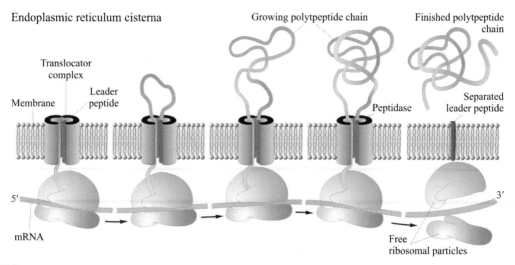

FIGURE 24.2 **Translocation of a protein through the membrane of the endoplasmic reticulum (ER).** The initial portion of the polypeptide chain, represented in *dark gray*, corresponds to the signal peptide or leader peptide (start transfer); this leads the nascent strand through the membrane into the cisternae of the ER. The leader peptide is separated from the rest by a peptidase.

glycosylation process in the ER, when they pass to the Golgi apparatus they enter through the *cis* network enclosed in vesicles whose membrane is derived from the ER membranes. This vesicles fuse with that of the Golgi, where it releases its content. Again, different alternatives are offered to the protein: it can remain in the Golgi, can be targeted to lysosomes or membranes, or it can be secreted.

N-glycosylation and disulfide bond formation in the ER are very important for the folding of membrane and secreted proteins.

Proteins destined to the lysosomes, predominantly hydrolases, are modified in the Golgi by an enzymatic process that adds phosphate at position 6 of a mannose molecule. Mannose-6-phosphate is recognized by a membrane receptor protein and binds to it. Vesicles released from the Golgi, targeted to lysosomes, fuse with their membrane. Finally, the content of the vesicle is released inside the organelle.

Cell I disease or *mucolipidosis II* is a genetic defect produced by a lack of the phosphotransferase, one of the enzymes that catalyzes

formation of the mannose-6-P. Hydrolytic enzymes cannot penetrate into the lysosomes and are excreted outside the cell, with accumulation of lipid complexes and glycans which cannot be catabolized.

Proteins "for export" follow the secretory pathway. Synthesized on membrane-bound ribosomes of the ER, they penetrate into the cisternae, where they undergo various transformations (signal peptide cleavage, disulfide bond formation between cysteines, and addition of carbohydrates). These proteins, enclosed in vesicles formed by evagination of the ER membrane, are sent to the *cis* Golgi sector, within which the content is released after membrane fusion. Within the Golgi, proteins undergo further modifications, such as carbohydrate remodeling of the chains and addition of new saccharides, a process known as complex glycosylation. After moving through the various cisternae, proteins are packaged into vesicles and emerge from the *trans* Golgi network. These charged vesicles are targeted selectively to the plasma membrane or any organelle, mainly lysosomes.

In the process described, several factors participate directing the formation of vesicles, ensuring recognition of the target site and promoting membrane fusion (see p. 246).

SUMMARY

Protein folding is essential for a polypeptide chain to acquire its proper structure and function. Protein folding is assisted by HSP called chaperones. Multimeric complexes that form hollow structures, called chaperonins, also participate in protein folding.

Other posttranslational modifications, critical to protein function, include elimination of the N-terminal formyl methionine residue, formation of disulfide bonds between cysteines, covalent modifications, hydroxylation, carboxylation, acetylation, methylation, amidation, deamidation, phosphorylation, ADP-ribosylation, addition of oligosaccharides, addition of prosthetic groups, sumoylation, and nucleotidylation.

Protein misfolding can lead to the formation of intra- and extracellular aggregates of fibrillar structure, which have damaging effects to the cell (as occurs in amyloidosis and neurodegenerative diseases). *Prion* protein PrP^c, can change to a misfolded conformation, PrPs^c, that behaves as infectious agents. These are responsible for different alterations, including "mad cow" disease).

Posttranslational modifications. These include disulfide bridge formation, isomerization of peptide bonds, hydrolysis of polypeptides, hydroxylation, carboxylation, methylation, phosphorylation, addition of carbohydrates, lipids, and prosthetic groups, ADP-ribosylation, acetylation, deacetylation (sirtuins), ubiquitination, and sumoylation.

Protein trafficking is required to send the protein to its final destination in the cell. Signals within the protein, dictated by the amino acid sequence are responsible for proper protein targeting. Proteins released in the cytosol can remain there or can be transferred to the nucleus, mitochondria, or peroxisomes. Cytosol proteins do not need signals to direct their final destination; others have specific sequences that direct them to their target location in the cell. Proteins introduced into the ER are guided by recognition particles. The polypeptide chain penetrates the ER lumen, where it undergoes modifications (glycosylation, disulphide bond formation). Some proteins eventually pass to the Golgi apparatus where further glycosylation and phosphorylation steps are performed. From the Golgi, proteins can move to lysosomes, the plasma membrane, or can be secreted into the extracellular space. This takes place by vesicular transport in which specific proteins are involved.

Bibliography

Choiti, F., Dobson, C.M., 2006. Protein misfolding, functional amyloid and human disease. Annu. Rev. Biochem. 75, 333–366.

DeArmond, S.J., Prusiner, S.B., 2003. Perspectives on prion biology, prion disease pathogenesis and pharmacologic approaches to treatment. Clin. Lab. Med. 23, 1–41.

Deuerling, E., Bukau, B., 2004. Chaperone-assisted folding of newly synthesized protein in the cytosol. Crit. Rev. Biochem. Mol. Biol. 39, 261–277.

Floho, A., Melchior, F., 2013. Sumoylation: a regulatory protein modifications in health and disease. Annu. Rev. Biochem. 82, 357–385.

Fridman, J., 2001. Folding of newly translated proteins in vivo. The role of molecular chaperones. Annu. Rev. Biochem. 70, 603–649.

García-Pino, A., Zenkin, N., Loris, R., 2014. The many faces of Fic: structural and functional aspects of Fic enzymes. Trends Biochem. Sci. 24, 171–178.

Geiss-Friedlander, R., Melchior, F., 2007. Concepts in sumoylation: a decade on. Nat. Rev. Mol. Cell Biol. 8, 947–956.

Gonfloni, S., Iannizzotto, V., Maiani, E., Bellusci, G., Ciccone, S., Diederich, M., 2014. P5e and Sirt1: routes of metabolism and genome stability. Biochem. Pharmacol. 92, 149–156.

Selkoe, D.J., 2003. Folding proteins in fatal ways. Nature 426, 900–904.

Smith, A., 2003. Protein misfolding. Nature 426, 883–909.

Westaway, D., Carlson, G.A., 2002. Mammalian prion proteins: enigma, variation and vaccination. Trends Biochem. Sci. 27, 301–307.

Yi, J., Luo, J., 2010. SIRT 1 and p53, effect on cancer, senescence and beyond. Biochim. Biophys. Acta 1804, 1684–1689.

Zhao, J., 2007. Sumoylation regulates diverse biological processes. Cell. Mol. Life Sci. 64, 3017–3033.

Biochemical Basis of Endocrinology (I) Receptors and Signal Transduction

Modulation of enzyme activity by allosteric effectors and regulation of gene activity in response to environmental stimuli are processes that ensure the adequate function of metabolism in unicellular organisms. However, those mechanisms are insufficient in multicellular organisms due to their higher complexity. The development of highly specialized tissues and organs required communication and "remote control" systems to regulate their function in an integrated manner that will serve the needs of each cell of the body. This integration is provided by the nervous and endocrine systems.

Despite their structural and functional differences, nervous and endocrine systems share a similar basic mechanism: the secretion of chemical intermediates or signaling molecules that trigger a particular response in the effector cells.

The endocrine system is constituted by a variety of cells, which are grouped in discrete glands that produce and secrete active substances, called *hormones*. In many cases, hormones are secreted into the circulation in response to specific stimuli. Transported by blood, they reach the target cells in which they will produce particular effects. Some of these secreted substances do not reach the blood, they act on the same cell that secretes them (autocrine mechanism)

or on adjacent cells (paracrine mechanism). Among these are eicosanoids, growth factors and cytokines (the latter will be considered in Chapter 30).

The nervous system serves as a fast communication network that connects different regions of the body. The interaction between neurons is based on the release to the synapsis of compounds called *neurotransmitters*.

However, neuronal cells can also secrete hormones. For example, neurons, whose bodies are found in hypothalamic centers, project their axons, and have their terminals in the posterior pituitary gland, where they release the hormones oxytocin and vasopressin. These hormones act on the mammary gland and kidney, respectively. Likewise, factors produced by neurons in the hypothalamus regulate the secretion of hormones of the anterior pituitary or adenohypophysis, and arrive at this gland from the hypothalamus by a portal venous system. Several peptides originally described as gastrointestinal hormones, are also produced in the nervous system. Epinephrine, hormone of the adrenal medulla and norepinephrine, are agents that, in addition to their role as hormones, also act as neurotransmitters in adrenergic terminals.

Fig. 25.1 presents some of the many relationships between the nervous and endocrine

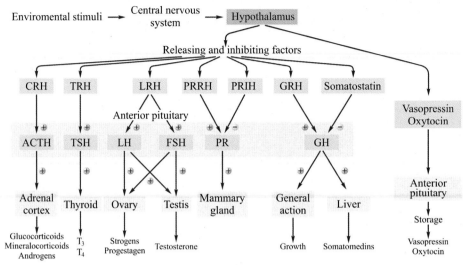

FIGURE 25.1   **Nervous system–endocrine system interactions.** In general, the hormone level in blood regulates the release of the corresponding tropin. Feedback regulation is not represented. *Hypothalamic releasing hormones*: CRH corticotropin, TRH thyrotropin, LRH gonadotropin hormones (luteinizing and follicle stimulating), PRRH prolactin, GRH somatotroph (growth hormone). *Inhibitor hormones*: PRIH prolactin, somatostatin, growth hormone. *Anterior pituitary hormones*: ACTH adrenocorticotropin (corticotropin), TSH thyroid stimulating (thyrotropin), LH luteinizing, FSH follicle stimulating PR prolactin, GH growth (somatotropin). *Thyroid hormones*: T_3 and T_4.

systems. Due to their tight interaction, they constitute what is called the *neuroendocrine system*.

In addition to the neuroendocrine interactions, these systems communicate with the immune system (see Chapter 30). Via the release of cytokines, the immune system can influence neuroendocrine functions. For this reason, hormones, growth factors, neurotransmitters, and cytokines constitute biological systems of communication, which integrate the function of different organs.

RECEPTORS

The specificity of action and the capacity of hormones, neurotransmitters, and other chemical messengers to identify their target cells are possible thanks to the presence of receptors on the effector cells. These receptors are proteins that selectively bind to hormones

by conformational changes that allow them to adapt to the ligand. Hormone (H) and receptor (R) form a complex (HR) with the following properties:

1. *Induced adaptation*. Like the substrate–enzyme binding, hormone binding to the receptor involves a reciprocal structural adjustment of both molecules.
2. *Saturability*. The number of receptors on a cell is limited and they can become saturated at high concentrations of hormone. If the amount of hormone receptors bound to hormone are plotted as a function of the hormone concentration, a hyperbolic curve is obtained.
3. *Reversibility*. The binding of the hormone to its receptor is reversible.

The concept is applicable to receptor macromolecules that selectively bind hormones, neurotransmitters, growth factors, cytokines, and other molecules that are able to induce a conformational change in the receptor.

The type of response induced depends on the functional specialization of the target cell. Sometimes a single hormone triggers different responses in different cells. For example, epinephrine produces activation of skeletal muscle glycogenolysis, but it stimulates lipolysis in adipocytes.

Agonists are compounds of similar structure to that of the physiological agent (hormone, neurotransmitter) with ability to bind to the receptor and trigger a response. This may be of equal, greater, or lesser intensity than that induced by the natural agent. *Antagonists* bind to the receptor but do not elicit responses. They function as competitive inhibitors.

Localization of receptors. Hormone receptors can be located inside the cell or on the cell plasma membrane. Hormones of slightly polar character, such as steroids, thyroid hormones, vitamin D metabolites, and retinoids, readily cross the membranes and bind to intracellular receptors. Those of protein or peptide nature and frankly polar small molecules cannot cross the lipid bilayer; they bind to receptors on the surface of the target cell. Eicosanoids, despite its solubility in lipids, bind to cell surface receptors. The membrane is not rigid but has a high degree of fluidity, and membrane-associated proteins, like some receptors, are free to move in all directions in the lipid bilayer plane; they are mobile receptors.

Some hormones bound to plasma membrane receptors are first internalized into the cell by endocytosis (p. 242).

Receptor number can be estimated by the use of hormones labeled with radioactive isotopes that detect the formation of HR complexes. The number of receptors on the surface of a cell can reach many thousands. The amount of intracellular receptors is usually much lower. It is not necessary that all of the cell receptors bind hormone to obtain a maximum response. Commonly, this occurs when approximately 20% of the receptors are occupied by the hormone. The rest is the so-called receptor reserve.

The amount of receptors for a given ligand varies in different physiological states. Generally, the concentration of hormone regulates the amount of its specific receptors on target cells. A sustained increase in the level of hormone decreases the receptor number; this down regulation mechanism is called "desensitization" of the receptor. On the contrary, deficiency of the specific ligand increases or upregulates the receptor number. In some cases, membrane receptor downregulation is produced by endocytosis and degradation of receptors. Conversely, upregulation may be due to increased synthesis and membrane insertion of receptors contained in intracellular vesicles. The decrease in receptor activity could be due, for example, to genetic alterations (mutations) that affect the receptor or any of the steps of the signal transduction system that regulates its function or plasma membrane targeting. In addition, autoimmune processes that produce antibodies against a particular receptor can impair receptor function.

Mechanism of action. To exert its action, the HR complex must interact with other cell structures. This involves different mechanisms depending whether the receptor is located inside the cell, or at the plasma membrane.

Intracellular Receptors

These receptors mediate the actions of steroid and thyroid hormones, active metabolites of vitamin D (p. 655) and retinoids (p. 652). Also, other compounds that can bind to these types of receptors are fatty acids, bile acids, xenobiotics, and less polar molecules that can enter cells by diffusion through membranes (or are introduced into the cell by facilitated transport). Some of these receptors are located in the cytoplasm, others are present in the nucleus. Both, cytoplasmic and nuclear receptors, have a similar molecular structure and, when stimulated by their specific ligand, they exert a direct action on the DNA-regulating transcriptional activity.

Intracellular receptors are grouped into two families: (1) *steroid receptors* (often found in the cytoplasm), which comprise the glucocorticoid (GR), mineralocorticoid (MR), progesterone (PR), and androgen (AR) receptors; and (2) *thyroid* or *nuclear receptors*, which include the estrogen (ER), thyroid hormone (TR), vitamin D metabolites [1,25-$(OH)_2$-D_3] (VDR), retinoids or retinoic acid (RAR and RXR), and *activated by peroxisome proliferator* (PPAR) receptors (pp. 551).

Steroid receptors. These are generally found in the cytoplasm, complexed with heat shock proteins (HSP 90, HSP 70, and HSP 56), which keep them inactive. Binding of the hormone with high affinity displaces HSP from the receptor. The receptor undergoes a conformational change, forms dimers (homodimers) that enter the nucleus and bind to defined sites of DNA, called *hormone response elements* (HRE), usually located "upstream" of the target gene promoter. Hormone response elements are inverted repeats of palindromic segments, separated by two to five base pairs. From its binding to the response element, the hormone-receptor complex interacts with transcription factors bound to the promoter site and influences the initiation complex that locate properly the polymerase II and starts transcription (Fig. 25.2).

Thyroid or nuclear receptors are localized in the nucleus, where they are linked to response elements in DNA in an inactive state. In the absence of the stimulating hormone, the estrogen receptor (ER) is bound to HSP, the other nuclear receptors are instead associated with a corepressor molecule that inhibits transcription. The corepressor (or HSP for the ER) is displaced upon formation of the hormone-receptor complex, which dimerizes and acquires the capacity to influence transcription. Only the ER is associated as homodimers; TR, VDR, RAR, PPAR, and RXR commonly form heterodimers.

Structure. Intracellular receptors belong to a superfamily of structurally homologous molecules, derived from a common ancestral gene.

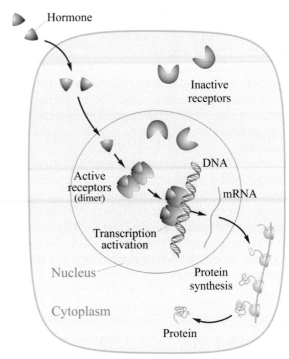

FIGURE 25.2 **Mechanism of action of steroidal hormones.**

They have three functional domains (Fig. 25.3): (1) a first segment of great variability among family members, both in length and in sequence; it is the hypervariable sector (HV). A portion of this domain is involved in transcription regulation. In glucocorticoid, mineralocorticoid, progesterone, and androgens receptors, the HV portion is longer than in estrogen, thyroid hormone, 1,25-$(OH)_2$-D_3, and retinoid receptors. (2) The central domain, containing two zinc

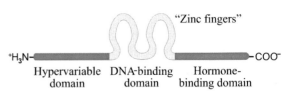

FIGURE 25.3 **Schematic representation of a steroidal receptor.**

fingers (p. 527), interacts with specific DNA sequences. This segment shows great homology among the different members of the superfamily (60%–95% identical amino acids). Many proteins that function as nuclear transcription factors possess the same structural motif. (3) The third domain, which includes the C-terminus, has the hormone-binding site. In the thyroid receptor family, this portion has high homology with the protein of the *c-erb A* proto-oncogen.

Peroxisome proliferator activated receptor (PPAR). The ligands of this type of nuclear receptor are products of the body's metabolism. PPAR is expressed as three subtypes (α, β/δ, and γ) that differ in tissue distribution; all of them are activated by polyunsaturated fatty acids and related compounds. They are mainly found in liver, muscle, and adipose tissue, but also are expressed in other organs. Hypolipidemic drugs of the fibrate type are good ligands of αPPARs.

PPARs function as a transcription factor, regulating gene activity. Ligand binding causes its dimerization with the retinoid receptor (RXR), which allows it to associate with response elements in regulatory regions of DNA in the target genes. Moreover, to stimulate transcription, PPAR also functions as a suppressor of the action of interfering agents, such as nuclear factor κB, STAT proteins, and others. The role of PPAR can be described as regulator or coordinator of metabolic pathways. One of the main actions of αPPARs is to stimulate fatty acid oxidation. Activation of δPPAR increases oxidation in adipose and muscle tissue. γPPARs are necessary in adipose tissue for the differentiation of adipocytes. Its action is exerted by controlling the expression of genes involved not only in adipogenesis and lipid metabolism, but also in general metabolism homeostasis and inflammation. PPAR stimulation improves glucose tolerance and decreases insulin resistance in patients with type 2 diabetes. Antidiabetic drugs that bind with high affinity to γPPAR improve insulin sensitivity.

Plasma Membrane Receptors

Receptors on the surface of cells are associated with different signal transduction systems. Binding of a hormone (the first messenger) causes conformational changes in the receptor, which are transmitted to effector proteins (enzymes or channels). As a result, small molecules (second messengers) are produced that rapidly diffuse in the cell and amplify transmission of the signal within the cell.

Different hormones provoke particular responses in different cell types, using the same or specific intermediaries. In general, the repertoire of signal transducer systems is smaller than the number of hormones. The structure and properties of the major types of membrane receptors follows.

G Protein Coupled Receptors

They comprise a large family (more than 800 members) of cell surface receptors that have seven transmembrane α helices of 22–26 hydrophobic residues each (Fig. 25.4); the amino acid sequence of these domains has been preserved without major modifications in all family members. The N-terminal, to which oligosaccharide chains are inserted, and the loops that connect the 2–3, 4–5, and 6–7 segments are extracellular, the hormone binds to the site of the receptor

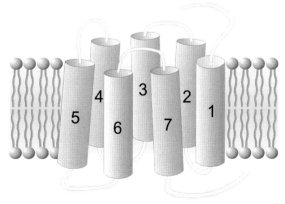

FIGURE 25.4 **Scheme of a G protein coupled receptor.**

formed by the outer ends of the transmembrane α helices. The loops between helices 3–4 and 5–6 and the C-terminal are in the cytosolic side and interact, when the receptor is activated by the hormone, with G proteins located on the inner face of the plasma membrane. Five or six different types of this receptor have been described that differ in amino acid sequence and length of the N-terminal extracellular segment. Most belong to three classes (A, B, and C).

Parathyroid hormone, luteinizing hormone, follicle stimulating, thyroid stimulating, ACTH, glucagon, vasopressin, angiotensin II, platelet activating factor, substance P, prostaglandins, α and β adrenergic, cholinergic muscarinic, serotoninergic, dopaminergic, and retinal rods (rhodopsin) receptors belong to this family.

The general mechanism of action, shown in Fig. 25.5 consists of the following steps: (1) Binding of the hormone to the receptor induces a conformational change that allows it to interact with a G protein on the membrane inner face. (2) The G protein, bound to GDP in its inactive state, exchanges GDP for GTP and becomes activated. (3) The activated G protein stimulates an enzyme localized in the membrane, which catalyzes the production of second messengers.

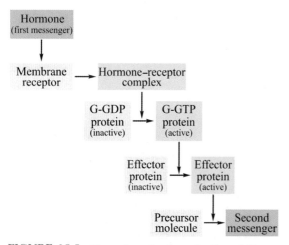

FIGURE 25.5 **General mechanism of action of G protein coupled receptors.**

(4) The second messenger continues the series of reactions in "cascade," causing changes in cellular proteins responsible for the final effect.

G Proteins

The name G protein derives from their property to bind guanine nucleotides (GDP or GTP). These proteins play an essential role in signal transduction systems, linking transmembrane receptors with effector proteins inside the cell.

Many circulating hormones or locally released neurotransmitters and other stimuli (photic, gustatory, or odorant) activate their specific receptors and initiate a signal chain in which G proteins are involved.

Structurally, G proteins are heterotrimers constituted by a α subunit with a molecular mass between 40 and 45 kDa, a β subunit of 37 kDa, and a γ of 8 kDa. The β and γ subunits are closely associated and function as a unit (dimer βγ). The heterotrimer is anchored to the inner or cytosolic face of the plasma membrane. The α subunit is bound to the double lipid layer by a myristoyl residue whose carboxyl binds to the amine group of the protein N-terminal glycine, forming an amide. Some types of α subunit are anchored by a palmitoyl residue. The hydrocarbon chain of the fatty acid is inserted within the hydrophobic membrane. The γ subunit is associated to the double layer by an isoprenoid chain (farnesyl or geranyl-geranyl) that forms a thioester bond with a cysteine residue near the C-terminal end of the polypeptide and then is inserted in the membrane (Fig. 25.6).

The α subunit has a site where guanine nucleotides (GDP or GTP) bind with high affinity. While bound to GDP, the αβγ assembly is maintained firmly associated and the heterotrimer is inactive. Hormone binding to the receptor modifies the orientation of transmembrane segments, enabling them to interact with the inactive G protein and to determine a change in its conformation. This change causes dissociation of the GDP bound to the α subunit and

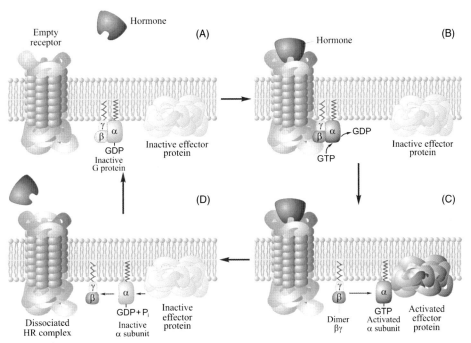

FIGURE 25.6 **General mechanism of action of membrane receptor proteins.** (A) The receptor is empty and the G protein (trimer $\alpha\beta\gamma$) is inactive. (B) Hormone binding to the extracellular portion of the receptor produces a conformation change that is transmitted to its cytosolic domain, which modulates the G protein. The GDP of the α subunit is replaced by GTP. (C) The activated GTP-α subunit dissociates from the $\beta\gamma$ dimer, allowing it to interact with a protein effector. (D) The ligand separates from the receptor, GTP is hydrolyzed to GDP and P_i, and the inactive $\alpha\beta\gamma$ trimer is reconstituted. The cycle is closed.

promotes binding of GTP, with higher cellular concentration than GDP. The α-GTP complex separates from the $\beta\gamma$ dimer and acquires modulating activity on the next effector protein in the signal system (Fig. 25.6).

The α subunit has GTPase activity, it is able to hydrolyze GTP to GDP and P_i. Then, the α subunit bound to GDP binds again with the $\beta\gamma$ dimer and the inactive heterotrimer is reconstituted.

A total of 20 different α subunits, 5 β and 12 γ polypeptides have been recognized, which creates the possibility of many different G heterotrimers. These proteins are grouped into different families, depending on the composition of their α subunit type. The G_s proteins have a α_s subunit and are found in almost all cells. In most cases, their action is expressed by activation of adenylate cyclase; others modulate ion channels (Ca^{2+} and Na^+). The G proteins with α_{olf} subunit present in the olfactory epithelium, linked to the transmission of odor stimuli, belong to this group.

The G_i proteins, with α_i subunits, are distributed in almost all cells; they exert inhibitory effect on adenylate cyclase and regulate the functioning of ion channels (K^+ and Ca^{2+}). Included in the G_i proteins group, are also G proteins containing α_o (found in brain), α_t (or transducing, present in retina rods and cones), α_g (expressed in taste buds), and α_z (present in brain, adrenal, and platelets). The subunits α_s and α_i are sensitive to pertussis and cholera toxins, respectively (see subsequent sections). Other G protein types, such as G_q and G_{12}, are not affected by the those toxins. The G_q proteins contain α_q subunits and are widely distributed in tissues. Their activation enables them to stimulate phospholipase C and the phosphatidyl-inositol-bisphosphate system, which will be considered later. In some tissues, other α subunits belonging to this group have been found (designated with numbers, α_{11} to α_{16}).

Experimental evidence indicates that the βγ dimer, to which initially no significant role was recognized, also serves as an intermediary in the signal system. It functions on different target proteins, including K^+ and Ca^{2+} channels, adenylate cyclase, phospholipase C, and various protein kinases.

Mutations responsible for pathological conditions in genes that control the synthesis of G protein coupled receptors, as well as in those encoding the α subunit, have been described.

Protein-Tyrosine Kinase Receptors

The G protein coupled receptors are required to link the changes of the receptor, brought by the ligand, with changes in the activity of members of the intracellular signaling system. Besides these system, there are receptors that have endogenous catalytic activity and others that are directly associated to enzymes.

Tyrosine kinase receptors with intrinsic activity. These receptors mediate the actions of, for example, insulin and growth factors. Their extracellular N-terminal end has the ligand-binding site; which contains numerous cysteine residues. The central portion consists of a α helix transmembrane domain that anchors the receptor to the plasma membrane. Finally, the cytosolic segment, corresponding to the C-terminal segment is the portion with tyrosine kinase activity (Fig. 25.7A).

The insulin receptor has a higher level of complexity; it is composed of two αβ heterodimers linked by disulfide bridges. The α subunits are extracellular proteins that form the hormone-binding site. Each of the β subunits has a transmembrane segment and a cytosolic tyrosine kinase domain (Fig. 25.7B).

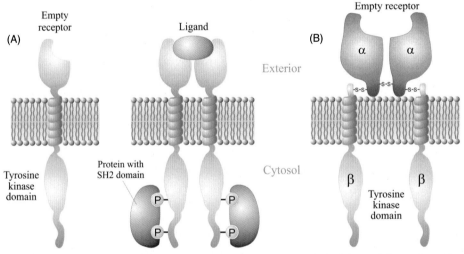

FIGURE 25.7 **Scheme of receptors with intrinsic tyrosine kinase activity.** (A) The left panel shows the epidermal growth factor receptor. The right panel shows the same receptor, but activated. Binding of the ligand to the extracellular portion of the receptor causes its dimerization and activates the tyrosine kinase that phosphorylates tyrosine residues. The presence of tyrosinylphosphate groups promotes binding of proteins with SH2 domains, which are phosphorylated and continue the transmission of signals in the system. (B) Insulin receptor composed by two αβ heterodimers linked by a disulfide bond.

Ligand binding to the extracellular subunit of these receptors induces their dimerization (except for the insulin receptor, which is constitutively formed by two heterodimers) and activation of tyrosine kinase activity. The receptor autophosphorylates, catalyzing the crossed phosphorylation from one chain to the other in several tyrosine residues of the cytosolic domain. As a result of these phosphorylations, the activity of the kinase further increases and promotes its association with signal transduction proteins that contain SH2 and SH3 domains. SH sequence domains (or Src homologous, due to their similarity with the oncogenic protein of *Src* Rous sarcoma virus consist of approximately 100 amino acid residues and are present in many proteins comprised in signal transduction systems. The protein containing the SH domain, attached to the activated receptor, is phosphorylated on tyrosine residues and in turn, it promotes the association of other proteins that also contain SH domains.

Receptor associated to serine-threonine kinases, have similar structure to that of tyrosine kinase receptors, but differ from them in that they phosphorylate serine or threonine residues in the target proteins. They mediate the action of a large number of ligands of the transforming growth factor (TGF) family of ligands. Forty two genes encoding receptors for cytokines of the TGFβ family have been described, including *activin* and *inhibin*, which modulate hormones related to reproduction.

The activated serine-threonine kinase receptors stimulate other proteins such as *Smad*, effector molecules in the signal chain. They also phosphorylate many receptor tyrosine kinases, including the insulin receptor, and inhibit them.

Receptors associated to extrinsic tyrosine kinase. There is another family of receptors without tyrosine kinase activity, but able to associate with protein tyrosine kinases in the cytoplasm. This group includes cytokines (p. 773) and some protein hormones receptors (e.g., growth hormone, prolactin, leptin). They are similar in structure

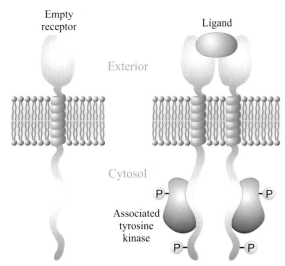

FIGURE 25.8 **Scheme of receptor associated to extrinsic tyrosine kinase.** On the left, the inactive receptor is shown. On the right, the effect of ligand binding is shown. The ligand (growth hormone, cytokine) induces receptor dimerization and modifies the cytosolic domain so that it can interact and activate a tyrosine kinase which phosphorylates tyrosine residues in the enzyme and the receptor.

to the tyrosine kinase receptors, although they do not have a catalytic site (Fig. 25.8). When the ligand binds to the extracellular domain and dimerization occurs, the cytosolic portion interacts with tyrosine kinase. The receptor is phosphorylated by the kinase in several tyrosine residues that facilitate the binding of proteins with SH2 domains. The result is essentially the same as that described for the activation of tyrosine kinase receptors.

These receptors can be distinguished into two main families. One includes the *Src* family of proteins. The others belong to the *JAK* family, associated to the *tyrosine kinase Janus*. The first are involved in signaling from cytokine receptors and antigens in B lymphocytes and T cells (see Chapter 30). *JAK* kinase coupled receptors phosphorylate a variety of target molecules in the cell.

Tyrosine phosphatase receptors. These receptors have been described as associated with protein-tyrosine phosphatases, such as CD45 on B lymphocytes and T cells. They remove the phosphate inserted by kinases and in this manner, they antagonize kinase action. The human genome encodes a large number of tyrosine phosphatases.

Guanylate cyclase associated receptors. Some hormones, and other peptidic agents, bind to membrane receptors whose cytosolic domain has guanylate cyclase activity, catalyzing the formation of cGMP from GTP. These guanylate cyclase receptors are composed of a polypeptide chain with an extracellular domain that binds the ligand, a transmembrane α helix domain, and a cytosolic portion with enzyme activity. Binding of the ligand stimulates the cyclase activity and generates cyclic GMP in the cytosol, which acts as a second messenger. Atrial natriuretic peptide receptors belong to this class of receptors.

Other types of guanylate cyclase receptors are not membrane receptors; they are cytosolic hemoproteins activated by small molecules that easily cross membranes, such as the paracrine messengers nitric oxide (NO) and carbon monoxide (CO).

The retina rods (rhodopsin) receptor, activated by photic stimuli transmitted through G protein is also related to cGMP, it reduces the 3′,5′-cyclic GMP concentration by activation of a specific phosphodiesterase.

Desensitization. As mentioned before, the continued exposure to an agonist results in decreased responsiveness of the receptors. This phenomenon of desensitization has been extensively studied in G protein coupled receptors. Receptor desensitization can be induced by changes in the receptor, including phosphorylation, by its internalization from the plasma membrane into the cell via endocytosis, or by decreasing the receptor's synthesis.

Phosphorylation of serine or threonine residues by specific kinases is the fastest way to produce the uncoupling of the receptor from the G protein. This occurs, for example, in the retinal photoreceptor (rhodopsin); however, for this receptor, phosphorylation alone is not sufficient to achieve complete receptor inactivation; the additional participation of a protein called *arrestin* is required. Arrestin is also found in other tissues, especially the nervous tissue.

SIGNAL TRANSDUCTION SYSTEMS

3′,5′-Cyclic-AMP System

The discovery of adenosine 3′,5′-monophosphate (cyclic AMP) in the late 1950s opened a new area of study, which advanced our understanding of the basic molecular mechanisms by which hormones elicit their action. 3′,5′-Cyclic monophosphate, or just cyclic AMP (cAMP) (Fig. 25.9) owes its name to the cyclic structure that phosphate forms between the hydroxyl groups of carbons 3′ and 5′ in ribose. Cyclic AMP is ubiquitous and found in all living organisms and in almost all mammalian cells. The exposure to certain hormones produces a rapid increase in cAMP concentration in cells.

cAMP is generated from ATP in a reaction catalyzed by adenylate cyclase, an enzyme that requires Mg^{2+} and is located in the plasma membrane.

$$ATP \xrightarrow[Mg^{2+}]{Adenylate\ cyclase} 3′,\ 5′\text{-cyclic AMP} + PP_i$$

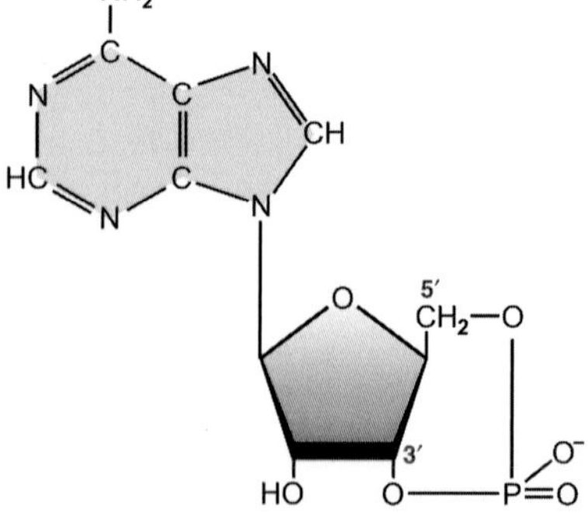

FIGURE 25.9   **3′,5′-cyclic-AMP.**

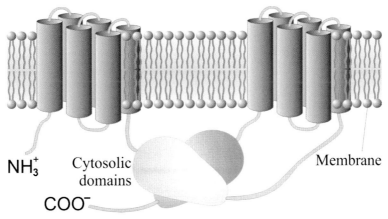

FIGURE 25.10 **Structure of adenylate cyclase.**

Adenylate cyclase. This is an integral plasma membrane protein with a mass of over 100 kDa. It consists of two equal portions, linked in tandem, each having an N-terminal intracellular segment, followed by six transmembrane α helices, and ending with a long cytosolic domain (Fig. 25.10) that contains the catalytic site. This structure resembles that of some membrane transporters or channels. However, adenylate cyclase does not have channel activity and it does not share sequence homology with membrane carriers. Nine different forms of adenylate cyclase have been identified in mammals; they are expressed in varying proportions in different tissues and have specific regulatory properties. Virtually all forms of adenylate cyclase are activated by α_s subunit of G_s protein; not all are inhibited by α_i. Ca^{2+}, Ca^{2+}-calmodulin and protein kinase C have modulatory effect only on some of them.

The system, shown schematically in Fig. 25.11, operates as follows:

1. Binding of a ligand is the first signal and causes a conformational change in the receptor, which is transmitted to the G protein. There are both stimulatory receptors and G proteins (R_s and G_s) and inhibitory G proteins (R_i and G_i). When the receptor occupied by the hormone is stimulatory (R_s type), it interacts with G_s protein. In contrast, the receptors coupled to G_i proteins transmit an inhibitory signal. While the receptor is empty, the G protein heterotrimer remains inactive, with its α subunit bound to GDP.

2. Formation of the hormone-receptor complex causes a conformation change in the G protein α subunit, which releases GDP and binds GTP from the cytosol. GTP binding dissociates the α subunit from the $\beta\gamma$ dimer. The α_s-GTP complex activates adenylate cyclase. In the case of receptors coupled to G_i proteins, α_i-GTP is formed, with inhibitory action on the enzyme. The $\beta\gamma$ dimer can also exert actions as a signal transmitter by itself.

3. Activated adenylate cyclase catalyzes the formation of cAMP from ATP and raises the concentration of this second messenger within the cell.

4. The α subunit of G protein has GTPase activity, it catalyzes the hydrolysis of GTP to GDP and reassociates with the $\beta\gamma$ dimer to reconstitute the inactive G protein.

The toxin produced by *Vibrio cholerae* causes profuse diarrhea, dehydration, and ion imbalance due to the increased level of cAMP in the

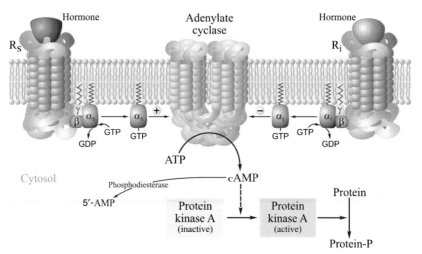

FIGURE 25.11   **Scheme of the 3′,5′-cyclic-AMP system.**

cells of the intestinal mucosa. The toxin pro-motes ADP-ribosylation (p. 538) of the α subunit of G$_s$ protein. The reaction uses NAD as a co-factor and ADP-ribose from NAD is transferred to an arginine residue. This covalent modifica-tion blocks the GTPase activity of the α subunit. Thus, adenylate cyclase is not deactivated and hence cAMP production is sustained.

The ADP-ribosylation mechanism is common in the action of various bacterial toxins. The tox-in produced by *Bordetella pertussis*, which causes whooping cough, acts by ADP-ribosylation of the subunit α$_i$ and keeps it bound to GDP and the βγ dimer preventing its inhibitory action on adenylate cyclase. The result is an increase in cAMP in the target cell.

Mode of action of cyclic AMP. cAMP diffuses inside the cell and stimulates protein kinase A. In the absence of cAMP, protein kinase A cAMP-dependent (PKA) is inactive, forming a tetramer consisting of two subunits called catalytic (C) and two subunits that are regulatory (R). When the level of cyclic AMP in the cell increases, two molecules of this nucleotide bind to specific binding sites in each of the regulatory subunits of PKA (Fig. 25.12), and produce a conformational

change that induces the separation of R from the C monomers. The free C subunits have en-zymatic activity. The process can be represented by the following equation:

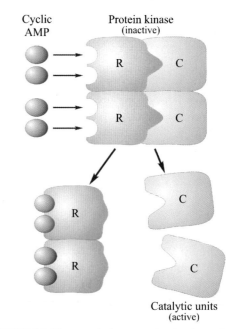

FIGURE 25.12   **Protein kinase activation by cyclic AMP.**

$$R_2C_2 + 4\,AMP_C \longrightarrow R_2\text{-}AMP_{C_4} + 2C$$

Inactive tetramer | Regulating units–cyclic AMP complex | Free catalytic units (active)

The catalytic unit of PKA transfers phosphate from ATP to serine or threonine residues in target proteins which, when phosphorylated, acquire their active state.

Cyclic AMP is a messenger that causes different responses depending on the cell type (Table 25.1). Although tissue response varies according to the tissue considered, in all cases the action of cAMP is carried out through "cascade" reactions initiated with activation of protein kinase A, and continuing with the activation of proteins that are target of PKA phosphorylation.

Different mechanisms of action of cAMP can be distinguished; these include:

1. *Enzyme regulation.* The phosphorylation promoted by PKA is an important mechanism that stimulates or inhibits enzyme activity; this is essential for the proper operation of metabolic pathways (see p. 430).

2. *Modulation of the activity of membrane transport systems.* Besides the actions mediated by protein kinase, cAMP can affect ion channels by mechanisms independent from phosphorylation. For example, host cells sensitive to odoriferous stimuli in the nasal mucosa are coupled to G proteins of the G_{olf} type. The α_{olf} subunit activates adenylate cyclase. The increased cAMP level directly causes opening of Na^+ channels, membrane depolarization, and generates nerve impulses.

3. *Transcriptional regulation.* In the nucleus, cAMP can influence transcription. There are specific DNA sequences, called cAMP-dependent response elements (CRE) that are sensitive to cAMP. The increase in cAMP

TABLE 25.1 Responses Mediated by Cyclic AMP

Hormone	Tissue	Main response
Epinephrine	Muscle	Glycogenolysis
	Adipose	Lipolysis
	Heart	Increase of frequency and strength of contractions
Glucagon	Liver	Glycogenolysis
	Adipose	Lipolysis
	Heart	Glycogenolysis
Adreno-corticotropin (ACTH)	Adrenals	Steroidogenesis
	Adipose	Lipolysis
Luteinizing (LH)	Ovary	Steroidogenesis
	Testis	Lipolysis
	Adipose	
Thyrostimulant (TSH)	Thyroid	Secretion thyroid hormones
	Adipose	Lipolysis
Parathyroid	Bone	Ca^{2+} reabsorption
	Kidney	Phosphate excretion
Vasopressin	Kidney	Water reabsorption
Prostaglandins	Fibroblasts	Growth inhibition
	Platelets	Inhibition of aggregation

in the cell stimulates protein kinase A and this phosphorylates a protein, designated CREB (for cAMP response element binding), which then binds to DNA CRE sequences and activates gene transcription. Phosphoenolpyruvate carboxykinase, tyrosine aminotransferase, and various cytochromes P_{450} are regulated by this mechanism.

Phosphodiesterase. Due to the potent activity of cyclic AMP as a regulator, cells must closely control its concentration. In most tissues, the enzyme phosphodiesterase catalyzes the hydrolysis of the phosphate bound to the 3′ carbon and converts cAMP to adenosine 5′-monophosphate, which is inactive. Methylxanthines, such as caffeine, theophylline, and aminophylline, inhibit the degradation of cAMP by phosphodiesterase, maintaining cAMP actions. Catecholamines binding to α_1 adrenergic receptors and insulin activate phosphodiesterase, reducing cAMP levels. The proteins phosphorylated by protein kinase are normally dephosphorylated by protein phosphatases, which are associated with transmembrane receptors, or free in the cytosol.

Phosphatidylinositol Bisphosphate System

Inositol phosphate (IP) is a component of cell membranes. Although it is less abundant than other membrane phospholipids, it has great functional significance. IP is part of the plasma membrane double lipid layer, located mainly on the inner leaflet, facing the cytoplasm. Phosphatidylinositol is phosphorylated at carbons 4 and 5 of inositol by transfer of phosphate from ATP to form phosphatidylinositol-4,5-bisphosphate (PIP_2) (Fig. 25.13). This substance is an intermediate in the signal transduction system.

The binding of a specific ligand to a seventransmembrane receptor produces a conformational change of its cytosolic portion, enabling it to interact with a G_q protein. The α_q subunit

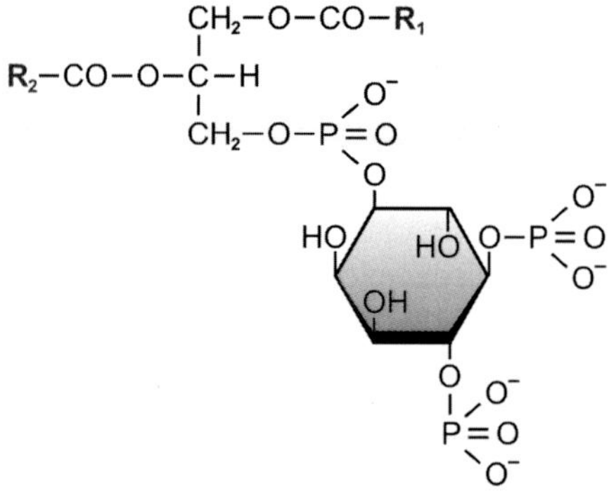

FIGURE 25.13   **Phosphatidylinositol-4,5-bisphosphate.**

in the G_q protein replaces its GDP for GTP and separates from the $\beta\gamma$ dimer. The α_q-GTP complex stimulates phospholipase C (β form). This enzyme catalyzes the hydrolysis of phosphatidylinositol-4,5-bisphosphate included in the membrane, to produce diacylglycerol (DAG) and inositol-1,4,5-trisphosphate (Fig. 25.14). Both molecules function as second messengers. Another phospholipase C, the γ form, is associated with protein tyrosine kinase receptors (see subsequent sections).

Fig. 25.15 shows a scheme of the phosphatidylinositol-4,5-bisphosphate system. Inositol-1,4,5-trisphosphate (IP_3) released in the cytosol binds to receptors in the membrane of the endoplasmic reticulum (ER). These receptors are tetrameric proteins with multiple transmembrane domains which form Ca^{2+} channels. IP_3 binding

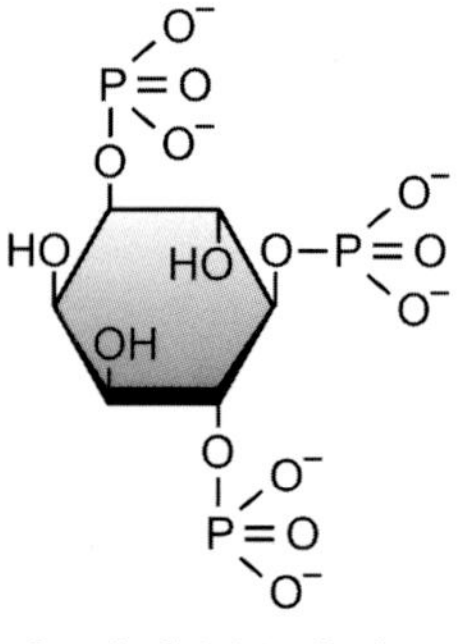

FIGURE 25.14   **Inositol-1,4,5-trisphosphate.**

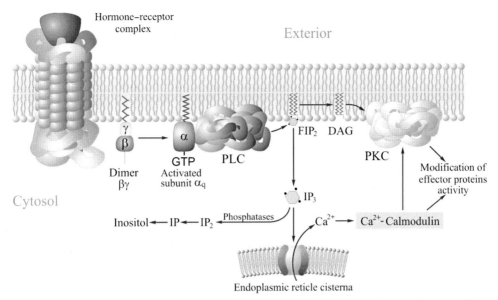

FIGURE 25.15 **Phosphatidylinositol bisphosphate system.** *DAG*, diacylglycerol; *PKC*, protein kinase C; *PLC*, phospholipase C.

opens the channel and induces the release of Ca^{2+} stored in the ER, causing a sudden increase of Ca^{2+} in the cytosol, which triggers diverse cellular responses.

The action of inositol-1,4,5,-trisphosphate is stopped by phosphatases. This is achieved by the dephosphorylation of IP_3 first to IP_2, and then to IP and inositol by specific IP phosphomonoesterases. IP hydrolysis to inositol is inhibited by lithium ions.

Inositol-3,4,5-trisphosphate is formed from inositol-4,5-bisphosphate (PI-4,5-P_2) by phosphatidilnositol 3-kinase (PI$_3$K). PI$_3$K is activated in response to stimulation of cell surface receptors by various hormones, neurotransmitters and growth factors. IP_3 induces activation of protein kinase B (PKB, also called *AKT*), important component of insulin-dependent and growth factors signal systems; it binds to proteins containing a domain called PH (for its homology with pleckstrin), including *AKT/PKB* and activates them. Highly phosphorylated inositol compounds such as inositol pentakisphosphate and inositol

hexakisphosphate (both IP_7) have inhibitory effect on the stimulation of *AKT/PKB* by IP_3.

The *DAG* remaining in the membrane after phosphatidylinositol bisphosphate hydrolysis also functions as a second messenger. It exerts its action through activation of protein kinase C, located in the membrane (Fig. 25.15). This enzyme phosphorylates proteins related to cell division and transcription factors. Table 25.2 presents responses mediated by this system. Some members of the kinase C protein family and DAG require Ca^{2+} for activation. The stimulation of protein kinase C by DAG is interrupted by hydrolysis of this compound. The reaction generally releases arachidonate, a precursor of eicosanoids.

The study of the tumorigenic action of phorbol esters has shown the importance of protein kinase C in the control of cell multiplication. Phorbol, a polycyclic alcohol, forms esters with fatty acids, whose structure resembles that of DAG. These compounds stimulate protein kinase C. Different from the activation produced

TABLE 25.2   Responses Mediated by Phosphatidylinositol Bisphosphate System

First messenger	Tissue	Main response
Thyrotropin releasing hormone (TRH)	Adenohypophysis	Prolactin secretion
Vasopressin	Liver	Glycogenolysis
Growth factors	Fibroblasts	DNA synthesis
Acetylcholine	Pancreas (exocrine) Parotid	Secretion of digestive enzymes (amylase, trypsinogen)
	Pancreas (β cells)	Secretion of insulin
	Smooth muscle (vascular, gastric)	Contraction
Thrombin	Platelets	Aggregation
Antigen	Mast cells	Secretion of histamine

by DAG, the action of phorbol esters is sustained because they are not degraded. Persistent stimulation of protein kinase C promotes tumor development.

Besides DAG generated by phosphatidylinositol-4,5-bisphosphate hydrolysis, other lipid derivatives can also act as second messengers.

Lipids in Signaling Systems

DAG produced from phosphatidylcholine. Another type of phospholipase, phospholipase D (PLD) is activated by various stimuli in the cell. PLD hydrolyzes phosphatidylcholine, which is abundant in the cell plasma membrane, producing phosphatidic acid and choline. Phosphatidic acid is hydrolyzed by phosphatidic acid hydrolase to release DAG and phosphate. This is a second pathway that generates DAG. While this intermediate is the product of the action of both PLC and PLD, cellular responses in both cases are usually not identical. This is probably explained by differences in the cellular localization of enzymes or the fatty acid composition of the DAG generated.

Ceramide. Some ligands, such as cytokines and 1,25-$(OH)_2$-D_3, activate sphingomyelinase, which converts sphingomyelin into ceramide and phosphocholine. Ceramide is another important intracellular messenger of lipid nature. Several functions have been assigned to

ceramide. One of the most important is its participation in the process of apoptosis or programmed cell death (Chapter 32).

Sphingosine-1-phosphate (S1P) binds with great affinity to G protein coupled receptors, generating multiple signals. It plays an important role in vascular development and endothelia integrity, and in the control of cardiac rhythm.

3′,5′-Cyclic-GMP

The observation that the intracellular level of guanosine 3′,5′-monophosphate (cyclic GMP, cGMP) changed after stimulation of cells with different factors, led to propose that it functions as a second messenger. Moreover, it was found that cGMP promotes actions that are opposed to those elicited by cyclic AMP. The 3′,5′-cyclic GMP (Fig. 25.16) is generated from GTP by the action of *guanylate cyclase* located in the membrane. It is also produced by the action of guanylate cyclase located in the cytosol. Guanylate cyclase can be activated by factors that easily pass through membranes, such as nitric oxide and carbon monoxide. The finding that these toxic gases are mediators of signal systems was completely unexpected.

Cyclic GMP is involved in various processes: (1) activation of protein kinases associated with the modulation of cell growth and proliferation, (2) smooth muscle relaxation and vasodilation, (3) stimulation of diuresis and natriuresis (the

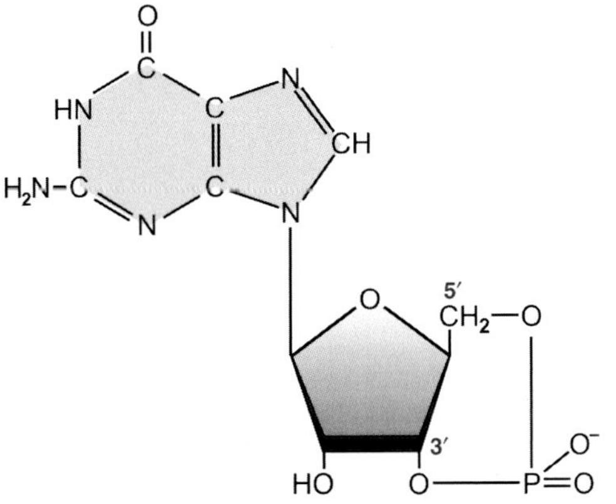

FIGURE 25.16   **3′,5′-cyclic-GMP.**

binding of atrial natriuretic peptide to its receptor in the plasma membrane activates guanylate cyclase and increases the level of cGMP in the cells), and (4) sensing of light and olfactory stimuli. Cyclic GMP participates as a second messenger that regulates the opening and closing of Na$^+$ channels. In the retina, the photoreceptor of rods has the basic structure of seven transmembrane α helices coupled to a G protein called *transducin* (with subunits α$_t$). Activated α$_t$ subunit stimulates cGMP phosphodiesterase and the cGMP level decreases. This causes Na$^+$ channel closure, and membrane hyperpolarization, generating nerve impulses responsible for visual sensations.

In the intestine, the increase of cGMP in cells of the intestinal mucosa causes diarrhea.

Ras and MAP Kinases System

This signaling pathway sets in motion a cascade of protein kinases very important in the regulation of numerous cell functions. All components of the system are proteins, small molecule second messengers are not generated.

When a protein-tyrosine kinase receptor or one associated to a protein tyrosine kinase is activated by its ligand, tyrosine residues are phosphorylated in the cytoplasmic domain, creating binding sites for proteins with SH2 and SH3, as the *growth factor receptor binding protein* (Grb), for example. Grb acts as an adapter protein that facilitates the binding of other proteins of this signaling system, that is, *Sos*, so named because of its similarity to that isolated in *Drosophila* (son of sevenless). The Grb–Sos complex bound to the receptor in a place close to the inner face of the plasma membrane, interacts with *Ras* (name that comes from its homology with rat sarcoma proteins). *Ras* is constitutively inactive; its stimulation depends on its association with *Grb–Sos* (Fig. 25.17).

Ras proteins. They are monomers of about 21 kDa, which bind guanine nucleotides (GDP or GTP) and have GTPase activity. They are anchored on the cytosolic face of the plasma membrane by an isoprenoid farnesyl chain bound to a cysteine residue near the C-terminal. *Ras* linked to GDP is inactive, the interaction with *Grb–Sos* promotes exchange of GDP for GTP and produces its activation. The same *Ras* molecule has the capacity of self-regulation, since it catalyzes the hydrolysis of GTP to GDP and P$_i$ and completes the activation–inactivation cycle. Note the similarity between *Ras* and G protein α subunit.

Ras stimulation initiates a cascade of phosphorylation events on serine-threonine residues of enzymes. These activated enzymes are generically called *MAPK* (*mitogen activated protein kinases*). The first MAPK phosphorylates and activates a second one, and this a third one, which is responsible for the phosphorylation of the effector protein. The final effect in the cell is proliferation and differentiation (Fig. 25.17). Actually, there are several pathways initiated by *Ras*. Each of the participating kinases is identified with different acronyms. The pathway begins with activation by *Ras*-GTP of *Raf* (a *MAP* kinase protein), which in turn phosphorylates and stimulates a second *MAP* kinase called *MEK* (*MAP* kinase-ERK). *MEK* is a dual kinase that transfers phosphate to serine-threonine and also to tyrosine residues. It activates members of the *MAP kinase family regulated by extracellular signals*, called *ERK* (extracellular regulated kinase).

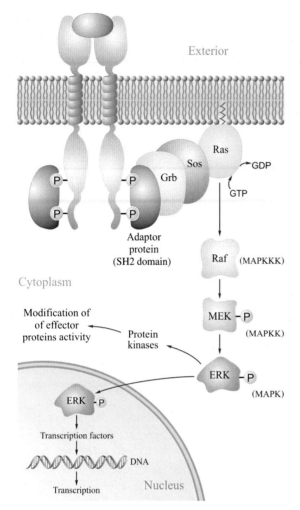

FIGURE 25.17   *Ras* and the *MAP* kinase cascade.

These kinases phosphorylate a variety of target proteins, including other protein kinases and transcription factors in the nucleus. Among the genes whose activity is induced by these pathways are immediate-early genes, most of which encode transcription factors that influence the expression of other genes.

Interest in *Ras* proteins increased substantially when it was known that about 30% of human cancers are associated with mutations in the *ras* gene. Normally, *Ras* is a proto-oncogen encoding a protein related to regulation of cell differentiation and proliferation. *Ras* mutations gives *Ras* oncogene activity that affect GTPase activity of the protein. These mutations prevent GTP hydrolysis, therefore *Ras* remains permanently activated, causing uncontrolled cell proliferation and malignant transformation.

Ras proteins belong to the superfamily of low molecular weight proteins that bind and hydrolyze GTP, also called *small GTPases*. At present more than a hundred, classified into five families, *Ras, Rho, Rab, Sar1/Arf*, and *Ran*, have been identified. Their function is related to regulation of gene expression, *Rho* modulates gene activity and cytoskeletal organization, *Rab* and *Sar1/Arf* modulate intracellular vesicular trafficking, *Ran* controls the transport between nucleus and cytoplasm during the G1, S, and G2 cell cycle phases and the reorganization of microtubules in M-phase.

JAK-STAT System

This pathway provides a more direct connection between the ligand-receptor and its downstream effectors than the previously described systems. *STAT* (*signal transducers and activators of transcription*) proteins contain SH2 domains. In unstimulated cells, *STAT* proteins are found in the cytosol. Binding of some hormones and cytokines to receptors associated to tyrosine kinase type Janus (*JAK*) activates the kinase, which phosphorylates tyrosine residues, to which *STAT* proteins bind by their SH2 domains. The subsequent phosphorylation of *STAT* promotes its dimerization and transport to the nucleus, where it stimulates transcription of its target genes (Fig. 25.18). Growth hormone, prolactin, various cytokines that control production of blood cells, and γ interferon, use this pathway.

TOR

TOR (*target of rapamycin*) is a serine/threonine kinase that functions in signaling pathways. It is a protein of a large mass (about 300 kDa),

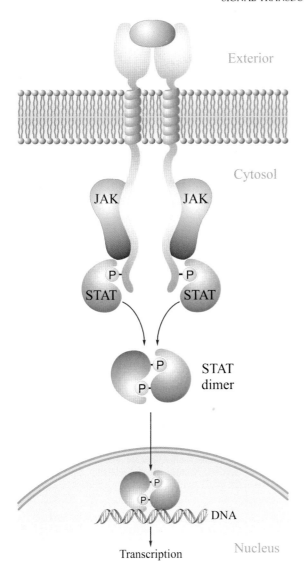

Exterior

Cytosol

JAK   JAK

P-   -P
STAT   STAT

P-   STAT
P-   dimer

P-
P-   DNA

Transcription   Nucleus

FIGURE 25.18 *JAK-STAT* System.

evolutionarily conserved from yeast to humans; in mammals, it is indicated by the letter *m* (*mTOR*).

mTOR binds to several proteins to form functional complexes (*mTORC*) in which *mTOR* is the catalytic center. Two complexes (*mTORC1* and *mTORC2*) have been recognized. They are constituted by different proteins, some of which are common to both groups and others are different, such as *Raptor* (*regulatory associated protein of mTOR*), specific for *mTORC1*, and *Rictor* (*rapamycin insensitive companion of mTOR*) exclusive for *mTORC2*. The mTORC complexes differ in their composition, properties, functions and sensitivity to inhibitors.

Rapamycin, a macrolide produced by *Streptomyces hygroscopicus*, is an inhibitor of *mTORC1* but not *mTORC2*; it has a powerful antifungal and immunosuppressive action. In some cell lines, prolonged treatment with rapamycin decreases also *mTORC2* activity.

mTORC1 detects and integrates intra- and extracellular signals through phosphorylation cascades triggered by various stimuli, including high supply of nutrients (mainly amino acids), high energy level in the cell (low value of the AMP/ATP ratio) and growth factors. Various signal pathways converge to activate *mTORC1*, including those activated by insulin and insulin-like growth factor (IGF-I), such as PI3K, PDK1 and AKT, those activated by epithelial growth factor (EGF), such as *MAPK*, and those regulated by the cell energy levels, such as AMP kinase (AMPK). These stimuli, with the exception of AMPK, activate *mTORC1* signaling, phosphorylating various substrates, among which are S6K1 and eIF4E, related to initiation of mRNA translation.

Stimulation of the mTOR pathway determines increased protein synthesis and ribosome production, cell proliferation and growth (higher number and cell mass), and increment of cellular metabolism. In contrast, dietary restriction (nutrient deficiency), the lack of growth factors, and cellular stress signals, depress *mTORC1* activity and promote autophagy. *mTORC1* also favors lipogenesis. The coordinated increase of protein and lipid synthesis *mTOR*-dependent, is a decisive factor promoting cell growth.

mTORC2 is activated by growth factors (e.g., IGF-I) and insulin, via PI3K. An important substrate of *mTORC2* is the serine/threonine kinase AKT or PKB. When AKT is stimulated by phosphorylation, it promotes cell proliferation, survival and migration, and modulates some metabolic processes. The best known function of *mTORC2* is to control the organization of the actin cytoskeleton.

In invertebrates, *TOR* signals pathway play an important role in determining their life span. Depression of *TOR* activity prolongs survival. This action, demonstrated in invertebrates has not been confirmed in mammals yet. Experiments in mice indicate that dietary restriction promotes longevity and delays the appearance of the organic impairments associated with aging. One mechanism that would explain this phenomenon is the depression of the *mTOR* activity.

In rodents, treatment with rapamycin decreases the incidence of tumors and cardiovascular changes that accompany aging. Rapamycin produces cell cycle arrest in G1 phase, inhibits cell proliferation, growth, and migration (in cancer, this action would reduce the production of metastasis) and increases autophagy. The mTOR signaling system has been implicated in pathological conditions, including obesity, type 2 diabetes, neurodegeneration and cancer. These results have encouraged human clinical trials; trials with rapamycin (in clinics also called *sirolimus*) and analogs in the treatment of certain types of malignancies are currently performed. Depression of *mTORC* activity to decrease the incidence and severity of cardiovascular disease has also been investigated.

Ca²⁺ Signal

Normally, the intracellular calcium ion concentration $[Ca^{2+}]_i$ is very low ($\sim 10^{-7}$ M or 0.1 µM), approximately 10,000 times lower than that of the extracellular space. Different stimuli can determine rapid changes in $[Ca^{2+}]_i$, which function as a signal that regulates cell function.

The calcium signal is linked to a myriad of vital physiological processes, from egg fertilization to embryo development and programmed cell death (Chapter 32). The mechanisms responsible for changes in $[Ca^{2+}]_i$ are complex and finely regulated (p. 723).

The $[Ca^{2+}]_i$ results from movements of this cation between the cell cytosol and the extracellular medium, and through its exchange with intracellular organelles that serve as Ca^{2+} stores. Transporters in cell membranes allow the rapid flow of Ca^{2+} into the cell and the production of transient Ca^{2+} "pulses" that function as signals to transmit messages. The structures responsible for the movement of calcium will be considered later (Chapter 29).

Ca^{2+} entry to the cell from the environment depends on various stimuli, such as ligand binding to cell surface receptors, depolarization of the plasma membrane, and opening of store-operated Ca^{2+} channels. The release of Ca^{2+} from internal stores (sarco-/ER, mitochondria, Golgi) is activated by calcium and by messengers, such as inositol-1,4,5-trisphosphate (I-1,4,5-P$_3$), cyclic ADP-ribose (cADPR), nicotinic acid adenine dinucleotide phosphate (NAADP), and sphingosine-1-phosphate. I-1,4,5-P$_3$ is produced by activation of phospholipase C, which consists of five isoforms. Each isoform is activated differently by specific stimuli. For example, G protein coupled receptors stimulate PLCβ, tyrosine kinase coupled receptors activate PLCγ, $[Ca^{2+}]_i$ increase stimulates PLCδ, activation of the *Ras–MAPK* system increases PLCε activity. In sperm, PLCζ induces the release of Ca^{2+} that is required for egg fertilization.

STIM (stromal interaction molecule) is a protein that senses and coordinates cellular Ca^{2+} signals. In the cytosol, calcium binds to various proteins. A large number of Ca^{2+} binding proteins have been identified. Among those are troponin C, synaptotagmin, S100 protein, annexin, calmodulin. They all have a common structural motif composed of two α helices held together by a loop where Ca^{2+} is linked to multiple coordination bonds. These α helices, designated E and F, are arranged with their axes perpendicular to each other. The domain has been termed "EF hand" because it resembles a hand with its thumb, and index fingers extended and the middle flexed into the palm. The thumb and index represent the E and F helices; the middle finger represents the loop that binds Ca^{2+}.

Calmodulin. It is one of the Ca^{2+}-binding proteins most widely distributed. This protein, of mass close to 17 kDa and acidic at the pH of the body, is found in all tissues. It has four binding sites for Ca^{2+}. At both ends, the molecule contains globular domains, each of which has two Ca^{2+}-binding motifs composed of "EF hand domains." In the middle of the molecule, there is a long α helical segment (Fig. 25.19). When bound to Ca^{2+}, calmodulin undergoes conformational changes and acquires the ability to regulate the activity of numerous target proteins, including various protein kinases that form part of phosphorylation cascades. These phosphorylation events modify the activity of enzymes, ion channels, and transcription factors. For example, the CREB transcription factor, protein kinase activated by cAMP, is also stimulated by Ca^{2+}-calmodulin. This shows that there is cross talk between pathways that use cAMP and Ca^{2+} as messengers (Table 25.3).

Importance of protein kinases. Protein kinases that transfer a phosphoryl group to serine-threonine or tyrosine residues in proteins are a common component of signal transduction systems. More than 1000 types of protein kinases exist in mammalian cells, which highlights the importance, diversity, and complexity of these enzymes in the cell. Some depend on cAMP, others on cGMP, DAG, Ca^{2+}, and tyrosine kinase receptors. They catalyze the phosphorylation of a variety of proteins, involved in the regulation of essential cell processes, such as membrane transport systems, cell division, nucleic acid, and protein synthesis.

Protein phosphatases. As important as protein kinases are protein phosphatases, which catalyze the dephosphorylation of proteins, enhancing, or inhibiting their activity. Protein phosphatases are essential components of signal transmission systems. There are several families of protein-tyrosine and protein-serine/threonine phosphatases, which display substrate specificity and are differentially regulated by a variety of various mechanisms, including phosphorylation–dephosphorylation.

Redox signaling. Superoxide anion, hydrogen peroxide, and nitric oxide are reactive oxygen species (ROS) that have deleterious action in cells

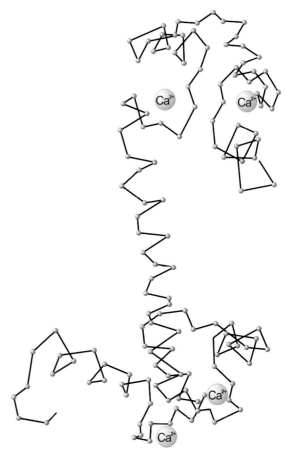

FIGURE 25.19 **Scheme of a calmodulin molecule with its four binding sites occupied by calcium.** Only the α carbons of the polypeptide chain are represented.

TABLE 25.3 Actions Mediated by Ca^{2+}–Calmodulin Complex

Activation of	Cyclic AMP phosphodiesterase
	Adenylcyclase
	Ca^{2+}-dependent ATPase
	Guanylate cyclase
	Phospholipase A
	Phosphorylase kinase
	NAD kinase
	Ca^{2+}-dependent protein kinase
	Kinase of myosin light chains
Deaggregation of microtubuli	
Liberation of neurotransmitters	

(see Chapter 10). However, in low levels, ROS have beneficial effects and function as messengers in signal transduction systems. In this role, they can, for example, influence the immune system, stem cell renewal, aging, and oncogenesis.

Wnt proteins. They belong to a family of secreted proteins, highly conserved along evolution. Nineteen Wnt proteins of 39–46 kDa each have been identified in humans; they are rich in cysteine residues and have a fatty acid (palmitate) covalently bound to their structure. They initiate signal transduction events that mainly have short distance effects (cell to cell). They induce the activation of cytoplasmic effector molecules that finally regulate target genes. One of the most important Wnt intermediates is *β-catenin* that, within the nucleus, acts as transcription factor.

The Wnt signaling pathway has an important role in embryo development. Also, acting on stem cells, they promote tissue regeneration in adults and regulate many other processes. In the embryonic nervous system, Wnt proteins control neural crest formation and the subsequent stages of nervous system conformation. Wnt proteins also regulate apoptosis.

Mutations in the genes coding for components of the Wnt pathway can cause degenerative diseases and cancer.

FOXO transcription factors are important regulators of many physiologic processes, including cell metabolism, cell cycle, cellular proliferation, apoptosis, inflammation and immunity, oxidative stress response, and longevity. They are modulated by posttranslational modifications (phosphorylation, acetylation, ubiquitylation). FOXO factors are activated through signal transduction pathways, such as phosphoinositide-3-kinase (AKT), or RAS/MAP kinases.

Cross Talk Between Signaling Systems

The coordination of signaling pathways is achieved through the communication between systems that ensure their integrated function. The relationship or *cross talk* between signal transduction systems allows them to coordinate the different functions of the cell. Just a few examples of cross talk between signal transduction pathways are the activation of the MAP kinase cascade by G proteins and independent from *Ras–Raf*; the inhibition of *Raf* by high levels of cAMP, the stimulation by stress conditions, or by the *Ras* pathway, of the transcription factor Jun, which is commonly activated by the specific kinase JNK (*c-Jun N-terminal kinase*).

Scaffold Proteins

There is a group of proteins, known as scaffold proteins, which role is to bring together and in close proximity other proteins involved in cell signal transduction events. To function as scaffolds, the involved proteins contain specific domains that allow them to engage in protein–protein interaction. Protein tethering in signaling complexes help localizing components of a signaling pathway to specific areas of a cell and they ensure that the sequence of the reactions in a signaling cascade occurs in the appropriate order. Scaffold proteins increase signaling specificity by preventing interactions of a particular signaling pathway with components of another. Also, they enhance the velocity and efficiency of response of the cell to a specific signal. Example of scaffold proteins include those that control the MAPK system that phosphorylate different target proteins and proteins that regulate the opening and closing of ion channels (especially Ca^{2+} channels). In addition, scaffold proteins participate in gene regulation and in the immune innate and adaptive systems responses (Chapter 30). Besides proteins, long noncoding RNA have been proposed to function as scaffolds (p. 531).

NEUROTRANSMITTER RECEPTORS

Neurotransmitter receptors are present in the plasma membrane of postsynaptic cells (in some cases also in the presynaptic terminal), which selectively bind the transmitter. They are integral membrane glycoproteins with multiple trans-membrane segments.

Pharmacological approaches, using compounds that bind to the receptor and reproduce (agonists) or block (antagonist) the action of the neurotransmitter, revealed the high complexity in function of these receptors. Several receptor subtypes have been described for each neurotransmitter. The advent of molecular biology methods revealed a high molecular diversity for these receptors. A description of some of the best-known neurotransmitter receptors follows.

Cholinergic receptors. At the beginning of the 20th century, shortly after the recognition of acetylcholine as a chemical neurotransmitter, it was discovered that it triggered two types of responses. Some actions were similar to those exerted by nicotine, while others could be reproduced by the compound muscarine. The first were blocked by D-tubocurarine and the second by atropine. These observations led to propose the existence of two types of acetylcholine receptors, designated nicotinic and muscarinic.

Nicotinic receptors. These are pentameric glycoproteins whose subunits are integral plasma membrane proteins of postsynaptic cells. In the neuromuscular junction, the receptor consists of four monomers: α, β, γ, and δ. The α subunit is repeated, so that the stoichiometry of the full receptor is $\alpha_2\beta\gamma\delta$, with a mass of approximately 50 kDa. The five subunits are arranged around a central channel (Fig. 25.20). A lateral view of the receptor shows a cylinder perpendicularly inserted in the plane of the membrane, which protrudes into the synaptic cleft more than into the cytoplasm. The nicotinic receptor in the central nervous system is composed by two α and three β subunits ($\alpha_2\beta_3$). There are several types of α and β subunits, which give the possibility of multiple $\alpha_2\beta_3$ combinations for the receptor.

The structure of each nicotinic receptor monomer is similar for all of them. They consists of a large hydrophilic domain at the N-terminal end facing the synaptic cleft, four hydrophobic transmembrane helix segments, and two hydrophilic cytoplasmic domains (Fig. 25.21). The

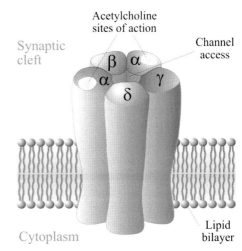

FIGURE 25.20 **Schematic representation of a neuromuscular junction nicotinic cholinergic receptor.** Five subunits (2 α, 1 β, 1 γ, and 1 δ) constitute the receptor forming a ion channel structure.

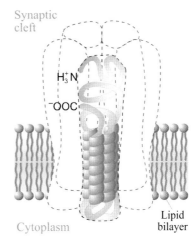

FIGURE 25.21 **Scheme of a nicotinic receptor subunit comprising four transmembrane α helices and a large N-terminal domain toward the synaptic cleft.** The larger of the two cytoplasmic loops is the one linking helices 3 and 4. GABAergic receptors units have similar structures.

acetylcholine-binding site is in the hydrophilic external domain of each α subunit. Therefore, it can bind two neurotransmitter molecules per receptor.

The nicotinic receptors function as ligand-gated ion channels. They belong to the category of ionotropic receptors. The binding of acetylcholine causes the opening of the central pore, permeable to Na$^+$ and K$^+$. Na$^+$ entry causes membrane depolarization and generates an excitatory postsynaptic potential. They are found in postsynaptic neurons in the central nervous system and ganglia and in the muscle endplate.

Muscarinic receptors. They are glycoproteins of 80 kDa with seven transmembrane helix segments (Fig. 25.22). The initial hydrophilic domain, glycosylated, and the small loops that connect helices 2–3, 4–5, and 6–7 emerge to the synaptic space. Three hydrophilic loops, of which the longest connects helices 5–6, and the C-terminal segment protrude into the cytoplasm.

There are other receptors with the same basic structure of muscarinic receptors; all associate to G proteins and belong to a superfamily probably originated from a common ancestral gene.

Pharmacological studies have helped to identify three subtypes of muscarinic receptors (M_1, M_2, and M_3), which show different location and specificity. Molecular cloning studies have identified a total of five receptors (m_1–m_5). M_1 to M_3 are the same as m_1 to m_3. M_1 is located in ganglia and exocrine cells, M_2 in smooth muscle and myocardium, and M_3 in smooth muscle and glands.

All muscarinic receptors exert their actions through G proteins. Receptors with odd number (m_1, m_3, and m_5) interact with G_q protein and activate phospholipase C, catalyzing the hydrolysis of phosphatidylinositol bisphosphate, and generating inositol trisphosphate (IP_3) and DAG. The m_2 and m_4 receptors couple with G_i, inhibit adenylate cyclase and decrease cAMP levels.

Adrenergic receptors. They are integral glycoproteins homologous to the muscarinic receptor with seven transmembrane helices (Fig. 25.22). The N-terminal portion faces the synaptic cleft and is glycosylated. The C-terminus is cytoplasmic and can be phosphorylated by protein kinases. The loop connecting helices 5–6 is the portion of the receptor that interacts with G proteins. In the extracellular side, the transmembrane segments form a niche where the neurotransmitter is bound.

Initially, two types of adrenergic receptors were distinguished, α and β. Later, several subtypes were recognized for each, which differ in anatomical distribution and affinity for different drugs. The α receptors includes two, $α_1$ and $α_2$; of which there are three variants. Three subtypes of β receptors were recognized: $β_1$, $β_2$, and $β_3$. All genes have been cloned.

The actions of β receptors are mediated by G_s protein. They activate adenylate cyclase and increase the level of cAMP in the cell. They also promote the opening of Ca^{2+} channels in the plasma membrane of cardiac and skeletal muscles.

Adrenergic $α_1$ receptors interact with G_q protein, activates phospholipase C and form the second messengers IP_3 and DAG. This results in the release of Ca^{2+} from intracellular stores and the activation of protein kinases.

Binding of the neurotransmitter to the $α_1$ adrenergic receptors couples them to G_i protein and inhibits adenylate cyclase; cAMP level is reduced and this inactivates protein kinases.

Dopaminergic receptors. These receptors belong to the superfamily comprising muscarinic and

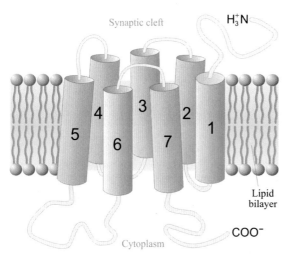

FIGURE 25.22 **Scheme of a muscarinic cholinergic receptor with seven transmembrane helical segments (represented as *cylinders*).** Dopaminergic, adrenergic, rhodopsin, and all G protein coupled receptors have a similar basic structure.

adrenergic receptors with seven transmembrane helices (Fig. 25.22). They are related to G protein; several subtypes have been demonstrated, D_1 interacts with the G_s protein while D_2 is inhibitory.

GABAergic receptors. They are pentameric structures homologous to nicotinic cholinergic receptors. Each of the constituent subunits has four transmembrane helices (Fig. 25.21). $GABA_A$ receptors are ionotropic; they form ion channels driven by the ligand. They allow the entry of Cl^- and hyperpolarize the membrane, having therefore an inhibitory action.

SUMMARY

Receptors (R) are present in cells and they can be located inside the cell or at the plasma membrane.

Intracellular receptors, located in the nucleus and cytosol, bind nonpolar or weakly polar molecules, which can easily cross membranes, such as steroid hormones, thyroid hormones, active metabolites of vitamin D and retinoids. The hormone-receptor complex (HR) dimerizes and binds to specific DNA sequences (*hormone responsive elements*), modifying the activity of transcription. Intracellular receptors belong to a family of homologous molecules that have three domains.

1. *Hypervariable*, participating in transcription regulatory actions.
2. *Central*, with two zinc fingers, which interacts with defined DNA sequences.
3. *Terminal*, which is the binding site of the hormone.

Membrane receptors are localized in the cell surface. Upon ligand binding, these receptors undergo conformational changes that are transmitted to other protein intermediate of a *signal cascade system*. There are several types:

1. G protein coupled receptors, which have seven transmembrane helices. G proteins are $\alpha\beta\gamma$ heterotrimers and under basal conditions they are inactive, with their α subunit bound to GDP. Binding of the ligand to the receptor causes the replacement of GDP for GTP, which frees the α-GTP subunit and allows it to activate downstream effectors. The $\beta\gamma$ dimer can also act as an intermediate in the signaling process. GTP is hydrolyzed to GDP and P_i, and the α subunit bound to GDP reconstitutes the inactive $\alpha\beta\gamma$ trimer.
2. Tyrosine kinase coupled receptor (TK) consists of an extracellular segment with the ligand-binding site, one transmembrane helix and a cytoplasmic portion containing the kinase activity. Formation of the ligand receptor complex promotes the dimerization of the recep-

tor, activates TK that autophosphorylates the receptor and promotes the attachment of SH2 domain proteins.
3. Receptor associated with extrinsic TK, are similar to the previous ones but without the catalytic site. Tyrosine kinase associates when HR is formed.
4. Receptor linked to guanilate cyclase catalyze the formation of 3',5'-cyclic-GMP from GTP.

PPAR is a nuclear receptor that functions as a transcription factor. Three types have been identified. They regulate metabolic pathways and the cell cycle.

Signal transduction systems in the cell include a series of different intermediaries. These include:

1. *3',5'-cyclic-AMP*, which is produced by *adenylate cyclase*. Once the hormone (first messenger) binds to a G protein coupled receptor (G_s or G_i proteins), the GTP α subunit interacts with *adenylate cyclase* that catalyzes the conversion of ATP to *3',5'-cyclic-AMP* (cAMP), considered second messenger in the system. There are stimulatory and inhibitory α subunits (α_s and α_i, respectively). When α_s is the subunit involved, cAMP level increases and activates *protein kinase A*. This enzyme is a tetramer in the inactive state, composed of two catalytic (C) and two regulatory (R) subunits. cAMP binds to the R subunits, which separate from the C subunit, which becomes activated. cAMP causes different responses in different cells. Inactivation of cAMP is performed by phosphodiesterase.
2. *Phosphatidylinositol-4,5-bisphosphate (FIP₂)* is produced after formation of the HR complex, associated to protein G_q.

    When activated, the α_q subunit of G_q binds GTP, it separates from the $\beta\gamma$ dimer and stimulates *phospholipase C* which hydrolyzes FIP_2 in the membrane to *inositol-1,4,5-trisphosphate* (IP_3) and DAG, both second messengers. IP_3 opens Ca^{2+} channels in the ER, raising Ca^{2+} in the cytosol; DAG activates protein kinase C.
3. *Ras-MAP kinase pathway* is activated by binding of a ligand to the TK receptor. Activation of the TK receptor induces its association with proteins containing SH2 domains, such as *Grb*, which bind *Sos*. The *Grb–Sos* complex activates *Ras*, which triggers the activation of *MAP* kinases (*Raf-MEK-ERK*). This finally activates other protein kinases in the cytosol or transcription factors in the cell nucleus. *Ras* proteins belong to the superfamily of *small GTPases*.
4. *JAK-STAT system* is activated when the receptor associates to *Janus type* tyrosine kinase (*JAK*). This kinase phosphorylates tyrosine residues to which *STAT* proteins (containing SH2 domains) bind. Phosphorylated *STATs* dimerize and penetrate into the nucleus, where they affect gene transcription.
5. TOR or *mTOR* in mammals, forms complexes with other proteins (*mTORC1* and *mTorcC2*). *TORC1* is acti-

vated by high levels of energy, nutrients, and growth factors. Through phosphorylation cascades TORC1 activates cell growth and proliferation. *TORC2* is activated by insulin and IGF1. One of its actions is the organization of the actin cytoskeleton.

6. *Ca^{2+}* serves as a messenger for signal transduction. Different stimuli can increase cytosolic Ca^{2+} levels, which regulate a variety of cellular functions. Ca^{2+} entering from the extracellular space or from intracellular stores binds to several proteins. One of the most widely distributed Ca^{2+}-binding proteins is *calmodulin*. When bound to Ca^{2+}, calmodulin undergoes conformational changes that allow the activation of effector proteins.

Nervous system receptors are glycoproteins with multiple transmembrane segments. There are different kinds:

Cholinergic receptors include different subgroups.

1. *Nicotinic* receptor is a pentamer in which the five constituent subunits form a channel. In muscle its composition is $\alpha_2\beta\gamma\delta$· Each α monomer has a binding site for acetylcholine. The binding of acetylcholine causes opening of the central pore, and increases the membrane permeability to Na$^+$ and K$^+$ (the receptor functions as a ligand-gated, ionotropic ion channel. Depolarization occurs and an excitatory postsynaptic potential is generated.

2. *Muscarinic* receptors have seven helix transmembrane segments. They exert their actions through G proteins. Five subtypes are found, which exert different effects. m_1, m_3, and m_5 types interact with G_q protein and activate *phospholipase C generating IP$_3$ and DAG*. m_2 and m_4 are coupled to G_i protein and *inhibit adenylate cyclase*.

Adrenergic receptors are homologous to the muscarinic receptor. Two types of these receptors, α and β exist, each of which present several variants. The α_1 receptors associate to G_q protein that activates *phospholipase C*. The α_2 receptor uses G_i proteins as mediators; it *reduces cAMP* levels. The β receptors couple to G_s. They elevate cAMP levels.

Dopaminergic receptors are homologous to muscarinic and adrenergic receptors. They also show different variants. D_1 interact with the G_s protein; in contrast, D_2 are inhibitory.

GABA receptors form ion channels that allow Cl$^-$ entry, producing hyperpolarization. They have inhibitory action.

Bibliography

Ahmadian, M., Suh, J.M., Hah, N., Liddle, E., Atkins, A.R., Downes, M., Evans, R.M., 2013. PPAR signaling and metabolism: the good, the bad and the future. Nat. Med. 19, 557–566.

Becker, K.L., 2002. Principles and Practice of Endocrinology and Metabolism, third ed. J.P. Lippincott, Philadelphia, PA.

Birnbaumer, L., 2007. The discovery of signal transduction by G proteins. Biochim. Biohys. Acta 1768, 756–771.

Coleman, M.L., Marshall, C.J., Olson, M.F., 2004. Ras and Rho GTPases in G1-phase cell cycle regulation. Nat. Rev. Mol. Cell Biol. 5, 355–366.

Escribá, P.V., 2007. G protein-coupled receptors, signaling mechanisms and pathophysiological relevance. Biochim. Biophys. Acta 1768, 747–898.

Foster, K.G., Fingar, D.C., 2010. Mammalian target of rapamycin (mTOR) conducting the cellular signaling symphony. J. Biol. Chem. 285, 14071–14077.

Gerbett, D., Bretscher, A., 2014. The surprising dynamics of scaffol proteins. Molecular Biol. Cell 25, 2315–2319.

Hancock, J.F., 2003. Ras proteins: different signals from different location. Nat. Rev. Mol. Cell Biol. 4, 373–385.

Holmström, K.H., Finkel, T., 2014. Cellular mechanisms and physiologic consequences of redox-dependent signaling. Nat. Rev. Mol. Cell. Biol. 15, 411–421.

Hubbard, S.R., Till, J.H., 2000. Protein tyrosine kinase structure and function. Annu. Rev. Biochem. 69, 373–398.

Klaus, A., Birchmeier, N., 2008. Wnt signaling and its impact on development and cancer. Nat. Rev. Cancer 8, 397–398.

Kronenberg, H.M., 2008. Williams Textbook of Endocrinology, eleventh ed. Saunders Elsevier, Philadelphia, PA.

Laplante, M., Sabatini, D.M., 2012. mTOR signaling in growth control and disease. Cell 149, 274–293.

Reddy, J.K., Hashimoto, T., 2001. Peroxisomal β-oxidation and peroxisome proliferator activated receptor. An adaptive metabolic system. Annu. Rev. Nutr. 21, 193–230.

Rockman, H.A., Koch, W.J., Lefkowitz, R.J., 2002. Seven transmembrane-spanning receptors and heart function. Nature 415, 206–212.

Rosen, H., Stevens, R.C., Hanson, M., Roberts, E., 2013. Sphingosine-1-phosphate and its receptor: structure, signaling, and influence. Annu. Rev. Biochem. 82, 637–662.

Stanfel, M.N., Shameh, L.S., Kaeberlein, M., Kennedy, B.K., 2009. The TOR pathway comes of age. Biochim. Biophys. Acta 1790, 1067–1074.

Takai, Y., Sasaki, T., Matosaki, T., 2001. Small GTP-binding proteins. Physiol. Rev. 81, 153–208.

Taleisnik, S., 2006. Receptores Celulares y la Transducción de Señales. Encuentro Grupo Editor, Córdoba.

Vondriska, T.M., Pass, J.M., Ping, P., 2004. Scaffold proteins and assembly of multiprotein signalin complexes. J. Mol. Cell Cardiol. 37, 391–397.

Wang, Y., Zhow, Y., Graves, D.T., 2014. FOXO transcription factors: their clinical significance and regulation. Biomed. Res. Int. 2014, 1–13.

Wilson, M.S., Livermore, T.M., Saiardi, A., 2013. Inositol pyrophosphates: between signalling and metabolism. Biochem. J. 452, 369–379.

Biochemical Bases of Endocrinology (II) Hormones and Other Chemical Intermediates

HORMONES: CHEMICAL NATURE

According to their chemical nature, hormones can be classified into five categories:

1. *Steroids*. These hormones are chemically related with the cyclopentaneperhydrophenanthrene, the chemical structure of cholesterol. Among this group of hormones are glucocorticoids, aldosterone, androgens from the adrenal cortex, estrogens and progesterone from the ovaries, testosterone from the testis, and 1,25-$(OH)_2$-D_3, the active metabolite of vitamin D_3. Owing to their poor polar character, these hormones readily cross cell membranes. In blood, they are transported in association with proteins.

2. *Amino acid derivatives*. This type of hormones include epinephrine or adrenaline and norepinephrine or noradrenaline (catecholamines) from the adrenal medulla, the tyrosine derivatives thyroxine and triiodothyronine produced by the thyroid, and the tryptophan derivative melatonin synthesized in the pineal gland. Catecholamines are soluble in aqueous media; they circulate in blood free or weakly bound to albumin and do not penetrate the target cells. Thyroid hormones are less polar and can cross cell membranes by diffusion.

3. *Fatty acid derivatives*. These include autocrine or paracrine compounds such as prostaglandins (PGs), thromboxanes (TXs), and leukotrienes (LTs) (eicosanoids). They originate from polyunsaturated fatty acids. Arachidonic acid is the most important precursor.

4. *Peptides*. Regulatory factors and hormones such as vasopressin and oxytocin from the hypothalamus belong to this group. Others include adrenocorticotropin hormone (ACTH), melanocyte-stimulating hormone (MSH) from the anterior pituitary, calcitonin made in the thyroid, glucagon from the pancreas, and gastrin, secretin, pancreozymin, and other hormones of the gastrointestinal tract.

5. *Proteins*. Among these hormones are prolactin (PR), follicle-stimulating hormone (FSH), luteinizing hormone (LH), growth hormone (GH), thyrotropin or thyroid-stimulating hormone (TSH). These all arise from the adenohypophysis. Other examples include parathyroid hormone

(PTH) and insulin, which are synthesized in the parathyroid gland and pancreas, respectively.

Protein and peptide hormones are synthesized in ribosomes associated with the rough endoplasmic reticulum (ER). Initially, a prohormone is formed. A signal peptidase removes a peptide from the N-terminus of the protein releasing the prohormone into the ER lumen. There another segment of the chain is removed to render the final "mature" hormone. Sometimes, the original chain precursor contains multiple active factors. This is the case of proopiomelanocortin (POMC), from which a variety of hormonal agents are generated by hydrolysis (p. 579). Protein and peptide hormones are stored in intracellular vesicles and are released by exocytosis upon different stimuli. They are soluble in aqueous media and, in most cases, circulate in blood free.

Types of Actions Promoted by Hormones

Despite the diversity and the variety of responses that each hormone triggers on a target organ, they can be grouped into three main categories according to the place where they exert their mechanisms of action. Thus, they can regulate (1) membrane transport, (2) the activity of enzymes present in the cell, and (3) protein synthesis.

1. *Cell membrane actions*. Some hormones modify the flow of metabolites or ions across membranes through their action on different transport systems of the cell plasma membrane.
2. *Enzyme activity modification*. In intact animals or in isolated tissues, changes in the activity of certain enzymes occur immediately after the treatment with hormones; this action is not only rapid, but also transient. It is mainly exerted at the level of regulatory enzymes whose activity is increased or decreased by covalent modification.
3. *Protein synthesis regulation*. Many hormones modulate the synthesis of enzymes and of

other proteins. They act predominantly by regulating gene transcription. This action is slower than direct enzyme regulation, but it has more sustained effects.

The same hormone can exert more than one of the mentioned mechanisms. For example, insulin promotes transport of certain metabolites through membranes and influences enzyme activity and protein synthesis.

All are closely related mechanisms and lead to interactions. For example, the effects on membrane transport systems can determine entry into cells of substances with the capacity to modulate enzyme activity or transcription in the nucleus. Also, a direct effect on enzymes determines changes in the availability of substrates for the operation of metabolic pathways or processes affecting protein synthesis and membrane transport. In some cases, this interplay between different mechanisms makes it difficult to pinpoint what the primary action of the hormone is.

General Properties of Hormones

Activity. Hormones exert their effects at very low concentrations. In this respect, they resemble biological catalysts. Usually, hormone levels in plasma are very low. For example, plasma concentrations of protein hormones range from 10^{-12} to 10^{-9} M, while steroidal hormones circulate at concentrations from 10^{-9} to 10^{-6} M.

Average life. Owing to their biological activity, hormones are degraded relatively rapidly and converted into inactive products because their accumulation in the body has damaging effects. The average duration of a hormone varies from seconds to days.

Rate and rhythm of secretion. In general, hormone secretion is not constant. It often responds to stimuli from the environment or the internal medium. For example, insulin secretion is promoted by increases in blood glucose concentration.

Secretion of many hormones presents cyclical variations, such as pituitary gonadotrophic hormones and ovarian hormones during the female sexual cycle, or adrenal cortex steroids during each 24-h period (circadian variation).

Specificity. One of hormones' most notable properties is their highly specific action. Hormones are released into the general circulation and reach all tissues; however, its action is exerted only on a limited number of cells and leads to a defined type of response. This specificity is due to the existence of receptors present in the target cells that allow the hormone to selectively bind to it.

Hormone Determination Methods

Many methods have been developed to determine the amount of hormone contained in plasma or other biological samples.

Bioassay. When a hormone is applied on an intact animal, an isolated organ, or cells in culture, the magnitude of the effect obtained is directly proportional to the hormone concentration in the preparation. These types of tests are valuable because they directly measure functional amounts of a hormone; however, they lack sensitivity. It is rarely used today.

Chemical assay. Current isolation and purification techniques, including chromatography, mass spectrometry, electrophoresis, and fractional extraction with solvents, can be used to determine the absolute amount of hormone in a sample. These methods have the disadvantage of being very laborious and not always applicable to protein hormones. Moreover, some of these techniques are not sensitive enough to detect very low concentrations of hormone. At present, the methods most frequently used are immunoassays, high performance chromatography, and mass spectrometry.

Immunoassay. This is an assay based on the competition between the hormone of interest in a sample and the same hormone, purified and labeled, for binding to a specific antibody. The

hormone can be labeled with a radioactive isotope (generally ^{125}I), in which case the assay is called *radioimmuno assay*, or with a compound which can be detected by colorimetric, fluorometric, or chemiluminescence assays.

The antibodies used for these assays can be monoclonal or polyclonal, made against protein hormones, which have excellent antigenicity, or small molecule hormones, which are less antigenic, but can be made more antigenic by binding them to the serum proteins that normally transport the hormone in blood. The assay mixture contains the sample with (1) the non labeled hormone, whose concentration is to be determined; (2) a known amount of the labeled hormone; and (3) the specific antibody in amounts lower than the total of hormone in the sample. The labeled hormone and the hormone in the sample compete for binding to the antibody. Radioactivity (or colorimetric indexes, fluorescence, chemiluminescence) in the free fraction remaining in the mixture is directly proportional to the concentration of the hormone in the sample. The method is extremely sensitive and allows for the determination of hormone concentrations of less than 1 ng (10^{-9} g)/mL. Previous calibrations are required with known amounts of hormone.

Enzyme-linked immunosorbent assay (ELISA) is another antibody-based assay, as sensitive as radioimmunoassay. In ELISA, the sample containing the hormone or the specific antibody coupled to an enzyme (peroxidase) is immobilized on a solid support (generally a polysterene microtiter plate). Formation of the hormone–antibody complex is directly proportional to the quantity of hormone present in the sample and can be detected by the development of color in the reaction catalyzed by peroxidase.

A variant of the ELISA method, also called *immunometric technique*, has been developed, which uses two monoclonal antibodies that recognize different epitopes in the hormone. This enhances the specificity of the method. The first antibody is attached to a solid matrix and must be in excess relative to the hormone in the sample. After allowing the hormone to bind to the antibody, a second antibody, conjugated to an enzyme for detection, is added. The hormone is bound by two antibodies, which has given this technique the name of "sandwich" method. The amount of the second antibody bound is proportional to the amount of the hormone in the sample, allowing its quantification.

Pituitary (Hypophysis)

From the embryological, anatomical, and functional points of view, the pituitary gland consists of two distinct parts: the anterior pituitary or adenohypophysis and the posterior pituitary or neurohypophysis. In lower vertebrates, an intermediate portion or lobe is also found.

The neurohypophysis is connected both anatomically and functionally with the base of the brain, more precisely with the hypothalamus through the infundibular stalk, which contains blood vessels of a portal system and nerve fibers.

Removal of the pituitary (hypophysectomy) in young animals produces growth arrest and lack of development of the sex glands. If hypophysectomy is performed in adult animals, it causes atrophy of sex organs and glands, involution of the thyroid and adrenal cortex, and alterations in general metabolism. This shows that the pituitary has widespread effects on the body and on other endocrine glands. The pituitary gland produces hormones (tropins) that stimulate the function of other endocrine glands. In turn, the secretion of pituitary hormones is controlled by regulatory factors produced in the hypothalamus.

HYPOTHALAMIC HORMONES

The anatomical relationship between the hypothalamus and the pituitary gland has great functional relevance. Blood vessels, which originate in the median eminence of the hypothalamus, arrive to the anterior pituitary via the infundibular stalk, where they again branch into capillaries, forming a portal system. These capillaries bring to the hypophysis factors generated in the hypothalamus, which regulate the release of hormones synthesized in the anterior pituitary. Neurons located in the hypothalamic center produce hormones that are taken via their axons in the infundibular stem to the nerve endings in the neurohypophysis.

Nine factors (regulatory anterior pituitary hormones) produced in the hypothalamus are known, all of which are peptides (except dopamine). They regulate the synthesis and release of tropins from the adenohypophysis. Some of these factors act on more than one pituitary hormone.

There are factors that stimulate the secretion of all the hormones produced in the pituitary. There are only three factors that are inhibitory; they suppress the release of the GH, prolactin, and MSH. Inhibitory factors for those three hormones are necessary because they are the only pituitary hormones not controlled by feedback mechanisms. In contrast, secretion of the remaining tropins is inhibited by increased blood levels of the hormone that they stimulate (negative feedback mechanism).

The secretion of releasing hormones is regulated by a group of neurons that innervate the hypothalamus, which explains why neural stimuli regulate endocrine activity and produce metabolic responses.

Fig. 26.1 presents the regulatory factors of the hypothalamus and the hormones of the anterior hypophysis that they control.

Corticotropin-releasing hormone (CRH). This hormone was the first hypothalamic regulatory factor discovered. It stimulates the release of adrenocorticotrophic hormone (ACTH) in the anterior pituitary and other products generated by the precursor molecule POMC (see subsequent section). It is a 41-amino acid polypeptide derived from a precursor polypeptide of 196 amino acids.

Hypothalamus hormones		Anterior pituitary hormones
ACTH releasing (CRH)	⊕ →	Adrenocorticotropin (ACTH) and β-endorphin
TSH releasing (TRH)	⊕ →	Thyrostimulant (TSH)
LH and FSH releasing (LHRH)	⊕ →	Luteinizing (LH) and follicle stimulant (FSH)
GH releasing (GHRH)	⊕ →	Growth hormone (GH)
GH inhibiting or somatostatin (GHIH)	⊖ →	Growth hormone (GH)
PR releasing (PRRH)	⊕ →	Prolactin (PR)
PR inhibiting (PRIH)	⊖ →	Prolactin (PR)
MSH releasing (MSRH)	⊕ →	Melanocyte stimulant (MSH)
MSH inhibiting (MSIH)	⊖ →	Melanocyte-stimulant (MSH)

FIGURE 26.1 **Hypothalamic regulatory factors and their effect on hormones of the adenohypophysis.**

Antidiuretic hormone (ADH) from the neurohypophysis and angiotensin II potentiate the action of CRH; oxytocin inhibits it. A CRH-binding protein has been described in serum and in several cell types. CRH is also secreted by the human placenta.

Thyrotropin-releasing hormone (TRH). This hormone is the main regulator of the secretion of TSH. It is a tripeptide (pyroglutamyl-histidyl-prolinamide) derived from a precursor molecule of 242 amino acids (29 kDa), which contains six copies of the Gln-His-Pro-Gly sequence from which TRH is generated by cyclization of glutamyl and amidation of the proline residue (Fig. 26.2).

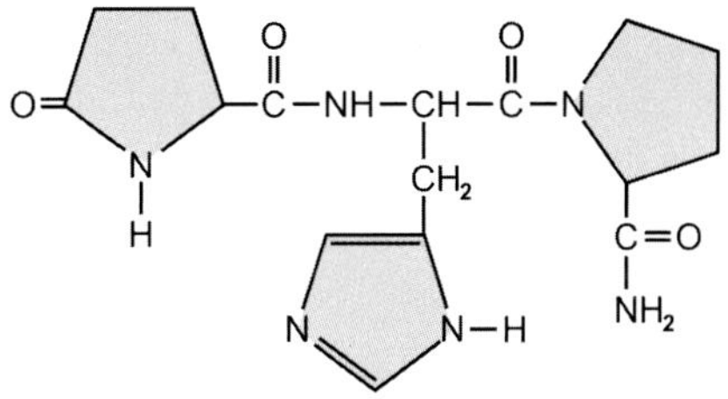

Gonadotropin-releasing hormone (GnRH). The function of this hormone is to stimulate the release of LH and FSH. The effect of LH and FSH explains the influence of nerve impulses in the control of gonadal function. It is a linear decapeptide that has a pyroglutamic moiety at the

FIGURE 26.2 **Thyrotropin-releasing hormone.**

N-terminal end and an amide function at the C-terminal end. It is derived from a precursor protein of 92 amino acids. The same luteinizing hormone–releasing hormone (LHRH) gene encodes a polypeptide of 56 amino acids, called GnRH-associated peptide, which exerts prolactin-releasing activity.

Prolactin-releasing hormone (PRRH) and prolactin inhibitory hormone (PRIH) have been proposed as regulators of the synthesis and secretion of prolactin in the adenohypophysis; however, at present, these hormones have not been yet isolated. The best known stimulating factors of PR release are the TRH, oxytocin, and vasoactive intestinal peptide (VIP). PRIH has been identified with dopamine. The neurotransmitter γ-aminobutyric acid (GABA) and cholinergic pathways also inhibit prolactin release.

Growth hormone–releasing hormone (GHRH) and growth hormone inhibitory hormone (GHIH). GHRH is a 44-amino acid peptide derived from a precursor protein of 108 amino acids in length. Its biological activity lies in the first 29 amino acids, which have homology with other peptides such as glucagon, secretin, and VIP. GHIH, also called *somatostatin*, is a peptide derived from a 116-amino acid precursor, processed to form a tetradecapeptide (somatostatin-14), and another 28-amino acid peptide (somatostatin-28), which differs from somatostatin-14 by an additional peptide sequence at the N-terminal end. It has a cyclic structure due to the presence of two cysteine residues that form a disulfide bridge. It belongs to a family of peptides derived from the same gene by alternative splicing of mRNA, and by posttranslational modifications of the propeptide. GHIH has been found not only in the hypothalamus, but also within δ cells of the Langerhans' islets in the pancreas, the gastrointestinal mucosa, and in thyroid C cells. In the hypothalamus, somatostain-14 is the main form, while in intestine somatostatin-28 is more important. Somatostatin has inhibitory action on the synthesis and secretion of GH in the adenohypophysis, and also exerts a wide range of inhibitory actions on the secretion of (1) insulin and glucagon in the pancreas, (2) TSH, and (3) gastrin and secretin.

Melanocyte-regulating hormones (MSRH and MSIH). A melanocyte-releasing hormone (MSRH) and a melanocyte inhibitory hormone (MSIH) are produced in the hypothalamus. The first is a pentapeptide with a sequence identical to the initial portion of oxytocin, a hypothalamic hormone released by the posterior pituitary gland. The MSIH is a tripeptide with the same sequence of the three amino acids of the oxytocin C-terminal end.

Mechanism of action of releasing factors. Releasing peptides bind to specific membrane receptors in the target cells. These receptors have seven transmembrane helices, coupled to G proteins such as G_s, which activates adenylate cyclase, increases levels of cAMP, and stimulates a protein kinase that phosphorylates proteins responsible for secretion of the corresponding hormone. The increase in cAMP also causes opening of Ca^{2+} channels and intracellular Ca^{2+} increase, which is another factor that stimulates hormone secretion (GnRH is released in this way).

TRH and LHRH bind to a G_q protein-coupled receptor which stimulates phospholipase C (PLC), and generate the second messengers inositoltrisphosphate (IP_3) and diacylglycerol (DAG), which elevate cytosolic Ca^{2+} concentration and activate protein kinase C. The hydrolysis of DAG produces arachidonate, which originates eicosanoids. PGs also trigger the release of GH and ACTH. Leukotrienes stimulate the

secretion of gonadotropins. Somatostatin and dopamine (inhibiting factors) act partly by depressing adenylate cyclase activity (G_i protein-coupled receptors).

The action of stimulating and inhibiting factors is very fast; its effects can be noted in less than a minute.

ANTERIOR PITUITARY (ADENOHYPOPHYSYS)

One of the most remarkable features of the anterior pituitary is the production of hormones or tropins, which stimulate the synthesis and secretion of hormones in other endocrine glands.

Different cell types have been identified in the anterior pituitary, including corticotroph, thyrotroph, gonadotroph, lactotroph, and somatotroph, which are responsible for the secretion of ACTH, thyrotropin or TSH, gonadotropins (LH and FSH), prolactin (PR), and GH. All of the anterior pituitary hormones are polypeptides.

Most tropin hormones are subjected to negative feedback inhibition and positive control by hypothalamic-releasing factors. The blood concentration of the hormone from the stimulated gland regulates the secretion of the corresponding tropin by acting on the pituitary or hypothalamus. An increase in plasma concentration of sex steroids, glucocorticoid, or thyroxine inhibits the release of gonadotropins, corticotropin, or TSH, respectively.

ADRENOCORTICOTROPIN HORMONE

Corticotroph cells produce a precursor protein, called pre-proopiomelanocortin (POMC), which is fragmented into several biologically active peptides.

POMC. The protein pre-POMC (280 amino acids in length) loses its initial portion (leader or

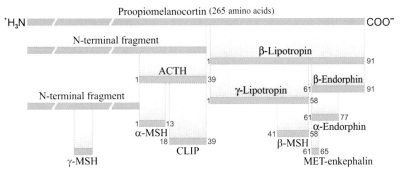

FIGURE 26.3   **Proopiomelanocortin and its products of hydrolysis.**

signal peptide) to give POMC, which undergoes hydrolysis at defined bonds in its chain, generating several molecules with hormonal activity. Some of these hormones undergo other posttranslational changes, such as acetylation, amidation, methylation, and glycosylation. POMC is hydrolyzed into two segments: the N-terminal polypeptide has a length of 170 amino acids and the C-terminal portion or β-lipotropin has a length of 91 amino acids (Fig. 26.3).

The N-terminal fragment undergoes additional hydrolysis to generate a 39-amino acid polypeptide corresponding to ACTH. In some species, two fragments derived from ACTH can be generated. One consists of amino acids 1–13 and is known as α-MSH; the other is a peptide called *corticotropin-like intermediate lobe peptide* (CLIP), which comprises amino acids 18–39 of ACTH. Both of these peptides are found in species with a developed intermediate lobe, but not in humans. The remaining portion of the N-terminal after removal of the ACTH fragment is hydrolyzed at two sites to give a small segment, γ-MSH.

The 91-amino acid β-lipotropin, generated from the hydrolysis of the POMC, is secreted by corticotrophic cells in equimolar amounts with ACTH. The β-lipotropin is again hydrolyzed, forming γ-lipotropin and β-endorphin (Fig. 26.3). The β- and γ-lipotropins act on adipose tissue, promoting lipolysis and fatty acid mobilization.

The β-endorphin (31 amino acids in length) is also cleaved in a segment containing the N-terminal 17 amino acids of α-endorphin, which can be further hydrolyzed at its N-terminus to give a pentapeptide called MET-enkephalin. Endorphins and enkephalins are peptides of the pituitary and central nervous system, which have potent analgesic action. They bind to the same receptors used by morphine and other opiates. They are related to pain perception.

The melanocortin system. It is composed of a group of small molecule hormones derived from the posttranslational cleavage of POMC, including α-, β-, and γ-melanocyte stimulatory factors and ACTH. This system is important in the regulation of energetic balance and other neuroendocrine processes, such as arterial pressure regulation, feeding, and pain sensation. Melanocyte stimulatory hormones (MSH) are important products of the pituitary *pars intermedia* in fish and batrachians, reason by which they are also designated *intermedins*. In these organisms, MSH acts on skin pigmented cells or melanocytes, promoting the darkening of teguments.

ACTH. The anterior pituitary produces a tropin that stimulates the adrenal cortex. ACTH is a polypeptide of 39 amino acids (~4500 Da). The initial 23 amino acids segment of the polypeptide, which is highly conserved among species, displays most of the activity of ACTH.

Variations between species reside in the remaining 16 amino acids, apparently unrelated to the activity of the hormone.

ACTH stimulates the synthesis and secretion of hormones from the adrenal cortex. It exerts its effects preferably on cells of the fascicular area, which are the main producers of glucocorticoids. Although the main action is related to the synthesis and secretion of glucocorticoids (cortisol), it also stimulates the synthesis and release of other corticosteroids, including mineralocorticoids (aldosterone) and androgenic corticoids.

Corticotropin increases not only the production of steroids, but also the synthesis of proteins involved in maintaining the structure of the adrenal gland. Steroidogenic acute regulatory (StAR) protein is among the proteins whose synthesis is stimulated by ACTH.

ACTH concentration in blood varies between 9 and 52 pg/mL (2–11 pmol/L). Maximum values occur in the morning (7:00–9:00 a.m.), and levels decrease during the day to reach a minimum at night. ACTH secretion is mediated through neural influences by several hormones, particularly CRH and also vasopressin, interleukin 1, catecholamines, angiotensin II, and serotonin. Production of ACTH is increased by stressors such as trauma, surgery, hypoxia, severe hypoglycemia, and exposure to cold. High levels of corticosteroids, particularly of cortisol, inhibit ACTH secretion by feedback mechanisms.

The primary action of ACTH is exerted via binding to membrane receptors coupled to G_s protein, adenylate cyclase activation, and cAMP increase.

Disorders related to ACTH secretion. High levels of ACTH are observed in Addison's disease, with deficiency in the production of cortisol, which suppresses ACTH feedback inhibition. There is also elevated ACTH in some pituitary functioning tumors. *Reduced levels* of ACTH are found in tumors of the adrenal cortex, which increase cortisol production.

THYROID-STIMULATING HORMONE

TSH is a glycoprotein of 28 kDa; 10%–15% of the molecule is represented by carbohydrates. Its structure is similar to that of gonadotropin, which is composed of two polypeptide chains: α and β. In the same species, the α subunit is identical for TSH and gonadotropins. The β chain determines the specific action for each of the hormones. The isolated subunit has little activity, only the αβ dimer exerts hormonal effect. The sequence of the β subunit has been highly conserved in mammals. The β subunit binds to high affinity receptors and stimulates thyroid iodine uptake as well as synthesis and release of thyroid hormones [3,5,3′,5′-tetraiodothyronine (T_4) and 3,5,3′-triiodothyronine (T_3)]. It also promotes synthesis of mRNA and protein, resulting in an increased size and vascularization of the gland. These actions are mediated by adenylate cyclase activation and increased cAMP levels in thyroid. Secretion of TSH is stimulated by TRH from the hypothalamus and modulated by the concentration of T_3 and T_4 in plasma.

GONADOTROPINS

These hormones are produced by the gonadotrophic cells of the anterior lobe of the pituitary gland. They act on the maturation and function of the ovaries and testes. Gonadotropins include FSH and LH, both of which are glycoproteins. FSH (~34 kDa) contains 10% carbohydrates, while LH (~25 kDa) has 15% carbohydrates. The hydrocarbon chains of both gonadotropins have mannose, galactose, fucose, glucosamine, galactosamine, and sialic acid. Gonadotropins consist of two different polypeptides, α and β. The α subunit is identical for both LH and FSH; the β chain is specific for each hormone. Although the activity depends on the β chain, the presence of the two subunits is necessary for the hormone's biological functions.

FSH actions. FSH induces the maturation and development of the Graafian follicle in the ovary. Together with LH, it stimulates estrogen and progesterone synthesis in the ovary. The highest levels of FSH in plasma are attained before ovulation.

In testes, FSH promotes the development of the seminiferous tubules and is one of the factors involved in the initiation of spermatogenesis. It targets Sertoli cells to stimulate the production of estrogens from androgens, and, together with testosterone, it induces the synthesis of androgen-binding proteins in these cells, which helps maintaining high levels of testosterone locally, necessary for spermatogenesis. In men plasma, FSH levels are very constant.

LH actions. In females, LH controls the development of the corpus luteum and stimulates estrogen and progesterone secretion. It induces StAR protein and the synthesis of enzymes involved in pregnenolone and other steroid hormone precursors. Although the Graafian follicle development is mainly controlled by FSH, estrogen production depends on both corticotropins.

In males, LH is also called interstitial cell (Leydig cells)-stimulating hormone (ICSH). It promotes the production and secretion of testosterone, which in turn helps to maintain spermatogenesis and development of secondary sexual organs.

FSH and LH activate adenylate cyclase in the target cells. Levels of LH and FSH vary with age; they are low before puberty and elevated in postmenopausal women. Increased LH in males and cyclic FSH secretion in girls precede the onset of puberty. LH and FSH vary during the menstrual cycle. In the initial phase of the cycle, LH increases slowly and has a marked increase at midcycle, which triggers ovulation. FSH increases slightly at first, then drops, and at midcycle, increases parallel to that of LH occurs. The concentrations of both hormones fall after ovulation.

Secretion is controlled by GnRH; circulating sex steroids influence GnRH secretion by feedback.

Other paracrine factors of ovarian origin are activin and inhibin, proteins that have opposite effects on the gonadotrophic cells, and follicle statin, a peptide that regulates activin and inhibin.

LACTOGENIC HORMONE OR PROLACTIN

Prolactin is a 199-amino acid protein with a mass of approximately 23 kDa. Along with GH and human placental lactogen, PR constitutes a family of proteins with high structural similarity. All three hormones are derived from a same ancestral gene that, by duplication and divergent evolution, resulted in the genes coding for PR, GH, and human placental lactogen. Owing to the structural similarities, these hormones give immune cross-reactions.

PR is secreted predominantly by lactotrophic cells of the anterior pituitary, but it is also expressed and secreted in the brain, hypothalamus, placenta, and the amnion. Basal PR levels in adults are highly variable, with an average of 13 ng/mL (0.6 nmol/L) in women and 5 ng/mL (0.23 nmol/L) in men. The prolactin receptors are plasma membrane proteins consisting of a single transmembrane domain. They belong to the cytokine receptor superfamily. Binding of the hormone to the receptor promotes its dimerization. The cytoplasmic domain of the receptor associates with Janus tyrosine kinase (*JAK-2*) and is activated. *JAK-2* autophosphorylates and phosphorylates tyrosine residues in the receptor, which then induces association with the SH2 domains of cytosolic signal transduction and activator of transcription (*STAT*) proteins (Fig. 24.18).

Active *STAT* proteins dimerize, move into the nucleus, and bind to the promoter of certain genes, such as those of β-lactoglobulin and casein (proteins of milk), activating their transcription.

The *JAK-STAT* pathway is activated by prolactin, GH, epidermal growth factor (EGF), erythropoietin, and cytokines.

Together with LH, prolactin promotes the formation of the *corpus luteum* and the production of progesterone. Its best-characterized function is stimulation of growth of the mammary gland during pregnancy and the postpartum lactation period. Although estrogen and progesterone contribute to these actions, they have depressant effects on lactogenesis. The decrease of these hormones after delivery allows the initiation of lactation, for which prolactin is the major factor. PR also activates immune function by stimulating lymphocyte proliferation. Prolactin secretion increases during pregnancy and decreases 8–10 weeks after delivery. High concentrations of PR inhibit secretion of GnRH, luteinizing and FSH. This reduces fertility and pregnancy during lactation.

GROWTH HORMONE OR SOMATOTROPIN

GH is synthesized as a prohormone in the anterior pituitary somatotroph cells. Before being secreted, a 26-amino acid peptide is removed from its N-terminal end. The "mature" GH is a single protein of 21.5 kDa, composed of 191 amino acids. It consists of a single chain, which has two disulfide bridges. Somatotropin has structural homology with prolactin and placental lactogen, all derived from a common ancestral gene.

GH function depends on a relatively small portion of the molecule; partial hydrolysis with chymotrypsin does not cancel the activity. Although there are great similarities between somatotropins of different species, there is a high degree of specificity. For example, bovine and porcine hormones have no activity in man, which responds only to the human and primate forms of the hormone.

GH secretion is under control of a set of neural, metabolic, and hormonal factors. Of these, the most important are somatotropin-releasing hormone (GHRH) which stimulates GH secretion, and somatostatin, which inhibits GH release. Ghrelin, a protein of 218 amino acids synthesized in various tissues, especially gastric mucosa and the hypothalamus, stimulates GH release, induces food intake, and promotes obesity.

Most of GH is transported in plasma by a specific protein. It has a half-life of 20–50 min, and its concentration in adults, early in the morning, is less than 2 ng/mL (90 pmol/L).

Somatropin is a multifunctional hormone; it promotes postnatal growth of skeletal and soft tissues through direct and indirect actions. Indirect effects are mediated by *insulin-like growth factors* (IGF).

The main function of somatotropin is to stimulate cell proliferation and maturation in many tissues, particularly bone, cartilage, and muscle. It has general effects on the metabolism of proteins, nucleic acids, lipids, and carbohydrates.

Proteins. GH stimulates protein anabolism; it increases amino acid uptake into cells by activation of transport systems, especially in muscle and liver, and decreases protein catabolism. It promotes the use of fat deposits contributing to reduce the consumption of protein for energy purposes.

Nucleic acids. GH increases the activity of DNA and RNA polymerases, accelerating mRNA transcription and translation.

Lipids. GH decreases lipogenesis; it exerts lipolytic action and increases the concentration of free fatty acids in blood. This stimulation of fatty acids mobilization is accompanied by activation of their degradation and conversion to acetyl-CoA.

Carbohydrates. GH decreases glucose utilization. In muscle, somatotropin depresses glucose transport and glycolytic activity; it has effects that are opposed to those of insulin. In the liver, GH increases gluconeogenesis from amino acids, and helps to increase glycogenesis. The

TABLE 26.1   Metabolic Actions of Growth Hormone

↑ Synthesis of DNA, RNA, and proteins
↑ Retention of nitrogen and phosphorus
↓ Free amino acids and urea in blood
↑ Lipolysis
↑ Free fatty acids in plasma
↓ Glucolysis
↑ Gluconeogenesis
↑ Glycemia
↑ Somatomedins
↑ Secretion of insulin
↑ Incorporation of sulfate to cartilage

decrease in peripheral glucose utilization and gluconeogenesis activation tends to raise the circulating levels of glucose. GH behaves as hyperglycemic (diabetogenic); it is a major determinant of insulin resistance.

GH is also secreted in other organs, such as gonads, placenta, and mammary gland, and in nervous system, where its acts in autocrine or paracrine ways. In gonads, GH is involved in sexual differentiation, pubertal maturation, steroidogenesis, gametogenesis, and ovulation. In the nervous system, GH favors development, cell proliferation, differentiation, and survival, and it participates in cognitive and behavioral functions.

Table 26.1 summarizes the metabolic actions of GH. These effects do not completely explain the actions of somatropin in promoting growth.

The function of GH is exerted in part through mediators produced in various tissues under the influence of GH. The existence of these mediators is supported by the discovery of a sulfation factor responsible for the incorporation of sulfate into chondroitin sulfate molecules of cartilage. These mediators are called *somatomedins* or insulin-like growth factors (IGF).

IGF. Two hormones of this family, IGF-I and IGF-II, are synthesized in various tissues. IGF-II is the major form in the fetus; it participates on the development of the fetus in utero. IGF-I is active at all stages of development, both

pre- and postnatally. However, GH deficiency does not significantly affect prenatal growth, indicating that IGF-I begins to be regulated by GH after birth.

Both somatomedins have structural similarity to insulin (IGF-I has 42% homology with proinsulin), which supports the idea of a common ancestral gene for all three molecules.

Some characteristics of IGF-I include the following:

1. its concentration in blood plasma is regulated by GH;
2. it stimulates sulfate and glucosamine incorporation into cartilage proteoglycans;
3. it increases the mitotic activity of fibroblasts;
4. it mimics the action of insulin in muscle and adipose tissue; and
5. it is activated via the *PI3K-AKT* pathway and by *TORC1* and *TORC2*.

Liver is the main site for the production of somatomedins; they have also been found in kidney and in cell cultures of lung fibroblast treated with GH.

IGF activity in plasma decreases under protein malnutrition. This effect is possibly due to the presence of inhibitors of somatomedins that limit growth when nutrition is deficient. Glucocorticoids and estrogens increase the production of these inhibitors.

Mechanisms of action of GH. The effects of GH are initiated by binding of the hormone to plasma membrane receptors on the target cells. The GH receptor belongs to a superfamily of receptors that includes the prolactin, EGF, and the cytokine receptors. It presents two forms: one is an integral membrane protein, and the other is a shorter soluble molecule, whose sequence is identical to the extracellular segment of the membrane receptor.

GH has binding sites for two separate receptors that upon binding form a dimer. Once activated, the receptor has tyrosine kinase activity and associates with *JAK-2* which autophosphorylates

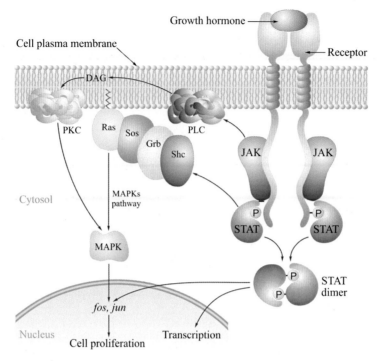

FIGURE 26.4   **Signal transduction pathways activated by growth hormone.**

and initiates a cascade of phosphorylation events. The actions of GH on gene transcription are mediated by *STAT* proteins that activate early response genes, such as *c-fos* and *c-jun*, and serine protease inhibitor (*Spi 2.1*). *JAK-STAT* is not the only pathway by which GH influences the transcriptional gene activation (Fig. 26.4). GH also activates the *MAP* protein kinase cascade and phosphatidylinositol-3-kinase (PI$_3$K). Also, protein kinase C is involved in some of GH actions. IGF-1 is one of the proteins whose synthesis is activated by GH.

Somatomedins, mediators of GH action, bind to their own specific receptor, a heterotetramer similar to that of insulin (Fig. 25.7B). They have the catalytic site of protein tyrosine kinases in the cytosolic domain. When activated by IGF, the receptor triggers a cascade of phosphorylating events that will result in stimulation of gene transcription. The insulin receptor substrates

(IRS) are among the phosphorylated proteins in the pathway activated by IGF.

Pathologic Conditions Associated with Abnormalities of Growth Hormone

GH excess. Hyperactivity of the pituitary gland during childhood and adolescence, before there is epiphyseal closure, results in *gigantism*. Long bones increase in length, making the affected individuals abnormally tall. When the hyperactivity of the pituitary gland occurs after epiphyseal closure and growth has ceased, the resulting defect is called *acromegaly*. Patients have characteristic facial changes, due to thickening of bones and cartilages; there is also enlargement of the hands, feet, and various organs.

The excess of GH blocks insulin actions, inhibiting glucose transport in muscle and adipose tissue. Patients show altered glucose tolerance tests and hyperglycemia.

GH deficiency. Low levels or lack of GH are observed in several hereditary conditions. Congenital inability to synthesize GH produces serious growth alterations, leading to *pituitary dwarfism*. In these cases, there is a proportionate reduction of body dimensions; intellectual development is not affected.

Because GH from other species is inactive in humans, these patients need to be treated with isolated human pituitary somatotropin. This represented in the past as a serious limitation in the treatment of these patients. Genetic engineering now allows the production of the hormone in vitro, in bacteria to which the GH gene has been incorporated.

In some cases the deficiency includes other pituitary hormones (panhypopituitarism), for example, a condition secondary to mutations in the gene encoding a transcription factor called *Pit-1*. Pit-1 is required for the transcription of several pituitary hormones and for the development of somatotroph, lactotroph, and tyrotroph cells.

There is another clinical condition, called *Laron dwarfism*, in which GH levels in blood are normal or elevated and treatment with GH has no effect. These patients present resistance to GH due to a congenital defect that prevents the expression of the somatropin receptor. In these cases, there is no production of IGF in response to GH.

African pygmies are another example of growth alteration due to partial defects in the GH receptor. Administration of hormone does not increase IGF-I. In contrast, IGF-II is normally produced.

MELANOCYTE STIMULATORY HORMONE

This hormone is secreted by the middle lobe (pars intermedia) of the pituitary gland of lower vertebrates; it is also designated *intermedin*. It acts on pigment cells or melanocytes, darkening the skin of fish and amphibians. It is also present in the adenohypophysis of higher animals; in humans it may target the central nervous system and influence attention and motivation processes.

Three peptides with MSH activity (α, β, and γ) have been isolated from the anterior pituitary, all derived from POMC, which is also a precursor of ACTH and endorphin lipotropins. α-MSH is a peptide of 13 amino acids identical to the amino acid sequence of the N-terminal portion of ACTH. Although the secretion of these hormones is controlled by releasing and inhibiting factors from the hypothalamus, the main inhibitory control is exerted by dopaminergic nerves.

PLACENTA

The placenta produces hormones with actions that are similar to those of some pituitary hormones.

CHORIONIC GONADOTROPIN HORMONE

This placental hormone is present in plasma and urine of pregnant women, where its identification can be used in the diagnostic tests for pregnancy. Chorionic gonadotropin hormone is a glycoprotein with biological and immunological properties similar to those of pituitary LH. It consists of two subunits (α and β). The α chain is almost identical to that of LH, FSH, and TSH.

PLACENTAL LACTOGEN

This hormone has physicochemical and immunological similarities with GH and prolactin. It has many activities, mainly somatotrophic (growth stimulation), mamotrophic (development of mammary gland alveoli and milk production), luteotrophic (maintenance of the corpus luteum), and erythropoietic. After binding to its receptor, placental lactogen activates adenylate cyclase and increases cAMP. Among its metabolic actions are (1) stimulation of lipolysis, with increased levels of circulating fatty acids; (2) inhibition of glucose utilization, resulting in hyperglycemia; and (3) increase of plasma total nitrogen. These metabolic changes elicited by placental lactogen ensure a good supply of glucose, fatty acids, and amino acids to the fetus.

POSTERIOR PITUITARY (NEUROHYPOPHYSIS)

The neurohypophysis produces two well characterized active substances: arginine vasopressin, with antidiuretic and pressor action, and oxytocin, which stimulates the smooth muscle of mammary gland and uterus. Vasopressin and oxytocin are synthesized by secretory neurons in the supraoptic and paraventricular nuclei of the hypothalamus. First, these hormones are synthesized as preprohormones. Then, the signal peptide is removed when the protein enters the ER and prohormones are formed. Pro-oxitocin comprises the hormone itself and an additional piece of approximately 10 kDa called *neurophysin I*. Provasopressin is associated with *neurophysin II*, also of approximately 10 kDa. Prohormones are stored in secretory granules or vesicles and transported along the neuronal axon from the cell body to the nerve endings in the neurohypophysis. Hydrolysis within the vesicle liberates the hormones from the neurophysins. Vasopressin and oxytocin secretion to the circulation is achieved by exocytosis of the vesicles containing the hormones. It is not yet known whether the released neurophysins accomplish any function besides being part of prohormones.

OXYTOCIN

It is a cyclic nonapeptide of a molecular mass of approximately 1000 Da. The first six amino acids (cysteine, tyrosine, isoleucine, glutamine, asparagine, and cysteine) form a cycle, maintained by a disulfide bond between the two cysteines. A three amino acid chain (proline, leucine, and glycinamide) is bound to the last cysteine (Fig. 26.5).

Oxytocin circulates in the blood free; it is rapidly degraded, mainly in liver and kidney, its half-life is 3–5 min.

Oxitocin is also synthesized in uterus, placenta, amnion, corpus luteum, testis, and heart.

The oxytocin receptor is coupled to G_q protein and activates PLC.

Actions. Oxytocin stimulates uterus contraction, an effect that depends on the presence of estrogen and progesterone. Estrogens increase the contractile effect, while progesterone inhibits it. Toward the end of pregnancy, there is a significant reduction of progesterone levels, and the sensitivity to oxytocin is therefore enhanced. Oxytocin increases in blood when labor begins, contributing decisively to facilitate uterine contractions. Although it is believed that oxytocin is not the factor that initiates contractions, pharmacological administration of oxytocin favors the induction of labor.

Oxitocin stimulates the contraction of the smooth muscle (myoepithelium) of mammary gland and helps with milk excretion. Oxytocin secretion is triggered by suction of the newborn on the nipple. Unlike PR, the oxytocin response

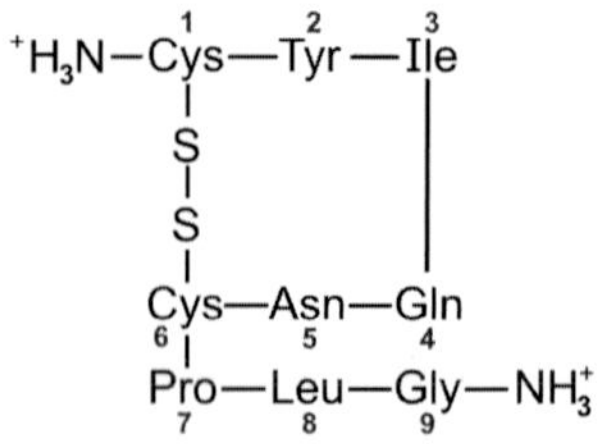

FIGURE 26.5 **Oxytocin.**

does not decline until 6 months postpartum, provided breastfeeding continues. Oxytocin has no action on milk production. Other functions are to stimulate follicle luteinization and steroidogenesis in the ovary. In men, oxytocin stimulates spontaneous erections.

ARGININE VASOPRESSIN

This hormone is also called antidiuretic hormone or ADH. It is a nonapeptide which differs from oxytocin in only two amino acids: isoleucine at position 3 and leucine at position 8 of oxytocin are replaced for phenylalanine and arginine in ADH (Fig. 26.6).

ADH circulates free in blood; its half-life is of approximately 15 min.

Actions. The main function of ADH is to control water reabsorption in the nephron distal tubules and collecting ducts. ADH also decreases glomerular filtration rate, which helps reducing diuresis. ADH along with other factors, such as the renin–angiotensin–aldosterone system, natriuretic peptides, and the sympathetic nervous system, collaborate in maintaining body water volume.

ADH binds to a membrane receptor called V_2 coupled to G_s protein in cells of distal convoluted tubules and medullar collecting ducts. It induces adenylate cyclase activation, increases cAMP level, and stimulates protein kinase A, which induces the phosphorylation and targeting of water channels (aquaporin 2) from intracellular stores to the apical membrane of the renal epithelial cells (p. 231). This allows increase of water absorption from the tubular lumen and the production of a urine with higher osmolality and lower volume.

Besides its antidiuretic action, vasopressin has powerful vasopressor action, increasing total peripheral resistance and blood pressure. This effect is the result of the binding of ADH to V_1 receptors coupled to G_q protein in vascular smooth muscle of blood vessels. ADH activates PLC, releases the second messengers DAG and IP_3, and causes an increase in intracellular Ca^{2+}, which mediates smooth muscle contraction.

Other actions of vasopressin via V_1 receptors are the stimulation of PG synthesis and the activation of liver glycogenolysis. In addition, ADH binds to a third type of receptors, V_3, associated with G_q proteins, which are localized in the pituitary gland. Acting on these receptors, ADH enhances the action of CRH and contributes to the release of ACTH.

Clinical conditions related to ADH. Lack of vasopressin secretion causes a disease called *diabetes insipidus*, which is characterized by excessive diuresis; the patients excrete a high volume of urine (15–20 L/day) with low osmolarity.

Tumors (i.e., craniopharyngioma) and other disorders of the hypothalamus (secondary to pituitary or hypothalamic tumors surgery) that cause diabetes insipidus can cause hypopituitarism.

Conditions have been described with the same symptoms, in which ADH production and hormone levels in blood are normal or even increased, but there is a lack of response to ADH in the kidney. This is called *nephrogenic diabetes insipidus*. It can be seen in chronic renal diseases, particularly those affecting the renal medulla and the collecting ducts. In addition, there are very rare, inherited forms of diabetes insipidus, associated with defects in the gene coding for the V_2 receptor, or with mutations in the gene for aquaporin 2. The inherited forms of diabetes insipidus can be recessive or dominant, and they start during childhood. Other forms are

$$^+H_3N-\overset{1}{Cys}-\overset{2}{Tyr}-\overset{3}{Phe}$$

FIGURE 26.6 **Antidiuretic hormone or arginine vasopressin.**

idiopathic or of unknown cause; they occur in adolescence or adulthood. They do not respond to ADH.

An *inappropriate ADH secretion syndrome* has been described, in which the mechanism of hormone secretion regulation is altered. It occurs in patients with lung cancer, pulmonary tuberculosis, severe trauma, and other conditions, in which the feedback control of ADH is lost and excessive quantities of ADH are secreted.

THYROID

The thyroid gland is composed of follicles with epithelial cells (follicle cells) surrounding a viscous material (colloid) containing the glycoprotein thyroglobulin, which contains within its structure the thyroid hormones. The thyroid hormones comprise thyroxine T_4 and T_3 (the thyroid gland also produces calcitonin, a hormone related to calcium homeostasis. It is discussed in another section of this chapter). Both T_4 and T_3 are iodinated tyrosine derivatives. Their basic structure is thyronine, constituted by two phenolic rings, one of which has an alanine chain (Fig. 26.7).

In T_4, four atoms of iodine are added at positions 3,5,3' and 5'. T_3 has iodine at positions

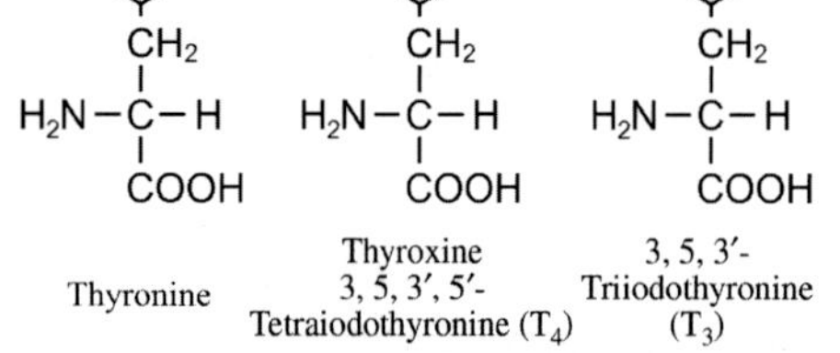

FIGURE 26.7 **Thyronine and thyroid hormones (T_4 and T_3).**

3,5 and 3'; it is eight times more active than T_4 and its action is much faster. The presence of iodine in positions 3 and 5 is essential for thyroid hormone activity. In tissues, 3,3',5'-triiodothyronine is found; this form is called *inverted or reverse T_3* (rT_3), which is devoid of activity.

Biosynthesis of Thyroid Hormones

Iodine is an essential constituent of thyroid hormones. The thyroid gland incorporates in thyroid hormones iodide that mainly comes as inorganic iodide from the diet. Iodide intake should be of 100 μg/day to ensure normal functioning of the thyroid gland.

Iodide absorption and distribution. Iodide from foods is absorbed in the duodenum and transported in blood bound to plasma proteins; iodide concentration in extracellular fluids ranges from 1.0 to 1.5 μg/dL. Approximately two-thirds of the absorbed iodide is excreted via the kidneys. The remaining third is taken up by the thyroid. Very small amounts are secreted by the salivary glands and intestine.

Iodide uptake. Iodide moves into the thyroid gland against its concentration gradient in a process that requires energy. Iodide transport through the basement membrane of the thyroid cells is mediated by a Na^+/I^- symporter, which depends on the activity of the Na^+,K^+-ATPase. Anions as pertechnate (TcO_4^-), perchlorate (ClO_4^-), and thiocyanate (SCN^-) are competitive inhibitors of iodine transport.

The thyroid capacity to "trap" iodide is remarkable; iodine concentration in the gland is 30–40 times higher than in plasma. Thyroid iodine uptake can be studied using radioactive iodine (^{125}I or ^{131}I). Iodide moves from the cells to the follicle lumen by a I/Cl^- exchanger in the apical membrane of the cells called *pendrin*.

Iodide oxidation. In the cell–colloid interface, iodide is oxidized (activated) by *thyroid peroxidase*, a hemoprotein that uses hydrogen peroxide

(H$_2$O$_2$) as oxygen source. The enzyme is present at the apical cell membrane, next to the follicle lumen. H$_2$O$_2$ is generated by an NADPH-dependent oxidase.

Apothyroglobulin. This is a glycoprotein of 660 kDa. It has tyrosine residues in its molecule, located in accessible positions for iodination. The protein is synthesized in the rough ER of thyroid cells, and it is glycosylated in the Golgi apparatus. Apothyroglobulin is secreted into the follicle lumen by exocytosis.

Iodine organification. This process is also catalyzed by thyroid peroxidase. Active iodine is attached in position 3 to the first phenyl ring of the apothyroglobulin tyrosine residues, to form 3-monoiodothyrosinyl (MIT). A second iodine is added at 5-position, to form 3,5-diiodothyrosinyl (DIT). Note that iodine is not bound to free tyrosine, but to those residues forming part of the polypeptide chain of apothyroglobulin thyrosinyl residues. The following step is the coupling of the phenolic group of DIT or MIT on a DIT in the chain, to form T$_4$ or T$_3$, respectively, which remains included as amino acid residues of the protein (Fig. 26.8). The coupling produces a kinol–ether intermediary that then gives iodothyronine. After iodination, apothyroglobulin becomes thyroglobulin with T$_4$, T$_3$, MIT, and DIT residues as constituents of its structure; there is 10 times more T$_4$ than T$_3$. Thyroglobulin is stored in the follicles, integrating the colloid.

The phenolic moiety is separated from the polypeptide, leaving an alanyl residue of the thyrosinyl donor as dehydroalanine. Some of the iodinated tyrosine residues are close in the thyroglobulin chain, which facilitates the transfer of iodinated phenolic moiety from one DIT to another.

Secretion of T$_3$ and T$_4$. Stimuli which trigger the release of TSH initiate the mobilization of thyroglobulin contained in the colloid. The luminal membrane of the thyroid cell projects membrane extensions that engulf colloid droplets and internalize them by pinocytosis. Fusion of the vesicles with lysosomes forms secondary lysosomes

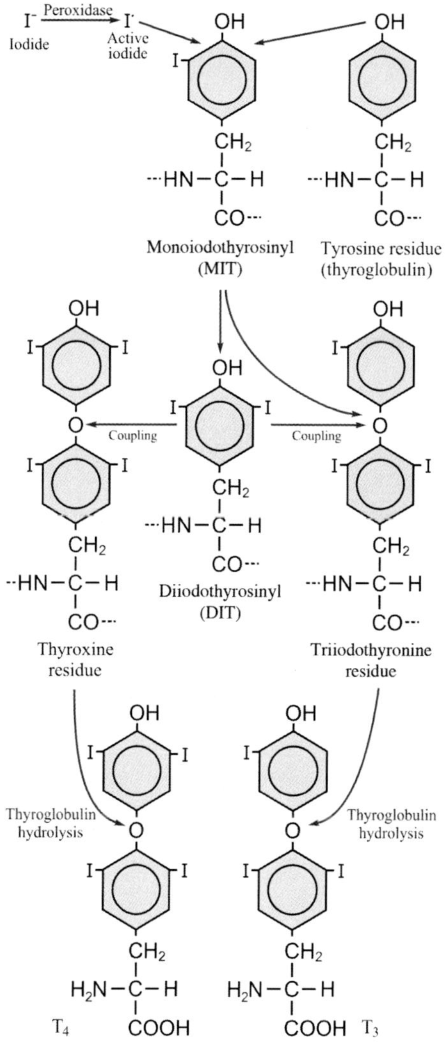

FIGURE 26.8 **Biosynthesis of thyroid hormones.**

where hydrolysis of thyroglobulin takes place. The products are free amino acids, including thyroxine T$_4$, T$_3$, DIT, and MIT. DIT and MIT do not have biological activity, if they were released from the cell, a large amount of iodine would be lost. This does not occur, thanks to the presence of *deiodinase* (there are three types with different cellular distribution), which separate iodine molecules from free MIT and DIT, to reuse it for the synthesis of thyroid hormone.

Free T_3 and T_4 are secreted through the cell basal membrane to reach the blood capillaries. All stages of thyroid hormone synthesis and secretion are activated by the hypophysis TSH. T_3 and T_4 levels in circulating blood have regulatory action on the pituitary gland via negative feedback.

Compounds belonging to the thiocarboamides, such as propylthiouracyl, are inhibitors of thyroid peroxidase; they block oxidation and I organification, and also act on peripheral tissues, where they inhibit 5'-monodeiodase that catalyzes conversion of T_4 to T_3, a more active product. Methimazole, and carbimazole, are other compounds with antithyroidal effect; they inhibit iodine oxidation and organification, but do not affect peripheral deiodination.

Excess iodine administration to normal individuals can inhibit organification and thyroid hormone synthesis. This phenomenon, called *Wolff–Chaicoff effect*, is only temporary. After several days, even if the high intake of iodide continues, thyroid hormone production is regained.

Transport in plasma. T_3 and T_4 are transported in plasma in association to blood plasma proteins. Two proteins act as specific thyroid hormone transporters. One is *thyroxine-binding globulin* (TBG), a 50-kDa glycoprotein with electrophoretic mobility intermediate between α_1 and α_2 globulins, and another is thyroxine binding prealbumin, also called *transthyretin* (TTR). TBG has a higher affinity for T_3 and T_4, binding 70%–75% of the total T_4 associated with plasma proteins; TTR instead binds only approximately 10% of circulating thyroid hormones. The affinity of TBG and TTR for T_3 is around 10 times higher than for T_4. Approximately 15% of T_4 and T_3 in plasma are also bound to albumin. This fraction rapidly dissociates and is a good source of free hormone for tissues. The concentration of T_4 in plasma is 8 µg/dL (100 nmol/L) and that of T_3 is 50 times lower, 0.12 µg/dL (0.18 nmol/L). Only a very small proportion of T_3 and T_4 are free, in dynamic equilibrium with the protein-bound hormones. Almost all of T_4 (99.4%) in plasma is

bound, only 0.04% is free. For T_3, the free hormone is 0.4%. The free form is readily available to the cells, and is the functionally important hormone. Homeostatic mechanisms regulate the free hormone and not the total. Estrogen stimulates the synthesis of TBG; during pregnancy, the TBG levels increase and total T_4 concentration may be duplicated without significant increase in the amount of free hormone. This should be taken into account to avoid wrong diagnoses of hyperthyroidism from total hormone measurements in plasma.

The half-life of T_4 in plasma is approximately 7 days, while the T_3 is 1 day.

Access to peripheral tissues. Although they are hydrophobic molecules, thyroid hormones poorly cross the plasma membrane of the target cells. Most of their uptake depends on plasma membrane transporters. Once in the cells, T_4 loses iodine 5' or 5 in a reaction catalyzed by *5' or 5 monodeiodase*, producing T_3 and T_3r (T_3 reversed, 3,3',5'-triiodothyronine). T_3 is significantly more active than T_4, and is considered the real thyroid hormone, while T_4 behaves as a prohormone. T_3r has virtually no activity. The rate at which T_4 is converted to T_3 or T_3r varies from one tissue to another, even in the same tissue it changes according to their metabolic state. The conversion of T_4 to T_3r (inactive) is proportionally greater in old age, malnutrition, fever, serious chronic diseases, cancer, and severe burns. Apparently, this represents a compensatory mechanism to reduce stimulation of catabolism promoted by thyroid hormones.

Degradation. The primary pathway for thyroid hormone degradation is deiodination, initiated by the selenoenzymes *5'-monodeiodase*, which produces T_3, the most potent compound, and 5-monodeiodase, which produces the inactive T_3r. T_3 is converted to T_2, T_1, and thyronine.

Some of the products of thyroid hormone metabolism previously considered to be inactive, have now been recognized to have activity, and have been named "nonclassical THs." For example, T_2 (3,5-diiodothyronine) in mammals induces the expression of several genes regulated by T_3; doses of T_2 100-fold greater than those of T_3 are required to obtain the same effects.

Another catabolic pathway for thyroid hormone occurs in liver, where T_4 and T_3 are conjugated with glucuronic acid and, to a lesser extent, with sulfate. In both cases, the phenolic hydroxyl group is esterified. The products are excreted with the bile. Part of the conjugated thyroid hormone is reabsorbed and transported to the kidney, where it can be excreted intact or free of iodine.

A third metabolic alternative involves the decarboxylation and deamination of T_4 and T_3 to form tetra-iodothyroacetic acid (*Tetrac*) and tri-iodothyroacetic (*Triac*), respectively. These products are considerably less active than their precursors.

Mechanism of action. Thyroid hormones cross the cell plasma membrane by diffusion, and mainly by facilitated transport. They reach the nucleus, where they bind to two specific receptors, α and β. The α receptor is the predominant type in brain, β is expressed in liver, and both are represented in similar proportions in the myocardial cells. They belong to the steroid receptors superfamily and have structural homology with them. Thyroxin receptors have an affinity of approximately 10 times greater for T_3 than for T_4. The receptors are found as homodimers or heterodimers, in association with the retinoid receptor (RXR). These receptors are attached to response elements in the DNA and to a corepressor that depresses gene transcription. When T_3 or T_4 binds to the RXR receptor, conformational changes in the receptor displace the corepressor and enable the hormone-receptor complex to activate transcription of defined genes. The effects on transcription are not immediate; they require hours or days to attain its maximum expression.

Actions. Thyroid hormone actions are so general that virtually no cell in the body is free from its influence. Among the effects of T_3 or T_4 are (1) activation of messenger, ribosomal and transfer RNA synthesis, and protein synthesis; (2) stimulation of cell growth; (3) increase in Na^+,K^+-ATPase activity in almost all cells (except brain, spleen, and testis). This effect is due to the increased number of Na^+,K^+-ATPase molecules inserted in the cell membrane, due to enhanced synthesis and targeting from intracellular stores, (4) augment in oxygen consumption in tissues and increase in glucose, lipid, and amino acid utilization, and (5) stimulation of cholesterol and LDL receptor synthesis, which results in more cholesterol being taken up from plasma.

Broadly speaking, at physiological concentrations thyroid hormones have anabolic effects. They increase the basal metabolic rate and induce synthesis of mitochondrial oxidoreductases, including glycerophosphate dehydrogenase, succinate dehydrogenase, and cytochrome oxidase. In contrast, at high doses, thyroid hormones have catabolic action, proteolysis is intensified, the nitrogen balance becomes negative, and carbohydrate and lipid degradation increases. Apparently, thyroid hormones exert an uncoupling effect, with increased oxygen consumption and heat production, decrease in ATP production, and reduction of the P/O ratio. The activity of superoxide dismutase is reduced, resulting in possible increase of superoxide anion radicals. This may contribute to the deleterious effects of hyperthyroidism. It has been proposed that the thermogenic action of thyroid hormones is due to increased activity of Na^+,K^+-ATPase, resulting in augmented energy expenditure in resting cells. Increased number of β-adrenergic receptors and operation of futile metabolic cycles can also contribute to heat release. Often, thyroid hormones activate simultaneously some metabolic pathways in both directions, anabolic and catabolic.

Actions on muscle. Thyroid hormones increase muscle contractility, through their combined effects on Na^+,K^+-ATPase, myosin ATPase and Ca^{2+}-ATPase activation, and Ca^{2+} increase in sarcoplasmic reticulum and in the cell. On heart muscle they have positive chronotropic and inotropic effects. Diastolic tone and number of β-adrenergic receptors increase.

Actions on adipose tissue. T_3 and T_4 have a role in the development and functioning of the white and brown adipose tissues. They induce adipocyte differentiation.

Actions on bone. Thyroid hormones increase bone resorption and, to a lesser extent, new bone formation.

Actions on the nervous system. T_3 and T_4 are essential for nervous system development, especially from the 6th month of pregnancy until 2 years of life. They regulate neuronal differentiation, myelin formation, neuronal growth, and synapses formation. Owing to the high content of 5-deiodinase in placenta, T_4 and T_3 are inactivated, and the amount of thyroid hormones is markedly reduced in fetal circulation. In the early weeks of gestation, this does not significantly affect development, but after the 11th week, the fetus is totally dependent on the secretion of its own thyroid hormone. In the absence of T_4 and T_3, there is reduced growth and development, with the brain and skeletal muscle being the most affected. This leads to *cretinism*, characterized by mental retardation and dwarfism.

Nongenomic actions. Although the main actions of T_3 and T_4 are mediated by nuclear receptors and influence gene expression, other effects have been described. Different from the genomic effects, T_4 and T_3 trigger responses in seconds or minutes, though signal transduction systems not yet well identified. Examples of these actions include increased transport of glucose and amino acids, activation of plasma membrane and sarco- and endoplasmic reticulum Ca^{2+}-ATPase, activation of adenylate cyclase and pyruvate kinase, and reduced activity of 5'-deiodinase. T_3 and T_4 depress cytochrome oxidase sensitivity to inhibition by ATP, which accounts for its rapid action on mitochondrial respiration. Table 26.2 presents a summary of the physiological actions of thyroid hormones, and the effects resulting from its deficiency and excess.

Hypothyroidism. There are congenital causes of hypothyroidism produced by alterations in the genes that encode enzymes involved in the biosynthesis of thyroid hormones. The insufficient development or congenital absence of thyroid activity produces *cretinism*, with growth disorders (dwarfism) and mental retardation. If

TABLE 26.2   Actions of Thyroid Hormones

	Hypothyroidism	Normal	Hyperthyroidism
Basal metabolism	↓ Basal metabolism	Normal	↑ Basal metabolism
Proteins	↓ Synthesis ↓ Degradation	Anabolic	↑ Degradation ↑ Creatinuria
Lipids	↓ Synthesis ↓ Degradation ↑ Serum cholesterol	↑ β-Oxidation ↑ Lipolysis ↑ Lipogenesis	↑ Synthesis ↑ Degradation ↓ Serum cholesterol
Glucose	↓ Glycolysis	Normal	↑ Glycolysis Abnormal test of glucose tolerance
Glycogen	↓ Glucogenolysis ↑ Glycogen store	Normal	↑ Glucogenolysis ↓ Glycogen store
Simpathetic nervous system			Mimmetizes effects of β-adrenergic stimulation
Cardiovascular system	↓ Amplitude ECG waves	↑ Cardiac frequency ↑ Volume minute ↑ Contractility	↑ Amplitude ECG waves

hypothyroidism is initiated in adulthood, one of the most striking symptoms is *myxedema*, which is produced by the accumulation of glycosaminoglycan and water in the subcutaneous tissue, with thickening and "pasty" appearance of the skin. Table 26.2 summarizes symptoms that accompany hypothyroidism.

Untreated congenital hypothyroidism produces severe and irreversible brain damage. The incidence of congenital hypothyroidism is 1 in 4000–5000 births. This requires early diagnosis through systematic screening programs in all newborns to start treatment within the first few months of life.

Simple *endemic goiter* or endemic hypothyroidism is a condition caused by inadequate supply of iodine in food. In some regions, usually far from the sea and surrounded by mountains, iodine content in soil and water is very poor and, therefore locally produced foods lack this element. Iodine deficiency in the diet depresses the synthesis of thyroid hormones and causes thyroid gland hypertrophy as a compensatory phenomenon, with T_3 and T_4 deficit. This disorder is also called *regional endemic goiter*.

It is estimated that 1 billion people in the world live in iodine deficient areas. When iodine lacks during pregnancy, there is irreversible damage to the fetus central nervous system that causes cretinism.

The addition of potassium iodide to table salt of common use is the most effective approach to eliminate iodine deficiency.

Another cause of hypothyroidism is caused by autoimmune disease, in which the patients produce antibodies against their own thyroid (*Hashimoto's disease*).

Hyperthyroidism. Table 26.2 shows the disturbances caused by excessive thyroid hormone activity. The hyperthyroid goiter presents with increased size of the thyroid gland and hyperfunction. In the so-called *toxic goiter* or *Graves' disease*, there is excessive stimulation of the thyroid gland, which results in thyroid hyperfunction. This is not due to increased pituitary TSH, which is actually markedly diminished by negative feedback, due to high levels of T_3 and T_4. The disease is autoimmune; affected patients produce thyroid-stimulating immunoglobulin G, which targets the TSH receptor and has an effect similar to that of TSH. Originally, these antibodies were designated as *long-acting thyroid stimulators*.

ADRENAL GLAND

The adrenal gland is composed of two distinct parts: the cortex, of mesodermal origin, which produces steroid hormones, and the medulla, functionally associated with the autonomic nervous system, which produces catecholamines identical to those released by the adrenergic system.

ADRENAL CORTEX HORMONES

Three distinct regions can be distinguished in the cortex: (1) external or *glomerular*, (2) media or *fascicular*, and (3) inner or *reticular*. Although these different portions produce different types of hormones, the fascicular and reticular zones can be considered as a unit. The adrenal cortex produces hormones which are all structurally related to the cyclopentaneperhydrophenanthrene. According to their general functions, the hormones of the adrenal cortex are grouped into three types:

1. *Glucocorticoids*, which exert primary action on carbohydrate, lipid, and protein metabolism. They are produced predominantly in the fascicular zone and to a lesser extent in the reticular zone.
2. *Mineralocorticoids*, which primarily function on electrolyte and water control. They are secreted by the cells of the glomerular zone.
3. *Androgenic steroids*, which are involved in development and maintenance of secondary sexual characters. They are preferentially

synthesized in the reticular zone and also in the fascicular zone.

This classification is based on the major functional properties of each hormone; however, although the hormones of each group exhibit a dominant specific effect, they also have secondary effects, which are shared with those of the corticoids from the other group.

The main steroids produced by the adrenal cortex include the glucocorticoids *cortisol* or cortisone and *hydrocortisone*, the mineralocorticoids *deoxycorticosterone* (DOC) and *aldosterone*, and the androgens *androstenedione* and *dehydroepiandrosterone*. Only cortisol, aldosterone, and corticosterone are found in circulating blood.

Glucocorticoids and mineralocorticoids have a common structural feature (Fig. 26.9), consisting

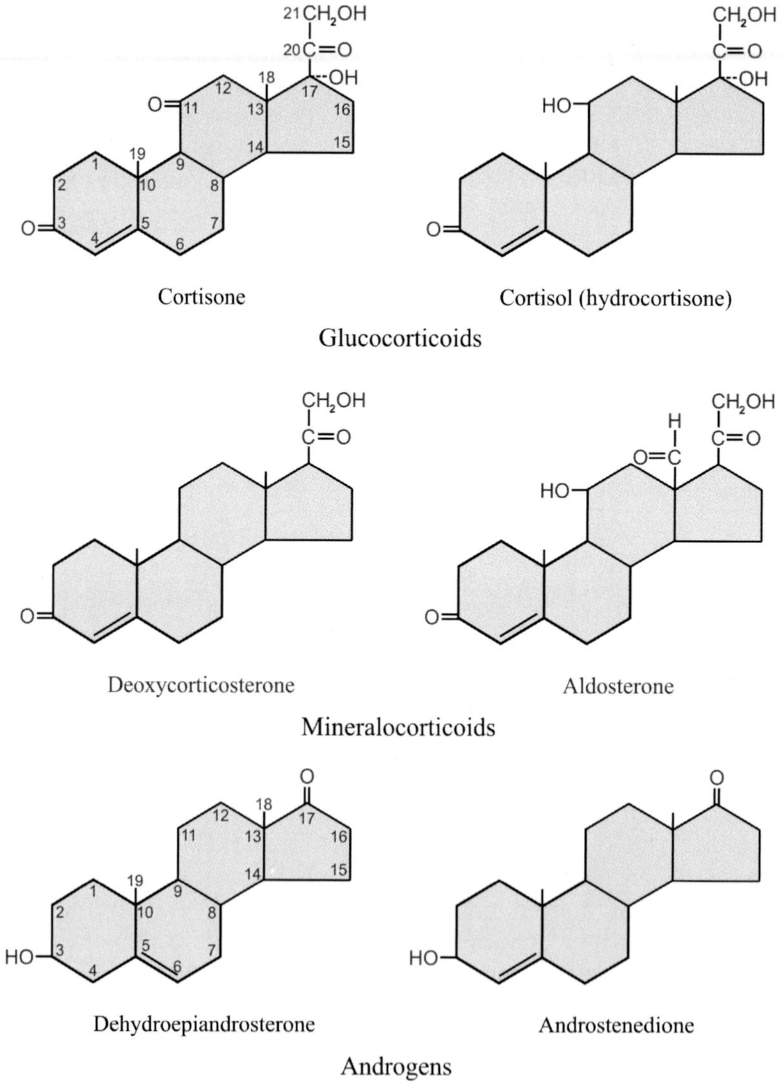

Cortisone   Cortisol (hydrocortisone)

Glucocorticoids

Deoxycorticosterone   Aldosterone

Mineralocorticoids

Dehydroepiandrosterone   Androstenedione

Androgens

FIGURE 26.9 **Hormones of the adrenal cortex.**

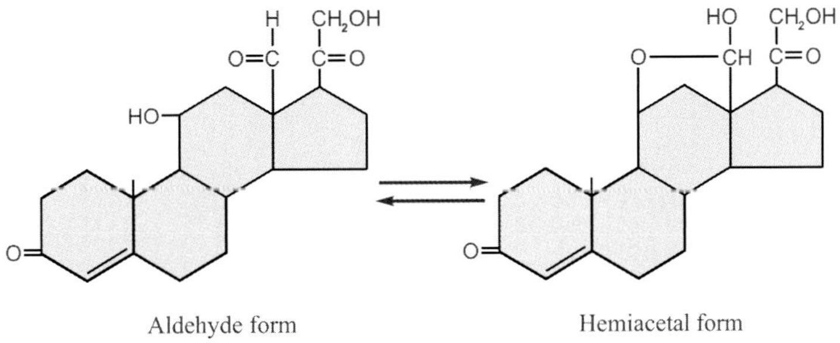

FIGURE 26.10 **Different chemical structures of aldosterone.**

of (1) a ketone function at C3, (2) a double bond between carbons 4 and 5, and (3) two-carbon side chains at C17. In both hormone types, this chain has a ketone and a primary alcohol function. The presence of a hydroxyl group at C17 is exclusive of glucocorticoids. Furthermore, glucocorticoids always have a hydroxyl or ketone group on C11.

Aldosterone has an aldehyde function in C18. Androgens do not have side chain; C17 has a ketone function. The double bond in the cycle is located between carbons 4 and 5 or 5 and 6 (Fig. 26.9).

Cortisol is the major glucocorticoid in humans. The most potent mineralocorticoid is *aldosterone*, which may adopt aldehyde or hemiacetal forms (Fig. 26.10). In solution, aldosterone is found as hemiacetal. The presence of a hydroxyl in C11 gives it some glucocorticoid activity.

Biosynthesis of Adrenal Cortex Hormones

The initial stage of steroid synthesis is the production of cholesterol from active acetates. The adrenal cortex is relatively rich in cholesterol, approximately 80% of the total cholesterol that it uses can be synthesized in the gland itself; the rest comes from plasma LDL and captured by the cells by LDL receptors on the cell plasma membrane. Different acute stimuli induce StAR protein, which mediates cholesterol transport in mitochondria from the outer membrane to the matrix.

In the early steps of steroids synthesis, cholesterol is hydroxylated at carbons 20 and 22;

later, side chains are cleaved by the action of *20,22-desmolase*. Desmolase is induced by ACTH in the fascicular and reticular and by angiotensin II in the glomerular zones of the gland. A six-carbon moiety (isocaproate) is separated, and a two-carbon residue with a ketone function is left on C17. A multienzyme complex located in mitochondria leads to the synthesis of *pregnenolone*.

Pregnenolone serves as an intermediary in the synthesis of all steroid hormones. The following transformations mostly involve dehydrogenation and hydroxylation steps, catalyzed by mixed function oxygenases, linked to cytochrome P_{450}. These enzymes require molecular oxygen and NADPH as coenzymes. A specific dehydrogenase catalyzes the oxidation of C3. An isomerase changes the double bond between carbons 5 and 6 to carbons 4 and 5. These reactions convert pregnenolone into progesterone.

Mineralocorticoids and glucocorticoids are synthesized from progesterone. The following reactions are catalyzed by specific hydroxylases that add hydroxyl groups in C11, C17, or C21. The enzymes involved in these biosynthetic pathways are found in mitochondria and ER. Hydroxylation at C21 occurs in all compounds with gluco- and mineralocorticoid activity.

The synthesis of aldosterone instead requires hydroxylation at C18. C18 hydroxylase activity is restricted to the glomerular layer cells.

The glomerular zone producing aldosterone lacks 17α-hydroxylase, so it cannot synthesize 17α-OH-pregnenolone or 17α-OH-progesterone,

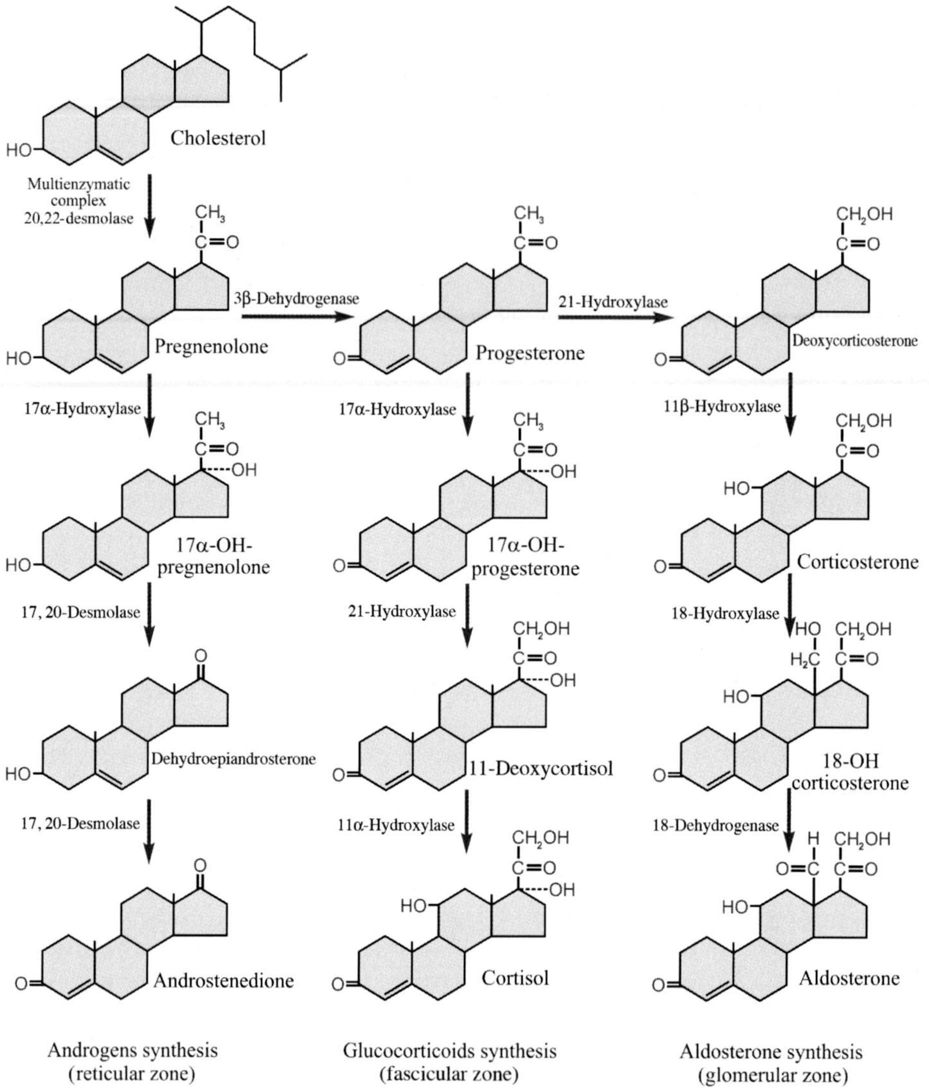

FIGURE 26.11   **Biosynthesis of adrenal steroids.**

which are cortisol and androgen precursors, respectively.

The primary adrenal androgen, *dehydroepiandrosterone*, is produced by side chain removal and oxidation in C17, with a ketone formation. Fig. 26.11 summarizes the synthetic routes of the main adrenal steroids.

Transport in blood plasma. Once synthesized, steroids pass immediately to the blood. A normal adult secretes daily between 5 and 30 mg of cortisol, 1 and 6 mg of corticosterone, and 50 and 100 µg of aldosterone.

Cortisol is the most abundant steroid in plasma, with a concentration from 5 to 15 µg/dL.

Approximately 10% of this amount is free and biologically active. Aldosterone concentration is much lower (0.02 µg/dL). Adrenocortical hormones are transported in part linked to plasma proteins. Binding to protein is stronger as the steroid polarity decreases. For example, androstenedione, with two ketone functions, forms a stronger complex with the transport proteins than the polar molecule of cortisol (with two ketones and three hydroxyl groups). The binding is noncovalent (hydrophobic interactions and hydrogen bonds) and is readily reversible under physiological conditions.

Transcortin (corticosteroid-binding globulin, CBG) and serum albumin bind steroids and carry them in blood. Cortisol and corticosterone preferentially bind to transcortin (CBG), 77% of blood cortisol is normally linked to CBG and 15% to albumin; approximately 7% remains free in plasma (1 µg/dL). Cortisol has a half-life of 70–120 min.

Aldosterone shows less binding to CBG, large part (30%–50% of the total) is free. Free hormone is the biologically active form and is in equilibrium with that bound to protein. Its half-life is 10–20 min.

Inactivation and excretion. Steroids in circulation are rapidly eliminated from the body; virtually all of them are excreted by 48 h. The liver is the main organ responsible for corticosteroids inactivation. They undergo reduction of the double bond, the ketone group at C3 is converted to an hydroxyl group, and conjugated with glucuronic acid or, to a lesser extent, with sulfate. Conjugated corticosteroids are excreted with the bile into the intestine, where they are partially reabsorbed and returned to the liver via the portal circulation (enterohepatic cycle). Main excretion is performed by urine. Androgenic steroids metabolites are excreted in urine as 17-ketosteroids, conjugated with glucuronic acid. Urinary 17-ketosteroids derive from adrenal androgens and from those produced in the gonads. In women and children, all 17-ketosteroids in urine come from the adrenal cortex. In the male, the excreted amount is greater because androgens from testicular hormones are added. The testes contribute to one-third of the total urinary 17-ketosteroids.

Synthesis and secretion regulation. The synthesis and secretion of adrenal steroids is stimulated by ACTH from the anterior pituitary. This hormone acts preferentially on the fascicular zone of the adrenal gland. ACTH secretion, in turn, is regulated by the CRH and arginine vasopressin produced in the hypothalamus, and by stressors. The level of circulating adrenal steroids also controls, by feedback, the secretion of ACTH. The levels of plasma ACTH and cortisol present parallel variations. Both have a circadian rhythm; they increase during night to reach a maximum in the last hours of sleep, and then gradually decline throughout the day until after dark. There are small increases produced by food ingestion or exercise. Stress, both physical and psychic, also raises ACTH and cortisol circulating levels. Adrenal stimulation by ACTH causes rapid decrease of cholesterol and ascorbic acid content in the gland. This phenomenon is an indication of activation of steroid synthesis. The role of ascorbic acid or vitamin C, abundant in adrenal cortex, is not exactly known. Its function may be related to its capacity of providing reducing equivalents for NADPH-dependent hydroxylation reactions.

Cortisol secretion is inhibited by somatostatin, heparin, atrial natriuretic factor, and dopamine. Aldosterone production in the adrenal gland is stimulated by angiotensin II (p. 628), potassium, and, to a lesser extent, ACTH. ACTH binds to membrane receptors of the adrenal cortex cells and causes activation of adenylate cyclase. Thus, its action is mediated by cAMP.

Mechanism of action. All steroids have primary action on the cell nucleus, stimulating gene transcription. Hormones of the adrenal cortex hormones cross the cell membrane by simple diffusion and bind to specific proteins of the steroid receptor family. These receptors are localized in the cytoplasm and the nucleus, forming inactive complexes with heat shock proteins (HSP90).

They are ~94 kDa proteins, which contain three main domains. The central domain has two "zinc finger" motifs that recognize and bind to response elements in the DNA, the N-terminal domain also interacts with DNA, and the C-terminal domain is the hormone binding site. Corticoid-receptor binding causes the separation of HSP from the receptor and receptor dimerization. This triggers the shift of the receptor to the nucleus, where it binds to defined sequences in the DNA (next to promoter or enhancer sites) and functions as a transcription factor (Fig. 25.2). The corticoid receptor interacts with other transcription factors such as nuclear factor κB, an important regulator of cytokine genes. It stimulates the synthesis of specific mRNAs and proteins. The metabolic effects of corticosteroids are largely dependent on changes in the synthesis of enzymes.

Two types of receptors for adrenal cortex hormones (for glucocorticoids and for mineralocorticoids) have been recognized. Glucocorticoid receptors are specific for the hormones of the group; however, the mineralocorticoid receptor (MR) has equal affinity for cortisol and aldosterone. Despite this relative lower specificity, and the concentration of cortisol being two to three orders of magnitude higher than that of mineralocorticoids, target tissues (kidney, colon, and salivary glands epithelia) respond to aldosterone. This is explained by the existence in those tissues of 11β-hydroxysteroid dehydrogenase, which converts cortisol into an inactive derivative, cortisone.

Metabolic actions (glucocorticoids). These hormones increase glucose, free fatty acids, and amino acids in plasma. In peripheral tissues (skin, adipose and connective tissue, and muscle), they depress the pathways for glucose utilization, mainly glycolysis, and stimulate protein degradation (they have catabolic action). High amounts of cortisol inhibit DNA synthesis and cell division; they decrease collagen synthesis in the skin and protein synthesis in connective tissue and muscle, which results in their atrophy.

In adipose tissue, they activate lipolysis, promoting the release and degradation of free fatty acids. Paradoxically, despite the enhanced lipolysis, a sustained glucocorticoid excess increases fat deposition. This is explained by the effects of glucocorticoids on the activation of adipocyte differentiation and the genes that control lipoprotein lipase, glycerol-3-phosphate dehydrogenase, and leptin synthesis. High levels of cortisol stimulate appetite and are accompanied by hyperinsulinemia.

In liver, glucocorticoids stimulate the synthesis of enzymes involved in amino acid metabolism, such as aminotransferases, tryptophan pyrrolase, and key enzymes in the regulation of gluconeogenesis (pyruvate carboxylase, phosphoenolpyruvate carboxykinase, fructose-1,6-bisphosphatase, and glucose-6-phosphatase). These effects result in activation of gluconeogenesis from amino acids. Stimulation of liver gluconeogenesis, decreased glucose uptake in peripheral tissues, and increased catabolism of lipids and proteins favor the elevation of circulating glucose level (hyperglycemia). There is also increased liver deposition of glycogen (Table 26.3).

Glucocorticoids have a permissive effect on other hormones, such as catecholamines and glucagon, which result in insulin resistance.

TABLE 26.3   Metabolic Actions of Glucocorticoids (Cortisol)

Peripheral tissues (muscle, adipose, lymphoid)

↓ Glycolysis and glucose consumption
↑ Protein degradation
↓ Protein synthesis

Adipose tissue

↑ Lipolysis
↓ Fatty acid and triacylglycerol syntheses

Liver

↑ RNA and protein syntheses
↑ Gluconeogenesis from amino acids
↑ Urea production

Blood pressure. Elevated concentrations of glucocorticoids increase vascular smooth muscle sensitivity to pressor agents, including catecholamines and angiotensin II, reduce the vasodilator effect of nitric oxide (NO), and increase the synthesis of angiotensinogen.

Inflammation and immunity. Glucocorticoid levels normally present in blood have inhibitory action on the activity of inflammatory cells and the immune system. At concentrations above the physiological level, glucocorticoids have a potent antiinflammatory effect and are immune response depressors. They stimulate apoptosis of peripheral blood lymphocytes, reduce the number of eosinophils, prevent the differentiation of monocytes into macrophages, and reduce macrophage phagocytic and cytotoxic activity.

These actions of glucocorticoids are exerted primarily by repressing genes that encode various cytokines (such as interleukin-1 and tumor necrosis factor) and proinflammatory enzymes (inducible isoform of cyclooxygenase [COX-2] and nitric oxide synthase, NOS). Also, glucocorticoids induce genes that direct the synthesis of antiinflammatory proteins and peptides (protease inhibitors, lipocortin 1, and annexin). The cortisol-receptor complex binds to nuclear factor κB and prevents the initiation of the inflammatory process.

Other effects of cortisol have been proposed. These include inhibition of phospholipase A_2, which decreases arachidonate release, an eicosanoid precursor that is essential for the production of PGs and related compounds. Cortisol also stabilizes lysosomal membranes and inhibits the release of enzymes that contribute to the inflammatory process.

High doses of cortisol inhibit the body defense mechanisms against infection, especially cellular immunity (decreases the number of circulating T lymphocytes). Antibody production by B lymphocytes is not greatly affected. Owing to their ability to inhibit the immune response, glucocorticoids are applied in clinics to reduce the consequences of unwanted or exaggerated immune reactions (allergic and anaphylactic reactions), in chronic inflammatory conditions (rheumatoid arthritis), or to depress the rejection reaction of transplanted organs.

Calcium homeostasis. Glucocorticoids decrease the intestinal absorption and renal reabsorption of calcium, which tends to reduce plasma calcium levels and stimulate the secretion of parathyroid hormone. As a result, glucocorticoids cause bone resorption, which explains the characteristic osteopenia and osteoporosis that accompany cortisol excess. Bone formation, fibroblast proliferation, and collagen formation are depressed.

Nongenomic actions. In general, the most important actions of steroid hormones are exerted through nuclear receptors, modifying gene activity. This requires time, and the response attains its maximum after several hours, or even days. Glucocorticoids can also exert some quick responses that are nongenomic, such as activation of nitric oxide synthesis, which results in vascular dilation.

Mineralocorticoids. The function of these hormones is to maintain normal body concentrations of Na^+ and K^+ and extracellular fluid volume. The response is mediated by mineralocorticoid receptors (MR), which have a tissue specific pattern of expression. For example, higher concentrations of these receptors are located in the distal nephron, colon, and hippocampus. The tissues with the lowest expression of MR are the myocardium and the gastrointestinal tract, except the colon.

Although mineralocorticoids increase Na^+ reabsorption in the renal tubules, they decrease Na^+ excretion in sweat cells, salivary glands, and gastrointestinal tract.

Aldosterone is the most potent corticosteroid acting on sodium retention. It is ~35 times more active than 11-DOC and ~1000 times more than cortisol. It acts on the distal nephron (last portion of the distal convoluted tubule and collecting ducts). Aldosterone induces the expression of Na^+ channels in the apical and of Na^+,K^+-ATPase in the basolateral membrane of renal tubular

TABLE 26.4   Metabolic Actions of Mineralocorticoid (Aldosterone)

↑ Sodium and chloride absorption in renal tubules
↑ Potassium urinary excretion
↑ Volume of extracellular fluid
↑ Synthesis of RNA and proteins

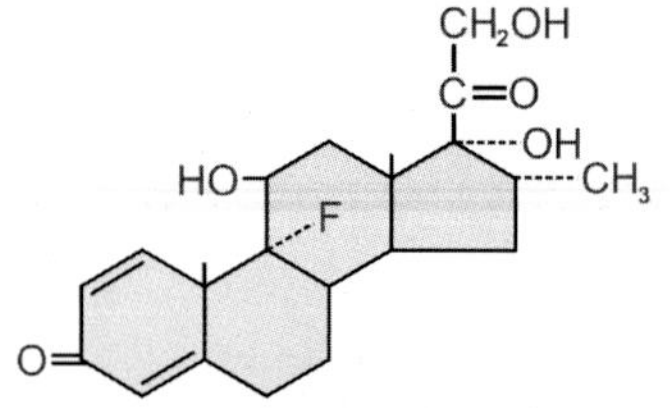

Dexamethasone

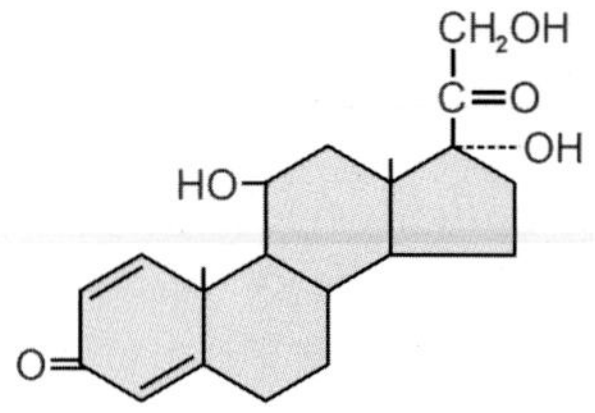

Prednisolone

FIGURE 26.12   **Synthetic corticosteroids.**

cells from the distal tubule and collecting ducts. As a result of the increased Na^+ absorption, negativity in the lumen increases, and K^+ secretion through K^+ channels is activated. The increase in Na,K-ATPase activity favors this K^+ secretion. Mineralocorticoids are involved in regulation of sodium and potassium distribution in the intra- and extracellular spaces. As the movement of ions is accompanied by water displacement, mineralocorticoids are an important factor in regulating body water volume (Table 26.4).

The lack of mineralocorticoids determines excessive loss of sodium, chloride, and water in the urine, with retention of potassium in the extracellular space. Hyponatremia with hyperkalemia occurs and acid-base balance is disturbed. The loss of sodium and water reduces plasma volume and blood pressure. The electrolyte disbalance can lead to coma and death.

Among the mineralocorticoids induced proteins are enzymes of the citric acid cycle, especially citrate synthase. Greater availability of enzymes activates tricarboxylic cycle operation and the production of ATP, which is used by the sodium pump.

Nongenomic actions. Aldosterone can bind to different receptors and activate signal transduction systems that increase cAMP and intracellular calcium concentration $[Ca^{2+}]_i$. These effects mediate the increase in contractility that aldosterone causes in skeletal muscle and its inotropic effects in the myocardium. These actions are manifested within minutes.

Androgenic steroids. The main steroids in this group are dehydroepiandrosterone and androstenedione (Fig. 26.11). They have weak androgenic and protein anabolic action and can be converted, in peripheral tissues, into testosterone and even estrogens. They do not have significant effect on Na^+, K^+, and water regulation. The adrenal cortex is the main source of androgens in women; when produced in excessive amounts, they can cause masculinization.

Synthetic corticosteroids. Synthetic steroids with higher activity than the natural hormones have been obtained. The addition of fluorine in C9, or a double bond between carbons 1 and 2, or a methyl in position 16, increases the affinity for the receptor. Several of these compounds are widely used in clinical practice. Fig. 26.12 shows dexamethasone and prednisolone, the first is ~30 times more potent than cortisol. Other synthetic derivatives function as competitive inhibitors. Spironolactone (aldactone) is an aldosterone antagonist; it binds to the aldosterone receptor inactivating the hormone-receptor complex.

Adrenocortical Hormones Alterations

Decreased function. Adrenocortical hormone deficiency produces Addison's disease. Patients exhibit sodium and chloride loss and potassium retention, with acid base disbalance (hyperkalemic acidosis). In addition, it is manifested by hypoglycemia, hypothermia, increased skin pigmentation, muscle weakness, and cachexia. Owing to the low corticosteroid level in blood of these patients, the adenohypophysis increases secretion of ACTH. This explains the hyperpigmentation because ACTH and an ACTH precursor molecule originate MSH.

Increased function. The excess of adrenocortical hormones is observed in hyperplasia of the gland or functioning tumors of the adrenal gland, and in cases of excessive secretion of ACTH (e.g., in adenohypophysis basophil cells tumor). Overproduction may be limited to glucocorticoids, mineralocorticoids, or androgens.

Cushing syndrome. This disorder is caused by glucocorticoid excess. It is characterized by hyperglycemia, impaired hydro mineral balance, adiposity ("full moon" face), osteoporosis, and muscle atrophy.

Primary aldosteronism. An aldosterone excess increases potassium urinary excretion, hypokalemia, and total body potassium reduction. A hereditary form of pseudoaldosteronism or *apparent mineralocorticoid excess* has been described. In this disease, 11β-hydroxysteroid dehydrogenase, that converts cortisol to cortisone, is inactive. Cortisol binds to mineralocorticoid receptors (MR), while cortisone has little affinity. When the enzyme is missing, there is a relative cortisol excess not inactivated to cortisone, which binds to MR and exerts its action.

Adrenogenital syndrome. The excessive production of androgens causes virilization in women (they develop male secondary sex characteristics) and early pseudo puberty in males. The defect is genetic. It affects enzymes implicated in adrenocortical hormones biosynthesis. Masculinization does not occur in all cases.

In *adrenocortical hyperplasia*, there is a deficiency in *21β-hydroxylase*. The affected patients cannot produce cortisol, deoxycortisol, corticosterone, or aldosterone in normal amounts. The decreased secretion of cortisol increases ACTH levels, which stimulates steroids synthesis. Metabolites from the reactions upstream of

21β-hydroxylase accumulate. During intrauterine life, this causes masculinization of the female fetuses.

If the abnormality affects the next metabolic step, catalyzed by 11β-hydroxylase, the mineralocorticoid DOC accumulates. Patients retain Na^+ and water and excrete K^+. There is also increase in androgens and masculinization. Lack of *3β-hydroxysteroid dehydrogenase* is lethal.

ADRENAL MEDULLA HORMONES

The adrenal medulla is considered a differentiated branch of the sympathetic nervous system. The chromaffin cells of this gland produce *catecholamines*, composed by catechol (ortodihydroxybenzene) with the addition of an amine group. They include epinephrine or adrenaline, norepinephrine or noradrenaline, and dopamine, all derived from the amino acid tyrosine (Fig. 26.13). The natural products are L isomers, compounds of the D series, and have little activity.

Epinephrine is mainly produced in the adrenal medulla; norepinephrine is further synthesized mainly in sympathetic system peripheral nerves. Dopamine, a precursor of both, is in peripheral sympathetic nerves and in specialized neurons of the central nervous system, sympathetic ganglia, and carotid body, and acts as a neurotransmitter.

Biosynthesis. Catecholamine synthesis from tyrosine has been described earlier (p. 388). The initial step is conversion of tyrosine to

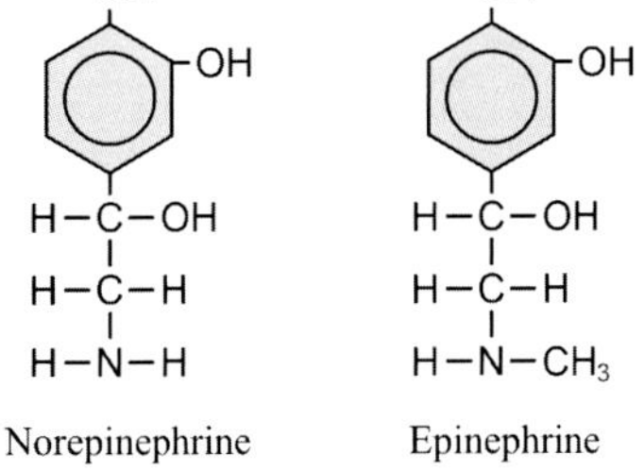

Norepinephrine   Epinephrine

FIGURE 26.13 **Adrenal medulla hormones.**

dihydroxyphenylalanine by *hydrotyrosine hydroxylase*. Dihydroxyphenylalanine is converted into dopamine by a *decarboxylase*. Dopamine is stored in vesicles (chromaffin granules) in cells of the adrenal medulla and in neurons of the sympathetic system. These vesicles contain *dopamine α-hydroxylase*, responsible for the conversion of dopamine to norepinephrine. The last step for the formation of noradrenaline is catalyzed by *phenylethanolamine-N-methyltransferase*, a cytosolic enzyme. Noradrenaline leaves the vesicles, is transformed into adrenaline, and reenters the secretory granules. Phenylethanolamine-N-methyltransferase expression is induced by cortisol.

Catecholamine synthesis is regulated by adrenaline and noradrenaline, which allosterically inhibit tyrosine hydroxylase.

$$\text{Tyrosine} \xrightarrow[\text{hydroxylase}]{\text{Tyrosine}} \text{DOPA} \xrightarrow{\text{Decarboxylase}} \text{Dopamine} \longrightarrow$$

$$\xrightarrow[\text{β-hydroxylase}]{\text{Dopamine}} \text{Norepinephrine} \xrightarrow{\text{Methyltransferase}} \text{Epinephine}$$

Most chromaffin cells in the adrenal medulla synthesize epinephrine; only in 10% of the cells the synthesis pathway ends with noradrenaline formation. These hormones are stored in vesicles, through a process that requires energy, provided by ATP. The molar ratio catecholamines/ATP is about 4:1; within the vesicles, ATP concentration is approximately 30 times greater than in the cytosol. Along with catecholamines, there are also peptides, including *chromogranins, enkephalin precursors, adrenomedullin, vasoactive intestinal peptide*, and *ACTH* in the vesicles.

Secretion. The release of the adrenal cell vesicle content is triggered by nerve impulses that cause Ca^{2+} entry into the cell. This is followed by the targeting and fusion of the vesicles to the plasma membrane, which discharges their content by exocytosis.

Each vesicle releases ~3 million catecholamine molecules, 800,000 ATP molecules, 5,000 chromogranins, and several thousand enkephalins or enkephalin precursors. In circulation, approximately half of the catecholamines are loosely bound to albumin. Its half-life is 10–100 s.

Inactivation. Secreted catecholamines are metabolized in tissues, mainly liver, by aminated chain oxidation and methylation of one of the phenolic groups. The main enzymes involved in these reactions are *monoamine oxidase* (MAO) and *catechol-O-methyltransferase* (COMT). MAO is located in the outer mitochondrial membrane and has broad substrate specificity. It catalyzes the oxidation of various monoamine side chains. COMT is a cytosolic enzyme responsible for the methylation reaction of the hydroxyl group in position 3; the methyl donor is S-adenosyl methionine. COMT metabolizes several types of catecholamines. The liver is particularly rich in MAO and COMT.

Catecholamine inactivation may start either by oxidation or methoxylation catalyzed by MAO or COMT, respectively (Fig. 26.14). Despite the order of these reactions, the final products are the same. The metabolites are excreted in urine, mostly conjugated with sulfate or glucuronate. One of the major urinary metabolites generated from epinephrine and norepinephrine is 4-hydroxy-3-methoxy-mandelic acid, also called *vanylmandelic* or *vanillylmandelic acid*. Another metabolite eliminated in urine in significant quantities is 3-metoxiadrenaline or *metanephrine*.

Approximately half of excreted metabolites correspond to metanephrine and vanillylmandelic acid, 10%–35% to conjugated catecholamines and other metabolites, and less than 5% correspond to free catecholamines.

Mechanism of action. Catecholamines bind to specific plasma membrane receptors on effector cells. There are three groups of receptors, all of which belong to the seven transmembrane helices family: α ($α_1$ and $α_2$), β ($β_1$, $β_2$, and $β_3$), and dopamine receptors (DA_1–DA_5).

The mechanism of action of catecholamines depends on the receptor they bind to. The $α_1$ receptors associate with G_q protein and activate the phosphatidylinositol bisphosphate

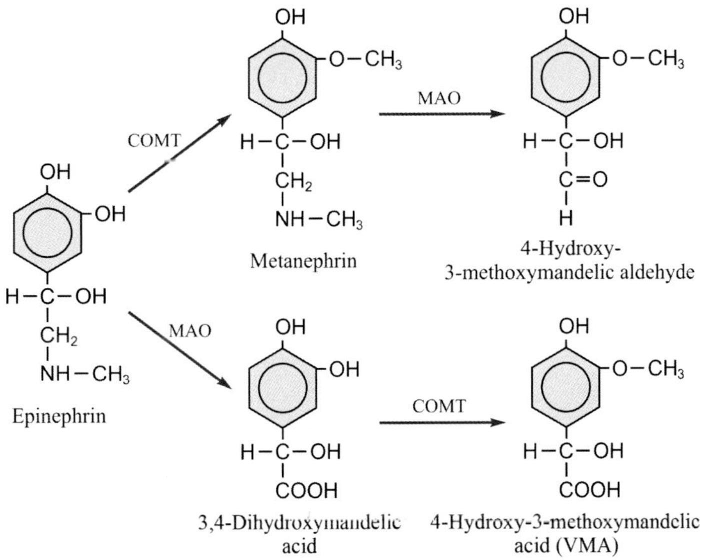

FIGURE 26.14   **Epinephrine inactivation.**

pathway. The α_2 receptors are coupled to G_i protein, which has inhibitory activity on adenylate cyclase and reduces cAMP levels. The β_1 and β_2 receptors couple with G_s proteins, which stimulate adenylate cyclase and increase intracellular cAMP concentration. These receptors are present in many tissues. The β_1 receptors are preferentially expressed in heart, while β_2 receptors are in liver, bronchial smooth muscle, and uterine arteries.

The α_1 receptors are postsynaptic, found in the smooth muscle of blood vessels, sphincters of the gastrointestinal tract, ureter, skin, iris, and in the seminal tract. Their stimulation produces constriction in blood vessels and blood pressure increase; sphincter contraction in the digestive tract, bladder, and ureter; pupil dilation (mydriasis); piloerection in skin; sweating in palms and feet; and ejaculation.

The α_2 receptors bind norepinephrine released by sympathetic postganglionic fibers and epinephrine from the adrenal medulla. Their activation inhibits noradrenaline release (by negative feedback). They are located on smooth muscle of some arterioles and veins; they lower blood pressure, decrease motility and stimulate sphincter contraction in the gastrointestinal tract. They inhibit insulin release in pancreatic β cell islets and induce glucagon secretion. They also promote platelet aggregation.

The β_1 receptors are located predominantly in heart. Their stimulation produces increase in heart rate (positive chronotropic effect), contractility (inotropic effect), and conductivity and cardiac automaticity (dromotropic effect). These effects enhance cardiac output. Other effects of β_1 receptor activation are the increase in renin release by the kidney juxtaglomerular apparatus and the stimulation of secretion in salivary glands.

The β_2 receptors are preferentially localized in liver, bronchial smooth muscle, and uterine arteries. They determine relaxation of the smooth muscle in the bronchi (bronchodilation), the gastrointestinal tract, bladder (inhibits micturition), and uterus. They also dilate arterioles in the heart, liver, and skeletal muscle. Among its metabolic effects are activation of glycogenolysis and gluconeogenesis in liver and glycogenolysis and lactate release in skeletal muscle.

TABLE 26.5   Epinephrine Metabolic Actions

↑ Glycogenolysis
↓ Glycogenesis
↑ Glycemia
↑ Production of lactate in muscle
↑ Gluconeogenesis
↑ Lipolysis in adipose tissue
↑ Free fatty acids in blood
↑ β-Oxidation of fatty acids in peripheral tissues

The β_3 receptors regulate energy expenditure and lipolysis; they promote lipolysis in adipose tissue and stimulate thermogenesis in brown fat.

Dopamine receptors (DR) are found in adrenergic presynaptic terminals of the central nervous system, in pituitary gland, heart, and renal and mesenteric vascular beds. The effects of DR are exerted by activation of adenylate cyclase. DR_1 predominate in brain, kidney, and mesenteric and coronary arteries, and cause vasodilation. DR_2 are located on presynaptic terminals, sympathetic ganglia, and brain. They inhibit norepinephrine release, ganglionic transmission, and prolactin release.

Adrenaline has a greater effect on β_2 receptor than noradrenaline; effects on β_1, α_1, and α_2 receptors are similar for both hormones. Table 26.5 presents some of the physiological responses triggered by adrenergic receptor activation. There are drugs that can bind to catecholamine receptors and function as selective agonists or antagonists. For example, phentolamine is an α antagonist; propranolol, atenolol, and metoprolol block β_1 receptors; isoproterenol, terbutaline, and albuterol are β_2 agonists.

Metabolic actions. Metabolic actions depend on the stimulation of β receptors in the target organs. Adrenaline exerts stronger metabolic effects than noradrenaline. In general, catecholamines increase oxygen consumption and heat production. In skeletal muscle, adrenaline promotes glycogen degradation mediated by phosphorylase. This effect determines that in muscle during exercise lactic acid is formed and released to the blood.

In adipose tissue, increased cAMP caused by catecholamines activates lipase and stimulates lipolysis, with the release of fatty acids into the circulation. These fatty acids are used as fuel by different tissues.

In liver, adrenaline stimulates glycogenolysis and inhibits glycogenesis by inactivating glycogen synthase. It also increases gluconeogenesis from lactate generated in muscles during high activity. Stimulation of hepatic glycogenolysis and gluconeogenesis increases glucose release to the blood and produces hyperglycemia. These effects are accentuated by the inhibitory action of catecholamines on insulin release. The gluconeogenic effect is mediated by α peroxisome proliferator-activated receptor (αPPAR) (pp. 551). αPPAR stimulation is due to increased cAMP, glucocorticoids, or a hepatic nuclear factor; it activates expression of key enzymes involved in gluconeogenesis. Increased fatty acid supply and oxidation in tissues, combined with lower glucose utilization, have ketogenic effect. Table 26.5 summarizes the metabolic actions of epinephrine.

The responses triggered by catecholamines, both cardiovascular and general, as well as metabolic, tend to prepare the body for stress or emergency situations that require extra energy expenditure.

Besides catecholamines, the adrenal medulla synthesizes other hormones, including the *adrenomedullin*, a peptide of 52 amino acids, whose action is exerted through a G_s protein-coupled receptor that causes cAMP increase in the effector cells. It is a potent vasodilator and natriuretic factor.

Chromogranins have been isolated in other endocrine glands and brain; their function is not clear. The enkephalin precursors (preproenkephalins) generate, by successive steps of hydrolysis, pentapeptides that behave like hormones or neurotransmitters.

Clinical alterations. No clinical symptoms are observed for epinephrine and norepinephrine deficiency; however, there is a pathological condition caused by the excessive production of these hormones. Pheochromocytoma, a tumor of chromaffin cells. Affected patients suffer from hypertension, hyperglycemia, and exacerbation of catecholamine metabolic responses. Vanillylmandelic acid and metanephrine urinary excretion are markedly increased.

PANCREAS

The endocrine function of the pancreas resides in the Langerhans' islets, which are clusters of epithelial cells scattered throughout the organ. These islets contain different types of cells, each responsible for producing a particular hormone. The α cells or A represent 15%–20% of the total islet cells and are responsible for *glucagon* synthesis. The β or B cells (60%–80% of total) produce *insulin*. The δ or D cells (5%–10% of total) secrete *gastrin* and *somatostatin*. Finally, the PP cells (<2% of total) produce *pancreatic polypeptide*. Somatostatin and pancreatic polypeptide have inhibitory actions on other endocrine and exocrine pancreatic secretions. There is also communication among α, β, and δ cells, which modulate each other via connexons or gap junctions.

INSULIN

Insulin is a protein that can be obtained highly purified from pancreatic extracts. It was the first protein whose amino acid sequence was determined. Insulin has 6000 Da and consists of two polypeptide chains. One of the chains, designated A, is composed of 21 amino acids; the other, or B chain, has 30 amino acids. Both polypeptides are linked together by two disulfide bridges extending from the cysteines in positions 7 and 20 of the A chain to cysteines 7 and 19 of the B chain. The A chain has a disulfide bond on cysteines located at positions 6 and 11 (Fig. 26.15). The presence of these disulfide

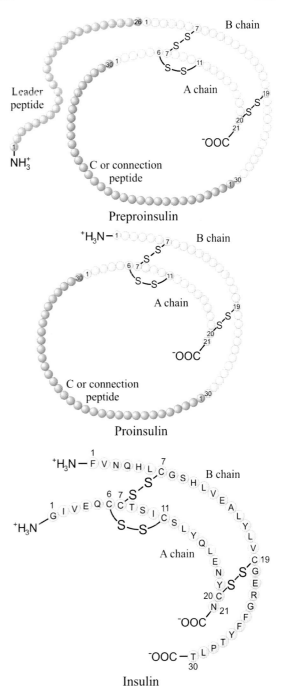

FIGURE 26.15 **Scheme of human preproinsulin, proinsulin, and insulin.**

bonds is essential for the biological activity of the hormone; their rupture with alkalis or reducing agents determines loss of the physiological properties of insulin. Also, the tertiary and quaternary structures of insulin are important for its activity.

The structure described earlier corresponds to a single insulin unit. In the pancreas, insulin is often found as polymers (dimers or hexamers) containing zinc. The hexamer consists of three dimers, associated to two Zn^{2+} ions.

There are differences in the primary structure of insulin from different species; however, all show the same activity when administered to individuals of a different species. For example, bovine, ovine, and equine insulin differ from the human hormone by three amino acids. Despite this difference, they are functional in humans, allowing their clinical use in patients with insulin deficiency. Sometimes, their prolonged use may induce the formation of antibodies against these exogenous insulin proteins.

Porcine insulin differs from that of humans by only one amino acid. Alanine 30 in the B chain C-terminus corresponds to threonine in human insulin. By enzymatic methods, it is possible to replace B chain alanine with threonine, and to obtain semisynthetic human insulin from porcine. It has been used in patients who have developed antibodies after previous treatments with heterologous hormone. Currently, human insulin is produced by recombinant DNA techniques, which solves the immunologic problems posed by the use of insulin from heterologous species.

Biosynthesis. Preproinsulin is a 111-amino acid insulin precursor synthesized by the rough ER of the pancreatic β cells. This protein enters the ER lumen and immediately loses the 26-amino acid leader peptide from its N-terminus. *Proinsulin* of 85 amino acids (~9000 Da) is formed, which has no hormonal activity. The first 30 amino acids of proinsulin correspond to those of the B chain of insulin, and the last 21 correspond to the A chain. A 34-amino acid piece, known as the *connection* or *C-peptide*, separates the A and B chains (Fig. 26.15). The spatial arrangement of proinsulin places the A and B chains in the proper orientation to form disulfide bridges.

Secretion. Proinsulin is transported within vesicles from the ER to the Golgi apparatus, where, after successive steps of hydrolysis (catalyzed by peptidases), it is converted into active insulin. Hydrolysis of proinsulin frees the connection peptide and two dipeptides located at the ends of the C-peptide. These remain in the secretory vesicles, along with the insulin dimers and hexamers. Secretion of the content of vesicles into the extracellular space is carried out by exocytosis, upon specific stimuli. Once in blood, insulin hexamers dissociate, rendering most plasma insulin as monomers.

Equimolar amounts of insulin and C-peptide are released; the connection peptide apparently does not perform any hormonal function. However, the determination of plasma C-peptide allows an estimation of the amount of endogenous insulin in patients treated with exogenous hormone.

One of the most powerful stimuli for the release of insulin and somatostatin in β and δ cells, respectively, is blood sugar increase. Instead, increase in free amino acids activates glucagon secretion (certain amino acids also stimulate insulin). The effect of food on the release of these hormones is dependent on the relationship between their carbohydrate and protein content. A food rich in carbohydrates depresses glucagon production and, if it is more abundant in protein, stimulates glucagon.

Glucose must be metabolized in the β cell to stimulate insulin secretion; it is phosphorylated by glucokinase and enters glycolysis and final oxidation to produce ATP. The intracellular increase in ATP/ADP ratio causes K^+ channels to close, which in turn depolarizes the cell plasma membrane and opens voltage sensitive Ca^{2+} channels. Ca^{2+} entry into the cell stimulates the transport and exocytosis of the insulin containing vesicles. The transport vesicles are guided to the plasma membrane by the cell cytoskeleton (microtubules) and driven by motor proteins (actin and myosin).

cAMP is one of the main modulators of insulin release; it mobilizes Ca^{2+} from the mitochondria and elevates its levels in the cytosol. Although $[Ca^{2+}]_i$ is an intracellular stimulus for secretion, its increase in the absence of glucose has no effect on insulin secretion.

Sulfonylureas, compounds used in the treatment of diabetes mellitus type II, induce K^+ channel closing in β cells and stimulate insulin secretion. Instead, diazoxide opens K^+ channels and inhibits insulin release.

Elevated levels of the amino acids, arginine and lysine, and free fatty acids stimulate insulin secretion. In addition, ingestion of food activates the secretion of gastrointestinal hormones [gastrin, cholecystokinin (CCK), secretin, enteroglucagon], which promote insulin release. For this reason, elevated blood insulin is higher after a meal than after the intravenous injection of glucose in amounts similar to those contained in the ingested food.

Other regulator of insulin release is glucagon. Vagal and β-adrenergic receptors stimulation increase insulin release. In contrast, activation of α-adrenergic receptors inhibits insulin secretion. Somatostatin depresses the release of both insulin and glucagon.

Under basal conditions, insulin concentration in plasma is 0.4–0.8 ng/mL (~61–122 pmol/L). Eight to 10 min after a meal rich in carbohydrates, insulin concentration begins to rise and reaches a maximum at 30–45 min. In normal individuals, insulin rarely increases above 4 ng/mL (610 pg/mol/L). Foods with low glycemic index (p. 317) produce fewer fluctuations in insulin secretion.

Degradation. Insulin has a short half-life, less than 10 min in humans. The hormone is degraded by insulinase in liver, kidney, and other organs and tissues. Approximately half of the insulin that passes through the liver is removed in a single step.

Mechanism of action. Insulin binds to a specific receptor (*insulin receptor*) in the plasma membrane of the target cells. This receptor is found in all mammalian tissues. Their number varies from 40 in red blood cell to many thousands in hepatocytes and adipocytes.

The *insulin receptor* is an integral membrane glycoprotein of approximately 400 kDa, which belongs to the family of growth factor receptors, with tyrosine kinase activity in its cytosolic portion. It is a heterotetramer of two α (135 kDa) and two β (95 kDa) subunits, which are glycosylated and linked by disulfide bonds (Figs. 25.7 and 26.16B). The α subunits are on the outside of the plasma membrane and are the subunits that bind insulin. One molecule of insulin binds to both α subunits. The β subunits are embedded in the double lipid layer and emerge on both sides of the plasma membrane. In the cytosolic portion is the protein tyrosine kinase site. Kinase activity is kept inactive while the receptor is empty; under these conditions, the α subunit functions as a β subunit allosteric inhibitor.

When insulin occupies the binding sites on the α subunits of the receptor, it produces a conformational change that is transmitted to the β subunits. This activates the tyrosine kinase activity in the β subunit. The activated receptor induces its autophosphorylation and acquires the capacity to phosphorylate tyrosine residues of proteins that contain SH2 domains, which then bind to the receptor. The best characterized of these proteins are the insulin receptor substrates (IRS), of which four different molecular forms have been identified. IRS-1 and IRS-2 are distributed in most tissues and are especially abundant in liver and muscle. IRS-3 is found in adipocytes, fibroblasts, and hepatocytes, and IRS-4 is present in embryonic kidney. At least other five proteins have been identified to bind to the insulin-activated receptor. They function as intermediates in signal transduction systems.

Once phosphorylated, the tyrosine residues in IRS bind to other proteins containing SH2 domains, including phosphoinositol-3-kinase

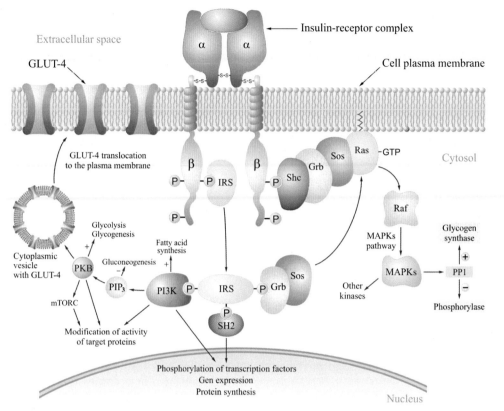

FIGURE 26.16 **Signal transduction systems activated by insulin.**

(PI3K) and *Grb-2*, which continue the cascade of signaling events in the cells. Fig. 26.16 shows some of the systems activated by insulin signals.

The PI3K is one of the main signal transduction systems activated by the insulin-receptor complex. PI3K catalyzes the addition of phosphate at position 3 of the inositol in inositol-4,5-bisphosphate (PI-4,5-P_2) to form inositol-3,4,5-trisphosphate (PI-3,4,5-P_3). This intermediate brings together and activates proteins containing the plekstrin homology domain. One of the most important enzymes of this pathway is stimulated by protein kinase B (PKB), serine–threonine kinase, also known as *Akt* (from its relation with the retroviral oncogene v-Akt).

Activated *PKB* or *Akt* phosphorylates various effector proteins, promoting changes in enzyme activity or gene expression. Some of the effects of insulin, mediated by Akt, include the following.

Glucose transport activation. Phosphorylation of GLT4, catalyzed by Akt/PKB, is the major factor determining translocation of this transporter from intracellular vesicles to the plasma membrane. This action, which increases glucose uptake, is particularly notable in muscle and adipocytes. In skeletal muscle, glucose uptake is further stimulated by muscle contraction and hypoxia, indicating that another pathway, independent from insulin, also promotes translocation of GLUT4 to the cell plasma membrane. This observation supports the benefit of physical activity for patients with diabetes insipidus.

Glycogen synthesis. Akt/PKB depresses glycogen synthase kinase 3 and stimulates protein phosphatase 1 (pp. 433), which helps to maintain glycogen synthase dephosphorylated and active.

Glycolysis. PKB activates phosphofructokinase 2, which catalyzes the conversion of fructose-2,6-bisphosphate (F-2,6-bisP) from fructose-6-P. The F-2,6-bisP is a powerful positive effector of phosphofructokinase 1, major regulatory enzyme of the glycolytic pathway.

Gluconeogenesis. PI-(3,4,5)-P_3 inhibits glucose-6-phosphatase, reducing the conversion of glucose-6-P into glucose.

Conversion of glucose into fatty acids. Inactivation of glycogen synthase kinase 3 by *Akt/PKB* maintains active ATP-citrate lyase, which stimulates the conversion of glucose into fatty acids. In addition, insulin, through the PI3K pathway, promotes the synthesis of fatty acids by activation of pyruvate dehydrogenase and acetyl-CoA carboxylase, and the production and secretion of VLDL.

Actions on gene activity. These are mediated via the PI3K pathway and include (1) increased expression of hexokinase in muscle, (2) repression of transcription of the gene coding for phosphoenolpyruvate carboxykinase, and (3) activation of glucose-6-P dehydrogenase synthesis, an enzyme of the pentose phosphate pathway.

Triacylglycerol metabolism. Insulin has antilipolytic effect, inhibiting lipase in adipose tissue. This effect is mediated by PKB-mediated activation of phosphodiesterase, which hydrolyzes and reduces cAMP levels.

Insulin enhances lipoprotein lipase activity in the vascular endothelium, which catalyzes the hydrolysis of plasma lipoproteins triacylglycerols, which will increase the offer of fatty acids for fat synthesis in adipocytes. In addition, the uptake and degradation of glucose promoted by insulin contributes to that synthesis increasing the availability of glycerol-3-phosphate.

Protein synthesis. The IRS-PI3K signaling system participates in the regulation of amino acids transport, gene transcription, and mRNA translation. The fast effects of insulin on protein synthesis are in great part mediated by increase in translation of preexistent mRNA. PKB activates initiation and elongation complexes that participate in mRNA translation, such as the eFI2b and eFI4E initiation factors.

In addition, insulin can activate other pathways, for example, it induces cell mitosis via IRS, stimulates Grb-2, Sops, and the Ras-MAP-kinases cascade and AKT, mTORC1, and mTORC2 signaling.

Insulin effects on gluconeogenesis and lipid metabolism are due, in part, to the action on the αPPAR.

Other pathway downstream of IRS includes phospholipase Cγ (PLCγ) and PKC. All of the examples described earlier show the remarkable extent and diversity and crosstalks of mechanisms by which insulin impacts cells and tissues.

Signal transduction initiated by insulin can be depressed by several regulators. Insulin-receptor autophosphorylation is inactivated by a protein tyrosine phosphatase 1B, also called PTPN1.

After it exerts its effects, the insulin-receptor complex is internalized by endocytosis. The receptor is dissociated from insulin and is sent back to the plasma membrane. Insulin is degraded in lysosomes. The receptor is also hydrolyzed by lysomal proteases as a mechanism to downregulate its function.

Metabolic Actions of Insulin

Insulin actions on metabolism are multiple and involve a variety of physiological mechanisms, including modification of the activity of membrane transport systems, regulation of RNA and protein synthesis, and modulation of the activity of enzymes. It is difficult to establish whether a given effect of insulin is direct or secondary to the changes produced at another level. Metabolic changes promoted by insulin in a tissue can cause changes in the supply of substrates or allosteric effectors, which in turn affect the function of enzymes of a particular metabolic pathway.

The main target tissues of insulin action are muscle, fat, and liver.

1. *Effects on glucose transport.* In muscle and adipose tissue, insulin stimulates cell uptake of glucose, amino acids, nucleosides, and phosphate. Mammalian cells have plasma membrane transporters (GLUT) responsible for facilitated diffusion of glucose (pp. 227, 285). Of the various types of glucose transporters, GLUT1 and GLUT3 are widely distributed in tissues; they ensure the basal supply of glucose to the cells even at low blood glucose levels. In contrast, GLUT4, present in muscle and adipose tissue, is regulated by insulin, which (via IRS-PI3K) induces the translocation of GLUT4 from intracellular vesicles to the plasma membrane. GLUT4 is therefore functional when plasma glucose levels increase. Once plasma glucose decreases, GLUT4 is internalized back by endocytosis. This recycling of transporters controlled by insulin regulates glucose uptake in muscle and adipose tissue as needed.

2. *Effects on carbohydrate metabolism.* Insulin is the main hypoglycemic hormone. This action is due to its capacity to stimulate glucose uptake and utilization in different tissues. *Liver.* Insulin significantly influences liver mechanisms for glucose utilization. It increases glucokinase activity, increasing the availability of glucose to different pathways, including glycogenesis, glycolysis, pentose phosphate pathway, glucose oxidation, and its conversion into lipids. These effects are due not only to a direct modulation of enzyme activity in glucose utilization pathways, but also to induction of enzyme synthesis. Thus, insulin stimulates the activity of glycolytic enzymes (glucokinase, phosphofructokinase, and pyruvate kinase). This action is accompanied by depression of glycogenolysis and gluconeogenesis, attained by direct inactivation of glycogen phosphorylase and phosphorylase kinase, and suppression of pyruvate carboxylase, phosphoenolpyruvate carboxykinase, fructose-1,6-bisphosphatase, and glucose-6-phosphatase synthesis.
Muscle and adipose tissue. Insulin stimulates glucose uptake in skeletal and cardiac muscle, and in adipocytes (via GLUT4 translocation to the plasma membrane). In muscle, insulin promotes glucose utilization or storage as glycogen.

3. *Effects on lipid metabolism.* Insulin stimulates fatty acid and triacylglycerol synthesis in the liver, adipose tissue, lactating mammary gland, and other tissues. Activation of glycolysis, glucose oxidative degradation, and pentose pathway increases acetyl-CoA and NADPH production, which are used in fatty acid biosynthesis; glycerophosphate and triacylglycerols precursors are generated. Lipogenesis is also favored because insulin activates pyruvate dehydrogenase, acetyl-CoA carboxylase, and glycerophosphate acyltransferase. When there is excess in food intake, acetyl-CoA is preferably directed to fatty acid synthesis. Insulin activates lipoprotein lipase bound to the membrane of peripheral capillaries, increasing fatty acid supply to tissues (preferably adipose) for triacylglycerol synthesis. Moreover, the hormone inhibits adipocytes lipase, depresses lipolysis, and reduces the levels of circulating free fatty acids. It also reduces ketogenesis.

4. *Effects on protein metabolism.* Insulin stimulates amino acid uptake by cells and their incorporation into proteins; it activates the transcription and translation processes. The anabolic effects of insulin are expressed by decrease in the production of urea in the liver. Table 26.6 summarizes the metabolic actions of insulin.

TABLE 26.6   Metabolic Actions of Insulin

↑ Glucose uptake by peripheral tissues
↑ Activity of all pathways consuming glucose
   ↑ Glycolysis
   ↑ Pentose pathway
   ↑ Total glucose oxidation
↑ Glycogenesis
↓ Glycogenolysis
↓ Gluconeogenesis
↓ Glycemia
↓ Lipolysis
↓ Free fatty acids in plasma
↑ Lipogenesis from glucose
↑ Amino acid uptake in tissues
↑ Protein synthesis
↓ Urea production

---

5. *Effects on potassium.* Insulin increases the activity of Na^+,K^+-ATPase. During insulin deficiency, potassium leaks out of the cell and into the extracellular compartment and is eliminated by the kidneys.

Paracrine actions. In addition to its systemic endocrine effects, insulin also exerts local paracrine actions. Together with somatostatin from δ pancreatic cells (p. 612), it inhibits glucagon secretion in neighboring α cells.

Clinical Pictures Related to Insulin Alterations

Hypoinsulinism. Absolute or relative insulin deficiency produces a very common clinical condition called *diabetes mellitus*, which will be described later in this chapter.

Hyperinsulinism. There are tumors of the pancreatic β cells which result in excessive production of insulin. Patients suffering from this condition undergo hypoglycemic crises, weakness, intense sweating, and dizziness. If the alteration is not treated by providing glucose, patients can suffer a hypoglycemic shock, with loss of consciousness, seizures, coma, and even death. The same symptoms may be produced by administration of insulin in high doses.

GLUCAGON

Glucagon is produced by the α or A cells of the pancreatic Langerhans' islets. It is a 29-amino acid peptide with a mass close to 3500 Da. Unlike insulin, it does not contain cysteines, but possesses methionine and tryptophan, which are absent in insulin. Glucagon is synthesized as a larger precursor molecule, preproglucagon, which is converted to proglucagon and glucagon by successive steps of hydrolyses. Proglucagon, which has 160 amino acids, originates *glicentin-related peptide* and *glucagon-like peptides 1* and *2* (GLP-1 and GLP-2). Glicentin is a 69-amino acid polypeptide composed by the glicentin-related peptide and glucagon. It is produced mainly in intestine (in greater proportion than in the pancreas), and it is also known as *enteroglucagon*. Another peptide, *incretin*, which results from removal of a six-amino acid piece to GLP-1, is secreted into the intestine, and stimulates insulin secretion, enhancing the action of glucose.

The concentration of immunoreactive glucagon in plasma during fasting is 75 pg/mL (25 pmol/L). Only 35% of this is pancreatic glucagon; the rest is represented by other molecules derived from glucagon or secreted by the intestine. Glucagon circulates in plasma free and has a short half-life (~6 min). The main regulator of glucagon secretion is the plasma glucose level. Increased glucose inhibits glucagon secretion, while hypoglycemia stimulates it. The mechanism by which blood glucose regulates glucagon secretion is not well known. It has been proposed that glucose does not act directly, but rather through the release of insulin and pancreatic somatostatin. Both of these hormones exert paracrine inhibitory action on α cells.

Increase in plasma amino acids, particularly arginine and alanine, stimulation of the sympathetic nervous system, catecholamines, gastrointestinal hormones, and glucocorticoids are also activators of glucagon secretion. High levels of fatty acids inhibit glucagon release. Leucine, which promotes insulin release, does not influence glucagon secretion.

Mechanism of action. In hepatocyte and adipocyte membranes, glucagon binds to specific receptors coupled to G_s proteins, activating adenylate cyclase, increasing cAMP concentration, and initiating a cascade of phosphorylation events that regulate the activity of various enzymes.

Effects. Glucagon favors the supply of energetic substrates to tissues in the intervals between meals, when there is not nutrients entrance from intestine.

1. *Effects on carbohydrate metabolism.* In hepatocytes, glucagon, via increase of cAMP and activation of protein kinase A, stimulates glycogen phosphorylase and inhibits glycogen synthase. This promotes glycogenolysis, with the release of glucose into the extracellular space; glycogenesis is repressed. Although glucagon has similar effects as epinephrine, it is active in liver and has little effect on muscle tissue, where adrenaline action predominates. Glucagon increases free fatty acids in plasma, which in turn depresses muscle glucose uptake. The hyperglycemic effect of glucagon is enhanced by its ability to increase gluconeogenesis.

2. *Effects on lipid metabolism.* In adipose tissue, glucagon increases cAMP, which activates hormone-sensitive lipase. This stimulates triacylglycerol degradation and the release of fatty acids into the blood. Glucagon increases fatty acid β oxidation, which reduces NAD levels in the cell and decreases the citric acid cycle capacity to oxidize glucose. The synthesis of acetyl-CoA and ketone bodies are increased.

3. *Effects on nitrogen metabolism.* Glucagon stimulates nitrogen catabolism, increases urea, creatinine, and uric acid excretion by the kidney, and favors a negative nitrogen balance. It activates gluconeogenesis from amino acids. Table 26.7 summarizes the metabolic actions of glucagon.

TABLE 26.7   Glucagon Metabolic Actions

On liver

↑ Glycogenolysis
↓ Glycogenesis
↑ Gluconeogenesis
↑ Glycemia
↑ Urea, creatinine, uric acid

On adipose tissue

↑ Lipolysis
↑ Free fatty acids in plasma

SOMATOSTATIN

Somatostatin, besides being produced in the hypothalamus, is also produced by the δ cells of the pancreas islets, enterocytes, and thyroid parafollicular C cells.

Alternative processing of the C-terminal 116 amino acids of the precursor *preprosomatostatin* generates *somatostatin 14*, a tetradecapeptide, and *somatostatin 28*, a polypeptide double in size. Both exert similar actions. Somatostatin 14 predominates in hypothalamus and pancreas, and somatostatin 28, in intestine. Both bind the same receptors, coupled to G_i proteins, which regulate intracellular cAMP and Ca^{2+} levels, to inhibit the secretion of several hormones. In addition to its remarkable inhibitory effect on pituitary GH, somatostatin also depresses insulin, glucagon, gastrin, secretin, and VIP secretion, and affects the thyrotrophs pituitary cells, enhancing thyroid hormones feedback inhibitory action. Somatostatins 14 and 28 also affect phosphotyrosine phosphatase and the MAP kinase pathway, which mediate the antiproliferative actions of somatostatin on tumor cells. Somatostatins also reduce the absorption of nutrients in the intestine. Almost all activators of insulin secretion also produce somatostatin release in pancreas.

GLUCOSE HOMEOSTASIS

Blood glucose concentration is maintained constant (between 70 and 110 mg/dL or 3.9 and 6.1 mM) by a stringent regulation, in which a group of hormones are involved. Plasma glucose levels are the result of a delicate balance between effects aimed at reducing the level of circulating glucose and those that elevate it.

The hypoglycemic actions include (1) facilitation of glucose uptake into cells, (2) stimulation of glycogenesis, (3) glucose degradation (via activation of glycolysis and the pentose phosphate pathway), and (4) glucose conversion to other substances, mainly lipids. Insulin is the main activator of all metabolic processes leading to hypoglycemia.

The hyperglycemic processes include (1) hepatic glycogenolysis, (2) gluconeogenesis, and (3) absorption of glucose in the intestine.

Hyperglycemic actions are exerted by: (1) *glucagon*, which activates hepatic glycogenolysis and gluconeogenesis; (2) *epinephrine*, which activates muscle and hepatic glycogenolysis. Only liver glucose is released in significant quantities to increase glycemia. In muscle, glucose degradation continues to lactate, which is converted into glucose by the liver. (3) *Glucocorticoids* (cortisol), which increase gluconeogenesis in liver and inhibit glucose utilization in extrahepatic tissues; and (4) *GH*, which decreases glucose uptake in muscle and glucose utilization in other tissues.

Normally, these actions are highly regulated and maintain fasting levels of plasma glucose within constant values. This is particularly important because it ensures a continuous supply of glucose, especially to tissues that absolutely depend on it.

DIABETES MELLITUS

Diabetes mellitus is a pathological condition characterized by persistent hyperglycemia, glycosuria, and a series of symptoms that are a consequence of an overall metabolic imbalance. Its most striking symptoms include (1) *hyperglycemia*, which reflects the decreased in insulin/glucagon ratio; (2) *glucosuria*, which results from the inability of the kidney to retain all of the glucose filtered; it occurs when the blood glucose level exceeds 160 mg/dL; (3) *polyuria* driven by the osmotic diuresis caused by the high levels of glucose in urine; (4) *dehydration* due to the loss of water in urine; (5) *polydipsia* due to the increase thirst to compensate the water loss; and (6) *polyphagia* due to increased appetite.

Diabetes mellitus can be caused by the absolute lack of insulin (inability to synthesize hormone), as in the *insulin-dependent* or *type 1* clinical form of the disease, also known as *juvenile diabetes*. Another form, in which insulin production is reduced, normal, or even increased, is known as *type 2 or noninsulin-dependent diabetes mellitus*. It is observed in adults, who present tissue resistance to insulin.

Type 1 diabetes mellitus. It is present in less than 10% of all diabetic patients. It becomes evident at 10–14 years of age. It is the most severe form of diabetes mellitus. This form must be treated with insulin administered parenterally. It is an autoimmune disease; the patients abnormally produce antibodies against the pancreas β cells, which become gradually lost. In patients with type 1 diabetes, the disease has been associated with alterations in the major histocompatibility complex (HLA) types DR3 and DR4 (Chapter 30).

Mutations in the gene encoding insulin have been identified, which produce an inactive hormone.

Metabolic disorders of type 1 diabetes mellitus. The main sign is fasting hyperglycemia, due to increased gluconeogenesis in liver. Glucose uptake by the tissues is normal at basal conditions, but after ingesting glucose, blood sugar increases excessively because glucose intake by peripheral tissues and its storage as glycogen are reduced; also, glucose release from the liver is not suppressed. Most oxaloacetate molecules are diverted to gluconeogenesis. The activity of the citric acid cycle is reduced. Inactivation of glycogen synthase and low production of ATP (low cell energy charge) contribute to reduce glycogen synthesis in all tissues.

The increase in blood glucose above the renal threshold produces glucosuria. The glucose in urine is necessarily accompanied by a corresponding amount of water, which explains the polyuria and dehydration that accompanies the disease.

Lipolysis and concentration of nonesterified fatty acids in blood increase. Malonyl-CoA formation, the first step of the synthesis of fatty acids, decreases. Malonyl-CoA is a carnitine acyltransferase I competitive inhibitor. When malonyl-CoA is reduced, carnitine acyltransferase I activity increases, a greater proportion of fatty acids enter β-oxidation in the mitochondrial matrix. This produces ketone bodies, at a rate higher than their metabolism, which leads to their accumulation in in the body. Ketone production is not only due to insulin deficiency, but also due to the action of glucagon. Decreased blood pH (diabetic ketoacidosis) is common in patients, who are poorly controlled. The high level of ketone bodies acetoacetate and β-OH-butyrate contribute to osmotic diuresis and dehydration.

Amino acid transport and utilization is reduced, which determines an elevation of free amino acids in blood. The production of urea is increased, protein synthesis is reduced, protein catabolism predominates, and nitrogen balance becomes negative. There is enhanced use of amino acids for glucose synthesis (gluconeogenesis).

Type 2 diabetes mellitus. This form comprises between 80% and 90% of all cases of diabetes and is generally observed in people over 40 years of age, most of whom have some degree of obesity. There is genetic (polygenic) predisposition to the disease, aggravated by factors such as unbalanced or excessive diet, sedentary lifestyle, abdominal and visceral obesity, and aging.

Patients with type 2 diabetes initially have no deficit in insulin production, usually it is even increased; however, there is a relative insufficiency of the hormone. The primary condition leading to this type of diabetes mellitus is the altered response of the pancreatic β cells to variations in glucose, and failure of glucose uptake in muscle and adipose tissue. Both defects tend to produce hyperglycemia and determine increase in insulin secretion, in an attempt to compensate for the peripheral tissues resistance to the hormone. Prolonged exposure to high levels of glucose produces gradual desensitization of β cells. The stress produced by the metabolic alterations progressively deteriorates β cell function, reduces synthesis and secretion of insulin, and finally causes apoptotic death of the β cells (Chapter 32).

These patients have very poor response to insulin (insulin resistance) that may be due to various factors, including dephosphorylation of IRS, defects in *Akt*/PKB target proteins, or loss of IRS.

Metabolic disorders of type II diabetes mellitus. The postprandial insulin-dependent glucose uptake is decreased, particularly in skeletal muscle. Glucose transport and utilization and glycogen synthesis are decreased. The free fatty acids and ketone bodies are slightly elevated in plasma. There is not a great tendency to ketosis. Protein metabolism is normal. Table 26.8 presents the most important metabolic alterations in diabetes mellitus.

Diabetes treatment can control many of the aforementioned metabolic disorders. However, it does not prevent the occurrence of nerve, kidney, retina, arteries, and capillaries system damages; only it retards them. These tissue alterations are due to the nonenzymatic glycosylation of proteins.

Insulin resistance. This alteration is often the result of the food overload that often accompanies obesity. Several factors, including accumulation of products of incomplete metabolism, along with changes in fatty acid uptake, hormonal, innate immunity, and inflammatory factors, can promote accumulation of lipid

TABLE 26.8 Metabolic Alterations in Diabetes Mellitus

↓ Glucose uptake by peripheral tissues

↓ Glucose consumption by peripheral tissues

↑ Hepatic glycogenolysis

↑ Hepatic gluconeogenesis

↓ Glycogenesis

↑ Glycemia

↑ Lipolysis in adipose tissue

↓ Lipogenesis in adipose tissue

↑ β-Oxidation of fatty acids

↑ Free fatty acids in plasma

↑ Production of ketones: ketonemia-ketonuria-ketoacidosis

↑ Acylglycerols and VLDL synthesis in liver

↑ Lipidemia

↑ Protein catabolism

↑ Free amino acids

↑ Urea production

The glucose load filtered in glomeruli surpasses the reabsorption capacity of the renal tubules, leading to glycosuria. Water and sodium are lost through urine (osmotic diuresis) resulting in poliuria, hypovolemia, and hyperosmolarity.

metabolites (DAGs and ceramides) in liver and muscle, and excessive fat deposition in different organs, which determine a state of impaired insulin signaling, insulin resistance, and eventually type 2 diabetes. The following are factors involved in insulin resistance:

Endocrine functions of adipose tissue. Adipocytes have a regulatory role in the development of insulin resistance. They produce hormones such as *leptin, adiponectin, resistin,* and inflammatory cytokines, particularly *interleukin-6* and *tumor necrosis factor* α. Leptin and adiponectin have antidiabetogenic action, decreasing triacylglycerol synthesis, stimulating fatty acid β oxidation, and favoring insulin action in the liver and skeletal muscle. Both hormones act stimulating AMP-activated protein kinase (AMPK). AMPK is a key factor in the regulation of energy balance; it responds to reduced ATP levels or increased AMP/ATP ratio in tissues, by activating the oxidation of glucose and fatty acids. In obese individuals, leptin levels increase, but tissues do not respond to it, while adiponectin levels decrease. The production of proinflammatory cytokines activates the recruitment of macrophages into tissues and increases the cell damage.

Effects of dietary overload. In liver, chronic and excessive food intake increases the production of malonyl-CoA, stimulating fatty acid synthesis and inhibiting carnitine-acyl transferase I. The fatty acids cannot be oxidized in mitochondria and the citric acid cycle, and the electron transport chain operation is reduced. Overnutrition imposes an anabolic overload to the endoplasmic reticulum, which undergoes alterations that affect its function, such as protein misfolding. Metabolic stress induces the expression of protein serine kinases that block insulin inhibitory action on gluconeogenesis. The liver continues releasing glucose into the circulation.

In skeletal muscle, increased uptake of fatty acids and stimulation by the α/δPPAR promotes β oxidation, but it is not coordinated with the citric acid cycle. Consequently, products of incomplete oxidation of fatty acids (acyl-carnitine, reactive oxygen species) accumulate in mitochondria. The subsequent stress activates serine kinases that suppress signal transduction initiated by insulin and prevents vesicles containing GLUT4 to transfer and insert it in the plasma membrane.

Exercise or physical activity counteracts the lipidic stress, activates the α/δPPAR pathway, and coordinates oxidative pathways and mitochondria operation, preventing insulin sensitivity.

Metabolic syndrome. This condition is characterized by the simultaneous presence of abdominal obesity (increased waist circumference), insulin resistance resulting in disruption of glucose homeostasis, hyperglycemia, hyperinsulinemia, dyslipidemia, intravascular clot formation due to inhibition of fibrinolysis, hypertension, and microalbuminuria. The excessive insulin increase produces sodium retention by the kidney, contributing to hypertension. The dyslipidemia shows the atherogenic profile, with high levels of VLDL and LDL (hypertriglyceridemia and hypercholesterolemia) and reduced HDL. These factors are a frequent cause of coronary heart disease and stroke.

Metabolic syndrome is a prevalent condition in the majority of the world population and can affect 40% of adults over 60 years. Affected individuals have an increased risk of type 2 diabetes mellitus and cardiovascular accidents.

Glucose tolerance test. This test has been used for the early detection of diabetes mellitus. The method consists in administering by mouth a solution of 1 g of glucose/kg body weight to an individual who has been fasting for approximately 10 h. Then, blood samples are drawn at the beginning (time 0) and every half an hour later, for up to 2–4 h. In normal individuals, the basal fasting blood glucose level is less than 110 mg/dL, half an hour after glucose intake, the maximum value is attained, which does not exceed 200 mg/dL. By the second hour, blood glucose level is lower than 140 mg/dL. In contrast, in patients with diabetes mellitus, fasting glucose levels exceed 140 mg/dL, and at least two of the subsequent determinations show values that surpass 200 mg/dL. Also, the elevation of blood glucose is maintained for a time longer than normal (Fig. 26.17).

Experimental diabetes. Diabetes mellitus can be produced by total pancreatectomy or parenteral administration of drugs such as streptozotocin, a nitrosourea and alloxan, a substance derived from pyrimidine. These compounds produce permanent diabetes in laboratory animals by inducing selective chemical injury of the pancreatic β cells.

Oral hypoglycemic agents. Several compounds have been developed that, when administered orally, can reduce blood glucose levels. These include sulfonylureas, which stimulate insulin secretion by pancreatic β cells. Obviously, they are effective only when there is production of insulin. Hypoglycemic biguanides function even in the absence of insulin; they increase glucose entry in cells and depress gluconeogenesis.

Thiazolidinediones are agonists of the nuclear receptor γPPAR; they enhance the tissue sensitivity to insulin, probably activating, via leptin and adiponectin, the AMPK. These compounds are useful in the treatment of type 2 diabetes.

Gonads

Besides producing germ cells (ovules and spermatozoa), ovary and testis secrete hormones that are essential for reproductive functions. These hormones are steroids whose actions (1) promote the development and function of reproductive organs and reproductive accessory glands, (2) determine sexual secondary characteristics, and (3) regulate sexual cycles production in women and spermatogenesis in men.

TESTIS ANDROGENS

Various steroids with androgenic activity are produced in the testis (Fig. 26.18). The main one is *testosterone*, which is synthesized in Leydig or interstitial cells. Besides testosterone, the testis secretes a more potent androgen [*dihydrotestosterone* (DHT)] and two weaker ones (*dehydroepiandrosterone* and *androstenedione*). In addition, Leydig cells produce small amounts of estradiol, estrone, pregnenolone, progesterone, 17 α-hydroxypregnenolone, and 17-α-hydroxyprogesterone. Estradiol and DHT are

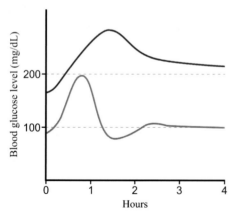

FIGURE 26.17 **Glucose tolerance curve.** Normal response is shown in *grey*; response from a diabetic patient is shown in *red*.

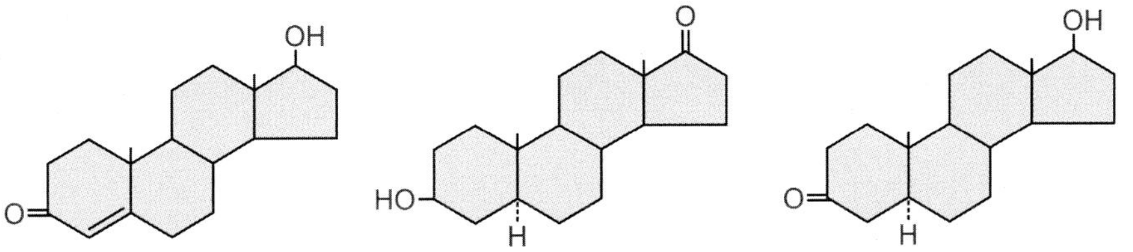

FIGURE 26.18 **Steroids with androgenic activity.**

produced not only in the gonads, but also in peripheral tissues by conversion of androgens and estrogen precursors secreted in testis and adrenal cortex.

Steroidogenesis starts in the fetus between 8 and 18 weeks of gestation, a period in which androgens play a critical role in the development of the male reproductive tract. Except a transient and slight increase at 2–3 months of age, testosterone levels fall after birth and remain low until puberty. At the onset of puberty, pituitary LH production is activated. LH binds to receptors on Leydig cells and stimulates testosterone synthesis and secretion. Testosterone levels are maintained during adult life, with a gradual decline at older ages.

Androgens and FSH target Sertoli cells and stimulate the production of various polypeptides, including *inhibin*, which represses, and *activin*, which stimulates FSH secretion in the pituitary gland, various *cytokines*, and the *androgen-binding protein* that carries sex hormones in plasma. The androgen-binding protein ensures maintenance of high testosterone levels within the testis. Sertoli cells have *aromatase* that catalyzes the conversion of testosterone to estrogens.

Testosterone biosynthesis. Testosterone is produced from cholesterol (Fig. 26.19). Almost half of the cholesterol in testis is synthesized de novo by the Leydig cells; the rest is taken up from plasma LDL by receptor-mediated endocytosis.

In a first step, cholesterol, of 27 carbons, is converted into pregnenolone (21 carbons), by the multienzyme complex *20,22-desmolase*, located in mitochondria. From pregnenolone there are two alternative pathways: the first leads to the formation of progesterone, 17-α-OH-progesterone, androstenedione, and testosterone. The second results in 17-α-OH-pregnenolone, dehydroepiandrosterone, androstenediol, and testosterone production. The stages from 17-α-OH-progesterone and 17-OH-pregnenolone to androstenedione and dehydroepiandrosterone, respectively, involve the loss of the two-carbon chain at position 17; the starting compounds, with 21 carbons, give 19 carbon products. The enzymes participating in these pathways are found in membranes of the smooth ER (Fig. 26.19). Most are mixed-function oxidases linked to cytochrome P_{450}. Three of them (20,22-desmolase, 3-β-hydroxysteroid dehydrogenase, and 17-α-hydroxylase) are the same as those participating in the adrenal androgens and glucocorticoids pathways. The StAR protein is one of the factors involved in the testosterone synthesis.

A normal adult testis secretes between 4 and 12 mg of testosterone/day. The hormone passes to the blood, where it is transported bound to proteins. Approximately 60% of circulating androgens bind with high affinity to a sex steroid-binding β globulin. Almost 40% of testosterone binds to other proteins, mainly albumin, with lower affinity, but larger transport capacity for the hormone. Only a small portion (less than 2%) of the blood androgen remains free and rapidly passes through cell membranes and is the biologically active form.

Testosterone synthesis and secretion is activated by the ICSH or LH of adenohypophysis.

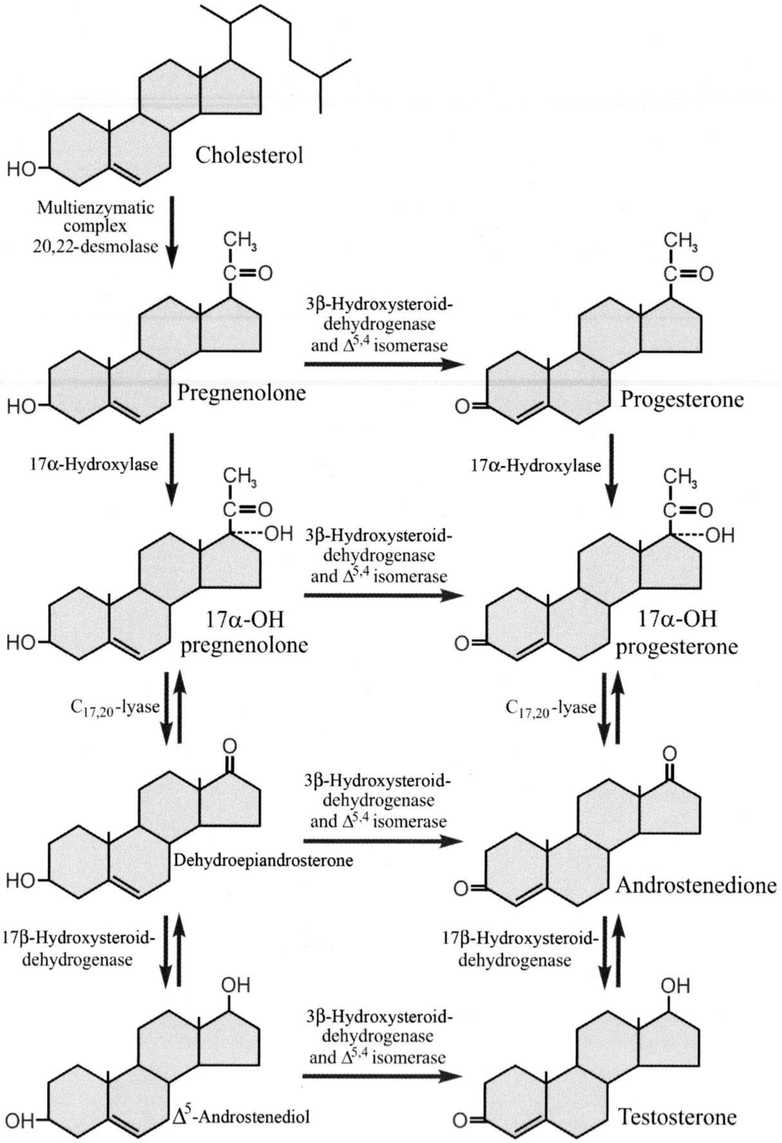

FIGURE 26.19   **Testosterone biosynthesis.**

FSH also has an important role in testicular function. It helps to increase the weight of the testis and stimulates spermatogenesis, but not androgen production. When blood testosterone level increases, there is an inhibitory effect on ICSH secretion by negative feedback.

Metabolism. Testosterone can be converted to DHT (Fig. 26.18) and estrogens. These transformations occur in the testis and, to a greater extent, in other tissues, catalyzed by *5α-reductase*, NADPH dependent. This enzyme is found mainly in liver, skin (hair follicles), and accessory reproductive organs; 8% of total secreted testosterone is metabolized to DHT, a product with more than twice the activity of testosterone, which is why it is considered the true hormone.

Production of estrogens (estradiol) from testosterone comprises a series of reactions (hydroxylation, oxidation, carbon 19 removal, and aromatization of the A ring catalyzed by *aromatase*, a complex inserted in ER membranes that includes NADPH-cytochrome P_{450} reductase). Aromatase is in Sertoli cells and in various tissues, particularly adipose.

Another transformation pathway for testosterone involves the reduction of the ketone group on carbon 3 and the saturation of the double bond of ring A of cyclopentanoperhydrophenanthrene. The resulting products, *androsterone* (Fig. 26.18) and *etiocholanolone*, are eliminated in the urine, conjugated with sulfate or glucuronate. The metabolites of testicular hormones represent about one-third of all 17-ketosteroids eliminated by urine; the remainder is derived from adrenocortical androgens.

Mechanism of action. Both testosterone and DHT freely cross cellular membranes and reach their receptor, a 110-kDa protein that belongs to the steroid receptor family, found in the target cells cytosol and nucleus. The same receptor binds testosterone and DHT; its affinity is higher for the latter. Before hormone binding, the receptor is inactive, associated with HSP. When the hormone-receptor complex is formed, a conformational change occurs, the HSP is released, the receptor dimerizes and binds to a specific response element in DNA, from which it regulates gene expression.

Androgen Effects

Androgens stimulate the development of reproductive organs and accessory glands. Testis development does not depend entirely on androgens; gonadotropins also play an important role. Activation of spermatogenesis requires the presence of testosterone, FSH, and other factors, mostly produced by Sertoli cells. Androgens are responsible for the development of male secondary sexual characteristics, maintain the libido, and promote bone density.

Metabolic effects. Androgens stimulate anabolism, particularly of proteins. They are responsible for nitrogen retention, muscle mass increase, lipid reduction, and sodium, potassium, calcium, phosphate, and sulfate retention. In liver and kidney, the protein anabolic effects of androgens are also apparent, although much less than in muscle.

The anabolic effect of androgens has led to their uncontrolled use, especially in athletes and individuals wishing to increase their muscle mass. All anabolic steroids retain a certain degree of androgenic activity and the abuse produces undesirable effects.

Androgens also act on bone tissue, where the response is dose dependent. Low doses, such as those circulating in blood before puberty, induce epiphyseal cartilage proliferation with increased synthesis of collagen and glycosaminoglycans. High androgen concentrations stimulate calcium uptake, promoting bone calcification and epiphyseal closure.

Androgens have stimulating effect on the synthesis of erythropoietin, which explains the higher hematocrit values that men have compared with women.

In the fetus, testosterone is important for masculinization. It promotes survival of Wolffian duct and its differentiation in epidydimides, ductus deferens, and seminal vesicles. DHT mediates differentiation and development of male urogenital structures.

In fetal and prepubertal testes, Sertoli cells produce *Müllerian-inhibiting substance*. This is a 150-kDa protein that causes degeneration of Müller ducts. It is also secreted in very low amounts in adults, acting in a paracrine manner directly on Leydig cells with inhibitory effects on androgen synthesis.

Insulin-like 3 (Insl-3), member of the insulin like peptides, is produced by Leydig cells after birth and increases markedly during puberty. Adult mice with a mutation in the gene coding Insl-3 show lack of testes descent from the abdomen to the scrotum (cryptorchidism) and spermatogenesis alterations. Although Insl-3 is not essential for gametogenesis, it has some effect on male fertility.

Androgen Alterations

Deficiency of 5α-reductase. The absence of 5α-reductase impairs the production of DHT. In males, genitals are incompletely developed and the individuals who suffer it appear as females at birth (male incomplete hermaphroditism).

Androgens insensitivity syndrome. This is a genetic alteration linked to the X chromosome in which the gene that controls testosterone and DHT synthesis is affected. Androgen deficiency could be total or partial; the phenotype is that of a female, presenting as a pseudohermaphroditism syndrome.

OVARY

The ovaries contain the female germ cells (oocytes) and produce hormones that are essential for reproduction and development of female sexual characteristics.

The major steroid hormones produced in the ovary are as follows: (1) follicular or estrogenic hormones, secreted by developing Graaf follicle cells; (2) progestational hormone or progesterone produced by the corpus luteum formed in ovary after follicle rupture at ovulation; and (3)

androgenic hormones, particularly androstenedione and testosterone.

Ovarian steroids synthesis is controlled by the pituitary gonadotropins LH and FSH.

In addition, the ovary produces peptidic hormones, *relaxin*, secreted by the corpus luteum during pregnancy, and *activin* and *inhibin*, which are found in the follicular fluid.

OVARIAN HORMONES

Estrogens. These hormones are synthesized in granulosa and antral follicular cells. They differ from all other steroids by an aromatic A ring and the absence of carbon 19 at C10 (C10 has a methyl group). The estrogens produced in ovary are *estradiol* and *estrone* (Fig. 26.20), both are interconvertible. Estradiol is the most important estrogen.

Progesterone. This hormone (Fig. 26.21) is produced in all steroidogenic ovary cells, especially in the corpus luteum that develops after the rupture of the follicle during ovulation. It is also produced in the placenta, particularly in late pregnancy, and in the adrenal cortex, as an intermediate metabolite in corticosteroids synthesis (Fig. 26.11).

Androgens. Androstenedione and testosterone are produced mainly in interstitial cells and thecal cells of the ovary.

Biosynthesis and metabolism. The precursor of all ovarian steroidal hormones is cholesterol. Although a small proportion of cholesterol used by the ovary is synthesized locally, most of it is taken up from the blood, through endocytosis of

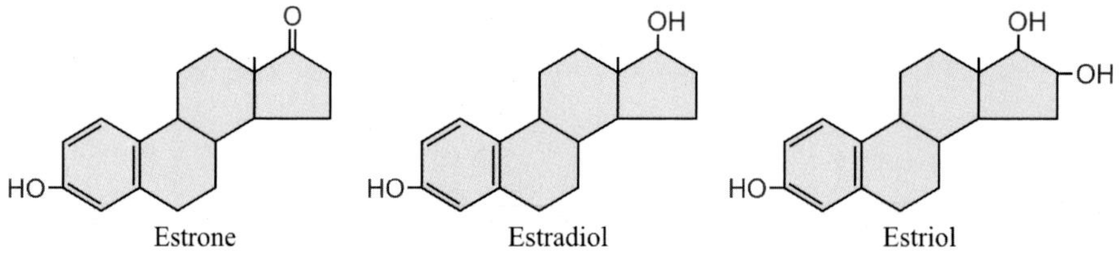

FIGURE 26.20   **Estrogens.**

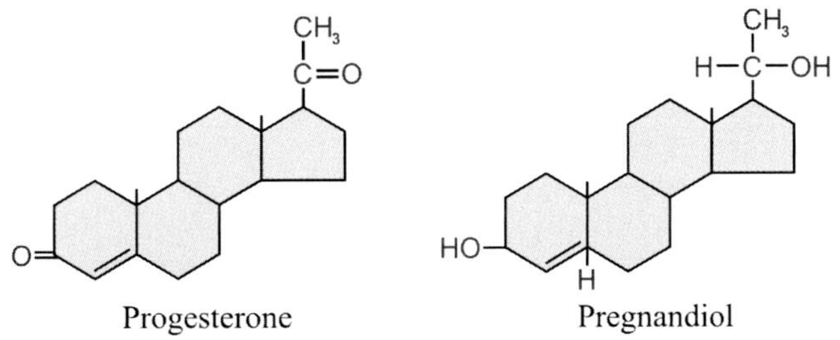

FIGURE 26.21 **Progestagens.**

LDL. Cholesterol released into the lysosomes is sent to the mitochondria by the StAR.

The synthesis of steroids in the ovary initially leads to the formation of androstenedione and testosterone, following reactions identical to those described for the testis. Androstenedione and testosterone are converted into estrone and estradiol, respectively, by *aromatase*, which catalyzes a complex series of reactions: hydroxylations, oxidations, C19 removal, and aromatization of the A ring. Except for the first stage, from cholesterol to pregnenolone, catalyzed by the mitochondrial complex 20,22-desmolase, all of the reactions of synthesis take place in the ER.

The enzymes involved in the synthesis, except 3β- and 17β-hydroxysteroid dehydrogenases, are mixed function oxidases that require molecular oxygen, NADPH, cytochrome P_{450}, and an electron transport system.

Progesterone and androgen are intermediary products in testosterone and estrogen synthesis pathways (Figs. 26.19 and 26.22).

Estrogen synthesis is stimulated by the gonadotropins of the pituitary gland (FSH and LH). Estrone and estradiol are released to blood, where they are transported bound to plasma proteins. Approximately 60% is bound to *sex steroid-binding globulin* and 20% to albumin. Bound estrogens are in equilibrium with the free fraction, which corresponds to 20% of the total hormone; this fraction is the biologically active form. Androgens bind to the same protein. Progesterone is bound with low affinity to transcortin and albumin.

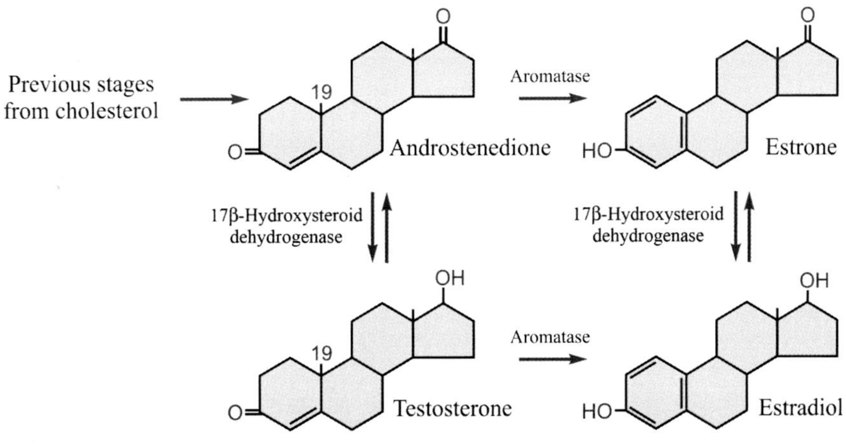

FIGURE 26.22 **Estrogen biosynthesis.** Previous stages, starting from cholesterol, are shown in Fig. 26.19.

Estrogens are primarily metabolized in the liver; the C17 ketone function is reduced to −OH and a hydroxyl group is added to C16. The resulting compound, called *estriol* (Fig. 25.20), is the main estrogen metabolite eliminated by urine, conjugated with sulfate or glucuronate.

The major progesterone metabolites are pregnandiol (Fig. 26.21) and pregnanetriol, excreted as glucuronide conjugates.

Mechanism of action. All steroids share similar mechanisms of action. They cross the plasma membrane and reach their receptors in the nucleus of the target cells. There steroids form hormone-receptor complexes, in which HSPs are also involved. Two types of estrogen receptors, α and β, exist. They differ in their tissue distribution and function. The β receptor can modulate the activity of the α receptor.

The association of estrogens to the receptor causes a conformational change that determines the release of the HSP and the dimerization of the receptor. The zinc finger domain of the receptor recognizes specific hormone response elements in the DNA, which binds transcription factors and modulates protein synthesis (Fig. 25.2).

Estrogen Actions

Effects on genital organs. Estrogens stimulate the development of the ovary, Fallopian tubes, vagina, and uterus. Estrogens stimulate uterine growth and prepare the uterine lining for the action of progestational hormones. They induce endometrial development and increase its vascularization. They produce characteristic changes in the Fallopian tubes and the vaginal epithelium, and are responsible for the development and maintenance of the female secondary sexual characteristics.

Metabolic effects. Estrogens have anabolic action in female genital organs. They increase the uptake of water, sodium, amino acids, and glucose by the myometrium. These actions are secondary to stimulation of protein synthesis. The anabolic action on protein metabolism is weak in muscle, liver, and kidney.

Estrogens antagonize the effects of insulin in peripheral tissues, decreasing glucose tolerance, reducing plasma cholesterol concentration, and increasing HDL levels. They stimulate the liver synthesis of transport proteins for thyroid and sex hormones, and transcortin. They promote bone growth and epiphyseal closure at puberty, and inhibit osteoclast activity. In adult women, estrogens participate in bone remodeling, and play a crucial role in the maintenance of bone mass. The rapid decrease in estrogen secretion, common in postmenopausal women, is an important factor that contributes to osteoporosis.

Nongenomic actions. Estradiol also exerts effects that are not mediated by the nuclear receptor. For example, estradiol causes rapid increase in intracellular calcium concentration $[Ca^{2+}]_i$ in endometrium, maturing oocytes, and granulosa cells. It also has direct action on the vascular system, and activates nitric oxide synthesis and produces vasodilation.

Progesterone Actions

Effects on genital organs. Progesterone appears in blood after ovulation. It favors the endometrium development and prepares the uterus to receive the embryo. It also suppresses estrous, ovulation, and production of LH, which stimulates the formation of the corpus luteum.

When fertilization occurs, the corpus luteum is maintained, and ovulation and menstruation are suspended. High levels of progesterone are important not only to facilitate implantation, but also to maintain pregnancy, stimulating uterine growth, and opposing the action of factors that favor myometrium contraction. In the mammary gland, progesterone activates the development of acini, preparing the gland for milk production. Progesterone is responsible for the rise in basal body temperature that occurs during the second half of the menstrual cycle. Along with estrogen, progesterone regulates the production of pituitary GnRHs.

Progesterone derivatives have been developed and are used as contraceptives.

Metabolic effects. Progestagens have nonspecific effects, mimicking those of glucocorticoids. Although this effect is minor, they mobilize tissue proteins for use in liver gluconeogenesis. Progesterone acts as a competitive inhibitor of aldosterone in the kidney and has natriuretic action.

Nongenomic actions. Most of these actions are exerted on the germ cells, oocytes, and spermatozoa. The sperm responds to progesterone initiating the acrosome reaction, a process that is important for the ability of sperm to fertilize the egg. This effect is secondary to changes in intracellular calcium concentration in sperm.

Oral contraceptives. These are combinations of estrogen and synthetic progestogens, which are used to prevent ovulation through the inhibition of gonadotropin secretion.

Estrogens repress FSH secretion, provide stability, and increase the number of progesterone receptors in the endometrium. Progestagens, in pharmacological doses, inhibit ovulation by suppressing GnRH pulses and LH release, prevent egg implantation, and induce production of very thick cervical mucus that hinders sperm penetration.

Variations of Ovary Hormones in Blood

FSH and LH undergo cyclic variations in their secretion. This secondarily leads to changes in estrogen and progesterone production and release, causing sexual cyclicity in women.

During the first half of the cycle, from menstruation to ovulation, FSH and LH stimulate the maturation of ovarian follicles, of which only one will reach full development. The maturing follicles synthesize estrogens, and their rapid increase in blood stimulates LH release. At the time of ovulation, the levels of FSH, LH, and estrogens reach maximum values. Then, the second phase of the cycle, or postovulatory period, begins; FSH, LH, and estrogens concentrations fall sharply

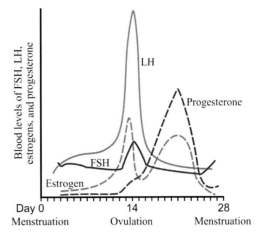

FIGURE 26.23 **Hormone changes in blood during the female cycle.**

(Fig. 26.23). The ruptured follicle becomes corpus luteum, which actively secretes progesterone and also estrogens, both of which attain high blood levels and inhibit FSH production. This prevents the maturation of new follicles.

If the ovule is not fertilized, the corpus luteum begins to regress and the levels of estrogens and progesterone abruptly decrease. This happens around the 26th day of the menstrual cycle and determines the typical menstrual hemorrhage. This is followed by the initiation of a new FSH and LH production cycle (Fig. 26.23).

If the egg is fertilized and pregnancy occurs, there is gradual increase in estrogen and progesterone during the first half of the gestation cycle.

The development of the placenta, an organ that synthesizes gonadotropins, estrogens, and progesterone, contributes significantly to increase the circulating levels of these hormones.

At the beginning of second half of pregnancy, estrogens and progesterone undergo a fast increase, reaching a maximum around the 6th month of gestation. This high level is maintained until immediately before labor, when both estrogen and progesterone suddenly drop. Throughout pregnancy, the changes in concentrations of follicular and progestational hormones follow a parallel course.

OVARIAN PEPTIDE HORMONES

Relaxin. This hormone is produced in the corpus luteum during pregnancy. It has also been found in decidual tissue. Its production is stimulated by chorionic gonadotropin. It is composed by two peptide chains linked by disulfide bridges. Its structure is similar to that of insulin. Relaxin inhibits uterine motility and favors reproductive tract structures relaxation during gestation.

Inhibins. These hormones were isolated from follicular fluid; they are produced predominantly in the preovulatory follicle and corpus luteum of primates. It is a heterodimer of α and β chains. Two forms have been characterized, both of 32 kDa. They exert inhibitory control of the secretion of FSH in the pituitary gland. Together with follistatin, they oppose to activin action. The balance activin/inhibin changes during follicle development; activin prevails during the earliest stages, while inhibin predominates at later stages.

Activins. These hormones are homodimers of β chains similar to those of inhibin, but with opposite functional properties. GnRH stimulates activin synthesis which, in turn, selectively activates FSH secretion. Primarily, activins exert paracrine action within the ovary, where it increases FSH-receptor gene expression. There is evidence suggesting a role of activin in cell differentiation processes.

Follistatins. These hormones are 40 kDa proteins produced in follicular cells that bind activin with high affinity and neutralize its biological activity.

PARATHYROID GLAND

The parathyroid glands are four small structures with the size of a rice grain, located behind the thyroid. They produce a hormone, whose function is to regulate the body calcium homeostasis.

PARATHYROID HORMONE

This hormone is a 84-amino acid protein of 9500 Da. Its partial hydrolysis releases the biological active portion of the hormone, which corresponds to the first 34 amino acids, responsible for binding to the parathyroid hormone receptor.

The gland synthesizes inactive parathyroid hormone precursors, larger than the hormone. The original gene product is a preprohormone of 115 amino acids and 13 kDa, which loses a segment of 25 amino acid residues at its N-terminal (leader or signal peptide) in the ER to form the prohormone of 90 amino acids. The prohormone passes to the Golgi, where a new hydrolysis occurs, catalyzed by an enzyme called *furin*, which separates the first six amino acids. This generates the active PTH, which moves to the cytosol, where it remains in secretory granules. Unlike other hormones, PTH is not stored in significant amounts and PTH reserve is usually small.

Secretion. PTH synthesis and secretion are both regulated by extracellular Ca^{2+}. The main stimulus for the secretion of PTH is the decrease in the concentration of circulating Ca^{2+}. In contrast, increase in serum calcium inhibits PTH release.

Normally, plasma calcium concentration is maintained with great constancy at 10 mg/dL (2.5 mM); 50% of the total calcium in plasma is present as Ca^{2+} (1.25 mM). The rest is bound to albumin (40%) or to nonionizable complexes (10%).

When serum calcium falls below 9 mg/dL ($[Ca^{2+}]$ ~1.12 mM), PTH secretion increases rapidly, reaching a maximum at 6 mg/dL ($[Ca^{2+}]$ ~0.75 mM). At lower plasma calcium levels, the release of PTH does not further increase. On the contrary, the increase in calcemia depresses hormone release. The maximum inhibitory effect is attained at plasma calcium values of 11 mg/dL ($[Ca^{2+}]$ ~1.37 mM); however, secretion is never completely suppressed.

The fluctuations in the extracellular Ca^{2+} concentration are monitored by a calcium "sensor" in the plasma membrane of the parathyroid cells, renal tubules, thyroid C, cells and in other tissues. This Ca^{2+} sensor is a 120-kDa protein with three domains: the first, extracellular, is rich in acidic amino acids probably involved in interaction with Ca^{2+} ions. The second portion consists of seven transmembrane helices characteristic of the G-protein-coupled receptors superfamily. The third domain includes the cytoplasmic C-terminus coupled to G_q proteins and probably also to G_i.

Under high extracellular Ca^{2+}, the calcium sensor undergoes a conformational change, which activates PLC and triggers the phosphatidylinositol bisphosphate signal system that causes the release of Ca^{2+} from intracellular stores and its uptake through channels in the plasma membrane. In addition, the Ca^{2+} sensor interacts with G_i proteins and depresses adenylate cyclase and cAMP levels. Plasma Ca^{2+} levels regulate not only PTH secretion, but also PTH synthesis, possibly by activation of gene transcription. This transcription is inhibited by high levels of 1,25-$(OH)_2$-D_3 vitamin.

Besides the effect of intracellular Ca^{2+}, cAMP reduction is another factor that suppresses PTH secretion by inhibiting fusion of the granules containing PTH with the plasma membrane.

In most cells, Ca^{2+} increase stimulates exocytosis. In contrast, PTH release is stimulated by the decrease in cytosolic Ca^{2+}. Mg^{2+} activates PTH secretion; however, hypermagnesemia inhibits PTH release. Catecholamines, acting through β-adrenergic receptors, and cAMP have a modest stimulatory effect on PTH secretion.

Degradation. PTH degradation occurs mainly in liver and kidney. PTH is cleaved in two segments: a N-terminal portion (containing the first 34 amino acids), which is still active, and the remaining of the molecule, which is inactive. The entire hormone and the cleaved segments can be detected in circulating blood. The method of choice for PTH determination is the immunoradiometric assay.

The half-life of the full length PTH and the active segment is of 2–4 min; both are completely degraded in the target tissues, kidney, and bone. The C-terminal inactive segment (amino acids 35–84) has a longer half-life (hours) and is eliminated in urine. It does not appear to play a physiological role.

Mechanism of action. The parathyroid hormone receptor is coupled to G_q and G_s proteins. Two types of PTH receptors are expressed in the target cells, PTH-1 and PTH-2. PTH-1 receptor recognizes PTH and PTHrP (see subsequent section), while PTH-2 is activated only by PTH. Binding of PTH or PTHrP to its receptors has the following effects: (1) activation of G_q protein and PLC, which leads to increased cytosolic Ca^{2+} and activation of protein kinase C, and (2) activation of G_s protein, which elevates cAMP concentration in the cells.

Actions. The main function of PTH is to contribute, along with other factors, to maintain the concentration of calcium constant in the extracellular compartment. It targets bone, renal tubules, and vitamin D metabolism (see p. 655). Table 26.9 summarizes the actions of PTH.

1. *Effects on bone.* PTH stimulates the mobilization of calcium from bone. PTH is

TABLE 26.9   Parathyroid Hormone Actions

On bone

↑ Resorption of mineral
↑ Citrate and other organic acid in osteocytes
↑ Ca^{2+} efflux to the extracellular fluid

On kidney

↑ Resorption of Ca^{2+} in distal tubules
↑ Urinary excretion of phosphate

On vitamin D_3

↑ Synthesis of 1,25-$(OH)_2$-D_3:
  ↑ Intestinal Ca^{2+} absorption

General

↑ Calcemia

involved in two processes: calcium release (osteolysis) and osteoclast resorption. PTH stimulates the Ca^{2+}-ATPase and the active transport of Ca^{2+} to the extracellular fluid. This response occurs within minutes and requires the presence of 1,25-$(OH)_2$-D_3, the active metabolite of vitamin D. PTH activates glycolysis and the citric acid cycle in osteocytes; it increases production of citrate and other organic acids, which favors bone mineral resorption. PTH stimulates the differentiation of osteoclast precursor cells and activates the production of cytokines, stimulators of bone resorption. This effect requires hours to days.

The end result of PTH effect is the removal of Ca^{2+} and demineralization of bone, which increases calcium movement to the extracellular space.

2. *Effect on kidney.* PTH stimulates Ca^{2+} reabsorption in the nephron distal convoluted tubules and decreases Ca^{2+} excretion in urine. Most of the renal Ca^{2+} reabsorption occurs predominantly in the proximal tubule (and in the lower portion of the loop of Henle and thick ascending limb), and is not regulated by PTH. Although PTH only controls 10% of the total calcium absorbed in the kidney, it is of high physiological relevance.

When the plasma phosphate concentration rises, PTH excretion is stimulated and causes a reduction of phosphate reabsorption in proximal tubules and, to a lesser extent, in the distal nephron.

Bone mineral mobilization increases the circulating levels of calcium and phosphate. PTH increases phosphaturia, which helps preventing calcium phosphate being deposited in soft tissues.

PTH decreases renal reabsorption of sodium, potassium, citrate, and bicarbonate. These actions tend to produce discrete acidosis. PTH also reduces excretion of uric acid in urine.

3. *Effect on vitamin D metabolism.* PTH administration increases intestinal calcium absorption. It is not a primary effect of the hormone, but secondary to the formation of 1,25-$(OH)_2$-D_3, active vitamin D_3 metabolite (p. 655), which is formed in kidney by hydroxylation of 25-OH-D_3, a vitamin D_3 metabolite. The hydroxylation reaction is activated by PTH.

PTH-related protein (PTHrP). This hormone is normally produced by fetal and adult tissues. The N-terminal portion has high homology with PTH and an affinity for binding to PTH-1 receptors on bone and kidney that is similar to that of PTH. PTHrP promotes hypercalcemia and decreases phosphate and cAMP concentrations in the kidney.

During intrauterine development, PTHrP functions as a regulator of chondrocyte proliferation and mineralization, and calcium transport in placenta. After birth, it contributes to the development and cell differentiation of mammary gland, skin, and pancreas. PTHrP has cytokine type effects during the inflammatory response, relaxes smooth muscle, and is neuroprotective during aging.

Under physiological conditions, PTHrP action is local (paracrine and autocrine). When used in higher concentrations, it reduces 1,25-$(OH)_2$-D_3 formation and uncouples bone resorption and new bone formation, resulting in severe loss of bone tissue.

Parathyroid Hormone Alterations

Hypoparathyroidism. This alteration may result from unintentional removal of the parathyroid glands during thyroidectomy. It is accompanied by decreased serum calcium and increased serum phosphorus. The increase in extracellular fluid phosphate promotes calcium phosphate deposition in bone, subtracting Ca^{2+} from the circulation and accentuating the hypocalcemia. Bone calcium mobilization decreases by reduction of osteolysis as well as osteoclastic resorption.

As 1,25-$(OH)_2$-D_3 synthesis is depressed, intestinal absorption of Ca^{2+} is poor. Renal Ca^{2+} reabsorption is reduced, but there is not increased urinary calcium because serum calcium is low. Excretion of bicarbonate is reduced and alkalosis may occur.

The most striking symptom of hypocalcemia is associated with neuromuscular hyperexcitability, manifested by paresthesia in fingers and toes, cramps, reduced myocardial contractility, prolonged QT interval, irritability, tetany, and convulsions.

Pseudohypoparathyroidism is a rare genetic disorder in which the PTH receptor is affected, leading to tissue resistance to the hormone.

Hyperparathyroidism. This disorder can be primary or secondary to other pathologies, including parathyroid functioning tumors (adenomas with high PTH production), kidney failure (which causes continuous loss of calcium in urine), or nutritional deficiencies with sustained calcium intake reduction.

Hyperparathyroidism is characterized by activation of calcium and phosphate mobilization from bone, increased plasma calcium level, hypercalciuria, and hyperphosphaturia. Often, calcium phosphate (stones) precipitates in the urinary tract or in soft tissues. Demineralization of bone can lead to generalized osteodystrophy.

High calcium levels affect neuromuscular excitability. Frequently, patients complain of fatigue; they have mental disorder, depression, and heart disfunction with a shortened cardiac QT interval in the electrocardiogram.

Some types of cancer, such as lung squamous cell carcinoma, breast, and kidney cancer, secrete PTHrP, producing a clinical disorder similar to that of hyperparathyroidism.

Therapeutic use of PTH in osteoporosis. Although hyperparathyroidism causes bone loss, the administration of PTH in daily injections to patients with osteoporosis has shown positive results, with reduction of bone reabsorption, restoration of bone mass, and lower incidence of fractures. This result, apparently paradoxical, suggests that there are differences in the action of sustained elevated levels and intermittent increases of PTH.

CALCITONIN

In mammals, calcitonin is synthesized and secreted by thyroid C cells (which are clear, with oval nucleus). These cells embryologically derive from the neural crest (fifth branchial pouch). Calcitonin is a 32-amino acid peptide with a mass of 3600 Da. Disulfide bridges generate in the molecule a six-amino acid cycle, analogous to those of vasopressin and oxytocin. The whole molecule is required for biological activity; loss of a single amino acid changes its functional capacity.

Its synthesis is directed by a gene that transcribes into different mRNAs by alternative splicing. C cells generate a precursor protein of 141 amino acids. In other tissues, particularly in the central nervous system, the same gene directs the synthesis of a 128-amino acid precursor from which a 37-amino acid peptide called *calcitonin gene-related peptide* is generated; it acts as potent vasodilator and neurotransmitter.

The primary stimulus for calcitonin secretion is the increase in extracellular calcium concentration. In addition, some gastrointestinal hormones (gastrin, secretin, and cholecystokinin) influence calcitonin secretion.

Actions. Calcitonin exerts its effects by binding to G_s protein-coupled receptors. It produces rapid decline in calcium and phosphate in blood. Administered at higher than physiological doses, it decreases the number and activity of osteoclasts in bone and calcium and phosphate tubular reabsorption in kidney.

The physiological role of calcitonin in humans is still unclear. It is a very important hormone in fish and birds, but as the surgical removal of calcitonin-producing cells in patients does not determine clinical disorders, some authors think that their role is not essential in humans, and that it just represents an evolutionary relic. Its relationship with gastrointestinal hormones suggests that it could prevent hypercalcemia produced by foods rich in calcium.

From a clinical point of view, it is of interest as a thyroid tumor "marker" and also as an inhibitor of osteoclasts activity and bone resorption.

Calcitonin has analgesic properties; it is effective in the treatment of some pain-related syndromes. As it increases calcium in bone, it has also been used in osteoporosis, but its administration should be limited to the lowest possible doses and for short periods of time because it was observed that a high proportion of patients could develop different types of cancer.

KIDNEY

In addition to maintaining the constancy of the extracellular fluid volume and composition of the body, the kidney also acts as an endocrine organ. It secretes hormones with systemic action, such as renin, erythropoietin, and calcitriol. Furthermore, similar to other organs, it synthesizes a variety of agents with autocrine or paracrine effects, such as prostaglandins, kinins, endothelin, urodilatin, and dopamine. These substances exert their effects locally, within the kidney.

RENIN

Renin is a protease of approximately 40 kDa. It is synthesized by the nephron granular juxtaglomerular cells as a 340-amino acid precursor or preprorenin, which is inactive. Preprorenin reaches the ER, where it loses a 23-amino acid segment from the amino terminus and becomes prorenin. This product traffics to the Golgi apparatus, where it is glycosylated and then moves to the cell cytoplasm, staying in vesicles loaded with secretory granules. Prorenin is processed in the same granules with loss of 20 amino acids to generate renin, which has proteolytic activity. Upon stimulation, renin is released into the circulation. Also, prorenin is secreted; between 50% and 90% of total renin in plasma is prorenin, whose function is unknown.

Renin release is controlled by several factors: (1) decreased systemic blood pressure and blood flow to the kidney, which is sensed by baroreceptors in the afferent glomerular arteriole; (2) Na^+ and Cl^- concentration in the tubular fluid, which is sensed at the macula densa in the juxtaglomerular apparatus; (3) sympathetic nervous system activity; (4) K^+ concentration (with hypokalemia stimulating and hyperkalemia depressing renin release); and (5) angiotensin and atrial natriuretic peptide (ANP).

In blood, renin acts on angiotensinogen, a plasma α_2 globulin of ~60 kDa, synthesized in the liver. Renin catalyzes the removal of a 10-amino acid segment from the N-terminus of angiotensinogen. The remaining decapeptide called *angiotensin I* is physiologically inactive. In plasma, a dipeptidyl carboxypeptidase, the *angiotensin-converting enzyme* (ACE) from the plasma membrane of vascular endothelial cells, excises two amino acids from the C-terminus of angiotensin I, converting it into *angiotensin II*. The octapeptide angiotensin II is active and binds to plasma membrane receptors of target cells. Two main types of angiotensin II receptors, AT1 and AT2, exist. AT1 is the mediator of almost all angiotensin II actions. AT2 apparently stimulates the release of bradykinin and nitric oxide (NO), which cause vasodilation; it is also involved in regulating cell growth and differentiation.

AT1 receptor is coupled to G_q protein and activates phosphatidylinositol bisphosphate signal system. The cell signaling pathway associated with AT2 receptors is still not clearly known.

AT1 activation by angiotensin II results in the following responses: (1) systemic arteriolar vasoconstriction, which raises blood pressure by increasing peripheral resistance. (2) Increased Na^+ reabsorption in the nephron proximal tubules and water, which increases plasma volume. (3) Activation of aldosterone secretion by the adrenal cortex. Aldosterone increases Na^+ reabsorption in the loop of Henle, thick ascending limb, distal tubules, and collecting ducts. (4) Negative feedback effect on renin secretion. In summary, angiotensin II actions tend to maintain fluid volume and pressure in the vascular system. The half-life of angiotensin II in circulation is less than 60 s.

Inhibitors of the ACE (ACE inhibitors) are widely used in clinics to prevent angiotensin II production. Also, AT1 receptor blockers have important therapeutic use. Both of these drugs produce a decrease in angiotensin II and aldosterone, which results in systemic vasodilation and reduced blood pressure.

The renin–angiotensin system is not only present in the circulation, but it is also found in various tissues (brain, heart, kidney, adrenal, ovary, and testis), where it produces angiotensin II locally.

Another form of ACE was discovered, known as ACE2, present only in heart, kidney, and testis. Like ACE1, ACE2 is a carboxypeptidase, but only separates one amino acid from angiotensin I instead of two. The product of ACE2 action is a nonapeptide (angiotensin I-9), which apparently has no physiological function. It has been suggested that ACE2 prevents angiotensin II formation.

ACE2 is also able to hydrolyze from angiotensin II a heptapeptide (angiotensin I-7), which has vasodilating action. This enzyme is thought to have a role in vasoconstriction and vasodilation balance in heart and kidney, regulating vascular function in these organs.

Another enzyme, aminopeptidase A, removes the aspartate residue at the amino-terminal end of angiotensin II and produces a heptapeptide named angiotensin III, with the same capacity to stimulate aldosterone secretion as angiotensin II.

ERYTHROPOIETIN

Erythropoietin is a glycoprotein of 166 amino acids (30.4 kDa), with a 60% protein and 40% carbohydrate content. In adults it is synthesized in the kidney and, to a small extent, in the liver. In the fetus and newborn, the primary site of erythropoietin origin is the liver. In kidney it is produced as a 193-amino acid precursor by mesangial cells located between the renal cortex and medulla.

Synthesis and secretion of erythropoietin are stimulated by reduction of the oxygen tension in the kidney. This may be secondary to hypoxia, anemia, or deficiencies in renal blood flow. It is believed that, by reducing the oxygen partial pressure, prostaglandins (PG) are released, and these are the mediators which stimulate erythropoietin release.

Erythropoietin targets the bone marrow and binds to receptors in erythroid progenitor cells. The receptors belong to the family associated with extrinsic tyrosine kinase (*JAK-2*). This phosphorylates *STAT*, an intracellular protein that dimerizes and enters the cell nucleus, where it activates the transcription of several genes, including *c-myc*. This stimulates mitotic activity and differentiation of erythroblasts precursor cells. The end result is the increased release of reticulocytes and red blood cells into the circulation, increasing the oxygen carrying capacity of blood.

Erythropoietin administration in adequate doses has beneficial effects on anemia of chronic renal patients. Currently, erythropoietin is produced by recombinant DNA techniques for therapeutic use.

BRADYKININ

Kinins are hormones of peptide nature. They are produced in the kidney and in other tissues from an inactive plasma precursor called kininogen. Nephron distal tubules cells produce *kallikrein*, a protease that hydrolyzes kininogen and liberates *bradykinin*, a powerful vasodilator. Bradykinin stimulates NO and PGs release and tends to counteract the reduction in renal blood flow induced by angiotensin II. It is rapidly inactivated by the ACE.

ENDOTHELIN

Endothelin is a 21-amino acid peptide. Three forms are known, encoded by different genes (ET-1, ET-2, and ET-3). ET-1 is the predominant form in kidney and is synthesized by renal endothelial, mesangial, and distal tubular cells.

Angiotensin II, epinephrine, and distension of vessel walls stimulate endothelin secretion.

Endothelin is a potent vasoconstrictor of efferent and afferent glomerular arterioles; it reduces renal blood flow, inhibiting renin secretion and increasing the release of prostaglandin PGE2.

Increased endothelin has been observed in patients with malignant hypertension, heart failure, pulmonary hypertension, and myocardial infarction. A rare type of tumor, hemangioendothelioma, secretes endothelin and produces hypertension.

URODILATIN

Urodilatin is a 32-amino acid peptide synthesized in renal distal tubule cells. It has homology with the atrial natriuretic peptide (ANP), differing from it by four additional amino acids at the N-terminal end of the polypeptide. Like ANP, urodilatin promotes sodium excretion.

CALCITRIOL

Calcitriol or 1,25-$(OH)_2$-D_3 is the active metabolite of vitamin D. The final hydroxylation that generates this hormone is produced in the kidney. Its synthesis and actions are considered in Chapter 27.

HEART

Atrial Natriuretic Peptide

The heart acts as an endocrine gland; it produces a hormone called *atrial natriuretic peptide* (ANP), which is involved in sodium and water excretion and to the regulation of extracellular fluid volume and blood pressure.

After the discovery of atrial natriuretic factor, similar peptides were found in other organs, all of which share a 17-amino acid segment that forms a cycle with a disulfide bridge. In humans, the major circulating natriuretic peptide (ANP) has 28 amino acids and a mass of 3060 Da. The others include natriuretic peptide B (32 amino acids), also produced in heart, and C (22 amino acids) isolated from brain and vascular endothelium. ANP is originally synthesized as an inactive preprohormone of 151 amino acids. The leader or signal peptide is a hydrophobic segment of 24 amino acids at the N-terminal. The prohormone is hydrolyzed to produce the active peptide.

Mechanism of action. ANP binds to specific receptors located in kidney, blood vessels, and the adrenal gland. Depending on the target tissue, activation of these receptors by ANP stimulates guanylate cyclase activity and increases cyclic GMP, or inhibits adenylate cyclase and reduces cAMP.

Actions. ANP has several effects: (1) *on the renin–angiotensin system*, it inhibits renin secretion and, therefore, aldosterone release. In addition, ANP can inhibit aldosterone secretion directly. (2) *On kidney*, ANP stimulates water and sodium excretion and increases glomerular filtration rate. (3) *On the vasculature*, ANP relaxes smooth muscle, especially in kidney arterioles, which lowers blood pressure and cardiac output. (4) *On the nervous system*, ANP locally regulates extracellular fluid volume and blood pressure and inhibits production of vasopressin.

Clinical alterations. In patients with heart or renal function failure and in patients with essential hypertension, elevated levels of ANPs in circulating blood are observed. This has prompted the use of natriuretic peptides as markers for heart function and as a pharmacological approach for the treatment of cardiovascular and renal disease.

GASTROINTESTINAL HORMONES

The neuroendocrine cells of the gastric and intestinal mucosa produce and secrete a series of peptides with hormonal activity in response to stimuli from the digestive tract lumen. The cells that synthesize these hormones are not grouped in discrete glandular structures but instead, they are widely distributed in between the exocrine and absorptive cells of the gastrointestinal mucosa and in other organs. These cells form a system, which has been called APUD for amine precursor uptake and decarboxylation, due to the ability of the cells to capture amine compounds and to decarboxilate them.

Some of the hormones of this system include the following.

GASTRIN

Gastrin is produced by the G cells of gastric antral region and duodenal bulb mucosa. A 101-amino acid precursor, preprogastrin, is first synthesized, which undergoes processing to produce various active molecular forms of 14, 17, and 34 amino acids in length. Its activity depends on a small segment at the C-terminal portion of the polypeptide. *Pentagastrin*, a synthetic peptide containing the last five amino acid residues of gastrin, is used to stimulate gastric secretion.

Peptides with gastrin immunoreactivity have been identified in the central and peripheric nervous system, pituitary and adrenal glands, and respiratory and genital tracts. The fetal pancreas produces amidated gastrin, suggesting a possible role of this hormone in pancreatic development.

Gastrin release normally occurs in response to gastric distension and to the presence of protein-rich foods in the stomach. Aromatic amino acids, small peptides, and calcium are potent stimuli for gastrin secretion.

The 17 and 34 amino acid amidated gastrins bind to G_q protein-coupled receptors, trigger the phosphatidylinositol bisphosphate signaling system, activate PLC, and stimulate the release of histamine. Histamine binds to H_2 receptors in the basolateral membrane of parietal cells and induces the production and secretion of hydrochloric acid. Furthermore, gastrin exerts genomic actions modulating gene expression in parietal cells. In addition, gastrin activates stomach smooth muscle contractility, accelerating gastric emptying.

Gastrin is also secreted by the β cells of the pancreas Langerhans' islets.

Gastrin-secreting tumors (gastrinoma) of the pancreas produce high amounts of gastrin. They cause the *Zollinger–Ellison syndrome*, which is characterized by gastrointestinal mucosal ulceration, chronic diarrhea, vomiting, and loss of weight and appetite. Gastrinomas could present as single tumors, or as multiple tumors in the pancreas, duodenum, lymph nodes, pituitary gland, and other organs, constituting the multiple endocrine neoplasia type I syndrome.

SECRETIN

Duodenal and intestinal mucosal cells synthesize a 27-amino acid polypeptide structurally homologous to glucagon that is known as secretin. The presence of acid in the duodenum, coming from the stomach acid chyme, is the stimulus for secretin release. Secretin binds with high affinity to G_s-coupled receptors, which mediate adenylate cyclase activation, and with low affinity to G_q-coupled receptors, which activates the phosphatidylinositol bisphosphate system.

Secretin stimulates the secretion of pancreatic juice, rich in bicarbonate and poor in enzymes; this action is potentiated by cholecystokinin (CCK). This contributes to neutralize the acid chyme reaching the intestine. Secretin also inhibits HCl secretion in the stomach and delays the emptying of stomach content.

CHOLECYSTOKININ

This hormone is synthesized in the mucosal endocrine cells of the duodenum and jejunum, in neurons of the myenteric plexus of the stomach, and the colon submucosa, where it functions as a neurotransmitter. In the central nervous system and other organs, several peptides that react with antibodies against CCK have been described. CCK is synthesized as a 115-amino acid precursor, which after posttranslational processing, generates multiple molecular CCK forms. In intestine, CCK forms of 33 and 39 amino acids have been identified, while in brain peptides of 8–58 amino acids were isolated. The last five amino acids of CCK are identical to those of gastrin. CKK is the same hormone that some authors have designated as *pancreozymin*.

CCK binds with high affinity to seven transmembrane domain receptors that are expressed in pancreatic acinar cells, gallbladder, gastric mucosa cells, and peripheral and central nervous system.

CCK secretion is stimulated by triacylglycerols, long chain fatty acids, aliphatic and aromatic amino acids, and proteins arriving to the intestinal tract. The main negative regulator of CCK is the presence of bile salts in the duodenum.

CCK is the main regulator of gallbladder motility; it produces gallbladder contraction and relaxation of Oddi's sphincter. CCK increases contraction of stomach antrum and pylorus, and stimulates the secretion of pancreatic juice rich in enzymes or their zymogens (pancreozymin effect). Acting on hypothalamic receptors, CCK inhibits appetite.

Different CCK peptides function as neurotransmitters in the central nervous system.

VASOACTIVE INTESTINAL PEPTIDE

The vasointestinal intestinal peptide (VIP) is a 28-amino acid molecule produced in different types of neurons of the spinal cord and brain, intestine, lungs, urogenital tract, and endocrine glands. VIP is structurally homologous to secretin and glucagon.

Receptors for VIP have been found in the intestinal epithelium, smooth muscle and arterioles, stomach, gall bladder, colon, exocrine pancreas, parotid gland, lymphocytes, and hepatocytes. These receptors are of two types: high affinity, coupled to G_s protein, and another low affinity, coupled to G_q.

VIP is a potent vasodilator and a neuromediator with a wide range of biological activities. It plays a role in regulating motility of intestinal smooth muscle, pancreatic and intestinal secretions, and gastrointestinal tract blood flow. It is a powerful relaxant of intestinal and vascular smooth muscle.

ENTEROGLUCAGON

Enteroglucagon is produced in intestinal cells as part of a precursor molecule which, after processing, renders several *glucagon-like peptides*. The first hydrolytic cleavage of the precursor generates *glicentin* (amino acid residues 1–69 of the precursor protein). A second cleavage of glicentin produces *oxyntomodulin* (amino acids 33–69 of the precursor), which includes enteroglucagon.

Enteroglucagon stimulates intestinal mucosa growth. Glucagon-type peptides and oxyntomodulin inhibit stomach acid secretion, delay gastric emptying, and have inhibitory effect on food intake.

It increases insulin secretion stimulated by glucose and reduces blood glucose.

GHRELIN

Oxyntic cells of the stomach synthesize and secrete the 28-amino acid peptide ghrelin. A small number of ghrelin-producing cells were also identified in the small and large intestine. Ghrelin is an orexigenic or a"hunger" hormone, involved in hunger sensation. Fasting increases the synthesis of ghrelin; its circulating levels are high before food ingestion and fall after a meal. Ghrelin binds to specific receptors in hypothalamus, pituitary gland, and other tissues. On the nervous system it acts on the vagal nerve and on appetite regulatory centers, including hypothalamus, hindbrain, and the mesolimbic system.

In addition to appetite stimulation, ghrelin has two- to threefold higher activity on GH release than hypothalamic GHRH. Ghrelin also contributes to the regulation of insulin sensitivity and liver release of glucose; it causes vasodilation and depresses proinflammatory responses.

Intravenous administration of ghrelin increases acid secretion in stomach and stimulates gastric motility.

AMYLIN

This hormone is a 37-amino acid peptide produced in β cells of the pancreatic islets and in cells scattered in the stomach and proximal intestine. It has high homology to calcitonin. Amylin exerts its action by interacting with a modifier protein and calcitonin receptor.

Amylin effects include the reduction of food intake and inhibition of gastric acid secretion. In pharmacological doses, it suppresses gastric emptying and glucagon secretion. It has been proposed that excessive production of amylin could be related to diabetes pathogenesis.

GALANIN

Galanin is expressed in the central and peripheral nervous system, pituitary gland, intestinal neurons, pancreas, and thyroid and adrenal glands. In intestine galanin is predominantly synthesized in neurons of the myenteric plexus, which innervates circular and longitudinal smooth muscle layers.

Galanin presents two molecular forms, one of 19 and the other of 30 amino acids. These hormones are released in response to intestinal distension and act through several types of receptors coupled to G-proteins, which are expressed in gastric and intestinal smooth muscle, pancreas, and central nervous system.

In the nervous system, galanin modulates neural transmission, pain sensation, and response to stress; it participates in cognitive and memory processes. It has also general effects on energy balance, innate immunity, inflammation, and cancer. In the digestive tract, galanin regulates intestinal motility and ion transport, gallbladder contraction, pancreas exocrine secretion, and moderates food ingestion.

Other peptides. Other peptides produced in the intestinal mucosa include *neuropeptide Y*, or *peptide YY*, produced in the intestinal mucosa, and *pancreatic polypeptide*, synthesized in the pancreas islets. These peptides are related to appetite control; neuropeptide Y is secreted primarily by neuronal cells of the central and peripheral nervous system; it is a powerful stimulant of food intake. Peptide YY and pancreatic polypeptide suppress appetite.

PINEAL GLAND

The pineal gland of vertebrates is the major site of *melatonin* production. Although in minimal amount, it is also synthesized in other tissues, especially retina.

Melatonin is an indolamine compound derived from tryptophan. Serotonin or

5-OH-tryptamine is one of the intermediate metabolites of melatonin synthesis; the final metabolic steps are catalyzed by *N*-acetyl and O-methyltransferase (see p. 392).

The synthesis of melatonin is highly regulated. A circadian rhythm of synthesis and secretion is regulated from the hypothalamic suprachiasmatic nucleus. In mammals, this nucleus is the "pacemaker" of circadian rhythms, including body temperature, sleep, and cortisol secretion regular changes. It is possible that the pineal gland, via melatonin, modulates the circadian system sensitivity to ambient light and helps to synchronize the "biological clock."

Melatonin is released from the pineal gland to the circulation during the dark period of the daily cycle. Its secretion is suppressed in individuals exposed to intense light. There are specific melatonin receptors in several areas of the mammalian brain, especially in the suprachiasmatic nucleus. Some of these receptors are coupled to G_i and others to G_q proteins. The first inhibits adenylate cyclase, and the second triggers the phosphatidylinositol bisphosphate system.

Multiple actions are attributed to melatonin, which have not been totally confirmed, such as sleep induction, antiaging effects, and sexual arousal. According to the available experimental data, it can be concluded that melatonin pulses serve as an internal signal to regulate light–dark cycles and to adjust the circadian rhythm of the body.

In humans, melatonin administration appears to be useful to reset the biological clock in individuals who undergo circadian rhythms shifts due to transmeridian travels, day/night schedule changes, or other causes.

Studies on the effectiveness of melatonin administration on sleep disorders have not provided a clear answer as to the benefits of melatonin. Actions on sexual function have been observed in species with seasonal reproduction. In sheep, for example, the extension of the daily dark periods, with high levels of melatonin, favors reproductive activity.

Some experimental evidence indicates that melatonin has antiapoptotic properties. This has led to propose that it could participate in tumor prevention and in modulation of immune reactions.

Independently of its hormonal action, melatonin has been shown to be an antioxidant agent in vivo.

EICOSANOIDS

Eicosanoids are a group of compounds that comprise prostaglandins, tromboxane, *prostacyclins* (PGI), and leukotrienes. They are synthesized from 20-carbon polyunsaturated fatty acids in all mammalian cells, except erythrocytes. They have very short half-life, but exert important biological effects. They function as autacoid hormones, with autocrine or paracrine action on the cells of origin or neighboring cells. The fatty acids precursors of these substances form part of phospholipids of cell membranes, which are released by phospholipase A_2. The most important of these fatty acids is arachidonic acid [20:4(5,8,11,14)].

The reactions which convert fatty acids into PGs, TXs, and PGI are initiated by *cyclooxygenase* (p. 352). The most important PGs are PGE_2, $PGF_{2\alpha}$, and PGD_2 (Fig. 26.24).

Thromboxane TXA_2 (Fig. 26.25) is mainly synthesized in platelets, and prostacyclin PGI_2 is formed in blood vessels endothelium.

Leukotrienes (Fig. 26.26) are derived from arachidonic acid by the *lipoxygenase* pathway (p. 352); they are produced by leukocytes.

Eicosanoids are rapidly metabolized; their modulatory actions are mostly paracrine.

Mechanism of action. Eicosanoids bind to specific receptors on the plasma membrane of the target cells. Some of these receptors are linked to G_s proteins, with cAMP as mediator. Others are coupled to G_i proteins. A third group of eicosanoids uses receptors linked to G_q proteins and activate PLC, an increase intracellular Ca^{2+} levels.

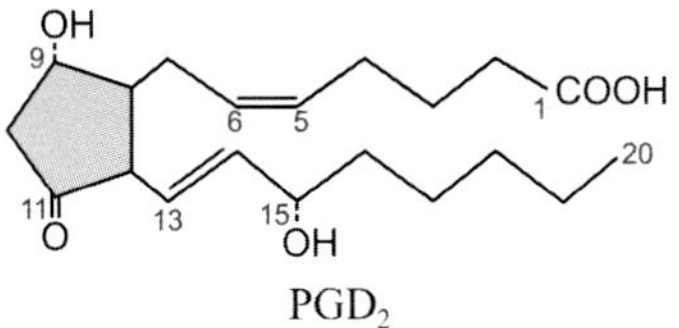

PGD$_2$

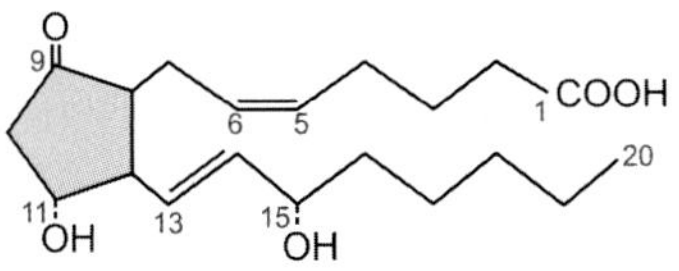

PGE$_2$

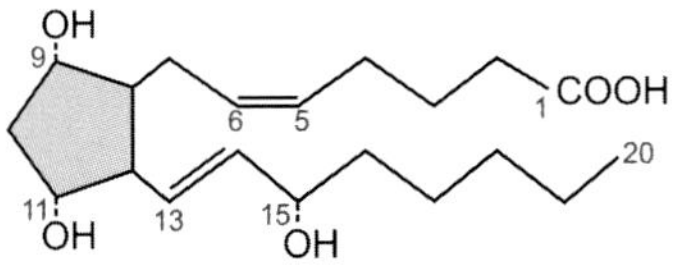

PGF$_{2\alpha}$

FIGURE 26.24 **Structure of the most common prostaglandins.**

Actions. Eicosanoids promote various activities on basically all body organs. The range of effects is very wide; only some examples will be mentioned.

Respiratory system. PGD$_2$ and PGF$_{2\alpha}$, TXA$_2$ and LTC$_4$, and LTD$_4$ and LTE$_4$ leukotrienes cause contraction of the smooth muscle of bronchi.

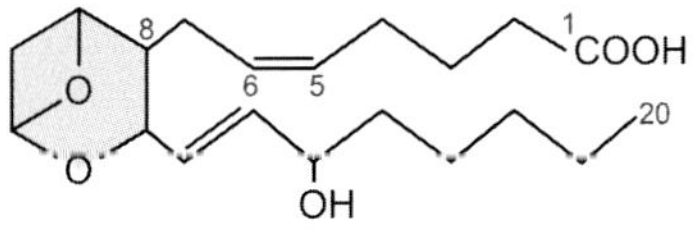

TXA$_2$

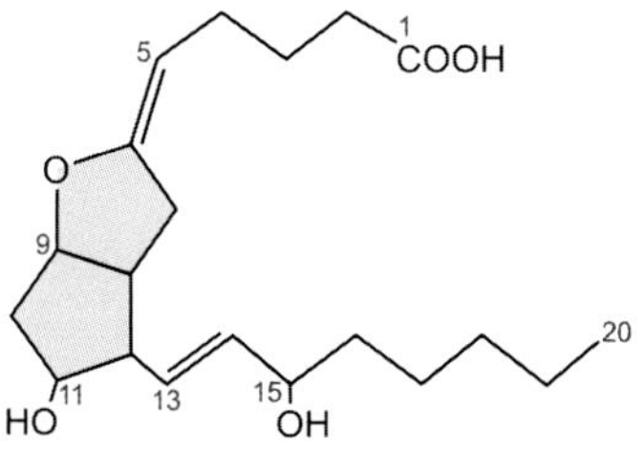

PGI$_2$

FIGURE 26.25 **Thromboxane and prostacyclin structure.**

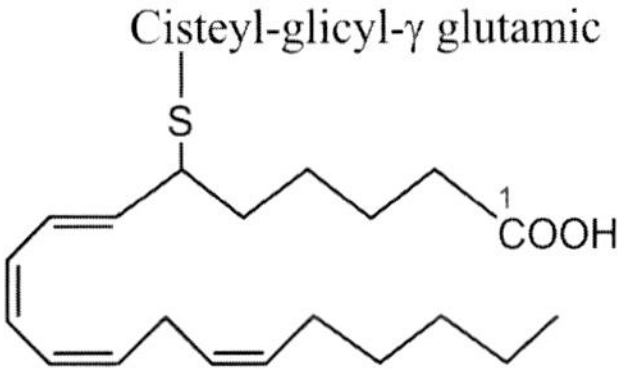

LTC$_4$

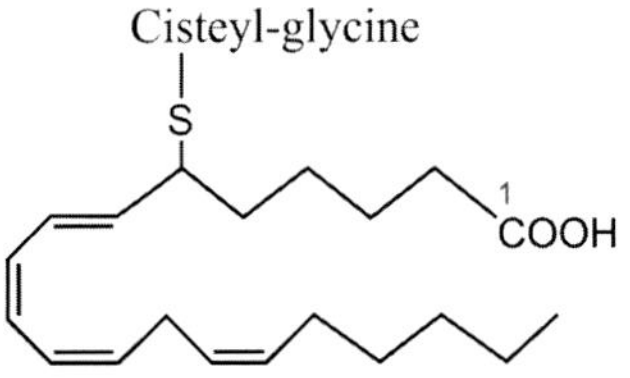

LTD$_4$

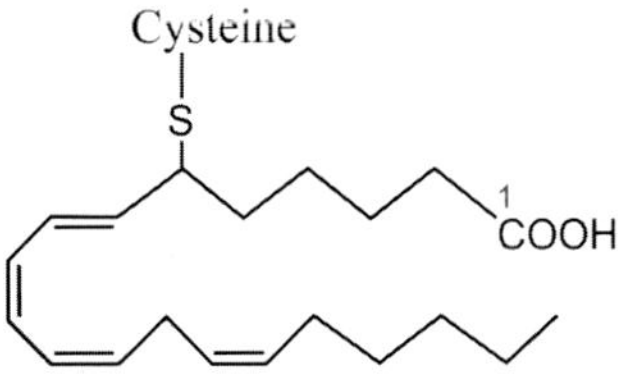

LTE$_4$

FIGURE 26.26 **Leukotriene structure.**

Most potent in this regard are LTs, whose effect is 100–1000 times stronger than histamine. Leukotrienes are the main triggers of asthma attacks. PGE, on the contrary, are potent bronchodilators.

Cardiovascular system. PGE and PGI$_2$ have vasodilator action; they reduce peripheral resistance and blood pressure, while thromboxane TXA$_2$ is a vasoconstrictor.

Prostacyclin (PGI$_2$) inhibits platelet aggregation; conversely, TXA$_2$ is a powerful platelet aggregating agent.

Digestive tract. PGE and PGI have inhibitory action on stomach secretion, decreasing gastric juice volume, acidity, and pepsin activity. They also have cytoprotective action, probably by stimulating mucus production. PGE favors gastric and intestinal peristalsis, increasing the contraction of longitudinal smooth muscle and relaxation of the circular smooth muscle. PGF$_{2\alpha}$

enhances the contraction of both circular and longitudinal smooth muscle fibers.

Reproduction. PGs, particularly PGE_2, acting on hypothalamus, are implicated in the regulation of secretion of GnRH; it also stimulates the release of LHRH. Increase in the synthesis of $PGF_{2\alpha}$ appears to be an important factor in labor initiation; it stimulates uterine muscle contraction. Oxytocin has a major effect in later stages of labor. PGs are involved in ovulation; they promote follicle rupture and have luteolytic action.

Autonomic nervous system. PGF facilitates neurotransmitter release. Instead, PGE_2 reduces norepinephrine release; it has inhibitory action on presynaptic terminals.

Inflammation. PGE and LTs promote the inflammatory process. It activates migration and aggregation of polymorphonuclear leukocytes and increases vascular permeability. The *slow-reacting substance of anaphylaxis*, important in mediating asthma accesses and hypersensitivity reactions, is a mixture of leukotrienes LTC_4, LTD_4, and LTE_4.

Some authors have shown that the antiinflammatory effects of corticosteroids are in part due to inhibition of phospholipase A_2, which reduces arachidonic acid release and eicosanoids synthesis.

GROWTH FACTORS

Development of a new organism from the egg requires intense mitotic activity. An adult human has $\sim 10^{18}$ cells, all generated by successive divisions of an original cell. At the same time, the newly formed cells need to differentiate into a wide variety of tissues with diverse functional capacity. Except for a few cell types, cell multiplication continues throughout the life of an individual. Growth and differentiation are complex processes that involve the selective activation and repression of multiple genes, which need to be exquisitely regulated. A series of substances, produced in different tissues, have

been identified, which modulate cellular proliferation. An important group of these substances include the *growth factors*, all of which are proteins. Growth factors can reach the circulation and exert actions at distant sites (endocrine action); they can be secreted to the extracellular space and influence neighboring cells (paracrine action), or they can bind to receptors on the same cell where they were originated (autocrine action).

Their receptors are plasma membrane tyrosine kinases, which trigger signaling events that regulate gene expression.

Some growth factors include the following.

Nerve growth factor (NGF). This is a 118-amino acid protein homodimer of 13,259 Da. It belongs to the family of neurotropins compounds, which includes the brain-derived neurotropin factor and neurotropins 3 and 4/5, generated from a 250-amino acid precursor.

NGF exerts its action through the *Trk* receptors, which have intrinsic tyrosine kinase activity. They activate PLCγ, PI3K, and the adapter protein *Shc*. TRK also stimulates the *Ras-MAPK* pathway. NGF activates neurite growth and has neuronal antiapoptotic effect. It stimulates division and differentiation of sympathetic and sensory neurons in the embryo.

Fibroblast growth factor (FGF). This growth factor includes a family of polypeptides of approximately 146 amino acids, which control proliferation and differentiation of various cell lineages (see next page). They are involved in the development of the bone and nervous system.

There are at least seven FGFs having autocrine and paracrine actions. They bind to four types of receptors differently expressed on the plasma membrane of the target tissues (FGFR1–FGFR4).

Achondroplasia is a type of dwarfism caused by a genetic defect in the FGFR3 receptor, which is involved in chondrocyte growth in bone growth plates.

Mutations in FGFR1–FGFR3 receptors produce *craniosynostosis*, a condition characterized by premature closure of the skull sutures.

Epidermal growth factor (EGF). This is a 53-amino acid polypeptide (6045 Da) with three disulfide bridges. EGF was originally found in relatively high concentration in the submaxillary gland of mice. EGF has also been identified with *β-urogastrone* a protein of human urine.

EGF is synthesized by cleavage of a precursor protein of 1168 amino acids. Owing to its large size, it is believed that this molecule should give rise to more than one active polypeptide.

EGF binds to a receptor on the plasma membrane of target cells, which is an integral membrane protein of 170 kDa. The outer portion or N-terminal segment of the receptor has the binding site for EGF, while the intracytoplasmic C-terminal domain possesses protein–tyrosine kinase activity. Binding of EGF to its receptor activates phosphorylation of tyrosine residues in the receptor itself (autophosphorylation) and in other proteins.

EGF stimulates proliferation of various types of epithelial cells. It also activates the expression of *Myc*, *Fos*, and *Jun* protooncogenes.

Platelet-derived growth factor (PDGF). It is a glycoprotein resistant to heat, composed by two homologous subunits, A of 125 and B of 160 amino acids, linked by disulfide bonds. It is found in α granules of platelets and is released during blood coagulation.

PDGF binding to its receptor (a 170-kDa protein) activates a protein–tyrosine kinase, which phosphorylates the receptor itself in its intracytoplasmic domain and other proteins. The PDGF-receptor complex is internalized by endocytosis and degraded in the cell.

PDGF is a powerful mitogen that stimulates proliferation in keratinocytes, fibroblasts, glial cells, monocytes, neutrophils, and smooth muscle cells. It favors cell division during wound healing. It is the main growth factor in human serum.

Other growth factors. There are several polypeptides that stimulate division and differentiation of stem cells from bone marrow. Some of these factors will be treated in the cytokine section (Chapter 30).

The hematopoietic growth factor or erythropoietin, involved in division and maturation of red blood cell precursors, and other growth factors, such as insulin IGF-I and IGF-II factors, have been described earlier.

The mechanism of action of EGF, PDGF, somatomedin, IGF-I, and IGF-II GH has great similarities with that of insulin. There is also striking homology between the receptors of these factors; they all belong to a family of proteins derived from a common ancestral gene.

There is close relationship between certain oncogenes and genes encoding growth factors and their receptors; this explains the capacity of oncogenes to modify cell proliferation.

NERVOUS SYSTEM CHEMICAL INTERMEDIARIES

Neurotransmitters

The nerve impulse is transmitted across synapses by chemical intermediates or neurotransmitters released from the presynaptic neuron axon terminal. Once in the synaptic cleft, the neurotransmitter binds to receptors in the postsynaptic neuron, or other cells innervated by the presynaptic neuron, where it will elicit a specific response.

To be considered a neurotransmitter, a compound must meet several conditions. These include (1) synthesis in neuronal cells; (2) store and release in the presynaptic terminal, in quantities sufficient to elicit a particular action; (3) to function in a manner similar to that of the endogenous neurotransmitter when administered exogenously; and (4) rapid degradation and removal from the synaptic cleft.

This last condition is essential for the normal transmission of the nerve impulse, which would not be possible if the chemical intermediate persists for long periods of time. Inactivation is achieved by hydrolysis or chemical modification of the neurotransmitter in the same space or by uptake through the presynaptic membrane.

According to the molecule size, two types of neurotransmitters are distinguished:

Small molecule neurotransmitters. These include acetylcholine, the amino acid glycine, glutamate and γ-aminobutyrate, and biogenic amines (amino acid derivatives) dopamine, norepinephrine, epinephrine, serotonin, and histamine. ATP and derivatives (adenosine, adenine) also function as neurotransmitters.

Acetylcholine. This neurotransmitter is stored in the synaptic vesicles and cytoplasm of cholinergic neuron axon terminals. It is released into the synaptic space in the presence of Ca^{2+} ions and binds to its receptors, promoting depolarization of the postsynaptic neuron.

Acetylcholine is the transmitter of spinal cord motor neurons, autonomic nervous system preganglionar neurons, and parasympathetic postganglionic neurons. It is also released in many brain synapses.

Acetylcholine degradation is produced in situ by *acetylcholinesterase*, which catalyzes its conversion to choline and acetate:

$$CH_3-COO-CH_2-CH_2-\overset{+}{N}\overset{CH_3}{\underset{CH_3}{-CH_3}} + H_2O \xrightarrow{\text{Acetylcholinesterase}}$$

Acetylcholine

$$\xrightarrow{\hspace{1cm}} CH_3-COO^- + HOH_2C-CH_2-\overset{+}{N}\overset{CH_3}{\underset{CH_3}{-CH_3}}$$

Acetate   Choline

Substances which inhibit cholinesterase prolong the neurotransmitter action and enhance the response. Some acetylcholinesterase inhibitors have reversible action, as eserine and neostigmine. Others behave as irreversible inhibitors. Among these are organophosphorus compounds, some of which are used in insect control.

To replenish acetylcholine content after nerve impulse, the presynaptic neuron resynthesizes acetylcholine by transferring acetate from

acetyl-CoA to a choline molecule, catalyzed by *choline acetyltransferase*.

Catecholamines. These include neurotransmitters of the adrenergic and dopaminergic systems (dopamine, epinephrine, and norepinephrine), which are synthesized from tyrosine (p. 388).

Dopamine is released primarily by neurons of the mesencephalon and diencephalon. In the central nervous system, norepinephrine is produced by cerulean nucleus neurons projecting to the cerebral cortex, cerebellum, and spinal cord. Norepinephrine is also the neurotransmitter of the sympathetic nervous system postganglionic neurons. Epinephrine is found in different neurons in the brain.

Catecholamines are taken up from the synaptic cleft into the neuron by active transport and inactivated in a process in which MAO and COMT (p. 602) are involved. Substances which inhibit catecholamine uptake, such as cocaine and amphetamines, enhance catecholamine-induced responses.

Serotonin. Also known as 5-hydroxytryptamine is synthesized in the synaptic terminals of neurons of the medium raphe nuclei that project to the whole brain and spinal cord. It is inactivated by MAO. The end product of its metabolism is 5-OH-indole acid (5-HIA).

Alteration of serotonin, dopamine, and norepinephrine levels results in serious pathological conditions. Depressive states improve with the administration of drugs that increase transmission at serotoninergic synapses. The deficient production of dopamine is involved in the development of Parkinson's disease. Alteration of dopaminergic transmission is also involved in schizophrenia.

Histamine. This neurotransmitter is produced by histidine decarboxylation, and it plays a major role in nerve activity in invertebrates and in hypothalamic neurons of vertebrates.

Amino acids. GABA and glycine inhibit nerve transmission and cause partial hyperpolarization or depolarization of neuronal cells. GABA is the main inhibitory neurotransmitter in brain neurons and spinal cord. It is also found in cerebellum basket and Purkinje cells, in olfactory bulb granulose cells, and in retina amacrine cells. GABA is produced by decarboxylation of glutamate. Drugs, such as benzodiazepines, exert their depressant action on the CNS by modulating the GABA-receptor complex. Baclofen, a muscle relaxant, produces GABA release.

Glycine is an inhibitory neurotransmitter in the spinal cord.

Glutamate produces neuronal depolarization. It is an excitatory neurotransmitter in many neurons of the central nervous system. The best known glutamate receptor is the N-methyl-D-aspartate (NMDA) receptor (a selective agonist for this type of receptor). Glutamate binding to NMDA receptors in the postsynaptic neuron opens Ca^{2+} channels, increases intracellular Ca^{2+} and its binding to calmodulin, and activates NOS. The NO produced is another intermediary for neuronal transmission.

Purines. In some neurons, ATP, adenosine, and adenine function as transmitters. Purinergic neurons are present in the sympathetic system (they innervate myocardium, intestinal smooth muscle, and vas deferens) and in dorsal root ganglia. The synaptic vesicles of these neurons are richer in ATP than in other neurons.

Neuropeptides. Peptides that exhibit strong physiological activity have been isolated in the central nervous system. Hypothalamic regulatory factors (p. 576) are examples of such substances. Unlike small molecule neurotransmitters, which are synthesized in the neuronal cytoplasm and synaptic terminal, the neuropeptides are produced in the cell body, processed in the ER and Golgi apparatus, and finally transported in secretory granules to the axonal terminal. Immediately after release into the synaptic cleft, they are degraded by serine proteases.

Over 50 active peptides have been described in nerve cells. Some of them have been identified as hormones with action outside the brain (i.e., oxytocin, vasopressin, somatostatin, hypothalamic-releasing factors, and pituitary gland hormones). Others were initially described as hormones not synthesized in the nervous system (i.e., gastrin, galanin, vasoactive intestinal polypeptide, glucagon, secretin, ANP, angiotensin, and calcitonin gene-related peptides). However, many of those also function as neurotransmitters. Only a few will be described here.

Substance P is a peptide consisting of 11 amino acids (Arg-Pro-Lys-Pro-Gln-Gln-Phe-Phe-Gly-Leu-Met-NH$_2$) that plays a neurotransmitter role in sensory neurons. It is related to transmission of pain sensation.

Endogenous opioids are a group of substances with morphine-like action. The existence of these substances in the nervous system was suspected after the discovery of morphine and opioid receptors in cells. Three opioid peptide families have been described: endorphins, enkephalins, and dynorphins, all isolated from the central nervous system. They function as neuromodulators. For example, in neurons transmitting pain sensation, they have analgesic effects by inhibiting the release of substance P. They also have other actions on behavior, appetite, and endocrine system.

Endorphins. The main substance in this group is β-endorphin, a 31-amino acid peptide, generated by cleavage of the POMC precursor molecule that also contains ACTH, lipotropin, and α-MSH (p. 579).

Enkephalins. Two pentapeptides were discovered, namely methionine-enkephalin or *MET-enkephalin* (Tyr-Gly-Gly-Phe-Met), and leucine-enkephalin or *leuenkephalin,* which differ from the MET-enkephalin by having a leucine replacing methionine (Tyr-Gly-Gly Phe-Leu). Both are derived from the same molecule, preproenkephalin, which, after cleavage of a 20–25-amino acid signal peptide, generates proenkephalin. This produces MET-enkephalin, leuencefalin, and E peptide, a 25-amino acid peptide with opioid activity.

Dynorphins. These compounds are synthesized from a precursor molecule, preprodinorphin. After losing a signal peptide, it becomes prodynorphin, which generates several products, including dynorphin, a peptide of 17 amino acids, and α-neoendorphin. Dynorphin is the most potent endogenous opioid peptide; it has analgesic effect, which is ~50 times stronger than β-endorphin, ~200 times greater than morphine, and ~700 times more powerful than leuencephalin.

Nitric oxide (NO). This is a free radical, which has high chemical reactivity due to the presence of an extra electron. It is very labile; its duration in the body does not exceed 6–8 s, after which it is converted into nitrates and nitrites. Despite these properties, apparently incompatible with useful physiological functions, NO participates as a chemical messenger in the nervous system and in other tissues.

In various cells, including some brain and cerebellar neurons, the existence of NO synthetase (NOS), which catalyzes the conversion of arginine to NO and citrulline, has been shown. NOS is a flavoprotein that contains FMN and FAD; it requires molecular oxygen and NADPH, and it is activated by Ca^{2+}-calmodulin (p. 567).

NO binds to the heme of guanylate cyclase and induces the convertion of GTP to cyclic GMP, which in turn activates protein kinases.

Although NO is not a neurotransmitter, it functions as an important modulator in some CNS synapses. NO has also been identified as chemical mediator in learning and memory processes. The ease with which the NO small molecule passes through cell membranes allows it to exert effects as a "retrograde messenger" in the phenomenon of long-term potentiation, which is considered to be the main mechanism of memory storage.

In stroke, hypoxia produces a great discharge of glutamate in brain neurons. This leads to increase the production of NO, which diffuses out of the neuron and exerts toxic effects on neighboring cells. It is believed that brain damage in such cases is predominantly due to NO, whose toxic action is accentuated on the hypoxia-stressed cells. This finding has been the basis for early use of NOS inhibitors or NMDA receptor blockers to prevent stroke damage.

NO also acts as a "second messenger" in processes that trigger vasodilation. Acetylcholine, released by nerve terminals at blood vessels, binds to receptors present in endothelial cells, activates NOS, and generates NO, which diffuses through the cell membrane, penetrates into the adjacent smooth muscle cells, and causes vessel relaxation. Although various organic nitrates have been used as coronary vasodilators for some time, their mechanisms of action through NO production have been only recently understood.

Other cells outside of the nervous system express NOS and, therefore, are capable of producing NO. Among those cells are macrophages. NO produced by activated macrophages allows free radical oxidation and lysis of tumor cells, bacteria, fungi, and other invading agents.

Carbon monoxide. This is a toxic gas produced in the reaction catalyzed by *heme oxygenase*. CO probably exerts in the CNS effects that are similar to those of NO. Both activate guanylate cyclase and elevate cGMP levels.

SUMMARY

Hormones, growth factors, cytokines, and *neurotransmitters* are chemical mediators, which communicate and integrate biological systems.

Hormones include multiple different substances. They can be classified, depending on their chemical nature, into:
1. steroids (cyclopentaneperhydrophenanthrene derivatives);
2. amino acid derivatives;
3. eicosanoids (derivatives of polyunsaturated fatty acids);
4. peptides; and
5. proteins.

Hormonal mechanisms of action are varied and include:
1. Cell plasma membrane effects, modulating the activity of transport systems.
2. Regulation of cell enzymes, which are involved in metabolism of different substrates and in cell energetics.
3. Genomic actions, which depend on activation or repression of DNA transcription and protein synthesis.

Specific properties characterize hormones. They operate in very low concentrations, have half-lives of seconds to days, and their secretion is regulated by a variety of stimuli that depend on the hormone type. Hormones exhibit high specificity, which is due to their binding to receptors in the target cells. Upon binding, they form a hormone-receptor complex, which has properties that resemble those of the enzyme–substrate complex (induced adaptation, saturability, and reversibility).

Depending on their site of origin, the following hormones can be described:

Hypothalamic hormones include various factors that stimulate the synthesis and secretion of all hormones produced in the pituitary gland, as well as inhibitory factors for GH, MSH, and prolactin. The hormone-releasing factors are as follows: (1) *CRH*, (2) *TRH*, (3) *LHRH* and *FSHRH*, (4) *PRRH*, (5) *GHRH*, and (6) *MSRH*. The inhibitory factors are as follows: (7) *PRIH*, (8) *GHIH*, also called *somatostatin*, and (9) *MSIH*.

The pituitary glands secrete different hormones, which are produced in its anterior portion or adenohypophysis and in the posterior of neurohypophysis.

Anterior pituitary or *adenohypophysis* hormones include:

1. *ACTH* is a polypeptide of 39 amino acids, originated from a 280-amino acid precursor protein known as *POMC*, which also generates α-MSH, β-MSH, γ-MSH, γ-lipotropin, β-endorphin, α-endorphin, and MET-enkephalin. The first 23 amino acids are responsible for the biological actions of ADH, which consist in stimulating the synthesis of adrenal gland hormones, primarily glucocorticoids.
2. *TSH* is a 30-kDa glycoprotein, integrated by α and β chains. Its function is the activation of thyroid hormone synthesis and release by the thyroid gland.
3. *Gonadotropins*, *FSH*, and *LH* are glycoproteins with α and β subunits. In *ovary*, FSH induces Graafian follicle maturation and development; along with LH, they stimulate ovarian estrogen production. LH controls the development of the corpus luteum and ovarian estrogen and progesterone secretion. In *testis*, LH stimulates the development of the seminiferous tubules and the production of testosterone.
4. *Lactogenic hormone* or *prolactin* (PR) is a simple protein, which stimulates corpus luteum formation and progesterone production. It promotes mammary gland development.
5. *MSH* consists of three peptides (α, β, and γ), all POMC derivatives. This hormone plays a role on pigment cells or melanocytes, darkening the skin of fish and amphibians. In humans, it may be involved on the central nervous system attention and motivation processes.
6. *Somatotropin* or *GH* is a 191-amino acid protein, which stimulates cell proliferation and growth. It has many actions on metabolism, including activation of protein synthesis, decrease in lipogenesis, activation of lipolysis, reduction of glucose transport and utilization in muscle, and increase in gluconeogenesis. It is a hormone that leads to hyperglycemia.

Somatomedins or *IGF* produced in the liver by GH action, stimulate glucosamine sulfate incorporation in cartilage proteoglycans, increase fibroblasts mitotic activity, mimic insulin action in muscle and adipose tissue.

Neurohypophysis

Hormones of the posterior hypophysis include:

1. *Oxytocin*, a nonapeptide, which stimulates uterus contraction.
2. *Vasopressin* or *ADH*, a nonapeptide that differs from oxytocin in two amino acids. It has vasopressor effects systemically and is involved in water reabsorption in kidney collecting ducts. It is essential to maintain body fluid volume.

The placenta produces hormones with actions similar to some of those secreted by the adenohypophysis: chorionic gonadotropins (GH) and placental lactogen.

The thyroid gland produces *thyroxine T_4* and T_3. Its synthesis requires iodine, which is uptaken from the blood and activated in the thyroid follicles by *thyroid peroxidase*. This enzyme conjugates iodine to apothyroglobulin at positions 3 and 5 of tyrosine residues, forming *MIT* and *DIT*, respectively. These couple to DIT residues to produce T_3 and T_4. After iodination, apothyroglobulin becomes thyroglobulin, which enters the cell by endocytosis. In the phagolysosome, T_4, T_3, DIT, and MIT are released; DIT and MIT are deiodinated by selenoenzymes. T_3 and T_4 are secreted and transported in blood bound to specific proteins (TGB and TTR). In the liver, T_3 and T_4 are conjugated with glucuronide and sulfate and excreted in the bile or deamidated and decarboxylated to give *Tetrac* and *Triac*, inactive metabolites. In the target cells, T_3 and T_4 attach to nuclear receptors that bind to DNA response elements to modulate gene expression.

T_3 and T_4 stimulate RNA and protein synthesis; they increase Na^+,K^+-ATPase activity, oxygen consumption, and promote glucose, lipid, and amino acid utilization. At physiological concentrations, they have anabolic effects. In contrast, high doses have catabolic action. T_3 and T_4 also exert nongenomic actions.

Calcitonin is also synthesized and secreted by the thyroid in the C cells of this gland. It is a 32-amino acid peptide. It binds to a membrane receptor and increases cAMP level in the target cells. Its function is to decrease of Ca^{2+} and phosphate levels in plasma.

The adrenal gland hormones include those produced in the cortex and those from the medulla.

Glucocorticoids (cortisol–cortisone), mineralocorticoids (aldosterone–deoxycorticosterone), and androgens (androstenedione–dehydroepiandrosterone) are all steroidal hormones synthesized in the adrenal cortex from cholesterol.

Corticosteroids are transported in the blood bound to *transcortin* and albumin. The free hormone, in equilibrium with the bound one, is the biologically active form. ACTH stimulates the synthesis and secretion of steroids, preferably corticosteroids. *Aldosterone* main stimulator is *angiotensin II*. Corticosteroids are rapidly degraded by the liver and secreted in the bile. They return to the circulation by absorption in intestine, and then they are excreted predominantly in the urine.

Corticosteroids bind receptors in the nucleus (or cytoplasm). The hormone-receptor complex in the nucleus binds response elements in DNA and modulates transcription.

The effect of glucocorticoids is to decrease glucose utilization and to stimulate protein degradation in extrahepatic tissues. In adipose tissue, they activate lipolysis and depress triacylglycerols synthesis. In liver, they activate gluconeogenesis from amino acids and increase protein synthesis. They tend to increase blood sugar and glycogen stores, and increase free fatty acids and amino acids in plasma. Glucocorticoids have antiinflammatory action; they depress immune responses.

Mineralocorticoids increase Na^+ and Cl^- reabsorption and K^+ excretion in renal tubules. They are important in electrolyte balance in the body.

Androgenic steroids exert anabolic effects on proteins.

Medulla of the adrenal gland produces the following hormones:

Adrenaline or *epinephrine* and *noradrenalin* or *norepinephrine*, both derived from tyrosine. They are inactivated by *MAO* and *COMT*. Metabolites are excreted in urine; the most important ones are *vanillylmandelic acid* and *metanephrine*.

Noradrenaline and *adrenaline* bind to various types of plasma membrane receptors. These include α1 (coupled to G_q protein, they activate the PIP_2 system), $α_2$ (coupled to G_i protein, inhibit adenylate cyclase), $β_1$ and $β_2$, which interact with G_s protein and $β_3$.

Noradrenaline and adrenaline stimulate glycogenolysis, inhibit glycogenesis, and increase gluconeogenesis; they are hyperglycemic. They prepare the body to cope with stress conditions.

The pancreatic hormones include *insulin, glucagon*, and *somatostatin*.

Insulin is a 6000-Da protein, which consists of an A chain (21 amino acids) and a B chain (30 amino acids), linked by two disulfide bridges. It is synthesized as *preproinsulin* by the β cells in the Langerhans' islets of the pancreas. Removal of a 30-amino acid connecting peptide (C-peptide) from proinsulin generates active insulin. The most effective stimulus for insulin secretion is the increase in plasma glucose concentration. Insulin binds to a specific plasma membrane receptor, which is a tetrameric glycoprotein composed of two α and two β sub-

units. Upon binding to the α subunits, insulin activates the *protein tyrosine kinase* cytosolic domain of the receptor's β subunit. This induces autophosphorylation of the receptor and the phosphorylation of other proteins in the cytosol, mainly the *IRS*, which trigger different signal transduction pathways in the target cells. Among the pathways activated by insulin are *phosphoinositide-3-kinase*, which form PI-(3,4,5)-P_3, *protein kinase B* or *Akt*, and the *Ras-MAP kinase system*. Insulin stimulates glucose uptake by moving GLUT4 from the cytoplasm to the plasma membrane, and uptake of amino acids, nucleosides, and phosphate in cells. Insulin activates all pathways of glucose utilization and depresses gluconeogenesis; it is the main hypoglycemic hormone. It stimulates fatty acid and triacylglycerol synthesis, reduces lipolysis in adipose tissue, and activates RNA and protein synthesis.

Glucagon is a 29-amino acid polypeptide produced by α cells of the Langerhans' islets of the pancreas. Its secretion is stimulated by decreased blood glucose. Glucagon binds to receptors on the membrane and activates adenylate cyclase. In hepatocytes, glucagon promotes glycogenolysis and represses glycogenesis, increases gluconeogenesis, activates lipase in adipocytes, and promotes release of fatty acids. It stimulates nitrogen catabolism.

Somatostatin exists as a tetradecapeptide (*Somatostatin 14*) and a polypeptide double the size (*Somatostatin 28*). They are produced from a 116 amino acid precursor by the δ cells of pancreas, hypothalamus, enterocytes, and thyroid C cells. Somatostatins bind to receptors coupled to G_i protein and activate *phosphotyrosine phosphatase* and *MAPK kinase*. They have inhibitory action on the synthesis and secretion of GH, insulin and glucagon, TSH, gastrin and secretin.

Glucose homeostasis is regulated by the interplay of several hormones. During fasting conditions, glycemia is maintained between 70 and 110 mg/dL (3.9–6.1 mM). *Insulin* is the main factor which reduces plasma glucose levels. Its actions include (1) activation of glucose uptake by cells, (2) stimulation of glycogenesis, (3) activation of glucose degradation (via glycolysis and the pentose phosphate pathway), and (4) stimulation of the conversion of glucose into other type of compounds (mainly lipids).

In contrast to insulin, *glucagon, epinephrin, cortisol*, and *GH* tend to increase glycemia. This is achieved by (1) activation of hepatic glycogenolysis, (2) stimulation of gluconeogenesis, and (3) increase in intestinal glucose absorption.

Diabetes mellitus is a pathologyc condition characterized by *hyperglycemia, glucosuria, polyuria, polydipsia*, and *a series of metabolic disturbances*. Two forms of the disease exist, *type 1* or *juvenile*, which is insulin dependent, and *type 2*, or noninsulin dependent. Although the first one is due to lack of insulin production or function, the second is due to insulin resistance in the target tissues, which abrogates the response to insulin.

Metabolic syndrome is a condition characterized by obesity, insulin resistance with the resultant alteration in glucose homeostasis, and arterial hypertension.

The testis produces testosterone, a steroidal compound that is converted into *dihydrotestosterone*, with greater activity. The main metabolites of testosterone are *androsterone* and *etiocholanolone*, which are part of the *17-ketosteroids* released in urine.

Androgens favor the development of reproductive organs, accessory glands, and secondary sex characteristics of males. They stimulate protein anabolism.

The ovary produces estrogens (estradiol and estrone) and progesterone.

Estradiol and *estrone* are steroidal hormones. They have anabolic action on the female genital organs, increase water, electrolytes, amino acids and glucose uptake by myometrial cells.

Progesterone is a *progestational hormone*. In general, steroids bind to nuclear receptors and activate transcription and synthesis of specific proteins.

The ovary also produces *relaxin, inhibin, activin,* and *folistastin.*

Parathyroid glands produce PTH, an 84-amino acid protein, synthesized as a preprohormone. Its biological activity resides in the first 34 amino acid segment. The main stimulus for secretion is decreased blood Ca^{2+}. PTH binds to a membrane receptor and stimulates *adenylate cyclase*. In bone, PTH promotes Ca^{2+} mobilization, *osteoclasts* activity, and RNA and protein synthesis; it enhances dissolution of bone mineral. In *osteocytes*, PTH activates glycolysis and the citric acid cycle; it activates Ca^{2+} ATPase, and increases Ca^{2+} passage from bone to extracellular fluid. In *kidney*, PTH promotes Ca^{2+} and phosphate reabsorption in distal tubules. It also increases Na^+, K^+, citrate, and bicarbonate urinary excretion and decreases elimination of H^+ and NH_4^+. PTH activates 1,25-$(OH)_2$-D_3 formation.

The *kidney* synthesizes various hormonal agents. These include the following:

Renin, which is synthesized by the juxtaglomerular apparatus as preprorenin, an inactive compound. Hydrolysis of the precursor produces active renin, which is secreted when blood pressure, renal blood flow, and Na^+ concentration in tubular fluid are low. In plasma, renin acts on *angiotensinogen*, separating a decapeptide and producing *angiotensin I*, which is inactive. Angiotensin convertin enzyme cleaves two amino acids of angiotensin I and generates the octapeptide *angiotensin II*, with biological activity. *Angiotensin II* binds to receptors on target cells (AT1 and AT2). *AT1*, coupled to G_q proteins, mediates arteriolar contraction, increases blood pressure, and retains Na^+ and water. It also stimulates aldosterone secretion in adrenal cortex.

Erythropoietin is secreted by mesangial cells of the renal cortex and medulla. Its release is stimulated by reduction of O_2 tension in the kidney. Erythropoietin binds to a receptor associated with tyrosine kinase. It stimulates mitotic activity and differentiation of erythropoietic cells and increases the number of circulating red blood cells and O_2 transport capacity.

Calcitriol, or 1,25-$(OH)_2$-D_3, the active metabolite of vitamin D, is formed in kidney.

Bradykinin is a kinin produced in the kidney which has vasodilator effect.

The heart synthesizes atrial natriuretic peptides. These are a family of peptides comprising 24–28 amino acids. They bind to the membrane receptors and activate *guanylyl cyclase*. They inhibit renin secretion, stimulate excretion of water and Na^+, decrease Na^+ reabsorption in renal tubules, and relax smooth muscle.

The gastrointestinal tract releases several polypeptides, among which are *gastrin, secretin, cholecystokinin, ghrelin, vasoactive intestinal peptide,* and *enteroglucagon*. These hormones regulate gastrointestinal contraction and secretions.

The pineal gland synthesizes melatonin, an indoleamine derived from tryptophan. Melatonin is released during the dark periods and serves as internal signal of light–dark cycles, adjusting the circadian rhythms.

Eicosanoids are a group of compounds expressed in most cells. They include three main type of substances: *PGs, TXs, PGI,* and *LTs*. They are derivatives of 20 carbons polyunsaturated fatty acids, mainly from arachidonic acid. They exert a variety of actions. In the respiratory system, they constrict the smooth muscle of bronchi. In the cardiovascular system, PG has vasodilator effects, while TX is vasoconstrictor. In the digestive tract, PG inhibits gastric juice production and favors intestinal peristalsis. In the reproductive system, PGE regulates the secretion of GnRH and LHRH. In the autonomic nervous system, PG modulates neurotransmitter release. PG and LTs also have effect on Inflammation.

Growth factors include a series of different compounds, all of which promote cell proliferation and tissue growth and development. They include:

Nerve growth factor, a homodimer of a 118-amino acid polypeptide. It stimulates the division and differentiation of sympathetic and sensory neurons in embryos and activates neurite outgrowth.

Epidermal growth factor is a 53-amino acid polypeptide that stimulates proliferation of various epithelial cells. It is formed by cleavage of a 1168-amino acid precursor.

Platelet derived growth factor consists of two homologous subunits of 18 amino acids each, linked by disulfide bonds. It is a powerful mitogen.

Fibroblasts growth factor is a polypeptide of ~146 amino acids. It is involved in proliferation and differentiation of various cell lineages.

Neurotransmitters are generally low mass substances, which are involved in synaptic transmission in different areas of the nervous system. They include *acetylcholine, catecholamines,* such as *dopamine, norepinephrine,* and *epinephrine, serotonin* synthesized from tryptophan, *biogenic amines, amino acids* (*GABA, glycine,* and *glutamate*), *purines,* and *neuropeptides,* such as *substance P, endogenous opioids, endorphins,* and *enkephalins.*

Bibliography

Aramburo, C., Alba-Betancourt, C., Luna, M., Harvey, S., 2014. Expression and function of growth hormone in the nervous system: a brief review. Gen. Comp. Endocrinol. 203, 35–42.

Becker, K.L., 2002. Principles and Practice of Endocrinology and Metabolism, third ed. J.P. Lippincott, Philadelphia, PA.

Carter-Su, C., Schwartz, J., Smith, L.S., 1996. Molecular mechanisms of growth hormone action. Annu. Rev. Physiol. 58, 187–207.

Cassone, V.M., 1998. Melaton: a role in vertebrate circadian rhythms. Chronobiol. Int. 15, 457–473.

Chew, S.L., Leslie, D., 2006. Clinical Endocrinology and Diabetes. Churchill Livingstone, Edinburgh.

Cohen, S., 2008. Origins of growth factors: NGF and EGF. J. Biol. Chem. 283, 33793–33797.

Donangelo, I., 2014. Thyroid hormone and central control of metabolic homeostasis. J. Endocrinol. Diabetes Obes. 2, 1047–1051.

Goffin, V., Binart, N., Touraine, P., Kelly, P.A., 2002. Prolactin: the new biology of an old hormone. Annu. Rev. Physiol. 64, 47–67.

Greenspan, F.S., Gardner, D.G., 2001. Basic and Clinical Endocrinology, seventh ed. Lange Medical Books, McGraw-Hill, NY.

Incerpi, S., 2002. Actions of thyroid hormone on ion transport. Curr. Opin. Endocrinol. Diabetes 9, 381–386.

Jeong, J.K., Kim, G., Lee, B.J., 2014. Participation of the central melanocortin system in metabolic regulation and energy homeostasis. Cell. Mol. Life Sci. 71, 3799–3803.

Kronenberg, H.M., 2008. Williams Textbook of Endocrinology, eleventh ed. Saunders Elsevier, Philadelphia, PA.

Lusis, A.J., Attie, A.D., Reve, K., 2008. Metabolic syndrome: from epidemiology to system biology. Nat. Rev. Genet. 9, 819–830.

Mackensie, R.W.A., Elliott, B.T., 2014. AKt/PKB activation and insulin signaling: a novel insulin signaling pathway in the treatment of type 2 diabetes. Diabetes Metab. Syndr. Obes. 7, 55–64.

Melmed, S., Conn, P.M. (Eds.), 2005. Endocrinology: Basic and Clinical Principles. second ed. Humana Press, Totowa, NJ.

Muoio, D.M., Newgard, C.B., 2008. Molecular and metabolic mechanisms of insulin resistance and β-cell failure in type 2 diabetes. Nat. Rev. Mol. Cell Biol. 9, 193–205.

Nef, S., Parada, L.F., 2015. Hormones in male sexual development. Genes Dev. 11, 3075–3086.

Nilsson, S., Gustafsson, J.A., 2002. Biological role of estrogen and estrogen receptors. Crit. Rev. Biochem. Mol. Biol. 37, 1–28.

Pearce, D., Bhargava, A., Cole, T.J., 2003. Aldosterone: its receptor, target genes and action. Vitam. Horm. 66, 29–76.

Ross, S.A., Gulve, E.A., Wang, M., 2004. Chemistry and biochemistry of type 2 diabetes. Chem. Rev. 104, 1255–1282.

Sato, T., Nakamura, Y., Shiimura, Y., Ohgusu, H., Kangawa, K., Kojima, M., 2012. Structure, regulation and function of ghrelin. J. Biochem. 151, 119–123.

Small, C.J., Bloom, S.R., 2004. Gut hormones and the control of appetite. Trends Endocrinol. Metab. 15, 259–263.

Varman, T.S., Shulman, G.I., 2012. Mechanisms of insulin resistance. Common threads and missing links. Cell 148, 852–871.

Weiss, M.D., 2009. Proinsulin and the genetics of diabetes mellitus. J. Biol. Chem. 284, 19159–19165.

White, M.F., 2003. Insulin signaling in health and disease. Science 302, 1710–1711.

27

Vitamins

In the late 19th century, it was believed that to ensure a good nutritional status, a complete diet needed proteins, carbohydrates, lipids, inorganic salts, and water. It was thought that by combining those compounds in quantity and quality identical to that found in food, it would be possible to obtain a balanced diet. Studies in animals showed that this was not the case and that such a diet resulted in severe alterations that lead even to death. It was apparent that food contained other substances that were essential, but that existed in such low concentrations that they could not be detected by the techniques used at the time. Although the nature of those substances was unknown, they were predicted and designated *dietary accessory factors*.

In the early 20th century, one of the essential dietary *accessory factors* was isolated; because this molecule had an amine group it was called "vital amine" and later *vitamin*. Since then, other nutritional accessory factors were discovered and, while most of them did not have an amine group, they were generically named *vitamins*.

General properties. Vitamins have the following characteristics:

1. They are organic compounds of varied and relatively simple chemical structure, different to carbohydrates, lipids, or proteins.
2. They are found in natural foods in very small concentrations.
3. They are essential for maintaining health and normal growth.
4. They must be supplied in the diet because they are not synthesized by the body.
5. Their deficiency results in pathological conditions (*avitaminosis*).

Functional role. Vitamins are not directly involved in providing structural or energetic components to the body. They are similar to enzymes, hormones, and other metabolic regulators, which act at very low concentrations. In fact, many vitamins integrate enzyme systems as coenzymes, and others accomplish their functions as hormones do.

Nomenclature. Initially, the existence of two different types of vitamins was recognized: one is soluble in organic solvents, and was called fat-soluble factor A (later known as *vitamin A*), and the other, the original "vital amine" of polar nature, was called hydrosoluble factor B or *vitamin B*. In the following years, other factors were discovered and were named with letters, following in alphabetical order (C, D, and E). An exception was vitamin K, which was named after its role in blood coagulation (Koagulation in Danish, language of its discoverer). Vitamin B turned out to be not one, but a group of compounds. As different members within this group were discovered, they were designated with the letter B and subscripts numbers (B_1, B_2, B_3, ..., B_{12}); the original vitamin B became the "vitamin B complex."

For some vitamins, more than one compound with similar activity has been isolated. These are called *vitamers*.

According to their solubility characteristics, vitamins are divided into lipid and water soluble. The vitamins soluble in lipids (A, D, E, and K) are found in the fatty component of foods. In general, when they are ingested in excess, lipid-soluble vitamins are stored in different tissues, mainly in the liver, and are eliminated very slowly. Intake of these vitamins in amounts higher than normal requirements and during long time produces undesirable effects.

Water-soluble vitamins include all components of the B complex and vitamin C. In general, they are not stored and must be provided everyday. For most of them, high doses do not produce toxic effects because the excess is easily eliminated by urine.

The separation of vitamins based on their solubility is justified from the nutritional point of view. Vitamins of the same group are commonly found together in foods, and their deficiency commonly comprises several vitamins with similar solubility.

Provitamins. Some vitamins come in foods as precursors or *provitamins*, which, when metabolized, generate the corresponding vitamin. For example, plant pigments called *carotenes* can be converted into vitamin A.

Antivitamins are substances with chemical structure similar to that of a vitamin that function as vitamin antagonists. Owing to their structural analogy, they occupy the place corresponding to the vitamin on enzyme systems or receptors and block their action.

Avitaminosis is a clinical condition caused by the lack of vitamins. The symptoms of lack of vitamin depend on the particular type of vitamin. Vitamin deficiency can be caused by (1) *poor diet*; the most common cause of hypo- or avitaminosis is the lack or insufficient quantity of fresh and varied food. (2) *Exclusive consumption of foods preserved or cooked at high temperatures*; cooking inactivates some vitamins. (3) *Deficient intestinal absorption*, which could be due to many different factors. (4) *Increased vitamin requirements*; there are physiological situations (pregnancy, lactation, and during active growth in children) and diseases (hyperthyroidism, fever) in which vitamin requirements are increased, and their normal intake is not enough to satisfy the body's requirements. (5) *Unbalanced diet*; excesive ingestion of carbohydrates increases the requirements for vitamin B_1. Excessive alcohol ingestion interferes with the absorption of several vitamins.

Vitamin requirement. A normal individual must have a diet containing all of the vitamins in the amount needed to maintain health.

From a nutritional point of view, *bioavailability* of a vitamin is important. This term refers to the percentage of the total vitamin present in foods which is effectively absorbed.

The amount of a vitamin needed varies with sex, age, and physiological conditions, such as pregnancy or lactation. *Estimated average requirement* is the lowest daily intake that, maintained over long time, prevents vitamin deficiency symptoms. *Recommended dietary allowance* is the daily intake sufficient to meet the needs of a healthy individual. Recommended dietary allowance values for all the vitamins are shown in Table 27.1.

Pharmacological actions. High doses of some vitamins can exert beneficial effects on certain pathological conditions. These effects are not due to the function of vitamin as a nutrient, but to the pharmacological effects of the vitamin.

FAT-SOLUBLE VITAMINS

Vitamin A

Synonymy. Retinol, axerophthol, antixerophtalmic vitamin.

Chemistry. Vitamin A or *retinol* is insoluble in water, and soluble in fats and organic solvents. It crystallizes in yellow prisms; its molecular formula is $C_{20}H_{30}O$; and it has a six-carbon ring

TABLE 27.1 Vitamins: Recommended Dietary Ingestion (per day)

Groups	A[a]	D[b]	E[c]	K[d]	B$_1$[e]	B$_2$[f]	B$_3$[g]	B$_5$[h]	B$_6$[i]	B$_7$[j]	B$_9$[k]	B$_{12}$[l]	C[m]
CHILDREN													
0–6 months	375	5	4	5	0.2	0.3	2	1.7	0.1	5	80	0.4	25
7–12 months	400	5	5	10	0.3	0.4	4	1.8	0.3	6	80	0.7	30
1–3 years	400	5	6	15	0.5	0.5	6	2.0	0.5	8	150	0.9	30
4–6 years	450	5	7	20	0.6	0.6	8	3.0	0.6	12	200	1.2	30
7–9 years	500	5	11	25	0.9	0.9	12	4.0	1.0	20	300	1.8	35
Adolescents													
10–18 years	600	5	15	45	1.2	1.2	16	5.0	1.3	25	400	2.4	40
ADULTS													
Women													
19–65 years	500	10	15	55	1.1	1.1	14	5.0	1.3	30	400	2.4	45
>65 years	600	15	15	55	1.1	1.1	14	5.0	1.5	30	400	2.4	45
Pregnants	800	5	15	55	1.4	1.4	18	6.0	1.9	30	600	2.6	55
Lactating	1300	5	19	55	1.5	1.6	17	7.0	2.0	35	500	2.8	70
Men													
19–65 years	600	10	15	65	1.2	1.3	16	5.0	1.3	30	400	2.4	45
>65 years	600	15	15	65	1.2	1.3	16	5.0	1.7	30	400	2.4	45

[a] *Activity equivalents for retinol.*
[b] *Micrograms for calcipherol.*
[c] *Milligrams for α-tocopherol.*
[d] *Micrograms for phylloquinone.*
[e] *Milligrams for B$_1$-thiamine.*
[f] *Milligrams for B$_2$-riboflavin.*
[g] *Milligrams for B$_3$-niacin.*
[h] *Milligrams for B$_5$-pantothenic acid.*
[i] *Milligrams for B$_6$-pyridoxin.*
[j] *Micrograms for B$_7$-biotin.*
[k] *Micrograms for B$_9$-folic acid.*
[l] *Micrograms for B$_{12}$-cobalamin.*
[m] *Milligrams for C-ascorbic acid.*

(β-ionone) with a side chain composed of two isoprene units, and a primary alcohol at position 6. The *trans* isomers in all double bonds (all-*trans*) of the chain (Fig. 27.1) have maximum activity.

Retinol decomposes when subjected to prolonged heating in contact with air.

Two vitamers of vitamin A exist (A$_1$ and A$_2$) (Fig. 27.1): A$_2$ predominates in mammalian tissues, and it has an additional double bond between carbons 3 and 4 of β-ionone. A$_2$ is more abundant in fish.

Retinaldehyde or *retinal* and *retinoic acid* are obtained by retinol oxidation, both possess

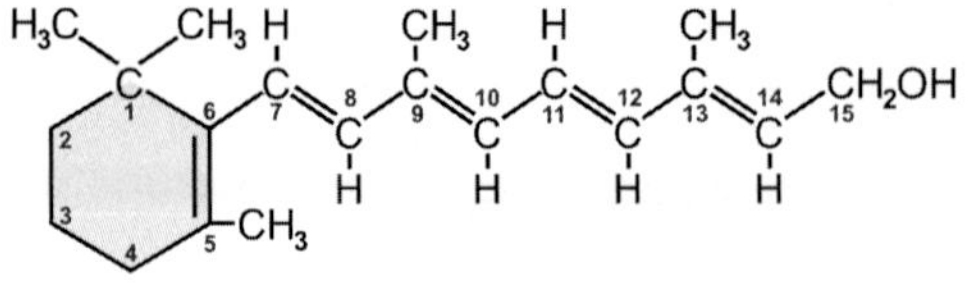

Vitamin A₁

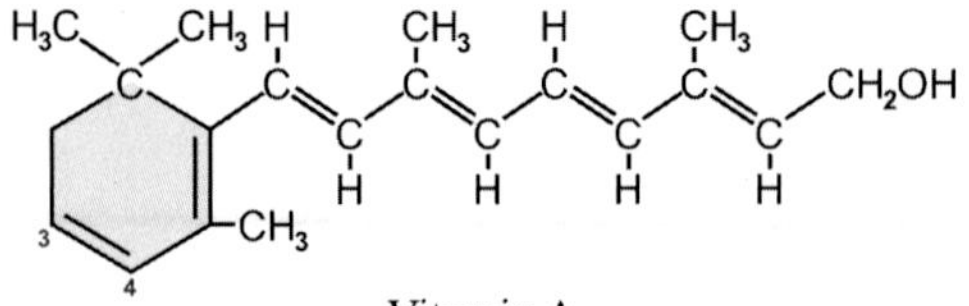

Vitamin A₂

FIGURE 27.1   **All-*trans* retinol.**

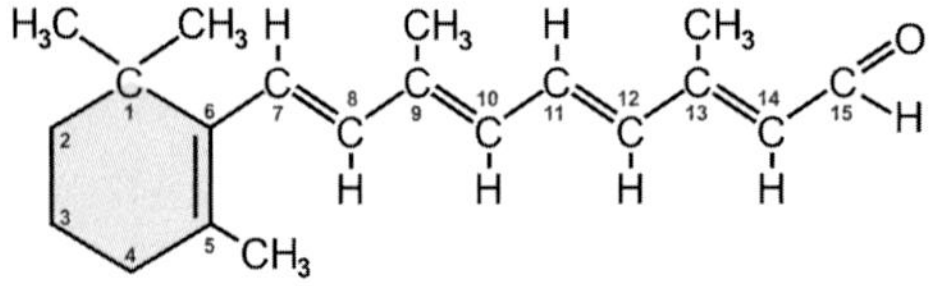

All-*trans*

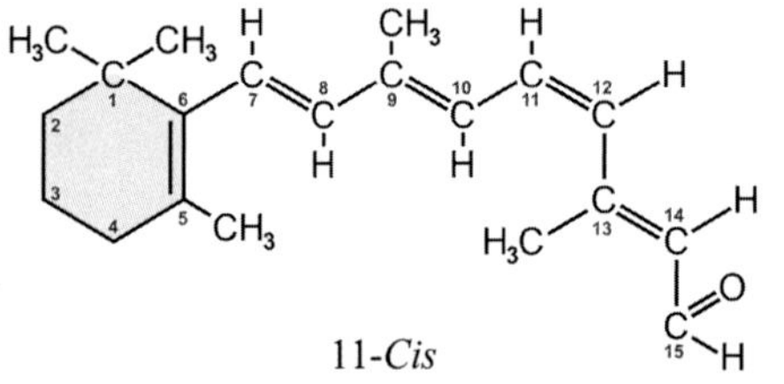

11-*Cis*

FIGURE 27.2   **Retinal.**

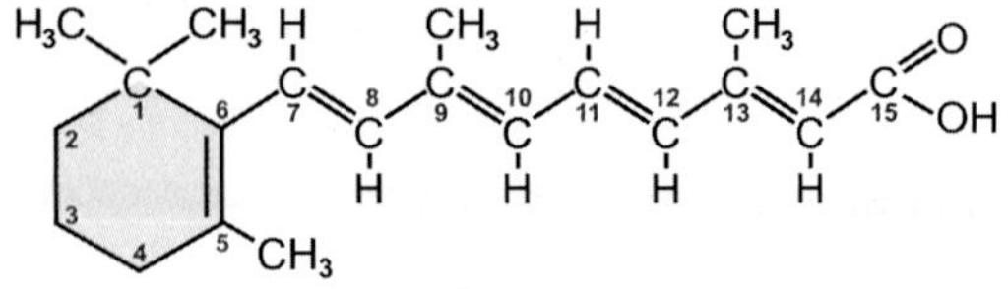

All-*trans*

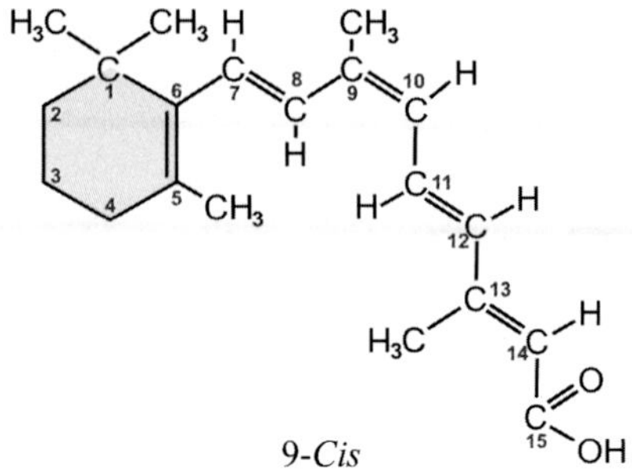

9-*Cis*

FIGURE 27.3   **Retinoic acid.**

biological activity. An isomer of retinal, 11-*cis* retinal (Fig. 27.2), is present in retina.

Retinoic acid is obtained from oxidation of retinal; there are all-*trans* and 9-*cis* isomers (Fig. 27.3).

Vitamin A includes compounds with similar activity to all-*trans* retinol; the term retinoid includes natural forms of vitamin A and analogs obtained by synthesis. In plants, there are pigments called *carotenes*, which are precursors or provitamins A; in the body, they generate vitamin A. The compounds of this group are called *carotenoids*. They form an isoprene chain with a β-ionone ring at one or both ends. The most common are β-carotene, α-carotene, and β-cryptoxanthin (Fig. 27.4). The all-*trans* isomers are the most abundant and stable.

β-Carotene, a symmetrical 40-carbon molecule, is the most widely distributed in nature. It has the highest vitamin A activity because it generates, by oxidation, two molecules of retinal. The other carotenes precursors of vitamin A generate only one. Carotenoids are not affected by cooking.

Natural sources. Vitamin A is found in foods of animal origin, such as milk, cheese, butter, eggs, and fish (tuna, sardines); kidney and meat contain smaller amounts. The liver, especially from some fish (shark, cod), is remarkably rich in vitamin A.

Carotenoids are found in pigmented vegetables (carrots, pumpkin, sweet potato, tomato, yellow corn, broccoli, peas, spinach, chard, and fruit). β-Carotene is the most abundant.

Intestinal absorption. Retinol, as well as its precursors, is associated with lipids, and as such they must be emulsified by bile salts to be absorbed. Any condition that prevents absorption of retinol (lack of bile acids or fat in food) reduces its intestinal absorption.

Retinol esters are hydrolyzed by pancreatic juice and intestinal brush border lipases and esterases. Thanks to its low polarity, retinol easily crosses the intestinal cell apical membrane, but their uptake mainly occurs by facilitated diffusion. Between 70% and 90% of retinol and 10%–20% of β-carotene foods is absorbed. In the intestinal brush border, a monooxygenase catalyzes the cleavage of the double bond between carbons 15 and 15′ of β-carotene and generates two molecules of retinal. Retinal is converted to retinol by retinal reductase or retinol dehydrogenase, and it is esterified with a long acyl chain. A great portion of carotenoids are not cleaved by the enterocytes; they move intact into the circulation, and are included into chylomicrons.

Metabolism and storage. Chylomicrons from the circulation are taken up by the liver. Retinol is stored in the liver, which plays a central role in maintaining vitamin A homeostasis. When retinol intake is poor, liver mobilizes that stored as reserve.

Retinol dehydrogenase, which is dependent on NAD⁺ or NADP⁺, converts retinol to retinal in a reversibly reaction. From retinal, retinoic acid is produced in an irreversible reaction catalyzed by *retinal dehydrogenase*. There are no enzymes that can reduce retinoic acid to retinal.

Carotenoids that fail to be transformed into retinal are incorporated into VLDL and sent to other tissues, particularly adipose tissue that acquires a yellowish tint.

Transport in blood. Retinol esters are hydrolyzed in the liver; all-*trans* retinol is sent to

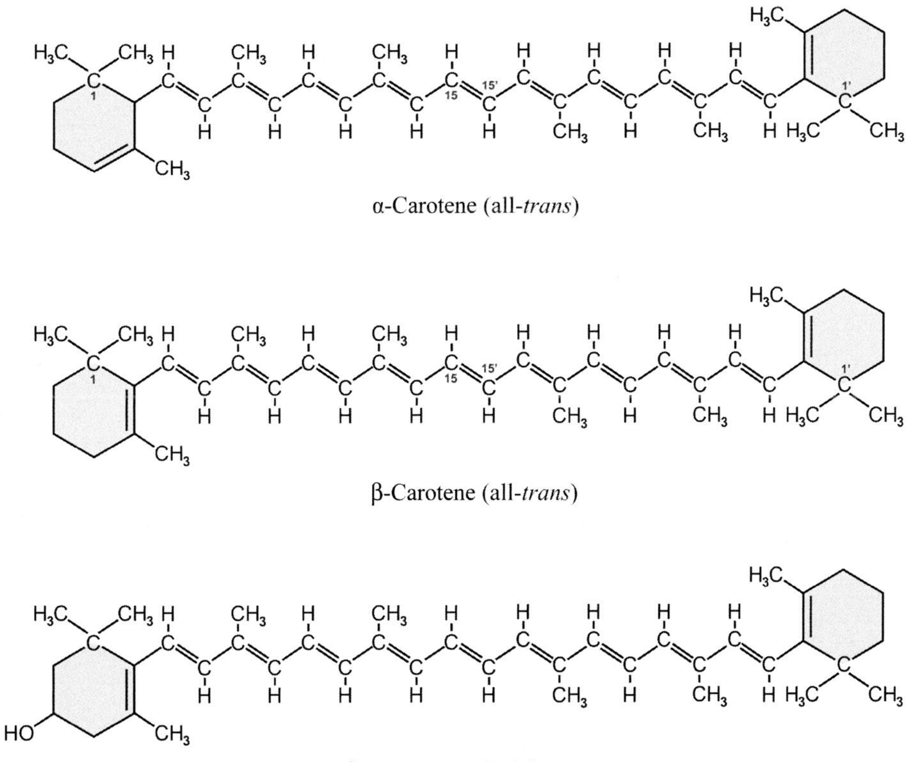

α-Carotene (all-*trans*)

β-Carotene (all-*trans*)

β-Cryptoxanthin (all-*trans*)

FIGURE 27.4 **Carotenes.**

the circulation bound to *retinol-binding protein* synthesized in the liver. In blood, the retinol–retinol-binding protein complex binds to *trans-thyretin* and forms a macromolecular group that prevents the loss of vitamin through the urine. Once reaching the target cells, retinoids are freed from their carrier protein and enter the cell cytoplasm. There are no carotenoids-specific transport proteins in plasma, those passing from liver into the blood are incorporated into VLDL.

Functional role. The main active form of vitamin A is retinoic acid, which favors functions, such as growth, development, reproduction, and immunity. Retinol is involved in reproduction, and retinal is essential for vision.

Mechanism of action. Most of retinoid actions, such as those of many hormones, are genomic, mediated by receptors in the cell nucleus, capable of binding to specific sites in DNA and to regulate the activity of genes encoding enzymes, transcription factors and other proteins, including some related to programmed cell death or apoptosis. It is estimated that retinoids modulate the expression of several hundreds of genes.

All-*trans* retinoic acid binds within cells to the cellular retinoic acid-binding protein which transports it to nuclear receptors. These receptors, designated RAR and RXR, belong to the steroid and thyroid hormone receptor superfamily. Retinoid binding causes a conformational change that induces formation of RAR–RXR heterodimers. These dimers bind to specific DNA sequences in the promoter response elements and influence transcription.

Cell differentiation and proliferation. Retinoids modulate the activity of genes involved in cell proliferation. They target epithelia (skin and gastrointestinal, respiratory, and urogenital tracts). Retinoic acid helps to maintain the structure and function of epithelial cells, promoting differentiation of keratinocytes in the epidermis, and mucus secretion in other epithelia. In addition, vitamin A participates in the synthesis of enzymes involved in the production of cell surface glycoproteins.

Experiments in vitro and with animals have demonstrated that retinoids can suppress the growth of tumors and protect cells from the effect of carcinogens.

Development and growth. During embryo development, gene expression is tightly regulated. Retinoic acid is involved in the induction of morphogenesis through controlling the *homeobox* family (Hox) genes, which guide body development. In late stages of pregnancy, retinol also contributes to these actions.

Reproduction. Retinoic acid is required for embryo implantation. Along with retinol, it stimulates progesterone production. In the male, retinoic acid and retinol are involved in the regulation of testicular function and spermatogenesis.

Immunity. Vitamin A plays a role in the immune system. In particular, retinoic acid stimulates the development of B- and T-helper lymphocytes, their phagocytic activity, and the production of cytokines. Also, the effects of vitamin A on maintaining skin and mucous membranes integrity serve as a mechanism of defense.

Vision. The role of vitamin A in vision is dual. Retinal, through nongenomic actions, participates in the process of light detection, and retinoic acid, by genomic mechanisms, is essential to maintain integrity and transparency of the corneal epithelium.

In the retina, rods are responsible for night vision or vision under poorly lit environments. Rhodopsin, a protein that belongs to the G protein-coupled family of receptors, is key for this process. Rhodopsin is composed by the apoprotein *opsin* and a prosthetic group, 11-*cis* retinal. In the retinal pigment epithelium, all-*trans* retinol is converted to 11-*cis* retinol by an *isomerase* and oxidized to 11-*cis* retinal by *11-cis retinol dehydrogenase*. 11-*cis* Retinal is carried by the *cellular retinaldehyde-binding protein* (CRALBP) to the gap that exists in the photoreceptor epithelium, and donates it to the *interstitial retinol-binding protein*. Once it reaches the rods, the

11-*cis* retinal is released and binds to opsin to form rhodopsin.

Light produces changes in the retina; retinal is converted from 11-*cis* to all-*trans*. This conversion is essential to detect the incoming light. The *cis–trans* isomerization of retinal determines its separation from the opsin, and conformational changes are transmitted to a G protein called *transducing*. This G protein decreases cGMP levels in the cell, closes Na^+ channels, hyperpolarizes the cells, and generates nerve impulses responsible for light perception.

Rhodopsin must be regenerated to continue functioning. All-*trans* retinal is reduced to all-*trans* retinol and transported from the photoreceptor to the pigment epithelium by interstitial retinol-binding protein. Rhodopsin is reconstituted by a metabolic cycle, diagrammed in Fig. 27.5.

The time required to restore rhodopsin explains the momentary impairment of vision after the incidence of a light beam on the retina, as occurs when passing from a bright to a dark room.

In normal individuals, rhodopsin regeneration cycle is accomplished within a very short time because the pigmented epithelial cells maintain a pool of retinyl esters. In vitamin A deficiency, the reduction in the precursor *cis* retinal slows down the cycle, causing "night blindness" or *nyctalopia*.

Requirements. Vitamin A activity is expressed in International Units (IU), which equal 0.3 μg of retinol (Table 27.1). At present, it is preferred to use the expression *retinol activity equivalent*, which is equal to 1 μg of retinol, 12 μg of β-carotene (14 μg for some authors), or 24–28 μg of other vitamin A carotenoid precursors. These equivalents take into consideration the less efficient absorption of carotenoids.

Toxic effects. Retinol is deposited in the liver and eliminated very slowly, so the risk exists for an excessive accumulation of this vitamin if it is administered in high doses. Prolonged treatment with higher than normally required amounts produces chronic poisoning. This results in loss of appetite, fatigue, skin, and mucosal changes

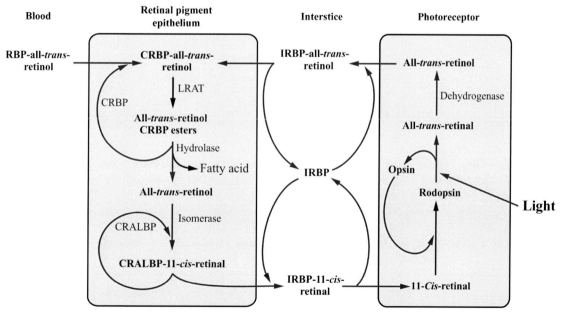

FIGURE 27.5 **Retinol cycle in the retina.**

(erythema, pruritus, desquamation), bone and joint pain, hepatomegaly, edema, neuromuscular, and psychological disorders. A single excessive dose (i.e., 300,000 retinol activity equivalents) can cause acute poisoning, abdominal pain, headache, vomits, and increase in intracranial pressure. Intake of more than 7500 μg of retinol in women during early pregnancy (preconception and first trimester) causes fetal abnormalities. Higher than recommended doses must not be consumed during pregnancy.

Carotenoids produce no toxic effects, mainly because their absorption in the intestine is limited.

Avitaminosis. Vitamin A deficiency affects mainly people of lower social class. It is estimated that over 30% of the world population is at risk of hypovitaminosis A, with a particular higher incidence for children under 6 years of age.

Early symptoms of vitamin A deficiency are related to visual function; discomfort for light or *photophobia* is characteristic. There is also an increase in time required to perceive images again after a sharp reduction in light intensity (night blindness or nyctalopia), due to the failure to regenerate 11-*cis* retinal.

The growth and differentiation of all epithelial cells are also affected; epidermal cells do not regenerate, flatten, and accumulate keratin. The skin becomes dry, with hyperkeratosis and severe scaling. Lacrimal glands reduce their secretion, resulting in erosions and ulcers in the mucosa and cornea. Eyelids become swollen with scaly edges; infections with purulent discharge enssues. If not treated on time, the spontaneous healing of the corneal ulcers causes its opacification and blindness, a condition called *xerophthalmia*.

The epithelium of the respiratory, gastrointestinal, and genitourinary tracts can also be severely compromised. Keratinization impairs the capacity of epithelia to serve as a barrier, and infections are common, especially in children. Systemic effects that compromise the innate and adaptive (humoral and cellular) immune mechanisms of resistance to infection (Chapter 30) are also evident. All these factors determine increased risk of morbidity and mortality to infectious diseases.

Alterations in epiphyseal cartilage with bone growth retardation and defects in bone structure are also apparent. Defects in the dental enamel cause tooth sprout retardation and cavities.

Administration of vitamin A to population at risk is effective in reducing the problems associated with the lack of retinoids in various regions of the world. Addition of vitamin A to milk is an effective measure.

Pharmacological actions. Numerous epidemiological studies indicate that vitamin A deficiency is often accompanied by increased risk of developing cancer. Administration of vitamin A has been shown to help the remission of acute promyelocytic leukemia.

All-*trans* retinoic acid and synthetic derivatives are used in the treatment of acne and psoriasis.

Administration of retinoids should be carefully monitored especially in women of childbearing age, due to its teratogenic effects.

Vitamin D

Synonymy. Calciferol, antirachitic vitamin.

Chemistry. Vitamin D is obtained as white crystals; it is soluble in fats and organic solvents and resistant to heat and oxidation. Structurally, it is derived from cyclopentaneperhydrophenanthrene. Two vitamers exist: vitamin D_2 or *ergocalciferol*, present in plants, and vitamin D_3 or *cholecalciferol*, present in animal tissues. Vitamers differ in their side chain (Fig. 27.6). Natural precursors are *ergosterol* (provitamin D_2) and *7-dehydrocholesterol* (provitamin D_3). Ultraviolet light opens ring B of cyclopentaneperhydrophenanthrene and generates vitamin D (Fig. 27.6).

Natural sources. Vitamin D exists in common foods in poor quantity. Egg yolk, whole milk, butter, liver, beef or calf meat, and vegetable oils

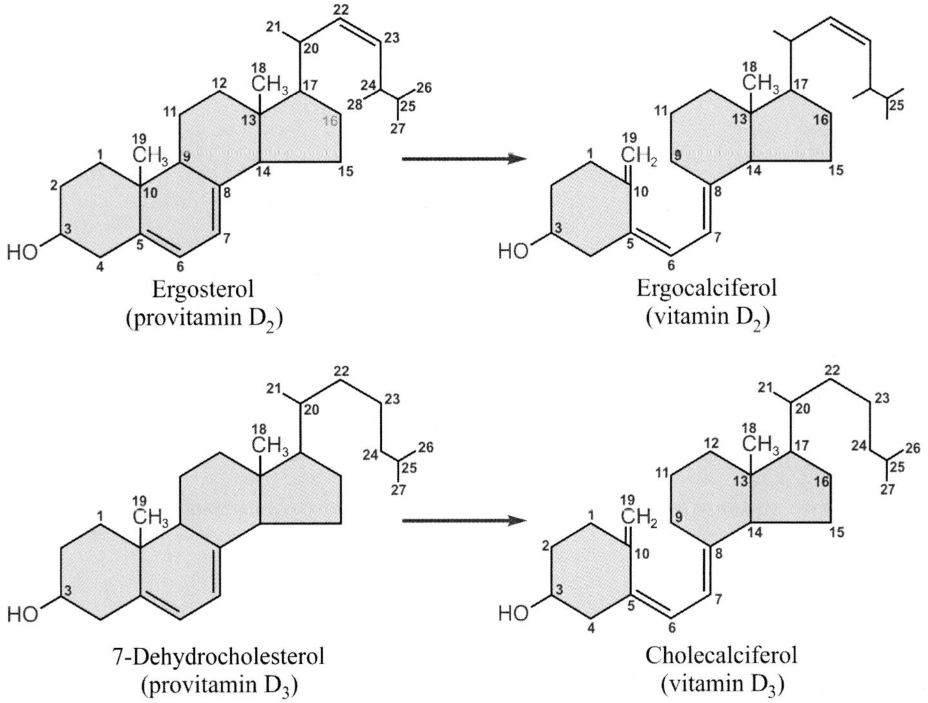

FIGURE 27.6 **Structure of precursors and natural forms (vitamers) of vitamin D.**

only contain small amount. Some fish (salmon, tuna, sardines, herring) are somewhat richer. Vegetables and fruits do not contain vitamin D. The vitamin D precursor ergosterol is found in plants, fungi, and yeasts; 7-dehydrocholesterol is widely distributed in the animal kingdom, and it is synthesized in the sebaceous glands and secreted to the skin surface, where it penetrates the epidermis. When skin is exposed to sunlight, ultraviolet B radiation (wavelength between 290 and 315 nm, with a peak action at 297 nm) converts 7-dehydrocholesterol to vitamin D_3 or cholecalciferol (Fig. 27.6).

Intestinal absorption. Some foods of animal origin contain small amounts of vitamin D esterified with fatty acids. The esters are hydrolyzed in the intestinal lumen, and the vitamin is absorbed in the upper two-thirds of the small intestine. The presence of bile salts and factors that ensure adequate fat digestion and absorption is required. Vitamin D, along with other lipids, passes through the luminal membrane by simple diffusion. In enterocytes, vitamin D is incorporated into chylomicrons and sent to the lymphatic system.

Transport. Chylomicrons enter the circulation, where much of the vitamin is transferred to a specific binding protein [vitamin D-binding protein (DBP)]. Vitamin D formed in the skin passes to the blood, binds to DBP, and follows the same path as that absorbed in the intestine. Both remain bound to DBP and are taken up by the liver.

Metabolism. The production and availability of vitamin D depend on exposure to ultraviolet light and to its absorbtion in the intestine. Vitamin D is considered a prohormone without biological activity; to acquire its functional

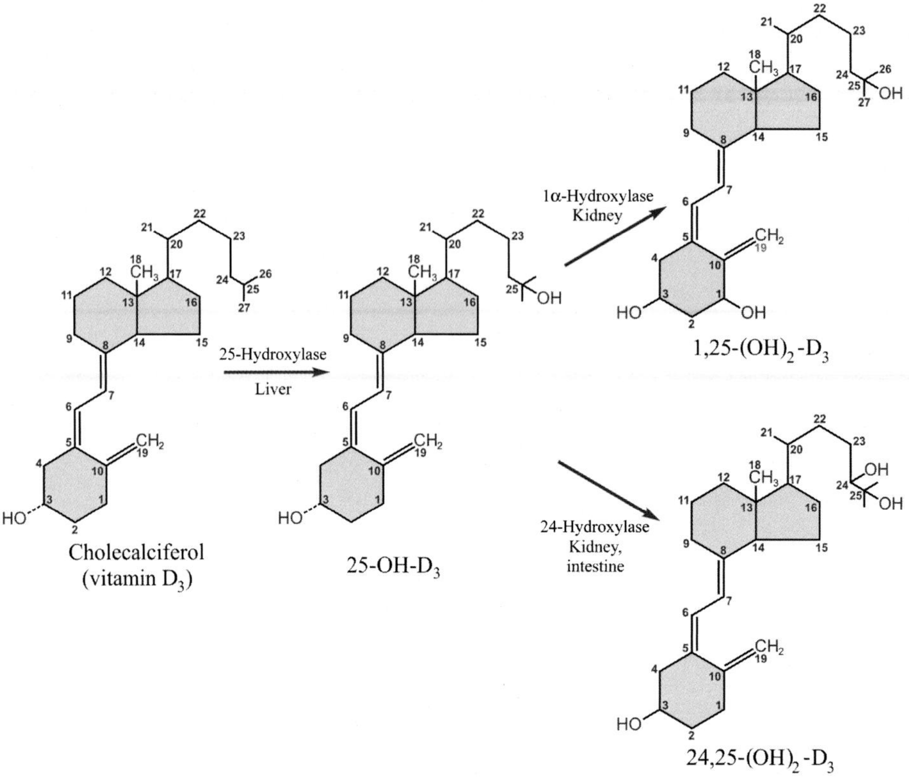

FIGURE 27.7   **Formation of the active metabolites of vitamin D₃.**

properties, it must undergo chemical changes. Cholecalciferol and ergocalciferol follow the same chemical transformations.

Formation of 25-OH-D₃. The first step in vitamin D activation is accomplished in liver. *Calcidiol*, 25-hydroxycholecalciferol (25-OH-D₃) (Fig. 27.7), is generated by action of *cholecalciferol-25-hydroxylase*. The amount of 25-OH-D₃ formed is proportional to the vitamin D₃ supplied.

Formation of 1,25-(OH)₂-D₃. The next stage in the activation of vitamin D takes place in the kidney; after the calcidiol–DBP complex has filtered through the glomeruli, 25-OH-D₃ traverses the brush border of the proximal tubules and is transported into the mitochondria of the tubular cells, where *25-OH-D₃-1α-hydroxylase* produces

dihydroxycholecalciferol (1,25-(OH)₂-D₃) or *calcitriol*. In blood, calcidiol and calcitriol are carried by the DBP.

Formation of 24,25-(OH)₂-D₃. The 25-OH-D₃-24-hydroxylase in kidney mitochondria adds an OH at carbon 24 of calcitriol and generates 24,25-dihydroxy cholecalciferol (24,25-(OH)₂-D₃).

The 1 and 24 hydroxylases are highly regulated factors that trigger the production of 1,25-(OH)₂-D₃, inhibit the formation of 24,25-(OH)₂-D₃, and vice versa. Increased levels of calcitriol induce the synthesis of 24-hydroxylase.

Functional role. The active metabolite of vitamin D regulates calcium and phosphate homeostasis. It increases the extracellular levels of calcium and phosphorus. Its main target organs are the intestinal mucosa, bone, and kidneys. It also has actions in other tissues, which are independent of its effects on calcium.

Mechanism of action. The most active derivative of vitamin D is $1,25\text{-}(OH)_2\text{-}D_3$; it exerts genomic and nongenomic actions. Much less active are $24,25\text{-}(OH)_2\text{-}D_3$ and $25\text{-}OH\text{-}D_3$.

Genomic actions. Calcitriol acts in the nucleus of effector cells, on DNA, in a way similar to that of steroidal, thyroid, and retinoid hormones. Calciferol crosses the plasma membrane without difficulty; in the cytosol and nucleus, there are specific receptors (VDR_n) that belong to the family of steroid receptors. VDR_n has higher affinity for $1,25\text{-}(OH)_2\text{-}D_3$ than for the other metabolites. VDR_n acts as a ligand-activated transcription factor. Binding of $1,25\text{-}(OH)_2\text{-}D_3$ induces a conformational change that enables the D–VDR complex to be transported to the nucleus and forms heterodimers with the retinoid receptor (RXR). The VDR portion of the calcitriol–VDR–RXR complex has a domain with two "zinc fingers" that specifically bind to *vitamin D response elements* in the promoter of the target genes, activating transcription. The best studied of the proteins whose synthesis is stimulated by $1,25\text{-}(OH)_2\text{-}D_3$ is the calcium-binding protein *calbindin*. Calcitriol also induces the synthesis of other proteins, such as alkaline phosphatase and Ca^{2+}-ATPase in the intestine, $25\text{-}OH\text{-}D_3\text{-}24\text{-}hy$-droxylase in the kidney and other organs, and is involved in the regulation of the synthesis of various proteins in the bone matrix (collagen I, osteocalcin, bone sialoprotein, osteopontin).

Nongenomic actions. Calcitriol can exert quick responses by binding to a membrane receptor (VDR_m). These cause calcium channel activation, enhanced phospholipid metabolism, increased membrane fluidity, and elevated intracellular Ca^{2+} concentration and prostaglandin production.

Vitamin D is essential in the intestine–bone–kidney–parathyroid maintenance of calcemia within narrow limits.

Effects on the intestine. The main effect of $1,25\text{-}(OH)_2\text{-}D_3$ on intestine is the increased absorption of calcium (see p. 720). It produces (1) nongenomic, fast-acting effects, such as calcium increase, which are secondary to the opening of Ca^{2+} channels, and (2) genomic, slower and more sustained effects, which are due to induction of calbindin synthesis and increased Ca^{2+}-ATPase. It also increases phosphate absorption in the distal jejunum and ileum.

Effects on bone. Vitamin D promotes mineralization and remodeling of bone: (1) it increases the local concentration of calcium and phosphate, and (2) it activates bone resorption. The synthesis of type I collagen and bone sialoprotein is regulated by several factors including $1,25\text{-}(OH)_2\text{-}D_3$. Expression of osteopontin and osteocalcin is stimulated by calcitriol. This, in addition to the increase in calcium and phosphate, promotes formation of the bone matrix.

Effects on the kidney. $1,25\text{-}(OH)_2\text{-}D_3$ activates reabsorption of calcium and phosphate. Calcitriol exerts a regulatory effect on itself by repressing the synthesis of $25\text{-}OH\text{-}D_3\text{-}1\alpha$-hydroxylase and inducing $25\text{-}OH\text{-}D_3\text{-}24$-hydroxylase.

Action on the parathyroid gland. Calcitriol–VDR is a powerful regulator of parathyroid gland function. It suppresses PTH synthesis and secretion, and modulates the parathyroid response to serum calcium levels.

Other actions. Calcitriol regulates the expression of over a hundred of genes, whose function exceed those related to calcium homeostasis. For example, it promotes cell differentiation and maturation, regulates apoptosis, inhibits cell proliferation, and enhances the differentiation of monocyte–macrophage antigen-presenting cells, dendritic cells, and lymphocytes, all of which explain the capacity of vitamin D to improve the response to infections and the tolerance to transplantations.

Functions of $24,25\text{-}(OH)_2\text{-}D_3$. Although the effects of $24,25\text{-}(OH)_2\text{-}D_3$ are not as clear as those of $1,25\text{-}(OH)_2\text{-}D_3$, experimental evidence shows that the presence of both metabolites is required to exert the actions of vitamin D. $24,25\text{-}(OH)_2\text{-}D_3$ is essential for bone fracture repair, acting by paracrine, autocrine, or systemic (endocrine) mechanisms.

Requirements. The amount of vitamin D is expressed in IU, which are equivalent to 0.025 µg of cholecalciferol or vitamin D_3 (1 µg = 40 IU); the use of weight units is preferred (Table 27.1).

The dietary requirements of vitamin D are not easy to estimate because the human body synthesizes provitamin 7-dehydrocholesterol and vitamin is generated when the skin is exposed to sunlight. It is considered that exposure of the skin areas that normally receive sunlight (face, neck, and hands) for 15 min/day is sufficient to produce the cholecalciferol needed by the body. However, to ensure an adequate supply, intake of vitamin D is recommended.

Toxic effects. Vitamin D hypervitaminosis occurs when very high amounts are administered. This causes loss of appetite, nausea, increased urine output, thirst, hypercalcemia, hyperphosphatemia, hypercalciuria, and hyperphosphaturia. Usually, calcification in soft tissues (kidney, lung) occurs.

An excess of sun exposure does not produce hypervitaminosis; thanks to the slow conversion of 7-dehydrocholesterol into cholecalciferol and the formation of inactive isomers.

Avitaminosis. Vitamin D deficiency in early life causes *rickets*, a disease known since ancient times. In adults, it produces *osteomalacia*. Insufficient exposure of skin to sunlight results in hypovitaminosis D. Even if nutrition is inadequate, no severe rickets is observed in children who have ample opportunity to receive sunlight. In regions far from the equator (above 45 degrees latitude), and in large cities where there are tall buildings and polluted air, the decreased sunshine disfavors vitamin D production.

Vitamin D deficiency results in faulty bone calcification. In young children, rickets delays closure and thickening of epiphyses and costochondral joints; the cartilage continues to grow, but is not calcified. When the child begins to walk, the poorly calcified leg bones bend due to the body's weight (*genus valgum*); both legs

become curved, with an outward convexity. Radiographically, bones are less dense than normal. Late sprout and poor development of teeth occur.

Deficiency of vitamin D and calcium in the adult result in bone demineralization, a disorder called osteomalacia. The main symptoms are muscle weakness and bone pain, especially in the hips. Skeletal deformities are rarely observed. In people over 75 years of age, the risk for vitamin D deficiency is higher. This is due to a decrease in the synthesis of provitamin 7-dehydrocholesterol in the skin and the lower exposure to sunlight. The reduced bone mineralization favors production of fractures.

Epidemiological studies have confirmed the existence of an association between prostate, breast and colon cancer, and vitamin D deficiency. Hypovitaminosis increases the risk of malignant cell transformation.

Renal insufficiency and hypoparathyroidism affect 1α-hydroxylation of 25-OH-D_3 and can be associated with rickets or osteomalacia. Genetic defects produce rickets (familial hypophosphatemic rickets and vitamin D-resistant rickets).

Pharmacological applications. In chronic renal failure and hypoparathyroidism, administration of high doses of vitamin D (250–500 µg/day) has proven to be beneficial. The therapeutic action depends on the large increase in 25-(OH)-D_3 in plasma, which follows the administration of calciferol, which mimics the action of calcitriol.

1,25-(OH)$_2$-D_3 and synthetic analogs are being used in the treatment of renal osteodystrophy and osteoporosis, psoriasis, and secondary hyperparathyroidism. Its benefits as immunosuppressive agent and antineoplastic agent in breast, colon, prostate, and leukemia are the matter of intense research.

Vitamin E

Synonym. Rat antiesterility vitamin.

Chemistry. Vitamin E is insoluble in water and soluble in fat, alcohol, and other organic

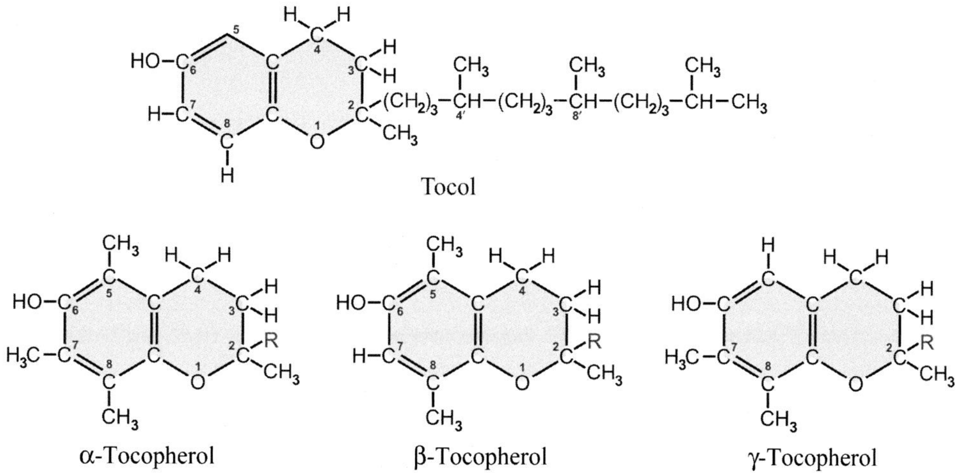

Tocol

α-Tocopherol   β-Tocopherol   γ-Tocopherol

FIGURE 27.8   **Tocopherols.**

solvents. It is relatively stable. There are two groups of compounds with vitamin E activity, *tocopherols* and *tocotrienols*, all methylated derivatives of a basic structure called *tocol*. Tocol has a chromane ring with a hydroxyl group at position 6 (6-hydroxychromane or 6-chromanol), and a methyl and a phytyl side chain of 15 carbons, formed by the union of three isoprene units, in position 2 (Fig. 27.8).

Tocopherols and tocotrienols differ in the number and position of their methyl groups, the degree of their chain saturation, and their biological activity. Those called α have three methyl groups in positions 5, 7, and 8 of chromanol.

Those named β have two methyl groups at 5 and 8. The ones known as γ have two methyl groups at positions 7 and 8. Finally, those named δ have only one methyl group in position 8 (Fig. 27.8).

Tocopherol's isoprenoid chain is saturated, while that of tocotrienols have double bonds between carbons 3'-4',7-,8' and 11'-12' (Fig. 27.9).

The antioxidant activity of vitamin D is due to the ease with which the −OH at position 6 donates its H.

Natural sources. Cereal and seeds rich in lipids are excellent sources of vitamin E. Soybean, corn, sunflower, and peanuts contain good amounts, as well as wheat germ, almonds, walnuts, and

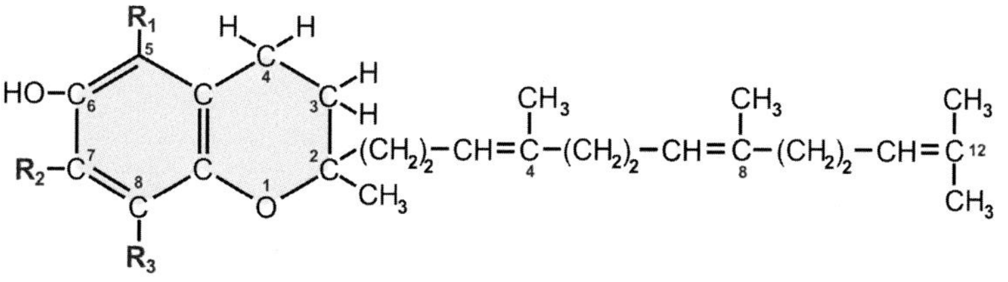

α-Tocotrienol: $R_1 = CH_3$; $R_2 = CH_3$; $R_3 = CH_3$

β-Tocotrienol: $R_1 = CH_3$; $R_2 = H$; $R_3 = CH_3$

γ-Tocotrienol: $R_1 = H$; $R_2 = CH_3$; $R_3 = CH_3$

FIGURE 27.9   **Tocotrienols.**

hazelnuts. There is very little vitamin E in meat, eggs, and dairy products. Foods differ not only in their content of vitamin E, but also in the relative amount of its vitamers.

Intestinal absorption. α- and γ-Tocopherols are the most common in the diet; they are free or esterified with fatty acids, which are hydrolyzed by enzymes from the pancreas and small intestine. Absorption of vitamin E takes place by passive diffusion, and like other fat-soluble vitamins, it requires the presence of factors that ensure normal digestion of fats. The proportion absorbed decreases as the amount ingested increases. In enterocytes, vitamin E is incorporated into chylomicrons.

Transport. In blood, chylomicrons become remnants. Part of the vitamin E is transferred to HDL that exchange it with VLDL and LDL. The vitamin E remaining in chylomicron remnants is taken up by the liver, incorporated into VLDL and released into the circulation.

Metabolism. After VLDL conversion into LDL, tocopherols enter into cells through LDL receptors. Excess vitamin is stored in various tissues, particularly in the adipose tissue.

Functional role. The most remarkable property of vitamin E is its antioxidant capacity. Its best known role is the protection of polyunsaturated fatty acids against oxidative damage (see Chapter 10). Vitamin E reacts with peroxyl radicals, forming hydroperoxides and a relatively stable radical form of vitamin E. Vitamin E also prevents oxidation of other membrane compounds and LDL.

After reaction with free radicals, vitamin E is regenerated to its initial state by antioxidants, such as vitamin C [ascorbic acid (AA)].

Other functions. Thanks to its saturated isoprenoid chain, α-tocopherol interacts with polyunsaturated fatty acids in the lipid bilayer and helps to stabilize the membrane structure.

Requirements. Presently it is preferred to use the expression *α-tocopherol equivalent* (αTE), which is equal to 1 mg of the compound (Table 27.1). The greater consumption of polyunsaturated fatty acids, the higher is the requirement of vitamin E. It is advisable to keep the ratio of 0.5 αTE per gram of polyunsaturated fatty acids.

The normal diet provides the required amount of vitamin E; in healthy people and additional administration is not required. No significant toxic effects were observed for vitamin E in humans.

Avitaminosis. Manifestations of vitamin E deficiency vary considerably depending on the species studied. In rats, mice, and pigs, vitamin E deficiency produces serious damage in the reproductive organs, with alteration of the germinal epithelium and infertility. In the female, fertility is impaired and if fecundation occurs, the embryos die. In males, there is atrophy of the seminiferous epithelium and male sterility. The role of vitamin E in reproduction suggested its use as an "antiesterility factor" that could be beneficial for the treatment of human infertility. Results in humans were negative and human infertility is not due to lack of tocopherol.

In most animals, avitaminosis E causes functional and structural alterations of cardiac, skeletal, and smooth muscles. In pig and rat, liver necrosis is observed. In chickens, there are symptoms of nervous system degeneration. Erythrocytes life shortening and hemolytic anemia are frequent. Most of the symptoms are expression of altered cell membranes.

Hypovitaminosis E is rare in humans. Deficiency symptoms occur in severe cases of fat malabsorption and abetalipoproteinemia (lack of LDL). Neuropathies and myopathies similar to those described in animals are observed. Large axons of peripheral sensory nerves, axonal dystrophy of posterior columns of the spinal cord, and dorsal and ventral spinocerebellar tracts are affected.

In malnourished children with low plasma tocopherol, half-life of erythrocytes is reduced. Prematures and low weight newborns deficient in vitamin E present anemia, reticulocytosis, thrombosis, and edema; artificial

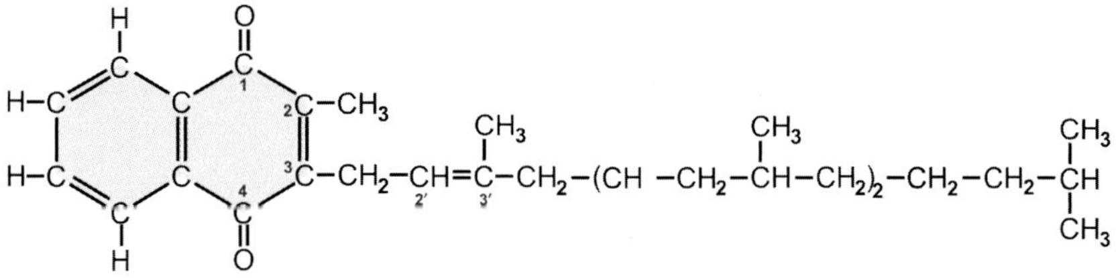

2-Mettyl-3-phytyl-1,4-naphthoquinone (phylloquinone)

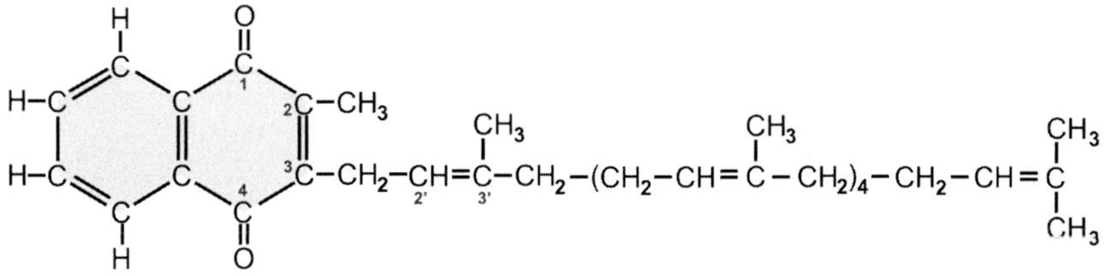

Menaquinone-6

FIGURE 27.10 **Vitamin K vitamers.**

feeding does not correct the deficiency. Continued lack of vitamin during the first two decades of life produces serious neurological disorders. Early treatment with tocopherol supplements is important to avoid the mentioned alterations.

Pharmacological actions. Given the reduced toxicity of vitamin E, studies have been conducted on a large number of individuals to establish the efficacy of high doses (~1000 mg/day) in the prevention of cardiovascular events, cancer, defects of the immune response, and aging, all processes that are related to the action of reactive oxygen species. Some authors report favorable effects; however, it is premature to draw conclusions at this time; additional research is needed before recommending vitamin E supplementation.

Vitamin K

Synonymy. Antihemorraghic vitamin.

Chemistry. There are several naturally occurring vitamers of vitamin K that are naphthoquinone derivatives; K_1 and K_2 are the most important. K_1, or *phylloquinone*, is synthesized by plants. It is the 2-methyl-3-phytyl-1,4-naphthoquinone, with a phytyl 20-carbon side chain that has a double bond between C2' and C3' (Fig. 27.10).

K_2 comprises a family of compounds synthesized by bacteria, called *menaquinones*. They consist of isoprenyl groups that differ in length in different members of this family of compounds. They are designated with the acronim MK and a figure indicating the number of isoprenyl units. The chain may consist of up to 15 units, the most common have 6–10. *Farnoquinone* or MK-6 is the 2-methyl-3-difarnesyl-1,4-naphthoquinone (Fig. 27.10); the side chain has difarnesyl (30C). There are no significant differences in activity between the different vitamers.

K vitamins are fat soluble, resistant to heat, and reducing agents, but sensitive to light. The activity is supressed by ultraviolet light, alkalis, strong acids, and oxidizing agents.

The synthetic compound 2-methyl-1,4-naphthoquinone, called *menadione* (Fig. 27.11),

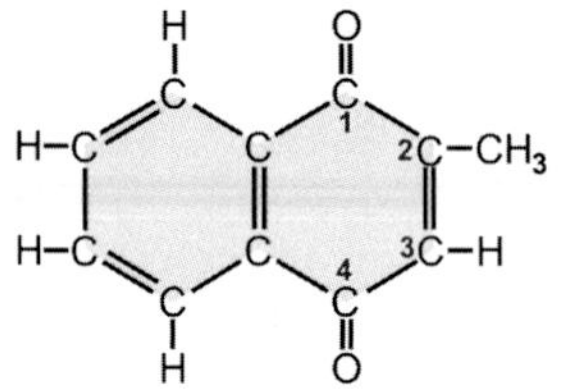

2-Methyl-1,4-naphthoquinone

FIGURE 27.11   **Menadione.**

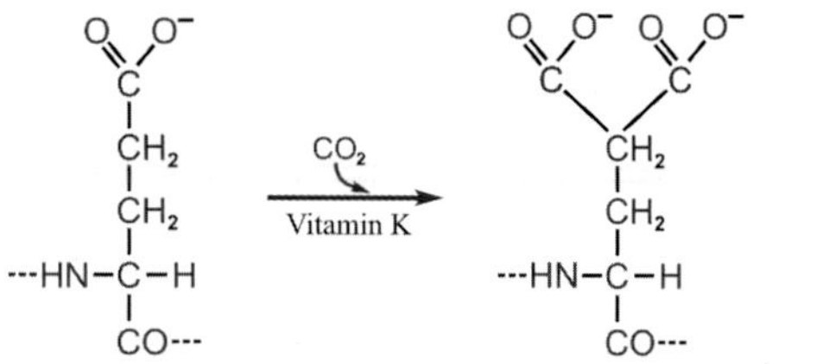

Glutamate residue   γ-Carboxyglutamate residue

FIGURE 27.12   **γ-Carboxylation of glutamate.**

is a provitamin K, converted into active MK-4 by addition of a chain of four isoprenyl units in position 3.

Natural sources. Cabbage, cauliflower, broccoli, tomatoes, spinach, and soybean, olive, peanut, corn, and sunflower oils are rich in phylloquinone. Breast milk is low in vitamin K. Menaquinones have a more restricted distribution in food. They are in fermented products, such as certain cheeses, and are also found in liver. Vitamin K_2 is synthesized by bacteria of the normal intestinal flora.

Intestinal absorption. Because it is fat soluble, vitamin K requires normal fat digestion and absorption. Menaquinones produced by the bacterial flora also contribute to supply vitamin K. They are absorbed by passive diffusion across the intestinal tract, including the colon, which is the primary site of bacterial vitamin K synthesis.

Transport. Within enterocytes, phylloquinone and menaquinones are incorporated into chylomicrons, which enter to the lymphatic system and then to the general circulation; the chylomicron remnants are taken up by the liver. Menadione is absorbed in the intestine and is carried to the liver via the portal vein. Liver tissue is the primary storage site. The reserve declines rapidly if the intake is insufficient. From the liver, the vitamin passes to the circulation with VLDL, which transfers it in part to HDL.

Metabolism. Vitamin K is metabolized in the liver. Phylloquinone is degraded slower than menaquinones. Menadione is converted into MK-4 in hepatocytes.

Functional role. Vitamin K participates as coenzyme in carboxylation reactions of various proteins called *Gla* proteins, including *prothrombin* (factor II), *proconvertin* (factor VII), *plasma thromboplastin component* (factor IX or Christmas), and *Stuart–Prower factor* (factor X), all procoagulant, and proteins S, C, and Z, with anticoagulant action (see Chapter 31), *osteocalcin* and *matrix bone tissue Gla protein*, and Gla proteins present in other tissues.

These proteins are subjected to posttranslational modification, such as carboxylation of glutamyl residues (Glu) γ carbon, which become γ-carboxyglutamyl (Gla) (Fig. 27.12).

The γ-carboxyglutamate residues chelate calcium ions (Fig. 27.13) and interact with phospholipids on the cell surface.

The γ-carboxylation of glutamate residues is catalyzed by vitamin K-dependent γ-carboxylase, which is localized in the endoplasmic reticulum

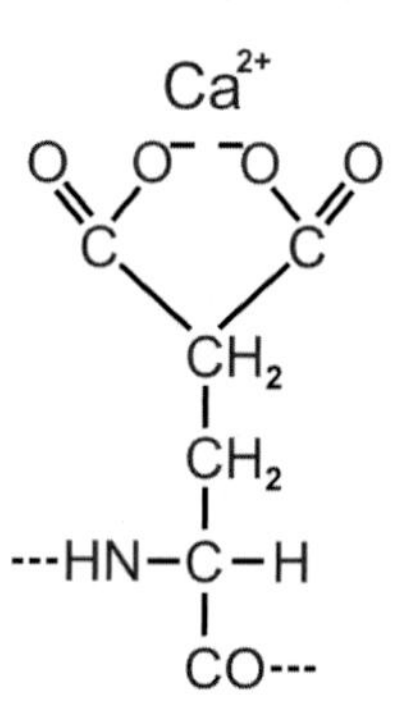

FIGURE 27.13   **Chelating action of α-carboxylate.**

membrane. The reduced form of vitamin K (KH$_2$, hydroquinone) acts as cofactor. CO_2 is added to carbon γ of glutamyl residues, and vitamin K-2,3-epoxide is formed. Hydroquinone is continuously regenerated in the *vitamin K cycle* (Fig. 27.14) with two steps catalyzed by reductases. In the first, the epoxide is reduced to quinone by *vitamin K-2,3-epoxide reductase*, and then vitamin K quinone reductase catalyzes the formation of hydroquinone.

Vitamin K and coagulation. Since its discovery, vitamin K was related to blood clotting, a process that involves more than a dozen different proteins, membrane phospholipids, Ca^{2+} ions (Chapter 31). Four of the procoagulant proteins are vitamin K-dependent (factor II prothrombin and factors VII, IX, and X). They are synthesized in the liver, γ-carboxylated, and secreted into the circulation as inactive zymogens. Other vitamin K-dependent proteins related to clotting are the S, C, and Z, which function as inhibitors of

coagulation. Thus, vitamin K is involved in the synthesis of activating and inhibitory proteins, but the procoagulant effect predominates. In vitamin K deficiency, incompletely carboxylated Gla *proteins induced by vitamin K absence* (PIVKA) appear in blood.

Initially, it was thought that the function of vitamin K was limited to γ-carboxylation of glutamate residues in clotting factors, but later the existence of Gla proteins in various tissues, such as bone, kidney, spleen, placenta and lung, was confirmed. When the function of these proteins becomes known, vitamin K will acquire a broader functional relevance.

Requirements. The need for vitamin K is shown in Table 27.1. It is not easy to establish with certainty the amount of vitamin K that needs to be ingested with the diet because the actual contribution of the bacterial flora is unknown. However, it is estimated that the amount of synthesized menaquinones absorbed in the intestine is

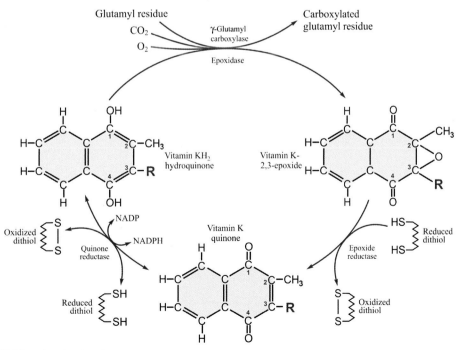

FIGURE 27.14 **Vitamin K cycle.**

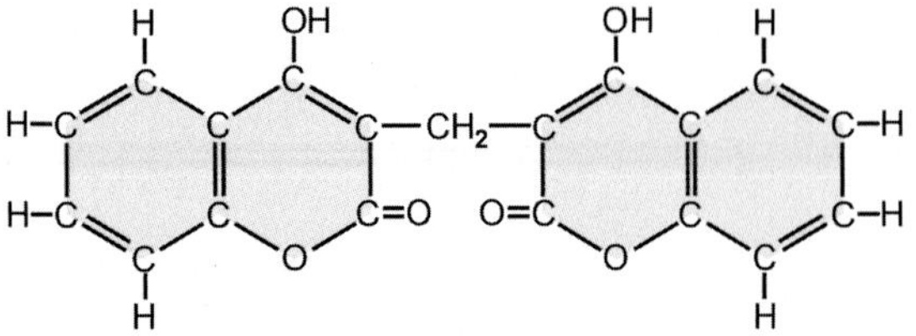

FIGURE 27.15 **Dicumarol.**

not sufficient to meet the normal requirements. No toxic effects have been observed by high doses of natural K vitamers. Instead, menadione (synthetic) and derivatives, in doses greater than 5 mg, produce hemolytic anemia, hyperbilirubinemia, and even kernicterus in infants.

Antivitamins. Dicumarol [3,3'-metylbis-(4-hydroxycoumarin)] (Fig. 27.15) is an antagonist of vitamin K, by interfering competitively its association with vitamin K-2,3-epoxide-reductase. It was the first drug applied clinically to prolong the clotting time. Currently, it is used to treat patients prone to intravascular clotting (thrombosis). One of the antagonists used in prolonged treatments is *warfarin* [(3-α-acetonylbenzyl)-4-hydroxycoumarin] (Fig. 27.16), also used as a poison to kill rats and mice.

Avitaminosis. Deficiency of vitamin K results in insufficient γ-carboxylation of Gla proteins.

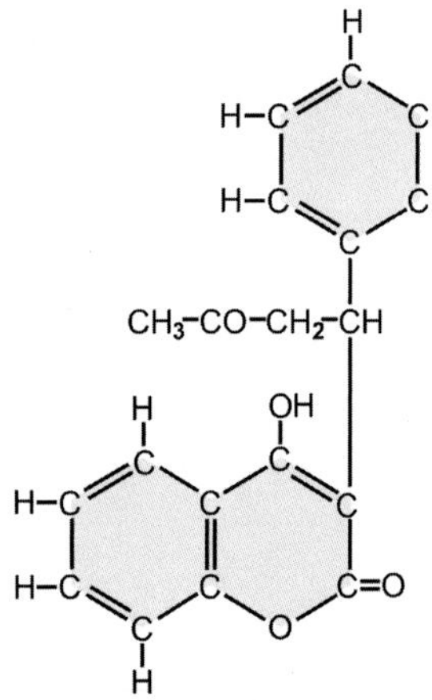

FIGURE 27.16 **Warfarin.**

The vitamin K-deficient patients have great tendency to bleed, even from little wounds. In adult humans, K vitamin deficiency is rare because of the relative abundance of vitamin in common foods and the additional contribution of menaquinones synthesized by bacteria in the intestine. Deficiencies can occur due to (1) poor absorption because of lack of bile salts or syndromes of malabsorption; (2) malnutrition, or patients undergoing prolonged treatment with antibiotics, which reduces the intestinal flora, and with this, the production of menaquinones. Some antibiotics, such as cephalosporins, have an effect that is similar to coumarin anticoagulants.

Several factors determine hypovitaminosis in infants during the first 6 days of life: (1) vitamin K passes with difficulty from mother to fetus, which has low levels of phylloquinone in blood and tissues, (2) the immature liver in babies has limited capacity to generate Gla proteins, (3) breast milk is low in vitamin K, and (4) the newborn intestine has not yet developed the flora that produces vitamin K. Prothrombinemia is low at birth and decreases further, reaching its minimum value at the third day of life. The risks are even higher in premature and low weight newborns.

Vitamin K deficiency in the newborn produces the so-called *hemorrhagic disease of the newborn*; currently, the name *bleeding due to vitamin K deficiency* is a preferred name. This consists in hematomas of head, chest and abdomen, gastrointestinal, and nasal bleeding. Although the incidence of this disease is low, preventive treatment with phylloquinone orally for several days after birth is indicated. Alternatively, one dose (1 mg) is recommended intramuscularly immediately after delivery.

Pharmacological actions. Some evidence suggested favorable effects of menaquinones administration on bone mineralization. In Japan, a study was conducted in a large number of postmenopausal women with osteoporosis. Good results have been described, with increased bone density and reduced incidence of fractures, but further studies are still required. A similar dosage of phylloquinone does not give the same results.

SOLUBLE VITAMINS

Vitamin B Complex

The "water-soluble factor B," initially considered a single substance, was later found to be a mixture of different components. As they were isolated, they were designated by the letter B and a subscript. However, currently, it is preferred to use names related to the chemical structure or the original source. Vitamin B complex includes *thiamine* (B_1), *riboflavin* (B_2), *niacin* (B_3), *pantothenic acid* (B_5), *pyridoxine* (B_6), *biotin* (B_7), *folic acid* (B_9), and *cobalamin* (B_{12}).

These vitamins are often found together in natural sources, deficiencies normally result from reduction of more than one factor. All members of the B complex are coenzymes or part of coenzymes.

When foods are boiled, these vitamins, which are water soluble, pass to the boiling water. This liquid should be used in the preparation of foods, so these vitamins are not lost.

Thiamine

Synonymy. Vitamin B_1, aneurin, antineuritic factor, or antiberiberi factor.

Chemistry. Thiamine is very soluble in water, slightly soluble in alcohol, insoluble in ether and chloroform, and stable to dry heat. It is a crystalline, colorless substance; in acid solution, it withstands temperatures from 120 to 130°C. However, heating in alkaline solution and in contact with air degrades the vitamin.

Thiamine is synthesized by plants in the presence of light and by various bacteria and yeast.

Structurally, it consists of a pyrimidine (4′-amino-2′-methylpyrimidine) attached to a

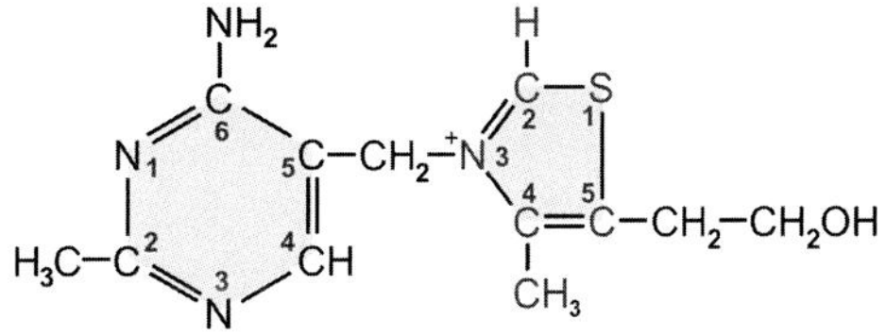

FIGURE 27.17 **Thiamine.**

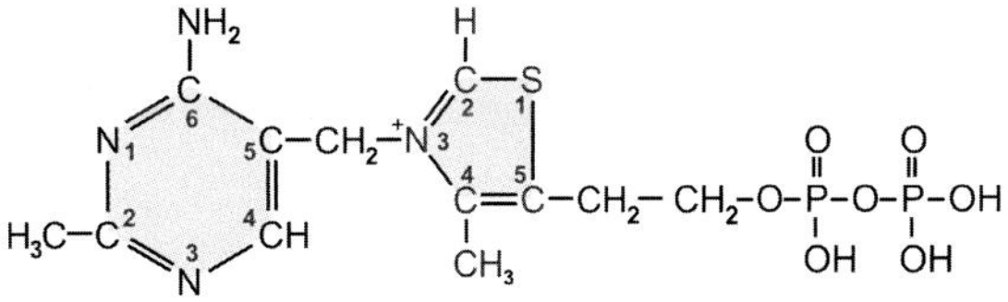

FIGURE 27.18 **Thiamine pyrophosphate.**

thiazole ring (4-methyl-5-hydroxyethylthiazol) by a methylene (—CH_2—) bridge. A two-carbon chain with a primary alcohol is bound to position 5 of thiazole (Fig. 27.17) forming esters with one, two, or three phosphates.

Monophosphate (TMP) is inactive; diphosphate or thiamine pyrophosphate (TPP) (Fig. 27.18) acts as a coenzyme, and it is the most abundant form. Triphosphate (TTP) is related to *nerve impulse* transmission.

Natural sources. Vitamin B_1 is found in plant and animal foods. The best sources are cereals (whole grains), yeast, and pork. Egg yolk, liver, kidney, walnuts, peanuts, and asparagus contain somewhat less. Milk and fruits are poor in this vitamin.

Grain milling separates the bran and with it more than 80% of the thiamin contained in the whole grain. Flour and white bread have very low thiamine, whole flour instead conserves it.

In animal tissues, more than 90% of thiamine is phosphorylated, predominantly as pyrophosphate, in cereals, most of it is free, nonphosphorylated.

Some seafood, plants (ferns), and bacteria contain *thiaminase*, an enzyme that hydrolyzes thiamine and inactivates it as a vitamin. Thiaminase is inactivated by cooking. The ingestion of raw seafood or fish may cause thiamine deficiency.

Intestinal absorption. Thiamine phosphates from food are hydrolyzed by phosphatases. The free thiamine is absorbed in the proximal small intestine. Tea and coffee contain polyphenols (flavonoids, tannic, and cinnamic acid) that interfere with thiamine absorption. In enterocytes, thiamine is phosphorylated to TMP and TPP, which "traps" the vitamin in the cell; it is not stored but dephosphorylated to thiamine and TMP, and sent to the bloodstream.

Transport. Thiamine reaches the liver via the portal vein, and then it is distributed to all tissues bound to albumin in blood.

Metabolism. In tissues, thiamine and TMP are phosphorylated to TPP by *thiamine pyrophosphokinase*, by using phosphate from ATP. In the nervous system, TPP is converted into TTP by *TPP ATP phosphoryltransferase*.

The amount of thiamine stored is small; when intake is insufficient, reduction in tissues is rapid.

TPP pyrophosphate is hydrolyzed by phosphatases. When the supply is normal, small amount of thiamine (free or esterified with sulfate) is excreted in urine; when thiamine is ingested above the normal requirements, the excess is rapidly eliminated unmodified.

Functional role. Thiamine plays a very important role in cell intermediary metabolism. The active form is TPP.

Oxidative decarboxylation reactions. TPP is a coenzyme of three multienzyme systems that catalyze oxidative decarboxylation of α-ketoacids. Two of these systems, pyruvate and α-ketoglutarate dehydrogenases, fulfill its role in pathways that provide energy; the third, branched-chain α-ketoacid dehydrogenase acts on α-ketoacids derived from leucine, isoleucine, and valine.

Transketolation reactions. TPP functions as a coenzyme of transketolases, which transfer two-carbon ketol group (—CO—CH₂OH) from a ketose donor to an aldose acceptor (see p. 307).

Alpha oxidation and cleavage of fatty acids in peroxisomes. Multienzyme systems in peroxisomes that require TPP have been described; they catalyze the oxidation and cleavage of 3-methyl fatty acids and the cleavage of straight chain fatty 2-hydroxyacids.

Regulatory functions. TPP is a major regulator of metabolic pathways. In noncoding mRNAs, there are segments that recognize and bind TPP (riboswitches) and disrupt transcription or translation.

Functions of TTP. Thiamine TTP has different function from TPP. TTP is in the nervous system and skeletal muscle, and it is related to nerve impulse transmission. The functional importance of TTP is evident when it is missing, leading to a severe disorder of the central nervous system, *Leigh disease* or *subacute necrotizing encephalomyelopathy*, in which the synthesis of TTP in the brain is impaired due to a genetic defect.

Requirements. The needs of thiamine depend on the diet. Excessive intake of carbohydrates increases the amount required (Table 27.1). It is advisable to provide 0.5 mg of thiamine per 1000 calories of food. In pathological conditions, such as hyperthyroidism and prolonged fever, the demand for thiamine is increased.

Avitaminosis. In humans, the effects of B₁ hypovitaminosis are anorexia, weight loss, gastrointestinal and cardiocirculatory disorders, peripheral and central neuropathies, muscular weakness, and peripheral symmetrical polyneuropathy. Severe deficiency causes a condition called *beriberi*. This disease was common in the past in the Far East, where many people ate almost exclusively rice devoid of its skin. The disease begins with tiredness and fatigue, headaches, insomnia, dizziness, anorexia, and tachycardia. On the basis of predominant symptoms, several conditions were distinguished: (1) *dry beriberi*, in which the dominant symptoms are neurological; (2) *wet beriberi*, in which the prevailing symptoms are circulatory disorders, lower limb edema, and effusions in serous cavities; (3) *severe acute beriberi*, the heart is the most affected organ; and (4) *mixed beriberi*, with symptoms corresponding to the first two categories. In almost all cases, there is dysfunction of the autonomic nervous system. In clinical practice a precise distinction of symptoms as described are not observed, reason by which there is a tendency to abandon this classification.

Beriberi also affects infants of B_1 deficient mothers. The most noticeable symptoms are body stiffness, constipation, oliguria, weakness, edema, enlarged heart, cyanosis, and tachicardia or irregular heartbeat.

Beriberi is usually accompanied by other symptoms of deficiency comprising various members of the vitamin B complex.

Plasma and urine of patients with beriberi present increased pyruvic and its derivative, lactic acid. Although this is harmful to the nervous system and myocardium, it is not the only factor responsible for the polyneuritis. Another metabolic consequence of thiamine deficit is the reduction of acetyl-CoA, acetylcholine, and other neurotransmitters. It is possible that TTP deficiency contributes to the nervous system disorders.

Severe vitamin deficiency is currently rare, but there are populations at risk. Elderly are a risk group; hypovitaminosis is a common finding in alcoholics; chronic alcohol abuse alters the absorption and utilization of thiamine; and in some of these patients, the problem is exacerbated by poor diet. In alcoholics, the most notable symptom is caused by lesions in the central nervous system that determines the *Wernicke–Korsakoff syndrome*, characterized also by psychiatric manifestations.

B_1 vitamin deficiencies are observed in cases of persistent vomiting and diarrhea, in chronic renal patients treated with dialysis, in intestinal malabsorption, and in patients with increased metabolic rate (hyperthyroidism, prolonged fevers).

Therapeutic uses. Administration of thiamine is justified only in cases of deficiency.

Riboflavin

Synonymy. Vitamin B_2, lactoflavin.

Chemistry. Riboflavin presents as orange yellow crystals, poorly soluble in water and insoluble in organic solvents. It behaves as a weak base, stable to heat, and degraded by exposure to light.

Riboflavin or vitamin B_2 is comprised by dimetylisoalloxazine (flavin tricyclic ring) attached to a D-ribitol, ribose-derived polyol; it is the 7,8-dimethyl-10-(1'-D-ribityl) isoalloxazine (Fig. 27.19).

The esterification with phosphate of ribitol primary alcohol (carbon 5') forms a monophosphate called *flavin mononucleotide* (FMN) (Fig. 27.20). Addition of adenosyl phosphate gives riboflavin-5'-adenosyldiphosphate, called *flavin adenine dinucleotide* (FAD) (Fig. 27.21).

Natural sources. Foods of animal origin are richer in riboflavin than vegetables. Milk and dairy products are the most important source and also are liver, kidney, meat, and egg yolk. Among plants, riboflavin is present in spinach, tomatoes, carrots, broccoli, asparagus, and legumes. Instead, grains, even whole, and fruits are very poor in riboflavin. In milk and eggs, an important amount of the vitamin is free; in other foods, it is as FMN and FAD bound to proteins.

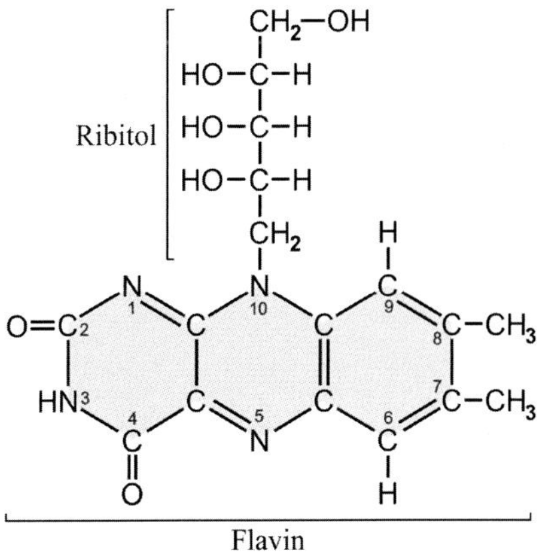

FIGURE 27.19 **Riboflavin.**

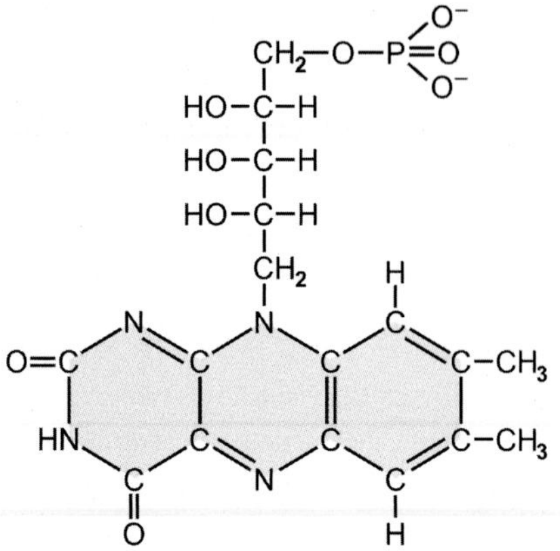

FIGURE 27.20   **Flavin mononucleotide (FMN).**

Intestinal absorption. The acidic environment of the stomach and proteases in the gastric and intestinal lumen release FAD and FMN from their bound proteins; then, they are hydrolyzed to free riboflavin. The intestinal bacterial flora synthesizes riboflavin. Absorption of riboflavin takes place in the small intestine.

In enterocytes, *flavin kinase* transfers phosphate from ATP and riboflavin becomes FMN, which is trapped in the cell. Then, FAD is formed by the action of *FAD synthetase* by transfer of adenosine monophosphate (AMP) from ATP.

After hydrolysis of FAD and FMN, riboflavin crosses the basolateral cell membrane by a transport mechanism similar to that present in the luminal membrane.

Metabolism. Riboflavin reaches the liver by the portal circulation, and then it is distributed to all tissues, combined with albumin and globulins, mainly immunoglobulins. In cells, it is converted back into FMN and FAD, prosthetic groups, which are incorporated into various enzyme systems. In most tissues, 80% or more is bound to enzymes as FAD; the reserve of vitamin B_2 is small.

Riboflavin circulating free in blood is filtered in the renal glomeruli and reabsorbed in the tubules. Only when there is excess in blood, it is excreted in urine. Approximately an hour after taking a supplement of 1.5–2 mg of vitamin B_2, it appears in the urine, which acquires an orange color.

Functional role. Riboflavin integrates coenzymes FMN and FAD, which function as electron carriers. The isoalloxazine is the hydrogen acceptor/donor (Fig. 9.7). Various flavoenzymes participate in dehydrogenation, hydroxylation, and oxidative decarboxylation reactions; they also act as coenzymes of oxidases and

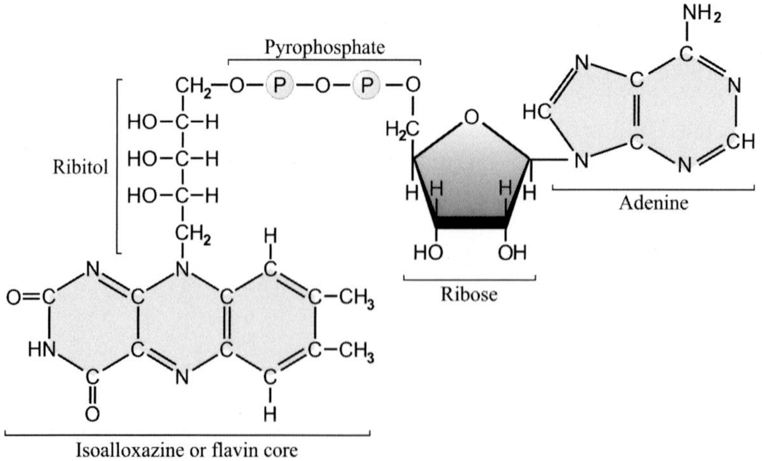

FIGURE 27.21   **Flavin adenine dinucleotide (*FAD*).**

oxygenases. The number of reactions dependent on FAD is greater than those in which FMN is involved.

There is a relationship between riboflavin and other B vitamins, pyridoxine, niacin, and folic acid. Flavins integrate mixed-function oxidases associated with cytochrome P_{450} and NADP, amino-N-oxidase and S-oxidases.

Requirements. The normal diet easily provides the required amount (Table 27.1). The reserve in the body is small and rapidly exhausted in cases of insufficient supply. No toxic effects of riboflavin have been observed.

Avitaminosis. Despite the fundamental role of riboflavin, vitamin B_2 deficiency is rarely lethal. There is not a specific disease caused by lack of riboflavin. Many of the symptoms of *arriboflavinosis* are also observed in other B vitamins deficiencies.

In humans, the symptoms include fatigue, edema of oral and pharyngeal mucosa, swollen tongue (glossitis) and lips (cheilitis), lesions at the corners of the lips (cheilosis) and nasolabial and retroauricular grooves, skin changes (seborrheic dermatitis, follicular keratosis), and ocular manifestations (conjunctivitis, photophobia, and inflammation of the cornea).

The main cause of hyporriboflavinosis is insufficient feeding, and also intestinal absorption failures. Excessive ingestion of alcohol interferes riboflavin absorption in the digestive tract.

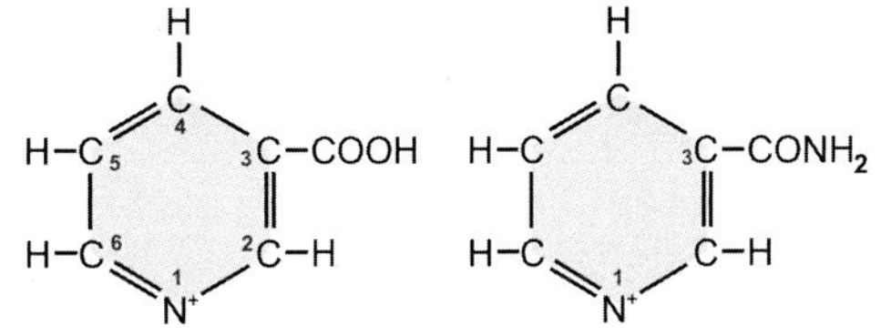

FIGURE 27.22 **Niacin vitamers.**

Niacin

Synonymy. Nicotinic acid, nicotinamide, pellagra preventive factor, PP factor. It is also called vitamin B_3, a name rarely used.

Chemistry. Niacin applies to both nicotinic acid (3-pyridinecarboxylic acid) and its amide, nicotinamide; they are vitamers derived from the pyridine ring (Fig. 27.22).

Nicotinic acid may be obtained by oxidation of nicotine. It crystallizes in colorless needles, and is poorly soluble in water or alcohol and insoluble in organic solvents; it is stable to heat, even in contact with air. Nicotinamide also crystallizes in colorless needles, and is highly water soluble and thermostable.

The functional forms are nicotinamide adenine dinucleotide (NAD) and NADP. NAD consists of nicotinamide and adenosyldiphosphate ribose linked by a β-glycosidic high energy bond from the C1' of ribose to N1 of nicotinamide (Fig. 27.23). NADP has an additional phosphate group added at C2' of ribose (Fig. 27.23). In the

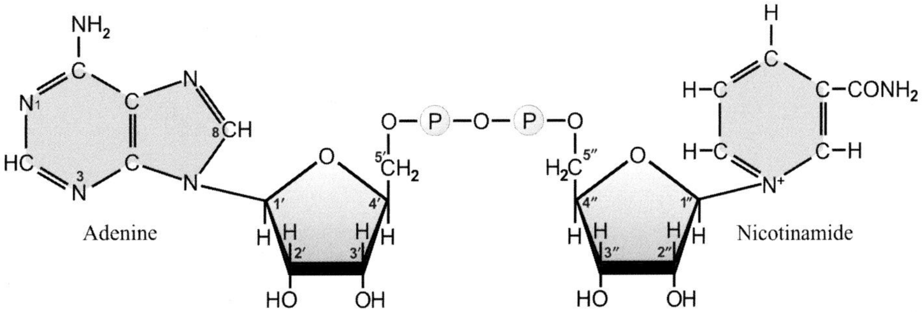

FIGURE 27.23 **Nicotinamide adenine dinucleotide (NAD).**

oxidized form, they are represented by the notation NAD^+ and $NADP^+$.

Natural sources. Liver and meat are rich in niacin; it is also found in eggs, whole grain cereals, legumes, peanuts, fruits, coffee, and tea. There are lesser amounts in milk. Whole wheat is an excellent source, but most of the vitamin is lost in milling. The contribution of niacin by plants in the usual diet is poor because it is attached to macromolecules (polysaccharides, glycopeptides) and is only available after exposure to alkaline medium. A vegan diet is often poor in niacin.

Tryptophan as nicotinic acid source. Nicotinic acid in liver is synthesized from the amino acid tryptophan (p. 392). Tryptophan-rich foods (meat, milk, eggs) ensure provision of the vitamin. Serious deficiencies are never observed when the diet includes animal protein. Most of these proteins contain approximately 1% tryptophan. Each 60 mg of tryptophan consumed produces approximately 1 mg of nicotinic acid.

Intestinal absorption. The nicotinamide nucleotide from food is hydrolyzed in the intestine to nicotinamide mononucleotide and AMP. The free nicotinamide and nicotinic acid quickly enter the intestinal mucosa. Within the enterocyte, niacin is converted to NAD or sent to the portal vein.

Metabolism. The most abundant form in blood is nicotinamide. Tissues take up both nicotinamide and nicotinic acid mainly by simple diffusion and rapidly form NAD. Most tissues synthesize NAD and NADP, preferentially from nicotinamide. The NAD, in a reaction requiring ATP, is phosphorylated at the ribose 2′ carbon corresponding to the adenine nucleotide, to give NADP. NAD concentration in the cells is higher than that of NADP, NAD is found predominantly in oxidized form (NAD^+); instead, NADPH is more abundant than $NADP^+$.

Functional role. Nicotinamide is the vitamer that integrates the NAD and NADP molecules.

Initially, the role assigned to these compounds was as coenzymes of oxidoreductases.

Oxidoreductions. NAD and NADP are hydrogen acceptors/donors. In dehydrogenations catalyzed by NAD- or NADP-dependent enzymes, two hydrogens are subtracted to the substrate and the coenzyme acquires its reduced form (NADH + H and NADPH + H) (Fig. 9.6). NAD is used by several enzymes, including oxidative decarboxylation multienzyme systems (pyruvate dehydrogenases, α-ketoglutarate, and branched-chain α-ketoacids). NADP acts as a coenzyme in two stages of the pentose phosphate pathway: in the reaction of the malic enzyme, which participates in the citrate shuttle, and in hydroxylations of mixed-function oxygenase associated with cytochrome P_{450}. In erythrocytes, NADPH helps to maintain hemoglobin in its reduced state.

NADH hydrogens are mainly transferred to the transport system of the respiratory chain; the energy liberated is used to generate ATP. Instead, H from NADPH is preferably utilized in reactions of biosynthesis (fatty acid, cholesterol, steroid hormones, deoxyribonucleotide, and other compounds); it also participates in regeneration of glutathione.

ADP-ribosylation. For a long time it was thought that all functions of niacin were related to redox reactions. Recently, it has been shown that NAD participates in processes of ADP-ribosylation. The ADP-ribosyl moiety separates from NAD, and binds to a variety of acceptors. Different types of reactions of ADP-ribosylation have been described (p. 538). Variants of ADP-ribosylation are cyclic ADP–ribose and nicotinic acid-ADP–ribosephosphate (NAADP) formation.

Cyclic ADP–ribose. A reaction catalyzed by ADP–ribose cyclase releases nicotinamide from NAD and establishes a glycosidic linkage between ribose and adenine N1 to form a cyclic structure (Fig. 27.24). It can be considered as an auto-ADP-ribosylation of ADP-ribosyl moiety, or in other words, an intramolecular ADP-ribosylation.

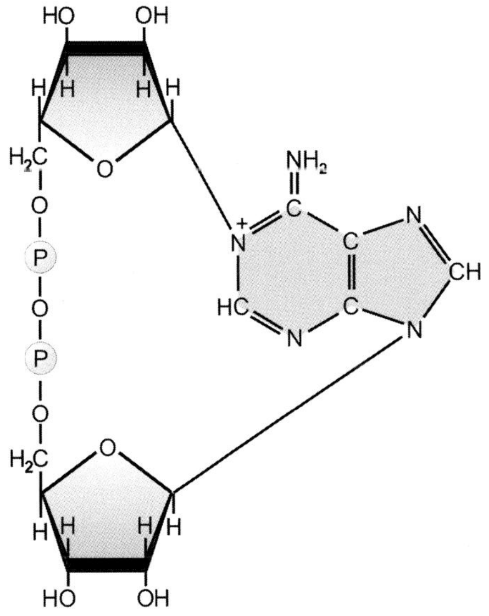

FIGURE 27.24 **Cyclic ADP–ribose.**

The leukocyte surface antigen CD38 is an ectoenzyme that has ADP–ribose cyclase and also contributes, as hydrolase, to open the ADP–ribose cycle.

Cyclic ADP–ribose mobilizes calcium from stores in the endoplasmic reticulum cisternae, promoting opening of ryanodine receptor channels type II and III.

NAADP. The NAADP (Fig. 27.25) is formed by reaction between nicotinic acid and NADP, which can also be considered a variant of ADP-ribosylation in which the acceptor is nicotinic

acid; nicotinamide is released. It is believed that the same enzyme that forms cyclic ADP–ribose is responsible for the production of NAADP.

NAADP is a potent agent for mobilization of calcium from intracellular stores. In some cells, it acts on ryanodine receptor channels, and in others it induces calcium efflux from acidic vesicles (lysosomes?). It has been shown that NAADP promotes release of calcium into pancreas β cells during glucose response, in the smooth muscle cells, in response to endothelin, and in the acinar cells of the exocrine pancreas in response to cholecystokinin.

Requirements. The needs for niacin are shown in Table 27.1. As niacin can be generated from tryptophan; the amount is expressed in equivalents, corresponding to 1 mg of niacin or 60 mg of tryptophan.

Avitaminosis. Lack of nicotinic acid or nicotinamide produces a disease known as *pellagra*. It occurs in situations in which the food supply is poor, particularly lacking high quality proteins. The condition is characterized by skin lesions, mainly in areas exposed to friction or sunlight. Lesions are symmetrical, in the face they invade both cheeks and nasal region ("butterfly" dermatitis), and also it affects the back of the hands, knees, elbows, and neck (Casal necklace). The skin is initially red, and then there is desquamation and hyperpigmentation. Also, the disorder is accompanied by alterations in the digestive tract, such as swollen tongue (glossitis) and buccal

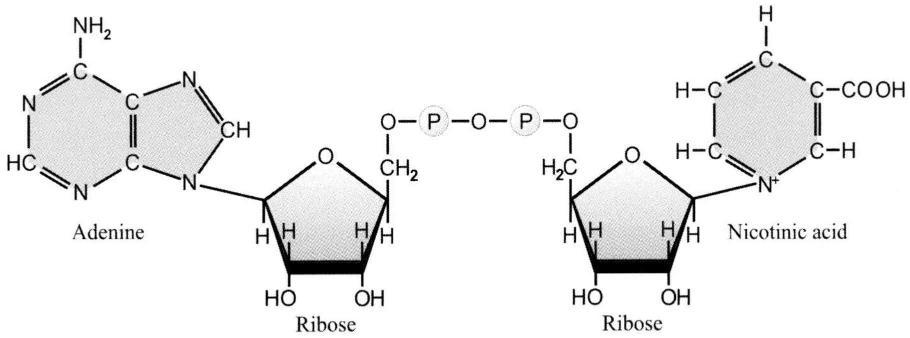

FIGURE 27.25 **Nicotinic acid–ADP–ribosephosphate (NAADP).**

mucosa inflammation (stomatitis), nausea, vomiting, achlorhydria, enteritis, and diarrhea. If the deficiency is not corrected, neurological and mental symptoms appear. Patients suffer headaches, apathy, insomnia, memory loss, depressive psychosis, delusions, hallucinations, and dementia. If untreated, pellagra leads to death.

The functional interrelation of niacin, riboflavin, and pyridoxine explains the production of "pellagroid" symptoms in deficiency of any of those vitamins. Deficiency of riboflavin or pyridoxine significantly reduces production of niacin from tryptophan.

When only the role of niacin as part of coenzymes was known, all this pathology was attributed to failures in redox reactions and energy generation caused by vitamin deficiency. However, some symptoms of pellagra, neurological and dermatological, may be related to failures in ADP-ribosylation and calcium mobilization in cells.

Pharmacological actions. Several studies have shown that administration of high doses of niacin (2–6 g/day) has beneficial effect on dyslipidemia and reduces mortality in patients with atherosclerosis at high risk of cardiovascular accidents. Treatment with high niacin lowers cholesterol and triacylglycerides in plasma, reduces VLDL, LDL, and lipoprotein(a), and increases HDL. Only nicotinic acid exerts these effects; nicotinamide has no action on lipemia.

High doses of nicotinamide also protect against cell death and inhibit the production of inflammatory mediators. They also enhance the cytotoxic effects of chemotherapy and radiotherapy. Excessively high doses of nicotinic acid are not well tolerated, causing intense vasodilation (flushing) with redness and burning of the skin of the upper body, nausea, and vomiting.

Pantothenic Acid

Synonymy. Chicken antidermatitis factor, vitamin B$_5$.

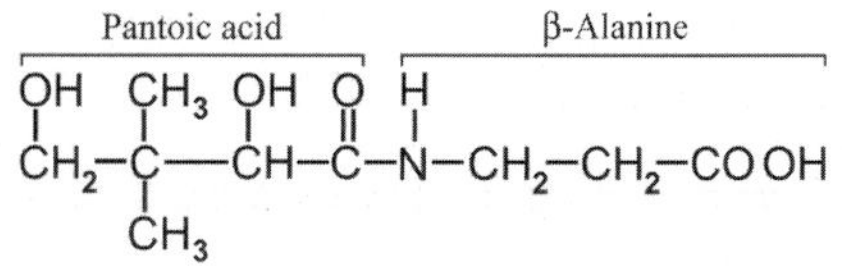

FIGURE 27.26 **Pantothenic acid.**

Chemistry. Pantothenic acid is a viscous yellow liquid, soluble in water, which is decomposed by acids, alkalis, and heat. It consists of β-alanine and pantoic acid (2,4-dihydroxy-3,3-dimethylbutyric acid), linked by an amide bond (Fig. 27.26). Humans are unable to synthesize pantoic acid.

Pantothenic acid integrates the active compounds 4′-phosphopantetheine and coenzyme A. The pantetheine is obtained by adding β-mercaptoethylamine to the free carboxyl of pantothenic acid (Fig. 27.27). CoA is formed by pantetheine and a pyrophosphate bridge to adenosine-3-monophosphate.

Natural sources. The designation pantothenic (from the Greek pantos: everywhere) indicates its wide distribution in animal and plant tissues. Liver, kidney, heart, bovine and chicken meat, egg yolk, peas, cabbage, sweet potatoes, broccoli, legumes, whole grains, peanuts, yeast, and fungi are rich in pantothenic acid. Potatoes, tomatoes, milk, and other animal and plant foods contain somewhat less. Its relative abundance in food makes it virtually impossible that even a restricted diet lacks the required amount of pantothenic acid.

The bacteria in the colon synthesize pantothenic acid; the amount absorbed coming from this source is unknown.

Intestinal absorption. The amount of free pantothenic acid in food is scarce, approximately 85% exists as CoA and the rest as 4′-phosphopantetheine, both must be hydrolyzed to release pantothenic acid. Once in the enterocytes, pantothenic acid moves into the portal circulation and is distributed to all tissues. In all cells, pantothenic acid is converted to 4′-phosphopantetheine. Coenzyme A synthase catalyzes a series of reactions which lead to the synthesis of CoA. 4′-Phosphopantetheine binds to a polypeptide

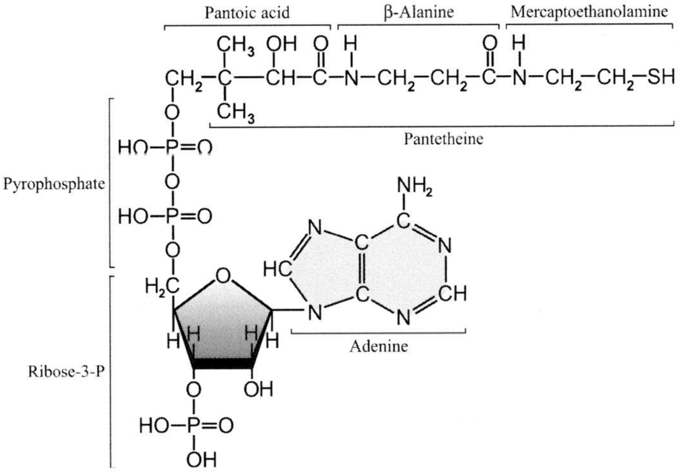

FIGURE 27.27 **Coenzyme A.**

of approximately 8600 Da to form the acyl carrier protein (ACP). ACP is incorporated as part of the fatty acid synthase complex (p. 345).

Functional role. CoA plays a central role in the metabolism of carbohydrates and lipids. ACP functions as a flexible arm, which transports the fatty acid chain being synthesized from one catalytic site of the fatty acid synthase complex to another. The combination of acyl groups with the SH group of coenzyme A or that of phosphopantetheine ACP is a thioester, which is a high energy chemical bond.

CoA forms part of oxidative decarboxylation systems of α-ketoacids (pyruvate, α-ketoglutarate, and branched-chain α-ketoacids dehydrogenases). In lipid metabolism, CoA is involved in fatty acid β-oxidation and in fatty acid, cholesterol and ketone synthesis. Acetyl-CoA is required for acetylation reactions. The transfer of long-chain acyls from CoA to carnitine is the mechanism by which the fatty acids enter into mitochondria to be β-oxidized. CoA bound to myristoyl, palmitoyl, or polyisoprenyls transfers these long hydrophobic chains to serve as anchor of proteins in cell membranes.

Requirements. The average normal diet provides the required amounts of panthotenic acid (Table 27.1).

High doses do not appear to have toxic effects.

Avitaminosis. The lack of pantothenic acid has been experimentally produced in various species. In young individuals, the earliest sign is growth retardation. In rat, pig, and chicken, lack of pantothenic acid causes gastrointestinal disorders (gastritis and enteritis), skin disorders (hyperkeratosis, skin desquamation, dermatitis, discoloration, graying, and loss of hair or feathers), anemia, and adrenal glands alterations. Also, disorders in the nervous, reproductive, and immune systems have been reported.

Thanks to its wide distribution in foods; the deficiency of pantothenic acid is rare in humans. It is manifested by restlessness, fatigue, hypotension, motor incoordination, headache, insomnia, anorexia, marked tachycardia during exercise, hyperactive tendon reflexes, loss of touch sense, burning sensation in hands and feet, and decreased antibody production.

Pharmacological actions. It has been claimed that high doses of pantothenic acid increase endurance in athletes, and relieve symptoms of rheumatoid arthritis patients, or that cosmetics added pantothenic acid or derivatives improve the appearance and vitality of hair. None of these results have been confirmed.

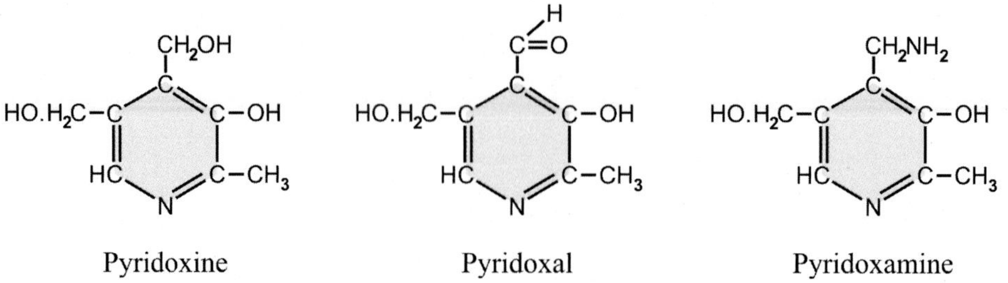

FIGURE 27.28   **Vitamin B$_6$ vitamers.**

Pyridoxine

Synonymy. Vitamin B$_6$ (widely used name), pyridoxol, adermin.

Chemistry. *Pyridoxine* (PN) is a pyridine derivative (2-methyl-3-hydroxy-4,5-dihydroxymethylpyridine). Oxidation generates pyridoxal (PL), with a formyl function (aldehyde) at position 4; addition of an amine group produces *pyridoxamine* (PM), with a methylamine function at carbon 4 (Fig. 27.28).

In foods, these three products are present in varying proportions. In the body, PN is readily converted to PL and PM. They are resistant to heat and sensitive to light; in alkaline media, PL is oxidized to 4′-pyridoxic acid, a metabolically inactive compound.

The phosphoric esters pyridoxine phosphate (PNP), pyridoxal phosphate (PLP), and pyridoxamine phosphate (PMP) (Fig. 27.29) are present in natural products. Phosphate esterification occurs at the 5-position of hydroxymethyl; 5′-O-β-glucoside-D-glucosyl PN is found in plants.

Natural sources. Free or phosphorylated vitamin B$_6$ is widely distributed in foods. In plants, there is a β-glucoside of PN (PNG) and PNP. Whole grains, cabbage, vegetables, and nuts are good sources. Same as thiamine, vitamin B$_6$ in cereals is lost when the bran is separated. Foods of animal origin, such as liver, pork and, to a lesser extent, milk, meat, fish, and egg yolk, contain PLP and PL, sometimes attached to proteins. Also, some PM and PMP is found in those foods. The bioavailability of PN varies depending on the source; the one coming from vegetable sources is lower than that of animal origin. This is due to the binding of vitamin B$_6$ as β-glucoside in vegetables, which decreases its utilization.

Intestinal absorption. The phosphorylated forms of vitamin B$_6$ are hydrolyzed by alkaline phosphatase in the small intestine. The hydrolysis of PNG is catalyzed by β-glucosidases in mucous membranes. PL, PM, and PN are absorbed by the intestinal mucosa, particularly in the jejunum. In the enterocytes, the three vitamers are phosphorylated by PL kinase again and become "trapped" in the cell.

Transport. Most of the absorbed vitamers are released to the portal circulation after dephosphorylation. In plasma PL, PM and PN are

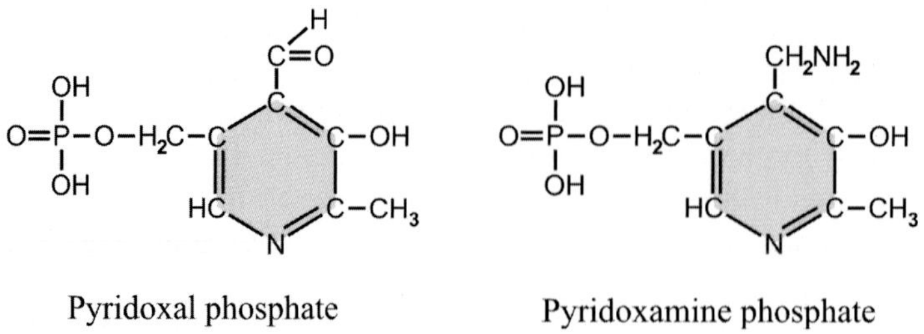

FIGURE 27.29   **Phosphoric esters of vitamin B$_6$.**

associated with albumin. Erythrocytes capture and store them as PLP, bound to hemoglobin.

Metabolism. To enter the tissues, PLP is dissociated from the albumin and dephosphorylated. Free vitamers enter by simple diffusion. In tissues, and mainly in liver, they are transformed into the corresponding phosphate (PLP, PMP, and PNP). The phosphates of PN and PM are oxidized to PL by PN 5'-phosphate oxidase, a flavoprotein with FMN. The human body contains approximately 250 mg of vitamin B_6, most of it (80%) is in muscle as PLP bound to glycogen phosphorylase. The liver contains approximately 10% of the total.

Functional role. The active form of vitamin B_6 is PLP. It functions as a coenzyme of 140 enzymes involved in the metabolism of amino acids, lipids, gluconeogenesis, heme synthesis, biosynthesis of neurotransmitters, and glycogenolysis. PLP is also a modulator of gene transcription.

The reactions of amino acid metabolism in which PLP-dependent enzymes participate follow the same general principle: the carbonyl of 4-formyl group forms a Schiff base (Fig. 16.3) with the amino acid α-amine group and weakens all substrate α carbon bonds. Reactions of transamination, decarboxylation, deamination, and interconversion of amino acids are thus possible.

PLP is also involved in the pathway of heme synthesis, in glycogenolysis (it appears to be essential to maintain the structure of glycogen phosphorylase), nucleic acid biosynthesis, synthesis of sphingolipids precursor, synthesis of carnitine and taurine.

PLP interacts with the steroid receptor into the cell nucleus and alters its binding to DNA. Through this action, it can modulate the expression of genes that respond to steroid hormones.

Vitamin B_6 influences the differentiation and production of helper T lymphocytes and reinforces its immunocompetence.

Requirements. PN requirement (Table 27.1) is related to the protein intake; 1 μg of PN per gram of protein is considered appropriate. A balanced diet contains adequate amounts of vitamin B_6.

Avitaminosis. In cases of very low supply, or factors that interfere with its absorption,

symptoms of vitamin B_6 deficiency appear. When diet is inadequate, not only PN, but other vitamins of the B complex are lacking, so specific symptoms of a single B vitamin are difficult to separate from one to another.

Skin (seborrheic dermatitis) and gastrointestinal disorders, nervous depression and mental confusion, decreased circulating lymphocytes, and sometimes hypochromic microcytic anemia (decrease in the size and hemoglobin concentration of erythrocytes) are described. In children, vitamin deficiency causes irritability, vomiting, diarrhea, and epileptiform seizures. These are related to the lowering of γ-aminobutyric acid in the central nervous system. Lack of vitamin B_6 causes alterations in the immune system; there is decreased production of lymphocytes and antibodies. Also, xanthurenic acid is excreted in urine and homocysteine in plasma is also observed.

The elderly are at risk of hypovitaminosis, absorbtion in the intestine decreases, and oxidation of PL to inactive pyridoxic acid increases with age. In chronic alcoholics, conversion of PN into PLP and PM is affected. Patients treated with the antituberculosis drug isonicotinic hydrazide acid may present deficiency symptoms (peripheral neuritis and convulsions).

Biotin

Synonymy. Vitamin H (from the German haut: skin), or vitamin B_7, which is a designation rarely used.

Chemistry. Biotin crystallizes as long needles, poorly soluble in water and ethanol, insoluble in ether and chloroform. It is acidic, stable to heat, and forms salts that are very soluble in water.

It consists of two heterocyclic rings, tetrahydrothiophene fused to ureid (urea derivative), which forms an imidazolidone cycle. The carbon in position 2 of the thiophene possesses a valeric or pentanoic acid side chain (Fig. 27.30).

In natural products, most of the biotin is associated with apoenzymes by an amide type bond. Protease action frees ε-N-biotinyl-L-lysine, a water-soluble compound called *biocytin*.

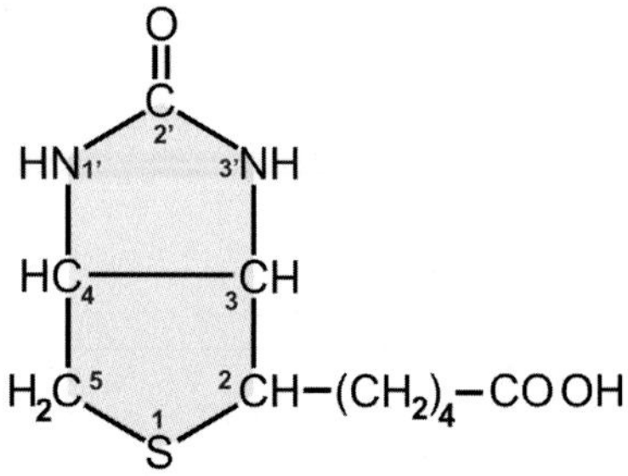

FIGURE 27.30 **Biotin.**

Natural sources. Biotin is widely distributed in foods. In plants there is a greater proportion of free biotin (liver, kidney, milk, egg yolk, tomatoes, and other vegetables, soy flour, cereals, peanuts, and yeast, are excellent sources). In human milk, almost all existing biotin is free.

Biotin is synthesized by the human intestine microbial flora; the amount absorbed is unknown.

Intestinal absorption. The ratio of free/bound (attached to protein) biotin on foods varies widely. Proteases hydrolyze the peptide bonds of these proteins, but not the binding of biotin with the ε-amine group of lysine residues or some links between neighboring amino acids, so that they release biocytin and peptidyl-biotinyl. A *biotinidase* present in pancreatic juice and brush border membrane releases biotin from biocytin.

In the egg, a glycoprotein called *avidin* binds to biotin to form a complex that is resistant to pancreatic proteases and prevents the absorption of this vitamin. Avidin is denatured by heating; therefore, eating cooked eggs does not affect biotin absorption.

Transport. In plasma, slightly over 80% of the biotin is free and most of the rest is carried by albumin. Biocytin in blood is cleaved by biotinidase into biotin and lysine. Biotin remains bound to biotinidase, which thus becomes a carrier of the vitamin.

Metabolism. Liver, lymphocytes, and other tissues (kidney and central nervous system) take biotin from blood. In kidney, biotin filtered in glomeruli is reabsorbed in the tubules.

Biotin binds to apoenzymes to form holocarboxylase. Once their life cycle is accomplished,

they are hydrolyzed, leaving peptidylbiotinyl and free biocytin, biotinidase releases biotin, used in the synthesis of new holocarboxylases. There is an active vitamin recycling. Besides acting as a biotin carrier in blood, biotinidase catalyzes transfer of biotin to acceptor molecules.

Functional role. In mammals biotin functions as the coenzyme for five carboxylases. In contrast to vitamin K-dependent carboxylations, which attach CO_2, biotin-dependent carboxylations use bicarbonate (HCO_3^-). The biotin moiety serves as an acceptor and donor of HCO_3^- (Fig. 27.31).

The binding of biotine to the apoenzymes is catalyzed by *holocarboxylase sinthethase*. The enzyme is responsible for the formation of the five biotin-dependent carboxylases present in humans. These are *acetyl-CoA* (α and β), *pyruvate, propionyl-CoA*, and *β-methyl-crotonyl carboxylases*. They catalyze the attachment of bicarbonate to organis acids. Acetyl-CoA α carboxylase is cytosolic; the others are mitochondrial.

Biotin transfer to histones. Biotinylation and debiotinylation catalyzed by holocarboxylase synthethase and biotinidase modify chromatin packing, and thus DNA transcription and replication.

Gene expression regulation. Participation of biotin in nonenzymatic processes has been demonstrated recently. Biotinyl-AMP regulate mRNA transcription and translation of genes codifying key enzymes in intermediary metabolism. Besides, they affect the expression of genes that encode for cytokines and their receptors.

Cellular multiplication. Biotin appears to be a factor needed to accomplish the cellular cycle.

Actions on immune system. In laboratory animals, biotin is necessary for the normal function

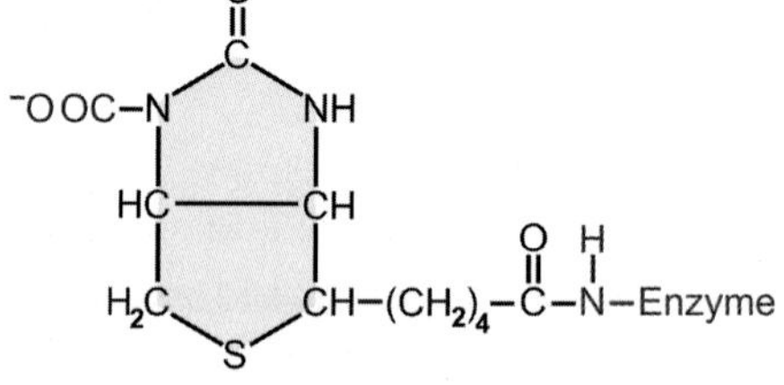

FIGURE 27.31 **CO₂–biotine–apoenzyme complex (activated CO₂).**

of different cells of the immune system (macrophages, T and B lymphocytes, and response of T-cytoxic cells). Biotin participates in regulation of immunoglobulin A synthesis.

Requirements. The needs for biotin are not know with certainty; see recommended daily amounts in Table 27.1.

Avitaminosis. Owing to the wide distribution of biotin in the meals and to its synthesis by the bacteria present in the intestine, lack of biotin is rarely observed. Biotin deficiency produces skin abnormalities, anemia, anorexia, nauseas, muscle pain, hypotonia, paresthesia, ataxia, and hallucinations. It also presents with abnormalities in the humoral and cellular immune responses. Lactic acidosis and organic aciduria may appear. During pregnancy, lack of biotin is teratogenic. In infants and children, there is growth retardation. Severe deficiency may produce metabolic alterations due to reduced activity of carboxylases. The biotinylation of histones and biotin dependent genetic regulation are affected.

Besides insufficient amounts of biotin in the body, failure in the synthesis of carboxylases can be due to genetic alterations. Deficiency of biotinidase is a condition in which biotin cannot be released from biocitin present in the food. The recovery of the vitamin from biocitin and peptidyl biotin and the biotinylation of proteins are blocked.

Folic Acid

Synonymy. Pteroylglutamic acid, folacin, or vitamin B_9, which is rarely used.

Chemistry. It is slightly soluble in water, stable to heat. The name is due to its presence in plant leaves. It consists of three components: a bicyclic core pteridine, paraaminobenzoic, and glutamic acids. From carbon 6 of the 2-amino-4-hydroxypteridine core, a methylene bridge (C9) extends to the amine group (N 10) of *p*-aminobenzoic acid to form *pteroic acid*; it binds to glutamic acid by a peptide bond type (Fig. 27.32).

There are pteroylpolyglutamates in nature; they are folic acid derivatives, formed by successive addition of glutamyl residues, linked through γ-peptide bonds. They are called folates; folic acid is the name reserved for monoglutamate. Pentaglutamate is abundant in the liver of most mammals and heptaglutamate is found in plants.

By reduction, folic acid gives 7,8-dihydrofolate (DHF) and 5,6,7,8-tetrahydrofolate (THF) (Fig. 27.33).

Functional derivatives of folate are polyglutamyl THF, with one-carbon unit at positions 5 or 10, forming a bridge. The one-carbon group added can be methyl ($-CH_3$), formyl ($-CH=O$), formimine ($-CH=NH$), methylene ($-CH_2-$), or methenyl ($-CH=$). The 5-formyl-THF, which is resistant to oxidation, is used in pharmaceutical preparations; it is called *folinic acid*.

Natural sources. Asparagus, beans, broccoli, strawberries, oranges, mushrooms, peanuts, liver, kidney, and yeast are good sources of folic acid. Vitamin is also found in lesser amounts in meat and wheat. Folates from natural products are a mixture of reduced folate polyglutamyl

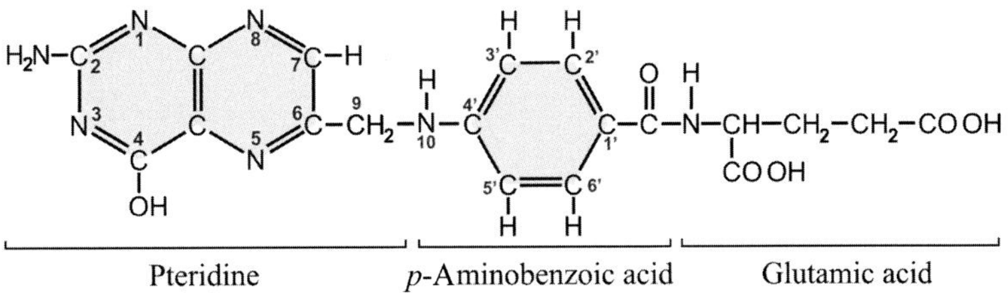

Pteridine | *p*-Aminobenzoic acid | Glutamic acid

FIGURE 27.32 **Folic acid.**

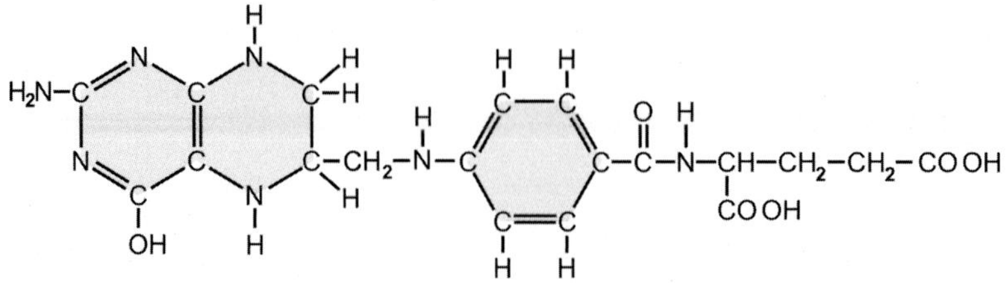

FIGURE 27.33   **Tetrahydrofolic acid.**

[THF(Glu)$_n$], especially 5-methyl-THF and 10-formyl-THF, bound to proteins and reserve polysaccharides.

Intestinal absorption. Polyglutamates must be converted into monoglutamates (THF) for absorption. Digestive proteases do not hydrolyze γ-peptide bonds, *polyglutamate hydrolase* or *folate conjugase*, from jejunal mucosa, and pancreatic juice liberates the terminal glutamate units one by one. THF is absorbed in the proximal part of the small intestine, mainly in the jejunum. The colon bacterial flora synthesizes folates that are partially absorbed.

Metabolism. In the enterocyte, folate is reduced by *dihydrofolate reductase*, NADPH-dependent, which transfers hydrogen to folic acid and gives DHF and THF. THF is methylated at N 5 of pteridine to form 5-methyl-THF and sent to the portal circulation. In cells, *folylpolyglutamate synthetase* demethylates 5-methyl-THF, which later forms polyglutamate.

Half of the folate total content in an adult (15–30 mg) is in the liver, most as pentaglutamate. If there is no further supply of folic acid, the reserves take 3 months to be consumed.

THF is partially methylated, secreted with bile, and reabsorbed in the intestine, approximately 100 µg of folate are recycled daily via enterohepatic circulation.

Functional role. Pteroilglutamates participate as coenzymes in one-carbon transfer reactions produced during interconversion and catabolism of amino acids, in purine and pyrimidine synthesis, and in the methylating agent S-adenosylmethionine (SAM) formation.

Formation of one-carbon donor moieties. The THF(Glu)$_n$ accepts methyl groups generated in serine, glycine, and choline catabolism, by reactions catalyzed by methyltransferases.

Examples of folate participation in metabolic pathways follow.

Conversion of pyrimidines. Thymidylate synthase catalyzes the transfer of a methyl group to deoxyuridine monophosphate and forms deoxythymidine monophosphate.

Purine synthesis. 10-formyl-THF, produced from 5,10-methylene-THF, participates in de novo synthesis of adenine and guanine.

All dividing cells require folate, but obviously those with higher mitotic activity are more dependent on it. This explains why one of the most noticeable signs of folate deficiencies is depression of erythropoiesis and of leukocytes and platelets formation. Cells of the intestinal mucosa and spermatogenic progeny also have an active turnover and, therefore, are sensitive to the reduction of folic acid.

Methionine formation. One of the pathways that generate homocysteine and methionine requires *methionine synthase*, 5-methyl-THF, and cobalamin. The 5-methyl-THF methyl is transferred to homocysteine.

There is strong evidence that folic acid plays an important role in the development of the central nervous system in the first weeks of gestation.

Requirements. The requirements of folate are shown in Table 27.1. Moderate doses of folate in acute or chronic treatments are not toxic. A tolerable maximum dose has been established, which is not due to toxic effects, but to the risk of masking other pathologic conditions, such as megaloblastic anemia.

Avitaminosis is observed in people with a poor diet and in patients with intestinal absorption disorders, or treated with antiinflammatory and anticonvulsant drugs.

Folate deficiency usually occurs in the elderly and in chronic alcoholics. Alcohol interferes with folate absorption, uptake by the liver and reabsorption in renal tubules.

The avitaminosis causes megaloblastic anemia, characterized by the presence of immature, larger than normal erythrocytes, and reduced number of leukocytes and platelets. Anemia causes tiredness and fatigue at the slightest exertion, pallor, headache, and irritability. There are alterations in tongue and buccal mucosa, with superficial ulceration. Serious deficiencies produce depression of spermatogenesis and intestinal disorders. Hyperhomocysteinemia, a risk factor for cardiovascular disease, is also observed.

Pregnant women are likely to have hypovitaminosis as folate demand increases during embryo development. The nervous system is particularly sensitive; lack of folic acid in the first weeks of pregnancy can cause abnormalities, especially in neural tube closure. Several studies have shown that supplements of folic acid (400 µg/day) immediately before and during early pregnancy (periconceptional period) decrease the incidence of malformations. A relationship between cancer incidence and folate deficiency has been found.

Genetic defects. Hereditary defects in genes related to folate metabolism have been described. One of the most common is a mutation in the gene for methylene-THF-reductase, which results in the synthesis of a labile and less active than normal enzyme. Homozygous women have the risk of having children with central nervous system developmental anomalies, primarily defects in neural tube closure (*spina bifida*). The risk increases if the diet is deficient in folic acid and decreases if high doses of the vitamin are administered a few months before and after conception.

Antivitamins. There are structural analogs of folic acid that act as potent metabolic antagonists or antivitamins. One is methotrexate (amethopterin or 4-amine-10-methylfolic), that binds to the active site and inhibits *dihydrofolate reductase*. Intracellular folate concentration decreases, and there is accumulation of DHF and 10-formyl-THF, with cytotoxic action. Aminopterin and 5-fluorouracil (4-aminefolic acid) are also antifolics. All these drugs block DNA synthesis, and reduce cell proliferation. They are used in the treatment of malignant tumors. These drugs also affect normal hematopoietic cells; close monitoring of treated patients is required.

Cobalamin

Synonymy. Vitamin B_{12}, antipernicious anemia factor, or extrinsic factor.

Chemistry. Cobalamin is the only natural organic compound that contains cobalt; it also has phosphorus and has a deep red color. A portion of the molecule is similar in structure to the porphine core; it consists of the *corrin* tetrapyrrole ring, with a cobalt atom at its center (like the Fe^{2+} in heme). Cobalt has six coordination sites, four of them bind to the nitrogen atoms of the pyrrol groups. A fifth site associates with a "nucleotide" formed by 5,6-dimethylbenzimidazole, ribose, phosphate, and aminopropanol, which is also attached to a side chain of corrin. The sixth coordination site can be occupied by different ligands: −OH in hydroxycobalamin, H_2O in aquocobalamin, or −CN in cyanocobalamin (Fig. 27.34), the latter is the most stable form.

The derivatives methylcobalamin and coenzyme B_{12} or 5′-deoxyadenosylcobalamin function as coenzymes. The first has a methyl, and the second has 5′-deoxyadenosine in the sixth coordination site of cobalt. Both are coupled to enzymes as cofactors.

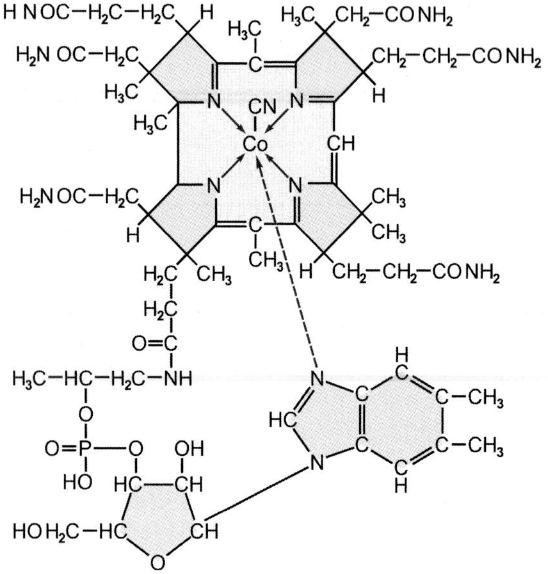

FIGURE 27.34   **Cyanocobalamin.**

Natural sources. Foods of animal origin are the only significant source of vitamin B_{12}. The liver is very rich; also kidney, heart, meat, seafood, milk, butter, cheese, and egg yolk have vitamin B_{12}. It is absent in plant foods, except for those that have fermented due to contamination with bacteria.

Vitamin B_{12} is essential for humans. Although bacteria in the large intestine flora synthesize cobalamin, very little of it is absorbed. Vitamin B_{12} is nontoxic, and maximal tolerable upper limits have not been established.

Intestinal absorption. The greater the supply of cobalamin, the lower is the proportion absorbed. Vitamin B_{12} in food is bound to protein. The acidic environment of the stomach and pepsin favor the release of methylcobalamin and adenosylcobalamin and binding to *haptocorrin*, protein secreted by the salivary glands and other exocrine glands. Haptocorrin is hydrolyzed in the duodenum by pancreatic proteases that release cobalamin, which binds very selectively to *intrinsic factor* (IF), a glycoprotein of 45 kDa, produced by the parietal cells of the gastric mucosa,

the same cells that secrete HCl. Vitamin B_{12} was called *extrinsic factor.* The complex formed by vitamin B_{12} and IF is resistant to hydrolysis by proteases. When the complex reaches the terminal ileum, it binds to specific receptors located in the mucosa that selectively recognize the IF. The cobalamin–FI–receptor complex enters the cells by endocytosis, and cobalamin is released within the enterocyte.

In the vicinity of the basolateral membrane, cobalamin binds to transcobalamin, a β-globulin of approximately 43 kDa, and as such it enters the portal circulation.

Metabolism. The transcobalamin–vitamin B_{12} complex binds to a specific receptor in plasma membrane of all cells; it is internalized into the cells by pinocytosis and incorporated into lysosomes, where cobalamine is released. Liver and kidney are the richest cells in these receptors. The body stores 3–5 mg of cobalamin, 80% of which is in the liver. This reserve is important in case of deficiencies in the diet. In liver and other tissues, vitamin B_{12} is converted into two coenzymes: methylcobalamin, which is generated in the cytoplasm by methyl transfer from 5-methyl-THF, and coenzyme B_{12} or 5′-deoxyadenosylcobalamin, which is formed in mitochondria using adenosine from ATP. Both methylcobalamin and coenzyme B_{12} are stored bound to the apoenzyme. Vitamin B_{12} is not degraded in the body, but it is excreted intact in the urine.

Functional role. In mammals there are two enzymes dependent on vitamin B_{12}, *methionine synthase,* which requires methylcobalamin, and *methylmalonyl-CoA mutase,* which uses 5′-deoxyadenosylcobalamin.

Conversion of homocysteine to methionine. A derivative of methionine, S-adenosyl-mehionine (SAM), is a methyl donor in numerous syntheses and in DNA, RNA, and protein methylation. In these reactions, SAM is converted into S-adenosylhomocysteine, and then it is hydrolyzed to homocysteine and adenosine. Homocysteine is remethylated to regenerate methionine by *methionine synthase,* dependent on methylcobalamin.

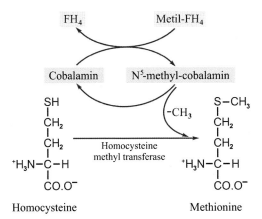

FIGURE 27.35 **Conversion of homocysteine to methionine.**

Methylcobalamin and 5-methyl-THF (Fig. 27.35) participate as coenzymes.

Transmethylation to cobalamin is the major route of THF regeneration, which can form 5,10-methylene-THF. When B_{12} lacks, methionine synthesis and formation of THF decrease, while homocysteine and 5-methyl-THF accumulate. The latter is not usable for other purposes (5-methyl-THF trap) and is excreted by urine.

Isomerization of L-methylmalonyl-CoA to succinyl-CoA. During methionine catabolism, threonine, valine, isoleucine, and β-oxidation of odd chain fatty acids, there is production of propionyl-CoA, which is carboxylated to methylmalonyl-CoA. The L-methylmalonyl-CoA is converted, by action of *L-methylmalonyl-CoA mutase*, in succinyl-CoA, intermediate of the citric acid cycle (Fig. 27.36). L-Methylmalonyl-CoA mutase uses 5'-deoxyadenosylcobalamin as coenzyme.

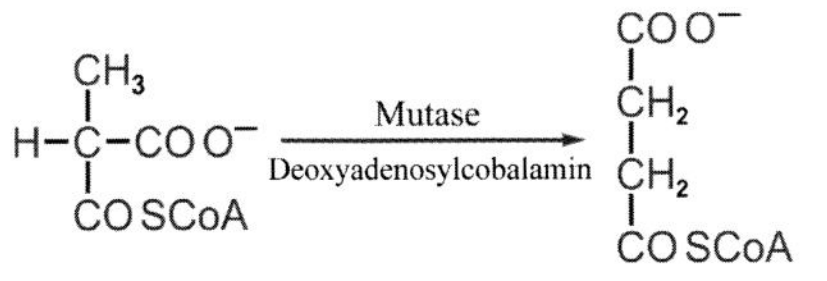

$$\underset{\text{L-Methylmalonyl-CoA}}{\overset{\displaystyle \overset{CH_3}{\underset{|}{\overset{|}{H-C-COO^-}}}}{\underset{\displaystyle COSCoA}{}}} \xrightarrow[\text{Deoxyadenosylcobalamin}]{\text{Mutase}} \underset{\text{Succinyl-CoA}}{\overset{\displaystyle \overset{COO^-}{\underset{|}{\overset{|}{\underset{|}{\underset{|}{CH_2}}}}}}{\underset{\displaystyle COSCoA}{}}}$$

FIGURE 27.36 **Isomerization of L-methylmalonyl-CoA to succinyl-CoA.**

Requirements. The needs for vitamin B_{12} are shown in Table 27.1. No toxic effects have been reported by administration of high doses of this vitamin.

Avitaminosis produces megaloblastic anemia similar to that described in the absence of folic acid. The functional relationship between vitamin B_{12} and folate explains why deficiency of any of them results in the same type of anemia. Cobalamin deficiency causes anemia due to secondary lack of folate. In blood, pancytopenia, immature red blood cells and hypersegmented granulocytes are observed. In bone marrow, immature cells accumulate and many of them die probably by apoptosis.

Vitamin B_{12} deficiency differs from folate avitaminosis by the greater severity of neurological and psychiatric disorders. There is demyelination in peripheral nerves, in posterior and lateral spinal cord columns, and in the central nervous system. Symptoms begin in feet and move up to legs, arms, and hands. There is weakness and painful paresthesia in the extremities, ataxia, loss of fine coordination, spasticity, and posture disorders. When demyelination reaches the cerebral hemispheres, neuropsychiatric symptoms appear (impaired memory, irritability, apathy, depression, and occasionally psychosis, dementia, and delirium). If left untreated, vitamin B_{12} deficiency leads to death. Once neurological damage has occurred, the lesions are irreversible and administration of vitamin B_{12} only stops the progression of the disease.

Infants with vitamin B_{12} deficiency experience irritability, abnormal nervous system development and growth, and megaloblastic anemia. In vitamin B_{12} deficiency there are increased levels of methylmalonic acid, homocysteine, and cystathionine in plasma and urine.

Among the causes of vitamin B_{12} is the alteration of its absorption. The absorption of cobalamine is reduced in total gastrectomy, surgical resection of more than 50 cm of ileum, hypochlorhydria with atrophic gastritis, and the prolonged use of histamine receptor blockers and inhibitors of proton pump (used in the treatment of peptic ulcer and hyperacidity). Deficiency due to malabsorption is common in elderly. Also, dietary deficiency due to lack of food from animal sources is a problem, whose incidence is increasing in several poor regions of the world. It is also an issue for individuals under a very strict vegetarian diet.

The most serious symptoms of vitamin B_{12} deficiency are observed in humans with malabsorption; the most important one is related to a deficiency in the intrinsic factor (IF). Failure to produce IF is the cause of *pernicious anemia*, which presents all the symptoms described earlier. Pernicious anemia is an autoimmune disease. Most of the symptoms are expressed in subjects older than 50 years. It is treated with vitamin B_{12}, which needs to be administered parenterally because if it is given orally, it would not be absorbed. Cobalamin is a therapeutic agent of extraordinary potency. With administration of only 1 mg/week, for 4–8 weeks, favorable responses are obtained; continued treatment with one injection per month maintains a satisfactory erythropoiesis. Folic acid can improve the symptoms of anemia, but has no action on the neurological disorders caused by lack of cobalamin.

Genetic disorders. Various genetic defects that affect the synthesis of proteins involved in the uptake and transport of cobalamin cause deficiency. These include mutations in the IF gene, which determine total lack of IF or production of a nonfunctional protein. It affects children aged 1–3 years, when the reserves of vitamin B_{12} transmitted by the mother are exhausted.

Pharmacological actions. The use of vitamin B_{12} in high doses has been proposed for treatment of neuralgia and neuritis or improvement of the general condition. There is no evidence to support these applications of cobalamin.

Ascorbic Acid

Synonymy. Vitamin C, ascorbate, and antiscurvy factor.

Chemistry. Vitamin C is a white crystal, very soluble in water and insoluble in organic solvents and lipids. It is a six-carbon α-keto lactone structure, which resembles that of hexoses. It adopts the furanose configuration (5-membered heterocycle). Unlike the vast majority of metabolizable carbohydrates of our organism, belonging to the D series, active ascorbic acis (AA) is the L-isomer (Fig. 27.37). It has acidic character due to the ionization of carbon 3 hydroxyl. At physiological pH, the molecule exists as a monovalent anion (ascorbate). It is a strong reducer, which easily releases two hydrogens: initially one electron is lost, forming the *ascorbyl* radical, relatively stable; the release of a second electron produces dehydroascorbic acid (DAA) (Fig. 27.37).

Antioxidant action. The interconversion of ascorbic into DAA takes place spontaneously and is important in redox reactions. In biological media, the concentration of AA is higher than that of dehydroascorbic and this is greater than ascorbyl radical.

When L-DAA is hydrated, it is converted to the inactive compound 2,3-diketogulonic acid

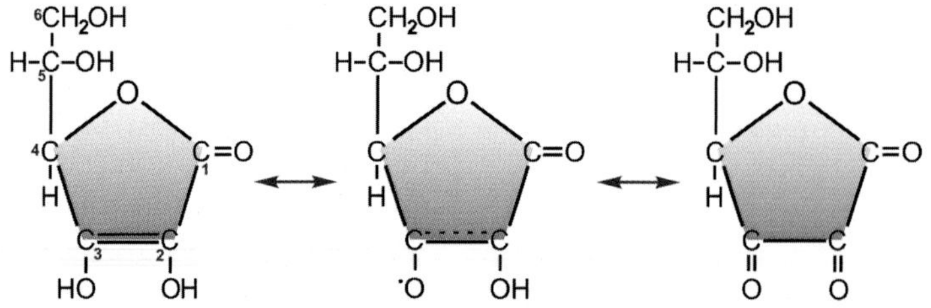

FIGURE 27.37 **Structures of ascorbic acid, ascorbyl radical, and dehydroascorbic acid.**

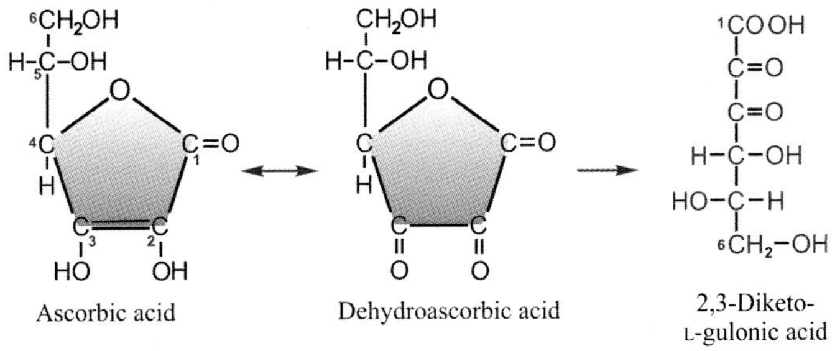

FIGURE 27.38 **Formation of 2,3-diketogulonic acid.**

(Fig. 27.38). In neutral or alkaline solutions, this hydration occurs spontaneously.

Natural sources. AA is found mainly in fresh plants. Fruits (lemon, orange, grapefruit strawberry, kiwi, and cantaloupe) and tomatoes are excellent sources. Spinach, broccoli, cabbage, cauliflower, peppers, potatoes, asparagus, peas, and beans have vitamin C, but some is lost during transportation, storage at room temperature, and mainly cooking.

AA is inactivated in dried vegetables. In citrus and tomato juices in contact with air, there is slow progressive loss of vitamin C. Liver is one of the richest sources of vitamin C among the foods of animal origin; kidney and heart have also AA. The small amount present in cow's milk disappears during pasteurization. Freezing does not affect the content of vitamin C in food. Breast milk provides the infant sufficient amounts of AA.

Intestinal absorption. In foods, 85% of the vitamin C is in its reduced form (AA), and the rest exists as DAA. Part of AA in the intestinal lumen oxidizes to reduce some components of the diet, such as iron. Both the reduced and acid forms of vitamin C enter through the intestinal mucosa, especially the ileum. If high doses are ingested, the percentage absorbed decreases. Within the enterocyte, the DAA is reduced to AA. In plasma, AA is free and distributed to all cells.

Metabolism. In the cells, AA is converted to dehydroascorbic by *ascorbate reductase*, which requires glutathione as a hydrogen acceptor/donor. The content of vitamin C in tissues is higher than that of other water-soluble vitamins. Indeed, in normal individuals there is a reserve of approximately 1500 mg, which allows tolerating a deficient diet for 2 or more months. The concentration of AA is high in some endocrine glands (adrenal and pituitary gland) and leukocytes; it is important in liver, kidney, pancreas, spleen, and brain. AA and DAA present in plasma filter through the renal glomeruli and are reabsorbed in the proximal collectors tubules. Approximately 25% is converted to oxalate and excreted in urine, the rest is oxidized to CO_2. When high doses (over 500 mg) are administered, almost 100% of the ingested dose is excreted intact in the urine. Oxalate excretion did not significantly vary even with doses greater than 5 g/day.

Functional role. Probably all of the AA functions are related to its reducing or antioxidant capacity. In mammals it is a cofactor for at least eight oxidoreductases.

Collagen formation. Vitamin C is essential for maintaining the normal structure of the collagen fibers by hydroxylating proline and lysine residues in collagen. In collagen, 4-hydroxyproline residues are important for stability of the molecule, and hydroxylysine are responsible for the interchain bridges that maintain collagen triple helix structure. *Carnitine synthesis* requires two AA-dependent dioxygenases, metalloenzymes with iron; ascorbate maintains the iron in the ferrous state.

Tyrosine synthesis and catabolism. Some reactions in these pathways require vitamin C (p. 389).

Neurotransmitters and hormones synthesis. Vitamin C is involved in the synthesis of catecholamines and serotonin and in posttranslational C-terminal amidation of some hormones.

Antioxidant action. Ascorbate not only acts as a cofactor in various reactions, but it has also nonenzymatic functions as a reducing agent in aqueous media, such as blood and the intracellular and extracellular spaces. It is committed to protective actions against reactive oxygen and nitrogen species (see Chapter 11).

Iron absorption. Much of the iron that reaches the intestine is oxidized to Fe^{3+} and must be reduced to ferrous iron (Fe^{2+}) to be absorbed. This action is accomplished by AA.

Requirements. Table 27.1 shows the needs for AA. In fever, infections, diarrhea, hyperthyroidism, trauma, and surgery, it is advisable to provide 200 mg or more per day.

Avitaminosis. Among mammals, only humans, primates, guinea pigs, and some species of bats present symptoms of vitamin C deficiency when they are subjected to AA-free diets. People that consume diets poor in vegetables and fresh fruits are at risk. Lack of vitamin C in the diet causes *scurvy*, a disease known since antique times. It was common among members of expeditions and military campaigns that consumed only dried foods during long time.

Lack of vitamin C is characterized by muscle fatigue and weakness, subcutaneous petechiae or hematoma in flexion or friction zones, swollen gums that bleed easily, losen teeth, subperiosteal and joint bleeding, and severe joint pain. All these symptoms are caused by extravasation of blood from the capillaries. Also, defects in calcification of bones and teeth, and anemia produced by poor absorption of iron, occur. In children, the changes are most noticeable at the sites of active bone growth. Complete lack of vitamin C is rarely observed; it leads to death if not treated promptly. Less serious cases of vitamin

C deficiency are observed, with delayed wound healing, decreased resistance to infections, and some of the disorders mentioned previously.

Pharmacological actions. Ingestion of high doses of vitamin C (1, 2, or more grams per day) has been proposed as a preventive of colds and other infections, complications of diabetes, cataract development, or to increase the wound healing capacity. So far, there is not solid evidence to support those claims.

Other Nutritional Factors

There is a group of compounds, important from the metabolic and nutritive points of view, which are not considered vitamins because they are synthesized in the organism. Some of these substances were originally classified as components of the vitamin B complex, a reason why they are included in this chapter.

Choline

Choline is a quaternary ammonium base (trimethylamine 2-hydroxyethyl) (Fig. 27.39), soluble in water and alcohol and insoluble in chloroform and ether; it is resistant to prolonged heating. Choline is not a vitamin because it is synthesized in the body. It is a constituent of complex lipids of cell membranes [phosphatidylcholine (PC) and sphingomyelin] and acetylcholine, a neurotransmitter of the nervous system.

Natural sources. Choline is widely distributed in foods, as PC. The richest sources are liver, egg yolk, meat, wheat germ, soybeans, cereals, cauliflower, beans, and peanuts; in soybean, 45% of the total is free choline.

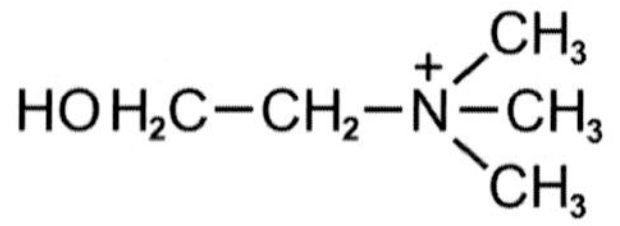

$$HOH_2C-CH_2-\overset{+}{N}\begin{smallmatrix} CH_3 \\ -CH_3 \\ CH_3 \end{smallmatrix}$$

FIGURE 27.39 **Choline.**

Absorption. PC (also called lecithin) is the main source of choline in the common diet; it is hydrolyzed by phospholipases. Sphingomyelin from food releases choline in the intestine. Choline is a positively charged molecule; it does not freely cross the plasma membrane; it is introduced into the enterocytes through a carrier. The main route of choline in tissues is the synthesis of PC.

Functions. Choline is an important contributor to the structure and properties of cell membranes. PC is the most abundant phospholipid of membranes, preferably located in the outer lipid layer. Sphingomyelin is also, to a lesser extent, in all membranes.

Transmission of nerve impulses. Acetylcholine is essential for the function of cholinergic neurons; it is synthesized by choline acetyltransferase and accumulates in synaptic vesicles. Membrane depolarization releases acetylcholine, which binds to receptors in the postsynaptic neuron. In the interneuronal synaptic space, choline is rapidly hydrolyzed by acetylcholinesterase.

Osmotic regulation. Cell volume depends on the concentration of osmolytes. Choline and its derivatives betaine and glycerophosphocholine are organic osmolytes.

One-carbon metabolism. Choline is a methyl donor related to the formation of methionine, which in turn generates SAM, the principal agent of methylations.

Others. The platelet-activating factor is a phospholipid with choline (1-O-alkyl-2-acetyl-glycerylphosphorylcholine). An important component of the surfactant factor (surface tension reducer) in lung is dipalmitoyl phosphatidylcholine.

Requirement. Choline is abundant in a normal mixed diet, and the use of supplements is not justified. In healthy people, no toxic effects have been detected by administration of high doses.

Choline deficiency. The administration of a choline-deficient diet for prolonged periods to laboratory animals affects growth and development, with alterations in liver, kidney, and pancreas.

Fatty liver occurs, probably due to reduction of phospholipid synthesis, essential for incorporation of triacylglycerols in VLDL. Choline administration to these animals mobilizes abnormal fat reserves; this action is designated *lipotropic effect*. Lack of choline and betaine osmolytes cause kidney alterations, including impaired urine concentrating ability, tubular necrosis, and interstitial hemorrhage.

Taurine

Taurine (2-aminesulfonic acid) (Fig. 27.40) is an amino acid found free in most mammal tissues, specially in skeletal muscle, heart, and nervous system.

It is generated by the catabolism of cysteine, which is formed during methionine degradation. A normal adult synthesizes 200–400 µmol/day. Taurine is present in foods from animal origin; a strict vegan diet has scarce amounts of taurine.

Functions. During long time, taurine was considered to be only a final product of the catabolism of sulfur containing amino acids; however, more recently, it has been recognized that taurine exerts various functions in the body. The best known one is the binding of hydrophobic compounds to produce acidic products that are more soluble in water and can be easily excreted. Biliary acids can be better absorbed in the intestine once they are conjugated with taurine and glycine. Taurine binds to other substances, such as retinoic acid and xenobiotics, facilitating their excretion. Taurine is related to various physiological processes; it has a neuromodulatory function, participates in stabilization of muscle membrane potential, has antioxidant and anti-inflammatory properties, and functions as an osmolyte.

$$\begin{array}{c} CH_2-NH_2 \\ | \\ CH_2-SO_3H \end{array}$$

FIGURE 27.40 **Taurine.**

Taurine synthesis is insufficient in new borns and especially in prematures. Breast milk contains approximately 400 μmol/L, ensuring an adequate supply. Formula that substitutes breast feeding contains very little taurine.

Uses. Commonly, taurine is included along with caffeine in "energizing" drinks, publicized as enhancers of physical performance. There is not evidence indicating that, in normal individuals, consumption of such type of products has any beneficial effects.

Carnitine

Carnitine (3-hydroxy-4N-trimethylaminobutiric) (Fig. 27.41) is an acid with a quaternary ammonium function. In living beings, it is found free or esterified with fatty acids. It is mainly synthesized in the liver (also in muscle) from lysine and methyl donated by SAM.

Foods of animal origin provide appropriate amounts, but fruits and vegetables are low in carnitine. It is absorbed in the intestine, passes to the blood, and is taken up by most tissues, mainly skeletal muscle, where its concentration is up to 50 times higher than in plasma.

Functions. The major role of carnitine is in metabolism, participating in the system that transfers fatty acids from the cytosol into the mitochondrial matrix for complete oxidation (p. 338).

Carnitine deficiency. Carnitine is not an essential nutrient in humans; the endogenous synthesis meets the needs. Deficiency has not been observed in adults. More vulnerable are newborns, particularly premature infants, whose ability to synthesize carnitine is poor, or young children given soy milk, which does not contain carnitine. In these cases, hypoglycemia, depression of the fatty acid oxidation, occurs. It is advisable to supplement breast milk substitutes with carnitine.

Uses. Addition of several grams of carnitine per day to the regular diet is believed to enhance the performance of elite athletes, or to increase muscular capacity of normal subjects. There is no scientific evidence to support such use.

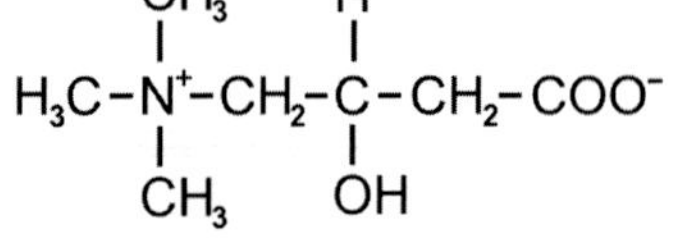

Inositol

It is a hexahydroxycyclohexane, which has nine isomers. The mesoinositol, also called myoinositol (Fig. 27.42), is the most important in nature and the only active isomer.

In all cells, inositol-1-phosphate is synthesized from glucose-6-phosphate. In plants there are significant amounts of inositol phosphate, including hexaphosphate or phytic acid, a chelator of calcium, iron, magnesium and zinc; it forms insoluble products that cannot be absorbed. It is widely distributed in natural foods, in fruits, meat, milk, nuts, legumes, and whole grains.

It integrates phospholipid molecules, such as phosphatidylinositol, phosphorylated on carbons 4 and 5; hydrolysis of this compound liberates inositol-4,5-bisphosphate (IP_2), member of a signal transduction system (p. 562).

In recent years, evidence has been collected for the role of compounds like inositol polyphosphates (IP_7, inositolhexaquisphosphate, inositolpentaquisphosphate) and inositol pyrophosphate. All of them have diverse actions in energy metabolism, exocytosis, response to infection and stress, and apoptosis; they inhibit cell response to inositol 3,4,5-trisphosphate (IP_3).

CH₃   H

H₃C–N⁺–CH₂–C–CH₂–COO⁻

CH₃   OH

FIGURE 27.41 **Carnitine.**

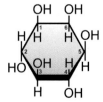

FIGURE 27.42 **Inositol.**

Another molecule with inositol is glycosylphosphatidylinositol, which serves to anchor proteins to the external face of cell membranes.

Consequences of inositol deficiency have been studied in laboratory animals. In mice, alopecia, lactation failure, and growth retardation are observed. Inositol, along with choline, has lipotropic action; it facilitates mobilization of fatty deposits in liver. Inositol is not an essential nutrient in humans because the endogenous production meets the normal requirements.

Lipoic Acid

Also known as thioctic acid, it is the 6,8-dithiooctanoic acid. Some authors include it among the members of the B complex, but it is not a dietary requirement in mammals. It is widely distributed in nature, in their oxidized and reduced forms (Fig. 27.43). In foods, lipoic acid is bound to lysine residues of proteins. Liver, kidney, heart, yeast, spinach, and peas are excellent sources.

Lipoic acid in foods is released from its protein binding in intestine; it is absorbed and converted into dihydrolipoic. It is one of the coenzyme members of multienzyme systems that catalyze oxidative decarboxylation of α-ketoacids (pyruvate, α-ketoglutarate, and branched-chain α-ketoacids dehydrogenases). Dihydrolipoic acid is an antioxidant agent, active against hydroxyl radicals, hypochlorous acid, and singlet oxygen. Dihydrolipoic acid can reduce the oxidized forms of vitamins C and E. No supplemental intake is necessary in healthy individuals.

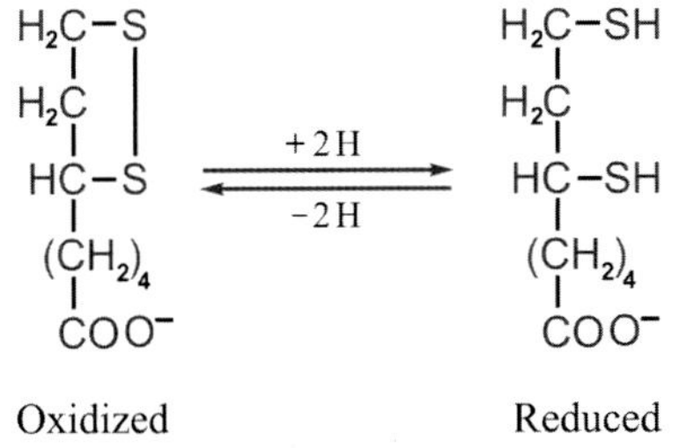

FIGURE 27.43 **Lipoic acid.**

SUMMARY

Vitamins are organic compounds essential for normal health. They are not synthesized in the body and must be supplied with the diet. Some vitamins form coenzymes, and others function as hormones. Their deficiency produces specific pathological conditions. Vitamins are divided into fat soluble (A, D, E, and K) and water soluble (B and C).

Fat-soluble vitamins include the following:

Vitamin A or *retinol* (*axerophthol*) and its active derivatives, *retinal* and retinoic acid. *Carotenes* are precursors or provitamin A compounds. Vitamin A is found in animal foods and carotenoids in pigmented plants. Retinol is oxidized to retinal and retinoic acid. In the cell nucleus, vitamin A associates with receptors (RAR and RXR), which dimerize and bind to specific DNA sites.

Avitaminosis or lack of vitamin A produces lesions in skin (hyperkeratosis, desquamation), eyes (xerophthalmia), and in tracheobronchial epithelium. The first symptom of deficiency is *night blindness*.

Functional role is to modulate the activity of various genes involved in the regulation of cell division and proliferation and synthesis of glycoproteins. It stimulates the development of B and T helper lymphocytes. Retinoic acid promotes cell differentiation and apoptosis. It also has nongenomic actions. Retinal is involved in the process of vision. Rods have rhodopsin, with 11-*cis* retinal as prosthetic group. The incidence of light converts the *cis* isomer in all-*trans* and triggers the light stimulus; the all-*trans* retinal is reduced to retinol. *Dehydrogenase* catalyzes the reaction, and the *isomerase* converts all-*trans* retinol into 11-*cis* retinal.

Vitamin D, *calciferol*, or antirachitic vitamin is a sterol derivative. It is scarce in natural foods. In the body there is a provitamin, *7-dehydrocholesterol*, which is converted to vitamin D by exposure of skin to sunlight (ultraviolet rays). Vitamin D must undergo some modifications to become active. In the liver it is hydroxylated to 25-OH-D$_3$; in the kidneys a new hydroxylation forms *1,25-(OH)$_2$-D$_3$* or *calcitriol*, the most active metabolite. In the nucleus, calcitriol binds receptors VDR and RXR that link to DNA.

Avitaminosis causes rickets in young children (failure in calcification, bone deformities, retarded, and abnormal tooth eruption). In adults, it causes *osteomalacia* (bone demineralization).

Functional role is to maintain Ca^{2+} homeostasis and also participates in phosphate balance. In intestine it stimulates Ca^{2+} absorption (via *calbindin* synthesis), in bone it increases deposition, and in kidney it activates

Ca^{2+} and phosphate tubular reabsorption. It is also a factor in the regulation of cell differentiation and maturation. Its actions are genomic and nongenomic.

Vitamin E or *tocopherol* is present in corn, peanut, and soybean oils, and in wheat.

Avitaminosis is rare in man, but when it occurs, it increases fragility of red blood cells and creatinuria. In children, lack of vitamin D causes anemia and decreased half-life of erythrocytes.

Functional role is to function as an antioxidant.

Vitamin K or *antihemorrhagic vitamin* is a derivative of naphthoquinone, sensitive to light. It is found in cabbage, cauliflower, spinach, tomato, cheese, egg, and liver. It is synthesized by intestinal flora bacteria.

Avitaminosis causes delayed blood coagulation.

Functional role is to be a factor of carboxylase, forming γ-carboxyglutamate in Gla proteins, which are essential for the formation of coagulation factors II (prothrombin), VII, IX, and X.

Water-soluble vitamins include the following:

Vitamin B complex, which is composed of thiamine, riboflavin, niacin, panthotenic acid, piridoxin, biotin, folic acid, and cobalamin.

Thiamin or *vitamin B$_1$* (aneurin) is decomposed by heating in humid environment. It is found in cereal whole grains, pork, liver, legumes, and brewer's yeast. The requirements are higher in alcoholics and in people who eat an excess of carbohydrates.

Avitaminosis causes *beriberi*, a disease characterized by weakness, fatigue, headaches, insomnia, dizziness, loss of appetite, tachycardia, neuritis, and paralysis. In some cases, neurological symptoms predominate, while in others cardiovascular alterations prevail.

Functional role is exerted by its active form (thiamine pyrophosphate, TPP), coenzyme of multienzyme systems catalyzing decarboxylation of α-ketoacids. It is also coenzyme of transketolase. *Thiamine TTP* is associated with nerve transmission.

Riboflavin or *vitamin B$_2$* is present in milk, liver, kidney, meat, fish, egg yolk, spinach, tomato, and carrot.

Avitaminosis causes glossitis, lesions in the lips corners (cheilosis), seborrhea, and conjunctivitis.

Functional role is to form part of FMN and FAD, which are coenzymes of *oxidoreductases*.

Niacin or *vitamin B$_3$* has two vitamers, *nicotinic acid* and *nicotinamide*, both of which are active. Niacin is present in liver, meat, eggs, and whole grains.

Avitaminosis causes *pellagra* (dermatitis, glossitis, stomatitis, nausea, vomiting, enteritis, diarrhea, neurological, and mental symptoms).

Functional role depends on its role in NAD and NADP, coenzymes of *oxidoreductases*. NAD is also involved in *ADP-ribosylation* of proteins, formation of *NAADP* and *cyclic ADP–ribose*, Ca^{2+}-mobilizing agents.

Pantothenic acid or *vitamin B$_5$* is widely distributed in all foods.

Avitaminosis is not observed in humans.

Functional role. Pantothenic acid integrates *coenzyme A* and *phosphopantetheine*. *Acetyl-CoA* is an important intermediate compound of different metabolic pathways. *Succinyl-CoA* is an intermediary in the citric acid cycle and in the synthesis of heme. *Phosphopantetheine*, attached to the acyl carrier protein, integrates the multifunctional *fatty acid synthase* protein.

Pyridoxine (PN) or *vitamin B$_6$* is oxidized to pyridoxal (PL), or binds to an amine group to give pyridoxamine (PM), both of which are active. PN is present in whole grains, cabbage, legumes, liver, pork, and cereals.

Avitaminosis is rare in man; it consists in skin and gastrointestinal disorders, anemia, nervous depression, and mental confusion.

Functional role depends on its active form, *PLP*, coenzyme in many reactions of amino acid metabolism (transamination, decarboxylation, deamination, etc.) and heme synthesis.

Biotin, vitamin H, or *B$_7$* is present in liver, kidney, milk, egg yolk, tomato, and yeast. It is synthesized by bacteria of the intestinal flora.

Avitaminosis is not observed in humans. Avidin of egg white prevents its absorption in the intestine.

Functional role is to function as a coenzyme of *carboxylases* (acetyl-CoA and pyruvate carboxylases).

Folic acid, pteroylglutamic acid, or *vitamin B$_9$* is present in legumes, liver, kidney, and yeast.

Avitaminosis causes megaloblastic anemia, cardiovascular changes, increased homocysteine in plasma and tissues. Deficiency in the mother produces alterations of the development of the nervous system in the embryo.

Functional role. As *tetrahydrofolic acid* acts as coenzyme for one-carbon transfer, it is important in the synthesis of purine, thymine, and in amino acid metabolism.

Cobalamin or *vitamin B$_{12}$* has cobalt. It is present in liver, kidney, meat, milk, and eggs, almost nonexistent in plants. Absorption in intestine requires intrinsic factor (IF), a glycoprotein secreted in gastric mucosa.

Avitaminosis causes megaloblastic anemia and nervous system disturbances. Lack of IF by autoimmune disease prevents absorption of vitamin B$_{12}$ and results in a serious condition called *pernicious anemia*.

Functional role. It forms part of a coenzyme as methylcobalamin. Along with methyl-THF, it participates in the conversion of homocysteine to methionine. As deoxyadenosylcobalamin, it is a coenzyme in the isomerization reaction of L-methylmalonyl-CoA to succinyl-CoA.

Vitamin C or *ascorbic acid (AA)* is an important reductor agent, easily inactivated in neutral or alkaline solutions. AA is abundant in citrus fruits, tomatoes, and vegetables. It is lost by cooking or food preservation.

Avitaminosis causes *scurvy* (joints pain, petechiae, anemia, hematomas, gum inflammation).

Functional role is related to its involvement in redox processes, in collagen synthesis (synthesis of OH-lysine and OH-proline), in catecholamines synthesis, and in the reduction of Fe^{3+} to Fe^{2+}, to allow its absorbtion in the intestine).

Bibliography

Ball, G.F.M., 2004. Vitamins. Their Role in the Human Body. Blackwell Publishing Ltd, Oxford.

Bender, D.A., 2003. Nutritional Biochemistry of the Vitamins, second ed. Cambridge University Press, Cambridge.

Blanco, A., 2009. Micronutrientes. Biomed, Buenos Aires.

Bowman, B.A., Russell, R.M. (Eds.), 2003. Present Knowledge in Nutrition. eighth ed. International Life Sciences Institute, Washington, DC.

Dusso, A., Brown, A.J., Slatopolsky, E., 2005. Vitamina D. Am. J. Physiol. 289, F8–F28.

Gropper, S.S., Smith, J.L., Groff, J.L., 2005. Advanced Nutrition and Human Metabolism, fourth ed. Thompson Wadsworth, Belmont, CA.

Shils, M.E., Shike, M., Ross, A.C., Caballero, B., Cousins, R.J. (Eds.), 2006. Modern Nutrition in Health and Disease. tenth ed. Lippincott Williams & Wilkins, Philadelphia, PA.

World Health Organization and FAO of the United Nations, 2004. Vitamin and Mineral Requirements in Human Nutrition: Report of a Joint FAO/WHO Expert Consultation. WHO, Bangkok.

Zempleni, J., Rucker, R.B., McCormick, D.B., Suttie, J.W. (Eds.), 2007. Handbook of Vitamins. fourth ed. CRC Press, Boca Raton, FL, pp. 559–570.

Water and Acid–Base Balance

Water and electrolyte balance has particular importance for normal body function. Water is the solvent in which almost all biological reactions take place. In all living organisms water carries the nutrients that cells need, compounds that are synthesized and metabolized by the cells from those nutrients, and dissolved minerals, many of which ionize and behave as electrolytes.

Water is quantitatively the most important component of living beings. In humans, it represents approximately 65 and 60% of the body weight of an adult male or female, respectively. This gender difference is mainly due to the relatively higher proportion of adipose tissue in women. Generally, the relative amount of water in the body decreases with age. In premature infants it comprises up to 80% of their weight, in normal newborns it reaches 77% and then slowly decreases to 70%–65% shortly after the first year of life. This is maintained throughout adulthood. From age 60 and above, the proportion of water is progressively reduced to values around 50% mainly due to the decrease in muscle mass.

The amount of water varies significantly depending on the tissue and organ considered. For example, skin has 72% water, muscle 75%, bone 22%, liver 68%, kidney 82%, intestine 74%, and adipose tissue 10%.

In a normal individual, total water content is stable within narrow limits, maintained by a series of regulatory mechanisms. Since total water content is in relation with the amount of adipose tissue, the proportion of the body weight corresponding to water decreases as the percentage of fat increases. Total water is relatively low in obese individuals.

Total water determination. Total water can be measured by techniques that are based on the dilution of solutes that easily cross cell membranes and distribute uniformly throughout the body. To achieve this, it is necessary to administer a known amount of the solute, wait until it equilibrates in all tissues, and determine its concentration in blood plasma. Total water volume (TV) in liters, is calculated using the following formula:

$$\text{TV} - \text{Administered amount(mg)}/\text{concentration(mg/L)}$$

Antipyrine, thiourea, and isotopes (deuterium or tritium oxides) are examples of solutes used for body water determination. A newer technique involves the measurement of the bioelectrical impedance of body tissues, which provides estimates of total water, fat free mass, and body fat. Besides determining adiposity, this approach is also used to calculate cell mass and total body water in different clinical conditions.

WATER DISTRIBUTION
IN THE BODY

The water in the body is distributed into different compartments. Two main body compartments include the intracellular and extracellular spaces.

1. The *intracellular* space (IC) is contained within the cells' plasma membrane. Its particular chemical composition compared to the interstitial fluid bathing the cells depends on the selective membrane permeability of the cells. The intracellular compartment contains two-thirds of total body water (about 27 L in a 70 kg adult male). There are no direct methods for measuring IC water; its amounts are obtained by the difference between total and extracellular water.

2. The *extracellular* space (EC), termed by Claude Bernard "internal milieu," is the medium in which all cells are immersed. The cells receive from the extracellular medium all the nutrients for their anabolic (synthetic) processes, and return to it their secretions and catabolic waste products. The ability to regulate the composition of the internal environment has been the result of evolutionary changes that have allowed living beings to become independent from the external medium. The overall regulation of physiological processes is what maintains the constancy of the internal environment.

The extracellular compartment comprises the *intravascular* and *interstitial* fluids. The intravascular fluid or plasma is confined to the vascular circulatory system. The interstitial fluid is in direct contact with the cells bathed in it. The EC space contains 20% of total water (about 12 L in a normal 70 kg adult). There is an active exchange between the intravascular and interstitial fluids through the capillary walls.

The relative distribution of body water among the different body compartments is shown in Table 28.1.

An additional compartment is represented by the *transcellular* space, which comprises the liquid in the lumen of the digestive, urinary, and respiratory tracts, the cerebrospinal, pleural, pericardial, and synovial fluids, and the aqueous humor of the eye. The fluid in all these transcellular compartments combined represents only 2.5% of the total body fluid (about 1 L).

The different body fluid spaces are separated by membranes that allow the passage of water and solutes. Small molecules (O_2, CO_2, urea) move freely between all compartments, but other substances are restricted in their movements. Plasma proteins, for example, are confined to the intravascular space; the capillary wall is not permeable to macromolecules.

The interstitial fluid is not physiologically uniform. It comprises a free phase that is in exchange with the other compartments, and a partially "sequestered" phase, which is located in the bone matrix and dense connective tissue. The mineralized bone and the connective tissue extracellular matrix, composed of anionic glycosaminoglycans polymers, selectively bind cations and retain water in an almost crystalline lattice. In this manner, a significant fraction of the total water ($\sim$10%) does not readily exchange with the other organic liquids.

TABLE 28.1   Distribution of Body Water in Humans

	Total water	Intracellular fluid	ECF	Interstitial fluid	Intravascular fluid
Man (adult)	65	50	15	10	5
Woman (adult)	60	45	15	10	5
Infant	75	48	27	22	5

Values are expressed as percent of total body weight. ECF, Extracellular fluid.

Chemical differences between fluid compartments can be summarized as follows: the intravascular fluid or plasma has many crystalloid substances of small molecular weight in true solution, and proteins (7 g/dL) in colloidal dispersion. The latter, of large molecular size, cannot pass through the capillary walls, which acts as a dialyzing membrane. The interstitial fluid composition is similar to that of plasma in terms of substances of small mass but its protein content is poor. The intracellular fluid differs significantly in composition from the fluid in other compartments.

Extracellular water determination. The water volume in the EC can be measured by dilution techniques, using solutes that diffuse only into this compartment, for example, inulin, ^{77}Br, or sulfate with ^{35}S. The latter gives more accurate results. The water volume of plasma can be estimated with Evans Blue dye, labeled albumin, ^{131}I or ^{125}I. In EC water measurements, the transcellular water volume is not included.

WATER BALANCE

Under adequate dietary intake conditions, the body weight of a normal adult remains constant despite the usual variations in water intake. This indicates that body water content is effectively regulated.

Water is incorporated orally with liquids and solids, which can contain 40% or more of their weight in water. In addition, the body generates water as a final product of metabolism. A mixed balanced diet produces about 12 g of water per each 100 calories.

The kidney is able to maintain body water volume constant, excreting diluted urine when there is water excess or highly concentrated urine when water needs to be retained. The mechanism of thirst is in great part responsible for the control of water ingestion. There is obliged water loss from skin by insensible perspiration and from lung because the exhaled air is saturated with water vapor. These losses increase in environments where temperature is high or the air

TABLE 28.2   Water Balance of a Normal Adult (in 24 h)

In (mL)		Out (mL)	
Drinks	1400	Obliged loss (skin and lungs)	850
Solid foods	800	Feces	150
Metabolic water	300	Urine	1500
Total	2500	Total	2500

is dry, or in some pathological conditions (i.e., high fever). Normally, the water loss through the stools is scarce (150 mL). In the gastrointestinal tract approximately 8 L of water is secreted per day: 1000 mL with saliva, 2000 mL with the gastric juice, 500 mL with the bile, 1500 mL with the pancreatic juice, and 3000 mL with the intestinal juice. Almost all this water is reabsorbed back into the body compartments.

Under pathological conditions that cause vomiting or diarrhea, the majority of the gastrointestinal fluid cannot be reabsorbed; the loss of water and electrolytes can be considerable, producing dehydration and ionic imbalances.

Table 28.2 shows the daily intake and loss of water in a normal adult.

IONIC COMPOSITION OF BODY FLUIDS

Although the separation of water body fluids into intra- and extracellular compartments is essentially anatomical, this distinction reflects functional and chemical differences, which have biological and clinical significance. This results from the selective permeability and activity of transport systems in the cell membrane. Even within the cell, the chemical composition of the fluids contained in the different subcellular organelles is different. For example, mitochondria and endoplasmic reticulum can concentrate certain ions.

Sodium is the major extracellular cation and potassium is the most abundant cation in the intracellular space. Among anions, chloride is preferably found extracellularly, and phosphates, proteins, and sulfates have higher concentration in the intracellular space. The fluid in each body compartment has the same concentration of anions and cations and is electroneutral. Water freely diffuses through the membranes, which is why the osmotic pressure in all the compartments is approximately the same.

Compared to other compartments of the body, the intracellular fluid has the highest concentration of anions and cations. The higher number of particles dissolved per liter in the cytoplasmic compartment results in higher osmotic pressure inside the cell compared to the interstitial fluid. Osmotic pressure depends on the number of particles dispersed in a predetermined volume, regardless of their electrical charge and chemical properties. The osmotic pressure difference between intra- and extracellular osmotic compartments is much lower than that suggested by the individual concentrations of each ion. The intracellular fluid has a high proportion of ions with charge greater than 1 and some intracellular ions are bound to different molecules and, therefore, they do not contribute to increase osmotic pressure.

Blood plasma is more accessible to analysis than the other compartments. The interstitial fluid is an ultrafiltrate of plasma; its composition is calculated from the values of plasma by applying a correction factor (see subsequent chapters). The protein concentration of the interstitial fluid is less than that of plasma.

Plasma contains approximately 92% of water and 8% of solids, which are represented mainly by proteins and lipids. The plasma protein concentration varies between 6 and 8 g/dL. Since 8% of plasma is represented by solids, the amount of solutes increases in that percentage when expressed per liter of plasma water. Thus, if the concentration of Na^+ in plasma is 142 mEq./L, it becomes 153 mEq./L when expressed in terms of plasma water. Na^+ is the predominant cation in plasma, and its concentration is maintained within narrow limits. Other cations, such as K^+, Ca^{2+}, and Mg^{2+}, are found in much smaller quantities. The main anions in plasma are chloride (Cl^-) and bicarbonate (HCO_3^-) (Table 28.3). The interstitial fluid is a filtrate of plasma through the capillary walls, which are highly permeable to water, electrolytes, and solutes of small mass.

TABLE 28.3　Ion Concentration in Blood Plasma, Interstitial, and Intracellular Fluids (in mEq./L)

Cations	Plasma	Plasma water	Interstitial fluid	Intracellular fluid (muscle)	Gastric juice	Pancreatic fluid	Sweat
Sodium	142.0	152.7	147.5	13.0	60	130	45
Potassium	4.5	5.0	5.0	150.0	7	7	5
Calcium	5.0	5.4	4.0	1×10^{-7}	—	—	—
Magnesium	1.7	1.9	1.5	26.0	—	—	—
Cations total	153.2	165.0	158.0	189.0	—	—	—
Anions							
Chloride	102.2	110.0	113.0	3.0	100	60	58
Bicarbonate	26.0	28.0	28.5	10.0	0	100	0
Phosphates	2.0	2.2	2.3	107.0	—	—	—
Sulfate	1.0	1.1	1.2	20.0	—	—	—
Organic acids	6.0	6.5	6.0	—	—	—	—
Proteins	16.0	17.2	7.0	49.0	—	—	—
Anions total	153.2	165.0	158.0	189.0	—	—	—

The amount of protein that crosses the capillary wall is very scarce. The composition of the lymph fluid is similar to that of interstitial fluid; its protein concentration is 1–3 g/dL.

The movement of fluid through the capillary wall depends on the balance between hydrostatic and osmotic forces (see p. 696). There is a slight difference between the concentrations of electrolytes in plasma and interstitial fluid, determined by the Gibbs–Donnan equilibrium. The concentration of diffusible cations is ~4% higher and that of the anions is ~4% lower in plasma than in the interstitial compartment.

The particles dissolved in each body fluid compartment are responsible for the osmotic force which drives water movements between compartments. While Na^+, Cl^-, and HCO_3^- are primarily responsible for maintaining water in the extracellular compartment, K^+, Mg^{2+}, phosphates, and different organic substances, are the osmotically active solutes that keep water inside the cells.

Although the IC fluid composition varies in different cell types, the concentration of K^+ in the IC is always higher (~150 mEq./L) than that of Na^+ (~10 mEq./L). Another cation present in the IC fluid in higher concentration than in the EC fluid is Mg^{2+} (~26 mEq./L compared to ~2 mEq./L, respectively). The anion composition between these compartments differs, Cl^- and HCO_3^- concentrations are lower and phosphates and sulfates are higher in the IC compared to the EC fluid.

Gibbs–Donnan equilibrium. At the pH of the biological fluids (~7.4) proteins behave as anions. Since proteins cannot pass through the semipermeable barriers of the cell membrane and the capillary walls, their presence on one side of these barriers drives an unequal distribution of diffusible ions across those barriers. This causes the following:

1. The total concentration of anions is equal to that of cations in both compartments.
2. In the compartment containing proteins, the concentration of diffusible anions is lower, and the concentration of diffusible cations

is greater, than in the compartment lacking proteins.
3. The osmotic pressure is slightly higher in the space that contains proteins.

Imagine two compartments of equal size (A and B in Fig. 28.1) separated by a semipermeable membrane. Compartment A contains a mixture of 5 mEq. protein and 5 mEq. sodium. In compartment B, an equal volume of solution with 10 mEq. chloride and 10 mEq. sodium is added. After a certain amount of time, the ion concentration in each compartment changes, until a balance between them is reached. On side A, besides the initial 5 mEq. of proteins and 5 mEq. of sodium, there will be an additional 4 mEq. of chloride (making a total of 9 mEq. of Cl) and 4 mEq. of sodium, which diffused from compartment B (now containing 6 mEq. of chloride and 6 mEq. of sodium).

At equilibrium, the concentrations of sodium and chloride in the A compartment containing protein ($[Na_A^+]$ and $[Cl_A^-]$) and in the B compartment, free of nondiffusible ions ($[Na_B^+]$ and $[Cl_B^-]$), attain the following relationship:

$$[Na_A^+] \cdot [Cl_A^-] = [Na_B^+] \cdot [Cl_B^-]$$

or

$$\frac{[Na_A^+]}{[Na_B^+]} = \frac{[Cl_B^-]}{[Cl_A^-]}$$

The charge on either side of the membrane is electrically neutral. Therefore, $[Na_B^+]$ is equal to $[Cl_B^-]$ on the compartment lacking protein. In the compartment containing the nondiffusible anionic proteins, electroneutrality is reached when

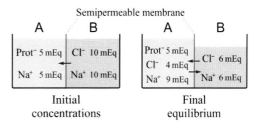

FIGURE 28.1 **Gibbs–Donnan equilibrium.**

there is enough Na^+ to balance the negative charges from proteins and Cl^-. In this manner:

$$[Na_A^+] > [Cl_A^-] \text{ and } [Na_A^+] + [Cl_A^-] > [Na_B^+] + [Cl_B^-]$$

The osmotic pressure results to be greater on the side containing proteins. For each diffusible ionic species there is a gradient of equal magnitude but opposite direction for cations and anions.

A similar phenomenon occurs between the blood plasma and the interstitial fluid compartments. To calculate the concentration of monovalent anions in the interstitial fluid, their concentration is determined in plasma and multiplied by the factor 1.05 (Donnan factor); for monovalent cations, the correction factor is 0.95.

The Gibbs–Donnan equilibrium explains why the amount of anions and cations is greater in the cell, richer in protein, than in the interstitial fluid. However, the uneven distribution of Na^+, K^+, or Ca^{2+} at both sides of the plasma membrane cannot be inferred from the Gibbs–Donnan theory. These ions are maintained by active transport mechanisms that pump Na^+ or Ca^{2+} out of the cell. Transport systems in cell membranes are responsible for the difference in ion concentrations between the intracellular and the interstitial space. Activity of Na^+,K^+-ATPase plays a key role in Na distribution between body compartments.

Osmotic pressure of body fluids. The osmotic pressure of a solution depends on the number of particles (ions and molecules) present in a given volume of solvent. As the volume of solvent varies with the temperature and, therefore, the concentration is modified by temperature changes, it is preferred to express concentration in number of dispersed particles per unit weight of solvent (molality). A solution containing 1 mole of particles ($6.022 \cdot 10^{23}$) in 1000 g of water (with a concentration equal to 1 molal or 1 m) develops an osmotic pressure of 22.4 atmospheres and freezes at $-1.86°C$ (cryoscopic descent).

One mole of particles also corresponds to a unit called *osmol*. An osmol of a substance is the amount corresponding to the Avogadro's number of particles of that substance dissolved in 1000 g of water (1 osmolal solution). Since glucose does not ionize, 1 mol of this hexose dissolved in 1000 g of water (1 m solution) produces 1 mol of particles with an osmolality of 1; NaCl ionizes completely in aqueous solution, forming two ions (or particles); therefore, 1 m NaCl solution is 2 osmolal. K_2SO_4 ionizes in solution giving three ions (one SO_4^{2-} and two K^+); then, a 1 m solution of K_2SO_4 is 3 osmolal.

One osmole of glucose is 1 mol; for NaCl 1 osmol is equal to 0.5 moles, and for K_2SO_4, 1 osmol is 0.333 moles. In practice, 1000th osmoles or milliosmole (mOsm) is commonly used.

Sometimes, osmolality and osmolarity are used indistinctly, which is incorrect. Osmolarity expresses osmoles concentration per liter of solution, while osmolality, expresses concentration of a solute per 1000 g of solvent (the same difference that exists between molarity and molality). At the temperature of the body, when solute concentrations are low, the difference between osmolarity and osmolality is not very important and both can be used without introducing a major error.

IC and EC compartments are in osmotic equilibrium, so that both have a similar osmolar concentration. Water shifts follow the osmotically active solutes concentration in either side of the membrane. In the extracellular fluid (ECF), practically all of the osmotic pressure is exerted by ions and low-mass molecules, which readily cross the capillary wall. Plasma determinations provide information regarding the extracellular osmolality, mainly due to the concentration of sodium and accompanying anions and, to a lesser extent, glucose and urea per kg of water.

The osmotic pressure of plasma can be measured directly by cryoscopy. If the freezing temperature of plasma is determined, it is possible to calculate its concentration in osmotically active particles (cryoscopic descent, one of the colligative properties, related to the concentration of particles). Normal plasma freezes at $-0.56°C$; therefore, its osmolality is $-0.56/1.86 = 0.301$ osmol/kg of water. In practice, no appreciable error is made if it is said that normal plasma osmolality is approximately 300 mOsm/L.

Determination of sodium, glucose, and urea in plasma, also allow to calculate osmolarity, by applying the following formula:

$$mOsm/kg = 2[Na] + \frac{[glucose]}{18} + \frac{[N \text{ of urea}]}{2.8}$$

Na^+ concentration (usually expressed in mEq./L) is multiplied by 2 to include the accompanying anions. The concentration of glucose and urea nitrogen (expressed in mg/dL) are divided by 18 and 2.8, respectively, to obtain their value in mEq./L (molecular weight of glucose is 180 and for urea is 28).

Nonelectrolytic substances contribute very little to the osmotic pressure of plasma, normally 5 mOsm/kg correspond to glucose. Only in the case of diabetes mellitus, at very high plasma glucose concentration, osmolarity from glucose becomes significant. Moreover, as urea traverses membranes and equilibrates rapidly in all compartments, it is not osmotically effective. Under normal conditions, no major error is introduced if the effective osmotic pressure of the ECF is inferred from the plasma sodium concentration.

$$\text{Effective osmotic pressure }(mOsm/kg) = 2[Na^+]$$

Effective osmolality is also designated as tonicity. Solutions with equal effective osmolality to that of the body fluids are considered *isotonic*, for example, a 0.9 g/dL NaCl solution (normal saline). Solutions with higher or lower osmolarity than that of plasma are referred to as hypertonic or hypotonic, respectively.

Oncotic pressure. A small fraction of the total osmotic pressure of plasma is due to proteins. This is known as colloid osmotic or oncotic pressure. Despite its small absolute value, oncotic pressure is of great practical importance since it influences the movement of diffusible ions across the cell membrane. Actually, the oncotic pressure from plasma proteins is greater than that corresponding to their concentration. The

difference is due in part to the Gibbs–Donnan equilibrium, as more particles exist in solution in the intravascular space than into the interstitial fluid. The Gibbs–Donnan equilibrium states the following:

$$[Na^+]_p \cdot [Cl^-]_p = [Na^+]_i \cdot [Cl^-]_i \quad (28.1)$$

where p and i subscripts correspond to plasma and interstice, respectively.

The concentration of Na^+ and Cl^- are similar to each other in the interstitial fluid. In contrast, in plasma, the difference between Na^+ and Cl^- is approximately 15 mEq./kg higher (at an average plasma protein concentration) than in the interstitial fluid. Assuming 145 mEq./kg for Na^+ and Cl in the interstitial compartment:

$$[Na^+]_i \cdot [Cl^-]_i = 145 \cdot 145$$

According to Eq. (28.1) this product must be equal to $[Na^+]_p \cdot [Cl^-]_p$

$$[Na^+]_p = 152.7 \text{ mEq./kg and } [Cl^-]_p$$
$$= 137.7 \text{ mEq./kg}$$

The net result is that the total concentration of Na^+ and Cl^- in plasma is higher than in the interstitial fluid ($145 + 145 = 290$; $152.7 + 137.7 = 290.4$). This difference of 0.4 mEq./kg or 0.4 mOsm/kg appears to be small but it is significant. The normal plasma protein concentration is ~0.9 mmol/kg; accordingly, the total osmotic effect of plasma increases from 0.9 to 1.3 mOsm/kg due to the Gibbs–Donnan effect. Since 1 mOsm/kg generates an osmotic pressure of 19.3 mmHg, the Gibbs–Donnan effect increases the oncotic pressure in the capillary by 25–26 mmHg ($0.9 \cdot 19.3 \text{ mmHg} = 17.4$; $1.3 \cdot 19.3 = 25$–26 mmHg).

The water exchange between plasma and interstitial fluid is regulated by the balance between the opposing forces represented by the hydrostatic pressure and oncotic pressures. An important function of plasma proteins is to serve in the regulation of fluid exchange between the circulating blood and the interstitial space. This exchange takes place at the level of the capillaries, whose walls are readily permeable to water and small molecular weight substances, but not to macromolecules, such as proteins. Therefore, the concentration of small molecular weight substances is similar in plasma and in the interstitial fluid; the osmotic pressure they exert is almost the same on either side of the capillary wall. However, the transcapillary difference in protein concentration determines a higher osmotic pressure in the vascular space, which allows water movement into the vessels. In contrast, the hydrostatic pressure in blood, drives the movement (filtration) of fluid toward the interstitium.

The relationship between fluid filtration from the capillary (driven by the blood hydrostatic force) and fluid uptake (driven by the oncotic pressure gradient across the wall) is expressed by the Starling law:

$$\text{Net filtration} = \text{LpS}(\Delta P_{cap} - \Delta P_{onc})$$

where Lp is the capillary wall permeability, S the area available for water movement, ΔP_{cap} the hydrostatic pressure difference between the capillary and interstitial space, and ΔP_{onc}, the oncotic pressure difference between capillary and interstitial space.

The following example will illustrate this mechanism. Due to the difference in protein concentration, the colloid osmotic pressure is approximately 30 mmHg in plasma and only 10 mmHg in the interstitial fluid (these values are approximate and are given only as an example). The difference of 20 mmHg forces water to move from the interstitium into the capillary.

The hydrostatic pressure in the capillary is opposed to the colloid osmotic force. In the proximal, or "arterial" segment of the capillary, the effective hydrostatic pressure that drives liquid out of the capillary is approximately 30 mmHg. As the blood flows along the capillary, this pressure decreases to a value of approximately 10 mmHg in the "venous" capillary side.

The difference between the hydrostatic and osmotic forces (~10 mmHg) in the arterial side of the capillary determines the passage of water and solutes to the interstitial space. In the capillary venous segment, the situation is reversed, with a 10 mmHg difference in favor of the oncotic pressure, which allows fluid to return to the bloodstream (Fig. 28.2).

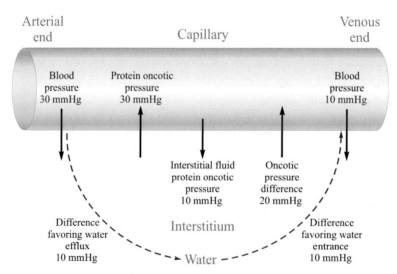

FIGURE 28.2 **Schematic representation of the Starling phenomenon.** Role of plasma proteins in the maintenance of blood volume.

Except for the oncotic pressure in the capillary, the other pressure values are difficult to measure accurately. Also, they are not uniform throughout the body; significant differences exist among different tissues and even between different zones of the same organ.

The described aqueous exchange is called *Starling equilibrium* and ensures the maintenance of the circulating fluid volume. Alterations of this mechanism can determine fluid accumulation in the interstitial space, a common clinical sign called edema.

ECF volume and osmolality regulation. The extracellular compartment has a variety of systems responsible for maintaining the constancy of its fluid volume and osmolality. Homeostatic mechanisms that regulate ECF osmolality are not identical to those responsible for maintaining ECF volume; however, they are closely related. Osmolality is dependent on the solute/water ratio, while the ECF volume is determined by the absolute amounts of Na^+ and water. The volume of the extracellular space depends on its osmolality and this, in large part, is determined by its sodium content. Due to this, many of the factors that regulate osmolality, operate on the concentration of sodium in the ECF.

The kidney, by excreting salt and water, plays a major role in the control of the ECF volume and osmolarity. The thirst mechanism, which controls water ingestion, is also essential in maintaining body water balance. Regulation of body fluid osmolality and volume are achieved through changes in water excretion in the kidney, or water uptake triggered by thirst. The mechanisms that control ECF volume primarily affect Na^+ urinary excretion.

The system monitoring osmolality of the ECF depends on osmoreceptors located in the hypothalamus. Pressure receptors or baroreceptors in the carotid sinus, aortic arch, kidney afferent arteries, large pulmonary veins, and atria are responsible for detecting changes in blood volume.

The signals generated in the osmoreceptors regulate the thirst center in the brain and stimulate the release of antidiuretic hormone (ADH) or vasopressin from the neurohypophysis. ADH stimulates water reabsorption in the kidney distal tubules and collecting ducts. Baroreceptors in the aorta and great veins also activate the production and release of ADH and, in addition, they stimulate aldosterone and atrial natriuretic peptide (ANP) release to the bloodstream. Aldosterone is secreted by the adrenal gland cortex, mainly as a response to the renin–angiotensin system activation. Aldosterone stimulates the reabsorption of Na^+ and excretion of K^+ in the kidney distal tubules and collecting ducts. ANP is released in atrial cardiomyocytes in response to heart baroreceptors, this peptide increases the rate of glomerular filtration and Na^+ excretion in urine.

The volume of the intravascular fluid depends on the plasma proteins as has been mentioned previously. The volume of the extracellular space depends on the sodium content in this space, which is the main solute responsible for the osmolality of the interstitium. Many of the factors that regulate body fluid osmolality are directed to maintain the sodium concentration in the ECF.

ALTERATIONS OF THE WATER BALANCE

Disorders of total body water content, either reduction (dehydration) or increase (hyperhydration), are accompanied by changes in the extracellular concentration of electrolytes. When ECF osmolality is unchanged, the disorder is *isotonic*, when osmolality increases or decreases, the alteration is hypertonic or hypotonic, respectively. As the ECF osmotic pressure depends mainly on the concentration of Na^+ and its accompanying anions (Cl^- and CO_3H^-), hypertonic or hypotonic changes are usually followed by variations in the concentration of these ions in plasma. Changes in plasma Na$^+$ (hypernatremia or hyponatremia) are used as important parameters to correct water balance disorders.

Dehydration. The loss of body water in excess to electrolytes results in dehydration. This happens if a person does not receive the required amount of water, or has excessive water loss due to intense sweating or hyperventilation. A particular type of dehydration occurs in patients with diabetes insipidus, where, due to lack of ADH or failure of the receptors for this hormone in the kidney, water is wasted not accompanied by electrolytes. This leads to *hypertonic dehydration*. In general, this condition is accompanied by hypernatremia, with concentrations in plasma Na$^+$ in excess of 150 mEq./L. The extracellular compartment becomes hypertonic compared to the intracellular space. To compensate, water moves from the cells into the extracellular space. While this mechanism tends to keep the circulating plasma volume, it results in a cell volume reduction. This affects the nervous system and is responsible for irritability, lethargy, and even seizures seen in advanced cases of hypertonic dehydration.

In individuals who perform intense physical labor in environments with high temperature, intense sweating leads to dehydration. Although the fluid lost is hypotonic with respect to plasma, there is also significant loss of electrolytes (mainly Na$^+$ and Cl$^-$). If only water is ingested to replenish the volume loss, the condition worsens. In these situations, fluids needs to be replenished with the addition of salts; such as a solution of 1–1.5 g/L of NaCl. In general, for any individual, during warm weather, it is advisable to consume fluids containing salts, to replenish the fluid lost by the increased perspiration and sweating. In acute cases of hypertonic dehydration, the intravascular compartment volume must be replaced with isotonic saline (0.9 g/dL NaCl solution).

Another common disorder in clinics is the parallel loss of water and electrolytes. This causes isotonic dehydration, in which plasma concentrations of Na$^+$ are within the normal range (between 130 and 145 mEq./L). This occurs in diarrhea, vomiting, and patients with fistulas, who are eliminating liquids from the digestive tract; or during hemorrhage and burns, in which important amounts of plasma are lost. Although in these cases there is decrease in the ECF, there are no changes in osmolarity and therefore, there is not movement of water without electrolytes from or to the cells. However, the loss of ECF causes movement of water and electrolytes from the cells, in an attempt to compensate for changes in the circulating plasma volume. Eventually, there is reduction in blood volume, depending on the severity of the case, with hypotension, and even hypovolemic shock if tissue perfusion is greatly compromised.

These cases of dehydration are treated by supplying isotonic saline solution, administered orally in mild cases, or parenterally in severe cases. If only water or glucose solutions are administered trying to correct dehydration, and electrolytes are not properly replenished, hypotonic expansion of ECF can be induced, with very serious consequences.

In the dehydration caused by vomiting, gastric juice is lost, predominantly composed of H$^+$ and Cl$^-$ and, to a lesser amount Na$^+$ and K$^+$. Diarrhea causes loss of Na$^+$, Cl$^-$, and HCO$_3^-$, major ions of the intestinal, pancreatic, and biliary secretions. Na$^+$ reduction is followed by K$^+$ exit from the cells. For this reason, in the treatment of these conditions, not only Na$^+$, but also other ions must be replaced.

Hypotonic dehydration is observed when electrolyte loss exceeds water loss. It occurs in chronic treatment with natriuretic compounds, or adrenal insufficiency and deficiency of aldosterone, in which there is increased Na$^+$ excretion in urine. Trying to counteract the fall in intravascular volume, there is increased secretion of ADH, which influences negatively, because this causes water reabsorption, which exacerbates the hyponatremia. This condition presents with Na$^+$ plasma concentrations lower than 130 mEq./L, which is accompanied by neurological signs, such as irritability and drowsiness. As in other cases of dehydration, the deficit in the extracellular space must be corrected by providing isotonic saline solutions. In addition, to correct the hyponatremia, the supply of water needs to be accompanied by electrolytes.

Hyperhydration. Excess fluid in the body may or may not occur with changes in osmolality of body fluids. Isotonic overhydration occurs when there is parallel accumulation of water and electrolytes in the extracellular space. This occurs in generalized edema caused by heart failure, nephrotic syndrome, marked hypoproteinemia, or excessive intravenous infusion of isotonic solutions. As there is no change in the osmolality of the extracellular space, there is no net movement of water between the extracellular and intravascular space. Treatment in these cases depends on the correction of the underlying cause. Salt intake should be reduced and diuretics should be administered to promote the excretion of Na⁺ and water.

Hypotonic hyperhydration results from excessive water intake, as in the excessive administration of fluids devoid of electrolytes (glucose solutions). A syndrome known as water intoxication occurs. This is common in psychological disorders that lead to excessive water ingestion (psychogenic polydipsia), in athletes who rehydrate by drinking water without electrolytes, or infants who have ingested too much water. Another cause of this type of overhydration is the excessive production of ADH. This is known as *syndrome of inappropriate ADH secretion* (SIADH). It occurs in patients with head traumas, by ingestion of certain drugs that stimulate ADH release, and lung carcinomas that secrete substances with ADH-like activity. In all these cases, there is an increase in total body water and, although the total amount of Na⁺ is normal, the dilution by excess water causes hyponatremia. Water moves into the cells, which increase in volume and decrease their osmolarity. This affects the brain, causing headaches, nausea and vomiting, changes in behavior, and even lethargy and coma. In cases of SIADH, urine is hyperosmotic compared to plasma. The treatment consists in correcting the disorder that causes electrolyte imbalance, and, simultaneously, to provide saline intravenously.

Hypertonic hyperhydration is observed in individuals treated with solutions of high osmolarity, in patients with Cushing's syndrome (increased secretion of glucocorticoids), in primary hyperaldosteronism, or in chronic treatment with corticosteroids. In addition to treating the cause, glucose solutions (5%) must be supplied to correct the hyperosmolality. Diuretics that stimulate natriuresis are important to reduce Na⁺ excess.

H⁺ CONCENTRATION
IN BODY FLUIDS

The ion composition of the ECF is maintained constant by efficient homeostatic mechanisms of control. Particularly critical for cellular functions is the hydrogen ion concentration. Acid or alkaline substances, coming with the diet, or generated by cell metabolism, tend to alter the hydrogen ion concentration of body fluids. However, the pH in plasma is maintained with great constancy at 7.4, corresponding to a $0.4 \cdot 10^{-7}$ Eq./L of H⁺, or $40 \cdot 10^{-6}$ mEq./L, or 40 nM of H⁺ (Table 28.4). [At the organism temperature (37°C), neutral pH is 6.7 ($2.1 \cdot 10^{-7}$ Eq./L), plasma is slightly alkaline. For a better understanding of the topics discussed in this section, it is advisable to review the concepts exposed in Chapter 2.] Under normal conditions, only very slight modifications, within a pH range of 7.35–7.45 (45–35 nM) are produced, which shows the existence of efficient regulatory systems, capable to preserve the acid–base balance of the body fluids.

Compared to the concentration of other ions in the ECF, the amount of H⁺ is very small. However, hydrogen ions are extremely important

TABLE 28.4   Relationship Between pH and [H⁺] Within the Physiological Range

pH	[H⁺] (nmol/L or nEq./L)
7.8	16
7.7	20
7.6	26
7.5	32
7.4	40
7.3	50
7.2	63
7.1	80
7.0	100
6.9	125
6.8	160

in biological processes. The organism is very sensible to changes in [H⁺]; pH values below 7.0 ($1 \cdot 10^{-7}$ Eq./L or $100 \cdot 10^{-6}$ mEq./L or 100 nM), or above pH 7.8 ($0.16 \cdot 10^{-7}$ Eq./L or $16 \cdot 10^{-6}$ mEq./L, or 16 nM) are incompatible with life.

Intracellular pH varies from one tissue to another and even from different organelles in a single cell; it is generally slightly lower than the ECF.

Some terms commonly used in clinic referring to acid–base balance in the body are not strictly correct from a chemical point of view, and this needs to be clarified. According to the Brønsted–Lowry concept, acids are compounds or ions that give up protons (H⁺) and bases are substances with the capacity to accept protons from the medium. In clinics, it is common to use the terms acids and bases for anions and cations, respectively, which obviously does not agree with the definition.

Anions that maintain their charge over a wide range of pH, for example, SO_4^{2-} or Cl^-, are called *fixed acids* and must necessarily be accompanied by cations to maintain electroneutrality. Other anions, generated by weak electrolytes, easily bind protons and lose their ionized state when

the hydrogen ion concentration in the medium increases. These anions integrate buffer systems (HCO_3^-, HPO_4^{2-}, protein⁻, Hb⁻). In contrast, *volatile acids* are those substances easily decomposed and eliminated by lung (i.e., carbonic acid).

Cations that maintain their positive charge over a wide range of pH, are called *fixed bases*.

While the body has the capacity to neutralize both, acids and alkalis, normal metabolic activity produces an excess of acids. An adult with a mixed diet generates about 20,000 mmol of CO_2 per day. This gas dissolves in body fluids and forms carbonic acid ($CO_2 + H_2O \rightarrow H_2CO_3$). Ionization of H_2CO_3 releases H⁺. Normally, the lung removes all of the CO_2 produced in the body (~300 L/day); this large amount of CO_2 moves through the body fluids leaving no acid equivalents.

The net effect of metabolism under a normal diet is the production of 50–100 mEq. of H⁺ per day, derived mainly from the oxidation of amino acids containing sulfur (methionine and cysteine), cationic amino acids (arginine and lysine), and phosphates coming in foods as $H_2PO_4^-$. Despite the excess of acid, the pH of the EC and IC fluids does not decrease. To cope with the continuous addition and removal of acids and bases, the body has several mechanisms. These include: (1) buffer systems existing in intra- and extracellular compartments, which provide an immediate defense against changes in the H⁺ concentration, (2) systems of pH control, including pulmonary ventilation and renal tubular activity. While the first provides a rapid response, the second one is slower.

REGULATION OF H⁺ CONCENTRATION

Buffer systems. Buffers in the body are generally composed of a weak acid and its salt. The buffering action is not dependent on a buffer in particular, but is the result of the joint action of all available systems. In the intracellular space, the systems formed by proteins (Prot⁻/HProt) and phosphates (HPO_4^{2-} / $H_2PO_4^-$) represent a significant proportion of the total buffering capacity of the body.

Within ECFs, blood buffers are particularly important because they are the first line of defense against pH fluctuations. This is important since blood is the medium through which acids or alkalis are distributed to the whole body. Among blood buffers the bicarbonate/carbonic acid system has particular physiological and clinical interest. Another important buffer is hemoglobin in its oxygenated and deoxygenated forms (HbO_2^-/$HHbO_2$ and Hb^-/HHb). Plasma proteins (Prot⁻/HProt) and phosphates (HPO_4^{2-} / $H_2PO_4^-$) also act as buffers, but their relative importance is much smaller than that of Hb.

Proteins-hemoglobin. Proteins generally behave as buffers thanks to the presence of the imidazole group of histidine in its molecule, which can accept or donate protons according to the pH in the medium (Fig. 28.3), pK of imidazole is 6.0.

At the pH of biological fluids, the carboxyl groups of the dicarboxylic amino acids, and the free amine groups of diaminated amino acids, are completely ionized (=CO.O⁻ and –NH₃⁺). The phenolic group of tyrosine and the sulfhydryl group of cysteine display no electric charge, because their pK_a values are far apart from the physiological pH. Consequently, the imidazole group of histidine is the only one capable to receive or transfer protons and gives proteins their buffer role.

In blood, oxyhemoglobin and deoxygenated hemoglobin are buffer systems (HbO_2^- / $HHbO_2$ and Hb^-/HHb), their role is more important than that of plasma proteins, not only because

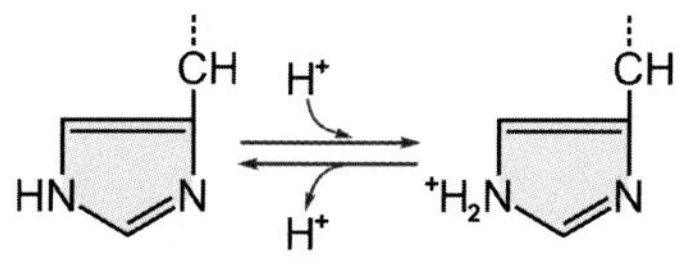

of the higher concentration of Hb (15g/dL vs. 7 g/dL for plasma proteins), but because its reversible conversion HbO_2 1Hb influences the bicarbonate/carbonic acid system (see subsequent sections).

Phosphates. Phosphoric acid (H_3PO_4) has three ionizable protons (triprotic acid) that are released in successive reactions:

1. $H_3PO_4 \leftrightarrow H_2PO_4^-$ $\qquad pK_{a1} = 2.1$

2. $H_2PO_4^- \leftrightarrow HPO_4^{2-} + H^+$ $\qquad pK_{a2} = 6.8$ (at 37°C)

3. $HPO_4^{2-} \leftrightarrow PO_4^{3-} + H^+$ $\qquad pK_{a3} = 12.0\ pK_3$

The pK_{a2} value is the closest to 7.4, indicating that at physiological pH, the dominant ionic species are HPO_4^{2-} and $H_2PO_4^-$; practically all phosphoric acid (H_3PO_4) or PO_4^{3-} ions are dissociated.

According to the Henderson–Hasselbalch equation:

$$pH = pK_a + \log\frac{[salt]}{[acid]}$$

In the HPO_4^{2-}/$H_2PO_4^-$ buffer system, in normal blood plasma (pH 7.4):

$$pH = 6.8 + \log[HPO_4^{2-}]/[H_2PO_4^-]$$

then

$$\log[HPO_4^{2-}]/[H_2PO_4^-] = 7.4 - 6.8 = 0.6$$

taking antilogarithm on both terms of the equation:

$$[HPO_4^{2-}]/[H_2PO_4^-] = 4$$

Normally, in plasma there is a 4 to 1 ratio between [HPO_4^{2-}] and [$H_2PO_4^-$], indicating that the plasma phosphate system is more efficient in buffering acids than bases. However, due to the low concentration of phosphate ions (~2 mEq./L), they play a minor role in plasma acid–base balance.

Bicarbonate/carbonic acid. CO_2, formed as a product of cell metabolism, dissolves in body

FIGURE 28.3 **Ionization of histidine imidazole.**

fluids and partly hydrates to give carbonic acid, which ionizes into bicarbonate ion and H⁺:

$$CO_2 + H_2O \leftrightarrow H_2CO_3 \leftrightarrow HCO_3^- + H^+$$

H_2CO_3 concentration is directly related to the amount of dissolved CO_2, indicated by the notation $[CO_2]$. It is possible to modify the Henderson–Hasselbalch equation for the HCO_3^-/H_2CO_3 system by replacing $[H_2CO_3]$ by $[CO_2]$. The pK value for the system is 6.1. Then, at the pH of normal plasma (7.4):

$$7.4 = 6.1 + \log [HCO_3^-]/[CO_2]$$

then

$$\log[HCO_3^-]/[CO_2] = 7.4 - 6.1 = 1.3$$

taking antilogarithms of both terms of the equation:

$$[HCO_3^-]/[CO_2] = 20$$

This high value for the [salt]/[acid] ratio indicates that the system is more efficient to buffer the acid than the alkali excess. The normal concentration of bicarbonate in plasma is approximately 25 mM, and that of carbonic acid, 1.25 mM, the ratio 25/1.25 = 20. Since the resulting pH depends on the value of this ratio, variations in absolute concentrations of bicarbonate and CO_2 do not affect the hydrogen ion concentration, provided that the value of 20 for the $[HCO_3^-]/[CO_2]$ ratio is maintained constant.

There is a close relationship between the bicarbonate/carbonic acid buffer system and the transport of CO_2 in blood. This topic will be discussed next.

CARBON DIOXIDE TRANSPORT

A gas in contact with a liquid dissolves in an amount proportional to its solubility constant and to the gas partial pressure (Henry's law). Carbon dioxide, formed as a result of cellular activity, diffuses and dissolves in body fluids according to Henry's law.

The CO_2 generated in the body exceeds the capacity of blood to physically dissolve it for transport. Only a small fraction is in solution; another fraction constitutes compounds that readily release CO_2 in the lungs. Most of the transported CO_2 (~90% of the total) is combined as bicarbonate; approximately 5% is bound to hemoglobin, forming carbamino hemoglobin, and the rest is physically dissolved.

Carbamino. Carbon dioxide may be transported in combination with the α-amine end groups of hemoglobin chains. These carbamino compounds form spontaneously in a reversible nonenzymatic reaction:

$$Hb\text{-}NH_2 + CO_2 \leftrightarrow Hb\text{-}NH\text{-}COO^- + H^+$$

The binding of CO_2 is influenced by the hemoglobin oxygenation status. In the lung, formation of oxyhemoglobin facilitates CO_2 release, while in the tissues, hemoglobin deoxygenation favors CO_2 fixation as carbamino hemoglobin. The conversion of Hb into HbO_2 also influences formation of bicarbonate in the blood. This action has consequences similar to those mentioned for the formation of carbamino, CO_2 is released in the lungs, and Hb deoxygenation in tissues favors its uptake. This phenomenon, called Haldane effect, appears to be opposed to the Bohr effect.

Bicarbonate. CO_2 that enters the blood is hydrated in red blood cells and forms carbonic acid.

$$CO_2 + H_2O \leftrightarrow H_2CO_3$$

This reaction occurs spontaneously but very slowly. In blood this conversion is very fast due to the existence in erythrocytes of the enzyme *carbonic anhydrase*, which catalyzes the reaction in both, the forward and reverse reactions.

The amount of carbonic acid formed is proportional to the gas partial pressure (P_{CO_2}). The acid ionizes to form bicarbonate and H⁺:

$$H_2CO_3 \leftrightarrow HCO_3^- + H^+$$

This reaction is significantly shifted to the right if there are hydrogen ion acceptors available and cations to neutralize the bicarbonate in the medium. In blood, the substance that most effectively performs this exchange is hemoglobin.

Hemoglobin oxygenation converts it to oxyhemoglobin, which is an acid able to retain or neutralize fixed bases. In the red cell, K^+ is the predominant fixed base present; HbO_2 neutralizes more K^+ ions than deoxygenated hemoglobin. When HbO_2 releases its oxygen, it loses some of its acidic character and accepts hydrogen ions from the dissociation of carbonic acid.

Fig. 28.4 shows the different capacity of HbO_2 and Hb to neutralize cations. Within the physiological pH range, oxyhemoglobin fixes cations (bases) in a greater extent than deoxyhemoglobin. Therefore, when the HbO_2 delivers oxygen in tissue capillaries, it frees bases usable to neutralize the bicarbonate formed. Thus, hemoglobin plays an important role in the formation of bicarbonate; the capacity of plasma alone to transport CO_2 and bicarbonate is much lower than that of whole blood.

Chloride shifts. Red blood cells undergo a series of ion changes associated to gas transport.

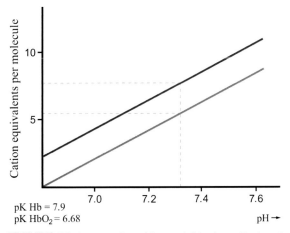

pK Hb = 7.9
pK HbO₂ = 6.68

pH →

FIGURE 28.4 **Capacity of hemoglobin (*gray line*) and oxyhemoglobin (*red line*) to neutralize cations.** Only the effect at pH values in the range compatible with life are shown.

These ion changes depend on the activity of transporters in the red blood cell plasma membrane. Ion exchange between erythrocytes and the medium varies depending on the blood territory considered (pulmonary alveoli or tissue capillaries).

In tissues, when the blood reaches the peripheral capillaries:

1. The CO_2 formed in cells as a result of metabolic activity, moves to plasma and into the erythrocytes, where CO_2 and H_2O, in a reaction catalyzed by carbonic anhydrase, form H_2CO_3 (which ionizes into bicarbonate and H^+).

2. Arterial oxyhemoglobin in tissue blood capillaries, with a low oxygen tension, releases oxygen and becomes deoxygenated hemoglobin, relatively less acidic than HbO_2.

3. Oxyhemoglobin deoxygenation releases some of the cations associated to it (mainly K^+) and accepts protons (HHb) formed by the dissociation of carbonic acid. Bicarbonate is neutralized by the K^+ released.

4. The diffusion of CO_2 into the erythrocytes increases the bicarbonate concentration within the cell. This enhances bicarbonate diffusion to plasma, where its level is lower. An electroneutral HCO_3^-/Cl^- exchanger moves bicarbonate out of the cell in exchange for Cl^- in a 1:1 stoichiometry. This antiporter, found in red blood cell membrane, is a protein known as *band 3* (Fig. 28.5).

In pulmonary alveoli, when blood returns to the lung capillaries, the shifts occur in opposite directions to those described in tissues.

1. The greater partial pressure of oxygen in the alveoli determines conversion of deoxygenated hemoglobin into oxyhemoglobin which is more acidic, releases protons and attracts free cations (K^+).

2. The protons liberated by Hb are captured by HCO_3^- ions to reconstitute carbonic acid. By action of carbonic anhydrase, the H_2CO_3

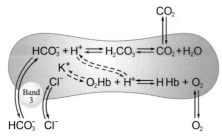

FIGURE 28.5 **Schematic representation of ion exchange between plasma and red blood cells (displacement of chloride or Zuntz–Hamburger phenomenon).** *Black arrows* indicate direction of the reactions that occur in pulmonary alveolar capillaries. *Red arrows* indicate reactions in tissues.

decomposes into H_2O and CO_2. This gas diffuses into the plasma and is eliminated through the alveolar epithelium.

3. As a result of those reactions, bicarbonate concentration in red blood cells decreases below that of the plasma. This determines HCO_3^- diffusion from plasma into erythrocytes, mediated by a HCO_3^-/Cl^- exchanger.

As a consequence of the ion movements, red blood cells in arterial blood have a lower chloride content than venous blood erythrocytes (Fig. 28.5).

The ion exchanges indicated previously, called *Zuntz–Hamburger phenomenon* or *chloride cycle*, explain the higher CO_2 transport capacity (in the form of bicarbonate) of plasma in contact with red cells (true plasma), compared with plasma alone (isolated plasma). The HCO_3^-/Cl^- exchange allows plasma to gain bases released from the erythrocytes during the conversion HbO_2–Hb.

Bone buffers. Bone has an important buffer role. In the presence of an acid overload, the bone releases Na^+ and K^+ from bone mineral matrix resorption, initially releasing to the EC space $NaHCO_3$ and $KHCO_3$, and later $CaCO_3$ and $CaHPO_4$. In general, acid load increases Ca^{2+} release from bone and favors its excretion in urine. When there is overload of bases, the bone increases carbonate deposition.

REGULATORY SYSTEMS

Bicarbonate, main form of CO_2 transported in the blood, is also a component of a very important buffer system. The blood pH is regulated by adjusting the ratio [bicarbonate]/[carbonic acid]. Both sides of this relationship may be modified independently: (1) the respiratory system is capable of modifying the P_{CO_2} and thus, the carbonic acid concentration in plasma; (2) kidney is responsible for modulating the bicarbonate concentration.

RESPIRATORY REGULATION

The respiratory center in the brainstem is sensitive to blood changes in pH and P_{CO_2}. A pH decrease or a P_{CO_2} increase enhance the frequency and depth of respiration to intensify gas exchange and CO_2 elimination. In contrast, a P_{CO_2} fall, or a pH elevation, reduces the frequency and amplitude of breathing, leading to CO_2 accumulation in blood. Therefore, respiration functions as a regulatory system to compensate for pH changes in body fluids, changing the value of the denominator of the [bicarbonate]/[CO_2] ratio.

According to the Henderson–Hasselbalch equation, the pH is 7.4 when the ratio [HCO_3^-]/[CO_2] is equal to 20, regardless the absolute values of HCO_3^- and CO_2. An excess of acid results in decreased blood bicarbonate concentration and thereby pH reduction. In such case, the respiratory system attempts to maintain the ratio [HCO_3^-]/[CO_2] = 20 increasing ventilation. This reduces blood P_{CO_2} and carbonic acid concentration in plasma. If there is excess of alkali, the numerator of the [HCO_3^-]/[CO_2] ratio increases. In this case, the respiratory system decreases ventilation, increasing blood P_{CO_2} and thus, carbonic acid concentration.

Respiratory response to acid–base disorders is rapid, it begins within minutes and reaches its maximum compensation within 12 h.

RENAL REGULATION

The body metabolism under a standard diet produces 50–100 mEq. of noncarbonic acid per day. This excess acid is removed by the kidneys, which maintain acid–base balance primarily by their ability to regulate H^+ excretion and to reabsorb bicarbonate filtered in the glomeruli. The CO_2 coming from the bloodstream, or produced in the renal tubular cells, hydrates to form carbonic acid, in a reaction catalyzed by carbonic anhydrase. Immediately, the H_2CO_3 ionizes into H^+ and bicarbonate:

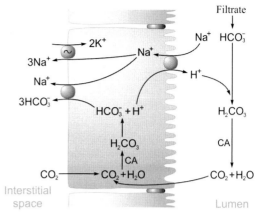

FIGURE 28.6 **Schematic representation of the bicarbonate reabsorption process in the nephron proximal tubule cells.** *CA*, Carbonic anhydrase.

$$CO_2 + H_2O \underset{\text{Carbonic anhydrase}}{\rightleftharpoons} H_2CO_3 \rightleftharpoons HCO_3^- + H^+$$

The H ions are secreted from the cell into the tubular lumen through transport mechanisms that operate in the nephron proximal and distal portions. These include the Na^+/H^+ exchanger in the apical membrane of proximal tubules, which moves H^+ from the cell into the tubular lumen in exchange for cytoplasmic Na^+. This is an electroneutral ion transporter with a Na^+:H^+ exchange of 1:1, which uses the gradient created by the Na^+,K^+-ATPase as driving force. The Na^+ coming into the tubular cells is actively pumped by the Na^+,K^+-ATPase to the peritubular interstitial space, from where it passes to the blood. The bicarbonate ions produced by ionization of H_2CO_3, move into the bloodstream via the Na^+/HCO_3^- exchanger (Fig. 28.6).

In the distal portions of the nephron (distal tubules and collecting ducts), H^+ secretion into the tubular lumen is driven by two apically located proton pumps, the H^+-ATPase and the H^+,K^+-ATPase. Bicarbonate returns to the circulation by a Cl^-/HCO_3^- exchanger in the basolateral membrane. Cl^- enters the cell favored by the gradient for this ion.

These basic renal mechanisms for H^+ secretion are associated to three processes that counteract acid excess: (1) bicarbonate reabsorption, (2) urine acidification, and (3) ammonium ions production.

1. *Bicarbonate reabsorption.* Bicarbonate is an ion with renal threshold. When the bicarbonate plasma level exceeds 28 mEq./L, it appears in urine. If the ion concentration is 27 mEq./L or less, the renal tubules reabsorb all the filtered bicarbonate and it is not excreted in urine, provided that the $[CO_2]$ is maintained within normal limits. When $[CO_2]$ decreases, the kidney eliminates HCO_3^-, even if its plasma level is less than 28 mEq./L. In this case, the requirement to maintain constant the $[HCO_3^-]/[CO_2]$ ratio predominates.

   The rate of bicarbonate reabsorption differs according to the nephron level under consideration. It predominates in the proximal convoluted tubules where it reaches a significant magnitude. A normal person with a glomerular filtration rate of 180 L/day and a HCO_3^- plasma concentration of 24 mEq./L, reabsorbs approximately 4300 mEq. of bicarbonate per day, 90% of which is reabsorbed in the proximal tubules.

The bicarbonate contained in the tubular fluid is combined with H^+ to give carbonic acid, which produces CO_2 and H_2O by action of the carbonic anhydrase in the apical brush border membrane of the tubular cells. CO_2 easily enters the tubular cells, where it is hydrated in a reaction catalyzed by cytosolic carbonic anhydrase. The carbonic acid formed produces H^+ and HCO_3^-. As mentioned, H^+ is secreted into the lumen by the Na^+/H^+ antiporter and the proton pump in the renal proximal and distal tubules, respectively. Meanwhile, bicarbonate returns to the circulation, mainly through the $Na^+/3HCO_3^-$ cotransporter in the basolateral membrane (Fig. 28.6). The $Na^+/3HCO_3^-$ symporter is a secondary electrogenic active transport system driven by the electrical gradient generated by the Na,K$^+$-ATPase. In nephron distal segments, bicarbonate is reabsorbed through a bicarbonate antiporter which exchanges $Cl^-/3HCO_3^-$.

2. *Urine acidification.* The kidney can produce a urine with pH ranging between extreme values of 4.5 and 7.9, according to the organism needs. The glomerular ultrafiltrate has the same pH as that of blood plasma; however, as it flows through the tubules, it is modified to achieve the final pH of the urine.

The simple change in the pH of urine is insufficient to offset the acid load from the body. In 1500 mL of urine, a volume normally eliminated per day, even at the lowest pH (4.5), the total amount of free H^+ that can be excreted is 0.00006 mEq. Obviously, the pH regulatory capacity of the kidney depends on other mechanisms.

The kidney capacity to remove free acids (nonionized) depends on the pK of the acid. When the pK_a is equal to the pH in urine, half of the acid is ionized (Henderson–Hasselbalch). When the pK_a is one unit higher than the pH of the urine, 9/10 parts

of the acid are not ionized, and only 1/10 must be neutralized by bases. In the case of strong acid radicals, such as Cl^- or SO_4^{2-}, they must be accompanied by bases, since they remain ionized, even at the lowest pH that the urine can attain.

The glomeruli filter various weak acids that form buffer systems in the urine. Thanks to its pK_a (6.8) and its relatively high rate of excretion, the major urinary system is the buffer $HPO_4^{2-}/H_2PO_4^-$. The following values give an idea of its importance: at the normal plasma pH (7.4), the $HPO_4^{2-}/H_2PO_4^-$ ratio is 4:1. Considering an excretion of 50 mmol phosphate in the filtrate, 40 exist as HPO_4^{2-} and 10 as $H_2PO_4^-$, if the pH of the tubular fluid drops to 6.8, the ratio $HPO_4^{2-}/H_2PO_4^-$ will be 1:1, and at pH 4.8, virtually all will remain as $H_2PO_4^-$. That is, the phosphate buffer may excrete approximately 40 mEq. H^+.

Carbonic anhydrase in the tubular cells is essential in the renal acidification of the urine, it generates HCO_3^- and H^+ ions. For each H^+ excreted, one HCO_3^- returns to the blood (Fig. 28.7).

The magnitude of the acidification produced by the kidney is determined by

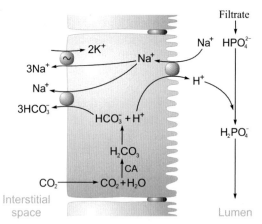

FIGURE 28.7 **Schematic representation of urine acidification in kidney tubules.** *CA*, Carbonic anhydrase.

titration of the urine with an alkali to bring the pH to the plasma value. The amount of alkali used in the tritation corresponds to the urine *titratable acidity*, and it is useful to assess the renal contribution to acid–base balance. Titratable acidity is primarily dependent on H^+ buffering by HPO_4^{2-} and, to a much lesser extent, by other buffers as creatinine ($pK_a = 4.97$) and uric acid ($pK_a = 5.75$). Significant H^+ excretion occurs in diabetic ketoacidosis, in which significant amounts of β-hydroxybutyric acid are excreted. This acid acts as urinary buffer. At pH 4.8, half of the β-hydroxybutyric acid binds protons and is nonionized.

3. *Production of ammonia.* The ability to excrete H^+ with phosphates is limited. To counteract acid overload, kidney still has another resource: secretion of ammonia.

The genesis of ammonia in the renal tubule is mainly from glutamine by the action of glutaminase. This enzyme is abundant in the renal proximal and distal tubules, and in collecting ducts. The amide bond of glutamine is hydrolyzed and glutamate and ammonia are produced. Ammonia is also formed by deamination of amino acids, especially glutamate, in a reaction catalyzed by glutamate dehydrogenase. Thus, one molecule of glutamine produces two molecules of ammonia, glutaminase first hydrolyzes glutamine to glutamate and this is converted to α-ketoglutarate by glutamate dehydrogenase. Metabolization of α-ketoglutarate in the citric acid cycle produces two CO_2 molecules.

The NH_3, polar small molecule, diffuses freely across membranes and can pass to the tubular lumen, where it forms ammonium by binding H^+ secreted through the antiporter Na^+/H^+. The membranes are impermeable to NH_4^+ and this ion cannot return to the cell.

Only part of the NH_3 generated in the proximal tubules moves as such to the lumen, a major proportion is converted to ammonium ion within the cell. The proton is originated by ionization

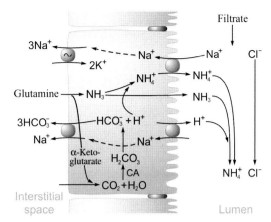

FIGURE 28.8 **Schematic representation of ammonium formation in renal proximal tubules.** *CA*, Carbonic anhydrase.

of the H_2CO_3 formed by action of carbonic anhydrase. The remaining HCO_3^- enters the circulation by the $Na^+/3HCO_3^-$ cotransporter in the basolateral membrane. The ammonium ion is excreted into the lumen by the Na^+/H^+ exchanger in the apical membrane, which also functions as Na^+/NH_4^+ antiporter (Fig. 28.8).

Excretion of H^+ as ammonia is very important in the regulation of renal acid–base, since the rate of production of NH_4^- can be adjusted to the body needs. The sum of titratable acidity plus ammonium released into the lumen, less bicarbonate in urine, represents the renal net acid excretion, a good index of the kidney total contribution to the process of $[H^+]$ excretion.

The renal compensatory mechanisms are slower than the pulmonary system; they attain their maximum capacity after 2 days. However, the kidney is the most efficient regulatory organ.

REGULATION OF INTRACELLULAR pH

While pH remains very constant in the EC fluid, the IC space shows different pH values depending on the cell type. For example, in skeletal muscle pH varies between 6.9 and 7.2, in brain it is 7.1, in liver 7.2. Even more significant are the differences between the various organelles of the same cell.

The viability and functional status of cells are highly dependent on their pH. The vast majority of cells regulate the cytoplasmic H^+ concentration primarily through proton transport systems present at their plasma membranes. The most important are the following:

Na$^+$/H$^+$ exchanger (NHE). In epithelial cells, this ion transporter is located both in the apical and basolateral membranes. Using the chemical energy from the transmembrane Na^+ gradient created by the sodium pump, NHE performs the electroneutral exchange of Na^+ for H^+ in a 1:1 stoichiometry.

HCO_3^-/Cl$^-$ exchanger independent of Na$^+$. This electroneutral antiporter, exchanges HCO_3^- for Cl^- in a 1:1 stoichiometry. Its function depends on the Cl^- and HCO_3^- gradients, which dictate the direction of flow in the forward or reverse modes. The best studied exchanger of this type is that of erythrocytes, identified as the *band 3* protein, which plays an important role in red blood cell CO_2 transport and plasma pH control.

HCO_3^-/Cl$^-$ exchanger dependent on Na$^+$. Present in many mammalian cells. It introduces HCO_3^- and Na^+ into the cell and releases H^+ and Cl^- in an electroneutral fashion and with a stoichiometry of 1:1:1:1. It is important in the normalization of intracellular pH after an acid overload. Its operation depends on the Na^+ gradient created by the Na^+,K^+-ATPase.

Na$^+$/HCO$_3^-$ exchanger. Present in the basolateral membrane of kidney proximal tubule cells, it exchanges three HCO_3^- per each Na^+ and it is important for bicarbonate reabsorption.

Proton ATPases. These are pumps that directly couple ATP hydrolysis to the transport of H^+. Some are electroneutral and others are electrogenic. Two proton transporting ATPases have been described. The H^+,K^+-ATPase is expressed in the apical membrane of the parietal cells of the gastric mucosa. It performs the electroneutral exchange of two H^+ ions into the gastric lumen (contributing to acidify it) for every two K^+ ions that are transported into the cell. The H^+,K^+-ATPase is also expressed in the intestinal and renal epithelium.

Another proton ATPase is the H^+-ATPase located in the apical membrane of renal tubular cells of the distal nephron segments. This transporter and H^+,K^+-ATPase contribute to acidify the urine. In subcellular organelles (endosomes, lysosomes, mitochondria, and Golgi) there are electrogenic pumps classified as V type ATPases. They acidify the lumen of lysosomes, endosomes, and Golgi. Proton pumping coupled to electron transport are important events in the alkalization of the mitochondrial matrix.

ACID–BASE BALANCE DISORDERS

Alterations in body acid–base balance are reflected in blood, so, these clinical disorders are defined according to the pH deviations observed in plasma. *Acidemia* occurs when the blood pH is lower than 7.35 ([H^+] = 45 nM). Conditions causing acidemia are designated with the term *acidosis*. *Alkalemia* consists in the increase in blood pH above 7.45 ([H^+] = 35 nM) and is produced by conditions that lead to *alkalosis*.

Clinically, the blood bicarbonate/carbonic acid buffer is a valuable index to assess the acid–base status of an individual. Logically, pH disorders equally affect all blood buffer systems, but the HCO_3^-/CO_2 system is the one that can be more easily determined. Moreover, body pH regulatory mechanisms directly impact on this buffer pair; the respiratory system controls body [CO_2], while kidneys regulate [HCO_3^-].

Depending on the primary change in the components of the bicarbonate/carbonic acid buffer, disorders of acid–base balance can be distinguished into four main groups: (1) respiratory acidosis, (2) metabolic acidosis, (3) respiratory alkalosis, and (4) metabolic alkalosis.

1. *Respiratory acidosis*. Reduction of CO_2 removal by the lung results in increase of the partial pressure of CO_2 P_{CO_2}, this results in a raise of carbonic acid concentration and a decrease in blood pH. Without adjustment by the body, this condition is called *noncompensated acidosis*.

   This situation is produced by diseases where pulmonary ventilation is compromised, which reduces gas exchange; for example, (1) hypoventilation by depression of the respiratory center (by drugs or other agents), (2) respiratory failure due to pulmonary diseases (emphysema, bronchial asthma, pneumonia, or other processes that disturb ventilation of large areas of the lung), and (3) circulatory deficiencies that decrease blood flow to the lung and slow lung gas exchange.

   To compensate for the imbalance, the kidneys will increase bicarbonate reabsorption and the urinary secretion of acid (H^+) and ammonia (enhancing urine acidity and net urine acid excretion). These effects increase bicarbonate concentration in plasma, tending to restore the value of the $[HCO_3^-]/[CO_2]$ ratio to normal levels (20 and a pH = 7.4). This situation is known as *compensated acidosis*.

2. *Metabolic acidosis*. This condition causes a primary reduction in plasma bicarbonate and it is produced by excessive loss of bicarbonate or the increased production, retention, or intake of acids. The reduction of $[HCO_3^-]$ results in reduced blood pH (*noncompensated acidosis*).

   This type of disturbance occurs in uncontrolled diabetes, starvation, and other situations in which there is enhanced production of ketone bodies (acetoacetate,

β-hydroxybutyrate); in severe diarrhea in which there is loss of digestive juices rich in bicarbonate; or in kidney failure in which tubular H^+ secretion mechanisms are disturbed.

   The immediate compensatory response in this type of alterations is the stimulation of the respiratory center to increase pulmonary ventilation, which tends to decrease P_{CO_2} and $[CO_2]$. If renal function is normal, the kidneys will respond with increased excretion of acid and production of ammonia (which increases urine acidity and net acid excretion). This restoration of body pH is known as *compensated acidosis*.

3. *Respiratory alkalosis*. This is a condition in which the primary alteration is the decrease in P_{CO_2}. This occurs during pulmonary hyperventilation do to: (1) abnormal stimulation of the respiratory center by diseases, such as encephalitis, (2) pronounced hypoxia, (3) salicylate intoxication, and (4) patients with anxiety and hysteria. The decrease in $[H_2CO_3]$ produces a raise in pH (*uncompensated alkalosis*). Compensation includes increased bicarbonate and reduced ammonia excretion by the kidneys (decreased titratable acidity and lower net acid excretion).

4. *Metabolic alkalosis*. This state is produced by a primary increase in bicarbonate, which elevates blood pH (*uncompensated alkalosis*). It can result from excessive intake of alkali ($NaHCO_3$) or by excessive removal of acids from the body. The most common example for this condition is continuous vomiting, caused by many different pathologies in which gastric juice, rich in hydrochloric acid, is lost.

Compensatory mechanisms for this condition include depression of ventilation, which decreases CO_2 elimination and the reduction of bicarbonate reabsorption and ammonia formation (net acid excretion decrease) by the kidneys.

If the blood pH is restored to normal, the condition is known as *compensated alkalosis*.

H⁺ EXCRETION REGULATION

All types of acidosis decrease the pH in the kidney tubular cells and increase the [H⁺] gradient in the apical membrane. This facilitates proton secretion to the tubular lumen. In alkalosis, the H⁺ gradient is reduced and H⁺ secretion is depressed. The increase in [H⁺] or P_{CO_2} in the cytoplasm have a double effect: (1) they enhance the H⁺ gradient that drives proton transport, and (2) they stimulate the activity and promote the expression and insertion of additional transporters in the apical membrane of renal tubular cells, involved in H⁺ and HCO_3^- transport. Thus, number of Na⁺/H⁺ exchangers in the proximal tubules luminal membrane, and H⁺-ATPase pumps in collecting ducts, increases in acidosis. Proton secretion into the tubular fluid tends to hyperpolarize the apical membrane of the tubular epithelial cells, which reduces the performance of the H⁺ and HCO_3^- carriers. Another limiting factor is the pH gradient in the collecting tubules, which affects H⁺ secretion when the urine reaches a pH of 4.5. The buffers present in the tubular fluid diminish this effect by capturing the secreted protons, to maintain tubular [H⁺] low which favors the H⁺ gradient driving proton secretion.

The concentration of bicarbonate in the glomerular filtrate depends on plasma bicarbonate levels. Factors that alter tubular fluid [HCO_3^-] modify proton secretion and thus, bicarbonate resorption or excretion.

In general, factors that regulate ECF volume and sodium balance (renin-angiotensin-aldosterone system) also affect the excretion of H⁺. For example, angiotensin II augments the activity of the Na⁺/H⁺ in the proximal tubules, which increases H⁺ secretion. Aldosterone stimulates the ATPases that excrete protons in the collecting ducts.

K⁺ balance can modify the movement of H⁺ and NH_4^+ in the proximal tubules. Secretion of H⁺ and NH_4^+ is reduced in hyperkalemia and increased in hypokalemia.

NH_4^+ production is stimulated when the cell pH is reduced. These depends on stimulation of the enzymes involved in glutamine metabolism. In alkalosis, NH_4^+ production is inhibited.

LABORATORY STUDIES IN ACID–BASE DISORDERS

Determination of plasma pH allows the diagnosis of uncompensated acidosis or alkalosis; however, it does not provide information regarding the primary cause of the disorder. Furthermore, in compensated situations, the pH is normal and this complicates the evaluation of the acid–base disorder.

Acid–base alterations are characterized by changes in the absolute values of HCO_3^- and CO_2, highlighting the clinical relevance of the HCO_3^-/CO_2 buffer system.

According to the Henderson–Hasselbalch equation:

$$pH = 6.1 + \log[HCO_3^-]/[CO_2] \qquad (28.1)$$

Two variables (pH and [CO_2]) out of the three present in the equation can be measured directly. The pH and P_{CO_2} are determined by potentiometric methods. The [CO_2] in millimoles per liter is obtained from the product of P_{CO_2} (in mmHg) and the solubility coefficient of CO_2 in plasma ($\alpha = 0.0301$ mM/L/mmHg):

$$[CO_2] = \alpha \cdot P_{CO_2} = 0.0301 \cdot P_{CO_2} \qquad (28.2)$$

Knowing two variables in the equation allows the third one to be calculated, [HCO_3^-] in this case:

$$[HCO_3^-] = (0.0301 \cdot P_{CO_2}) \cdot \text{antilog} \, (pH - 6.1)$$

There are volumetric methods, at present rarely used, to measure plasma total CO_2 amount.

Total CO_2 represents the sum of the physically dissolved CO_2 and bicarbonate:

$$Total\ CO_2 = [HCO_3^-] + [CO_2]$$

from where:

$$[HCO_3^-] = Total\ CO_2 - [CO_2]$$

substituting the value of $[CO_2]$ (Eq. 28.2):

$$[HCO_3^-] = Total\ CO_2 \cdot (0.0301\ P_{CO_2})$$

and substituting in Eq. (28.1):

$$pH = 6.1 + \log \frac{CO_2\ total - (0.0301 \cdot P_{CO_2})}{(0.0301 \cdot P_{CO_2})}$$

Nomograms are available that allow obtaining the value of the remaining variable when two of them are known. It should be noted that in clinics the cases are frequently complex. Compensatory mechanisms and alterations added to the primary cause of the pH disturbance, sometimes complicate the interpretation of the pathological condition.

Another parameter used to evaluate pH disturbances is the evaluation of the metabolic component ($[HCO_3^-]$), which is referred to as the *base excess*. This is calculated from the values of pH, P_{CO_2}, and hemoglobin concentration using an equation developed by Siggaard–Andersen. Nomograms have been developed which provide the value of the base excess directly from the pH, P_{CO_2}, and hemoglobin values.

Under normal conditions (pH 7.4, P_{CO_2} 40 mmHg, Hb 15 g/dL, 37°C), the value for the base excess is 0. The increase in bicarbonate gives a positive base excess, while its reduction gives a "negative base excess," which actually means a base deficit. In metabolic acidosis, there is a negative base excess, and in metabolic alkalosis the base excess is positive. In respiratory disorders, the base excess is initially 0.

Knowing the value of the base excess is useful to help guiding the treatment of patients with acid–base disorders. It should be noted that due to pH compensatory processes, serial measurements are needed to better estimate the base excess of the patient during the disease.

In addition to pH and total CO_2, diagnosis of pH alterations requires the determination of plasma anion and cation concentrations. Usually Na^+, K^+, and Cl^- are measured. The remaining anions (sulfates, phosphates, proteins) and cations (Ca^{2+}, Mg^{2+}) are not measured in routine studies, but are estimated indirectly. As the body fluids always maintain electrical neutrality, the sum of positive and negative charges in plasma is equal ([cation] = [anion]), even if the concentration of one of them can vary.

The Na^+ and K^+ in plasma represent some 95% of all cations, while Cl^- and HCO_3^- comprise 85% of anions. The remaining negative charges, known as *plasma anion gap* is approximately 15% of the total and is represented mainly by phosphates, sulfates, and proteins. The concentration of these anions can be estimated by the following equation:

$$Plasma\ anion\ gap = ([Na^+] + [K^+]) - ([Cl^-] + [HCO_3^-])$$

The normal value for the anion gap is between 10 and 15 mEq./L. Determination of the plasma anion gap is useful in the determination of the cause of metabolic acidosis. Its level is increased in metabolic acidosis (i.e., lactic acidosis).

SUMMARY

Water represents a significant amount of the total body weight (~65% in an adult man, ~60% in an adult woman, and ~77% in infants). Water is distributed in the *intracellular* (IC) and an *extracellular* (EC) compartments. The ECF is called *internal environment*. In adult men, 45% of the water is in the ICF and 15% in the ECF. This last includes the intravascular space (10%), and the interstitial fluid (5%). Daily water ingestion is approximately 2500 mL/day (1400 mL represented by beverages, 800 mL by solid foods, and 300 mL generated through the reactions of body metabolism). Daily water eliminated include 1500 mL excreted with the urine, 150 mL with the feces, and 850 mL lost through skin and lungs.

The main ECF cation is Na^+, the principal cation in the ECF is K^+ in the ICF. Among the anions, Cl^- predominates in the ECF and phosphates, sulfates, and proteins in the ICF.

Gibbs–Donnan equilibrium. The presence of proteins (which behave as anions at the pH of biological fluids) in the cell and in spaces surrounded by semipermeable membranes determines an unequal distribution of diffusible ions. This phenomenon, called the Gibbs–Donnan equilibrium causes the following effects:

1. The total concentration of anions is equal to that of cations in each side of the membrane.
2. The concentration of diffusible anions is lower and that of cations is higher in the space containing the proteins than in the space devoid of proteins.
3. The osmotic pressure in the compartment with proteins is slightly higher than in the one lacking proteins.

Body fluid osmolarity depends on the osmotically active solutes present in those fluids. The osmolarity of plasma is approximately 300 mOsm/L, corresponding mainly to low molecular weight electrolytes. Na^+ and K^+ play an important role in this respect and their movement across membranes is conditioned by ion transport systems, such as the Na^+,K^+-ATPase. Regulation of ECF osmolarity and volume mainly depends on renal activity and its regulation by factors, such as the *ADH*, *aldosterone*, and *ANP*. Also the mechanism of thirst influences ECF water balance.

Dehydration and *hyperhydration* are alterations in body fluid volume. They can present without changes in osmolarity (*isotonic* with respect to normal plasma), or can be accompanied by increased or reduced osmolarity (*hypotonic* or *hypertonic*, respectively).

Oncotic pressure is the osmotic pressure driven by the proteins present in body fluids and is of primary importance for liquid exchange between body compartments. The difference in oncotic pressure between blood and the interstitial space and its balance with the hydrostatic pressures in the capillaries creates the gradient that helps maintain fluid exchange in the interstitial space.

Acid–base balance is essential to cope with the continuous changes in pH due to intake and production of acids and bases in the body.

Buffer systems in control of body pH. In the intracellular space proteins ($Prot^-/ProtH$) and phosphates ($HPO_4^{2-}/H_2PO_4^-$) function as buffer systems. In the intravascular space, blood *proteins*, in particularly *Hb*, has buffer capacity (imidazole of histidines can bind or donate H^+, as its pK_a is near the pH of biological fluids). The most important buffer system in the ECF is the *bicarbonate/carbonic acid* pair. CO_2, produced by cellular activity, dissolves in body fluids and is hydrated in a reaction catalyzed by *carbonic anhydrase* to give carbonic acid, which rapidly ionizes into bicarbonate and H^+. The H_2CO_3 concentration depends on the amount of CO_2 dissolved and P_{CO_2}. At normal pH (7.4), the ratio $[HCO_3^-]/[CO_2]$ value is 20. Most of the CO_2 (~90%) is transported in blood as bicarbonate, 5% as carbamino hemoglobin, and the rest is physically dissolved. Whole blood, due to the presence of Hb, has greater capacity to transport CO_2 than plasma alone. HbO_2 is a relatively stronger acid than Hb. This causes ionic exchange to be associated with the transport of gases (*Zuntz–Hamburger phenomenon*).

Bicarbonate/carbonic acid regulation is achieved by close control of the HCO_3^-/CO_2 system. This takes place at two main levels:

Respiratory regulation depends on the effects of blood pH and P_{CO_2} changes on the CNS respiratory center. A decrease in pH or increase in P_{CO_2} stimulates the frequency and depth of breathing, augmenting CO_2 removal. P_{CO_2} reduction or pH increase depress pulmonary ventilation and promotes CO_2 accumulation in blood. This mechanisms allow keeping the $[HCO_3^-]/[CO_2]$ ratio = 20 and body pH close to normal. Respiratory compensatory response is rapid (within 12–24 h).

Renal regulation controls body pH via three main mechanisms: bicarbonate reabsorption, urine acidification, and production of ammonium ions. Urine titratable acidity plus the amount of ammonium, less than that of bicarbonate, gives an estimate of the *net acid excretion* by the kidney, which is an index of the kidney's contribution to $[H^+]$ regulation when there is an acid excess. Changes in pH in the renal tubular cells modify the activity and expression of the mechanisms involved in H^+ and HCO_3^- transport. Increased H^+ secretion enhances HCO_3^- reabsorption and acid excretion. Renal compensation of pH is slower than respiratory regulation; it reaches its maximum capacity after 2–4 days.

Disorders of acid–base balance are characterized by modifications of the HCO_3^-/CO_2 system. Depending on the alteration, four types of acid–base alterations can be distinguished:

Respiratory acidosis consists of a decrease in pH originated by alterations in respiration, which reduce CO_2 removal and increase the partial pressure of P_{CO_2}.

Metabolic acidosis is the drop in pH caused by a primary reduction in plasma bicarbonate and it is produced by excessive loss of bicarbonate or the increased production, retention, or intake of acids.

Respiratory alkalosis is the condition characterized by the increase in pH in which the primary alteration is the decrease in P_{CO_2}, due to situations that produce pulmonary hyperventilation.

Metabolic alkalosis is the state caused by a primary increase in bicarbonate, which elevates blood pH, due to excessive intake of alkali or the excessive removal of acids from the body.

Respiratory and metabolic acidosis and alkalosis may be compensated or uncompensated. Nomograms have been developed, which allow characterizing the pH disbalance through measurement of blood pH and P_{CO_2} values. If Hb concentration is determined, one can calculate the *base excess*. *Anion gap* is important in the differential diagnosis of acidosis and it is obtained from the algebraic sum $([Na^+]+[K^+])-([Cl^-]+[HCO_3^-])$. Its normal value ranges from 10 to 15 mEq./L.

Bibliography

Arieff, A.I., De Frunzo, R.A., 1995. Fluid, Electrolyte and Acid–Base Disorders, second ed. Churchill Livingston, New York, NY.

Mota Hernández, F., Velázques Jones, L., 2004. Agua y Electrólitos. McGraw Hill Interamericana, México.

Rose, B.D., 2000. Clinical Physiology of Acid–Base and Electrolyte Disorders, fifth ed. McGraw Hill, Inc., New York, NY.

Essential Minerals

MACROMINERALS

Calcium, phosphorus, potassium, sodium, chloride, magnesium, and iron are, overall, only 3.2%–3.3% of a normal adult body weight. However, despite these relatively low levels, these minerals play essential physiological roles.

Sodium (Na)

A normal adult has 60 mEq. of sodium per kilogram of body weight, a total amount of approximately 4200 mEq. (~100 g); 55% of the sodium is in extracellular fluids (ECFs), 40% is in bone, and the rest is found in the intracellular compartment. The body contains two types of sodium deposits, one that can be easily exchanged between body compartments and another that is not readily available for exchange. The first corresponds to the sodium present in the intra- and extracellular spaces and almost half of that contained in bone. The second deposit is less dynamic and is confined to the bone matrix.

Sodium and chloride are the major electrolytes of the extracellular compartment and the main determinants of the water volume in that space. Sodium is maintained at relatively high and potassium at relatively low concentrations in the extracellular space compared to the intracellular space, due to the activity of the Na^+,K^+-ATPase.

Natural sources. Sodium concentration in biological media tends to parallel that of chloride. Milk, meat, eggs, and vegetables provide only 10% of the sodium consumed daily. The salt added to food is the main source. The 43% of NaCl mass corresponds to sodium.

Sodium intake varies between 100 and 200 mEq./day (6–12 g of NaCl); to maintain the balance, a similar amount is eliminated. Sodium intake is influenced by salt appetite and it is regulated by the renin–angiotensin system; the mechanism of thirst also plays a significant role in sodium uptake and dilution.

Absorption. Practically all of the ingested sodium is absorbed in the intestine. Less than 5% is excreted with the feces. Sodium enters the enterocytes by several different mechanisms driven by the Na^+ gradient created by the Na^+,K^+-ATPase. A small portion can use the paracellular pathway, through the tight junctions between enterocytes. From the brush border, Na^+ passes to the basolateral membrane, where the Na^+,K^+-ATPase transports it into the interstitial space and from there it reaches the blood.

Plasma sodium concentration is normally maintained between 135 and 145 mEq./L (average 320 mg/dL); in interstitial fluid it is slightly lower due to the Gibbs–Donnan equilibrium.

In the intracellular space the Na^+ level is much lower, with an average of 10 mEq./L. This great difference in concentrations on either side of the plasma membrane is maintained by the activity of the Na^+,K^+-ATPase. Sodium is responsible for half of the osmotic pressure in the ECF compartment.

Excretion. Na^+ is excreted mainly by urine (more than 90% of total). Na^+ is also eliminated through skin with sweat, which has a concentration of approximately 50 mEq./L. The liquid evaporated by insensible perspiration is practically free of Na^+. Only when the temperature is high, under strong physical activity or fever, Na^+ and water loss via sweating becomes important.

Excretion of Na^+ also takes place in the intestine. The amount of Na^+ lost via this route is very small (not more than 2 mEq./day), but increases markedly in cases of severe diarrhea or via catheters placed in the digestive tract.

Renal handling of sodium. Regulation of Na^+ excretion by the kidney is essential in maintaining body salt and water balance. In normal adults, the renal glomeruli filter approximately 27,000 mmol of Na^+ per day. More than 99% is reabsorbed back by the renal tubules; only 150 mmol are excreted with the urine. Excretion of Na^+ in urine is adjusted to the intake to maintain Na^+ balance. If the amount of sodium in foods is completely suppressed, the Na^+ excreted in urine falls to undetectable levels after 24–48 h. Most of the sodium filtered in glomeruli is reabsorbed in proximal convoluted tubules (~60%–70%), the rest in the loop of Henle (~25%) and distal and collecting tubules (~5%–10%). Sodium reabsorption in the tubules is controlled by several hormones:

Angiotensin II regulates Na^+ reabsorption in the proximal convoluted tubule. It also promotes Na^+ and water absorption in proximal intestine. This effect contributes to increase plasma volume and blood pressure. Other effects of angiotensin II are secondary to its ability to activate aldosterone synthesis and secretion.

Aldosterone is responsible for 5%–10% of total Na^+ reabsorption. It acts on the distal nephron (principal cells of the collecting duct), where it stimulates the expression of new Na^+ channel molecules and increases the activity of preexisting Na^+ channel units in the apical membrane of the cells. It also stimulates the Na^+,K^+-ATPase in the basolateral membrane of the cells, which in addition to enhance Na^+ reabsorption, promotes K^+ secretion to the tubular fluid. Due to the stimulatory action of the proton ATPase, aldosterone favors the renal excretion of H^+. Aldosterone also promotes Na^+ absorption in the colon.

Atrial natriuretic peptide (ANP) stimulates Na^+ excretion when there is an expansion of the ECF. ANP increases glomerular filtration pressure and reduces Na^+ reabsorption in the renal collecting duct.

Na^+ daily requirements. An intake of 500 mg/day is the minimum amount of Na^+ recommended. Ingestion of larger amounts is not advisable; in patients with hypertension and cardiac failure, Na^+ from the diet must be reduced to diminish the effects on the volume of the ECF.

Alterations of Na^+ homeostasis. Failure of the mechanisms that control Na^+ balance results in an alteration of body salt and fluid volume homeostasis. Total Na^+ content in the body can vary due to changes in either its intake or output. There are no effective mechanisms to adjust sodium intake in response to changes in the vascular volume and Na^+ regulation is mainly achieved by changes in salt handling by the kidney.

Disorders resulting in sodium accumulation in the body tend to increase the interstitial space fluid volume, with production of edema. This situation occurs in congestive heart failure, hypoalbuminemia, renal diseases that affect sodium excretion, and hyperaldosteronism.

On the other hand, Na^+ depletion occurs in severe diarrhea, fistulas or drainage of intestinal fluids, excessive sweating without salt replacement, extensive burns that cause fluid and electrolytes loss through the skin, inability to reabsorb sodium by kidney tubules, or hypoaldosteronism.

The plasma Na^+ concentration does not always reflect changes in total body Na^+ content. Sodium retention may not be manifested by hypernatremia. Hypernatremia occurs when there is relative excess of sodium with respect to water, or when there is proportionately greater loss of water than sodium. Hypernatremia can appear even when there is a sodium deficit.

Hyponatremia occurs when sodium loss exceeds water loss, or when water is in relative excess compared to sodium. This leads to hyponatremia due to fluid dilution, which can occur even if the total body content of Na^+ is normal or increased. Its symptoms include cramps, nausea, vomit, dizziness, shock, and even coma.

Given the amount of sodium added to foods, deficiency of dietary Na^+ is extremely rare. It can result from profuse sweating (of 3% or more of body weight).

Sodium and hypertension. Increased Na^+, whatever is the cause that originates it, results in increase in effective vascular volume and blood pressure elevation.

Salt restriction is effective in lowering blood pressure. A reduction of 4 g daily in salt intake causes a drop from 5 to 3 mmHg in systolic and diastolic pressures in hypertensive patients, and from 2 to 1 mmHg in normotensive patients.

Potassium (K)

A normal adult has approximately 50 mEq. of potassium per kilogram of body weight, a total of 3500 mEq. or 136 g. A 98% of the K^+ is located in the intracellular space, where it represents the most abundant cation. Its concentration in this compartment is ~150 mEq./L. In blood plasma and ECF its concentration ranges from 4.0 to 5.5 mEq./L. The wide difference for K^+ across the plasma membrane is maintained by the activity of the Na^+,K^+-ATPase, which introduces K^+ from the extracellular space into the cell.

Potassium has two main functions: (1) it plays an important role in cell metabolism, for example, it is essential for protein and glycogen synthesis. (2) The relationship between the K^+ extracellular and intracellular concentrations is the major determinant of plasma membrane resting potential, required for the generation of action potentials, which are critical for nerve and muscle function.

Natural sources. Potassium is widely distributed in foods, especially fruits, plants, and meats. Therefore, it is virtually impossible to find potassium deficits through the diet. Fruits and vegetables, especially bananas, citrus, apricots, melons, tomatoes, potatoes have a high K^+ content. Polished rice and wheat flour are poor sources of potassium. In general, meat and fish have a higher K^+ content than nuts and cereals.

The daily intake in a mixed diet is approximately 4 g (102 mEq.). This amount is similar to all the potassium present in the ECF.

Absorption. Over 90% of ingested potassium is absorbed in the small intestine and colon. The mechanisms involved in K^+ uptake are not known as well as those for Na^+. Once in the intestinal cells, K^+ diffuses into the interstitial space and blood through potassium channels.

Excretion. The kidney is the main organ responsible for K^+ excretion and balance modulation. A total of 90%–95% of the ingested K^+ is eliminated by the kidneys in a tightly regulated manner. The rest is excreted through the feces and sweat, which is not subjected to regulation. The loss of K^+ by the stool is more or less constant, usually less than 10% of the intake, but can be very large in patients with severe diarrhea. There is also significant loss in cases of severe vomiting.

Renal handling of potassium. Under a standard diet, approximately 15% of the total K^+ filtered in the renal glomeruli is excreted in the urine. This shows that K^+ is highly reabsorbed in the tubules. If the diet contains very little or no K^+, less K^+ appears in urine; however, there is always some K^+ present in urine (at least 1% of the amount filtered). Therefore, an individual with a diet low in K^+ could develop hypokalemia. The amount of potassium excreted in urine can be greater than that filtered, which indicates that K^+ is also secreted by the kidneys. Plasma potassium concentration (or kalemia) and aldosterone are the main factors that regulate K^+ secretion, ADH has a lesser effect.

Increased extracellular [K^+] (hyperkalemia) stimulates aldosterone production and secretion to the bloodstream, which activates K^+ secretion in the renal collecting ducts. The hormone increases the number and activity of Na^+,K^+-ATPase molecules in the basolateral membrane of the renal tubular cells, increasing the intracellular K^+. The passage of K^+ to the tubular lumen is favored by the action of aldosterone in stimulating the activity of K^+ channels in the apical membrane of renal tubular cells.

Regulation of K^+ intracellular and extracellular concentrations. After its intake, the changes in K^+ in blood are quickly compensated by its movement to the cells. The total amount of K^+ in ICF is so large compared to that in ECF, that the cation displacement from the IC compartment can correct changes in plasma [K^+] with insignificant modification of the intracellular concentration.

The uptake of K^+ by cells is controlled by several hormones.

Insulin promotes K^+ entry, particularly in skeletal muscle and liver. It produces stimulation of Na^+,K^+-ATPase. This action depends on the transfer of previously synthesized Na^+,K^+-ATPase from intracellular stores to the cell plasma membrane. Also, insulin action is secondary to the activation of the Na^+/H^+ exchanger in the target cells; Na^+ entrance into the cells is an important activator of the activity of Na^+,K^+-ATPase.

Catecholamines exert different effects, depending on their action in β_2 or α adrenergic receptors. β_2 favor K^+ entry into cells by activating Na^+,K^+-ATPase activity. The mechanism involves the increase in cyclic AMP, activation of protein kinase A, and phosphorylation of Na^+,K^+-ATPase. By stimulating the adrenergic α receptors, epinephrine promotes K^+ efflux from cells. Catecholamines also have an indirect action; by stimulating glycogenolysis, they produce a rise in blood glucose and insulin release, which activates K^+ entry into cells through the mechanisms listed previously.

Aldosterone favors K^+ entry into the cells when there is hyperkalemia. It has little or no effect if plasma K^+ concentration is normal.

Alterations of K^+ homeostasis. Maintenance of K^+ homeostasis is critical since variations in this cation significantly influences heart, skeletal, and smooth muscle cell contractility, and affects nervous system excitability. It is also important in electrolyte and hydrogen ion balance.

In general, K^+ excess occurs when there is failure in its renal excretion, produced by hypoaldosteronism or renal failure. Also, increase in plasma K^+ is usually observed in metabolic acidosis (diabetic keto acidosis), in which there is K^+ leakage from the cells into the ECF. The increase of K^+ in blood enhances its filtration and secretion in the nephron, which augments urine K^+ output. In contrast to acidosis, alkalosis activates the movement of K^+ from the ECF to the cells.

Potassium deficiency is common during severe diarrhea.

Changes in plasma K^+ concentration do not solely occur due to changes in the total K^+ content of the body; they can be caused by K^+ shifts between intracellular and extracellular compartments. In diabetic keto acidosis, for example, hyperkalemia may appear even when total K^+ is decreased. Given the large concentration difference between IC and EC spaces, a relatively small shift of intracellular K^+ markedly increases serum K^+.

Hypokalemia causes alterations in ventricular repolarization, which is manifested by changes in heart rate and a characteristic trace in the electrocardiogram. Cell membrane potential tends to increase in hypokalemia and decrease in hyperkalemia.

Hyperkalemia is manifested by neuromuscular symptoms, such as muscle weakness, mental confusion, numbness, and even paralysis. Arrhythmias and electrocardiographic abnormality appear when K plasma concentration is higher than 5.5 mEq./L. Above 6 mEq./L levels, K^+ can cause cardiac arrest.

Recommended K^+ intake. It is almost impossible to produce hyperkalemia by administration of K^+ with the diet in a normal individual; this is due to the effective body control mechanisms for K^+. Nutritional hypokalemia is also rare, except when the ingestion of food is suppressed; the abundance of K^+ in foods ensures an adequate intake even with poor diets. In any case, ingestion of 3.5 g/day, which is easily provided by a well-balanced diet, is advised.

Potassium and high blood pressure. Numerous studies have shown that a diet rich in K^+ lowers blood pressure. Eating fruits and vegetables provides good amounts of potassium and is recommended in the prevention and control of hypertension. An increase of 40 mEq./day of K^+ in the diet reduces blood pressure more than a decrease of 60–80 mEq./day in sodium intake. Probably the mechanism of this action of K^+ on blood pressure is due to stimulation of diuresis.

Chloride (Cl^-)

The main anion in the ECF is chloride. A normal adult has approximately 30 mEq. Cl^- per kilogram of body weight (a total of 2.100 mEq. or 75 g). About 88% of Cl^- is in the ECF and the rest is in the ICF. The differences in electrical potential across the plasma membrane (cell interior electronegative with respect to the outside) restrict Cl^- entry to the cells.

Natural sources. Most of the Cl^- ingested with food is associated with sodium. Chloride is also found in eggs and meats. The average daily intake of Cl^- in an adult varies between 100 and 200 mEq.

Absorption. The intake of Cl^- parallels that of Na^+, mostly because it enters as NaCl. Chloride is almost entirely absorbed in the small intestine, coupled to Na^+, via the Na^+/Cl^- symporter. Cl^- enters the cell against its electrochemical gradient, driven by the energy derived from the Na^+ gradient and in exchange for HCO_3^- through the Cl^-/HCO_3^- exchanger. From the enterocytes, Cl^- is transferred to the ECF via Cl^- selective channels and the K^+/Cl^- symporter driven by the K^+ gradient. In addition, Cl^- crosses the intestinal epithelium via the paracellular pathway, between the cells.

In the gastric mucosa Cl^- is secreted into the lumen via specific channels and the activity of a H^+,K^+-ATPase.

Chloride in plasma. Plasma Cl^- concentration is 102 mEq./L (365 mg/dL). This value is dependent on the hydration state of the body. Cl^- content may be decreased and yet its plasma concentration be normal or even higher if there is an imbalance in chloride and water loss. Due to the Gibbs–Donnan equilibrium, and opposite to what occurs for Na^+ and K^+, Cl^- level in the interstitial fluid is higher than in plasma.

Excretion. Cl^- is excreted in urine, skin, and feces, with the kidney being the main organ for its elimination.

Renal handling of chloride. Under normal conditions, the renal tubules reabsorb nearly all the chloride present in the glomerular filtrate. Cl^- reabsorption is mostly associated with Na^+ throughout the nephron; the excretion rate is usually similar for both ions and is influenced by the same factors.

Chloride balance disorders. Variations in body chloride content follow those of sodium. Generally, Na^+ excesses or deficits are accompanied by similar changes in Cl^-. Metabolic alkalosis may be an exception, since in this disorder, Cl^- loss is greater than that of Na^+ (i.e., alkalosis caused by vomiting). Chloride deficiencies can cause seizures.

Calcium (Ca)

Calcium is the fifth element in order of abundance in the body, after O, C, H, and N and is the major divalent cation in the ECF. A normal 70-kg adult male has approximately 1.4 kg of calcium; a 60-kg woman, around 1.1 kg; and a newborn at term close to 25 g. Calcium (99%) is found in bone; 1% is distributed in the intravascular, interstitial, and intracellular fluids.

Calcium plays an essential physiological role in the body. Its salts are the minerals that constitute bone. In addition, Ca^{2+} dissolved in the body fluids is essential to many biochemical processes, including nerve excitability, muscle contraction, hormone secretion, regulation of enzymes, and blood clotting.

The Ca^{2+} level in plasma is maintained very constant at 10 mg/dL (2.5 mM or 5 mEq./L). Normal daily Ca^{2+} fluctuations in serum are less than 3%, even when calcium intake varies significantly. Calcium concentration remains unchanged throughout life; only higher values are observed in fetal cord blood (11.6 mg/dL or 2.9 mM).

Plasma calcium is found in three different forms: (1) as ion (Ca^{2+}), which represents approximately 50% of the total calcium. (2) Bound to proteins, primarily albumin (~40% of the total), which is in dynamic equilibrium with the plasma Ca^{2+}. (3) Nonionic (or diffusible), associated with anions, such as citrate or lactate.

Ionized calcium is the physiologically active form of Ca^{2+}; its level varies between 4.6 and 5.2 mg/dL (1.25 and 1.3 mM). The balance between ionic and protein-bound calcium is affected by changes in blood pH. When the pH in blood increases (alkalemia), Ca^{2+} decreases and the opposite occurs when plasma pH decreases. For every 0.1 unit of pH increase, calcium ion level is reduced by 0.4%.

Obviously, the fraction of Ca^{2+} bound to protein is related to the amount of protein present, therefore, in alterations that affect protein levels

(increased in dehydration; decreased in chronic renal disease, liver failure, and severe protein malnutrition) there are changes in the amount of protein-bound calcium. An albumin decrease of 1 g/dL in plasma, produces a 0.8-mg/dL decrease in plasma calcium.

Determinations of plasma total calcium do not provide direct information on the amount of free Ca^{2+}, which is the physiologically relevant form of calcium. Especially when there are changes in blood pH or proteinemia, a patient can be hyper- or hypocalcemic, with a total calcium (free plus bound) concentration within the normal range (9.5–10.5 mg/dL). Direct measurement of ionic calcium can be performed by electrometric methods.

The constancy of calcium concentration in ECF indicates the existence of a remarkably efficient homeostatic mechanism; parathyroid hormone, calcitriol and probably calcitonin, are factors involved in its regulation. Furthermore, bone is a vast reservoir of calcium, which stores it or releases it to the ECF as needed.

Natural sources. Milk and dairy products, especially cheese and yogurt, are the richest in calcium. A liter of cow's milk provides more than 1 g. Other good sources of calcium are salmon, sardines, clams, and oysters and among plants, broccoli, cauliflower, turnip, and cabbage. Meats, grains (beans), and nuts are low in calcium. The bioavailability of calcium in a standard diet is approximately 30% of the total calcium contained in the ingested food.

Intestinal absorption. Calcium intake depends on intestinal absorption. With a normal diet, an adult absorbs around 30% of the 800–900 mg of calcium that come in the diet per day.

Calcium absorption in the intestine is carried out by two different mechanisms: active and saturable transcellular transport, and paracellular passive diffusion. The main active site of absorption is the duodenum; paracellular diffusion occurs throughout the small intestine. Although absorption is more efficient in the duodenum and jejunum, the amount absorbed is higher in the ileum, where the intestinal content stays for a longer time.

When calcium in the intestinal lumen is high and the active transport system is saturated, an increasing proportion of Ca^{2+} is taken up by passive diffusion.

The transcellular transport of Ca^{2+} comprises its passage through the apical membrane via Ca^{2+} channels, following the electrochemical gradient. Once in the cell, calcium binds *calbindin*, a 9-kDa protein in mammals (Ca9k), which can attach two calcium ions per molecule, serving as an intracellular carrier for Ca^{2+}. Also by binding Ca^{2+} calbindin protects the cell from the effects of high free Ca^{2+}. In the basolateral membrane, Ca^{2+} is moved into the extracellular space, against its electrochemical gradient, by two active mechanisms, the Ca^{2+}-ATPase that uses adenosine triphosphate (ATP) and the Na^+/Ca^{2+} exchanger, which uses the electrochemical Na^+ gradient maintained by the Na^+,K^+-ATPase. The Na^+/Ca^{2+} exchanger countertransports one Ca^{2+} out of the cell for every three Na^+ that are brought in, in an electrogenic fashion.

The paracellular diffusion of Ca^{2+} takes place through tight junctions between mucosal cells. This is a passive process that is not saturable.

Ca^{2+} absorption is regulated according to body needs and it depends on the amounts of Ca^{2+} in the diet. A diet low in calcium increases the uptake efficiency. There are additional dietary factors that influence calcium influx. Lactose, other sugars, and alcohols derived from monosaccharides, such as xylitol, favor Ca^{2+} absorption. Phytate (myoinositol hexaphosphate) present in legumes, nuts, and grains binds Ca^{2+} and reduces its bioavailability. Oxalates found in spinach, beet, eggplant, strawberry, walnut, peanut, tea, and chocolate form insoluble, nonabsorbable calcium salts that are excreted with the feces. When free fatty acids are abundant in intestine, they form insoluble calcium soaps, which prevents Ca^{2+} uptake. This becomes important in cases of steatorrhea.

Calcemia. The Ca^{2+} absorbed reaches the plasma, where its concentration is maintained, with little variation, in 10 mg/dL. Calcium circulates in blood in different forms: (1) nondialyzable or bound to proteins (4.5 mg/dL), (2) as free ion, with high filterability (5.0 mg/dL), and nondissociable, combined with citrate, phosphate, and bicarbonate (0.5 mg/dL).

Excretion. Calcium is excreted in urine, feces and, to a lesser degree, in sweat. Total excretion in an adult is approximately 300 mg/day. A high protein intake contributes to increased calcium output in urine. During lactation, there is a considerable amount of calcium that is secreted through the mammary gland.

The calcium eliminated in feces (~100 mg/day) corresponds to that unabsorbed plus the Ca^{2+} that comes from the mucosal cells that are released in the intestine and from the digestive secretions. The loss of Ca^{2+} in sweat is approximately 25 mg/day. It is also eliminated through recycled skin cells, hair, and nails.

Renal handling of calcium. Most of the Ca^{2+} is excreted by the kidneys. A 55%–60% of the total calcium in plasma is diffusible and filters in the glomeruli. The remaining 40% is bound to proteins and cannot pass the glomerular membrane. Normally, almost 10 g of calcium per day (500 mEq. or 250 moles) reach the renal tubules; 98% of this amount is reabsorbed and only 2% is excreted in urine (~200 mg/day). Transcellular movement of Ca^{2+} through the apical membrane of the renal tubular cells is driven by the large electrochemical gradient (intracellular calcium concentration $[Ca^{2+}]_i$ is about 10,000 times lower than the concentration in the tubular fluid). In the cells, a 28-kDa calbindin (Ca28k), different from the intestinal Ca9k binds Ca^{2+}. The transport of Ca^{2+} from the renal epithelial cells to the interstitial space is carried out in the basolateral membrane by active processes: a Ca^{2+}-ATPase and the $3Na^+/Ca^{2+}$ exchanger, similar to that present in the intestine.

The distal tubules and collecting ducts receive approximately 12% of the total calcium filtered

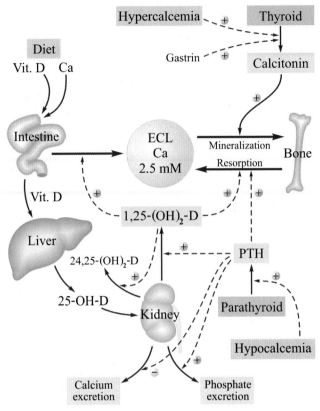

FIGURE 29.1   **Factors controlling calcium homeostasis.** *PTH,* Parathyroid hormone.

in the glomeruli and reabsorb over two-thirds of that amount. In these and all nephron sections, calcium is reabsorbed through the transcellular pathway. The mechanisms that regulate Ca^{2+} reabsorption and excretion take place in the distal nephron.

Calcium homeostasis. Calcium homeostasis is controlled by endocrine factors, mainly PTH and 1,25-$(OH)_2$-D_3 (Fig. 29.1).

Parathyroid hormone (PTH) is secreted when calcium concentration in the EC fluid drops below its normal limits. PTH exerts direct action on bone and kidney, and has indirect effect on intestine. The skeleton is the main reservoir for Ca^{2+} and this store is used when there is little or no exogenous calcium from the diet. PTH promotes rapid mobilization of calcium from bone

tissue by activation of osteolysis, with calcium release into the ECF. In the kidney, PTH stimulates Ca^{2+} reabsorption and decreases phosphate uptake (stimulates phosphaturia). In the intestine, PTH has no direct action. It induces 1α-hydroxylase in the kidney, which is involved in the synthesis of 1,25-$(OH)_2$-D_3. This active form of vitamin D stimulates calcium absorption in the intestine and increases calcium in the ECF. The overall effect of PTH is the increase in calcemia without elevation of phosphatemia.

Vitamin D exerts its main actions after conversion to 1,25-$(OH)_2$-D_3 or calcitriol. Vitamin D increases Ca^{2+} absorption in the intestine activating all the mechanisms for the transport of this cation. In bone, vitamin D metabolites stimulate both mineralization and resorption. The absorption of calcium and its release into the ECF requires the joint action of $1,25\text{-}(OH)_2\text{-}D_3$ and PTH.

Calcitonin. The role of this hormone is not entirely clarified in humans. When administered pharmacologically it inhibits bone resorption and decreases renal calcium and phosphate reabsorption. This reduces the level of calcium and inorganic phosphate in plasma.

Fig. 29.1 summarizes the factors involved in Ca^{2+} homeostasis.

Other hormones. Growth hormones, thyroid hormones, glucocorticoids, and androgens are also involved in bone formation, and therefore exert effects on calcium homeostasis. Estrogens inhibit bone resorption mediated by PTH and depress the secretion of cytokines (interleukins 1 and 6) (p. 773) and prostaglandin E_2, factors that favor Ca^{2+} resorption. The decline in estrogen production contributes to osteoporosis, frequently observed in postmenopausal women.

Calcium balance. The overall balance of calcium in the body is positive during periods of bone formation and growth. When development is completed, a balance is reached and, although there is continuous remodeling of bone, calcium entry and exit remain equal. After 45 years of age, Ca^{2+} balance begins to be negative, there is decrease in the mineral matrix of bone, at a rate of ~5% of the total every 10 years. In postmenopausal women, the rate of demineralization is faster than in men. Osteoporosis is a consequence of prolonged calcium loss.

Alterations in Ca^{2+} homeostasis.

Hypocalcemia. The reduction in plasma calcium may be due to various causes: (1) hypoparathyroidism (congenital or acquired due to resistance of the effector organs), (2) vitamin D deficiency (or alterations of its metabolism), (3) renal tubular defects (with decreased renal reabsorption), and (4) severe and prolonged dietary deficiencies. When the plasma level falls below normal, Ca^{2+} is mobilized from bone to normalize calcemia, which compromises bone formation. The fall of extracellular Ca^{2+} below 3.5 mg per 100 mL (~0.83 mM) causes neuromuscular

hyperexcitability and can lead to tetany (sustained muscle contraction).

Hypercalcemia occurs in hyperparathyroidism, vitamin D intoxication, sarcoidosis, and some cancers. High Ca^{2+} in the body affects the neuromuscular system, increasing the threshold for depolarization. In the heart, it has inotropic and chronotropic effect, with changes in the ECG, and it could even lead to cardiac arrest. The excess Ca^{2+} can deposit in different tissues, especially in kidney, leading to kidney stones and nephrocalcinosis. The increase in Ca^{2+} can also result in psychiatric alterations.

Regulation of intracellular calcium. Calcium, in addition to being an important component of bone and teeth, plays a variety of functions related to the nervous system activity, muscle contraction, cell motility, and it is involved in hormonal actions. While homeostatic mechanisms ensure the maintenance of calcium levels in the ECF, the multiple functions that Ca^{2+} exerts depend on rapid variations in its concentration, which require close monitoring and fast regulation. The calcium concentration in cytoplasm is approximately 0.1 μM, about 10,000 times lower than that in ECF. The large gradient across the plasma membrane, facilitates the production of sudden changes in the cation intracellular concentration. The cytoplasmic Ca^{2+} concentration can be increased 10 or more times in just milliseconds by rapid uptake from the extracellular space, or by release of the cation from intracellular stores, such as the endoplasmic (or sarcoplasmic) reticulum, mitochondria, and other organelles. This increase is the signal for a multitude of physiological actions.

Ca^{2+} entrance. Different stimuli acting on plasma membrane receptors promote Ca^{2+} entry and sharply raise its concentration in the cytosol. This raise in Ca^{2+} functions as a "messenger" of signaling systems used by hormones and other agents. Ca^{2+} entry from the ECF is driven by the electrochemical gradient that favors its uptake into the cell. Several Ca^{2+} channel types are involved in moving this cation into the cells: (1) *voltage-operated channels* (VOC), which are found

mainly in excitable cells, generate rapid Ca^{2+} influx and activate processes, such as muscle contraction and cell exocytosis; (2) *receptor-operated channels* (ROC), which respond to different agents coming from the extracellular medium (i.e., neurotransmitters) or from the cell cytosol (i.e., second messengers); and (3) *store-operated channels* (SOC), which are controlled by the Ca^{2+} levels in cell intracellular stores.

Ca^{2+} exit. To maintain the IC calcium concentration $[Ca^{2+}]_i$ constant, cytosolic Ca^{2+} entering the cell must be rapidly extruded. The main system responsible for this function is the cell plasma membrane Ca^{2+}-ATPase which, due to its high affinity for Ca^{2+}, becomes activated at small elevations in $[Ca^+]_i$. Also, the plasma membrane Na^+/Ca^+ exchanger contributes to Ca^{2+} extrusion from the cell.

Calcium movement from intracellular deposits. Ca^{2+} is stored by active processes into organelles from which it is released to the cytosol by ligand-activated channels.

Sarcoplasmic or endoplasmic reticulum (ER). A Ca^{2+}-ATPase operates (sarcoplasmic/endoplasmic Ca^{2+}-ATPase, or SERCA) that introduces Ca^{2+} in the ER lumen. Calcium release from the ER takes place through two channels that are both sensitive to $[Ca^{2+}]$. One, called *inositol 1,4,5-trisphosphate receptor* (I-1,4,5-P$_3$R), is activated by Ca^{2+} and by inositol 1,4,5-trisphosphate. The other, designated *ryanodine receptor* (RyR) opens in the presence of low concentrations of cytosolic Ca^{2+} and cyclic adenylribose (ADPc). RYR is inhibited by calmodulin and β adrenergic stimulation.

Mitochondria. These organelles import calcium from the cytosol through a uniporter driven by the negative potential of the mitochondrial matrix. The release of Ca^{2+} from mitochondria is mediated by the Na^+/Ca^{2+} exchanger. Mitochondria can minimize changes in $[Ca^{2+}]_i$. When there is a cytosolic raise in Ca^{2+}, the Ca^{2+} uniporter is activated and Ca^{2+} is accumulated in the matrix. When $[Ca^{2+}]$ in the cytosol diminishes, mitochondria calcium is released to maintain $[Ca^{2+}]_i$

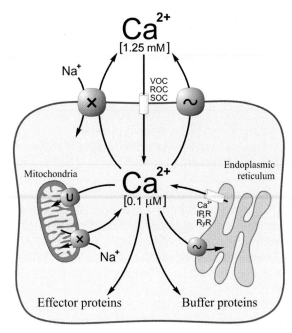

FIGURE 29.2 **Intracellular Ca^{2+} regulation.** *IP3R*, Inositol 1,4,5-trisphosphate receptor; *ROC*, receptor-operated channel; *RyR*, ryanodine receptor; *SOC*, store-operated channel; *U*, uniporter; *VOC*, voltage-operated channel; *X*, Na$^+$/Ca^{2+} exchanger; ~, Ca^{2+}-ATPase.

at basal levels. Fig. 29.2 schematically summarizes the buffering action that mitochondria exert on Ca^{2+}.

Golgi apparatus. A Ca^{2+}-ATPase imports Ca^{2+} and the I-1,4,5-P$_3$R-activated channel releases it from the Golgi cisternae.

Ca^{2+} binding to specific proteins. Approximately 200 proteins have been recognized that specifically bind Ca^{2+} in cells. These proteins can be distinguished into those that play a calcium buffer role, and those that function as effectors. While the Ca^{2+} buffer proteins regulate the calcium signal, Ca^{2+} effectors undergo conformational changes upon Ca^{2+} binding, which enable them to act on a substrate, sometimes with catalytic action. One of the most widely distributed Ca^{2+} effector proteins is *calmodulin*, which contains four Ca^{2+}-binding sites (p. 567). Fig. 29.2 shows the factors that regulate intracellular Ca^{2+}.

Ca²⁺ requirement. Ca^{2+} needs vary at different times of life. In the first 2 years, 120 mg of calcium per day are required to maintain normal growth. Breast-fed children require an intake of 300 mg/day, which is the Ca^{2+} content in the average daily secretion from the mammary gland. As the bioavailability of Ca^{2+} from cow's milk is lower, 400 mg must be provided when mother's milk is replaced by cow's milk. At years 2–9, the recommended daily intake of Ca^{2+} is 220 mg, which requires an intake of 600 mg from the diet. In adolescents (10–17 years of age), the marked increase in bone mineralization, which occurs earlier in girls than in boys requires approximately 440 mg of Ca^{2+} per day, which can be met with an intake of 1300 mg. In adults, the Ca^{2+} requirements are approximately 320 mg, provided by an intake of 1000 mg daily. Adults over age 45 have reduced absorption, which is why the intake should be of 1300 mg. In menopausal women there is increased Ca^{2+} loss in urine, which requires additional dietary calcium. During pregnancy, the Ca^{2+} needs increase, especially in the last quarter; during this period, the fetus retains about 240 mg/day. Pregnant women absorb calcium more efficiently; the recommended intake is 1200 mg/day. Women who are breastfeeding secrete approximately 36 mg of Ca^{2+} per 100 mL of milk; considering a daily production of 750 mL, this represents a total of 270 mg of Ca^{2+} per day. If urinary and other losses are added, with maximum absorption efficiency, the daily requirement is 1040 mg and the recommended intake is 1300 mg.

An intake of up to 2.5 g of calcium per day does not cause harmful effects and are considered the tolerable limit. The use of larger amounts may cause hypercalcemia and Ca^{2+} deposits in soft tissues. In patients with hypercalciuria, excessive Ca intake increases the risk of kidney stone formation.

Ca²⁺ deficiency may be due to inadequate Ca^{2+} absorption or excessive Ca^{2+} loss. The risks of deficiency are higher in the first 2 years of life, during puberty and adolescence; in pregnant women, especially in their third trimester; and in lactating and postmenopausal women. Males over 65 years are another population group at risk.

Decreased levels of plasma ionized calcium (hypocalcemia) may cause muscle pain, paresthesia, and tetany, characterized by sustained muscle contraction, especially in hands, arms, and legs. Calcium deficiency in children can produce clinical manifestations similar to those of rickets (Chapter 26). In adults, osteoporosis occurs, with loss of bone mass. This increases bone fragility and the risk of fractures.

Phosphorus (P)

Phosphorus integrates a series of body compounds that are essential for the formation of cellular structures and are involved in numerous metabolic processes in the body. An average male adult contains approximately 700 g of phosphorus (~1% of body weight), an amount that is slightly lower in women. From this, 85% of phosphorus is found in bone, primarily in the form of *hydroxyapatite* [$Ca_{10}(PO_4)_6(OH)_2$], the crystal structure that constitutes the mineral portion of bone. The remaining 15% is found in soft tissues and only a 0.3% is present in ECFs as phosphates. The phosphate in cells and in the ECF, exists as organic phosphate esters, or integrating nucleic acids, phospholipids, and phosphoproteins. At the pH of the body, inorganic phosphate (P_i) forms di- and monovalent anions HPO_4^{2-}, $H_2PO_4^{-}$.

Natural sources. Foods rich in phosphate include meat, poultry, fish, eggs, milk, and milk derivatives. Nuts, legumes, and grains contain phosphorus, but its bioavailability is higher in animal products. Coffee and tea contain some phosphate. Cola drinks also contain phosphorus as phosphoric acid.

Phosphate from foods is not found free, but forming organic and inorganic compounds. In its organic form phosphate is bound to proteins, sugars, and lipids. The relative amount of organic and inorganic phosphorus varies depending on the diet. In milk, for example, one-third of the total phosphate is inorganic phosphate (P_i). More than 80% of the phosphorus in the diet is in cereals (wheat, maize, and rice). In legumes and nuts, it is found as phytate (inositol hexaphosphate).

Absorption. Due to its binding to organic molecules, phosphate needs to be separated before it can be absorbed by the intestine. Phospholipase C catalyzes the separation of phosphorus from glycerophosphate of phospholipids, while alkaline phosphatase in the brush border of enterocytes, stimulated by calcitriol, releases phosphate from organic compounds. The bioavailability of phosphorus from the diet varies between 50% and 70%.

Inorganic phosphate is absorbed throughout the small intestine, particularly duodenum and jejunum, by passive diffusion and by saturable transport. The passive paracellular absorption is performed when the luminal concentration of phosphorus exceeds 1.5 mM, which is the usual amount after a meal.

The transport across the apical membrane of the enterocytes is mediated by a Na^+/P_i symporter driven by the sodium gradient, and stimulated by calcitriol. The movement of P_i to the interstitial space is carried out by a Na^+-independent transporter, driven by the P_i electrochemical gradient.

The proper absorption of P_i is linked to that of calcium and a ratio of 2:1 Ca:P_i ratio is optimal for absorption. If P_i exceeds Ca, it tends to form Ca phosphate compounds that are insoluble and are eliminated with the feces.

Phosphatemia. Phosphate appears in blood 1 h after ingestion. Plasma P_i (70%) is bound to organic compounds, such as phospholipids in lipoproteins. Most of the remaining 30% P_i circulates as HPO_4^{2-} and $H_2PO_4^-$; at normal plasma pH, the relationship $HPO_4^{2-}:H_2PO_4^-$ is 4:1. The plasma level of phosphorus (expressed as weight of elemental phosphorus) is 2.5–4.8 mg/dL (1.45–2.78 mEq./L) in adults and 4.0–7.0 mg/dL (2.37–4.06 mEq./L) in children during the first year of life.

Excretion. Phosphorus (70%–90%) is excreted in urine in its inorganic form. The remaining 10–30% is eliminated with the feces.

Renal handling of phosphate. The renal proximal tubule reabsorb ~85% of the phosphate filtered in the glomeruli, less than 5% is reabsorbed in the thick ascending limb of Henle's loop, the distal convoluted tubules, and collecting ducts. Approximately 10% of the total phosphate filtered is excreted in urine. Phosphate renal reabsorption takes place by the transcellular pathway. In the luminal membrane there is a $2Na^+/HPO_4^{2-}$ symporter system that transports phosphate from the lumen. The transfer to the interstitial space is performed by an anion/phosphate basolateral exchanger.

Calcitriol, growth hormone, insulin-like growth factor I, insulin and thyroid hormones stimulate phosphate reabsorption in the renal tubules. In contrast, calcitonin, at high doses, reduces it. Parathyroid hormone stimulates renal excretion.

Homeostasis. In a normal adult, the balance between intake and loss of P_i is zero. Phosphate levels in the body are controlled at different levels, including intestinal absorption, kidney reabsorption, and bone exchange. Moreover, these processes are regulated by several factors, such as parathyroid hormone, 1,25-$(OH)_2$-D_3, glucocorticoids, dopamine, and insulin. The existence of several peptides generically called *phosphaturic phosphatonins* has been demonstrated. The best one studied is the *fibroblast growth factor 23* (FGF-23), produced in bone tissue and released when the level of phosphate in the ECF increases. FGF-23 inhibits the expression of the Na^+/P_i transporter in the renal tubules, reduces phosphate reabsorption and increases phosphaturia. Also, FGF-23 inhibits 1α-hydroxylase activity in the kidney, decreasing the production of calcitriol, and therefore, the absorption of phosphate in the intestine and kidney tubules.

Function of phosphate in the body. Most of the functions of phosphate are exerted in bone, where it contributes to the structural and functional characteristics of this tissue. In cells and in the extracellular spaces in general, it is an essential component of many different molecules. Some of the processes in which P_i participates are the following:

Bone structure. The role of P_i in bone is to constitute the hydroxyapatite deposited on collagen for bone mineralization.

Nucleic acids. Phosphate is an important component of DNA and RNA. As part of free nucleotides it participates in the intermediary metabolism, storage, and energy transfer compounds, such as adenosine triphosphate (ATP), creatine phosphate, and UTP and UDP, which have a role in different metabolic reactions.

Intracellular second messengers. As a constituent of cyclic AMP (cAMP) and inositol trisphosphate (IP_3), P_i is involved in intracellular signaling systems.

Phosphoproteins. P_i plays a major role in regulation of protein function. Thus, activation of protein kinases promotes the phosphorylation of specific proteins (enzymes and others), modifying their activity.

Membrane structure. Phospholipids are essential components of cell membranes. Their amphiphilic properties allow them to form the double lipid layer, which is essential to form the basic structure and support functionality of cell membranes.

Acid–base balance. Within cells, phosphate is part of body buffers. In the kidney, Na_2HPO_4 from the filtrate binds H^+ and releases sodium, contributing to the elimination of hydrogen ions by urine.

Transport of O_2 by hemoglobin. In erythrocytes, P_i is part of 2,3-bisphosphoglycerate, which binds to and decreases the affinity of hemoglobin for oxygen. This promotes O_2 release to the tissues.

Recommended Intake. The dietary amounts of P_i are approximately the same as those of calcium. For an adult, 800 mg/day is recommended. Children require 200–800 mg/day with a progressive increase until 9 years of age. Adolescents, pregnant women, and lactating mothers need 1250 mg/day. Drinking approximately 500 mL of milk per day for adults and 1 L for teens and women during pregnancy and lactation meets the dietary needs of P_i, as well as Ca^{2+}.

Deficiency. Hypophosphatemia is a relatively rare condition, since the usual diet provides adequate amounts of P_i. Deficits can be found when the renal excretion of phosphate is increased, the intestinal absorption is decreased, or there is an important loss via the digestive tract. Pathological conditions that lead to acidosis and stimulation of phosphate excretion can cause hypophosphatemia. Low phosphate is also found in individuals who ingest large amounts of antacids containing Al, Mg, or Ca, which reduce the intestinal absorption of P_i. Starved or malnourished patients and those subjected to refeeding by oral gavage or parenterally, can become hypophosphatemic if P_i is not provided in adequate amounts. This condition is usually known as "refeeding syndrome."

As P_i is involved in numerous biological processes, its deficiency affects all cells. Clinical symptoms become evident when phosphatemia falls below 1.5 mg/dL and include anorexia, bone mineral loss, muscle weakness, neurological disorders (ataxia, paresthesia). These symptoms are the expression of a generalized metabolic disorder, with low ATP levels in cells. The decrease in 2,3-bisphosphoglycerate in erythrocytes alters the release of oxygen from Hb and this leads to hypoxia. Hypophosphatemia is a potentially lethal condition.

Magnesium (Mg)

Magnesium is found in all cells and body fluids. It is the second most abundant cation in the intracellular fluid (after potassium) and the fourth most abundant one, when all body cations are considered (after Na, K, and Ca). A 70 kg adult contains between 25 and 30 g of magnesium; more than half of this amount is found in bone, combined with calcium and phosphate. Mg that is not part of bone ($\sim$25%) is distributed in soft tissues, mainly muscle, liver, and kidney.

Magnesium is an essential element in many enzymatic reactions, especially in those in which ATP, GTP, or UTP are involved, and in those catalyzed by pyrophosphatase. In some cases, Mg^{2+} can be replaced by Mn^{2+}.

Magnesium is predominantly an intracellular cation (only 1% of total body magnesium is extracellular); the concentration inside the cell is 10 or more times greater than in plasma. The intracellular/extracellular relationship between Mg^{+2} and Ca^{+2} is somewhat similar to that of K^+ and Na^+.

Natural sources. Whole grains and legumes (beans, soy) are rich in Mg^{2+}. Ca^{2+} forms part of chlorophyll; therefore, it is present in all green leaf vegetables. It is also found in nuts, fruits, meats, dairy products, chocolate, coffee, and tea. The processing and preparation of food, as well as grinding grain with bran removal, reduces 80% of the original Mg.

Absorption. Mg is absorbed in the small intestine, mainly in jejunum and ileum, and part also in colon. In normal adults, Mg bioavailability is 30%–70% of the total Mg present in food. The percentage of Mg absorbed is inversely related to the amount offered; it increases when Mg^{2+} in the diet is low. Several factors influence intestinal absorption; phytate and oxalate interfere with intestinal Mg^{2+} uptake. Unabsorbed fatty acids (steatorrhea) form insoluble soaps with Mg, which are excreted with the feces. Calcium, phosphate, and other minerals, such as zinc, reduce Mg absorption. In contrast, some carbohydrates (fructose and lactose) enhance Mg absorption. Active vitamin D [1,25-(OH)$_2$-D$_3$] has no effect at physiological doses, but activates intestinal Mg absorption at high doses. Mg^{2+} absorption in the intestinal epithelium occurs through Mg^{2+}channels.

Magnesemia. Mg from the diet appears in blood 1 h after its ingestion. Normal blood plasma concentration of Mg^{2+} is 0.625–1.25 mM or 1.25–2.5 mEq./L (1.5–3.0 mg/dL); 30% of this amount is bound to protein, primarily albumin, 55% is free, in ionic state, and the rest forms complexes with different anions.

Excretion. The kidney is the main organ for Mg excretion. Other routes for Mg elimination are the intestine and skin (15 mg/day).

Renal handling of Mg^{2+}. One-third of the ingested Mg is eliminated in urine. A 70% of total plasma Mg, including ionic and constituting diffusible complexes, filter in the glomeruli ($\sim$100 mmol or 2.400 mg/day), a large proportion is reabsorbed in the tubules. The reabsorption in the distal tubules and collecting ducts (5%–10% of total filtered) is carried out through Mg^{2+} channels, following the Mg electrochemical gradient.

Homeostasis. The balance of Mg in the body is regulated by different actions on intestinal absorption, renal reabsorption, and bone exchange. Parathyroid hormone stimulates Mg^{2+} reabsorption in the loop of Henle. Also, aldosterone is involved in regulating Mg renal reabsorption. Bone is an important Mg reservoir, containing approximately 55% of the body Mg. A 70% of that amount is associated with phosphate and calcium in the bone mineral matrix. The rest is on the bone surface, in an amorphous form, which can be easily mobilized and exchanged with plasma Mg.

Functions. Mg is associated with phospholipids of the plasma and organelles' membranes. It also forms part of proteins; Mg is involved in more than 300 reactions, either as a structural cofactor, or an allosteric activator of enzymes. Mg constitutes nucleic acids; up to 90% of Mg is bound to nucleotides, especially ADP and ATP, forming complexes, such as $Mg^{2+}ATP^{3-}$ or $Mg^{2+}ADP^{2-}$ in which Mg assists in the transfer of the phosphate group.

Some of the pathways in which Mg participates are: (1) glycolytic pathway (hexokinase and phosphofructokinase); (2) pentose phosphate pathway (transketolase); (3) citric acid cycle (oxidative decarboxylation); (4) formation of creatine (creatine kinase); (5) initiation of β-oxidation of fatty acids (acyl-CoA synthetase); (6) synthesis of nucleic acids; (7) DNA transcription; (8) protein synthesis; (9) amino acid activation; (10) contractility and excitability of skeletal, cardiac, and smooth muscles; (11) reactions catalyzed by alkaline phosphatase and pyrophosphatase; (12) cyclic AMP formation (adenylate cyclase); (13) blood clotting; (14) regulation of ion channels, especially K channels; (15) vitamin D hydroxylation in the liver; and (16) calcium homeostasis (secretion and actions of PTH in bone, kidney, and intestine, and activation of Ca^{2+}-ATPase).

Recommended intake levels. The dietary requirement of Mg is estimated in 400 mg/day for males 19–30 years in age, and 420 mg for those over 30 years of age. Women 19–30 years of age need 310 mg/day and over 30, 320 mg/day. During pregnancy and lactation, an intake of 350 mg/day is recommended.

Normally, the excess Mg is rapidly eliminated in the urine, so that no significant increases in plasma are observed. A high intake of Mg is not toxic, except in patients with renal insufficiency.

Homeostasis alterations. Magnesium deficiency is rare in the general population because of its abundance in foods. Mg deficits can be seen during postoperative complications, after massive transfusions (chelating effect of the added citrate), refeeding syndrome in patients with starvation, diabetic ketoacidosis, extensive burns, and in a high percentage of alcoholics. Hypomagnesemia is asymptomatic unless the levels are lower than 0.5 mmol/L (1.2 mg/dL). With these levels, symptoms, such as nausea, vomiting, anorexia, muscle weakness, personality changes, hallucinations, spasms, tremors, and even tetany can occur. Another effect associated with low Mg is hypocalcemia, which may cause hyperexcitability, arrhythmias, and cardiac arrest. Hypermagnesemia has depressant effects on the central nervous system.

Iron (Fe)

Iron represents a very small proportion of the total mass of the body (~0.0057% of body weight in an adult human). Most authors include it among the trace elements. Despite its low amount, iron is a component of great importance. In adult women the total Fe content is approximately 3 g; in men, 4 g. Quantitative variations are observed with body weight, age, and physiological conditions, such as pregnancy.

Distribution. Iron has a key role in the body as part of molecules that play vital activities. This Fe has been called *functional* Fe and represents ~80% of the total Fe. A large part (65%, 2.0 g in women and 2.6 g in men) is found in hemoglobin, where it is involved in the transport of oxygen in blood; 10% (0.3 g in women and 0.4 g in men) is found in myoglobin, which carries and stores oxygen in muscle; 2.5% (0.075 and 0.10 g in female and male, respectively) corresponds to cytochromes of the mitochondrial respiratory chain, the cytochrome P_{450} enzymes, and numerous oxygenases, such as catalase, peroxidases, and others involved in redox reactions. All the mentioned proteins are hemoproteins; they contain a heme as the prosthetic group. In heme, the Fe is at the center of a tetrapyrrole protoporphyrin IX and is the key element for the functional properties of hemoproteins (Fig. 3.26).

There are other proteins associated with nonheme iron. They represent approximately 2.0% (0.06–0.08 g) of the total Fe. Some of these molecules have iron–sulfur clusters (Fe–S, 2Fe–2S, 3Fe–3S, and 4Fe–4S) including the I, II, and III complexes of the mitochondrial respiratory chain, and metalloenzymes, such as aconitase (enzyme of the Krebs, cycle), ferrochelatase (enzyme of heme synthesis last step), oxidases (ribonucleotide reductase), and glycerolphosphate dehydrogenase.

Approximately 20% of the total Fe is found in body reserves (iron deposits). These reserves, present in all cells, contain Fe complexed with proteins (*ferritin* and *hemosiderin*). These stores are dynamic and they accumulate Fe when it is in excess and release it when required.

Finally, 1% of Fe is found in blood, bound to *transferrin*.

Fe toxicity. Body iron is found in two oxidation states, ferrous (Fe^{2+}) and ferric (Fe^{3+}). Free Fe, especially Fe^{2+}, can cause tissue damage; for this reason, it is always associated to cell and plasma proteins. The toxicity of free iron is due to its ability to form free radicals.

Natural sources. Foods rich in iron include liver, heart, meat of any kind, seafood, tomatoes, beans, cauliflower. Although in lesser amounts, it is also found in green vegetables, cereals, legumes, and nuts. Milk and dairy products are very poor in iron. In foods of animal origin, approximately 60% of the iron is in hemoproteins and the rest in nonheme proteins. All iron in vegetables is associated with nonheme proteins.

Iron balance. The normal organism is very efficient in handling iron. Fe is constantly recycled with very little loss; an example is Fe in hemoglobin: red blood cells, after 120 days, reach the end of their useful life, are lysed, and their hemoglobin is degraded in the spleen. A total of 8–9 g of Hb are catabolized per day, which represents 25 mg of iron. From this, approximately 24 mg is reused for the synthesis of new hemoglobin. This Fe recycling is very active; only 12 h after erythrocytes lysis, 60% of their Fe content is already incorporated into new Hb.

A normal adult loses an average of 1 mg of iron per day, via epithelial cells desquamated from the gastrointestinal mucosa, skin, and urinary tract; and the small amount of blood that eventually is eliminated with stool (occult blood, less than 0.35 mg Fe per day), negligible amounts are excreted in the bile, urine, or normal perspiration. Women during reproductive age have the additional loss caused by menstrual bleeding, which represents an average loss of approximately 40 mL of blood or more than 20 mg of Fe per month; there are individual variations, in some women the loss can reach up to 100 mL of blood per month.

A normal diet contains between 10 and 20 mg of Fe (per day); the intestine absorbs 8%–10% of this amount per day (1.0–1.5 mg/day). The amount absorbed varies according to the organism needs.

Digestion. Foods of animal origin contain iron predominantly associated with heme proteins. Fe associated to heme is more bioavailable than nonheme Fe. The protein portion of hemoproteins is degraded by digestive proteases, releasing heme, which is soluble in the intestinal environment, especially in the presence of protein digestion products (amino acids and small peptides).

For proteins containing nonheme Fe, which is mostly in ferric state, the acid medium of the stomach (pH 1.5) maintains both Fe^{2+} and Fe^{3+} in soluble form. Some of the Fe^{3+} can be converted into Fe^{2+}, especially if there is ascorbic acid (vitamin C) in the lumen. In the duodenum and jejunum, where the pH is higher, Fe^{2+} remains in solution at a pH of 7.5–8.0, but at a pH above 5.0, Fe^{3+} forms ferric hydroxide [$Fe(OH)_3$], which precipitates and is eliminated in the stool.

Absorption. Both, heme and nonheme iron, are absorbed in the proximal small intestine, the first one is more efficiently absorbed than the latter.

Intestinal absorption of Fe can be divided into several stages: (1) uptake in the brush border of enterocytes, and transport across the apical membrane, (2) transport in the enterocyte cytoplasm to the basolateral membrane, and (3) transfer to the circulation after crossing the basolateral membrane of the enterocytes.

1. *Fe absorption in the enterocyte brush border.* Heme is uptaken by the enterocytes at the apical membrane through the *heme transport protein 1*, which transfers heme intact into the cytoplasm. A *hemoxigenase* associated to the cell membrane separates free iron from heme; protoporphyrin continues its catabolism independently to form bilirubin. The iron separated from nonheme proteins is mostly Fe^{3+}, poorly soluble and not readily available

to the cell. Fe transport in the cell requires that specific transporters and enzymes change its oxidation state from Fe^{3+} to Fe^{2+}.

The amount of Fe present in the diet influences its absorption; Fe uptake increases when dietary Fe increases; however, the overall efficiency of transport diminishes. Another factor that influences intestinal Fe absorption is the diet composition. Fe bioavailability can be greatly reduced by the presence of agents that interfere with its intestinal uptake, such as phytate in legumes, nuts, and cereals and some polyphenols as tannins in tea. A mixed diet provides factors that counteract the action of these agents and favor Fe absorption.

Lack of meat in the diet is usually one of the main causes of iron deficiency.

Once inside the enterocyte, Fe from heme and nonheme proteins, becomes part of the same Fe pool from which Fe can be transferred to the bloodstream, stored, or used for the synthesis of iron-containing molecules.

2. *Fe transport to the basolateral membrane and circulation* is performed by binding it to a protein called *mobilferrin* (56 kDa). This complexation with proteins avoids the potentially harmful effects of free Fe. The movement of Fe from the enterocytes to the blood, and across the basolateral membrane, is mediated by the transmembrane transport protein, *ferroportin*. Before transport, Fe must be oxidized from Fe^{2+} to Fe^{3+}. This is catalyzed by the ferroxidase *hephaestin*, a protein associated to ferroportin. This same action is exerted by plasma *ceruloplasmin*, another ferroxidase, which, similar to hephaestin, contains copper in its structure.

3. *Storage in cells*. Iron can remain in cells bound to *apoferritin*, a large protein of 480 kDa and 24 subunits that form a hollow sphere whose central cavity can accommodate up to 4500 Fe atoms. By capturing Fe, apoferritin

becomes *ferritin* (holoferritin), the main form of iron storage, not only in the intestine, but also in other cells, particularly liver and the *mononuclear phagocyte system*. Iron absorption in intestine is regulated and all factors involved in Fe absorption are upregulated when Fe reserves decrease. In cases of increased demand (increased erythropoiesis subsequent to bleeding, severe hemolysis, hypoxic conditions) Fe uptake and transfer to the blood are incremented.

Fig. 29.3 shows a scheme of the process of Fe uptake in enterocytes.

Blood transport. In the normal adult (both male and female), Fe concentration in plasma is 60–150 µg/dL (10–30 µmol/L). Serum Fe is higher in the newborn and rapidly decreases after the fourth month of life. There is a significant circadian variation of plasma Fe, the highest values are found in the early morning and the lowest occur late in the afternoon.

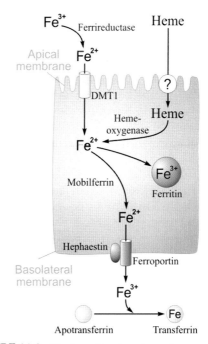

FIGURE 29.3   **Fe absorption in enterocytes.**

The Fe transferred to the plasma is transported by *apotransferrin*, also called *siderofilin*. This is a glycosylated carrier protein of 80 kDa, which is synthesized primarily in liver. It can bind two ferric atoms per molecule and becomes *transferrin* (Tf) (holotransferrin). The presence of Tf in plasma and ferritin in cells maintain almost all Fe in a complexed state, preventing the toxic effects of free inorganic iron. Not all plasma transferrin is saturated with Fe. Less than one-third is bound to two Fe^{3+} atoms (diferric Tf), the rest contains one atom (monoferric Tf) or has no Fe (apotransferrin). An indirect measure of the amount of Tf in plasma can be obtained by measuring the plasma *total iron-binding capacity*. Normal values range between 250 and 410 µg/dL. The difference between total iron-binding capacity and serum Fe is known as unsaturated iron-binding capacity, the normal range is 140–280 µg/dL. The total iron-binding capacity is elevated in pregnancy and chronic Fe deficiency, it decreases in liver failure (liver synthesizes Tf) and during protein malnutrition.

Transfer to the tissues. Transferrin binds to receptors located in the plasma membrane of cells. All cells, except mature red blood cells, have transferrin receptors. Erythrocyte precursor cells are the richest in these receptors. The transferrin–receptor complex at the cell surface is introduced into the cells by endocytosis. The decrease in pH within the endosome promotes release of Fe from Tf. This is accompanied by the reduction of Fe^{3+} to Fe^{2+}, which is transported to the cytosol through the endosomal membrane by a carrier identical to that in intestinal mucosa.

Apotransferrin remains bound to the receptor in the membrane of the endosome and it is returned to the cell plasma membrane. The apotransferrin–receptor complex, exposed to the extracellular space where the pH is 7.4, dissociates and apotransferrin is released. Once in plasma, apotransferrin binds Fe again repeating the cycle.

Haptoglobin. Cells can capture Fe by mechanisms different from that involving the transferrin receptor. One is by direct uptake of hemoglobin or heme from plasma. Normally, lysis of erythrocytes releases a small amount of hemoglobin in the bloodstream. This free hemoglobin filters in the renal glomerulus and it would be eliminated. However, hemoglobin is bound by plasma *haptoglobin*, the macromolecule formed cannot pass the glomerular barrier and it is retained.

Hemopexin. When there is free heme is in plasma, it is bound by another protein called *hemopexin*.

The Fe released into the cell cytosol can be stored in ferritin, exported to the circulation by ferroportin, or used for the synthesis of compounds containing Fe. One of the main destinations of Fe in cells, especially in the erythroblastic progeny, is the mitochondrion, which contains many enzymes associated with Fe.

Macrophages. Another mechanism for Fe uptake operates in macrophages of the mononuclear phagocyte system, particularly in the spleen. This system is in charge of lysing erythrocytes that have completed their life cycle. Old or damaged red blood cells are lysed and phagocytosed by macrophages, hemoglobin is degraded, and iron, in part, is stored in the macrophage as ferritin. A significant portion of Fe is exported via ferroportin to reload the circulating apotransferrin.

Erythroblasts. Erythrocyte precursor cells have a high affinity for binding Fe and have many transferrin receptors in their outer membrane. One of the main destinations of Fe is the mitochondria, where the last step of heme synthesis takes place.

Storage. Most of the iron in the cells (particularly in liver, bone marrow, and spleen) binds to apoferritin and is stored as ferritin. The liver contains approximately 60% of the iron deposited in the body. The rest is mainly divided into cells of the mononuclear phagocyte system of spleen, bone marrow, and enterocytes.

Plasma ferritin. A small portion of ferritin passes to plasma, where its concentration is related to the amount of stored Fe.

Hemosiderin. When there is too much iron, ferritin is partially denatured by lysosomal action, and forms granules containing ferritin aggregates, in which iron comprises a total of up to 50% of its mass. Ferritin and hemosiderin are storage forms usable when the body requires Fe. When the body needs iron in larger amounts than usual, for example, after a hemorrhage, deposits in liver and spleen are mobilized. After these reserves are consumed, the iron deposited as ferritin in cells of the intestinal mucosa is used. This decreases the Fe saturated apoferritin in the intestine, and Fe absorption is activated.

Fe release from cells. In a normal adult, the amount of iron coming into the cells is balanced by that released. The Fe is exported from the cells through ferroportin.

Ceruloplasmin. Fe^{2+} that reaches the circulation is oxidized to Fe^{3+} by the ferroxidase *ceruloplasmin*. Ceruloplasmin, as enterocyte basolateral membrane hephaestin, is a metalloenzyme with copper as the electron acceptor.

Fig. 29.4 summarizes these processes.

Iron homeostasis. Since Fe excretion cannot be regulated, Fe homeostasis depends on the modulation of its absorption in the intestine.

Bone marrow erythrocyte precursor cells, which actively synthesize hemoglobin, consume the largest amounts of Fe. Macrophages of the mononuclear phagocyte system are the main Fe suppliers. In addition, hepatocytes have great capacity to store Fe and transfer it as needed. There is a tight coordination between all these tissues to ensure that the requirements are satisfied and that excessive accumulation of Fe does not occur.

Hepcidin, a 25 amino acid peptide hormone, is the main regulator of all tissues involved in Fe homeostasis. It is synthesized primarily in the liver and released into the circulation when Fe reserves in hepatocytes are high. Its action is mainly exerted on enterocytes and macrophages. In the intestine, hepcidin decreases Fe absorption. In macrophages, it depresses Fe

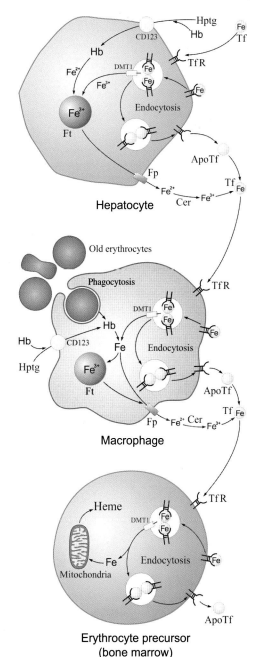

FIGURE 29.4 **Different routes for Fe in cells with active iron metabolism.** *CD123* and *DMT1*, Carriers; *Cer*, ceruloplasmin; *Fp*, ferroportin; *Ft*, ferritin; *Hptg*, Haptoglobin; *Tf*, transferrin; *TfR*, transferrrin receptor.

recycling, which favors accumulation of Fe as ferritin.

Excretion. Fe is removed from the body via the intestine in a nonregulated manner. In feces, Fe is released from the desquamated epithelial cells of the gastrointestinal tract, eventually from small amounts of blood lost, and from the unabsorbed Fe from food. Due to the association of Fe in macromolecular complexes (transferrin or hemoglobin–haptoglobin), it is hardly excreted in urine.

Requirements. Only a small proportion, about 10% of the Fe in the diet is absorbed and reaches the circulation, the remainder is excreted in the feces. That percentage may increase or decrease depending on the body needs; 10 mg/day of Fe are recommended for an adult male. During growth, needs are greater; in children 3 months to 3 years of age, 15 mg/day must be provided. The normal newborn has a reserve for 3–6 months, which were stored in the liver during fetal development. Fe supplementation is needed in infants, since milk is very low in this element. Adolescents require 18 mg/day, women 12–50 years of age need 18 mg/day, to compensate for menstrual losses, pregnant women should receive more than 18 mg/day.

Deficiency. Reduced Fe amounts is a common medical problem and the most frequent nutritional deficiency. The number of people who suffer it in the world is estimated to reach over one billion, and includes individuals of lower economic class. The most common cause is inadequate dietary intake of Fe for extended periods of time. As heme iron is absorbed the best, when the meat is absent in ordinary food, the availability of Fe is significantly reduced. Other factors that lead to Fe deficiency include parasitic diseases that produce continuous blood loss, such as hookworm and schistosomiasis, common in some countries. There are stages of life and physiological conditions in which Fe requirements are significantly increased and in these the risk for Fe deficiency increases. For example, during pregnancy there is an increase

in the volume of plasma and erythrocyte dilution with a relative reduction in the concentration of Fe in blood. In addition, Fe is transferred from mother to fetus in amounts that can reach 3 or 4 mg/day in the third quarter. Premature infants and very low–weight newborns, who have poor reserves, are prone to Fe deficiency if supplements are not administered. In women, menstrual blood loss can cause deficiency if the iron is not replaced with adequate food or supplements.

There are pathological conditions that determine alterations in Fe intestinal absorption and Fe deficiency, as for example: lack of hydrochloric acid in the gastric secretion (achlorhydria), gastrectomy, excessive consumption of antacids, intestinal malabsorption syndromes, accelerated intestinal transit, and avitaminosis C. Important hemorrhage, or small amounts of blood loss, maintained for a long time, as occurs in chronic infections and in parasitic diseases, chronic infectious or inflammatory processes (lasting more than 2 months), *Helicobacter pylori* infections, associated with gastric ulcers, neoplasias, and lead poisoning.

Anemia due to iron deficiency. When Fe is reduced and its stores are depleted, hemoglobin synthesis and erythropoiesis is impaired, leading to anemia. The concentration of hemoglobin in blood decreases (values under 12 g/dL in women and 13 g/dL in men are considered abnormal). The hematocrit and mean corpuscular volume of RBCs is reduced. Erythrocytes are smaller than normal and poor in hemoglobin (*hypochromic microcytic anemia*). Patients are pale, their capacity for work is reduced, and they experience fatigue at the slightest effort. Lack of Fe during the first 2 years of life produces irreversible alterations in the central nervous system that do not improve with later supplementation treatments. These children suffer from behavioral problems, learning difficulties and reduced cognitive ability.

Iron overload. Excessive accumulation of Fe results from excessive Fe absorption in the intestine, hematological conditions requiring repeated transfusions (i.e., thalassemia), impaired heme synthesis that prevents the use of iron, genetic faults affecting factors involved in income regulatory processes.

Initially, the Fe accumulation is as ferritin, if the overload persists, hemosiderin is formed and deposited in liver, pancreas, skin, and joints (hemosiderosis). These deposits alter the function of the organs where Fe accumulates and are the cause of a condition called *hemochromatosis*. Also, several genetic syndromes that cause iron overload have been described.

TRACE ELEMENTS

These constitute a group of elements (also called oligoelements) that present in the body in concentrations ranging from picograms to nanograms per gram of wet tissue. Although found as traces, they are essential for maintaining normal metabolism. They are described in the following sections.

Zinc (Zn)

Zinc is an essential element present in all body tissues. A normal adult contains approximately 2.5 g of zinc, mainly as divalent ion (Zn^{2+}). Zinc forms part of proteins and enzymes and participates in catalytic, structural, and regulatory functions.

Natural sources. Zn present in foods is associated with proteins, peptides, and nucleic acids. Beef, chicken, and fish meats are the richest sources. Lower amounts are found in milk and dairy products. Whole grain cereals, legumes, and vegetables, contain Zn, but the presence of phytate in them reduces Zn bioavailability. Fruits and refined grains are very poor in Zn. Animal foods provide 40%–70% of Zn in a mixed diet.

Absorption. Zn is released from proteins and nucleic acids by the hydrochloric acid in the stomach and the proteases and nucleases present in the gastrointestinal tract. Pancreatic secretions are rich in Zn.

Zn is absorbed through the small intestine, mainly in the duodenum. It is transported through the apical membrane of the cells as free Zn^{2+} by a saturable active carrier. Zn absorption is influenced by various factors; it is favored by small ligands, such as citrus juice acids, picolinic acid (intermediate of the pathway from tryptophan to niacin), prostaglandins, glutathione, and free amino acids (principally histidine, cysteine, and glycine). In contrast, phytates from whole grains; oxalate from spinach, chocolate, and tea and polyphenols (i.e., tannins) from whole grains, fruits, tea, and vegetable fiber, form nonabsorbable complexes with Zn. The presence of animal foods lowers the inhibitory action of these products. Other divalent cations (Fe^{2+}, Cu^{2+}, and Ca^{2+}) compete with Zn for binding to its carrier, reducing Zn absorption. Usually, Zn absorption transports 35% of the total Zn contained in the intestinal lumen and reaches 3 or 4 mg (45–60 μmoles)/day; this amount is adjusted to the organism needs.

Within the enterocyte, Zn has several destinations; it can be used by the same cell, it can be transported to the basolateral membrane to be released in plasma, or stored bound to proteins. *Thionein* is a small protein (10 kDa) rich in cysteine residues, which serves as the main ligand for Zn and other minerals, such as copper, forming metallothionein. Zn bound to metallothionein is lost when cells are shed from the villi in the intestinal lumen. Diets that are rich in Zn induce the synthesis of metallothionein.

Transport. In the portal vein, Zn binds mainly to albumin; it is captured by the liver and it is sent back to the circulation bound to albumin, α_2-microglobulin, and immunoglobulin G. Albumin carries approximately 70% of the total Zn present in plasma, where its concentration is maintained within narrow limits, at 12 μg/dL (10 μM). In whole blood, 70%–80% remains in the intracellular space, with higher amount in leukocytes than in erythrocytes.

Storage. There is no specific organ where Zn is deposited, it is found in all tissues, especially liver, kidney, muscle, skin, blood, and in prostate secretions.

Excretion. Feces are the main routes for Zn elimination. The Zn unabsorbed from food, and that contained in pancreatic secretions and cells that shed from the gastrointestinal mucosa, are the sources for the Zn that is lost. The amount of Zn released through the urine is very small (less than 1 mg/day).

Functions. Zinc is an essential component of many enzymes (at least 200), more than any other trace mineral. In many of these enzymes, Zn forms part of the catalytic site; in others, it ensures the structural stability of the enzyme. Some of the enzymes containing Zn are: carbonic anhydrase, superoxide dismutase, alcohol dehydrogenase, alkaline phosphatase, aminopeptidase, carboxypeptidase, δ-aminolevulinic acid dehydratase, collagenase, phospholipase C, polyglutamate hydrolase, transcriptases.

Other functions. Zn is also important as a structural component of transcription factors that contain "zinc finger" domains, which bind to response elements in the promoter of a series of genes, activating or repressing their activity. Particularly relevant are the steroid, thyroid, calcitriol, and retinoid hormone receptors.

Redox functions. Zn is involved in redox reactions mainly catalyzed by superoxide dismutase.

Regulation of apoptosis. Zn influences some of the steps that trigger the process of programmed cell death (Chapter 32).

Control of cell proliferation. Zn has direct effect on hormones, receptors, and genes responsible for cell growth.

Membrane stabilization. Zn functions in stabilizing the structure of membrane proteins and enzymes.

Immune system. Zn improves the body defense against infections.

Pancreas. Zn is associated with insulin deposits in the pancreatic β cells and also promotes insulin secretion.

In saliva, a zinc metalloprotein, called *gustin*, is related to taste sensations.

Deficiency. Insufficient intake, intestinal malabsorption, loss or disturbances in utilization are causes of Zn insufficiency. It is estimated that approximately half of the world's population is at risk of deficiency. Among the groups at highest risk are those that require higher amounts, such as children, adolescents, pregnant, and lactating women. Alcoholics with liver cirrhosis often excrete high amounts of Zn in urine and have decreased plasma levels of Zn. Diabetic and patients with infections are also at risk of Zn deficiency.

Reduction of Zn in the body is a condition that develops slowly due to the effectiveness of the adaptive mechanisms that regulate homeostasis and control income and loss. Zn deficiency in human presents with skin lesions, decreased appetite, altered taste and smell, growth defects, delayed puberty, hypogonadism that can be explained by the lack of Zn in the androgen and estrogen receptors, which reduces sensitivity of target tissues to these hormones. In addition, Zn deficiency causes increased susceptibility to infections.

Copper (Cu)

Copper is a widely distributed element in nature. It has great functional importance, mainly as a component of metalloenzymes, in which copper functions as an electron acceptor changing from Cu^{2+} to Cu^{+}. A normal adult contains approximately 150 mg of copper, distributed mainly in liver, muscle, red cells, and plasma. Dietary deficiency of Cu is rare; however, some studies have shown that marginal copper deficiency is a relatively common condition.

Natural sources. Liver, seafood, whole grain cereals, nuts, raisins, potatoes, fruits, and meats are foods rich in Cu. Lower concentrations are found in vegetables, poultry, and fish. Milk and dairy products have very small amount. The average diet provides 2.5–5.0 mg/day. Most of the

Cu present in food is bound to proteins, primarily as Cu^{2+}.

Absorption. Generally, the bioavailability of Cu is 50% of that coming in the diet. The percentage absorbed increases when the intake is less than 1 mg/day. Pepsin and hydrochloric acid in the stomach facilitate Cu release from proteins and the action of the proteases of the small intestine facilitates Cu absorption. The main region of the intestine involved in Cu absorption is the duodenum, which uptakes it through a saturable active transport system.

Cu absorption is influenced by various dietary factors. It is favored by free amino acids, which bind to Cu and carry it along as they are transported across the intestinal epithelium by different amino acid transport systems. Some organic acids (citric, lactic, malic, acetic, and gluconic) enhance Cu absorption. Phytates hinder the entry of Cu, and also inhibit Fe, Ca, and Zn absorption. Cu absorption is regulated; when the uptake by enterocytes is high, the excess is stored and eliminated with cells that shed from the mucosa.

Once inside the enterocyte, Cu can follow different routes: (1) it can be used by the same cell to synthesize cuproteins, (2) it can be transferred to blood by active transport systems, and (3) can be bound to *thionein*, a metal-fixing apoprotein that becomes *metallothionein*.

Binding of Cu to proteins, amino acids, or glutathione prevents the generation of free Cu radicals that can cause cell damage.

Transport. Cu circulates in blood bound to albumin; it can also be linked to free amino acids, histidine, and the transport protein transcuprein. From blood, Cu moves into tissues, mainly liver, where it is incorporated into cuproteins. One of these proteins is *ceruloplasmin*, a α_2-globulin of ~160 kDa that binds six Cu atoms per molecule. From the liver, ceruloplasmin passes to the general circulation. More than 80% of the Cu present in plasma is bound to ceruloplasmin, which is also involved in the metabolism of iron catalyzing the oxidation of Fe^{2+} to Fe^{3+}. The rest of plasma Cu is attached to albumin, transcuprein, and histidine.

Excretion. Copper is excreted in the bile and eliminated with the feces. This is the main route for Cu elimination and helps to regulate the body Cu content. The amount being excreted in urine is very small (~30 µg/day) as well as its elimination through sweat and desquamated skin cells.

Functions. Copper is an essential cofactor for many enzyme reactions, acting as an intermediary compound in electron transfer. In several proteins and enzymes it also is a structural component.

Among the most important molecules containing Cu are: ceruloplasmin, superoxide dismutase, cytochrome *c* oxidase, monoamine oxidase, *p*-OH-phenylpyruvate hydroxylase, dopamine monooxygenase, lysyl oxidase, tyrosinase, α-amidating peptidylglycine monooxygenase. Other proteins containing Cu include *prion* proteins (p. 536), important for normal function of the nervous system and *factors V* and *VIII*, which are involved in the process of blood clotting (Chapter 31).

Deficiency. Dietary copper deficiency is rare, but it can be observed in prematures, malnourished children, patients under chronic parenteral nutrition, intestinal malabsorption, and people consuming high amounts of antacids.

The most frequent symptoms of Cu deficit are: hypochromic microcytic anemia similar to that of iron deficiency, skin and hair depigmentation, bone demineralization, impaired immune function, poor myelination. In these patients, the synthesis and degradation of serotonin is decreased, which can compromise nerve excitability. There are genetic defects related to alteration of copper metabolism, such as Wilson's disease or hepatolenticular degeneration and Menkes' disease.

Iodine (I)

Iodine is required for the synthesis of thyroid hormone. A normal adult has 20–50 mg of iodine distributed in the thyroid gland, muscle, and other tissues. In thyroid gland the concentration is much higher than that present in muscle and blood plasma. Molecular iodine presents as I_2, in nature it is often found as iodides (I^-) and iodates IO_3^-.

Natural sources. The concentration of iodine in foods is variable because it depends on its amounts in the soil of the particular geographic region considered. Sea water is relatively rich in iodine. Salt water fish and seaweed contain it in significant quantity.

Absorption. Iodine from foods comes mostly as iodides or iodates, linked to free amino acids. The amount in a standard diet is 100–200 µg/day.

In the digestive tract, iodate is reduced to iodide in reactions in which glutathione is usually involved. Iodide is readily absorbed through the digestive tract, including the stomach. The bioavailability of iodide is virtually 100%. Thyroxine (T_4) and triiodothyronine (T_3) are absorbed without change in the intestinal mucosa with a bioavailability of 75%. Therefore, it is possible to treat hypothyroidism with thyroid hormone administered orally.

Transport. Iodide enters the circulation by the portal vein system and is distributed throughout the body. Blood plasma contains 4–8 µg of iodide per deciliter, predominantly bound to plasma proteins (proteic iodine). Approximately one-third of the total iodide absorbed is taken up and concentrated by the thyroid gland. Iodide excess is eliminated mainly by urine. The iodide thyroid/plasma ratio is normally 30–40. In the thyroid, iodides pass from the cell to the lumen of the gland's follicles.

Metabolism. Iodide metabolism is considered in p. 588.

Functions. So far, the only known physiological role of iodine is to form part of thyroid hormones.

Deficiency. Since iodine is essential for the synthesis of triiodothyronine and thyroxine, chronic deficiency of this element in the diet results in reduced production of these hormones and causes hypothyroidism. Iodine deficiency affects all ages. However, the groups most at risk are pregnant women, infants, and children under 3 years of age.

Iodide deficiency results in a reduction of iodide reserves in the thyroid gland and in T_3 and T_4 synthesis. The reduction of T_4 in plasma increases, through a positive feedback mechanism, the secretion of TSH and its releasing hormone, which results in thyroid hyperstimulation and *simple goiter*. This is commonly due to insufficient iodine intake and is characterized by an enlarged thyroid gland.

Ocean water contains iodine; therefore, areas close to the coasts have adequate iodine levels in the soil, water, and foods. In contrast, regions far from the sea or surrounded by mountains that block the winds coming from the ocean, are poor in iodine and the incidence of simple goiter with deficient production of thyroid hormones and hypertrophy of the gland is high (greater than 10%); this is known as *endemic goiter*. The deficiency can be compensated by adding iodine to the diet; the most practical method to do this is the addition of iodine to table salt in a ratio of 1 part in 100,000. In many countries iodide addition to salt is required by law.

Deficiency of iodide in the fetus results in *cretinism*, which is characterized by growth failure (dwarfism) and serious neurological damage, mental retardation, hearing loss, and motor disorders (spasticity and muscle rigidity).

Failure of the thyroid gland (hypothyroidism) is manifested by various symptoms: decreased basal metabolism, reduced synthesis and degradation of proteins and lipids, increased serum cholesterol, decreased glycolysis and glycogenolysis, decreased amplitude of electrocardiogram waves. Early treatment with iodine improves these symptoms, although prenatal brain damage is irreversible.

If hypothyroidism begins in adulthood, one of the striking symptoms is *myxedema*, due to accumulation of glycosaminoglycans and water in the subcutaneous tissue, with thickening and "pasty" skin appearance.

There are also congenital forms of hypothyroidism caused by genetic alterations affecting enzymes involved in the biosynthesis of thyroid hormones. Early diagnosis is important because the lack of T_4 and T_3 in the newborn produces severe developmental disorders, especially of the nervous system.

Manganese (Mn)

Manganese is essential to many animal species, including humans. It is important for the activity of various enzymes and is a component of several metalloenzymes. A normal adult contains approximately 20 mg of Mn.

Natural sources. Mn is supplied in the diet mainly through plant foods. Nuts, whole grains, vegetables, cereals, and tea are good sources; white flour has much less than whole wheat. Meat, fish, and dairy products contain lower amounts. The average diet provides 2–9 mg of Mn per day, which covers the normal requirements of an adult.

Absorption. Mn is absorbed throughout the small intestine. Low molecular weight ligands, (histidine and citrate) increase its absorption. Fiber, phytates, and oxalate form nonabsorbable manganese precipitates.

Transport. From the intestine Mn passes to the portal vein, where it can remain free as Mn^{2+} or can be bound to α_2-macroglobulin and albumin and be taken up by the liver. In hepatocytes, Mn can be excreted in the bile, incorporated to enzymes, or secreted back to the circulation. Little is known about the mechanism by which Mn enters the cells.

Mn is not stored in the body, but its concentration is higher in bone, which contains about 25% of the total body Mn, mainly associated with hydroxyapatite. Other organs containing Mn are liver, kidney, pancreas, and hair.

Excretion. Almost all of the excess Mn is eliminated with the feces. This includes the Mn excreted in the bile or that not absorbed in the intestine.

Functions. The functional relevance of Mn is related to its role in the activation of enzymes or as a constituent of metalloenzymes. Enzymes associated to Mn include: oxidoreductases, hydrolases, mutases, transferases, lyases, and ligases. The activation of enzymes by Mn is not specific and it can be replaced by other divalent cations, such as Mg^{2+}. The only enzymes that specifically depend on Mn are glycosyl transferases, glutamine synthetase, and farnesyl pyrophosphate synthetase. Mn metalloenzymes are mitochondrial superoxide dismutase, arginase, and pyruvate carboxylase.

Deficiency. The deficit in Mn is rarely seen and only occurs when there is malnutrition. Low Mn can cause nausea, vomiting, dermatitis, decreased growth of hair and nails, change in hair pigmentation. Alterations in the reproductive functions, glucose tolerance, lipid metabolism, and plasma cholesterol levels have also been described. Also, failures in the synthesis of oligosaccharides, glycoproteins, and proteoglycans produced by reduced activity of the glucosyl transferase activity can occur.

Selenium (Se)

Selenium is an essential trace element in humans. Due to its similarity with sulfur, it can replace sulfur in methionine, cysteine, and cystine and form seleno-amino acids. These can be used for the synthesis of various proteins, including enzymes. The normal adult contains from 3 to 15 mg, depending on the geographic area of residence.

Natural sources. Se content in food depends on the existing in the soil and varies greatly in different regions. In general, products of animal origin have more Se than plants. Se is present in organic compounds, mainly proteins, either as selenocysteine, which predominates in foods from animals, or as selenomethionine, more abundant in plants. It is also in inorganic compounds, including selenite SeO_3^{2-} or selenate SeO_4^{2-}. The best sources of Se are salt water fish, shellfish, and Brazil nuts (cashew).

Absorption. After digestion of proteins containing seleno-amino acids, Se is transported into the enterocytes by the same carriers that transport the corresponding sulfur amino acids.

Transport. Se is transported in circulation bound to plasma low and very low density lipoproteins (LDL and VLDL). Most of it (60%–80%) is bound to *selenoprotein P*. The liver contains over 30% of the total body Se, muscle has 30%, kidneys 15%, and 10% circulates in the plasma.

Metabolism. The seleno-amino acids may remain in the cell as part of the general amino acid pool. Selenomethionine, mainly from vegetables, is not distinguished from its sulfur analog by mammals; it is metabolized and incorporated into protein by the same ways as methionine. Free selenium is converted to hydrogen selenide (H_2Se) with H donated by glutathione, and seleno-phosphate by the seleno-phosphate synthetase, which transfers phosphate from ATP. Seleno-phosphate is intermediate in the biosynthesis of selenoproteins, containing predominantly selenocysteine in animals. Selenoprotein biosynthesis uses a special tRNA that initially binds serine. By a process that requires seleno-phosphate, the serine bound to tRNA is converted to selenocysteine. The UGA codon, which usually indicates termination of protein synthesis, when it coexists with a specific mRNA sequence on the 3′ untranslated side, induces insertion of tRNA loaded with selenocysteine.

Excretion. Se is excreted in urine and feces. Urinary excretion is the major mechanism for the homeostatic control of Se.

Functions. Se is a component of proteins, many of them with enzymatic activity. Twenty-five selenoprotein genes have been identified in the human genome, but only 15 of them have been characterized. The concentration of selenoproteins in endocrine glands (thyroid, adrenal, pituitary, testis, and ovary) is higher than in other organs. The best known selenoproteins are: (1) *glutathione peroxidase* (GSPx), which presents five isozymes that catalyze the removal of hydrogen peroxide and other peroxides from different compounds; GSPx forms part of the antioxidant system of the body (see Chapter 10); (2) *thioredoxin reductase*, a NADPH-dependent flavoenzyme with selenocysteine in its catalytic site; and (3) *iodothyronine deiodinases*, which are selenoenzymes involved in the metabolism of thyroid hormones (p. 588). In addition, *selenophosphate synthetase, selenoprotein P* and *sperm selenoprotein* that results from the transformation of one of the glutathione peroxidases.

Deficiency. The deficit has been described in patients maintained with parenteral nutrition for prolonged periods of time. Symptoms of Se deficiency are pain and muscle weakness, loss of hair, and skin pigmentation. In children, growth retardation and development are due to failures in thyroid hormone metabolism. Some studies have shown that selenium deficiency is correlated with an increased incidence of neoplasias, especially prostate cancer.

Molybdenum (Mo)

Molybdenum is a transition element that in the body is usually linked to S or O in two oxidation states, Mo^{4+} and Mo^{6+}. Mo is an essential element; it is biologically inactive unless attached to a pterin core forming a cofactor similar to folic acid, *molybdopterin*, indispensable for the activity of several enzymes.

Natural sources. Mo is widely distributed in common foods, but its content varies according to the Mo concentration in the soil. Legumes, whole grains, cereals, chicken, liver, and kidney are good sources of Mo. The content in beef depends on the diet that cattle are fed with.

Absorption. Mo is absorbed throughout the gastrointestinal tract, in higher amounts in the proximal portions of the small intestine.

Transport. From enterocytes Mo enters the circulation, where it is transported as molybdate (MoO_4^{2-}) bound to albumin and α_2-macroglobulin, and also loosely bound to erythrocytes. Liver and kidneys are the organs that concentrate Mo at the highest amounts. Livestock, grazing on soil contaminated with Mo from industrial or mining areas, shows copper deficiency.

Excretion. Most Mo is excreted in the urine as molybdate in quantities that are directly related to the intake. It is also eliminated with the bile. Mo homeostasis depends on the regulation of its excretion more than its absorption.

Functions. Mo in molybdopterin, binds to the active site of three hydroxylases: *sulfite oxidase*, *xanthine oxidase/xanthine dehydrogenase*, and *aldehyde oxidase* and it is important for their activity.

Deficiency. Mo deficit is very rare in man, it could be produced by prolonged ingestion of diets containing high concentrations of sulfate, copper, or tungsten. For people living in areas with very low content of Mo in soil, increased incidence of esophageal cancer has been described.

Fluorine (F)

Although fluorine is not an essential element in humans, it has, at very low concentrations, beneficial effects, protecting against demineralization of calcified tissues. At present there is no evidence to include it as an essential factor in the diet. It is harmful if its intake exceeds 10 mg/day.

Natural sources. F is present in small amounts in foods, in concentrations that vary depending on the geographical region they come from. It is commonly found as fluorides bound to metals, nonmetals, or organic compounds, including proteins. Saltwater fish, consumed with its bone, and tea are rich sources of F. In some countries, fluoridation of tap water is performed, with addition of 37–63 mM (0.7–1.2 mg/L) of F. In this case, the daily intake of F can reach 1.4–3.4 mg/day.

Absorption. F is absorbed throughout the gastrointestinal tract.

Functions. F contributes to the mineralization of bones and teeth. F promotes the precipitation of calcium and phosphate and becomes itself incorporated into bone hydroxyapatite. The protein in the bone matrix has a high affinity for fluoride. Osteoporosis, bone decalcification common in older individuals, improves with consumption of moderate amounts of F. Fluorohydroxyapatite, which is less acid soluble than hydroxyapatite, makes teeth more resistant to decay. Moreover, it has inhibitory action of bacteria that proliferate in the dental plaque. Water fluoridation, at a concentration of 1 part per million, has a preventive effect on tooth decay, favoring the mineralization of the dental enamel.

Toxicity. Chronic ingestion of larger amounts of F than those indicated previously, produces *fluorosis*. In some areas, drinking water contains more than 4 mg/L of fluoride, which can cause toxic effects. Fluorosis is characterized by alterations in bone, kidney, muscles, the nervous system, and skin, where it leads to lesions that could favor skin cancer. In children, fluorosis causes white spots and brown streaks in newly formed teeth, which become brittle. F has inhibitory action on various enzymes, such as enolase, which results in inhibition of glycolysis.

Cobalt (Co)

Cobalt is not an essential element in humans, it is a component of cobalamin or vitamin B_{12} (p. 677), and is provided preformed in the food. Free Co is not used by the organism.

SUMMARY

Sodium is the main cation of the ECF (55% of the total Na^+ is in the ECF); the rest is located mainly in bone (40%). Plasma Na^+ concentration is 135–145 mEq./L, while inside the cells, it only reaches 10 mEq./L. Na^+ contributes to half of the osmolarity of ECF and drives fluid movement between different body compartments. The major route for Na^+ excretion is the kidney. Aldosterone, the main Na^+ regulator, activates Na^+ reabsorption in the renal tubules. Atrial natriuretic peptide also contributes to Na^+ homeostasis by favoring its urine excretion.

Potassium is the main cation of the intracellular space (90% of the total K^+ is in the ICF), with a concentration of

~150 mEq./L. In plasma, K^+ circulates in concentrations of 4–5.5 mEq./L. The uneven distribution of K^+ and Na^+ between intra- and extracellular compartments is due to the activity of the Na^+,K^+-ATPase of the cell plasma membrane. K^+ amounts are regulated by its elimination in the urine, via aldosterone-induced secretion in the renal distal tubules.

Chloride is the main anion in ECF, corresponding to 88% of the total body Cl^-. Plasma Cl^- concentration is 102 mEq./L. Since Cl^- is Na^+ main attending anion, its amounts follow those of Na^+. Cl^- contributes importantly to plasma osmolarity. Its regulation in the body follows that of Na^+.

Calcium is abundant in bone, which contains 99% of the total body Ca^{2+}. The plasma concentration of Ca^{2+} is 10 mg/dL (2.5 mM or 5 mEq./L), 50% of which exists as free Ca^{2+} and the rest is bound to proteins or forms nonionizable compounds. Ca^{2+} is the physiologically active form. When the pH increases (alkalosis) Ca^{2+} decreases and the opposite occurs in acidosis. Adults and children 1–10 years of age need to ingest ~800 mg of Ca^{2+} per day; infants require 350–550 mg/day; and women during pregnancy and lactation, 1200 mg/day. Milk and dairy products are the best sources of Ca^{2+}. Ca^{2+} homeostasis is primarily under the control of parathyroid hormone and 1,25-$(OH)_2$-D_3. The intracellular Ca^{2+} concentration is very low (0.1 μM). Various stimuli acting via membrane receptors promote Ca^{2+} entrance into the cell, via Ca^{2+} channels opening. This increase in Ca^{2+} serves as a signal to regulate many cell processes. Ca^{2+}-ATPase and Na^+/Ca^{2+} exchanger in the plasma membrane extrude Ca^{2+} from the cell, maintaining its cytoplasmic levels low. Cytosolic Ca^{2+} also varies by movement of this cation from intracellular stores (endoplasmic reticulum and mitochondria). Reduction of extracellular Ca^{2+} (below 3.5 mg/dL) causes increase in neuromuscular excitability (tetany). Prolonged loss of Ca^{2+} causes osteoporosis.

Phosphorus is mainly located in bone (over 80% of the total P) forming part of *hydroxyapatite*. In the ECF, P exists partly as phosphate anions $HPO_4^{2-}/H_2PO_4^-$; the plasma concentration of P is 2.5–4.8 mEq./L. P is absorbed in the jejunum and ileum, regulated by 1,25-$(OH)_2$-D_3. Phosphate levels are also maintained by controlling reabsorption in the kidney tubules, by mechanisms that are activated by calcitriol and inhibited by parathyroid hormone.

Magnesium is an essential cation, involved in many enzymatic reactions, neuromuscular excitability. It forms complexes with ATP and other triphosphate nucleotides. The plasma concentration of Mg is 1.5–3.0 mg/dL (0.625–1.25 mM or 1.25–2.5 mEq./L), of which 30% is bound to proteins.

Iron is mainly bound to proteins in the body. An adult contains a total of 4.0–4.5 g; 2.6–3.0 of which are in Hb and 1.0–1.5 g bound to *ferritin* and *hemosiderin* in intracellular stores. A normal adult man loses ~1 mg Fe per day; women have a higher loss, due to menstrual bleeding. Animal foods contain heme proteins whose Fe is more easily absorbed than Fe from plants. The presence of HCl and ascorbic acid in gastric juice is important for Fe absorption. The recommended amounts of Fe for an adult are 10 mg/day; higher amounts are required for pregnant women, children, and adolescents. Fe is absorbed in the intestine in its ferrous state. Fe transport across the intestinal mucosa is proportional to the body needs. Once in the enterocyte, Fe is transferred to the basolateral membrane and to the circulation, or stored in the cell as *ferritin*. Passage of Fe to blood is mediated by *ferroportin*; this requires the prior oxidation of Fe^{2+} to Fe^{3+}, which is catalyzed by *hephaestin* and *ceruloplasmin*. In plasma, Fe is carried by *transferrin*. The normal plasma level of Fe is 60–150 μg/dL. Transferrin transfers Fe to cells by binding to cell membrane receptors, which are internalized by cells via endocytosis. Fe is released in the cytosol, stored as ferritin, and the transferrin receptor is recycled after secretion from the cell. Fe is being continuously recycled in the body, mainly by the lysis and production of erythrocytes and Hb. Fe is eliminated mainly by the intestine. An overload of Fe results in formation of *hemosiderin* granules and hemosiderosis. Lack of Fe in the diet produces *hypochromic microcytic anemia*.

Trace elements include the following:

Zinc is present in foods of animal origin. It is absorbed mainly in the jejunum. In the enterocyte it is linked to metallothionein and in plasma, it is transported bound to albumin. When in excess, Zn is excreted in pancreatic juice. Zn is a constituent of many enzymes. Its deficiency causes delayed development, anemia, and hypogonadism.

Copper amounts in the body reach a total of 150 mg; 50% of Cu is found in muscle and bone, 10% is found in liver. Cu is part of enzymes, such as cytochrome oxidase, superoxide dismutase, monoamine oxidase, tyrosinase, and ceruloplasmin. Cu deficiency is mainly characterized by anemia.

Iodine is almost exclusively required for the synthesis of thyroid hormones. It is absorbed in the intestine as iodide. Thyroid captures a third of the total iodide absorbed, peroxidase oxidizes it and incorporates it in tyrosine residues into thyroglobulin, from which the T_3 and T_4 hormones are released. Chronic iodine deficiency causes hypothyroidism (*endemic goiter*).

Manganese is mainly provided by vegetables in the diet. It is required as a cofactor for some enzymes.

Molybdenum is an important component of several metalloenzymes.

Selenium is associated to several amino acids. Selenomethionine and selenocysteine form part of different proteins, such as glutathione peroxidase, iodotyrosine deiodinases, thioredoxin reductase, and synthetase selenophosphate.

Bibliography

Blanco, A., 2009. Micronutrientes. Ed. Biomed, Buenos Aires.

Bowman, B.A., Russell, R.M. (Eds.), 2003. Present Knowledge in Nutrition. eighth ed. International Life Sciences Institute, Washington, DC.

DiSilvestro, R.A., 2005. Handbook of Minerals as Nutritional Supplements. CRC Press, Boca Raton, FL.

Gropper, S.S., Smith, J.L., Groff, J.L., 2005. Advanced Nutrition and Human Metabolism, fourth ed. Thomson Wadsworth, Belmont, CA, pp. 402–407.

Preuss, H.G., 2003. Sodium, chloride and potassium. In: Bowman, B.A., Russell, R.M. (Eds.), Present Knowledge in Nutrition. eighth ed. International Life Sciences Institute, Washington, DC, pp. 330–339.

Shils, M.E., Shike, M., Ross, A.C., Caballero, B., Coussins, R.J. (Eds.), 2006. Modern Nutrition in Health and Disease. tenth ed. Lippincott Williams & Wilkins, Philadelphia, PA, pp. 194–210.

World Health Organization, Food and Agricultural Organization of the United Nations, 2004. Vitamin and Mineral Requirements in Human Nutrition, second ed. WHO, FAO, Geneva, pp. 59–93.

Molecular Basis of Immunity

IMMUNE SYSTEM

The body is constantly exposed to a variety of agents (viruses, bacteria, protozoa, and fungi); that can potentially cause disease. Against this, the body has a series of defense mechanisms, which constitute what is called the immune system. This system surveys the body and reacts against any incoming pathogen. The immune system has two main branches: (1) the *innate* system, which is present at birth, is nonspecific, and does not require previous contact with the invading agent to be turned on; and (2) the *adaptive* system, which evolved in vertebrates, is specific, adaptable, and requires a first encounter with the invading agent to become activated. In addition, the adaptive system has "memory" capacity. After the first contact with the invading agent, the system remembers it, and in subsequent encounters, will recognize it, responding with a faster, stronger, and more effective response.

INNATE SYSTEM

The components of this system are the skin and the epithelia lining the respiratory and gastrointestinal tracts. Their integrity is an essential barrier against the invasion of germs and molecules foreign to the body. If the incoming agents can overcome these barriers, a second line of defense, which involves immune cells, chemical mediators, and microbicides, will try to eliminate the invader.

Cells of the innate system include neutrophils, macrophages, natural killer (NK) cells, and dendritic cells.

Neutrophils and macrophages. These phagocytes express on their surface receptors that recognize molecules that are commonly found in many microorganisms. The most important among these receptors are the *toll-like* receptors (TLR).

TLR. Eleven different TLRs have been described in humans. They are expressed not only in neutrophils and macrophages, but also in epithelial, endothelial, and dendritic cells. TLRs have a first segment that binds extracellular ligands, a transmembrane α helix, which anchors the receptor to the cell plasma membrane, and a third cytoplasmic domain that is associated to signal transduction systems.

The ligands for TLR receptors are lipopeptides and lipopolysaccharides from the wall of many bacteria, double-stranded RNA from some virus, unmethylated DNA from endogenous cells, and formylmethionine residues from bacterial proteins.

Binding of a ligand to the TLR receptor causes its dimerization and induces a conformational change that enables the receptor to interact with adapter proteins, including MyD88 (*myeloid differentiation factor 88*) and TRIF (*TLR-associated activator of interferon*). Attachment of these proteins activates protein kinases and initiates a series of phosphorylation events, which will ultimately induce *cytokine* synthesis. The MyD88-dependent pathway activates *nuclear transcription factor κB* (NF-κB) which determines the release of inflammatory cytokines. The main cytokines of the innate immune system include tumor necrosis factor (TNF), IL-1, and chemokines (see subsequent sections). In some cases, these agents promote and maintain the inflammatory process which will contribute to eliminate the invading microorganism. In others, they determine production of interferon (IFN) (p. 774), which counteracts the action of viruses.

The TRIF pathway activates the transcription factor IRF-3, which controls the production of βIFN. This, in turn, activates *STAT* proteins (p. 564) and stimulates cytokine production. Mitogen-activated protein kinases (MAPK) (p. 563) also participate in facilitating cytokine release.

Binding of certain ligands to TLR may produce nonseptic (without infection) inflammation, which explains the production of autoimmune diseases not dependent on autoantibodies or autoreactive T cells, such as rheumatoid arthritis, ankylosing spondylitis, psoriasis, and others.

Lectins are molecules that serve for the recognition of carbohydrates (mainly mannose) commonly present on the surface of many microorganisms.

NK cells participate in eliminating cells of the body that have been infected. NK cells recognize the target cells through surface receptors. Also, they induce production of IFNγ, activate macrophages and promote destruction of those cells.

Dendritic cells are derived from precursor cells in the bone marrow and are related to mononuclear phagocytes. They have receptors that recognize microbes or virally infected cells, leading to cytokine production, including interferons (IFN).

Chemokines are among the cytokines released, particularly, by macrophages. The chemotactic effects on circulating leukocytes attract and recruit them to the area of infection. Chemokines promote the adhesion to the vascular endothelium walls, and stimulate their migration to the site of infection through the vascular walls.

Convergence of all these cell types and factors at the site of infection initiates the inflammatory reaction. Phagocytes engulf germs and foreign particles and degrade them via hydrolytic enzymes in lysosomes and by oxygen and nitrogen reactive species released during the "respiratory burst" that occurs on activated cells.

Circulating molecules of the family of *pentraxins*, such as *C reactive protein*, *collectins*, which act as lectins and *ficolins*, also have microbicidal action. Complement proteins (see p. 760) can also be considered as part of the innate system, especially those of the alternative pathway.

Several of the factors related to the innate system are also involved in the adaptive system. In many cases, the activation of the innate system contributes to enhance the response of the adaptive system, which greatly increases the effectiveness of the response. If the invading agent overcomes the barriers of the innate system, the adaptive system is the next mechanism of body defense.

ADAPTIVE SYSTEM

Humoral Immunity

The major components of the adaptive immune system include the lymphocytes, which specifically recognize and respond to the presence of invading agents (microorganisms, particles, or molecules). Two different types of lymphocytes exist, B and T. They circulate in the blood and lymph, they concentrate in lymphoid organs, and they are located in virtually all other tissues.

B and T lymphocytes. B lymphocytes are involved in the production of antibodies or immunoglobulins. Since antibodies are released to the extracellular space and are present in different body fluids, it is said that B lymphocytes constitute the *humoral immunity*. T lymphocytes, in association with other cells, eliminate the invading agent, and therefore, they are involved in *cellular immunity*.

Both types of lymphocytes are originated in the bone marrow from the lymphoblastic cell progeny. B cells remain in the bone marrow until they are fully mature; instead, T cells precursors pass to the thymus, where they undergo a process of differentiation that converts them into mature T cells. Once maturation is complete, both B and T lymphocytes enter the circulation and reach the spleen, lymph nodes, tonsils, and other lymphoid tissues. It is in these organs where immune responses are initiated. During differentiation, lymphocytes have distinctive surface proteins. A group of these proteins, the *cluster of differentiation* (CD) are differentially expressed at defined stages of their ontogenesis. CD proteins can therefore be used as markers to identify lymphocyte types and their stage of maturity.

Antigen. This is a term used for any agent that, recognized by the adaptive system (B or T lymphocytes), is able to trigger an immune response. The immune system recognizes as antigens agents that have molecular structural characteristics different from those of the invaded organism, which allows it to discriminate between own and foreign structures. Aggressors, such as viruses, bacteria, protozoa, and fungi are immunogenic. Also, cells and macromolecules different from those of the host individual (from another species, or even from a different individual of the same species), and tumor cells from the same individual behave as antigens.

Epitope. To identify it as an antigen, the immune system does not need to recognize the entire invading organism, cell or macromolecule. Recognition of relatively small specific portions of the intruder agent or its molecules is sufficient to identify it as foreign to the body. These regions are known as *antigenic determinants* or *epitopes*. Specific domains within proteins and heteropolysaccharides behave as epitopes. Microorganisms contain multiple of these epitopes, each of which can trigger the immune response that will serve to identify and react against them.

Kinetics of the Immune Response

When the immune system has its first contact with an antigen, the *primary response* is evident approximately 5 days after the exposure by the appearance in plasma of *antibodies* of the IgM type (p. 751) that specifically recognize the antigen. A few days later, antibodies of the IgG type, with the same antigen specificity, also appear. The IgM concentration decreases rapidly, while the IgG continues increasing until they reach a maximum 10–15 days after first contact with the antigen. The antibody level is maintained for 15–20 days, after which it begins to slowly descend. After ~4 months, the concentration of specific Ig in blood plasma is reduced to low values (Fig. 30.1A).

If after several months or years from the original exposure, there is a second invasion by the same antigen, the immune system responds with a *secondary response*. This response is different from the previous one in various aspects: (1) the appearance of antibodies occurs more rapidly; it is evident on the third day, (2) the Ig concentration in plasma reaches levels up to 100 times higher and lasts longer than that of the primary response (Fig. 30.1B), and (3) the predominant immunoglobulins in the secondary response are of the IgG type, which have higher affinity for the antigen than those of the primary response.

The causes of these changes in levels of plasma antibodies and the differences between primary and secondary response will be discussed later.

Structure of Immunoglobulins

All classes of immunoglobulins (IgG, IgA, IgM, IgD, and IgE) have a similar structure. They are composed of four polypeptide chains symmetrically arranged. These polypeptides are assembled resembling a letter Y (Fig. 30.2). While this is the basic shape of all Ig, there are

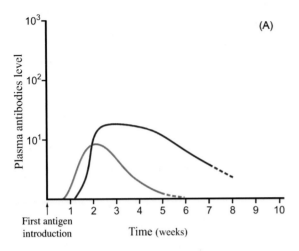

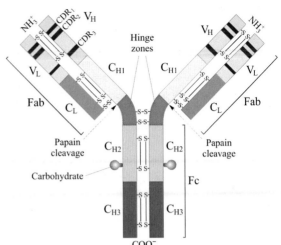

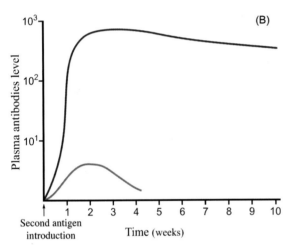

FIGURE 30.1 **Kinetics for the appearance of immunoglobulins (IgM** *gray,* **IgG** *red***) in blood plasma in response to an antigen.** The levels of antibody are plotted against the time elapsed since inoculation of the foreign agent. (A) First contact of the antigen with the immune system (primary response). (B) Second encounter with the antigen, 6 months after the initial contact (secondary response).

FIGURE 30.2 **Scheme of the structure of an IgG unit.** The polypeptide arrangement is symmetrical, containing four chains [two light (L) and two heavy (H)], bound together by disulfide bridges. Each subunit has a variable domain (V_L for light, V_H for heavy). *Transversal red areas* in the variable regions (CDR_1, CDR_2, and CDR_3) are hypervariable segments. The constant and variable domains (C_L for light; C_{H1}, C_{H2}, and C_{H3} for heavy) have S—S intrachain bonds. Between C_{H1} and C_{H2} there is a *hinge region*. Carbohydrates are linked to the C_{H2} domains. Papain treatment separates three fragments, two Fab fragments, each consisting of a complete L and an H chain piece containing V_H and C_{H1} domains. L and H pieces are linked by a disulfide bridge. The N-terminal portion of the Fab fragment is the antigen-binding site. The remaining fragment, named Fc, consists of the C_{H2}, C_{H3}, and both hinge portions linked by S—S bonds.

differences among different types of antibodies. Table 30.1 summarizes the characteristics of all five classes of immunoglobulins. The following description corresponds to IgG.

Two of the chains are larger and are designated *heavy* or H. They are identical to each other, have 440 amino acid residues each and a mass of approximately 50 kDa. The other two subunits, called *light* or L, are also indistinguishable from each other, they consist of 220 amino acids, and have a weight of approximately 25 kDa.

The four chains of immunoglobulins are held together by disulfide bonds (—S—S— interchain bonds) and noncovalent interactions (Fig. 30.2). Each of the subunits has domains of similar secondary and tertiary structure, consisting in ~110 amino acid residues, with intrachain disulfide bridges between cysteine residues separated by ~60 amino acids. Two of these domains are found in the light chains and four in the heavy chains (Figs. 30.2 and 30.3).

TABLE 30.1 Immunoglobulins

| Class | Mass (kDa) | Sedimentation coefficient (S) | Plasma concentration | | Light chains | Heavy chains | Structure |
			(mg/dL)	(%)			
IgG	150	7	1.250	75.6	κ o λ	γ	$κ_2γ_2$ or $λ_2γ_2$
IgA	150–600	7–13	250	15.1	κ o λ	α	$(κ_2α_2)_n$ or $(λ_2α_2)_n$
IgM	950	18–20	130	7.9	κ o λ	μ	$(κ_2 μ_2)_5$ or $(λ_2 μ_2)_5$
IgD	180	7	20	1.2	κ o λ	δ	$κ_2δ_2$ or $λ_2δ_2$
IgE	180	8	3	0.2	κ o λ	ε	$κ_2ε_2$ or $λ_2ε_2$

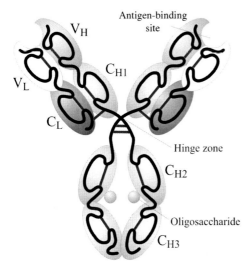

FIGURE 30.3 **Scheme of a structural unit of IgG.** The domains of the heavy (H) and light (L) subunits are represented. *Pink,* heavy chains variable domains (V_H); *light gray,* heavy chains constant domains (C_H); *white,* light chains variable domain (V_L); *dark gray,* light chains constant domains (C_L). *Red lines* indicate intra- and interchain disulfide bridges. *Pink spheres* indicate carbohydrates.

Variable domains. The V initial domain of each chain comprises half of the total length of the light and one-fourth of the heavy chain. This protein region presents differences in the amino acid sequence among different antibodies and is called the *variable region*. It is distinguished with the notation V_L for variable region of the light chain and V_H for the heavy chain. Within

these variable regions there are at least three segments in which the sequence differences among different antibodies are even greater. These are called *hypervariable* segments. Hypervariable sequences represent approximately 25% of the total length of the V_L and V_H domains.

Antigen-binding site. The variable region of a heavy and a light chain form a niche, which can bind the antigen. This region, called *paratope*, provides the Ig its specificity. The conformation that the paratope adopts allows the antibody to recognize a particular antigen. When there is structural complementarity between the binding site of the Ig and the antigen epitope, noncovalent bonds (hydrogen bonds, hydrophobic and electrostatic attractions, and van der Waals forces) are established between the epitope and paratope. The largest differences in the amino acid sequence are found in the hypervariable segments, they are primarily responsible for the specificity of the antibodies. Those segments are also called *complementarity determining regions* (CDR).

An immunoglobulin G molecule has two binding sites and can fix two antigens. Both antigen binding sites of an Ig structural unit have identical configuration and equal specificity. The affinity of the antibody for the antigen depends on the degree of complementarity and reciprocal interactions between epitope and paratope. The strength with which an antibody binds to the antigen is called *avidity* and depends on antigen and antibody's affinity.

When antigens that possess two or more identical epitopes bind to an antibody, they form a net in which the Ig appears as a bridge between two antigens, which usually precipitates.

In the case of some bacterial toxins or viral antigens, binding to the antibody is sufficient to neutralize and prevent their deleterious cell effects; however, in general, the elimination of many immunogens requires their interaction with other cells or molecules (see subsequent sections).

Constant domains. The segment of the H and L chains, from the end of the variable region to the C-terminus, has the same primary structure for all subunits of the same type and is called the constant region. Each light chain has one constant domain, designated C_L. In contrast, heavy chains have three constant zones (C_{H1} to C_{H3}). A 10–15 amino acids segment between the C_{H1} and C_{H2} domains corresponds to the *hinge* region (Fig. 29.2). This area contains several cysteine and proline residues. The presence of proline gives flexibility to this segment of the molecule, modifying the angle between the two arms of the immunoglobulin, which facilitates approximation and binding of the antigen. Carbohydrates bind to the immunoglobulin C_{H2} domain. This N-glycosidation takes place between an asparagine and *N*-acetyl-glucosamine.

Fab and Fc fragments. If an immunoglobulin is treated with papain or papaya proteinase I (a proteolytic enzyme from papaya), the complex is hydrolyzed at the hinge region of both heavy chains and originate three segments: two of them, identical to each other, are constituted by a complete light chain bound through disulfide bonds to a portion of heavy chain comprising the V_H and C_{H1} domains. They are called *Fab* segments and contain one of the antigen-binding sites. The third fragment consists of the remaining constant portions of both heavy chains, from the beginning of the hinge region to the C-terminus, linked by S—S bridges. This segment is called *Fc* (Fig. 30.2).

The *Fc* segment is responsible for nonspecific interactions of the antigen–antibody complex with other molecules and receptors on the surface of cells. The Fc fragment participates in different aspects of the immune defense. Examples of its role includes the following: (1) Antibody binding to the surface of invading bacteria is not enough to stop their multiplication. However, the antibody promotes the uptake of the bacteria by phagocytic cells that will lyse them. Phagocytes can recognize bacteria by binding to the Fc regions of the Ig bound to the bacteria. The process of coating of pathogens or foreign particles by antibodies is called *opsonization*. (2) Activation of the complement system (see subsequent sections) is initiated when one of its components binds to the Fc portion of antigen–antibody complexes. This activation triggers lysis or phagocytosis of the invading agent. (3) Immune protection of the newborn depends on IgG transferred from maternal blood during fetal life. The Fc segment of antibodies allows the transport of Ig across the placenta. (4) A specific type of immunoglobulin, IgE is involved in the defense against parasites and in the allergic response. The Fc portion of IgE binds to specific receptors on eosinophils, platelets, and neutrophils, to mediate toxic action against parasites. IgE can also bind to basophils, mast cells, and macrophages, which triggers hypersensitivity phenomena related to the release of leucotrienes, prostaglandins, histamine, and lysosomal enzymes, all responsible for allergic and anaphylactic reactions. (5) IgA is a secreted form of immunoglobulin, which is released by submucosal plasma cells. The transport of IgA across epithelia depends on binding of the Fc portion of IgA to a receptor on epithelial cells. Moreover, the complex interactions of immunoglobulins with different cells are mediated through these cell surface receptors.

Heavy Chain Isotypes

The five types of immunoglobulins (IgG, IgA, IgM, IgD, and IgE) are distinguished by the

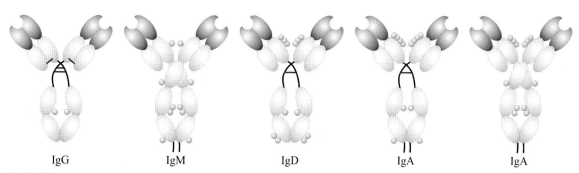

FIGURE 30.4 **Basic structure of different types of Ig.** *Red lines* indicate interchain disulfide bridges. *Pink*, variable domains, *gray*, constant domains. *Pink spheres*, oligosaccharides.

isotype or class of heavy chain. Five different variants of Ig heavy subunits are expressed in the body, designated by Greek letters: μ (mu), γ (gamma), α (alpha), δ (delta), and ε (epsilon). IgM chains have μ chains, of which there are four different subclasses, μ_1 to μ_4. IgG has γ chains, with four subclasses (γ_1 to γ_4), γ_3 has a much longer hinge region than the other IgGs. IgA has two α chains (α_1 and α_2), IgD, has δ and IgE, ε chain types. For all immunoglobulins, the light chain may be either κ (kappa) or λ (lambda). Approximately 65% of circulating antibodies have kappa light chains and the rest have lambda chains. In an Ig molecule, both light and heavy chains are always of the same class, that is, there are not Ig having κ in one light subunit and λ in the other, or heavy chains with μ in one subunit and γ in the other.

The heavy chains of IgG (γ), IgA (α), and IgD (δ) have three constant domains (C_{H1}, C_{H2}, and C_{H3}) in addition to the variable (V_H) domain. The heavy chains of IgM and IgE (μ and ε, respectively) have four constant domains (C_{H1} to C_{H4}) and lack the hinge region. The α and μ chains have a C-terminal piece of 18 amino acids following the final domain (C_{H3} and C_{H4}, respectively). The amount of oligosaccharide chains and their place of insertion vary depending on the immunoglobulin considered (Fig. 30.4).

Immunoglobulin G, D, and E are formed by a single structural unit as the one described. Each has two antigen-binding sites. IgM are

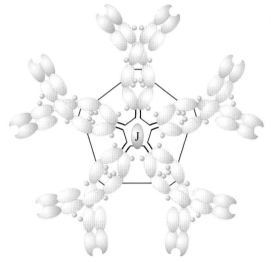

FIGURE 30.5 **IgM pentamer.** *Red lines* indicate disulfide bridges. *J*, linking chain.

pentamers that have a total of 10 heavy chains, 10 light chains, and 10 antigen-binding sites. The pentamer units are linked by disulfide bonds, placed between two adjacent C_{H3} domains and between the 18 amino acid C terminal pieces. An additional chain of 15 kDa, designated J, is bound by S—S to two of these segments and promotes the polymerization of the pentamer (Fig. 30.5). The C-terminal ends of all heavy chains in IgM are directed toward the center of the complex and antigen-binding sites radiate toward the periphery of the molecule (Fig. 30.5).

IgA is found in plasma as a monomer; however, in secretions (saliva, tears, and milk), and in epithelia from the respiratory, genitourinary, and gastrointestinal tracts, where it is more abundant, it exists as dimers. The IgA$_1$ subtype predominates in serum, nasal secretions, tears, and milk. IgA$_2$ is most abundant in colon. Both units of the IgA dimer are associated by a J chain; another chain of 70 kDa, known as the *secretory component*, facilitates the transport of IgA and protects it from the digestion by proteases (Fig. 30.6).

Membrane Immunoglobulins

B lymphocytes have both IgM and IgD attached to their plasma membrane. Membrane-bound immunoglobulins (Ig$_m$) have an additional piece of 29 amino acids in the C-terminal portion of the heavy chains, which is absent in the corresponding plasma Ig. This additional segment consists of hydrophobic amino acids that cross the membrane lipid bilayer and serves to anchor the Ig to the B-cell surface. The last residues, of hydrophilic character, project toward the cytoplasmic side. Membrane IgM is

a monomer; it does not form pentamers as the IgM from plasma.

The antigen-binding sites of membrane immunoglobulins face the external side of the cell and this allows them to function as receptors. The antigen specificity of membrane immunoglobulins is the same in a given lymphocyte. However, there is a great variety with respect to antigen specificity among different B lymphocytes in an individual. This diversity confers the body with receptors that can virtually recognize any invading antigenic molecule. The origin of this variability will be discussed later.

IgM and IgD embedded in the membrane of B cells are associated to two protein complexes, which are composed by a heterodimer of α and β subunits (Fig. 30.7). Each α and β subunit has a transmembrane segment and a cytoplasmic domain which can be phosphorylated by tyrosine kinases; this allows them to initiate signal transduction signals in the cell.

Clonal Selection

The immune response begins when an antigen that entered the body binds to the Ig on a

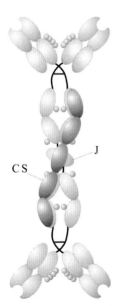

FIGURE 30.6 **IgA dimer.** *Red lines*, disulfide bridges; *J*, linking chain; *CS*, secretory component *(dark gray)*.

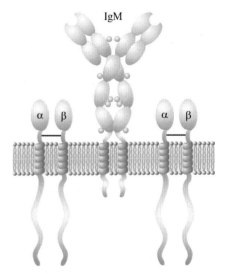

FIGURE 30.7 **IgM–receptor complex of the B-cell membrane.**

B lymphocyte surface. Obviously, only cells having receptors with paratopes complementary to the antigen epitope participate in the response and are then activated. One of the T-cell types also plays an important role in this process.

The activation of the B lymphocytes after association with the antigen (see subsequent section) triggers the division and differentiation of the B cells into *plasmatic cells*. These cells synthesize and secrete antibodies (designated as Ig$_s$ for secretory) with the same specificity as those present in the plasma membrane of the B cells. The multiplication of plasma cells from a single B cell represents a *clonal selection* of these cells.

Since an invading agent may possess multiple antigenic determinants, usually many plasmatic cell clones are generated, and thus many different antibodies are produced. This is referred to as a *polyclonal response*.

Memory cells. In the process of B-cell differentiation, in addition to the production of plasma cells, *memory cells* are produced that also synthesize and secrete antibodies with the same specificity of those present in the original B lymphocyte, except that they have a much longer half-life. While the original B lymphocytes last several weeks, memory cells remain in circulation for months or years. Formation of these memory cells explains why, during a second antigen invasion, the body responds more rapidly and intensely than during the first exposure. The second response produces a greater number of B cells. Subsequent activation of these cells exponentially multiplies the immune response.

Characteristics of the Antigen

Antigens or immunogens that trigger the humoral immune response are generally large molecules, usually proteins or heteropolysaccharides. The antigenic determinant or epitope is only a portion of these macromolecules and must be exposed; otherwise, it could not interact directly with the immunoglobulin-binding site.

Lipids and nucleic acids are ineffective as antigens, but increase their immunogenicity when bound to protein. Some small molecules that may not be immunogenic when free, become strong antigens when attached to appropriate macromolecules. These small molecules are called *haptens*.

The fit between antibody and antigen is not rigid (as in the "key and lock" model); instead there is a mutual molecular adaptation, similar to that occurring between an enzyme and its substrate. Antigenic determinants with more plasticity have better chances of becoming accommodated into the Ig-binding site, which is also plastic. Folded portions of the immunogen with certain degree of flexibility represent good antigenic determinants. For antigens of protein nature, an epitope is commonly constituted by approximately 15 amino acids, not necessarily contiguous along the polypeptide chain. In this manner, the epitope results from the clustering of amino acids in a particular region of the molecule as a result of the folding and three-dimensional arrangement of the polypeptide chain. If the tertiary and secondary structures of the protein are disrupted by denaturing agents, the protein may no longer be recognized by the antibody, since the epitope may be lost.

While the most striking characteristic of an antibody is its specificity, sometimes an Ig can bind to various antigens. This event is called *cross-reactivity* and occurs when different immunogens present common or similar epitopes, as is the case for homologous proteins from different species.

GENETIC DIVERSITY OF IMMUNOGLOBULINS

Antibody Heterogeneity

It is estimated that the body is able to produce over 10^7 immunoglobulins with different antigen specificity. This ensures that, although the variety of antigens to which an individual may be exposed throughout life is very large, an antibody will be produced that will have the ability to recognize and bind to the incoming antigen.

This remarkable structural diversity of immunoglobulins primarily resides in the differences in amino acid sequence in the hypervariable segments of the V_H and V_L polypeptide chains. The rest of the immunoglobulin remains very constant. Antibody diversity was a phenomenon difficult to explain from a genetic standpoint because the number of genes encoding for immunoglobulin-related proteins is much smaller than the number of different antibodies that can be synthesized in an individual. Advances in the field of molecular biology helped clarifying this phenomenon. The key discovery was the finding that the genes encoding for immunoglobulins, cluster in three loci of different chromosomes, and they undergo a DNA rearrangement. During maturation of B lymphocytes, at each of the immunoglobulin loci, DNA segments are randomly selected and combined to provide a unique genetic template from which distinct immunoglobulins can be synthesized. This provides each B cell with the capacity to generate antibodies different from those made by other B cells, highly enhancing the antibody diversity that the body needs to react against the many possible incoming antigens.

Arrangement of the Ig Heavy-Chain Genes

In germ cells, the variable region of the heavy chain (V_H) is encoded by three sets of genes. Listed from the 5' to the 3' end, these gene sets include the V (or variable) genes, the D (or diversity) genes, and the J (or joining) genes. These genes are separated by noncoding DNA sequences (Fig. 29.8A).

The genomic V region contains 51 genes, each of which is composed of two exons. They encode a leader or signal peptide, required for the passage of the polypeptide chain through the endoplasmic reticulum (ER) membrane, and a 98–amino acid polypeptide corresponding to the variable region. The genomic D region includes a group of 25 genes that encode amino acids 99–105 of the V_H polypeptide. The J gene set includes six genes, which encode the last amino acids of the V_H domain.

Near the J genes are the genes that encode for the constant zone (C_H). Unlike the variable regions that have multiple genes, C_H possesses only one gene for each of the H chains and its isotypes. These genes are arranged from the 5' to the 3' end, in the following manner: μ, δ, α_1, γ_2, γ_4, ϵ, and α_2. Each is composed of multiple exons for the constant domains (three or four depending on the class). In some Ig isotypes, between exons for the C_{H1} and C_{H2} domains, there is a smaller exon, which encodes for the Ig hinge region. At the end of the constant domain exons, the gene corresponding to the intramembrane IgM terminal portion is found (Fig. 30.11).

Arrangement of Light-Chain Genes

Kappa chains. The variable region of the κ light chains is encoded by two groups of genes (V and J). The set of D genes corresponding to the heavy chains is missing. The V genes include approximately 40 genes, each of which encodes the leader peptide and amino acids 1–95 of the variable domain. The J group of genes has 5 genes which encode for amino acids 96–108. Downstream from this region is a gene for the constant domain (Fig. 30.8B).

Lambda chains. Two sets of genes encode for the variable region of the λ light chain. The estimated number for the V genes is 41 and for J is 4. The constant region of the H and L_κ chains are encoded by only one gene. In contrast, the L_λ chain has at least six genes for domain C. The arrangement of these genes is shown in Fig. 30.8C.

Rearrangement of Ig Genes

During maturation of B lymphocytes in the bone marrow, a recombination and removal of the genes encoding the variable domains occurs. The process begins with the rearrangement of the genes corresponding to the variable region

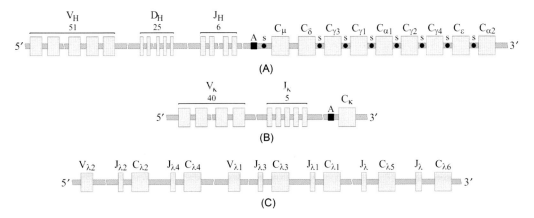

FIGURE 30.8 **Arrangement of the immunoglobulin genes in human germ cells (schematic).** (A) Heavy chain genes (located on chromosome 14). Variable domain genes are separated into three groups, V_H comprises 51 functional genes, D_H 25 genes, and J_H 6 genes. Each gene of the V group consists of two exons (not shown here); one encodes the leader or signal sequence and the other, the rest of the protein. The constant region has only one gene for each isotype. Each C gene is composed of multiple exons and the corresponding introns (not shown). There are exons for each of the C_H domains, for the hinge region and for the Ig_m intramembrane portion. The *black squares* in front of the C genes (A) represent the enhancer sequence. The *black circles* in front of each H chain gene (except the δ) indicate a sequence controlling the isotype switch. (B) κ light chain genes (located on chromosome 2). Variable domain genes are separated into two groups: C_κ, 40 genes, and J_κ, 5 genes. There is only one constant region gene, preceded by the enhancer sequence (A). (C) λ light chains genes (located on chromosome 22). Two sets of variable genes exist (including 31 V_λ and 4 J_λ) and 6 constant domain genes (genes C_λ intercalated with J).

of the H chains. One of the D genes becomes close and binds to a J gene to form a DJ segment. Subsequently, a V gene, selected randomly joins the DJ piece to form a VDJ gene block. The DNA segments between the V, D, and J genes are cleaved and removed. The resulting *rearranged configuration* encodes the variable domain of the heavy chain (Fig. 30.9).

For the light chain locus, the rearrangement involves the approach and binding of one V gene to a J gene to form the VJ block that will be transcribed (Fig. 30.10).

An enzyme system catalyzes the splicing of DJ, VDJ, or VJ segments, removing material between the selected pieces. Noncoding sequences at the 3′ end of the V and D genes and the 5′ end of the D and J genes are complementary. This allows their pairing, forming a loop, which will be cleaved, to allow their removal from the B cell genome. This DNA rearrangement only takes place in the lymphocyte germline; it is not present in other eukaryotic cells.

The DNA rearrangement of B cells allows the synthesis of a wide number of immunoglobulins. It is estimated that DNA rearrangement can generate 7650 different possible VDJ combinations for the H chain (51 V × 25 D × 6 J). For the L_κ chain, the number of different VJ assemblies is 200 (40 V × 5 J) and for the L_λ chain 124 (31 V × 4 J). Therefore, a total of 324 different light chains (κ + λ) can be produced.

Once the H and L chains are formed, there are no restrictions for their association into full Ig molecules. The H chains resulting from any VDJ assembly can associate with L chains in any possible configuration. The amount of different antigen-binding sites that can be formed between H and L (κ and λ) chains are close to 2.5×10^6. DNA rearrangements occur at random; the possibility of obtaining identical assemblies VDJ–VJ in two different cells is very low.

Other factors that can increase immunoglobulin variability. In addition to DNA rearrangements,

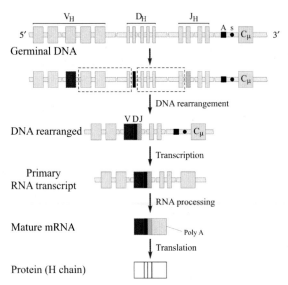

FIGURE 30.9 Rearrangement of the variable region of the H gene in human B lymphocytes. Genes of the variable chain are indicated in *red* for V, in *black* for D and in *pink* for J. For the constant region, only Cμ is shown. Surrounded by *dashed line rectangles* are intermediate portions of DNA which are eliminated, leaving a VDJ block encoding for the variable region. Those portions of the V and J genes are removed from the primary transcript during RNA splicing and the remaining constant region is added to complete the "mature" mRNA that will be translated into the corresponding region of the H chain. In the 3′ end, a poly-A tail is inserted. The *square* (A) and the *black circle* indicate *enhancer* and *switch* control regions, respectively.

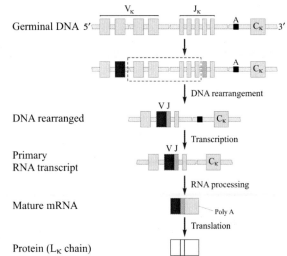

FIGURE 30.10 Rearrangement of the variable region of the Lκ chain gene in B lymphocytes. The DNA intermediate portion, surrounded by a *dashed rectangle* corresponds to the DNA segment eliminated. The rearranged DNA forms a VJ block coding for the variable region of the immunoglobulin. During RNA splicing, regions of the V and J nonselected genes are removed and the constant region gene transcript is added. In the 3′ end, the poly-A tail is added. The *black square* (A) indicates the enhancer. *Red,* Selected V genes; *pink,* J genes.

other mechanisms contribute to increase the variability of immunoglobulins. These include:

1. Differences in splicing of the VDJ and VJ regions.
2. Addition of nucleotides to the 3′ end of the VDJ and VJ segments catalyzed by terminal deoxynucleotidyl transferases.

Both of these processes contribute to modify the nucleotide sequence of the immunoglobulin gene, generating codons that were not present in the original genes, or altering the DNA reading frame. These mechanisms take place during the gene rearrangement event in the lymphocyte B precursor cells and can significantly increase immunoglobulin diversity. If the nucleotide change

results in a stop codon, the DNA rearrangement is abortive and nonviable. Another factor that increases the diversity of the immunoglobulin variable region is *somatic hypermutation*. This event takes place after the antibody chains have been assembled (see subsequent sections).

Allelic exclusion. When a VDJ assembly is formed, a phenomenon called allelic exclusion takes place, which consists in the inhibition of recombination in the other allele of the same cell. (All diploid cells have one set of maternal and another set of paternal chromosomes. Genes of each set occupying the same locus in homologous chromosomes are called alleles.) Allelic exclusion ensures that synthesis of only one immunoglobulin arrangement occurs in each B cell. Only when there is an abortive rearrangement in one allele, the immunoglobulin assembly continues on the other. If this second attempt does not obtain a functioning

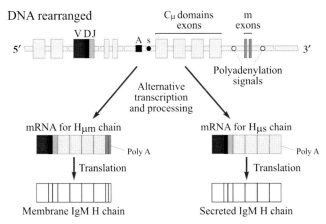

FIGURE 30.11 **Scheme of membrane and secreted IgM transcription and processing.** Only the exons of isotype μ and the genes of the rearranged variable region are represented. Downstream of the exons encoding the constant domain (C_{H1} to C_{H4}), two small exons corresponding to the transmembrane domain *(dark gray)*, followed by polyadenylation signals are found. Transcription of Ig_m or Ig_s depends on these exons. Since the VDJ block does not change, both Ig have identical antigen specificity.

VDJ block, the cell does not continue its differentiation.

Once a productive VDJ piece is obtained, the assembly of the genes for the variable region of the κ light chain begins. If a viable VJ is obtained, there is allelic exclusion, and the other allele for the V domain of the κ chain does not undergo rearrangement. In addition, the rearrangement of the λ chain gene is also inhibited. This is the reason why there is never simultaneous presence of subunits for the κ and λ chains in the same cell. Only if assembly in both κ alleles has failed, the VJ for the λ chain is rearranged.

Membrane Ig synthesis. Mature B lymphocytes synthesize membrane immunoglobulins (Ig_m), mostly IgM and IgD, which function as plasma membrane receptors. Ig_m synthesis involves transcription and rearrangement of the H genes (to obtain the VDJ block) and of the constant region of the μ and δ genes, including two exons encoding the transmembrane segment. During RNA splicing, the selection of the exons corresponding to the constant domains of the μ and δ chains takes place. Addition of the transmembrane domain is controlled by the polyadenylation signal in the 3' end of the corresponding exons. Although the generated Igs have different heavy chains, their antigen specificity is identical, because the same rearranged VDJ is used (Fig. 30.11).

Synthesis of Different Ig Isotypes

A lymphocyte activated by binding to the antigen gives rise to a large number of plasma cells, which synthesize antibodies of the same specificity as that of the original lymphocyte (clonal selection). The Ig synthesized by plasma cells and secreted into the medium (Ig_s) do not possess the intramembrane domain at the C-terminus of the H chains. Selective Ig_m or Ig_s synthesis is regulated by polyadenylation signals flanking the exons for the intramembrane section (Fig. 30.11).

Once activation of the B cell has started, IgD are no longer produced. Plasma cells first generate IgM and, after a few days, they produce IgG and other Ig types. The switch in heavy chain isotype requires a new DNA rearrangement (Fig. 30.12). The VDJ block approaches to a gene different from μ; the intermediate segment between VDJ and the new C gene is removed. Some DNA pieces of conserved sequence located in the noncoding segment before each C gene (indicated by s in Fig. 30.12), play an important role in the Ig isotype switching. The δ gene is the only one that does not have this sequence.

Despite the change in isotype, the specificity of the Ig synthesized is not modified because the same VDJ is always used.

Regulatory Sequences in Ig Genes

Ig synthesis only takes place in mature B cells and in plasma cells derived from them.

Regulation of the Ig genes primarily depends on a DNA region located between the sequences of the J gene and the genes encoding the constant segments of the Ig chains. These are enhancer sequences, responsible for activating transcription and expression of the Ig genes. When the isotype switch occurs, the enhancer is moved along with the VDJ block, immediate to other genes encoding constant Ig regions (Fig. 30.12).

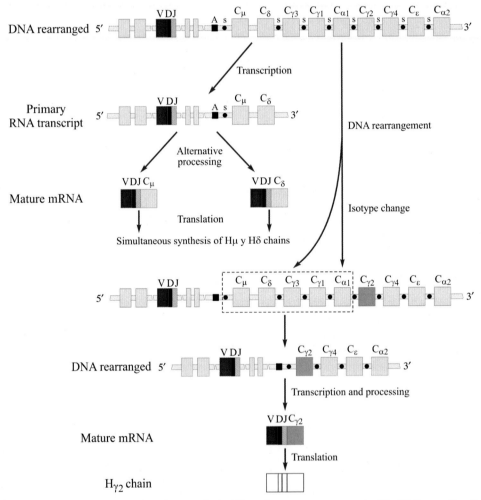

FIGURE 30.12 **Switch of Ig heavy chain isotype.** In the B lymphocyte, rearrangement of the DNA encoding the variable region of the α and δ chains are alternatively transcribed and processed. The switch in Ig isotype requires a new DNA rearrangement (the figure provides an example for the switch to γ₂). C genes other than that selected are removed (shown in the *dashed line rectangle*). *Black circles* (s) before each C gene indicate switch sequences. Independent from the isotype chosen, the VDJ block used is always the same; therefore, all synthesized H chains have the same specificity.

The enhancer sequences are not only involved in activating Ig synthesis in plasma cells, but also in the neoplastic transformation that these cells undergo when an oncogene is inserted at a position close to an Ig enhancer (i.e., Burkitt's lymphoma).

Memory Cells

A fraction of the lymphocytes generated by division of the original B cell, serve as *memory cells*, which can remain in circulation for long time (months to years). Each memory cell has the same antigen binding specificity of that of the original cell. Rearrangement of the VDJ and VJ genes in these cells has a high rate for generating mutations (10^{-3} per nucleotide per cell generation). This phenomenon, called *somatic hypermutation*, causes amino acid changes in the variable regions of Ig. This allows increasing the affinity of the antibody for a particular epitope, improving the capacity of the Ig to recognize and bind the antigen. After the first exposure, and in following encounters with the antigen, the cells that had undergone a favorable mutation react preferentially, generating new memory cells. This phenomenon explains why the response to repeated encounters with the antigen is stronger than the previous one. In addition, the new exposure to the antigen finds a much larger number of B lymphocytes that can recognize it and from which clonal expansion selects those which will be best adapted to bind the antigen.

Monoclonal Antibodies

These are antibodies synthesized by plasma cells that belong to the same clone and are all derived from a single original lymphocyte. These monoclonal antibodies contrast with polyclonal antibodies, which are made by different B cells. Monoclonal antibodies are monospecific; they have identical specificity for the epitope in a particular antigen. Köhler and Milstein were the first to develop methods to produce monoclonal antibodies that could recognize an antigen of interest. The method is based in the hybridization

of lymphocytes (isolated from the spleen of a mouse immunized with an antigen of interest) and malignant myeloma cells containing defects that render them incapable of growing in a culture medium with a special composition. The hybridization of the spleen and myeloma cells can be carried out in vitro by using a fusogenic agent such as propylene glycol. These myeloma cells have lost their capacity to synthesize hypoxanthine guanine phosphoribosyl transferase, an enzyme involved in the synthesis of nucleic acids. This defect makes the cells susceptible to the folic acid analog aminopterin and thymidine, which when added to the culture medium will prevent the survival of the cells. The unfused spleen cells cannot grow in culture indefinitely, because they are not immortal. In contrast, the lymphocyte-myeloma hybrid cells can selectively and indefinitely grow in the aminopterin and thymidine–containing medium. This is because the hybrid cells have recovered hypoxanthine guanine phosphoribosyl transferase from the spleen cells and have gained immortality through the myeloma cells. The hybrid cells generated can be grown indefinitely and produces the desired antibodies, all identical from those of the plasma cell from which they originated. These antibodies can be easily isolated and purified from the culture medium in which the hybrid cells are growing. Expansion of the cells allows the production of high quantities of antibody.

An alternative approach includes growing the hybrid cells in vivo, after inoculating them into a mouse, frequently in the peritoneal cavity. Under this condition, the cells produce tumors called *hybridomas*, which produce the desired antibody. The Ig can then be recovered from blood plasma, or the peritoneal fluid of the mice.

Monoclonal antibodies have many different applications in medical research, mainly related with diagnostic tests. For example, they can be applied to identify tumor antigens, allowing early diagnosis of tumors and metastases, and to recognize pathogenic microorganisms or histocompatibility antigens. This includes different immunochemical assays, including immunoblot, immunohistochemistry, and immunocytochemistry. Monoclonal antibodies can also be used to purify their target antigen, using immunoprecipitation methods. They have also been used as therapeutic agents, particularly in the treatment of cancer and rheumatoid arthritis.

COMPLEMENT

The complement system consists of a series of plasma proteins that are involved in body defense, promoting the lysis or phagocytosis of foreign cells, bacteria, and viruses. The complement cooperates with and enhances the capacity of antibodies and phagocytic cells. Some of these proteins circulate in an inactive form, as zymogens or proenzymes and are activated by a sequential cascade of reactions that involve proteases. The final action of complement activation is to attract and stimulate phagocytes and to turn on the cell killing membrane attack complex, which will destroy the attacking cells, bacteria and viruses.

Two pathways for complement activation are known:

1. *Classical pathway*, which is activated by the formation of antigen–antibody complexes. This system helps to eliminate cells and microorganisms previously recognized by immunoglobulins.
2. *Alternative pathway* does not require the presence of immune complexes for its activation.

Classical Pathway

The components of the classical pathway are all plasma glycoproteins designated with a number preceded by the letter C (C1 to C9) (Table 30.2). The C1, C2, and C4 are unique components of this pathway, in contrast, C3 is common to the classical and alternative path-

ways. C5 to C9 compose the *lytic attack complex* or *membrane attack complex*, which is formed in both pathways.

C1 structure. The C1 component is a complex of three different glycoproteins: C1q, C1r, and C1s. C1q consists of 18 polypeptide chains of 217 amino acids each, corresponding to three chains (A, B, and C) grouped into six subunits. The first 81 amino acids in these polypeptides are similar to those of collagen, containing a glycine residue every three amino acids and several hydroxyproline residues. The A, B, and C chains are associated in a collagen-like helix (p. 47). C1q is elongated, fibrillar in its initial portion and globular at the end. The fibrillar portion is bent at the middle, forming an obtuse angle (Fig. 30.13A). In the C1q molecule, the six subunits are arranged with their parallel fibrillary stems bundled in the initial portion, diverging radially in their middle portion, and ending in their globular regions; they look like flowers in a bouquet (Fig. 30.13B).

The six globular heads of C1q have binding sites for the Fc region of immunoglobulins. Only immunoglobulins that are bound to their corresponding antigen can associate with C1q; free antibodies do not bind to C1q. Conformational changes in the Ig due to antigen binding allow recognition by C1q. Antibody–C1q binding capacity increases in the presence of multiple Fc regions close to each other, as occurs in immune aggregates.

Not all immunoglobulins have the same ability to bind to C1q. IgG_3 has the highest affinity, followed by IgG_1 and IgG_2; IgG_4 does not bind C1q. When IgM are attached to large antigens, they bind firmly to C1q. The other Ig types do not associate with C1q.

Other proteins of the C1 complex are C1r and C1s. These molecules are elongated, with two globular ends that are different in size. Both are zymogens or proenzymes with the catalytic site located on the larger end. In plasma, C1r and C1s form linear associations $C1r_2$–$C1s_2$ (Fig. 30.13C), which locate in the space that exists between the

TABLE 30.2   Components of the Complement System

	Mass (kDa)	No. of chains before activation	Plasma concentration (mg/dL)	Substrate hydrolyzed by the active form	Products of hydrolysis
CLASSIC PATHWAY					
C1q	462	18 (6A, 6B, 6C)	8.0	—	—
C1r[a]	83	1	5.0	C1r, C1s	C1r
C1s[a]	83	1	5.0	C4, C2	C1s
C4	205	3 (α, β, γ)	60.0	—	C4a-C4b
C2	102	1	2.0	C3, C5	C2a-C3b
C3	185	2 (α, β)	130.0	—	C3a-C3b
ALTERNATIVE PATHWAY					
Factor D[a]	24	1	0.1	Factor B	—
Factor B[a]	92	1	21.0	C3, C5	Activated factor B
Properdin	220		2.0	—	—
C3	185	2 (α, β)	130.0	—	C3a-C3b
TERMINAL COMPONENTS					
C5	190	2 (α, β)	7.0	—	C5a-C5b
C6	120	1	6.4	—	—
C7	110	1	5.6	—	—
C8	150	3 (α, β, γ)	5.5	—	—
C9	71	1	5.9	—	—

[a]These components are zymogens or proenzymes.

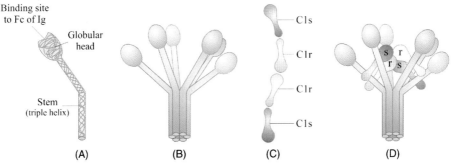

FIGURE 30.13   **Structure of the complement C1 protein complex.** (A) Diagram of a C1q subunit, composed of the association of three polypeptide chains (A, B, and C). These proteins associate in a triple helix structure, with the fibrillar portion bending at an obtuse angle. The C-terminal ends of the three chains present a globular conformation, forming a "head" where the binding site to the Ig Fc portion is located. (B) C1q molecule, with its six subunits arranged in a bundle. (C) Linear association of two subunits C1r and two C1s (C1r$_2$–C1s$_2$ complex). The catalytic site of these proenzymes is at the major globular end of C1r and C1s. (D) Complete C1 component. The C1r$_2$–C1s$_2$ complex folds between C1q subunits with the major globular C1r and C1s heads oriented inward.

stems of C1q, folded so that the larger globular end of C1r and C1s (which possess the catalytic site) are facing the inner part of the molecule (Fig. 30.13D). Formation of this complex requires Ca^{2+}.

C1 activation. Normally C1 is in an inactive state, maintained by association with a plasma protein C1 inhibitor (*C1-inh*) that prevents activation of the C1r zymogen. This effect of the C1 inhibitor ends when C1 binds to the antigen–antibody complex. The interaction of two or more globular heads of C1q with Fc regions of the antibody induces a conformational change in C1, which suppresses the inhibition. Then, C1r self-activates, displaying its protease activity. Activated C1r induces C1s zymogen activity, which initiates the cascade of proteolytic reactions.

The mechanism of C1r and C1s activation is similar to that of digestive proteases, zymogens are converted to active enzyme after undergoing hydrolysis of specific peptide bonds. Activated C1r and C1s, as well as other enzymes of the complement are serine proteases.

The steps of the classical pathway are presented in Fig. 30.14.

C2 activation. C1s catalyzes the hydrolysis of the C4 component, which splits into two fragments, a smaller one, C4a, and a larger one, C4b. The latter covalently links to neighboring molecules, preferably antigens bound to the antibodies that initiated C1 activation. C4b interacts with another component of the complement, proenzyme C2. Binding of C4b and C2 depends on Mg^{2+}. C1s also acts on C2 and divides it into C2a and C2b. C2a has serine protease activity, C2b fragment is released, leaving on the plasma membrane the C4bC2a complex, called *C3 convertase.*

C3 activation. The C3 member of the complement is a relatively abundant protein in plasma (130 mg/dL), which has a central position in the pathway; it is the point where the classical and alternative pathways converge.

In the classical pathway, the C4bC2a complex binds to and activates C3 by hydrolytic cleavage.

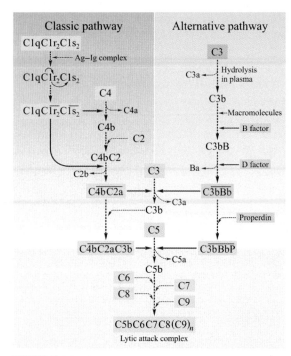

FIGURE 30.14 **Classical and alternative pathways for complement activation.** In *red background* are the components with proteolytic activity. *Solid arrows* indicate hydrolytic effects.

A small fragment, C3a is released, C3b associates to C4b forming the complex C4bC2aC3b. The hydrolysis of C3 leaves exposed in the rest of the molecule (C3b) a thioester group, which can readily bind to hydroxyl or amine groups to form esters or amides, respectively. This allows C3b to covalently associate with molecules in nearby cell surfaces.

C5 activation. Similar to the other member of the complement, C5 is activated by hydrolysis, by the C4bC2C3b complex, also known as *C5 convertase.* C5 is cleaved into C5a and C5b, which is the first component of the lytic membrane attack complex.

Membrane Attack Complex

The five glycoproteins C5b to C9 form a complex close to 2×10^6 Da, which is responsible for

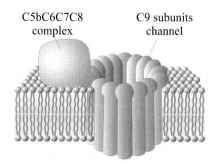

C5bC6C7C8 complex

C9 subunits channel

FIGURE 30.15 **Scheme of a membrane lytic attack complex.** The C5bC6C7C8(C9)$_n$ complex is shown with the C9 subunits forming the transmembrane channel.

the final lytic action on cells. The complex is the result of successive assembly of its components. C5b associates to C6, later to C7, which binds to the membrane of the target cells; finally, C8 is added to the complex.

Once in the membrane, C5bC6C7C8 functions by creating transmembrane pores or channels which will help lyse the cell where the complex is attached. These channels are made up by polymerization of several subunits (up to 18) of the C9 component (Fig. 30.15). The complex C5bC6C7C8(C9)$_n$ is included as an integral membrane protein complex (membrane attack complex). The free flow of small molecules and ions through the formed pores causes osmotic cell lysis.

Alternative Pathway

The members of this pathway are three proteins, factors B, D, and *properdin* (P). Activation does not depend on the presence of antibodies, but of surface macromolecules on cells or particles. Generally, C3 hydrolysis occurs in plasma, and very small amounts of C3b are formed continuously. This C3b is rapidly inactivated by various agents, including factors H and I, which are regulatory plasma proteins. The equilibrium between production and inactivation of C3b can be altered by the presence of macromolecules (i.e., polysaccharides and lipopolysaccharides)

on some microorganisms or cells to which C3b fixes, allowing C3b to escape inactivation by factors H and I. Under these conditions, factor B, a plasma proenzyme, binds to C3b. In the presence of factor D, a protease, B is hydrolyzed in two pieces, Ba and Bb. The latter is a serine protease that, bound to C3b, form the C3bBb complex or *C3 convertase*, with identical action to that of C3 convertase of the classical pathway (C4bC2a). Properdin (P) binds to C3b in the complex, stabilizing it and prolonging its effects. C3bBbP also acts as C5 convertase.

Both pathways converge to C3b, from which the following steps (leading to formation of the membrane attack complex) are the same as those of the alternative and classical pathways (Fig. 30.14).

Lectin pathway. A third pathway, known as the lectin pathway, exists for activation of the complement. This is homologous to the classical pathway and can be activated independently of antibodies. It is initiated by *mannan-binding protein* (MBP) or *mannan-binding lectin*, structurally homologous to C1q, present in blood plasma, associated to serine proteases similar to C1r and C1s. MBP can join terminal mannose groups on the surface of bacteria and activate accompanying proteolytic enzymes; they act on C4 and C2 and generate C3 convertase as in the classical pathway.

Regulatory Factors

Pathways in cascade have a multiplier effect, one initial molecule activates several molecules of the proenzyme immediately downstream, progressively amplifying the response. These reactions must be regulated; otherwise the process would only stops once the components are depleted. Regulation is accomplished by several inhibitory proteins, whose action is exerted at different levels. As mentioned previously, C1 inhibitor (C1-inh) functions to control the pathway. In addition, H and I are factors that acting on C3b, and other inhibitory proteins of the C3bBb formation, control the alternative pathway. The membrane attack complex is regulated by several agents which affects its assembly; among these regulators protein S is the most important one.

Functions of the Complement System

The common effector of both the classic and alternative pathways is the membrane attack complex, whose intense cytolytic action leads to the destruction of the foreign agent. This action is performed only in membranes adjacent to the activation site, either where the antigen to which Ig is attached (classical pathway) or at the place which has the activator macromolecules (alternative route). The C3b fragments unbound to these membranes remain free in the liquid phase, where different factors quickly suppress their capacity to establish covalent bonds with other molecules. In this manner, the cells of the host remain undamaged.

An important role of the complement is to eliminate antigen–antibody complexes and to prevent their accumulation. Individuals with genetic defects that reduce or totally inhibit the synthesis of C1, C2, C3, and C4 can develop disorders caused by deposition of antigen–antibody aggregates (i.e., glomerulonephritis and rheumatoid arthritis).

Other actions of the complement are mediated by intermediate products of the activation pathways. Thus, the C3b and C4b fragments are able to covalently bind to particles or cells and coat their surface, a phenomenon that is called *opsonization*. Various phagocytic cells (macrophages, monocytes, and polymorphonuclear leukocytes) have receptors for C3b and C4b. The presence of these molecules on the surface of particles or cells facilitates the adhesion of phagocytes, which will endocytose and eliminate the pathogen.

The C3a, C4a, and C5a fragments, polypeptides of 9 kDa structurally homologous to each other, are designated *anaphylatoxins*. They bind to specific receptors on mast cells, basophils, platelets, and smooth muscle cells, and trigger the release of histamine, leukotrienes, and other chemical mediators. These mediators cause local vasodilatation, increased vascular permeability, facilitate the production of edema, promote the release of lysosomal enzymes, and cause smooth muscle contraction. In addition, C5a has *chemotactic* action, inducing the migration of leukocytes to the site where activation of the complement occurs. The mentioned effects are characteristic of the inflammatory reactions.

Anaphylatoxins are inactivated by removal of an arginine from their C-terminus. This is catalyzed by *carboxypeptidase N*, an enzyme circulating in plasma.

The importance of complement is manifested in patients with deficiencies in this system. Defects in complement activation are the underlying cause of many inflammatory, degenerative, and autoimmune diseases. For example, mutations in the genes coding H, I, B factors, or CD46 have been related to atypical hemolytic uremic syndrome, systemic erythematous lupus, angioneurotic edema, and paroxysmal nocturnal hemoglobinuria.

CELLULAR IMMUNITY

T Lymphocytes

T lymphocytes are the effectors of cell-mediated immunity. They contain receptors that selectively recognize the antigen and bind to it. Unlike immunoglobulin-binding sites, which bind free antigens in the extracellular medium, T-cell receptors (TCRs) detect the antigen only when it is presented by another cell.

Two types of T cells with different functions can be distinguished:

1. *T helper cells* (T_H). These cells, when stimulated, secrete factors that activate the proliferation and differentiation of other cells. A group of T cells interact with B cells which have bound antigen, inducing their proliferation and differentiation and the synthesis and secretion of antibodies.

Another group of T_H cells is related to phagocytes and enhance their capacity to destroy foreign agents.

2. *T killer cells* (T_K). These cells are responsible for the destruction of cells infected by viruses or other pathogens.

T_H and T_K cells recognize the antigen only when it is bound to specific proteins on the surface of other cells.

T suppressor cells (T_S) have a regulatory role and are able to inhibit the immune response. They are not considered a group of T cells.

Approximately 15% of circulating lymphocytes do not belong to the B or T type, but they exert defense functions. They are the natural killer (NK) *cells*, with nonspecific cytotoxic properties. They can recognize and kill tumor and virus-infected cells.

T cells arise from lymphoblastic precursor cells in the bone marrow, they migrate to the thymus, where they differentiate and develop into mature T lymphocytes. Once they complete their differentiation in the thymus, T cells enter the circulation and stay in secondary lymphoid organs, where there are also B lymphocytes.

During the stages of maturation, T cells express different proteins on their surface, which can be identified and used as markers to determine their stage of differentiation and cell type. Among these markers are CD3, CD4, and CD8. CD3 is a complex of six polypeptide chains present in the plasma membrane of all mature T cells; it colocalizes with several specific receptors of T cells and is necessary for their expression in the cell surface. CD4 and CD8 allow distinguishing T cells with different functions. Generally, the T helper cells (T_H) have CD4, which is why these type of cells are known as T_HCD4 or simply T4. T killer cells (T_K) have CD8 on their plasma membrane and they are named T_KTD8 or T8 cells.

T-Cell Receptor

The vast majority of circulating T lymphocytes expresses a receptor composed of two different chains named α and β, which are glycoproteins of 46 and 41 kDa, respectively. They have two extracellular globular domains of approximately 110 amino acids each, followed by a transmembrane segment that crosses the plasma membrane once. One of the globular domains shows differences in the amino acid sequence when compared to the corresponding domain of other T cells, and is called the *variable region*. The second globular domain is more conserved; it is called the *constant region*. The variable domains of the α and β subunits are in front of each other and they form the antigen recognition site. The variable and constant domains contain a disulfide bond. Also, there are S—S bridges linking cysteines in the α and β subunits near the plasma membrane. The extracellular portion of the receptor shares structural homology with the Fab fragment of immunoglobulins. The transmembrane domain is a hydrophobic segment of 22 amino acid residues and spans the plasma membrane once. A hydrophilic segment of 12 residues at the C-terminal faces the cytoplasm (Fig. 30.16). The $\alpha\beta$ receptor is expressed in ~95% of peripheral T cells and in the majority of thymocytes. Some T cells have a receptor that consists of γ and δ chains, which have a structure that is similar to that of the α and β subunits. T cells containing $\gamma\delta$ receptors are in the thymus and secondary lymphoid organs of intestinal epithelium, tongue, and uterus.

The great sequence variability of these receptors is due to a set of genes that undergo genetic rearrangements similar to those described for the immunoglobulins. As occurs for the Ig genes, the genes encoding the variable regions of the TCRs are arranged in separate sets (V and J for the α chain; and V, D and J for the β chain). The rearrangement of these genes takes place during differentiation of T cells in the thymus. Similar to immunoglobulins, the structural diversity of the $\alpha\beta$ receptors is increased by the lack of precision in the D–J, V–DJ and V–J assemblies, and by the addition of nucleotides catalyzed by terminal deoxynucleotidyl transferase. The number of TCRs with different specificity is as large as the number of immunoglobulins expressed.

The TCRs, αβ or γδ are always associated to the CD3 complex, a group of polypeptides (γ, δ, ε, and ζ) with highly conserved structure. All polypeptides of the CD3 complex have a transmembrane piece and a cytoplasmic domain of 40 or more amino acids (Fig. 30.16), which can be phosphorylated by tyrosine kinases. The CD3 γ, δ, and ε polypeptides, are members of the Ig superfamily of proteins, with an external domain containing an intrachain disulfide bond. The ζ chains have a smaller extracellular segment. The TCRs polypeptides associate into two αβ heterodimers with their variable and constant domains facing the extracellular space. The variable portions of each dimer form the antigen-binding site and provide specificity to the receptor. The stoichiometry of the complex is (αβ)$_2$, γ, δ, ε$_2$, ζ$_2$ (Fig. 30.16). CD3 is involved in signal transduction events that are triggered by binding of an antigen to the receptor.

CD4 and CD8 play an important role in the differential recognition of proteins of the major histocompatibility complex (MHC) (see next section). During recognition of the antigen, CD4 and CD8 associate to components of T-lymphocyte receptor. CD4 is a polypeptide chain with four extracellular Ig-like domains, a transmembrane segment and a cytoplasmic domain that can be phosphorylated by tyrosine kinase (Fig. 30.17). It participates in signal transduction within the lymphocyte. CD8 is a heterodimer of α and β chains linked by a disulfide bridge. Each subunit has an Ig-type domain in its external portion, a transmembrane segment and a cytosolic domain, which can bind a tyrosine kinase (Fig. 30.17). CD4 and CD8 function together as coreceptors.

Major Histocompatibility Complex

T cells cannot recognize an antigen if this is not bound to specific proteins in other accessory cells. These specific proteins are encoded by genes of the MHC. In humans, these proteins are also designated with the acronym HLA (*human leucocyte antigen*). There are three classes of proteins synthesized under control of the complex (I, II, and III). The first two will be described, the third comprises a set of diverse proteins that includes some components of the complement system and others related to antigen processing.

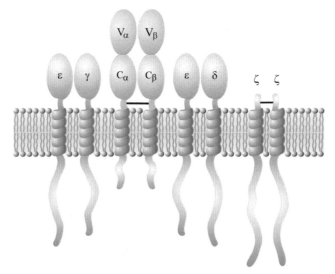

FIGURE 30.16 **Scheme of the T-cell receptor (TCR) and the CD3 complex.**

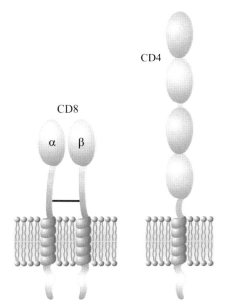

FIGURE 30.17   **Scheme of the CD8 and CD4 TCRs.**

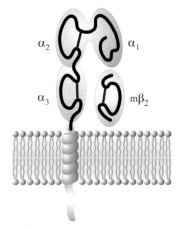

FIGURE 30.18   **Scheme of protein major histocompatibility (MHC, HLA) class I complex.**

Class I proteins. They are formed by two polypeptide chains (α- and β_2-microglobulin). The α chain is an integral membrane protein of 45 kDa, linked to carbohydrates. It is found in all nucleated cells. It has three domains of approximately 90 amino acid residues each, named α_1, α_2, and α_3. The amino acid sequence differences among the different class I proteins are mainly in α_1 and α_2 domains; the α_3 is the most conserved. Following the α_3 domain, there is a transmembrane segment of 25 amino acids, predominantly hydrophobic, and an intracellular C-terminal hydrophilic stretch of 30–40 amino acid residues (Fig. 30.18), which can be phosphorylated by tyrosine kinases.

The α chain is noncovalently associated to the β_2-microglobulin (β_{2m}), which contains 99 amino acids (12 kDa) that are not inserted into the membrane. The presence of β_{2m} is essential for expression of MHC (or HLA) class I on the cell surface. The α_2, α_3, and β_{2m} domains have a disulfide bridge established between cysteine residues separated by ~60 amino acid residues.

X-ray crystallographic studies have provided a three-dimensional image of MHC class I protein. The antigen-binding site resides within the α_1 and α_2 domains; which creates a niche, closed at the bottom by eight antiparallel β sheets that accommodate the antigen (Fig. 3.22).

In humans, there are three different subclasses of HLA class I protein, encoded by three different loci HLA-A, HLA-B, and HLA-C [in addition, the existence of additional genes (E, F, G) has been described]. These subgroups differ in the sequence of their α_1 and α_2 domains; however, they share the same β_2-microglobulin. HLA-A, HLA-B, and HLA-C are expressed simultaneously in all cells except in erythrocytes and the trophoblast syncytium.

There is a marked polymorphism of the HLA proteins in the population; numerous alleles have been identified for each of them. Given the great variability, most individuals are heterozygous for the three loci. As the alleles are codominant, an individual can express up to six different class I proteins, two of each of the A, B, and C subclasses.

Class II proteins. These are integral membrane glycoproteins, formed by one α (33 kDa) and one β (28 kDa) chains. Both subunits are noncovalently associated and show structural homology. They possess two extracellular domains, a transmembrane segment of ~30 hydrophobic residues, and an intracytoplasmic hydrophilic

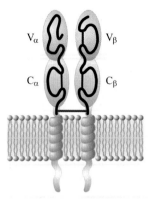

FIGURE 30.19 **Scheme of class II major histocompatibility complex (MHC).**

segment of ~14 amino acids (Fig. 30.19). The α and β chains have an intrachain disulfide bridge close to the external side of the membrane.

There is a marked degree of homology between TCRs and the MHC class I and II proteins, in particular with respect to the S—S bridges, which are repeated in all these molecules. It is possible that the genes responsible for their synthesis have been generated by successive duplications of a common ancestral gene. They belong to a superfamily class of genes.

The α_1 and β_1 domains of Class 2 MHC form a niche similar to that of α_1 and β_2 domains of class I MHC proteins, flanked by two α helices on a background of antiparallel β sheets. Antigenic peptides bind to this niche. Class II proteins can accommodate peptides longer than those accepted by MHC I. Unlike class I MHC proteins, present in all cells (except red blood cells), class II are only expressed on the surface of macrophages, monocytes, B cells, and dendritic cells, which are called *antigen-presenting cells* (APCs).

There are three different subclasses of class II HLA, DR, DQ, and DP. As for class I proteins, they also show a pronounced polymorphism in the population, with the existence of six or more MHC class II proteins. Class II MHC are encoded by three groups of genes (DR, DQ, and DP), each of which contains information for at least one α chain and one β chain. It is common to find more

than one functional gene for those polypeptides, especially β, which allows the synthesis of more than six MHC class II proteins. Given the genetic variability in the CMH, with the exception of identical twins, it is very rare to find two individuals with identical HLA proteins.

MHC genotypes can be identified by serological reactions. MHC determination is important for organ donation. Tissue transplantation whose HLA proteins are recognized as foreign by the recipient patient will induce rejection reactions in the host. The main effectors of this response are the T lymphocytes

MHC (HLA) protein functions. The function of the MHC proteins is to present antigenic peptides to the T lymphocytes. HLA proteins function as guiding molecules for T cells in the recognition of the antigen that is bound to them. The antigen is only "seen" by T cells when associated with MHC proteins of presenting cells from the same individual. This phenomenon is called *MHC restriction.* The antigen bound to HLA cells other than those of the subject is not recognized.

According to the class of proteins which recognize MHC, T cells are divided into functionally distinct subpopulations; in general, T helper cells (T_H) recognize antigens bound to HLA class II, whereas the killer (T_K), identify antigens linked to HLA class I. TCRs must not only recognize the antigen but also the MHC protein carrier. CD4 or CD8 coreceptors assist in the recognition of the MHC invariant portion.

Antigen Recognition

Detection of the antigen by T cells differs substantially from that of B cells and antibodies. While the recognition of the antigen by B cells or secreted antibodies is the result of a direct ligand–receptor interaction, T cells can only bind antigens when they are associated to the MHC. The T cell must form a receptor–antigen–MHC protein ternary complex.

The characteristics of the antigenic determinants or epitopes recognized by immunoglobulins and T lymphocytes also differ. In general, antibodies bind to epitopes exposed on the surface of the antigen; it is therefore very important to maintain the antigen native conformation to obtain a humoral immune response. In contrast, T cells can recognize denatured proteins or better, small pieces of the molecule, often inaccessible in the native protein. The tertiary structure of the antigen, important for antibody binding, is not critical for the T receptor, this usually binds short peptides of 8–15 amino acids, of extended linear structure.

Antigen Processing

T cells are activated by binding to an antigen processed by the presenting cell and associated to MHC molecules. Two pathways for the processing of antigens are known; an endogenous pathway for antigens produced in the presenting cell and an exogenous pathway for antigens that invade the cell.

Endogenous or biosynthetic pathway. This pathway is triggered by proteins different from those of the individual and processed within the host cell, as for example, those synthesized by virus-infected cells. Once produced, these proteins are targeted to proteasomes (p. 369), where after hydrolysis they render peptides 5–15 amino acids long. These peptides, released into the cytosol, are transported to the ER and introduced into its cisternae by TAP1 and TAP2, ABC type carriers (p. 238).

The newly synthesized α and β_{2m} chains also enter the ER, where the MHC class I protein is assembled. Within the ER cisternae, the antigenic peptides associate with the MHC. This process is assisted by chaperones such as *calnexin*. The MHC I–Ag complex passes to the Golgi system. Finally it is released in vesicles that are transferred to the plasma membrane of the presenting cell, where it is exposed and can interact specifically with the T_C-cell receptor. Fig. 30.20 schematically shows the process.

Exogenous or endocytic pathway. In this pathway, the antigen originates outside the presenting cell. It can be captured by IgM or IgD of a B-lymphocyte membrane, or it can be directly taken from the medium by other cells. In both

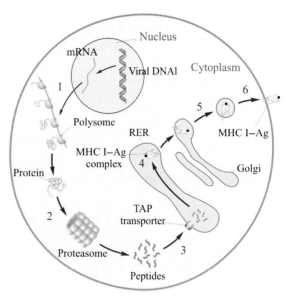

FIGURE 30.20 **Antigen (Ag) processing in the endogenous pathway.** The scheme refers to a virus-infected cell. 1, The cell transcribes the viral DNA incorporated into its genome and produces mRNA, which passes to the cytosol. There, the mRNA serves as a template to synthesize viral protein. 2, The protein released in the cytosol is introduced into the proteasome and hydrolyzed to small peptides. 3, The peptides enter the rough endoplasmic reticulum (RER) via TAP transporters. 4, Immediately after synthesis in the cytosol, α and β_{2m} chains are incorporated into the RER where they form the MHC class I that binds to the antigenic peptide, aided by the chaperone calnexin. 5, MHC I–Ag complex passes to the Golgi system, and is released by vesicle-mediated transport. 6, The vesicle fuses with the plasma membrane and the MHC I–Ag complex is exposed to the extracellular space.

cases, the antigen enters the cell by endocytosis or pinocytosis (p. 242) and is carried in vesicles (endosomes). The endosome fuses with the lysosome loaded with hydrolytic enzymes that degrade the antigenic protein into small peptides. The α and β chains of the MHC are introduced into the rough endoplasmic reticulum (RER) where, after forming the MHC class II, bind to a polypeptide called *invariant chain* (Ii). The complex then moves to the Golgi and finally to the endosome–lysosome compartment, to meet the foreign protein that has been degraded. MHC II

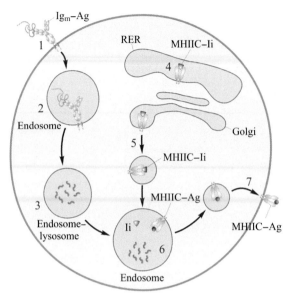

FIGURE 30.21 **Antigen processing in the exogenous pathway.** An antigen (Ag) presented by a B lymphocyte is shown as an example. 1, An antigenic protein epitope is attached to the paratope of Ig_m inserted into the lymphocyte membrane. 2, The Ig–Ag complex is internalized by endocytosis. 3, The formed endosome fuses with a lysosome and the antigen is degraded into small peptides. 4, The α and β2m chains move into the RER and form class II MHC that binds the invariant chain (Ii). 5, The MHC II–Ii complex passes to the RER and the Golgi apparatus and from there it is transported via vesicle-mediated transport with the endosome–lysosome system. 6, In the endosome, Ii separates and an antigenic peptide recognized by MHC II is bound. 7, The MHC II–Ag complex is inserted into the plasma membrane and exposed to the extracellular space.

releases the Ii chain and binds to one of the antigen peptides, containing the appropriate structure. The MHC II–Ag complex is placed in the plasma membrane and exposed to the extracellular space (Fig. 30.21).

Binding of the processed peptide to MHC does not require a degree of structural selectivity as stringent as those of the T cell or Ig receptors; a single type of HLA molecule may present different antigens. Therefore, a relatively small number of HLA molecules are sufficient to process and present a wide variety of antigens. However, not all HLA variants are equally

efficient, some have poor antigen-presenting capacity and this decreases the immune response of individuals who possess them.

The CD4 and CD8 accessory molecules, associated to nonvariable regions of MHC, participate in the interaction of the receptor with the antigen–HLA. In general, CD4 present in T_H lymphocytes cells interacts with HLA class II molecules, and CD8 in T_K cells, associates with class I proteins.

After the receptor has recognized the antigen, clonal proliferation and differentiation depend on substances released into the medium by the presenting cells or by the T lymphocytes themselves. Those substances are the *cytokines*, which are described later. Stimulation of cell proliferation generates helper, killer, and memory T cells, all with a receptor of identical specificity to that of the original T cell.

Activation of T Lymphocytes

After completing their development in the thymus, T cells enter the circulation, where they eventually find a presenting immune cell containing an antigen that the T cell can specifically recognize. The T cell is then stimulated and starts multiplying and differentiating into effector T cells. These cells can react when faced with MHC–Ag complexes on other cells.

There are three classes of effector T cells with the capacity to detect antigens and to participate in immune reactions, the T_K CD8[+], T_H1 CD4[+], and T_H2 CD4[+] cells.

Antigens derived from pathogens that multiply within the cells are processed and exposed on the surface of presenting cells by MHC (or HLA) class I. Then, they are detected by killer T cells (T_K CD8[+]) that destroy the affected cell.

Antigens admitted into the cell by endocytosis, phagocytosis or pinocytosis, are processed and inserted into the plasma membrane bound to MHC class II molecules and recognized by helper T cells (T_H CD4[+]). These cells can be differentiated into two types: (1) T_H1 or *inflammatory* lymphocytes, which activate infected macrophages that will destroy the invading pathogen, and (2) T_H2, which stimulate B cells to multiply and differentiate into antibody-producing plasma cells. They also potentiate allergic reactions.

The first contact of the T lymphocytes with the antigen, results in a primary response that generates memory cells. These cells provide protection in subsequent invasions of the same pathogen.

Activation. The activation of T lymphocytes requires the recognition of an antigenic peptide fragment bound to a MHC (HLA) molecule and also a series of stimulatory signals. Only the APCs (macrophages, dendritic cells and B lymphocytes) are capable to express MHC and other molecules related to signaling systems that enhance the clonal expansion of T lymphocytes and their differentiation into effector cells.

While MHC–Ag complex recognition by the TCR is specific, there are costimulatory molecules that can establish nonspecific interactions. These molecules, which mediate the adhesion between presenting cells and T lymphocytes, include the *selectins, integrins, addressins,* and other proteins of the immunoglobulin superfamily. In addition to the CD4, CD8, and CD19, coreceptors associated to nonvariable MHC domains, the *intercellular adhesion molecules* (ICAM) that bind integrins LFA (lymphocyte function-associated to antigen), and glycoprotein B7 or B80 (which binds to CD28), also belong to the Ig family of intercellular adhesion molecules.

These molecules promote synthesis and secretion into the medium of substances, known as cytokines, which function in the activation of lymphocytes. Fig. 30.22 shows the molecules involved in lymphocyte–APCs and cytokines responsible for the stimulation of T cells.

B-lymphocyte activation. Mature B lymphocytes use IgM and IgD membrane receptors. These cells circulate for several weeks and eventually die unless they find a related antigen. The first stimulus to their proliferation and differentiation occurs when a suitable antigen binds to a B cell Ig_m. Molecules associated with the membrane IgM and IgD ($\alpha\beta$ heterodimers) activate

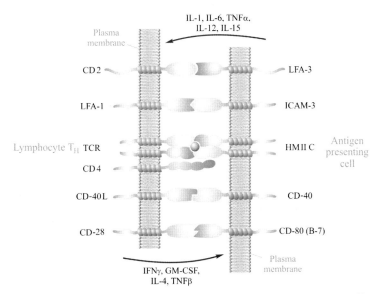

FIGURE 30.22 **Molecules participating in the interaction of T lymphocyte and presenting cell and cytokines released in response to the signals generated by this protein–protein interaction.**

the lymphocytes. In most cases, this stimulation is not enough, but also requires interaction with a helper T cell (T_H).

B cells internalize the antigen bound to their Ig_m by endocytosis and degrade it into small peptides as described in the previous section. Peptides bind to HLA class II molecules and the complex is exposed on the surface of the lymphocytes (Fig. 30.21). A TCR, which recognizes the MHC II–Ag complex, binds to it. Such interaction with the T cell is also required to trigger the clonal expansion of B cells, since the T cell releases to the medium cytokines that induce B-lymphocyte proliferation and differentiation into plasma and memory cells.

The specific interaction between the MHC class II–antigen complex and the T_H receptor (TCR) takes place on the surface of the B lymphocytes. The CD3 and CD4 molecules associated to the receptor, as well as MHC proteins of presenting cells are involved in the transmission of signals within the respective cells. The cytoplasmic domains of the CD3 complex subunits γ, δ, ε, and ζ, and a CD4-associated tyrosine kinase, initiate phosphorylation events that finally determine the activation of transcription factors and the synthesis and release of cytokines. Phospholipase C is also activated, producing inositol trisphosphate and diacylglycerol as second messengers.

The stimulus is enhanced by the presence of other molecules on the surface of both cells, including adhesion molecules (ICAM), CD40 and CD80 on B lymphocytes and LFA, CD40L (CD40 ligand), and CD28 on TH cells.

The activation of T and B lymphocytes shares common basic mechanisms. In both cell types, when the receptor recognizes the antigen, there is a receptor mediated response which initiates signal transduction events in the cell. Also, T and B lymphocytes respond with other nonspecific signals through costimulatory molecules. These signals are responsible for the release of activating agents by presenting or T-helper cells.

Lytic action of T_K lymphocytes. T_K cells use a variety of mechanisms to destroy their targets. These include cell to cell direct actions and indirect mechanisms through signal transduction systems and cytokine release. A killer lymphocyte activated by binding to a cell with the HLA class I–Ag complex, releases granules that have lytic action. The granules contain a 70-kDa protein called *perforin*, which inserts and polymerizes in the membrane of the target cell to form pores of about 10 nm in diameter, similar to those described for the C9 component of the complement system (Fig. 30.23). These pores allow the free flow of ions and small molecules with subsequent osmotic swelling and lysis of the attacked cells.

Perforin polymerization depends on the presence of Ca^{2+}. The released perforin units can also be assembled in the liquid phase, but the pores formed cannot insert into the lipid bilayer; the membrane only accepts one unit at a time, which functions as a protective mechanism for the neighboring cells not directly associated to the T_K lymphocyte. The membrane adjacent to the T_K lymphocyte is the only one receiving, one after the other, the perforin units, which will assemble the active pore.

Besides perforin, the granules contain *granzyme* (granule enzymes) or *fragmentins*, which are serine proteases. These enzymes are introduced into the target cell through the perforin channels and degrade critical proteins in the target cell.

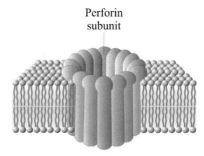

Perforin subunit

FIGURE 30.23　**Scheme of the perforin pore in the cell membrane.**

The lytic action is also completed by "programmed cell death." T_K lymphocyte surface molecules acting as ligands of Fas receptor on the target cell, activate the cascade of reactions characteristic of apoptosis (Chapter 32).

Macrophage activation. Macrophages that have phagocytized a pathogen are stimulated by signals generated from inflammatory T_H CD4$^+$ (T_H1). When macrophages are activated, they undergo changes in oxygen consumption, nitric oxide production, phagocytic and cytotoxic capacity, expression of MHC class II protein, and secretion of IFNγ cytokine and other proteins (IL-1, TNF, GM-CSF, and M-CSF) with high biological activity.

CYTOKINES

In experiments in vitro with antigen-activated lymphocytes, it was observed that substances appeared in the culture medium that were able to influence the behavior of immune system cells. The existence of these mediators and their multiple biological actions were confirmed by numerous subsequent studies. As these substances were found to be produced by lymphocytes, they were called *lymphokines*. Later, other factors secreted by mononuclear phagocytes were found which were designated *monokines*. Many were originally labeled with names that suggested their biological in vitro activity, for example, the T-cell growth factor, T-cell migration inhibitory factor, B-cell activating factor, thymocyte mitogenic factor, B-cell differentiation factor, etc. Those that presented the property of attracting leukocytes (chemotaxis) to areas of infection were called *chemokines*. The finding that, in addition to lymphocytes and monocytes, epithelial, endothelial, and other cell types can also produce these proteins with immune action, led to name them all with the term *cytokines*. Many cytokines synthesized by leukocytes, macrophages or T cells, act on other leukocytes, which prompt to call them *interleukins* (IL). As new cytokines were

characterized, they were designated with the acronym IL followed by a number (IL-1, IL-2, ..., IL-27, etc.). The names cytokines and interleukins have widespread use today; however, for some cytokines, names related to their principal action are also used; this include, for example, *granulocyte and macrophage colonies stimulating factor* (GM-CSF), and TNF.

More than 100 cytokines have been described, most of which are glycoproteins of 10–80 kDa. The genes encoding most of the known cytokines have been cloned and have been produced using recombinant DNA techniques. This has allowed the study of their molecular structure and biological activity.

General properties of cytokines. Cytokines are responsible for many of the actions of both the innate and the adaptive system. They mediate and regulate immune and inflammatory reactions. Their action is usually local, autocrine or paracrine; only when secreted in large amounts, can they move to the circulation and have distant effects (endocrine action).

The same cytokines are produced by various cell types, and a given cytokine can act on different cells. All exert their effects by binding to specific receptors on the membrane of the target cell and activate signal transduction systems. The responses to most cytokines consist in modification of gene expression that results in changes, such as B- and T-lymphocyte differentiation and proliferation.

The chemokines represent a large family of cytokines with approximately a total of 50 members. Their function is to attract and recruit leukocytes at the site of infection, to promote their adhesion to the walls of the vessel and their migration from the blood into the interstitial space.

Cytokine receptors. These are classified according to their structure and the transduction pathway that they use. Most consist of two or more subunits. They activate different signaling pathways such as *Jak-STAT* (p. 564), *TLR/IL-1* (toll like/IL-1), and *TRAF* (TNF receptor–associated factor), phosphorylation-mediated cascades that ultimately regulate gene transcription. Chemokine receptors consist of seven transmembrane polypeptides associated to G proteins.

Cytokine functions. Cytokines fulfill critical functions in the defense against pathogens. They stimulate lymphocyte growth and differentiation, regulating the magnitude of immune responses and increasing its efficiency; they also contribute to coordinate the interaction between the innate and adaptive systems. While some cytokines function in both systems, others predominantly participate in one of them.

The cytokines involved with the innate immunity are primarily synthesized by mononuclear phagocytes in response to infectious agents. NK cells also produce cytokines. These cells have TLR receptors that bind, for example, to lipopolysaccharides from the wall of Gram negative bacteria or to double-stranded viral RNA. Upon binding to these antigens, NK cells activate production and secretion of cytokines, most of which stimulate the inflammatory response and the antimicrobial activity. Some, such as TNF, interleukin 1, 12 and 18 (IL-1, IL-12, and IL-18), promote the production of IFNγ with antiviral action. IL-23 and IL-27 regulate the inflammatory response, IL-10 inhibits macrophages.

Cytokines of the adaptive immune system are generated mainly in T lymphocytes in response to specific antigen recognition. The major mediators are interleukin 2, 3, 4, 5 (IL-2, IL-3, IL-4, IL-5), and IFNγ. They regulate growth, differentiation, and activation of different lymphocyte populations. IL-12, IL-18, IL-23, and IL-27, regulate the adaptive immune response that follows the innate immune response.

Other functions of cytokines. In addition to their role in the immune system, cytokines exert other important functions. Some behave as growth factors, others are regulators of red blood cell maturation. IL-3 and GM-CSF are hematopoietic factors of wide specificity, they influence pluripotent bone marrow cells and promote the differentiation of blood cells. Others, such as macrophage colony stimulating factor (M-CSF),

granulocyte colony stimulating factor (G-CSF), and IL-5 (eosinophil proliferation factor), have more specific effects on certain cell lines in later stages of their differentiation. TNF and lymphotoxin (LT) have strong cytotoxic action on some malignant tumors.

In cases of excessive cytokine production, their raise in blood can cause pathological effects, such as fever, stimulation of liver production of acute phase proteins in response to inflammation, thrombosis, metabolic disorders, decreased blood glucose, and reduction in myocardial contractility. A severe complication of serious infections is septic shock.

Interferon

Cells that have been invaded by a virus synthesize and secrete small molecular weight (~18 kDa) glycoproteins, of the cytokine family, called *interferons* (IFNs). Two types of IFNs, I and II have been isolated.

Type I IFNs are mediators of the initial response of the innate system to viral infections. Humans produce several IFNs, designated by Greek letters (αIFN, βIFN, εIFN, κIFN, and ωIFN). The most common forms of type I IFN are αIFN and βIFN; αIFN is primarily synthesized by dendritic cells, βIFN mainly by fibroblasts. The most potent stimuli for the release of these cytokines are viral nucleic acids (RNA, double or single-stranded, and unmethylated DNA).

Once IFNs are released, they bind to their receptors in nearby cells, which are associated to tyrosine kinases (TyK2), *Janus* (Jak) and *STAT* proteins 1 and 2. The activated *STAT* moves to the nucleus, where it binds to DNA response elements to induce gene transcription and translation.

The primary function of type I IFNs is to counteract viral infections. This is accomplished by: (1) promoting the immune response mediated by cells affected by intracellular microorganisms, (2) inhibiting virus replication by inducing the expression of proteins that block the transcription of viral RNA or DNA, and favor the translation of enzymes that degrade these nucleic acids, (3) stimulating the expression of MHC class I and T-helper cell development, and (4) favoring the recruitment and activation of lymphocytes in lymph nodes.

The action of IFNs is primarily paracrine; infected cells secrete IFN that protect neighboring uninfected cells. Although they can also have an autocrine action, IFNs are more effective in preventing viral infection in cells still not infected, than in stopping viral replication in the already invaded cells.

Type II IFNs. The main IFN in this class is IFNγ. It is secreted by T helper and NK cells; it activates macrophages, increases production of reactive oxygen species and nitric oxide in these and other cells, which increase their capacity to eliminate microbes. IFNγ modulates the immunoglobulin isotype switch in B lymphocytes, increasing the synthesis of IgG and reducing IgE. IgG activates the complement and, therefore, stimulates phagocytosis of opsonized microbes. Another function of IFNγ is the stimulation of expression of MHC proteins class I and II. Overall, the main role of IFNγ is to promote macrophage inflammatory reactions and to inhibit the eosinophil-dependent reaction of IgE.

Although IFNs are nonspecific with respect to viruses, they are species specific, being only active in individuals of the same species that have produced them.

Besides their antiviral action, IFNs possess other biological activities, such as inhibition of cell proliferation. This action sparked interest for their potential application in the treatment of cancer.

Until recently, the use of IFN for therapeutic purposes was limited by the inability to obtain adequate amounts of these substances. Currently, genetic engineering has allowed to highly increase their production for therapeutic purposes. It is used to treat different viral infections and some malignancies.

Pathologies Related to the Immune System

The effective defense against infectious agents requires the normal functioning of both innate and adaptive immune systems. Failures of components of these systems (lymphocytes, phagocytes, signal transduction pathways, and mediating factors) produce serious disturbances, some of which are lethal.

In general, these alterations are known as immune deficiencies; they can be of congenital (primary) or acquired (secondary) origin. All of them are accompanied by excessive susceptibility to infections. Its severity depends on the affected component and the magnitude of the disturbance.

Congenital immunodeficiencies. These could primarily affect the innate or adaptive immune systems.

Innate system. There are different types of defects that impair the function of phagocytes. These include: alterations in the production of reactive oxygen species, lysosomal function, TLR receptors, the adhesion and migration of leukocytes, or the complement alternative pathway. All these anomalies reduce the efficiency of the innate system as a first line of defense.

Adaptive system. Congenital immunodeficiencies can affect B lymphocytes, T lymphocytes, or both.

Humoral immunity. There is a group of diseases known as *common variable immunodeficiencies*, which comprise B-cell failures, with incapacity to respond to an antigen and to produce antibodies, or defects in B–T cell interaction.

Abnormalities in the development of B lymphocytes disrupt antibody synthesis with reduction (hypogammaglobulinemia), or total lack (agammaglobulinemia) of immunoglobulins. Rarely, even hypergammaglobulinemia may result from B cells developmental defects.

Hipogammaglobulinemias. Among the most common ones is the selective deficiency of IgA.

Patients suffer from recurrent infections of the respiratory and genitourinary tracts due to lack of mucosal secretory IgA. Even more rarely, there are also cases of selective deficiencies of other immunoglobulins (such as IgG_2 and IgG_4) or involving several types of Ig. Some inherited agammaglobulinemias are linked to the X chromosome and others are autosomal recessive. Usually, they are accompanied by reduction in the number of circulating B lymphocytes. An X-linked IgM hypergammaglobulinemia has been described. Patients with this disease present, in addition to increased IgM, low levels of IgG, IgA, and IgE. Commonly due to a defect in Ig isotype switch. Immunodeficiencies affecting B lymphocytes reduce the resistance to pyogenic (pus producing) bacterial infections. The patients are prone to otitis, pneumonia, meningitis, and osteomyelitis. They also suffer recurrent infections by enteric bacteria or some parasites.

Normal children 2–5 months of age have hypogammaglobulinemia that can be considered physiological. The newborn has immunoglobulins transferred from maternal blood through the placenta. The level of these Ig in plasma progressively decreases from birth until reaching minimum values at 3 months. The deficiency affects mainly IgG, but the Ig concentration is restored in the following months, as the child acquires ability to synthesize antibodies. In premature infants, this transient hypogammaglobulinemia takes a longer period of time.

Cellular immunity. Defects in the activation of T lymphocytes can also alter the immune response. Some are produced by mutations in the genes encoding the TCRs. Normal function of the cellular immune system is also disturbed by failures in the expression of the MHC.

Conditions that affect both systems, cellular and humoral, are called *severe combined immunodeficiency* syndromes. Some are due to abnormalities in cytokine production or the transduction pathways that they activate (e.g., mutations in the cytokine receptors). Other defects are caused by alterations in the development of the thymus. The dysfunction of T cells can be extended to B and NK cells. Approximately 50% of severe combined immunodeficiencies have autosomal recessive inheritance, many of these are due to lack of adenine deaminase, which leads to an increase in deoxyadenosine and dATP, especially in T lymphocytes. Accumulation of these compounds has toxic effects that cause cellular destruction and DNA synthesis block. It is a lethal condition that leads to death from infections at early age. Transplantation of bone marrow or stem cells and gene therapy are approaches to the treatment of this condition.

Acquired immunodeficiencies. Protein-caloric malnutrition seriously compromises protein synthesis in general and the production of antibodies and other molecules of the immune system. Especially children with severe malnutrition are susceptible to infections. Malignant tumors, mainly those that produce bone marrow metastasis and leukemia, are accompanied by immunodeficiency. Immunosuppressing therapy in patients of organ transplantation or autoimmune diseases also compromises the individual's resistance to infections. Infection with certain viruses causes immune system disorders. The *acquired immunodeficiency syndrome* (AIDS) is an example. This disease is caused by the human immunodeficiency virus (HIV) of sexual, blood, and transplacental transmission. The virus causes reduction of helper T lymphocytes and alterations in macrophages, disabling the capacity of the individual to appropriately respond to infections and increasing the development of malignant tumors.

Autoimmune diseases. Throughout this chapter it has been suggested that immune mechanisms come into play to remove foreign body invaders. The T receptors and antibodies discriminate with remarkable accuracy between the self and foreign components of the body. However, there are pathological conditions in which the system abnormally reacts against the individual's own structures. These alterations are known as *autoimmune diseases.*

Normally, there are mechanisms that protect against the development of autoreactive leukocytes. These are known as mechanisms of *tolerance*. *Central tolerance* is exerted in the bone marrow and thymus, where B and T cells bearing receptors that recognize self-antigens are removed before they mature. Eventually, autoreactive leukocytes can escape to the circulation, and are deactivated in the peripheral lymphoid organs (*peripheral tolerance*). When these tolerance processes fail, the activated B and T cells are able to generate autoimmune reactions and damage the own tissues. In some cases, the damage is caused by antibodies, in others, by T lymphocytes.

Examples of autoimmune diseases due to antibodies that react against body structures are: *myasthenia gravis*, in which antibodies against the acetylcholine receptor are synthesized; *thrombocytopenic purpura*, which affects platelet membrane proteins; *Graves' disease*, which presents antibodies that bind to thyroid stimulating hormone receptors and activate T_3 and T_4 production (hyperthyroidism); *pernicious anemia*, which has autoantibodies against the gastric parietal cells that produce intrinsic factor, necessary for absorption of vitamin B_{12}.

Sometimes, antibody–autoantigen complexes are deposited in large quantity on the walls of arterioles, activating the complement and promoting the initiation of inflammatory processes that damage the vessels and adjacent tissues. This occurs, for example, in *lupus erythematosus*, which is characterized by systemic alterations due to the synthesis of antibodies that react against multiple structures of the body, including DNA, nuclear proteins, renal glomeruli membranes, and erythrocyte membranes. Among those produced mainly by T cells are *type 1 diabetes mellitus* (insulin dependent), in which antibodies against the β cells of the pancreas are made; *multiple sclerosis*, with autoantibodies against myelin; some cases of *rheumatoid arthritis*, in which the target proteins of auto-antibodies are in the joints.

Autoimmune diseases have simple Mendelian inheritance, they are mostly polygenic, which determines, in some individuals, predisposition to allergies. Among the genes associated with these diseases are those of the MHC class II.

Viral or bacterial infections sometimes contribute to the development or aggravation of autoimmunity. Some promote the release of factors that increase the immune response, others induce the production of proteins that mimic those of the individual, causing the synthesis of antibodies with cross-reactivity to the own body proteins.

SUMMARY

The immune system is actively involved in the protection of the body against invading foreign agents. It consists of two different systems, innate and adaptive.

The innate system defends the body against the invasion of pathogens. It is unspecific and does not require the prior exposure of the foreign agent to respond. The skin and mucous membranes are primary barrier of this system. Phagocytic cells with toll-like receptors recognize molecules in microorganisms, such as lipopolysaccharides, lipopeptides, and double-stranded RNA, and initiate the inflammatory process or promote the release of IFN.

The adaptive system is specific, its response is enhanced by new encounters with the invading agent and it can develop memory against the aggressor. It is composed of *B* and *T lymphocytes*, located in lymphoid organs and circulating blood. The B cells are involved in the production of *immunoglobulins* (Ig) or antibodies (*humoral immunity*); the T lymphocytes are effector of the cell-mediated immunity (*cellular immunity*).

Antigens (Ag) are agents (microorganisms, particles, cells, or molecules) that can trigger the immune response. Small portions of the antigen (antigenic determinants or *epitopes*), recognized as foreign by the immune system, are responsible for the immune reaction.

Humoral immunity depends on the production of antibodies. In response to an antigen, a B lymphocyte multiplies and differentiates (clonal selection) into plasma cells, which produce antibodies of the same type as those of original B lymphocyte.

Immunoglobulins are composed of a structural unit of four polypeptide chains linked together by disulfide bridges. Two of the polypeptides are larger (50 kDa, 440 amino acids) and are known as the *heavy or H chains*. The other two (of 25 kDa, 220 amino acids), are the *light or L chains*. The first 110 amino acid residues in both chains constitute the *variable domains* (V_H and V_L), with at least three hypervariable segments that present the greatest differences in amino acid sequence among the different Ig. They are called the *complementarity determinant regions* (CDR). V_H and V_L domains face each other and form a niche where the antigen epitope binds. Each antibody has two identical antigen binding sites. The antibody–antigen binding is similar to that of enzyme–substrate. From the end of the variable region to the C-terminal region of each Ig, are the *constant domains* (C_H and C_L). The L chains have one constant domain (C_L) and the H chains have three to four domains, depending on the type of heavy chain considered. In some Ig, there is a *hinge region* between the C_{H1} and C_{H2} domains. When an Ig is treated with the enzyme papain, three fragments are produced, two are called the Fab fragments and one the Fc fragment. The Fab fragment contains the binding site for the antigen, while the Fc portion performs the nonspecific functions of Ig.

Five classes or isotypes of Ig exists that differ by their heavy chain. IgG, IgA, IgM, IgD, and IgE contain γ, α, μ, δ, and ε chains, respectively. Light chains can be either of two types, κ or λ. IgG, IgD, and IgE are monomers, IgM are pentamers, and IgA are dimers of the basic Ig structure. B cells have both IgM and IgD inserted in their membrane, which function as receptors that bind the corresponding antigen.

Antigens (Ag) are in general macromolecules. The most common antigens are proteins and heteropolysaccharides. Lipids, nucleic acids, and small molecules (*haptens*) have low antigenicity, but their antigenicity can be enhanced when they are bound to proteins. The antigenic determinant or *epitope* of a protein comprises ~12–20 amino acid residues in the protein surface which converge to a small area of the molecular surface thanks to the tertiary structure. A denatured protein is generally not recognized by an antibody specific for the native molecule.

Antibodies are characterized by their diversity. An individual is capable of synthesizing more than 10^7 Igs with different specificities. Such heterogeneity in Ig synthesis is achieved through rearrangement of the genes encoding the different polypeptides that form the antibody. Three set of genes control Ig synthesis, one group for the H chain, one for the L_κ, and another for L_λ chains. The genes of the *H chain* are arranged in three sets for the variable V region (a total of 51 genes), 25 genes for the D domain, 6 genes for the J polypeptide, and 1 gene for each of the isotypes of the constant regions. The genes encoding the L_κ *chain* are in two groups for the variable V domain (total of 40 genes), the J segment (5 genes) and one for the constant region. L_λ *chain genes* comprise 31 V genes and 4 J genes for the variable region and 6 genes for the constant.

During the maturation of B lymphocytes, a rearrangement of Ig genes takes place, beginning with the H variable region genes. One of the genes in each group are randomly selected and rearranged to form a continuous VDJ block that encodes for the variable region. There are at least 7650 possibilities of forming different VDJ combinations for H chain genes. Then, the genes encoding the variable regions of the Lκ chain are reordered. A VJ block is produced from any of the V and J genes. If a viable chain rearrangement fails to assemble L_κ, a VJ block for L_λ is arranged. The assembly of VDJ or VJ gene requires the cleavage and remotion of the DNA sequences that originally separated the selected genes. The Ig gene reordering is only performed in one of the alleles in the locus (*allelic exclusion*) to ensure monospecific Ig synthesis in each B cell.

The variety of Ig produced is increased even more because of the existence of inaccuracy in gene splicing and the addition of bases to the 3′ end of the segments. In the rearranged VDJ and VJ gene blocks there is a high incidence of point mutations (*somatic hypermutation*), which may allow synthesizing Ig with higher affinity to bind the antigen and to enhance the response in subsequent exposures of the antibody. Isotype synthesis of the heavy chain requires a new DNA rearrangement for the constant region genes.

Monoclonal antibodies are those synthesized by plasma cells derived from the same clone. All these antibodies have identical antigen specificity.

The complement includes a series of plasma proteins that are involved in body defense, promoting the lysis or phagocytosis of foreign cells, bacteria and viruses. The complement cooperates with and enhances the capacity of antibodies and phagocytic cells. These proteins circulate as zymogens, which are activated by a sequential cascade of reactions that involve proteases. The final action of complement activation is to attract and stimulate phagocytes, and to form the cell killing membrane attack complex, which will destroy the invading agent. Two pathways for complement activation are known:

Classical pathway. This pathway is activated by antigen–antibody complexes. All components are glycoproteins. C1 is composed by three proteins, C1q, C1r, and C1s. C1 binds to the Fc portion of Ig–Ag complexes and activates C1r, which initiates a series of proteolytic reactions. C1r catalyzes hydrolysis of C1s. C1s catalyzes C4 separation into two pieces, C4a and C4b. C4b binds to nearby surfaces and to C2. C1s promotes hydrolysis of C2 into C2a and C2b. C2a is a protease, bound to C4b, which forms the C4bC2a complex or *C3 convertase* that divides C3 into C3a and C3b fragments, forming the C4bC2aC3b complex or *C5 convertase*. C4bC2aC3b catalyzes the hydrolysis of C5 into C5a and C5b. The latter initiates the formation of the *membrane attack* or *lytic complex*, which subsequently binds C6, C7, and C8. The C5bC6C7C8 complex binds to the membrane of cells and promotes the polymerization of C9 units, which form pores or channels in the lipid bilayer. The *lytic attack complex* $C5bC6C7C8(C9)_n$ is responsible for the destruction of cells or microorganisms.

Alternative pathway. This pathway is activated by macromolecules of the surface of microorganisms. Plasma factor D hydrolyzes factor B and produces Ba and Bb. Bb is a protease that binds to C3b and forms the C3bBb complex or *C3 convertase*, which stimulates the formation of more C3b from C3. Then, it joins *properdin* to give C3bBbP or *C5 convertase*, which acts on C5 and splits it into C5a and C5b. C5b initiates the formation of the *lytic attack complex*.

Besides destroying invading microorganisms, the complement system has other functions. It prevents formation of clusters of antigen–antibody complexes; it facilitates macrophage adhesion and subsequent phagocytosis; and acts on mast cells, basophils, platelets, and smooth muscle cells, promoting the release of chemical mediators of inflammation and smooth muscle contraction.

Cellular immunity is carried out by *effector T cells*, of which there are several types: *helper* (T_H), *killer* (T_K), and *suppressor* (T_S) cells. The T_H cells have CD4 proteins in the membrane (CD4$^+$ T_H or T4 cells), T_K cells, protein CD8 (CD8$^+$ T_K or T8 cells). These cells express receptors (TCRs), which consist of two polypeptide chains (α and β). Each of these polypeptides has two domains, a variable (V) and a constant (C) domain, and a segment that crosses the membrane once. Both V regions form the Ag-binding site. There is an enormous diversity of TCRs, whose synthesis is controlled by rearrangement of genes similar to that of the Ig. TCR is composed of two $\alpha\beta$ heterodimers responsible for their specificity, a coreceptor (formed by CD4 or CD8) and the CD3 complex, composed of six polypeptides (γ, δ, two ε and two ζ). TCRs recognize the Ag only if it is bound to the *major histocompatibility complex* (MHC or HLA in humans) of the surface of accessory cells. There are two classes of MHC proteins: I and II.

The MHC consists of two classes of proteins:

Class I MHC proteins comprise a α chain with three domains (α_1, α_2, and α_3) and a transmembrane segment, and a β_2 microglobulin, which has only one domain. MHC class I is expressed in all cells except in red blood cells and in the trophoblast syncytium. There are three subclasses of class I MHC proteins (A, B and C) that have high genetic polymorphism in the human population. Most individuals are heterozygous at the three loci and express six different MHC class I proteins.

Class II MHC proteins consist of two chains (α and β). Both have two domains and a transmembrane portion and are expressed only on antigen-presenting cells (macrophages, monocytes, B lymphocytes, dendritic cells). There are three subclasses of MHC class II proteins (DR, DQ, and DP).

T cells detect the presence of foreign MHC proteins and initiate an immune response aimed at eliminating the cells that possess them. MHC is also involved in cell recognition and is the cause of incompatible organ grafts rejection. MHC proteins present the Ag to TCR, serving as guiding molecules for immune recognition. The TCRs detect the Ag only if it is presented in association with MHC proteins of the same individual (MHC restriction). T_H cell receptors recognize Ag bound to class II MHC proteins and T_K receptors, recognize Ag bound to MHC class I.

Antigen recognition depends on two different pathways:

Exogenous pathway. This pathway is followed by proteins from the environment. The antigen is admitted by endocytosis, degraded to small peptides in endo-lysosomal compartment, and is then bound to MHC-II proteins. The MHC II–Ag complex is exposed on the plasma membrane, to be recognized by T cells.

Endogenous pathway. This pathway is followed by endogenous proteins synthesized by cells of the individual, viral proteins produced by virus-infected cells or abnormal proteins produced by cancer cells. These proteins are introduced into proteasomes, where they are hydrolyzed to small peptides, transported into the cisternae of the ER, and bound to MHC class I protein. MHC I–Ag complex then passes to the Golgi and to the surface of the host cell, where it is exposed to be recognized by the T cells.

T_K cells recognize antigen bound to MHC class I and exert their lytic action through the secretion of *perforin*, a protein that forms channels or transmembrane pores, similar to those generated by the complement, which will lyse the cell. Furthermore, the lytic action is enhanced by proteases, called *granzymes*, and the stimulation of *apoptosis*.

Cytokines are proteins secreted by cells of the immune system with paracrine or autocrine action. The name *interleukin* (IL) is also used for such substances. T lymphocytes are activated by a double stimulation: one specific, produced by Ag binding to MHC proteins and other, nonspecific, induced by chemical mediators (cytokines). CD4$^+$ T_H cells bind to antigens attached to the MHC class II complex on the surface of presenting cells, which, as well as the activated T_H cells, secrete various cytokines, induce the expression of IL-2 receptor and activate the clonal expansion of T lymphocytes. T_H cells assist in the activation of B lymphocytes. When an Ig$_m$ on the surface of a B cell binds an Ag, it is endocytosed, degraded and the resulting peptides are bound to MHC class II, exposing the complex on its surface. A CD4$^+$ T$_H$ cell binds to the MHC class II–Ag and secretes cytokines that stimulate proliferation and differentiation of B lymphocytes into plasma and memory cells.

IFNs are low molecular mass glycoproteins of the cytokine family, which are synthesized by cells that have been invaded by viruses. Type I IFNs mediate the initial response of the innate system to viral infections. Type II IFNs are secreted by T helper and NK cells. The general role of IFN is to promote the inflammatory response of macrophages.

Bibliography

Abbas, A.K., Lichtman, A.H., Pillai, S., 2007. Cellular and Molecular Immunology, sixth ed. Saunders Elsevier, Philadelphia, PA.

Beutler, B., 2004. Inferences, questions and possibilities in toll-like receptor signalling. Nature 430, 257–263.

Infection and Immunity, 2004. Serie de Artículos. Nature 430, 241–271.

Kindt, T.J., Goldsby, R.A., Osborne, B.A., 2007. Inmunology, sixth ed. McGraw Hill, New York, NY.

Murphy, K.M., Travers, P., Walfort, M., 2007. Janeway's Immunobiology: The Immune System in Health and Disease, seventh ed. Garland Publishing, London.

31

Hemostasis

Hemostasis is a process that limits blood loss from a blood vessel that has been injured. Hemostasis consists of several events that help seal back the damaged vessel, including local vasoconstriction, platelet accumulation, and blood clotting (which is the change of the blood from liquid to gel, also known as coagulation).

Platelet aggregation. Immediately following the rupture of a vessel, or an injury of the endothelium, platelets adhere to the subendothelial collagen. This is accomplished through protein–protein interaction between an adhesion protein expressed on the surface of platelets, the integrin *glycoprotein Ib* (GpIb) and the *von Willebrand factor* (vWf). This factor is a multimeric protein synthesized by endothelial cells and secreted into the subendothelium, where it binds to collagen. Once attached to the subendothelium, platelets become activated and release substances stored in their cytoplasm in granules (serotonin, ADP). These compounds promote further adhesion and activation of neighbor platelets, which will form a plug that will close the opening of the vessel's wall. If platelet aggregation is not controlled, it will continue and may completely occlude the vessel. To avoid this, there are platelet antiaggregating agents, such as prostacyclin (prostaglandin I$_2$), nitric oxide (previously known as endothelium-derived vascular relaxation factor), and ADPase released by the endothelial cells. Platelet aggregation can also be inhibited by pharmaceutical agents, among which is acetylsalicylic acid (aspirin).

BLOOD COAGULATION

Blood clotting comprises a series of enzymatic reactions in which more than 12 proteins, cell membrane phospholipids, and Ca^{2+} participate (Table 31.1). The existence of specific factors involved in blood coagulation was discovered from clinical observations in patients who suffered genetically determined bleeding disorders. Most of these factors are designated with Roman numerals, according to the order in which they were discovered. Activated factors are denoted by adding the letter *a* to their names.

Formation of the clot is due to the conversion of *fibrinogen* (factor I), which is a soluble protein present in plasma, into an insoluble polymer, *fibrin*. Fibrin forms a net that traps the cells circulating in blood. Transformation of fibrinogen to fibrin is the last stage of a series of a cascade of proteolytic reactions (Fig. 31.1). The factors that initiate coagulation circulate in plasma in very low concentrations; it is the amplification capacity of the cascade reaction mechanisms, which ensures an efficient response.

Seven of the coagulation factors, including prekallikrein, prothrombin (factor II), proconvertin (VII), Christmas factor (IX), Stuart–Prower factor (X), plasma thromboplastin (XI), and Hageman factor (XII), are zymogens that are converted into active proteases by selective hydrolysis of one or more peptide bonds. The enzymes belong to the family of proteases, among which are trypsin, chymotrypsin, and elastase of

TABLE 31.1   Blood Clotting Factors

No.	Name	Molecular mass (kDa)	Plasma concentration (mg/dL and molar)	Function
I	Fibrinogen	340	250–400 (7 μM)	It is converted to fibrin by thrombin, forming a net that traps the blood cells.
II	Prothrombin	72	10–14 (1.4 μM)	It is activated to thrombin by factor Xa.
III	Thromboplastin	44	—	It is released when there is cellular damage. It participates in factor X activation.
IV	Ca^{2+}	—	4–5 (1.25 μM)	It mediates union of factors IX, X, VII, and II with phospholipids.
V	Proaccelerin	350	(120 nM)	It potentiates Xa action on prothrombin.
VI	It does not exist	—	—	—
VII	Proconvertin	45–54	0.05 (10 nM)	It forms the complex with factor III and Ca^{2+} that activates factor X (extrinsic path).
VIII	Antihemophilic factor (VIII:C)	285	0.1–0.2 (0.7 nM)	Associated to factor IXa it activates factor X. Its absence causes hemophilia A.
	von Willebrand factor (VIII:R)	Polymer > 10,000	—	It mediates the union of platelets to collagen in vessels subendothelium. Its absence causes von Willebrand disease.
IX	Christmas factor	57	0.3 (90 nM)	It is converted to IXa by XIa. Its absence causes hemophilia B.
X	Stuart–Prower factor	59	1 (170 nM)	It is converted to Xa by IXa–VII–Ca^{2+} (intrinsic path), by VIIa–III–Ca^{2+} (extrinsic).
XI	Plasma thromboplastin	160	0.5 (30 nM)	It is converted to XIa by factor XIIa and activates factor IX.
XII	Hageman factor	76	—	Activated by kallikrein and kininogen in contact with strange surfaces, it converts XI in XIa.
XIII	Laki–Lorand factor	320	1–2 (70 nM)	Activated to XIIIa by thrombin, it forms cross-bridges in fibrin filaments.
XIV	Fletcher factor	—	—	Activated to kallikrein it converts XII in XIIa together with kininogen.
XV	Kininogen of high molecular mass	—	—	Together with kallikrein, they activate factor XII.
XVI	Protein C	62	(60 nM)	Activated to Ca it binds to protein S and hydrolyzes Va and VIII inhibitor
XVII	TFPI	30	(2.5 nM)	In the presence of Xa it blocks complex VIIa and TF in extrinsic pathway.
XVIII	Antithrombin III	58	0.15–0.40 (2.5 nM)	Inhibitor of thrombin and factors IXa and Xa.
XIX	Thrombomodulin	100	—	Bound to α-thrombin it activates C to Ca.
XX	Protein S	—	—	Bound to Ca it facilitates inhibition of thrombin generation.

pancreatic juice. All these enzymes have an essential serine in their active site.

Factors II, VII, IX, and X are proteins synthesized in the liver. They have γ-carboxyglutamate (Gla) residues in the initial portions of the molecule. The γ-carboxyglutamate residues are Ca^{2+} chelators, which favor the binding of the proteins that contain them to phospholipids. The formation of Gla residues by posttranslational glutamate carboxylation, requires vitamin K as cofactor (p. 660). Avitaminosis K produces coagulation disorders because factors II, VII, IX, and X cannot be properly carboxylated and remain inactive. Factor XIII is a zymogen that, after activation, is transformed into transamidase (transglutaminase or factor XIIIa).

Stages of Blood Coagulation

The final stage of blood clotting, the conversion of fibrinogen into fibrin, is catalyzed by a protease called *thrombin* (or factor IIa), which results from activation of the precursor *prothrombin* (or factor II), by proteolysis, catalyzed by factor Xa. Two different pathways converge to the production of factor Xa, the *intrinsic* and the *extrinsic* pathways. All the factors of the intrinsic pathway are circulating in blood. In contrast, the extrinsic pathway utilizes a tissue factor, originated in the damaged tissues.

The whole process can be divided into three stages: the first leads to the activation of factor Xa, the second comprises thrombin activation, and the third, the production of fibrin (Fig. 31.1).

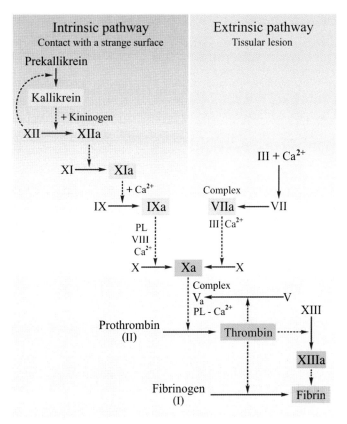

FIGURE 31.1 **Pathways that mediate blood coagulation.** *Solid arrows* indicate conversion of zymogen into active product; *dashed line arrows* show proteolytic action; *red boxes* indicate activated factors.

First Stage: Factor Xa Formation

1. *Intrinsic pathway.* This coagulation pathway comprises three steps:

   a. *XIa formation.* This starts when the blood comes in contact with a surface that is different from the vascular endothelium. In cases of a lesion in the vessel wall, the damage exposes the basal lamina or collagen fibers of the vessels, which are substrates for the initiation of coagulation. In general, polyanionic substrates, even inorganic surfaces, such as glass or kaolin, can trigger the coagulation reaction. The factors involved in this process include *prekallikrein*, high molecular weight *kininogen*, *Hageman factor* (XII), and *plasma thromboplastin antecedent*. These four factors are adsorbed to the surface. Factor XII is a zymogen; however, it has weak proteolytic activity and catalyzes the conversion of prekallikrein into kallikrein. The activity of this protease is potentiated by kininogen, both of which activate factor XII. Active protease XII (XIIa) converts the plasma thromboplastin antecedent (XI) into active factor XIa. Different from the reactions that follow, this step of coagulation does not require Ca^{2+}. Deficiency of factor XI, resulting from a genetic disease not linked to chromosome X, causes *hemophilia C*. This condition is milder than common forms of hemophilia; hemorrhages are infrequent.

   b. *IXa formation.* Factor IX (Christmas factor) is a plasma zymogen. In the presence of Ca^{2+}, factor XIa catalyzes the hydrolysis of a peptide bond in factor IX, releasing a 10-kDa glycopeptide and activating factor IX into IXa. Factor IX is a glycoprotein encoded by a chromosome X gen. Its deficiency produces hemophilia B, characterized by severe coagulation disorders.

   c. *Xa formation.* Factors IXa, X, and VIII via interaction of their γ-carboxyglutamate residues with phospholipids, bind and form a complex on the plasma membrane of platelets. IXa and X are the first factors to become in contact with the platelets, then factor VIII is attached. Factor VIII functions as a cofactor for the activation of factor X. The IXa–X–VIII complex is called *intrinsic tenase*. The gene that encodes for factor VIII is located in chromosome X. Mutations in this gene causes *hemophilia A*, the most frequent genetic disease of coagulation. Heterozygote women for the defect are asymptomatic because the normal gene compensates for the mutated allele. Males that receive an abnormal allele from the mother suffer the disease, since they have only the maternal X chromosome. These patients require frequent transfusions and periodic administration of concentrated plasma rich in factor VIII. Due to the repeated transfusion, patients with hemophilia A have a higher risk for viral infections, such as AID and hepatitis. The gene of factor VIII has been cloned, which offers the possibility of producing and administering it, avoiding blood transfusions.

   Factor VIII associates in plasma to the von Willebrand factor (vWF). This polymeric glycoprotein, formed by 225 kDa subunits favors the interaction of intrinsic tenase with platelets. vWF has structural homology with other adhesive proteins, such as fibronectin. Hereditary coagulation disorders due to vWF absence have been described. Once the IXa–X–VIII complex is formed, factor IXa activates factor X into Xa with protease activity.

2. *Extrinsic pathway.* When blood contacts damaged tissues or it is mixed with tissue extracts, factor Xa is rapidly generated. The mechanism of factor X activation via this route consists in the formation of a complex in which factor VII, Ca^{2+} and tissue factor

(thromboplastin or factor III) are involved. Tissue factor is a lipoprotein present in all tissues, but it is present in highest amounts in lung, brain, and placenta. The site of tissue factor synthesis appears to be subendothelial cells. Factor VII is activated by its association with the phospholipid domain of tissue factor, via γ-carboxyglutamate residues. The VIIa–tissue factor–Ca^{2+}, also called *extrinsic tenase*, converts factor X into active factor Xa.

The extrinsic pathway is set on in few seconds, while the intrinsic pathway takes several minutes. Both pathways converge in the formation of factor Xa (Fig. 31.1). The stages that follow are common to both pathways.

Patients with deficiencies in factors VIII or IX suffer from severe hemorrhage even when the extrinsic pathway is normal. This suggested that the extrinsic pathway was of less importance. On the other hand, the observations that deficiency in factors of the intrinsic pathway (prekallikrein, kininogen, XII) does not cause coagulation problems, the fact that lack of factor VII is accompanied by hemorrhagic episodes, and the finding of a *tissue factor pathway inhibitor* (TFPI), prompted researchers to revise the pathways involved in coagulation. A model that integrates both extrinsic and intrinsic pathways developed.

The coagulation process starts when damage of a blood vessel exposes blood to the tissue factor (TF or III). In plasma, small amounts of VIIa are circulating. Binding of VIIa to the exposed tissue factor forms the *extrinsic tenase complex* (TF–VIIa), which markedly increases the catalytic activity of VIIa. This activates factor Xa, which is essential for thrombin formation. Extrinsic tenase also promotes IXa formation, but its action is soon limited by the presence of TFPI, which binds to Xa and then to Tf–VII to give the Xa–TFPI–TF–VIIa quaternary complex. This complex has no coagulating

activity. Despite inhibition by the Xa–TFPI–TF–VIIa complex, the reaction cascade continues because factor IXa–VIII (intrinsic tenase complex) activates factor X. Besides, formation of IXa initially promoted by extrinsic tenase, is also catalyzed by α-thrombin. In this pathway, the first step of the intrinsic pathway, comprising the activation by contact factors, is functionally less significant.

Second Stage: Thrombin Formation

The protease *α-thrombin* (factor IIa) is generated by hydrolysis of the zymogen *prothrombin* (factor II), a glycoprotein of 582 amino acids containing 12 disulfide bridges. Lysis of an arginine–threonine bond at the N-terminal end of prothrombin, catalyzed by Xa, releases a 32-kDa fragment. A second hydrolysis of an arginine–leucine peptide bond generates other two segments that remain linked by an S—S bridge (Fig. 31.2). This structure of 36.7 kDa is α-thrombin, a serine protease homologous to trypsin, but with a higher proteolytic selectivity; it hydrolyzes only arginine bonds at defined positions.

Prothrombin conversion into α-thrombin is catalyzed by factor Xa, in a reaction that is enhanced by a complex with proaccelerin (factor V), platelets membrane phospholipids, and Ca^{2+} (*prothrombinase complex*). Factor Xa and prothrombin bind to phospholipids through γ-carboxyglutamate residues and Ca^{2+}. Formation of the prothrombinase complex favors the interactions between the molecules. Factor V is activated to accelerin (Va) by α-thrombin. In turn, factor Va accelerates the proteolysis of prothrombin catalyzed by Xa.

Third Stage: Fibrin Formation

Fibrinogen (factor I) is a glycoprotein composed by six polypeptide chains (two Aα, two Bβ, and two γ) linked by disulfide bonds. The

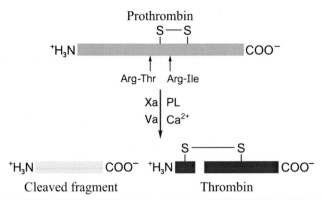

FIGURE 31.2 **Formation of α-thrombin.** Prothrombin is hydrolyzed at two peptidic bonds (Arg-Thr and Arg-Ile), indicated by *arrows*, in a reaction catalyzed by Xa and enhanced by Va of the prothrombinase complex linked to membrane phospholipids (PL). The N-terminal fragment of prothrombin is released, the other two segments remain bound by a disulfide bridge.

molecule extends longitudinally and is symmetric, with three globular domains connected by fibrillar segments (Fig. 31.3). Two of the globular domains are located at the ends of the molecule, while the third globular segment is in the middle. Each half of the molecule is formed by association of three polypeptide chains (Aα, Bβ, and γ). These chains are wrapped around the longitudinal axis forming a triple helix. The amino-terminal extremes of all six protein chains are in the central domain, those of Aα and Bβ emerge as free ends in the middle zone of the molecule (Fig. 31.3). These ends comprise the A and B portions of Aα and Bβ, respectively, rich in aspartate

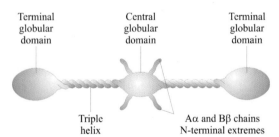

FIGURE 31.3 **Fibrinogen.** The scheme represents a molecule with its globular ends and a minor central domain, where the Aα and Bβ N-terminal extremes emerge. The central domain is connected to the globular ends by a triple helix segment formed by Aα, Bβ, and two γ chains.

and glutamate residues. The Bβ terminal ends contain tyrosine-O-sulfate residues, formed posttranslationally. These groups contribute to provide an electronegative region in the central part of the molecule and are responsible for the mutual repulsion of different fibrinogen molecules that help maintaining them in solution.

α-Thrombin hydrolyses arginyl–glycine bonds at the free N-terminal segments of Aα and Bβ chains, separating 4 peptides of 18 amino acids from each of the Aα chains and 1 of 20 amino acids from each of the Bβ chains (Fig. 31.4). These peptides are the *fibrinopeptides*; the remaining of the fibrin molecule is a monomer $(\alpha\beta\gamma)_2$. With the removal of fibrinopeptides rich in electronegative groups, fibrin loses its intermolecular repulsion, monomers tend to polymerize spontaneously, and highly ordered molecular associations are formed. The monomers are oriented one after another, head against head, in long linear threads, probably paired by electrostatic attractions and/or H bonds between two neighbor globular heads from one strand and a central domain of an adjacent thread (Fig. 31.5).

Fibrin stabilization. The fibrin bundles constitute a loose net, and cannot form a stable clot unless they are reinforced by the addition of

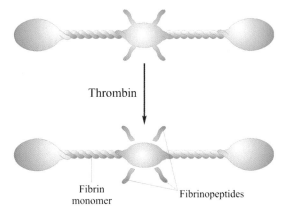

adjacent strands. NH_4^+ is released in the reaction (Fig. 31.6).

Regulation of Coagulation

Commonly, clot formation is limited to the area of the blood vessel injured. This suggests the existence of regulatory mechanisms that control coagulation and avoid massive vessels blockage (thrombosis).

A series of mechanisms are involved in the regulation of clot formation. Some of them reduce the levels of coagulation factors, others inhibit their function. These include the following:

1. Normal blood flow in the vessels drags clotting factors in excess at the site of injury and subsequently dilutes them in the bloodstream. When there is an obstruction to blood flow (i.e., venous stasis), thrombus formation is favored.
2. The liver uptakes and degrades active factors from circulation.

FIGURE 31.4 **Fibrinogen hydrolysis by α-thrombin.** Below is a fibrin monomer after cleavage of four fibrinopeptides, two A and two B.

covalent bonds between threads. These bonds are generated by a transamidase (factor XIIIa or *transglutaminase*), formed from factor XIII by α-thrombin. Transglutaminase catalyzes the formation of an amidic link between fibrin monomers glutamine and lysine residues in

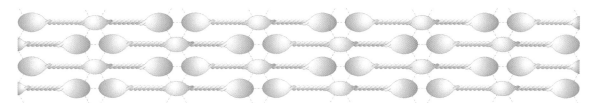

FIGURE 31.5 **Scheme of the bundle formed by polymerization of fibrin monomers, arranged regularly in parallel strands.** Two neighbor heads from one thread establish attractions with the central domain of the adjacent strand.

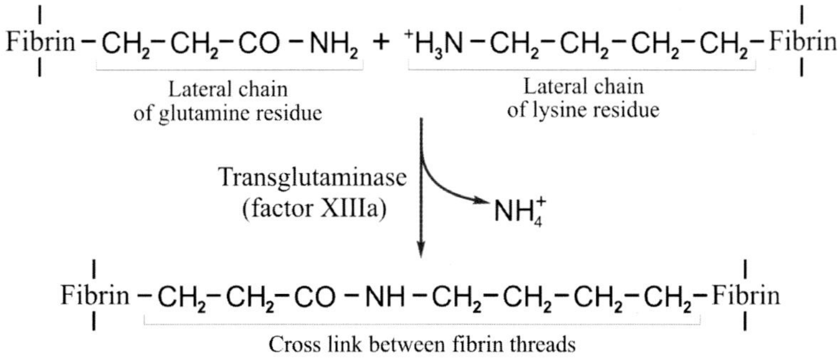

$$Fibrin-CH_2-CH_2-CO-NH_2 \; + \; {}^+H_3N-CH_2-CH_2-CH_2-CH_2-Fibrin$$

Lateral chain of glutamine residue   Lateral chain of lysine residue

Transglutaminase (factor XIIIa) $\searrow NH_4^+$

$$Fibrin-CH_2-CH_2-CO-NH-CH_2-CH_2-CH_2-CH_2-Fibrin$$

Cross link between fibrin threads

FIGURE 31.6 **Formation of cross links between adjacent fibrin strands.**

3. Factors that promote coagulation can be removed by binding of α-thrombin to *thrombomodulin* (TM), a receptor present on the surface of endothelial cells surface. The α-thrombin–TM complex catalyzes the activation of a plasma zymogen, C protein (CP). This protein depends on the presence of vitamin K, in its active form (ACP) is a serine protease that hydrolyzes VIIIa and Va of the intrinsic tenase and prothrombinase complexes, respectively. Hydrolytic activity of ACP is stimulated by protein S, which has Gla domains. It is important to note that α-thrombin plays a dual role; it catalyzes fibrin formation and, once the process is underway, it exerts actions to stop it, promoting the removal of clotting factors.

4. Plasma contains inhibitors of active clotting factors. The main one is antithrombin III, a 60 kDa glycoprotein that irreversibly inactivates thrombin and other factors, such as kallikrein, IXa, Xa, Xia, and XIIa. Patients with antithrombin III deficiency often suffer from thrombosis and embolism. Antithrombin III action is markedly stimulated by the heteropolysaccharides heparin and heparan sulfate, compounds found in mast cells and in the vascular endothelium, respectively. They have anticoagulant action that favor formation of complexes between antithrombin III and activated clotting factors. Table 31.1 presents the factors involved in blood coagulation.

Other plasma antiproteases, such as α_2-macroglobulin and α_1-antiprotease, also inhibit thrombin, although more weakly than antithrombin III.

Fibrinolysis

After the clot has been established, repair of the damaged tissues starts, with the formation of a scar. At that time, the blood clot is no longer necessary and it is removed by rupture of the fibrin net. This is catalyzed by *plasmin*, a serine protease that hydrolyses peptide bonds in the triple helix regions of the fibrin monomers. Plasmin is generated from an inactive precursor, *plasminogen*, by the action of intrinsic and extrinsic factors. These intrinsic factors include prekallikrein and factors XI and XII. The extrinsic factors are produced in the vascular endothelium. The most potent one is the *plasminogen tissue activator*. The gene encoding this protein has been cloned, synthesized by recombinant DNA, and it is currently utilized in clinics to dissolve blood clots.

SUMMARY

Hemostasis is the process by which blood loss from vessels that have been damaged is stopped.

Blood clotting is achieved by a cascade of enzymatic reactions, which involve a series of factors. Among them are the zymogens *prekallikrein, prothrombin* (II), *factors VII, IX, X, XI,* and *XII*, which are converted to active proteases by hydrolysis. Factors II, VII, IX, and X are dependent on vitamin K, which acts as a cofactor of carboxylations (formation γ-carboxyglutamate, Ca^{2+} chelator) serves to fix the protein to membrane phospholipids).

Two main pathways mediate blood coagulation:

The *intrinsic pathway* is triggered when blood comes in contact with a foreign surface. This causes the adsorbtion of prekallikrein, kininogen, XI, and XII. Prekallikrein is activated to kallikrein; this, together with kininogen, convert factor XII in the active protease XIIa. XIIa hydrolyses factor XI to factor XIa, which in the presence of Ca^{2+} transforms zymogen IX into IXa. The complex IX–X–VIII and Ca^{2+} is formed and this activates factor X to Xa.

The *extrinsic pathway*, which is activated when blood comes in contact with damaged tissue. This generates factor Xa, in a process that is catalyzed by the factor VII, Ca^{2+} and thromboplastin (III) complex.

Both the intrinsic and extrinsic pathways converge in the formation of factor Xa. The stages that follow, including thrombin and fibrin formation, are common to both pathways.

Thrombin formation is mediated by factor Xa, which converts prothrombin into α-thrombin. This reaction is accelerated by factor Va. Phospholipids and Ca^{2+} are required in the process.

Fibrin formation is catalyzed by α-thrombin, which hydrolyses four small fibrinopeptides from fibrinogen to form fibrin. Fibrin monomers polymerize in fiber bundles to generate the clot. The fibrin network is stabilized by factor XIII, activated by thrombin, which acts as transamidase, forming bridges between glutamine and lysine residues of adjacent fibrin strands.

Regulation of coagulation is achieved by removal of the factors involved in blood clotting, or by inactivating them. The α-thrombin–thrombomodulin complex activates plasma protein C, which degrades factors Va and VIIIa. Antithrombin III, a protease inhibitor, inactivates thrombin, kallikrein, IXa, Xa, XIa, and XIIa. Antithrombin III action is potentiated by heparin.

Fibrinolysis is the process that leads to clot dissolution. It is produced by hydrolysis of fibrin. This is catalyzed by *plasmin*, which is generated from plasminogen. The conversion factor for plasminogen to plasmin includes factors XI and XII (intrinsic factors) and plasminogen tissue activator (extrinsic factor).

Bibliography

Broze, G.T., 1997. Tissue factor pathway inhibitor and the revised theory of coagulation. Annu. Rev. Med. 46, 103–112.

Chapin, J.C., Hajjar, K.A., 2015. Fibrinolysis and the control of blood coagulation. Blood Rev. 29, 17–24.

Kalafatis, M., Egan, J.O., van't Veer, C., Cawthern, K.M., Mann, K.G., 1997. The regulation of clotting factors. Crit. Rev. Eukaryot. Gene Expr. 7, 241–280.

Scully, M.F., 1992. The biochemistry of blood clotting: the digestion of a liquid to form a solid. Essays Biochem. 27, 17–36.

Apoptosis

Under normal conditions, cells are continuously removed from body tissues. This can take place through a controlled process known as *programmed cell death* or *apoptosis* (from the Greek, "falling off" of petals from a flower or "dropping off" leaves from a tree).

Apoptosis serves several important functions. These include:

1. Elimination of unnecessary or damaged cells, or cells carrying alterations that make them potentially dangerous to the body (i.e., cells which have been invaded by viruses or cancer cells).
2. Atrophy of certain tissues in organs that undergo cyclic changes induced by hormones (i.e., mammary gland at the end of lactation).
3. Removal of cells from the immune system that did not reach a satisfactory gene recombination or whose antigen specificity makes them potentially reactive against own tissue structures (Chapter 30).
4. Regression of embryonic structures, which should not persist in postnatal life (i.e., interdigital membranes), or cells produced in excess (it is estimated that 50% of brain cells are removed before birth).

In a normal adult, it is estimated that more than 10 billion cells are destroyed by apoptosis each day. Cells sentenced to programmed cell death show a number of characteristic morphological changes. Cell size is reduced due to cleavage of cytoskeletal filament fibers, cell cytoplasm, and nucleus condense, DNA becomes fragmented by the action of endonucleases into pieces of 200 (or multiples of 200) base pairs, cell organelles (Golgi apparatus, endoplasmic reticulum, and mitochondria) also become fragmented, the cell plasma membrane displays a "blistered" like appearance, cells become round and separate from neighboring cells. Finally, the cell is reduced to vesicular remnants that are phagocytosed by neighboring cells and macrophages.

One of the biochemical changes during apoptosis is the translocation of phosphatidylserine from one to the other side of the cell plasma membrane. Normally phosphatidylserine is located only in the inner leaflet of the lipid bilayer, upon apoptosis the phospholipid moves (via flip-flop) to the outer membrane layer, where it is exposed and serves as a marker that can be recognized by receptors on phagocytic cells. In addition, translocation of phosphatidylserine is responsible for blocking inflammatory factors that usually accompany phagocytosis. In addition to phosphatidylserine, other molecules also become exposed to the surface of the apoptotic cell. Among them are annexin I and calreticulin. The shift of these molecules from one to the other leaflet of the plasma membrane is catalyzed by enzymes named floppases and flippases (see p. 218).

The changes that cells undergo during apoptosis are very different from those occurring during cell necrosis. Necrosis is secondary to severe cell damage; it leads to cell plasma membrane breakdown and the loss of the cell content to the interstitial space, which causes an intense inflammatory reaction.

The discovery of apoptosis was the result of studies in a small nematode (*Caenorhabditis elegans*). It was observed that during embryo development of this organism, 131 out of 1090 cells were selectively eliminated, leaving 959 cells, which will constitute the tissues of the adult nematode. Three genes were found to be directly involved in the process. Two of them, *ced-3* and *ced-4*, encode executing cell death proteins (CED-3 and CED-4), and a third, *ced-9*, is responsible for the synthesis of an apoptosis inhibitory protein (CED-9).

The findings in the nematode encouraged the search for homologous genes in mammals. Genes, as well as proteins, involved in programmed cell death were found in several species, including humans. These proteins have been highly conserved during evolution.

Proapoptotic proteins. Proteins homologous to the *C. elegans* CED-3 have been demonstrated in humans. They comprise a family of proteases called *caspases*. All have a critical cysteine residue in the active site and they selectively hydrolyze peptide bonds that involve the carboxyl group of aspartic acid (this gives them their name, c, cysteine; asp, aspartate). They were designated by numbers, reflecting the order in which they were discovered. At present, more than 14 different caspases have been identified.

Caspases are found in cells as zymogens and are activated by hydrolysis. Caspases are cleaved into three fragments, a small piece from the N-terminal end of the protein and two polypeptides of different size from the rest of the molecule, which bind in a heterodimer. Association of two of these heterodimers forms the enzyme, which has two active sites.

Caspases can be distinguished into two groups, the *initiator* and the *effector* caspases. Initiator caspases are activated by autohydrolysis and once active, they hydrolyze other caspases, and finally the effector or executor caspases, which specifically lyse molecules that are essential for cell survival. Once apoptosis is started, the process is irreversible.

Not all caspases are involved in cell apoptosis. The first protein identified in mammals that was homologous to the nematode CED-3 was *interleukin 1b–converting enzyme* (ICE), which activates the inflammatory cytokine IL-1b. Subsequently, other members of the caspase family were isolated. Caspases 2, 8, 9, and 10 are involved in programmed cell death and function as initiator caspases. On the other hand, caspases 3, 6, and 7 function as effector caspases. Caspases 1, 4, 5, and 12 participate in the processing of proinflammatory cytokines (p. 773), caspase 13 is specific to bovine and caspase 14 is expressed only in embryonic tissues.

It is estimated that approximately 400 proteins can be the substrates of caspases. Among these proteins are cell adhesion molecules as cadherins, cytoskeleton proteins, such as actin, spectrin, gelsolin, microtubule tubulin, dynein, and laminin of the nuclear envelope sheet. Caspases also hydrolyze several transcription factors, initiation factors, and ribosomal proteins, suppressing the transcription and translation processes. A protein responsible for maintaining endonuclease inactive is also target of caspases.

Proteins homologous to *C. elegans* CED-4, which promote activation of the apoptotic protease, have also been recognized in mammals. They are designated *apoptotic protease activating factor* (APAF).

Antiapoptotic proteins. Proteins have been identified that protect cells from apoptosis. These correspond to the nematode CED-9 protein, and the mammalian homologs, the Bcl family of proteins, among which is Bcl-2.

Pathways leading to apoptosis. Programmed cell death is triggered by external stimuli, which activate the extrinsic signaling pathway of apoptosis, or by stimuli coming from the same cell, which activate the intrinsic pathway.

EXTRINSIC PATHWAY

This pathway is activated by the binding of a ligand to a cell receptor. Usually the ligand is a cytokine of the *tumor necrosis factor* (TNF) family (p. 774), which binds to the tumor necrosis factor receptor (TNFR), also called *Fas*. This receptor is a homotrimer of a protein that has an extracellular domain, which accommodates the ligand, a transmembrane segment, and an intracytoplasmic or *death domain* (DD). Binding of the ligand produces a conformational change of the DD domain in the receptor that allows it to associate with adaptor proteins, forming the receptor–adaptor protein-procaspase complex or *death-inducing signaling complex* (DISC). The adaptor proteins include TNFR-associated death domain (TRADD) and Fas-associated death domain (FADD). These proteins contain a death effector domain (DED) that interacts with the initiator procaspases (8–10). These proenzymes undergo hydrolysis and initiate a proteolytic cascade, which stimulate effector caspases 3 and 7, with the consequent degradation of cell proteins.

Activation of other pathways contributes to enhance cell apoptosis. For example, increase in *ceramide* through the *Ras/MAP* kinase system, release of Ca^{2+} from the endoplasmic reticulum, and the activation of the extrinsic pathway of apoptosis also promote programmed cell death.

INTRINSIC OR MITOCHONDRIAL PATHWAY

Apoptosis can be activated by stimuli coming within the cell, including cell stressors, such as hypoxia or lack of nutrients, and agents that cause damage of DNA or other cell structures.

In vertebrates this pathway is initiated by the release of apoptosis mediators from the mitochondrial intermembrane space, when certain intracellular signals permeabilize the outer membrane of the mitochondria.

A major protein in this process is *cytochrome c*, component of the electron transport chain on the external face of the inner mitochondrial membrane. When released into the cytosol, cytochrome c binds to APAF-1, which oligomerizes to form a complex or *apoptosome*. Once formed this complex, comprising APAF-1–cytochrome c–procaspase 9, triggers the activation of procaspase 9 to caspase 9. This initiator protease starts the hydrolytic cascade that stimulates the effector caspases (3, 6, and 7).

In addition to cytochrome c, other molecules are released from mitochondria. These include *Diablo* or second mitochondria-derived activator of caspases (SMAC), which by blocking inhibitor of apoptosis proteins induces programmed cell death; *Omi*, which is a stress-regulated endoprotease that activates caspases; and endonuclease G, which is responsible for DNA degradation, chromatin condensation, and DNA fragmentation.

Other apoptotic pathways. A third pathway leading to apoptosis is specific of cytotoxic T lymphocytes (CTL) and natural killer cells (NK) (Chapter 30). A serine protease designated *granzyme B* (Gra-B) functions as caspases, hydrolyzing peptide bonds at aspartate residues. Gra-B enters the cells through channels formed by perforin. Gra-B also activates procaspases 3 and 10 and can stimulate the mitochondrial pathway.

A fourth pathway for caspase activation is triggered by cell organelle–mediated cell death. Apoptosis can be initiated by stressors that affect the integrity of the nucleus, endoplasmic reticulum, Golgi, and lysosomes.

Regulation of apoptosis. Due to the irreversible consequence of apoptosis, both the intrinsic and extrinsic pathways are under tight control. The major regulator of apoptosis is the *Bcl* (from B cell lymphoma) family of proteins. The members of this group of proteins include proapoptotic and antiapoptotic agents, which control the release of cytochrome c and other mitochondrial proteins to the cytosol, regulating downstream caspase activation. The Bcl family of proteins have BH (Bcl-2 homology) domains, which gave them the designation BH1234. Bcl-2 has four BH domains and was initially identified as a protein encoded by the oncogene that stimulates the development of B cells lymphomas. Unlike other oncogenes, such as *Ras*, that stimulate cell proliferation, Bcl-2 is an inhibitor of programmed cell death, promoting the survival of cancer cells, which continue their uncontrolled growth. Other Bcl proteins with antiapoptotic function are Bcl-xL, Bcl-W, and other proteins that contain BH1234 homologous domains.

Not all the members of the Bcl family of proteins are antiapoptotic. *Bcl-2 associated X protein* (*Bax*) and *Bcl-2 antagonist killer* (*Bak*) have three BH domains instead of four (designated BH123). They promote programmed cell death. When the intrinsic pathway is activated, Bax and Bak form oligomers on the cytosolic side of the outer mitochondrial membrane, creating pores that permeabilize the mitochondrial membrane, allowing the release of cytochrome c and other intermembrane proapoptotic proteins, including *Diablo* and *Omi*, to the cytosol.

The fate of a cell depends on the balance between pro- and antiapoptotic proteins, which antagonize each other. Bcl-2 interferes with Bax and Bak proteins, disrupting the balance in favor of cell survival.

The intrinsic and extrinsic pathways are both regulated by the apoptosis inhibitor protein (IAP). IAP directly interacts with caspases and suppresses apoptosis either by inhibiting them, or labeling them for ubiquitination and subsequent degradation in the proteasome (p. 368). The Diablo and Omi proteins released from the intermembrane space, are proapoptotic. Diablo opposes to Bcl and Omi

stimulates caspase activity by interfering the action of AIP. Fig. 32.1 is a scheme of the factors involved in apoptosis.

An important function of programmed cell death is the elimination of damaged cells, especially those that have undergone changes in their DNA. These cells are dangerous as they can accumulate mutations and become carcinogenic. Many neoplastic cells show an upregulation of the expression of antiapoptotic members of the Bcl-2 family. In mammalian cells there are mechanisms that prevent malignant transformation of cells. In response to DNA damage, these mechanisms are mediated by the transcription factor of p53 (tumor suppressor protein) that induces expression of cyclin inhibitors and stops cell cycle progression. In addition to this action, p53 also triggers apoptosis by inducing expression of protein BH3 of the intrinsic pathway.

Medical importance. Many diseases are associated with disturbances in the regulation of apoptosis. For example, in Alzheimer's disease and other neurodegenerative diseases, abnormal apoptosis leads to exacerbated neuronal destruction in specific regions of the brain. There is also increased apoptotic activity in T lymphocytes in the acquired immunodeficiency syndrome (AIDS).

Other examples of increased cell destruction are heart attacks and strokes. In these cases, the necrosis resulting from the lack of oxygen supply predominates, but some cells affected by hypoxia undergo apoptosis. It is expected that the development of caspase inhibitor drugs will be useful in these conditions.

Mutations in the genes encoding factors associated with programmed cell death result in serious alterations. A failure of the gene that directs the synthesis of Fas death receptors, reduces cell removal, leading to cell accumulation in spleen and lymph nodes, which is a cause of autoimmune diseases.

Elimination of unwanted cells is impaired in some types of tumors. In lymphoma, there is

Extrinsic pathway

Intrinsic pathway

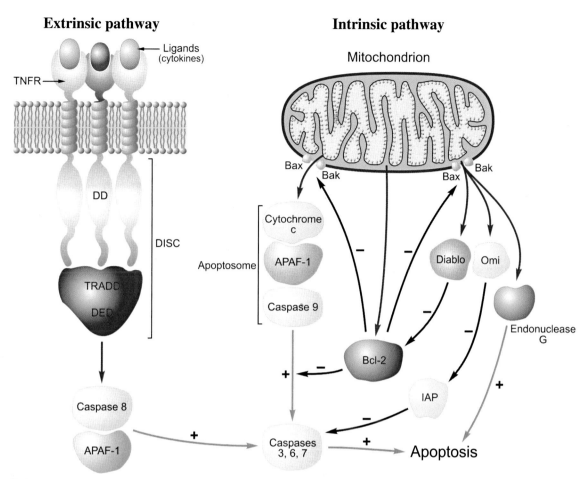

FIGURE 32.1   **Scheme of the factors involved in the apoptotic process.** *APAF-1*, Apoptosis protease activator factor-1; *Bak*, Bcl-associated antagonist killer; *Bax*, Bcl-associated X protein; *Bcl-2*, B cell lymphoma-2; *DD*, death domain; *DED*, death effector domain; *IAP*, inhibitor apoptosis protein; *TNFR*, tumor necrosis factor receptor; *TRADD*, TNFR-associated death domain.

a chromosomal translocation that determines excessive production of Bcl-2 protein and inhibition of apoptosis. In 50% of human cancers, mutations in the p53 gene have been shown. Some of the drugs used in the treatment of neoplasias induce apoptosis and cell cycle arrest by a mechanism dependent on p53. In cases in which the p53 gene is mutated, the cancer cells are not sensitive to chemotherapy.

SUMMARY

Apoptosis or programmed cell death is a physiological process by which cells are removed from tissues in a controlled manner. Cells sentenced to apoptosis show characteristic morphological changes, including condensation of the cell cytoplasm and nucleus, DNA fragmentation, dissociation of cell organelles, and disruption of the cell plasma membrane. Finally, the cell is reduced to vesicular remnants that are phagocytosed by neighboring cells and macrophages. Among the biochemical changes in apoptosis are the translocation of phosphatidylserine to the outer side of the cell plasma membrane, the exposure of anexin I and calreticulin to the cell surface, and the increase in cytochrome c in the cell cytoplasm.

Caspases are proapoptotic proteins found in cells as zymogens and activated by hydrolysis. Once activated, they function as cysteine proteases to activate other caspases. Caspases can be distinguished into two groups, the *initiator* caspases, which start the process and the *effector* caspases, which specifically lyse molecules that are essential for cell survival.

Apoptosis can be triggered by stimuli coming from the outside or inside the cell. These activate the apoptotic *extrinsic* and *intrinsic pathways, respectively.*

Apoptotic extrinsic pathway. It is activated by the binding of a cytokine (TNF) to its plasma membrane receptor (TNFR) in the cell. This stimulates the association of the receptor to adapter proteins (FADD or TRADD) interacting with initiator procaspases to form the *DISC complex.* Activated procaspase initiates the proteolytic cascade, which will activate effector caspases and induce the digestion of essential cell proteins.

Intrinsic pathway. It is activated by cell stressors, which will cause the release of proteins (*cytochrome c* and others) from the mitochondria intermembrane space into the cytosol. Cytochrome c binds to APAF-1 and procaspase 9, forming the *apoptosome* complex. This triggers the activation of procaspase 9 to caspase 9 and the protease hydrolysis cascade.

Regulation of apoptosis is achieved by proteins of the *Bcl* family of proteins. The members of this group of proteins include proapoptotic and antiapoptotic agents. Among the proapoptotic members are *Bcl-2 associ-* *ated X protein (Bax)* and *Bcl-2 antagonist killer (Bak).* Bcl proteins with four BH domains are antiapoptotic. The fate of a cell depends on the balance between pro- and antiapoptotic proteins, which antagonize each other.

Bibliography

Danial, N.N., Korsmeyer, S.J., 2004. Cell death: critical control points. Cell 116, 205–219.

Delhalle, S., Duvoix, A., Schnekenburger, M., Morceau, F., Dicato, M., Diederich, M., 2003. An introduction to the molecular mechanisms of apoptosis. Ann. N Y Acad. Sci. 1010, 1–8.

Elmore, S., 2007. Apoptosis: a review of programmed cell death. Toxicol. Pathol. 35, 495–516.

Hassan, M., Watari, H., Almaaty, A.A., Ohba, Y., Sakuragi, N., 2014. Apoptosis and molecular targeting therapy in cancer. Biomed. Res. Int. 2014, 1–23.

Moldoveanu, T., Follis, A.V., Kriwacki, R.N., Green, D.R., 2014. Many players in BCL-2 family affairs. Trends Biochem. Sci. 39, 101–111.

Newmeyer, D.D., 2003. Releasing power for life and unleashing the machineries of death. Cell 112, 481–490.

Pop, C., Salvesen, G.S., 2009. Human caspases: activation, specificity and regulation. J. Biol. Chem. 284, 21777–21781.

Shiozaki, E.N., Shi, Y., 2004. Caspases, IAC and Smac/DIABLO: mechanisms from structural biology. Trends Biochem. Sci. 29, 486–494.

Taylor, R.C., Cullen, S.P., Martin, S.J., 2008. Apoptosis: controlled demolition at the cellular level. Nat. Rev. Mol. Cell Biol. 9, 231–241.

Varfolomeev, E.E., Ashkenazi, A., 2004. Tumor necrosis factor: an apoptosis juNKie? Cell 116, 491–497.

Alphabetic Index

A

ABC transporters, 238
Absorption in intestine, 267
  amino acids, 269
  carbohydrates, 267
  lipids, 268
  proteins, 269
  water and electrolytes, 271
Acetylcholine, 568, 637
  esterase, 637
Acid-base balance, 704
  disorders, 708
  laboratory studies, 710
  regulation, 704
Acidosis, 709
Acids and bases, 11
  strenght, 12
  titration curve, 15
ACTH, 578
Activation energy, 150
Active transport, 233
Acylglycerols, 104
  chemical properties, 106
  physical properties, 105
Adenilate/s, 139
  cyclase, 139, 557
  regulatory role, 443
  adenine, 122
Adenosine phosphate, 139
ADP, 139
  AMP, 139
  Cyclic, 139
ATP, 138
ADP ribosylation, 538
Adrenal, 593
  cortex, 593
    hyper- and hypofunction, 600
    corticoids, 593
    biosynthesis, 595
  medulla, 601
    hormones, 601
Adrenocorticotrpin hormone, 578
Aerobic work, 458
Alanine, 24

Albumin, 60
Alcalosis, 711
Aldosterone, 599
α-helix, 39
Allosteric enzymes, 55
Amino acids, 22
  acid-base properties, 27
  catabolism, 373
  classification, 24
  chemical properties, 30
  decarboxylation, 384
  essential, 370
  fate, 372
  in biotransformation, 393
  metabolism, 367
    inborn errors, 390, 396
    regulation, 441
    role of various organs, 381
  titration curve, 28
  transport, 372
Aminosugars, 82
Ammonia, 376
  metabolism, 376
  toxicity, 381
Amphipathic compounds, 11
Amylase, 253, 259
Amylin, 632
Amylopeptin, 84
Amylose, 84
Anaerobic work, 456
Androgenic corticosteroids, 593
Anemia, 53
Anterior hypophysis, 578
Antiapoptotic proteins, 792
Antibiotic action on
        DNA, 508
Antibodies, 747
  monoclonal, 759
  variability, 753
Antidiuretic hormone, 567
Antigen, 747
  nature of, 753
  processing, 769
  recognition, 769

Antioxidants, 210
  nutrients, 211
Antiporters, 226
Antivitamins, 646
Apoenzyme, 186
Apoptosis, 791
  medical importance, 794
  pathways, 793
    extrinsic, 793
    mitochondrial, 793
    others, 793
    regulation, 793
Aquaporins, 231
Arginine, 24
Arginine-vasopressin, 587
Asparagine, 25
Aspartic acid, 25
ATP, 147, 178
  -ADP transport, 196
  synthase, 191
  sinthesis, 195
ATPases, 233
  F-type, 237
  P-type, 233
  V-type, 237
Atrial natriuretic peptide, 630
  actions, 630
Autoimmune diseases, 776
Autophagia, 368
Avitaminosis, 645

B

Bacterial putrefaction, 392
Bacteriophage, 138
Beriberi, 664
β-lymphocytes, 746
β-sheet, 40
Bile, 261
  acids, 261
  composition, 261
  bile pigments, 263
  cholesterol, 263
  phospholipids, 263
  gallstones, 263

Bilirubin, 263
  direct and indirect, 408
  transport, 406
Biliverdin, 406
2,3-bisphosphoglycerate, 55
Biogenic elements, 1
Bioenergetics, 177
Biological amines, 384
  biosynthesis, 384
Biological oxidations, 181
Biotin, 674
  avitaminosis, 675
  chemistry, 674
  functional role, 674
  intestinal absorption, 674
  metabolism, 674
  natural sources, 674
  requirements, 675
  synonymy, 674
  transport, 674
Biotransformation, 450
Blood coagulation or clotting, 781
  extrinsic pathway, 783
  intrinsic pathway, 783
  regulation, 787
  stages, 783
Blood glucose, 317
  homeostasis, 314, 612
Blood lipids, 326
Blood groups, 94
Blood plasma proteins, 58
B lymphocytes, 746
  activation, 771
Body fluids, 690
  H* concentration, 699
    regulation, 700
      renal, 705
      respiratory, 704
  ionic composition, 691
  osmotic pressure, 694
Bohr effect, 55
Bradikinin, 629
Branched chain amino acids, 24
  metabolic fate, 383
Brown fat, 199
Brush border enzymes, 260
Buffers, 13
  mechanism of action, 13

C
Ca^{2+}-ATPase, 236
Cadherins, 248
Calbindin, 721
Calcitonin, 627
  actions, 627

Calcitriol, 630, 654
Calcemia, 720
Calcium, 720
  absorption, 720
  deficiency, 723
  excretion, 721
  homeostasis, 722
    alterations, 723
  intracellular concentration, 723
    regulation, 722
  natural sources, 720
  renal handling, 721
  requirement, 725
Calmodulin, 567
Calorie, 142
Capsid, 137
Capsomer, 137
Carbohydrates, 73
  isomerism, 74
  metabolism, 283
Carbon dioxide transport, 55, 702
Carboxyhemoglobin, 56
Carboxypeptidase, 259
Cardiolipin, 110
Carnitine, 338, 684
  deficiency, 684
  functions, 684
Carotene, 648
Carotenoids, 649
Ca* signal, 566
Caspases, 792
Catalase, 210
Ceramide, 112
Catecholamines, 388, 601, 638
  biosynthesis, 388
Catepsins, 368
Caveolin, 245
Cell
  adhesion, 247
  as open system, 146
  cycle, 465
Cellulose, 87
Cerebrosides, 113
Channels, 228
  amiloride sensitive Na, 230
  aquaporins, 231
  chloride, 230
  extracellular ligands, 230
  ion, 228
  ligand gated, 229
  voltage-gated, 229
Chemical composition of
  human tissues, 2
  living organisms, 1
Chemical kinetics, 148

Chemical reactions, 144
  direction of, 144
  equilibrium, 144
  order, 149
Chilomicrons, 328
Chimotrypsin, 259
Chitin, 87
Chloride, 719
  absorption, 719
  excretion, 719
  balance alterations, 719
  natural sources, 719
  renal handling, 719
Cholecalciferol, 652
Cholecystokinin, 631
Cholesterol, 115
  biosynthesis, 358
    regulation, 440
  catabolism, 361
  metabolism, 358
    regulation, 361
  plasma, 360
Choline, 355, 682
  absorption, 683
  deficiency, 683
  functions, 683
  natural sources, 682
  requirements, 683
Chondroitin sulfate, 88
Chromatin, 130
Chromatosome, 131
Cistron, 495
Citric acid cycle, 301
  energy balance, 305
  functional role, 305
  genetic defects, 303
  regulation, 438
Clathrin, 243
Cobalt, 741
Codon, 495
Coenzyme, 156
Coenzyme A, 671
Coenzyme Q, 187
Collagen, 47
Complement, 760
  alternative pathway, 763
  classical pathway, 761
  lectin pathway, 763
  components, 761
  functions, 764
  membrane attack complex, 762
  regulatory factors, 763
Complex lipids degradation, 356
  congenital disorders, 356
Conjugated linoleic acids, 107

Conjugated proteins, 46
Connexin, 233
Copper, 736
Cori's cycle, 311
Corticosteroids, 593
    synthetic, 600
Cortisol, 599
Creatin, 394
    synthesis, 394
3',5'-cyclic AMP, 139
    signal transduction system, 556
Cyanocobalamin, 677
Cyclic ADP ribose, 668
Cyclooxigenase, 350
Cyclopentaneperhydro-
    phenantrene, 115
Cystic fibrosis, 238
    conductance, 238
Cytochrome, 187
Cysteine, 25
Cystine, 27
Cytokines, 773
    functions, 774
    general properties, 773
    receptors, 773
Dalton, 31
Deamination, 375
Dehydration, 698
7-Dehydrocholesterol, 117
Deoxyribonucleic acid, 124
Deoxiribose, 122
Deoxisugars, 81
Dermatansulfate, 88
Desensitization, 554
Dextrans, 87
Dextrins, 86
Diabetes insipidus, 587
Diabetes mellitus, 613
Dialysis, 35
Dicumarol, 662
Dietary fiber, 89, 264
Diffusion, 224
    facilitated, 226
Digestion, 251
Digestive process, 264
    carbohydrates, 264
    lipids, 265
    nucleic acids, 267
    proteins, 266
1,25-$(OH)_2$-D_3, 653
Disaccharides, 83
DNA, 124
    antibiotic action on, 508
    base sequence, 481
    circular, 132

conformations, 128
denaturation, 129
electrophoresis, 481
fingerprinting, 487
microarrays, 514
mitochoncrial, 133, 497
molecular structure, 125
nuclear, 495
polymerase/s, 468
reverse transcriptase, 480
recombinant, 482
renaturation, 130
repair, 472
    abnormalities, 473
repetitive sequences, 130
replication, 465
    mechanism, 469
satellite, 132
Dolichol, 115

E
Eicosanoids, 349, 634
    actions, 350
    biosynthesis, 349
Elastase, 259
Electrochemical cell, 180
Electron transport, 184
    transport chain, 184
    other systems for, 202
    inhibitors, 189
Electrophoresis, 33
Embden-Meyerhof pathway, 291
Enantiomers, 22
Endocitosis, 242
Endogenous opioids, 639
Endoplasmic reticulum, 223
Endothelin, 629
Energy, 141
    free, 143
    in chemical reactions, 142
    standard free, 145
    coupled reactions, 148
Enkephalins, 32
Enteroglucagon, 632
Enthalpy, 142
Entropy, 143
Enzymatic catalysis, 157
Enzymes, 153
    active site, 158
    activity, 161
        determination, 161, 173
        regulation, 168
        factors that affect, 161
    nomenclature, 153
    adaptive or induced fit, 159

constitutive, 170
covalent modification, 170
feedback inhibition, 169
in blood plasma, 173
inducible, 170
processes in cascade, 171
inhibitors, 165
    anticompetitive, 168
    competitive, 165
    irreversible, 165
    noncompetitive, 167
    reversible, 165
Epigenetics, 521
Epinephrin, 388, 603
Epitope, 747
Equilibrium constant, 9
    of water ionization, 10
Equivalent, 17
Ergosterol, 117
Erythropoietin, 629
Essential minerals, 715
Estrogens, 620
    actions, 622
    biosynthesis, 621
Ethanol metabolism, 452
Exchangers, 226
    Na^+Ca^{2+}, 240
    Na^+H^+, 240
Exocytosis, 245
Exons, 495
Exosomes, 245
Expression of concentrations, 16
Extracellular water, 691
Extracellular fluid, 690
    volune, 694
    osmolality, 694
        regulation, 694

F
Facilitated diffussion, 227
Fats, 104
    metabolism, 334
    nutitional importance, 106
Fatty acids, 99
    biosynthesis, 343
        of unsaturated, 348
    β-oxidation, 336
        genetic defects, 338
    catabolism, 335
    chemical properties, 102
    essential, 104
    metabolism, 335
        regulation, 439
    oxidation, 336
        energy balance, 339

Fatty acids (*cont.*)
    in peroxisomes, 339
    of unsaturated, 339
    physical properties, 101
    synthase, 345
Fibrin, 785
Fibrinogen, 785
Fibrinolysis, 788
Fick's law, 224
Flavinadeninedinucleotide, 186
Flavinmononucleotide, 186, 666
Flavoproteins, 186
Fluorine, 941
Folic acid, 675
    antivitamins, 677
    avitaminosis, 677
    chemistry, 675
    functional role, 676
    intestinal absorption, 676
    metabolism, 676
    natural sources, 675
    requirements, 676
    sinonymy, 675
Free energy, 143
Free radicals, 205
Fructose, 77
    metabolism, 312

G

Galactose, 77
    metabolism, 313
Galanin, 633
Gamma-aminobutyric acid, 385
Gamma-glutamyl cycle, 395
Gangliosides, 113
Gap junction, 232
Gastric juice, 254
    analysis, 257
    digestive action, 256
    hydrochloric acid, 254
    intrinsic factor, 257
    lipase, 257
    mucin, 257
    pepsin, 256
Gastrin, 630
Gastrointestinal hormones, 630
Gene/s, 493
    cloning, 488
    repeated, 495
    tumor suppressor, 520
Genetic
    code, 493
    information, 465
    mutations, 510
Genome, 133

Ghrelin, 354, 632
Gibbs-Donnan equilibrium, 693
Gla proteins, 660
Globin, 51
Globosides, 113
Globulin, 60
Glucagon, 611
    metabolic actions, 611
Glucocorticoids, 593
Glucokinase, 286
Gluconeogenesis, 308
    energy cost, 312
    regulation, 432
Glucose, 75
    -alanine cycle, 442
    cyclic structure, 75
    -fatty acid cycle, 442
    oxidation energy balance, 306
    phosphorylation, 285
    tolerance test, 616
    transport, 227, 285
    uptake in cells, 284
Glutamate deamination, 375
Glutamine, 25
    formation, 376
Glutathione, 32, 395
    peroxidase, 211
Glycemia, 314
    homeostasis, 314
Glycemic impact, 315
Glycemic index, 315
Glycerol metabolism, 334
Glycerophospholipids, 108
    Properties, 110
Glycine, 24
Glycocalyx, 222
Glycogen, 86
    genetic diseases related to, 291
Glycogenesis, 287
Glycogenesis, 287
    energy requirement, 289
    regulation, 431
Glycogenolysis, 279
    regulation, 429
Glycolipids, 112
Glycolysis, 291
    energy balance, 296
    functional role, 298
Glycogenin, 288
Glycogenosis, 291
Glycoproteins, 91
    degradation, 321
    oligosaccharide synthesis, 316
Glycosaminoglycans, 88
Glycosides, 80

Golgi complex, 223
Gonadotropins, 580
Gonads, 616
Gout, 418
G proteins, 552
Granzyme, 772
Growth factors, 635
Growth hormone, 582
    pathologic conditions, 584
Guanine, 122
Guanylate cyclase receptor, 562
Guldberg and Waage law, 144
Gums, 90

H

Haptens, 753
Haworth's formulas, 78
H^+ dependent transporters, 241
Heart hormone, 610
Helicase, 468
Heme, 50
    biosynthesis, 401
    catabolism, 405
    metabolism, 401
Hemicelluloses, 90
Hemoglobin, 49
    abnormal, 56
    A_{1c}, 56
    S, 57
    derivatives, 56
    functions, 52
    oxygen dissociation curve, 53
    structure, 51, 52
Hemostasia, 781
Henderson-Hasselbalch, 14
Heparan sulfate, 89
Heparin, 89
Heteropolysaccharides, 88
H^* excretion regulation, 710
Hexokinases, 286
Hexose-phosphate pathway, 307
Hexose oxidation products, 81
Hexose reduction products, 80
High density lipoproteins, 330
High-energy compounds, 146
Histamine, 384, 638
Histidine, 26
Histones, 131
Holoenzyme, 156
Homocysteine, 386
Homopolysaccharides, 84
Homoserine, 27
Hormones, 573
    chemical nature, 573
    determination methods, 575

general properties, 574
types of action, 574
Human genome project, 496
Hyaluronic acid, 88
Hydrogen bond, 5
[H⁺] in body fluids, 699
  regulation, 700
Hydrogen shuttle systems, 199
Hydrolases, 155
Hydroxiapatite, 725
Hydroxyl radical, 207
Hypercholesterolemia, 333
Hyperhydration, 699
Hypophysis, 576
Hypothalamus hormones, 577

I

Immune response, 747
Immune system, 745
  adaptive, 746
  innate, 745
  related pathology, 775
Immunity, 745
  cellular, 764
  humoral, 746
  molecular basis, 745
Immunodeficiencies, 775
Immunoglobulins, 747
  clonal selection, 752
  genes, 754
    alledlic exclusion, 756
    rearrangement, 754
    regulatory sequences, 758
  genetic diversity, 753
  heavy chains, 748
    arrangement, 754
    isotypes, 750
    synthesis, 757
  light chains, 748
    arrangement, 754
  membrane, 752
    synthesis, 758
  monoclonal, 759
  structure, 747
  superfamily, 248
Imprinting, 522
Inositol, 109, 684
Insulin, 605
  biosynthesis, 606
  hyper- and hypofunction, 611
  like growth factors, 583
  mechanism of action, 607
  metabolic actions, 608, 609
  receptor, 607
  resistance, 615

secretion, 606
Integrins, 248
Interconversion carbohydrates, lipids,
  amino acids, 425
Interferon, 774
Interleukins, 773
Intestinal mucosa, 260
  brush border enzymes, 260
  disaccharidases, 260
  endopeptidases, 260
  exopeptidases, 260
  nucleases, 261
  phosphatases, 261
Intracellular pH regulation, 707
Intracellular receptors, 549, 690
Intrinsic factor, 678
Introns, 495
Inulin, 87
Iodine, 738
Ionic compounds, 7
Ions of body fluids, 691
Ionofores, 231
Iron, 729
  absorption, 730
  balance, 730
  deficiency, 734
  excretion, 734
  homeostasis, 733
    alterations, 734
  natural sources, 730
  overload, 734
  requirement, 734
  storage, 731
  transfer to tissues, 732
  transport in blood, 731
Iron-sulfur centers, 186
Isoelecric point, 28
Isoforms, 172
Isoleucine, 24
Isomerases, 155
Isoprene, 114
Isozymes, 172

J

Jaundice, 408
Joule, 142

K

Keratansulfate, 89
Keratins, 49
Kernicterus, 407
Ketogenesis, 340
Ketone bodies, 340
Ketosis, 342
Kidney, 627

hormones, 628
Km, 163
Krebs cycle, 301

L

Lab ferment, 256
Lactic acidosis, 300
Lactogenic hormone, 581
Lactose, 83
Lectins, 94, 223
Leucine, 24
Lecithin, 109
Leptin, 354
Leucotrienes, 349
Ligases, 155
Lignin, 90
Lineaweaver-Burk equation, 163
Lipids, 99
  metabolism, 325
Lipoic acid, 299, 685
Liprotein (a), 331
Lipoproteins, 114
  and atherosclerosis, 332
  receptors, 331
Liposomes, 216
Low density lipoproteins, 330
Lyases, 155
Lysine, 26
Lysosomes, 223, 368

M

Macrophage activation, 773
Magnesium, 727
Maltose, 83
Melanine synthesis, 390
Major histocompatibility, 766
  class I proteins, 767
  class II proteins, 767
  functions, 768
Manganese, 739
Mannose, 77
Melanin synthesis, 390
Melanocortin system, 579
Melanocyte stimulant, 585
Melatonin, 392, 633
  Synthesis, 392
Membranes, 215
  transport across, 224
  carbohydrates, 222
  constitution, 215
  lipids, 215
  proteins, 220
  structure, 217
Memory cells, 755, 761
Menadione, 662

Menaquinone, 661
Metabolic pathways, 277
Metabolic syndrome, 617
Metabolism, 277
  compartmentalization, 282
  integration, 427
  in some tissues, 449
    adipose tissue, 462
    heart, 461
    liver, 449
    nervous, 463
    skeletal muscle, 455
      effect of training, 459
  interactions, 442
  of amino acids, 367
  of carbohydrates, 283
  of ethanol, 452
  of heme, 405
  of lipids, 325
  regulation, 279
  studies, 277
    systems used, 278
Metabolome, 279
Metalloenzymes, 157
Methemoglobin, 56
Methionine, 25
Michaelis constant, 163
Microarrays, 514
Mineralocorticoids, 593
Mitochondria, 183
Molality, 17
Molarity, 16
Molecular biology
  methods used, 481
Molibdenum, 740
Monocarbon groups, 386
  transfer, 386
Monosaccharides, 73
  derivatives, 81
  phosphoric esters, 82
Mucilages, 90
Mucopolysaccharidoses, 320
Multidrug resistance protein, 238
Multienzyme systems, 162
Multimolecular complexes, 43
Muscle contraction, 67
Muscle proteins, 54
  actin, 66
  myosin, 65
  nebulin, 67
  phospholamban, 67
  tropomyosin, 67
  troponin, 67
Mutagenic agents, 511
Mutarotation, 75

Mutation, 510
Myoglobin, 53

N
NAADP, 669
NAD and NADP, 180
Na^+ dependent transport, 240
  amino acid, 240
  glucose, 240
  Na^+K^+Chloride, 240
Na^+Ca^{2+} exchanger, 240
Na^+H^+ exchanger, 240
Na^+ independent transporters, 240
Na^+K^+-ATPase, 233
Nerve impulse, 241
Neuraminic acid, 82
Neuroendocrine system, 548
Neurohypophysis, 586
Neuropeptides, 638
Neurophysin, 586
Neurotransmitters, 568, 637
Niacin, 667
  avitaminosis, 669
  chemistry, 667
  functional role, 668
  intestinal absorption, 668
  metabolism, 668
  natural sources, 668
  pharmacological actions, 670
  requirements, 669
  sinonymy, 667
Nicotineamide, 185, 667
Nicotinic acid, 667
  biosynthesis, 392
Nictalopia (night blindness), 651
Nitric oxide, 207, 639
  synthesis, 393
Nitrogenous bases, 122
Non polar compounds, 5
Norepinephrin, 388, 603
Nuclear magnetic resonance
  spectroscopy, 279
Nucleic acids, 123
Nucleosome, 131
Nucleotides, 121
  biosynthesis, 421
  free, 138
  sugars, 315

O
Obesity, 353
Oil-water partition, 224
Okasaki fragment, 470
Oligoelements, 735
Oligosaccharides, 92

biosynthesis, 316
Oncogene, 518
Oncotic pressure, 695
Open reading frame, 549
Opioids, 639
Optical activity, 23
Optical isomerism, 22
Ornithine, 27
Osteomalacia, 656
Ovary, 620
  hormones, 620
    variations of levels, 523
  peptide hormones, 623
Oxidation-reduction, 178
Oxidative phosphorylation, 190
  ATP yield, 196
  inhibitors, 197
  regulation, 198, 442
Oxidoreductases, 154
Oxigen debt, 311, 457
Oxytocin, 586

P
Pancreas hormones, 604
Pancreatic juice, 258
  digestive actions, 258
  inorganic components, 258
Pantothenic acid, 570
Parathyroid, 624
  hormone, 624
    actions, 625
    alterations, 626
    related protein, 626
    secretion, 624
Passive transport, 226
Pasteur effect, 436
Pectins, 90
Pellagra, 669
Pentoses, 78
Pentose phosphate pathway, 307
Pepsin, 256
Peptides, 30
  acid-base properties, 31
Peptide bond, 30, 38
Peptidoglycans, 91
Pernicious anemia, 680
Peroxinitrite, 208
Peroxide radicals, 207
Peroxisome, 223
  proliferator activated recep, 551
Personalized medicine, 516
pH, 12
Phagocytosis, 242
Phenylalanine, 25
  Metabolism, 387

Phosphatemia, 726
Phosphatidic acid, 106
Phosphatidylcholine, 107
Phosphatidylglycerol, 110
Phosphatidyl inositolbis-P, 109
  signal transduction system, 560
Phosphodiesterase, 560
Phospholipids, 108
  biosynthesis, 354
4'-Phosphopantetheine, 671
Phosphorus, 725
Phosphrylase, 289
Phosphorylation
  at substrate level, 201
  oxidative, 190
Phylloquinone, 659
Pineal gland, 633
Pinocytosis, 243
Pituitary, 576
  anterior (adenohypophisis), 578
  posterior (neurohypophys.), 586
Placenta, 585
  chorionic gonatotropin, 585
  lactogen, 585
Plasmalogens, 110
Plasma proteins, 60
  general functions, 64
  lipoproteins, 326
    metabolism, 328
Plasmids, 133
Platelet aggregation, 781
Polar compounds, 5
Polarized light, 22
Polyamines, 385
Polymerase chain reaction, 485
Polypeptides, 30
Polysaccharides, 84
Polysomes, 507
Porphin, 50
Porphobilinogen, 402
Porphyrias, 403
Porphyrin, 50
Potassium, 717
  absorption, 717
  and hypertension, 719
  homeostasis alterations, 718
  excretion, 717
  natural sources, 717
  recommended intake, 719
  regulation of concentration, 718
  renal handling, 717
Potential phosphoryl transfer, 147
Precursor messenger RNA, 477
  synthesis, 477
Proapoptotic proteins, 792

Progesterone, 620
  actions, 622
  biosynthesis, 621
  metabolism, 621
Programmed cell death, 791
Prolactin, 581
Proline, 26
Pro-opiomelanocortin, 579
Prostacyclins, 632
Prosthetic group, 46
Proteasome, 368
Proteins, 33
  acid-base properties, 33
  biosynthesis, 493
    mechanism, 502
  classification, 46
  degradation, 368
  denaturation, 45
  domains, 44
  folding, 535
  half life, 368
  heat shokc, 535
  in nutrition, 47
  kinases, 554, 567
  molecular mass, 34
  molecular shape, 36
  phosphatases, 567
  random arrangement, 41
  solubility, 34
  structure, 36, 43
    primary, 36
    quaternary, 42
    secondary, 38
    tertiary, 41
  traffic in the cell, 542
  tyrosine kinase receptors, 554
Proteoglycans, 90
  degradation, 320
Proteome, 515
Prothrombin, 787
Proto-oncogene, 519
Provitamins, 646
Pteroylglutamic acid, 675
Pseudogenes, 496
Ptyalin, 253
P type ATPases, 233
Purine, 122
  biosynthesis, 413
    inhibitors, 422
  catabolism, 416
    genetic defects, 416, 417
  metabolism, 413
  salvage pathway, 416
Pyridoxal, 672
Pyridoxamine, 672

Pyridoxine, 672
Pyrimidine, 122
  biosynthesis, 419
    genetic defects, 421
  catabolism, 421
  metabolism, 421
Pyruvate, 294
  oxidative decarboxylation, 298
  regulation, 437

Q
Q cycle, 194

R
Ras proteins, 563
Reaction order, 153
Reactive species, 205
  defense mechanisms, 210
    endogenous systems, 210
  harmful effects, 208
  related pathology, 208
Reactive oxygen species, 206
Reactive nitrogen species, 207
Real time PCR, 487
Receptors, 548
  adrenergic, 590
  cholinergic, 568
  dopaminergic, 570
  GABAergic, 570
  G proteins coupled, 551
  guanylate cyclase, 556
  intracellular, 549
  membrane, 551
  muscarinic, 569
  neurotransmitter, 568
  nicotinic, 569
  peroxisome proliferator, 551
  toll like, 745
  tyrosine kinase, 554
Recombinant DNA, 482
Redox signaling, 568
Reduction equivalent transfer, 184
Reduction potential, 179
Relationship pH and [H*], 12
Releasing hormones, 577
Renin, 628
Respiratory chain, 184
  components, 185
  complexes, 189
  genetic deficiencies, 198
Respiratory control, 198
Respiratory exchange ratio, 462
Restriction endonucleases, 474
Retinal, 647
Retinol, 647

Retinoic acid, 647
Reverse cholesterol transport, 330
Riboflavin, 665
   avitaminosis, 667
   chemistry, 665
   functional role, 666
   intestinal absorption, 666
   mechanism of action, 666
   metabolism, 666
   natural sources, 665
   requirements, 667
   sinonymy, 665
Ribonucleic acid, 134
Ribose, 122
Ribosome, 136
Rickets, 656
Ribozymes, 156, 501
RNA, 134
   long non-coding, 531
   messenger, 134, 497
      mature, 499
      processing, 497
         differential, 499
      synthesis, 475
   ribosomal, 134
      synthesis, 480
   small nuclear, 136
   transfer, 135, 500
      synthesis, 479
Rodopsin, 650

S
Saccharose, 83
Saliva, 251
   composition, 251
   digestive action, 253
   other functions, 254
Sarcosine, 27
Scaffold proteins, 568
Scurvy, 682
Secondary active transport, 239
Secretin, 631
Selenium, 739
Seleno amino acids, 26
Serine, 24
Serotonin, 391, 638
   biosynthesis, 391
Sialic acids, 82
Signal transduction systems, 556
   ceramide in, 562
   crosstalk, 568
   3',5'-cyclicAMP, 556
   3;5'-cyclicGMP, 362
   JAK=STAT, 564

   lipids in, 562
   Ras and MAP kinase, 563
   redox signaling, 568
   sphingosine-1-phosphate, 562
   TOR, 564
Singlet oxygen, 207
Sirtuins, 541
Sodium, 715
   absorption, 715
   and hypertension, 717
   homeostasis alterations, 716
   excretion, 716
   natural sources, 715
   renal handling, 716
   requirements, 716
Somatostatin, 612
Somatotropin hormone, 582
Southern blotting, 484
Sphyngolipidoses, 356
Sphingomyelin, 112
Sphingophospholipids, 112
Sphingosine, 112
Splicing, 498
   alternative, 499
   alterations, 499
Starch, 84
Starling equilibrium, 696
STAT proteins, 564
Stercobilin, 408
Steroids, 115
   Receptors, 550
Subcellular organelles, 223
Sulphonylurea, 616
Superoxide anion, 206
Superoxide dismutase, 210

T
Taurine, 683
   functions, 683
T cells receptors, 765
Telomerase, 473
Terpenes, 114
Testis, 616
   hormone alterations, 620
Testosterone, 616
   actions, 519
   biosynthesis, 617
   metabolism, 618
Tetrahydrofolic acid, 675
Thalassemia, 57
Thermodinamics, 141
   first law, 141
   second law, 141
Threonine, 24

Thiamine, 663
   avitaminosis, 664
   chemistry, 663
   functional role, 664
   intestinal absorption, 664
   mechanism of action, 664
   metabolism, 664
   natural sources, 663
   requirenents, 664
   synonymy, 663
   therapeutic uses, 665
   transport, 664
Thrombin, 785
Thromboxane, 349
Thyramine, 385
Thryptamine, 385
Thymine, 122
Thyroid hormones, 588
   hyper- and hypo, 592
   mechanism of action, 591
   secretion, 589
   synthesis, 380
Thyroid stimulating hormone, 580
Thyrotropin releasing, 577
Tioctic acid, 685
T lymphocytes, 764
   activation, 770
   antigen recognition, 768
   antigen processing, 769
   receptor, 765
T_K lymphocytes lytic action, 772
Tocopherol, 657
Tocotrienol, 657
Topoisomerases, 468
Total water determination, 689
Trace elements, 735
Transamination, 373
Transcription, 475
   Factors, 477
   FOXO, 568
   in eukaryotes, 477
   in prokaryotes, 475
Transcriptome, 500
Transfer RNA, 135
   Synthesis, 479
Transferases, 154
Transgenesis, 513
Translation, 502
Translational medicine, 493
Transport of CO_2, 55
Transposons, 512
Triacylglycerols, 104
   biosynthesis, 353
   metabolism, 334

Tricarboxylic acid cycle, 301
Tripsin, 259
Tryptophan, 25
   metabolism, 391
     serotonin biosynthesis, 391
Tumor supressor genes, 520
Tyronine, 588
Tyrosine, 25
   metabolism, 387
     inborn errors, 390
Tyroxine, 588

U

Ubiquinone, 187
Ubiquitin, 368
Ultrafiltration, 35, 541
Uncoupler oxidat. phosphor., 197
Uracil, 122
Urea, 377
   cycle, 377
     inborn errors, 380
Uric acid, 418
Urobilin, 408
Urodilatin, 630

V

Valine, 24
Variome, 513
Vasoactive intestinal peptide, 632
Vasopressin, 587
Very low density lipoproteins, 329
Vesicular transport, 246
Virion, 137
Virus, 136
   mechanism of action, 516
Visual retinol cycle, 651
Vitamers, 646
Vitamin A, 646
   avitaminosis, 652
   chemistry, 646
   functional role, 650
   intestinal absorption, 648
   mechanism of action, 650
   metabolism, 649
   natural sources, 648
   pharmacological actions, 652
   requirement, 679
   sinonymy, 677
Vitamin B_1 (*See* thiamine)
Vitamin B_2 (see riboflavine)

Vitamin B_3 (*See* niacin)
Vitamin B_5 (*See* pantothenic acid)
Vitamin B_6 (*See* pyridoxine)
Vitamin B_7 (*See* biotin)
Vitamin B_9 (pteroyl glutamic acid)
Vitamin B_{12} (cobalamine), 677
   avitaminosis, 679
   chemistry, 677
   functional role, 678
   intestinal absorption, 678
   mechanism of action, 678
   metabolism, 678
   natural sources, 678
   pharmacological actions, 680
   sinonymy, 677
   Vitamin B complex, 663
Vitamin C, 680
   avitaminosis, 682
   chemistry, 680
   functional role, 681
   intestinal absorption, 681
   metabolism, 681
   natural sources, 682
   pharmacological actions, 682
   requirement, 682
   sinonymy, 680
Vitamin D, 652
   avitaminosis, 656
   chemistry, 652
   functional role, 654
   intestinal absorption, 653
   mechanism of action, 655
   metabolism, 653
   natural sources, 652
   pharmacological actions, 656
   requirement, 656
   sinonymy, 652
   toxic effects, 656
   transport in blood, 653
Vitamin E, 656
   avitaminosis, 658
   chemistry, 656
   functional role, 658
   intestinal absorption, 658
   mechanism of action, 658
   metabolism, 658
   natural sources, 657
   requirement, 658
   sinonymy, 656
   transport in blood, 658

Vitamin K, 659
   antivitamins, 662
   avitaminosis, 662
   chemistry, 659
   cycle, 661
   functional role, 660
   intestinal absorption, 660
   mechanism of action, 660
   metabolism, 660
   natural sources, 660
   pharmacological actions, 662
   requirement, 661
   sinonymy, 659
   transport in blood, 660
Vitamins, 645
   fat soluble, 646
   functional role, 645
   general properties, 645
   hydrosoluble, 646
   liposoluble, 646
   nomenclature, 645
   pharmacological actions, 646
   requirements, 646

W

Warfarin, 662
Water, 5
   as electrolyte, 8
   as solvent, 7
   balance, 691
     alterations, 697
   channels, 231
   determination, 689
   distribution in the body, 690
   ion product, 10
   molecular structure, 5
Warburg effect, 436
Waxes, 108
Wnt proteins, 568

X

Xenobiotics, 450
Xerophthalmia, 652

Z

Zinc, 735
Zuntz-Hamburger, 704
Zwitterion, 27
Zymogens, 160

Printed in the United States
By Bookmasters